DISTANCE AND MIDPOINT FORMULAS

Distance between $P_1(x_1, y_1)$ and $P_2(x_2, y_2)$:

$$d = \sqrt{(x_2 - x_1)^2 + (y_2 - y_1)^2}$$

Midpoint of P_1P_2: $\left(\dfrac{x_1 + x_2}{2}, \dfrac{y_1 + y_2}{2} \right)$

LINES

Slope of line through $P_1(x_1, y_1)$ and $P_2(x_2, y_2)$

$m = \dfrac{y_2 - y_1}{x_2 - x_1}$

Point-slope equation of line through $P_1(x_1, y_1)$ with slope m

$y - y_1 = m(x - x_1)$

Slope-intercept equation of line with slope m and y-intercept b

$y = mx + b$

Two-intercept equation of line with x-intercept a and y-intercept b

$\dfrac{x}{a} + \dfrac{y}{b} = 1$

LOGARITHMS

$y = \log_a x$ means $a^y = x$

$\log_a a^x = x$ $\qquad$ $a^{\log_a x} = x$

$\log_a 1 = 0$ $\qquad$ $\log_a a = 1$

$\log x = \log_{10} x$ $\qquad$ $\ln x = \log_e x$

$\log_a xy = \log_a x + \log_a y$ $\qquad$ $\log_a\left(\dfrac{x}{y}\right) = \log_a x - \log_a y$

$\log_a x^b = b \log_a x$ $\qquad$ $\log_b x = \dfrac{\log_a x}{\log_a b}$

EXPONENTIAL AND LOGARITHMIC FUNCTIONS

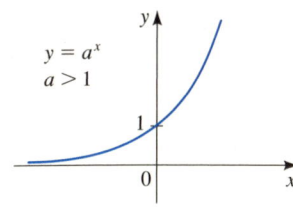

$y = a^x$
$a > 1$

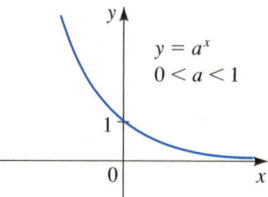

$y = a^x$
$0 < a < 1$

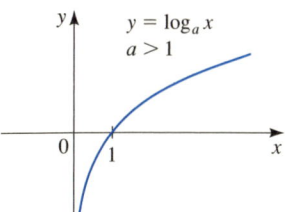

$y = \log_a x$
$a > 1$

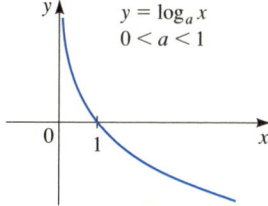

$y = \log_a x$
$0 < a < 1$

GRAPHS OF FUNCTIONS

Linear functions: $f(x) = mx + b$

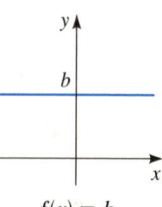

$f(x) = b$

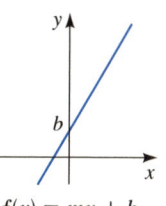

$f(x) = mx + b$

Power functions: $f(x) = x^n$

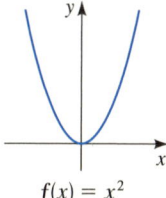

$f(x) = x^2$

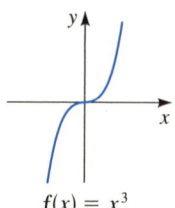

$f(x) = x^3$

Root functions: $f(x) = \sqrt[n]{x}$

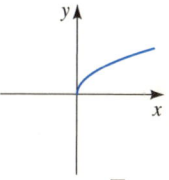

$f(x) = \sqrt{x}$

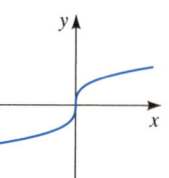

$f(x) = \sqrt[3]{x}$

Reciprocal functions: $f(x) = 1/x^n$

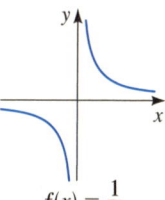

$f(x) = \dfrac{1}{x}$

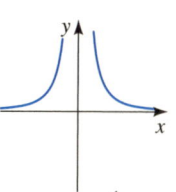

$f(x) = \dfrac{1}{x^2}$

Absolute value function $\qquad$ Greatest integer function

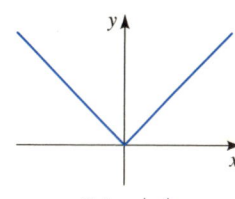

$f(x) = |x|$

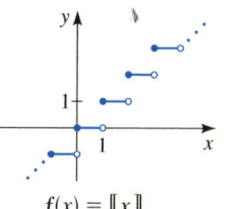

$f(x) = [\![x]\!]$

COMPLEX NUMBERS

For the complex number $z = a + bi$

the **conjugate** is $\bar{z} = a - bi$

the **modulus** is $|z| = \sqrt{a^2 + b^2}$

the **argument** is θ, where $\tan \theta = b/a$

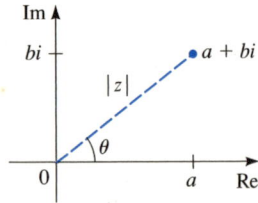

Polar form of a complex number

For $z = a + bi$, the **polar form** is

$$z = r(\cos \theta + i \sin \theta)$$

where $r = |z|$ is the modulus of z and θ is the argument of z

DeMoivre's Theorem

$$z^n = [r(\cos \theta + i \sin \theta)]^n = r^n(\cos n\theta + i \sin n\theta)$$

$$\sqrt[n]{z} = [r(\cos \theta + i \sin \theta)]^{1/n}$$

$$= r^{1/n}\left(\cos \frac{\theta + 2k\pi}{n} + i \sin \frac{\theta + 2k\pi}{n}\right)$$

where $k = 0, 1, 2, \ldots, n - 1$

ROTATION OF AXES

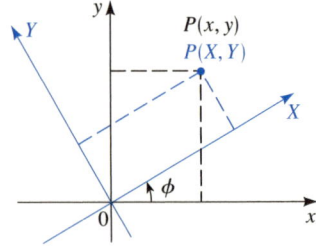

Rotation of axes formulas

$$x = X \cos \phi - Y \sin \phi$$
$$y = X \sin \phi + Y \cos \phi$$

Angle-of-rotation formula for conic sections

$$\cot 2\phi = \frac{A - C}{B}$$

POLAR COORDINATES

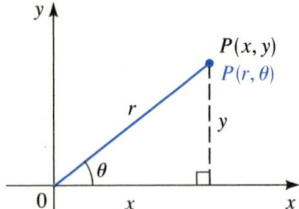

$$x = r \cos \theta$$
$$y = r \sin \theta$$
$$r^2 = x^2 + y^2$$
$$\tan \theta = \frac{y}{x}$$

SUMS OF POWERS OF INTEGERS

$$\sum_{k=1}^{n} 1 = n$$

$$\sum_{k=1}^{n} k = \frac{n(n + 1)}{2}$$

$$\sum_{k=1}^{n} k^2 = \frac{n(n + 1)(2n + 1)}{6}$$

$$\sum_{k=1}^{n} k^3 = \frac{n^2(n + 1)^2}{4}$$

THE DERIVATIVE

The **average rate of change** of f between a and b is

$$\frac{f(b) - f(a)}{b - a}$$

The **derivative** of f at a is

$$f'(a) = \lim_{x \to a} \frac{f(x) - f(a)}{x - a}$$

$$f'(a) = \lim_{h \to 0} \frac{f(a + h) - f(a)}{h}$$

AREA UNDER THE GRAPH OF f

The **area under the graph of** f on the interval $[a, b]$ is the limit of the sum of the areas of approximating rectangles

$$A = \lim_{n \to \infty} \sum_{k=1}^{n} f(x_k)\Delta x$$

where

$$\Delta x = \frac{b - a}{n}$$

$$x_k = a + k \, \Delta x$$

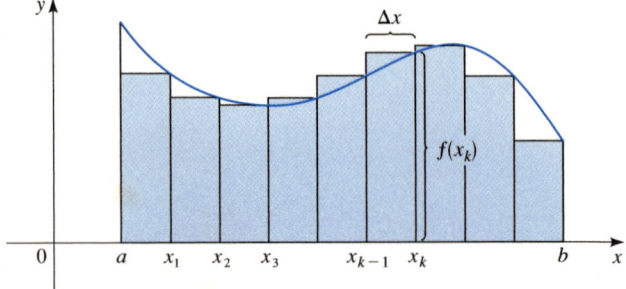

www.brookscole.com

www.brookscole.com is the World Wide Web site for Brooks/Cole and is your direct source to dozens of online resources.

At *www.brookscole.com* you can find out about supplements, demonstration software, and student resources. You can also send email to many of our authors and preview new publications and exciting new technologies.

www.brookscole.com
Changing the way the world learns®

www.brookscole.com

www.brookscole.com is the World Wide Web site for Brooks/Cole and is your direct source to dozens of online resources.

At *www.brookscole.com* you can find out about supplements, demonstration software, and student resources. You can also send email to many of our authors and preview new publications and exciting new technologies.

www.brookscole.com
Changing the way the world learns®

PRECALCULUS
Mathematics for Calculus

FIFTH EDITION

THOMSON

BROOKS/COLE

Precalculus: Mathematics for Calculus, Fifth Edition
Stewart, Redlin, Watson

Publisher: Bob Pirtle
Assistant Editor: Stacy Green
Editorial Assistant: Katherine Cook
Technology Project Manager: Earl Perry
Senior Marketing Manager: Karin Sandberg
Marketing Assistant: Jennifer Velasquez
Marketing Communications Manager: Bryan Vann
Senior Project Manager, Editorial Production: Janet Hill
Senior Art Director: Vernon Boes
Print/Media Buyer: Judy Inouye
Production Service: Martha Emry

Text Designer: John Edeen
Photo Researcher: Stephen Forsling
Copy Editor: Luana Richards
Art Editor: Martha Emry
Illustrator: Jade Myers, Matrix Art Services
Cover Designer: Roy E. Neuhaus
Cover Image: Bill Ralph
Cover Printer: Phoenix Color Corp
Compositor: G & S Book Services
Printer: R.R. Donnelley/Willard

Thomson Higher Education
10 Davis Drive
Belmont, CA 94002-3098
USA

For more information about our products, contact us at:
Thomson Learning Academic Resource Center
1-800-423-0563

For permission to use material from this text or product,

submit a request online at **http://www.thomsonrights.com.**
Any additional questions about permissions can be submitted
by e-mail to **thomsonrights@thomson.com.**

Library of Congress Control Number: 2005926064

Student Edition: ISBN 0-534-49277-0

PRECALCULUS
Mathematics for Calculus

FIFTH EDITION

James Stewart
McMaster University

Lothar Redlin
The Pennsylvania State University

Saleem Watson
California State University, Long Beach

THOMSON

BROOKS/COLE

Australia • Brazil • Canada • Mexico • Singapore • Spain
United Kingdom • United States

About the Cover

The art on the cover was created by Bill Ralph, a mathematician who uses modern mathematics to produce visual representations of "dynamical systems." Examples of dynamical systems in nature include the weather, blood pressure, the motions of the planets, and other phenomena that involve continual change. Such systems, which tend to be unpredictable and even chaotic at times, are modeled mathematically using the concepts of composition and iteration of functions (see Section 2.7 and the Discovery Project on pages 223–224). The basic idea is to start with a particular function and evaluate it at some point in its domain, yielding a new number. The function is then evaluated at the new number. Repeating this process produces a sequence of numbers called *iterates* of the function. The original domain is "painted" by assigning a color to each starting point; the color is determined by certain properties of its sequence of iterates and the mathematical concept of "dimension." The result is a picture that reveals the complex patterns of the dynamical system. In a sense, these pictures allow us to look, through the lens of mathematics, at exotic little universes that have never been seen before.

Professor Ralph teaches at Brock University in Canada. He can be contacted by e-mail at bralph@spartan.ac.brocku.ca.

About the Authors

James Stewart was educated at the University of Toronto and Stanford University, did research at the University of London, and now teaches at McMaster University. His research field is harmonic analysis.

He is the author of a best-selling calculus textbook series published by Brooks/Cole, including *Calculus, 5th Ed., Calculus: Early Transcendentals, 5th Ed.,* and *Calculus: Concepts and Contexts, 3rd Ed.,* as well as a series of high-school mathematics textbooks.

Lothar Redlin grew up on Vancouver Island, received a Bachelor of Science degree from the University of Victoria, and a Ph.D. from McMaster University in 1978. He subsequently did research and taught at the University of Washington, the University of Waterloo, and California State University, Long Beach.

He is currently Professor of Mathematics at The Pennsylvania State University, Abington College. His research field is topology.

Saleem Watson received his Bachelor of Science degree from Andrews University in Michigan. He did graduate studies at Dalhousie University and McMaster University, where he received his Ph.D. in 1978. He subsequently did research at the Mathematics Institute of the University of Warsaw in Poland. He also taught at The Pennsylvania State University.

He is currently Professor of Mathematics at California State University, Long Beach. His research field is functional analysis.

The authors have also published *College Algebra, Fourth Edition* (Brooks/Cole, 2004), *Algebra and Trigonometry, Second Edition* (Brooks/Cole, 2007), and *Trigonometry* (Brooks/Cole, 2003).

**To our students,
from whom we have learned so much.**

Contents

3 ▸ Polynomial and Rational Functions   248

4 ▸ Exponential and Logarithmic Functions   326

Contents

Preface

What do students really need to know to be prepared for calculus? What tools do instructors really need to assist their students in preparing for calculus? These two questions have motivated the writing of this book.

To be prepared for calculus a student needs not only technical skill but also a clear understanding of concepts. Indeed, *conceptual understanding* and *technical skill* go hand in hand, each reinforcing the other. A student also needs to gain an appreciation for the power and utility of mathematics in *modeling* the real-world. Every feature of this textbook is devoted to fostering these goals.

We are keenly aware that good teaching comes in many different forms, and that each instructor brings unique strengths and imagination to the classroom. Some instructors use *technology* to help students become active learners; others use the *rule of four*, "topics should be presented geometrically, numerically, algebraically, and verbally," to promote conceptual reasoning; some use an expanded emphasis on *applications* to promote an appreciation for mathematics in everyday life; still others use *group learning, extended projects*, or *writing exercises* as a way of encouraging students to explore their own understanding of a given concept; and all present mathematics as a *problem-solving* endeavor. In this book we have included all these methods of teaching precalculus as enhancements to a central core of fundamental skills. These methods are tools to be utilized by instructors and their students to navigate their own course of action in preparing for calculus.

In writing this fifth edition our purpose was to further enhance the utility of the book as an instructional tool. The main change in this edition is an expanded emphasis on modeling and applications: In each section the applications exercises have been expanded and are grouped together under the heading *Applications*, and each chapter (except Chapter 1) now ends with a *Focus on Modeling* section. We have also made some organizational changes, including dividing the chapter on analytic trigonometry into two chapters, each of more manageable size. There are numerous other smaller changes—as we worked through the book we sometimes realized that an additional example was needed, or an explanation could be clarified, or a section could benefit from different types of exercises. Throughout these changes, however, we have retained the overall structure and the main features that have contributed to the success of this book.

Many of the changes in this edition have been drawn from our own experience in teaching, but, more importantly, we have listened carefully to the users of the current edition, including many of our closest colleagues. We are also grateful to the many letters and e-mails we have received from users of this book, instructors as well as students, recommending changes and suggesting additions. Many of these have helped tremendously in making this edition even more user-friendly.

Special Features

EXERCISE SETS   The most important way to foster conceptual understanding and hone technical skill is through the problems that the instructor assigns. To that end we have provided a wide selection of exercises.

- **Exercises**   Each exercise set is carefully graded, progressing from basic conceptual exercises and skill-development problems to more challenging problems requiring synthesis of previously learned material with new concepts.

- **Applications Exercises**   We have included substantial applied problems that we believe will capture the interest of students. These are integrated throughout the text in both examples and exercises. In the exercise sets, applied problems are grouped together under the heading, *Applications*. (See, for example, pages 127, 156, 314, and 451.)

- **Discovery, Writing, and Group Learning**   Each exercise set ends with a block of exercises called *Discovery•Discussion*. These exercises are designed to encourage students to experiment, preferably in groups, with the concepts developed in the section, and then to write out what they have learned, rather than simply look for "the answer." (See, for example, pages 232 and 369.)

A COMPLETE REVIEW CHAPTER   We have included an extensive review chapter primarily as a handy reference for the student to revisit basic concepts in algebra and analytic geometry.

- **Chapter 1**   This is the review chapter; it contains the fundamental concepts a student needs to begin a precalculus course. As much or as little of this chapter can be covered in class as needed, depending on the background of the students.

- **Chapter 1 Test**   The test at the end of Chapter 1 is intended as a diagnostic instrument for determining what parts of this review chapter need to be taught. It also serves to help students gauge exactly what topics they need to review.

FLEXIBLE APPROACH TO TRIGONOMETRY   The trigonometry chapters of this text have been written so that either the right triangle approach or the unit circle approach may be taught first. Putting these two approaches in different chapters, each with its relevant applications, helps clarify the purpose of each approach. The chapters introducing trigonometry are as follows:

- **Chapter 5: Trigonometric Functions of Real Numbers**   This chapter introduces trigonometry through the unit circle approach. This approach emphasizes that the trigonometric functions are functions of real numbers, just like the polynomial and exponential functions with which students are already familiar.

- **Chapter 6: Trigonometric Functions of Angles**   This chapter introduces trigonometry through the right triangle approach. This approach builds on the foundation of a conventional high-school course in trigonometry.

Another way to teach trigonometry is to intertwine the two approaches. Some instructors teach this material in the following order: Sections 5.1, 5.2, 6.1, 6.2, 6.3, 5.3, 5.4, 6.4, 6.5. Our organization makes it easy to do this without obscuring the fact that the two approaches involve distinct representations of the same functions.

GRAPHING CALCULATORS AND COMPUTERS   Calculator and computer technology extends in a powerful way our ability to calculate and visualize mathematics. The availability of graphing calculators makes it not less important, but far *more* important to understand the concepts that underlie what the calculator produces. Accordingly, all our calculator-oriented subsections are preceded by sections in which students must graph or calculate by hand, so that they can understand precisely what the calculator is doing when they later use it to simplify the routine, mechanical part of their work. The graphing calculator sections, subsections, examples, and exercises, all marked with the special symbol 🖩, are optional and may be omitted without loss of continuity. We use the following capabilities of the calculator:

- **Graphing Calculators**   The use of the graphing calculator is integrated throughout the text to graph and analyze functions, families of functions, and sequences, to calculate and graph regression curves, to perform matrix algebra, to graph linear inequalities, and other powerful uses.

- **Simple Programs**   We exploit the programming capabilities of a graphing calculator to simulate real-life situations, to sum series, or to compute the terms of a recursive sequence. (See, for instance, pages 702, 825, and 829.)

FOCUS ON MODELING   The "modeling" theme has been used throughout to unify and clarify the many applications of precalculus. We have made a special effort, in these modeling sections and subsections, to clarify the essential process of translating problems from English into the language of mathematics. (See pages 204 or 647.)

- **Constructing Models**   There are numerous applied problems throughout the book where students are given a model to analyze (see, for instance, page 200). But the material on modeling, where students are required to *construct* mathematical models for themselves, has been organized into clearly defined sections and subsections (see, for example, pages 203, 369, 442, and 848).

- **Focus on Modeling**   Each chapter concludes with a *Focus on Modeling* section. The first such section, after Chapter 2, introduces the basic idea of modeling a real-life situation by fitting lines to data (linear regression). Other sections present ways in which polynomial, exponential, logarithmic, and trigonometric functions, and systems of inequalities can all be used to model familiar phenomena from the sciences and from everyday life (see, for example, pages 320, 386, or 459). Chapter 1 concludes with a section entitled *Focus on Problem Solving*.

DISCOVERY PROJECTS   One way to engage students and make them active learners is to have them work (perhaps in groups) on extended projects that give a feeling of substantial accomplishment when completed. Each chapter contains one or more *Discovery Projects* (see the table of contents); these provide a challenging but accessible set of activities that enable students to explore in greater depth an interesting aspect of the topic they have just learned. (See, for instance, pages 223, 432, or 700.)

MATHEMATICAL VIGNETTES   Throughout the book we make use of the margins to provide historical notes, key insights, or applications of mathematics in the modern world. These serve to enliven the material and show that mathematics is an important, vital activity, and that even at this elementary level it is fundamental to everyday life.

- **Mathematical Vignettes**   These vignettes include biographies of interesting mathematicians and often include a key insight that the mathematician discovered and which is relevant to precalculus. (See, for instance, the vignettes on Viète, page 49; coordinates as addresses, page 88; and radiocarbon dating, page 360.)

- **Mathematics in the Modern World**   This is a series of vignettes that emphasizes the central role of mathematics in current advances in technology and the sciences. (See pages 256, 656, and 746, for example).

CHECK YOUR ANSWER   The *Check Your Answer* feature is used wherever possible to emphasize the importance of looking back to check whether an answer is reasonable. (See, for instance, page 363.)

REVIEW SECTIONS AND CHAPTER TESTS   Each chapter ends with an extensive review section, including a *Chapter Test* designed to help the students gauge their progress. Brief answers to odd-numbered exercises in each section (including the review exercises), and answers to all questions in the Chapter Tests, are given in the back of the book.

The review material in each chapter begins with a *Concept Check*, designed to get the students to think about and explain in their own words the ideas presented in the chapter. These can be used as writing exercises, in a classroom discussion setting, or for personal study.

Major Changes for the Fifth Edition

- More than 20 percent of the exercises are new. New exercises have been chosen to provide more practice with basic concepts, as well as to explore ideas that we do not have space to cover in the discussion and examples in the text itself. Many new applied exercises have been added.

- Each chapter now begins with a *Chapter Overview* that introduces the main themes of the chapter and explains why the material is important.

- Six new *Focus on Modeling* sections have been added, with topics ranging from Mapping the World (Chapter 8) to Traveling and Standing Waves (Chapter 7).

- Five new *Discovery Projects* have been added, with topics ranging from the uses of vectors in sailing (see page 626) to the uses of conics in architecture (see page 771).

- A few more mathematical vignettes have been added (see for example the vignette on splines, page 252, and the one on Maria Agnesi, page 802.)

- We have moved the section on variation from Chapter 2 to Chapter 1, thus focusing Chapter 2 more clearly on the essential concept of a function.

- In Chapter 5, Trigonometric Functions of Real Numbers, we have incorporated the material on harmonic motion as a new section. The *Focus on Modeling* section is now about fitting sinusoidal curves to data.

- In Chapter 7, Analytic Trigonometry, we now include only the material on trigonometric identities and equations. This change was done at the request of users.

- Chapter 8, Polar Coordinates and Vectors, is a new chapter, incorporating material that was previously in other chapters. The topics in this chapter, which also include the polar representation of complex numbers, are united by the theme of using the trigonometric functions to locate the coordinates of a point or describe the components of a vector.

- In Chapter 9, Systems of Equations and Inequalities, we have put the section on graphing of inequalities as the last section, so it now immediately precedes the material on linear programming in the *Focus on Modeling* section.

- Chapter 10, Analytic Geometry, now includes only the conic sections and parametric equations. The material on polar coordinates is in the new Chapter 8.

- In Chapter 11, Sequence and Series, we have expanded the material on recursive sequences by adding a *Focus on Modeling* section on the use of such sequences in modeling real-world phenomena.

Acknowledgments

We thank the following reviewers for their thoughtful and constructive comments.

REVIEWERS FOR THE FOURTH EDITION   Michelle Benedict, Augusta State University; Linda Crawford, Augusta State University; Vivian G. Kostyk, Inver Hills Community College; and Heather C. McGilvray, Seattle University.

REVIEWERS FOR THE FIFTH EDITION   Kenneth Berg, University of Maryland; Elizabeth Bowman, University of Alabama at Huntsville; William Cherry, University of North Texas; Barbara Cortzen, DePaul University; Gerry Fitch, Louisiana State University; Lana Grishchenko, Cal Poly State University, San Luis Obispo; Bryce Jenkins, Cal Poly State University, San Luis Obispo; Margaret Mary Jones, Rutgers University; Victoria Kauffman, University of New Mexico; Sharon Keener, Georgia Perimeter College; YongHee Kim-Park, California State University Long Beach; Mangala Kothari, Rutgers University; Andre Mathurin, Bellarmine College Prep; Donald Robertson, Olympic College; Jude Socrates, Pasadena City College; Enefiok Umana, Georgia Perimeter College; Michele Wallace, Washington State University; and Linda Waymire, Daytona Beach Community College.

We have benefited greatly from the suggestions and comments of our colleagues who have used our books in previous editions. We extend special thanks in this regard to Linda Byun, Bruce Chaderjian, David Gau, Daniel Hernandez, YongHee Kim-Park, Daniel Martinez, David McKay, Robert Mena, Kent Merryfield, Florence Newberger, Viet Ngo, Marilyn Oba, Alan Safer, Angelo Segalla, Robert Valentini, and Derming Wang, from California State University, Long Beach; to Karen Gold, Betsy Gensamer, Cecilia McVoy, Mike McVoy, Samir Ouzomgi, and Ralph Rush, of the Pennsylvania State University, Abington College; to Gloria Dion, of Educational Testing Service, Princeton, New Jersey; to Mark Ashbaugh and Nakhlé Asmar of the University of Missouri, Columbia; to Fred Safier, of the City College of San Francisco; and Steve Edwards, of Southern Polytechnic State University in Marietta, Georgia. We have also received much invaluable advice from our students, especially Devaki Shah and Ellen Newman.

We especially thank Martha Emry, manager of our production service, for her excellent work and her tireless attention to quality and detail. Her energy, devotion, experience, and intelligence were essential components in the creation of this book. We are grateful to Luana Richards, our copy editor, who over the years has expertly shaped the language and style of all of our books. At Matrix Art Services we thank Jade Meyers for his elegant graphics. We thank the team at G & S Book Services for the high quality and consistency brought to the page composition. Our special thanks to Phyllis Panman-Watson for her dedication and care in creating the answer section. At Brooks/Cole our thanks go to the book team: Assistant Editor Stacy Green; Editorial Assistant Katherine Cook; Senior Marketing Manager Karin Sandberg; Marketing Assistant Jennifer Velasquez; Marketing Communications Project Manager Bryan Vann; Senior Project Production Manager Janet Hill; Senior Art Director Vernon Boes; and Technology Project Manager Earl Perry.

We are particularly grateful to our publisher Bob Pirtle for guiding this book through every stage of writing and production. His support and editorial insight when crucial decisions had to be made were invaluable.

For the Instructor

LECTURE PREPARATION

Instructor's Guide 0-534-49300-9

This helpful teaching companion is written by Doug Shaw, author of the Instructor Guides for the Stewart calculus texts. It contains points to stress, suggested time to allot, text discussion topics, core materials for lecture, workshop/discussion suggestions, group work exercises in a form suitable for handout, solutions to group work exercises, and suggested homework problems.

Complete Solutions Manual 0-534-49316-5

This manual provides worked-out solutions to all of the problems in the text.

Solution Builder CD 0-495-10820-0

This CD is an electronic version of the complete solutions manual. It provides instructors with an efficient method for creating solution sets to homework or exams that can then be printed or posted.

PRESENTATION TOOLS

Text Specific DVDs 0-534-49309-2

This set of DVDs is available free upon adoption of the text. Each chapter of the text is broken down into 10- to 20-minute problem-solving lessons that cover each section of the chapter.

JoinIn™ on TurningPoint® 0-495-10754-9

With this book-specific **JoinIn**™ content for electronic response systems, and Microsoft® PowerPoint® slides of your own lecture, you can transform your classroom and assess your students' progress with instant in-class quizzes and polls.

TESTING TOOLS

iLrn™ Assessment

iLrn Assessment is a powerful and fully integrated teaching and course management system that can be used for homework, quizzing, or testing purposes. Easy to use, it offers you complete control when creating assessments; you can draw from the wealth of exercises provided or create your own questions.

A real timesaver, **iLrn Assessment** offers automatic grading of text-specific homework, quizzes, and tests with results flowing directly into the gradebook. The auto-enrollment feature also saves time with course set up as students self-enroll into the course gradebook. A wide range of problem types provides greater variety and more diverse challenges in your tests.

iLrn Assessment provides seamless integration with Blackboard® and WebCT®.

ExamView® Computerized Testing 0-495-01994-1

Create, deliver, and customize tests (both print and online) in minutes with this easy-to-use assessment system. *ExamView is a registered trademark of FSCreations, Inc. Used herein under license.*

Test Bank 0-534-49299-1

The **Test Bank consists of two parts.** Part 1 includes six tests per chapter including 3 final exams. Part 2 of the Test Bank contains test questions broken down by section. Question types are free response and multiple choice.

For the Student

HOMEWORK TOOLS

Student Solutions Manual 0-534-49290-8

The student solutions manual provides worked out solutions to the odd-numbered problems in the text.

iLrn™ Homework and Tutorial Student Version

iLrn Tutorial supports students with active text examples, explanations from the text, step-by-step problem solving help, unlimited practice, video lessons, quizzes, and more.

In addition, a personalized study plan can be generated by course-specific diagnostics built into **iLrn Tutorial**.

iLrn Homework Assigned homework can be done online. Immediate feedback lets the student know where additional study is needed.

vMentor™ provides additional assistance within iLrn via live online tutoring.

LEARNING TOOLS

Interactive Video Skillbuilder CD ROM 0-534-49287-8

Included with the text! The **Interactive Video Skillbuilder CD-ROM** contains hours of video instruction. To help students evaluate their progress, each section contains a 10-question web quiz and each chapter contains a chapter test, with answers to each problem on each test. Also includes **MathCue Tutorial**, dual-platform software that presents and scores problems and tutors students by displaying annotated, step-by-step solutions. Problem sets may be customized as desired.

Study Guide 0-534-49289-4

Contains detailed explanations, worked-out examples, practice problems, and key ideas to master. Each section of the main text has a corresponding section in the **Study Guide**.

ADDITIONAL RESOURCES

Book Companion Website
http://mathematics.brookscole.com
This outstanding site features chapter-by-chapter online tutorial quizzes, a sample final exam, chapter outlines, chapter review, chapter-by-chapter web links, flashcards, and more! Plus, the Brooks/Cole Mathematics Resource Center features historical notes, math news, and career information.

To the Student

This textbook was written for you to use as a guide to mastering precalculus mathematics. Here are some suggestions to help you get the most out of your course.

First of all, you should read the appropriate section of text *before* you attempt your homework problems. Reading a mathematics text is quite different from reading a novel, a newspaper, or even another textbook. You may find that you have to reread a passage several times before you understand it. Pay special attention to the examples, and work them out yourself with pencil and paper as you read. With this kind of preparation you will be able to do your homework much more quickly and with more understanding.

Don't make the mistake of trying to memorize every single rule or fact you may come across. Mathematics doesn't consist simply of memorization. Mathematics is a *problem-solving art*, not just a collection of facts. To master the subject you must solve problems—lots of problems. Do as many of the exercises as you can. Be sure to write your solutions in a logical, step-by-step fashion. Don't give up on a problem if you can't solve it right away. Try to understand the problem more clearly—reread it thoughtfully and relate it to what you have learned from your teacher and from the examples in the text. Struggle with it until you solve it. Once you have done this a few times you will begin to understand what mathematics is really all about.

Answers to the odd-numbered exercises, as well as all the answers to each chapter test, appear at the back of the book. If your answer differs from the one given, don't immediately assume that you are wrong. There may be a calculation that connects the two answers and makes both correct. For example, if you get $1/(\sqrt{2} - 1)$ but the answer given is $1 + \sqrt{2}$, your answer *is* correct, because you can multiply both numerator and denominator of your answer by $\sqrt{2} + 1$ to change it to the given answer.

The symbol 🚫 is used to warn against committing an error. We have placed this symbol in the margin to point out situations where we have found that many of our students make the same mistake.

The Interactive Video Skillbuilder CD-ROM bound into the cover of this book contains video instruction designed to further help you understand the material of this course. The symbol 📼 points to topics for which additional examples and explanations can be found on the CD-ROM.

Calculators and Calculations

Calculators are essential in most mathematics and science subjects. They free us from performing routine tasks, so we can focus more clearly on the concepts we are studying. Calculators are powerful tools but their results need to be interpreted with care. In what follows, we describe the features that a calculator suitable for a precalculus course should have, and we give guidelines for interpreting the results of its calculations.

Scientific and Graphing Calculators

For this course you will need a *scientific* calculator—one that has, as a minimum, the usual arithmetic operations $(+, -, \times, \div)$ as well as exponential, logarithmic, and trigonometric functions (e^x, 10^x, ln, log, sin, cos, tan). In addition, a memory and at least some degree of programmability will be useful.

Your instructor may recommend or require that you purchase a *graphing* calculator. This book has optional subsections and exercises that require the use of a graphing calculator or a computer with graphing software. These special subsections and exercises are indicated by the symbol ⊞. Besides graphing functions, graphing calculators can also be used to find functions that model real-life data, solve equations, perform matrix calculations (which are studied in Chapter 9), and help you perform other mathematical operations. All these uses are discussed in this book.

It is important to realize that, because of limited resolution, a graphing calculator gives only an *approximation* to the graph of a function. It plots only a finite number of points and then connects them to form a *representation* of the graph. In Section 1.9, we give guidelines for using a graphing calculator and interpreting the graphs that it produces.

Calculations and Significant Figures

Most of the applied examples and exercises in this book involve approximate values. For example, one exercise states that the moon has a radius of 1074 miles. This does not mean that the moon's radius is exactly 1074 miles but simply that this is the radius rounded to the nearest mile.

One simple method for specifying the accuracy of a number is to state how many **significant digits** it has. The significant digits in a number are the ones from the first nonzero digit to the last nonzero digit (reading from left to right). Thus, 1074 has four significant digits, 1070 has three, 1100 has two, and 1000 has one significant digit. This rule may sometimes lead to ambiguities. For example, if a distance is 200 km to the nearest kilometer, then the number 200 really has three significant digits, not just one. This ambiguity is avoided if we use scientific notation—that is, if we express the number as a multiple of a power of 10:

$$2.00 \times 10^2$$

When working with approximate values, students often make the mistake of giving a final answer with *more* significant digits than the original data. This is incorrect because you cannot "create" precision by using a calculator. The final result can be no more accurate than the measurements given in the problem. For example, suppose we are told that the two shorter sides of a right triangle are measured to be 1.25 and 2.33 inches long. By the Pythagorean Theorem, we find, using a calculator, that the hypotenuse has length

$$\sqrt{1.25^2 + 2.33^2} \approx 2.644125564 \text{ in.}$$

But since the given lengths were expressed to three significant digits, the answer cannot be any more accurate. We can therefore say only that the hypotenuse is 2.64 in. long, rounding to the nearest hundredth.

In general, the final answer should be expressed with the same accuracy as the *least*-accurate measurement given in the statement of the problem. The following rules make this principle more precise.

Rules for Working with Approximate Data

1. When multiplying or dividing, round off the final result so that it has as many *significant digits* as the given value with the fewest number of significant digits.

2. When adding or subtracting, round off the final result so that it has its last significant digit in the *decimal place* in which the least-accurate given value has its last significant digit.

3. When taking powers or roots, round off the final result so that it has the same number of *significant digits* as the given value.

As an example, suppose that a rectangular table top is measured to be 122.64 in. by 37.3 in. We express its area and perimeter as follows:

Area = length × width = 122.64 × 37.3 ≈ 4570 in²   *Three significant digits*

Perimeter = 2(length + width) = 2(122.64 + 37.3) ≈ 319.9 in.   *Tenths digit*

Note that in the formula for the perimeter, the value 2 is an exact value, not an approximate measurement. It therefore does not affect the accuracy of the final result. In general, if a problem involves only exact values, we may express the final answer with as many significant digits as we wish.

Note also that to make the final result as accurate as possible, *you should wait until the last step to round off your answer*. If necessary, use the memory feature of your calculator to retain the results of intermediate calculations.

Abbreviations

cm	centimeter		**mg**	milligram
dB	decibel		**MHz**	megahertz
F	farad		**mi**	mile
ft	foot		**min**	minute
g	gram		**mL**	milliliter
gal	gallon		**mm**	millimeter
h	hour		**N**	Newton
H	henry		**qt**	quart
Hz	Hertz		**oz**	ounce
in.	inch		**s**	second
J	Joule		**Ω**	ohm
kcal	kilocalorie		**V**	volt
kg	kilogram		**W**	watt
km	kilometer		**yd**	yard
kPa	kilopascal		**yr**	year
L	liter		**°C**	degree Celsius
lb	pound		**°F**	degree Fahrenheit
lm	lumen		**K**	Kelvin
M	mole of solute per liter of solution		$\Rightarrow$	implies
m	meter		$\Leftrightarrow$	is equivalent to

Mathematical Vignettes

MATHEMATICS IN THE MODERN WORLD

PRECALCULUS
Mathematics for Calculus

FIFTH EDITION

1 Fundamentals

Chapter Overview

In this first chapter we review the real numbers, equations, and the coordinate plane. You are probably already familiar with these concepts, but it is helpful to get a fresh look at how these ideas work together to solve problems and model (or describe) real-world situations.

Let's see how all these ideas are used in the following real-life situation: Suppose you get paid $8 an hour at your part-time job. We are interested in how much money you make.

To describe your pay we use *real numbers*. In fact, we use real numbers every day—to describe how tall we are, how much money we have, how cold (or warm) it is, and so on. In algebra, we express properties of the real numbers by using letters to stand for numbers. An important property is the distributive property:

$$A(B + C) = AB + AC$$

To see that this property makes sense, let's consider your pay if you work 6 hours one day and 5 hours the next. Your pay for those two days can be calculated in two different ways: $8(6 + 5)$ or $8 \cdot 6 + 8 \cdot 5$, and both methods give the same answer. This and other properties of the real numbers constitute the rules for working with numbers, or the rules of algebra.

We can also model your pay for any number of hours by a formula. If you work x hours then your pay is y dollars, where y is given by the algebraic formula

$$y = 8x$$

So if you work 10 hours, your pay is $y = 8 \cdot 10 = 80$ dollars.

An *equation* is a sentence written in the language of algebra that expresses a fact about an unknown quantity x. For example, how many hours would you need to work to get paid 60 dollars? To answer this question we need to solve the equation

$$60 = 8x$$

We use the rules of algebra to find x. In this case we divide both sides of the equation by 8, so $x = \frac{60}{8} = 7.5$ hours.

The *coordinate plane* allows us to sketch a graph of an equation in two variables. For example, by graphing the equation $y = 8x$ we can "see" how pay increases with hours worked. We can also solve the equation $60 = 8x$ graphically by finding the value of x at which the graphs of $y = 8x$ and $y = 60$ intersect (see the figure).

In this chapter we will see many examples of how the real numbers, equations, and the coordinate plane all work together to help us solve real-life problems.

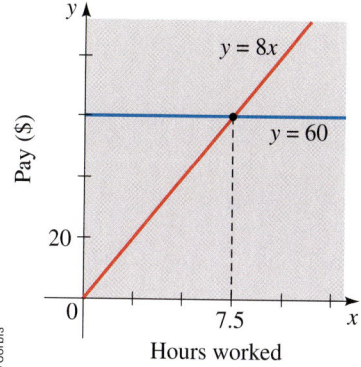

Hours worked

1

1.1   Real Numbers

Let's review the types of numbers that make up the real number system. We start with the **natural numbers**:

$$1, 2, 3, 4, \ldots$$

The **integers** consist of the natural numbers together with their negatives and 0:

$$\ldots, -3, -2, -1, 0, 1, 2, 3, 4, \ldots$$

We construct the **rational numbers** by taking ratios of integers. Thus, any rational number r can be expressed as

$$r = \frac{m}{n}$$

where m and n are integers and $n \neq 0$. Examples are:

$$\frac{1}{2} \qquad -\frac{3}{7} \qquad 46 = \frac{46}{1} \qquad 0.17 = \frac{17}{100}$$

(Recall that division by 0 is always ruled out, so expressions like $\frac{3}{0}$ and $\frac{0}{0}$ are undefined.) There are also real numbers, such as $\sqrt{2}$, that cannot be expressed as a ratio of integers and are therefore called **irrational numbers**. It can be shown, with varying degrees of difficulty, that these numbers are also irrational:

$$\sqrt{3} \qquad \sqrt{5} \qquad \sqrt[3]{2} \qquad \pi \qquad \frac{3}{\pi^2}$$

The set of all real numbers is usually denoted by the symbol $\mathbb{R}$. When we use the word *number* without qualification, we will mean "real number." Figure 1 is a diagram of the types of real numbers that we work with in this book.

The different types of real numbers were invented to meet specific needs. For example, natural numbers are needed for counting, negative numbers for describing debt or below-zero temperatures, rational numbers for concepts like "half a gallon of milk," and irrational numbers for measuring certain distances, like the diagonal of a square.

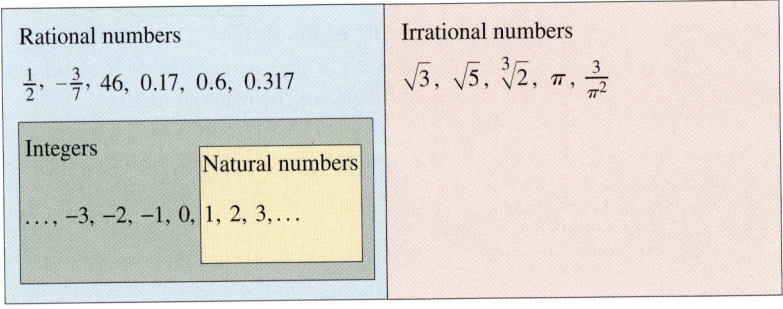

Figure 1
The real number system

Every real number has a decimal representation. If the number is rational, then its corresponding decimal is repeating. For example,

$$\frac{1}{2} = 0.5000\ldots = 0.5\overline{0} \qquad\qquad \frac{2}{3} = 0.66666\ldots = 0.\overline{6}$$

$$\frac{157}{495} = 0.3171717\ldots = 0.3\overline{17} \qquad\qquad \frac{9}{7} = 1.285714285714\ldots = 1.\overline{285714}$$

(The bar indicates that the sequence of digits repeats forever.) If the number is irrational, the decimal representation is nonrepeating:

$$\sqrt{2} = 1.414213562373095\ldots \qquad\qquad \pi = 3.141592653589793\ldots$$

A repeating decimal such as

$$x = 3.5474747\ldots$$

is a rational number. To convert it to a ratio of two integers, we write

$$1000x = 3547.47474747\ldots$$
$$\underline{10x = 35.47474747\ldots}$$
$$990x = 3512.0$$

Thus, $x = \frac{3512}{990}$. (The idea is to multiply x by appropriate powers of 10, and then subtract to eliminate the repeating part.)

If we stop the decimal expansion of any number at a certain place, we get an approximation to the number. For instance, we can write

$$\pi \approx 3.14159265$$

where the symbol $\approx$ is read "is approximately equal to." The more decimal places we retain, the better our approximation.

Properties of Real Numbers

We all know that $2 + 3 = 3 + 2$ and $5 + 7 = 7 + 5$ and $513 + 87 = 87 + 513$, and so on. In algebra, we express all these (infinitely many) facts by writing

$$a + b = b + a$$

where a and b stand for any two numbers. In other words, "$a + b = b + a$" is a concise way of saying that "when we add two numbers, the order of addition doesn't matter." This fact is called the *Commutative Property* for addition. From our experience with numbers we know that the properties in the following box are also valid.

Properties of Real Numbers

Property	Example	Description
Commutative Properties		
$a + b = b + a$	$7 + 3 = 3 + 7$	When we add two numbers, order doesn't matter.
$ab = ba$	$3 \cdot 5 = 5 \cdot 3$	When we multiply two numbers, order doesn't matter.
Associative Properties		
$(a + b) + c = a + (b + c)$	$(2 + 4) + 7 = 2 + (4 + 7)$	When we add three numbers, it doesn't matter which two we add first.
$(ab)c = a(bc)$	$(3 \cdot 7) \cdot 5 = 3 \cdot (7 \cdot 5)$	When we multiply three numbers, it doesn't matter which two we multiply first.
Distributive Property		
$a(b + c) = ab + ac$	$2 \cdot (3 + 5) = 2 \cdot 3 + 2 \cdot 5$	When we multiply a number by a sum of two numbers, we get the same result as multiplying the number by each of the terms and then adding the results.
$(b + c)a = ab + ac$	$(3 + 5) \cdot 2 = 2 \cdot 3 + 2 \cdot 5$	

The Distributive Property applies whenever we multiply a number by a sum. Figure 2 explains why this property works for the case in which all the numbers are positive integers, but the property is true for any real numbers a, b, and c.

The Distributive Property is crucial because it describes the way addition and multiplication interact with each other.

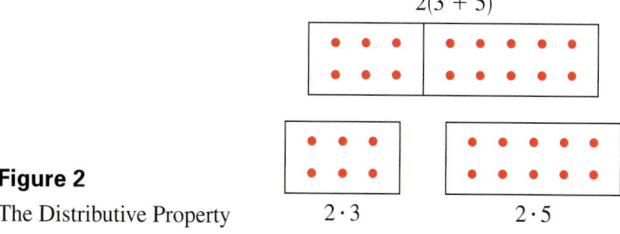

Figure 2

The Distributive Property

Example 1   Using the Distributive Property

(a) $2(x + 3) = 2 \cdot x + 2 \cdot 3$ Distributive Property

$\qquad\qquad = 2x + 6$ Simplify

(b) $(a + b)(x + y) = (a + b)x + (a + b)y$ Distributive Property

$\qquad\qquad\qquad = (ax + bx) + (ay + by)$ Distributive Property

$\qquad\qquad\qquad = ax + bx + ay + by$ Associative Property of Addition

In the last step we removed the parentheses because, according to the Associative Property, the order of addition doesn't matter.   ■

⊘ Don't assume that $-a$ is a negative number. Whether $-a$ is negative or positive depends on the value of a. For example, if $a = 5$, then $-a = -5$, a negative number, but if $a = -5$, then $-a = -(-5) = 5$ (Property 2), a positive number.

The number 0 is special for addition; it is called the **additive identity** because $a + 0 = a$ for any real number a. Every real number a has a **negative**, $-a$, that satisfies $a + (-a) = 0$. **Subtraction** is the operation that undoes addition; to subtract a number from another, we simply add the negative of that number. By definition

$$a - b = a + (-b)$$

To combine real numbers involving negatives, we use the following properties.

Properties of Negatives

Property	Example
1. $(-1)a = -a$	$(-1)5 = -5$
2. $-(-a) = a$	$-(-5) = 5$
3. $(-a)b = a(-b) = -(ab)$	$(-5)7 = 5(-7) = -(5 \cdot 7)$
4. $(-a)(-b) = ab$	$(-4)(-3) = 4 \cdot 3$
5. $-(a + b) = -a - b$	$-(3 + 5) = -3 - 5$
6. $-(a - b) = b - a$	$-(5 - 8) = 8 - 5$

Property 6 states the intuitive fact that $a - b$ and $b - a$ are negatives of each other. Property 5 is often used with more than two terms:

$$-(a + b + c) = -a - b - c$$

Example 2   Using Properties of Negatives

Let x, y, and z be real numbers.

(a) $-(x + 2) = -x - 2$ Property 5: $-(a + b) = -a - b$

(b) $-(x + y - z) = -x - y - (-z)$ Property 5: $-(a + b) = -a - b$

$\qquad\qquad\qquad = -x - y + z$ Property 2: $-(-a) = a$ ■

The number 1 is special for multiplication; it is called the **multiplicative identity** because $a \cdot 1 = a$ for any real number a. Every nonzero real number a has an **inverse**, $1/a$, that satisfies $a \cdot (1/a) = 1$. **Division** is the operation that undoes multiplication; to divide by a number, we multiply by the inverse of that number. If $b \neq 0$, then, by definition,

$$a \div b = a \cdot \frac{1}{b}$$

We write $a \cdot (1/b)$ as simply a/b. We refer to a/b as the **quotient** of a and b or as the **fraction** a over b; a is the **numerator** and b is the **denominator** (or **divisor**). To combine real numbers using the operation of division, we use the following properties.

Properties of Fractions

Property	Example	Description
1. $\dfrac{a}{b} \cdot \dfrac{c}{d} = \dfrac{ac}{bd}$	$\dfrac{2}{3} \cdot \dfrac{5}{7} = \dfrac{2 \cdot 5}{3 \cdot 7} = \dfrac{10}{21}$	When **multiplying fractions**, multiply numerators and denominators.
2. $\dfrac{a}{b} \div \dfrac{c}{d} = \dfrac{a}{b} \cdot \dfrac{d}{c}$	$\dfrac{2}{3} \div \dfrac{5}{7} = \dfrac{2}{3} \cdot \dfrac{7}{5} = \dfrac{14}{15}$	When **dividing fractions**, invert the divisor and multiply.
3. $\dfrac{a}{c} + \dfrac{b}{c} = \dfrac{a + b}{c}$	$\dfrac{2}{5} + \dfrac{7}{5} = \dfrac{2 + 7}{5} = \dfrac{9}{5}$	When **adding fractions** with the **same denominator**, add the numerators.
4. $\dfrac{a}{b} + \dfrac{c}{d} = \dfrac{ad + bc}{bd}$	$\dfrac{2}{5} + \dfrac{3}{7} = \dfrac{2 \cdot 7 + 3 \cdot 5}{35} = \dfrac{29}{35}$	When **adding fractions** with **different denominators**, find a common denominator. Then add the numerators.
5. $\dfrac{ac}{bc} = \dfrac{a}{b}$	$\dfrac{2 \cdot 5}{3 \cdot 5} = \dfrac{2}{3}$	**Cancel** numbers that are **common factors** in numerator and denominator.
6. If $\dfrac{a}{b} = \dfrac{c}{d}$, then $ad = bc$	$\dfrac{2}{3} = \dfrac{6}{9}$, so $2 \cdot 9 = 3 \cdot 6$	**Cross multiply.**

When adding fractions with different denominators, we don't usually use Property 4. Instead we rewrite the fractions so that they have the smallest possible common denominator (often smaller than the product of the denominators), and then we use Property 3. This denominator is the **L**east **C**ommon **D**enominator (LCD) described in the next example.

Example 3   Using the LCD to Add Fractions

Evaluate: $\dfrac{5}{36} + \dfrac{7}{120}$

Solution   Factoring each denominator into prime factors gives

$$36 = 2^2 \cdot 3^2 \qquad \text{and} \qquad 120 = 2^3 \cdot 3 \cdot 5$$

We find the least common denominator (LCD) by forming the product of all the factors that occur in these factorizations, using the highest power of each factor.

Thus, the LCD is $2^3 \cdot 3^2 \cdot 5 = 360$. So

$$\frac{5}{36} + \frac{7}{120} = \frac{5 \cdot 10}{36 \cdot 10} + \frac{7 \cdot 3}{120 \cdot 3} \qquad \textcolor{teal}{\textit{Use common denominator}}$$

$$= \frac{50}{360} + \frac{21}{360} = \frac{71}{360} \qquad \textcolor{teal}{\textit{Property 3: Adding fractions with the same denominator}} \qquad \blacksquare$$

The Real Line

The real numbers can be represented by points on a line, as shown in Figure 3. The positive direction (toward the right) is indicated by an arrow. We choose an arbitrary reference point O, called the **origin**, which corresponds to the real number 0. Given any convenient unit of measurement, each positive number x is represented by the point on the line a distance of x units to the right of the origin, and each negative number $-x$ is represented by the point x units to the left of the origin. The number associated with the point P is called the coordinate of P, and the line is then called a **coordinate line**, or a **real number line**, or simply a **real line**. Often we identify the point with its coordinate and think of a number as being a point on the real line.

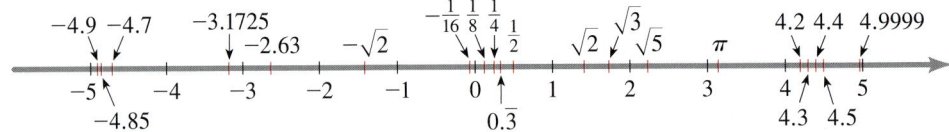

Figure 3   The real line

The real numbers are *ordered*. We say that ***a* is less than *b*** and write $a < b$ if $b - a$ is a positive number. Geometrically, this means that a lies to the left of b on the number line. Equivalently, we can say that ***b* is greater than *a*** and write $b > a$. The symbol $a \le b$ (or $b \ge a$) means that either $a < b$ or $a = b$ and is read "a is less than or equal to b." For instance, the following are true inequalities (see Figure 4):

$$7 < 7.4 < 7.5 \qquad -\pi < -3 \qquad \sqrt{2} < 2 \qquad 2 \le 2$$

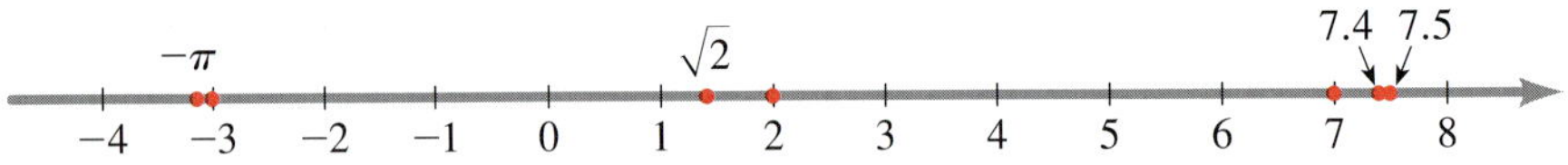

Figure 4

Sets and Intervals

A **set** is a collection of objects, and these objects are called the **elements** of the set. If S is a set, the notation $a \in S$ means that a is an element of S, and $b \notin S$ means that b is not an element of S. For example, if Z represents the set of integers, then $-3 \in Z$ but $\pi \notin Z$.

Some sets can be described by listing their elements within braces. For instance, the set A that consists of all positive integers less than 7 can be written as

$$A = \{1, 2, 3, 4, 5, 6\}$$

We could also write A in **set-builder notation** as

$$A = \{x \mid x \text{ is an integer and } 0 < x < 7\}$$

which is read "A is the set of all x such that x is an integer and $0 < x < 7$."

If S and T are sets, then their **union** $S \cup T$ is the set that consists of all elements that are in S *or* T (or in both). The **intersection** of S and T is the set $S \cap T$ consisting of all elements that are in both S *and* T. In other words, $S \cap T$ is the common part of S and T. The **empty set**, denoted by $\varnothing$, is the set that contains no element.

Example 4 Union and Intersection of Sets

If $S = \{1, 2, 3, 4, 5\}$, $T = \{4, 5, 6, 7\}$, and $V = \{6, 7, 8\}$, find the sets $S \cup T$, $S \cap T$, and $S \cap V$.

Solution

$$S \cup T = \{1, 2, 3, 4, 5, 6, 7\} \qquad \text{\color{blue}All elements in S or T}$$

$$S \cap T = \{4, 5\} \qquad \text{\color{blue}Elements common to both S and T}$$

$$S \cap V = \varnothing \qquad \text{\color{blue}S and V have no element in common} \qquad ■$$

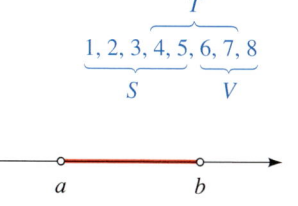

Certain sets of real numbers, called **intervals**, occur frequently in calculus and correspond geometrically to line segments. If $a < b$, then the **open interval** from a to b consists of all numbers between a and b and is denoted (a, b). The **closed interval** from a to b includes the endpoints and is denoted $[a, b]$. Using set-builder notation, we can write

$$(a, b) = \{x \mid a < x < b\} \qquad [a, b] = \{x \mid a \le x \le b\}$$

Figure 5

The open interval (a, b)

Figure 6

The closed interval $[a, b]$

Note that parentheses ( ) in the interval notation and open circles on the graph in Figure 5 indicate that endpoints are *excluded* from the interval, whereas square brackets [ ] and solid circles in Figure 6 indicate that the endpoints are *included*. Intervals may also include one endpoint but not the other, or they may extend infinitely far in one direction or both. The following table lists the possible types of intervals.

The symbol ∞ ("infinity") does not stand for a number. The notation (a, ∞), for instance, simply indicates that the interval has no endpoint on the right but extends infinitely far in the positive direction.

Notation	Set description	Graph
(a, b)	$\{x \mid a < x < b\}$	 a b (open, open)
$[a, b]$	$\{x \mid a \le x \le b\}$	 a b (closed, closed)
$[a, b)$	$\{x \mid a \le x < b\}$	 a b (closed, open)
$(a, b]$	$\{x \mid a < x \le b\}$	 a b (open, closed)
(a, ∞)	$\{x \mid a < x\}$	 a (open)
$[a, \infty)$	$\{x \mid a \le x\}$	 a (closed)
$(-\infty, b)$	$\{x \mid x < b\}$	 b (open)
$(-\infty, b]$	$\{x \mid x \le b\}$	b (closed)
$(-\infty, \infty)$	$\mathbb{R}$ (set of all real numbers)	

No Smallest or Largest Number in an Open Interval

Any interval contains infinitely many numbers—every point on the graph of an interval corresponds to a real number. In the closed interval $[0, 1]$, the smallest number is 0 and the largest is 1, but the open interval $(0, 1)$ contains no smallest or largest number. To see this, note that 0.01 is close to zero, but 0.001 is closer, 0.0001 closer yet, and so on. So we can always find a number in the interval $(0, 1)$ closer to zero than any given number. Since 0 itself is not in the interval, the interval contains no smallest number. Similarly, 0.99 is close to 1, but 0.999 is closer, 0.9999 closer yet, and so on. Since 1 itself is not in the interval, the interval has no largest number.

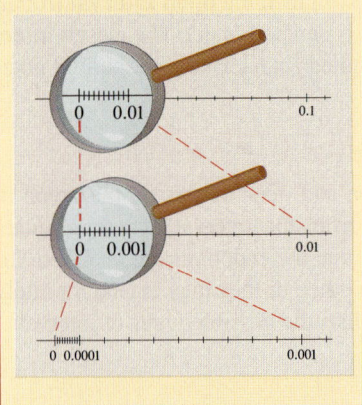

Example 5   Graphing Intervals

Express each interval in terms of inequalities, and then graph the interval.

(a) $[-1, 2) = \{x \mid -1 \le x < 2\}$

(b) $[1.5, 4] = \{x \mid 1.5 \le x \le 4\}$

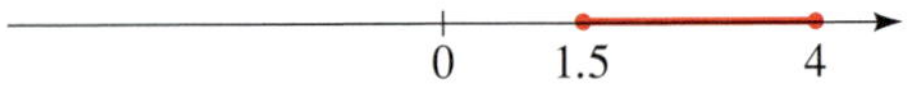

(c) $(-3, \infty) = \{x \mid -3 < x\}$

Example 6   Finding Unions and Intersections of Intervals

Graph each set.

(a) $(1, 3) \cap [2, 7]$ 　　　　(b) $(1, 3) \cup [2, 7]$

Solution

(a) The intersection of two intervals consists of the numbers that are in both intervals. Therefore

$$(1, 3) \cap [2, 7] = \{x \mid 1 < x < 3 \text{ and } 2 \le x \le 7\}$$
$$= \{x \mid 2 \le x < 3\} = [2, 3)$$

This set is illustrated in Figure 7.

(b) The union of two intervals consists of the numbers that are in either one interval or the other (or both). Therefore

$$(1, 3) \cup [2, 7] = \{x \mid 1 < x < 3 \text{ or } 2 \le x \le 7\}$$
$$= \{x \mid 1 < x \le 7\} = (1, 7]$$

This set is illustrated in Figure 8.

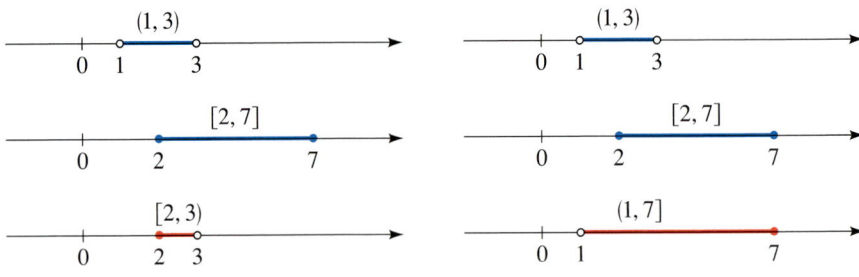

Figure 7
$(1, 3) \cap [2, 7] = [2, 3)$

Figure 8
$(1, 3) \cup [2, 7] = (1, 7]$

Absolute Value and Distance

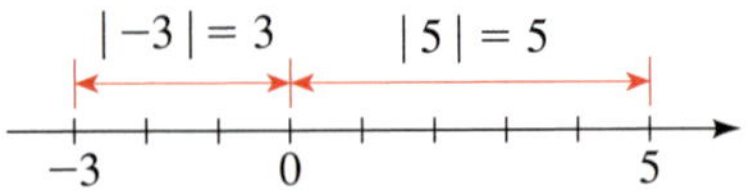

The **absolute value** of a number a, denoted by $|a|$, is the distance from a to 0 on the real number line (see Figure 9). Distance is always positive or zero, so we have $|a| \ge 0$ for every number a. Remembering that $-a$ is positive when a is negative, we have the following definition.

Figure 9

Definition of Absolute Value

If a is a real number, then the **absolute value** of a is

$$|a| = \begin{cases} a & \text{if } a \geq 0 \\ -a & \text{if } a < 0 \end{cases}$$

Example 7   Evaluating Absolute Values of Numbers

(a) $|3| = 3$

(b) $|-3| = -(-3) = 3$

(c) $|0| = 0$

(d) $|3 - \pi| = -(3 - \pi) = \pi - 3$ (since $3 < \pi \;\Rightarrow\; 3 - \pi < 0$)   ∎

When working with absolute values, we use the following properties.

Properties of Absolute Value

Property	Example	Description
1. $|a| \geq 0$	$|-3| = 3 \geq 0$	The absolute value of a number is always positive or zero.
2. $|a| = |-a|$	$|5| = |-5|$	A number and its negative have the same absolute value.
3. $|ab| = |a||b|$	$|-2 \cdot 5| = |-2||5|$	The absolute value of a product is the product of the absolute values.
4. $\left|\dfrac{a}{b}\right| = \dfrac{|a|}{|b|}$	$\left|\dfrac{12}{-3}\right| = \dfrac{|12|}{|-3|}$	The absolute value of a quotient is the quotient of the absolute values.

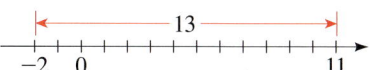

Figure 10

What is the distance on the real line between the numbers -2 and 11? From Figure 10 we see that the distance is 13. We arrive at this by finding either $|11 - (-2)| = 13$ or $|(-2) - 11| = 13$. From this observation we make the following definition (see Figure 11).

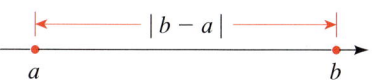

Figure 11

Length of a line segment $= |b - a|$

Distance between Points on the Real Line

If a and b are real numbers, then the **distance** between the points a and b on the real line is

$$d(a, b) = |b - a|$$

From Property 6 of negatives it follows that $|b - a| = |a - b|$. This confirms that, as we would expect, the distance from a to b is the same as the distance from b to a.

Example 8   Distance between Points on the Real Line

The distance between the numbers -8 and 2 is

$$d(a, b) = |-8 - 2| = |-10| = 10$$

We can check this calculation geometrically, as shown in Figure 12.

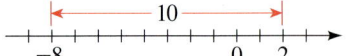

Figure 12

1.1   Exercises

1–2 ■ List the elements of the given set that are

(a) natural numbers

(b) integers

(c) rational numbers

(d) irrational numbers

1. $\{0, -10, 50, \frac{22}{7}, 0.538, \sqrt{7}, 1.2\overline{3}, -\frac{1}{3}, \sqrt[3]{2}\}$

2. $\{1.001, 0.333\ldots, -\pi, -11, 11, \frac{13}{15}, \sqrt{16}, 3.14, \frac{15}{3}\}$

3–10 ■ State the property of real numbers being used.

3. $7 + 10 = 10 + 7$

4. $2(3 + 5) = (3 + 5)2$

5. $(x + 2y) + 3z = x + (2y + 3z)$

6. $2(A + B) = 2A + 2B$

7. $(5x + 1)3 = 15x + 3$

8. $(x + a)(x + b) = (x + a)x + (x + a)b$

9. $2x(3 + y) = (3 + y)2x$

10. $7(a + b + c) = 7(a + b) + 7c$

11–14 ■ Rewrite the expression using the given property of real numbers.

11. Commutative Property of addition, $x + 3 =$ ▢

12. Associative Property of multiplication, $7(3x) =$ ▢

13. Distributive Property, $4(A + B) =$ ▢

14. Distributive Property, $5x + 5y =$ ▢

15–20 ■ Use properties of real numbers to write the expression without parentheses.

15. $3(x + y)$

16. $(a - b)8$

17. $4(2m)$

18. $\frac{4}{3}(-6y)$

19. $-\frac{5}{2}(2x - 4y)$

20. $(3a)(b + c - 2d)$

21–26 ■ Perform the indicated operations.

21. (a) $\frac{3}{10} + \frac{4}{15}$ **(b)** $\frac{1}{4} + \frac{1}{5}$

22. (a) $\frac{2}{3} - \frac{3}{5}$ **(b)** $1 + \frac{5}{8} - \frac{1}{6}$

23. (a) $\frac{2}{3}\left(6 - \frac{3}{2}\right)$ **(b)** $0.25\left(\frac{8}{9} + \frac{1}{2}\right)$

24. (a) $\left(3 + \frac{1}{4}\right)\left(1 - \frac{4}{5}\right)$ **(b)** $\left(\frac{1}{2} - \frac{1}{3}\right)\left(\frac{1}{2} + \frac{1}{3}\right)$

25. (a) $\dfrac{\frac{2}{3} - \frac{2}{3}}{2}$ **(b)** $\dfrac{\frac{1}{12}}{\frac{1}{8} - \frac{1}{9}}$

26. (a) $\dfrac{2 - \frac{3}{4}}{\frac{1}{2} - \frac{1}{3}}$ **(b)** $\dfrac{\frac{2}{5} + \frac{1}{2}}{\frac{1}{10} + \frac{3}{15}}$

27–28 ■ Place the correct symbol ($<$, $>$, or $=$) in the space.

27. (a) 3 ▢ $\frac{7}{2}$ **(b)** -3 ▢ $-\frac{7}{2}$ **(c)** 3.5 ▢ $\frac{7}{2}$

28. (a) $\frac{2}{3}$ ▢ 0.67 **(b)** $\frac{2}{3}$ ▢ -0.67 **(c)** $|0.67|$ ▢ $|-0.67|$

29–32 ■ State whether each inequality is true or false.

29. (a) $-6 < -10$ **(b)** $\sqrt{2} > 1.41$

30. (a) $\dfrac{10}{11} < \dfrac{12}{13}$ **(b)** $-\dfrac{1}{2} < -1$

31. (a) $-\pi > -3$ **(b)** $8 \leq 9$

32. (a) $1.1 > 1.\overline{1}$ **(b)** $8 \leq 8$

33–34 ■ Write each statement in terms of inequalities.

33. (a) x is positive

    (b) t is less than 4

    (c) a is greater than or equal to π

    (d) x is less than $\frac{1}{3}$ and is greater than -5

    (e) The distance from p to 3 is at most 5

34. (a) y is negative

    (b) z is greater than 1

    (c) b is at most 8

(d) w is positive and is less than or equal to 17

(e) y is at least 2 units from π

35–38 ■ Find the indicated set if

$$A = \{1, 2, 3, 4, 5, 6, 7\} \qquad B = \{2, 4, 6, 8\}$$
$$C = \{7, 8, 9, 10\}$$

35. (a) $A \cup B$ **(b)** $A \cap B$

36. (a) $B \cup C$ **(b)** $B \cap C$

37. (a) $A \cup C$ **(b)** $A \cap C$

38. (a) $A \cup B \cup C$ **(b)** $A \cap B \cap C$

39–40 ■ Find the indicated set if

$$A = \{x \mid x \geq -2\} \qquad B = \{x \mid x < 4\}$$
$$C = \{x \mid -1 < x \leq 5\}$$

39. (a) $B \cup C$ **(b)** $B \cap C$

40. (a) $A \cap C$ **(b)** $A \cap B$

41–46 ■ Express the interval in terms of inequalities, and then graph the interval.

41. $(-3, 0)$ **42.** $(2, 8]$

43. $[2, 8)$ **44.** $\left[-6, -\frac{1}{2}\right]$

45. $[2, \infty)$ **46.** $(-\infty, 1)$

47–52 ■ Express the inequality in interval notation, and then graph the corresponding interval.

47. $x \leq 1$ **48.** $1 \leq x \leq 2$

49. $-2 < x \leq 1$ **50.** $x \geq -5$

51. $x > -1$ **52.** $-5 < x < 2$

53–54 ■ Express each set in interval notation.

53. (a)
$$\xleftarrow{\hspace{1cm}} \underset{-3}{\bullet} \quad \underset{0}{+} \quad \underset{5}{\bullet} \xrightarrow{\hspace{1cm}}$$

     (b)
$$\xleftarrow{\hspace{1cm}} \underset{-3}{\circ} \quad \underset{0}{+} \quad \underset{5}{\bullet} \xrightarrow{\hspace{1cm}}$$

54. (a)
$$\xleftarrow{\hspace{1cm}} \underset{0}{\bullet} \quad \underset{2}{\circ} \xrightarrow{\hspace{1cm}}$$

     (b)
$$\xleftarrow{\hspace{1cm}} \underset{-2}{\circ} \quad \underset{0}{\bullet} \xrightarrow{\hspace{1cm}}$$

55–60 ■ Graph the set.

55. $(-2, 0) \cup (-1, 1)$ **56.** $(-2, 0) \cap (-1, 1)$

57. $[-4, 6] \cap [0, 8)$ **58.** $[-4, 6) \cup [0, 8)$

59. $(-\infty, -4) \cup (4, \infty)$ **60.** $(-\infty, 6] \cap (2, 10)$

61–66 ■ Evaluate each expression.

61. (a) $|100|$ **(b)** $|-73|$

62. (a) $|\sqrt{5} - 5|$ **(b)** $|10 - \pi|$

63. (a) $\big| \, |-6| - |-4| \, \big|$ **(b)** $\dfrac{-1}{|-1|}$

64. (a) $\big| \, 2 - |-12| \, \big|$ **(b)** $-1 - \big| \, 1 - |-1| \, \big|$

65. (a) $|(-2) \cdot 6|$ **(b)** $\left| \left(-\frac{1}{3}\right)(-15) \right|$

66. (a) $\left| \dfrac{-6}{24} \right|$ **(b)** $\left| \dfrac{7 - 12}{12 - 7} \right|$

67–70 ■ Find the distance between the given numbers.

67.
$$\xleftarrow{\hspace{0.3cm}}\underset{-3}{+}\ \underset{-2}{\bullet}\ \underset{-1}{+}\ \underset{0}{+}\ \underset{1}{+}\ \underset{2}{\bullet}\ \underset{3}{+}\xrightarrow{\hspace{0.3cm}}$$

68.
$$\xleftarrow{\hspace{0.3cm}}\underset{-3}{+}\ \underset{-2}{\bullet}\ \underset{-1}{+}\ \underset{0}{+}\ \underset{1}{+}\ \underset{2}{\bullet}\ \underset{3}{+}\xrightarrow{\hspace{0.3cm}}$$

69. (a) 2 and 17

     (b) -3 and 21

     (c) $\frac{11}{8}$ and $-\frac{3}{10}$

70. (a) $\frac{7}{15}$ and $-\frac{1}{21}$

     (b) -38 and -57

     (c) -2.6 and -1.8

71–72 ■ Express each repeating decimal as a fraction. (See the margin note on page 2.)

71. (a) $0.\overline{7}$ **(b)** $0.2\overline{8}$ **(c)** $0.\overline{57}$

72. (a) $5.\overline{23}$ **(b)** $1.3\overline{7}$ **(c)** $2.1\overline{35}$

Applications

73. Area of a Garden   Mary's backyard vegetable garden measures 20 ft by 30 ft, so its area is $20 \times 30 = 600 \text{ ft}^2$. She decides to make it longer, as shown in the figure, so that the area increases to $A = 20(30 + x)$. Which property of real numbers tells us that the new area can also be written $A = 600 + 20x$?

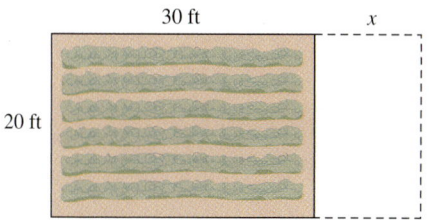

74. Temperature Variation   The bar graph shows the daily high temperatures for Omak, Washington, and Geneseo, New York, during a certain week in June. Let T_O represent the temperature in Omak and T_G the temperature in Geneseo. Calculate $T_O - T_G$ and $|T_O - T_G|$ for each day shown.

Which of these two values gives more information?

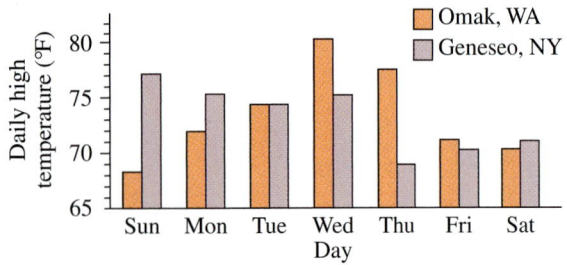

75. Mailing a Package   The post office will only accept packages for which the length plus the "girth" (distance around) is no more than 108 inches. Thus, for the package in the figure, we must have

$$L + 2(x + y) \le 108$$

(a) Will the post office accept a package that is 6 in. wide, 8 in. deep, and 5 ft long? What about a package that measures 2 ft by 2 ft by 4 ft?

(b) What is the greatest acceptable length for a package that has a square base measuring 9 in. by 9 in?

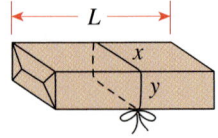

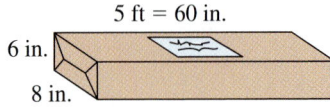

5 ft = 60 in.

6 in.

8 in.

Discovery • Discussion

76. Signs of Numbers   Let a, b, and c be real numbers such that $a > 0$, $b < 0$, and $c < 0$. Find the sign of each expression.

(a) $-a$ (b) $-b$ (c) bc

(d) $a - b$ (e) $c - a$ (f) $a + bc$

(g) $ab + ac$ (h) $-abc$ (i) ab^2

77. Sums and Products of Rational and Irrational Numbers   Explain why the sum, the difference, and the product of two rational numbers are rational numbers. Is the product of two irrational numbers necessarily irrational? What about the sum?

78. Combining Rational Numbers with Irrational Numbers   Is $\frac{1}{2} + \sqrt{2}$ rational or irrational? Is $\frac{1}{2} \cdot \sqrt{2}$ rational or irrational? In general, what can you say about the sum of a rational and an irrational number? What about the product?

79. Limiting Behavior of Reciprocals   Complete the tables. What happens to the size of the fraction $1/x$ as x gets large? As x gets small?

x	$1/x$
1	
2	
10	
100	
1000	

x	$1/x$
1.0	
0.5	
0.1	
0.01	
0.001	

80. Irrational Numbers and Geometry   Using the following figure, explain how to locate the point $\sqrt{2}$ on a number line. Can you locate $\sqrt{5}$ by a similar method? What about $\sqrt{6}$? List some other irrational numbers that can be located this way.

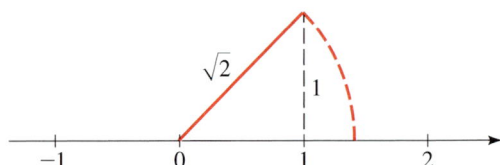

81. Commutative and Noncommutative Operations   We have seen that addition and multiplication are both commutative operations.

(a) Is subtraction commutative?

(b) Is division of nonzero real numbers commutative?

1.2   Exponents and Radicals

In this section we give meaning to expressions such as $a^{m/n}$ in which the exponent m/n is a rational number. To do this, we need to recall some facts about integer exponents, radicals, and nth roots.

Integer Exponents

A product of identical numbers is usually written in exponential notation. For example, $5 \cdot 5 \cdot 5$ is written as 5^3. In general, we have the following definition.

Exponential Notation

If a is any real number and n is a positive integer, then the **nth power** of a is

$$a^n = \underbrace{a \cdot a \cdot \cdots \cdot a}_{n \text{ factors}}$$

The number a is called the **base** and n is called the **exponent**.

Example 1   Exponential Notation

(a) $\left(\frac{1}{2}\right)^5 = \left(\frac{1}{2}\right)\left(\frac{1}{2}\right)\left(\frac{1}{2}\right)\left(\frac{1}{2}\right)\left(\frac{1}{2}\right) = \frac{1}{32}$

(b) $(-3)^4 = (-3) \cdot (-3) \cdot (-3) \cdot (-3) = 81$

(c) $-3^4 = -(3 \cdot 3 \cdot 3 \cdot 3) = -81$ ∎

> ⊘ Note the distinction between $(-3)^4$ and -3^4. In $(-3)^4$ the exponent applies to -3, but in -3^4 the exponent applies only to 3.

We can state several useful rules for working with exponential notation. To discover the rule for multiplication, we multiply 5^4 by 5^2:

$$5^4 \cdot 5^2 = \underbrace{(5 \cdot 5 \cdot 5 \cdot 5)}_{4 \text{ factors}}\underbrace{(5 \cdot 5)}_{2 \text{ factors}} = \underbrace{5 \cdot 5 \cdot 5 \cdot 5 \cdot 5 \cdot 5}_{6 \text{ factors}} = 5^6 = 5^{4+2}$$

It appears that *to multiply two powers of the same base, we add their exponents*. In general, for any real number a and any positive integers m and n, we have

$$a^m a^n = \underbrace{(a \cdot a \cdot \cdots \cdot a)}_{m \text{ factors}}\underbrace{(a \cdot a \cdot \cdots \cdot a)}_{n \text{ factors}} = \underbrace{a \cdot a \cdot a \cdot \cdots \cdot a}_{m + n \text{ factors}} = a^{m+n}$$

Thus $a^m a^n = a^{m+n}$.

We would like this rule to be true even when m and n are 0 or negative integers. For instance, we must have

$$2^0 \cdot 2^3 = 2^{0+3} = 2^3$$

But this can happen only if $2^0 = 1$. Likewise, we want to have

$$5^4 \cdot 5^{-4} = 5^{4+(-4)} = 5^{4-4} = 5^0 = 1$$

and this will be true if $5^{-4} = 1/5^4$. These observations lead to the following definition.

Zero and Negative Exponents

If $a \neq 0$ is any real number and n is a positive integer, then

$$a^0 = 1 \qquad \text{and} \qquad a^{-n} = \frac{1}{a^n}$$

Example 2   Zero and Negative Exponents

(a) $\left(\frac{4}{7}\right)^0 = 1$

(b) $x^{-1} = \frac{1}{x^1} = \frac{1}{x}$

(c) $(-2)^{-3} = \frac{1}{(-2)^3} = \frac{1}{-8} = -\frac{1}{8}$ ∎

Familiarity with the following rules is essential for our work with exponents and bases. In the table the bases a and b are real numbers, and the exponents m and n are integers.

Laws of Exponents

Law	Example	Description
1. $a^m a^n = a^{m+n}$	$3^2 \cdot 3^5 = 3^{2+5} = 3^7$	To multiply two powers of the same number, add the exponents.
2. $\dfrac{a^m}{a^n} = a^{m-n}$	$\dfrac{3^5}{3^2} = 3^{5-2} = 3^3$	To divide two powers of the same number, subtract the exponents.
3. $(a^m)^n = a^{mn}$	$(3^2)^5 = 3^{2\cdot 5} = 3^{10}$	To raise a power to a new power, multiply the exponents.
4. $(ab)^n = a^n b^n$	$(3 \cdot 4)^2 = 3^2 \cdot 4^2$	To raise a product to a power, raise each factor to the power.
5. $\left(\dfrac{a}{b}\right)^n = \dfrac{a^n}{b^n}$	$\left(\dfrac{3}{4}\right)^2 = \dfrac{3^2}{4^2}$	To raise a quotient to a power, raise both numerator and denominator to the power.

■ **Proof of Law 3**   If m and n are positive integers, we have

$$(a^m)^n = \underbrace{(a \cdot a \cdots \cdot a)}_{m \text{ factors}}{}^{n}$$

$$= \underbrace{\underbrace{(a \cdot a \cdots \cdot a)}_{m \text{ factors}}\underbrace{(a \cdot a \cdots \cdot a)}_{m \text{ factors}} \cdots \underbrace{(a \cdot a \cdots \cdot a)}_{m \text{ factors}}}_{n \text{ groups of factors}}$$

$$= \underbrace{a \cdot a \cdots \cdot a}_{mn \text{ factors}} = a^{mn}$$

The cases for which $m \le 0$ or $n \le 0$ can be proved using the definition of negative exponents.   ■

■ **Proof of Law 4**   If n is a positive integer, we have

$$(ab)^n = \underbrace{(ab)(ab) \cdots (ab)}_{n \text{ factors}} = \underbrace{(a \cdot a \cdots \cdot a)}_{n \text{ factors}} \cdot \underbrace{(b \cdot b \cdots \cdot b)}_{n \text{ factors}} = a^n b^n$$

Here we have used the Commutative and Associative Properties repeatedly. If $n \le 0$, Law 4 can be proved using the definition of negative exponents.   ■

You are asked to prove Laws 2 and 5 in Exercise 88.

Example 3   Using Laws of Exponents

(a) $x^4 x^7 = x^{4+7} = x^{11}$ Law 1: $a^m a^n = a^{m+n}$

(b) $y^4 y^{-7} = y^{4-7} = y^{-3} = \dfrac{1}{y^3}$ Law 1: $a^m a^n = a^{m+n}$

(c) $\dfrac{c^9}{c^5} = c^{9-5} = c^4$ Law 2: $a^m / a^n = a^{m-n}$

(d) $(b^4)^5 = b^{4\cdot5} = b^{20}$ Law 3: $(a^m)^n = a^{mn}$

(e) $(3x)^3 = 3^3x^3 = 27x^3$ Law 4: $(ab)^n = a^n b^n$

(f) $\left(\dfrac{x}{2}\right)^5 = \dfrac{x^5}{2^5} = \dfrac{x^5}{32}$ Law 5: $(a/b)^n = a^n/b^n$ ■

Example 4   Simplifying Expressions with Exponents

Simplify:

(a) $(2a^3b^2)(3ab^4)^3$ (b) $\left(\dfrac{x}{y}\right)^3\left(\dfrac{y^2x}{z}\right)^4$

Solution

(a) $(2a^3b^2)(3ab^4)^3 = (2a^3b^2)[3^3a^3(b^4)^3]$ Law 4: $(ab)^n = a^n b^n$

$= (2a^3b^2)(27a^3b^{12})$ Law 3: $(a^m)^n = a^{mn}$

$= (2)(27)a^3a^3b^2b^{12}$ Group factors with the same base

$= 54a^6b^{14}$ Law 1: $a^m a^n = a^{m+n}$

(b) $\left(\dfrac{x}{y}\right)^3\left(\dfrac{y^2x}{z}\right)^4 = \dfrac{x^3}{y^3}\dfrac{(y^2)^4x^4}{z^4}$ Laws 5 and 4

$= \dfrac{x^3}{y^3}\dfrac{y^8x^4}{z^4}$ Law 3

$= (x^3x^4)\left(\dfrac{y^8}{y^3}\right)\dfrac{1}{z^4}$ Group factors with the same base

$= \dfrac{x^7y^5}{z^4}$ Laws 1 and 2   ■

When simplifying an expression, you will find that many different methods will lead to the same result; you should feel free to use any of the rules of exponents to arrive at your own method. We now give two additional laws that are useful in simplifying expressions with negative exponents.

Laws of Exponents

Law	Example	Description
6. $\left(\dfrac{a}{b}\right)^{-n} = \left(\dfrac{b}{a}\right)^{n}$	$\left(\dfrac{3}{4}\right)^{-2} = \left(\dfrac{4}{3}\right)^{2}$	To raise a fraction to a negative power, invert the fraction and change the sign of the exponent.
7. $\dfrac{a^{-n}}{b^{-m}} = \dfrac{b^m}{a^n}$	$\dfrac{3^{-2}}{4^{-5}} = \dfrac{4^5}{3^2}$	To move a number raised to a power from numerator to denominator or from denominator to numerator, change the sign of the exponent.

■ **Proof of Law 7**   Using the definition of negative exponents and then Property 2 of fractions (page 5), we have

$$\frac{a^{-n}}{b^{-m}} = \frac{1/a^n}{1/b^m} = \frac{1}{a^n}\cdot\frac{b^m}{1} = \frac{b^m}{a^n}$$ ■

You are asked to prove Law 6 in Exercise 88.

Example 5 Simplifying Expressions with Negative Exponents

Eliminate negative exponents and simplify each expression.

(a) $\dfrac{6st^{-4}}{2s^{-2}t^2}$ (b) $\left(\dfrac{y}{3z^3}\right)^{-2}$

Solution

(a) We use Law 7, which allows us to move a number raised to a power from the numerator to the denominator (or vice versa) by changing the sign of the exponent.

> t^{-4} moves to denominator and becomes t^4.

$$\frac{6st^{-4}}{2s^{-2}t^2} = \frac{6ss^2}{2t^2t^4} \qquad \text{Law 7}$$

> s^{-2} moves to numerator and becomes s^2.

$$= \frac{3s^3}{t^6} \qquad \text{Law 1}$$

(b) We use Law 6, which allows us to change the sign of the exponent of a fraction by inverting the fraction.

$$\left(\frac{y}{3z^3}\right)^{-2} = \left(\frac{3z^3}{y}\right)^2 \qquad \text{Law 6}$$

$$= \frac{9z^6}{y^2} \qquad \text{Laws 5 and 4}$$

Scientific Notation

Exponential notation is used by scientists as a compact way of writing very large numbers and very small numbers. For example, the nearest star beyond the sun, Proxima Centauri, is approximately 40,000,000,000,000 km away. The mass of a hydrogen atom is about 0.00000000000000000000000166 g. Such numbers are difficult to read and to write, so scientists usually express them in *scientific notation*.

Scientific Notation

A positive number x is said to be written in **scientific notation** if it is expressed as follows:

$$x = a \times 10^n \qquad \text{where } 1 \le a < 10 \text{ and } n \text{ is an integer}$$

For instance, when we state that the distance to the star Proxima Centauri is 4×10^{13} km, the positive exponent 13 indicates that the decimal point should be moved 13 places to the *right*:

$$4 \times 10^{13} = 40,000,000,000,000$$

> Move decimal point 13 places to the right.

When we state that the mass of a hydrogen atom is 1.66×10^{-24} g, the exponent -24 indicates that the decimal point should be moved 24 places to the *left*:

$$1.66 \times 10^{-24} = 0.00000000000000000000000166$$

Move decimal point 24 places to the left.

Example 6   Writing Numbers in Scientific Notation

(a) $\underbrace{327,900}_{\text{5 places}} = 3.279 \times 10^5$ (b) $\underbrace{0.000627}_{\text{4 places}} = 6.27 \times 10^{-4}$ ∎

To use scientific notation on a calculator, press the key labeled $\boxed{\texttt{EE}}$ or $\boxed{\texttt{EXP}}$ or $\boxed{\texttt{EEX}}$ to enter the exponent. For example, to enter the number 3.629×10^{15} on a TI-83 calculator, we enter

$$3.629 \; \boxed{\texttt{2ND}} \; \boxed{\texttt{EE}} \; 15$$

and the display reads

$$\texttt{3.629E15}$$

Scientific notation is often used on a calculator to display a very large or very small number. For instance, if we use a calculator to square the number 1,111,111, the display panel may show (depending on the calculator model) the approximation

$$\boxed{\texttt{1.234568   12}} \quad \text{or} \quad \boxed{\texttt{1.23468 E12}}$$

Here the final digits indicate the power of 10, and we interpret the result as

$$1.234568 \times 10^{12}$$

Example 7   Calculating with Scientific Notation

If $a \approx 0.00046$, $b \approx 1.697 \times 10^{22}$, and $c \approx 2.91 \times 10^{-18}$, use a calculator to approximate the quotient ab/c.

Solution   We could enter the data using scientific notation, or we could use laws of exponents as follows:

$$\frac{ab}{c} \approx \frac{(4.6 \times 10^{-4})(1.697 \times 10^{22})}{2.91 \times 10^{-18}}$$

$$= \frac{(4.6)(1.697)}{2.91} \times 10^{-4+22+18}$$

$$\approx 2.7 \times 10^{36}$$

We state the answer correct to two significant figures because the least accurate of the given numbers is stated to two significant figures.   ∎

Radicals

We know what 2^n means whenever n is an integer. To give meaning to a power, such as $2^{4/5}$, whose exponent is a rational number, we need to discuss radicals.

The symbol $\sqrt{}$ means "the positive square root of." Thus

$$\boxed{\sqrt{a} = b \quad \text{means} \quad b^2 = a \quad \text{and} \quad b \geq 0}$$

It is true that the number 9 has two square roots, 3 and -3, but the notation $\sqrt{9}$ is reserved for the *positive* square root of 9 (sometimes called the *principal square root* of 9). If we want the negative root, we must write $-\sqrt{9}$, which is -3.

Since $a = b^2 \geq 0$, the symbol $\sqrt{a}$ makes sense only when $a \geq 0$. For instance,

$$\sqrt{9} = 3 \quad \text{because} \quad 3^2 = 9 \quad \text{and} \quad 3 \geq 0$$

Square roots are special cases of nth roots. The nth root of x is the number that, when raised to the nth power, gives x.

Definition of nth Root

If n is any positive integer, then the **principal nth root** of a is defined as follows:

$$\sqrt[n]{a} = b \quad \text{means} \quad b^n = a$$

If n is even, we must have $a \geq 0$ and $b \geq 0$.

Thus

$$\sqrt[4]{81} = 3 \qquad \text{because} \qquad 3^4 = 81 \qquad \text{and} \qquad 3 \geq 0$$

$$\sqrt[3]{-8} = -2 \qquad \text{because} \qquad (-2)^3 = -8$$

But $\sqrt{-8}$, $\sqrt[4]{-8}$, and $\sqrt[6]{-8}$ are not defined. (For instance, $\sqrt{-8}$ is not defined because the square of every real number is nonnegative.)

Notice that

$$\sqrt{4^2} = \sqrt{16} = 4 \qquad \text{but} \qquad \sqrt{(-4)^2} = \sqrt{16} = 4 = |-4|$$

So, the equation $\sqrt{a^2} = a$ is not always true; it is true only when $a \geq 0$. However, we can always write $\sqrt{a^2} = |a|$. This last equation is true not only for square roots, but for any even root. This and other rules used in working with nth roots are listed in the following box. In each property we assume that all the given roots exist.

Properties of nth Roots

Property	Example
1. $\sqrt[n]{ab} = \sqrt[n]{a}\sqrt[n]{b}$	$\sqrt[3]{-8 \cdot 27} = \sqrt[3]{-8}\sqrt[3]{27} = (-2)(3) = -6$
2. $\sqrt[n]{\dfrac{a}{b}} = \dfrac{\sqrt[n]{a}}{\sqrt[n]{b}}$	$\sqrt[4]{\dfrac{16}{81}} = \dfrac{\sqrt[4]{16}}{\sqrt[4]{81}} = \dfrac{2}{3}$
3. $\sqrt[m]{\sqrt[n]{a}} = \sqrt[mn]{a}$	$\sqrt{\sqrt[3]{729}} = \sqrt[6]{729} = 3$
4. $\sqrt[n]{a^n} = a$ if n is odd	$\sqrt[3]{(-5)^3} = -5, \quad \sqrt[5]{2^5} = 2$
5. $\sqrt[n]{a^n} = |a|$ if n is even	$\sqrt[4]{(-3)^4} = |-3| = 3$

Example 8   Simplifying Expressions Involving nth Roots

(a) $\sqrt[3]{x^4} = \sqrt[3]{x^3 x}$ Factor out the largest cube

$\qquad\quad = \sqrt[3]{x^3}\sqrt[3]{x}$ Property 1: $\sqrt[3]{ab} = \sqrt[3]{a}\sqrt[3]{b}$

$\qquad\quad = x\sqrt[3]{x}$ Property 4: $\sqrt[3]{a^3} = a$

(b) $\sqrt[4]{81x^8y^4} = \sqrt[4]{81}\,\sqrt[4]{x^8}\,\sqrt[4]{y^4}$ Property 1: $\sqrt[4]{abc} = \sqrt[4]{a}\,\sqrt[4]{b}\,\sqrt[4]{c}$

$\qquad\qquad = 3\sqrt[4]{(x^2)^4}\,|y|$ Property 5: $\sqrt[4]{a^4} = |a|$

$\qquad\qquad = 3x^2|y|$ Property 5: $\sqrt[4]{a^4} = |a|,\ |x^2| = x^2$ ∎

It is frequently useful to combine like radicals in an expression such as $2\sqrt{3} + 5\sqrt{3}$. This can be done by using the Distributive Property. Thus

$$2\sqrt{3} + 5\sqrt{3} = (2 + 5)\sqrt{3} = 7\sqrt{3}$$

The next example further illustrates this process.

Example 9 Combining Radicals

 Avoid making the following error:

$$\sqrt{a + b} \ne \sqrt{a} + \sqrt{b}$$

For instance, if we let $a = 9$ and $b = 16$, then we see the error:

$$\sqrt{9 + 16} \stackrel{?}{=} \sqrt{9} + \sqrt{16}$$

$$\sqrt{25} \stackrel{?}{=} 3 + 4$$

$$5 \stackrel{?}{=} 7 \quad \text{Wrong!}$$

(a) $\sqrt{32} + \sqrt{200} = \sqrt{16\cdot 2} + \sqrt{100\cdot 2}$ Factor out the largest squares

$\qquad\qquad\quad = \sqrt{16}\sqrt{2} + \sqrt{100}\sqrt{2}$ Property 1: $\sqrt{ab} = \sqrt{a}\sqrt{b}$

$\qquad\qquad\quad = 4\sqrt{2} + 10\sqrt{2} = 14\sqrt{2}$ Distributive Property

(b) If $b > 0$, then

$$\sqrt{25b} - \sqrt{b^3} = \sqrt{25}\sqrt{b} - \sqrt{b^2}\sqrt{b} \qquad \text{Property 1: } \sqrt{ab} = \sqrt{a}\sqrt{b}$$

$$\qquad\qquad\quad = 5\sqrt{b} - b\sqrt{b} \qquad\qquad \text{Property 5, } b > 0$$

$$\qquad\qquad\quad = (5 - b)\sqrt{b} \qquad\qquad \text{Distributive Property} \quad ∎$$

Rational Exponents

To define what is meant by a *rational exponent* or, equivalently, a *fractional exponent* such as $a^{1/3}$, we need to use radicals. In order to give meaning to the symbol $a^{1/n}$ in a way that is consistent with the Laws of Exponents, we would have to have

$$(a^{1/n})^n = a^{(1/n)n} = a^1 = a$$

So, by the definition of nth root,

$$a^{1/n} = \sqrt[n]{a}$$

In general, we define rational exponents as follows.

Definition of Rational Exponents

For any rational exponent m/n in lowest terms, where m and n are integers and $n > 0$, we define

$$a^{m/n} = (\sqrt[n]{a})^m \qquad \text{or equivalently} \qquad a^{m/n} = \sqrt[n]{a^m}$$

If n is even, then we require that $a \ge 0$.

With this definition it can be proved that *the Laws of Exponents also hold for rational exponents.*

Example 10   Using the Definition of Rational Exponents

(a) $4^{1/2} = \sqrt{4} = 2$

(b) $8^{2/3} = (\sqrt[3]{8})^2 = 2^2 = 4$ Alternative solution: $8^{2/3} = \sqrt[3]{8^2} = \sqrt[3]{64} = 4$

(c) $125^{-1/3} = \dfrac{1}{125^{1/3}} = \dfrac{1}{\sqrt[3]{125}} = \dfrac{1}{5}$

(d) $\dfrac{1}{\sqrt[3]{x^4}} = \dfrac{1}{x^{4/3}} = x^{-4/3}$ ∎

Example 11   Using the Laws of Exponents with Rational Exponents

(a) $a^{1/3}a^{7/3} = a^{8/3}$ Law 1: $a^m a^n = a^{m+n}$

(b) $\dfrac{a^{2/5}a^{7/5}}{a^{3/5}} = a^{2/5+7/5-3/5} = a^{6/5}$ Law 1, Law 2: $\dfrac{a^m}{a^n} = a^{m-n}$

(c) $(2a^3b^4)^{3/2} = 2^{3/2}(a^3)^{3/2}(b^4)^{3/2}$ Law 4: $(abc)^n = a^n b^n c^n$

   $= (\sqrt{2})^3 a^{3(3/2)} b^{4(3/2)}$ Law 3: $(a^m)^n = a^{mn}$

   $= 2\sqrt{2}a^{9/2}b^6$

(d) $\left(\dfrac{2x^{3/4}}{y^{1/3}}\right)^3 \left(\dfrac{y^4}{x^{-1/2}}\right) = \dfrac{2^3(x^{3/4})^3}{(y^{1/3})^3} \cdot (y^4 x^{1/2})$ Laws 5, 4, and 7

   $= \dfrac{8x^{9/4}}{y} \cdot y^4 x^{1/2}$ Law 3

   $= 8x^{11/4}y^3$ Laws 1 and 2   ∎

Example 12   Simplifying by Writing Radicals as Rational Exponents

(a) $(2\sqrt{x})(3\sqrt[3]{x}) = (2x^{1/2})(3x^{1/3})$ Definition of rational exponents

   $= 6x^{1/2+1/3} = 6x^{5/6}$ Law 1

(b) $\sqrt{x\sqrt{x}} = (xx^{1/2})^{1/2}$ Definition of rational exponents

   $= (x^{3/2})^{1/2}$ Law 1

   $= x^{3/4}$ Law 3   ∎

Rationalizing the Denominator

It is often useful to eliminate the radical in a denominator by multiplying both numerator and denominator by an appropriate expression. This procedure is called **rationalizing the denominator**. If the denominator is of the form $\sqrt{a}$, we multiply numerator and denominator by $\sqrt{a}$. In doing this we multiply the given quantity by 1, so we do not change its value. For instance,

$$\frac{1}{\sqrt{a}} = \frac{1}{\sqrt{a}} \cdot 1 = \frac{1}{\sqrt{a}} \cdot \frac{\sqrt{a}}{\sqrt{a}} = \frac{\sqrt{a}}{a}$$

Note that the denominator in the last fraction contains no radical. In general, if the denominator is of the form $\sqrt[n]{a^m}$ with $m < n$, then multiplying the numerator and denominator by $\sqrt[n]{a^{n-m}}$ will rationalize the denominator, because (for $a > 0$)

$$\sqrt[n]{a^m}\sqrt[n]{a^{n-m}} = \sqrt[n]{a^{m+n-m}} = \sqrt[n]{a^n} = a$$

Example 13 Rationalizing Denominators

(a) $\dfrac{2}{\sqrt{3}} = \dfrac{2}{\sqrt{3}} \cdot \dfrac{\sqrt{3}}{\sqrt{3}} = \dfrac{2\sqrt{3}}{3}$

(b) $\dfrac{1}{\sqrt[3]{x^2}} = \dfrac{1}{\sqrt[3]{x^2}} \dfrac{\sqrt[3]{x}}{\sqrt[3]{x}} = \dfrac{\sqrt[3]{x}}{\sqrt[3]{x^3}} = \dfrac{\sqrt[3]{x}}{x}$

(c) $\sqrt[7]{\dfrac{1}{a^2}} = \dfrac{1}{\sqrt[7]{a^2}} = \dfrac{1}{\sqrt[7]{a^2}} \dfrac{\sqrt[7]{a^5}}{\sqrt[7]{a^5}} = \dfrac{\sqrt[7]{a^5}}{\sqrt[7]{a^7}} = \dfrac{\sqrt[7]{a^5}}{a}$ ∎

1.2 Exercises

1–8 ■ Write each radical expression using exponents, and each exponential expression using radicals.

	Radical expression	Exponential expression
1.	$\dfrac{1}{\sqrt{5}}$	
2.	$\sqrt[3]{7^2}$	
3.		$4^{2/3}$
4.		$11^{-3/2}$
5.	$\sqrt[5]{5^3}$	
6.		$2^{-1.5}$
7.		$a^{2/5}$
8.	$\dfrac{1}{\sqrt{x^5}}$	

9–18 ■ Evaluate each expression.

9. (a) -3^2 (b) $(-3)^2$ (c) $(-3)^0$

10. (a) $5^2 \cdot \left(\frac{1}{5}\right)^3$ (b) $\dfrac{10^7}{10^4}$ (c) $\dfrac{3}{3^{-2}}$

11. (a) $\dfrac{4^{-3}}{2^{-8}}$ (b) $\dfrac{3^{-2}}{9}$ (c) $\left(\frac{1}{4}\right)^{-2}$

12. (a) $\left(\frac{2}{3}\right)^{-3}$ (b) $\left(\frac{3}{2}\right)^{-2} \cdot \frac{9}{16}$ (c) $\left(\frac{1}{2}\right)^4 \cdot \left(\frac{5}{2}\right)^{-2}$

13. (a) $\sqrt{16}$ (b) $\sqrt[4]{16}$ (c) $\sqrt[4]{1/16}$

14. (a) $\sqrt{64}$ (b) $\sqrt[3]{-64}$ (c) $\sqrt[5]{-32}$

15. (a) $\sqrt[3]{\dfrac{8}{27}}$ (b) $\sqrt[3]{\dfrac{-1}{64}}$ (c) $\dfrac{\sqrt[5]{-3}}{\sqrt[5]{96}}$

16. (a) $\sqrt{7}\sqrt{28}$ (b) $\dfrac{\sqrt{48}}{\sqrt{3}}$ (c) $\sqrt[4]{24}\sqrt[4]{54}$

17. (a) $\left(\frac{4}{9}\right)^{-1/2}$ (b) $(-32)^{2/5}$ (c) $-32^{2/5}$

18. (a) $1024^{-0.1}$ (b) $\left(-\frac{27}{8}\right)^{2/3}$ (c) $\left(\frac{25}{64}\right)^{-3/2}$

19–22 ■ Evaluate the expression using $x = 3$, $y = 4$, and $z = -1$.

19. $\sqrt{x^2 + y^2}$ 20. $\sqrt[4]{x^3 + 14y + 2z}$

21. $(9x)^{2/3} + (2y)^{2/3} + z^{2/3}$ 22. $(xy)^{2z}$

23–26 ■ Simplify the expression.

23. $\sqrt{32} + \sqrt{18}$ 24. $\sqrt{75} + \sqrt{48}$

25. $\sqrt[5]{96} + \sqrt[5]{3}$ 26. $\sqrt[4]{48} - \sqrt[4]{3}$

27–44 ■ Simplify the expression and eliminate any negative exponent(s).

27. $a^9 a^{-5}$ 28. $(3y^2)(4y^5)$

29. $(12x^2y^4)(\frac{1}{2}x^5y)$ 30. $(6y)^3$

31. $\dfrac{x^9(2x)^4}{x^3}$ 32. $\dfrac{a^{-3}b^4}{a^{-5}b^5}$

33. $b^4(\frac{1}{3}b^2)(12b^{-8})$ 34. $(2s^3t^{-1})(\frac{1}{4}s^6)(16t^4)$

35. $(rs)^3(2s)^{-2}(4r)^4$ 36. $(2u^2v^3)^3(3u^3v)^{-2}$

37. $\dfrac{(6y^3)^4}{2y^5}$ 38. $\dfrac{(2x^3)^2(3x^4)}{(x^3)^4}$

39. $\dfrac{(x^2y^3)^4(xy^4)^{-3}}{x^2y}$ 40. $\left(\dfrac{c^4d^3}{cd^2}\right)\left(\dfrac{d^2}{c^3}\right)^3$

41. $\dfrac{(xy^2z^3)^4}{(x^3y^2z)^3}$

42. $\left(\dfrac{xy^{-2}z^{-3}}{x^2y^3z^{-4}}\right)^{-3}$

43. $\left(\dfrac{q^{-1}rs^{-2}}{r^{-5}sq^{-8}}\right)^{-1}$

44. $(3ab^2c)\left(\dfrac{2a^2b}{c^3}\right)^{-2}$

45–52 ■ Simplify the expression. Assume the letters denote any real numbers.

45. $\sqrt[4]{x^4}$

46. $\sqrt[5]{x^{10}}$

47. $\sqrt[4]{16x^8}$

48. $\sqrt[3]{x^3y^6}$

49. $\sqrt{a^2b^6}$

50. $\sqrt[3]{a^2b}\sqrt[3]{a^4b}$

51. $\sqrt[3]{\sqrt{64x^6}}$

52. $\sqrt[4]{x^4y^2z^2}$

53–70 ■ Simplify the expression and eliminate any negative exponent(s). Assume that all letters denote positive numbers.

53. $x^{2/3}x^{1/5}$

54. $(2x^{3/2})(4x)^{-1/2}$

55. $(-3a^{1/4})(9a)^{-3/2}$

56. $(-2a^{3/4})(5a^{3/2})$

57. $(4b)^{1/2}(8b^{2/5})$

58. $(8x^6)^{-2/3}$

59. $(c^2d^3)^{-1/3}$

60. $(4x^6y^8)^{3/2}$

61. $(y^{3/4})^{2/3}$

62. $(a^{2/5})^{-3/4}$

63. $(2x^4y^{-4/5})^3(8y^2)^{2/3}$

64. $(x^{-5}y^3z^{10})^{-3/5}$

65. $\left(\dfrac{x^6y}{y^4}\right)^{5/2}$

66. $\left(\dfrac{-2x^{1/3}}{y^{1/2}z^{1/6}}\right)^4$

67. $\left(\dfrac{3a^{-2}}{4b^{-1/3}}\right)^{-1}$

68. $\dfrac{(y^{10}z^{-5})^{1/5}}{(y^{-2}z^3)^{1/3}}$

69. $\dfrac{(9st)^{3/2}}{(27s^3t^{-4})^{2/3}}$

70. $\left(\dfrac{a^2b^{-3}}{x^{-1}y^2}\right)^3\left(\dfrac{x^{-2}b^{-1}}{a^{3/2}y^{1/3}}\right)$

71–72 ■ Write each number in scientific notation.

71. (a) 69,300,000   (b) 7,200,000,000,000

(c) 0.000028536   (d) 0.0001213

72. (a) 129,540,000   (b) 7,259,000,000

(c) 0.0000000014   (d) 0.0007029

73–74 ■ Write each number in decimal notation.

73. (a) 3.19×10^5 (b) 2.721×10^8

(c) 2.670×10^{-8} (d) 9.999×10^{-9}

74. (a) 7.1×10^{14} (b) 6×10^{12}

(c) 8.55×10^{-3} (d) 6.257×10^{-10}

75–76 ■ Write the number indicated in each statement in scientific notation.

75. (a) A light-year, the distance that light travels in one year, is about 5,900,000,000,000 mi.

(b) The diameter of an electron is about 0.0000000000004 cm.

(c) A drop of water contains more than 33 billion billion molecules.

76. (a) The distance from the earth to the sun is about 93 million miles.

(b) The mass of an oxygen molecule is about 0.000000000000000000000053 g.

(c) The mass of the earth is about 5,970,000,000,000,000,000,000,000 kg.

77–82 ■ Use scientific notation, the Laws of Exponents, and a calculator to perform the indicated operations. State your answer correct to the number of significant digits indicated by the given data.

77. $(7.2 \times 10^{-9})(1.806 \times 10^{-12})$

78. $(1.062 \times 10^{24})(8.61 \times 10^{19})$

79. $\dfrac{1.295643 \times 10^9}{(3.610 \times 10^{-17})(2.511 \times 10^6)}$

80. $\dfrac{(73.1)(1.6341 \times 10^{28})}{0.0000000019}$

81. $\dfrac{(0.0000162)(0.01582)}{(594,621,000)(0.0058)}$

82. $\dfrac{(3.542 \times 10^{-6})^9}{(5.05 \times 10^4)^{12}}$

83–86 ■ Rationalize the denominator.

83. (a) $\dfrac{1}{\sqrt{10}}$ (b) $\sqrt{\dfrac{2}{x}}$ (c) $\sqrt{\dfrac{x}{3}}$

84. (a) $\sqrt{\dfrac{5}{12}}$ (b) $\sqrt{\dfrac{x}{6}}$ (c) $\sqrt{\dfrac{y}{2z}}$

85. (a) $\dfrac{2}{\sqrt[3]{x}}$ (b) $\dfrac{1}{\sqrt[4]{y^3}}$ (c) $\dfrac{x}{y^{2/5}}$

86. (a) $\dfrac{1}{\sqrt[4]{a}}$ (b) $\dfrac{a}{\sqrt[3]{b^2}}$ (c) $\dfrac{1}{c^{3/7}}$

87. Let a, b, and c be real numbers with $a > 0$, $b < 0$, and $c < 0$. Determine the sign of each expression.

(a) b^5 (b) b^{10} (c) ab^2c^3

(d) $(b - a)^3$ (e) $(b - a)^4$ (f) $\dfrac{a^3c^3}{b^6c^6}$

88. Prove the given Laws of Exponents for the case in which m and n are positive integers and $m > n$.

(a) Law 2   (b) Law 5   (c) Law 6

Applications

89. Distance to the Nearest Star   Proxima Centauri, the star nearest to our solar system, is 4.3 light-years away. Use the information in Exercise 75(a) to express this distance in miles.

90. Speed of Light   The speed of light is about 186,000 mi/s. Use the information in Exercise 76(a) to find how long it takes for a light ray from the sun to reach the earth.

91. Volume of the Oceans   The average ocean depth is 3.7×10^3 m, and the area of the oceans is 3.6×10^{14} m². What is the total volume of the ocean in liters? (One cubic meter contains 1000 liters.)

92. National Debt   As of November 2004, the population of the United States was 2.949×10^8, and the national debt was 7.529×10^{12} dollars. How much was each person's share of the debt?

93. Number of Molecules   A sealed room in a hospital, measuring 5 m wide, 10 m long, and 3 m high, is filled with pure oxygen. One cubic meter contains 1000 L, and 22.4 L of any gas contains 6.02×10^{23} molecules (Avogadro's number). How many molecules of oxygen are there in the room?

94. How Far Can You See?   Due to the curvature of the earth, the maximum distance D that you can see from the top of a tall building of height h is estimated by the formula

$$D = \sqrt{2rh + h^2}$$

where $r = 3960$ mi is the radius of the earth and D and h are also measured in miles. How far can you see from the observation deck of the Toronto CN Tower, 1135 ft above the ground?

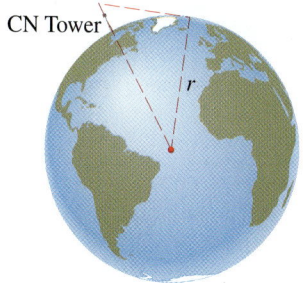

CN Tower

95. Speed of a Skidding Car   Police use the formula $s = \sqrt{30fd}$ to estimate the speed s (in mi/h) at which a car is traveling if it skids d feet after the brakes are applied suddenly. The number f is the coefficient of friction of the road, which is a measure of the "slipperiness" of the road. The table gives some typical estimates for f.

	Tar	Concrete	Gravel
Dry	1.0	0.8	0.2
Wet	0.5	0.4	0.1

(a) If a car skids 65 ft on wet concrete, how fast was it moving when the brakes were applied?

(b) If a car is traveling at 50 mi/h, how far will it skid on wet tar?

96. Distance from the Earth to the Sun   It follows from **Kepler's Third Law** of planetary motion that the average distance from a planet to the sun (in meters) is

$$d = \left(\frac{GM}{4\pi^2}\right)^{1/3} T^{2/3}$$

where $M = 1.99 \times 10^{30}$ kg is the mass of the sun, $G = 6.67 \times 10^{-11}$ N · m²/kg² is the gravitational constant, and T is the period of the planet's orbit (in seconds). Use the fact that the period of the earth's orbit is about 365.25 days to find the distance from the earth to the sun.

97. Flow Speed in a Channel   The speed of water flowing in a channel, such as a canal or river bed, is governed by the **Manning Equation**

$$V = 1.486 \, \frac{A^{2/3} S^{1/2}}{p^{2/3} n}$$

Here V is the velocity of the flow in ft/s; A is the cross-sectional area of the channel in square feet; S is the downward slope of the channel; p is the wetted perimeter in feet (the distance from the top of one bank, down the side of the channel, across the bottom, and up to the top of the other bank); and n is the roughness coefficient (a measure of the roughness of the channel bottom). This equation is used to predict the capacity of flood channels to handle runoff from heavy rainfalls. For the canal shown in the figure, $A = 75$ ft^2, $S = 0.050$, $p = 24.1$ ft, and $n = 0.040$.

(a) Find the speed with which water flows through this canal.

(b) How many cubic feet of water can the canal discharge per second? [*Hint:* Multiply V by A to get the volume of the flow per second.]

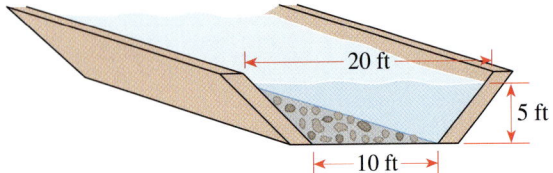

Discovery • Discussion

98. How Big Is a Billion? If you have a million (10^6) dollars in a suitcase, and you spend a thousand (10^3) dollars each day, how many years would it take you to use all the money? Spending at the same rate, how many years would it take you to empty a suitcase filled with a *billion* (10^9) dollars?

99. Easy Powers That Look Hard Calculate these expressions in your head. Use the Laws of Exponents to help you.

(a) $\dfrac{18^5}{9^5}$

(b) $20^6 \cdot (0.5)^6$

100. Limiting Behavior of Powers Complete the following tables. What happens to the nth root of 2 as n gets large? What about the nth root of $\frac{1}{2}$?

n	$2^{1/n}$
1	
2	
5	
10	
100	

n	$\left(\frac{1}{2}\right)^{1/n}$
1	
2	
5	
10	
100	

Construct a similar table for $n^{1/n}$. What happens to the nth root of n as n gets large?

101. Comparing Roots Without using a calculator, determine which number is larger in each pair.

(a) $2^{1/2}$ or $2^{1/3}$

(b) $\left(\frac{1}{2}\right)^{1/2}$ or $\left(\frac{1}{2}\right)^{1/3}$

(c) $7^{1/4}$ or $4^{1/3}$

(d) $\sqrt[3]{5}$ or $\sqrt{3}$

1.3   Algebraic Expressions

A **variable** is a letter that can represent any number from a given set of numbers. If we start with variables such as x, y, and z and some real numbers, and combine them using addition, subtraction, multiplication, division, powers, and roots, we obtain an **algebraic expression**. Here are some examples:

$$2x^2 - 3x + 4 \qquad \sqrt{x} + 10 \qquad \frac{y - 2z}{y^2 + 4}$$

A **monomial** is an expression of the form ax^k, where a is a real number and k is a nonnegative integer. A **binomial** is a sum of two monomials and a **trinomial** is a sum of three monomials. In general, a sum of monomials is called a *polynomial*. For example, the first expression listed above is a polynomial, but the other two are not.

Polynomials

A **polynomial** in the variable x is an expression of the form

$$a_n x^n + a_{n-1} x^{n-1} + \cdots + a_1 x + a_0$$

where $a_0, a_1, \ldots, a_n$ are real numbers, and n is a nonnegative integer. If $a_n \neq 0$, then the polynomial has **degree n**. The monomials $a_k x^k$ that make up the polynomial are called the **terms** of the polynomial.

Note that the degree of a polynomial is the highest power of the variable that appears in the polynomial.

Polynomial	Type	Terms	Degree
$2x^2 - 3x + 4$	trinomial	$2x^2, -3x, 4$	2
$x^8 + 5x$	binomial	$x^8, 5x$	8
$3 - x + x^2 - \frac{1}{2}x^3$	four terms	$-\frac{1}{2}x^3, x^2, -x, 3$	3
$5x + 1$	binomial	$5x, 1$	1
$9x^5$	monomial	$9x^5$	5
6	monomial	6	0

Combining Algebraic Expressions

Distributive Property

$ac + bc = (a + b)c$

We **add** and **subtract** polynomials using the properties of real numbers that were discussed in Section 1.1. The idea is to combine **like terms** (that is, terms with the same variables raised to the same powers) using the Distributive Property. For instance,

$$5x^7 + 3x^7 = (5 + 3)x^7 = 8x^7$$

 In subtracting polynomials we have to remember that if a minus sign precedes an expression in parentheses, then the sign of every term within the parentheses is changed when we remove the parentheses:

$$-(b + c) = -b - c$$

[This is simply a case of the Distributive Property, $a(b + c) = ab + ac$, with $a = -1$.]

Example 1 Adding and Subtracting Polynomials

(a) Find the sum $(x^3 - 6x^2 + 2x + 4) + (x^3 + 5x^2 - 7x)$.

(b) Find the difference $(x^3 - 6x^2 + 2x + 4) - (x^3 + 5x^2 - 7x)$.

Solution

(a) $(x^3 - 6x^2 + 2x + 4) + (x^3 + 5x^2 - 7x)$

$\qquad = (x^3 + x^3) + (-6x^2 + 5x^2) + (2x - 7x) + 4 \qquad$ Group like terms

$\qquad = 2x^3 - x^2 - 5x + 4 \qquad$ Combine like terms

(b) $(x^3 - 6x^2 + 2x + 4) - (x^3 + 5x^2 - 7x)$

$\qquad = x^3 - 6x^2 + 2x + 4 - x^3 - 5x^2 + 7x \qquad$ Distributive Property

$\qquad = (x^3 - x^3) + (-6x^2 - 5x^2) + (2x + 7x) + 4 \qquad$ Group like terms

$\qquad = -11x^2 + 9x + 4 \qquad$ Combine like terms ∎

To find the **product** of polynomials or other algebraic expressions, we need to use the Distributive Property repeatedly. In particular, using it three times on the product of two binomials, we get

$$(a + b)(c + d) = a(c + d) + b(c + d) = ac + ad + bc + bd$$

This says that we multiply the two factors by multiplying each term in one factor by each term in the other factor and adding these products. Schematically we have

$$(a + b)(c + d) = ac + ad + bc + bd$$
$$\uparrow \quad \uparrow \quad \uparrow \quad \uparrow$$
$$F \quad O \quad I \quad L$$

The acronym **FOIL** helps us remember that the product of two binomials is the sum of the products of the **F**irst terms, the **O**uter terms, the **I**nner terms, and the **L**ast terms.

In general, we can multiply two algebraic expressions by using the Distributive Property and the Laws of Exponents.

Example 2   Multiplying Algebraic Expressions

(a) $(2x + 1)(3x - 5) = 6x^2 - 10x + 3x - 5$ Distributive Property
$$\uparrow \qquad \uparrow \qquad \uparrow \quad \uparrow$$
$$F \qquad O \qquad I \quad L$$

$$= 6x^2 - 7x - 5 \qquad \text{Combine like terms}$$

(b) $(x^2 - 3)(x^3 + 2x + 1) = x^2(x^3 + 2x + 1) - 3(x^3 + 2x + 1)$ Distributive Property

$$= x^5 + 2x^3 + x^2 - 3x^3 - 6x - 3 \qquad \text{Distributive Property}$$

$$= x^5 - x^3 + x^2 - 6x - 3 \qquad \text{Combine like terms}$$

(c) $(1 + \sqrt{x})(2 - 3\sqrt{x}) = 2 - 3\sqrt{x} + 2\sqrt{x} - 3(\sqrt{x})^2$ Distributive Property

$$= 2 - \sqrt{x} - 3x \qquad \text{Combine like terms}$$

Certain types of products occur so frequently that you should memorize them. You can verify the following formulas by performing the multiplications.

See the Discovery Project on page 34 for a geometric interpretation of some of these formulas.

Special Product Formulas

If A and B are any real numbers or algebraic expressions, then

1. $(A + B)(A - B) = A^2 - B^2$ — Sum and product of same terms

2. $(A + B)^2 = A^2 + 2AB + B^2$ — Square of a sum

3. $(A - B)^2 = A^2 - 2AB + B^2$ — Square of a difference

4. $(A + B)^3 = A^3 + 3A^2B + 3AB^2 + B^3$ — Cube of a sum

5. $(A - B)^3 = A^3 - 3A^2B + 3AB^2 - B^3$ — Cube of a difference

---

The key idea in using these formulas (or any other formula in algebra) is the **Principle of Substitution**: We may substitute any algebraic expression for any letter in a formula. For example, to find $(x^2 + y^3)^2$ we use Product Formula 2, substituting x^2 for A and y^3 for B, to get

$$(x^2 + y^3)^2 = (x^2)^2 + 2(x^2)(y^3) + (y^3)^2$$

$$(A + B)^2 \;=\; A^2 \;+\; 2AB \;+\; B^2$$

Example 3   Using the Special Product Formulas

Use the Special Product Formulas to find each product.

(a) $(3x + 5)^2$ (b) $(x^2 - 2)^3$ (c) $(2x - \sqrt{y})(2x + \sqrt{y})$

Solution

(a) Substituting $A = 3x$ and $B = 5$ in Product Formula 2, we get

$$(3x + 5)^2 = (3x)^2 + 2(3x)(5) + 5^2 = 9x^2 + 30x + 25$$

(b) Substituting $A = x^2$ and $B = 2$ in Product Formula 5, we get

$$(x^2 - 2)^3 = (x^2)^3 - 3(x^2)^2(2) + 3(x^2)(2)^2 - 2^3$$
$$= x^6 - 6x^4 + 12x^2 - 8$$

(c) Substituting $A = 2x$ and $B = \sqrt{y}$ in Product Formula 1, we get

$$(2x - \sqrt{y})(2x + \sqrt{y}) = (2x)^2 - (\sqrt{y})^2$$
$$= 4x^2 - y$$ ∎

Factoring

We use the Distributive Property to expand algebraic expressions. We sometimes need to reverse this process (again using the Distributive Property) by **factoring** an expression as a product of simpler ones. For example, we can write

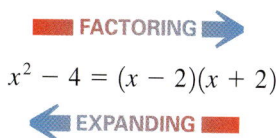

$$x^2 - 4 = (x - 2)(x + 2)$$

We say that $x - 2$ and $x + 2$ are **factors** of $x^2 - 4$.

The easiest type of factoring occurs when the terms have a common factor.

Example 4   Factoring Out Common Factors

Factor each expression.

(a) $3x^2 - 6x$ (b) $8x^4y^2 + 6x^3y^3 - 2xy^4$

(c) $(2x + 4)(x - 3) - 5(x - 3)$

Solution

(a) The greatest common factor of the terms $3x^2$ and $-6x$ is $3x$, so we have

$$3x^2 - 6x = 3x(x - 2)$$

(b) We note that

8, 6, and -2 have the greatest common factor 2

x^4, x^3, and x have the greatest common factor x

y^2, y^3, and y^4 have the greatest common factor y^2

So the greatest common factor of the three terms in the polynomial is $2xy^2$, and we have

$$8x^4y^2 + 6x^3y^3 - 2xy^4 = (2xy^2)(4x^3) + (2xy^2)(3x^2y) + (2xy^2)(-y^2)$$
$$= 2xy^2(4x^3 + 3x^2y - y^2)$$

Check Your Answer

Multiplying gives

$$3x(x - 2) = 3x^2 - 6x \quad ✓$$

Check Your Answer

Multiplying gives

$$2xy^2(4x^3 + 3x^2y - y^2) =$$
$$8x^4y^2 + 6x^3y^3 - 2xy^4 \quad ✓$$

(c) The two terms have the common factor $x - 3$.

$$(2x + 4)(x - 3) - 5(x - 3) = [(2x + 4) - 5](x - 3) \qquad \text{Distributive Property}$$
$$= (2x - 1)(x - 3) \qquad \text{Simplify} \qquad \blacksquare$$

To factor a trinomial of the form $x^2 + bx + c$, we note that

$$(x + r)(x + s) = x^2 + (r + s)x + rs$$

so we need to choose numbers r and s so that $r + s = b$ and $rs = c$.

Example 5   Factoring $x^2 + bx + c$ by Trial and Error

Factor: $x^2 + 7x + 12$

Solution   We need to find two integers whose product is 12 and whose sum is 7. By trial and error we find that the two integers are 3 and 4. Thus, the factorization is

$$x^2 + 7x + 12 = (x + 3)(x + 4)$$
$$\text{factors of 12}$$

$\blacksquare$

Check Your Answer

Multiplying gives

$(x + 3)(x + 4) = x^2 + 7x + 12$ ✓

To factor a trinomial of the form $ax^2 + bx + c$ with $a \neq 1$, we look for factors of the form $px + r$ and $qx + s$:

$$ax^2 + bx + c = (px + r)(qx + s) = pqx^2 + (ps + qr)x + rs$$

Therefore, we try to find numbers p, q, r, and s such that $pq = a$, $rs = c$, $ps + qr = b$. If these numbers are all integers, then we will have a limited number of possibilities to try for p, q, r, and s.

$$\overset{\text{factors of } a}{\underset{\text{factors of } c}{ax^2 + bx + c = (px + r)(qx + s)}}$$

Example 6   Factoring $ax^2 + bx + c$ by Trial and Error

Factor: $6x^2 + 7x - 5$

Solution   We can factor 6 as $6 \cdot 1$ or $3 \cdot 2$, and -5 as $-25 \cdot 1$ or $5 \cdot (-1)$. By trying these possibilities, we arrive at the factorization

$$\overset{\text{factors of 6}}{\underset{\text{factors of } -5}{6x^2 + 7x - 5 = (3x + 5)(2x - 1)}}$$

$\blacksquare$

Check Your Answer

Multiplying gives

$(3x + 5)(2x - 1) = 6x^2 + 7x - 5$ ✓

Example 7   Recognizing the Form of an Expression

Factor each expression.

(a) $x^2 - 2x - 3$ (b) $(5a + 1)^2 - 2(5a + 1) - 3$

Solution

(a) $x^2 - 2x - 3 = (x - 3)(x + 1)$ Trial and error

(b) This expression is of the form

$$\blacksquare^2 - 2 \,\blacksquare\, - 3$$

where ▢ represents $5a + 1$. This is the same form as the expression in part (a), so it will factor as $(▢ - 3)(▢ + 1)$.

$$(5a + 1)^2 - 2(5a + 1) - 3 = [(5a + 1) - 3][(5a + 1) + 1]$$
$$= (5a - 2)(5a + 2)$$ ∎

Some special algebraic expressions can be factored using the following formulas. The first three are simply Special Product Formulas written backward.

Special Factoring Formulas

Formula	Name
1. $A^2 - B^2 = (A - B)(A + B)$	Difference of squares
2. $A^2 + 2AB + B^2 = (A + B)^2$	Perfect square
3. $A^2 - 2AB + B^2 = (A - B)^2$	Perfect square
4. $A^3 - B^3 = (A - B)(A^2 + AB + B^2)$	Difference of cubes
5. $A^3 + B^3 = (A + B)(A^2 - AB + B^2)$	Sum of cubes

Example 8 Factoring Differences of Squares

Factor each polynomial.

(a) $4x^2 - 25$ (b) $(x + y)^2 - z^2$

Solution

(a) Using the Difference of Squares Formula with $A = 2x$ and $B = 5$, we have

$$4x^2 - 25 = (2x)^2 - 5^2 = (2x - 5)(2x + 5)$$

$$A^2 - B^2 = (A - B)(A + B)$$

(b) We use the Difference of Squares Formula with $A = x + y$ and $B = z$.

$$(x + y)^2 - z^2 = (x + y - z)(x + y + z)$$ ∎

Example 9 Factoring Differences and Sums of Cubes

Factor each polynomial.

(a) $27x^3 - 1$ (b) $x^6 + 8$

Solution

(a) Using the Difference of Cubes Formula with $A = 3x$ and $B = 1$, we get

$$27x^3 - 1 = (3x)^3 - 1^3 = (3x - 1)[(3x)^2 + (3x)(1) + 1^2]$$
$$= (3x - 1)(9x^2 + 3x + 1)$$

(b) Using the Sum of Cubes Formula with $A = x^2$ and $B = 2$, we have

$$x^6 + 8 = (x^2)^3 + 2^3 = (x^2 + 2)(x^4 - 2x^2 + 4)$$ ■

A trinomial is a perfect square if it is of the form

$$A^2 + 2AB + B^2 \qquad \text{or} \qquad A^2 - 2AB + B^2$$

So, we **recognize a perfect square** if the middle term ($2AB$ or $-2AB$) is plus or minus twice the product of the square roots of the outer two terms.

Example 10   Recognizing Perfect Squares

Factor each trinomial.

(a) $x^2 + 6x + 9$ (b) $4x^2 - 4xy + y^2$

Solution

(a) Here $A = x$ and $B = 3$, so $2AB = 2 \cdot x \cdot 3 = 6x$. Since the middle term is $6x$, the trinomial is a perfect square. By the Perfect Square Formula, we have

$$x^2 + 6x + 9 = (x + 3)^2$$

(b) Here $A = 2x$ and $B = y$, so $2AB = 2 \cdot 2x \cdot y = 4xy$. Since the middle term is $-4xy$, the trinomial is a perfect square. By the Perfect Square Formula, we have

$$4x^2 - 4xy + y^2 = (2x - y)^2$$ ■

When we factor an expression, the result can sometimes be factored further. In general, *we first factor out common factors*, then inspect the result to see if it can be factored by any of the other methods of this section. We repeat this process until we have factored the expression completely.

Example 11   Factoring an Expression Completely

Factor each expression completely.

(a) $2x^4 - 8x^2$ (b) $x^5y^2 - xy^6$

Solution

(a) We first factor out the power of x with the smallest exponent.

$$2x^4 - 8x^2 = 2x^2(x^2 - 4) \qquad \text{Common factor is } 2x^2$$

$$= 2x^2(x - 2)(x + 2) \qquad \text{Factor } x^2 - 4 \text{ as a difference of squares}$$

(b) We first factor out the powers of x and y with the smallest exponents.

$$x^5y^2 - xy^6 = xy^2(x^4 - y^4) \qquad \text{Common factor is } xy^2$$

$$= xy^2(x^2 + y^2)(x^2 - y^2) \qquad \text{Factor } x^4 - y^4 \text{ as a difference of squares}$$

$$= xy^2(x^2 + y^2)(x + y)(x - y) \qquad \text{Factor } x^2 - y^2 \text{ as a difference of squares}$$ ■

In the next example we factor out variables with fractional exponents. This type of factoring occurs in calculus.

Example 12 **Factoring Expressions with Fractional Exponents**

Factor each expression.

(a) $3x^{3/2} - 9x^{1/2} + 6x^{-1/2}$ (b) $(2 + x)^{-2/3}x + (2 + x)^{1/3}$

Solution

(a) Factor out the power of x with the *smallest exponent*, that is, $x^{-1/2}$.

$$3x^{3/2} - 9x^{1/2} + 6x^{-1/2} = 3x^{-1/2}(x^2 - 3x + 2) \qquad \text{Factor out } 3x^{-1/2}$$
$$= 3x^{-1/2}(x - 1)(x - 2) \qquad \begin{array}{l}\text{Factor the quadratic}\\ x^2 - 3x + 2\end{array}$$

To factor out $x^{-1/2}$ from $x^{3/2}$, we *subtract* exponents:

$$x^{3/2} = x^{-1/2}(x^{3/2 - (-1/2)})$$
$$= x^{-1/2}(x^{3/2 + 1/2})$$
$$= x^{-1/2}(x^2)$$

(b) Factor out the power of $2 + x$ with the *smallest exponent*, that is, $(2 + x)^{-2/3}$.

$$(2 + x)^{-2/3}x + (2 + x)^{1/3} = (2 + x)^{-2/3}[x + (2 + x)] \qquad \text{Factor out } (2 + x)^{-2/3}$$
$$= (2 + x)^{-2/3}(2 + 2x) \qquad \text{Simplify}$$
$$= 2(2 + x)^{-2/3}(1 + x) \qquad \text{Factor out } 2 \qquad \blacksquare$$

Check Your Answer

To see that you have factored correctly, multiply using the Laws of Exponents.

(a) $3x^{-1/2}(x^2 - 3x + 2)$

$\quad = 3x^{3/2} - 9x^{1/2} + 6x^{-1/2}$ ✔

(b) $(2 + x)^{-2/3}[x + (2 + x)]$

$\quad = (2 + x)^{-2/3}x + (2 + x)^{1/3}$ ✔

Polynomials with at least four terms can sometimes be factored by grouping terms. The following example illustrates the idea.

Example 13 **Factoring by Grouping**

Factor each polynomial.

(a) $x^3 + x^2 + 4x + 4$ (b) $x^3 - 2x^2 - 3x + 6$

Solution

(a) $x^3 + x^2 + 4x + 4 = (x^3 + x^2) + (4x + 4) \qquad \text{Group terms}$

$\qquad = x^2(x + 1) + 4(x + 1) \qquad \text{Factor out common factors}$

$\qquad = (x^2 + 4)(x + 1) \qquad \text{Factor out } x + 1 \text{ from each term}$

(b) $x^3 - 2x^2 - 3x + 6 = (x^3 - 2x^2) - (3x - 6) \qquad \text{Group terms}$

$\qquad = x^2(x - 2) - 3(x - 2) \qquad \text{Factor out common factors}$

$\qquad = (x^2 - 3)(x - 2) \qquad \begin{array}{l}\text{Factor out } x - 2 \text{ from each}\\ \text{term}\end{array} \qquad \blacksquare$

1.3   Exercises

1–6 ■ Complete the following table by stating whether the polynomial is a monomial, binomial, or trinomial; then list its terms and state its degree.

Polynomial	Type	Terms	Degree
1. $x^2 - 3x + 7$			
2. $2x^5 + 4x^2$			
3. -8			
4. $\frac{1}{2}x^7$			
5. $x - x^2 + x^3 - x^4$			
6. $\sqrt{2}x - \sqrt{3}$			

7–42 ■ Perform the indicated operations and simplify.

7. $(12x - 7) - (5x - 12)$ **8.** $(5 - 3x) + (2x - 8)$

9. $(3x^2 + x + 1) + (2x^2 - 3x - 5)$

10. $(3x^2 + x + 1) - (2x^2 - 3x - 5)$

11. $(x^3 + 6x^2 - 4x + 7) - (3x^2 + 2x - 4)$

12. $3(x - 1) + 4(x + 2)$

13. $8(2x + 5) - 7(x - 9)$

14. $4(x^2 - 3x + 5) - 3(x^2 - 2x + 1)$

15. $2(2 - 5t) + t^2(t - 1) - (t^4 - 1)$

16. $5(3t - 4) - (t^2 + 2) - 2t(t - 3)$

17. $\sqrt{x}(x - \sqrt{x})$

18. $x^{3/2}(\sqrt{x} - 1/\sqrt{x})$

19. $(3t - 2)(7t - 5)$

20. $(4x - 1)(3x + 7)$

21. $(x + 2y)(3x - y)$

22. $(4x - 3y)(2x + 5y)$

23. $(1 - 2y)^2$

24. $(3x + 4)^2$

25. $(2x^2 + 3y^2)^2$

26. $\left(c + \dfrac{1}{c}\right)^2$

27. $(2x - 5)(x^2 - x + 1)$

28. $(1 + 2x)(x^2 - 3x + 1)$

29. $(x^2 - a^2)(x^2 + a^2)$

30. $(x^{1/2} + y^{1/2})(x^{1/2} - y^{1/2})$

31. $\left(\sqrt{a} - \dfrac{1}{b}\right)\left(\sqrt{a} + \dfrac{1}{b}\right)$

32. $(\sqrt{h^2 + 1} + 1)(\sqrt{h^2 + 1} - 1)$

33. $(1 + a^3)^3$

34. $(1 - 2y)^3$

35. $(x^2 + x - 1)(2x^2 - x + 2)$

36. $(3x^3 + x^2 - 2)(x^2 + 2x - 1)$

37. $(1 + x^{4/3})(1 - x^{2/3})$

38. $(1 - b)^2(1 + b)^2$

39. $(3x^2y + 7xy^2)(x^2y^3 - 2y^2)$

40. $(x^4y - y^5)(x^2 + xy + y^2)$

41. $(x + y + z)(x - y - z)$

42. $(x^2 - y + z)(x^2 + y - z)$

43–48 ■ Factor out the common factor.

43. $-2x^3 + 16x$

44. $2x^4 + 4x^3 - 14x^2$

45. $y(y - 6) + 9(y - 6)$

46. $(z + 2)^2 - 5(z + 2)$

47. $2x^2y - 6xy^2 + 3xy$

48. $-7x^4y^2 + 14xy^3 + 21xy^4$

49–54 ■ Factor the trinomial.

49. $x^2 + 2x - 3$

50. $x^2 - 6x + 5$

51. $8x^2 - 14x - 15$

52. $6y^2 + 11y - 21$

53. $(3x + 2)^2 + 8(3x + 2) + 12$

54. $2(a + b)^2 + 5(a + b) - 3$

55–60 ■ Use a Special Factoring Formula to factor the expression.

55. $9a^2 - 16$

56. $(x + 3)^2 - 4$

57. $27x^3 + y^3$

58. $8s^3 - 125t^6$

59. $x^2 + 12x + 36$

60. $16z^2 - 24z + 9$

61–66 ■ Factor the expression by grouping terms.

61. $x^3 + 4x^2 + x + 4$

62. $3x^3 - x^2 + 6x - 2$

63. $2x^3 + x^2 - 6x - 3$

64. $-9x^3 - 3x^2 + 3x + 1$

65. $x^3 + x^2 + x + 1$

66. $x^5 + x^4 + x + 1$

67–70 ■ Factor the expression completely. Begin by factoring out the lowest power of each common factor.

67. $x^{5/2} - x^{1/2}$

68. $x^{-3/2} + 2x^{-1/2} + x^{1/2}$

69. $(x^2 + 1)^{1/2} + 2(x^2 + 1)^{-1/2}$

70. $2x^{1/3}(x - 2)^{2/3} - 5x^{4/3}(x - 2)^{-1/3}$

71–100 ■ Factor the expression completely.

71. $12x^3 + 18x$

72. $5ab - 8abc$

73. $x^2 - 2x - 8$

74. $y^2 - 8y + 15$

75. $2x^2 + 5x + 3$

76. $9x^2 - 36x - 45$

77. $6x^2 - 5x - 6$

78. $r^2 - 6rs + 9s^2$

79. $25s^2 - 10st + t^2$

80. $x^2 - 36$

81. $4x^2 - 25$

82. $49 - 4y^2$

83. $(a + b)^2 - (a - b)^2$

84. $\left(1 + \dfrac{1}{x}\right)^2 - \left(1 - \dfrac{1}{x}\right)^2$

85. $x^2(x^2 - 1) - 9(x^2 - 1)$

86. $(a^2 - 1)b^2 - 4(a^2 - 1)$

87. $8x^3 + 125$

88. $x^6 + 64$

89. $x^6 - 8y^3$

90. $27a^3 - b^6$

91. $x^3 + 2x^2 + x$

92. $3x^3 - 27x$

93. $y^3 - 3y^2 - 4y + 12$

94. $x^3 + 3x^2 - x - 3$

95. $2x^3 + 4x^2 + x + 2$

96. $3x^3 + 5x^2 - 6x - 10$

97. $(x - 1)(x + 2)^2 - (x - 1)^2(x + 2)$

98. $y^4(y + 2)^3 + y^5(y + 2)^4$

99. $(a^2 + 1)^2 - 7(a^2 + 1) + 10$

100. $(a^2 + 2a)^2 - 2(a^2 + 2a) - 3$

101–104 ■ Factor the expression completely. (This type of expression arises in calculus when using the "product rule.")

101. $5(x^2 + 4)^4(2x)(x - 2)^4 + (x^2 + 4)^5(4)(x - 2)^3$

102. $3(2x - 1)^2(2)(x + 3)^{1/2} + (2x - 1)^3(\tfrac{1}{2})(x + 3)^{-1/2}$

103. $(x^2 + 3)^{-1/3} - \tfrac{2}{3}x^2(x^2 + 3)^{-4/3}$

104. $\tfrac{1}{2}x^{-1/2}(3x + 4)^{1/2} - \tfrac{3}{2}x^{1/2}(3x + 4)^{-1/2}$

105. (a) Show that $ab = \tfrac{1}{2}[(a + b)^2 - (a^2 + b^2)]$.
   (b) Show that $(a^2 + b^2)^2 - (a^2 - b^2)^2 = 4a^2b^2$.
   (c) Show that
$$(a^2 + b^2)(c^2 + d^2) = (ac + bd)^2 + (ad - bc)^2$$
   (d) Factor completely: $4a^2c^2 - (a^2 - b^2 + c^2)^2$.

106. Verify Special Factoring Formulas 4 and 5 by expanding their right-hand sides.

Applications

107. Volume of Concrete   A culvert is constructed out of large cylindrical shells cast in concrete, as shown in the figure. Using the formula for the volume of a cylinder given on the inside back cover of this book, explain why the volume of the cylindrical shell is

$$V = \pi R^2 h - \pi r^2 h$$

Factor to show that

$$V = 2\pi \cdot \text{average radius} \cdot \text{height} \cdot \text{thickness}$$

Use the "unrolled" diagram to explain why this makes sense geometrically.

108. Mowing a Field   A square field in a certain state park is mowed around the edges every week. The rest of the field is kept unmowed to serve as a habitat for birds and small animals (see the figure). The field measures b feet by b feet, and the mowed strip is x feet wide.

(a) Explain why the area of the mowed portion is $b^2 - (b - 2x)^2$.

(b) Factor the expression in (a) to show that the area of the mowed portion is also $4x(b - x)$.

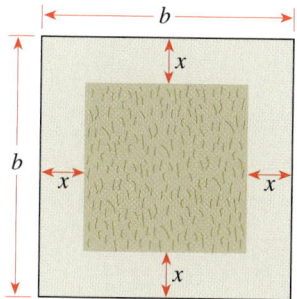

Discovery • Discussion

109. Degrees of Sums and Products of Polynomials   Make up several pairs of polynomials, then calculate the sum and product of each pair. Based on your experiments and observations, answer the following questions.

(a) How is the degree of the product related to the degrees of the original polynomials?

(b) How is the degree of the sum related to the degrees of the original polynomials?

110. The Power of Algebraic Formulas   Use the Difference of Squares Formula to factor $17^2 - 16^2$. Notice that it is easy to calculate the factored form in your head, but not so easy to calculate the original form in this way. Evaluate each expression in your head:

(a) $528^2 - 527^2$ **(b)** $122^2 - 120^2$ **(c)** $1020^2 - 1010^2$

Now use the Special Product Formula

$$(A + B)(A - B) = A^2 - B^2$$

to evaluate these products in your head:

(d) $79 \cdot 51$ **(e)** $998 \cdot 1002$

111. Differences of Even Powers

(a) Factor the expressions completely: $A^4 - B^4$ and $A^6 - B^6$.

(b) Verify that $18{,}335 = 12^4 - 7^4$ and that $2{,}868{,}335 = 12^6 - 7^6$.

(c) Use the results of parts (a) and (b) to factor the integers 18,335 and 2,868,335. Then show that in both of these factorizations, all the factors are prime numbers.

112. Factoring $A^n - 1$ Verify these formulas by expanding and simplifying the right-hand side.

$$A^2 - 1 = (A - 1)(A + 1)$$
$$A^3 - 1 = (A - 1)(A^2 + A + 1)$$
$$A^4 - 1 = (A - 1)(A^3 + A^2 + A + 1)$$

Based on the pattern displayed in this list, how do you think $A^5 - 1$ would factor? Verify your conjecture. Now generalize the pattern you have observed to obtain a factoring formula for $A^n - 1$, where n is a positive integer.

113. Factoring $x^4 + ax^2 + b$ A trinomial of the form $x^4 + ax^2 + b$ can sometimes be factored easily. For example, $x^4 + 3x^2 - 4 = (x^2 + 4)(x^2 - 1)$. But $x^4 + 3x^2 + 4$ cannot be factored in this way. Instead, we can use the following method.

$$x^4 + 3x^2 + 4 = (x^4 + 4x^2 + 4) - x^2 \qquad \text{Add and subtract } x^2$$
$$= (x^2 + 2)^2 - x^2 \qquad \text{Factor perfect square}$$
$$= [(x^2 + 2) - x][(x^2 + 2) + x] \qquad \text{Difference of squares}$$
$$= (x^2 - x + 2)(x^2 + x + 2)$$

Factor the following using whichever method is appropriate.

(a) $x^4 + x^2 - 2$

(b) $x^4 + 2x^2 + 9$

(c) $x^4 + 4x^2 + 16$

(d) $x^4 + 2x^2 + 1$

DISCOVERY PROJECT

Visualizing a Formula

Many of the Special Product Formulas that we learned in this section can be "seen" as geometrical facts about length, area, and volume. For example, the figure shows how the formula for the square of a binomial can be interpreted as a fact about areas of squares and rectangles.

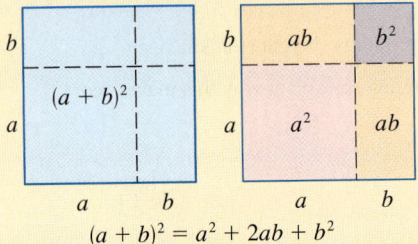

$$(a + b)^2 = a^2 + 2ab + b^2$$

In the figure, a and b represent lengths, a^2, b^2, ab, and $(a + b)^2$ represent areas. The ancient Greeks always interpreted algebraic formulas in terms of geometric figures as we have done here.

1. Explain how the figure verifies the formula $a^2 - b^2 = (a + b)(a - b)$.

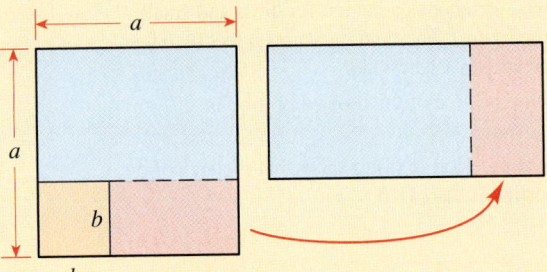

2. Find a figure that verifies the formula $(a - b)^2 = a^2 - 2ab + b^2$.

3. Explain how the figure verifies the formula
   $(a + b)^3 = a^3 + 3a^2b + 3ab^2 + b^3$.

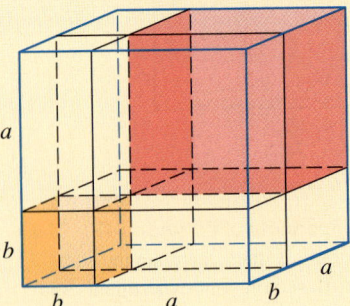

4. Is it possible to draw a geometric figure that verifies the formula for $(a + b)^4$? Explain.

5. (a) Expand $(a + b + c)^2$.

   (b) Make a geometric figure that verifies the formula you found in part (a).

| **1.4** | **Rational Expressions** |

A quotient of two algebraic expressions is called a **fractional expression**. Here are some examples:

$$\frac{2x}{x-1} \qquad \frac{\sqrt{x}+3}{x+1} \qquad \frac{y-2}{y^2+4}$$

A **rational expression** is a fractional expression where both the numerator and denominator are polynomials. For example, the following are rational expressions:

$$\frac{2x}{x-1} \qquad \frac{x}{x^2+1} \qquad \frac{x^3-x}{x^2-5x+6}$$

In this section we learn how to perform algebraic operations on rational expressions.

The Domain of an Algebraic Expression

Expression	Domain
$\dfrac{1}{x}$	$\{x \mid x \neq 0\}$
$\sqrt{x}$	$\{x \mid x \geq 0\}$
$\dfrac{1}{\sqrt{x}}$	$\{x \mid x > 0\}$

In general, an algebraic expression may not be defined for all values of the variable. The **domain** of an algebraic expression is the set of real numbers that the variable is permitted to have. The table in the margin gives some basic expressions and their domains.

Example 1   Finding the Domain of an Expression

Find the domains of the following expressions.

(a) $2x^2 + 3x - 1$ (b) $\dfrac{x}{x^2-5x+6}$ (c) $\dfrac{\sqrt{x}}{x-5}$

Solution

(a) This polynomial is defined for every x. Thus, the domain is the set $\mathbb{R}$ of real numbers.

(b) We first factor the denominator.

$$\frac{x}{x^2-5x+6} = \frac{x}{(x-2)(x-3)}$$

Denominator would be 0 if $x = 2$ or $x = 3$.

Since the denominator is zero when $x = 2$ or 3, the expression is not defined for these numbers. The domain is $\{x \mid x \neq 2 \text{ and } x \neq 3\}$.

(c) For the numerator to be defined, we must have $x \geq 0$. Also, we cannot divide by zero, so $x \neq 5$.

Must have $x \geq 0$ to take square root.

$$\frac{\sqrt{x}}{x-5}$$

Denominator would be 0 if $x = 5$.

Thus, the domain is $\{x \mid x \geq 0 \text{ and } x \neq 5\}$. ∎

Simplifying Rational Expressions

To **simplify rational expressions**, we factor both numerator and denominator and use the following property of fractions:

$$\frac{AC}{BC} = \frac{A}{B}$$

This allows us to **cancel** common factors from the numerator and denominator.

Example 2   Simplifying Rational Expressions by Cancellation

Simplify: $\dfrac{x^2 - 1}{x^2 + x - 2}$

Solution

 We can't cancel the x^2's in

$\dfrac{x^2 - 1}{x^2 + x - 2}$ because x^2 is not a factor.

$$\frac{x^2 - 1}{x^2 + x - 2} = \frac{(x - 1)(x + 1)}{(x - 1)(x + 2)} \qquad \textit{Factor}$$

$$= \frac{x + 1}{x + 2} \qquad \textit{Cancel common factors} \quad \blacksquare$$

Multiplying and Dividing Rational Expressions

To **multiply rational expressions**, we use the following property of fractions:

$$\frac{A}{B} \cdot \frac{C}{D} = \frac{AC}{BD}$$

This says that to multiply two fractions we multiply their numerators and multiply their denominators.

Example 3   Multiplying Rational Expressions

Perform the indicated multiplication and simplify: $\dfrac{x^2 + 2x - 3}{x^2 + 8x + 16} \cdot \dfrac{3x + 12}{x - 1}$

Solution   We first factor.

$$\frac{x^2 + 2x - 3}{x^2 + 8x + 16} \cdot \frac{3x + 12}{x - 1} = \frac{(x - 1)(x + 3)}{(x + 4)^2} \cdot \frac{3(x + 4)}{x - 1} \qquad \textit{Factor}$$

$$= \frac{3(x - 1)(x + 3)(x + 4)}{(x - 1)(x + 4)^2} \qquad \textit{Property of fractions}$$

$$= \frac{3(x + 3)}{x + 4} \qquad \textit{Cancel common factors} \quad \blacksquare$$

To **divide rational expressions**, we use the following property of fractions:

$$\frac{A}{B} \div \frac{C}{D} = \frac{A}{B} \cdot \frac{D}{C}$$

This says that to divide a fraction by another fraction we invert the divisor and multiply.

Example 4   Dividing Rational Expressions

Perform the indicated division and simplify: $\dfrac{x-4}{x^2-4} \div \dfrac{x^2-3x-4}{x^2+5x+6}$

Solution

$$\frac{x-4}{x^2-4} \div \frac{x^2-3x-4}{x^2+5x+6} = \frac{x-4}{x^2-4} \cdot \frac{x^2+5x+6}{x^2-3x-4} \qquad \text{Invert and multiply}$$

$$= \frac{(x-4)(x+2)(x+3)}{(x-2)(x+2)(x-4)(x+1)} \qquad \text{Factor}$$

$$= \frac{x+3}{(x-2)(x+1)} \qquad \text{Cancel common factors} \quad \blacksquare$$

Adding and Subtracting Rational Expressions

⊘ Avoid making the following error:

$$\frac{A}{B+C} \neq \frac{A}{B} + \frac{A}{C}$$

For instance, if we let $A = 2$, $B = 1$, and $C = 1$, then we see the error:

$$\frac{2}{1+1} \overset{?}{=} \frac{2}{1} + \frac{2}{1}$$

$$\frac{2}{2} \overset{?}{=} 2 + 2$$

$$1 \overset{?}{=} 4 \qquad \text{Wrong!}$$

To **add or subtract rational expressions**, we first find a common denominator and then use the following property of fractions:

$$\boxed{\frac{A}{C} + \frac{B}{C} = \frac{A+B}{C}}$$

Although any common denominator will work, it is best to use the **least common denominator** (LCD) as explained in Section 1.1. The LCD is found by factoring each denominator and taking the product of the distinct factors, using the highest power that appears in any of the factors.

Example 5   Adding and Subtracting Rational Expressions

Perform the indicated operations and simplify:

(a) $\dfrac{3}{x-1} + \dfrac{x}{x+2}$ (b) $\dfrac{1}{x^2-1} - \dfrac{2}{(x+1)^2}$

Solution

(a) Here the LCD is simply the product $(x-1)(x+2)$.

$$\frac{3}{x-1} + \frac{x}{x+2} = \frac{3(x+2)}{(x-1)(x+2)} + \frac{x(x-1)}{(x-1)(x+2)} \qquad \text{Write fractions using LCD}$$

$$= \frac{3x+6+x^2-x}{(x-1)(x+2)} \qquad \text{Add fractions}$$

$$= \frac{x^2+2x+6}{(x-1)(x+2)} \qquad \text{Combine terms in numerator}$$

(b) The LCD of $x^2 - 1 = (x - 1)(x + 1)$ and $(x + 1)^2$ is $(x - 1)(x + 1)^2$.

$$\frac{1}{x^2 - 1} - \frac{2}{(x + 1)^2} = \frac{1}{(x - 1)(x + 1)} - \frac{2}{(x + 1)^2} \qquad \text{Factor}$$

$$= \frac{(x + 1) - 2(x - 1)}{(x - 1)(x + 1)^2} \qquad \text{Combine fractions using LCD}$$

$$= \frac{x + 1 - 2x + 2}{(x - 1)(x + 1)^2} \qquad \text{Distributive Property}$$

$$= \frac{3 - x}{(x - 1)(x + 1)^2} \qquad \text{Combine terms in numerator}$$

Compound Fractions

A **compound fraction** is a fraction in which the numerator, the denominator, or both, are themselves fractional expressions.

Example 6   Simplifying a Compound Fraction

Simplify: $\dfrac{\dfrac{x}{y} + 1}{1 - \dfrac{y}{x}}$

Solution 1   We combine the terms in the numerator into a single fraction. We do the same in the denominator. Then we invert and multiply.

$$\frac{\dfrac{x}{y} + 1}{1 - \dfrac{y}{x}} = \frac{\dfrac{x + y}{y}}{\dfrac{x - y}{x}} = \frac{x + y}{y} \cdot \frac{x}{x - y}$$

$$= \frac{x(x + y)}{y(x - y)}$$

Solution 2   We find the LCD of all the fractions in the expression, then multiply numerator and denominator by it. In this example the LCD of all the fractions is xy. Thus

$$\frac{\dfrac{x}{y} + 1}{1 - \dfrac{y}{x}} = \frac{\dfrac{x}{y} + 1}{1 - \dfrac{y}{x}} \cdot \frac{xy}{xy} \qquad \text{Multiply numerator and denominator by } xy$$

$$= \frac{x^2 + xy}{xy - y^2} \qquad \text{Simplify}$$

$$= \frac{x(x + y)}{y(x - y)} \qquad \text{Factor}$$

The next two examples show situations in calculus that require the ability to work with fractional expressions.

Example 7 Simplifying a Compound Fraction

Simplify: $\dfrac{\dfrac{1}{a+h}-\dfrac{1}{a}}{h}$

Solution   We begin by combining the fractions in the numerator using a common denominator.

$$\dfrac{\dfrac{1}{a+h}-\dfrac{1}{a}}{h}=\dfrac{\dfrac{a-(a+h)}{a(a+h)}}{h} \qquad \text{\textit{Combine fractions in the numerator}}$$

$$=\dfrac{a-(a+h)}{a(a+h)}\cdot\dfrac{1}{h} \qquad \text{\textit{Property 2 of fractions (invert divisor and multiply)}}$$

$$=\dfrac{a-a-h}{a(a+h)}\cdot\dfrac{1}{h} \qquad \text{\textit{Distributive Property}}$$

$$=\dfrac{-h}{a(a+h)}\cdot\dfrac{1}{h} \qquad \text{\textit{Simplify}}$$

$$=\dfrac{-1}{a(a+h)} \qquad \text{\textit{Property 5 of fractions (cancel common factors)}}$$

■

Example 8 Simplifying a Compound Fraction

Simplify: $\dfrac{(1+x^2)^{1/2}-x^2(1+x^2)^{-1/2}}{1+x^2}$

Solution 1   Factor $(1+x^2)^{-1/2}$ from the numerator.

Factor out the power of $1+x^2$ with the *smallest* exponent, in this case $(1+x^2)^{-1/2}$.

$$\dfrac{(1+x^2)^{1/2}-x^2(1+x^2)^{-1/2}}{1+x^2}=\dfrac{(1+x^2)^{-1/2}[(1+x^2)-x^2]}{1+x^2}$$

$$=\dfrac{(1+x^2)^{-1/2}}{1+x^2}=\dfrac{1}{(1+x^2)^{3/2}}$$

Solution 2   Since $(1+x^2)^{-1/2}=1/(1+x^2)^{1/2}$ is a fraction, we can clear all fractions by multiplying numerator and denominator by $(1+x^2)^{1/2}$.

$$\dfrac{(1+x^2)^{1/2}-x^2(1+x^2)^{-1/2}}{1+x^2}=\dfrac{(1+x^2)^{1/2}-x^2(1+x^2)^{-1/2}}{1+x^2}\cdot\dfrac{(1+x^2)^{1/2}}{(1+x^2)^{1/2}}$$

$$=\dfrac{(1+x^2)-x^2}{(1+x^2)^{3/2}}=\dfrac{1}{(1+x^2)^{3/2}}$$

■

Rationalizing the Denominator or the Numerator

If a fraction has a denominator of the form $A + B\sqrt{C}$, we may rationalize the denominator by multiplying numerator and denominator by the **conjugate radical** $A - B\sqrt{C}$. This is effective because, by Special Product Formula 1 in Section 1.3, the product of the denominator and its conjugate radical does not contain a radical:

$$(A + B\sqrt{C})(A - B\sqrt{C}) = A^2 - B^2C$$

Example 9   Rationalizing the Denominator

Rationalize the denominator: $\dfrac{1}{1 + \sqrt{2}}$

Solution   We multiply both the numerator and the denominator by the conjugate radical of $1 + \sqrt{2}$, which is $1 - \sqrt{2}$.

Special Product Formula 1
$(a + b)(a - b) = a^2 - b^2$

$$\frac{1}{1 + \sqrt{2}} = \frac{1}{1 + \sqrt{2}} \cdot \frac{1 - \sqrt{2}}{1 - \sqrt{2}}$$ *Multiply numerator and denominator by the conjugate radical*

$$= \frac{1 - \sqrt{2}}{1^2 - (\sqrt{2})^2}$$ *Special Product Formula 1*

$$= \frac{1 - \sqrt{2}}{1 - 2} = \frac{1 - \sqrt{2}}{-1} = \sqrt{2} - 1$$ ∎

Example 10   Rationalizing the Numerator

Rationalize the numerator: $\dfrac{\sqrt{4 + h} - 2}{h}$

Solution   We multiply numerator and denominator by the conjugate radical $\sqrt{4 + h} + 2$.

$$\frac{\sqrt{4 + h} - 2}{h} = \frac{\sqrt{4 + h} - 2}{h} \cdot \frac{\sqrt{4 + h} + 2}{\sqrt{4 + h} + 2}$$ *Multiply numerator and denominator by the conjugate radical*

Special Product Formula 1
$(a + b)(a - b) = a^2 - b^2$

$$= \frac{(\sqrt{4 + h})^2 - 2^2}{h(\sqrt{4 + h} + 2)}$$ *Special Product Formula 1*

$$= \frac{4 + h - 4}{h(\sqrt{4 + h} + 2)}$$

$$= \frac{h}{h(\sqrt{4 + h} + 2)} = \frac{1}{\sqrt{4 + h} + 2}$$ *Property 5 of fractions (cancel common factors)*   ∎

Avoiding Common Errors

 Don't make the mistake of applying properties of multiplication to the operation of addition. Many of the common errors in algebra involve doing just that. The following table states several properties of multiplication and illustrates the error in applying them to addition.

Correct multiplication property	Common error with addition
$(a \cdot b)^2 = a^2 \cdot b^2$	$(a + b)^2 \neq a^2 + b^2$
$\sqrt{a \cdot b} = \sqrt{a}\,\sqrt{b} \quad (a, b \geq 0)$	$\sqrt{a + b} \neq \sqrt{a} + \sqrt{b}$
$\sqrt{a^2 \cdot b^2} = a \cdot b \quad (a, b \geq 0)$	$\sqrt{a^2 + b^2} \neq a + b$
$\dfrac{1}{a} \cdot \dfrac{1}{b} = \dfrac{1}{a \cdot b}$	$\dfrac{1}{a} + \dfrac{1}{b} \neq \dfrac{1}{a + b}$
$\dfrac{ab}{a} = b$	$\dfrac{a + b}{a} \neq b$
$a^{-1} \cdot b^{-1} = (a \cdot b)^{-1}$	$a^{-1} + b^{-1} \neq (a + b)^{-1}$

To verify that the equations in the right-hand column are wrong, simply substitute numbers for a and b and calculate each side. For example, if we take $a = 2$ and $b = 2$ in the fourth error, we find that the left-hand side is

$$\frac{1}{a} + \frac{1}{b} = \frac{1}{2} + \frac{1}{2} = 1$$

whereas the right-hand side is

$$\frac{1}{a + b} = \frac{1}{2 + 2} = \frac{1}{4}$$

Since $1 \neq \frac{1}{4}$, the stated equation is wrong. You should similarly convince yourself of the error in each of the other equations. (See Exercise 97.)

1.4   Exercises

1–6 ■ Find the domain of the expression.

1. $4x^2 - 10x + 3$

2. $-x^4 + x^3 + 9x$

3. $\dfrac{2x + 1}{x - 4}$

4. $\dfrac{2t^2 - 5}{3t + 6}$

5. $\sqrt{x + 3}$

6. $\dfrac{1}{\sqrt{x - 1}}$

7–16 ■ Simplify the rational expression.

7. $\dfrac{3(x + 2)(x - 1)}{6(x - 1)^2}$

8. $\dfrac{4(x^2 - 1)}{12(x + 2)(x - 1)}$

9. $\dfrac{x - 2}{x^2 - 4}$

10. $\dfrac{x^2 - x - 2}{x^2 - 1}$

11. $\dfrac{x^2 + 6x + 8}{x^2 + 5x + 4}$

12. $\dfrac{x^2 - x - 12}{x^2 + 5x + 6}$

13. $\dfrac{y^2 + y}{y^2 - 1}$

14. $\dfrac{y^2 - 3y - 18}{2y^2 + 5y + 3}$

15. $\dfrac{2x^3 - x^2 - 6x}{2x^2 - 7x + 6}$

16. $\dfrac{1 - x^2}{x^3 - 1}$

17–30 ■ Perform the multiplication or division and simplify.

17. $\dfrac{4x}{x^2 - 4} \cdot \dfrac{x + 2}{16x}$

18. $\dfrac{x^2 - 25}{x^2 - 16} \cdot \dfrac{x + 4}{x + 5}$

19. $\dfrac{x^2 - x - 12}{x^2 - 9} \cdot \dfrac{3 + x}{4 - x}$

20. $\dfrac{x^2 + 2x - 3}{x^2 - 2x - 3} \cdot \dfrac{3 - x}{3 + x}$

21. $\dfrac{t - 3}{t^2 + 9} \cdot \dfrac{t + 3}{t^2 - 9}$

22. $\dfrac{x^2 - x - 6}{x^2 + 2x} \cdot \dfrac{x^3 + x^2}{x^2 - 2x - 3}$

23. $\dfrac{x^2 + 7x + 12}{x^2 + 3x + 2} \cdot \dfrac{x^2 + 5x + 6}{x^2 + 6x + 9}$

24. $\dfrac{x^2 + 2xy + y^2}{x^2 - y^2} \cdot \dfrac{2x^2 - xy - y^2}{x^2 - xy - 2y^2}$

25. $\dfrac{2x^2 + 3x + 1}{x^2 + 2x - 15} \div \dfrac{x^2 + 6x + 5}{2x^2 - 7x + 3}$

26. $\dfrac{4y^2 - 9}{2y^2 + 9y - 18} \div \dfrac{2y^2 + y - 3}{y^2 + 5y - 6}$

27. $\dfrac{\dfrac{x^3}{x + 1}}{\dfrac{x}{x^2 + 2x + 1}}$

28. $\dfrac{\dfrac{2x^2 - 3x - 2}{x^2 - 1}}{\dfrac{2x^2 + 5x + 2}{x^2 + x - 2}}$

29. $\dfrac{x/y}{z}$

30. $\dfrac{x}{y/z}$

31–50 ■ Perform the addition or subtraction and simplify.

31. $2 + \dfrac{x}{x + 3}$

32. $\dfrac{2x - 1}{x + 4} - 1$

33. $\dfrac{1}{x + 5} + \dfrac{2}{x - 3}$

34. $\dfrac{1}{x + 1} + \dfrac{1}{x - 1}$

35. $\dfrac{1}{x + 1} - \dfrac{1}{x + 2}$

36. $\dfrac{x}{x - 4} - \dfrac{3}{x + 6}$

37. $\dfrac{x}{(x + 1)^2} + \dfrac{2}{x + 1}$

38. $\dfrac{5}{2x - 3} - \dfrac{3}{(2x - 3)^2}$

39. $u + 1 + \dfrac{u}{u + 1}$

40. $\dfrac{2}{a^2} - \dfrac{3}{ab} + \dfrac{4}{b^2}$

41. $\dfrac{1}{x^2} + \dfrac{1}{x^2 + x}$

42. $\dfrac{1}{x} + \dfrac{1}{x^2} + \dfrac{1}{x^3}$

43. $\dfrac{2}{x + 3} - \dfrac{1}{x^2 + 7x + 12}$

44. $\dfrac{x}{x^2 - 4} + \dfrac{1}{x - 2}$

45. $\dfrac{1}{x + 3} + \dfrac{1}{x^2 - 9}$

46. $\dfrac{x}{x^2 + x - 2} - \dfrac{2}{x^2 - 5x + 4}$

47. $\dfrac{2}{x} + \dfrac{3}{x - 1} - \dfrac{4}{x^2 - x}$

48. $\dfrac{x}{x^2 - x - 6} - \dfrac{1}{x + 2} - \dfrac{2}{x - 3}$

49. $\dfrac{1}{x^2 + 3x + 2} - \dfrac{1}{x^2 - 2x - 3}$

50. $\dfrac{1}{x + 1} - \dfrac{2}{(x + 1)^2} + \dfrac{3}{x^2 - 1}$

51–60 ■ Simplify the compound fractional expression.

51. $\dfrac{\dfrac{x}{y} - \dfrac{y}{x}}{\dfrac{1}{x^2} - \dfrac{1}{y^2}}$

52. $x - \dfrac{y}{\dfrac{x}{y} + \dfrac{y}{x}}$

53. $\dfrac{1 + \dfrac{1}{c - 1}}{1 - \dfrac{1}{c - 1}}$

54. $1 + \dfrac{1}{1 + \dfrac{1}{1 + x}}$

55. $\dfrac{\dfrac{5}{x - 1} - \dfrac{2}{x + 1}}{\dfrac{x}{x - 1} + \dfrac{1}{x + 1}}$

56. $\dfrac{\dfrac{a - b}{a} - \dfrac{a + b}{b}}{\dfrac{a - b}{b} + \dfrac{a + b}{a}}$

57. $\dfrac{x^{-2} - y^{-2}}{x^{-1} + y^{-1}}$

58. $\dfrac{x^{-1} + y^{-1}}{(x + y)^{-1}}$

59. $\dfrac{1}{1 + a^n} + \dfrac{1}{1 + a^{-n}}$

60. $\dfrac{\left(a + \dfrac{1}{b}\right)^m \left(a - \dfrac{1}{b}\right)^n}{\left(b + \dfrac{1}{a}\right)^m \left(b - \dfrac{1}{a}\right)^n}$

61–66 ■ Simplify the fractional expression. (Expressions like these arise in calculus.)

61. $\dfrac{\dfrac{1}{a + h} - \dfrac{1}{a}}{h}$

62. $\dfrac{(x + h)^{-3} - x^{-3}}{h}$

63. $\dfrac{\dfrac{1 - (x + h)}{2 + (x + h)} - \dfrac{1 - x}{2 + x}}{h}$

64. $\dfrac{(x + h)^3 - 7(x + h) - (x^3 - 7x)}{h}$

65. $\sqrt{1 + \left(\dfrac{x}{\sqrt{1 - x^2}}\right)^2}$

66. $\sqrt{1 + \left(x^3 - \dfrac{1}{4x^3}\right)^2}$

67–72 ■ Simplify the expression. (This type of expression arises in calculus when using the "quotient rule.")

67. $\dfrac{3(x + 2)^2(x - 3)^2 - (x + 2)^3(2)(x - 3)}{(x - 3)^4}$

68. $\dfrac{2x(x + 6)^4 - x^2(4)(x + 6)^3}{(x + 6)^8}$

69. $\dfrac{2(1 + x)^{1/2} - x(1 + x)^{-1/2}}{x + 1}$

70. $\dfrac{(1 - x^2)^{1/2} + x^2(1 - x^2)^{-1/2}}{1 - x^2}$

71. $\dfrac{3(1 + x)^{1/3} - x(1 + x)^{-2/3}}{(1 + x)^{2/3}}$

72. $\dfrac{(7 - 3x)^{1/2} + \frac{3}{2}x(7 - 3x)^{-1/2}}{7 - 3x}$

73–78 ■ Rationalize the denominator.

73. $\dfrac{1}{2 - \sqrt{3}}$

74. $\dfrac{2}{3 - \sqrt{5}}$

75. $\dfrac{2}{\sqrt{2} + \sqrt{7}}$

76. $\dfrac{1}{\sqrt{x} + 1}$

77. $\dfrac{y}{\sqrt{3} + \sqrt{y}}$

78. $\dfrac{2(x - y)}{\sqrt{x} - \sqrt{y}}$

79–84 ■ Rationalize the numerator.

79. $\dfrac{1 - \sqrt{5}}{3}$

80. $\dfrac{\sqrt{3} + \sqrt{5}}{2}$

81. $\dfrac{\sqrt{r} + \sqrt{2}}{5}$

82. $\dfrac{\sqrt{x} - \sqrt{x + h}}{h\sqrt{x}\sqrt{x + h}}$

83. $\sqrt{x^2 + 1} - x$

84. $\sqrt{x + 1} - \sqrt{x}$

85–92 ■ State whether the given equation is true for all values of the variables. (Disregard any value that makes a denominator zero.)

85. $\dfrac{16 + a}{16} = 1 + \dfrac{a}{16}$

86. $\dfrac{b}{b - c} = 1 - \dfrac{b}{c}$

87. $\dfrac{2}{4 + x} = \dfrac{1}{2} + \dfrac{2}{x}$

88. $\dfrac{x + 1}{y + 1} = \dfrac{x}{y}$

89. $\dfrac{x}{x + y} = \dfrac{1}{1 + y}$

90. $2\left(\dfrac{a}{b}\right) = \dfrac{2a}{2b}$

91. $\dfrac{-a}{b} = -\dfrac{a}{b}$

92. $\dfrac{1 + x + x^2}{x} = \dfrac{1}{x} + 1 + x$

Applications

93. Electrical Resistance   If two electrical resistors with resistances R_1 and R_2 are connected in parallel (see the figure), then the total resistance R is given by

$$R = \dfrac{1}{\dfrac{1}{R_1} + \dfrac{1}{R_2}}$$

(a) Simplify the expression for R.

(b) If $R_1 = 10$ ohms and $R_2 = 20$ ohms, what is the total resistance R?

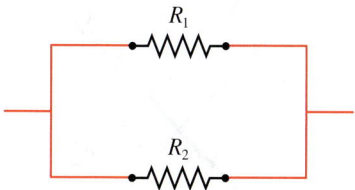

94. Average Cost   A clothing manufacturer finds that the cost of producing x shirts is $500 + 6x + 0.01x^2$ dollars.

(a) Explain why the average cost per shirt is given by the rational expression

$$A = \dfrac{500 + 6x + 0.01x^2}{x}$$

(b) Complete the table by calculating the average cost per shirt for the given values of x.

x	Average cost
10	
20	
50	
100	
200	
500	
1000	

Discovery • Discussion

95. Limiting Behavior of a Rational Expression   The rational expression

$$\dfrac{x^2 - 9}{x - 3}$$

is not defined for $x = 3$. Complete the tables and determine what value the expression approaches as x gets closer and closer to 3. Why is this reasonable? Factor the numerator of the expression and simplify to see why.

x	$\dfrac{x^2 - 9}{x - 3}$	x	$\dfrac{x^2 - 9}{x - 3}$
2.80		3.20	
2.90		3.10	
2.95		3.05	
2.99		3.01	
2.999		3.001	

96. Is This Rationalization?   In the expression $2/\sqrt{x}$ we would eliminate the radical if we were to square both numerator and denominator. Is this the same thing as rationalizing the denominator?

97. Algebraic Errors   The left-hand column in the table lists some common algebraic errors. In each case, give an example using numbers that show that the formula is not valid. An example of this type, which shows that a statement is false, is called a *counterexample*.

Algebraic error	Counterexample
$\dfrac{1}{a} + \dfrac{1}{b} \;\overset{\times}{=}\; \dfrac{1}{a+b}$	$\dfrac{1}{2} + \dfrac{1}{2} \neq \dfrac{1}{2+2}$
$(a+b)^2 \;\overset{\times}{=}\; a^2 + b^2$	
$\sqrt{a^2 + b^2} \;\overset{\times}{=}\; a + b$	
$\dfrac{a+b}{a} \;\overset{\times}{=}\; b$	
$(a^3 + b^3)^{1/3} \;\overset{\times}{=}\; a + b$	
$a^m/a^n \;\overset{\times}{=}\; a^{m/n}$	
$a^{-1/n} \;\overset{\times}{=}\; \dfrac{1}{a^n}$	

98. The Form of an Algebraic Expression   An algebraic expression may look complicated, but its "form" is always simple; it must be a sum, a product, a quotient, or a power. For example, consider the following expressions:

$$(1 + x^2)^2 + \left(\frac{x+2}{x+1}\right)^3 \qquad (1 + x)\left(1 + \frac{x+5}{1+x^4}\right)$$

$$\frac{5 - x^3}{1 + \sqrt{1 + x^2}} \qquad \sqrt{\frac{1+x}{1-x}}$$

With appropriate choices for A and B, the first has the form $A + B$, the second AB, the third A/B, and the fourth $A^{1/2}$. Recognizing the form of an expression helps us expand, simplify, or factor it correctly. Find the form of the following algebraic expressions.

(a) $x + \sqrt{1 + \dfrac{1}{x}}$ **(b)** $(1 + x^2)(1 + x)^3$

(c) $\sqrt[3]{x^4(4x^2 + 1)}$ **(d)** $\dfrac{1 - 2\sqrt{1 + x}}{1 + \sqrt{1 + x^2}}$

1.5 Equations

An equation is a statement that two mathematical expressions are equal. For example,

$$3 + 5 = 8$$

is an equation. Most equations that we study in algebra contain variables, which are symbols (usually letters) that stand for numbers. In the equation

$$4x + 7 = 19$$

the letter x is the variable. We think of x as the "unknown" in the equation, and our goal is to find the value of x that makes the equation true. The values of the unknown that make the equation true are called the **solutions** or **roots** of the equation, and the process of finding the solutions is called **solving the equation**.

Two equations with exactly the same solutions are called **equivalent equations**. To solve an equation, we try to find a simpler, equivalent equation in which the variable stands alone on one side of the "equal" sign. Here are the properties that we use to solve an equation. (In these properties, A, B, and C stand for any algebraic expressions, and the symbol $\Leftrightarrow$ means "is equivalent to.")

$x = 3$ is a solution of the equation $4x + 7 = 19$, because substituting $x = 3$ makes the equation true:

$$x = 3$$

$$4(3) + 7 = 19 \qquad \checkmark$$

Properties of Equality

Property	Description
1. $A = B \Leftrightarrow A + C = B + C$	Adding the same quantity to both sides of an equation gives an equivalent equation.
2. $A = B \Leftrightarrow CA = CB \quad (C \neq 0)$	Multiplying both sides of an equation by the same nonzero quantity gives an equivalent equation.

These properties require that you *perform the same operation on both sides of an equation* when solving it. Thus, if we say "*add* -7" when solving an equation, that is just a short way of saying "*add* -7 to each side of the equation."

Linear Equations

The simplest type of equation is a *linear equation*, or first-degree equation, which is an equation in which each term is either a constant or a nonzero multiple of the variable.

Linear Equations

A **linear equation** in one variable is an equation equivalent to one of the form

$$ax + b = 0$$

where a and b are real numbers and x is the variable.

Here are some examples that illustrate the difference between linear and nonlinear equations.

Linear equations	Nonlinear equations	
$4x - 5 = 3$	$x^2 + 2x = 8$	Not linear; contains the square of the variable
$2x = \frac{1}{2}x - 7$	$\sqrt{x} - 6x = 0$	Not linear; contains the square root of the variable
$x - 6 = \dfrac{x}{3}$	$\dfrac{3}{x} - 2x = 1$	Not linear; contains the reciprocal of the variable

Example 1 Solving a Linear Equation

Solve the equation $7x - 4 = 3x + 8$.

Solution We solve this by changing it to an equivalent equation with all terms that have the variable x on one side and all constant terms on the other.

$$
\begin{aligned}
7x - 4 &= 3x + 8 && \text{Given equation} \\
(7x - 4) + 4 &= (3x + 8) + 4 && \text{Add 4} \\
7x &= 3x + 12 && \text{Simplify} \\
7x - 3x &= (3x + 12) - 3x && \text{Subtract 3x} \\
4x &= 12 && \text{Simplify} \\
\tfrac{1}{4} \cdot 4x &= \tfrac{1}{4} \cdot 12 && \text{Multiply by } \tfrac{1}{4} \\
x &= 3 && \text{Simplify}
\end{aligned}
$$
∎

Because it is important to CHECK YOUR ANSWER, we do this in many of our examples. In these checks, LHS stands for "left-hand side" and RHS stands for "right-hand side" of the original equation.

Check Your Answer

$x = 3$:

$x = 3$	$x = 3$
LHS $= 7(3) - 4$	RHS $= 3(3) + 8$
$= 17$	$= 17$

LHS = RHS ✓

Many formulas in the sciences involve several variables, and it is often necessary to express one of the variables in terms of the others. In the next example we solve for a variable in Newton's Law of Gravity.

Example 2   Solving for One Variable in Terms of Others

Solve for the variable M in the equation

$$F = G\frac{mM}{r^2}$$

This is Newton's Law of Gravity. It gives the gravitational force F between two masses m and M that are a distance r apart. The constant G is the universal gravitational constant.

Solution   Although this equation involves more than one variable, we solve it as usual by isolating M on one side and treating the other variables as we would numbers.

$$F = \left(\frac{Gm}{r^2}\right)M \qquad \text{Factor M from RHS}$$

$$\left(\frac{r^2}{Gm}\right)F = \left(\frac{r^2}{Gm}\right)\left(\frac{Gm}{r^2}\right)M \qquad \text{Multiply by reciprocal of } \frac{Gm}{r^2}$$

$$\frac{r^2F}{Gm} = M \qquad \text{Simplify}$$

The solution is $M = \dfrac{r^2F}{Gm}$. ∎

Example 3   Solving for One Variable in Terms of Others

The surface area A of the closed rectangular box shown in Figure 1 can be calculated from the length l, the width w, and the height h according to the formula

$$A = 2lw + 2wh + 2lh$$

Solve for w in terms of the other variables in this equation.

Solution   Although this equation involves more than one variable, we solve it as usual by isolating w on one side, treating the other variables as we would numbers.

$$A = (2lw + 2wh) + 2lh \qquad \text{Collect terms involving } w$$

$$A - 2lh = 2lw + 2wh \qquad \text{Subtract } 2lh$$

$$A - 2lh = (2l + 2h)w \qquad \text{Factor } w \text{ from RHS}$$

$$\frac{A - 2lh}{2l + 2h} = w \qquad \text{Divide by } 2l + 2h$$

The solution is $w = \dfrac{A - 2lh}{2l + 2h}$. ∎

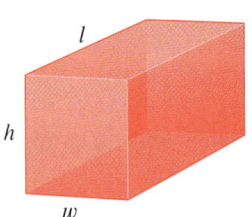

Figure 1

A closed rectangular box

Quadratic Equations

Linear equations are first-degree equations like $2x + 1 = 5$ or $4 - 3x = 2$. Quadratic equations are second-degree equations like $x^2 + 2x - 3 = 0$ or $2x^2 + 3 = 5x$.

Quadratic Equations

$$x^2 - 2x - 8 = 0$$

$$3x + 10 = 4x^2$$

$$\tfrac{1}{2}x^2 + \tfrac{1}{3}x - \tfrac{1}{6} = 0$$

Quadratic Equations

A **quadratic equation** is an equation of the form

$$ax^2 + bx + c = 0$$

where a, b, and c are real numbers with $a \neq 0$.

Some quadratic equations can be solved by factoring and using the following basic property of real numbers.

Zero-Product Property

$$AB = 0 \qquad \text{if and only if} \qquad A = 0 \quad \text{or} \quad B = 0$$

 This means that if we can factor the left-hand side of a quadratic (or other) equation, then we can solve it by setting each factor equal to 0 in turn. This method works only when the right-hand side of the equation is 0.

Example 4   Solving a Quadratic Equation by Factoring

Solve the equation $x^2 + 5x = 24$.

Solution   We must first rewrite the equation so that the right-hand side is 0.

Check Your Answers

$x = 3$:

$(3)^2 + 5(3) = 9 + 15 = 24$ ✓

$x = -8$:

$(-8)^2 + 5(-8) = 64 - 40 = 24$ ✓

$$x^2 + 5x = 24$$

$$x^2 + 5x - 24 = 0 \qquad \text{Subtract 24}$$

$$(x - 3)(x + 8) = 0 \qquad \text{Factor}$$

$$x - 3 = 0 \quad \text{or} \quad x + 8 = 0 \qquad \text{Zero-Product Property}$$

$$x = 3 \qquad\qquad x = -8 \qquad \text{Solve}$$

The solutions are $x = 3$ and $x = -8$. ■

Do you see why one side of the equation must be 0 in Example 4? Factoring the equation as $x(x + 5) = 24$ does not help us find the solutions, since 24 can be factored in infinitely many ways, such as $6 \cdot 4$, $\tfrac{1}{2} \cdot 48$, $\left(-\tfrac{2}{5}\right) \cdot (-60)$, and so on.

A quadratic equation of the form $x^2 - c = 0$, where c is a positive constant, factors as $(x - \sqrt{c})(x + \sqrt{c}) = 0$, and so the solutions are $x = \sqrt{c}$ and $x = -\sqrt{c}$. We often abbreviate this as $x = \pm\sqrt{c}$.

Solving a Simple Quadratic Equation

The solutions of the equation $x^2 = c$ are $x = \sqrt{c}$ and $x = -\sqrt{c}$.

Example 5   Solving Simple Quadratics

Solve each equation.

(a) $x^2 = 5$ (b) $(x - 4)^2 = 5$

Solution

(a) From the principle in the preceding box, we get $x = \pm\sqrt{5}$.

(b) We can take the square root of each side of this equation as well.

$$(x - 4)^2 = 5$$

$$x - 4 = \pm\sqrt{5} \qquad \textit{Take the square root}$$

$$x = 4 \pm \sqrt{5} \qquad \textit{Add 4}$$

The solutions are $x = 4 + \sqrt{5}$ and $x = 4 - \sqrt{5}$.   ∎

See page 30 for how to recognize when a quadratic expression is a perfect square.

As we saw in Example 5, if a quadratic equation is of the form $(x \pm a)^2 = c$, then we can solve it by taking the square root of each side. In an equation of this form the left-hand side is a *perfect square*: the square of a linear expression in x. So, if a quadratic equation does not factor readily, then we can solve it using the technique of **completing the square**. This means that we add a constant to an expression to make it a perfect square. For example, to make $x^2 - 6x$ a perfect square we must add 9, since $x^2 - 6x + 9 = (x - 3)^2$.

Completing the Square

Area of blue region is

$$x^2 + 2\left(\frac{b}{2}\right)x = x^2 + bx$$

Add a small square of area $(b/2)^2$ to "complete" the square.

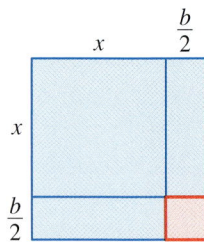

> **Completing the Square**
>
> To make $x^2 + bx$ a perfect square, add $\left(\dfrac{b}{2}\right)^2$, the square of half the coefficient of x. This gives the perfect square
>
> $$x^2 + bx + \left(\frac{b}{2}\right)^2 = \left(x + \frac{b}{2}\right)^2$$

Example 6   Solving Quadratic Equations by Completing the Square

Solve each equation.

(a) $x^2 - 8x + 13 = 0$ (b) $3x^2 - 12x + 6 = 0$

🚫 When completing the square, make sure the coefficient of x^2 is 1. If it isn't, you must factor this coefficient from both terms that contain x:

$$ax^2 + bx = a\left(x^2 + \frac{b}{a}x\right)$$

Then complete the square inside the parentheses. Remember that the term added inside the parentheses is multiplied by a.

Solution

(a) $x^2 - 8x + 13 = 0 \qquad \textit{Given equation}$

$$x^2 - 8x = -13 \qquad \textit{Subtract 13}$$

$$x^2 - 8x + 16 = -13 + 16 \qquad \textit{Complete the square: add } \left(\frac{-8}{2}\right)^2 = 16$$

$$(x - 4)^2 = 3 \qquad \textit{Perfect square}$$

$$x - 4 = \pm\sqrt{3} \qquad \textit{Take square root}$$

$$x = 4 \pm \sqrt{3} \qquad \textit{Add 4}$$

(b) After subtracting 6 from each side of the equation, we must factor the coefficient of x^2 (the 3) from the left side to put the equation in the correct form for completing the square.

$$3x^2 - 12x + 6 = 0 \qquad \text{\small\color{blue}Given equation}$$

$$3x^2 - 12x = -6 \qquad \text{\small\color{blue}Subtract 6}$$

$$3(x^2 - 4x) = -6 \qquad \text{\small\color{blue}Factor 3 from LHS}$$

Now we complete the square by adding $(-2)^2 = 4$ *inside* the parentheses. Since everything inside the parentheses is multiplied by 3, this means that we are actually adding $3 \cdot 4 = 12$ to the left side of the equation. Thus, we must add 12 to the right side as well.

$$3(x^2 - 4x + 4) = -6 + 3 \cdot 4 \qquad \text{\small\color{blue}Complete the square: add 4}$$

$$3(x - 2)^2 = 6 \qquad \text{\small\color{blue}Perfect square}$$

$$(x - 2)^2 = 2 \qquad \text{\small\color{blue}Divide by 3}$$

$$x - 2 = \pm\sqrt{2} \qquad \text{\small\color{blue}Take square root}$$

$$x = 2 \pm \sqrt{2} \qquad \text{\small\color{blue}Add 2} \qquad \blacksquare$$

We can use the technique of completing the square to derive a formula for the roots of the general quadratic equation $ax^2 + bx + c = 0$.

The Quadratic Formula

The roots of the quadratic equation $ax^2 + bx + c = 0$, where $a \neq 0$, are

$$x = \frac{-b \pm \sqrt{b^2 - 4ac}}{2a}$$

■ **Proof** First, we divide each side of the equation by a and move the constant to the right side, giving

$$x^2 + \frac{b}{a}x = -\frac{c}{a} \qquad \text{\small\color{blue}Divide by } a$$

We now complete the square by adding $(b/2a)^2$ to each side of the equation:

$$x^2 + \frac{b}{a}x + \left(\frac{b}{2a}\right)^2 = -\frac{c}{a} + \left(\frac{b}{2a}\right)^2 \qquad \text{\small\color{blue}Complete the square: Add } \left(\frac{b}{2a}\right)^2$$

$$\left(x + \frac{b}{2a}\right)^2 = \frac{-4ac + b^2}{4a^2} \qquad \text{\small\color{blue}Perfect square}$$

$$x + \frac{b}{2a} = \pm\frac{\sqrt{b^2 - 4ac}}{2a} \qquad \text{\small\color{blue}Take square root}$$

$$x = \frac{-b \pm \sqrt{b^2 - 4ac}}{2a} \qquad \text{\small\color{blue}Subtract } \frac{b}{2a} \qquad \blacksquare$$

The quadratic formula could be used to solve the equations in Examples 4 and 6. You should carry out the details of these calculations.

François Viète (1540–1603) had a successful political career before taking up mathematics late in life. He became one of the most famous French mathematicians of the 16th century. Viète introduced a new level of abstraction in algebra by using letters to stand for *known* quantities in an equation. Before Viète's time, each equation had to be solved on its own. For instance, the quadratic equations

$$3x^2 + 2x + 8 = 0$$

$$5x^2 - 6x + 4 = 0$$

had to be solved separately by completing the square. Viète's idea was to consider all quadratic equations at once by writing

$$ax^2 + bx + c = 0$$

where a, b, and c are known quantities. Thus, he made it possible to write a *formula* (in this case, the quadratic formula) involving a, b, and c that can be used to solve all such equations in one fell swoop.

Viète's mathematical genius proved quite valuable during a war between France and Spain. To communicate with their troops, the Spaniards used a complicated code that Viète managed to decipher. Unaware of Viète's accomplishment, the Spanish king, Philip II, protested to the Pope, claiming that the French were using witchcraft to read his messages.

Example 7   **Using the Quadratic Formula**

Find all solutions of each equation.

(a) $3x^2 - 5x - 1 = 0$ (b) $4x^2 + 12x + 9 = 0$ (c) $x^2 + 2x + 2 = 0$

Solution

(a) In this quadratic equation $a = 3$, $b = -5$, and $c = -1$.

$$b = -5$$

$$3x^2 - 5x - 1 = 0$$

$$a = 3 \qquad c = -1$$

By the quadratic formula,

$$x = \frac{-(-5) \pm \sqrt{(-5)^2 - 4(3)(-1)}}{2(3)} = \frac{5 \pm \sqrt{37}}{6}$$

If approximations are desired, we can use a calculator to obtain

$$x = \frac{5 + \sqrt{37}}{6} \approx 1.8471 \qquad \text{and} \qquad x = \frac{5 - \sqrt{37}}{6} \approx -0.1805$$

Another Method

$$4x^2 + 12x + 9 = 0$$

$$(2x + 3)^2 = 0$$

$$2x + 3 = 0$$

$$x = -\tfrac{3}{2}$$

(b) Using the quadratic formula with $a = 4$, $b = 12$, and $c = 9$ gives

$$x = \frac{-12 \pm \sqrt{(12)^2 - 4 \cdot 4 \cdot 9}}{2 \cdot 4} = \frac{-12 \pm 0}{8} = -\frac{3}{2}$$

This equation has only one solution, $x = -\tfrac{3}{2}$.

(c) Using the quadratic formula with $a = 1$, $b = 2$, and $c = 2$ gives

$$x = \frac{-2 \pm \sqrt{2^2 - 4 \cdot 2}}{2} = \frac{-2 \pm \sqrt{-4}}{2} = \frac{-2 \pm 2\sqrt{-1}}{2} = -1 \pm \sqrt{-1}$$

Since the square of any real number is nonnegative, $\sqrt{-1}$ is undefined in the real number system. The equation has no real solution. ∎

In Section 3.4 we study the complex number system, in which the square roots of negative numbers do exist. The equation in Example 7(c) does have solutions in the complex number system.

The quantity $b^2 - 4ac$ that appears under the square root sign in the quadratic formula is called the *discriminant* of the equation $ax^2 + bx + c = 0$ and is given the symbol D. If $D < 0$, then $\sqrt{b^2 - 4ac}$ is undefined, and the quadratic equation has no real solution, as in Example 7(c). If $D = 0$, then the equation has only one real solution, as in Example 7(b). Finally, if $D > 0$, then the equation has two distinct real solutions, as in Example 7(a). The following box summarizes these observations.

The Discriminant

The **discriminant** of the general quadratic $ax^2 + bx + c = 0$ $(a \neq 0)$ is $D = b^2 - 4ac$.

1. If $D > 0$, then the equation has two distinct real solutions.

2. If $D = 0$, then the equation has exactly one real solution.

3. If $D < 0$, then the equation has no real solution.

Example 8 Using the Discriminant

Use the discriminant to determine how many real solutions each equation has.

(a) $x^2 + 4x - 1 = 0$ (b) $4x^2 - 12x + 9 = 0$ (c) $\frac{1}{3}x^2 - 2x + 4 = 0$

Solution

(a) The discriminant is $D = 4^2 - 4(1)(-1) = 20 > 0$, so the equation has two distinct real solutions.

(b) The discriminant is $D = (-12)^2 - 4 \cdot 4 \cdot 9 = 0$, so the equation has exactly one real solution.

(c) The discriminant is $D = (-2)^2 - 4(\frac{1}{3})4 = -\frac{4}{3} < 0$, so the equation has no real solution. ∎

Now let's consider a real-life situation that can be modeled by a quadratic equation.

Example 9 The Path of a Projectile

An object thrown or fired straight upward at an initial speed of v_0 ft/s will reach a height of h feet after t seconds, where h and t are related by the formula

$$h = -16t^2 + v_0 t$$

Suppose that a bullet is shot straight upward with an initial speed of 800 ft/s. Its path is shown in Figure 2.

(a) When does the bullet fall back to ground level?

(b) When does it reach a height of 6400 ft?

(c) When does it reach a height of 2 mi?

(d) How high is the highest point the bullet reaches?

This formula depends on the fact that acceleration due to gravity is constant near the earth's surface. Here we neglect the effect of air resistance.

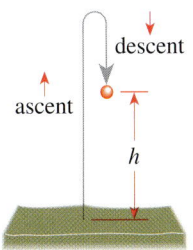

Figure 2

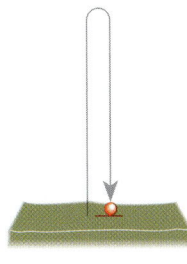

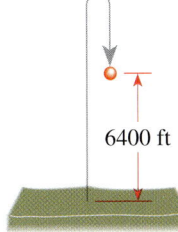

Solution Since the initial speed in this case is $v_0 = 800$ ft/s, the formula is

$$h = -16t^2 + 800t$$

(a) Ground level corresponds to $h = 0$, so we must solve the equation

$$0 = -16t^2 + 800t \qquad \text{Set } h = 0$$
$$0 = -16t(t - 50) \qquad \text{Factor}$$

Thus, $t = 0$ or $t = 50$. This means the bullet starts ($t = 0$) at ground level and returns to ground level after 50 s.

(b) Setting $h = 6400$ gives the equation

$$6400 = -16t^2 + 800t \qquad \text{Set } h = 6400$$
$$16t^2 - 800t + 6400 = 0 \qquad \text{All terms to LHS}$$
$$t^2 - 50t + 400 = 0 \qquad \text{Divide by 16}$$
$$(t - 10)(t - 40) = 0 \qquad \text{Factor}$$
$$t = 10 \quad \text{or} \quad t = 40 \qquad \text{Solve}$$

The bullet reaches 6400 ft after 10 s (on its ascent) and again after 40 s (on its descent to earth).

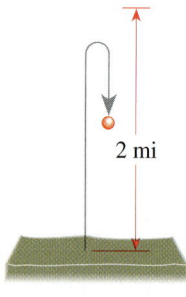

2 mi

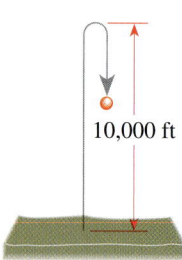

10,000 ft

(c) Two miles is $2 \times 5280 = 10{,}560$ ft.

$$10{,}560 = -16t^2 + 800t \qquad \text{Set } h = 10{,}560$$

$$16t^2 - 800t + 10{,}560 = 0 \qquad \text{All terms to LHS}$$

$$t^2 - 50t + 660 = 0 \qquad \text{Divide by 16}$$

The discriminant of this equation is $D = (-50)^2 - 4(660) = -140$, which is negative. Thus, the equation has no real solution. The bullet never reaches a height of 2 mi.

(d) Each height the bullet reaches is attained twice, once on its ascent and once on its descent. The only exception is the highest point of its path, which is reached only once. This means that for the highest value of h, the following equation has only one solution for t:

$$h = -16t^2 + 800t$$

$$16t^2 - 800t + h = 0 \qquad \text{All terms to LHS}$$

This in turn means that the discriminant D of the equation is 0, and so

$$D = (-800)^2 - 4(16)h = 0$$

$$640{,}000 - 64h = 0$$

$$h = 10{,}000$$

The maximum height reached is 10,000 ft. ∎

Other Types of Equations

So far we have learned how to solve linear and quadratic equations. Now we study other types of equations, including those that involve higher powers, fractional expressions, and radicals.

Example 10   An Equation Involving Fractional Expressions

Solve the equation $\dfrac{3}{x} + \dfrac{5}{x+2} = 2$.

Solution   We eliminate the denominators by multiplying each side by the lowest common denominator.

$$\left(\frac{3}{x} + \frac{5}{x+2}\right)x(x+2) = 2x(x+2) \qquad \text{Multiply by LCD } x(x+2)$$

$$3(x+2) + 5x = 2x^2 + 4x \qquad \text{Expand}$$

$$8x + 6 = 2x^2 + 4x \qquad \text{Expand LHS}$$

$$0 = 2x^2 - 4x - 6 \qquad \text{Subtract } 8x + 6$$

$$0 = x^2 - 2x - 3 \qquad \text{Divide both sides by 2}$$

$$0 = (x-3)(x+1) \qquad \text{Factor}$$

$$x - 3 = 0 \quad \text{or} \quad x + 1 = 0 \qquad \text{Zero-Product Property}$$

$$x = 3 \qquad\qquad x = -1 \qquad \text{Solve}$$

We must check our answers because multiplying by an expression that contains the variable can introduce extraneous solutions. From *Check Your Answers* we see that the solutions are $x = 3$ and -1.   ∎

When you solve an equation that involves radicals, you must be especially careful to check your final answers. The next example demonstrates why.

Example 11   An Equation Involving a Radical

Solve the equation $2x = 1 - \sqrt{2 - x}$.

Solution   To eliminate the square root, we first isolate it on one side of the equal sign, then square.

$$2x - 1 = -\sqrt{2 - x} \qquad \textcolor{teal}{\text{Subtract 1}}$$

$$(2x - 1)^2 = 2 - x \qquad \textcolor{teal}{\text{Square each side}}$$

$$4x^2 - 4x + 1 = 2 - x \qquad \textcolor{teal}{\text{Expand LHS}}$$

$$4x^2 - 3x - 1 = 0 \qquad \textcolor{teal}{\text{Add } -2 + x}$$

$$(4x + 1)(x - 1) = 0 \qquad \textcolor{teal}{\text{Factor}}$$

$$4x + 1 = 0 \quad \text{or} \quad x - 1 = 0 \qquad \textcolor{teal}{\text{Zero-Product Property}}$$

$$x = -\tfrac{1}{4} \qquad\qquad x = 1 \qquad \textcolor{teal}{\text{Solve}}$$

The values $x = -\tfrac{1}{4}$ and $x = 1$ are only potential solutions. We must check them to see if they satisfy the original equation. From *Check Your Answers* we see that $x = -\tfrac{1}{4}$ is a solution but $x = 1$ is not. The only solution is $x = -\tfrac{1}{4}$.   ∎

When we solve an equation, we may end up with one or more **extraneous solutions**, that is, potential solutions that do not satisfy the original equation. In Example 11, the value $x = 1$ is an extraneous solution. Extraneous solutions may be introduced when we square each side of an equation because the operation of squaring can turn a false equation into a true one. For example, $-1 \neq 1$, but $(-1)^2 = 1^2$. Thus, the squared equation may be true for more values of the variable than the original equation. That is why you must always check your answers to make sure that each satisfies the original equation.

An equation of the form $aW^2 + bW + c = 0$, where W is an algebraic expression, is an equation of **quadratic type**. We solve equations of quadratic type by substituting for the algebraic expression, as we see in the next two examples.

Example 12   A Fourth-Degree Equation of Quadratic Type

Find all solutions of the equation $x^4 - 8x^2 + 8 = 0$.

Solution   If we set $W = x^2$, then we get a quadratic equation in the new variable W:

$$(x^2)^2 - 8x^2 + 8 = 0 \qquad \textcolor{teal}{\text{Write } x^4 \text{ as } (x^2)^2}$$

$$W^2 - 8W + 8 = 0 \qquad \textcolor{teal}{\text{Let } W = x^2}$$

$$W = \frac{-(-8) \pm \sqrt{(-8)^2 - 4 \cdot 8}}{2} = 4 \pm 2\sqrt{2} \qquad \textcolor{teal}{\text{Quadratic formula}}$$

$$x^2 = 4 \pm 2\sqrt{2} \qquad \textcolor{teal}{W = x^2}$$

$$x = \pm\sqrt{4 \pm 2\sqrt{2}} \qquad \textcolor{teal}{\text{Take square roots}}$$

Check Your Answers

$x = -\tfrac{1}{4}$:

$$\text{LHS} = 2\left(-\tfrac{1}{4}\right) = -\tfrac{1}{2}$$

$$\text{RHS} = 1 - \sqrt{2 - \left(-\tfrac{1}{4}\right)}$$

$$= 1 - \sqrt{\tfrac{9}{4}}$$

$$= 1 - \tfrac{3}{2} = -\tfrac{1}{2}$$

$$\text{LHS} = \text{RHS} \qquad ✓$$

$x = 1$:

$$\text{LHS} = 2(1) = 2$$

$$\text{RHS} = 1 - \sqrt{2 - 1}$$

$$= 1 - 1 = 0$$

$$\text{LHS} \neq \text{RHS} \qquad ✗$$

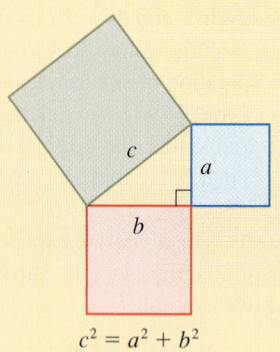
So, there are four solutions:

$$\sqrt{4 + 2\sqrt{2}}, \qquad \sqrt{4 - 2\sqrt{2}}, \qquad -\sqrt{4 + 2\sqrt{2}}, \qquad -\sqrt{4 - 2\sqrt{2}}$$

Using a calculator, we obtain the approximations $x \approx 2.61, 1.08, -2.61, -1.08$. ∎

Example 13   An Equation Involving Fractional Powers

Find all solutions of the equation $x^{1/3} + x^{1/6} - 2 = 0$.

Solution   This equation is of quadratic type because if we let $W = x^{1/6}$, then $W^2 = (x^{1/6})^2 = x^{1/3}$.

$$x^{1/3} + x^{1/6} - 2 = 0$$
$$W^2 + W - 2 = 0 \qquad \text{Let } W = x^{1/6}$$
$$(W - 1)(W + 2) = 0 \qquad \text{Factor}$$

$W - 1 = 0$	or	$W + 2 = 0$	Zero-Product Property
$W = 1$		$W = -2$	Solve
$x^{1/6} = 1$		$x^{1/6} = -2$	$W = x^{1/6}$
$x = 1^6 = 1$		$x = (-2)^6 = 64$	Take the 6th power

From *Check Your Answers* we see that $x = 1$ is a solution but $x = 64$ is not. The only solution is $x = 1$. ∎

Check Your Answers

$x = 1$:

LHS $= 1^{1/3} + 1^{1/6} - 2 = 0$

RHS $= 0$

LHS $=$ RHS ✓

$x = 64$:

LHS $= 64^{1/3} + 64^{1/6} - 2$

$= 4 + 2 - 2 = 4$

RHS $= 0$

LHS $\neq$ RHS ✗

When solving equations that involve absolute values, we usually take cases.

Example 14   An Absolute Value Equation

Solve the equation $|2x - 5| = 3$.

Solution   By the definition of absolute value, $|2x - 5| = 3$ is equivalent to

$$2x - 5 = 3 \qquad \text{or} \qquad 2x - 5 = -3$$
$$2x = 8 \qquad\qquad\qquad 2x = 2$$
$$x = 4 \qquad\qquad\qquad x = 1$$

The solutions are $x = 1$, $x = 4$. ∎

1.5 Exercises

1–4 ■ Determine whether the given value is a solution of the equation.

1. $4x + 7 = 9x - 3$
  (a) $x = -2$ (b) $x = 2$

2. $1 - [2 - (3 - x)] = 4x - (6 + x)$
  (a) $x = 2$ (b) $x = 4$

3. $\dfrac{1}{x} - \dfrac{1}{x-4} = 1$
  (a) $x = 2$ (b) $x = 4$

4. $\dfrac{x^{3/2}}{x-6} = x - 8$
  (a) $x = 4$ (b) $x = 8$

5–22 ■ The given equation is either linear or equivalent to a linear equation. Solve the equation.

5. $2x + 7 = 31$

6. $5x - 3 = 4$

7. $\frac{1}{2}x - 8 = 1$

8. $3 + \frac{1}{3}x = 5$

9. $-7w = 15 - 2w$

10. $5t - 13 = 12 - 5t$

11. $\frac{1}{2}y - 2 = \frac{1}{3}y$

12. $\dfrac{z}{5} = \dfrac{3}{10}z + 7$

13. $2(1 - x) = 3(1 + 2x) + 5$

14. $\dfrac{2}{3}y + \dfrac{1}{2}(y - 3) = \dfrac{y+1}{4}$

15. $x - \frac{1}{3}x - \frac{1}{2}x - 5 = 0$

16. $2x - \dfrac{x}{2} + \dfrac{x+1}{4} = 6x$

17. $\dfrac{1}{x} = \dfrac{4}{3x} + 1$

18. $\dfrac{2x-1}{x+2} = \dfrac{4}{5}$

19. $\dfrac{3}{x+1} - \dfrac{1}{2} = \dfrac{1}{3x+3}$

20. $\dfrac{4}{x-1} + \dfrac{2}{x+1} = \dfrac{35}{x^2-1}$

21. $(t - 4)^2 = (t + 4)^2 + 32$

22. $\sqrt{3}x + \sqrt{12} = \dfrac{x+5}{\sqrt{3}}$

23–36 ■ Solve the equation for the indicated variable.

23. $PV = nRT$; for R

24. $F = G\dfrac{mM}{r^2}$; for m

25. $\dfrac{1}{R} = \dfrac{1}{R_1} + \dfrac{1}{R_2}$; for R_1

26. $P = 2l + 2w$; for w

27. $\dfrac{ax+b}{cx+d} = 2$; for x

28. $a - 2[b - 3(c - x)] = 6$; for x

29. $a^2x + (a - 1) = (a + 1)x$; for x

30. $\dfrac{a+1}{b} = \dfrac{a-1}{b} + \dfrac{b+1}{a}$; for a

31. $V = \frac{1}{3}\pi r^2 h$; for r

32. $F = G\dfrac{mM}{r^2}$; for r

33. $a^2 + b^2 = c^2$; for b

34. $A = P\left(1 + \dfrac{i}{100}\right)^2$; for i

35. $h = \frac{1}{2}gt^2 + v_0 t$; for t

36. $S = \dfrac{n(n+1)}{2}$; for n

37–44 ■ Solve the equation by factoring.

37. $x^2 + x - 12 = 0$

38. $x^2 + 3x - 4 = 0$

39. $x^2 - 7x + 12 = 0$

40. $x^2 + 8x + 12 = 0$

41. $4x^2 - 4x - 15 = 0$

42. $2y^2 + 7y + 3 = 0$

43. $3x^2 + 5x = 2$

44. $6x(x - 1) = 21 - x$

45–52 ■ Solve the equation by completing the square.

45. $x^2 + 2x - 5 = 0$

46. $x^2 - 4x + 2 = 0$

47. $x^2 + 3x - \frac{7}{4} = 0$

48. $x^2 = \frac{3}{4}x - \frac{1}{8}$

49. $2x^2 + 8x + 1 = 0$

50. $3x^2 - 6x - 1 = 0$

51. $4x^2 - x = 0$

52. $-2x^2 + 6x + 3 = 0$

53–68 ■ Find all real solutions of the quadratic equation.

53. $x^2 - 2x - 15 = 0$

54. $x^2 + 30x + 200 = 0$

55. $x^2 + 3x + 1 = 0$

56. $x^2 - 6x + 1 = 0$

57. $2x^2 + x - 3 = 0$

58. $3x^2 + 7x + 4 = 0$

59. $2y^2 - y - \frac{1}{2} = 0$

60. $\theta^2 - \frac{3}{2}\theta + \frac{9}{16} = 0$

61. $4x^2 + 16x - 9 = 0$

62. $w^2 = 3(w - 1)$

63. $3 + 5z + z^2 = 0$

64. $x^2 - \sqrt{5}x + 1 = 0$

65. $\sqrt{6}x^2 + 2x - \sqrt{3/2} = 0$

66. $3x^2 + 2x + 2 = 0$

67. $25x^2 + 70x + 49 = 0$

68. $5x^2 - 7x + 5 = 0$

69–74 ■ Use the discriminant to determine the number of real solutions of the equation. Do not solve the equation.

69. $x^2 - 6x + 1 = 0$

70. $3x^2 = 6x - 9$

71. $x^2 + 2.20x + 1.21 = 0$

72. $x^2 + 2.21x + 1.21 = 0$

73. $4x^2 + 5x + \frac{13}{8} = 0$

74. $x^2 + rx - s = 0$ $(s > 0)$

75–98 ■ Find all real solutions of the equation.

75. $\dfrac{1}{x-1} + \dfrac{1}{x+2} = \dfrac{5}{4}$

76. $\dfrac{10}{x} - \dfrac{12}{x-3} + 4 = 0$

77. $\dfrac{x^2}{x+100} = 50$

78. $\dfrac{1}{x-1} - \dfrac{2}{x^2} = 0$

79. $\dfrac{x + 5}{x - 2} = \dfrac{5}{x + 2} + \dfrac{28}{x^2 - 4}$ **80.** $\dfrac{x}{2x + 7} - \dfrac{x + 1}{x + 3} = 1$

81. $\sqrt{2x + 1} + 1 = x$ **82.** $\sqrt{5 - x} + 1 = x - 2$

83. $2x + \sqrt{x + 1} = 8$ **84.** $\sqrt{\sqrt{x - 5} + x} = 5$

85. $x^4 - 13x^2 + 40 = 0$ **86.** $x^4 - 5x^2 + 4 = 0$

87. $2x^4 + 4x^2 + 1 = 0$ **88.** $x^6 - 2x^3 - 3 = 0$

89. $x^{4/3} - 5x^{2/3} + 6 = 0$ **90.** $\sqrt{x} - 3\sqrt[4]{x} - 4 = 0$

91. $4(x + 1)^{1/2} - 5(x + 1)^{3/2} + (x + 1)^{5/2} = 0$

92. $x^{1/2} + 3x^{-1/2} = 10x^{-3/2}$

93. $x^{1/2} - 3x^{1/3} = 3x^{1/6} - 9$ **94.** $x - 5\sqrt{x} + 6 = 0$

95. $|2x| = 3$ **96.** $|3x + 5| = 1$

97. $|x - 4| = 0.01$ **98.** $|x - 6| = -1$

Applications

99–100 ■ Falling-Body Problems   Suppose an object is dropped from a height h_0 above the ground. Then its height after t seconds is given by $h = -16t^2 + h_0$, where h is measured in feet. Use this information to solve the problem.

99. If a ball is dropped from 288 ft above the ground, how long does it take to reach ground level?

100. A ball is dropped from the top of a building 96 ft tall.

  (a) How long will it take to fall half the distance to ground level?

  (b) How long will it take to fall to ground level?

101–102 ■ Falling-Body Problems   Use the formula $h = -16t^2 + v_0 t$ discussed in Example 9.

101. A ball is thrown straight upward at an initial speed of $v_0 = 40$ ft/s.

  (a) When does the ball reach a height of 24 ft?

  (b) When does it reach a height of 48 ft?

  (c) What is the greatest height reached by the ball?

  (d) When does the ball reach the highest point of its path?

  (e) When does the ball hit the ground?

102. How fast would a ball have to be thrown upward to reach a maximum height of 100 ft? [*Hint:* Use the discriminant of the equation $16t^2 - v_0 t + h = 0$.]

103. Shrinkage in Concrete Beams   As concrete dries, it shrinks—the higher the water content, the greater the shrinkage. If a concrete beam has a water content of w kg/m^3, then it will shrink by a factor

$$S = \dfrac{0.032w - 2.5}{10,000}$$

where S is the fraction of the original beam length that disappears due to shrinkage.

  (a) A beam 12.025 m long is cast in concrete that contains 250 kg/m^3 water. What is the shrinkage factor S? How long will the beam be when it has dried?

  (b) A beam is 10.014 m long when wet. We want it to shrink to 10.009 m, so the shrinkage factor should be $S = 0.00050$. What water content will provide this amount of shrinkage?

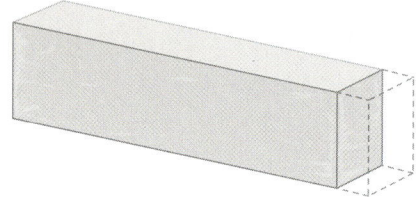

104. The Lens Equation   If F is the focal length of a convex lens and an object is placed at a distance x from the lens, then its image will be at a distance y from the lens, where F, x, and y are related by the *lens equation*

$$\dfrac{1}{F} = \dfrac{1}{x} + \dfrac{1}{y}$$

Suppose that a lens has a focal length of 4.8 cm, and that the image of an object is 4 cm closer to the lens than the object itself. How far from the lens is the object?

105. Fish Population   The fish population in a certain lake rises and falls according to the formula

$$F = 1000(30 + 17t - t^2)$$

Here F is the number of fish at time t, where t is measured in years since January 1, 2002, when the fish population was first estimated.

  (a) On what date will the fish population again be the same as on January 1, 2002?

  (b) By what date will all the fish in the lake have died?

106. Fish Population   A large pond is stocked with fish. The fish population P is modeled by the formula $P = 3t + 10\sqrt{t} + 140$, where t is the number of days since the fish were first introduced into the pond. How many days will it take for the fish population to reach 500?

107. Profit   A small-appliance manufacturer finds that the profit P (in dollars) generated by producing x microwave ovens per week is given by the formula $P = \frac{1}{10}x(300 - x)$ provided that $0 \le x \le 200$. How many ovens must be manufactured in a given week to generate a profit of $1250?

108. Gravity   If an imaginary line segment is drawn between the centers of the earth and the moon, then the net gravitational force F acting on an object situated on this line segment is

$$F = \frac{-K}{x^2} + \frac{0.012K}{(239 - x)^2}$$

where $K > 0$ is a constant and x is the distance of the object from the center of the earth, measured in thousands of miles. How far from the center of the earth is the "dead spot" where no net gravitational force acts upon the object? (Express your answer to the nearest thousand miles.)

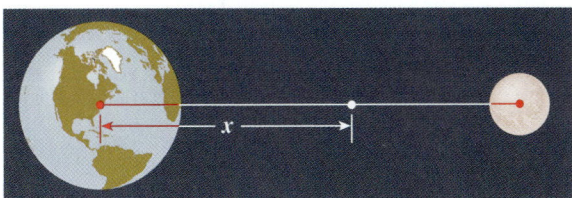

109. Depth of a Well   One method for determining the depth of a well is to drop a stone into it and then measure the time it takes until the splash is heard. If d is the depth of the well (in feet) and t_1 the time (in seconds) it takes for the stone to fall, then $d = 16t_1^2$, so $t_1 = \sqrt{d}/4$. Now if t_2 is the time it takes for the sound to travel back up, then $d = 1090t_2$ because the speed of sound is 1090 ft/s. So $t_2 = d/1090$. Thus, the total time elapsed between dropping the stone and hearing the splash is

$$t_1 + t_2 = \frac{\sqrt{d}}{4} + \frac{d}{1090}$$

How deep is the well if this total time is 3 s?

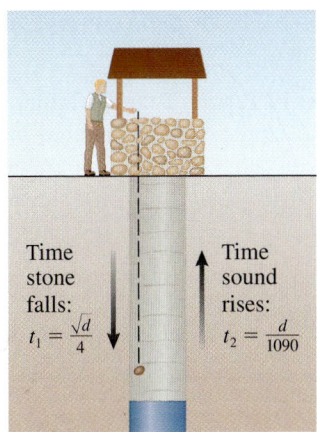

Time stone falls: $t_1 = \frac{\sqrt{d}}{4}$

Time sound rises: $t_2 = \frac{d}{1090}$

Discovery • Discussion

110. A Family of Equations   The equation

$$3x + k - 5 = kx - k + 1$$

is really a **family of equations**, because for each value of k, we get a different equation with the unknown x. The letter k is called a **parameter** for this family. What value should we pick for k to make the given value of x a solution of the resulting equation?

(a) $x = 0$ (b) $x = 1$ (c) $x = 2$

111. Proof That $0 = 1$?   The following steps appear to give equivalent equations, which seem to prove that $1 = 0$. Find the error.

$x = 1$	*Given*
$x^2 = x$	*Multiply by x*
$x^2 - x = 0$	*Subtract x*
$x(x - 1) = 0$	*Factor*
$\dfrac{x(x - 1)}{x - 1} = \dfrac{0}{x - 1}$	*Divide by x − 1*
$x = 0$	*Simplify*
$1 = 0$	*Given x = 1*

112. Volumes of Solids   The sphere, cylinder, and cone shown here all have the same radius r and the same volume V.

(a) Use the volume formulas given on the inside front cover of this book, to show that

$$\tfrac{4}{3}\pi r^3 = \pi r^2 h_1 \quad \text{and} \quad \tfrac{4}{3}\pi r^3 = \tfrac{1}{3}\pi r^2 h_2$$

(b) Solve these equations for h_1 and h_2.

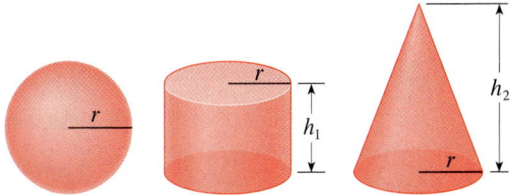

113. Relationship between Roots and Coefficients
The quadratic formula gives us the roots of a quadratic equation from its coefficients. We can also obtain the coefficients from the roots. For example, find the roots of the equation $x^2 - 9x + 20 = 0$ and show that the product of the roots is the constant term 20 and the sum of the roots is 9, the negative of the coefficient of x. Show that the same relationship between roots and coefficients holds for the following equations:

$$x^2 - 2x - 8 = 0$$

$$x^2 + 4x + 2 = 0$$

Use the quadratic formula to prove that in general, if the equation $x^2 + bx + c = 0$ has roots r_1 and r_2, then $c = r_1 r_2$ and $b = -(r_1 + r_2)$.

114. Solving an Equation in Different Ways   We have
learned several different ways to solve an equation in this
section. Some equations can be tackled by more than one
method. For example, the equation $x - \sqrt{x} - 2 = 0$ is of
quadratic type: We can solve it by letting $\sqrt{x} = u$ and
$x = u^2$, and factoring. Or we could solve for $\sqrt{x}$, square
each side, and then solve the resulting quadratic equation.

Solve the following equations using both methods
indicated, and show that you get the same final answers.

(a) $x - \sqrt{x} - 2 = 0$ quadratic type; solve for the
radical, and square

(b) $\dfrac{12}{(x-3)^2} + \dfrac{10}{x-3} + 1 = 0$ quadratic type; multiply
by LCD

1.6   Modeling with Equations

Many problems in the sciences, economics, finance, medicine, and numerous other
fields can be translated into algebra problems; this is one reason that algebra is so
useful. In this section we use equations as mathematical models to solve real-life
problems.

Guidelines for Modeling with Equations

We will use the following guidelines to help us set up equations that model situations
described in words. To show how the guidelines can help you set up equations, we
note them in the margin as we work each example in this section.

Guidelines for Modeling with Equations

1. **Identify the Variable.**   Identify the quantity that the problem asks you to
   find. This quantity can usually be determined by a careful reading of the
   question posed at the end of the problem. Then **introduce notation** for the
   variable (call it x or some other letter).

2. **Express All Unknown Quantities in Terms of the Variable.**   Read each
   sentence in the problem again, and express all the quantities mentioned in the
   problem in terms of the variable you defined in Step 1. To organize this information, it is sometimes helpful to **draw a diagram** or **make a table**.

3. **Set Up the Model.**   Find the crucial fact in the problem that gives a relationship between the expressions you listed in Step 2. **Set up an equation** (or
   model) that expresses this relationship.

4. **Solve the Equation and Check Your Answer.**   Solve the equation, check
   your answer, and express it as a sentence that answers the question posed in
   the problem.

The following example illustrates how these guidelines are used to translate a
"word problem" into the language of algebra.

Example 1   Renting a Car

A car rental company charges $30 a day and 15¢ a mile for renting a car. Helen rents a car for two days and her bill comes to $108. How many miles did she drive?

Solution   We are asked to find the number of miles Helen has driven. So we let

$$x = \text{number of miles driven}$$

Identify the variable

Then we translate all the information given in the problem into the language of algebra.

In Words	In Algebra
Number of miles driven	x
Mileage cost (at $0.15 per mile)	$0.15x$
Daily cost (at $30 per day)	$2(30)$

Express all unknown quantities in terms of the variable

Now we set up the model.

$$\boxed{\begin{array}{c}\text{mileage} \\ \text{cost}\end{array}} + \boxed{\begin{array}{c}\text{daily} \\ \text{cost}\end{array}} = \boxed{\text{total cost}}$$

Set up the model

$$0.15x + 2(30) = 108$$

Solve

$$0.15x = 48 \qquad \textit{Substract 60}$$

$$x = \frac{48}{0.15} \qquad \textit{Divide by 0.15}$$

$$x = 320 \qquad \textit{Calculator}$$

Helen drove her rental car 320 miles.   ∎

Check Your Answer

total cost = mileage cost + daily cost

$= 0.15(320) + 2(30)$

$= 108$ ✓

Constructing Models

In the examples and exercises that follow, we construct equations that model problems in many different real-life situations.

Example 2   Interest on an Investment

Mary inherits $100,000 and invests it in two certificates of deposit. One certificate pays 6% and the other pays $4\frac{1}{2}\%$ simple interest annually. If Mary's total interest is $5025 per year, how much money is invested at each rate?

Solution   The problem asks for the amount she has invested at each rate. So we let

Identify the variable

$$x = \text{the amount invested at 6\%}$$

Since Mary's total inheritance is $100,000, it follows that she invested $100,000 - x$ at $4\frac{1}{2}\%$. We translate all the information given into the language of algebra.

In Words	In Algebra
Amount invested at 6%	x
Amount invested at $4\frac{1}{2}\%$	$100{,}000 - x$
Interest earned at 6%	$0.06x$
Interest earned at $4\frac{1}{2}\%$	$0.045(100{,}000 - x)$

> **Express all unknown quantities in terms of the variable**

We use the fact that Mary's total interest is $5025 to set up the model.

> **Set up the model**

$$\boxed{\text{interest at } 6\%} \; + \; \boxed{\text{interest at } 4\tfrac{1}{2}\%} \; = \; \boxed{\text{total interest}}$$

> **Solve**

$$0.06x + 0.045(100{,}000 - x) = 5025$$

$$0.06x + 4500 - 0.045x = 5025 \qquad \text{Multiply}$$

$$0.015x + 4500 = 5025 \qquad \text{Combine the x-terms}$$

$$0.015x = 525 \qquad \text{Subtract 4500}$$

$$x = \frac{525}{0.015} = 35{,}000 \qquad \text{Divide by 0.015}$$

So Mary has invested $35,000 at 6% and the remaining $65,000 at $4\frac{1}{2}\%$. ∎

Check Your Answer

$$\text{total interest} = 6\% \text{ of } \$35{,}000 + 4\tfrac{1}{2}\% \text{ of } \$65{,}000$$
$$= \$2100 + \$2925 = \$5025 \quad \checkmark$$

Example 3   Dimensions of a Poster

> In a problem such as this, which involves geometry, it is essential to draw a diagram like the one shown in Figure 1.

A poster has a rectangular printed area 100 cm by 140 cm, and a blank strip of uniform width around the four edges. The perimeter of the poster is $1\frac{1}{2}$ times the perimeter of the printed area. What is the width of the blank strip, and what are the dimensions of the poster?

Solution   We are asked to find the width of the blank strip. So we let

$$x = \text{the width of the blank strip}$$

Then we translate the information in Figure 1 into the language of algebra:

> **Identify the variable**

In Words	In Algebra
Width of blank strip	x
Perimeter of printed area	$2(100) + 2(140) = 480$
Width of poster	$100 + 2x$
Length of poster	$140 + 2x$
Perimeter, of poster	$2(100 + 2x) + 2(140 + 2x)$

> **Express all unknown quantities in terms of the variable**

Now we use the fact that the perimeter of the poster is $1\frac{1}{2}$ times the perimeter of the printed area to set up the model.

$$\text{perimeter of poster} = \tfrac{3}{2} \cdot \text{ perimeter of printed area}$$

$$2(100 + 2x) + 2(140 + 2x) = \tfrac{3}{2} \cdot 480$$

$$480 + 8x = 720 \qquad \text{Expand and combine like terms on LHS}$$

$$8x = 240 \qquad \text{Subtract 480}$$

$$x = 30 \qquad \text{Divide by 8}$$

The blank strip is 30 cm wide, so the dimensions of the poster are

$$100 + 30 + 30 = 160 \text{ cm wide}$$

by

$$140 + 30 + 30 = 200 \text{ cm long}$$

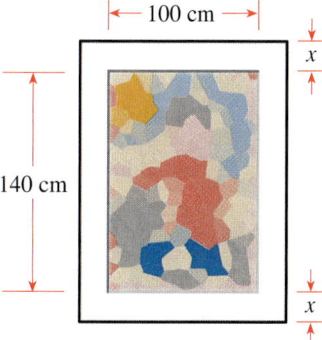

Figure 1 ∎

Example 4   Dimensions of a Building Lot

A rectangular building lot is 8 ft longer than it is wide and has an area of 2900 ft². Find the dimensions of the lot.

Solution   We are asked to find the width and length of the lot. So let

$$w = \text{width of lot}$$

Then we translate the information given in the problem into the language of algebra (see Figure 2 on page 62).

In Words	In Algebra
Width of lot	w
Length of lot	$w + 8$

Now we set up the model.

Set up the model		$\dfrac{\text{width}}{\text{of lot}} \cdot \dfrac{\text{length}}{\text{of lot}} = \dfrac{\text{area}}{\text{of lot}}$

$$w(w + 8) = 2900$$

Solve	$w^2 + 8w = 2900$	Expand
	$w^2 + 8w - 2900 = 0$	Subtract 2900
	$(w - 50)(w + 58) = 0$	Factor
	$w = 50 \quad \text{or} \quad w = -58$	Zero-Product Property

Since the width of the lot must be a positive number, we conclude that $w = 50$ ft. The length of the lot is $w + 8 = 50 + 8 = 58$ ft.

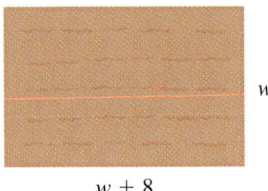

w

$w + 8$

Figure 2

Example 5   Determining the Height of a Building Using Similar Triangles

A man 6 ft tall wishes to find the height of a certain four-story building. He measures its shadow and finds it to be 28 ft long, while his own shadow is $3\frac{1}{2}$ ft long. How tall is the building?

Solution   The problem asks for the height of the building. So let

Identify the variable

$$h = \text{the height of the building}$$

We use the fact that the triangles in Figure 3 are similar. Recall that for any pair of similar triangles the ratios of corresponding sides are equal. Now we translate these observations into the language of algebra.

Express all unknown quantities in terms of the variable

In Words	In Algebra
Height of building	h
Ratio of height to base in large triangle	$\frac{h}{28}$
Ratio of height to base in small triangle	$\frac{6}{3.5}$

Since the large and small triangles are similar, we get the equation

Set up the model

$$\dfrac{\text{ratio of height to}}{\text{base in large triangle}} = \dfrac{\text{ratio of height to}}{\text{base in small triangle}}$$

$$\frac{h}{28} = \frac{6}{3.5}$$

Solve

$$h = \frac{6 \cdot 28}{3.5} = 48$$

The building is 48 ft tall.

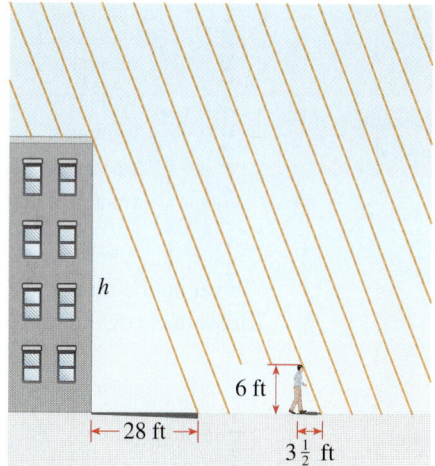

Figure 3

Example 6 Mixtures and Concentration

A manufacturer of soft drinks advertises their orange soda as "naturally flavored," although it contains only 5% orange juice. A new federal regulation stipulates that to be called "natural" a drink must contain at least 10% fruit juice. How much pure orange juice must this manufacturer add to 900 gal of orange soda to conform to the new regulation?

Solution   The problem asks for the amount of pure orange juice to be added. So let

Identify the variable

x = the amount (in gallons) of pure orange juice to be added

In any problem of this type—in which two different substances are to be mixed—drawing a diagram helps us organize the given information (see Figure 4).

Figure 4

We now translate the information in the figure into the language of algebra.

In Words	In Algebra
Amount of orange juice to be added	x
Amount of the mixture	$900 + x$
Amount of orange juice in the first vat	$0.05(900) = 45$
Amount of orange juice in the second vat	$1 \cdot x = x$
Amount of orange juice in the mix ture	$0.10(900 + x)$

Express all unknown quantities in terms of the variable

To set up the model, we use the fact that the total amount of orange juice in the mixture is equal to the orange juice in the first two vats.

Set up the model

$$\begin{pmatrix} \text{amount of} \\ \text{orange juice} \\ \text{in first vat} \end{pmatrix} + \begin{pmatrix} \text{amount of} \\ \text{orange juice} \\ \text{in second vat} \end{pmatrix} = \begin{pmatrix} \text{amount of} \\ \text{orange juice} \\ \text{in mixture} \end{pmatrix}$$

Solve

$$45 + x = 0.1(900 + x) \quad \text{From Figure 4}$$
$$45 + x = 90 + 0.1x \quad \text{Multiply}$$
$$0.9x = 45 \quad \text{Subtract 0.1x and 45}$$
$$x = \frac{45}{0.9} = 50 \quad \text{Divide by 0.9}$$

The manufacturer should add 50 gal of pure orange juice to the soda.   ■

Check Your Answer

$$\text{amount of juice before mixing} = 5\% \text{ of } 900 \text{ gal} + 50 \text{ gal pure juice}$$
$$= 45 \text{ gal} + 50 \text{ gal} = 95 \text{ gal}$$
$$\text{amount of juice after mixing} = 10\% \text{ of } 950 \text{ gal} = 95 \text{ gal}$$

Amounts are equal.   ✓

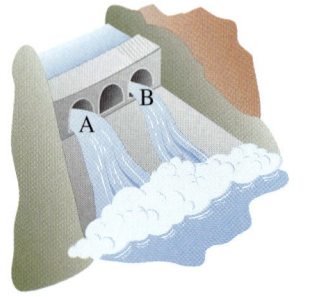

Example 7   Time Needed to Do a Job

Because of an anticipated heavy rainstorm, the water level in a reservoir must be lowered by 1 ft. Opening spillway A lowers the level by this amount in 4 hours, whereas opening the smaller spillway B does the job in 6 hours. How long will it take to lower the water level by 1 ft if both spillways are opened?

Solution   We are asked to find the time needed to lower the level by 1 ft if both spillways are open. So let

$$x = \text{the time (in hours) it takes to lower the water level}$$
$$\text{by 1 ft if both spillways are open}$$

Identify the variable

Finding an equation relating x to the other quantities in this problem is not easy.

The building is 48 ft tall.

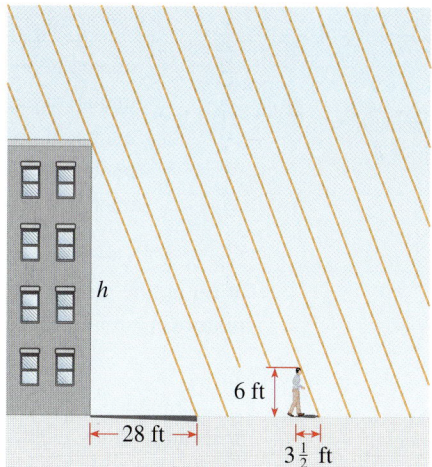

Figure 3

Example 6   Mixtures and Concentration

A manufacturer of soft drinks advertises their orange soda as "naturally flavored," although it contains only 5% orange juice. A new federal regulation stipulates that to be called "natural" a drink must contain at least 10% fruit juice. How much pure orange juice must this manufacturer add to 900 gal of orange soda to conform to the new regulation?

Solution   The problem asks for the amount of pure orange juice to be added. So let

Identify the variable

x = the amount (in gallons) of pure orange juice to be added

In any problem of this type—in which two different substances are to be mixed—drawing a diagram helps us organize the given information (see Figure 4).

	5% juice	100% juice	10% juice
Volume	900 gallons	x gallons	$900 + x$ gallons
Amount of orange juice	5% of 900 gallons = 45 gallons	100% of x gallons = x gallons	10% of $900 + x$ gallons = $0.1(900 + x)$ gallons

Figure 4

We now translate the information in the figure into the language of algebra.

In Words	In Algebra
Amount of orange juice to be added	x
Amount of the mixture	$900 + x$
Amount of orange juice in the first vat	$0.05(900) = 45$
Amount of orange juice in the second vat	$1 \cdot x = x$
Amount of orange juice in the mix ture	$0.10(900 + x)$

Express all unknown quantities in terms of the variable

To set up the model, we use the fact that the total amount of orange juice in the mixture is equal to the orange juice in the first two vats.

Set up the model

amount of orange juice in first vat	+	amount of orange juice in second vat	=	amount of orange juice in mixture

Solve

$$45 + x = 0.1(900 + x) \qquad \text{From Figure 4}$$
$$45 + x = 90 + 0.1x \qquad \text{Multiply}$$
$$0.9x = 45 \qquad \text{Subtract 0.1x and 45}$$
$$x = \frac{45}{0.9} = 50 \qquad \text{Divide by 0.9}$$

The manufacturer should add 50 gal of pure orange juice to the soda. ∎

Check Your Answer

$$\text{amount of juice before mixing} = 5\% \text{ of } 900 \text{ gal} + 50 \text{ gal pure juice}$$
$$= 45 \text{ gal} + 50 \text{ gal} = 95 \text{ gal}$$
$$\text{amount of juice after mixing} = 10\% \text{ of } 950 \text{ gal} = 95 \text{ gal}$$
Amounts are equal.   ✓

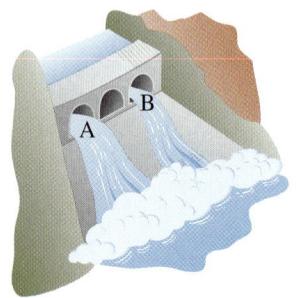

Example 7   Time Needed to Do a Job

Because of an anticipated heavy rainstorm, the water level in a reservoir must be lowered by 1 ft. Opening spillway A lowers the level by this amount in 4 hours, whereas opening the smaller spillway B does the job in 6 hours. How long will it take to lower the water level by 1 ft if both spillways are opened?

Solution   We are asked to find the time needed to lower the level by 1 ft if both spillways are open. So let

$$x = \text{the time (in hours) it takes to lower the water level}$$
$$\text{by 1 ft if both spillways are open}$$

Identify the variable

Finding an equation relating x to the other quantities in this problem is not easy.

Certainly x is not simply $4 + 6$, because that would mean that together the two spillways require longer to lower the water level than either spillway alone. Instead, *we look at the fraction of the job that can be done in one hour by each spillway.*

In Words	In Algebra
Time it takes to lower level 1 ft with A and B together	x h
Distance A lowers level in 1 h	$\frac{1}{4}$ ft
Distance B lowers level in 1 h	$\frac{1}{6}$ ft
Distance A and B together lower levels in 1 h	$\frac{1}{x}$ ft

Express all unknown quantities in terms of the variable

Now we set up the model.

Set up the model

fraction done by A   +   fraction done by B   =   fraction done by both

$$\frac{1}{4} + \frac{1}{6} = \frac{1}{x}$$

Solve

$$3x + 2x = 12 \qquad \text{Multiply by the LCD, } 12x$$

$$5x = 12 \qquad \text{Add}$$

$$x = \frac{12}{5} \qquad \text{Divide by 5}$$

It will take $2\frac{2}{5}$ hours, or 2 h 24 min to lower the water level by 1 ft if both spillways are open. ∎

The next example deals with distance, rate (speed), and time. The formula to keep in mind here is

$$\text{distance} = \text{rate} \times \text{time}$$

where the rate is either the constant speed or average speed of a moving object. For example, driving at 60 mi/h for 4 hours takes you a distance of $60 \cdot 4 = 240$ mi.

Example 8   A Distance-Speed-Time Problem

A jet flew from New York to Los Angeles, a distance of 4200 km. The speed for the return trip was 100 km/h faster than the outbound speed. If the total trip took 13 hours, what was the jet's speed from New York to Los Angeles?

Solution   We are asked for the speed of the jet from New York to Los Angeles. So let

Identify the variable

$$s = \text{speed from New York to Los Angeles}$$

Then $s + 100 = $ speed from Los Angeles to New York

Now we organize the information in a table. We fill in the "Distance" column first, since we know that the cities are 4200 km apart. Then we fill in the "Speed" column, since we have expressed both speeds (rates) in terms of the variable s. Finally, we calculate the entries for the "Time" column, using

$$\text{time} = \frac{\text{distance}}{\text{rate}}$$

	Distance (km)	Speed (km/h)	Time (h)
N.Y. to L.A.	4200	s	$\dfrac{4200}{s}$
L.A. to N.Y.	4200	$s + 100$	$\dfrac{4200}{s + 100}$

Express all unknown quantities in terms of the variable

The total trip took 13 hours, so we have the model

Set up the model

$$\begin{array}{c} \text{time from} \\ \text{N.Y. to L.A.} \end{array} + \begin{array}{c} \text{time from} \\ \text{L.A. to N.Y.} \end{array} = \begin{array}{c} \text{total} \\ \text{time} \end{array}$$

$$\frac{4200}{s} + \frac{4200}{s + 100} = 13$$

Multiplying by the common denominator, $s(s + 100)$, we get

$$4200(s + 100) + 4200s = 13s(s + 100)$$

$$8400s + 420{,}000 = 13s^2 + 1300s$$

$$0 = 13s^2 - 7100s - 420{,}000$$

Although this equation does factor, with numbers this large it is probably quicker to use the quadratic formula and a calculator.

Solve

$$s = \frac{7100 \pm \sqrt{(-7100)^2 - 4(13)(-420{,}000)}}{2(13)}$$

$$= \frac{7100 \pm 8500}{26}$$

$$s = 600 \qquad \text{or} \qquad s = \frac{-1400}{26} \approx -53.8$$

Since s represents speed, we reject the negative answer and conclude that the jet's speed from New York to Los Angeles was 600 km/h. ∎

Example 9 Energy Expended in Bird Flight

Ornithologists have determined that some species of birds tend to avoid flights over large bodies of water during daylight hours, because air generally rises over land and falls over water in the daytime, so flying over water requires more energy. A bird is released from point A on an island, 5 mi from B, the nearest point on a straight shoreline. The bird flies to a point C on the shoreline and then flies along the shoreline to its nesting area D, as shown in Figure 5. Suppose the bird has 170 kcal of energy reserves. It uses 10 kcal/mi flying over land and 14 kcal/mi flying over water.

(a) Where should the point C be located so that the bird uses exactly 170 kcal of energy during its flight?

(b) Does the bird have enough energy reserves to fly directly from A to D?

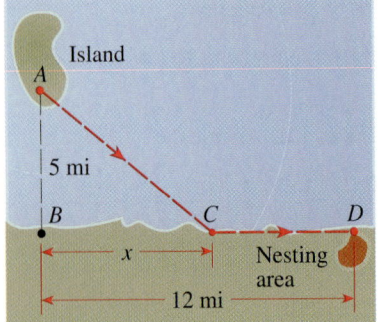

Figure 5

Solution

(a) We are asked to find the location of C. So let

Identify the variable

$$x = \text{ distance from } B \text{ to } C$$

From the figure, and from the fact that

$$\text{energy used } = \text{ energy per mile } \times \text{ miles flown}$$

we determine the following:

In Words	In Algebra	
Distance from B to C	x	
Distance flown over water (from A to C)	$\sqrt{x^2 + 25}$	Pythagorean Theorem
Distance flown over land (from C to D)	$12 - x$	
Energy used over water	$14\sqrt{x^2 + 25}$	
Energy used over land	$10(12 - x)$	

Express all unknown quantities in terms of the variable

Now we set up the model.

Set up the model

$$\begin{array}{c} \text{total energy} \\ \text{used} \end{array} = \begin{array}{c} \text{energy used} \\ \text{over water} \end{array} + \begin{array}{c} \text{energy used} \\ \text{over land} \end{array}$$

$$170 = 14\sqrt{x^2 + 25} + 10(12 - x)$$

To solve this equation, we eliminate the square root by first bringing all other terms to the left of the equal sign and then squaring each side.

Solve

$$170 - 10(12 - x) = 14\sqrt{x^2 + 25}$$ Isolate square-root term on RHS

$$50 + 10x = 14\sqrt{x^2 + 25}$$ Simplify LHS

$$(50 + 10x)^2 = (14)^2(x^2 + 25)$$ Square each side

$$2500 + 1000x + 100x^2 = 196x^2 + 4900$$ Expand

$$0 = 96x^2 - 1000x + 2400$$ All terms to RHS

This equation could be factored, but because the numbers are so large it is easier to use the quadratic formula and a calculator:

$$x = \frac{1000 \pm \sqrt{(-1000)^2 - 4(96)(2400)}}{2(96)}$$

$$= \frac{1000 \pm 280}{192} = 6\tfrac{2}{3} \quad \text{or} \quad 3\tfrac{3}{4}$$

Point C should be either $6\tfrac{2}{3}$ mi or $3\tfrac{3}{4}$ mi from B so that the bird uses exactly 170 kcal of energy during its flight.

(b) By the Pythagorean Theorem (see page 54), the length of the route directly from A to D is $\sqrt{5^2 + 12^2} = 13$ mi, so the energy the bird requires for that route is $14 \times 13 = 182$ kcal. This is more energy than the bird has available, so it can't use this route. ∎

1.6   Exercises

1–12 ■ Express the given quantity in terms of the indicated variable.

1. The sum of three consecutive integers; $n = $ first integer of the three

2. The sum of three consecutive integers; $n = $ middle integer of the three

3. The average of three test scores if the first two scores are 78 and 82; $s = $ third test score

4. The average of four quiz scores if each of the first three scores is 8; $q = $ fourth quiz score

5. The interest obtained after one year on an investment at $2\frac{1}{2}\%$ simple interest per year; $x = $ number of dollars invested

6. The total rent paid for an apartment if the rent is $795 a month; $n = $ number of months

7. The area (in ft^2) of a rectangle that is three times as long as it is wide; $w = $ width of the rectangle (in ft)

8. The perimeter (in cm) of a rectangle that is 5 cm longer than it is wide; $w = $ width of the rectangle (in cm)

9. The distance (in mi) that a car travels in 45 min; $s = $ speed of the car (in mi/h)

10. The time (in hours) it takes to travel a given distance at 55 mi/h; $d = $ given distance (in mi)

11. The concentration (in oz/gal) of salt in a mixture of 3 gal of brine containing 25 oz of salt, to which some pure water has been added; $x = $ volume of pure water added (in gal)

12. The value (in cents) of the change in a purse that contains twice as many nickels as pennies, four more dimes than nickels, and as many quarters as dimes and nickels combined; $p = $ number of pennies

Applications

13. **Number Problem**   Find three consecutive integers whose sum is 156.

14. **Number Problem**   Find four consecutive odd integers whose sum is 416.

15. **Number Problem**   Find two numbers whose sum is 55 and whose product is 684.

16. **Number Problem**   The sum of the squares of two consecutive even integers is 1252. Find the integers.

17. **Investments**   Phyllis invested $12,000, a portion earning a simple interest rate of $4\frac{1}{2}\%$ per year and the rest earning a rate of 4% per year. After one year the total interest earned

on these investments was $525. How much money did she invest at each rate?

18. **Investments**   If Ben invests $4000 at 4% interest per year, how much additional money must he invest at $5\frac{1}{2}\%$ annual interest to ensure that the interest he receives each year is $4\frac{1}{2}\%$ of the total amount invested?

19. **Investments**   What annual rate of interest would you have to earn on an investment of $3500 to ensure receiving $262.50 interest after one year?

20. **Investments**   Jack invests $1000 at a certain annual interest rate, and he invests another $2000 at an annual rate that is one-half percent higher. If he receives a total of $190 interest in one year, at what rate is the $1000 invested?

21. **Salaries**   An executive in an engineering firm earns a monthly salary plus a Christmas bonus of $8500. If she earns a total of $97,300 per year, what is her monthly salary?

22. **Salaries**   A woman earns 15% more than her husband. Together they make $69,875 per year. What is the husband's annual salary?

23. **Inheritance**   Craig is saving to buy a vacation home. He inherits some money from a wealthy uncle, then combines this with the $22,000 he has already saved and doubles the total in a lucky investment. He ends up with $134,000, just enough to buy a cabin on the lake. How much did he inherit?

24. **Overtime Pay**   Helen earns $7.50 an hour at her job, but if she works more than 35 hours in a week she is paid $1\frac{1}{2}$ times her regular salary for the overtime hours worked. One week her gross pay was $352.50. How many overtime hours did she work that week?

25. **Labor Costs**   A plumber and his assistant work together to replace the pipes in an old house. The plumber charges $45 an hour for his own labor and $25 an hour for his assistant's labor. The plumber works twice as long as his assistant on this job, and the labor charge on the final bill is $4025. How long did the plumber and his assistant work on this job?

26. **Career Home Runs**   During his major league career, Hank Aaron hit 41 more home runs than Babe Ruth hit during his career. Together they hit 1459 home runs. How many home runs did Babe Ruth hit?

27. **A Riddle**   A movie star, unwilling to give his age, posed the following riddle to a gossip columnist. "Seven years ago, I was eleven times as old as my daughter. Now I am four times as old as she is." How old is the star?

28. **A Riddle**   A father is four times as old as his daughter. In 6 years, he will be three times as old as she is. How old is the daughter now?

29. Value of Coins   A change purse contains an equal number of pennies, nickels, and dimes. The total value of the coins is $1.44. How many coins of each type does the purse contain?

30. Value of Coins   Mary has $3.00 in nickels, dimes, and quarters. If she has twice as many dimes as quarters and five more nickels than dimes, how many coins of each type does she have?

31. Law of the Lever   The figure shows a lever system, similar to a seesaw that you might find in a children's playground. For the system to balance, the product of the weight and its distance from the fulcrum must be the same on each side; that is

$$w_1 x_1 = w_2 x_2$$

This equation is called the **law of the lever**, and was first discovered by Archimedes (see page 748).

   A woman and her son are playing on a seesaw. The boy is at one end, 8 ft from the fulcrum. If the son weighs 100 lb and the mother weighs 125 lb, where should the woman sit so that the seesaw is balanced?

32. Law of the Lever   A plank 30 ft long rests on top of a flat-roofed building, with 5 ft of the plank projecting over the edge, as shown in the figure. A worker weighing 240 lb sits on one end of the plank. What is the largest weight that can be hung on the projecting end of the plank if it is to remain in balance? (Use the law of the lever stated in Exercise 31.)

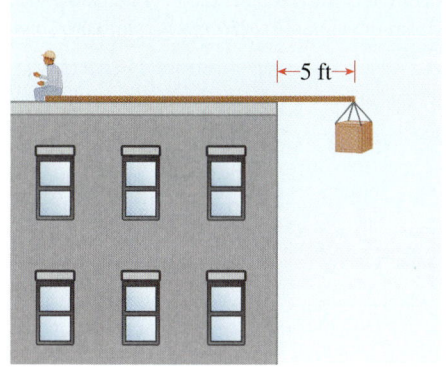

33. Length and Area   Find the length x in the figure. The area of the shaded region is given.

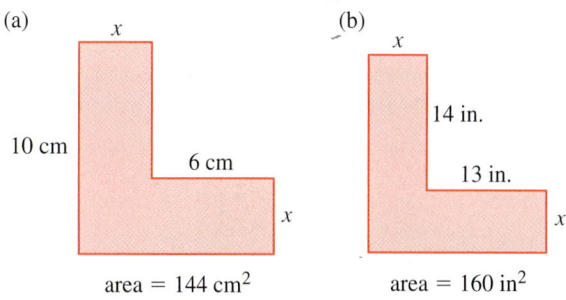

(a)
area = 144 cm²

(b)
area = 160 in²

34. Length and Area   Find the length y in the figure. The area of the shaded region is given.

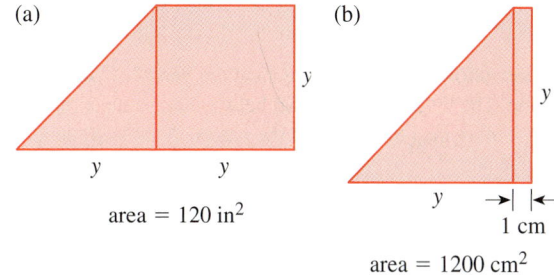

(a)
area = 120 in²

(b)
area = 1200 cm²

35. Length of a Garden   A rectangular garden is 25 ft wide. If its area is 1125 ft², what is the length of the garden?

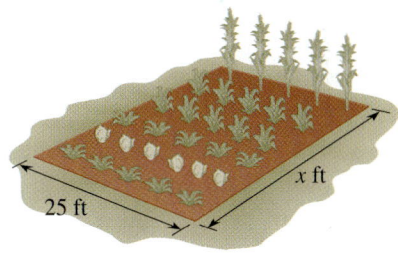

36. Width of a Pasture   A pasture is twice as long as it is wide. Its area is 115,200 ft². How wide is the pasture?

37. Dimensions of a Lot   A square plot of land has a building 60 ft long and 40 ft wide at one corner. The rest of the land outside the building forms a parking lot. If the parking lot has area 12,000 ft², what are the dimensions of the entire plot of land?

38. Dimensions of a Lot   A half-acre building lot is five times as long as it is wide. What are its dimensions? [*Note:* 1 acre = 43,560 ft².]

39. Dimensions of a Garden   A rectangular garden is 10 ft longer than it is wide. Its area is 875 ft². What are its dimensions?

40. Dimensions of a Room   A rectangular bedroom is 7 ft longer than it is wide. Its area is 228 ft². What is the width of the room?

41. Dimensions of a Garden   A farmer has a rectangular garden plot surrounded by 200 ft of fence. Find the length and width of the garden if its area is 2400 ft².

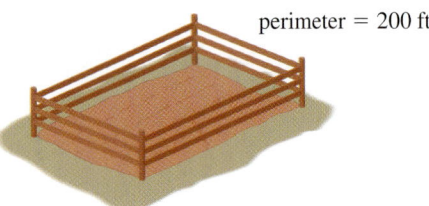

perimeter = 200 ft

42. Dimensions of a Lot   A parcel of land is 6 ft longer than it is wide. Each diagonal from one corner to the opposite corner is 174 ft long. What are the dimensions of the parcel?

43. Dimensions of a Lot   A rectangular parcel of land is 50 ft wide. The length of a diagonal between opposite corners is 10 ft more than the length of the parcel. What is the length of the parcel?

44. Dimensions of a Track   A running track has the shape shown in the figure, with straight sides and semicircular ends. If the length of the track is 440 yd and the two straight parts are each 110 yd long, what is the radius of the semicircular parts (to the nearest yard)?

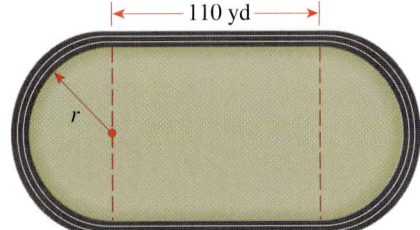

110 yd

r

45. Framing a Painting   Al paints with watercolors on a sheet of paper 20 in. wide by 15 in. high. He then places this sheet on a mat so that a uniformly wide strip of the mat shows all around the picture. The perimeter of the mat is 102 in. How wide is the strip of the mat showing around the picture?

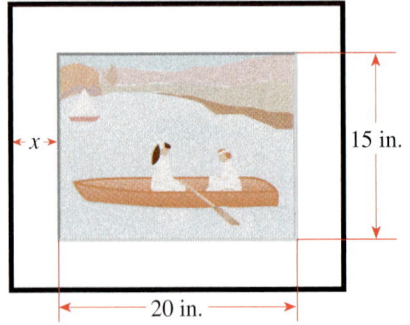

x   15 in.   20 in.

46. Width of a Lawn   A factory is to be built on a lot measuring 180 ft by 240 ft. A local building code specifies that a lawn of uniform width and equal in area to the factory must surround the factory. What must the width of this lawn be, and what are the dimensions of the factory?

47. Reach of a Ladder   A $19\frac{1}{2}$-foot ladder leans against a building. The base of the ladder is $7\frac{1}{2}$ ft from the building. How high up the building does the ladder reach?

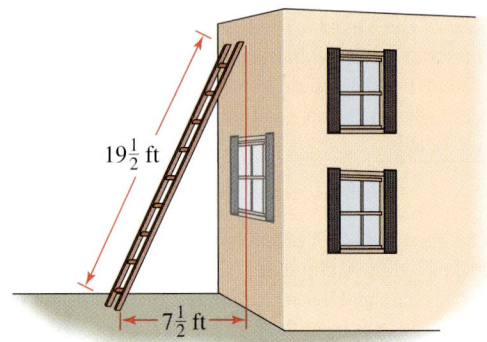

$19\frac{1}{2}$ ft

$7\frac{1}{2}$ ft

48. Height of a Flagpole   A flagpole is secured on opposite sides by two guy wires, each of which is 5 ft longer than the pole. The distance between the points where the wires are fixed to the ground is equal to the length of one guy wire. How tall is the flagpole (to the nearest inch)?

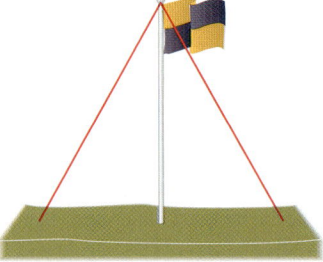

49. Length of a Shadow   A man is walking away from a lamppost with a light source 6 m above the ground. The man is 2 m tall. How long is the man's shadow when he is 10 m from the lamppost? [*Hint:* Use similar triangles.]

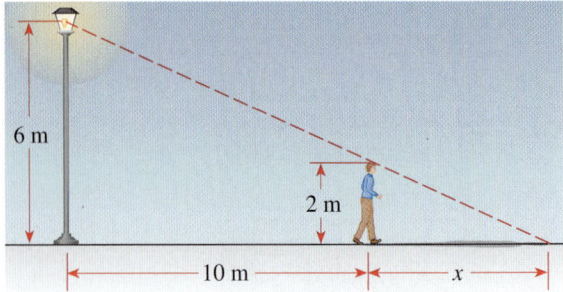

6 m   2 m   10 m   x

50. Height of a Tree   A woodcutter determines the height of a tall tree by first measuring a smaller one 125 ft away, then moving so that his eyes are in the line of sight along the tops of the trees, and measuring how far he is standing from the small tree (see the figure). Suppose the small tree is 20 ft tall, the man is 25 ft from the small tree, and his eye level is 5 ft above the ground. How tall is the taller tree?

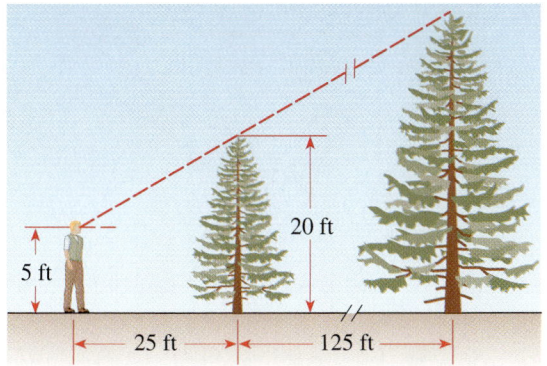

51. Buying a Cottage   A group of friends decides to buy a vacation home for $120,000, sharing the cost equally. If they can find one more person to join them, each person's contribution will drop by $6000. How many people are in the group?

52. Mixture Problem   What quantity of a 60% acid solution must be mixed with a 30% solution to produce 300 mL of a 50% solution?

53. Mixture Problem   A jeweler has five rings, each weighing 18 g, made of an alloy of 10% silver and 90% gold. He decides to melt down the rings and add enough silver to reduce the gold content to 75%. How much silver should he add?

54. Mixture Problem   A pot contains 6 L of brine at a concentration of 120 g/L. How much of the water should be boiled off to increase the concentration to 200 g/L?

55. Mixture Problem   The radiator in a car is filled with a solution of 60% antifreeze and 40% water. The manufacturer of the antifreeze suggests that, for summer driving, optimal cooling of the engine is obtained with only 50% antifreeze. If the capacity of the radiator is 3.6 L, how much coolant should be drained and replaced with water to reduce the antifreeze concentration to the recommended level?

56. Mixture Problem   A health clinic uses a solution of bleach to sterilize petri dishes in which cultures are grown. The sterilization tank contains 100 gal of a solution of 2% ordinary household bleach mixed with pure distilled water. New research indicates that the concentration of bleach should be 5% for complete sterilization. How much of the solution should be drained and replaced with bleach to increase the bleach content to the recommended level?

57. Mixture Problem   A bottle contains 750 mL of fruit punch with a concentration of 50% pure fruit juice. Jill drinks 100 mL of the punch and then refills the bottle with an equal amount of a cheaper brand of punch. If the concentration of juice in the bottle is now reduced to 48%, what was the concentration in the punch that Jill added?

58. Mixture Problem   A merchant blends tea that sells for $3.00 a pound with tea that sells for $2.75 a pound to produce 80 lb of a mixture that sells for $2.90 a pound. How many pounds of each type of tea does the merchant use in the blend?

59. Sharing a Job   Candy and Tim share a paper route. It takes Candy 70 min to deliver all the papers, and it takes Tim 80 min. How long does it take the two when they work together?

60. Sharing a Job   Stan and Hilda can mow the lawn in 40 min if they work together. If Hilda works twice as fast as Stan, how long does it take Stan to mow the lawn alone?

61. Sharing a Job   Betty and Karen have been hired to paint the houses in a new development. Working together the women can paint a house in two-thirds the time that it takes Karen working alone. Betty takes 6 h to paint a house alone. How long does it take Karen to paint a house working alone?

62. Sharing a Job   Next-door neighbors Bob and Jim use hoses from both houses to fill Bob's swimming pool. They know it takes 18 h using both hoses. They also know that Bob's hose, used alone, takes 20% less time than Jim's hose alone. How much time is required to fill the pool by each hose alone?

63. Sharing a Job   Henry and Irene working together can wash all the windows of their house in 1 h 48 min. Working alone, it takes Henry $1\frac{1}{2}$ h more than Irene to do the job. How long does it take each person working alone to wash all the windows?

64. Sharing a Job   Jack, Kay, and Lynn deliver advertising flyers in a small town. If each person works alone, it takes Jack 4 h to deliver all the flyers, and it takes Lynn 1 h longer than it takes Kay. Working together, they can deliver all the flyers in 40% of the time it takes Kay working alone. How long does it take Kay to deliver all the flyers alone?

65. Distance, Speed, and Time   Wendy took a trip from Davenport to Omaha, a distance of 300 mi. She traveled part of the way by bus, which arrived at the train station just in time for Wendy to complete her journey by train. The bus averaged 40 mi/h and the train 60 mi/h. The entire trip took $5\frac{1}{2}$ h. How long did Wendy spend on the train?

66. Distance, Speed, and Time   Two cyclists, 90 mi apart, start riding toward each other at the same time. One cycles twice as fast as the other. If they meet 2 h later, at what average speed is each cyclist traveling?

67. Distance, Speed, and Time A pilot flew a jet from Montreal to Los Angeles, a distance of 2500 mi. On the return trip the average speed was 20% faster than the outbound speed. The round-trip took 9 h 10 min. What was the speed from Montreal to Los Angeles?

68. Distance, Speed, and Time A woman driving a car 14 ft long is passing a truck 30 ft long. The truck is traveling at 50 mi/h. How fast must the woman drive her car so that she can pass the truck completely in 6 s, from the position shown in figure (a) to the position shown in figure (b)? [*Hint:* Use feet and seconds instead of miles and hours.]

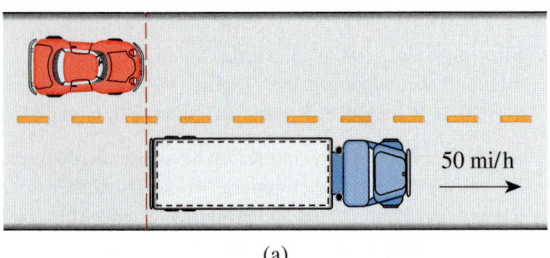

(a)

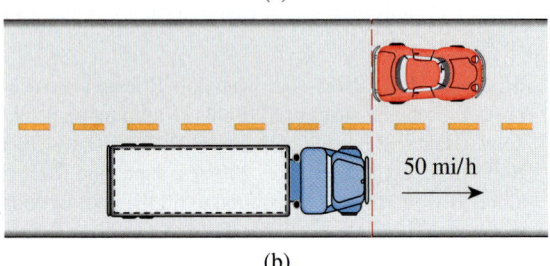

(b)

69. Distance, Speed, and Time A salesman drives from Ajax to Barrington, a distance of 120 mi, at a steady speed. He then increases his speed by 10 mi/h to drive the 150 mi from Barrington to Collins. If the second leg of his trip took 6 min more time than the first leg, how fast was he driving between Ajax and Barrington?

70. Distance, Speed, and Time Kiran drove from Tortula to Cactus, a distance of 250 mi. She increased her speed by 10 mi/h for the 360-mi trip from Cactus to Dry Junction. If the total trip took 11 h, what was her speed from Tortula to Cactus?

71. Distance, Speed, and Time It took a crew 2 h 40 min to row 6 km upstream and back again. If the rate of flow of the stream was 3 km/h, what was the rowing speed of the crew in still water?

72. Speed of a Boat Two fishing boats depart a harbor at the same time, one traveling east, the other south. The

eastbound boat travels at a speed 3 mi/h faster than the southbound boat. After two hours the boats are 30 mi apart. Find the speed of the southbound boat.

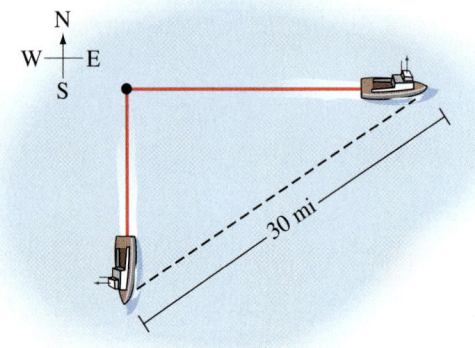

73. Dimensions of a Box A large plywood box has a volume of 180 ft³. Its length is 9 ft greater than its height, and its width is 4 ft less than its height. What are the dimensions of the box?

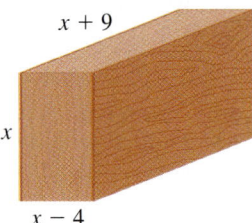

74. Radius of a Sphere A jeweler has three small solid spheres made of gold, of radius 2 mm, 3 mm, and 4 mm. He decides to melt these down and make just one sphere out of them. What will the radius of this larger sphere be?

75. Dimensions of a Box A box with a square base and no top is to be made from a square piece of cardboard by cutting 4-in. squares from each corner and folding up the sides, as shown in the figure. The box is to hold 100 in³. How big a piece of cardboard is needed?

 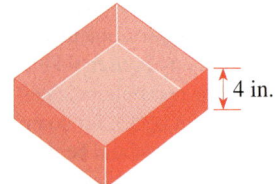

76. Dimensions of a Can A cylindrical can has a volume of 40π cm^3 and is 10 cm tall. What is its diameter? [*Hint:* Use the volume formula listed on the inside back cover of this book.]

10 cm

77. Radius of a Tank A spherical tank has a capacity of 750 gallons. Using the fact that one gallon is about 0.1337 ft^3, find the radius of the tank (to the nearest hundredth of a foot).

78. Dimensions of a Lot A city lot has the shape of a right triangle whose hypotenuse is 7 ft longer than one of the other sides. The perimeter of the lot is 392 ft. How long is each side of the lot?

79. Construction Costs The town of Foxton lies 10 mi north of an abandoned east-west road that runs through Grimley, as shown in the figure. The point on the abandoned road closest to Foxton is 40 mi from Grimley. County officials are about to build a new road connecting the two towns. They have determined that restoring the old road would cost $100,000 per mile, whereas building a new road would cost $200,000 per mile. How much of the abandoned road should be used (as indicated in the figure) if the officials intend to spend exactly $6.8 million? Would it cost less than this amount to build a new road connecting the towns directly?

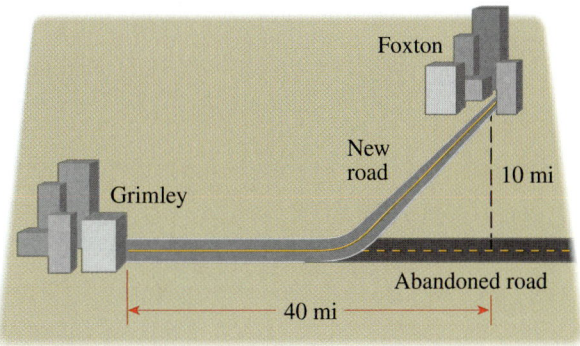

80. Distance, Speed, and Time A boardwalk is parallel to and 210 ft inland from a straight shoreline. A sandy beach lies between the boardwalk and the shoreline. A man is standing on the boardwalk, exactly 750 ft across the sand from his beach umbrella, which is right at the shoreline. The man walks 4 ft/s on the boardwalk and 2 ft/s on the sand. How far should he walk on the boardwalk before veering off onto the sand if he wishes to reach his umbrella in exactly 4 min 45 s?

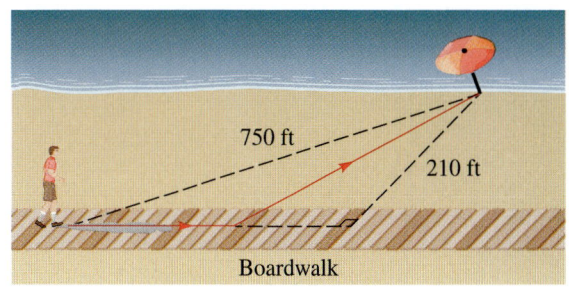

81. Volume of Grain Grain is falling from a chute onto the ground, forming a conical pile whose diameter is always three times its height. How high is the pile (to the nearest hundredth of a foot) when it contains 1000 ft^3 of grain?

82. TV Monitors Two television monitors sitting beside each other on a shelf in an appliance store have the same screen height. One has a conventional screen, which is 5 in. wider than it is high. The other has a wider, high-definition screen, which is 1.8 times as wide as it is high. The diagonal measure of the wider screen is 14 in. more than the diagonal measure of the smaller. What is the height of the screens, correct to the nearest 0.1 in.?

83. Dimensions of a Structure A storage bin for corn consists of a cylindrical section made of wire mesh, surmounted by a conical tin roof, as shown in the figure. The height of the roof is one-third the height of the entire structure. If the total volume of the structure is 1400π ft^3 and its radius is 10 ft, what is its height? [*Hint:* Use the volume formulas listed on the inside front cover of this book.]

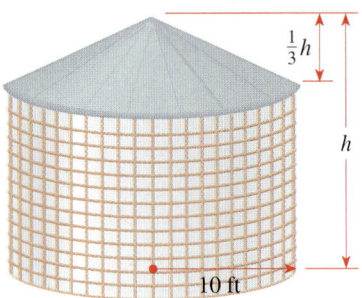

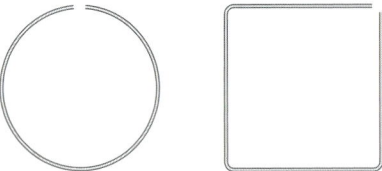

84. Comparing Areas A wire 360 in. long is cut into two pieces. One piece is formed into a square and the other into a circle. If the two figures have the same area, what are the lengths of the two pieces of wire (to the nearest tenth of an inch)?

85. An Ancient Chinese Problem This problem is taken from a Chinese mathematics textbook called *Chui-chang suan-shu*, or *Nine Chapters on the Mathematical Art*, which was written about 250 B.C.

A 10-ft-long stem of bamboo is broken in such a way that its tip touches the ground 3 ft from the base of the stem, as shown in the figure. What is the height of the break?

[*Hint:* Use the Pythagorean Theorem.]

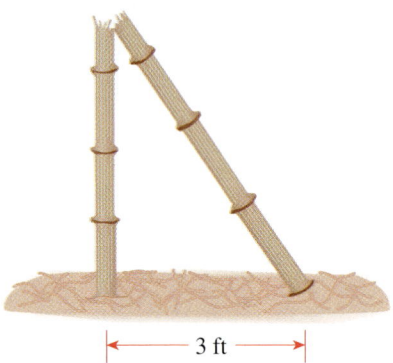

Discovery • Discussion

86. Historical Research Read the biographical notes on Pythagoras (page 54), Euclid (page 532), and Archimedes (page 748). Choose one of these mathematicians and find out more about him from the library or on the Internet. Write a short essay on your findings. Include both biographical information and a description of the mathematics for which he is famous.

87. A Babylonian Quadratic Equation The ancient Babylonians knew how to solve quadratic equations. Here is a problem from a cuneiform tablet found in a Babylonian school dating back to about 2000 B.C.

> I have a reed, I know not its length. I broke from it one cubit, and it fit 60 times along the length of my field. I restored to the reed what I had broken off, and it fit 30 times along the width of my field. The area of my field is 375 square nindas. What was the original length of the reed?

Solve this problem. Use the fact that 1 ninda = 12 cubits.

DISCOVERY PROJECT

The British Museum

Equations through the Ages

Equations have been used to solve problems throughout recorded history, in every civilization. (See, for example, Exercise 85 on page 74.) Here is a problem from ancient Babylon (ca. 2000 B.C.).

> I found a stone but did not weigh it. After I added a seventh, and then added an eleventh of the result, I weighed it and found it weighed 1 mina. What was the original weight of the stone?

The answer given on the cuneiform tablet is $\frac{2}{3}$ mina, 8 sheqel, and $22\frac{1}{2}$ se, where 1 mina = 60 sheqel, and 1 sheqel = 180 se.

In ancient Egypt, knowing how to solve word problems was a highly prized secret. The Rhind Papyrus (ca. 1850 B.C.) contains many such problems (see page 716). Problem 32 in the Papyrus states

> A quantity, its third, its quarter, added together become 2. What is the quantity?

The answer in Egyptian notation is $1 + \overline{4} + \overline{76}$, where the bar indicates "reciprocal," much like our notation 4^{-1}.

The Greek mathematician Diophantus (ca. 250 A.D., see page 20) wrote the book *Arithmetica*, which contains many word problems and equations. The Indian mathematician Bhaskara (12th century A.D., see page 144) and the Chinese mathematician Chang Ch'iu-Chien (6th century A.D.) also studied and wrote about equations. Of course, equations continue to be important today.

1. Solve the Babylonian problem and show that their answer is correct.

2. Solve the Egyptian problem and show that their answer is correct.

3. The ancient Egyptians and Babylonians used equations to solve practical problems. From the examples given here, do you think that they may have enjoyed posing and solving word problems just for fun?

4. Solve this problem from 12th-century India.

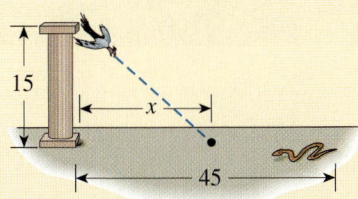

> A peacock is perched at the top of a 15-cubit pillar, and a snake's hole is at the foot of the pillar. Seeing the snake at a distance of 45 cubits from its hole, the peacock pounces obliquely upon the snake as it slithers home. At how many cubits from the snake's hole do they meet, assuming that each has traveled an equal distance?

5. Consider this problem from 6th-century China.

> If a rooster is worth 5 coins, a hen 3 coins, and three chicks together one coin, how many roosters, hens, and chicks, totaling 100, can be bought for 100 coins?

This problem has several answers. Use trial and error to find at least one answer. Is this a practical problem or more of a riddle? Write a short essay to support your opinion.

6. Write a short essay explaining how equations affect your own life in today's world.

1.7   Inequalities

x	$4x + 7 \le 19$
1	$11 \le 19$ ✓
2	$15 \le 19$ ✓
3	$19 \le 19$ ✓
4	$23 \le 19$ ✗
5	$27 \le 19$ ✗

Some problems in algebra lead to **inequalities** instead of equations. An inequality looks just like an equation, except that in the place of the equal sign is one of the symbols, $<$, $>$, $\le$, or $\ge$. Here is an example of an inequality:

$$4x + 7 \le 19$$

The table in the margin shows that some numbers satisfy the inequality and some numbers don't.

To **solve** an inequality that contains a variable means to find all values of the variable that make the inequality true. Unlike an equation, an inequality generally has infinitely many solutions, which form an interval or a union of intervals on the real line. The following illustration shows how an inequality differs from its corresponding equation:

		Solution	Graph
Equation:	$4x + 7 = 19$	$x = 3$	
Inequality:	$4x + 7 \le 19$	$x \le 3$	

To solve inequalities, we use the following rules to isolate the variable on one side of the inequality sign. These rules tell us when two inequalities are *equivalent* (the symbol $\Leftrightarrow$ means "is equivalent to"). In these rules the symbols A, B, and C stand for real numbers or algebraic expressions. Here we state the rules for inequalities involving the symbol $\le$, but they apply to all four inequality symbols.

Rules for Inequalities

Rule	Description
1. $A \le B \quad \Leftrightarrow \quad A + C \le B + C$	**Adding** the same quantity to each side of an inequality gives an equivalent inequality.
2. $A \le B \quad \Leftrightarrow \quad A - C \le B - C$	**Subtracting** the same quantity from each side of an inequality gives an equivalent inequality.
3. If $C > 0$, then $A \le B \quad \Leftrightarrow \quad CA \le CB$	**Multiplying** each side of an inequality by the same *positive* quantity gives an equivalent inequality.
4. If $C < 0$, then $A \le B \quad \Leftrightarrow \quad CA \ge CB$	**Multiplying** each side of an inequality by the same *negative* quantity *reverses the direction* of the inequality.
5. If $A > 0$ and $B > 0$, then $A \le B \quad \Leftrightarrow \quad \dfrac{1}{A} \ge \dfrac{1}{B}$	**Taking reciprocals** of each side of an inequality involving *positive* quantities *reverses the direction* of the inequality.
6. If $A \le B$ and $C \le D$, then $A + C \le B + D$	Inequalities can be added.

Pay special attention to Rules 3 and 4. Rule 3 says that we can multiply (or divide) each side of an inequality by a *positive* number, but Rule 4 says that if we multiply each side of an inequality by a *negative* number, then we reverse the direction of the inequality. For example, if we start with the inequality

$$3 < 5$$

and multiply by 2, we get

$$6 < 10$$

but if we multiply by -2, we get

$$-6 > -10$$

Linear Inequalities

An inequality is **linear** if each term is constant or a multiple of the variable.

Example 1   Solving a Linear Inequality

Solve the inequality $3x < 9x + 4$ and sketch the solution set.

Solution

$$3x < 9x + 4$$

$$3x - 9x < 9x + 4 - 9x \qquad \text{Subtract } 9x$$

$$-6x < 4 \qquad \text{Simplify}$$

Multiplying by the negative number $-\frac{1}{6}$ *reverses* the direction of the inequality.

$$\left(-\tfrac{1}{6}\right)(-6x) > \left(-\tfrac{1}{6}\right)(4) \qquad \text{Multiply by } -\tfrac{1}{6} \text{ (or divide by } -6\text{)}$$

$$x > -\tfrac{2}{3} \qquad \text{Simplify}$$

The solution set consists of all numbers greater than $-\frac{2}{3}$. In other words the solution of the inequality is the interval $\left(-\frac{2}{3}, \infty\right)$. It is graphed in Figure 1. ∎

Figure 1

Example 2   Solving a Pair of Simultaneous Inequalities

Solve the inequalities $4 \le 3x - 2 < 13$.

Solution   The solution set consists of all values of x that satisfy both of the inequalities $4 \le 3x - 2$ and $3x - 2 < 13$. Using Rules 1 and 3, we see that the following inequalities are equivalent:

$$4 \le 3x - 2 < 13$$

$$6 \le 3x < 15 \qquad \text{Add 2}$$

$$2 \le x < 5 \qquad \text{Divide by 3}$$

Therefore, the solution set is $[2, 5)$, as shown in Figure 2. ∎

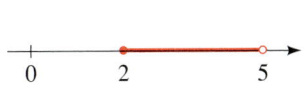

Figure 2

Nonlinear Inequalities

To solve inequalities involving squares and other powers of the variable, we use factoring, together with the following principle.

The Sign of a Product or Quotient

If a product or a quotient has an *even* number of *negative* factors, then its value is *positive*.

If a product or a quotient has an *odd* number of *negative* factors, then its value is *negative*.

Example 3   A Quadratic Inequality

Solve the inequality $x^2 - 5x + 6 \leq 0$.

Solution   First we factor the left side.

$$(x - 2)(x - 3) \leq 0$$

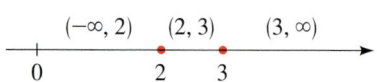

Figure 3

We know that the corresponding equation $(x - 2)(x - 3) = 0$ has the solutions 2 and 3. As shown in Figure 3, the numbers 2 and 3 divide the real line into three intervals: $(-\infty, 2)$, $(2, 3)$, and $(3, \infty)$. On each of these intervals we determine the signs of the factors using **test values**. We choose a number inside each interval and check the sign of the factors $x - 2$ and $x - 3$ at the value selected. For instance, if we use the test value $x = 1$ for the interval $(-\infty, 2)$ shown in Figure 4, then substitution in the factors $x - 2$ and $x - 3$ gives

$$x - 2 = 1 - 2 = -1 < 0$$

and
$$x - 3 = 1 - 3 = -2 < 0$$

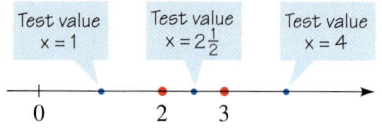

Figure 4

So both factors are negative on this interval. (The factors $x - 2$ and $x - 3$ change sign only at 2 and 3, respectively, so they maintain their signs over the length of each interval. That is why using a single test value on each interval is sufficient.)

Using the test values $x = 2\frac{1}{2}$ and $x = 4$ for the intervals $(2, 3)$ and $(3, \infty)$ (see Figure 4), respectively, we construct the following sign table. The final row of the table is obtained from the fact that the expression in the last row is the product of the two factors.

Interval	$(-\infty, 2)$	$(2, 3)$	$(3, \infty)$
Sign of $x - 2$	−	+	+
Sign of $x - 3$	−	−	+
Sign of $(x - 2)(x - 3)$	+	−	+

If you prefer, you can represent this information on a real number line, as in the following sign diagram. The vertical lines indicate the points at which the real line is divided into intervals:

	2	3

		2		3	
Sign of $x - 2$	$-$		$+$		$+$
Sign of $x - 3$	$-$		$-$		$+$
Sign of $(x - 2)(x - 3)$	$+$		$-$		$+$

Figure 5

We read from the table or the diagram that $(x - 2)(x - 3)$ is negative on the interval $(2, 3)$. Thus, the solution of the inequality $(x - 2)(x - 3) \leq 0$ is

$$\{x \mid 2 \leq x \leq 3\} = [2, 3]$$

We have included the endpoints 2 and 3 because we seek values of x such that the product is either less than *or equal to* zero. The solution is illustrated in Figure 5.   ∎

Example 3 illustrates the following guidelines for solving an inequality that can be factored.

Guidelines for Solving Nonlinear Inequalities

1. **Move All Terms to One Side.**   If necessary, rewrite the inequality so that all nonzero terms appear on one side of the inequality sign. If the nonzero side of the inequality involves quotients, bring them to a common denominator.

2. **Factor.**   Factor the nonzero side of the inequality.

3. **Find the Intervals.**   Determine the values for which each factor is zero. These numbers will divide the real line into intervals. List the intervals determined by these numbers.

4. **Make a Table or Diagram.**   Use test values to make a table or diagram of the signs of each factor on each interval. In the last row of the table determine the sign of the product (or quotient) of these factors.

5. **Solve.**   Determine the solution of the inequality from the last row of the sign table. Be sure to check whether the inequality is satisfied by some or all of the endpoints of the intervals (this may happen if the inequality involves $\leq$ or $\geq$).

   The factoring technique described in these guidelines works only if all nonzero terms appear on one side of the inequality symbol. If the inequality is not written in this form, first rewrite it, as indicated in Step 1. This technique is illustrated in the examples that follow.

🚫 It is tempting to multiply both sides of the inequality by $1 - x$ (as you would if this were an *equation*). But this doesn't work because we don't know if $1 - x$ is positive or negative, so we can't tell if the inequality needs to be reversed. (See Exercise 110.)

Example 4   An Inequality Involving a Quotient

Solve: $\dfrac{1 + x}{1 - x} \geq 1$

Solution   First we move all nonzero terms to the left side, and then we simplify using a common denominator.

$$\frac{1 + x}{1 - x} \geq 1$$

Terms to one side

$$\frac{1 + x}{1 - x} - 1 \geq 0 \qquad \text{Subtract 1 (to move all terms to LHS)}$$

$$\frac{1 + x}{1 - x} - \frac{1 - x}{1 - x} \geq 0 \qquad \text{Common denominator } 1 - x$$

$$\frac{1 + x - 1 + x}{1 - x} \geq 0 \qquad \text{Combine the fractions}$$

$$\frac{2x}{1 - x} \geq 0 \qquad \text{Simplify}$$

The numerator is zero when $x = 0$ and the denominator is zero when $x = 1$, so we construct the following sign diagram using these values to define intervals on the real line.

Make a diagram

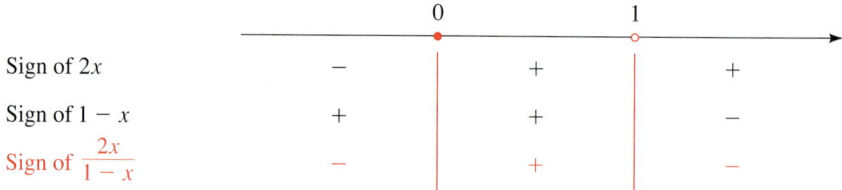

		0		1	
Sign of $2x$	$-$		$+$		$+$
Sign of $1 - x$	$+$		$+$		$-$
Sign of $\dfrac{2x}{1 - x}$	$-$		$+$		$-$

Solve

From the diagram we see that the solution set is $\{x \mid 0 \leq x < 1\} = [0, 1)$. We include the endpoint 0 because the original inequality requires the quotient to be greater than *or equal to* 1. However, we do not include the other endpoint 1, since 🚫 the quotient in the inequality is not defined at 1. Always check the endpoints of solution intervals to determine whether they satisfy the original inequality.

The solution set $[0, 1)$ is illustrated in Figure 6.   ∎

Figure 6

Example 5   Solving an Inequality with Three Factors

Solve the inequality $x < \dfrac{2}{x - 1}$.

Solution   After moving all nonzero terms to one side of the inequality, we use a common denominator to combine the terms.

Terms to one side

$$x - \frac{2}{x - 1} < 0 \qquad \text{Subtract } \frac{2}{x - 1}$$

$$\frac{x(x - 1)}{x - 1} - \frac{2}{x - 1} < 0 \qquad \text{Common denominator } x - 1$$

$$\frac{x^2 - x - 2}{x - 1} < 0 \qquad \text{Combine fractions}$$

Factor

$$\frac{(x + 1)(x - 2)}{x - 1} < 0 \qquad \text{Factor numerator}$$

The factors in this quotient change sign at -1, 1, and 2, so we must examine the intervals $(-\infty, -1), (-1, 1), (1, 2),$ and $(2, \infty)$. Using test values, we get the following sign diagram.

Find the intervals

	-1		1	2
Sign of $x + 1$	$-$	$+$	$+$	$+$
Sign of $x - 2$	$-$	$-$	$-$	$+$
Sign of $x - 1$	$-$	$-$	$+$	$+$
Sign of $\dfrac{(x+1)(x-2)}{x-1}$	$-$	$+$	$-$	$+$

Make a diagram

Since the quotient must be negative, the solution is

$$(-\infty, -1) \cup (1, 2)$$

as illustrated in Figure 7.

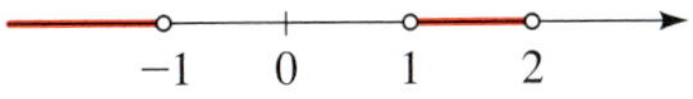

Figure 7

Absolute Value Inequalities

We use the following properties to solve inequalities that involve absolute value.

Properties of Absolute Value Inequalities

These properties hold when x is replaced by any algebraic expression. (In the figures we assume that $c > 0$.)

Inequality	Equivalent form
1. $|x| < c$	$-c < x < c$
2. $|x| \leq c$	$-c \leq x \leq c$
3. $|x| > c$	$x < -c$ or $c < x$
4. $|x| \geq c$	$x \leq -c$ or $c \leq x$

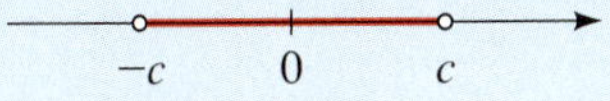

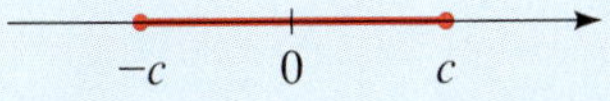

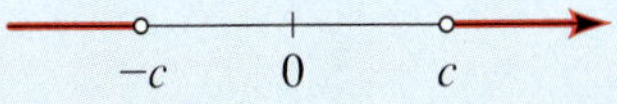

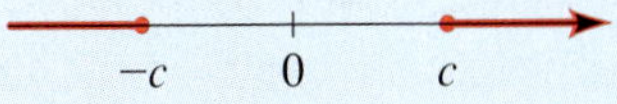

These properties can be proved using the definition of absolute value. To prove Property 1, for example, note that the inequality $|x| < c$ says that the distance from x to 0 is less than c, and from Figure 8 you can see that this is true if and only if x is between $-c$ and c.

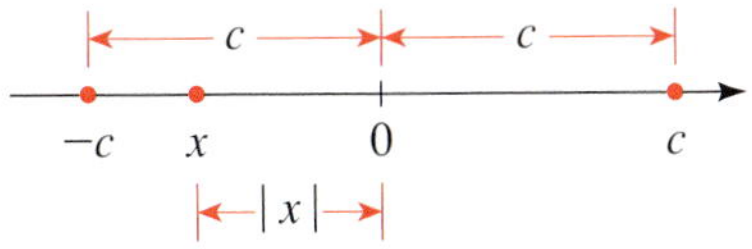

Figure 8

Example 6 Solving an Absolute Value Inequality

Solve the inequality $|x - 5| < 2$.

Solution 1 The inequality $|x - 5| < 2$ is equivalent to

$$-2 < x - 5 < 2 \quad \text{Property 1}$$
$$3 < x < 7 \quad \text{Add 5}$$

The solution set is the open interval $(3, 7)$.

Solution 2 Geometrically, the solution set consists of all numbers x whose distance from 5 is less than 2. From Figure 9 we see that this is the interval $(3, 7)$.

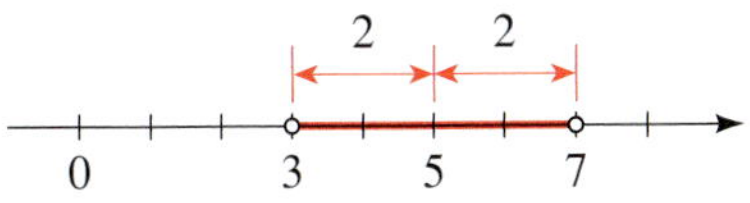

Figure 9

Example 7   Solving an Absolute Value Inequality

Solve the inequality $|3x + 2| \geq 4$.

Solution   By Property 4 the inequality $|3x + 2| \geq 4$ is equivalent to

$$3x + 2 \geq 4 \qquad \text{or} \qquad 3x + 2 \leq -4$$

$$3x \geq 2 \qquad\qquad\qquad 3x \leq -6 \qquad \text{Subtract 2}$$

$$x \geq \tfrac{2}{3} \qquad\qquad\qquad x \leq -2 \qquad \text{Divide by 3}$$

So the solution set is

$$\{x \mid x \leq -2 \quad \text{or} \quad x \geq \tfrac{2}{3}\} = (-\infty, -2] \cup [\tfrac{2}{3}, \infty)$$

The set is graphed in Figure 10.

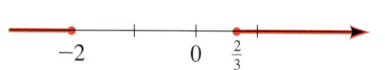

Figure 10

Modeling with Inequalities

Modeling real-life problems frequently leads to inequalities because we are often interested in determining when one quantity is more (or less) than another.

Example 8   Carnival Tickets

A carnival has two plans for tickets.

Plan A: $5 entrance fee and 25¢ each ride

Plan B: $2 entrance fee and 50¢ each ride

How many rides would you have to take for plan A to be less expensive than plan B?

Solution   We are asked for the number of rides for which plan A is less expensive than plan B. So let

Identify the variable

$$x = \text{number of rides}$$

The information in the problem may be organized as follows.

In Words	In Algebra
Number of rides	x
Cost with plan A	$5 + 0.25x$
Cost with plan B	$2 + 0.50x$

Express all unknown quantities in terms of the variable

Now we set up the model.

Set up the model

$$\begin{array}{ccc} \text{cost with} \\ \text{plan A} \end{array} < \begin{array}{ccc} \text{cost with} \\ \text{plan B} \end{array}$$

$$5 + 0.25x < 2 + 0.50x$$

Solve

$$3 + 0.25x < 0.50x \qquad \text{Subtract 2}$$

$$3 < 0.25x \qquad \text{Subtract 0.25x}$$

$$12 < x \qquad \text{Divide by 0.25}$$

So if you plan to take *more than* 12 rides, plan A is less expensive.

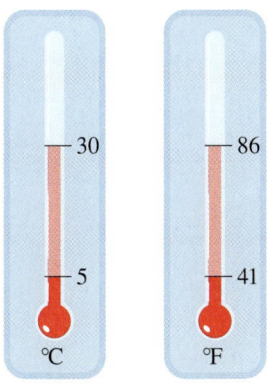

°C   °F

Example 9   Fahrenheit and Celsius Scales

The instructions on a box of film indicate that the box should be stored at a temperature between 5 °C and 30 °C. What range of temperatures does this correspond to on the Fahrenheit scale?

Solution   The relationship between degrees Celsius (C) and degrees Fahrenheit (F) is given by the equation $C = \frac{5}{9}(F - 32)$. Expressing the statement on the box in terms of inequalities, we have

$$5 < C < 30$$

So the corresponding Fahrenheit temperatures satisfy the inequalities

$$5 < \tfrac{5}{9}(F - 32) < 30$$

$$\tfrac{9}{5} \cdot 5 < F - 32 < \tfrac{9}{5} \cdot 30 \qquad \text{Multiply by } \tfrac{9}{5}$$

$$9 < F - 32 < 54 \qquad \text{Simplify}$$

$$9 + 32 < F < 54 + 32 \qquad \text{Add 32}$$

$$41 < F < 86 \qquad \text{Simplify}$$

The film should be stored at a temperature between 41 °F and 86 °F.   ■

Example 10   Concert Tickets

A group of students decide to attend a concert. The cost of chartering a bus to take them to the concert is $450, which is to be shared equally among the students. The concert promoters offer discounts to groups arriving by bus. Tickets normally cost $50 each but are reduced by 10¢ per ticket for each person in the group (up to the maximum capacity of the bus). How many students must be in the group for the total cost per student to be less than $54?

Solution   We are asked for the number of students in the group. So let

$$x = \text{number of students in the group}$$

Identify the variable

The information in the problem may be organized as follows.

In Words	In Algebra
Number of students in group	x
Bus cost per student	$\dfrac{450}{x}$
Ticket cost per student	$50 - 0.10x$

Express all unknown quantities in terms of the variable

Now we set up the model.

Set up the model

$$\begin{pmatrix} \text{bus cost} \\ \text{per student} \end{pmatrix} + \begin{pmatrix} \text{ticket cost} \\ \text{per student} \end{pmatrix} < 54$$

$$\frac{450}{x} + (50 - 0.10x) < 54$$

Solve

$$\frac{450}{x} - 4 - 0.10x < 0 \qquad \text{Subtract 54}$$

$$\frac{450 - 4x - 0.10x^2}{x} < 0 \qquad \text{Common denominator}$$

$$\frac{4500 - 40x - x^2}{x} < 0 \qquad \text{Multiply by 10}$$

$$\frac{(90 + x)(50 - x)}{x} < 0 \qquad \text{Factor numerator}$$

	-90		0		50	
Sign of $90 + x$	$-$		$+$		$+$	$+$
Sign of $50 - x$	$+$		$+$		$+$	$-$
Sign of x	$-$		$-$		$+$	$+$
Sign of $\dfrac{(90 + x)(50 - x)}{x}$	$+$		$-$		$+$	$-$

The sign diagram shows that the solution of the inequality is $(-90, 0) \cup (50, \infty)$. Because we cannot have a negative number of students, it follows that the group must have more than 50 students for the total cost per person to be less than \$54. ∎

1.7 Exercises

1–6 ■ Let $S = \{-2, -1, 0, \frac{1}{2}, 1, \sqrt{2}, 2, 4\}$. Determine which elements of S satisfy the inequality.

1. $3 - 2x \le \frac{1}{2}$

2. $2x - 1 \ge x$

3. $1 < 2x - 4 \le 7$

4. $-2 \le 3 - x < 2$

5. $\frac{1}{x} \le \frac{1}{2}$

6. $x^2 + 2 < 4$

7–28 ■ Solve the linear inequality. Express the solution using interval notation and graph the solution set.

7. $2x - 5 > 3$

8. $3x + 11 < 5$

9. $7 - x \ge 5$

10. $5 - 3x \le -16$

11. $2x + 1 < 0$

12. $0 < 5 - 2x$

13. $3x + 11 \le 6x + 8$

14. $6 - x \ge 2x + 9$

15. $\frac{1}{2}x - \frac{2}{3} > 2$

16. $\frac{2}{5}x + 1 < \frac{1}{5} - 2x$

17. $\frac{1}{3}x + 2 < \frac{1}{6}x - 1$

18. $\frac{2}{3} - \frac{1}{2}x \ge \frac{1}{6} + x$

19. $4 - 3x \le -(1 + 8x)$

20. $2(7x - 3) \le 12x + 16$

21. $2 \le x + 5 < 4$

22. $5 \le 3x - 4 \le 14$

23. $-1 < 2x - 5 < 7$

24. $1 < 3x + 4 \le 16$

25. $-2 < 8 - 2x \le -1$

26. $-3 \le 3x + 7 \le \frac{1}{2}$

27. $\frac{1}{6} < \frac{2x - 13}{12} \le \frac{2}{3}$

28. $-\frac{1}{2} \le \frac{4 - 3x}{5} \le \frac{1}{4}$

29–62 ■ Solve the nonlinear inequality. Express the solution using interval notation and graph the solution set.

29. $(x + 2)(x - 3) < 0$

30. $(x - 5)(x + 4) \ge 0$

31. $x(2x + 7) \ge 0$

32. $x(2 - 3x) \le 0$

33. $x^2 - 3x - 18 \le 0$

34. $x^2 + 5x + 6 > 0$

35. $2x^2 + x \ge 1$

36. $x^2 < x + 2$

37. $3x^2 - 3x < 2x^2 + 4$

38. $5x^2 + 3x \ge 3x^2 + 2$

39. $x^2 > 3(x + 6)$

40. $x^2 + 2x > 3$

41. $x^2 < 4$

42. $x^2 \ge 9$

43. $-2x^2 \le 4$

44. $(x + 2)(x - 1)(x - 3) \le 0$

45. $x^3 - 4x > 0$

46. $16x \le x^3$

47. $\frac{x - 3}{x + 1} \ge 0$

48. $\frac{2x + 6}{x - 2} < 0$

49. $\frac{4x}{2x + 3} > 2$

50. $-2 < \frac{x + 1}{x - 3}$

51. $\dfrac{2x + 1}{x - 5} \le 3$ **52.** $\dfrac{3 + x}{3 - x} \ge 1$

53. $\dfrac{4}{x} < x$ **54.** $\dfrac{x}{x + 1} > 3x$

55. $1 + \dfrac{2}{x + 1} \le \dfrac{2}{x}$ **56.** $\dfrac{3}{x - 1} - \dfrac{4}{x} \ge 1$

57. $\dfrac{6}{x - 1} - \dfrac{6}{x} \ge 1$ **58.** $\dfrac{x}{2} \ge \dfrac{5}{x + 1} + 4$

59. $\dfrac{x + 2}{x + 3} < \dfrac{x - 1}{x - 2}$ **60.** $\dfrac{1}{x + 1} + \dfrac{1}{x + 2} \le 0$

61. $x^4 > x^2$ **62.** $x^5 > x^2$

63–76 ■ Solve the absolute value inequality. Express the answer using interval notation and graph the solution set.

63. $|x| \le 4$ **64.** $|3x| < 15$

65. $|2x| > 7$ **66.** $\frac{1}{2}|x| \ge 1$

67. $|x - 5| \le 3$ **68.** $|x + 1| \ge 1$

69. $|2x - 3| \le 0.4$ **70.** $|5x - 2| < 6$

71. $\left| \dfrac{x - 2}{3} \right| < 2$ **72.** $\left| \dfrac{x + 1}{2} \right| \ge 4$

73. $|x + 6| < 0.001$ **74.** $3 - |2x + 4| \le 1$

75. $8 - |2x - 1| \ge 6$ **76.** $7|x + 2| + 5 > 4$

77–80 ■ A phrase describing a set of real numbers is given. Express the phrase as an inequality involving an absolute value.

77. All real numbers x less than 3 units from 0

78. All real numbers x more than 2 units from 0

79. All real numbers x at least 5 units from 7

80. All real numbers x at most 4 units from 2

81–86 ■ A set of real numbers is graphed. Find an inequality involving an absolute value that describes the set.

81.

82.

83.

84.

85.

86.

87–90 ■ Determine the values of the variable for which the expression is defined as a real number.

87. $\sqrt{16 - 9x^2}$ **88.** $\sqrt{3x^2 - 5x + 2}$

89. $\left(\dfrac{1}{x^2 - 5x - 14} \right)^{1/2}$ **90.** $\sqrt[4]{\dfrac{1 - x}{2 + x}}$

91. Solve the inequality for x, assuming that a, b, and c are positive constants.

  (a) $a(bx - c) \ge bc$ **(b)** $a \le bx + c < 2a$

92. Suppose that a, b, c, and d are positive numbers such that

$$\frac{a}{b} < \frac{c}{d}$$

Show that $\dfrac{a}{b} < \dfrac{a + c}{b + d} < \dfrac{c}{d}$

Applications

93. Temperature Scales   Use the relationship between C and F given in Example 9 to find the interval on the Fahrenheit scale corresponding to the temperature range $20 \le C \le 30$.

94. Temperature Scales   What interval on the Celsius scale corresponds to the temperature range $50 \le F \le 95$?

95. Car Rental Cost   A car rental company offers two plans for renting a car.

  Plan A: $30 per day and 10¢ per mile

  Plan B: $50 per day with free unlimited mileage

  For what range of miles will plan B save you money?

96. Long-Distance Cost   A telephone company offers two long-distance plans.

  Plan A: $25 per month and 5¢ per minute

  Plan B: $5 per month and 12¢ per minute

  For how many minutes of long-distance calls would plan B be financially advantageous?

97. Driving Cost   It is estimated that the annual cost of driving a certain new car is given by the formula

$$C = 0.35m + 2200$$

where m represents the number of miles driven per year and C is the cost in dollars. Jane has purchased such a car, and decides to budget between $6400 and $7100 for next year's driving costs. What is the corresponding range of miles that she can drive her new car?

98. Gas Mileage   The gas mileage g (measured in mi/gal) for a particular vehicle, driven at v mi/h, is given by the formula $g = 10 + 0.9v - 0.01v^2$, as long as v is between 10 mi/h and 75 mi/h. For what range of speeds is the vehicle's mileage 30 mi/gal or better?

99. Gravity   The gravitational force F exerted by the earth on an object having a mass of 100 kg is given by the equation

$$F = \frac{4,000,000}{d^2}$$

where d is the distance (in km) of the object from the center of the earth, and the force F is measured in newtons (N). For what distances will the gravitational force exerted by the earth on this object be between 0.0004 N and 0.01 N?

100. Bonfire Temperature   In the vicinity of a bonfire, the temperature T in °C at a distance of x meters from the center of the fire was given by

$$T = \frac{600,000}{x^2 + 300}$$

At what range of distances from the fire's center was the temperature less than 500°C?

101. Stopping Distance   For a certain model of car the distance d required to stop the vehicle if it is traveling at v mi/h is given by the formula

$$d = v + \frac{v^2}{20}$$

where d is measured in feet. Kerry wants her stopping distance not to exceed 240 ft. At what range of speeds can she travel?

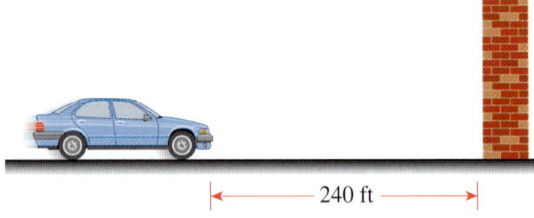

← 240 ft →

102. Manufacturer's Profit   If a manufacturer sells x units of a certain product, his revenue R and cost C (in dollars) are given by:

$$R = 20x$$

$$C = 2000 + 8x + 0.0025x^2$$

Use the fact that

$$\text{profit} = \text{revenue} - \text{cost}$$

to determine how many units he should sell to enjoy a profit of at least $2400.

103. Air Temperature   As dry air moves upward, it expands and in so doing cools at a rate of about 1 °C for each 100-meter rise, up to about 12 km.

   (a) If the ground temperature is 20 °C, write a formula for the temperature at height h.

   (b) What range of temperatures can be expected if a plane takes off and reaches a maximum height of 5 km?

104. Airline Ticket Price   A charter airline finds that on its Saturday flights from Philadelphia to London, all 120 seats will be sold if the ticket price is $200. However, for each $3 increase in ticket price, the number of seats sold decreases by one.

   (a) Find a formula for the number of seats sold if the ticket price is P dollars.

   (b) Over a certain period, the number of seats sold for this flight ranged between 90 and 115. What was the corresponding range of ticket prices?

105. Theater Tour Cost   A riverboat theater offers bus tours to groups on the following basis. Hiring the bus costs the group $360, to be shared equally by the group members. Theater tickets, normally $30 each, are discounted by 25¢ times the number of people in the group. How many members must be in the group so that the cost of the theater tour (bus fare plus theater ticket) is less than $39 per person?

106. Fencing a Garden   A determined gardener has 120 ft of deer-resistant fence. She wants to enclose a rectangular vegetable garden in her backyard, and she wants the area enclosed to be at least 800 ft². What range of values is possible for the length of her garden?

107. Thickness of a Laminate   A company manufactures industrial laminates (thin nylon-based sheets) of thickness 0.020 in, with a tolerance of 0.003 in.

   (a) Find an inequality involving absolute values that describes the range of possible thickness for the laminate.

   (b) Solve the inequality you found in part (a).

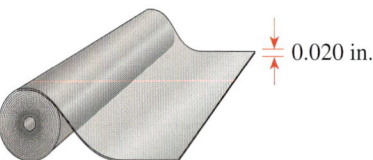

0.020 in.

108. Range of Height   The average height of adult males is 68.2 in, and 95% of adult males have height h that satisfies the inequality

$$\left| \frac{h - 68.2}{2.9} \right| \leq 2$$

Solve the inequality to find the range of heights.

Coordinates as Add

The coordinates of
xy-plane unique
cation. We c
nates as t
In Sal
dres

Discovery • Discussion

109. Do Powers Preserve Order? If $a < b$, is $a^2 < b^2$?
(Check both positive and negative values for a and b.) If
$a < b$, is $a^3 < b^3$? Based on your observations, state a
general rule about the relationship between a^n and b^n when
$a < b$ and n is a positive integer.

110. What's Wrong Here? It is tempting to try to solve an
inequality like an equation. For instance, we might try to
solve $1 < 3/x$ by multiplying both sides by x, to get $x < 3$,
so the solution would be $(-\infty, 3)$. But that's wrong; for

example, x
the origina
work (thin
correctly.

**111. Using Dis
ties** Rec
on the nur
and $|x -$
inequality
if $a < b$, x
$|x - a|$

1.8   Coordinate Geometry

The *coordinate plane* is the link between algebra and geometry. In the coordinate plane
we can draw graphs of algebraic equations. The graphs, in turn, allow us to "see" the
relationship between the variables in the equation. In this section we study the coordinate plane.

The Coordinate Plane

The Cartesian plane is named in honor
of the French mathematician René
Descartes (1596–1650), although
another Frenchman, Pierre Fermat
(1601–1665), also invented the principles of coordinate geometry at the
same time. (See their biographies on
pages 112 and 652.)

Just as points on a line can be identified with real numbers to form the coordinate line,
points in a plane can be identified with ordered pairs of numbers to form the **coordinate plane** or **Cartesian plane**. To do this, we draw two perpendicular real lines that
intersect at 0 on each line. Usually one line is horizontal with positive direction to the
right and is called the **x-axis**; the other line is vertical with positive direction upward
and is called the **y-axis**. The point of intersection of the x-axis and the y-axis is the
origin O, and the two axes divide the plane into four **quadrants**, labeled I, II, III, and
IV in Figure 1. (The points *on* the coordinate axes are not assigned to any quadrant.)

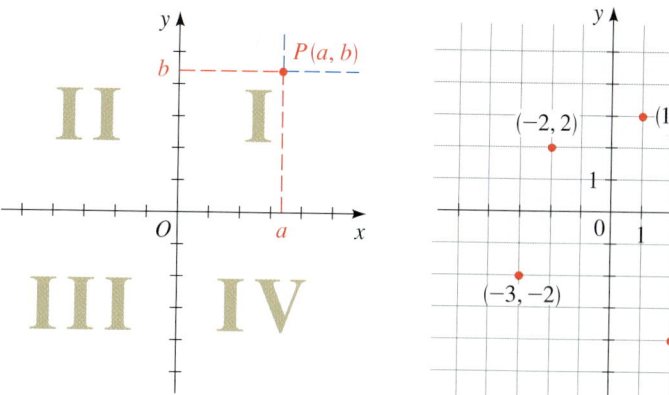

Figure 1   **Figure 2**

Although the notation for a point (a, b)
is the same as the notation for an open
interval (a, b), the context should make
clear which meaning is intended.

Any point P in the coordinate plane can be located by a unique **ordered pair** of
numbers (a, b), as shown in Figure 1. The first number a is called the **x-coordinate**
of P; the second number b is called the **y-coordinate** of P. We can think of the coordinates of P as its "address," because they specify its location in the plane. Several
points are labeled with their coordinates in Figure 2.

a point in the
y determine its lo-
think of the coordi-
"address" of the point.
Lake City, Utah, the ad-
ses of most buildings are in
ct expressed as coordinates. The
city is divided into quadrants with
Main Street as the vertical (North-
South) axis and S. Temple Street as
the horizontal (East-West) axis. An
address such as

1760 W   2100 S

indicates a location 17.6 blocks
west of Main Street and 21 blocks
south of S. Temple Street. (This is
the address of the main post office
in Salt Lake City.) With this logical
system it is possible for someone
unfamiliar with the city to locate
any address immediately, as easily
as one locates a point in the coordinate plane.

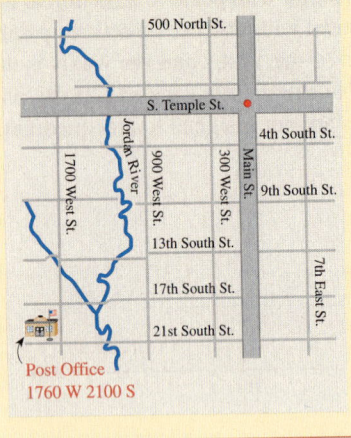

Example 1   Graphing Regions in the Coordinate Plane

Describe and sketch the regions given by each set.

(a) $\{(x, y) \mid x \geq 0\}$ (b) $\{(x, y) \mid y = 1\}$ (c) $\{(x, y) \mid |y| < 1\}$

Solution

(a) The points whose x-coordinates are 0 or positive lie on the y-axis or to the right of it, as shown in Figure 3(a).

(b) The set of all points with y-coordinate 1 is a horizontal line one unit above the x-axis, as in Figure 3(b).

(c) Recall from Section 1.7 that

$$|y| < 1 \qquad \text{if and only if} \qquad -1 < y < 1$$

So the given region consists of those points in the plane whose y-coordinates lie between -1 and 1. Thus, the region consists of all points that lie between (but not on) the horizontal lines $y = 1$ and $y = -1$. These lines are shown as broken lines in Figure 3(c) to indicate that the points on these lines do not lie in the set.

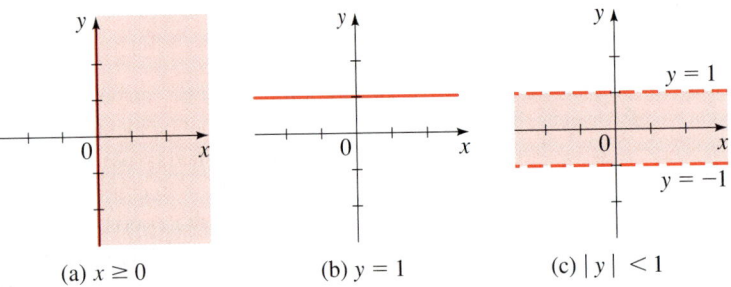

(a) $x \geq 0$ (b) $y = 1$ (c) $|y| < 1$

Figure 3

The Distance and Midpoint Formulas

We now find a formula for the distance $d(A, B)$ between two points $A(x_1, y_1)$ and $B(x_2, y_2)$ in the plane. Recall from Section 1.1 that the distance between points a and b on a number line is $d(a, b) = |b - a|$. So, from Figure 4 we see that the distance between the points $A(x_1, y_1)$ and $C(x_2, y_1)$ on a horizontal line must be $|x_2 - x_1|$, and the distance between $B(x_2, y_2)$ and $C(x_2, y_1)$ on a vertical line must be $|y_2 - y_1|$.

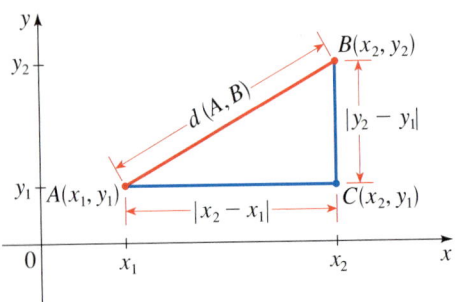

Figure 4

Since triangle ABC is a right triangle, the Pythagorean Theorem gives

$$d(A, B) = \sqrt{|x_2 - x_1|^2 + |y_2 - y_1|^2} = \sqrt{(x_2 - x_1)^2 + (y_2 - y_1)^2}$$

Distance Formula

The distance between the points $A(x_1, y_1)$ and $B(x_2, y_2)$ in the plane is

$$d(A, B) = \sqrt{(x_2 - x_1)^2 + (y_2 - y_1)^2}$$

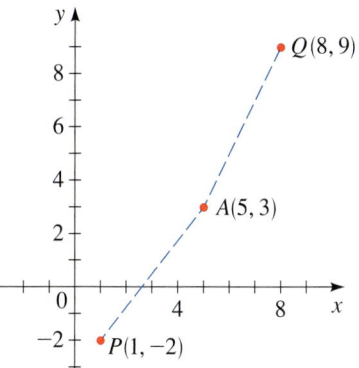

Figure 5

Example 2   Applying the Distance Formula

Which of the points $P(1, -2)$ or $Q(8, 9)$ is closer to the point $A(5, 3)$?

Solution   By the Distance Formula, we have

$$d(P, A) = \sqrt{(5 - 1)^2 + [3 - (-2)]^2} = \sqrt{4^2 + 5^2} = \sqrt{41}$$

$$d(Q, A) = \sqrt{(5 - 8)^2 + (3 - 9)^2} = \sqrt{(-3)^2 + (-6)^2} = \sqrt{45}$$

This shows that $d(P, A) < d(Q, A)$, so P is closer to A (see Figure 5).   ∎

Now let's find the coordinates (x, y) of the midpoint M of the line segment that joins the point $A(x_1, y_1)$ to the point $B(x_2, y_2)$. In Figure 6 notice that triangles APM and MQB are congruent because $d(A, M) = d(M, B)$ and the corresponding angles are equal.

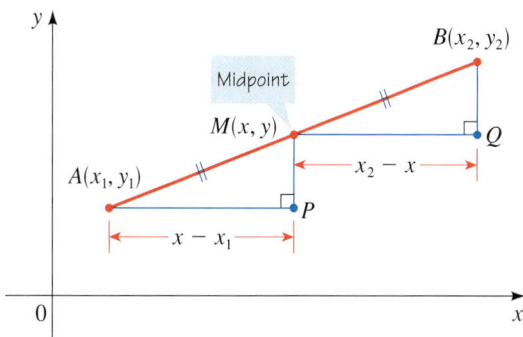

Figure 6

It follows that $d(A, P) = d(M, Q)$ and so

$$x - x_1 = x_2 - x$$

Solving this equation for x, we get $2x = x_1 + x_2$, and so $x = \dfrac{x_1 + x_2}{2}$. Similarly, $y = \dfrac{y_1 + y_2}{2}$.

Midpoint Formula

The midpoint of the line segment from $A(x_1, y_1)$ to $B(x_2, y_2)$ is

$$\left(\frac{x_1 + x_2}{2}, \frac{y_1 + y_2}{2} \right)$$

Example 3   Applying the Midpoint Formula

Show that the quadrilateral with vertices $P(1, 2)$, $Q(4, 4)$, $R(5, 9)$, and $S(2, 7)$ is a parallelogram by proving that its two diagonals bisect each other.

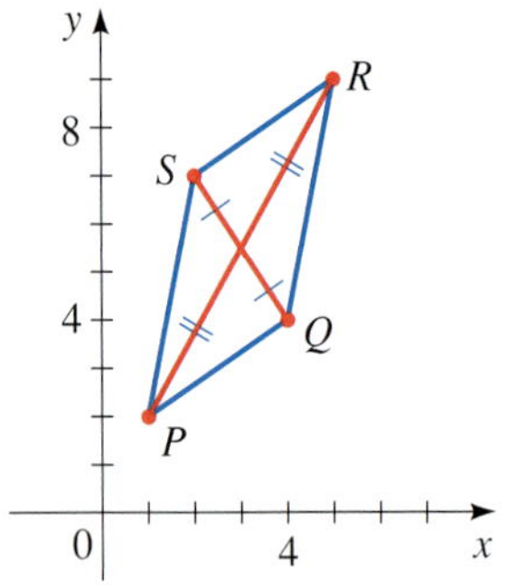

Solution   If the two diagonals have the same midpoint, then they must bisect each other. The midpoint of the diagonal PR is

$$\left(\frac{1 + 5}{2}, \frac{2 + 9}{2} \right) = \left(3, \frac{11}{2} \right)$$

and the midpoint of the diagonal QS is

$$\left(\frac{4 + 2}{2}, \frac{4 + 7}{2} \right) = \left(3, \frac{11}{2} \right)$$

so each diagonal bisects the other, as shown in Figure 7. (A theorem from elementary geometry states that the quadrilateral is therefore a parallelogram.)   ∎

Figure 7

Graphs of Equations in Two Variables

An **equation in two variables**, such as $y = x^2 + 1$, expresses a relationship between two quantities. A point (x, y) **satisfies** the equation if it makes the equation true when the values for x and y are substituted into the equation. For example, the point $(3, 10)$ satisfies the equation $y = x^2 + 1$ because $10 = 3^2 + 1$, but the point $(1, 3)$ does not, because $3 \neq 1^2 + 1$.

Fundamental Principle of Analytic Geometry

A point (x, y) lies on the graph of an equation if and only if its coordinates satisfy the equation.

The Graph of an Equation

The **graph** of an equation in x and y is the set of all points (x, y) in the coordinate plane that satisfy the equation.

The graph of an equation is a curve, so to graph an equation we plot as many points as we can, then connect them by a smooth curve.

Example 4   Sketching a Graph by Plotting Points

Sketch the graph of the equation $2x - y = 3$.

Solution   We first solve the given equation for y to get

$$y = 2x - 3$$

This helps us calculate the y-coordinates in the following table.

x	$y = 2x - 3$	(x, y)
-1	-5	$(-1, -5)$
0	-3	$(0, -3)$
1	-1	$(1, -1)$
2	1	$(2, 1)$
3	3	$(3, 3)$
4	5	$(4, 5)$

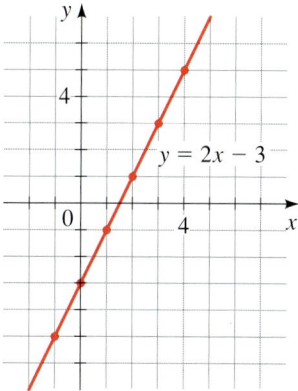

Figure 8

Of course, there are infinitely many points on the graph, and it is impossible to plot all of them. But the more points we plot, the better we can imagine what the graph represented by the equation looks like. We plot the points we found in Figure 8; they appear to lie on a line. So, we complete the graph by joining the points by a line. (In Section 1.10 we verify that the graph of this equation is indeed a line.) ■

Example 5 Sketching a Graph by Plotting Points

Sketch the graph of the equation $y = x^2 - 2$.

Solution We find some of the points that satisfy the equation in the following table. In Figure 9 we plot these points and then connect them by a smooth curve. A curve with this shape is called a *parabola*.

A detailed discussion of parabolas and their geometric properties is presented in Chapter 10.

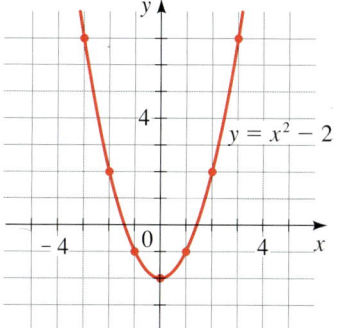

Figure 9

x	$y = x^2 - 2$	(x, y)
-3	7	$(-3, 7)$
-2	2	$(-2, 2)$
-1	-1	$(-1, -1)$
0	-2	$(0, -2)$
1	-1	$(1, -1)$
2	2	$(2, 2)$
3	7	$(3, 7)$

■

Example 6 Graphing an Absolute Value Equation

Sketch the graph of the equation $y = |x|$.

Solution We make a table of values:

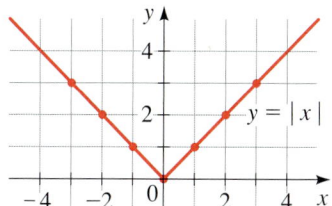

Figure 10

x	$y = |x|$	(x, y)
-3	3	$(-3, 3)$
-2	2	$(-2, 2)$
-1	1	$(-1, 1)$
0	0	$(0, 0)$
1	1	$(1, 1)$
2	2	$(2, 2)$
3	3	$(3, 3)$

In Figure 10 we plot these points and use them to sketch the graph of the equation.

■

Intercepts

The *x*-coordinates of the points where a graph intersects the *x*-axis are called the **x-intercepts** of the graph and are obtained by setting $y = 0$ in the equation of the graph. The *y*-coordinates of the points where a graph intersects the *y*-axis are called the **y-intercepts** of the graph and are obtained by setting $x = 0$ in the equation of the graph.

Definition of Intercepts

Intercepts	How to find them	Where they are on the graph
x-intercepts: The *x*-coordinates of points where the graph of an equation intersects the *x*-axis	Set $y = 0$ and solve for x	
y-intercepts: The *y*-coordinates of points where the graph of an equation intersects the *y*-axis	Set $x = 0$ and solve for y	

Example 7 Finding Intercepts

Find the *x*- and *y*-intercepts of the graph of the equation $y = x^2 - 2$.

Solution To find the *x*-intercepts, we set $y = 0$ and solve for x. Thus

$$0 = x^2 - 2 \qquad \text{Set } y = 0$$
$$x^2 = 2 \qquad \text{Add 2 to each side}$$
$$x = \pm\sqrt{2} \qquad \text{Take the square root}$$

The *x*-intercepts are $\sqrt{2}$ and $-\sqrt{2}$.

To find the *y*-intercepts, we set $x = 0$ and solve for y. Thus

$$y = 0^2 - 2 \qquad \text{Set } x = 0$$
$$y = -2$$

The *y*-intercept is -2.

The graph of this equation was sketched in Example 5. It is repeated in Figure 11 with the *x*- and *y*-intercepts labeled. ∎

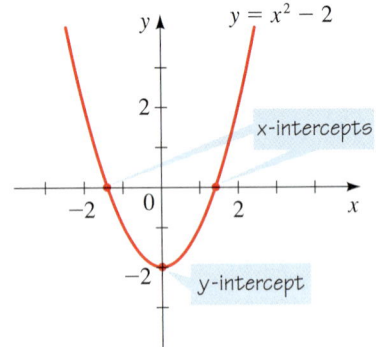

Figure 11

Circles

So far we have discussed how to find the graph of an equation in *x* and *y*. The converse problem is to find an equation of a graph, that is, an equation that represents a

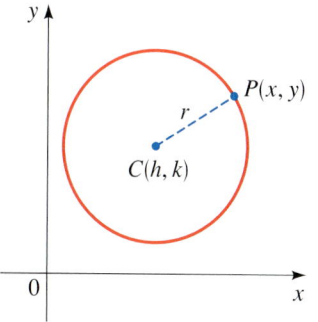

Figure 12

given curve in the xy-plane. Such an equation is satisfied by the coordinates of the points on the curve and by no other point. This is the other half of the fundamental principle of analytic geometry as formulated by Descartes and Fermat. The idea is that if a geometric curve can be represented by an algebraic equation, then the rules of algebra can be used to analyze the curve.

As an example of this type of problem, let's find the equation of a circle with radius r and center (h, k). By definition, the circle is the set of all points $P(x, y)$ whose distance from the center $C(h, k)$ is r (see Figure 12). Thus, P is on the circle if and only if $d(P, C) = r$. From the distance formula we have

$$\sqrt{(x - h)^2 + (y - k)^2} = r$$
$$(x - h)^2 + (y - k)^2 = r^2 \qquad \textcolor{blue}{\textit{Square each side}}$$

This is the desired equation.

Equation of a Circle

An equation of the circle with center (h, k) and radius r is

$$(x - h)^2 + (y - k)^2 = r^2$$

This is called the **standard form** for the equation of the circle. If the center of the circle is the origin $(0, 0)$, then the equation is

$$x^2 + y^2 = r^2$$

Example 8   Graphing a Circle

Graph each equation.

(a) $x^2 + y^2 = 25$ \qquad (b) $(x - 2)^2 + (y + 1)^2 = 25$

Solution

(a) Rewriting the equation as $x^2 + y^2 = 5^2$, we see that this is an equation of the circle of radius 5 centered at the origin. Its graph is shown in Figure 13.

(b) Rewriting the equation as $(x - 2)^2 + (y + 1)^2 = 5^2$, we see that this is an equation of the circle of radius 5 centered at $(2, -1)$. Its graph is shown in Figure 14.

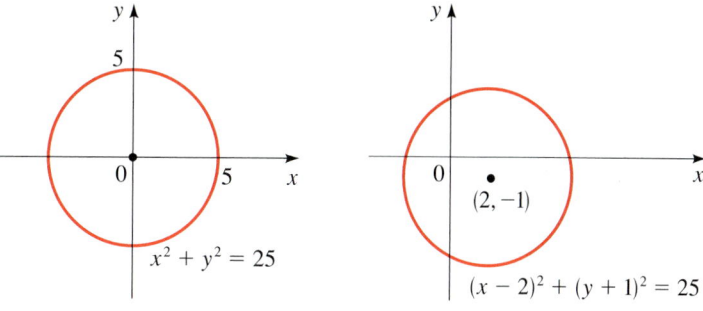

Figure 13 \qquad **Figure 14**

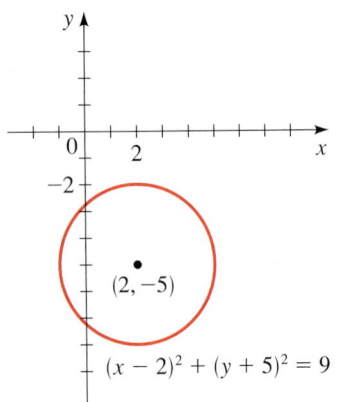

Figure 15

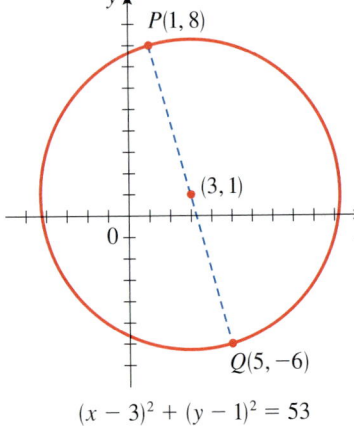

$(x - 3)^2 + (y - 1)^2 = 53$

Figure 16

Completing the square is used in many contexts in algebra. In Section 1.5 we used completing the square to solve quadratic equations.

Example 9 **Finding an Equation of a Circle**

(a) Find an equation of the circle with radius 3 and center $(2, -5)$.

(b) Find an equation of the circle that has the points $P(1, 8)$ and $Q(5, -6)$ as the endpoints of a diameter.

Solution

(a) Using the equation of a circle with $r = 3$, $h = 2$, and $k = -5$, we obtain

$$(x - 2)^2 + (y + 5)^2 = 9$$

The graph is shown in Figure 15.

(b) We first observe that the center is the midpoint of the diameter PQ, so by the Midpoint Formula the center is

$$\left(\frac{1 + 5}{2}, \frac{8 - 6}{2} \right) = (3, 1)$$

The radius r is the distance from P to the center, so by the Distance Formula

$$r^2 = (3 - 1)^2 + (1 - 8)^2 = 2^2 + (-7)^2 = 53$$

Therefore, the equation of the circle is

$$(x - 3)^2 + (y - 1)^2 = 53$$

The graph is shown in Figure 16.   ∎

Let's expand the equation of the circle in the preceding example.

$$(x - 3)^2 + (y - 1)^2 = 53 \qquad \text{Standard form}$$
$$x^2 - 6x + 9 + y^2 - 2y + 1 = 53 \qquad \text{Expand the squares}$$
$$x^2 - 6x + y^2 - 2y = 43 \qquad \text{Subtract 10 to get expanded form}$$

Suppose we are given the equation of a circle in expanded form. Then to find its center and radius we must put the equation back in standard form. That means we must reverse the steps in the preceding calculation, and to do that we need to know what to add to an expression like $x^2 - 6x$ to make it a perfect square—that is, we need to complete the square, as in the next example.

Example 10 **Identifying an Equation of a Circle**

Show that the equation $x^2 + y^2 + 2x - 6y + 7 = 0$ represents a circle, and find the center and radius of the circle.

Solution   We first group the x-terms and y-terms. Then we complete the square within each grouping. That is, we complete the square for $x^2 + 2x$ by adding $(\frac{1}{2} \cdot 2)^2 = 1$, and we complete the square for $y^2 - 6y$ by adding $[\frac{1}{2} \cdot (-6)]^2 = 9$.

$$(x^2 + 2x \quad) + (y^2 - 6y \quad) = -7 \qquad \text{Group terms}$$

$$(x^2 + 2x + 1) + (y^2 - 6y + 9) = -7 + 1 + 9 \qquad \text{Complete the square by adding 1 and 9 to each side}$$

$$(x + 1)^2 + (y - 3)^2 = 3 \qquad \text{Factor and simplify}$$

⊘ We must add the same numbers to *each side* to maintain equality.

Comparing this equation with the standard equation of a circle, we see that $h = -1$, $k = 3$, and $r = \sqrt{3}$, so the given equation represents a circle with center $(-1, 3)$ and radius $\sqrt{3}$. ∎

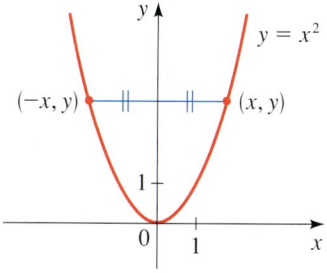

Figure 17

Symmetry

Figure 17 shows the graph of $y = x^2$. Notice that the part of the graph to the left of the y-axis is the mirror image of the part to the right of the y-axis. The reason is that if the point (x, y) is on the graph, then so is $(-x, y)$, and these points are reflections of each other about the y-axis. In this situation we say the graph is **symmetric with respect to the y-axis**. Similarly, we say a graph is **symmetric with respect to the x-axis** if whenever the point (x, y) is on the graph, then so is $(x, -y)$. A graph is **symmetric with respect to the origin** if whenever (x, y) is on the graph, so is $(-x, -y)$.

Definition of Symmetry

Type of symmetry	How to test for symmetry	What the graph looks like (figures in this section)	Geometric meaning
Symmetry with respect to the x-axis	The equation is unchanged when y is replaced by $-y$	(Figures 13, 18)	Graph is unchanged when reflected in the x-axis
Symmetry with respect to the y-axis	The equation is unchanged when x is replaced by $-x$	(Figures 9, 10, 11, 13, 17)	Graph is unchanged when reflected in the y-axis
Symmetry with respect to the origin	The equation is unchanged when x is replaced by $-x$ and y by $-y$	(Figures 13, 19)	Graph is unchanged when rotated 180° about the origin

The remaining examples in this section show how symmetry helps us sketch the graphs of equations.

Example 11   Using Symmetry to Sketch a Graph

Test the equation $x = y^2$ for symmetry and sketch the graph.

Solution   If y is replaced by $-y$ in the equation $x = y^2$, we get

$$x = (-y)^2 \qquad \textit{Replace y by } -y$$

$$x = y^2 \qquad \textit{Simplify}$$

and so the equation is unchanged. Therefore, the graph is symmetric about the x-axis. But changing x to $-x$ gives the equation $-x = y^2$, which is not the same as the original equation, so the graph is not symmetric about the y-axis.

We use the symmetry about the x-axis to sketch the graph by first plotting points just for $y > 0$ and then reflecting the graph in the x-axis, as shown in Figure 18.

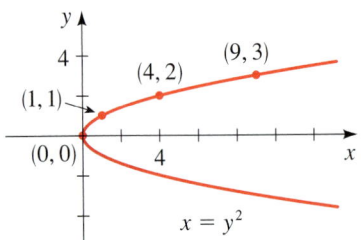

Figure 18

y	$x = y^2$	(x, y)
0	0	$(0, 0)$
1	1	$(1, 1)$
2	4	$(4, 2)$
3	9	$(9, 3)$

Example 12   Using Symmetry to Sketch a Graph

Test the equation $y = x^3 - 9x$ for symmetry and sketch its graph.

Solution   If we replace x by $-x$ and y by $-y$ in the equation, we get

$$-y = (-x)^3 - 9(-x) \qquad \textit{Replace x by } -x \textit{ and y by } -y$$

$$-y = -x^3 + 9x \qquad \textit{Simplify}$$

$$y = x^3 - 9x \qquad \textit{Multiply by } -1$$

and so the equation is unchanged. This means that the graph is symmetric with respect to the origin. We sketch it by first plotting points for $x > 0$ and then using symmetry about the origin (see Figure 19).

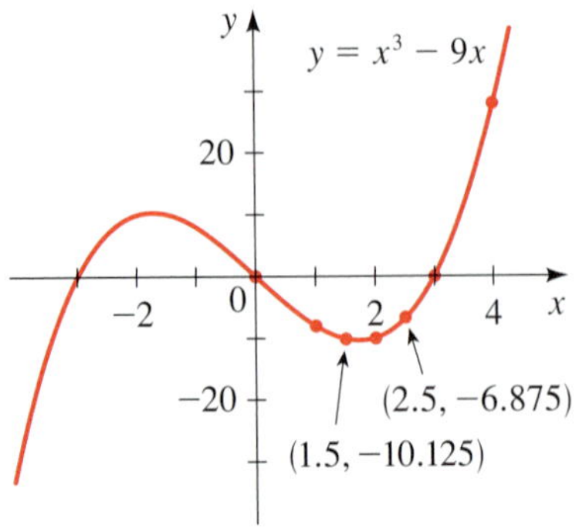

Figure 19

x	$y = x^3 - 9x$	(x, y)
0	0	$(0, 0)$
1	-8	$(1, -8)$
1.5	-10.125	$(1.5, -10.125)$
2	-10	$(2, -10)$
2.5	-6.875	$(2.5, -6.875)$
3	0	$(3, 0)$
4	28	$(4, 28)$

1.8 Exercises

1. Plot the given points in a coordinate plane:

$(2, 3), (-2, 3), (4, 5), (4, -5), (-4, 5), (-4, -5)$

2. Find the coordinates of the points shown in the figure.

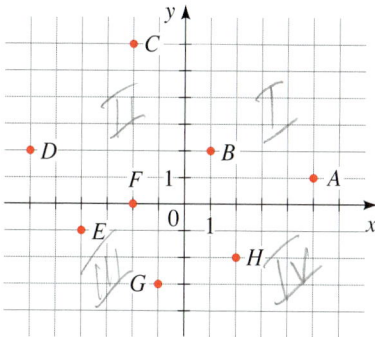

3–6 ■ A pair of points is graphed.

(a) Find the distance between them.

(b) Find the midpoint of the segment that joins them.

3.

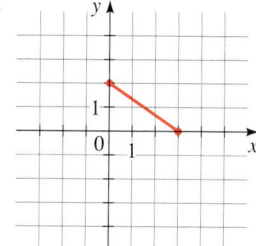

4.

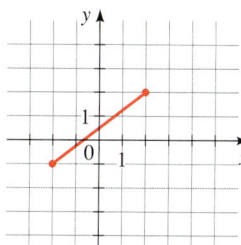

5.

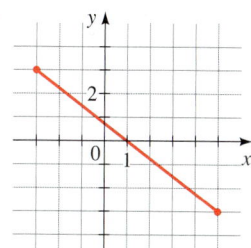

6.

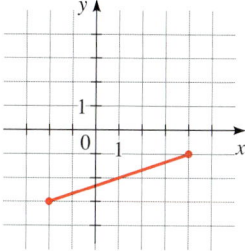

7–12 ■ A pair of points is graphed.

(a) Plot the points in a coordinate plane.

(b) Find the distance between them.

(c) Find the midpoint of the segment that joins them.

7. $(0, 8), (6, 16)$

8. $(-2, 5), (10, 0)$

9. $(-3, -6), (4, 18)$

10. $(-1, -1), (9, 9)$

11. $(6, -2), (-6, 2)$

12. $(0, -6), (5, 0)$

13. Draw the rectangle with vertices $A(1, 3)$, $B(5, 3)$, $C(1, -3)$, and $D(5, -3)$ on a coordinate plane. Find the area of the rectangle.

14. Draw the parallelogram with vertices $A(1, 2)$, $B(5, 2)$, $C(3, 6)$, and $D(7, 6)$ on a coordinate plane. Find the area of the parallelogram.

15. Plot the points $A(1, 0)$, $B(5, 0)$, $C(4, 3)$, and $D(2, 3)$, on a coordinate plane. Draw the segments AB, BC, CD, and DA. What kind of quadrilateral is $ABCD$, and what is its area?

16. Plot the points $P(5, 1)$, $Q(0, 6)$, and $R(-5, 1)$, on a coordinate plane. Where must the point S be located so that the quadrilateral $PQRS$ is a square? Find the area of this square.

17–26 ■ Sketch the region given by the set.

17. $\{(x, y) \mid x \geq 3\}$

18. $\{(x, y) \mid y < 3\}$

19. $\{(x, y) \mid y = 2\}$

20. $\{(x, y) \mid x = -1\}$

21. $\{(x, y) \mid 1 < x < 2\}$

22. $\{(x, y) \mid 0 \leq y \leq 4\}$

23. $\{(x, y) \mid |x| > 4\}$

24. $\{(x, y) \mid |y| \leq 2\}$

25. $\{(x, y) \mid x \geq 1 \text{ and } y < 3\}$

26. $\{(x, y) \mid |x| \leq 2 \text{ and } |y| \leq 3\}$

27. Which of the points $A(6, 7)$ or $B(-5, 8)$ is closer to the origin?

28. Which of the points $C(-6, 3)$ or $D(3, 0)$ is closer to the point $E(-2, 1)$?

29. Which of the points $P(3, 1)$ or $Q(-1, 3)$ is closer to the point $R(-1, -1)$?

30. (a) Show that the points $(7, 3)$ and $(3, 7)$ are the same distance from the origin.

(b) Show that the points (a, b) and (b, a) are the same distance from the origin.

31. Show that the triangle with vertices $A(0, 2)$, $B(-3, -1)$, and $C(-4, 3)$ is isosceles.

32. Find the area of the triangle shown in the figure.

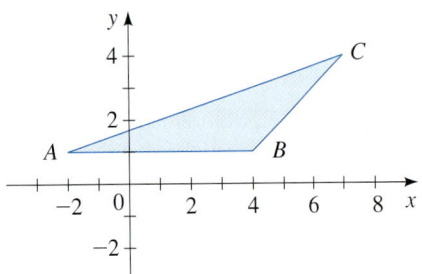

33. Refer to triangle ABC in the figure.

(a) Show that triangle ABC is a right triangle by using the converse of the Pythagorean Theorem (see page 54).

(b) Find the area of triangle ABC.

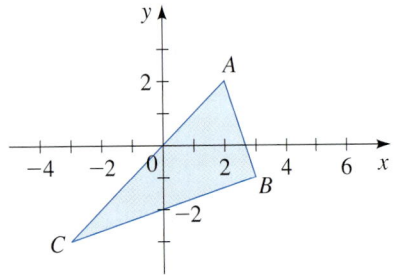

34. Show that the triangle with vertices $A(6, -7)$, $B(11, -3)$, and $C(2, -2)$ is a right triangle by using the converse of the Pythagorean Theorem. Find the area of the triangle.

35. Show that the points $A(-2, 9)$, $B(4, 6)$, $C(1, 0)$, and $D(-5, 3)$ are the vertices of a square.

36. Show that the points $A(-1, 3)$, $B(3, 11)$, and $C(5, 15)$ are collinear by showing that $d(A, B) + d(B, C) = d(A, C)$.

37. Find a point on the y-axis that is equidistant from the points $(5, -5)$ and $(1, 1)$.

38. Find the lengths of the medians of the triangle with vertices $A(1, 0)$, $B(3, 6)$, and $C(8, 2)$. (A *median* is a line segment from a vertex to the midpoint of the opposite side.)

39. Plot the points $P(-1, -4)$, $Q(1, 1)$, and $R(4, 2)$, on a coordinate plane. Where should the point S be located so that the figure $PQRS$ is a parallelogram?

40. If $M(6, 8)$ is the midpoint of the line segment AB, and if A has coordinates $(2, 3)$, find the coordinates of B.

41. (a) Sketch the parallelogram with vertices $A(-2, -1)$, $B(4, 2)$, $C(7, 7)$, and $D(1, 4)$.

(b) Find the midpoints of the diagonals of this parallelogram.

(c) From part (b) show that the diagonals bisect each other.

42. The point M in the figure is the midpoint of the line segment AB. Show that M is equidistant from the vertices of triangle ABC.

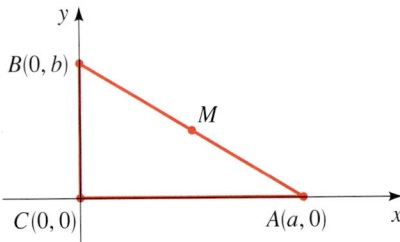

43–46 ■ Determine whether the given points are on the graph of the equation.

43. $x - 2y - 1 = 0$; $(0, 0), (1, 0), (-1, -1)$

44. $y(x^2 + 1) = 1$; $(1, 1), \left(1, \frac{1}{2}\right), \left(-1, \frac{1}{2}\right)$

45. $x^2 + xy + y^2 = 4$; $(0, -2), (1, -2), (2, -2)$

46. $x^2 + y^2 = 1$; $(0, 1), \left(\frac{1}{\sqrt{2}}, \frac{1}{\sqrt{2}}\right), \left(\frac{\sqrt{3}}{2}, \frac{1}{2}\right)$

47–50 ■ An equation and its graph are given. Find the x- and y-intercepts.

47. $y = 4x - x^2$

48. $\dfrac{x^2}{9} + \dfrac{y^2}{4} = 1$

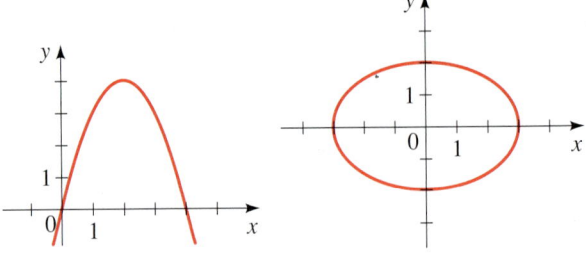

49. $x^4 + y^2 - xy = 16$ **50.** $x^2 + y^3 - x^2y^2 = 64$

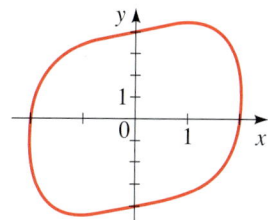   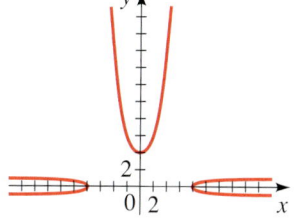

51–70 ■ Make a table of values and sketch the graph of the equation. Find the x- and y-intercepts and test for symmetry.

51. $y = -x + 4$ **52.** $y = 3x + 3$

53. $2x - y = 6$ **54.** $x + y = 3$

55. $y = 1 - x^2$ **56.** $y = x^2 + 2$

57. $4y = x^2$ **58.** $8y = x^3$

59. $y = x^2 - 9$ **60.** $y = 9 - x^2$

61. $xy = 2$ **62.** $y = \sqrt{x + 4}$

63. $y = \sqrt{4 - x^2}$ **64.** $y = -\sqrt{4 - x^2}$

65. $x + y^2 = 4$ **66.** $x = y^3$

67. $y = 16 - x^4$ **68.** $x = |y|$

69. $y = 4 - |x|$ **70.** $y = |4 - x|$

71–76 ■ Test the equation for symmetry.

71. $y = x^4 + x^2$ **72.** $x = y^4 - y^2$

73. $x^2y^2 + xy = 1$ **74.** $x^4y^4 + x^2y^2 = 1$

75. $y = x^3 + 10x$ **76.** $y = x^2 + |x|$

77–80 ■ Complete the graph using the given symmetry property.

77. Symmetric with respect   **78.** Symmetric with respect
to the y-axis   to the x-axis

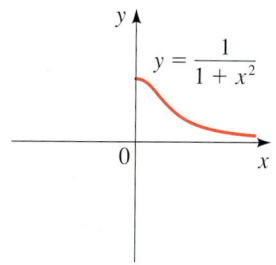   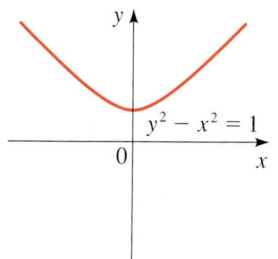

79. Symmetric with respect   **80.** Symmetric with respect
to the origin   to the origin

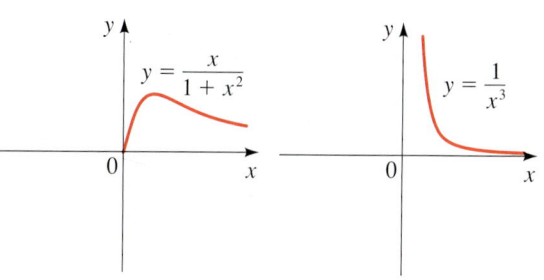

81–86 ■ Find an equation of the circle that satisfies the given conditions.

81. Center $(2, -1)$;   radius 3

82. Center $(-1, -4)$;   radius 8

83. Center at the origin;   passes through $(4, 7)$

84. Endpoints of a diameter are $P(-1, 1)$ and $Q(5, 9)$

85. Center $(7, -3)$;   tangent to the x-axis

86. Circle lies in the first quadrant, tangent to both x-and y-axes; radius 5

87–88 ■ Find the equation of the circle shown in the figure.

87.   **88.**

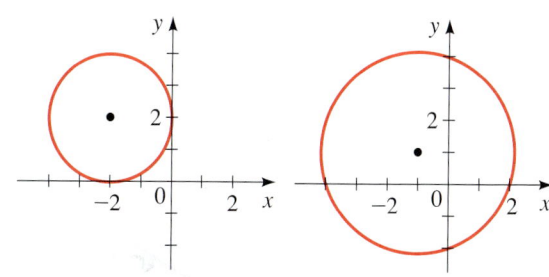

89–94 ■ Show that the equation represents a circle, and find the center and radius of the circle.

89. $x^2 + y^2 - 4x + 10y + 13 = 0$

90. $x^2 + y^2 + 6y + 2 = 0$

91. $x^2 + y^2 - \frac{1}{2}x + \frac{1}{2}y = \frac{1}{8}$

92. $x^2 + y^2 + \frac{1}{2}x + 2y + \frac{1}{16} = 0$

93. $2x^2 + 2y^2 - 3x = 0$

94. $3x^2 + 3y^2 + 6x - y = 0$

95–96 ■ Sketch the region given by the set.

95. $\{(x, y) \mid x^2 + y^2 \le 1\}$

96. $\{(x, y) \mid x^2 + y^2 > 4\}$

97. Find the area of the region that lies outside the circle $x^2 + y^2 = 4$ but inside the circle

$$x^2 + y^2 - 4y - 12 = 0$$

98. Sketch the region in the coordinate plane that satisfies both the inequalities $x^2 + y^2 \le 9$ and $y \ge |x|$. What is the area of this region?

Applications

99. Distances in a City   A city has streets that run north and south, and avenues that run east and west, all equally spaced. Streets and avenues are numbered sequentially, as shown in the figure. The *walking* distance between points A and B is 7 blocks—that is, 3 blocks east and 4 blocks north. To find the *straight-line* distances d, we must use the Distance Formula.

(a) Find the straight-line distance (in blocks) between A and B.

(b) Find the walking distance and the straight-line distance between the corner of 4th St. and 2nd Ave. and the corner of 11th St. and 26th Ave.

(c) What must be true about the points P and Q if the walking distance between P and Q equals the straight-line distance between P and Q?

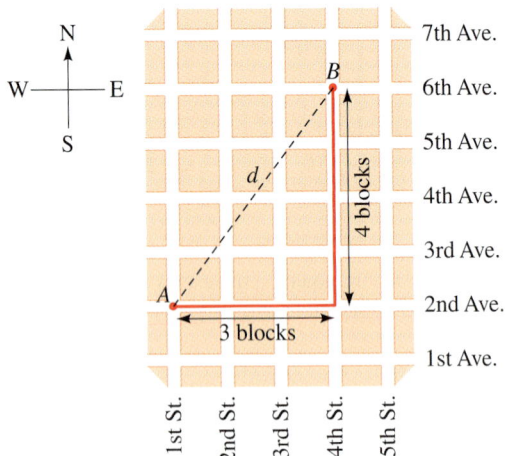

100. Halfway Point   Two friends live in the city described in Exercise 99, one at the corner of 3rd St. and 7th Ave., the other at the corner of 27th St. and 17th Ave. They frequently meet at a coffee shop halfway between their homes.

(a) At what intersection is the coffee shop located?

(b) How far must each of them walk to get to the coffee shop?

101. Orbit of a Satellite   A satellite is in orbit around the moon. A coordinate plane containing the orbit is set up with the center of the moon at the origin, as shown in the graph, with distances measured in megameters (Mm). The equation of the satellite's orbit is

$$\frac{(x-3)^2}{25} + \frac{y^2}{16} = 1$$

(a) From the graph, determine the closest and the farthest that the satellite gets to the center of the moon.

(b) There are two points in the orbit with y-coordinates 2. Find the x-coordinates of these points, and determine their distances to the center of the moon.

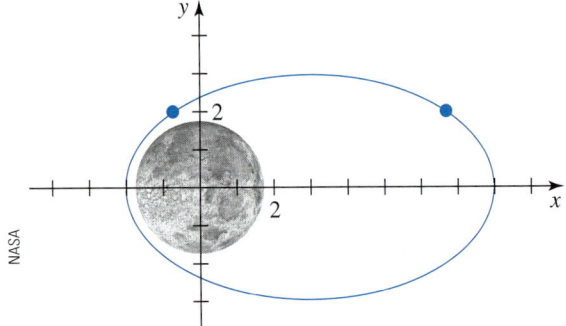

NASA

Discovery • Discussion

102. Shifting the Coordinate Plane   Suppose that each point in the coordinate plane is shifted 3 units to the right and 2 units upward.

(a) The point $(5, 3)$ is shifted to what new point?

(b) The point (a, b) is shifted to what new point?

(c) What point is shifted to $(3, 4)$?

(d) Triangle ABC in the figure has been shifted to triangle $A'B'C'$. Find the coordinates of the points A', B', and C'.

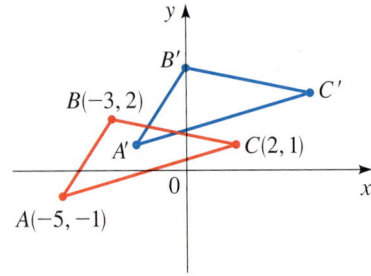

103. Reflecting in the Coordinate Plane   Suppose that the y-axis acts as a mirror that reflects each point to the right of it into a point to the left of it.

(a) The point $(3, 7)$ is reflected to what point?

(b) The point (a, b) is reflected to what point?

(c) What point is reflected to $(-4, -1)$?

(d) Triangle ABC in the figure is reflected to triangle $A'B'C'$. Find the coordinates of the points A', B', and C'.

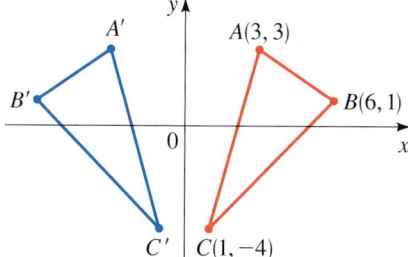

104. Completing a Line Segment   Plot the points $M(6, 8)$ and $A(2, 3)$ on a coordinate plane. If M is the midpoint of the line segment AB, find the coordinates of B. Write a brief description of the steps you took to find B, and your reasons for taking them.

105. Completing a Parallelogram   Plot the points $P(0, 3)$, $Q(2, 2)$, and $R(5, 3)$ on a coordinate plane. Where should the point S be located so that the figure $PQRS$ is a parallelogram? Write a brief description of the steps you took and your reasons for taking them.

106. Circle, Point, or Empty Set?   Complete the squares in the general equation $x^2 + ax + y^2 + by + c = 0$ and simplify the result as much as possible. Under what conditions on the coefficients a, b, and c does this equation represent a circle? A single point? The empty set? In the case that the equation does represent a circle, find its center and radius.

107. Do the Circles Intersect?

(a) Find the radius of each circle in the pair, and the distance between their centers; then use this information to determine whether the circles intersect.

   (i) $(x - 2)^2 + (y - 1)^2 = 9$;
   $(x - 6)^2 + (y - 4)^2 = 16$

   (ii) $x^2 + (y - 2)^2 = 4$;
   $(x - 5)^2 + (y - 14)^2 = 9$

   (iii) $(x - 3)^2 + (y + 1)^2 = 1$;
   $(x - 2)^2 + (y - 2)^2 = 25$

(b) How can you tell, just by knowing the radii of two circles and the distance between their centers, whether the circles intersect? Write a short paragraph describing how you would decide this and draw graphs to illustrate your answer.

108. Making a Graph Symmetric   The graph shown in the figure is not symmetric about the x-axis, the y-axis, or the origin. Add more line segments to the graph so that it exhibits the indicated symmetry. In each case, add as little as possible.

(a) Symmetry about the x-axis

(b) Symmetry about the y-axis

(c) Symmetry about the origin

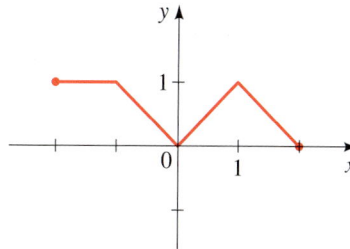

 1.9   # Graphing Calculators; Solving Equations and Inequalities Graphically

In Sections 1.5 and 1.7 we solved equations and inequalities algebraically. In the preceding section we learned how to sketch the graph of an equation in a coordinate plane. In this section we use graphs to solve equations and inequalities. To do this, we must first draw a graph using a graphing device. So, we begin by giving a few guidelines to help us use graphing devices effectively.

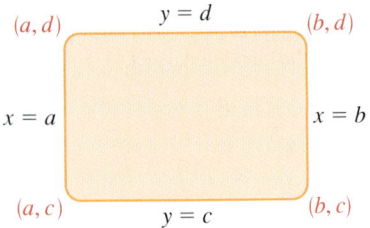

Figure 1
The viewing rectangle $[a, b]$ by $[c, d]$

Using a Graphing Calculator

A graphing calculator or computer displays a rectangular portion of the graph of an equation in a display window or viewing screen, which we call a **viewing rectangle**. The default screen often gives an incomplete or misleading picture, so it is important to choose the viewing rectangle with care. If we choose the x-values to range from a minimum value of $\mathtt{Xmin} = a$ to a maximum value of $\mathtt{Xmax} = b$ and the y-values to range from a minimum value of $\mathtt{Ymin} = c$ to a maximum value of $\mathtt{Ymax} = d$, then the displayed portion of the graph lies in the rectangle

$$[a, b] \times [c, d] = \{(x, y) \mid a \le x \le b, c \le y \le d\}$$

as shown in Figure 1. We refer to this as the $[a, b]$ by $[c, d]$ viewing rectangle.

The graphing device draws the graph of an equation much as you would. It plots points of the form (x, y) for a certain number of values of x, equally spaced between a and b. If the equation is not defined for an x-value, or if the corresponding y-value lies outside the viewing rectangle, the device ignores this value and moves on to the next x-value. The machine connects each point to the preceding plotted point to form a representation of the graph of the equation.

Example 1   Choosing an Appropriate Viewing Rectangle

Graph the equation $y = x^2 + 3$ in an appropriate viewing rectangle.

Solution   Let's experiment with different viewing rectangles. We'll start with the viewing rectangle $[-2, 2]$ by $[-2, 2]$, so we set

$$\mathtt{Xmin} = -2 \qquad \mathtt{Ymin} = -2$$
$$\mathtt{Xmax} = 2 \qquad \mathtt{Xmax} = 2$$

The resulting graph in Figure 2(a) is blank! This is because $x^2 \ge 0$, so $x^2 + 3 \ge 3$ for all x. Thus, the graph lies entirely above the viewing rectangle, so this viewing rectangle is not appropriate. If we enlarge the viewing rectangle to $[-4, 4]$ by $[-4, 4]$, as in Figure 2(b), we begin to see a portion of the graph.

Now let's try the viewing rectangle $[-10, 10]$ by $[-5, 30]$. The graph in Figure 2(c) seems to give a more complete view of the graph. If we enlarge the viewing rectangle even further, as in Figure 2(d), the graph doesn't show clearly that the y-intercept is 3.

So, the viewing rectangle $[-10, 10]$ by $[-5, 30]$ gives an appropriate representation of the graph.

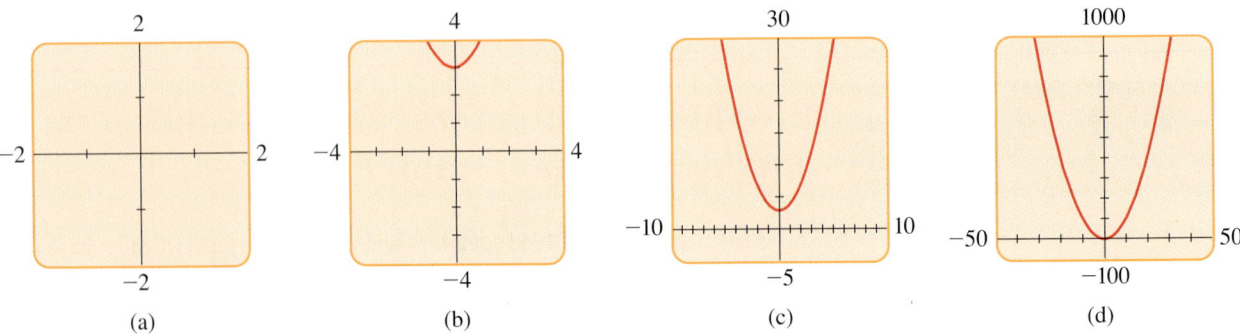

Figure 2   Graphs of $y = x^2 + 3$

Alan Turing (1912–1954) was at the center of two pivotal events of the 20th century—World War II and the invention of computers. At the age of 23 Turing made his mark on mathematics by solving an important problem in the foundations of mathematics that was posed by David Hilbert at the 1928 International Congress of Mathematicians (see page 708). In this research he invented a theoretical machine, now called a Turing machine, which was the inspiration for modern digital computers. During World War II Turing was in charge of the British effort to decipher secret German codes. His complete success in this endeavor played a decisive role in the Allies' victory. To carry out the numerous logical steps required to break a coded message, Turing developed decision procedures similar to modern computer programs. After the war he helped develop the first electronic computers in Britain. He also did pioneering work on artificial intelligence and computer models of biological processes. At the age of 42 Turing died of poisoning after eating an apple that had mysteriously been laced with cyanide.

Example 2   Two Graphs on the Same Screen

Graph the equations $y = 3x^2 - 6x + 1$ and $y = 0.23x - 2.25$ together in the viewing rectangle $[-1, 3]$ by $[-2.5, 1.5]$. Do the graphs intersect in this viewing rectangle?

Solution   Figure 3(a) shows the essential features of both graphs. One is a parabola and the other is a line. It looks as if the graphs intersect near the point $(1, -2)$. However, if we zoom in on the area around this point as shown in Figure 3(b), we see that although the graphs almost touch, they don't actually intersect.

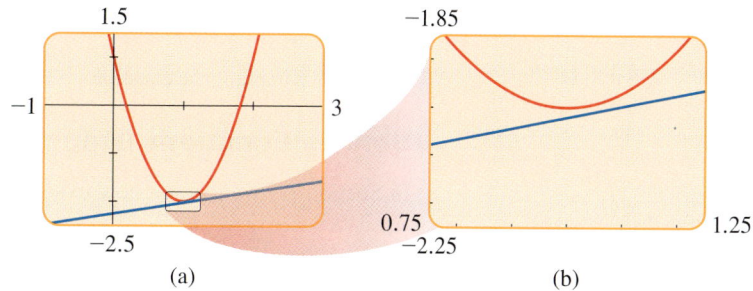

(a)   (b)

Figure 3

You can see from Examples 1 and 2 that the choice of a viewing rectangle makes a big difference in the appearance of a graph. If you want an overview of the essential features of a graph, you must choose a relatively large viewing rectangle to obtain a global view of the graph. If you want to investigate the details of a graph, you must zoom in to a small viewing rectangle that shows just the feature of interest.

Most graphing calculators can only graph equations in which y is isolated on one side of the equal sign. The next example shows how to graph equations that don't have this property.

Example 3   Graphing a Circle

Graph the circle $x^2 + y^2 = 1$.

Solution   We first solve for y, to isolate it on one side of the equal sign.

$$y^2 = 1 - x^2 \qquad \text{Subtract } x^2$$

$$y = \pm\sqrt{1 - x^2} \qquad \text{Take square roots}$$

Therefore, the circle is described by the graphs of *two* equations:

$$y = \sqrt{1 - x^2} \qquad \text{and} \qquad y = -\sqrt{1 - x^2}$$

The first equation represents the top half of the circle (because $y \geq 0$), and the second represents the bottom half of the circle (because $y \leq 0$). If we graph the first equation in the viewing rectangle $[-2, 2]$ by $[-2, 2]$, we get the semicircle shown in Figure 4(a). The graph of the second equation is the semicircle in Figure 4(b). Graphing these semicircles together on the same viewing screen, we get the full circle in Figure 4(c).

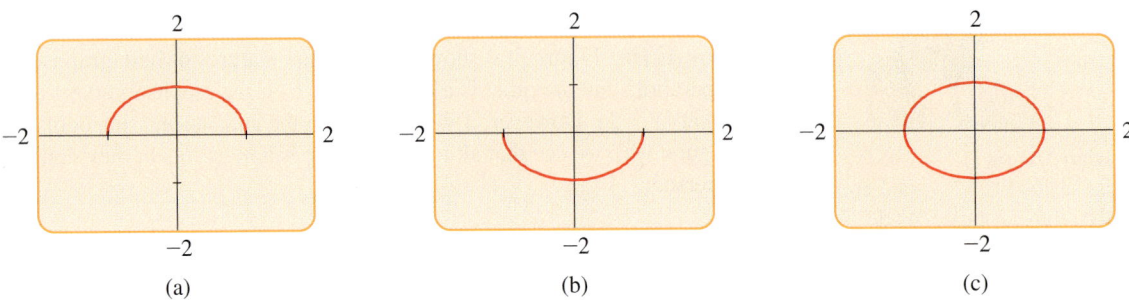

(a) (b) (c)

The graph in Figure 4(c) looks somewhat flattened. Most graphing calculators allow you to set the scales on the axes so that circles really look like circles. On the TI-82 and TI-83, from the `ZOOM` menu, choose `ZSquare` to set the scales appropriately. (On the TI-86 the command is `Zsq`.)

Figure 4 Graphing the equation $x^2 + y^2 = 1$

Solving Equations Graphically

In Section 1.5 we learned how to solve equations. To solve an equation like

$$3x - 5 = 0$$

we used the **algebraic method**. This means we used the rules of algebra to isolate x on one side of the equation. We view x as an *unknown* and we use the rules of algebra to hunt it down. Here are the steps in the solution:

$$3x - 5 = 0$$
$$3x = 5 \qquad \text{Add 5}$$
$$x = \tfrac{5}{3} \qquad \text{Divide by 3}$$

So the solution is $x = \tfrac{5}{3}$.

We can also solve this equation by the **graphical method**. In this method we view x as a *variable* and sketch the graph of the equation

$$y = 3x - 5$$

Different values for x give different values for y. Our goal is to find the value of x for which $y = 0$. From the graph in Figure 5 we see that $y = 0$ when $x \approx 1.7$. Thus, the solution is $x \approx 1.7$. Note that from the graph we obtain an approximate solution.

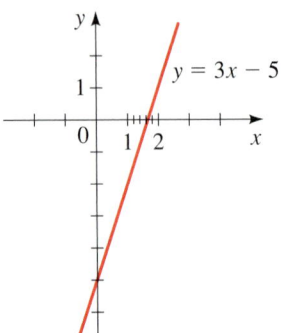

Figure 5

We summarize these methods in the following box.

Solving an Equation

Algebraic method

Use the rules of algebra to isolate the unknown x on one side of the equation.

Example: $2x = 6 - x$

$$3x = 6 \qquad \text{Add } x$$

$$x = 2 \qquad \text{Divide by 3}$$

The solution is $x = 2$.

Graphical method

Move all terms to one side and set equal to y. Sketch the graph to find the value of x where $y = 0$.

Example: $2x = 6 - x$

$$0 = 6 - 3x$$

Set $y = 6 - 3x$ and graph.

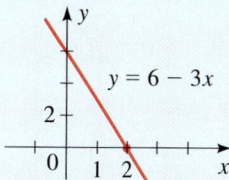

From the graph the solution is $x \approx 2$.

The Discovery Project on page 283 describes a numerical method for solving equations.

The advantage of the algebraic method is that it gives exact answers. Also, the process of unraveling the equation to arrive at the answer helps us understand the algebraic structure of the equation. On the other hand, for many equations it is difficult or impossible to isolate x.

The graphical method gives a numerical approximation to the answer. This is an advantage when a numerical answer is desired. (For example, an engineer might find an answer expressed as $x \approx 2.6$ more immediately useful than $x = \sqrt{7}$.) Also, graphing an equation helps us visualize how the solution is related to other values of the variable.

Example 4   Solving a Quadratic Equation Algebraically and Graphically

Solve the quadratic equations algebraically and graphically.

(a) $x^2 - 4x + 2 = 0$ (b) $x^2 - 4x + 4 = 0$ (c) $x^2 - 4x + 6 = 0$

Solution 1:   Algebraic

We use the quadratic formula to solve each equation.

The quadratic formula is discussed on page 49.

(a) $x = \dfrac{-(-4) \pm \sqrt{(-4)^2 - 4 \cdot 1 \cdot 2}}{2} = \dfrac{4 \pm \sqrt{8}}{2} = 2 \pm \sqrt{2}$

There are two solutions, $x = 2 + \sqrt{2}$ and $x = 2 - \sqrt{2}$.

(b) $x = \dfrac{-(-4) \pm \sqrt{(-4)^2 - 4 \cdot 1 \cdot 4}}{2} = \dfrac{4 \pm \sqrt{0}}{2} = 2$

There is just one solution, $x = 2$.

(c) $x = \dfrac{-(-4) \pm \sqrt{(-4)^2 - 4 \cdot 1 \cdot 6}}{2} = \dfrac{4 \pm \sqrt{-8}}{2}$

There is no real solution.

Solution 2:   Graphical

We graph the equations $y = x^2 - 4x + 2$, $y = x^2 - 4x + 4$, and $y = x^2 - 4x + 6$ in Figure 6. By determining the x-intercepts of the graphs, we find the following solutions.

(a) $x \approx 0.6$ and $x \approx 3.4$

(b) $x = 2$

(c) There is no x-intercept, so the equation has no solution.

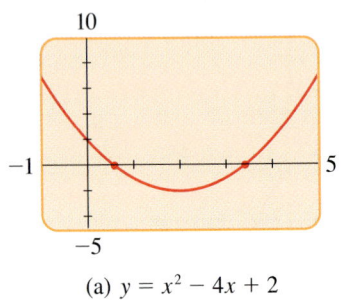

(a) $y = x^2 - 4x + 2$

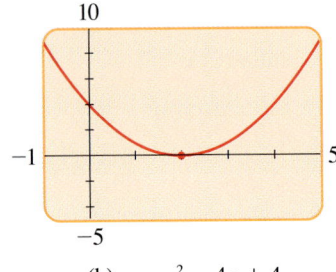

(b) $y = x^2 - 4x + 4$

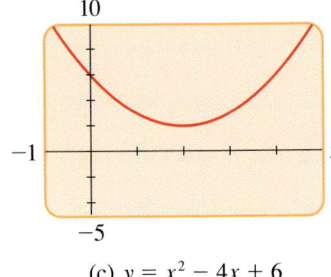

(c) $y = x^2 - 4x + 6$

Figure 6

The graphs in Figure 6 show visually why a quadratic equation may have two solutions, one solution, or no real solution. We proved this fact algebraically in Section 1.5 when we studied the discriminant.

Example 5   **Another Graphical Method**

Solve the equation algebraically and graphically: $5 - 3x = 8x - 20$

Solution 1:   Algebraic

$$5 - 3x = 8x - 20$$
$$-3x = 8x - 25 \qquad \text{Subtract 5}$$
$$-11x = -25 \qquad \text{Subtract 8x}$$
$$x = \frac{-25}{-11} = 2\tfrac{3}{11} \qquad \text{Divide by } -11 \text{ and simplify}$$

Solution 2:   Graphical

We could move all terms to one side of the equal sign, set the result equal to y, and graph the resulting equation. But to avoid all this algebra, we graph two equations instead:

$$y_1 = 5 - 3x \qquad \text{and} \qquad y_2 = 8x - 20$$

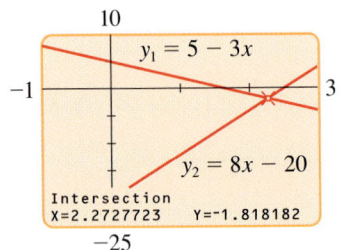

Figure 7

The solution of the original equation will be the value of x that makes y_1 equal to y_2; that is, the solution is the x-coordinate of the intersection point of the two graphs. Using the $\boxed{\text{TRACE}}$ feature or the `intersect` command on a graphing calculator, we see from Figure 7 that the solution is $x \approx 2.27$.

In the next example we use the graphical method to solve an equation that is extremely difficult to solve algebraically.

Example 6 Solving an Equation in an Interval

Solve the equation

$$x^3 - 6x^2 + 9x = \sqrt{x}$$

in the interval $[1, 6]$.

Solution We are asked to find all solutions x that satisfy $1 \le x \le 6$, so we will graph the equation in a viewing rectangle for which the x-values are restricted to this interval.

$$x^3 - 6x^2 + 9x = \sqrt{x}$$

$$x^3 - 6x^2 + 9x - \sqrt{x} = 0 \qquad \text{Subtract } \sqrt{x}$$

We can also use the `zero` command to find the solutions, as shown in Figures 8(a) and 8(b).

Figure 8 shows the graph of the equation $y = x^3 - 6x^2 + 9x - \sqrt{x}$ in the viewing rectangle $[1, 6]$ by $[-5, 5]$. There are two x-intercepts in this viewing rectangle; zooming in we see that the solutions are $x \approx 2.18$ and $x \approx 3.72$.

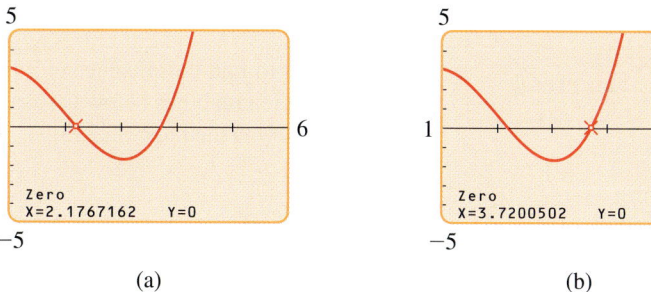

Figure 8   (a)   (b)

The equation in Example 6 actually has four solutions. You are asked to find the other two in Exercise 57.

Example 7 Intensity of Light

Two light sources are 10 m apart. One is three times as intense as the other. The light intensity L (in lux) at a point x meters from the weaker source is given by

$$L = \frac{10}{x^2} + \frac{30}{(10 - x)^2}$$

(See Figure 9.) Find the points at which the light intensity is 4 lux.

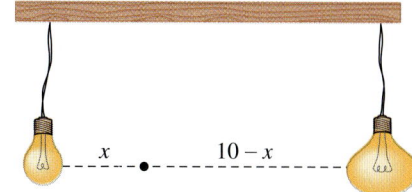

Figure 9

Solution We need to solve the equation

$$4 = \frac{10}{x^2} + \frac{30}{(10 - x)^2}$$

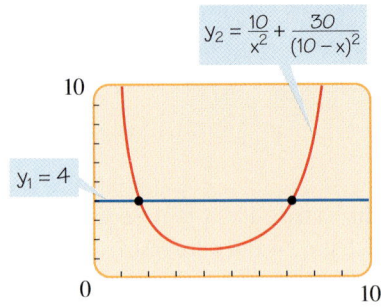

$$y_2 = \frac{10}{x^2} + \frac{30}{(10-x)^2}$$

$y_1 = 4$

Figure 10

The graphs of

$$y_1 = 4 \qquad \text{and} \qquad y_2 = \frac{10}{x^2} + \frac{30}{(10-x)^2}$$

are shown in Figure 10. Zooming in (or using the `intersect` command) we find two solutions, $x \approx 1.67431$ and $x \approx 7.1927193$. So the light intensity is 4 lux at the points that are 1.67 m and 7.19 m from the weaker source. ∎

Solving Inequalities Graphically

Inequalities can be solved graphically. To describe the method we solve

$$x^2 - 5x + 6 \le 0$$

This inequality was solved algebraically in Section 1.7, Example 3. To solve the inequality graphically, we draw the graph of

$$y = x^2 - 5x + 6$$

Our goal is to find those values of x for which $y \le 0$. These are simply the x-values for which the graph lies below the x-axis. From Figure 11 we see that the solution of the inequality is the interval $[2, 3]$.

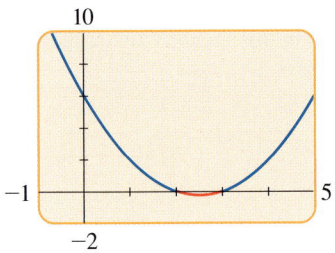

Figure 11
$x^2 - 5x + 6 \le 0$

Example 8 Solving an Inequality Graphically

Solve the inequality $3.7x^2 + 1.3x - 1.9 \le 2.0 - 1.4x$.

Solution We graph the equations

$$y_1 = 3.7x^2 + 1.3x - 1.9 \qquad \text{and} \qquad y_2 = 2.0 - 1.4x$$

in the same viewing rectangle in Figure 12. We are interested in those values of x for which $y_1 \le y_2$; these are points for which the graph of y_2 lies on or above the graph of y_1. To determine the appropriate interval, we look for the x-coordinates of points where the graphs intersect. We conclude that the solution is (approximately) the interval $[-1.45, 0.72]$. ∎

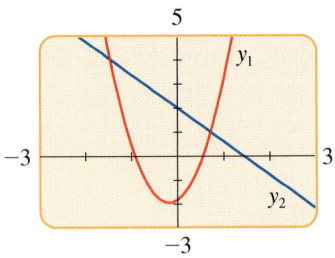

y_1

y_2

Figure 12
$y_1 = 3.7x^2 + 1.3x - 1.9$
$y_2 = 2.0 - 1.4x$

Example 9 Solving an Inequality Graphically

Solve the inequality $x^3 - 5x^2 \ge -8$.

Solution We write the inequality as

$$x^3 - 5x^2 + 8 \ge 0$$

and then graph the equation

$$y = x^3 - 5x^2 + 8$$

in the viewing rectangle $[-6, 6]$ by $[-15, 15]$, as shown in Figure 13. The solution of the inequality consists of those intervals on which the graph lies on or above the x-axis. By moving the cursor to the x-intercepts we find that, correct to one decimal place, the solution is $[-1.1, 1.5] \cup [4.6, \infty)$. ∎

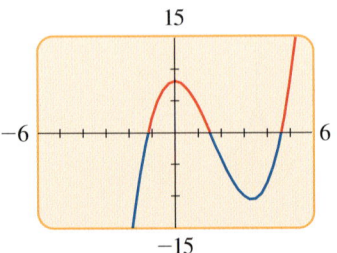

Figure 13
$x^3 - 5x^2 + 8 \ge 0$

1.9   Exercises

1–6 ■ Use a graphing calculator or computer to decide which viewing rectangle (a)–(d) produces the most appropriate graph of the equation.

1. $y = x^4 + 2$

  (a) $[-2, 2]$ by $[-2, 2]$

  (b) $[0, 4]$ by $[0, 4]$

  (c) $[-8, 8]$ by $[-4, 40]$

  (d) $[-40, 40]$ by $[-80, 800]$

2. $y = x^2 + 7x + 6$

  (a) $[-5, 5]$ by $[-5, 5]$

  (b) $[0, 10]$ by $[-20, 100]$

  (c) $[-15, 8]$ by $[-20, 100]$

  (d) $[-10, 3]$ by $[-100, 20]$

3. $y = 100 - x^2$

  (a) $[-4, 4]$ by $[-4, 4]$

  (b) $[-10, 10]$ by $[-10, 10]$

  (c) $[-15, 15]$ by $[-30, 110]$

  (d) $[-4, 4]$ by $[-30, 110]$

4. $y = 2x^2 - 1000$

  (a) $[-10, 10]$ by $[-10, 10]$

  (b) $[-10, 10]$ by $[-100, 100]$

  (c) $[-10, 10]$ by $[-1000, 1000]$

  (d) $[-25, 25]$ by $[-1200, 200]$

5. $y = 10 + 25x - x^3$

  (a) $[-4, 4]$ by $[-4, 4]$

  (b) $[-10, 10]$ by $[-10, 10]$

  (c) $[-20, 20]$ by $[-100, 100]$

  (d) $[-100, 100]$ by $[-200, 200]$

6. $y = \sqrt{8x - x^2}$

  (a) $[-4, 4]$ by $[-4, 4]$

  (b) $[-5, 5]$ by $[0, 100]$

  (c) $[-10, 10]$ by $[-10, 40]$

  (d) $[-2, 10]$ by $[-2, 6]$

7–18 ■ Determine an appropriate viewing rectangle for the equation and use it to draw the graph.

7. $y = 100x^2$

8. $y = -100x^2$

9. $y = 4 + 6x - x^2$

10. $y = 0.3x^2 + 1.7x - 3$

11. $y = \sqrt[4]{256 - x^2}$

12. $y = \sqrt{12x - 17}$

13. $y = 0.01x^3 - x^2 + 5$

14. $y = x(x + 6)(x - 9)$

15. $y = x^4 - 4x^3$

16. $y = \dfrac{x}{x^2 + 25}$

17. $y = 1 + |x - 1|$

18. $y = 2x - |x^2 - 5|$

19. Graph the circle $x^2 + y^2 = 9$ by solving for y and graphing two equations as in Example 3.

20. Graph the circle $(y - 1)^2 + x^2 = 1$ by solving for y and graphing two equations as in Example 3.

21. Graph the equation $4x^2 + 2y^2 = 1$ by solving for y and graphing two equations corresponding to the negative and positive square roots. (This graph is called an *ellipse*.)

22. Graph the equation $y^2 - 9x^2 = 1$ by solving for y and graphing the two equations corresponding to the positive and negative square roots. (This graph is called a *hyperbola*.)

23–26 ■ Do the graphs intersect in the given viewing rectangle? If they do, how many points of intersection are there?

23. $y = -3x^2 + 6x - \frac{1}{2}$, $y = \sqrt{7 - \frac{7}{12}x^2}$; $[-4, 4]$ by $[-1, 3]$

24. $y = \sqrt{49 - x^2}$, $y = \frac{1}{5}(41 - 3x)$; $[-8, 8]$ by $[-1, 8]$

25. $y = 6 - 4x - x^2$, $y = 3x + 18$; $[-6, 2]$ by $[-5, 20]$

26. $y = x^3 - 4x$, $y = x + 5$; $[-4, 4]$ by $[-15, 15]$

27–36 ■ Solve the equation both algebraically and graphically.

27. $x - 4 = 5x + 12$

28. $\frac{1}{2}x - 3 = 6 + 2x$

29. $\dfrac{2}{x} + \dfrac{1}{2x} = 7$

30. $\dfrac{4}{x + 2} - \dfrac{6}{2x} = \dfrac{5}{2x + 4}$

31. $x^2 - 32 = 0$

32. $x^3 + 16 = 0$

33. $16x^4 = 625$

34. $2x^5 - 243 = 0$

35. $(x - 5)^4 - 80 = 0$

36. $6(x + 2)^5 = 64$

37–44 ■ Solve the equation graphically in the given interval. State each answer correct to two decimals.

37. $x^2 - 7x + 12 = 0$; $[0, 6]$

38. $x^2 - 0.75x + 0.125 = 0$; $[-2, 2]$

39. $x^3 - 6x^2 + 11x - 6 = 0$; $[-1, 4]$

40. $16x^3 + 16x^2 = x + 1$; $[-2, 2]$

41. $x - \sqrt{x + 1} = 0$; $[-1, 5]$

42. $1 + \sqrt{x} = \sqrt{1 + x^2}$; $[-1, 5]$

43. $x^{1/3} - x = 0$; $[-3, 3]$

44. $x^{1/2} + x^{1/3} - x = 0$; $[-1, 5]$

45–48 ■ Find all real solutions of the equation, correct to two decimals.

45. $x^3 - 2x^2 - x - 1 = 0$

46. $x^4 - 8x^2 + 2 = 0$

47. $x(x - 1)(x + 2) = \frac{1}{6}x$

48. $x^4 = 16 - x^3$

49–56 ■ Find the solutions of the inequality by drawing appropriate graphs. State each answer correct to two decimals.

49. $x^2 - 3x - 10 \leq 0$

50. $0.5x^2 + 0.875x \leq 0.25$

51. $x^3 + 11x \leq 6x^2 + 6$

52. $16x^3 + 24x^2 > -9x - 1$

53. $x^{1/3} < x$

54. $\sqrt{0.5x^2 + 1} \leq 2|x|$

55. $(x + 1)^2 < (x - 1)^2$

56. $(x + 1)^2 \leq x^3$

57. In Example 6 we found two solutions of the equation $x^3 - 6x^2 + 9x = \sqrt{x}$, the solutions that lie between 1 and 6. Find two more solutions, correct to two decimals.

Applications

58. Estimating Profit   An appliance manufacturer estimates that the profit y (in dollars) generated by producing x cooktops per month is given by the equation

$$y = 10x + 0.5x^2 - 0.001x^3 - 5000$$

where $0 \leq x \leq 450$.

   (a) Graph the equation.

   (b) How many cooktops must be produced to begin generating a profit?

   (c) For what range of values of x is the company's profit greater than $15,000?

59. How Far Can You See?   If you stand on a ship in a calm sea, then your height x (in ft) above sea level is related to the farthest distance y (in mi) that you can see by the equation

$$y = \sqrt{1.5x + \left(\frac{x}{5280}\right)^2}$$

   (a) Graph the equation for $0 \leq x \leq 100$.

   (b) How high up do you have to be to be able to see 10 mi?

Discovery • Discussion

60. Equation Notation on Graphing Calculators   When you enter the following equations into your calculator, how does what you see on the screen differ from the usual way of writing the equations? (Check your user's manual if you're not sure.)

   (a) $y = |x|$

   (b) $y = \sqrt[5]{x}$

   (c) $y = \dfrac{x}{x - 1}$

   (d) $y = x^3 + \sqrt[3]{x + 2}$

61. Enter Equations Carefully   A student wishes to graph the equations

$$y = x^{1/3} \qquad \text{and} \qquad y = \frac{x}{x + 4}$$

on the same screen, so he enters the following information into his calculator:

$$Y_1 = X \wedge 1/3 \qquad Y_2 = X/X + 4$$

The calculator graphs two lines instead of the equations he wanted. What went wrong?

62. Algebraic and Graphical Solution Methods   Write a short essay comparing the algebraic and graphical methods for solving equations. Make up your own examples to illustrate the advantages and disadvantages of each method.

63. How Many Solutions?   This exercise deals with the family of equations

$$x^3 - 3x = k$$

   (a) Draw the graphs of

$$y_1 = x^3 - 3x \qquad \text{and} \qquad y_2 = k$$

in the same viewing rectangle, in the cases $k = -4$, $-2, 0, 2$, and 4. How many solutions of the equation $x^3 - 3x = k$ are there in each case? Find the solutions correct to two decimals.

   (b) For what ranges of values of k does the equation have one solution? two solutions? three solutions?

1.10   Lines

In this section we find equations for straight lines lying in a coordinate plane. The equations will depend on how the line is inclined, so we begin by discussing the concept of slope.

The Slope of a Line

We first need a way to measure the "steepness" of a line, or how quickly it rises (or falls) as we move from left to right. We define *run* to be the distance we move to the right and *rise* to be the corresponding distance that the line rises (or falls). The *slope* of a line is the ratio of rise to run:

$$\text{slope} = \frac{\text{rise}}{\text{run}}$$

Figure 1 shows situations where slope is important. Carpenters use the term *pitch* for the slope of a roof or a staircase; the term *grade* is used for the slope of a road.

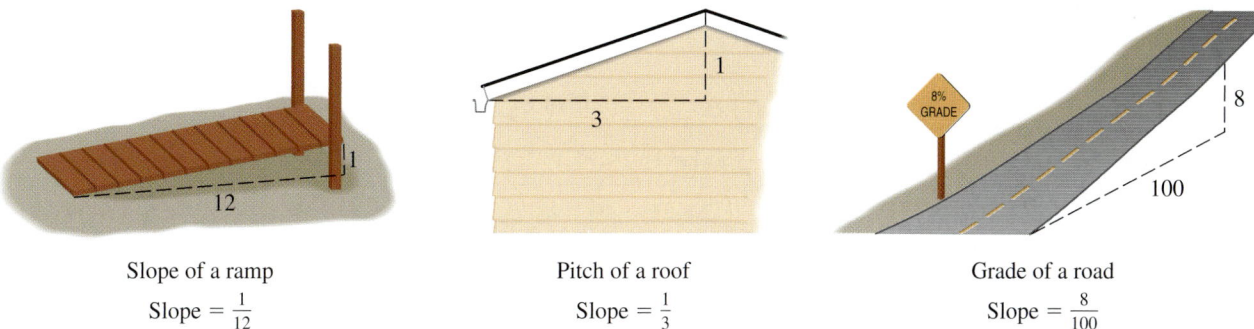

Slope of a ramp	Pitch of a roof	Grade of a road
Slope $= \frac{1}{12}$	Slope $= \frac{1}{3}$	Slope $= \frac{8}{100}$

Figure 1

If a line lies in a coordinate plane, then the **run** is the change in the *x*-coordinate and the **rise** is the corresponding change in the *y*-coordinate between any two points on the line (see Figure 2). This gives us the following definition of slope.

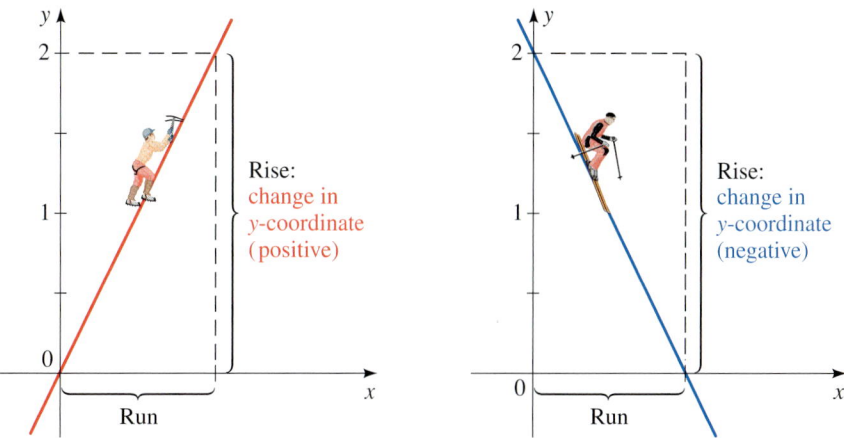

Figure 2

Slope of a Line

The **slope** m of a nonvertical line that passes through the points $A(x_1, y_1)$ and $B(x_2, y_2)$ is

$$m = \frac{\text{rise}}{\text{run}} = \frac{y_2 - y_1}{x_2 - x_1}$$

The slope of a vertical line is not defined.

The slope is independent of which two points are chosen on the line. We can see that this is true from the similar triangles in Figure 3:

$$\frac{y_2 - y_1}{x_2 - x_1} = \frac{y_2' - y_1'}{x_2' - x_1'}$$

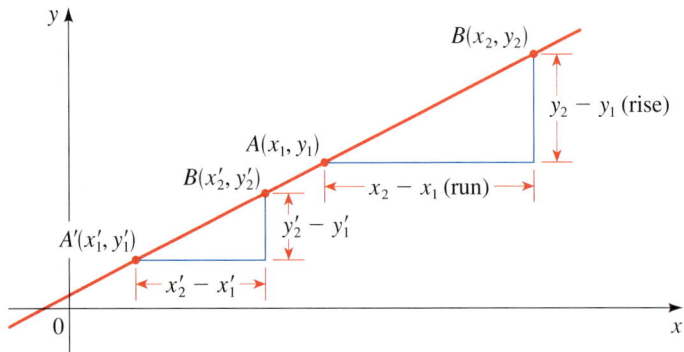

Figure 3

Figure 4 shows several lines labeled with their slopes. Notice that lines with positive slope slant upward to the right, whereas lines with negative slope slant downward to the right. The steepest lines are those for which the absolute value of the slope is the largest; a horizontal line has slope zero.

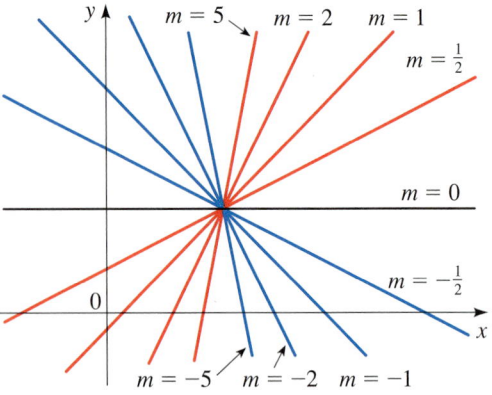

Figure 4
Lines with various slopes

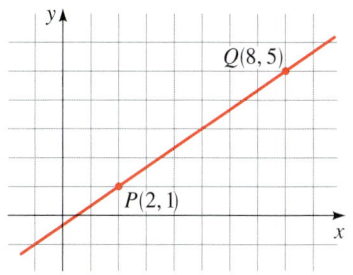

Figure 5

Example 1 Finding the Slope of a Line through Two Points

Find the slope of the line that passes through the points $P(2, 1)$ and $Q(8, 5)$.

Solution Since any two different points determine a line, only one line passes through these two points. From the definition, the slope is

$$m = \frac{y_2 - y_1}{x_2 - x_1} = \frac{5 - 1}{8 - 2} = \frac{4}{6} = \frac{2}{3}$$

This says that for every 3 units we move to the right, the line rises 2 units. The line is drawn in Figure 5. ∎

Equations of Lines

Now let's find the equation of the line that passes through a given point $P(x_1, y_1)$ and has slope m. A point $P(x, y)$ with $x \neq x_1$ lies on this line if and only if the slope of the line through P_1 and P is equal to m (see Figure 6), that is,

$$\frac{y - y_1}{x - x_1} = m$$

This equation can be rewritten in the form $y - y_1 = m(x - x_1)$; note that the equation is also satisfied when $x = x_1$ and $y = y_1$. Therefore, it is an equation of the given line.

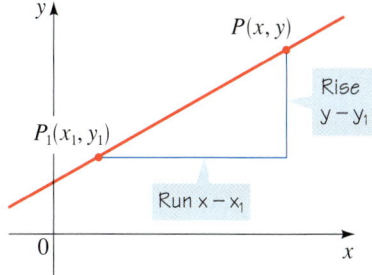

Figure 6

Point-Slope Form of the Equation of a Line

An equation of the line that passes through the point (x_1, y_1) and has slope m is

$$y - y_1 = m(x - x_1)$$

Example 2 Finding the Equation of a Line with Given Point and Slope

(a) Find an equation of the line through $(1, -3)$ with slope $-\frac{1}{2}$.

(b) Sketch the line.

Solution

(a) Using the point-slope form with $m = -\frac{1}{2}$, $x_1 = 1$, and $y_1 = -3$, we obtain an equation of the line as

$$y + 3 = -\tfrac{1}{2}(x - 1) \qquad \text{From point-slope equation}$$

$$2y + 6 = -x + 1 \qquad \text{Multiply by 2}$$

$$x + 2y + 5 = 0 \qquad \text{Rearrange}$$

(b) The fact that the slope is $-\frac{1}{2}$ tells us that when we move to the right 2 units, the line drops 1 unit. This enables us to sketch the line in Figure 7. ∎

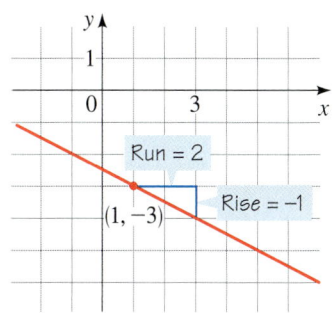

Figure 7

Example 3   Finding the Equation of a Line through Two Given Points

Find an equation of the line through the points $(-1, 2)$ and $(3, -4)$.

Solution   The slope of the line is

$$m = \frac{-4 - 2}{3 - (-1)} = -\frac{6}{4} = -\frac{3}{2}$$

Using the point-slope form with $x_1 = -1$ and $y_1 = 2$, we obtain

$$y - 2 = -\tfrac{3}{2}(x + 1) \qquad \text{From point-slope equation}$$

$$2y - 4 = -3x - 3 \qquad \text{Multiply by 2}$$

$$3x + 2y - 1 = 0 \qquad \text{Rearrange} \qquad \blacksquare$$

We can use *either* point, $(-1, 2)$ *or* $(3, -4)$, in the point-slope equation. We will end up with the same final answer.

Suppose a nonvertical line has slope m and y-intercept b (see Figure 8). This means the line intersects the y-axis at the point $(0, b)$, so the point-slope form of the equation of the line, with $x = 0$ and $y = b$, becomes

$$y - b = m(x - 0)$$

This simplifies to $y = mx + b$, which is called the **slope-intercept form** of the equation of a line.

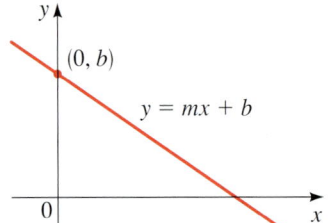

Figure 8

Slope-Intercept Form of the Equation of a Line

An equation of the line that has slope m and y-intercept b is

$$y = mx + b$$

Example 4   Lines in Slope-Intercept Form

(a) Find the equation of the line with slope 3 and y-intercept -2.

(b) Find the slope and y-intercept of the line $3y - 2x = 1$.

Solution

(a) Since $m = 3$ and $b = -2$, from the slope-intercept form of the equation of a line we get

$$y = 3x - 2$$

(b) We first write the equation in the form $y = mx + b$:

$$3y - 2x = 1$$

$$3y = 2x + 1 \qquad \text{Add } 2x$$

$$y = \tfrac{2}{3}x + \tfrac{1}{3} \qquad \text{Divide by 3}$$

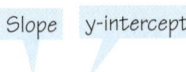

$$y = \tfrac{2}{3}x + \tfrac{1}{3}$$

From the slope-intercept form of the equation of a line, we see that the slope is $m = \tfrac{2}{3}$ and the y-intercept is $b = \tfrac{1}{3}$. $\qquad \blacksquare$

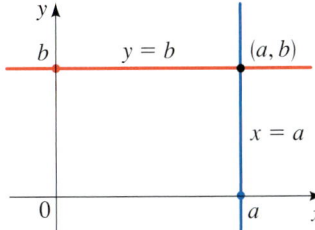

Figure 9

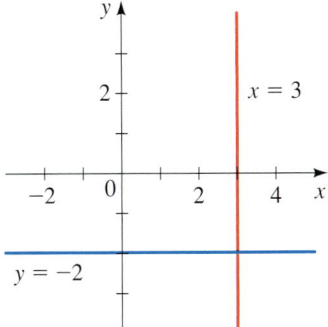

Figure 10

If a line is horizontal, its slope is $m = 0$, so its equation is $y = b$, where b is the y-intercept (see Figure 9). A vertical line does not have a slope, but we can write its equation as $x = a$, where a is the x-intercept, because the x-coordinate of every point on the line is a.

Vertical and Horizontal Lines

An equation of the vertical line through (a, b) is $x = a$.

An equation of the horizontal line through (a, b) is $y = b$.

Example 5   Vertical and Horizontal Lines

(a) The graph of the equation $x = 3$ is a vertical line with x-intercept 3.

(b) The graph of the equation $y = -2$ is a horizontal line with y-intercept -2.

The lines are graphed in Figure 10.   ■

A **linear equation** is an equation of the form

$$Ax + By + C = 0$$

where A, B, and C are constants and A and B are not both 0. The equation of a line is a linear equation:

■ A nonvertical line has the equation $y = mx + b$ or $-mx + y - b = 0$, which is a linear equation with $A = -m$, $B = 1$, and $C = -b$.

■ A vertical line has the equation $x = a$ or $x - a = 0$, which is a linear equation with $A = 1$, $B = 0$, and $C = -a$.

Conversely, the graph of a linear equation is a line:

■ If $B \neq 0$, the equation becomes

$$y = -\frac{A}{B}x - \frac{C}{B}$$

and this is the slope-intercept form of the equation of a line (with $m = -A/B$ and $b = -C/B$).

■ If $B = 0$, the equation becomes

$$Ax + C = 0$$

or $x = -C/A$, which represents a vertical line.

We have proved the following.

General Equation of a Line

The graph of every **linear equation**

$$Ax + By + C = 0 \qquad (A, B \text{ not both zero})$$

is a line. Conversely, every line is the graph of a linear equation.

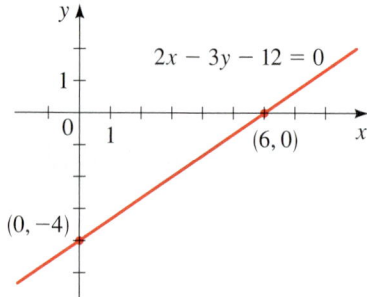

Figure 11

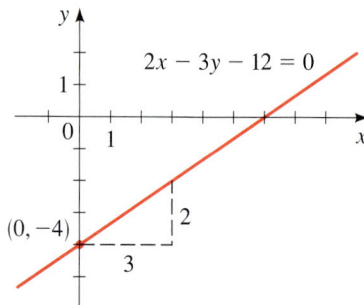

Figure 12

Example 6 Graphing a Linear Equation

Sketch the graph of the equation $2x - 3y - 12 = 0$.

Solution 1 Since the equation is linear, its graph is a line. To draw the graph, it is enough to find any two points on the line. The intercepts are the easiest points to find.

x-intercept: Substitute $y = 0$, to get $2x - 12 = 0$, so $x = 6$

y-intercept: Substitute $x = 0$, to get $-3y - 12 = 0$, so $y = -4$

With these points we can sketch the graph in Figure 11.

Solution 2 We write the equation in slope-intercept form:

$$2x - 3y - 12 = 0$$

$$2x - 3y = 12 \qquad \text{Add 12}$$

$$-3y = -2x + 12 \qquad \text{Subtract } 2x$$

$$y = \tfrac{2}{3}x - 4 \qquad \text{Divide by } -3$$

This equation is in the form $y = mx + b$, so the slope is $m = \tfrac{2}{3}$ and the y-intercept is $b = -4$. To sketch the graph, we plot the y-intercept, and then move 3 units to the right and 2 units up as shown in Figure 12. ■

Parallel and Perpendicular Lines

Since slope measures the steepness of a line, it seems reasonable that parallel lines should have the same slope. In fact, we can prove this.

Parallel Lines

Two nonvertical lines are parallel if and only if they have the same slope.

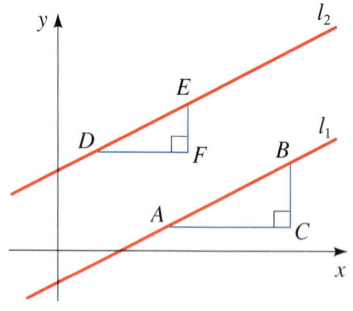

Figure 13

■ **Proof** Let the lines l_1 and l_2 in Figure 13 have slopes m_1 and m_2. If the lines are parallel, then the right triangles ABC and DEF are similar, so

$$m_1 = \frac{d(B, C)}{d(A, C)} = \frac{d(E, F)}{d(D, F)} = m_2$$

Conversely, if the slopes are equal, then the triangles will be similar, so $\angle BAC = \angle EDF$ and the lines are parallel. ■

Example 7 Finding the Equation of a Line Parallel to a Given Line

Find an equation of the line through the point $(5, 2)$ that is parallel to the line $4x + 6y + 5 = 0$.

Solution First we write the equation of the given line in slope-intercept form.

$$4x + 6y + 5 = 0$$

$$6y = -4x - 5 \qquad \text{Subtract } 4x + 5$$

$$y = -\tfrac{2}{3}x - \tfrac{5}{6} \qquad \text{Divide by } 6$$

So the line has slope $m = -\frac{2}{3}$. Since the required line is parallel to the given line, it also has slope $m = -\frac{2}{3}$. From the point-slope form of the equation of a line, we get

$$y - 2 = -\tfrac{2}{3}(x - 5) \qquad \text{Slope } m = -\tfrac{2}{3}, \text{ point } (5, 2)$$

$$3y - 6 = -2x + 10 \qquad \text{Multiply by 3}$$

$$2x + 3y - 16 = 0 \qquad \text{Rearrange}$$

Thus, the equation of the required line is $2x + 3y - 16 = 0$. ∎

The condition for perpendicular lines is not as obvious as that for parallel lines.

Perpendicular Lines

Two lines with slopes m_1 and m_2 are perpendicular if and only if $m_1 m_2 = -1$, that is, their slopes are negative reciprocals:

$$m_2 = -\frac{1}{m_1}$$

Also, a horizontal line (slope 0) is perpendicular to a vertical line (no slope).

■ **Proof** In Figure 14 we show two lines intersecting at the origin. (If the lines intersect at some other point, we consider lines parallel to these that intersect at the origin. These lines have the same slopes as the original lines.)

If the lines l_1 and l_2 have slopes m_1 and m_2, then their equations are $y = m_1 x$ and $y = m_2 x$. Notice that $A(1, m_1)$ lies on l_1 and $B(1, m_2)$ lies on l_2. By the Pythagorean Theorem and its converse (see page 54), $OA \perp OB$ if and only if

$$[d(O, A)]^2 + [d(O, B)]^2 = [d(A, B)]^2$$

By the Distance Formula, this becomes

$$(1^2 + m_1^2) + (1^2 + m_2^2) = (1 - 1)^2 + (m_2 - m_1)^2$$

$$2 + m_1^2 + m_2^2 = m_2^2 - 2m_1 m_2 + m_1^2$$

$$2 = -2m_1 m_2$$

$$m_1 m_2 = -1 \qquad ∎$$

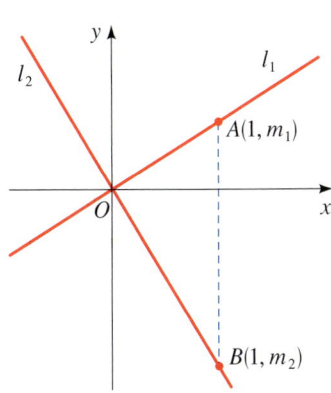

Figure 14

Example 8   Perpendicular Lines

Show that the points $P(3, 3)$, $Q(8, 17)$, and $R(11, 5)$ are the vertices of a right triangle.

Solution   The slopes of the lines containing PR and QR are, respectively,

$$m_1 = \frac{5 - 3}{11 - 3} = \frac{1}{4} \qquad \text{and} \qquad m_2 = \frac{5 - 17}{11 - 8} = -4$$

Since $m_1 m_2 = -1$, these lines are perpendicular and so PQR is a right triangle. It is sketched in Figure 15.

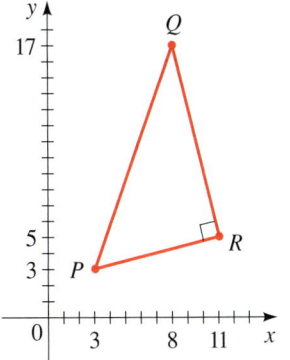

Figure 15

Example 9   Finding an Equation of a Line Perpendicular to a Given Line

Find an equation of the line that is perpendicular to the line $4x + 6y + 5 = 0$ and passes through the origin.

Solution   In Example 7 we found that the slope of the line $4x + 6y + 5 = 0$ is $-\frac{2}{3}$. Thus, the slope of a perpendicular line is the negative reciprocal, that is, $\frac{3}{2}$. Since the required line passes through $(0,0)$, the point-slope form gives

$$y - 0 = \tfrac{3}{2}(x - 0)$$

$$y = \tfrac{3}{2}x$$ ∎

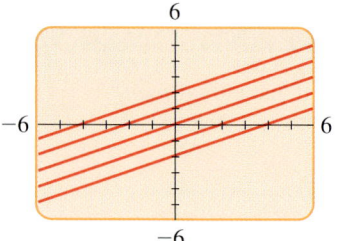

Figure 16
$y = 0.5x + b$

Example 10   Graphing a Family of Lines

Use a graphing calculator to graph the family of lines

$$y = 0.5x + b$$

for $b = -2, -1, 0, 1, 2$. What property do the lines share?

Solution   The lines are graphed in Figure 16 in the viewing rectangle $[-6, 6]$ by $[-6, 6]$. The lines all have the same slope, so they are parallel.   ∎

Applications: Slope as Rate of Change

When a line is used to model the relationship between two quantities, the slope of the line is the **rate of change** of one quantity with respect to the other. For example, the graph in Figure 17(a) gives the amount of gas in a tank that is being filled. The slope between the indicated points is

$$m = \frac{6 \text{ gallons}}{3 \text{ minutes}} = 2 \text{ gal/min}$$

The slope is the *rate* at which the tank is being filled, 2 gallons per minute. In Figure 17(b), the tank is being drained at the *rate* of 0.03 gallon per minute, and the slope is -0.03.

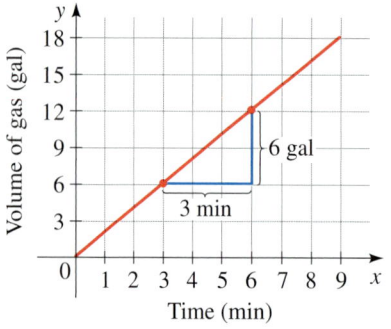

(a) Tank filled at 2 gal/min
Slope of line is 2

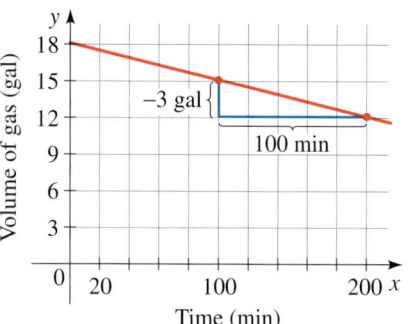

(b) Tank drained at 0.03 gal/min
Slope of line is −0.03

Figure 17

The next two examples give other situations where the slope of a line is a rate of change.

Example 11   Slope as Rate of Change

A dam is built on a river to create a reservoir. The water level w in the reservoir is given by the equation

$$w = 4.5t + 28$$

where t is the number of years since the dam was constructed, and w is measured in feet.

(a) Sketch a graph of this equation.

(b) What do the slope and w-intercept of this graph represent?

Solution

(a) This equation is linear, so its graph is a line. Since two points determine a line, we plot two points that lie on the graph and draw a line through them.

When $t = 0$, then $w = 4.5(0) + 28 = 28$, so $(0, 28)$ is on the line.

When $t = 2$, then $w = 4.5(2) + 28 = 37$, so $(2, 37)$ is on the line.

The line determined by these points is shown in Figure 18.

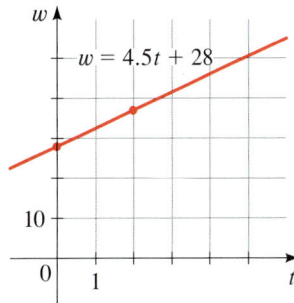

Figure 18

(b) The slope is $m = 4.5$; it represents the rate of change of water level with respect to time. This means that the water level *increases* 4.5 ft per year. The w-intercept is 28, and occurs when $t = 0$, so it represents the water level when the dam was constructed. ∎

Example 12   Linear Relationship between Temperature and Elevation

(a) As dry air moves upward, it expands and cools. If the ground temperature is 20°C and the temperature at a height of 1 km is 10°C, express the temperature T (in °C) in terms of the height h (in kilometers). (Assume that the relationship between T and h is linear.)

(b) Draw the graph of the linear equation. What does its slope represent?

(c) What is the temperature at a height of 2.5 km?

Solution

(a) Because we are assuming a linear relationship between T and h, the equation must be of the form

$$T = mh + b$$

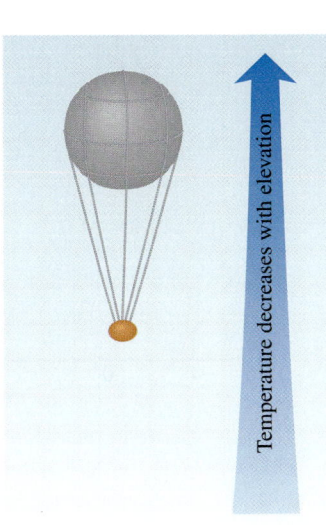

Temperature decreases with elevation

where m and b are constants. When $h = 0$, we are given that $T = 20$, so

$$20 = m(0) + b$$
$$b = 20$$

Thus, we have

$$T = mh + 20$$

When $h = 1$, we have $T = 10$ and so

$$10 = m(1) + 20$$
$$m = 10 - 20 = -10$$

The required expression is

$$T = -10h + 20$$

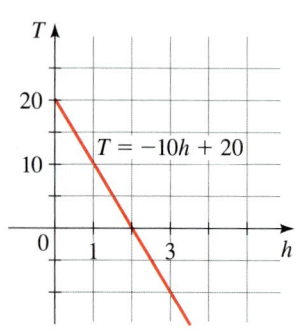

Figure 19

(b) The graph is sketched in Figure 19. The slope is $m = -10°C/km$, and this represents the rate of change of temperature with respect to distance above the ground. So the temperature *decreases* $10°C$ per kilometer of height.

(c) At a height of $h = 2.5$ km, the temperature is

$$T = -10(2.5) + 20 = -25 + 20 = -5°C$$

1.10   Exercises

1–8 ■ Find the slope of the line through P and Q.

1. $P(0,0)$, $Q(4,2)$

2. $P(0,0)$, $Q(2,-6)$

3. $P(2,2)$, $Q(-10,0)$

4. $P(1,2)$, $Q(3,3)$

5. $P(2,4)$, $Q(4,3)$

6. $P(2,-5)$, $Q(-4,3)$

7. $P(1,-3)$, $Q(-1,6)$

8. $P(-1,-4)$, $Q(6,0)$

9. Find the slopes of the lines l_1, l_2, l_3, and l_4 in the figure below.

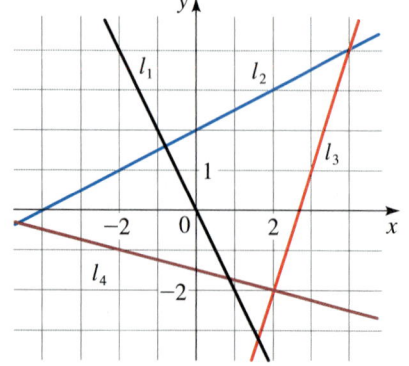

10. (a) Sketch lines through $(0,0)$ with slopes $1, 0, \frac{1}{2}, 2$, and -1.

(b) Sketch lines through $(0,0)$ with slopes $\frac{1}{3}, \frac{1}{2}, -\frac{1}{3}$, and 3.

11–14 ■ Find an equation for the line whose graph is sketched.

11.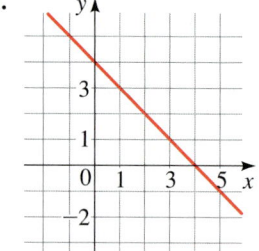

12.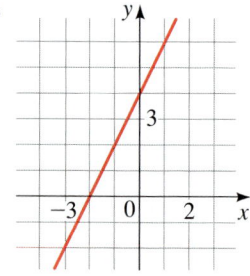

13.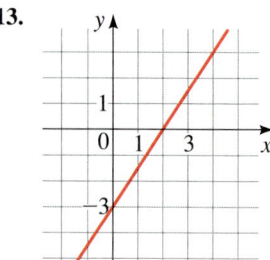

14.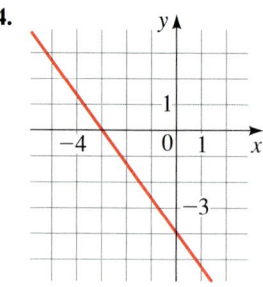

15–34 ■ Find an equation of the line that satisfies the given conditions.

15. Through $(2, 3)$;   slope 1

16. Through $(-2, 4)$;   slope -1

17. Through $(1, 7)$;   slope $\frac{2}{3}$

18. Through $(-3, -5)$;   slope $-\frac{7}{2}$

19. Through $(2, 1)$ and $(1, 6)$

20. Through $(-1, -2)$ and $(4, 3)$

21. Slope 3; y-intercept -2

22. Slope $\frac{2}{3}$; y-intercept 4

23. x-intercept 1; y-intercept -3

24. x-intercept -8; y-intercept 6

25. Through $(4, 5)$;   parallel to the x-axis

26. Through $(4, 5)$;   parallel to the y-axis

27. Through $(1, -6)$;   parallel to the line $x + 2y = 6$

28. y-intercept 6;   parallel to the line $2x + 3y + 4 = 0$

29. Through $(-1, 2)$;   parallel to the line $x = 5$

30. Through $(2, 6)$;   perpendicular to the line $y = 1$

31. Through $(-1, -2)$;   perpendicular to the line $2x + 5y + 8 = 0$

32. Through $(\frac{1}{2}, -\frac{2}{3})$;   perpendicular to the line $4x - 8y = 1$

33. Through $(1, 7)$;   parallel to the line passing through $(2, 5)$ and $(-2, 1)$

34. Through $(-2, -11)$;   perpendicular to the line passing through $(1, 1)$ and $(5, -1)$

35. (a) Sketch the line with slope $\frac{3}{2}$ that passes through the point $(-2, 1)$.

(b) Find an equation for this line.

36. (a) Sketch the line with slope -2 that passes through the point $(4, -1)$.

(b) Find an equation for this line.

 37–40 ■ Use a graphing device to graph the given family of lines in the same viewing rectangle. What do the lines have in common?

37. $y = -2x + b$ for $b = 0, \pm 1, \pm 3, \pm 6$

38. $y = mx - 3$ for $m = 0, \pm 0.25, \pm 0.75, \pm 1.5$

39. $y = m(x - 3)$ for $m = 0, \pm 0.25, \pm 0.75, \pm 1.5$

40. $y = 2 + m(x + 3)$ for $m = 0, \pm 0.5, \pm 1, \pm 2, \pm 6$

41–52 ■ Find the slope and y-intercept of the line and draw its graph.

41. $x + y = 3$ **42.** $3x - 2y = 12$

43. $x + 3y = 0$ **44.** $2x - 5y = 0$

45. $\frac{1}{2}x - \frac{1}{3}y + 1 = 0$ **46.** $-3x - 5y + 30 = 0$

47. $y = 4$ **48.** $4y + 8 = 0$

49. $3x - 4y = 12$ **50.** $x = -5$

51. $3x + 4y - 1 = 0$ **52.** $4x + 5y = 10$

53. Use slopes to show that $A(1, 1)$, $B(7, 4)$, $C(5, 10)$, and $D(-1, 7)$ are vertices of a parallelogram.

54. Use slopes to show that $A(-3, -1)$, $B(3, 3)$, and $C(-9, 8)$ are vertices of a right triangle.

55. Use slopes to show that $A(1, 1)$, $B(11, 3)$, $C(10, 8)$, and $D(0, 6)$ are vertices of a rectangle.

56. Use slopes to determine whether the given points are collinear (lie on a line).

(a) $(1, 1), (3, 9), (6, 21)$

(b) $(-1, 3), (1, 7), (4, 15)$

57. Find an equation of the perpendicular bisector of the line segment joining the points $A(1, 4)$ and $B(7, -2)$.

58. Find the area of the triangle formed by the coordinate axes and the line

$$2y + 3x - 6 = 0$$

59. (a) Show that if the x- and y-intercepts of a line are nonzero numbers a and b, then the equation of the line can be written in the form

$$\frac{x}{a} + \frac{y}{b} = 1$$

This is called the **two-intercept form** of the equation of a line.

(b) Use part (a) to find an equation of the line whose x-intercept is 6 and whose y-intercept is -8.

60. (a) Find an equation for the line tangent to the circle $x^2 + y^2 = 25$ at the point $(3, -4)$. (See the figure.)

(b) At what other point on the circle will a tangent line be parallel to the tangent line in part (a)?

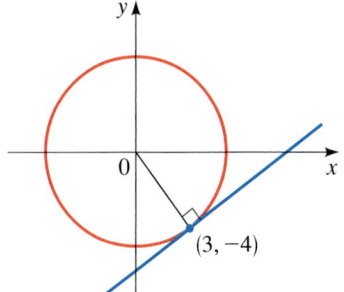

Applications

61. Grade of a Road   West of Albuquerque, New Mexico, Route 40 eastbound is straight and makes a steep descent toward the city. The highway has a 6% grade, which means that its slope is $-\frac{6}{100}$. Driving on this road you notice from elevation signs that you have descended a distance of 1000 ft. What is the change in your horizontal distance?

62. Global Warming   Some scientists believe that the average surface temperature of the world has been rising steadily. The average surface temperature is given by

$$T = 0.02t + 8.50$$

where T is temperature in °C and t is years since 1900.

(a) What do the slope and T-intercept represent?

(b) Use the equation to predict the average global surface temperature in 2100.

63. Drug Dosages   If the recommended adult dosage for a drug is D (in mg), then to determine the appropriate dosage c for a child of age a, pharmacists use the equation

$$c = 0.0417D(a + 1)$$

Suppose the dosage for an adult is 200 mg.

(a) Find the slope. What does it represent?

(b) What is the dosage for a newborn?

64. Flea Market   The manager of a weekend flea market knows from past experience that if she charges x dollars for a rental space at the flea market, then the number y of spaces she can rent is given by the equation $y = 200 - 4x$.

(a) Sketch a graph of this linear equation. (Remember that the rental charge per space and the number of spaces rented must both be nonnegative quantities.)

(b) What do the slope, the y-intercept, and the x-intercept of the graph represent?

65. Production Cost   A small-appliance manufacturer finds that if he produces x toaster ovens in a month his production cost is given by the equation

$$y = 6x + 3000$$

(where y is measured in dollars).

(a) Sketch a graph of this linear equation.

(b) What do the slope and y-intercept of the graph represent?

66. Temperature Scales   The relationship between the Fahrenheit (F) and Celsius (C) temperature scales is given by the equation $F = \frac{9}{5}C + 32$.

(a) Complete the table to compare the two scales at the given values.

(b) Find the temperature at which the scales agree. [*Hint:* Suppose that a is the temperature at which the scales agree. Set $F = a$ and $C = a$. Then solve for a.]

C	F
$-30°$	
$-20°$	
$-10°$	
$0°$	
	$50°$
	$68°$
	$86°$

67. Crickets and Temperature   Biologists have observed that the chirping rate of crickets of a certain species is related to temperature, and the relationship appears to be very nearly linear. A cricket produces 120 chirps per minute at 70°F and 168 chirps per minute at 80°F.

(a) Find the linear equation that relates the temperature t and the number of chirps per minute n.

(b) If the crickets are chirping at 150 chirps per minute, estimate the temperature.

68. Depreciation   A small business buys a computer for $4000. After 4 years the value of the computer is expected to be $200. For accounting purposes, the business uses *linear depreciation* to assess the value of the computer at a given time. This means that if V is the value of the computer at time t, then a linear equation is used to relate V and t.

(a) Find a linear equation that relates V and t.

(b) Sketch a graph of this linear equation.

(c) What do the slope and V-intercept of the graph represent?

(d) Find the depreciated value of the computer 3 years from the date of purchase.

69. Pressure and Depth   At the surface of the ocean, the water pressure is the same as the air pressure above the water, 15 lb/in². Below the surface, the water pressure increases by 4.34 lb/in² for every 10 ft of descent.

(a) Find an equation for the relationship between pressure and depth below the ocean surface.

(b) Sketch a graph of this linear equation.

(c) What do the slope and y-intercept of the graph represent?

(d) At what depth is the pressure 100 lb/in²?

Water pressure increases with depth

70. Distance, Speed, and Time Jason and Debbie leave Detroit at 2:00 P.M. and drive at a constant speed, traveling west on I-90. They pass Ann Arbor, 40 mi from Detroit, at 2:50 P.M.

(a) Express the distance traveled in terms of the time elapsed.

(b) Draw the graph of the equation in part (a).

(c) What is the slope of this line? What does it represent?

71. Cost of Driving The monthly cost of driving a car depends on the number of miles driven. Lynn found that in May her driving cost was $380 for 480 mi and in June her cost was $460 for 800 mi. Assume that there is a linear relationship between the monthly cost C of driving a car and the distance driven d.

(a) Find a linear equation that relates C and d.

(b) Use part (a) to predict the cost of driving 1500 mi per month.

(c) Draw the graph of the linear equation. What does the slope of the line represent?

(d) What does the y-intercept of the graph represent?

(e) Why is a linear relationship a suitable model for this situation?

72. Manufacturing Cost The manager of a furniture factory finds that it costs $2200 to manufacture 100 chairs in one day and $4800 to produce 300 chairs in one day.

(a) Assuming that the relationship between cost and the number of chairs produced is linear, find an equation that expresses this relationship. Then graph the equation.

(b) What is the slope of the line in part (a), and what does it represent?

(c) What is the y-intercept of this line, and what does it represent?

Discovery • Discussion

73. What Does the Slope Mean? Suppose that the graph of the outdoor temperature over a certain period of time is a line. How is the weather changing if the slope of the line is positive? If it's negative? If it's zero?

74. Collinear Points Suppose you are given the coordinates of three points in the plane, and you want to see whether they lie on the same line. How can you do this using slopes? Using the Distance Formula? Can you think of another method?

| 1.11 | **Modeling Variation** |

Mathematical models are discussed in more detail in Focus on Modeling, which begins on page 239.

When scientists talk about a mathematical model for a real-world phenomenon, they often mean an equation that describes the relationship between two quantities. For instance, the model may describe how the population of an animal species varies with time or how the pressure of a gas varies as its temperature changes. In this section we study a kind of modeling called *variation*.

Direct Variation

Two types of mathematical models occur so often that they are given special names. The first is called *direct variation* and occurs when one quantity is a constant multiple of the other, so we use an equation of the form $y = kx$ to model this dependence.

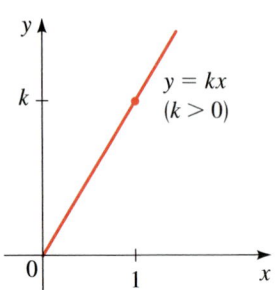

Figure 1

Direct Variation

If the quantities x and y are related by an equation

$$y = kx$$

for some constant $k \neq 0$, we say that y **varies directly as** x, or y is **directly proportional to** x, or simply y **is proportional to** x. The constant k is called the **constant of proportionality**.

Recall that the graph of an equation of the form $y = mx + b$ is a line with slope m and y-intercept b. So the graph of an equation $y = kx$ that describes direct variation is a line with slope k and y-intercept 0 (see Figure 1).

Example 1   Direct Variation

During a thunderstorm you see the lightning before you hear the thunder because light travels much faster than sound. The distance between you and the storm varies directly as the time interval between the lightning and the thunder.

(a) Suppose that the thunder from a storm 5400 ft away takes 5 s to reach you. Determine the constant of proportionality and write the equation for the variation.

(b) Sketch the graph of this equation. What does the constant of proportionality represent?

(c) If the time interval between the lightning and thunder is now 8 s, how far away is the storm?

Solution

(a) Let d be the distance from you to the storm and let t be the length of the time interval. We are given that d varies directly as t, so

$$d = kt$$

where k is a constant. To find k, we use the fact that $t = 5$ when $d = 5400$. Substituting these values in the equation, we get

$$5400 = k(5) \qquad \text{Substitute}$$

$$k = \frac{5400}{5} = 1080 \qquad \text{Solve for } k$$

Substituting this value of k in the equation for d, we obtain

$$d = 1080t$$

as the equation for d as a function of t.

(b) The graph of the equation $d = 1080t$ is a line through the origin with slope 1080 and is shown in Figure 2. The constant $k = 1080$ is the approximate speed of sound (in ft/s).

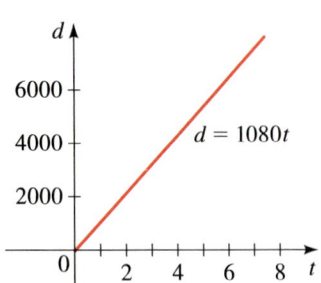

Figure 2

(c) When $t = 8$, we have

$$d = 1080 \cdot 8 = 8640$$

So, the storm is 8640 ft $\approx$ 1.6 mi away.   ∎

Inverse Variation

Another equation that is frequently used in mathematical modeling is $y = k/x$, where k is a constant.

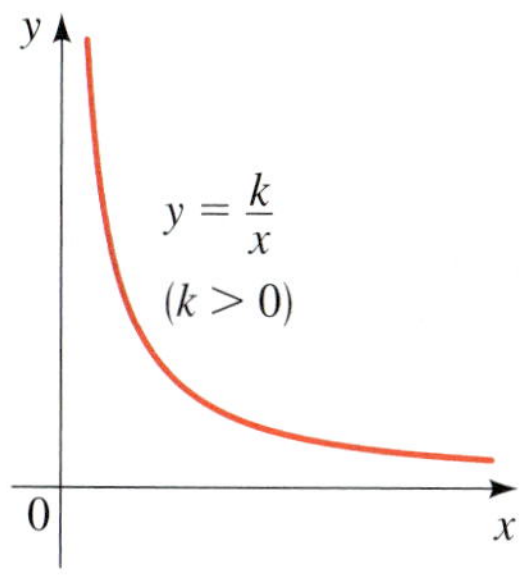

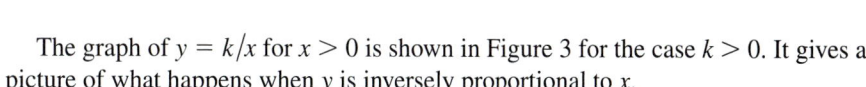

Inverse Variation

If the quantities x and y are related by the equation

$$y = \frac{k}{x}$$

for some constant $k \neq 0$, we say that y **is inversely proportional to** x, or y **varies inversely as** x.

y

$y = \dfrac{k}{x}$

$(k > 0)$

0   x

Figure 3

Inverse variation

The graph of $y = k/x$ for $x > 0$ is shown in Figure 3 for the case $k > 0$. It gives a picture of what happens when y is inversely proportional to x.

Example 2 Inverse Variation

Boyle's Law states that when a sample of gas is compressed at a constant temperature, the pressure of the gas is inversely proportional to the volume of the gas.

(a) Suppose the pressure of a sample of air that occupies 0.106 m³ at 25 °C is 50 kPa. Find the constant of proportionality, and write the equation that expresses the inverse proportionality.

(b) If the sample expands to a volume of 0.3 m³, find the new pressure.

Solution

(a) Let P be the pressure of the sample of gas and let V be its volume. Then, by the definition of inverse proportionality, we have

$$P = \frac{k}{V}$$

where k is a constant. To find k we use the fact that $P = 50$ when $V = 0.106$. Substituting these values in the equation, we get

$$50 = \frac{k}{0.106} \qquad \textcolor{blue}{\textit{Substitute}}$$

$$k = (50)(0.106) = 5.3 \qquad \textcolor{blue}{\textit{Solve for } k}$$

Putting this value of k in the equation for P, we have

$$P = \frac{5.3}{V}$$

(b) When $V = 0.3$, we have

$$P = \frac{5.3}{0.3} \approx 17.7$$

So, the new pressure is about 17.7 kPa.   ■

Joint Variation

A physical quantity often depends on more than one other quantity. If one quantity is proportional to two or more other quantities, we call this relationship *joint variation*.

Joint Variation

If the quantities x, y, and z are related by the equation

$$z = kxy$$

where k is a nonzero constant, we say that z **varies jointly as** x and y, or z **is jointly proportional to** x and y.

In the sciences, relationships between three or more variables are common, and any combination of the different types of proportionality that we have discussed is possible. For example, if

$$z = k\frac{x}{y}$$

we say that z **is proportional to** x and **inversely proportional to** y.

Example 3   Newton's Law of Gravitation

Newton's Law of Gravitation says that two objects with masses m_1 and m_2 attract each other with a force F that is jointly proportional to their masses and inversely proportional to the square of the distance r between the objects. Express Newton's Law of Gravitation as an equation.

Solution   Using the definitions of joint and inverse variation, and the traditional notation G for the gravitational constant of proportionality, we have

$$F = G\frac{m_1 m_2}{r^2}$$ ■

If m_1 and m_2 are fixed masses, then the gravitational force between them is $F = C/r^2$ (where $C = Gm_1m_2$ is a constant). Figure 4 shows the graph of this equation for $r > 0$ with $C = 1$. Observe how the gravitational attraction decreases with increasing distance.

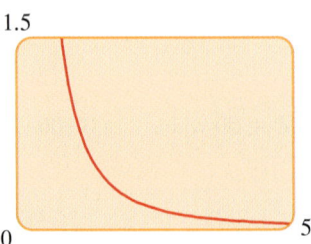

Figure 4
Graph of $F = \dfrac{1}{r^2}$

1.11 Exercises

1–12 ■ Write an equation that expresses the statement.

1. T varies directly as x.

2. P is directly proportional to w.

3. v is inversely proportional to z.

4. w is jointly proportional to m and n.

5. y is proportional to s and inversely proportional to t.

6. P varies inversely as T.

7. z is proportional to the square root of y.

8. A is proportional to the square of t and inversely proportional to the cube of x.

9. V is jointly proportional to l, w, and h.

10. S is jointly proportional to the squares of r and θ.

11. R is proportional to i and inversely proportional to P and t.

12. A is jointly proportional to the square roots of x and y.

13–22 ■ Express the statement as an equation. Use the given information to find the constant of proportionality.

13. y is directly proportional to x. If $x = 6$, then $y = 42$.

14. z varies inversely as t. If $t = 3$, then $z = 5$.

15. M varies directly as x and inversely as y. If $x = 2$ and $y = 6$, then $M = 5$.

16. S varies jointly as p and q. If $p = 4$ and $q = 5$, then $S = 180$.

17. W is inversely proportional to the square of r. If $r = 6$, then $W = 10$.

18. t is jointly proportional to x and y and inversely proportional to r. If $x = 2$, $y = 3$, and $r = 12$, then $t = 25$.

19. C is jointly proportional to l, w, and h. If $l = w = h = 2$, then $C = 128$.

20. H is jointly proportional to the squares of l and w. If $l = 2$ and $w = \frac{1}{3}$, then $H = 36$.

21. s is inversely proportional to the square root of t. If $s = 100$, then $t = 25$.

22. M is jointly proportional to a, b, and c, and inversely proportional to d. If a and d have the same value, and if b and c are both 2, then $M = 128$.

Applications

23. **Hooke's Law** Hooke's Law states that the force needed to keep a spring stretched x units beyond its natural length is directly proportional to x. Here the constant of proportional-

ity is called the **spring constant**.

(a) Write Hooke's Law as an equation.

(b) If a spring has a natural length of 10 cm and a force of 40 N is required to maintain the spring stretched to a length of 15 cm, find the spring constant.

(c) What force is needed to keep the spring stretched to a length of 14 cm?

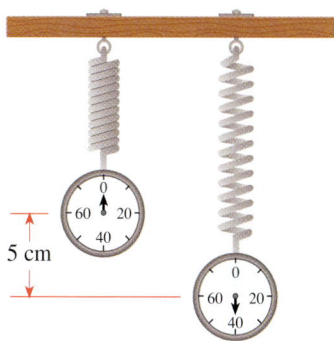

24. **Law of the Pendulum** The period of a pendulum (the time elapsed during one complete swing of the pendulum) varies directly with the square root of the length of the pendulum.

(a) Express this relationship by writing an equation.

(b) In order to double the period, how would we have to change the length l?

25. **Printing Costs** The cost C of printing a magazine is jointly proportional to the number of pages p in the magazine and the number of magazines printed m.

(a) Write an equation that expresses this joint variation.

(b) Find the constant of proportionality if the printing cost is $60,000 for 4000 copies of a 120-page magazine.

(c) How much would the printing cost be for 5000 copies of a 92-page magazine?

26. Boyle's Law   The pressure P of a sample of gas is directly proportional to the temperature T and inversely proportional to the volume V.

(a) Write an equation that expresses this variation.

(b) Find the constant of proportionality if 100 L of gas exerts a pressure of 33.2 kPa at a temperature of 400 K (absolute temperature measured on the Kelvin scale).

(c) If the temperature is increased to 500 K and the volume is decreased to 80 L, what is the pressure of the gas?

27. Power from a Windmill   The power P that can be obtained from a windmill is directly proportional to the cube of the wind speed s.

(a) Write an equation that expresses this variation.

(b) Find the constant of proportionality for a windmill that produces 96 watts of power when the wind is blowing at 20 mi/h.

(c) How much power will this windmill produce if the wind speed increases to 30 mi/h?

28. Power Needed to Propel a Boat   The power P (measured in horse power, hp) needed to propel a boat is directly proportional to the cube of the speed s. An 80-hp engine is needed to propel a certain boat at 10 knots. Find the power needed to drive the boat at 15 knots.

29. Loudness of Sound   The loudness L of a sound (measured in decibels, dB) is inversely proportional to the square of the distance d from the source of the sound. A person 10 ft from a lawn mower experiences a sound level of 70 dB; how loud is the lawn mower when the person is 100 ft away?

30. Stopping Distance   The stopping distance D of a car after the brakes have been applied varies directly as the square of the speed s. A certain car traveling at 50 mi/h can

stop in 240 ft. What is the maximum speed it can be traveling if it needs to stop in 160 ft?

31. A Jet of Water   The power P of a jet of water is jointly proportional to the cross-sectional area A of the jet and to the cube of the velocity v. If the velocity is doubled and the cross-sectional area is halved, by what factor will the power increase?

32. Aerodynamic Lift   The lift L on an airplane wing at take-off varies jointly as the square of the speed s of the plane and the area A of its wings. A plane with a wing area of 500 ft^2 traveling at 50 mi/h experiences a lift of 1700 lb. How much lift would a plane with a wing area of 600 ft^2 traveling at 40 mi/h experience?

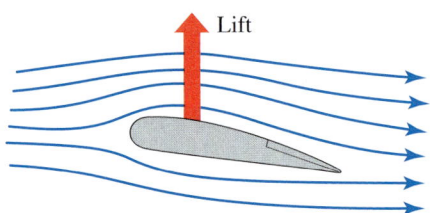

Lift

33. Drag Force on a Boat   The drag force F on a boat is jointly proportional to the wetted surface area A on the hull and the square of the speed s of the boat. A boat experiences a drag force of 220 lb when traveling at 5 mi/h with a wetted surface area of 40 ft^2. How fast must a boat be traveling if it has 28 ft^2 of wetted surface area and is experiencing a drag force of 175 lb?

34. Skidding in a Curve   A car is traveling on a curve that forms a circular arc. The force F needed to keep the car from skidding is jointly proportional to the weight w of the car and the square of its speed s, and is inversely proportional to the radius r of the curve.

(a) Write an equation that expresses this variation.

(b) A car weighing 1600 lb travels around a curve at 60 mi/h. The next car to round this curve weighs 2500 lb and requires the same force as the first car to keep from skidding. How fast is the second car traveling?

35. Electrical Resistance   The resistance R of a wire varies directly as its length L and inversely as the square of its diameter d.

  (a) Write an equation that expresses this joint variation.

  (b) Find the constant of proportionality if a wire 1.2 m long and 0.005 m in diameter has a resistance of 140 ohms.

  (c) Find the resistance of a wire made of the same material that is 3 m long and has a diameter of 0.008 m.

36. Kepler's Third Law   Kepler's Third Law of planetary motion states that the square of the period T of a planet (the time it takes for the planet to make a complete revolution about the sun) is directly proportional to the cube of its average distance d from the sun.

  (a) Express Kepler's Third Law as an equation.

  (b) Find the constant of proportionality by using the fact that for our planet the period is about 365 days and the average distance is about 93 million miles.

  (c) The planet Neptune is about 2.79×10^9 mi from the sun. Find the period of Neptune.

37. Radiation Energy   The total radiation energy E emitted by a heated surface per unit area varies as the fourth power of its absolute temperature T. The temperature is 6000 K at the surface of the sun and 300 K at the surface of the earth.

  (a) How many times more radiation energy per unit area is produced by the sun than by the earth?

  (b) The radius of the earth is 3960 mi and the radius of the sun is 435,000 mi. How many times more total radiation does the sun emit than the earth?

38. Value of a Lot   The value of a building lot on Galiano Island is jointly proportional to its area and the quantity of water produced by a well on the property. A 200 ft by 300 ft lot has a well producing 10 gallons of water per minute, and is valued at $48,000. What is the value of a 400 ft by 400 ft lot if the well on the lot produces 4 gallons of water per minute?

39. Growing Cabbages   In the short growing season of the Canadian arctic territory of Nunavut, some gardeners find it possible to grow gigantic cabbages in the midnight sun. Assume that the final size of a cabbage is proportional to the amount of nutrients it receives, and inversely proportional to the number of other cabbages surrounding it. A cabbage that received 20 oz of nutrients and had 12 other cabbages around it grew to 30 lb. What size would it grow to if it received 10 oz of nutrients and had only 5 cabbage "neighbors"?

40. Heat of a Campfire   The heat experienced by a hiker at a campfire is proportional to the amount of wood on the fire, and inversely proportional to the cube of his distance from the fire. If he is 20 ft from the fire, and someone doubles the amount of wood burning, how far from the fire would he have to be so that he feels the same heat as before?

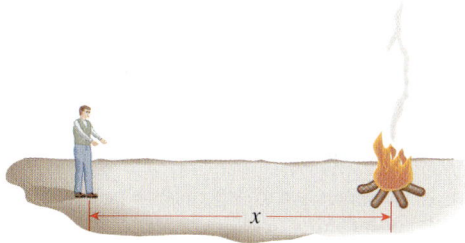

41. Frequency of Vibration   The frequency f of vibration of a violin string is inversely proportional to its length L. The constant of proportionality k is positive and depends on the tension and density of the string.

  (a) Write an equation that represents this variation.

  (b) What effect does doubling the length of the string have on the frequency of its vibration?

42. Spread of a Disease   The rate r at which a disease spreads in a population of size P is jointly proportional to the number x of infected people and the number $P - x$ who are not infected. An infection erupts in a small town with population $P = 5000$.

  (a) Write an equation that expresses r as a function of x.

  (b) Compare the rate of spread of this infection when 10 people are infected to the rate of spread when 1000 people are infected. Which rate is larger? By what factor?

  (c) Calculate the rate of spread when the entire population is infected. Why does this answer make intuitive sense?

Discovery • Discussion

43. Is Proportionality Everything?   A great many laws of physics and chemistry are expressible as proportionalities. Give at least one example of a function that occurs in the sciences that is *not* a proportionality.

1	**Review**

Concept Check

1. Define each term in your own words. (Check by referring to the definition in the text.)

    (a) An integer   **(b)** A rational number

    (c) An irrational number   **(d)** A real number

2. State each of these properties of real numbers.

    (a) Commutative Property

    (b) Associative Property

    (c) Distributive Property

3. What is an open interval? What is a closed interval? What notation is used for these intervals?

4. What is the absolute value of a number?

5. (a) In the expression a^x, which is the base and which is the exponent?

    (b) What does a^x mean if $x = n$, a positive integer?

    (c) What if $x = 0$?

    (d) What if x is a negative integer: $x = -n$, where n is a positive integer?

    (e) What if $x = m/n$, a rational number?

    (f) State the Laws of Exponents.

6. (a) What does $\sqrt[n]{a} = b$ mean?

    (b) Why is $\sqrt{a^2} = |a|$?

    (c) How many real nth roots does a positive real number have if n is odd? If n is even?

7. Explain how the procedure of rationalizing the denominator works.

8. State the Special Product Formulas for $(a + b)^2$, $(a - b)^2$, $(a + b)^3$, and $(a - b)^3$.

9. State each Special Factoring Formula.

    (a) Difference of squares   **(b)** Difference of cubes

    (c) Sum of cubes

10. What is a solution of an equation?

11. How do you solve an equation involving radicals? Why is it important to check your answers when solving equations of this type?

12. How do you solve an equation

    (a) algebraically?   **(b)** graphically?

13. Write the general form of each type of equation.

    (a) A linear equation   **(b)** A quadratic equation

14. What are the three ways to solve a quadratic equation?

15. State the Zero-Product Property.

16. Describe the process of completing the square.

17. State the quadratic formula.

18. What is the discriminant of a quadratic equation?

19. State the rules for working with inequalities.

20. How do you solve

    (a) a linear inequality?

    (b) a nonlinear inequality?

21. (a) How do you solve an equation involving an absolute value?

    (b) How do you solve an inequality involving an absolute value?

22. (a) Describe the coordinate plane.

    (b) How do you locate points in the coordinate plane?

23. State each formula.

    (a) The Distance Formula

    (b) The Midpoint Formula

24. Given an equation, what is its graph?

25. How do you find the x-intercepts and y-intercepts of a graph?

26. Write an equation of the circle with center (h, k) and radius r.

27. Explain the meaning of each type of symmetry. How do you test for it?

    (a) Symmetry with respect to the x-axis

    (b) Symmetry with respect to the y-axis

    (c) Symmetry with respect to the origin

28. Define the slope of a line.

29. Write each form of the equation of a line.

   (a) The point-slope form

   (b) The slope-intercept form

30. (a) What is the equation of a vertical line?

   (b) What is the equation of a horizontal line?

31. What is the general equation of a line?

32. Given lines with slopes m_1 and m_2, explain how you can tell if the lines are

   (a) parallel   **(b)** perpendicular

33. Write an equation that expresses each relationship.

   (a) y is directly proportional to x.

   (b) y is inversely proportional to x.

   (c) z is jointly proportional to x and y.

Exercises

1–4 ■ State the property of real numbers being used.

1. $3x + 2y = 2y + 3x$

2. $(a + b)(a - b) = (a - b)(a + b)$

3. $4(a + b) = 4a + 4b$

4. $(A + 1)(x + y) = (A + 1)x + (A + 1)y$

5–6 ■ Express the interval in terms of inequalities, and then graph the interval.

5. $[-2, 6)$ **6.** $(-\infty, 4]$

7–8 ■ Express the inequality in interval notation, and then graph the corresponding interval.

7. $x \geq 5$ **8.** $-1 < x \leq 5$

9–18 ■ Evaluate the expression.

9. $\big| 3 - |-9| \big|$ **10.** $1 - \big| 1 - |-1| \big|$

11. $2^{-3} - 3^{-2}$ **12.** $\sqrt[3]{-125}$

13. $216^{-1/3}$ **14.** $64^{2/3}$

15. $\dfrac{\sqrt{242}}{\sqrt{2}}$ **16.** $\sqrt[4]{4}\,\sqrt[4]{324}$

17. $2^{1/2} 8^{1/2}$ **18.** $\sqrt{2}\,\sqrt{50}$

19–28 ■ Simplify the expression.

19. $\dfrac{x^2(2x)^4}{x^3}$ **20.** $(a^2)^{-3}(a^3b)^2(b^3)^4$

21. $(3xy^2)^3 \left(\tfrac{2}{3}x^{-1}y\right)^2$ **22.** $\left(\dfrac{r^2 s^{4/3}}{r^{1/3}s}\right)^6$

23. $\sqrt[3]{(x^3 y)^2 y^4}$ **24.** $\sqrt{x^2 y^4}$

25. $\left(\dfrac{9x^3 y}{y^{-3}}\right)^{1/2}$

26. $\left(\dfrac{x^{-2}y^3}{x^2 y}\right)^{-1/2}\left(\dfrac{x^3 y}{y^{1/2}}\right)^2$

27. $\dfrac{8r^{1/2}s^{-3}}{2r^{-2}s^4}$

28. $\left(\dfrac{ab^2 c^{-3}}{2a^3 b^{-4}}\right)^{-2}$

29. Write the number $78{,}250{,}000{,}000$ in scientific notation.

30. Write the number 2.08×10^{-8} in ordinary decimal notation.

31. If $a \approx 0.00000293$, $b \approx 1.582 \times 10^{-14}$, and $c \approx 2.8064 \times 10^{12}$, use a calculator to approximate the number ab/c.

32. If your heart beats 80 times per minute and you live to be 90 years old, estimate the number of times your heart beats during your lifetime. State your answer in scientific notation.

33–48 ■ Factor the expression completely.

33. $12x^2 y^4 - 3xy^5 + 9x^3 y^2$ **34.** $x^2 - 9x + 18$

35. $x^2 + 3x - 10$ **36.** $6x^2 + x - 12$

37. $4t^2 - 13t - 12$ **38.** $x^4 - 2x^2 + 1$

39. $25 - 16t^2$ **40.** $2y^6 - 32y^2$

41. $x^6 - 1$ **42.** $y^3 - 2y^2 - y + 2$

43. $x^{-1/2} - 2x^{1/2} + x^{3/2}$ **44.** $a^4 b^2 + ab^5$

45. $4x^3 - 8x^2 + 3x - 6$ **46.** $8x^3 + y^6$

47. $(x^2 + 2)^{5/2} + 2x(x^2 + 2)^{3/2} + x^2\sqrt{x^2 + 2}$

48. $3x^3 - 2x^2 + 18x - 12$

49–64 ■ Perform the indicated operations and simplify.

49. $(2x + 1)(3x - 2) - 5(4x - 1)$

50. $(2y - 7)(2y + 7)$

51. $(1 + x)(2 - x) - (3 - x)(3 + x)$

52. $\sqrt{x}(\sqrt{x} + 1)(2\sqrt{x} - 1)$

53. $x^2(x - 2) + x(x - 2)^2$

54. $\dfrac{x^2 - 2x - 3}{2x^2 + 5x + 3}$

55. $\dfrac{x^2 + 2x - 3}{x^2 + 8x + 16} \cdot \dfrac{3x + 12}{x - 1}$

56. $\dfrac{t^3 - 1}{t^2 - 1}$

57. $\dfrac{x^2 - 2x - 15}{x^2 - 6x + 5} \div \dfrac{x^2 - x - 12}{x^2 - 1}$

58. $\dfrac{2}{x} + \dfrac{1}{x - 2} + \dfrac{3}{(x - 2)^2}$

59. $\dfrac{1}{x - 1} - \dfrac{2}{x^2 - 1}$

60. $\dfrac{1}{x + 2} + \dfrac{1}{x^2 - 4} - \dfrac{2}{x^2 - x - 2}$

61. $\dfrac{\dfrac{1}{x} - \dfrac{1}{2}}{x - 2}$

62. $\dfrac{\dfrac{1}{x} - \dfrac{1}{x + 1}}{\dfrac{1}{x} + \dfrac{1}{x + 1}}$

63. $\dfrac{\sqrt{6}}{\sqrt{3} + \sqrt{2}}$ (rationalize the denominator)

64. $\dfrac{\sqrt{x + h} - \sqrt{x}}{h}$ (rationalize the numerator)

65–80 ■ Find all real solutions of the equation.

65. $7x - 6 = 4x + 9$

66. $8 - 2x = 14 + x$

67. $\dfrac{x + 1}{x - 1} = \dfrac{3x}{3x - 6}$

68. $(x + 2)^2 = (x - 4)^2$

69. $x^2 - 9x + 14 = 0$

70. $x^2 + 24x + 144 = 0$

71. $2x^2 + x = 1$

72. $3x^2 + 5x - 2 = 0$

73. $4x^3 - 25x = 0$

74. $x^3 - 2x^2 - 5x + 10 = 0$

75. $3x^2 + 4x - 1 = 0$

76. $\dfrac{1}{x} + \dfrac{2}{x - 1} = 3$

77. $\dfrac{x}{x - 2} + \dfrac{1}{x + 2} = \dfrac{8}{x^2 - 4}$

78. $x^4 - 8x^2 - 9 = 0$

79. $|x - 7| = 4$

80. $|2x - 5| = 9$

81. The owner of a store sells raisins for \$3.20 per pound and nuts for \$2.40 per pound. He decides to mix the raisins and nuts and sell 50 lb of the mixture for \$2.72 per pound. What quantities of raisins and nuts should he use?

82. Anthony leaves Kingstown at 2:00 P.M. and drives to Queensville, 160 mi distant, at 45 mi/h. At 2:15 P.M. Helen leaves Queensville and drives to Kingstown at 40 mi/h. At what time do they pass each other on the road?

83. A woman cycles 8 mi/h faster than she runs. Every morning she cycles 4 mi and runs $2\frac{1}{2}$ mi, for a total of one hour of exercise. How fast does she run?

84. The hypotenuse of a right triangle has length 20 cm. The sum of the lengths of the other two sides is 28 cm. Find the lengths of the other two sides of the triangle.

85. Abbie paints twice as fast as Beth and three times as fast as Cathie. If it takes them 60 min to paint a living room with all three working together, how long would it take Abbie if she works alone?

86. A homeowner wishes to fence in three adjoining garden plots, one for each of her children, as shown in the figure. If each plot is to be 80 ft^2 in area, and she has 88 ft of fencing material at hand, what dimensions should each plot have?

87–94 ■ Solve the inequality. Express the solution using interval notation and graph the solution set on the real number line.

87. $3x - 2 > -11$

88. $-1 < 2x + 5 \le 3$

89. $x^2 + 4x - 12 > 0$

90. $x^2 \le 1$

91. $\dfrac{x - 4}{x^2 - 4} \le 0$

92. $\dfrac{5}{x^3 - x^2 - 4x + 4} < 0$

93. $|x - 5| \le 3$

94. $|x - 4| < 0.02$

95–98 ■ Solve the equation or inequality graphically.

95. $x^2 - 4x = 2x + 7$

96. $\sqrt{x + 4} = x^2 - 5$

97. $4x - 3 \geq x^2$

98. $x^3 - 4x^2 - 5x > 2$

99–100 ■ Two points P and Q are given.

(a) Plot P and Q on a coordinate plane.

(b) Find the distance from P to Q.

(c) Find the midpoint of the segment PQ.

(d) Sketch the line determined by P and Q, and find its equation in slope-intercept form.

(e) Sketch the circle that passes through Q and has center P, and find the equation of this circle.

99. $P(2, 0)$, $Q(-5, 12)$ **100.** $P(7, -1)$, $Q(2, -11)$

101–102 ■ Sketch the region given by the set.

101. $\{(x, y) \mid -4 < x < 4 \text{ and } -2 < y < 2\}$

102. $\{(x, y) \mid x \geq 4 \text{ or } y \geq 2\}$

103. Which of the points $A(4, 4)$ or $B(5, 3)$ is closer to the point $C(-1, -3)$?

104. Find an equation of the circle that has center $(2, -5)$ and radius $\sqrt{2}$.

105. Find an equation of the circle that has center $(-5, -1)$ and passes through the origin.

106. Find an equation of the circle that contains the points $P(2, 3)$ and $Q(-1, 8)$ and has the midpoint of the segment PQ as its center.

107–110 ■ Determine whether the equation represents a circle, a point, or has no graph. If the equation is that of a circle, find its center and radius.

107. $x^2 + y^2 + 2x - 6y + 9 = 0$

108. $2x^2 + 2y^2 - 2x + 8y = \frac{1}{2}$

109. $x^2 + y^2 + 72 = 12x$

110. $x^2 + y^2 - 6x - 10y + 34 = 0$

111–118 ■ Test the equation for symmetry and sketch its graph.

111. $y = 2 - 3x$

112. $2x - y + 1 = 0$

113. $x + 3y = 21$

114. $x = 2y + 12$

115. $y = 16 - x^2$

116. $8x + y^2 = 0$

117. $x = \sqrt{y}$

118. $y = -\sqrt{1 - x^2}$

119–122 ■ Use a graphing device to graph the equation in an appropriate viewing rectangle.

119. $y = x^2 - 6x$

120. $y = \sqrt{5 - x}$

121. $y = x^3 - 4x^2 - 5x$

122. $\dfrac{x^2}{4} + y^2 = 1$

123. Find an equation for the line that passes through the points $(-1, -6)$ and $(2, -4)$.

124. Find an equation for the line that passes through the point $(6, -3)$ and has slope $-\frac{1}{2}$.

125. Find an equation for the line that has x-intercept 4 and y-intercept 12.

126. Find an equation for the line that passes through the point $(1, 7)$ and is perpendicular to the line $x - 3y + 16 = 0$.

127. Find an equation for the line that passes through the origin and is parallel to the line $3x + 15y = 22$.

128. Find an equation for the line that passes through the point $(5, 2)$ and is parallel to the line passing through $(-1, -3)$ and $(3, 2)$.

129–130 ■ Find equations for the circle and the line in the figure.

129.

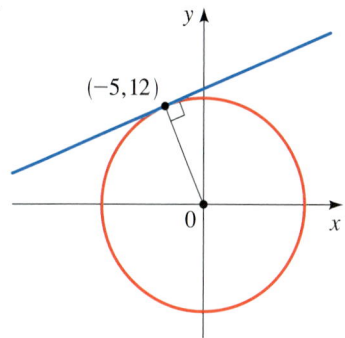

$(-5, 12)$

130.

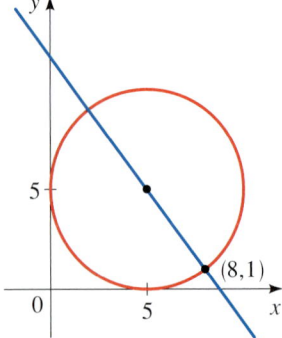

131. Hooke's Law states that if a weight w is attached to a hanging spring, then the stretched length s of the spring is linearly related to w. For a particular spring we have

$$s = 0.3w + 2.5$$

where s is measured in inches and w in pounds.

(a) What do the slope and s-intercept in this equation represent?

(b) How long is the spring when a 5-lb weight is attached?

132. Margarita is hired by an accounting firm at a salary of $60,000 per year. Three years later her annual salary has increased to $70,500. Assume her salary increases linearly.

(a) Find an equation that relates her annual salary S and the number of years t that she has worked for the firm.

(b) What do the slope and S-intercept of her salary equation represent?

(c) What will her salary be after 12 years with the firm?

133. Suppose that M varies directly as z, and $M = 120$ when $z = 15$. Write an equation that expresses this variation.

134. Suppose that z is inversely proportional to y, and that $z = 12$ when $y = 16$. Write an equation that expresses z in terms of y.

135. The intensity of illumination I from a light varies inversely as the square of the distance d from the light.

(a) Write this statement as an equation.

(b) Determine the constant of proportionality if it is known that a lamp has an intensity of 1000 candles at a distance of 8 m.

(c) What is the intensity of this lamp at a distance of 20 m?

136. The frequency of a vibrating string under constant tension is inversely proportional to its length. If a violin string 12 inches long vibrates 440 times per second, to what length must it be shortened to vibrate 660 times per second?

137. The terminal velocity of a parachutist is directly proportional to the square root of his weight. A 160-lb parachutist attains a terminal velocity of 9 mi/h. What is the terminal velocity for a parachutist weighing 240 lb?

138. The maximum range of a projectile is directly proportional to the square of its velocity. A baseball pitcher throws a ball at 60 mi/h, with a maximum range of 242 ft. What is his maximum range if he throws the ball at 70 mi/h?

1 Test

1. (a) Graph the intervals $(-5, 3]$ and $(2, \infty)$ on the real number line.
   (b) Express the inequalities $x \le 3$ and $-1 \le x < 4$ in interval notation.
   (c) Find the distance between -7 and 9 on the real number line.

2. Evaluate each expression.
   (a) $(-3)^4$ (b) -3^4 (c) 3^{-4} (d) $\dfrac{5^{23}}{5^{21}}$ (e) $\left(\dfrac{2}{3}\right)^{-2}$ (f) $16^{-3/4}$

3. Write each number in scientific notation.
   (a) $186{,}000{,}000{,}000$ (b) 0.0000003965

4. Simplify each expression. Write your final answer without negative exponents.
   (a) $\sqrt{200} - \sqrt{32}$ (b) $(3a^3b^3)(4ab^2)^2$ (c) $\left(\dfrac{3x^{3/2}y^3}{x^2y^{-1/2}}\right)^{-2}$

   (d) $\dfrac{x^2 + 3x + 2}{x^2 - x - 2}$ (e) $\dfrac{x^2}{x^2 - 4} - \dfrac{x+1}{x+2}$ (f) $\dfrac{\dfrac{y}{x} - \dfrac{x}{y}}{\dfrac{1}{y} - \dfrac{1}{x}}$

5. Rationalize the denominator and simplify: $\dfrac{\sqrt{10}}{\sqrt{5} - 2}$

6. Perform the indicated operations and simplify.
   (a) $3(x + 6) + 4(2x - 5)$ (b) $(x + 3)(4x - 5)$ (c) $(\sqrt{a} + \sqrt{b})(\sqrt{a} - \sqrt{b})$
   (d) $(2x + 3)^2$ (e) $(x + 2)^3$

7. Factor each expression completely.
   (a) $4x^2 - 25$ (b) $2x^2 + 5x - 12$ (c) $x^3 - 3x^2 - 4x + 12$
   (d) $x^4 + 27x$ (e) $3x^{3/2} - 9x^{1/2} + 6x^{-1/2}$ (f) $x^3y - 4xy$

8. Find all real solutions.
   (a) $x + 5 = 14 - \frac{1}{2}x$ (b) $\dfrac{2x}{x + 1} = \dfrac{2x - 1}{x}$ (c) $x^2 - x - 12 = 0$
   (d) $2x^2 + 4x + 1 = 0$ (e) $\sqrt{3} - \sqrt{x + 5} = 2$ (f) $x^4 - 3x^2 + 2 = 0$
   (g) $3|x - 4| = 10$

9. Mary drove from Amity to Belleville at a speed of 50 mi/h. On the way back, she drove at 60 mi/h. The total trip took $4\frac{2}{5}$ h of driving time. Find the distance between these two cities.

10. A rectangular parcel of land is 70 ft longer than it is wide. Each diagonal between opposite corners is 130 ft. What are the dimensions of the parcel?

11. Solve each inequality. Write the answer using interval notation, and sketch the solution on the real number line.
    (a) $-4 < 5 - 3x \le 17$ (b) $x(x - 1)(x + 2) > 0$
    (c) $|x - 4| < 3$ (d) $\dfrac{2x - 3}{x + 1} \le 1$

12. A bottle of medicine is to be stored at a temperature between 5 °C and 10 °C. What range does this correspond to on the Fahrenheit scale? [*Note:* Fahrenheit (F) and Celsius (C) temperatures satisfy the relation $C = \frac{5}{9}(F - 32)$.]

13. For what values of x is the expression $\sqrt{6x - x^2}$ defined as a real number?

14. Solve the equation and the inequality graphically.

(a) $x^3 - 9x - 1 = 0$ (b) $x^2 - 1 \le |x + 1|$

15. (a) Plot the points $P(0, 3)$, $Q(3, 0)$, and $R(6, 3)$ in the coordinate plane. Where must the point S be located so that $PQRS$ is a square?

(b) Find the area of $PQRS$.

16. (a) Sketch the graph of $y = x^2 - 4$.

(b) Find the x- and y-intercepts of the graph.

(c) Is the graph symmetric about the x-axis, the y-axis, or the origin?

17. Let $P(-3, 1)$ and $Q(5, 6)$ be two points in the coordinate plane.

(a) Plot P and Q in the coordinate plane.

(b) Find the distance between P and Q.

(c) Find the midpoint of the segment PQ.

(d) Find the slope of the line that contains P and Q.

(e) Find the perpendicular bisector of the line that contains P and Q.

(f) Find an equation for the circle for which the segment PQ is a diameter.

18. Find the center and radius of each circle and sketch its graph.

(a) $x^2 + y^2 = 25$ (b) $(x - 2)^2 + (y + 1)^2 = 9$ (c) $x^2 + 6x + y^2 - 2y + 6 = 0$

19. Write the linear equation $2x - 3y = 15$ in slope-intercept form, and sketch its graph. What are the slope and y-intercept?

20. Find an equation for the line with the given property.

(a) It passes through the point $(3, -6)$ and is parallel to the line $3x + y - 10 = 0$.

(b) It has x-intercept 6 and y-intercept 4.

21. A geologist uses a probe to measure the temperature T (in °C) of the soil at various depths below the surface, and finds that at a depth of x cm, the temperature is given by the linear equation $T = 0.08x - 4$.

(a) What is the temperature at a depth of one meter (100 cm)?

(b) Sketch a graph of the linear equation.

(c) What do the slope, the x-intercept, and T-intercept of the graph of this equation represent?

22. The maximum weight M that can be supported by a beam is jointly proportional to its width w and the square of its height h, and inversely proportional to its length L.

(a) Write an equation that expresses this proportionality.

(b) Determine the constant of proportionality if a beam 4 in. wide, 6 in. high, and 12 ft long can support a weight of 4800 lb.

(c) If a 10-ft beam made of the same material is 3 in. wide and 10 in. high, what is the maximum weight it can support?

If you had difficulty with any of these problems, you may wish to review the section of this chapter indicated below.

If you had trouble with this test problem	Review this section
1	Section 1.1
2, 3, 4(a), 4(b), 4(c)	Section 1.2
4(d), 4(e), 4(f), 5	Section 1.4
6, 7	Section 1.3
8	Section 1.5
9, 10	Section 1.6
11, 12, 13	Section 1.7
14	Section 1.9
15, 16, 17(a), 17(b)	Section 1.8
17(c), 17(d)	Section 1.10
17(e), 17(f), 18	Section 1.8
19, 20, 21	Section 1.10
22	Section 1.11

Focus on Problem Solving
General Principles

Stanford University News Service

George Polya (1887–1985) is famous among mathematicians for his ideas on problem solving. His lectures on problem solving at Stanford University attracted overflow crowds whom he held on the edges of their seats, leading them to discover solutions for themselves. He was able to do this because of his deep insight into the psychology of problem solving. His well-known book *How To Solve It* has been translated into 15 languages. He said that Euler (see page 288) was unique among great mathematicians because he explained *how* he found his results. Polya often said to his students and colleagues, "Yes, I see that your proof is correct, but how did you discover it?" In the preface to *How To Solve It*, Polya writes, "A great discovery solves a great problem but there is a grain of discovery in the solution of any problem. Your problem may be modest; but if it challenges your curiosity and brings into play your inventive faculties, and if you solve it by your own means, you may experience the tension and enjoy the triumph of discovery."

There are no hard and fast rules that will ensure success in solving problems. However, it is possible to outline some general steps in the problem-solving process and to give principles that are useful in solving certain problems. These steps and principles are just common sense made explicit. They have been adapted from George Polya's insightful book *How To Solve It*.

1. Understand the Problem

The first step is to read the problem and make sure that you understand it. Ask yourself the following questions:

> *What is the unknown?*
> *What are the given quantities?*
> *What are the given conditions?*

For many problems it is useful to

> *draw a diagram*

and identify the given and required quantities on the diagram.

Usually it is necessary to

> *introduce suitable notation*

In choosing symbols for the unknown quantities, we often use letters such as a, b, c, m, n, x, and y, but in some cases it helps to use initials as suggestive symbols, for instance, V for volume or t for time.

2. Think of a Plan

Find a connection between the given information and the unknown that enables you to calculate the unknown. It often helps to ask yourself explicitly: "How can I relate the given to the unknown?" If you don't see a connection immediately, the following ideas may be helpful in devising a plan.

■ **Try to recognize something familiar**

Relate the given situation to previous knowledge. Look at the unknown and try to recall a more familiar problem that has a similar unknown.

■ **Try to recognize patterns**

Certain problems are solved by recognizing that some kind of pattern is occurring. The pattern could be geometric, or numerical, or algebraic. If you can see regularity or repetition in a problem, then you might be able to guess what the pattern is and then prove it.

■ **Use analogy**

Try to think of an analogous problem, that is, a similar or related problem, but one that is easier than the original. If you can solve the similar, simpler problem, then it might give you the clues you need to solve the original, more difficult one. For instance, if a problem involves very large numbers, you could first try a similar problem with smaller numbers. Or if the problem is in three-dimensional geometry, you could look for something similar in two-dimensional geometry. Or if the problem you start with is a general one, you could first try a special case.

▪ Introduce something extra

You may sometimes need to introduce something new—an auxiliary aid—to make the connection between the given and the unknown. For instance, in a problem for which a diagram is useful, the auxiliary aid could be a new line drawn in the diagram. In a more algebraic problem the aid could be a new unknown that relates to the original unknown.

▪ Take cases

You may sometimes have to split a problem into several cases and give a different argument for each case. For instance, we often have to use this strategy in dealing with absolute value.

▪ Work backward

Sometimes it is useful to imagine that your problem is solved and work backward, step by step, until you arrive at the given data. Then you may be able to reverse your steps and thereby construct a solution to the original problem. This procedure is commonly used in solving equations. For instance, in solving the equation $3x - 5 = 7$, we suppose that x is a number that satisfies $3x - 5 = 7$ and work backward. We add 5 to each side of the equation and then divide each side by 3 to get $x = 4$. Since each of these steps can be reversed, we have solved the problem.

▪ Establish subgoals

In a complex problem it is often useful to set subgoals (in which the desired situation is only partially fulfilled). If you can attain or accomplish these subgoals, then you may be able to build on them to reach your final goal.

▪ Indirect reasoning

Sometimes it is appropriate to attack a problem indirectly. In using **proof by contradiction** to prove that P implies Q, we assume that P is true and Q is false and try to see why this cannot happen. Somehow we have to use this information and arrive at a contradiction to what we absolutely know is true.

▪ Mathematical induction

In proving statements that involve a positive integer n, it is frequently helpful to use the Principle of Mathematical Induction, which is discussed in Section 11.5.

3. Carry Out the Plan

In Step 2, a plan was devised. In carrying out that plan, you must check each stage of the plan and write the details that prove each stage is correct.

4. Look Back

Having completed your solution, it is wise to look back over it, partly to see if any errors have been made and partly to see if you can discover an easier way to solve the problem. Looking back also familiarizes you with the method of solution, and this may be useful for solving a future problem. Descartes said, "Every problem that I solved became a rule which served afterwards to solve other problems."

We illustrate some of these principles of problem solving with an example. Further illustrations of these principles will be presented at the end of selected chapters.

Problem Average Speed

A driver sets out on a journey. For the first half of the distance she drives at the leisurely pace of 30 mi/h; during the second half she drives 60 mi/h. What is her average speed on this trip?

■ **Thinking about the problem**

It is tempting to take the average of the speeds and say that the average speed for the entire trip is

$$\frac{30 + 60}{2} = 45 \text{ mi/h}$$

But is this simple-minded approach really correct?

Try a special case   Let's look at an easily calculated special case. Suppose that the total distance traveled is 120 mi. Since the first 60 mi is traveled at 30 mi/h, it takes 2 h. The second 60 mi is traveled at 60 mi/h, so it takes one hour. Thus, the total time is $2 + 1 = 3$ hours and the average speed is

$$\frac{120}{3} = 40 \text{ mi/h}$$

So our guess of 45 mi/h was wrong.

Solution We need to look more carefully at the meaning of average speed. It is *Understand the problem* defined as

$$\text{average speed} = \frac{\text{distance traveled}}{\text{time elapsed}}$$

Introduce notation   Let d be the distance traveled on each half of the trip. Let t_1 and t_2 be the times taken for the first and second halves of the trip. Now we can write down the information we have been given. For the first half of the trip, we have

State what is given   (1) $$30 = \frac{d}{t_1}$$

and for the second half, we have

(2) $$60 = \frac{d}{t_2}$$

Identify the unknown   Now we identify the quantity we are asked to find:

$$\text{average speed for entire trip} = \frac{\text{total distance}}{\text{total time}} = \frac{2d}{t_1 + t_2}$$

Connect the given with
the unknown

To calculate this quantity, we need to know t_1 and t_2, so we solve Equations 1 and 2 for these times:

$$t_1 = \frac{d}{30} \qquad t_2 = \frac{d}{60}$$

Now we have the ingredients needed to calculate the desired quantity:

$$\text{average speed} = \frac{2d}{t_1 + t_2} = \frac{2d}{\dfrac{d}{30} + \dfrac{d}{60}}$$

$$= \frac{60(2d)}{60\left(\dfrac{d}{30} + \dfrac{d}{60}\right)}$$

Multiply numerator and
denominator by 60

$$= \frac{120d}{2d + d} = \frac{120d}{3d} = 40$$

So, the average speed for the entire trip is 40 mi/h.   ■

Problems

Bettmann /Corbis

Don't feel bad if you don't solve these problems right away. Problems 2 and 6 were sent to Albert Einstein by his friend Wertheimer. Einstein (and his friend Bucky) enjoyed the problems and wrote back to Wertheimer. Here is part of his reply:

Your letter gave us a lot of amusement. The first intelligence test fooled both of us (Bucky and me). Only on working it out did I notice that no time is available for the downhill run! Mr. Bucky was also taken in by the second example, but I was not. Such drolleries show us how stupid we are!

(See *Mathematical Intelligencer*, Spring 1990, page 41.)

1. **Distance, Time, and Speed**   A man drives from home to work at a speed of 50 mi/h. The return trip from work to home is traveled at the more leisurely pace of 30 mi/h. What is the man's average speed for the round-trip?

2. **Distance, Time, and Speed**   An old car has to travel a 2-mile route, uphill and down. Because it is so old, the car can climb the first mile—the ascent—no faster than an average speed of 15 mi/h. How fast does the car have to travel the second mile—on the descent it can go faster, of course—in order to achieve an average speed of 30 mi/h for the trip?

3. **A Speeding Fly**   A car and a van are parked 120 mi apart on a straight road. The drivers start driving toward each other at noon, each at a speed of 40 mi/h. A fly starts from the front bumper of the van at noon and flies to the bumper of the car, then immediately back to the bumper of the van, back to the car, and so on, until the car and the van meet. If the fly flies at a speed of 100 mi/h, what is the total distance it travels?

4. **Comparing Discounts**   Which price is better for the buyer, a 40% discount or two successive discounts of 20%?

5. **Cutting up a Wire**   A piece of wire is bent as shown in the figure. You can see that one cut through the wire produces four pieces and two parallel cuts produce seven pieces. How many pieces will be produced by 142 parallel cuts? Write a formula for the number of pieces produced by n parallel cuts.

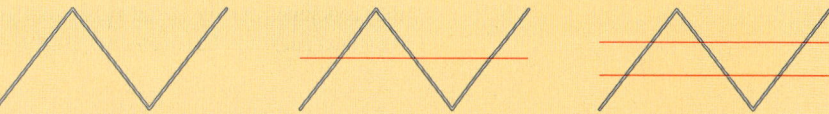

6. **Amoeba Propagation**   An amoeba propagates by simple division; each split takes 3 minutes to complete. When such an amoeba is put into a glass container with a nutrient fluid, the container is full of amoebas in one hour. How long would it take for the container to be filled if we start with not one amoeba, but two?

7. **Running Laps**   Two runners start running laps at the same time, from the same starting position. George runs a lap in 50 s; Sue runs a lap in 30 s. When will the runners next be side by side?

8. **Batting Averages**   Player A has a higher batting average than player B for the first half of the baseball season. Player A also has a higher batting average than player B for the second half of the season. Is it necessarily true that player A has a higher batting average than player B for the entire season?

9. **Coffee and Cream**   A spoonful of cream is taken from a pitcher of cream and put into a cup of coffee. The coffee is stirred. Then a spoonful of this mixture is put into the pitcher of cream. Is there now more cream in the coffee cup or more coffee in the pitcher of cream?

10. **A Melting Ice Cube**   An ice cube is floating in a cup of water, full to the brim, as shown in the sketch. As the ice melts, what happens? Does the cup overflow, or does the water level drop, or does it remain the same? (You need to know Archimedes' Principle: A floating object displaces a volume of water whose weight equals the weight of the object.)

11. **Wrapping the World**   A red ribbon is tied tightly around the earth at the equator. How much more ribbon would you need if you raised the ribbon 1 ft above the equator everywhere? (You don't need to know the radius of the earth to solve this problem.)

12. **Irrational Powers**   Prove that it's possible to raise an irrational number to an irrational power and get a rational result. [*Hint:* The number $a = \sqrt{2}^{\sqrt{2}}$ is either rational or irrational. If a is rational, you are done. If a is irrational, consider $a^{\sqrt{2}}$.]

13. **Babylonian Square Roots**   The ancient Babylonians developed the following process for finding the square root of a number N. First they made a guess at the square root—let's call this first guess r_1. Noting that

$$r_1 \cdot \left(\frac{N}{r_1}\right) = N$$

they concluded that the actual square root must be somewhere between r_1 and N/r_1, so their next guess for the square root, r_2, was the average of these two numbers:

$$r_2 = \frac{1}{2}\left(r_1 + \frac{N}{r_1}\right)$$

Continuing in this way, their next approximation was given by

$$r_3 = \frac{1}{2}\left(r_2 + \frac{N}{r_2}\right)$$

and so on. In general, once we have the nth approximation to the square root of N, we find the $(n + 1)$st using

$$r_{n+1} = \frac{1}{2}\left(r_n + \frac{N}{r_n}\right)$$

Use this procedure to find $\sqrt{72}$, correct to two decimal places.

14. **A Perfect Cube**   Show that if you multiply three consecutive integers and then add the middle integer to the result, you get a perfect cube.

15. **Number Patterns**   Find the last digit in the number 3^{459}. [*Hint:* Calculate the first few powers of 3, and look for a pattern.]

16. **Number Patterns**   Use the techniques of solving a simpler problem and looking for a pattern to evaluate the number

$$3999999999999^2$$

17. **Right Triangles and Primes**   Prove that every prime number is the leg of exactly one right triangle with integer sides. (This problem was first stated by Fermat; see page 652.)

18. **An Equation with No Solution**   Show that the equation $x^2 + y^2 = 4z + 3$ has no solution in integers. [*Hint:* Recall that an even number is of the form $2n$ and an odd number is of the form $2n + 1$. Consider all possible cases for x and y even or odd.]

19. **Ending Up Where You Started**   A woman starts at a point P on the earth's surface and walks 1 mi south, then 1 mi east, then 1 mi north, and finds herself back at P, the starting point. Describe all points P for which this is possible (there are infinitely many).

20. **Volume of a Truncated Pyramid**   The ancient Egyptians, as a result of their pyramid-building, knew that the volume of a pyramid with height h and square base of side length a is $V = \frac{1}{3}ha^2$. They were able to use this fact to prove that the volume of a truncated pyramid is $V = \frac{1}{3}h(a^2 + ab + b^2)$, where h is the height and b and a are the lengths of the sides of the square top and bottom, as shown in the figure. Prove the truncated pyramid volume formula.

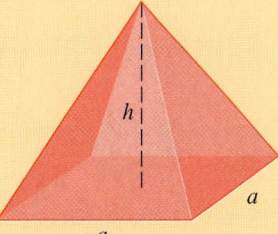 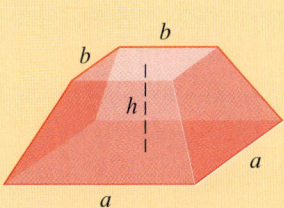

21. **Area of a Ring**   Find the area of the region between the two concentric circles shown in the figure.

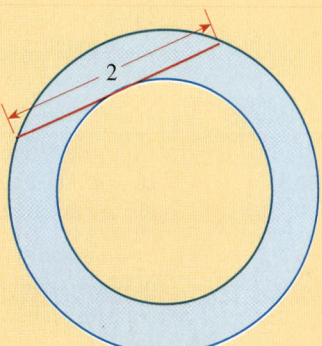

Bhaskara (born 1114) was an Indian mathematician, astronomer, and astrologer. Among his many accomplishments was an ingenious proof of the Pythagorean Theorem (see Problem 22). His important mathematical book *Lilavati* [*The Beautiful*] consists of algebra problems posed in the form of stories to his daughter Lilavati. Many of the problems begin "Oh beautiful maiden, suppose . . ." The story is told that using astrology, Bhaskara had determined that great misfortune would befall his daughter if she married at any time other than at a certain hour of a certain day. On her wedding day, as she was anxiously watching the water clock, a pearl fell unnoticed from her headdress. It stopped the flow of water in the clock, causing her to miss the opportune moment for marriage. Bhaskara's *Lilavati* was written to console her.

Entrance =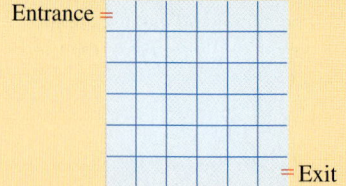
= Exit

22. Bhaskara's Proof   The Indian mathematician Bhaskara sketched the two figures shown here and wrote below them, "Behold!" Explain how his sketches prove the Pythagorean Theorem.

 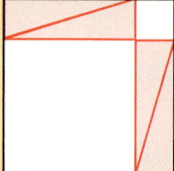

23. An Interesting Integer   The number 1729 is the smallest positive integer that can be represented in two different ways as the sum of two cubes. What are the two ways?

24. Simple Numbers

(a) Use a calculator to find the value of the expression

$$\sqrt{3 + 2\sqrt{2}} - \sqrt{3 - 2\sqrt{2}}$$

The number looks very simple. Show that the calculated value is correct.

(b) Use a calculator to evaluate

$$\frac{\sqrt{2} + \sqrt{6}}{\sqrt{2 + \sqrt{3}}}$$

Show that the calculated value is correct.

25. The Impossible Museum Tour   A museum is in the shape of a square with six rooms to a side; the entrance and exit are at diagonally opposite corners, as shown in the figure to the left. Each pair of adjacent rooms is joined by a door. Some very efficient tourists would like to tour the museum by visiting each room *exactly* once. Can you find a path for such a tour? Here are examples of attempts that failed.

Oops! Missed this room.

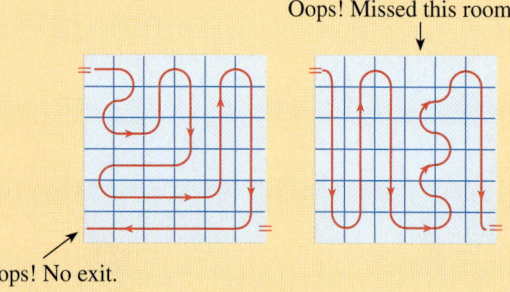

Oops! No exit.

Here is how you can prove that the museum tour is not possible. Imagine that the rooms are colored black and white like a checkerboard.

(a) Show that the room colors alternate between white and black as the tourists walk through the museum.

(b) Use part (a) and the fact that there are an even number of rooms in the museum to conclude that the tour cannot end at the exit.

26. Coloring the Coordinate Plane   Suppose that each point in the coordinate plane is colored either red or blue. Show that there must always be two points of the same color that are exactly one unit apart.

27. The Rational Coordinate Forest   Suppose that each point (x, y) in the plane, both of whose coordinates are rational numbers, represents a tree. If you are standing at the point $(0, 0)$, how far could you see in this forest?

28. **A Thousand Points**   A thousand points are graphed in the coordinate plane. Explain why it is possible to draw a straight line in the plane so that half of the points are on one side of the line and half are on the other. [*Hint:* Consider the slopes of the lines determined by each *pair* of points.]

29. **Graphing a Region in the Plane**   Sketch the region in the plane consisting of all points (x, y) such that

$$|x| + |y| \leq 1$$

30. **The Graph of an Equation**   Graph the equation

$$x^2y - y^3 - 5x^2 + 5y^2 = 0$$

[*Hint:* Factor.]

2 Functions

Chapter Overview

Perhaps the most useful mathematical idea for modeling the real world is the concept of *function*, which we study in this chapter. To understand what a function is, let's look at an example.

If a rock climber drops a stone from a high cliff, what happens to the stone? Of course the stone falls; how far it has fallen at any given moment depends upon how long it has been falling. That's a general description, but it doesn't tell us exactly when the stone will hit the ground.

 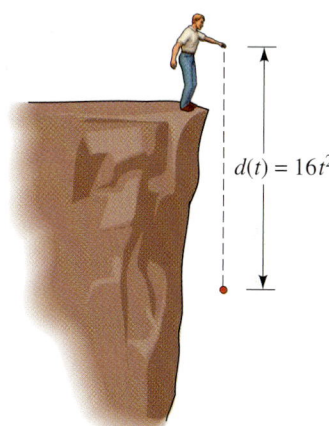

$d(t) = 16t^2$

General description: The stone falls.   **Function:** In t seconds the stone falls $16t^2$ ft.

What we need is a *rule* that relates the position of the stone to the time it has fallen. Physicists know that the rule is: In t seconds the stone falls $16t^2$ feet. If we let $d(t)$ stand for the distance the stone has fallen at time t, then we can express this rule as

$$d(t) = 16t^2$$

This "rule" for finding the distance in terms of the time is called a *function*. We say that distance is a *function* of time. To understand this rule or function better, we can make a table of values or draw a graph. The graph allows us to easily visualize how far and how fast the stone falls.

Time t	Distance $d(t)$
0	0
1	16
2	64
3	144
4	256

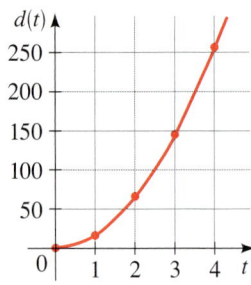

You can see why functions are important. For example, if a physicist finds the "rule" or function that relates distance fallen to elapsed time, then she can predict when a missile will hit the ground. If a biologist finds the function or "rule" that relates the number of bacteria in a culture to the time, then he can predict the number of bacteria for some future time. If a farmer knows the function or "rule" that relates the yield of apples to the number of trees per acre, then he can decide how many trees per acre to plant to maximize the yield.

In this chapter we will learn how functions are used to model real-world situations and how to find such functions.

2.1   What Is a Function?

In this section we explore the idea of a function and then give the mathematical definition of function.

Functions All Around Us

In nearly every physical phenomenon we observe that one quantity depends on another. For example, your height depends on your age, the temperature depends on the date, the cost of mailing a package depends on its weight (see Figure 1). We use the term *function* to describe this dependence of one quantity on another. That is, we say the following:

- Height is a function of age.
- Temperature is a function of date.
- Cost of mailing a package is a function of weight.

The U.S. Post Office uses a simple rule to determine the cost of mailing a package based on its weight. But it's not so easy to describe the rule that relates height to age or temperature to date.

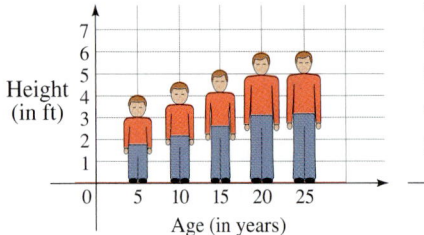

Height is a function of age.

Temperature is a function of date.

w (ounces)	Postage (dollars)
$0 < w \le 1$	0.37
$1 < w \le 2$	0.60
$2 < w \le 3$	0.83
$3 < w \le 4$	1.06
$4 < w \le 5$	1.29
$5 < w \le 6$	1.52

Postage is a function of weight.

Figure 1

Can you think of other functions? Here are some more examples:

- The area of a circle is a function of its radius.
- The number of bacteria in a culture is a function of time.
- The weight of an astronaut is a function of her elevation.
- The price of a commodity is a function of the demand for that commodity.

The rule that describes how the area A of a circle depends on its radius r is given by the formula $A = \pi r^2$. Even when a precise rule or formula describing a function is not available, we can still describe the function by a graph. For example, when you turn on a hot water faucet, the temperature of the water depends on how long the water has been running. So we can say

- Temperature of water from the faucet is a function of time.

Figure 2 shows a rough graph of the temperature T of the water as a function of the time t that has elapsed since the faucet was turned on. The graph shows that the initial temperature of the water is close to room temperature. When the water from the hot water tank reaches the faucet, the water's temperature T increases quickly. In the next phase, T is constant at the temperature of the water in the tank. When the tank is drained, T decreases to the temperature of the cold water supply.

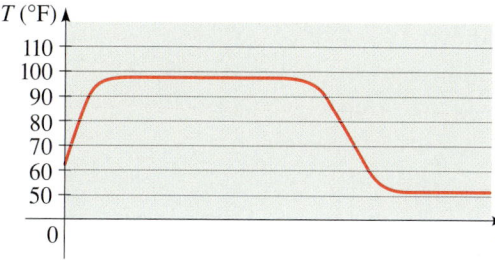

Figure 2

Graph of water temperature T as a function of time t

Definition of Function

We have previously used letters to stand for numbers. Here we do something quite different. We use letters to represent *rules*.

A function is a rule. In order to talk about a function, we need to give it a name. We will use letters such as $f, g, h, \ldots$ to represent functions. For example, we can use the letter f to represent a rule as follows:

$$\text{``}f\text{''} \qquad \text{is the rule} \qquad \text{``square the number''}$$

When we write $f(2)$, we mean "apply the rule f to the number 2." Applying the rule gives $f(2) = 2^2 = 4$. Similarly, $f(3) = 3^2 = 9$, $f(4) = 4^2 = 16$, and in general $f(x) = x^2$.

Definition of Function

A **function** f is a rule that assigns to each element x in a set A exactly one element, called $f(x)$, in a set B.

We usually consider functions for which the sets A and B are sets of real numbers. The symbol $f(x)$ is read "f of x" or "f at x" and is called the **value of f at x**, or the **image of x under f**. The set A is called the **domain** of the function. The **range** of f is the set of all possible values of $f(x)$ as x varies throughout the domain, that is,

$$\text{range of } f = \{f(x) \mid x \in A\}$$

The symbol that represents an arbitrary number in the domain of a function f is called an **independent variable**. The symbol that represents a number in the range of f is called a **dependent variable**. So if we write $y = f(x)$, then x is the independent variable and y is the dependent variable.

It's helpful to think of a function as a **machine** (see Figure 3). If x is in the domain of the function f, then when x enters the machine, it is accepted as an **input** and the machine produces an **output** $f(x)$ according to the rule of the function. Thus, we can think of the domain as the set of all possible inputs and the range as the set of all possible outputs.

> The $\boxed{\sqrt{}}$ key on your calculator is a good example of a function as a machine. First you input x into the display. Then you press the key labeled $\boxed{\sqrt{}}$. (On most *graphing* calculators, the order of these operations is reversed.) If $x < 0$, then x is not in the domain of this function; that is, x is not an acceptable input and the calculator will indicate an error. If $x \geq 0$, then an approximation to $\sqrt{x}$ appears in the display, correct to a certain number of decimal places. (Thus, the $\boxed{\sqrt{}}$ key on your calculator is not quite the same as the exact mathematical function f defined by $f(x) = \sqrt{x}$.)

Figure 3
Machine diagram of f x → f → $f(x)$
input   output

Another way to picture a function is by an **arrow diagram** as in Figure 4. Each arrow connects an element of A to an element of B. The arrow indicates that $f(x)$ is associated with x, $f(a)$ is associated with a, and so on.

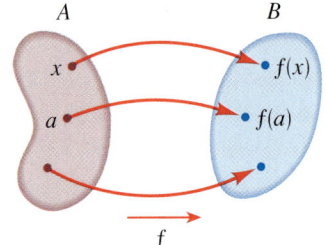

Figure 4
Arrow diagram of f

Example 1   The Squaring Function

The squaring function assigns to each real number x its square x^2. It is defined by

$$f(x) = x^2$$

(a) Evaluate $f(3)$, $f(-2)$, and $f(\sqrt{5})$.

(b) Find the domain and range of f.

(c) Draw a machine diagram for f.

Solution

(a) The values of f are found by substituting for x in $f(x) = x^2$.

$$f(3) = 3^2 = 9 \qquad f(-2) = (-2)^2 = 4 \qquad f(\sqrt{5}) = (\sqrt{5})^2 = 5$$

(b) The domain of f is the set $\mathbb{R}$ of all real numbers. The range of f consists of all values of $f(x)$, that is, all numbers of the form x^2. Since $x^2 \geq 0$ for all real numbers x, we can see that the range of f is $\{y \mid y \geq 0\} = [0, \infty)$.

(c) A machine diagram for this function is shown in Figure 5.   ∎

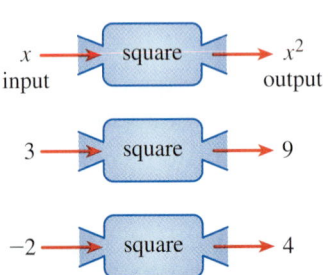

Figure 5

Machine diagram

Evaluating a Function

In the definition of a function the independent variable x plays the role of a "place-holder." For example, the function $f(x) = 3x^2 + x - 5$ can be thought of as

$$f(\quad) = 3 \cdot \quad^2 + \quad - 5$$

To evaluate f at a number, we substitute the number for the placeholder.

Example 2 Evaluating a Function

Let $f(x) = 3x^2 + x - 5$. Evaluate each function value.

(a) $f(-2)$ (b) $f(0)$ (c) $f(4)$ (d) $f\left(\frac{1}{2}\right)$

Solution To evaluate f at a number, we substitute the number for x in the definition of f.

(a) $f(-2) = 3 \cdot (-2)^2 + (-2) - 5 = 5$

(b) $f(0) = 3 \cdot 0^2 + 0 - 5 = -5$

(c) $f(4) = 3 \cdot 4^2 + 4 - 5 = 47$

(d) $f\left(\frac{1}{2}\right) = 3 \cdot \left(\frac{1}{2}\right)^2 + \frac{1}{2} - 5 = -\frac{15}{4}$ ■

Example 3 A Piecewise Defined Function

A cell phone plan costs $39 a month. The plan includes 400 free minutes and charges 20¢ for each additional minute of usage. The monthly charges are a function of the number of minutes used, given by

$$C(x) = \begin{cases} 39 & \text{if } 0 \le x \le 400 \\ 39 + 0.2(x - 400) & \text{if } x > 400 \end{cases}$$

Find $C(100), C(400),$ and $C(480)$.

Solution Remember that a function is a rule. Here is how we apply the rule for this function. First we look at the value of the input x. If $0 \le x \le 400$, then the value of $C(x)$ is 39. On the other hand, if $x > 400$, then the value of $C(x)$ is $39 + 0.2(x - 400)$.

A piecewise-defined function is defined by different formulas on different parts of its domain. The function C of Example 3 is piecewise defined.

Since $100 \le 400$, we have $C(100) = 39$.

Since $400 \le 400$, we have $C(400) = 39$.

Since $480 > 400$, we have $C(480) = 39 + 0.2(480 - 400) = 55$.

Thus, the plan charges $39 for 100 minutes, $39 for 400 minutes, and $55 for 480 minutes.   ■

Expressions like the one in part (d) of Example 4 occur frequently in calculus; they are called *difference quotients*, and they represent the average change in the value of f between $x = a$ and $x = a + h$.

Example 4 Evaluating a Function

If $f(x) = 2x^2 + 3x - 1$, evaluate the following.

(a) $f(a)$ (b) $f(-a)$

(c) $f(a + h)$ (d) $\dfrac{f(a + h) - f(a)}{h}, \quad h \neq 0$

Solution

(a) $f(a) = 2a^2 + 3a - 1$

(b) $f(-a) = 2(-a)^2 + 3(-a) - 1 = 2a^2 - 3a - 1$

(c) $f(a + h) = 2(a + h)^2 + 3(a + h) - 1$

$$= 2(a^2 + 2ah + h^2) + 3(a + h) - 1$$

$$= 2a^2 + 4ah + 2h^2 + 3a + 3h - 1$$

(d) Using the results from parts (c) and (a), we have

$$\frac{f(a + h) - f(a)}{h} = \frac{(2a^2 + 4ah + 2h^2 + 3a + 3h - 1) - (2a^2 + 3a - 1)}{h}$$

$$= \frac{4ah + 2h^2 + 3h}{h} = 4a + 2h + 3 \qquad \blacksquare$$

The weight of an object on or near the earth is the gravitational force that the earth exerts on it. When in orbit around the earth, an astronaut experiences the sensation of "weightlessness" because the centripetal force that keeps her in orbit is exactly the same as the gravitational pull of the earth.

Example 5   The Weight of an Astronaut

If an astronaut weighs 130 pounds on the surface of the earth, then her weight when she is h miles above the earth is given by the function

$$w(h) = 130\left(\frac{3960}{3960 + h}\right)^2$$

(a) What is her weight when she is 100 mi above the earth?

(b) Construct a table of values for the function w that gives her weight at heights from 0 to 500 mi. What do you conclude from the table?

Solution

(a) We want the value of the function w when $h = 100$; that is, we must calculate $w(100)$.

$$w(100) = 130\left(\frac{3960}{3960 + 100}\right)^2 \approx 123.67$$

So at a height of 100 mi, she weighs about 124 lb.

(b) The table gives the astronaut's weight, rounded to the nearest pound, at 100-mile increments. The values in the table are calculated as in part (a).

h	$w(h)$
0	130
100	124
200	118
300	112
400	107
500	102

The table indicates that the higher the astronaut travels, the less she weighs. $\qquad \blacksquare$

The Domain of a Function

Recall that the *domain* of a function is the set of all inputs for the function. The domain of a function may be stated explicitly. For example, if we write

$$f(x) = x^2, \quad 0 \le x \le 5$$

then the domain is the set of all real numbers x for which $0 \le x \le 5$. If the function is given by an algebraic expression and the domain is not stated explicitly, then by convention *the domain of the function is the domain of the algebraic expression—that is, the set of all real numbers for which the expression is defined as a real number.* For example, consider the functions

Domains of algebraic expressions are discussed on page 35.

$$f(x) = \frac{1}{x - 4} \qquad\qquad g(x) = \sqrt{x}$$

The function f is not defined at $x = 4$, so its domain is $\{x \mid x \ne 4\}$. The function g is not defined for negative x, so its domain is $\{x \mid x \ne 0\}$.

Example 6 Finding Domains of Functions

Find the domain of each function.

(a) $f(x) = \dfrac{1}{x^2 - x}$ (b) $g(x) = \sqrt{9 - x^2}$ (c) $h(t) = \dfrac{t}{\sqrt{t + 1}}$

Solution

(a) The function is not defined when the denominator is 0. Since

$$f(x) = \frac{1}{x^2 - x} = \frac{1}{x(x - 1)}$$

we see that $f(x)$ is not defined when $x = 0$ or $x = 1$. Thus, the domain of f is

$$\{x \mid x \ne 0, x \ne 1\}$$

The domain may also be written in interval notation as

$$(\infty, 0) \cup (0, 1) \cup (1, \infty)$$

(b) We can't take the square root of a negative number, so we must have $9 - x^2 \ge 0$. Using the methods of Section 1.7, we can solve this inequality to find that $-3 \le x \le 3$. Thus, the domain of g is

$$\{x \mid -3 \le x \le 3\} = [-3, 3]$$

(c) We can't take the square root of a negative number, and we can't divide by 0, so we must have $t + 1 > 0$, that is, $t > -1$. So the domain of h is

$$\{t \mid t > -1\} = (-1, \infty) \qquad\qquad\qquad \blacksquare$$

Four Ways to Represent a Function

To help us understand what a function is, we have used machine and arrow diagrams. We can describe a specific function in the following four ways:

- verbally (by a description in words)
- algebraically (by an explicit formula)

- visually (by a graph)
- numerically (by a table of values)

A single function may be represented in all four ways, and it is often useful to go from one representation to another to gain insight into the function. However, certain functions are described more naturally by one method than by the others. An example of a verbal description is

$$P(t) \quad \text{is} \quad \text{``the population of the world at time } t\text{''}$$

The function P can also be described numerically by giving a table of values (see Table 1 on page 386). A useful representation of the area of a circle as a function of its radius is the algebraic formula

$$A(r) = \pi r^2$$

The graph produced by a seismograph (see the box) is a visual representation of the vertical acceleration function $a(t)$ of the ground during an earthquake. As a final example, consider the function $C(w)$, which is described verbally as "the cost of mailing a first-class letter with weight w." The most convenient way of describing this function is numerically—that is, using a table of values.

We will be using all four representations of functions throughout this book. We summarize them in the following box.

Four Ways to Represent a Function

Verbal

Using words:

$P(t)$ is "the population of the world at time t"

Relation of population P and time t

Algebraic

Using a formula:

$$A(r) = \pi r^2$$

Area of a circle

Visual

Using a graph:

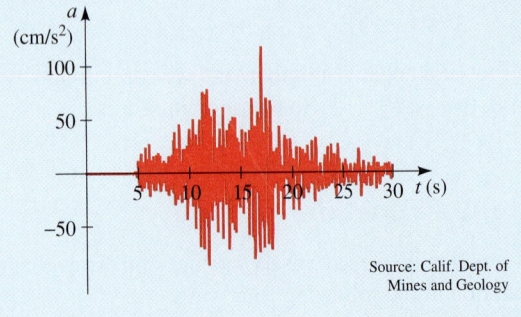

Source: Calif. Dept. of Mines and Geology

Vertical acceleration during an earthquake

Numerical

Using a table of values:

w (ounces)	$C(w)$ (dollars)
$0 < w \le 1$	0.37
$1 < w \le 2$	0.60
$2 < w \le 3$	0.83
$3 < w \le 4$	1.06
$4 < w \le 5$	1.29
$\vdots$	$\vdots$

Cost of mailing a first-class letter

2.1   Exercises

1–4 ■ Express the rule in function notation. (For example, the rule "square, then subtract 5" is expressed as the function $f(x) = x^2 - 5$.)

1. Add 3, then multiply by 2

2. Divide by 7, then subtract 4

3. Subtract 5, then square

4. Take the square root, add 8, then multiply by $\frac{1}{3}$

5–8 ■ Express the function (or rule) in words.

5. $f(x) = \dfrac{x - 4}{3}$

6. $g(x) = \dfrac{x}{3} - 4$

7. $h(x) = x^2 + 2$

8. $k(x) = \sqrt{x + 2}$

9–10 ■ Draw a machine diagram for the function.

9. $f(x) = \sqrt{x - 1}$

10. $f(x) = \dfrac{3}{x - 2}$

11–12 ■ Complete the table.

11. $f(x) = 2(x - 1)^2$

12. $g(x) = |2x + 3|$

x	$f(x)$
-1	
0	
1	
2	
3	

x	$g(x)$
-3	
-2	
0	
1	
3	

13–20 ■ Evaluate the function at the indicated values.

13. $f(x) = 2x + 1$;

$f(1), f(-2), f(\frac{1}{2}), f(a), f(-a), f(a + b)$

14. $f(x) = x^2 + 2x$;

$f(0), f(3), f(-3), f(a), f(-x), f\left(\dfrac{1}{a}\right)$

15. $g(x) = \dfrac{1 - x}{1 + x}$;

$g(2), g(-2), g(\frac{1}{2}), g(a), g(a - 1), g(-1)$

16. $h(t) = t + \dfrac{1}{t}$;

$h(1), h(-1), h(2), h(\frac{1}{2}), h(x), h\left(\dfrac{1}{x}\right)$

17. $f(x) = 2x^2 + 3x - 4$;

$f(0), f(2), f(-2), f(\sqrt{2}), f(x + 1), f(-x)$

18. $f(x) = x^3 - 4x^2$;

$f(0), f(1), f(-1), f(\frac{3}{2}), f\left(\dfrac{x}{2}\right), f(x^2)$

19. $f(x) = 2|x - 1|$;

$f(-2), f(0), f(\frac{1}{2}), f(2), f(x + 1), f(x^2 + 2)$

20. $f(x) = \dfrac{|x|}{x}$;

$f(-2), f(-1), f(0), f(5), f(x^2), f\left(\dfrac{1}{x}\right)$

21–24 ■ Evaluate the piecewise defined function at the indicated values.

21. $f(x) = \begin{cases} x^2 & \text{if } x < 0 \\ x + 1 & \text{if } x \geq 0 \end{cases}$

$f(-2), f(-1), f(0), f(1), f(2)$

22. $f(x) = \begin{cases} 5 & \text{if } x \leq 2 \\ 2x - 3 & \text{if } x > 2 \end{cases}$

$f(-3), f(0), f(2), f(3), f(5)$

23. $f(x) = \begin{cases} x^2 + 2x & \text{if } x \leq -1 \\ x & \text{if } -1 < x \leq 1 \\ -1 & \text{if } x > 1 \end{cases}$

$f(-4), f(-\frac{3}{2}), f(-1), f(0), f(25)$

24. $f(x) = \begin{cases} 3x & \text{if } x < 0 \\ x + 1 & \text{if } 0 \leq x \leq 2 \\ (x - 2)^2 & \text{if } x > 2 \end{cases}$

$f(-5), f(0), f(1), f(2), f(5)$

25–28 ■ Use the function to evaluate the indicated expressions and simplify.

25. $f(x) = x^2 + 1$; $f(x + 2), f(x) + f(2)$

26. $f(x) = 3x - 1$; $f(2x), 2f(x)$

27. $f(x) = x + 4$; $f(x^2), (f(x))^2$

28. $f(x) = 6x - 18$; $f\left(\dfrac{x}{3}\right), \dfrac{f(x)}{3}$

29–36 ■ Find $f(a)$, $f(a + h)$, and the difference quotient $\dfrac{f(a + h) - f(a)}{h}$, where $h \neq 0$.

29. $f(x) = 3x + 2$

30. $f(x) = x^2 + 1$

31. $f(x) = 5$

32. $f(x) = \dfrac{1}{x + 1}$

33. $f(x) = \dfrac{x}{x + 1}$

34. $f(x) = \dfrac{2x}{x - 1}$

35. $f(x) = 3 - 5x + 4x^2$

36. $f(x) = x^3$

37–58 ■ Find the domain of the function.

37. $f(x) = 2x$

38. $f(x) = x^2 + 1$

39. $f(x) = 2x, \quad -1 \le x \le 5$

40. $f(x) = x^2 + 1, \quad 0 \le x \le 5$

41. $f(x) = \dfrac{1}{x - 3}$

42. $f(x) = \dfrac{1}{3x - 6}$

43. $f(x) = \dfrac{x + 2}{x^2 - 1}$

44. $f(x) = \dfrac{x^4}{x^2 + x - 6}$

45. $f(x) = \sqrt{x - 5}$

46. $f(x) = \sqrt[4]{x + 9}$

47. $f(t) = \sqrt[3]{t - 1}$

48. $g(x) = \sqrt{7 - 3x}$

49. $h(x) = \sqrt{2x - 5}$

50. $G(x) = \sqrt{x^2 - 9}$

51. $g(x) = \dfrac{\sqrt{2 + x}}{3 - x}$

52. $g(x) = \dfrac{\sqrt{x}}{2x^2 + x - 1}$

53. $g(x) = \sqrt[4]{x^2 - 6x}$

54. $g(x) = \sqrt{x^2 - 2x - 8}$

55. $f(x) = \dfrac{3}{\sqrt{x - 4}}$

56. $f(x) = \dfrac{x^2}{\sqrt{6 - x}}$

57. $f(x) = \dfrac{(x + 1)^2}{\sqrt{2x - 1}}$

58. $f(x) = \dfrac{x}{\sqrt[4]{9 - x^2}}$

Applications

59. Production Cost   The cost C in dollars of producing x yards of a certain fabric is given by the function

$$C(x) = 1500 + 3x + 0.02x^2 + 0.0001x^3$$

   (a) Find $C(10)$ and $C(100)$.

   (b) What do your answers in part (a) represent?

   (c) Find $C(0)$. (This number represents the *fixed costs*.)

60. Area of a Sphere   The surface area S of a sphere is a function of its radius r given by

$$S(r) = 4\pi r^2$$

   (a) Find $S(2)$ and $S(3)$.

   (b) What do your answers in part (a) represent?

61. How Far Can You See?   Due to the curvature of the earth, the maximum distance D that you can see from the top of a tall building or from an airplane at height h is given by the function

$$D(h) = \sqrt{2rh + h^2}$$

where $r = 3960$ mi is the radius of the earth and D and h are measured in miles.

   (a) Find $D(0.1)$ and $D(0.2)$.

   (b) How far can you see from the observation deck of Toronto's CN Tower, 1135 ft above the ground?

   (c) Commercial aircraft fly at an altitude of about 7 mi. How far can the pilot see?

62. Torricelli's Law   A tank holds 50 gallons of water, which drains from a leak at the bottom, causing the tank to empty in 20 minutes. The tank drains faster when it is nearly full because the pressure on the leak is greater. **Torricelli's Law** gives the volume of water remaining in the tank after t minutes as

$$V(t) = 50\left(1 - \frac{t}{20}\right)^2 \quad 0 \le t \le 20$$

   (a) Find $V(0)$ and $V(20)$.

   (b) What do your answers to part (a) represent?

   (c) Make a table of values of $V(t)$ for $t = 0, 5, 10, 15, 20$.

63. Blood Flow   As blood moves through a vein or an artery, its velocity v is greatest along the central axis and decreases as the distance r from the central axis increases (see the figure). The formula that gives v as a function of r is called the **law of laminar flow**. For an artery with radius 0.5 cm, we have

$$v(r) = 18{,}500(0.25 - r^2) \quad 0 \le r \le 0.5$$

   (a) Find $v(0.1)$ and $v(0.4)$.

   (b) What do your answers to part (a) tell you about the flow of blood in this artery?

   (c) Make a table of values of $v(r)$ for $r = 0, 0.1, 0.2, 0.3, 0.4, 0.5$.

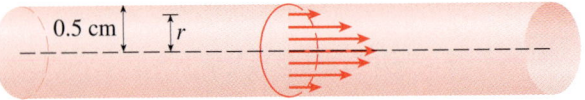

0.5 cm r

64. Pupil Size   When the brightness x of a light source is increased, the eye reacts by decreasing the radius R of the pupil. The dependence of R on x is given by the function

$$R(x) = \sqrt{\frac{13 + 7x^{0.4}}{1 + 4x^{0.4}}}$$

(a) Find $R(1)$, $R(10)$, and $R(100)$.

(b) Make a table of values of $R(x)$.

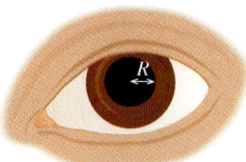

65. Relativity   According to the Theory of Relativity, the length L of an object is a function of its velocity v with respect to an observer. For an object whose length at rest is 10 m, the function is given by

$$L(v) = 10\sqrt{1 - \frac{v^2}{c^2}}$$

where c is the speed of light.

(a) Find $L(0.5c)$, $L(0.75c)$, and $L(0.9c)$.

(b) How does the length of an object change as its velocity increases?

66. Income Tax   In a certain country, income tax T is assessed according to the following function of income x:

$$T(x) = \begin{cases} 0 & \text{if } 0 \le x \le 10{,}000 \\ 0.08x & \text{if } 10{,}000 < x \le 20{,}000 \\ 1600 + 0.15x & \text{if } 20{,}000 < x \end{cases}$$

(a) Find $T(5{,}000)$, $T(12{,}000)$, and $T(25{,}000)$.

(b) What do your answers in part (a) represent?

67. Internet Purchases   An Internet bookstore charges $15 shipping for orders under $100, but provides free shipping for orders of $100 or more. The cost C of an order is a function of the total price x of the books purchased, given by

$$C(x) = \begin{cases} x + 15 & \text{if } x < 100 \\ x & \text{if } x \ge 100 \end{cases}$$

(a) Find $C(75)$, $C(90)$, $C(100)$, and $C(105)$.

(b) What do your answers in part (a) represent?

68. Cost of a Hotel Stay   A hotel chain charges $75 each night for the first two nights and $50 for each additional night's stay. The total cost T is a function of the number of nights x that a guest stays.

(a) Complete the expressions in the following piecewise defined function.

$$T(x) = \begin{cases} \rule{1.5cm}{0pt} & \text{if } 0 \le x \le 2 \\ \rule{1.5cm}{0pt} & \text{if } x > 2 \end{cases}$$

(b) Find $T(2)$, $T(3)$, and $T(5)$.

(c) What do your answers in part (b) represent?

69. Speeding Tickets   In a certain state the maximum speed permitted on freeways is 65 mi/h and the minimum is 40. The fine F for violating these limits is $15 for every mile above the maximum or below the minimum.

(a) Complete the expressions in the following piecewise defined function, where x is the speed at which you are driving.

$$F(x) = \begin{cases} \rule{1.5cm}{0pt} & \text{if } 0 < x < 40 \\ \rule{1.5cm}{0pt} & \text{if } 40 \le x \le 65 \\ \rule{1.5cm}{0pt} & \text{if } x > 65 \end{cases}$$

(b) Find $F(30)$, $F(50)$, and $F(75)$.

(c) What do your answers in part (b) represent?

70. Height of Grass   A home owner mows the lawn every Wednesday afternoon. Sketch a rough graph of the height of the grass as a function of time over the course of a four-week period beginning on a Sunday.

71. Temperature Change   You place a frozen pie in an oven and bake it for an hour. Then you take it out and let it cool before eating it. Sketch a rough graph of the temperature of the pie as a function of time.

72. Daily Temperature Change   Temperature readings T (in °F) were recorded every 2 hours from midnight to noon in Atlanta, Georgia, on March 18, 1996. The time t was measured in hours from midnight. Sketch a rough graph of T as a function of t.

t	T
0	58
2	57
4	53
6	50
8	51
10	57
12	61

73. Population Growth   The population P (in thousands) of San Jose, California, from 1988 to 2000 is shown in the table. (Midyear estimates are given.) Draw a rough graph of P as a function of time t.

t	P
1988	733
1990	782
1992	800
1994	817
1996	838
1998	861
2000	895

Discovery • Discussion

74. Examples of Functions   At the beginning of this section we discussed three examples of everyday, ordinary functions: Height is a function of age, temperature is a function of date, and postage cost is a function of weight. Give three other examples of functions from everyday life.

75. Four Ways to Represent a Function   In the box on page 154 we represented four different functions verbally, algebraically, visually, and numerically. Think of a function that can be represented in all four ways, and write the four representations.

2.2   # Graphs of Functions

The most important way to visualize a function is through its graph. In this section we investigate in more detail the concept of graphing functions.

Graphing Functions

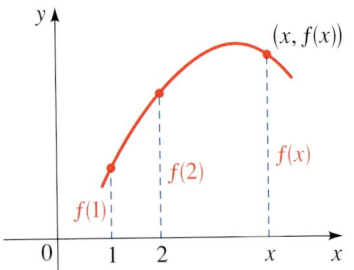

Figure 1
The height of the graph above the point x is the value of $f(x)$.

The Graph of a Function

If f is a function with domain A, then the **graph** of f is the set of ordered pairs

$$\{(x, f(x)) \mid x \in A\}$$

In other words, the graph of f is the set of all points (x, y) such that $y = f(x)$; that is, the graph of f is the graph of the equation $y = f(x)$.

The graph of a function f gives a picture of the behavior or "life history" of the function. We can read the value of $f(x)$ from the graph as being the height of the graph above the point x (see Figure 1).

A function f of the form $f(x) = mx + b$ is called a **linear function** because its graph is the graph of the equation $y = mx + b$, which represents a line with slope m and y-intercept b. A special case of a linear function occurs when the slope is $m = 0$. The function $f(x) = b$, where b is a given number, is called a **constant function** because all its values are the same number, namely, b. Its graph is the horizontal line $y = b$. Figure 2 shows the graphs of the constant function $f(x) = 3$ and the linear function $f(x) = 2x + 1$.

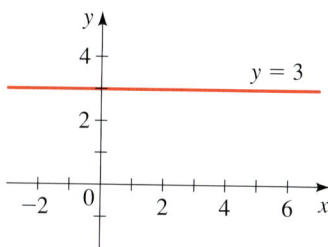

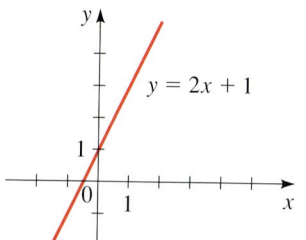

Figure 2   The constant function $f(x) = 3$ The linear function $f(x) = 2x + 1$

Example 1   Graphing Functions

Sketch the graphs of the following functions.

(a) $f(x) = x^2$ (b) $g(x) = x^3$ (c) $h(x) = \sqrt{x}$

Solution   We first make a table of values. Then we plot the points given by the table and join them by a smooth curve to obtain the graph. The graphs are sketched in Figure 3.

x	$f(x) = x^2$
0	0
$\pm\frac{1}{2}$	$\frac{1}{4}$
± 1	1
± 2	4
± 3	9

x	$g(x) = x^3$
0	0
$\frac{1}{2}$	$\frac{1}{8}$
1	1
2	8
$-\frac{1}{2}$	$-\frac{1}{8}$
-1	-1
-2	-8

x	$h(x) = \sqrt{x}$
0	0
1	1
2	$\sqrt{2}$
3	$\sqrt{3}$
4	2
5	$\sqrt{5}$

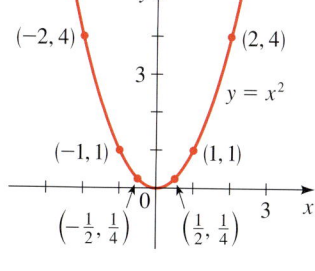

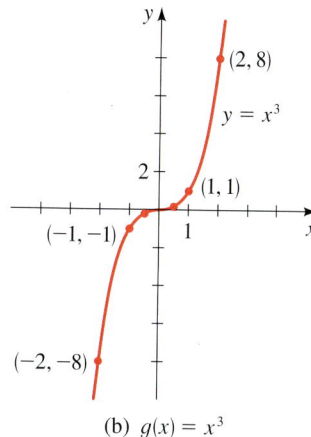

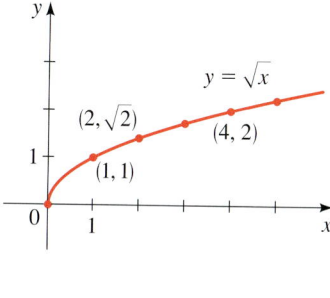

Figure 3   (a) $f(x) = x^2$ (b) $g(x) = x^3$ (c) $h(x) = \sqrt{x}$ ■

A convenient way to graph a function is to use a graphing calculator, as in the next example.

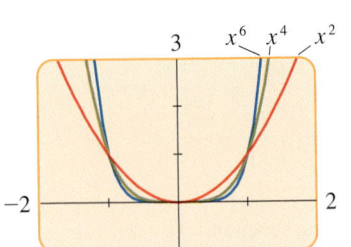

(a) Even powers of x

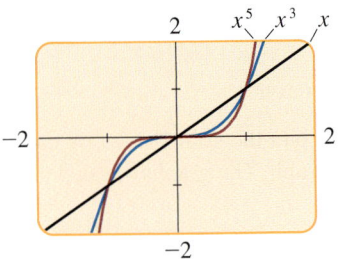

(b) Odd powers of x

Figure 4

A family of power functions $f(x) = x^n$

Example 2   A Family of Power Functions

(a) Graph the functions $f(x) = x^n$ for $n = 2, 4$, and 6 in the viewing rectangle $[-2, 2]$ by $[-1, 3]$.

(b) Graph the functions $f(x) = x^n$ for $n = 1, 3$, and 5 in the viewing rectangle $[-2, 2]$ by $[-2, 2]$.

(c) What conclusions can you draw from these graphs?

Solution   The graphs for parts (a) and (b) are shown in Figure 4.

(c) We see that the general shape of the graph of $f(x) = x^n$ depends on whether n is even or odd.

> If n is even, the graph of $f(x) = x^n$ is similar to the parabola $y = x^2$.
> If n is odd, the graph of $f(x) = x^n$ is similar to that of $y = x^3$. ■

Notice from Figure 4 that as n increases the graph of $y = x^n$ becomes flatter near 0 and steeper when $x > 1$. When $0 < x < 1$, the lower powers of x are the "bigger" functions. But when $x > 1$, the higher powers of x are the dominant functions.

Getting Information from the Graph of a Function

The values of a function are represented by the height of its graph above the x-axis. So, we can read off the values of a function from its graph.

Example 3   Find the Values of a Function from a Graph

The function T graphed in Figure 5 gives the temperature between noon and 6 P.M. at a certain weather station.

(a) Find $T(1)$, $T(3)$, and $T(5)$.

(b) Which is larger, $T(2)$ or $T(4)$?

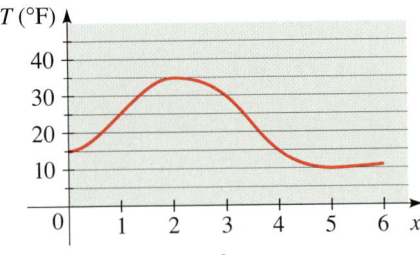

Figure 5

Temperature function

Hours from noon

Solution

(a) $T(1)$ is the temperature at 1:00 P.M. It is represented by the height of the graph above the x-axis at $x = 1$. Thus, $T(1) = 25$. Similarly, $T(3) = 30$ and $T(5) = 10$.

(b) Since the graph is higher at $x = 2$ than at $x = 4$, it follows that $T(2)$ is larger than $T(4)$. ■

The graph of a function helps us picture the domain and range of the function on the x-axis and y-axis as shown in Figure 6.

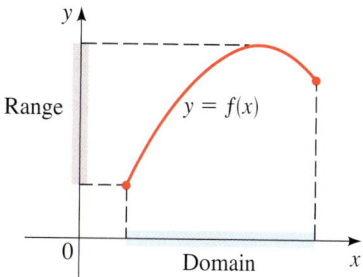

Figure 6

Domain and range of f

Example 4 **Finding the Domain and Range from a Graph**

(a) Use a graphing calculator to draw the graph of $f(x) = \sqrt{4 - x^2}$.

(b) Find the domain and range of f.

Solution

(a) The graph is shown in Figure 7.

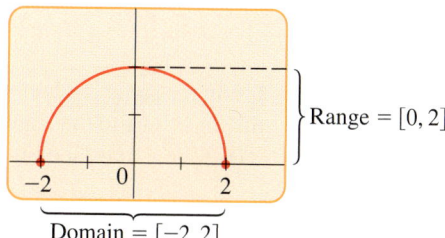

Figure 7

Graph of $f(x) = \sqrt{4 - x^2}$

(b) From the graph in Figure 7 we see that the domain is $[-2, 2]$ and the range is $[0, 2]$. ∎

Graphing Piecewise Defined Functions

A piecewise defined function is defined by different formulas on different parts of its domain. As you might expect, the graph of such a function consists of separate pieces.

Example 5 **Graph of a Piecewise Defined Function**

Sketch the graph of the function.

$$f(x) = \begin{cases} x^2 & \text{if } x \le 1 \\ 2x + 1 & \text{if } x > 1 \end{cases}$$

Solution If $x \le 1$, then $f(x) = x^2$, so the part of the graph to the left of $x = 1$ coincides with the graph of $y = x^2$, which we sketched in Figure 3. If $x > 1$, then $f(x) = 2x + 1$, so the part of the graph to the right of $x = 1$ coincides with the

On many graphing calculators the graph in Figure 8 can be produced by using the logical functions in the calculator. For example, on the TI-83 the following equation gives the required graph:

$$Y_1 = (X \le 1)X^2 + (X > 1)(2X + 1)$$

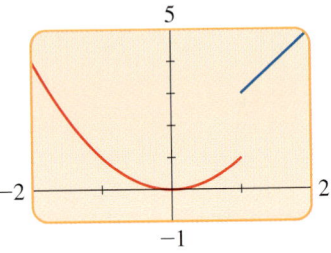

(To avoid the extraneous vertical line between the two parts of the graph, put the calculator in **Dot** mode.)

line $y = 2x + 1$, which we graphed in Figure 2. This enables us to sketch the graph in Figure 8.

The solid dot at $(1, 1)$ indicates that this point is included in the graph; the open dot at $(1, 3)$ indicates that this point is excluded from the graph.

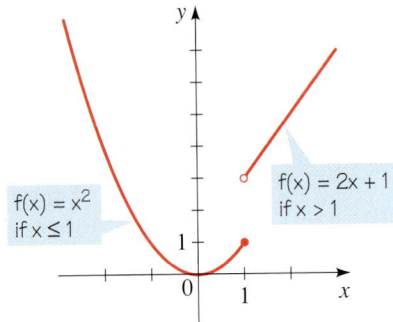

Figure 8

$$f(x) = \begin{cases} x^2 & \text{if } x \le 1 \\ 2x + 1 & \text{if } x > 1 \end{cases}$$

■

Example 6   Graph of the Absolute Value Function

Sketch the graph of the absolute value function $f(x) = |x|$.

Solution   Recall that

$$|x| = \begin{cases} x & \text{if } x \ge 0 \\ -x & \text{if } x < 0 \end{cases}$$

Using the same method as in Example 5, we note that the graph of f coincides with the line $y = x$ to the right of the y-axis and coincides with the line $y = -x$ to the left of the y-axis (see Figure 9).

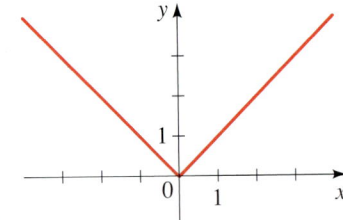

Figure 9
Graph of $f(x) = |x|$

■

The **greatest integer function** is defined by

$$[\![x]\!] = \text{greatest integer less than or equal to } x$$

For example, $[\![2]\!] = 2$, $[\![2.3]\!] = 2$, $[\![1.999]\!] = 1$, $[\![0.002]\!] = 0$, $[\![-3.5]\!] = -4$, $[\![-0.5]\!] = -1$.

Example 7   Graph of the Greatest Integer Function

Sketch the graph of $f(x) = [\![x]\!]$.

Solution   The table shows the values of f for some values of x. Note that $f(x)$ is constant between consecutive integers so the graph between integers is a horizontal line segment as shown in Figure 10.

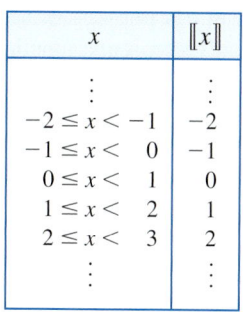

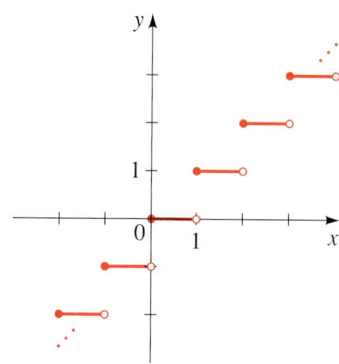

Figure 10
The greatest integer function, $y = [\![x]\!]$

The greatest integer function is an example of a **step function**. The next example gives a real-world example of a step function.

Example 8   **The Cost Function for Long-Distance Phone Calls**

The cost of a long-distance daytime phone call from Toronto to Mumbai, India, is 69 cents for the first minute and 58 cents for each additional minute (or part of a minute). Draw the graph of the cost C (in dollars) of the phone call as a function of time t (in minutes).

Solution   Let $C(t)$ be the cost for t minutes. Since $t > 0$, the domain of the function is $(0, \infty)$. From the given information, we have

$$C(t) = 0.69 \qquad\qquad\qquad \text{if } 0 < t \le 1$$
$$C(t) = 0.69 + 0.58 = 1.27 \qquad \text{if } 1 < t \le 2$$
$$C(t) = 0.69 + 2(0.58) = 1.85 \qquad \text{if } 2 < t \le 3$$
$$C(t) = 0.69 + 3(0.58) = 2.43 \qquad \text{if } 3 < t \le 4$$

and so on. The graph is shown in Figure 11.

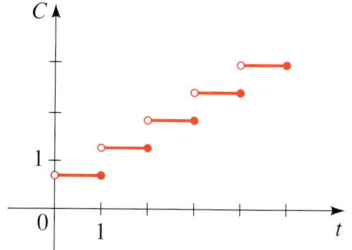

Figure 11
Cost of a long-distance call

The Vertical Line Test

The graph of a function is a curve in the xy-plane. But the question arises: Which curves in the xy-plane are graphs of functions? This is answered by the following test.

The Vertical Line Test

A curve in the coordinate plane is the graph of a function if and only if no vertical line intersects the curve more than once.

We can see from Figure 12 why the Vertical Line Test is true. If each vertical line $x = a$ intersects a curve only once at (a, b), then exactly one functional value is defined by $f(a) = b$. But if a line $x = a$ intersects the curve twice, at (a, b) and at (a, c), then the curve can't represent a function because a function cannot assign two different values to a.

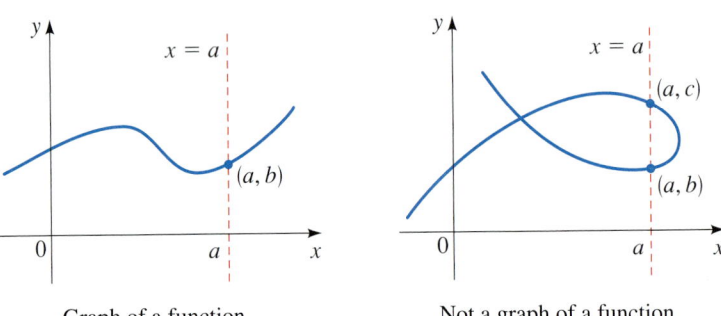

Graph of a function   Not a graph of a function

Figure 12
Vertical Line Test

Example 9   Using the Vertical Line Test

Using the Vertical Line Test, we see that the curves in parts (b) and (c) of Figure 13 represent functions, whereas those in parts (a) and (d) do not.

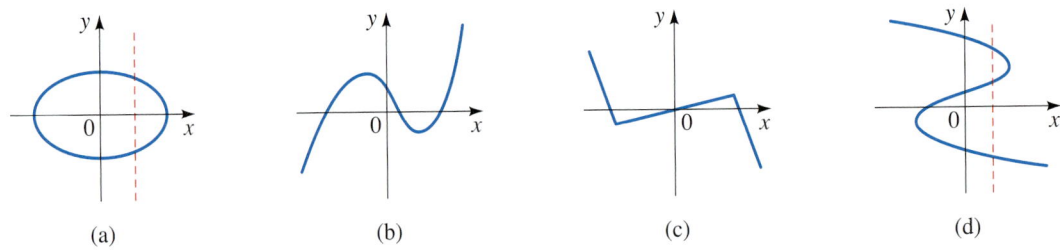

(a)   (b)   (c)   (d)

Figure 13

Equations That Define Functions

Any equation in the variables x and y defines a relationship between these variables. For example, the equation

$$y - x^2 = 0$$

defines a relationship between y and x. Does this equation define y as a *function* of x? To find out, we solve for y and get

$$y = x^2$$

We see that the equation defines a rule, or function, that gives one value of y for each

Donald Knuth was born in Milwaukee in 1938 and is Professor Emeritus of Computer Science at Stanford University. While still a graduate student at Caltech, he started writing a monumental series of books entitled *The Art of Computer Programming*. President Carter awarded him the National Medal of Science in 1979. When Knuth was a high school student, he became fascinated with graphs of functions and laboriously drew many hundreds of them because he wanted to see the behavior of a great variety of functions. (Today, of course, it is far easier to use computers and graphing calculators to do this.) Knuth is famous for his invention of TEX, a system of computer-assisted typesetting. This system was used in the preparation of the manuscript for this textbook. He has also written a novel entitled *Surreal Numbers: How Two Ex-Students Turned On to Pure Mathematics and Found Total Happiness*.

Dr. Knuth has received numerous honors, among them election as an associate of the French Academy of Sciences, and as a Fellow of the Royal Society.

value of x. We can express this rule in function notation as

$$f(x) = x^2$$

But not every equation defines y as a function of x, as the following example shows.

Example 10   Equations That Define Functions

Does the equation define y as a function of x?

(a) $y - x^2 = 2$

(b) $x^2 + y^2 = 4$

Solution

(a) Solving for y in terms of x gives

$$y - x^2 = 2$$
$$y = x^2 + 2 \qquad \text{Add } x^2$$

The last equation is a rule that gives one value of y for each value of x, so it defines y as a function of x. We can write the function as $f(x) = x^2 + 2$.

(b) We try to solve for y in terms of x:

$$x^2 + y^2 = 4$$
$$y^2 = 4 - x^2 \qquad \text{Subtract } x^2$$
$$y = \pm\sqrt{4 - x^2} \qquad \text{Take square roots}$$

The last equation gives two values of y for a given value of x. Thus, the equation does not define y as a function of x. ∎

The graphs of the equations in Example 10 are shown in Figure 14. The Vertical Line Test shows graphically that the equation in Example 10(a) defines a function but the equation in Example 10(b) does not.

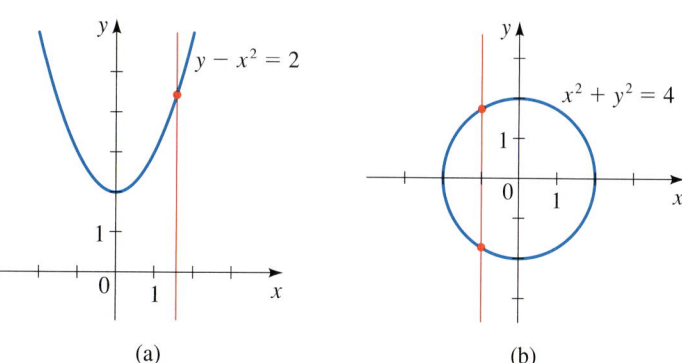

Figure 14   (a)   (b)

The following table shows the graphs of some functions that you will see frequently in this book.

Some Functions and Their Graphs

Linear functions
$f(x) = mx + b$

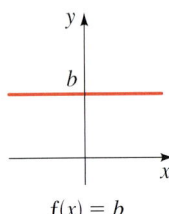

$f(x) = b$

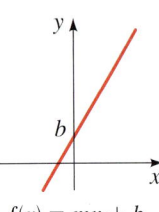

$f(x) = mx + b$

Power functions
$f(x) = x^n$

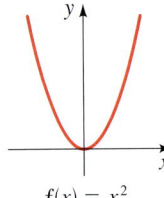

$f(x) = x^2$

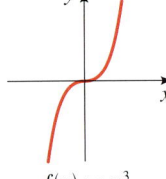

$f(x) = x^3$

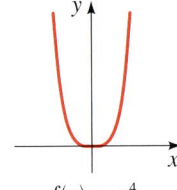

$f(x) = x^4$

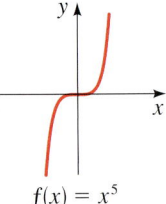

$f(x) = x^5$

Root functions
$f(x) = \sqrt[n]{x}$

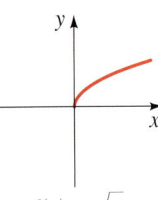

$f(x) = \sqrt{x}$

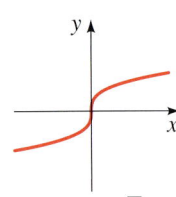

$f(x) = \sqrt[3]{x}$

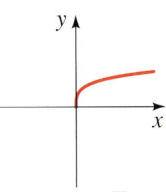

$f(x) = \sqrt[4]{x}$

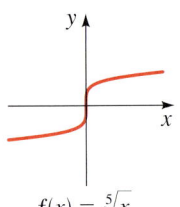

$f(x) = \sqrt[5]{x}$

Reciprocal functions
$f(x) = 1/x^n$

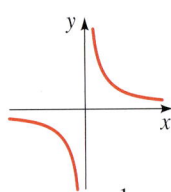

$f(x) = \dfrac{1}{x}$

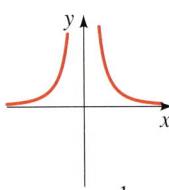

$f(x) = \dfrac{1}{x^2}$

Absolute value function
$f(x) = |x|$

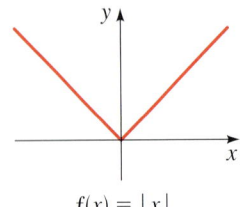

$f(x) = |x|$

Greatest integer function
$f(x) = [\![x]\!]$

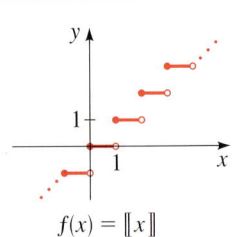

$f(x) = [\![x]\!]$

2.2   Exercises

1–22 ■ Sketch the graph of the function by first making a table of values.

1. $f(x) = 2$
2. $f(x) = -3$

3. $f(x) = 2x - 4$
4. $f(x) = 6 - 3x$

5. $f(x) = -x + 3$, $-3 \le x \le 3$

6. $f(x) = \dfrac{x-3}{2}$, $0 \le x \le 5$

7. $f(x) = -x^2$
8. $f(x) = x^2 - 4$

9. $g(x) = x^3 - 8$
10. $g(x) = 4x^2 - x^4$

11. $g(x) = \sqrt{x+4}$
12. $g(x) = \sqrt{-x}$

13. $F(x) = \dfrac{1}{x}$
14. $F(x) = \dfrac{1}{x+4}$

15. $H(x) = |2x|$
16. $H(x) = |x+1|$

17. $G(x) = |x| + x$
18. $G(x) = |x| - x$

19. $f(x) = |2x - 2|$
20. $f(x) = \dfrac{x}{|x|}$

21. $g(x) = \dfrac{2}{x^2}$
22. $g(x) = \dfrac{|x|}{x^2}$

23. The graph of a function h is given.
  (a) Find $h(-2)$, $h(0)$, $h(2)$, and $h(3)$.
  (b) Find the domain and range of h.

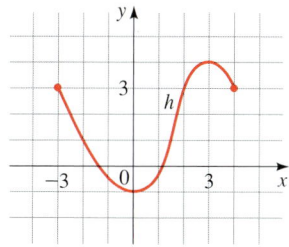

24. The graph of a function g is given.
  (a) Find $g(-4)$, $g(-2)$, $g(0)$, $g(2)$, and $g(4)$.
  (b) Find the domain and range of g.

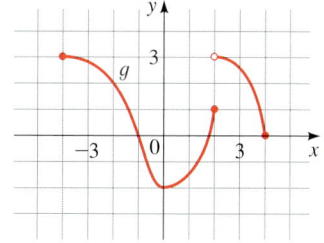

25. Graphs of the functions f and g are given.
  (a) Which is larger, $f(0)$ or $g(0)$?
  (b) Which is larger, $f(-3)$ or $g(-3)$?
  (c) For which values of x is $f(x) = g(x)$?

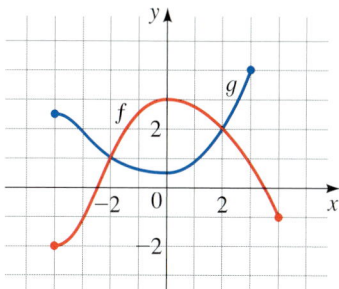

26. The graph of a function f is given.
  (a) Estimate $f(0.5)$ to the nearest tenth.
  (b) Estimate $f(3)$ to the nearest tenth.
  (c) Find all the numbers x in the domain of f for which $f(x) = 1$.

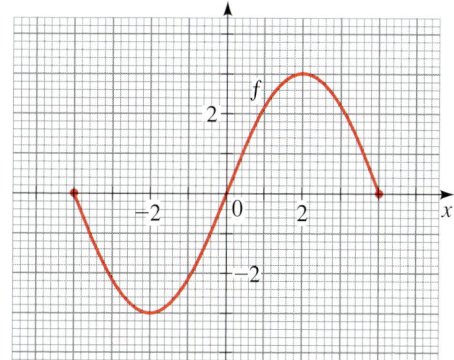

27–36 ■ A function f is given.
  (a) Use a graphing calculator to draw the graph of f.
  (b) Find the domain and range of f from the graph.

27. $f(x) = x - 1$
28. $f(x) = 2(x + 1)$

29. $f(x) = 4$
30. $f(x) = -x^2$

31. $f(x) = 4 - x^2$
32. $f(x) = x^2 + 4$

33. $f(x) = \sqrt{16 - x^2}$
34. $f(x) = -\sqrt{25 - x^2}$

35. $f(x) = \sqrt{x - 1}$
36. $f(x) = \sqrt{x + 2}$

37–50 ■ Sketch the graph of the piecewise defined function.

37. $f(x) = \begin{cases} 0 & \text{if } x < 2 \\ 1 & \text{if } x \ge 2 \end{cases}$

38. $f(x) = \begin{cases} 1 & \text{if } x \le 1 \\ x + 1 & \text{if } x > 1 \end{cases}$

39. $f(x) = \begin{cases} 3 & \text{if } x < 2 \\ x - 1 & \text{if } x \ge 2 \end{cases}$

40. $f(x) = \begin{cases} 1 - x & \text{if } x < -2 \\ 5 & \text{if } x \ge -2 \end{cases}$

41. $f(x) = \begin{cases} x & \text{if } x \le 0 \\ x + 1 & \text{if } x > 0 \end{cases}$

42. $f(x) = \begin{cases} 2x + 3 & \text{if } x < -1 \\ 3 - x & \text{if } x \ge -1 \end{cases}$

43. $f(x) = \begin{cases} -1 & \text{if } x < -1 \\ 1 & \text{if } -1 \le x \le 1 \\ -1 & \text{if } x > 1 \end{cases}$

44. $f(x) = \begin{cases} -1 & \text{if } x < -1 \\ x & \text{if } -1 \le x \le 1 \\ 1 & \text{if } x > 1 \end{cases}$

45. $f(x) = \begin{cases} 2 & \text{if } x \le -1 \\ x^2 & \text{if } x > -1 \end{cases}$

46. $f(x) = \begin{cases} 1 - x^2 & \text{if } x \le 2 \\ x & \text{if } x > 2 \end{cases}$

47. $f(x) = \begin{cases} 0 & \text{if } |x| \le 2 \\ 3 & \text{if } |x| > 2 \end{cases}$

48. $f(x) = \begin{cases} x^2 & \text{if } |x| \le 1 \\ 1 & \text{if } |x| > 1 \end{cases}$

49. $f(x) = \begin{cases} 4 & \text{if } x < -2 \\ x^2 & \text{if } -2 \le x \le 2 \\ -x + 6 & \text{if } x > 2 \end{cases}$

50. $f(x) = \begin{cases} -x & \text{if } x \le 0 \\ 9 - x^2 & \text{if } 0 < x \le 3 \\ x - 3 & \text{if } x > 3 \end{cases}$

51–52 ■ Use a graphing device to draw the graph of the piecewise defined function. (See the margin note on page 162.)

51. $f(x) = \begin{cases} x + 2 & \text{if } x \le -1 \\ x^2 & \text{if } x > -1 \end{cases}$

52. $f(x) = \begin{cases} 2x - x^2 & \text{if } x > 1 \\ (x - 1)^3 & \text{if } x \le 1 \end{cases}$

53–54 ■ The graph of a piecewise defined function is given. Find a formula for the function in the indicated form.

53.

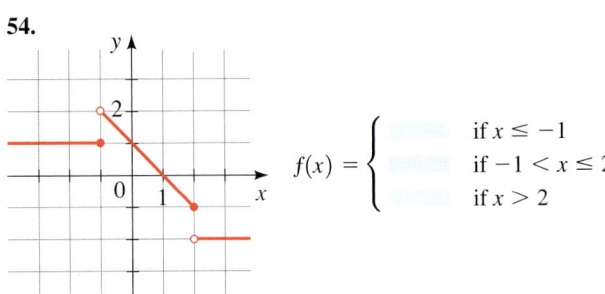

$$f(x) = \begin{cases} \rule{1cm}{0.4pt} & \text{if } x < -2 \\ \rule{1cm}{0.4pt} & \text{if } -2 \le x \le 2 \\ \rule{1cm}{0.4pt} & \text{if } x > 2 \end{cases}$$

54.

$$f(x) = \begin{cases} \rule{1cm}{0.4pt} & \text{if } x \le -1 \\ \rule{1cm}{0.4pt} & \text{if } -1 < x \le 2 \\ \rule{1cm}{0.4pt} & \text{if } x > 2 \end{cases}$$

55–56 ■ Determine whether the curve is the graph of a function of x.

55. (a)   (b)

(c)   (d)

56. (a)   (b)

(c)   (d)

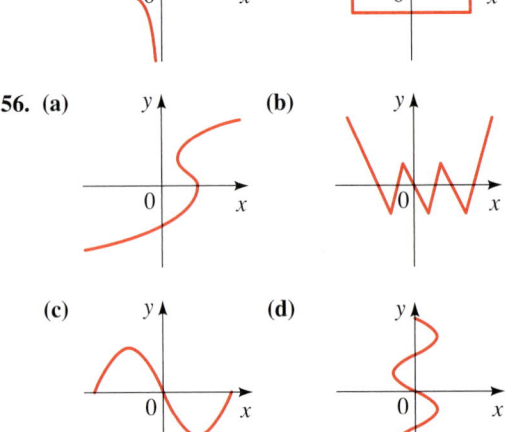

57–60 ■ Determine whether the curve is the graph of a function *x*. If it is, state the domain and range of the function.

57.

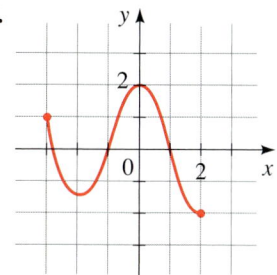

58.

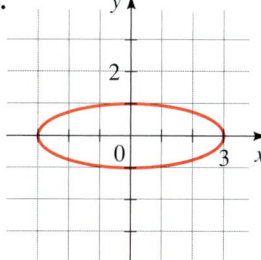

59.

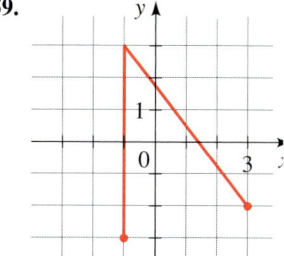

60.

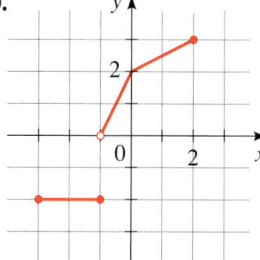

61–72 ■ Determine whether the equation defines *y* as a function of *x*. (See Example 10.)

61. $x^2 + 2y = 4$

62. $3x + 7y = 21$

63. $x = y^2$

64. $x^2 + (y-1)^2 = 4$

65. $x + y^2 = 9$

66. $x^2 + y = 9$

67. $x^2 y + y = 1$

68. $\sqrt{x} + y = 12$

69. $2|x| + y = 0$

70. $2x + |y| = 0$

71. $x = y^3$

72. $x = y^4$

73–78 ■ A family of functions is given. In parts (a) and (b) graph all the given members of the family in the viewing rectangle indicated. In part (c) state the conclusions you can make from your graphs.

73. $f(x) = x^2 + c$
(a) $c = 0, 2, 4, 6;\quad [-5, 5]$ by $[-10, 10]$
(b) $c = 0, -2, -4, -6;\quad [-5, 5]$ by $[-10, 10]$
(c) How does the value of *c* affect the graph?

74. $f(x) = (x - c)^2$
(a) $c = 0, 1, 2, 3;\quad [-5, 5]$ by $[-10, 10]$
(b) $c = 0, -1, -2, -3;\quad [-5, 5]$ by $[-10, 10]$
(c) How does the value of *c* affect the graph?

75. $f(x) = (x - c)^3$
(a) $c = 0, 2, 4, 6;\quad [-10, 10]$ by $[-10, 10]$
(b) $c = 0, -2, -4, -6;\quad [-10, 10]$ by $[-10, 10]$
(c) How does the value of *c* affect the graph?

76. $f(x) = cx^2$
(a) $c = 1, \frac{1}{2}, 2, 4;\quad [-5, 5]$ by $[-10, 10]$
(b) $c = 1, -1, -\frac{1}{2}, -2;\quad [-5, 5]$ by $[-10, 10]$
(c) How does the value of *c* affect the graph?

77. $f(x) = x^c$
(a) $c = \frac{1}{2}, \frac{1}{4}, \frac{1}{6};\quad [-1, 4]$ by $[-1, 3]$
(b) $c = 1, \frac{1}{3}, \frac{1}{5};\quad [-3, 3]$ by $[-2, 2]$
(c) How does the value of *c* affect the graph?

78. $f(x) = 1/x^n$
(a) $n = 1, 3;\quad [-3, 3]$ by $[-3, 3]$
(b) $n = 2, 4;\quad [-3, 3]$ by $[-3, 3]$
(c) How does the value of *n* affect the graph?

79–82 ■ Find a function whose graph is the given curve.

79. The line segment joining the points $(-2, 1)$ and $(4, -6)$

80. The line segment joining the points $(-3, -2)$ and $(6, 3)$

81. The top half of the circle $x^2 + y^2 = 9$

82. The bottom half of the circle $x^2 + y^2 = 9$

Applications

83. Weight Function The graph gives the weight of a certain person as a function of age. Describe in words how this person's weight has varied over time. What do you think happened when this person was 30 years old?

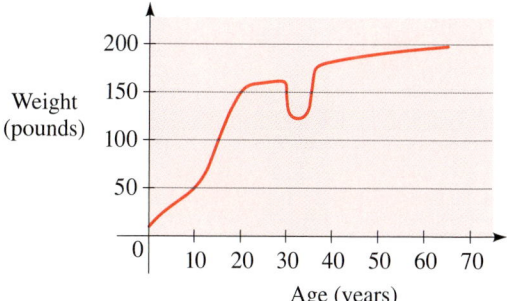

84. Distance Function The graph gives a salesman's distance from his home as a function of time on a certain day. Describe in words what the graph indicates about his travels on this day.

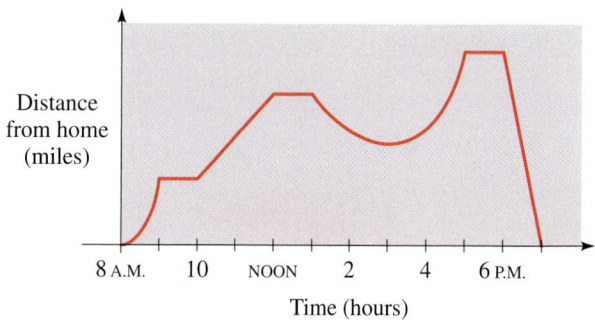

85. Hurdle Race   Three runners compete in a 100-meter hurdle race. The graph depicts the distance run as a function of time for each runner. Describe in words what the graph tells you about this race. Who won the race? Did each runner finish the race? What do you think happened to runner B?

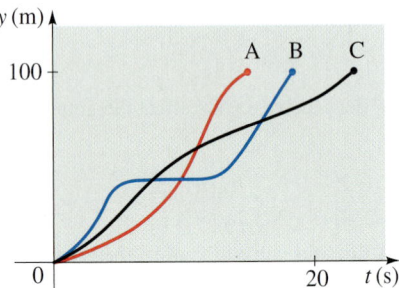

86. Power Consumption   The figure shows the power consumption in San Francisco for September 19, 1996 (P is measured in megawatts; t is measured in hours starting at midnight).

(a) What was the power consumption at 6 A.M.? At 6 P.M.?

(b) When was the power consumption the lowest?

(c) When was the power consumption the highest?

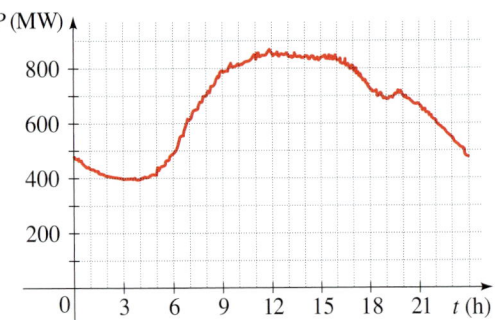

Source: Pacific Gas & Electric

87. Earthquake   The graph shows the vertical acceleration of the ground from the 1994 Northridge earthquake in Los Angeles, as measured by a seismograph. (Here t represents the time in seconds.)

(a) At what time t did the earthquake first make noticeable movements of the earth?

(b) At what time t did the earthquake seem to end?

(c) At what time t was the maximum intensity of the earthquake reached?

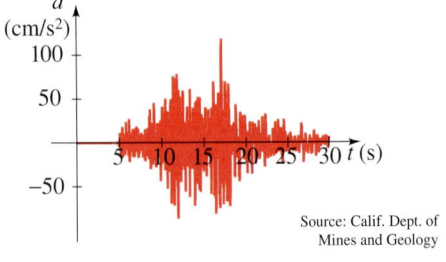

Source: Calif. Dept. of Mines and Geology

88. Utility Rates   Westside Energy charges its electric customers a base rate of $6.00 per month, plus 10¢ per kilowatt-hour (kWh) for the first 300 kWh used and 6¢ per kWh for all usage over 300 kWh. Suppose a customer uses x kWh of electricity in one month.

(a) Express the monthly cost E as a function of x.

(b) Graph the function E for $0 \le x \le 600$.

89. Taxicab Function   A taxi company charges $2.00 for the first mile (or part of a mile) and 20 cents for each succeeding tenth of a mile (or part). Express the cost C (in dollars) of a ride as a function of the distance x traveled (in miles) for $0 < x < 2$, and sketch the graph of this function.

90. Postage Rates   The domestic postage rate for first-class letters weighing 12 oz or less is 37 cents for the first ounce (or less), plus 23 cents for each additional ounce (or part of an ounce). Express the postage P as a function of the weight x of a letter, with $0 < x \le 12$, and sketch the graph of this function.

Discovery • Discussion

91. When Does a Graph Represent a Function?   For every integer n, the graph of the equation $y = x^n$ is the graph of a function, namely $f(x) = x^n$. Explain why the graph of $x = y^2$ is *not* the graph of a function of x. Is the graph of $x = y^3$ the graph of a function of x? If so, of what function of x is it the graph? Determine for what integers n the graph of $x = y^n$ is the graph of a function of x.

92. Step Functions   In Example 8 and Exercises 89 and 90 we are given functions whose graphs consist of horizontal line segments. Such functions are often called *step functions*, because their graphs look like stairs. Give some other examples of step functions that arise in everyday life.

93. Stretched Step Functions   Sketch graphs of the functions $f(x) = [\![x]\!]$, $g(x) = [\![2x]\!]$, and $h(x) = [\![3x]\!]$ on separate graphs. How are the graphs related? If n is a positive integer, what does the graph of $k(x) = [\![nx]\!]$ look like?

94. Graph of the Absolute Value of a Function

(a) Draw the graphs of the functions $f(x) = x^2 + x - 6$ and $g(x) = |x^2 + x - 6|$. How are the graphs of f and g related?

(b) Draw the graphs of the functions $f(x) = x^4 - 6x^2$ and $g(x) = |x^4 - 6x^2|$. How are the graphs of f and g related?

(c) In general, if $g(x) = |f(x)|$, how are the graphs of f and g related? Draw graphs to illustrate your answer.

DISCOVERY PROJECT

Relations and Functions

A function f can be represented as a set of ordered pairs (x, y) where x is the input and $y = f(x)$ is the output. For example, the function that squares each natural number can be represented by the ordered pairs $\{(1, 1), (2, 4), (3, 9), \ldots\}$.

A **relation** is *any* collection of ordered pairs. If we denote the ordered pairs in a relation by (x, y) then the set of x-values (or inputs) is the **domain** and the set of y-values (or outputs) is the **range**. With this terminology a **function** is a relation where for each x-value there is *exactly one* y-value (or for each input there is *exactly one* output). The correspondences in the figure below are relations—the first is a function but the second is not because the input 7 in A corresponds to two different outputs, 15 and 17, in B.

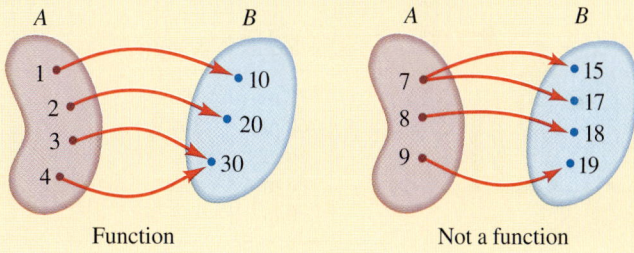

Function   Not a function

We can describe a relation by listing all the ordered pairs in the relation or giving the rule of correspondence. Also, since a relation consists of ordered pairs we can sketch its graph. Let's consider the following relations and try to decide which are functions.

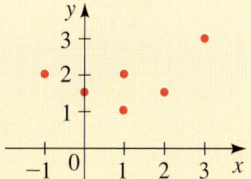

(a) The relation that consists of the ordered pairs $\{(1, 1), (2, 3), (3, 3), (4, 2)\}$.

(b) The relation that consists of the ordered pairs $\{(1, 2), (1, 3), (2, 4), (3, 2)\}$.

(c) The relation whose graph is shown to the left.

(d) The relation whose input values are days in January 2005 and whose output values are the maximum temperature in Los Angeles on that day.

(e) The relation whose input values are days in January 2005 and whose output values are the persons born in Los Angeles on that day.

The relation in part (a) is a function because each input corresponds to exactly one output. But the relation in part (b) is not, because the input 1 corresponds to two different outputs (2 and 3). The relation in part (c) is not a function because the input 1 corresponds to two different outputs (1 and 2). The relation in (d) is a function because each day corresponds to exactly one maximum temperature. The relation in (e) is not a function because many persons (not just one) were born in Los Angeles on most days in January 2005.

1. Let $A = \{1, 2, 3, 4\}$ and $B = \{-1, 0, 1\}$. Is the given relation a function from A to B?

(a) $\{(1, 0), (2, -1), (3, 0), (4, 1)\}$

(b) $\{(1, 0), (2, -1), (3, 0), (3, -1), (4, 0)\}$

2. Determine if the correspondence is a function.

(a) 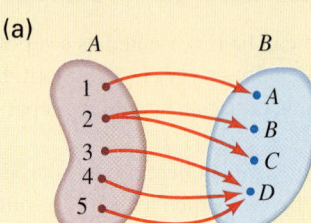   (b)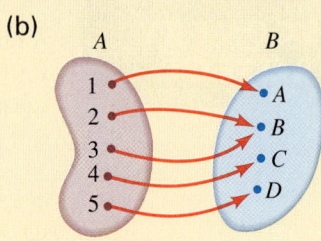

3. The following data were collected from members of a college precalculus class. Is the set of ordered pairs (x, y) a function?

(a)

x Height	y Weight
72 in.	180 lb
60 in.	204 lb
60 in.	120 lb
63 in.	145 lb
70 in.	184 lb

(b)

x Age	y ID Number
19	82-4090
21	80-4133
40	66-8295
21	64-9110
21	20-6666

(c)

x Year of graduation	y Number of graduates
2005	2
2006	12
2007	18
2008	7
2009	1

4. An equation in x and y defines a relation, which may or may not be a function (see page 164). Decide whether the relation consisting of all ordered pairs of real numbers (x, y) satisfying the given condition is a function.

(a) $y = x^2$ (b) $x = y^2$ (c) $x \leq y$ (d) $2x + 7y = 11$

5. In everyday life we encounter many relations which may or may not define functions. For example, we match up people with their telephone number(s), baseball players with their batting averages, or married men with their wives. Does this last correspondence define a function? In a society in which each married man has exactly one wife the rule is a function. But the rule is not a function. Which of the following everyday relations are functions?

(a) x is the daughter of y (x and y are women in the United States)

(b) x is taller than y (x and y are people in California)

(c) x has received dental treatment from y (x and y are millionaires in the United States)

(d) x is a digit (0 to 9) on a telephone dial and y is a corresponding letter

2.3   Increasing and Decreasing Functions; Average Rate of Change

Functions are often used to model changing quantities. In this section we learn how to determine if a function is increasing or decreasing, and how to find the rate at which its values change as the variable changes.

Increasing and Decreasing Functions

It is very useful to know where the graph of a function rises and where it falls. The graph shown in Figure 1 rises, falls, then rises again as we move from left to right: It rises from A to B, falls from B to C, and rises again from C to D. The function f is said to be *increasing* when its graph rises and *decreasing* when its graph falls.

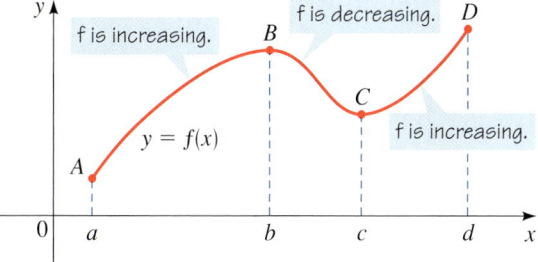

Figure 1

f is increasing on $[a, b]$ and $[c, d]$.
f is decreasing on $[b, c]$.

We have the following definition.

Definition of Increasing and Decreasing Functions

f is **increasing** on an interval I if $f(x_1) < f(x_2)$ whenever $x_1 < x_2$ in I.

f is **decreasing** on an interval I if $f(x_1) > f(x_2)$ whenever $x_1 < x_2$ in I.

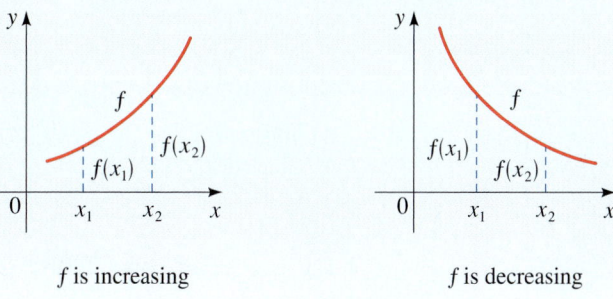

f is increasing f is decreasing

Example 1   Intervals on which a Function Increases and Decreases

The graph in Figure 2 gives the weight W of a person at age x. Determine the intervals on which the function W is increasing and on which it is decreasing.

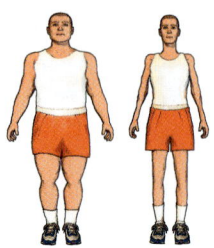

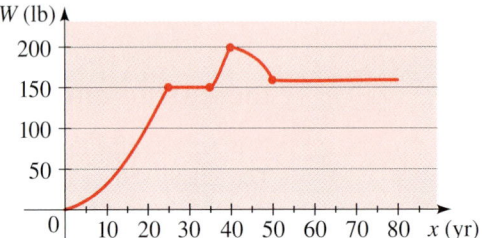

Figure 2

Weight as a function of age

Solution   The function is increasing on $[0, 25]$ and $[35, 40]$. It is decreasing on $[40, 50]$. The function is constant (neither increasing nor decreasing) on $[25, 35]$ and $[50, 80]$. This means that the person gained weight until age 25, then gained weight again between ages 35 and 40. He lost weight between ages 40 and 50.   ∎

Example 2   Using a Graph to Find Intervals where a Function Increases and Decreases

(a) Sketch the graph of the function $f(x) = x^{2/3}$.

(b) Find the domain and range of the function.

(c) Find the intervals on which f increases and decreases.

Solution

(a) We use a graphing calculator to sketch the graph in Figure 3.

(b) From the graph we observe that the domain of f is $\mathbb{R}$ and the range is $[0, \infty)$.

(c) From the graph we see that f is decreasing on $(-\infty, 0]$ and increasing on $[0, \infty)$.   ∎

Some graphing calculators, such as the TI-82, do not evaluate $x^{2/3}$ [entered as x^(2/3)] for negative x. To graph a function like $f(x) = x^{2/3}$, we enter it as $y_1 = (x^{\wedge}(1/3))^{\wedge}2$ because these calculators correctly evaluate powers of the form x^(1/n). Newer calculators, such as the TI-83 and TI-86, do not have this problem.

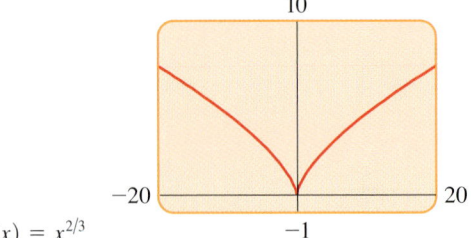

Figure 3

Graph of $f(x) = x^{2/3}$

Average Rate of Change

We are all familiar with the concept of speed: If you drive a distance of 120 miles in 2 hours, then your average speed, or rate of travel, is $\frac{120 \text{ mi}}{2 \text{ h}} = 60$ mi/h.

Now suppose you take a car trip and record the distance that you travel every few minutes. The distance s you have traveled is a function of the time t:

$$s(t) = \text{total distance traveled at time } t$$

We graph the function s as shown in Figure 4. The graph shows that you have traveled a total of 50 miles after 1 hour, 75 miles after 2 hours, 140 miles after 3 hours, and so on. To find your *average* speed between any two points on the trip, we divide the distance traveled by the time elapsed.

Let's calculate your average speed between 1:00 P.M. and 4:00 P.M. The time elapsed is $4 - 1 = 3$ hours. To find the distance you traveled, we subtract the distance at 1:00 P.M. from the distance at 4:00 P.M., that is, $200 - 50 = 150$ mi. Thus, your average speed is

$$\text{average speed} = \frac{\text{distance traveled}}{\text{time elapsed}} = \frac{150 \text{ mi}}{3 \text{ h}} = 50 \text{ mi/h}$$

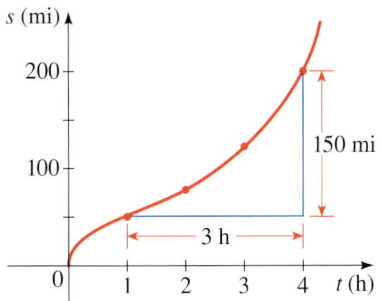

Figure 4

Average speed

The average speed we have just calculated can be expressed using function notation:

$$\text{average speed} = \frac{s(4) - s(1)}{4 - 1} = \frac{200 - 50}{3} = 50 \text{ mi/h}$$

Note that the average speed is different over different time intervals. For example, between 2:00 P.M. and 3:00 P.M. we find that

$$\text{average speed} = \frac{s(3) - s(2)}{3 - 2} = \frac{140 - 75}{1} = 65 \text{ mi/h}$$

Finding average rates of change is important in many contexts. For instance, we may be interested in knowing how quickly the air temperature is dropping as a storm approaches, or how fast revenues are increasing from the sale of a new product. So we need to know how to determine the average rate of change of the functions that model these quantities. In fact, the concept of average rate of change can be defined for any function.

Average Rate of Change

The **average rate of change** of the function $y = f(x)$ between $x = a$ and $x = b$ is

$$\text{average rate of change} = \frac{\text{change in } y}{\text{change in } x} = \frac{f(b) - f(a)}{b - a}$$

The average rate of change is the slope of the **secant line** between $x = a$ and $x = b$ on the graph of f, that is, the line that passes through $(a, f(a))$ and $(b, f(b))$.

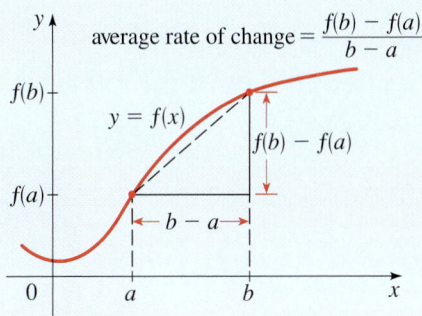

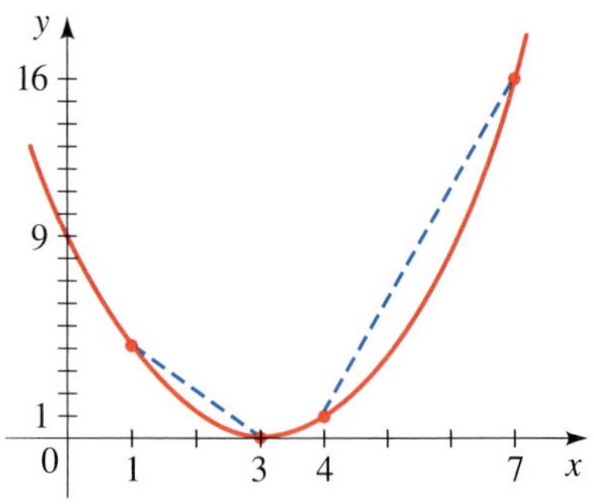

Figure 5
$f(x) = (x - 3)^2$

Example 3   Calculating the Average Rate of Change

For the function $f(x) = (x - 3)^2$, whose graph is shown in Figure 5, find the average rate of change between the following points:

(a) $x = 1$ and $x = 3$ (b) $x = 4$ and $x = 7$

Solution

(a) Average rate of change $= \dfrac{f(3) - f(1)}{3 - 1}$ Definition

$= \dfrac{(3 - 3)^2 - (1 - 3)^2}{3 - 1}$ Use $f(x) = (x - 3)^2$

$= \dfrac{0 - 4}{2} = -2$

(b) Average rate of change $= \dfrac{f(7) - f(4)}{7 - 4}$ Definition

$= \dfrac{(7 - 3)^2 - (4 - 3)^2}{7 - 4}$ Use $f(x) = (x - 3)^2$

$= \dfrac{16 - 1}{3} = 5$

Example 4   Average Speed of a Falling Object

If an object is dropped from a tall building, then the distance it has fallen after t seconds is given by the function $d(t) = 16t^2$. Find its average speed (average rate of change) over the following intervals:

(a) Between 1 s and 5 s (b) Between $t = a$ and $t = a + h$

Solution

(a) Average rate of change $= \dfrac{d(5) - d(1)}{5 - 1}$ Definition

$= \dfrac{16(5)^2 - 16(1)^2}{5 - 1}$ Use $d(t) = 16t^2$

$= \dfrac{400 - 16}{4} = 96 \text{ ft/s}$

(b) Average rate of change $= \dfrac{d(a + h) - d(a)}{(a + h) - a}$ Definition

$= \dfrac{16(a + h)^2 - 16(a)^2}{(a + h) - a}$ Use $d(t) = 16t^2$

$= \dfrac{16(a^2 + 2ah + h^2 - a^2)}{h}$ Expand and factor 16

$= \dfrac{16(2ah + h^2)}{h}$ Simplify numerator

$= \dfrac{16h(2a + h)}{h}$ Factor h

$= 16(2a + h)$ Simplify

The average rate of change calculated in Example 4(b) is known as a *difference quotient*. In calculus we use difference quotients to calculate *instantaneous* rates of change. An example of an instantaneous rate of change is the speed shown on the speedometer of your car. This changes from one instant to the next as your car's speed changes.

Example 5   Average Rate of Temperature Change

Time	Temperature (°F)
8:00 A.M.	38
9:00 A.M.	40
10:00 A.M.	44
11:00 A.M.	50
12:00 NOON	56
1:00 P.M.	62
2:00 P.M.	66
3:00 P.M.	67
4:00 P.M.	64
5:00 P.M.	58
6:00 P.M.	55
7:00 P.M.	51

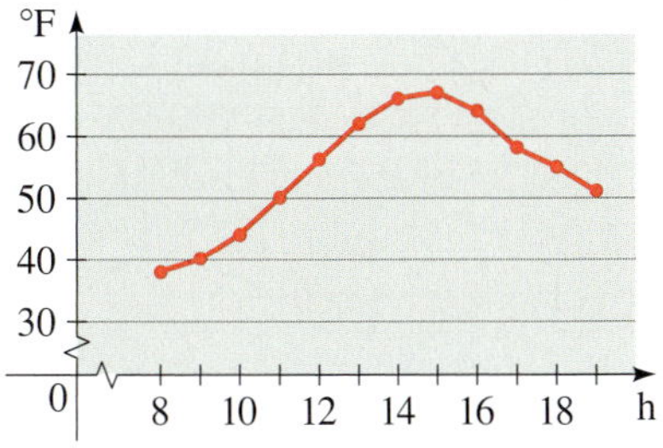

The table gives the outdoor temperatures observed by a science student on a spring day. Draw a graph of the data, and find the average rate of change of temperature between the following times:

(a) 8:00 A.M. and 9:00 A.M.

(b) 1:00 P.M. and 3:00 P.M.

(c) 4:00 P.M. and 7:00 P.M.

Solution   A graph of the temperature data is shown in Figure 6. Let t represent time, measured in hours since midnight (so that 2:00 P.M., for example, corresponds to $t = 14$). Define the function F by

$$F(t) = \text{temperature at time } t$$

Figure 6

(a) Average rate of change $= \dfrac{\text{temperature at 9 A.M.} - \text{temperature at 8 A.M.}}{9 - 8}$

$$= \frac{F(9) - F(8)}{9 - 8}$$

$$= \frac{40 - 38}{9 - 8} = 2$$

The average rate of change was $2\,°F$ per hour.

(b) Average rate of change $= \dfrac{\text{temperature at 3 P.M.} - \text{temperature at 1 P.M.}}{15 - 13}$

$$= \frac{F(15) - F(13)}{15 - 13}$$

$$= \frac{67 - 62}{2} = 2.5$$

The average rate of change was $2.5\,°F$ per hour.

(c) Average rate of change $= \dfrac{\text{temperature at 7 P.M.} - \text{temperature at 4 P.M.}}{19 - 16}$

$$= \frac{F(19) - F(16)}{19 - 16}$$

$$= \frac{51 - 64}{3} \approx -4.3$$

The average rate of change was about $-4.3\,°F$ per hour during this time interval. The negative sign indicates that the temperature was dropping. ■

The graphs in Figure 7 show that if a function is increasing on an interval, then the average rate of change between any two points is positive, whereas if a function is decreasing on an interval, then the average rate of change between any two points is negative.

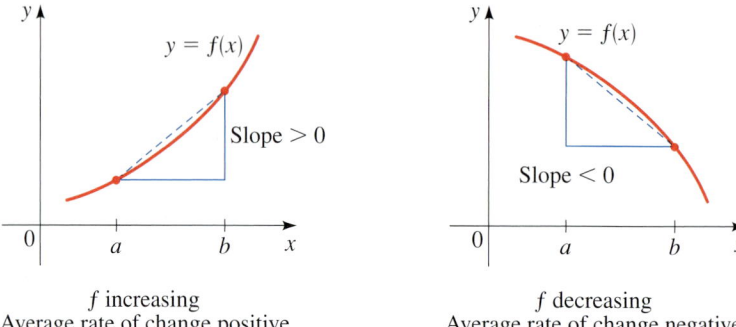

f increasing	f decreasing
Average rate of change positive	Average rate of change negative

Figure 7

Example 6 Linear Functions Have Constant Rate of Change

Let $f(x) = 3x - 5$. Find the average rate of change of f between the following points.

(a) $x = 0$ and $x = 1$ (b) $x = 3$ and $x = 7$ (c) $x = a$ and $x = a + h$

What conclusion can you draw from your answers?

Solution

(a) Average rate of change $= \dfrac{f(1) - f(0)}{1 - 0} = \dfrac{(3 \cdot 1 - 5) - (3 \cdot 0 - 5)}{1}$

$= \dfrac{(-2) - (-5)}{1} = 3$

(b) Average rate of change $= \dfrac{f(7) - f(3)}{7 - 3} = \dfrac{(3 \cdot 7 - 5) - (3 \cdot 3 - 5)}{4}$

$= \dfrac{16 - 4}{4} = 3$

(c) Average rate of change $= \dfrac{f(a + h) - f(a)}{(a + h) - a} = \dfrac{[3(a + h) - 5] - [3a - 5]}{h}$

$= \dfrac{3a + 3h - 5 - 3a + 5}{h} = \dfrac{3h}{h} = 3$

It appears that the average rate of change is always 3 for this function. In fact, part (c) proves that the rate of change between any two arbitrary points $x = a$ and $x = a + h$ is 3. ∎

As Example 6 indicates, for a linear function $f(x) = mx + b$, the average rate of change between any two points is the slope m of the line. This agrees with what we learned in Section 1.10, that the slope of a line represents the rate of change of y with respect to x.

2.3   Exercises

1–4 ■ The graph of a function is given. Determine the intervals on which the function is **(a)** increasing and **(b)** decreasing.

1.
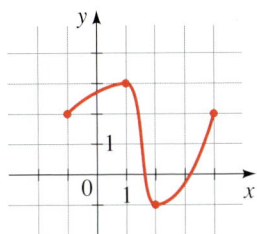

2.
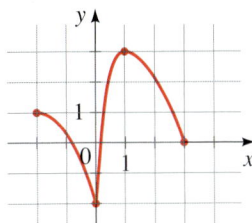

3.
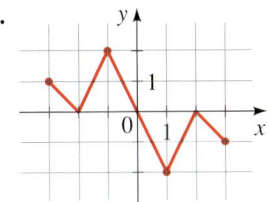

4.
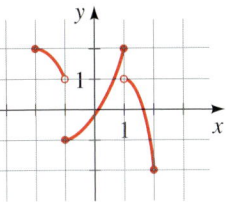

5–12 ■ A function f is given.

(a) Use a graphing device to draw the graph of f.

(b) State approximately the intervals on which f is increasing and on which f is decreasing.

5. $f(x) = x^{2/5}$

6. $f(x) = 4 - x^{2/3}$

7. $f(x) = x^2 - 5x$

8. $f(x) = x^3 - 4x$

9. $f(x) = 2x^3 - 3x^2 - 12x$

10. $f(x) = x^4 - 16x^2$

11. $f(x) = x^3 + 2x^2 - x - 2$

12. $f(x) = x^4 - 4x^3 + 2x^2 + 4x - 3$

13–16 ■ The graph of a function is given. Determine the average rate of change of the function between the indicated values of the variable.

13.
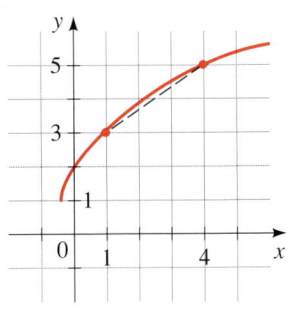

14.
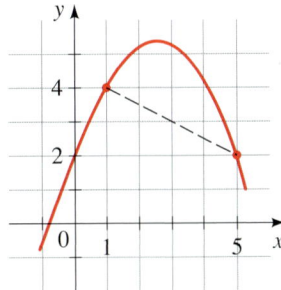

15.
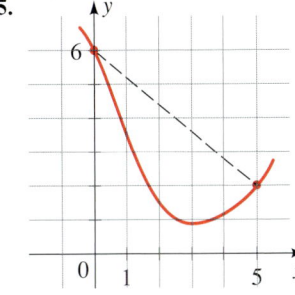

16.
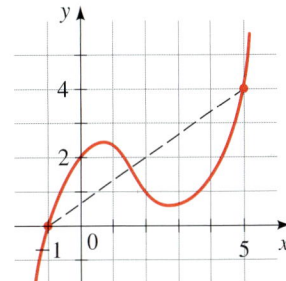

17–28 ■ A function is given. Determine the average rate of change of the function between the given values of the variable.

17. $f(x) = 3x - 2$; $x = 2$, $x = 3$

18. $g(x) = 5 + \frac{1}{2}x$; $x = 1$, $x = 5$

19. $h(t) = t^2 + 2t$; $t = -1$, $t = 4$

20. $f(z) = 1 - 3z^2$; $z = -2$, $z = 0$

21. $f(x) = x^3 - 4x^2$; $x = 0$, $x = 10$

22. $f(x) = x + x^4$; $x = -1$, $x = 3$

23. $f(x) = 3x^2$; $x = 2$, $x = 2 + h$

24. $f(x) = 4 - x^2$; $x = 1$, $x = 1 + h$

25. $g(x) = \dfrac{1}{x}$; $x = 1$, $x = a$

26. $g(x) = \dfrac{2}{x + 1}$; $x = 0$, $x = h$

27. $f(t) = \dfrac{2}{t};\quad t = a, t = a + h$

28. $f(t) = \sqrt{t};\quad t = a, t = a + h$

29–30 ■ A linear function is given.

(a) Find the average rate of change of the function between $x = a$ and $x = a + h$.

(b) Show that the average rate of change is the same as the slope of the line.

29. $f(x) = \frac{1}{2}x + 3$ **30.** $g(x) = -4x + 2$

Applications

31. Changing Water Levels   The graph shows the depth of water W in a reservoir over a one-year period, as a function of the number of days x since the beginning of the year.

(a) Determine the intervals on which the function W is increasing and on which it is decreasing.

(b) What was the average rate of change of W between $x = 100$ and $x = 200$?

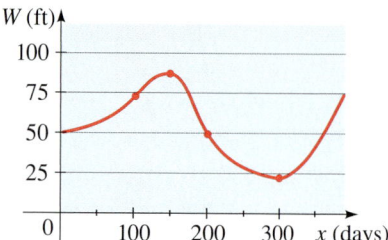

32. Population Growth and Decline   The graph shows the population P in a small industrial city from 1950 to 2000. The variable x represents the number of years since 1950.

(a) Determine the intervals on which the function P is increasing and on which it is decreasing.

(b) What was the average rate of change of P between $x = 20$ and $x = 40$?

(c) Interpret the value of the average rate of change that you found in part (b).

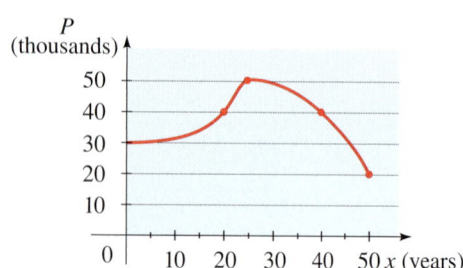

33. Population Growth and Decline   The table gives the population in a small coastal community for the period 1997–2006. Figures shown are for January 1 in each year.

(a) What was the average rate of change of population between 1998 and 2001?

(b) What was the average rate of change of population between 2002 and 2004?

(c) For what period of time was the population increasing?

(d) For what period of time was the population decreasing?

Year	Population
1997	624
1998	856
1999	1,336
2000	1,578
2001	1,591
2002	1,483
2003	994
2004	826
2005	801
2006	745

34. Running Speed   A man is running around a circular track 200 m in circumference. An observer uses a stopwatch to record the runner's time at the end of each lap, obtaining the data in the following table.

(a) What was the man's average speed (rate) between 68 s and 152 s?

(b) What was the man's average speed between 263 s and 412 s?

(c) Calculate the man's speed for each lap. Is he slowing down, speeding up, or neither?

Time (s)	Distance (m)
32	200
68	400
108	600
152	800
203	1000
263	1200
335	1400
412	1600

35. CD Player Sales   The table shows the number of CD players sold in a small electronics store in the years 1993–2003.

(a) What was the average rate of change of sales between 1993 and 2003?

(b) What was the average rate of change of sales between 1993 and 1994?

(c) What was the average rate of change of sales between 1994 and 1996?

(d) Between which two successive years did CD player sales *increase* most quickly? *Decrease* most quickly?

Year	CD players sold
1993	512
1994	520
1995	413
1996	410
1997	468
1998	510
1999	590
2000	607
2001	732
2002	612
2003	584

36. Book Collection   Between 1980 and 2000, a rare book collector purchased books for his collection at the rate of 40 books per year. Use this information to complete the following table. (Note that not every year is given in the table.)

Year	Number of books
1980	420
1981	460
1982	
1985	
1990	
1992	
1995	
1997	
1998	
1999	
2000	1220

Discovery • Discussion

37. 100-meter Race   A 100-m race ends in a three-way tie for first place. The graph shows distance as a function of time for each of the three winners.

(a) Find the average speed for each winner.

(b) Describe the differences between the way the three runners ran the race.

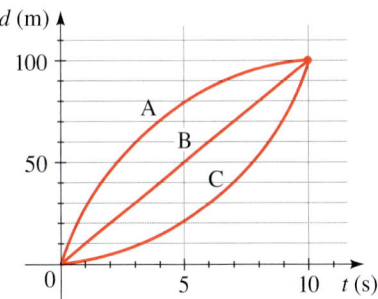

38. Changing Rates of Change: Concavity   The two tables and graphs give the distances traveled by a racing car during two different 10-s portions of a race. In each case, calculate the average speed at which the car is traveling between the observed data points. Is the speed increasing or decreasing? In other words, is the car *accelerating* or *decelerating* on each of these intervals? How does the shape of the graph tell you whether the car is accelerating or decelerating? (The first graph is said to be *concave up* and the second graph *concave down*.)

(a)

Time (s)	Distance (ft)
0	0
2	34
4	70
6	196
8	490
10	964

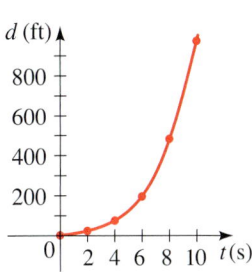

(b)

Time (s)	Distance (ft)
30	5208
32	5734
34	6022
36	6204
38	6352
40	6448

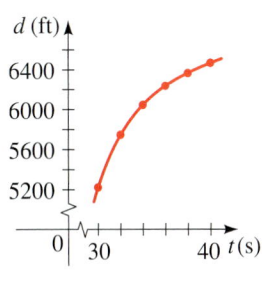

39. Functions That Are Always Increasing or Decreasing   Sketch rough graphs of functions that are defined for all real numbers and that exhibit the indicated behavior (or explain why the behavior is impossible).

(a) f is always increasing, and $f(x) > 0$ for all x

(b) f is always decreasing, and $f(x) > 0$ for all x

(c) f is always increasing, and $f(x) < 0$ for all x

(d) f is always decreasing, and $f(x) < 0$ for all x

2.4   Transformations of Functions

In this section we study how certain transformations of a function affect its graph. This will give us a better understanding of how to graph functions. The transformations we study are shifting, reflecting, and stretching.

Vertical Shifting

Adding a constant to a function shifts its graph vertically: upward if the constant is positive and downward if it is negative.

Example 1   Vertical Shifts of Graphs

Use the graph of $f(x) = x^2$ to sketch the graph of each function.

(a) $g(x) = x^2 + 3$ (b) $h(x) = x^2 - 2$

Solution   The function $f(x) = x^2$ was graphed in Example 1(a), Section 2.2. It is sketched again in Figure 1.

(a) Observe that

$$g(x) = x^2 + 3 = f(x) + 3$$

So the y-coordinate of each point on the graph of g is 3 units above the corresponding point on the graph of f. This means that to graph g we shift the graph of f upward 3 units, as in Figure 1.

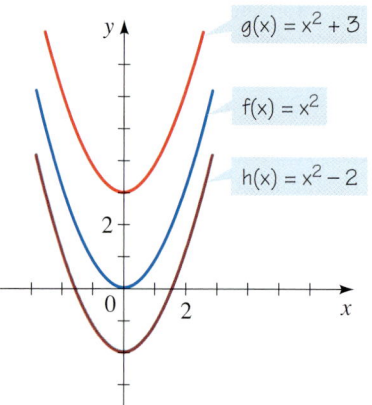

Figure 1

(b) Similarly, to graph h we shift the graph of f downward 2 units, as shown.   ∎

In the margin:

Recall that the graph of the function f is the same as the graph of the equation $y = f(x)$.

In general, suppose we know the graph of $y = f(x)$. How do we obtain from it the graphs of

$$y = f(x) + c \qquad \text{and} \qquad y = f(x) - c \qquad (c > 0)$$

The y-coordinate of each point on the graph of $y = f(x) + c$ is c units above the y-coordinate of the corresponding point on the graph of $y = f(x)$. So, we obtain the graph of $y = f(x) + c$ simply by shifting the graph of $y = f(x)$ upward c units. Similarly, we obtain the graph of $y = f(x) - c$ by shifting the graph of $y = f(x)$ downward c units.

Vertical Shifts of Graphs

Suppose $c > 0$.

    To graph $y = f(x) + c$, shift the graph of $y = f(x)$ upward c units.

    To graph $y = f(x) - c$, shift the graph of $y = f(x)$ downward c units.

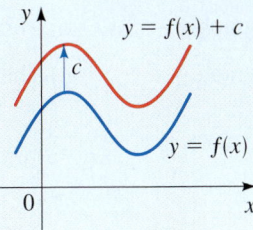 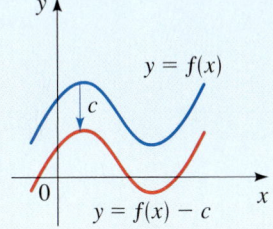

Example 2   Vertical Shifts of Graphs

Use the graph of $f(x) = x^3 - 9x$, which was sketched in Example 12, Section 1.8, to sketch the graph of each function.

(a) $g(x) = x^3 - 9x + 10$ (b) $h(x) = x^3 - 9x - 20$

Solution   The graph of f is sketched again in Figure 2.

(a) To graph g we shift the graph of f upward 10 units, as shown.

(b) To graph h we shift the graph of f downward 20 units, as shown.

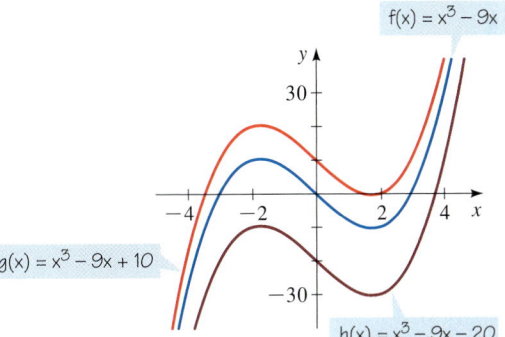

Figure 2

Horizontal Shifting

Suppose that we know the graph of $y = f(x)$. How do we use it to obtain the graphs of

$$y = f(x + c) \qquad \text{and} \qquad y = f(x - c) \qquad (c > 0)$$

The value of $f(x - c)$ at x is the same as the value of $f(x)$ at $x - c$. Since $x - c$ is c units to the left of x, it follows that the graph of $y = f(x - c)$ is just the graph of

$y = f(x)$ shifted to the right c units. Similar reasoning shows that the graph of $y = f(x + c)$ is the graph of $y = f(x)$ shifted to the left c units. The following box summarizes these facts.

Horizontal Shifts of Graphs

Suppose $c > 0$.

 To graph $y = f(x - c)$, shift the graph of $y = f(x)$ to the right c units.

 To graph $y = f(x + c)$, shift the graph of $y = f(x)$ to the left c units.

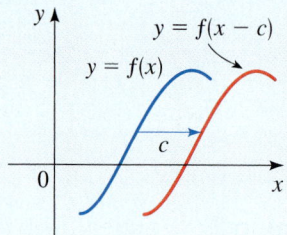
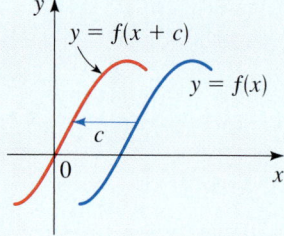

Example 3   Horizontal Shifts of Graphs

Use the graph of $f(x) = x^2$ to sketch the graph of each function.

(a) $g(x) = (x + 4)^2$ (b) $h(x) = (x - 2)^2$

Solution

(a) To graph g, we shift the graph of f to the left 4 units.

(b) To graph h, we shift the graph of f to the right 2 units.

The graphs of g and h are sketched in Figure 3.

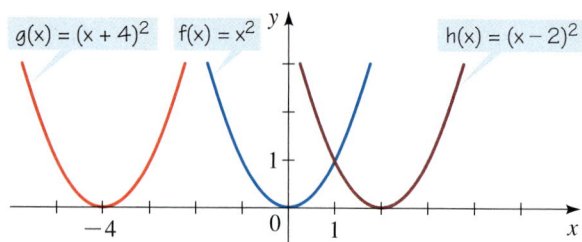

Figure 3

Example 4   Combining Horizontal and Vertical Shifts

Sketch the graph of $f(x) = \sqrt{x - 3} + 4$.

Solution We start with the graph of $y = \sqrt{x}$ (Example 1(c), Section 2.2) and shift it to the right 3 units to obtain the graph of $y = \sqrt{x - 3}$. Then we shift the resulting graph upward 4 units to obtain the graph of $f(x) = \sqrt{x-3} + 4$ shown in Figure 4.

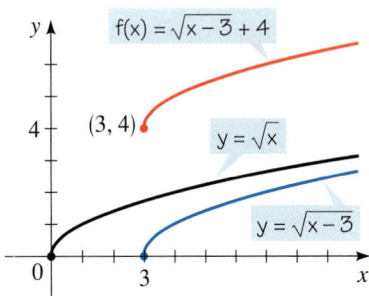

Figure 4

Reflecting Graphs

Suppose we know the graph of $y = f(x)$. How do we use it to obtain the graphs of $y = -f(x)$ and $y = f(-x)$? The y-coordinate of each point on the graph of $y = -f(x)$ is simply the negative of the y-coordinate of the corresponding point on the graph of $y = f(x)$. So the desired graph is the reflection of the graph of $y = f(x)$ in the x-axis. On the other hand, the value of $y = f(-x)$ at x is the same as the value of $y = f(x)$ at $-x$ and so the desired graph here is the reflection of the graph of $y = f(x)$ in the y-axis. The following box summarizes these observations.

Reflecting Graphs

To graph $y = -f(x)$, reflect the graph of $y = f(x)$ in the x-axis.

To graph $y = f(-x)$, reflect the graph of $y = f(x)$ in the y-axis.

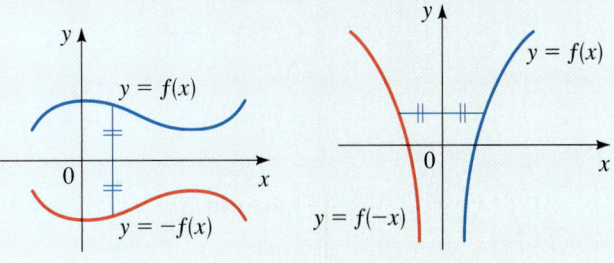

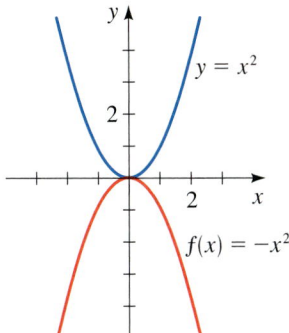

Figure 5

Example 5 Reflecting Graphs

Sketch the graph of each function.

(a) $f(x) = -x^2$ (b) $g(x) = \sqrt{-x}$

Solution

(a) We start with the graph of $y = x^2$. The graph of $f(x) = -x^2$ is the graph of $y = x^2$ reflected in the x-axis (see Figure 5).

(b) We start with the graph of $y = \sqrt{x}$ (Example 1(c) in Section 2.2). The graph of $g(x) = \sqrt{-x}$ is the graph of $y = \sqrt{x}$ reflected in the y-axis (see Figure 6). Note that the domain of the function $g(x) = \sqrt{-x}$ is $\{x \mid x \leq 0\}$.

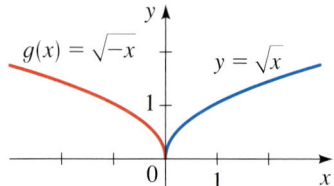

Figure 6

Vertical Stretching and Shrinking

Suppose we know the graph of $y = f(x)$. How do we use it to obtain the graph of $y = cf(x)$? The y-coordinate of $y = cf(x)$ at x is the same as the corresponding y-coordinate of $y = f(x)$ multiplied by c. Multiplying the y-coordinates by c has the effect of vertically stretching or shrinking the graph by a factor of c.

Vertical Stretching and Shrinking of Graphs

To graph $y = cf(x)$:

If $c > 1$, stretch the graph of $y = f(x)$ vertically by a factor of c.

If $0 < c < 1$, shrink the graph of $y = f(x)$ vertically by a factor of c.

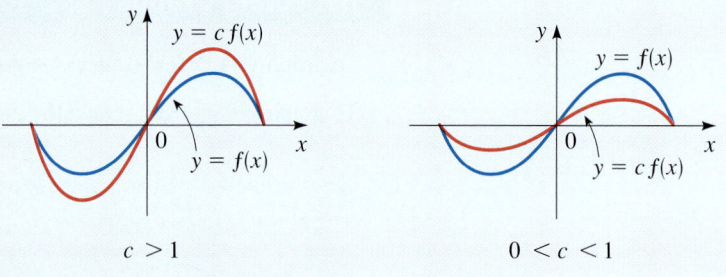

Example 6   Vertical Stretching and Shrinking of Graphs

Use the graph of $f(x) = x^2$ to sketch the graph of each function.

(a) $g(x) = 3x^2$ (b) $h(x) = \frac{1}{3}x^2$

Solution

(a) The graph of g is obtained by multiplying the y-coordinate of each point on the graph of f by 3. That is, to obtain the graph of g we stretch the graph of f vertically by a factor of 3. The result is the narrower parabola in Figure 7.

(b) The graph of h is obtained by multiplying the y-coordinate of each point on the graph of f by $\frac{1}{3}$. That is, to obtain the graph of h we shrink the graph of f vertically by a factor of $\frac{1}{3}$. The result is the wider parabola in Figure 7.

We illustrate the effect of combining shifts, reflections, and stretching in the following example.

Figure 7

Example 7   **Combining Shifting, Stretching, and Reflecting**

Sketch the graph of the function $f(x) = 1 - 2(x - 3)^2$.

Solution   Starting with the graph of $y = x^2$, we first shift to the right 3 units to get the graph of $y = (x - 3)^2$. Then we reflect in the x-axis and stretch by a factor of 2 to get the graph of $y = -2(x - 3)^2$. Finally, we shift upward 1 unit to get the graph of $f(x) = 1 - 2(x - 3)^2$ shown in Figure 8.

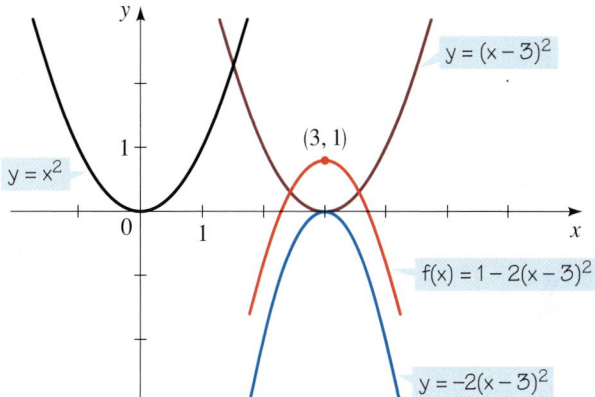

Figure 8

Horizontal Stretching and Shrinking

Now we consider horizontal shrinking and stretching of graphs. If we know the graph of $y = f(x)$, then how is the graph of $y = f(cx)$ related to it? The y-coordinate of $y = f(cx)$ at x is the same as the y-coordinate of $y = f(x)$ at cx. Thus, the x-coordinates in the graph of $y = f(x)$ correspond to the x-coordinates in the graph of $y = f(cx)$ multiplied by c. Looking at this the other way around, we see that the x-coordinates in the graph of $y = f(cx)$ are the x-coordinates in the graph of $y = f(x)$ multiplied by $1/c$. In other words, to change the graph of $y = f(x)$ to the graph of $y = f(cx)$, we must shrink (or stretch) the graph horizontally by a factor of $1/c$, as summarized in the following box.

Horizontal Shrinking and Stretching of Graphs

To graph $y = f(cx)$:

If $c > 1$, shrink the graph of $y = f(x)$ horizontally by a factor of $1/c$.

If $0 < c < 1$, stretch the graph of $y = f(x)$ horizontally by a factor of $1/c$.

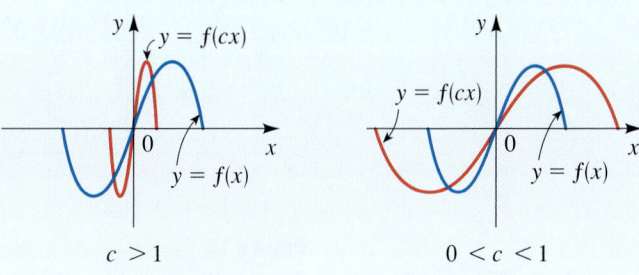

Sonya Kovalevsky (1850–1891) is considered the most important woman mathematician of the 19th century. She was born in Moscow to an aristocratic family. While a child, she was exposed to the principles of calculus in a very unusual fashion—her bedroom was temporarily wallpapered with the pages of a calculus book. She later wrote that she "spent many hours in front of that wall, trying to understand it." Since Russian law forbade women from studying in universities, she entered a marriage of convenience, which allowed her to travel to Germany and obtain a doctorate in mathematics from the University of Göttingen. She eventually was awarded a full professorship at the University of Stockholm, where she taught for eight years before dying in an influenza epidemic at the age of 41. Her research was instrumental in helping put the ideas and applications of functions and calculus on a sound and logical foundation. She received many accolades and prizes for her research work.

Example 8 Horizontal Stretching and Shrinking of Graphs

The graph of $y = f(x)$ is shown in Figure 9. Sketch the graph of each function.

(a) $y = f(2x)$ (b) $y = f(\frac{1}{2}x)$

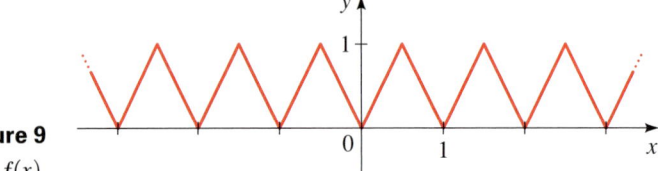

Figure 9
$y = f(x)$

Solution   Using the principles described in the preceding box, we obtain the graphs shown in Figures 10 and 11.

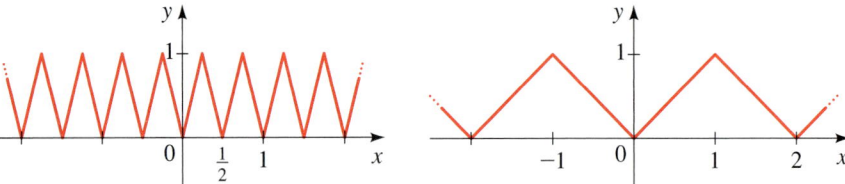

Figure 10
$y = f(2x)$

Figure 11
$y = f(\frac{1}{2}x)$

Even and Odd Functions

If a function f satisfies $f(-x) = f(x)$ for every number x in its domain, then f is called an **even function**. For instance, the function $f(x) = x^2$ is even because

$$f(-x) = (-x)^2 = (-1)^2 x^2 = x^2 = f(x)$$

The graph of an even function is symmetric with respect to the y-axis (see Figure 12). This means that if we have plotted the graph of f for $x \geq 0$, then we can obtain the entire graph simply by reflecting this portion in the y-axis.

If f satisfies $f(-x) = -f(x)$ for every number x in its domain, then f is called an **odd function**. For example, the function $f(x) = x^3$ is odd because

$$f(-x) = (-x)^3 = (-1)^3 x^3 = -x^3 = -f(x)$$

The graph of an odd function is symmetric about the origin (see Figure 13). If we have plotted the graph of f for $x \geq 0$, then we can obtain the entire graph by rotating

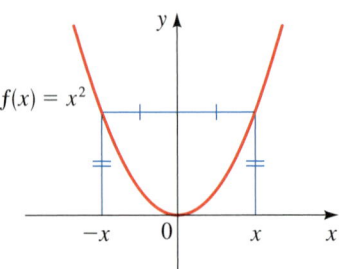

Figure 12
$f(x) = x^2$ is an even function.

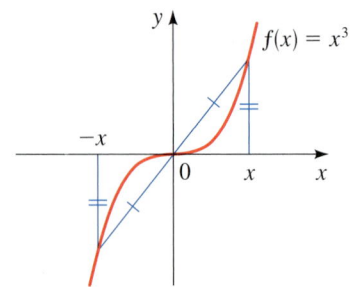

Figure 13
$f(x) = x^3$ is an odd function.

this portion through 180° about the origin. (This is equivalent to reflecting first in the x-axis and then in the y-axis.)

Even and Odd Functions

Let f be a function.

f is **even** if $f(-x) = f(x)$ for all x in the domain of f.

f is **odd** if $f(-x) = -f(x)$ for all x in the domain of f

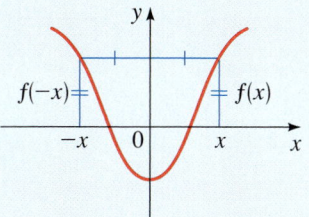   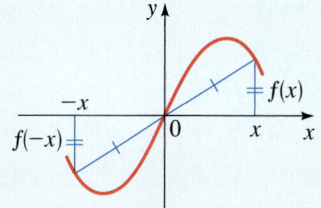

The graph of an even function is symmetric with respect to the y-axis.

The graph of an odd function is symmetric with respect to the origin.

Example 9   Even and Odd Functions

Determine whether the functions are even, odd, or neither even nor odd.

(a) $f(x) = x^5 + x$ (b) $g(x) = 1 - x^4$ (c) $h(x) = 2x - x^2$

Solution

(a) $f(-x) = (-x)^5 + (-x)$

$\qquad\quad = -x^5 - x = -(x^5 + x)$

$\qquad\quad = -f(x)$

Therefore, f is an odd function.

(b) $g(-x) = 1 - (-x)^4 = 1 - x^4 = g(x)$

So g is even.

(c) $h(-x) = 2(-x) - (-x)^2 = -2x - x^2$

Since $h(-x) \neq h(x)$ and $h(-x) \neq -h(x)$, we conclude that h is neither even nor odd. ■

The graphs of the functions in Example 9 are shown in Figure 14. The graph of f is symmetric about the origin, and the graph of g is symmetric about the y-axis. The graph of h is not symmetric either about the y-axis or the origin.

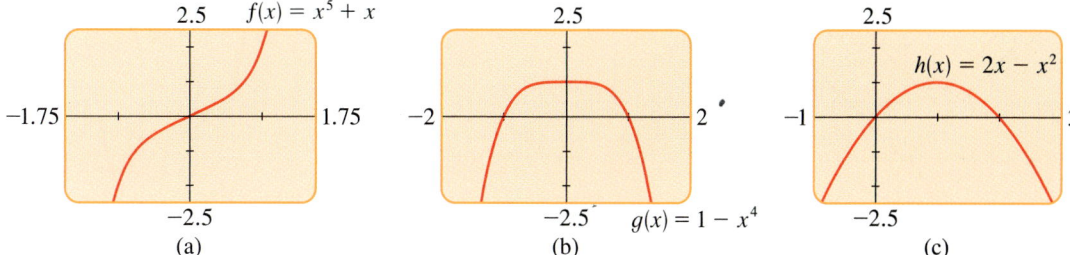

Figure 14

(a)   (b)   (c)

2.4 Exercises

1–10 ■ Suppose the graph of f is given. Describe how the graph of each function can be obtained from the graph of f.

1. (a) $y = f(x) - 5$ (b) $y = f(x - 5)$

2. (a) $y = f(x + 7)$ (b) $y = f(x) + 7$

3. (a) $y = f(x + \frac{1}{2})$ (b) $y = f(x) + \frac{1}{2}$

4. (a) $y = -f(x)$ (b) $y = f(-x)$

5. (a) $y = -2f(x)$ (b) $y = -\frac{1}{2}f(x)$

6. (a) $y = -f(x) + 5$ (b) $y = 3f(x) - 5$

7. (a) $y = f(x - 4) + \frac{3}{4}$ (b) $y = f(x + 4) - \frac{3}{4}$

8. (a) $y = 2f(x + 2) - 2$ (b) $y = 2f(x - 2) + 2$

9. (a) $y = f(4x)$ (b) $y = f(\frac{1}{4}x)$

10. (a) $y = -f(2x)$ (b) $y = f(2x) - 1$

11–16 ■ The graphs of f and g are given. Find a formula for the function g.

11.

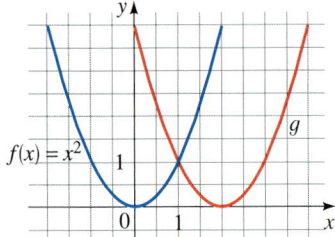

12.

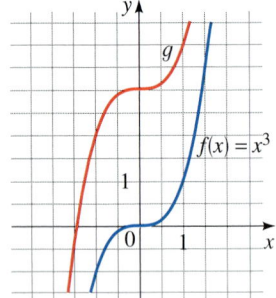

13.

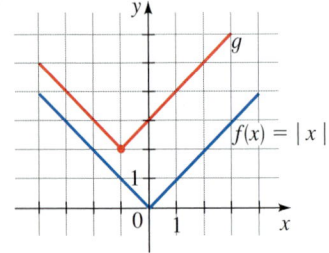

14.

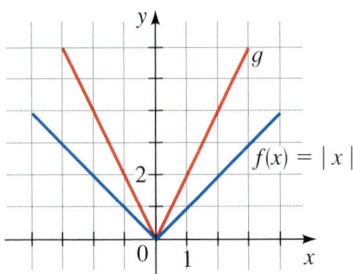

15.

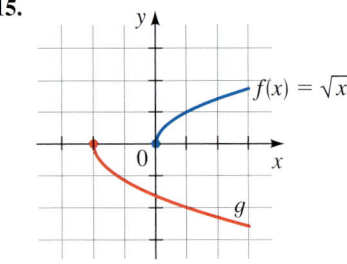

16.

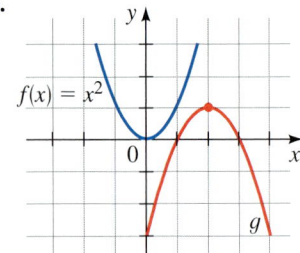

17–18 ■ The graph of $y = f(x)$ is given. Match each equation with its graph.

17. (a) $y = f(x - 4)$ (b) $y = f(x) + 3$

     (c) $y = 2f(x + 6)$ (d) $y = -f(2x)$

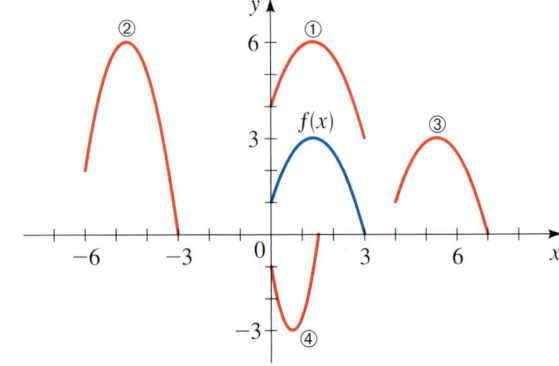

18. (a) $y = \frac{1}{3}f(x)$ **(b)** $y = -f(x + 4)$
   (c) $y = f(x - 4) + 3$ **(d)** $y = f(-x)$

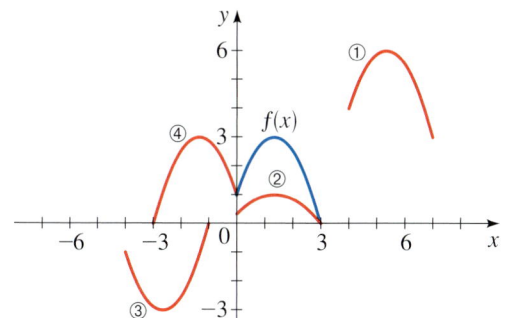

19. The graph of f is given. Sketch the graphs of the following functions.

 (a) $y = f(x - 2)$ **(b)** $y = f(x) - 2$
 (c) $y = 2f(x)$ **(d)** $y = -f(x) + 3$
 (e) $y = f(-x)$ **(f)** $y = \frac{1}{2}f(x - 1)$

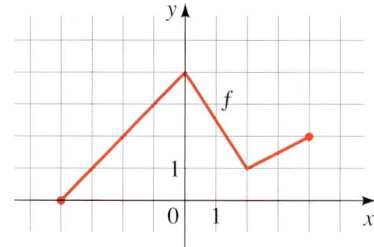

20. The graph of g is given. Sketch the graphs of the following functions.

 (a) $y = g(x + 1)$ **(b)** $y = -g(x + 1)$
 (c) $y = g(x - 2)$ **(d)** $y = g(x) - 2$
 (e) $y = -g(x) + 2$ **(f)** $y = 2g(x)$

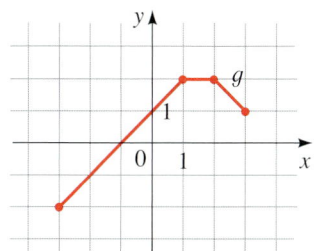

21. (a) Sketch the graph of $f(x) = \dfrac{1}{x}$ by plotting points.

   (b) Use the graph of f to sketch the graphs of the following functions.

      (i) $y = -\dfrac{1}{x}$ **(ii)** $y = \dfrac{1}{x - 1}$

      (iii) $y = \dfrac{2}{x + 2}$ **(iv)** $y = 1 + \dfrac{1}{x - 3}$

22. (a) Sketch the graph of $g(x) = \sqrt[3]{x}$ by plotting points.

   (b) Use the graph of g to sketch the graphs of the following functions.

      (i) $y = \sqrt[3]{x} - 2$ **(ii)** $y = \sqrt[3]{x + 2} + 2$
      (iii) $y = 1 - \sqrt[3]{x}$ **(iv)** $y = 2\sqrt[3]{x}$

23–26 ■ Explain how the graph of g is obtained from the graph of f.

23. (a) $f(x) = x^2$, $g(x) = (x + 2)^2$
    (b) $f(x) = x^2$, $g(x) = x^2 + 2$

24. (a) $f(x) = x^3$, $g(x) = (x - 4)^3$
    (b) $f(x) = x^3$, $g(x) = x^3 - 4$

25. (a) $f(x) = \sqrt{x}$, $g(x) = 2\sqrt{x}$
    (b) $f(x) = \sqrt{x}$, $g(x) = \frac{1}{2}\sqrt{x - 2}$

26. (a) $f(x) = |x|$, $g(x) = 3|x| + 1$
    (b) $f(x) = |x|$, $g(x) = -|x + 1|$

27–32 ■ A function f is given, and the indicated transformations are applied to its graph (in the given order). Write the equation for the final transformed graph.

27. $f(x) = x^2$; shift upward 3 units and shift 2 units to the right

28. $f(x) = x^3$; shift downward 1 unit and shift 4 units to the left

29. $f(x) = \sqrt{x}$; shift 3 units to the left, stretch vertically by a factor of 5, and reflect in the x-axis

30. $f(x) = \sqrt[3]{x}$; reflect in the y-axis, shrink vertically by a factor of $\frac{1}{2}$, and shift upward $\frac{3}{5}$ unit

31. $f(x) = |x|$; shift to the right $\frac{1}{2}$ unit, shrink vertically by a factor of 0.1, and shift downward 2 units

32. $f(x) = |x|$; shift to the left 1 unit, stretch vertically by a factor of 3, and shift upward 10 units

33–48 ■ Sketch the graph of the function, not by plotting points, but by starting with the graph of a standard function and applying transformations.

33. $f(x) = (x - 2)^2$ **34.** $f(x) = (x + 7)^2$

35. $f(x) = -(x + 1)^2$ **36.** $f(x) = 1 - x^2$

37. $f(x) = x^3 + 2$ **38.** $f(x) = -x^3$

39. $y = 1 + \sqrt{x}$ **40.** $y = 2 - \sqrt{x + 1}$

41. $y = \frac{1}{2}\sqrt{x + 4} - 3$ **42.** $y = 3 - 2(x - 1)^2$

43. $y = 5 + (x + 3)^2$ **44.** $y = \frac{1}{3}x^3 - 1$

45. $y = |x| - 1$ **46.** $y = |x - 1|$

47. $y = |x + 2| + 2$ **48.** $y = 2 - |x|$

49–52 ■ Graph the functions on the same screen using the given viewing rectangle. How is each graph related to the graph in part (a)?

49. Viewing rectangle $[-8, 8]$ by $[-2, 8]$

(a) $y = \sqrt[4]{x}$ (b) $y = \sqrt[4]{x + 5}$
(c) $y = 2\sqrt[4]{x + 5}$ (d) $y = 4 + 2\sqrt[4]{x + 5}$

50. Viewing rectangle $[-8, 8]$ by $[-6, 6]$

(a) $y = |x|$ (b) $y = -|x|$
(c) $y = -3|x|$ (d) $y = -3|x - 5|$

51. Viewing rectangle $[-4, 6]$ by $[-4, 4]$

(a) $y = x^6$ (b) $y = \frac{1}{3}x^6$
(c) $y = -\frac{1}{3}x^6$ (d) $y = -\frac{1}{3}(x - 4)^6$

52. Viewing rectangle $[-6, 6]$ by $[-4, 4]$

(a) $y = \dfrac{1}{\sqrt{x}}$ (b) $y = \dfrac{1}{\sqrt{x + 3}}$

(c) $y = \dfrac{1}{2\sqrt{x + 3}}$ (d) $y = \dfrac{1}{2\sqrt{x + 3}} - 3$

53. The graph of g is given. Use it to graph each of the following functions.

(a) $y = g(2x)$ (b) $y = g(\frac{1}{2}x)$

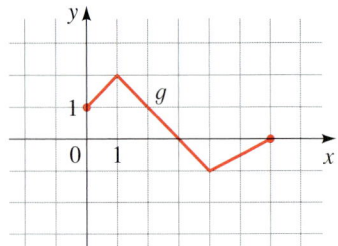

54. The graph of h is given. Use it to graph each of the following functions.

(a) $y = h(3x)$ (b) $y = h(\frac{1}{3}x)$

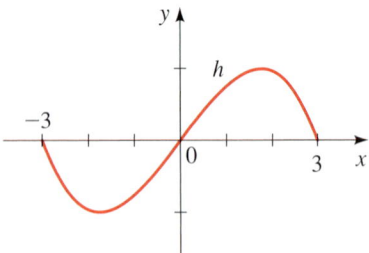

55–56 ■ The graph of a function defined for $x \geq 0$ is given. Complete the graph for $x < 0$ to make

(a) an even function

(b) an odd function

55.

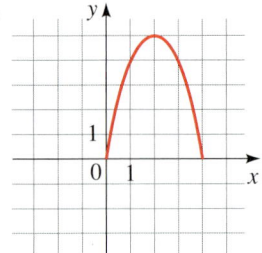

56.

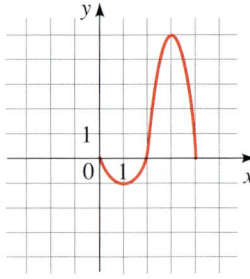

57–58 ■ Use the graph of $f(x) = [\![x]\!]$ described on pages 162–163 to graph the indicated function.

57. $y = [\![2x]\!]$ **58.** $y = [\![\frac{1}{4}x]\!]$

 59. If $f(x) = \sqrt{2x - x^2}$, graph the following functions in the viewing rectangle $[-5, 5]$ by $[-4, 4]$. How is each graph related to the graph in part (a)?

(a) $y = f(x)$ (b) $y = f(2x)$ (c) $y = f(\frac{1}{2}x)$

60. If $f(x) = \sqrt{2x - x^2}$, graph the following functions in the viewing rectangle $[-5, 5]$ by $[-4, 4]$. How is each graph related to the graph in part (a)?

(a) $y = f(x)$ (b) $y = f(-x)$ (c) $y = -f(-x)$
(d) $y = f(-2x)$ (e) $y = f(-\frac{1}{2}x)$

61–68 ■ Determine whether the function f is even, odd, or neither. If f is even or odd, use symmetry to sketch its graph.

61. $f(x) = x^{-2}$ **62.** $f(x) = x^{-3}$

63. $f(x) = x^2 + x$ **64.** $f(x) = x^4 - 4x^2$

65. $f(x) = x^3 - x$ **66.** $f(x) = 3x^3 + 2x^2 + 1$

67. $f(x) = 1 - \sqrt[3]{x}$ **68.** $f(x) = x + \dfrac{1}{x}$

69. The graphs of $f(x) = x^2 - 4$ and $g(x) = |x^2 - 4|$ are shown. Explain how the graph of g is obtained from the graph of f.

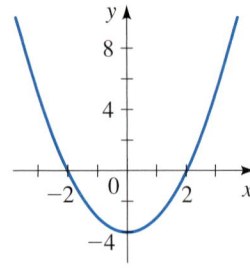

$f(x) = x^2 - 4$

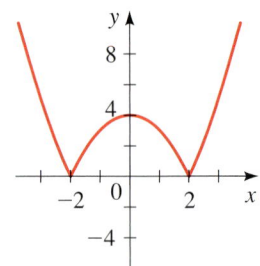

$g(x) = |x^2 - 4|$

70. The graph of $f(x) = x^4 - 4x^2$ is shown. Use this graph to sketch the graph of $g(x) = |x^4 - 4x^2|$.

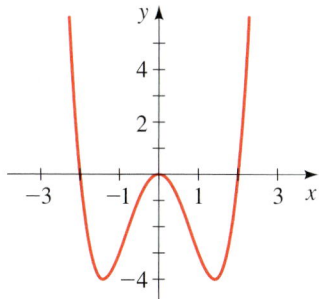

71–72 ■ Sketch the graph of each function.

71. (a) $f(x) = 4x - x^2$ **(b)** $g(x) = |4x - x^2|$

72. (a) $f(x) = x^3$ **(b)** $g(x) = |x^3|$

Applications

73. Sales Growth   The annual sales of a certain company can be modeled by the function $f(t) = 4 + 0.01t^2$, where t represents years since 1990 and $f(t)$ is measured in millions of dollars.

   (a) What shifting and shrinking operations must be performed on the function $y = t^2$ to obtain the function $y = f(t)$?

   (b) Suppose you want t to represent years since 2000 instead of 1990. What transformation would you have to apply to the function $y = f(t)$ to accomplish this? Write the new function $y = g(t)$ that results from this transformation.

74. Changing Temperature Scales   The temperature on a certain afternoon is modeled by the function

$$C(t) = \tfrac{1}{2}t^2 + 2$$

where t represents hours after 12 noon ($0 \le t \le 6$), and C is measured in °C.

   (a) What shifting and shrinking operations must be performed on the function $y = t^2$ to obtain the function $y = C(t)$?

   (b) Suppose you want to measure the temperature in °F instead. What transformation would you have to apply to the function $y = C(t)$ to accomplish this? (Use the fact that the relationship between Celsius and Fahrenheit degrees is given by $F = \tfrac{9}{5}C + 32$.) Write the new function $y = F(t)$ that results from this transformation.

Discovery • Discussion

75. Sums of Even and Odd Functions   If f and g are both even functions, is $f + g$ necessarily even? If both are odd, is their sum necessarily odd? What can you say about the sum if one is odd and one is even? In each case, prove your answer.

76. Products of Even and Odd Functions   Answer the same questions as in Exercise 75, except this time consider the *product* of f and g instead of the sum.

77. Even and Odd Power Functions   What must be true about the integer n if the function

$$f(x) = x^n$$

is an even function? If it is an odd function? Why do you think the names "even" and "odd" were chosen for these function properties?

| 2.5 | **Quadratic Functions; Maxima and Minima** |

A maximum or minimum value of a function is the largest or smallest value of the function on an interval. For a function that represents the profit in a business, we would be interested in the maximum value; for a function that represents the amount of material to be used in a manufacturing process, we would be interested in the minimum value. In this section we learn how to find the maximum and minimum values of quadratic and other functions.

Graphing Quadratic Functions Using the Standard Form

A **quadratic function** is a function f of the form

$$f(x) = ax^2 + bx + c$$

where a, b, and c are real numbers and $a \neq 0$.

In particular, if we take $a = 1$ and $b = c = 0$, we get the simple quadratic function $f(x) = x^2$ whose graph is the parabola that we drew in Example 1 of Section 2.2. In fact, the graph of any quadratic function is a **parabola**; it can be obtained from the graph of $f(x) = x^2$ by the transformations given in Section 2.4.

Standard Form of a Quadratic Function

A quadratic function $f(x) = ax^2 + bx + c$ can be expressed in the **standard form**

$$f(x) = a(x - h)^2 + k$$

by completing the square. The graph of f is a parabola with **vertex** (h, k); the parabola opens upward if $a > 0$ or downward if $a < 0$.

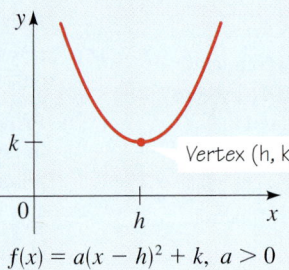

$f(x) = a(x - h)^2 + k, \ a > 0$

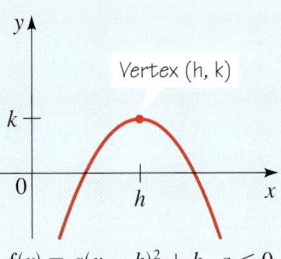

$f(x) = a(x - h)^2 + k, \ a < 0$

Example 1   Standard Form of a Quadratic Function

Let $f(x) = 2x^2 - 12x + 23$.

(a) Express f in standard form.

(b) Sketch the graph of f.

Solution

Completing the square is discussed in Section 1.5.

(a) Since the coefficient of x^2 is not 1, we must factor this coefficient from the terms involving x before we complete the square.

$$
\begin{aligned}
f(x) &= 2x^2 - 12x + 23 \\
&= 2(x^2 - 6x) + 23 && \text{Factor 2 from the x-terms} \\
&= 2(x^2 - 6x + 9) + 23 - 2 \cdot 9 && \text{Complete the square: Add 9 inside} \\
& && \text{parentheses, subtract } 2 \cdot 9 \text{ outside} \\
&= 2(x - 3)^2 + 5 && \text{Factor and simplify}
\end{aligned}
$$

$f(x) = 2(x - 3)^2 + 5$

Vertex is $(3, 5)$

The standard form is $f(x) = 2(x - 3)^2 + 5$.

(b) The standard form tells us that we get the graph of f by taking the parabola $y = x^2$, shifting it to the right 3 units, stretching it by a factor of 2, and moving it upward 5 units. The vertex of the parabola is at $(3, 5)$ and the parabola opens upward. We sketch the graph in Figure 1 after noting that the y-intercept is $f(0) = 23$.

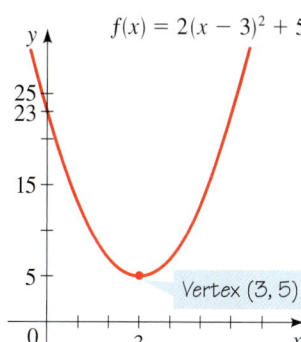

$f(x) = 2(x - 3)^2 + 5$

Vertex $(3, 5)$

Figure 1

Maximum and Minimum Values of Quadratic Functions

If a quadratic function has vertex (h, k), then the function has a minimum value at the vertex if it opens upward and a maximum value at the vertex if it opens downward. For example, the function graphed in Figure 1 has minimum value 5 when $x = 3$, since the vertex $(3, 5)$ is the lowest point on the graph.

Maximum or Minimum Value of a Quadratic Function

Let f be a quadratic function with standard form $f(x) = a(x - h)^2 + k$. The maximum or minimum value of f occurs at $x = h$.

If $a > 0$, then the **minimum value** of f is $f(h) = k$.

If $a < 0$, then the **maximum value** of f is $f(h) = k$.

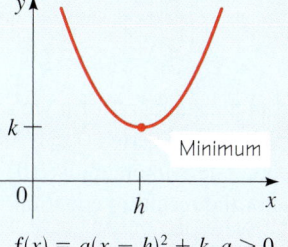

$f(x) = a(x - h)^2 + k,\ a > 0$

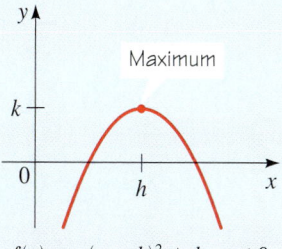

$f(x) = a(x - h)^2 + k,\ a < 0$

Example 2   Minimum Value of a Quadratic Function

Consider the quadratic function $f(x) = 5x^2 - 30x + 49$.

(a) Express f in standard form.

(b) Sketch the graph of f.

(c) Find the minimum value of f.

Solution

(a) To express this quadratic function in standard form, we complete the square.

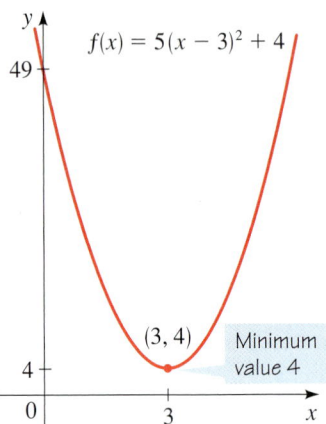

$f(x) = 5(x - 3)^2 + 4$

Figure 2

$$f(x) = 5x^2 - 30x + 49$$

$$= 5(x^2 - 6x) + 49 \qquad \text{Factor 5 from the x-terms}$$

$$= 5(x^2 - 6x + 9) + 49 - 5 \cdot 9 \qquad \begin{array}{l}\text{Complete the square: Add 9 inside} \\ \text{parentheses, subtract } 5 \cdot 9 \text{ outside}\end{array}$$

$$= 5(x - 3)^2 + 4 \qquad \text{Factor and simplify}$$

(b) The graph is a parabola that has its vertex at $(3, 4)$ and opens upward, as sketched in Figure 2.

(c) Since the coefficient of x^2 is positive, f has a minimum value. The minimum value is $f(3) = 4$. ∎

Example 3   Maximum Value of a Quadratic Function

Consider the quadratic function $f(x) = -x^2 + x + 2$.

(a) Express f in standard form.

(b) Sketch the graph of f.

(c) Find the maximum value of f.

Solution

(a) To express this quadratic function in standard form, we complete the square.

$$y = -x^2 + x + 2$$

$$= -(x^2 - x) + 2 \qquad \text{Factor } -1 \text{ from the x-terms}$$

$$= -\left(x^2 - x + \tfrac{1}{4}\right) + 2 - (-1)\tfrac{1}{4} \qquad \begin{array}{l}\text{Complete the square: Add } \tfrac{1}{4} \\ \text{inside parentheses, subtract} \\ (-1)\tfrac{1}{4} \text{ outside}\end{array}$$

$$= -\left(x - \tfrac{1}{2}\right)^2 + \tfrac{9}{4} \qquad \text{Factor and simplify}$$

(b) From the standard form we see that the graph is a parabola that opens downward and has vertex $\left(\tfrac{1}{2}, \tfrac{9}{4}\right)$. As an aid to sketching the graph, we find the intercepts. The y-intercept is $f(0) = 2$. To find the x-intercepts, we set $f(x) = 0$ and factor the resulting equation.

$$-x^2 + x + 2 = 0$$

$$-(x^2 - x - 2) = 0$$

$$-(x - 2)(x + 1) = 0$$

Thus, the x-intercepts are $x = 2$ and $x = -1$. The graph of f is sketched in Figure 3.

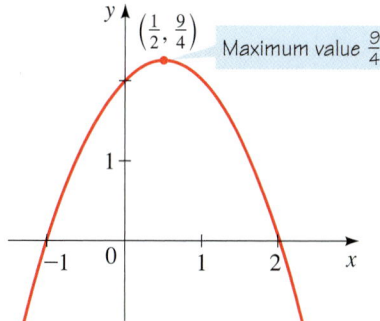

Figure 3
Graph of $f(x) = -x^2 + x + 2$

(c) Since the coefficient of x^2 is negative, f has a maximum value, which is $f(\frac{1}{2}) = \frac{9}{4}$. ∎

Expressing a quadratic function in standard form helps us sketch its graph as well as find its maximum or minimum value. If we are interested only in finding the maximum or minimum value, then a formula is available for doing so. This formula is obtained by completing the square for the general quadratic function as follows:

$$f(x) = ax^2 + bx + c$$

$$= a\left(x^2 + \frac{b}{a}x\right) + c \qquad \text{Factor } a \text{ from the x-terms}$$

$$= a\left(x^2 + \frac{b}{a}x + \frac{b^2}{4a^2}\right) + c - a\left(\frac{b^2}{4a^2}\right) \qquad \begin{array}{l}\text{Complete the square:}\\\text{Add } \frac{b^2}{4a^2} \text{ inside parentheses,}\\\text{subtract } a\left(\frac{b^2}{4a^2}\right) \text{ outside}\end{array}$$

$$= a\left(x + \frac{b}{2a}\right)^2 + c - \frac{b^2}{4a} \qquad \text{Factor}$$

This equation is in standard form with $h = -b/(2a)$ and $k = c - b^2/(4a)$. Since the maximum or minimum value occurs at $x = h$, we have the following result.

Maximum or Minimum Value of a Quadratic Function

The maximum or minimum value of a quadratic function $f(x) = ax^2 + bx + c$ occurs at

$$x = -\frac{b}{2a}$$

If $a > 0$, then the **minimum value** is $f\left(-\dfrac{b}{2a}\right)$.

If $a < 0$, then the **maximum value** is $f\left(-\dfrac{b}{2a}\right)$.

Example 4 **Finding Maximum and Minimum Values of Quadratic Functions**

Find the maximum or minimum value of each quadratic function.

(a) $f(x) = x^2 + 4x$ (b) $g(x) = -2x^2 + 4x - 5$

Solution

(a) This is a quadratic function with $a = 1$ and $b = 4$. Thus, the maximum or minimum value occurs at

$$x = -\frac{b}{2a} = -\frac{4}{2 \cdot 1} = -2$$

Since $a > 0$, the function has the *minimum* value

$$f(-2) = (-2)^2 + 4(-2) = -4$$

(b) This is a quadratic function with $a = -2$ and $b = 4$. Thus, the maximum or minimum value occurs at

$$x = -\frac{b}{2a} = -\frac{4}{2 \cdot (-2)} = 1$$

Since $a < 0$, the function has the *maximum* value

$$f(1) = -2(1)^2 + 4(1) - 5 = -3 \qquad \blacksquare$$

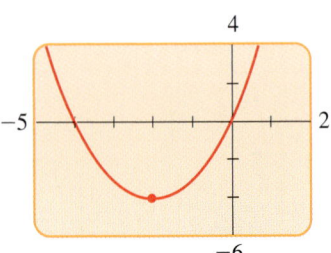

The minimum value occurs at $x = -2$.

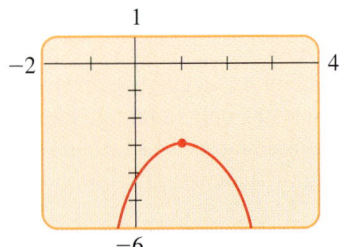

The maximum value occurs at $x = 1$.

Many real-world problems involve finding a maximum or minimum value for a function that models a given situation. In the next example we find the maximum value of a quadratic function that models the gas mileage for a car.

Example 5 **Maximum Gas Mileage for a Car**

Most cars get their best gas mileage when traveling at a relatively modest speed. The gas mileage M for a certain new car is modeled by the function

$$M(s) = -\frac{1}{28}s^2 + 3s - 31, \qquad 15 \le s \le 70$$

where s is the speed in mi/h and M is measured in mi/gal. What is the car's best gas mileage, and at what speed is it attained?

Solution The function M is a quadratic function with $a = -\frac{1}{28}$ and $b = 3$. Thus, its maximum value occurs when

$$s = -\frac{b}{2a} = -\frac{3}{2\left(-\frac{1}{28}\right)} = 42$$

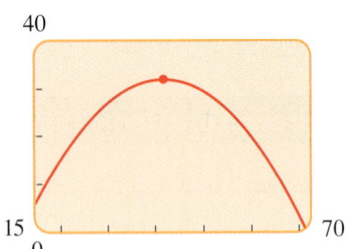

The maximum gas mileage occurs at 42 mi/h.

The maximum is $M(42) = -\frac{1}{28}(42)^2 + 3(42) - 31 = 32$. So the car's best gas mileage is 32 mi/gal, when it is traveling at 42 mi/h. $\blacksquare$

Using Graphing Devices to Find Extreme Values

The methods we have discussed apply to finding extreme values of quadratic functions only. We now show how to locate extreme values of any function that can be graphed with a calculator or computer.

If there is a viewing rectangle such that the point $(a, f(a))$ is the highest point on the graph of f *within* the viewing rectangle (not on the edge), then the number $f(a)$ is called a **local maximum value** of f (see Figure 4). Notice that $f(a) \geq f(x)$ for all numbers x that are close to a.

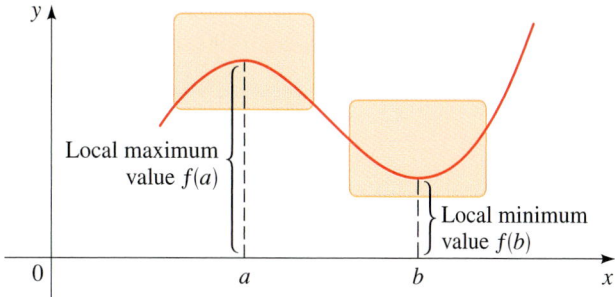

Figure 4

Similarly, if there is a viewing rectangle such that the point $(b, f(b))$ is the lowest point on the graph of f within the viewing rectangle, then the number $f(b)$ is called a **local minimum value** of f. In this case, $f(b) \leq f(x)$ for all numbers x that are close to b.

Example 6 Finding Local Maxima and Minima from a Graph

Find the local maximum and minimum values of the function $f(x) = x^3 - 8x + 1$, correct to three decimals.

Solution   The graph of f is shown in Figure 5. There appears to be one local maximum between $x = -2$ and $x = -1$, and one local minimum between $x = 1$ and $x = 2$.

Let's find the coordinates of the local maximum point first. We zoom in to enlarge the area near this point, as shown in Figure 6. Using the TRACE feature on the graphing device, we move the cursor along the curve and observe how the y-coordinates change. The local maximum value of y is 9.709, and this value occurs when x is -1.633, correct to three decimals.

We locate the minimum value in a similar fashion. By zooming in to the viewing rectangle shown in Figure 7, we find that the local minimum value is about -7.709, and this value occurs when $x \approx 1.633$.

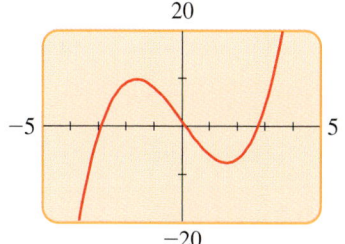

Figure 5

Graph of $f(x) = x^3 - 8x + 1$

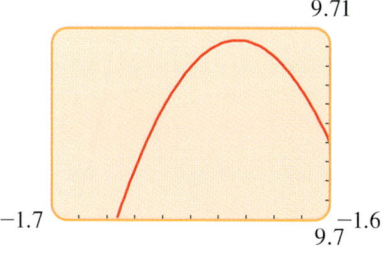

Figure 6

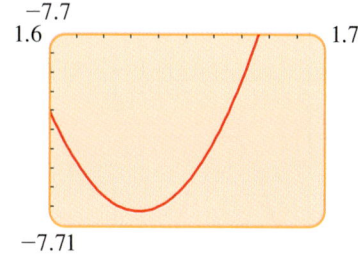

Figure 7

The `maximum` and `minimum` commands on a TI-82 or TI-83 calculator provide another method for finding extreme values of functions. We use this method in the next example.

Example 7   A Model for the Food Price Index

A model for the food price index (the price of a representative "basket" of foods) between 1990 and 2000 is given by the function

$$I(t) = -0.0113t^3 + 0.0681t^2 + 0.198t + 99.1$$

where t is measured in years since midyear 1990, so $0 \le t \le 10$, and $I(t)$ is scaled so that $I(3) = 100$. Estimate the time when food was most expensive during the period 1990–2000.

Solution   The graph of I as a function of t is shown in Figure 8(a). There appears to be a maximum between $t = 4$ and $t = 7$. Using the `maximum` command, as shown in Figure 8(b), we see that the maximum value of I is about 100.38, and it occurs when $t \approx 5.15$, which corresponds to August 1995.

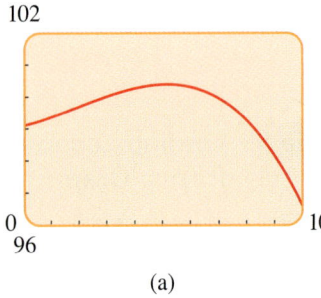

(a)

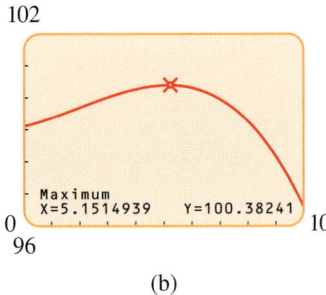

(b)

Figure 8

2.5   Exercises

1–4 ■ The graph of a quadratic function f is given.

(a) Find the coordinates of the vertex.

(b) Find the maximum or minimum value of f.

1. $f(x) = -x^2 + 6x - 5$ **2.** $f(x) = -\frac{1}{2}x^2 - 2x + 6$ **3.** $f(x) = 2x^2 - 4x - 1$ **4.** $f(x) = 3x^2 + 6x - 1$

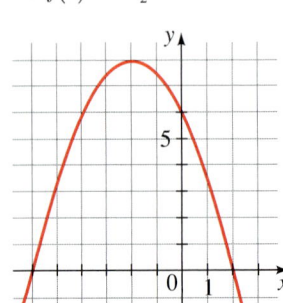
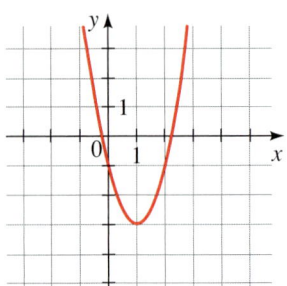
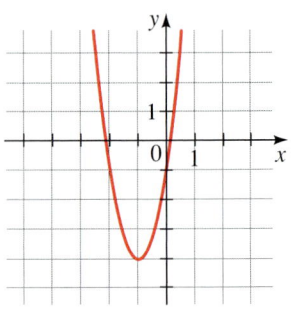

5–18 ■ A quadratic function is given.

(a) Express the quadratic function in standard form.

(b) Find its vertex and its x- and y-intercept(s).

(c) Sketch its graph.

5. $f(x) = x^2 - 6x$

6. $f(x) = x^2 + 8x$

7. $f(x) = 2x^2 + 6x$

8. $f(x) = -x^2 + 10x$

9. $f(x) = x^2 + 4x + 3$

10. $f(x) = x^2 - 2x + 2$

11. $f(x) = -x^2 + 6x + 4$

12. $f(x) = -x^2 - 4x + 4$

13. $f(x) = 2x^2 + 4x + 3$

14. $f(x) = -3x^2 + 6x - 2$

15. $f(x) = 2x^2 - 20x + 57$

16. $f(x) = 2x^2 + x - 6$

17. $f(x) = -4x^2 - 16x + 3$

18. $f(x) = 6x^2 + 12x - 5$

19–28 ■ A quadratic function is given.

(a) Express the quadratic function in standard form.

(b) Sketch its graph.

(c) Find its maximum or minimum value.

19. $f(x) = 2x - x^2$

20. $f(x) = x + x^2$

21. $f(x) = x^2 + 2x - 1$

22. $f(x) = x^2 - 8x + 8$

23. $f(x) = -x^2 - 3x + 3$

24. $f(x) = 1 - 6x - x^2$

25. $g(x) = 3x^2 - 12x + 13$

26. $g(x) = 2x^2 + 8x + 11$

27. $h(x) = 1 - x - x^2$

28. $h(x) = 3 - 4x - 4x^2$

29–38 ■ Find the maximum or minimum value of the function.

29. $f(x) = x^2 + x + 1$

30. $f(x) = 1 + 3x - x^2$

31. $f(t) = 100 - 49t - 7t^2$

32. $f(t) = 10t^2 + 40t + 113$

33. $f(s) = s^2 - 1.2s + 16$

34. $g(x) = 100x^2 - 1500x$

35. $h(x) = \frac{1}{2}x^2 + 2x - 6$

36. $f(x) = -\frac{x^2}{3} + 2x + 7$

37. $f(x) = 3 - x - \frac{1}{2}x^2$

38. $g(x) = 2x(x - 4) + 7$

39. Find a function whose graph is a parabola with vertex $(1, -2)$ and that passes through the point $(4, 16)$.

40. Find a function whose graph is a parabola with vertex $(3, 4)$ and that passes through the point $(1, -8)$.

41–44 ■ Find the domain and range of the function.

41. $f(x) = -x^2 + 4x - 3$

42. $f(x) = x^2 - 2x - 3$

43. $f(x) = 2x^2 + 6x - 7$

44. $f(x) = -3x^2 + 6x + 4$

45–46 ■ A quadratic function is given.

(a) Use a graphing device to find the maximum or minimum value of the quadratic function f, correct to two decimal places.

(b) Find the exact maximum or minimum value of f, and compare with your answer to part (a).

45. $f(x) = x^2 + 1.79x - 3.21$

46. $f(x) = 1 + x - \sqrt{2}x^2$

47–50 ■ Find all local maximum and minimum values of the function whose graph is shown.

47.

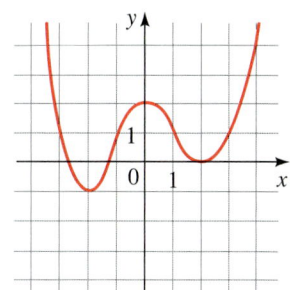

48.

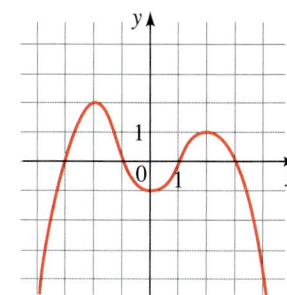

49.

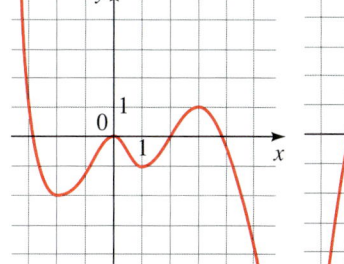

50.

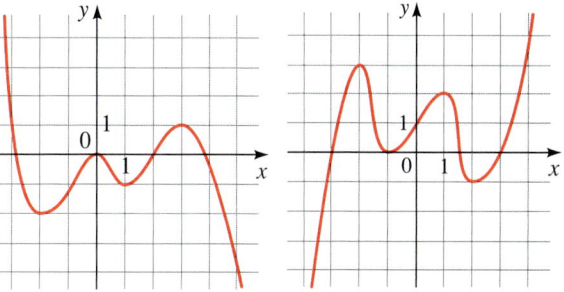

51–58 ■ Find the local maximum and minimum values of the function and the value of x at which each occurs. State each answer correct to two decimal places.

51. $f(x) = x^3 - x$

52. $f(x) = 3 + x + x^2 - x^3$

53. $g(x) = x^4 - 2x^3 - 11x^2$

54. $g(x) = x^5 - 8x^3 + 20x$

55. $U(x) = x\sqrt{6 - x}$

56. $U(x) = x\sqrt{x - x^2}$

57. $V(x) = \frac{1 - x^2}{x^3}$

58. $V(x) = \frac{1}{x^2 + x + 1}$

Applications

59. Height of a Ball If a ball is thrown directly upward with a velocity of 40 ft/s, its height (in feet) after t seconds is given by $y = 40t - 16t^2$. What is the maximum height attained by the ball?

60. Path of a Ball A ball is thrown across a playing field. Its path is given by the equation $y = -0.005x^2 + x + 5$, where x is the distance the ball has traveled horizontally, and y is its height above ground level, both measured in feet.

(a) What is the maximum height attained by the ball?

(b) How far has it traveled horizontally when it hits the ground?

61. Revenue A manufacturer finds that the revenue generated by selling x units of a certain commodity is given by the function $R(x) = 80x - 0.4x^2$, where the revenue $R(x)$ is measured in dollars. What is the maximum revenue, and how many units should be manufactured to obtain this maximum?

62. Sales A soft-drink vendor at a popular beach analyzes his sales records, and finds that if he sells x cans of soda pop in one day, his profit (in dollars) is given by

$$P(x) = -0.001x^2 + 3x - 1800$$

What is his maximum profit per day, and how many cans must he sell for maximum profit?

63. Advertising The effectiveness of a television commercial depends on how many times a viewer watches it. After some experiments an advertising agency found that if the effectiveness E is measured on a scale of 0 to 10, then

$$E(n) = \tfrac{2}{3}n - \tfrac{1}{90}n^2$$

where n is the number of times a viewer watches a given commercial. For a commercial to have maximum effectiveness, how many times should a viewer watch it?

64. Pharmaceuticals When a certain drug is taken orally, the concentration of the drug in the patient's bloodstream after t minutes is given by $C(t) = 0.06t - 0.0002t^2$, where $0 \le t \le 240$ and the concentration is measured in mg/L. When is the maximum serum concentration reached, and what is that maximum concentration?

65. Agriculture The number of apples produced by each tree in an apple orchard depends on how densely the trees are planted. If n trees are planted on an acre of land, then each tree produces $900 - 9n$ apples. So the number of apples produced per acre is

$$A(n) = n(900 - 9n)$$

How many trees should be planted per acre in order to obtain the maximum yield of apples?

66. Migrating Fish A fish swims at a speed v relative to the water, against a current of 5 mi/h. Using a mathematical model of energy expenditure, it can be shown that the total energy E required to swim a distance of 10 mi is given by

$$E(v) = 2.73v^3 \frac{10}{v - 5}$$

Biologists believe that migrating fish try to minimize the total energy required to swim a fixed distance. Find the value of v that minimizes energy required.

NOTE This result has been verified; migrating fish swim against a current at a speed 50% greater than the speed of the current.

67. Highway Engineering A highway engineer wants to estimate the maximum number of cars that can safely travel a particular highway at a given speed. She assumes that each car is 17 ft long, travels at a speed s, and follows the car in front of it at the "safe following distance" for that speed. She finds that the number N of cars that can pass a given point per minute is modeled by the function

$$N(s) = \frac{88s}{17 + 17\left(\dfrac{s}{20}\right)^2}$$

At what speed can the greatest number of cars travel the highway safely?

68. Volume of Water Between 0°C and 30°C, the volume V (in cubic centimeters) of 1 kg of water at a temperature T is given by the formula

$$V = 999.87 - 0.06426T + 0.0085043T^2 - 0.0000679T^3$$

Find the temperature at which the volume of 1 kg of water is a minimum.

69. Coughing When a foreign object lodged in the trachea (windpipe) forces a person to cough, the diaphragm thrusts upward causing an increase in pressure in the lungs. At the same time, the trachea contracts, causing the expelled air to move faster and increasing the pressure on the foreign object. According to a mathematical model of coughing, the velocity v of the airstream through an average-sized person's trachea is related to the radius r of the trachea (in centimeters) by the function

$$v(r) = 3.2(1 - r)r^2, \qquad \tfrac{1}{2} \le r \le 1$$

Determine the value of r for which v is a maximum.

Discovery • Discussion

70. Maxima and Minima In Example 5 we saw a real-world situation in which the maximum value of a function is important. Name several other everyday situations in which a maximum or minimum value is important.

71. Minimizing a Distance When we seek a minimum or maximum value of a function, it is sometimes easier to work with a simpler function instead.

(a) Suppose $g(x) = \sqrt{f(x)}$, where $f(x) \ge 0$ for all x. Explain why the local minima and maxima of f and g occur at the same values of x.

(b) Let $g(x)$ be the distance between the point $(3,0)$ and the point (x, x^2) on the graph of the parabola $y = x^2$. Express g as a function of x.

(c) Find the minimum value of the function g that you found in part (b). Use the principle described in part (a) to simplify your work.

72. Maximum of a Fourth-Degree Polynomial Find the maximum value of the function

$$f(x) = 3 + 4x^2 - x^4$$

[*Hint:* Let $t = x^2$.]

2.6 Modeling with Functions

Many of the processes studied in the physical and social sciences involve understanding how one quantity varies with respect to another. Finding a function that describes the dependence of one quantity on another is called *modeling*. For example, a biologist observes that the number of bacteria in a certain culture increases with time. He tries to model this phenomenon by finding the precise function (or rule) that relates the bacteria population to the elapsed time.

In this section we will learn how to find models that can be constructed using geometric or algebraic properties of the object under study. (Finding models from *data* is studied in the *Focus on Modeling* at the end of this chapter.) Once the model is found, we use it to analyze and predict properties of the object or process being studied.

Modeling with Functions

We begin with a simple real-life situation that illustrates the modeling process.

Example 1 Modeling the Volume of a Box

A breakfast cereal company manufactures boxes to package their product. For aesthetic reasons, the box must have the following proportions: Its width is 3 times its depth and its height is 5 times its depth.

(a) Find a function that models the volume of the box in terms of its depth.

(b) Find the volume of the box if the depth is 1.5 in.

(c) For what depth is the volume 90 in^3?

(d) For what depth is the volume greater than 60 in^3?

■ Thinking About the Problem

Let's experiment with the problem. If the depth is 1 in, then the width is 3 in. and the height is 5 in. So in this case, the volume is $V = 1 \times 3 \times 5 = 15$ in^3. The table gives other values. Notice that all the boxes have the same shape, and the greater the depth the greater the volume.

Depth	Volume
1	$1 \times 3 \times 5 = 15$
2	$2 \times 6 \times 10 = 120$
3	$3 \times 9 \times 15 = 405$
4	$4 \times 12 \times 20 = 960$

Solution

(a) To find the function that models the volume of the box, we use the following steps.

■ Express the Model in Words

We know that the volume of a rectangular box is

$$\text{volume} = \text{depth} \times \text{width} \times \text{height}$$

■ Choose the Variable

There are three varying quantities—width, depth, and height. Since the function we want depends on the depth, we let

$$x = \text{depth of the box}$$

Then we express the other dimensions of the box in terms of x.

In Words	In Algebra
Depth	x
Width	$3x$
Height	$5x$

■ Set up the Model

The model is the function V that gives the volume of the box in terms of the depth x.

$$\text{volume} = \text{depth} \times \text{width} \times \text{height}$$

$$V(x) = x \cdot 3x \cdot 5x$$

$$V(x) = 15x^3$$

The volume of the box is modeled by the function $V(x) = 15x^3$. The function V is graphed in Figure 1.

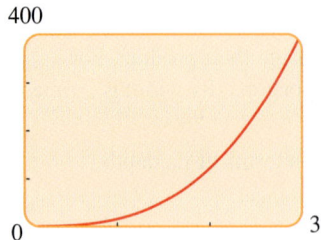

Figure 1

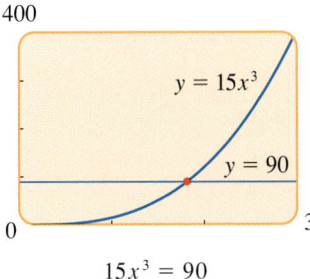

400

$y = 15x^3$

$y = 90$

0   3

$15x^3 = 90$

Figure 2

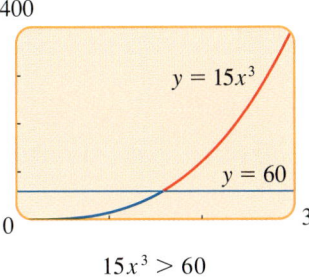

400

$y = 15x^3$

$y = 60$

0   3

$15x^3 > 60$

Figure 3

■ **Use the Model**

We use the model to answer the questions in parts (b), (c), and (d).

(b) If the depth is 1.5 in., the volume is $V(1.5) = 15(1.5)^3 = 50.625$ in^3.

(c) We need to solve the equation $V(x) = 90$ or

$$15x^3 = 90$$

$$x^3 = 6$$

$$x = \sqrt[3]{6} \approx 1.82 \text{ in.}$$

The volume is 90 in^3 when the depth is about 1.82 in. (We can also solve this equation graphically, as shown in Figure 2.)

(d) We need to solve the inequality $V(x) > 60$ or

$$15x^3 > 60$$

$$x^3 > 4$$

$$x > \sqrt[3]{4} \approx 1.59$$

The volume will be greater than 60 in^3 if the depth is greater than 1.59 in. (We can also solve this inequality graphically, as shown in Figure 3.) ■

The steps in Example 1 are typical of how we model with functions. They are summarized in the following box.

Guidelines for Modeling with Functions

1. **Express the Model in Words.** Identify the quantity you want to model and express it, in words, as a function of the other quantities in the problem.

2. **Choose the Variable.** Identify all the variables used to express the function in Step 1. Assign a symbol, such as x, to one variable and express the other variables in terms of this symbol.

3. **Set up the Model.** Express the function in the language of algebra by writing it as a function of the single variable chosen in Step 2.

4. **Use the Model.** Use the function to answer the questions posed in the problem. (To find a maximum or a minimum, use the algebraic or graphical methods described in Section 2.5.)

Example 2 Fencing a Garden

A gardener has 140 feet of fencing to fence in a rectangular vegetable garden.

(a) Find a function that models the area of the garden she can fence.

(b) For what range of widths is the area greater than or equal to 825 ft^2?

(c) Can she fence a garden with area 1250 ft^2?

(d) Find the dimensions of the largest area she can fence.

■ **Thinking About the Problem**

If the gardener fences a plot with width 10 ft, then the length must be 60 ft, because $10 + 10 + 60 + 60 = 140$. So the area is

$$A = \text{width} \times \text{length} = 10 \cdot 60 = 600 \text{ ft}^2$$

The table shows various choices for fencing the garden. We see that as the width increases, the fenced area increases, then decreases.

Width	Length	Area
10	60	600
20	50	1000
30	40	1200
40	30	1200
50	20	1000
60	10	600

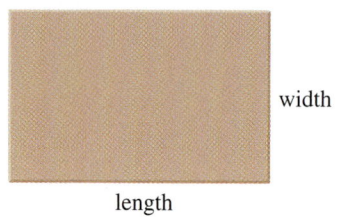

width

length

Solution

(a) The model we want is a function that gives the area she can fence.

■ **Express the Model in Words**

We know that the area of a rectangular garden is

$$\text{area} = \text{width} \times \text{length}$$

■ **Choose the Variable**

There are two varying quantities—width and length. Since the function we want depends on only one variable, we let

$$x = \text{width of the garden}$$

Then we must express the length in terms of x. The perimeter is fixed at 140 ft, so the length is determined once we choose the width. If we let the length be l as in Figure 4, then $2x + 2l = 140$, so $l = 70 - x$. We summarize these facts.

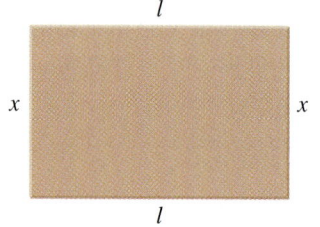

Figure 4

In Words	In Algebra
Width	x
Length	$70 - x$

■ **Set up the Model**

The model is the function A that gives the area of the garden for any width x.

$$\text{area} = \text{width} \times \text{length}$$

$$A(x) = x(70 - x)$$
$$A(x) = 70x - x^2$$

The area she can fence is modeled by the function $A(x) = 70x - x^2$.

■ **Use the Model**

We use the model to answer the questions in parts (b)–(d).

(b) We need to solve the inequality $A(x) \geq 825$. To solve graphically, we graph $y = 70x - x^2$ and $y = 825$ in the same viewing rectangle (see Figure 5). We see that $15 \leq x \leq 55$.

Maximum values of quadratic functions are discussed on page 195.

(c) From Figure 6 we see that the graph of $A(x)$ always lies below the line $y = 1250$, so an area of 1250 ft^2 is never attained.

(d) We need to find the maximum value of the function $A(x) = 70x - x^2$. Since this is a quadratic function with $a = -1$ and $b = 70$, the maximum occurs at

$$x = -\frac{b}{2a} = -\frac{70}{2(-1)} = 35$$

So the maximum area that she can fence has width 35 ft and length $70 - 35 = 35$ ft.

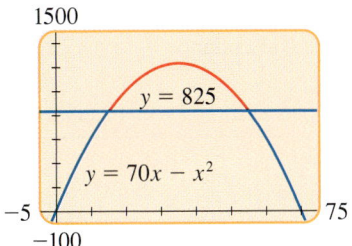

Figure 5

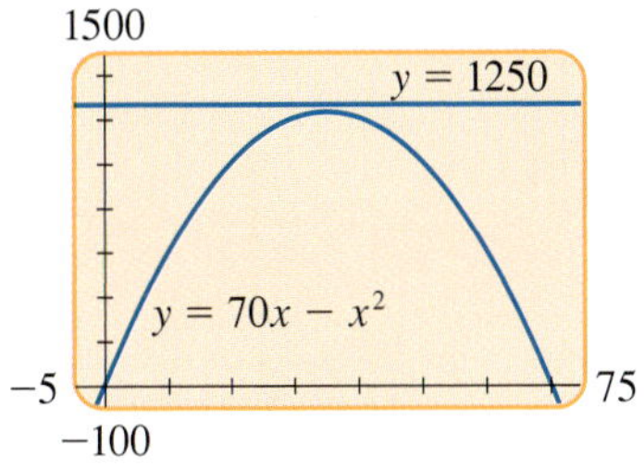

Figure 6

Example 3   Maximizing Revenue from Ticket Sales

A hockey team plays in an arena with a seating capacity of 15,000 spectators. With the ticket price set at $14, average attendance at recent games has been 9500. A market survey indicates that for each dollar the ticket price is lowered, the average attendance increases by 1000.

(a) Find a function that models the revenue in terms of ticket price.
(b) What ticket price is so high that no one attends, and hence no revenue is generated?
(c) Find the price that maximizes revenue from ticket sales.

■ **Thinking About the Problem**

With a ticket price of $14, the revenue is $9500 \times \$14 = \$133,000$. If the ticket price is lowered to $13, attendance increases to $9500 + 1000 = 10,500$, so the revenue becomes $10,500 \times \$13 = \$136,500$. The table shows the revenue for several ticket prices. Note that if the ticket price is lowered, revenue increases, but if the ticket price is lowered too much, revenue decreases.

Price	Attendance	Revenue
$15	8,500	$127,500
$14	9,500	$133,500
$13	10,500	$136,500
$12	11,500	$138,500
$11	12,500	$137,500
$10	13,500	$135,500
$9	14,500	$130,500

Solution

(a) The model we want is a function that gives the revenue for any ticket price.

■ **Express the Model in Words**

We know that

$$\boxed{\text{revenue}} \; = \; \boxed{\text{ticket price}} \; \times \; \boxed{\text{attendance}}$$

■ **Choose the Variable**

There are two varying quantities—ticket price and attendance. Since the function we want depends on price, we let

$$x = \text{ticket price}$$

Next, we must express the attendance in terms of x.

In Words	In Algebra
Ticket price	x
Amount ticket price is lowered	$14 - x$
Increase in attendance	$1000(14 - x)$
Attendance	$9500 + 1000(14 - x) = 23{,}500 - 1000x$

■ **Set up the Model**

The model is the function R that gives the revenue for a given ticket price x.

$$\boxed{\text{revenue}} \; = \; \boxed{\text{ticket price}} \; \times \; \boxed{\text{attendance}}$$

$$R(x) = x(23{,}500 - 1000x)$$

$$R(x) = 23{,}500x - 1000x^2$$

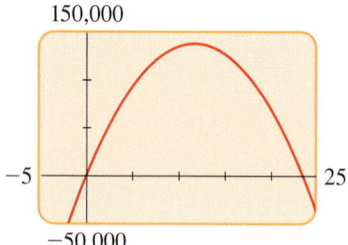

Figure 7

Maximum values of quadratic functions are discussed on page 195.

■ **Use the Model**

We use the model to answer the questions in parts (b) and (c).

(b) We want to find the ticket price x for which $R(x) = 23{,}500x - 1000x^2 = 0$. We can solve this quadratic equation algebraically or graphically. From the graph in Figure 7 we see that $R(x) = 0$ when $x = 0$ or $x = 23.5$. So, according to our model, the revenue would drop to zero if the ticket price is \$23.50 or higher. (Of course, revenue is also zero if the ticket price is zero!)

(c) Since $R(x) = 23{,}500x - 1000x^2$ is a quadratic function with $a = -1000$ and $b = 23{,}500$, the maximum occurs at

$$x = -\frac{b}{2a} = -\frac{23{,}500}{2(-1000)} = 11.75$$

So a ticket price of \$11.75 yields the maximum revenue. At this price the revenue is

$$R(11.75) = 23{,}500(11.75) - 1000(11.75)^2 = \$138{,}062.50$$

■

Example 4   Minimizing the Metal in a Can

A manufacturer makes a metal can that holds 1 L (liter) of oil. What radius minimizes the amount of metal in the can?

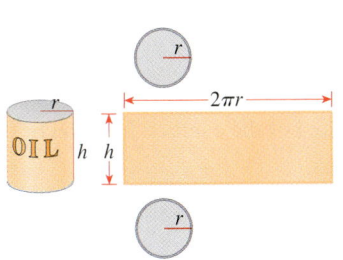

Figure 8

■ **Thinking About the Problem**

To use the least amount of metal, we must minimize the surface area of the can, that is, the area of the top, bottom, and the sides. The area of the top and bottom is $2\pi r^2$ and the area of the sides is $2\pi rh$ (see Figure 8), so the surface area of the can is

$$S = 2\pi r^2 + 2\pi rh$$

The radius and height of the can must be chosen so that the volume is exactly 1 L, or 1000 cm³. If we want a small radius, say $r = 3$, then the height must be just tall enough to make the total volume 1000 cm³. In other words, we must have

$$\pi(3)^2 h = 1000 \qquad \text{\color{teal}Volume of the can is } \pi r^2 h$$

$$h = \frac{1000}{9\pi} \approx 35.4 \text{ cm} \qquad \text{\color{teal}Solve for } h$$

Now that we know the radius and height, we can find the surface area of the can:

$$\text{surface area} = 2\pi(3)^2 + 2\pi(3)(35.4) \approx 729.1 \text{ cm}^3$$

If we want a different radius, we can find the corresponding height and surface area in a similar fashion.

Solution   The model we want is a function that gives the surface area of the can.

■ **Express the Model in Words**

We know that for a cylindrical can

$$\text{surface area} \;=\; \text{area of top and bottom} \;+\; \text{area of sides}$$

■ **Choose the Variable**

There are two varying quantities—radius and height. Since the function we want depends on the radius, we let

$$r = \text{radius of can}$$

Next, we must express the height in terms of the radius r. Since the volume of a cylindrical can is $V = \pi r^2 h$ and the volume must be 1000 cm³, we have

$$\pi r^2 h = 1000 \qquad \text{\color{teal}Volume of can is 1000 cm}^3$$

$$h = \frac{1000}{\pi r^2} \qquad \text{\color{teal}Solve for } h$$

We can now express the areas of the top, bottom, and sides in terms of r only.

In Words	In Algebra
Radius of can	r
Height of can	$\dfrac{1000}{\pi r^2}$
Area of top and bottom	$2\pi r^2$
Area of sides ($2\pi rh$)	$2\pi r\left(\dfrac{1000}{\pi r^2}\right)$

■ **Set up the Model**

The model is the function S that gives the surface area of the can as a function of the radius r.

$$\text{surface area} \;=\; \text{area of top and bottom} \;+\; \text{area of sides}$$

$$S(r) = 2\pi r^2 + 2\pi r\left(\frac{1000}{\pi r^2}\right)$$

$$S(r) = 2\pi r^2 + \frac{2000}{r}$$

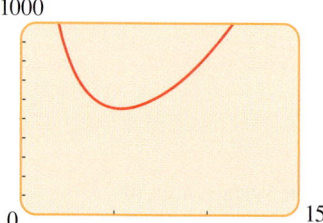

Figure 9

$$S = 2\pi r^2 + \frac{2000}{r}$$

■ **Use the Model**

We use the model to find the minimum surface area of the can. We graph S in Figure 9 and zoom in on the minimum point to find that the minimum value of S is about 554 cm² and occurs when the radius is about 5.4 cm.   ■

2.6   Exercises

1–18 ■ In these exercises you are asked to find a function that models a real-life situation. Use the guidelines for modeling described in the text to help you.

1. Area   A rectangular building lot is three times as long as it is wide. Find a function that models its area A in terms of its width w.

2. Area   A poster is 10 inches longer than it is wide. Find a function that models its area A in terms of its width w.

3. Volume   A rectangular box has a square base. Its height is half the width of the base. Find a function that models its volume V in terms of its width w.

4. Volume   The height of a cylinder is four times its radius. Find a function that models the volume V of the cylinder in terms of its radius r.

5. Area   A rectangle has a perimeter of 20 ft. Find a function that models its area A in terms of the length x of one of its sides.

6. Perimeter   A rectangle has an area of 16 m². Find a function that models its perimeter P in terms of the length x of one of its sides.

7. Area   Find a function that models the area A of an equilateral triangle in terms of the length x of one of its sides.

8. Area   Find a function that models the surface area S of a cube in terms of its volume V.

9. Radius   Find a function that models the radius r of a circle in terms of its area A.

10. Area   Find a function that models the area A of a circle in terms of its circumference C.

11. Area   A rectangular box with a volume of 60 ft³ has a square base. Find a function that models its surface area S in terms of the length x of one side of its base.

12. Length   A woman 5 ft tall is standing near a street lamp that is 12 ft tall, as shown in the figure. Find a function that models the length L of her shadow in terms of her distance d from the base of the lamp.

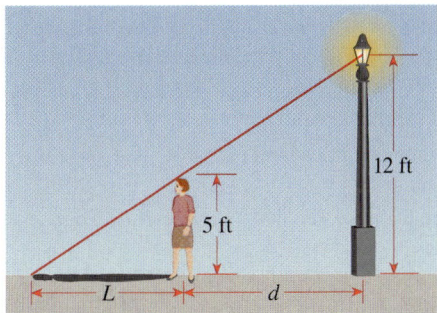

13. Distance Two ships leave port at the same time. One sails south at 15 mi/h and the other sails east at 20 mi/h. Find a function that models the distance D between the ships in terms of the time t (in hours) elapsed since their departure.

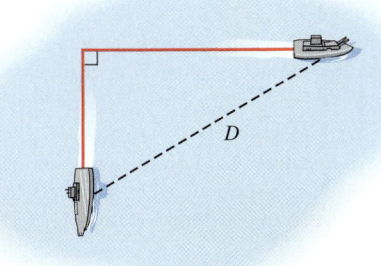

14. Product The sum of two positive numbers is 60. Find a function that models their product P in terms of x, one of the numbers.

15. Area An isosceles triangle has a perimeter of 8 cm. Find a function that models its area A in terms of the length of its base b.

16. Perimeter A right triangle has one leg twice as long as the other. Find a function that models its perimeter P in terms of the length x of the shorter leg.

17. Area A rectangle is inscribed in a semicircle of radius 10, as shown in the figure. Find a function that models the area A of the rectangle in terms of its height h.

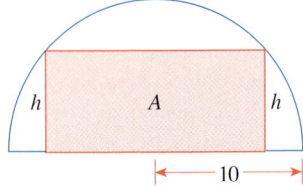

18. Height The volume of a cone is 100 in^3. Find a function that models the height h of the cone in terms of its radius r.

19–36 ■ In these problems you are asked to find a function that models a real-life situation, and then use the model to answer questions about the situation. Use the guidelines on page 205 to help you.

19. Maximizing a Product Consider the following problem: Find two numbers whose sum is 19 and whose product is as large as possible.

(a) Experiment with the problem by making a table like the one below, showing the product of different pairs of numbers that add up to 19. Based on the evidence in your table, estimate the answer to the problem.

First number	Second number	Product
1	18	18
2	17	34
3	16	48
⋮	⋮	⋮

(b) Find a function that models the product in terms of one of the two numbers.

(c) Use your model to solve the problem, and compare with your answer to part (a).

20. Minimizing a Sum Find two positive numbers whose sum is 100 and the sum of whose squares is a minimum.

21. Maximizing a Product Find two numbers whose sum is -24 and whose product is a maximum.

22. Maximizing Area Among all rectangles that have a perimeter of 20 ft, find the dimensions of the one with the largest area.

23. Fencing a Field Consider the following problem: A farmer has 2400 ft of fencing and wants to fence off a rectangular field that borders a straight river. He does not need a fence along the river (see the figure). What are the dimensions of the field of largest area that he can fence?

(a) Experiment with the problem by drawing several diagrams illustrating the situation. Calculate the area of each configuration, and use your results to estimate the dimensions of the largest possible field.

(b) Find a function that models the area of the field in terms of one of its sides.

(c) Use your model to solve the problem, and compare with your answer to part (a).

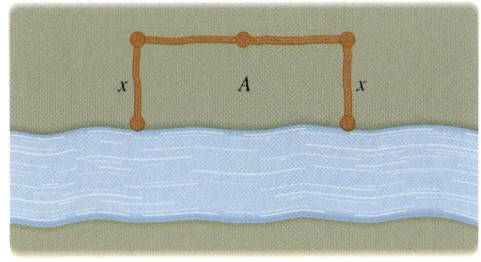

24. Dividing a Pen A rancher with 750 ft of fencing wants to enclose a rectangular area and then divide it into four pens with fencing parallel to one side of the rectangle (see the figure).

(a) Find a function that models the total area of the four pens.

(b) Find the largest possible total area of the four pens.

25. Fencing a Garden Plot A property owner wants to fence a garden plot adjacent to a road, as shown in the figure. The fencing next to the road must be sturdier and costs $5 per foot, but the other fencing costs just $3 per foot. The garden is to have an area of 1200 ft².

(a) Find a function that models the cost of fencing the garden.

(b) Find the garden dimensions that minimize the cost of fencing.

(c) If the owner has at most $600 to spend on fencing, find the range of lengths he can fence along the road.

26. Maximizing Area A wire 10 cm long is cut into two pieces, one of length x and the other of length $10 - x$, as shown in the figure. Each piece is bent into the shape of a square.

(a) Find a function that models the total area enclosed by the two squares.

(b) Find the value of x that minimizes the total area of the two squares.

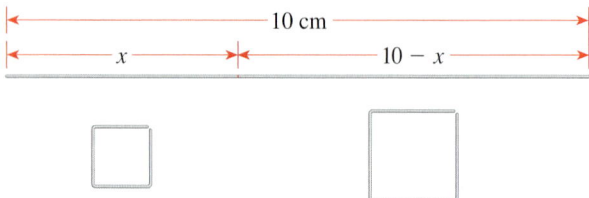

27. Stadium Revenue A baseball team plays in a stadium that holds 55,000 spectators. With the ticket price at $10, the average attendance at recent games has been 27,000. A market survey indicates that for every dollar the ticket price is lowered, attendance increases by 3000.

(a) Find a function that models the revenue in terms of ticket price.

(b) What ticket price is so high that no revenue is generated?

(c) Find the price that maximizes revenue from ticket sales.

28. Maximizing Profit A community bird-watching society makes and sells simple bird feeders to raise money for its conservation activities. The materials for each feeder cost $6, and they sell an average of 20 per week at a price of $10 each. They have been considering raising the price, so they conduct a survey and find that for every dollar increase they lose 2 sales per week.

(a) Find a function that models weekly profit in terms of price per feeder.

(b) What price should the society charge for each feeder to maximize profits? What is the maximum profit?

29. Light from a Window A Norman window has the shape of a rectangle surmounted by a semicircle, as shown in the figure. A Norman window with perimeter 30 ft is to be constructed.

(a) Find a function that models the area of the window.

(b) Find the dimensions of the window that admits the greatest amount of light.

30. Volume of a Box A box with an open top is to be constructed from a rectangular piece of cardboard with dimensions 12 in. by 20 in. by cutting out equal squares of side x at each corner and then folding up the sides (see the figure).

(a) Find a function that models the volume of the box.

(b) Find the values of x for which the volume is greater than 200 in³.

(c) Find the largest volume that such a box can have.

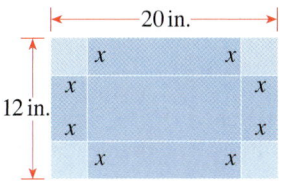

20 in.
12 in.

31. Area of a Box   An open box with a square base is to have a volume of 12 ft³.

   (a) Find a function that models the surface area of the box.

   (b) Find the box dimensions that minimize the amount of material used.

32. Inscribed Rectangle   Find the dimensions that give the largest area for the rectangle shown in the figure. Its base is on the x-axis and its other two vertices are above the x-axis, lying on the parabola $y = 8 - x^2$.

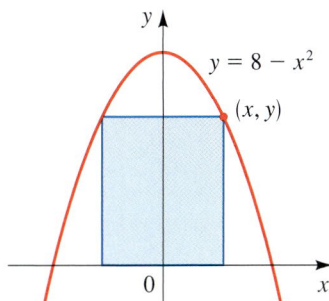

$y = 8 - x^2$
(x, y)

33. Minimizing Costs   A rancher wants to build a rectangular pen with an area of 100 m².

   (a) Find a function that models the length of fencing required.

   (b) Find the pen dimensions that require the minimum amount of fencing.

34. Minimizing Time   A man stands at a point A on the bank of a straight river, 2 mi wide. To reach point B, 7 mi downstream on the opposite bank, he first rows his boat to point P on the opposite bank and then walks the remaining distance x to B, as shown in the figure. He can row at a speed of 2 mi/h and walk at a speed of 5 mi/h.

   (a) Find a function that models the time needed for the trip.

(b) Where should he land so that he reaches B as soon as possible?

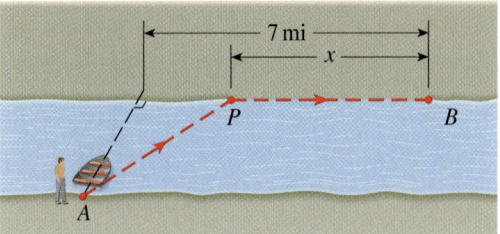

7 mi
x
P
B
A

35. Bird Flight   A bird is released from point A on an island, 5 mi from the nearest point B on a straight shoreline. The bird flies to a point C on the shoreline, and then flies along the shoreline to its nesting area D (see the figure). Suppose the bird requires 10 kcal/mi of energy to fly over land and 14 kcal/mi to fly over water (see Example 9 in Section 1.6).

   (a) Find a function that models the energy expenditure of the bird.

   (b) If the bird instinctively chooses a path that minimizes its energy expenditure, to what point does it fly?

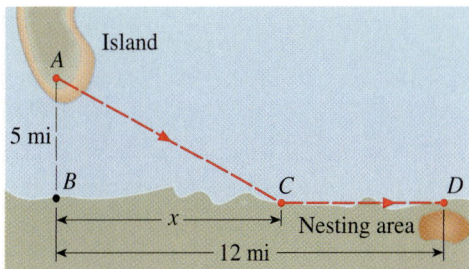

Island
A
5 mi
B
C
D
Nesting area
x
12 mi

36. Area of a Kite   A kite frame is to be made from six pieces of wood. The four pieces that form its border have been cut to the lengths indicated in the figure. Let x be as shown in the figure.

   (a) Show that the area of the kite is given by the function

$$A(x) = x(\sqrt{25 - x^2} + \sqrt{144 - x^2})$$

   (b) How long should each of the two crosspieces be to maximize the area of the kite?

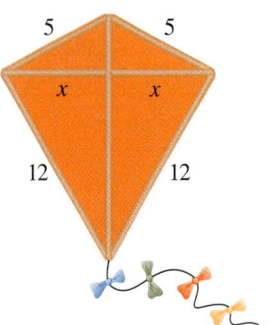

5   5
x x
12   12

2.7	**Combining Functions**

In this section we study different ways to combine functions to make new functions.

Sums, Differences, Products, and Quotients

Two functions f and g can be combined to form new functions $f + g$, $f - g$, fg, and f/g in a manner similar to the way we add, subtract, multiply, and divide real numbers. For example, we define the function $f + g$ by

$$(f + g)(x) = f(x) + g(x)$$

The sum of f and g is defined by

$$(f + g)(x) = f(x) + g(x)$$

The name of the new function is "$f + g$." So this $+$ sign stands for the operation of addition of *functions*. The $+$ sign on the right side, however, stands for addition of the *numbers* $f(x)$ and $g(x)$.

The new function $f + g$ is called the **sum** of the functions f and g; its value at x is $f(x) + g(x)$. Of course, the sum on the right-hand side makes sense only if both $f(x)$ and $g(x)$ are defined, that is, if x belongs to the domain of f and also to the domain of g. So, if the domain of f is A and the domain of g is B, then the domain of $f + g$ is the intersection of these domains, that is, $A \cap B$. Similarly, we can define the **difference** $f - g$, the **product** fg, and the **quotient** f/g of the functions f and g. Their domains are $A \cap B$, but in the case of the quotient we must remember not to divide by 0.

Algebra of Functions

Let f and g be functions with domains A and B. Then the functions $f + g$, $f - g$, fg, and f/g are defined as follows.

$$(f + g)(x) = f(x) + g(x) \qquad \text{Domain } A \cap B$$

$$(f - g)(x) = f(x) - g(x) \qquad \text{Domain } A \cap B$$

$$(fg)(x) = f(x)g(x) \qquad \text{Domain } A \cap B$$

$$\left(\frac{f}{g}\right)(x) = \frac{f(x)}{g(x)} \qquad \text{Domain } \{x \in A \cap B \mid g(x) \neq 0\}$$

Example 1   Combinations of Functions and Their Domains

Let $f(x) = \dfrac{1}{x - 2}$ and $g(x) = \sqrt{x}$.

(a) Find the functions $f + g$, $f - g$, fg, and f/g and their domains.

(b) Find $(f + g)(4)$, $(f - g)(4)$, $(fg)(4)$, and $(f/g)(4)$.

Solution

(a) The domain of f is $\{x \mid x \neq 2\}$ and the domain of g is $\{x \mid x \geq 0\}$. The intersection of the domains of f and g is

$$\{x \mid x \geq 0 \text{ and } x \neq 2\} = [0, 2) \cup (2, \infty)$$

Thus, we have

To divide fractions, invert the denominator and multiply:

$$\frac{1/(x-2)}{\sqrt{x}} = \frac{1/(x-2)}{\sqrt{x}/1}$$

$$= \frac{1}{x-2} \cdot \frac{1}{\sqrt{x}}$$

$$= \frac{1}{(x-2)\sqrt{x}}$$

$$(f+g)(x) = f(x) + g(x) = \frac{1}{x-2} + \sqrt{x} \qquad \text{Domain } \{x \mid x \geq 0 \text{ and } x \neq 2\}$$

$$(f-g)(x) = f(x) - g(x) = \frac{1}{x-2} - \sqrt{x} \qquad \text{Domain } \{x \mid x \geq 0 \text{ and } x \neq 2\}$$

$$(fg)(x) = f(x)g(x) = \frac{\sqrt{x}}{x-2} \qquad \text{Domain } \{x \mid x \geq 0 \text{ and } x \neq 2\}$$

$$\left(\frac{f}{g}\right)(x) = \frac{f(x)}{g(x)} = \frac{1}{(x-2)\sqrt{x}} \qquad \text{Domain } \{x \mid x > 0 \text{ and } x \neq 2\}$$

Note that in the domain of f/g we exclude 0 because $g(0) = 0$.

(b) Each of these values exist because $x = 4$ is in the domain of each function.

$$(f+g)(4) = f(4) + g(4) = \frac{1}{4-2} + \sqrt{4} = \frac{5}{2}$$

$$(f-g)(4) = f(4) - g(4) = \frac{1}{4-2} - \sqrt{4} = -\frac{3}{2}$$

$$(fg)(4) = f(4)g(4) = \left(\frac{1}{4-2}\right)\sqrt{4} = 1$$

$$\left(\frac{f}{g}\right)(4) = \frac{f(4)}{g(4)} = \frac{1}{(4-2)\sqrt{4}} = \frac{1}{4} \qquad \blacksquare$$

The graph of the function $f + g$ can be obtained from the graphs of f and g by **graphical addition**. This means that we add corresponding y-coordinates, as illustrated in the next example.

Example 2   Using Graphical Addition

The graphs of f and g are shown in Figure 1. Use graphical addition to graph the function $f + g$.

Solution   We obtain the graph of $f + g$ by "graphically adding" the value of $f(x)$ to $g(x)$ as shown in Figure 2. This is implemented by copying the line segment PQ on top of PR to obtain the point S on the graph of $f + g$.

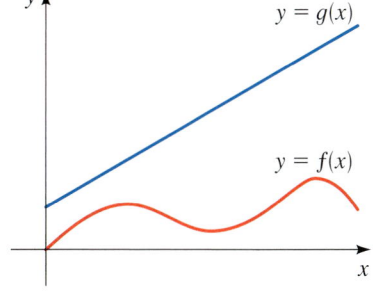

Figure 1

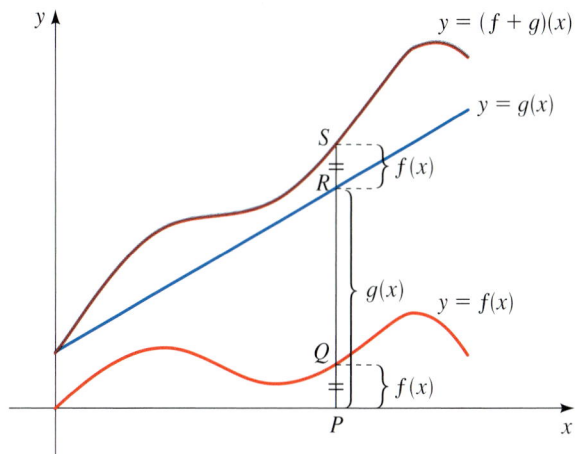

Figure 2

Graphical addition

Composition of Functions

Now let's consider a very important way of combining two functions to get a new function. Suppose $f(x) = \sqrt{x}$ and $g(x) = x^2 + 1$. We may define a function h as

$$h(x) = f(g(x)) = f(x^2 + 1) = \sqrt{x^2 + 1}$$

The function h is made up of the functions f and g in an interesting way: Given a number x, we first apply to it the function g, then apply f to the result. In this case, f is the rule "take the square root," g is the rule "square, then add 1," and h is the rule "square, then add 1, then take the square root." In other words, we get the rule h by applying the rule g and then the rule f. Figure 3 shows a machine diagram for h.

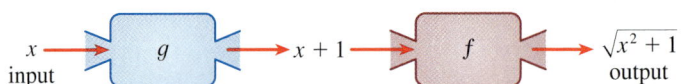

Figure 3

The h machine is composed of the g machine (first) and then the f machine.

In general, given any two functions f and g, we start with a number x in the domain of g and find its image $g(x)$. If this number $g(x)$ is in the domain of f, we can then calculate the value of $f(g(x))$. The result is a new function $h(x) = f(g(x))$ obtained by substituting g into f. It is called the *composition* (or *composite*) of f and g and is denoted by $f \circ g$ ("f composed with g").

Composition of Functions

Given two functions f and g, the **composite function** $f \circ g$ (also called the **composition** of f and g) is defined by

$$(f \circ g)(x) = f(g(x))$$

The domain of $f \circ g$ is the set of all x in the domain of g such that $g(x)$ is in the domain of f. In other words, $(f \circ g)(x)$ is defined whenever both $g(x)$ and $f(g(x))$ are defined. We can picture $f \circ g$ using an arrow diagram (Figure 4).

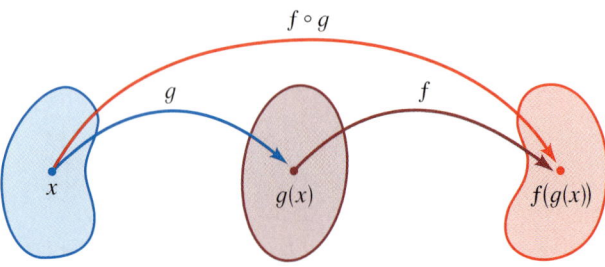

Figure 4

Arrow diagram for $f \circ g$

Example 3 Finding the Composition of Functions

Let $f(x) = x^2$ and $g(x) = x - 3$.

(a) Find the functions $f \circ g$ and $g \circ f$ and their domains.

(b) Find $(f \circ g)(5)$ and $(g \circ f)(7)$.

Solution

In Example 3, f is the rule "square" and g is the rule "subtract 3." The function $f \circ g$ *first* subtracts 3 and *then* squares; the function $g \circ f$ *first* squares and *then* subtracts 3.

(a) We have

$$(f \circ g)(x) = f(g(x)) \qquad \text{Definition of } f \circ g$$
$$= f(x - 3) \qquad \text{Definition of } g$$
$$= (x - 3)^2 \qquad \text{Definition of } f$$

and

$$(g \circ f)(x) = g(f(x)) \qquad \text{Definition of } g \circ f$$
$$= g(x^2) \qquad \text{Definition of } f$$
$$= x^2 - 3 \qquad \text{Definition of } g$$

The domains of both $f \circ g$ and $g \circ f$ are $\mathbb{R}$.

(b) We have

$$(f \circ g)(5) = f(g(5)) = f(2) = 2^2 = 4$$
$$(g \circ f)(7) = g(f(7)) = g(49) = 49 - 3 = 46$$ ∎

You can see from Example 3 that, in general, $f \circ g \neq g \circ f$. Remember that the notation $f \circ g$ means that the function g is applied first and then f is applied second.

Example 4 Finding the Composition of Functions

If $f(x) = \sqrt{x}$ and $g(x) = \sqrt{2 - x}$, find the following functions and their domains.

(a) $f \circ g$ (b) $g \circ f$ (c) $f \circ f$ (d) $g \circ g$

Solution

(a)
$$(f \circ g)(x) = f(g(x)) \qquad \text{Definition of } f \circ g$$
$$= f(\sqrt{2 - x}) \qquad \text{Definition of } g$$
$$= \sqrt{\sqrt{2 - x}} \qquad \text{Definition of } f$$
$$= \sqrt[4]{2 - x}$$

The domain of $f \circ g$ is $\{x \mid 2 - x \geq 0\} = \{x \mid x \leq 2\} = (-\infty, 2]$.

(b)
$$(g \circ f)(x) = g(f(x)) \qquad \text{Definition of } g \circ f$$
$$= g(\sqrt{x}) \qquad \text{Definition of } f$$
$$= \sqrt{2 - \sqrt{x}} \qquad \text{Definition of } g$$

For $\sqrt{x}$ to be defined, we must have $x \geq 0$. For $\sqrt{2 - \sqrt{x}}$ to be defined, we

The graphs of f and g of Example 4, as well as $f \circ g$, $g \circ f$, $f \circ f$, and $g \circ g$, are shown below. These graphs indicate that the operation of composition can produce functions quite different from the original functions.

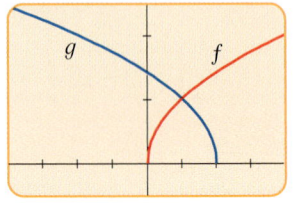

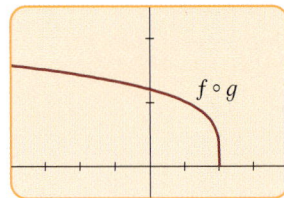

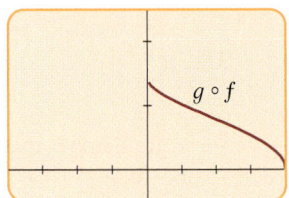

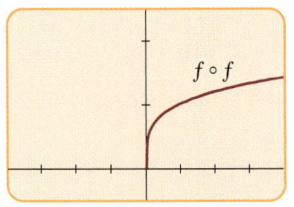

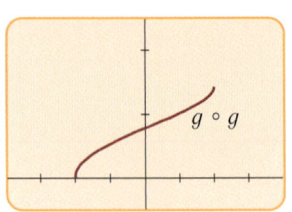

must have $2 - \sqrt{x} \geq 0$, that is, $\sqrt{x} \leq 2$, or $x \leq 4$. Thus, we have $0 \leq x \leq 4$, so the domain of $g \circ f$ is the closed interval $[0, 4]$.

(c)

$$(f \circ f)(x) = f(f(x)) \qquad \text{Definition of } f \circ f$$
$$= f(\sqrt{x}) \qquad \text{Definition of } f$$
$$= \sqrt{\sqrt{x}} \qquad \text{Definition of } f$$
$$= \sqrt[4]{x}$$

The domain of $f \circ f$ is $[0, \infty)$.

(d)

$$(g \circ g)(x) = g(g(x)) \qquad \text{Definition of } g \circ g$$
$$= g(\sqrt{2 - x}) \qquad \text{Definition of } g$$
$$= \sqrt{2 - \sqrt{2 - x}} \qquad \text{Definition of } g$$

This expression is defined when both $2 - x \geq 0$ and $2 - \sqrt{2 - x} \geq 0$. The first inequality means $x \leq 2$, and the second is equivalent to $\sqrt{2 - x} \leq 2$, or $2 - x \leq 4$, or $x \geq -2$. Thus, $-2 \leq x \leq 2$, so the domain of $g \circ g$ is $[-2, 2]$.   ■

It is possible to take the composition of three or more functions. For instance, the composite function $f \circ g \circ h$ is found by first applying h, then g, and then f as follows:

$$(f \circ g \circ h)(x) = f(g(h(x)))$$

Example 5   A Composition of Three Functions

Find $f \circ g \circ h$ if $f(x) = x/(x + 1)$, $g(x) = x^{10}$ and $h(x) = x + 3$.

Solution

$$(f \circ g \circ h)(x) = f(g(h(x))) \qquad \text{Definition of } f \circ g \circ h$$
$$= f(g(x + 3)) \qquad \text{Definition of } h$$
$$= f((x + 3)^{10}) \qquad \text{Definition of } g$$
$$= \frac{(x + 3)^{10}}{(x + 3)^{10} + 1} \qquad \text{Definition of } f \qquad ■$$

So far we have used composition to build complicated functions from simpler ones. But in calculus it is useful to be able to "decompose" a complicated function into simpler ones, as shown in the following example.

Example 6   Recognizing a Composition of Functions

Given $F(x) = \sqrt[4]{x + 9}$, find functions f and g such that $F = f \circ g$.

Solution   Since the formula for F says to first add 9 and then take the fourth root, we let

$$g(x) = x + 9 \qquad \text{and} \qquad f(x) = \sqrt[4]{x}$$

Then

$$(f \circ g)(x) = f(g(x)) \qquad \text{Definition of } f \circ g$$

$$= f(x + 9) \qquad \text{Definition of } g$$

$$= \sqrt[4]{x + 9} \qquad \text{Definition of } f$$

$$= F(x)$$

Example 7   An Application of Composition of Functions

A ship is traveling at 20 mi/h parallel to a straight shoreline. The ship is 5 mi from shore. It passes a lighthouse at noon.

(a) Express the distance s between the lighthouse and the ship as a function of d, the distance the ship has traveled since noon; that is, find f so that $s = f(d)$.

(b) Express d as a function of t, the time elapsed since noon; that is, find g so that $d = g(t)$.

(c) Find $f \circ g$. What does this function represent?

Solution   We first draw a diagram as in Figure 5.

(a) We can relate the distances s and d by the Pythagorean Theorem. Thus, s can be expressed as a function of d by

$$s = f(d) = \sqrt{25 + d^2}$$

(b) Since the ship is traveling at 20 mi/h, the distance d it has traveled is a function of t as follows:

$$d = g(t) = 20t$$

(c) We have

$$(f \circ g)(t) = f(g(t)) \qquad \text{Definition of } f \circ g$$

$$= f(20t) \qquad \text{Definition of } g$$

$$= \sqrt{25 + (20t)^2} \qquad \text{Definition of } f$$

The function $f \circ g$ gives the distance of the ship from the lighthouse as a function of time.

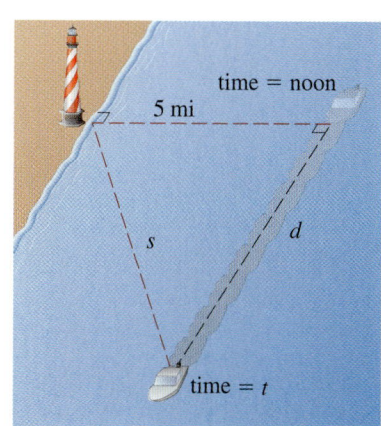

Figure 5

distance = rate $\times$ time

2.7   Exercises

1–6 ■ Find $f + g$, $f - g$, fg, and f/g and their domains.

1. $f(x) = x - 3$, $g(x) = x^2$

2. $f(x) = x^2 + 2x$, $g(x) = 3x^2 - 1$

3. $f(x) = \sqrt{4 - x^2}$, $g(x) = \sqrt{1 + x}$

4. $f(x) = \sqrt{9 - x^2}$, $g(x) = \sqrt{x^2 - 4}$

5. $f(x) = \dfrac{2}{x}$, $g(x) = \dfrac{4}{x + 4}$

6. $f(x) = \dfrac{2}{x + 1}$, $g(x) = \dfrac{x}{x + 1}$

7–10 ■ Find the domain of the function.

7. $f(x) = \sqrt{x} + \sqrt{1 - x}$

8. $g(x) = \sqrt{x + 1} - \dfrac{1}{x}$

9. $h(x) = (x - 3)^{-1/4}$

10. $k(x) = \dfrac{\sqrt{x + 3}}{x - 1}$

11–12 ■ Use graphical addition to sketch the graph of $f + g$.

11.

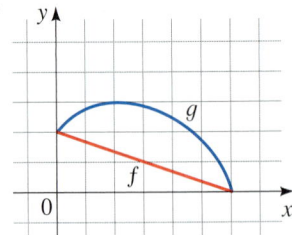

12.

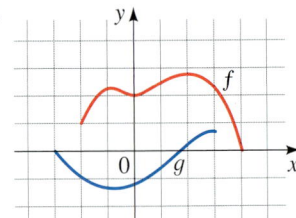

13–16 ■ Draw the graphs of f, g, and $f + g$ on a common screen to illustrate graphical addition.

13. $f(x) = \sqrt{1 + x}$, $g(x) = \sqrt{1 - x}$

14. $f(x) = x^2$, $g(x) = \sqrt{x}$

15. $f(x) = x^2$, $g(x) = \frac{1}{3}x^3$

16. $f(x) = \sqrt[4]{1 - x}$, $g(x) = \sqrt{1 - \dfrac{x^2}{9}}$

17–22 ■ Use $f(x) = 3x - 5$ and $g(x) = 2 - x^2$ to evaluate the expression.

17. (a) $f(g(0))$ **(b)** $g(f(0))$

18. (a) $f(f(4))$ **(b)** $g(g(3))$

19. (a) $(f \circ g)(-2)$ **(b)** $(g \circ f)(-2)$

20. (a) $(f \circ f)(-1)$ **(b)** $(g \circ g)(2)$

21. (a) $(f \circ g)(x)$ **(b)** $(g \circ f)(x)$

22. (a) $(f \circ f)(x)$ **(b)** $(g \circ g)(x)$

23–28 ■ Use the given graphs of f and g to evaluate the expression.

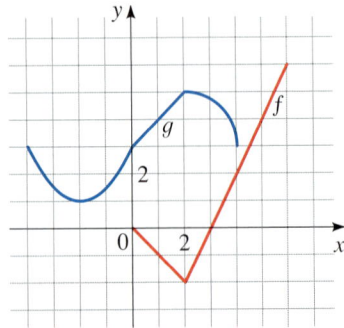

23. $f(g(2))$

24. $g(f(0))$

25. $(g \circ f)(4)$

26. $(f \circ g)(0)$

27. $(g \circ g)(-2)$

28. $(f \circ f)(4)$

29–40 ■ Find the functions $f \circ g$, $g \circ f$, $f \circ f$, and $g \circ g$ and their domains.

29. $f(x) = 2x + 3$, $g(x) = 4x - 1$

30. $f(x) = 6x - 5$, $g(x) = \dfrac{x}{2}$

31. $f(x) = x^2$, $g(x) = x + 1$

32. $f(x) = x^3 + 2$, $g(x) = \sqrt[3]{x}$

33. $f(x) = \dfrac{1}{x}$, $g(x) = 2x + 4$

34. $f(x) = x^2$, $g(x) = \sqrt{x - 3}$

35. $f(x) = |x|$, $g(x) = 2x + 3$

36. $f(x) = x - 4$, $g(x) = |x + 4|$

37. $f(x) = \dfrac{x}{x + 1}$, $g(x) = 2x - 1$

38. $f(x) = \dfrac{1}{\sqrt{x}}$, $g(x) = x^2 - 4x$

39. $f(x) = \sqrt[3]{x}$, $g(x) = \sqrt[4]{x}$

40. $f(x) = \dfrac{2}{x}$, $g(x) = \dfrac{x}{x + 2}$

41–44 ■ Find $f \circ g \circ h$.

41. $f(x) = x - 1$, $g(x) = \sqrt{x}$, $h(x) = x - 1$

42. $f(x) = \dfrac{1}{x}$, $g(x) = x^3$, $h(x) = x^2 + 2$

43. $f(x) = x^4 + 1$, $g(x) = x - 5$, $h(x) = \sqrt{x}$

44. $f(x) = \sqrt{x}$, $g(x) = \dfrac{x}{x - 1}$, $h(x) = \sqrt[3]{x}$

45–50 ■ Express the function in the form $f \circ g$.

45. $F(x) = (x - 9)^5$

46. $F(x) = \sqrt{x} + 1$

47. $G(x) = \dfrac{x^2}{x^2 + 4}$

48. $G(x) = \dfrac{1}{x + 3}$

49. $H(x) = |1 - x^3|$

50. $H(x) = \sqrt{1 + \sqrt{x}}$

51–54 ■ Express the function in the form $f \circ g \circ h$.

51. $F(x) = \dfrac{1}{x^2 + 1}$

52. $F(x) = \sqrt[3]{\sqrt{x} - 1}$

53. $G(x) = (4 + \sqrt[3]{x})^9$

54. $G(x) = \dfrac{2}{(3 + \sqrt{x})^2}$

Applications

55–56 ■ **Revenue, Cost, and Profit** A print shop makes bumper stickers for election campaigns. If x stickers are ordered (where $x < 10{,}000$), then the price per sticker is $0.15 - 0.000002x$ dollars, and the total cost of producing the order is $0.095x - 0.0000005x^2$ dollars.

55. Use the fact that

$$\text{revenue} = \text{price per item} \times \text{number of items sold}$$

to express $R(x)$, the revenue from an order of x stickers, as a product of two functions of x.

56. Use the fact that

$$\text{profit} = \text{revenue} - \text{cost}$$

to express $P(x)$, the profit on an order of x stickers, as a difference of two functions of x.

57. Area of a Ripple A stone is dropped in a lake, creating a circular ripple that travels outward at a speed of 60 cm/s.

(a) Find a function g that models the radius as a function of time.

(b) Find a function f that models the area of the circle as a function of the radius.

(c) Find $f \circ g$. What does this function represent?

58. Inflating a Balloon A spherical balloon is being inflated. The radius of the balloon is increasing at the rate of 1 cm/s.

(a) Find a function f that models the radius as a function of time.

(b) Find a function g that models the volume as a function of the radius.

(c) Find $g \circ f$. What does this function represent?

59. Area of a Balloon A spherical weather balloon is being inflated. The radius of the balloon is increasing at the rate of 2 cm/s. Express the surface area of the balloon as a function of time t (in seconds).

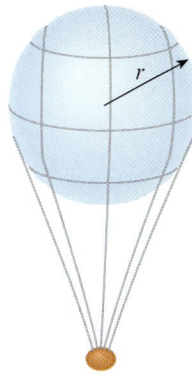

60. Multiple Discounts You have a $50 coupon from the manufacturer good for the purchase of a cell phone. The store where you are purchasing your cell phone is offering a 20% discount on all cell phones. Let x represent the regular price of the cell phone.

(a) Suppose only the 20% discount applies. Find a function f that models the purchase price of the cell phone as a function of the regular price x.

(b) Suppose only the $50 coupon applies. Find a function g that models the purchase price of the cell phone as a function of the sticker price x.

(c) If you can use the coupon and the discount, then the purchase price is either $f \circ g(x)$ or $g \circ f(x)$, depending on the order in which they are applied to the price. Find both $f \circ g(x)$ and $g \circ f(x)$. Which composition gives the lower price?

61. Multiple Discounts An appliance dealer advertises a 10% discount on all his washing machines. In addition, the manufacturer offers a $100 rebate on the purchase of a washing machine. Let x represent the sticker price of the washing machine.

(a) Suppose only the 10% discount applies. Find a function f that models the purchase price of the washer as a function of the sticker price x.

(b) Suppose only the $100 rebate applies. Find a function g that models the purchase price of the washer as a function of the sticker price x.

(c) Find $f \circ g$ and $g \circ f$. What do these functions represent? Which is the better deal?

62. Airplane Trajectory An airplane is flying at a speed of 350 mi/h at an altitude of one mile. The plane passes directly above a radar station at time $t = 0$.

(a) Express the distance s (in miles) between the plane and the radar station as a function of the horizontal distance d (in miles) that the plane has flown.

(b) Express d as a function of the time t (in hours) that the plane has flown.

(c) Use composition to express s as a function of t.

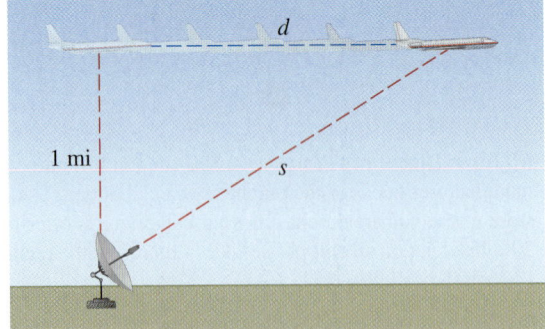

Discovery • Discussion

63. Compound Interest A savings account earns 5% interest compounded annually. If you invest x dollars in such an account, then the amount $A(x)$ of the investment after one year is the initial investment plus 5%; that is, $A(x) = x + 0.05x = 1.05x$. Find

$$A \circ A$$
$$A \circ A \circ A$$
$$A \circ A \circ A \circ A$$

What do these compositions represent? Find a formula for what you get when you compose n copies of A.

64. Composing Linear Functions The graphs of the functions

$$f(x) = m_1 x + b_1$$
$$g(x) = m_2 x + b_2$$

are lines with slopes m_1 and m_2, respectively. Is the graph of $f \circ g$ a line? If so, what is its slope?

65. Solving an Equation for an Unknown Function Suppose that

$$g(x) = 2x + 1$$
$$h(x) = 4x^2 + 4x + 7$$

Find a function f such that $f \circ g = h$. (Think about what operations you would have to perform on the formula for g to end up with the formula for h.) Now suppose that

$$f(x) = 3x + 5$$
$$h(x) = 3x^2 + 3x + 2$$

Use the same sort of reasoning to find a function g such that $f \circ g = h$.

66. Compositions of Odd and Even Functions Suppose that

$$h = f \circ g$$

If g is an even function, is h necessarily even? If g is odd, is h odd? What if g is odd and f is odd? What if g is odd and f is even?

DISCOVERY PROJECT

Iteration and Chaos

The **iterates** of a function f at a point x_0 are $f(x_0)$, $f(f(x_0))$, $f(f(f(x_0)))$, and so on. We write

$$x_1 = f(x_0) \qquad \text{The first iterate}$$

$$x_2 = f(f(x_0)) \qquad \text{The second iterate}$$

$$x_3 = f(f(f(x_0))) \qquad \text{The third iterate}$$

For example, if $f(x) = x^2$, then the iterates of f at 2 are $x_1 = 4$, $x_2 = 16$, $x_3 = 256$, and so on. (Check this.) Iterates can be described graphically as in Figure 1. Start with x_0 on the x-axis, move vertically to the graph of f, then horizontally to the line $y = x$, then vertically to the graph of f, and so on. The x-coordinates of the points on the graph of f are the iterates of f at x_0.

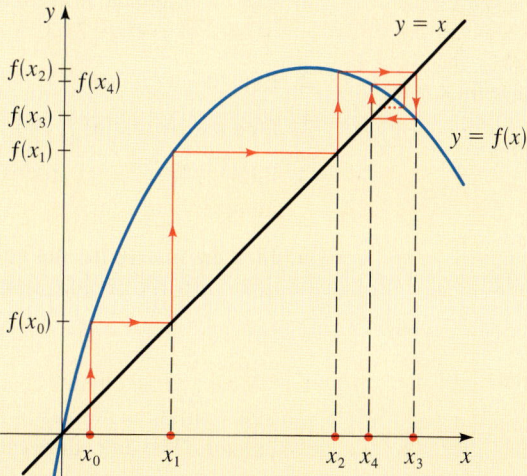

Figure 1

Iterates are important in studying the **logistic function**

$$f(x) = kx(1 - x)$$

n	x_n
0	0.1
1	0.234
2	0.46603
3	0.64700
4	0.59382
5	0.62712
6	0.60799
7	0.61968
8	0.61276
9	0.61694
10	0.61444
11	0.61595
12	0.61505

which models the population of a species with limited potential for growth (such as rabbits on an island or fish in a pond). In this model the maximum population that the environment can support is 1 (that is, 100%). If we start with a fraction of that population, say 0.1 (10%), then the iterates of f at 0.1 give the population after each time interval (days, months, or years, depending on the species). The constant k depends on the rate of growth of the species being modeled; it is called the **growth constant**. For example, for $k = 2.6$ and $x_0 = 0.1$ the iterates shown in the table to the left give the population of the species for the first 12 time intervals. The population seems to be stabilizing around 0.615 (that is, 61.5% of maximum).

In the three graphs in Figure 2, we plot the iterates of f at 0.1 for different values of the growth constant k. For $k = 2.6$ the population appears to stabilize at a value 0.615 of maximum, for $k = 3.1$ the population appears to oscillate between two values, and for $k = 3.8$ no obvious pattern emerges. This latter situation is described mathematically by the word **chaos**.

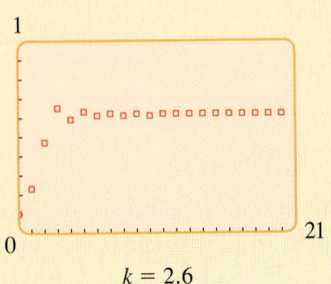

$k = 2.6$

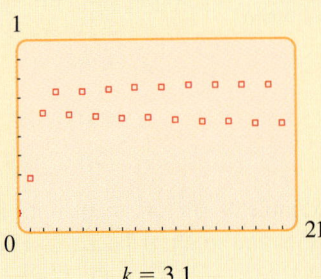

$k = 3.1$

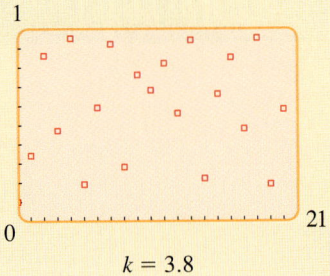

$k = 3.8$

Figure 2

The following TI-83 program draws the first graph in Figure 2. The other graphs are obtained by choosing the appropriate value for K in the program.

```
PROGRAM:ITERATE
:ClrDraw
:2.6→K
:0.1→X
:For(N,1,20)
:K*X*(1-X)→Z
:Pt-On(N,Z,2)
:Z→X
:End
```

1. Use the graphical procedure illustrated in Figure 1 to find the first five iterates of $f(x) = 2x(1 - x)$ at $x = 0.1$.

2. Find the iterates of $f(x) = x^2$ at $x = 1$.

3. Find the iterates of $f(x) = 1/x$ at $x = 2$.

4. Find the first six iterates of $f(x) = 1/(1 - x)$ at $x = 2$. What is the 1000th iterate of f at 2?

5. Find the first 10 iterates of the logistic function at $x = 0.1$ for the given value of k. Does the population appear to stabilize, oscillate, or is it chaotic?

   (a) $k = 2.1$ (b) $k = 3.2$ (c) $k = 3.9$

6. It's easy to find iterates using a graphing calculator. The following steps show how to find the iterates of $f(x) = kx(1 - x)$ at 0.1 for $k = 3$ on a TI-83 calculator. (The procedure can be adapted for any graphing calculator.)

$Y_1 = K * X * (1 - X)$	Enter f as Y_1 on the graph list
$3 → K$	Store 3 in the variable K
$0.1 → X$	Store 0.1 in the variable X
$Y_1 → X$	Evaluate f at X and store result back in X
0.27	Press ENTER and obtain first iterate
0.5913	Keep pressing ENTER to re-execute the
0.72499293	command and obtain successive iterates
0.59813454435	

You can also use the program in the margin to graph the iterates and study them visually.

Use a graphing calculator to experiment with how the value of k affects the iterates of $f(x) = kx(1 - x)$ at 0.1. Find several different values of k that make the iterates stabilize at one value, oscillate between two values, and exhibit chaos. (Use values of k between 1 and 4.) Can you find a value of k that makes the iterates oscillate between *four* values?

2.8 One-to-One Functions and Their Inverses

The *inverse* of a function is a rule that acts on the output of the function and produces the corresponding input. So, the inverse "undoes" or reverses what the function has done. Not all functions have inverses; those that do are called *one-to-one*.

One-to-One Functions

Let's compare the functions f and g whose arrow diagrams are shown in Figure 1. Note that f never takes on the same value twice (any two numbers in A have different images), whereas g does take on the same value twice (both 2 and 3 have the same image, 4). In symbols, $g(2) = g(3)$ but $f(x_1) \neq f(x_2)$ whenever $x_1 \neq x_2$. Functions that have this latter property are called *one-to-one*.

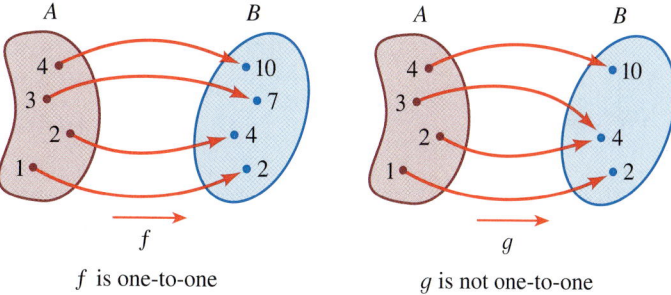

Figure 1 f is one-to-one g is not one-to-one

Definition of a One-to-one Function

A function with domain A is called a **one-to-one function** if no two elements of A have the same image, that is,

$$f(x_1) \neq f(x_2) \qquad \text{whenever } x_1 \neq x_2$$

An equivalent way of writing the condition for a one-to-one function is this:

$$\text{If } f(x_1) = f(x_2), \text{ then } x_1 = x_2.$$

If a horizontal line intersects the graph of f at more than one point, then we see from Figure 2 that there are numbers $x_1 \neq x_2$ such that $f(x_1) = f(x_2)$. This means that f is not one-to-one. Therefore, we have the following geometric method for determining whether a function is one-to-one.

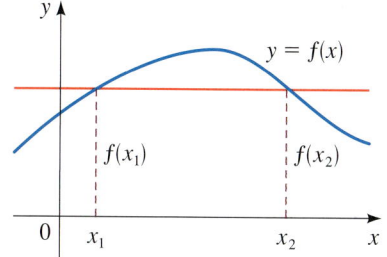

Figure 2

This function is not one-to-one because $f(x_1) = f(x_2)$.

Horizontal Line Test

A function is one-to-one if and only if no horizontal line intersects its graph more than once.

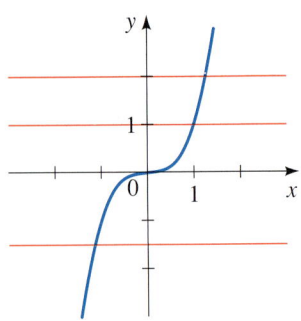

Figure 3

$f(x) = x^3$ is one-to-one.

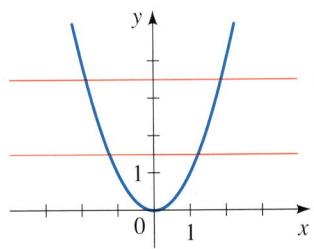

Figure 4

$f(x) = x^2$ is not one-to-one.

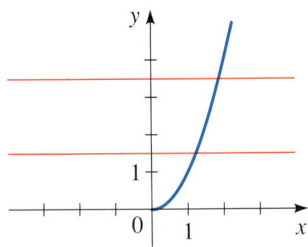

Figure 5

$f(x) = x^2$ $(x \geq 0)$ is one-to-one.

Example 1 **Deciding whether a Function Is One-to-One**

Is the function $f(x) = x^3$ one-to-one?

Solution 1 If $x_1 \neq x_2$, then $x_1^3 \neq x_2^3$ (two different numbers cannot have the same cube). Therefore, $f(x) = x^3$ is one-to-one.

Solution 2 From Figure 3 we see that no horizontal line intersects the graph of $f(x) = x^3$ more than once. Therefore, by the Horizontal Line Test, f is one-to-one. ■

Notice that the function f of Example 1 is increasing and is also one-to-one. In fact, it can be proved that *every increasing function and every decreasing function is one-to-one*.

Example 2 **Deciding whether a Function Is One-to-One**

Is the function $g(x) = x^2$ one-to-one?

Solution 1 This function is not one-to-one because, for instance,

$$g(1) = 1 \qquad \text{and} \qquad g(-1) = 1$$

and so 1 and -1 have the same image.

Solution 2 From Figure 4 we see that there are horizontal lines that intersect the graph of g more than once. Therefore, by the Horizontal Line Test, g is not one-to-one. ■

Although the function g in Example 2 is not one-to-one, it is possible to restrict its domain so that the resulting function is one-to-one. In fact, if we define

$$h(x) = x^2, \qquad x \geq 0$$

then h is one-to-one, as you can see from Figure 5 and the Horizontal Line Test.

Example 3 **Showing That a Function Is One-to-One**

Show that the function $f(x) = 3x + 4$ is one-to-one.

Solution

Suppose there are numbers x_1 and x_2 such that $f(x_1) = f(x_2)$. Then

$$3x_1 + 4 = 3x_2 + 4 \qquad \textcolor{blue}{\text{Suppose } f(x_1) = f(x_2)}$$

$$3x_1 = 3x_2 \qquad \textcolor{blue}{\text{Subtract 4}}$$

$$x_1 = x_2 \qquad \textcolor{blue}{\text{Divide by 3}}$$

Therefore, f is one-to-one. ■

The Inverse of a Function

One-to-one functions are important because they are precisely the functions that possess inverse functions according to the following definition.

⊘ Don't mistake the -1 in f^{-1} for an exponent.

$$f^{-1} \quad \textit{does not mean} \quad \frac{1}{f(x)}$$

The reciprocal $1/f(x)$ is written as $(f(x))^{-1}$.

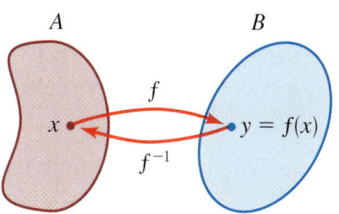

Figure 6

Definition of the Inverse of a Function

Let f be a one-to-one function with domain A and range B. Then its **inverse function** f^{-1} has domain B and range A and is defined by

$$f^{-1}(y) = x \quad \Leftrightarrow \quad f(x) = y$$

for any y in B.

This definition says that if f takes x into y, then f^{-1} takes y back into x. (If f were not one-to-one, then f^{-1} would not be defined uniquely.) The arrow diagram in Figure 6 indicates that f^{-1} reverses the effect of f. From the definition we have

$$\text{domain of } f^{-1} = \text{range of } f$$

$$\text{range of } f^{-1} = \text{domain of } f$$

Example 4   Finding f^{-1} for Specific Values

If $f(1) = 5$, $f(3) = 7$, and $f(8) = -10$, find $f^{-1}(5)$, $f^{-1}(7)$, and $f^{-1}(-10)$.

Solution   From the definition of f^{-1} we have

$$f^{-1}(5) = 1 \qquad \text{because} \qquad f(1) = 5$$
$$f^{-1}(7) = 3 \qquad \text{because} \qquad f(3) = 7$$
$$f^{-1}(-10) = 8 \qquad \text{because} \qquad f(8) = -10$$

Figure 7 shows how f^{-1} reverses the effect of f in this case.

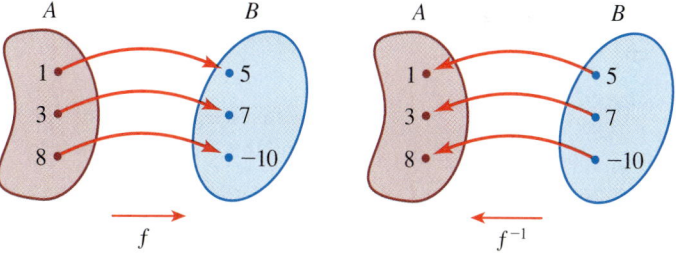

Figure 7

By definition the inverse function f^{-1} undoes what f does: If we start with x, apply f, and then apply f^{-1}, we arrive back at x, where we started. Similarly, f undoes what f^{-1} does. In general, any function that reverses the effect of f in this way must be the inverse of f. These observations are expressed precisely as follows.

Inverse Function Property

Let f be a one-to-one function with domain A and range B. The inverse function f^{-1} satisfies the following cancellation properties.

$$f^{-1}(f(x)) = x \qquad \text{for every } x \text{ in } A$$

$$f(f^{-1}(x)) = x \qquad \text{for every } x \text{ in } B$$

Conversely, any function f^{-1} satisfying these equations is the inverse of f.

These properties indicate that f is the inverse function of f^{-1}, so we say that f and f^{-1} are *inverses of each other*.

Example 5   Verifying That Two Functions Are Inverses

Show that $f(x) = x^3$ and $g(x) = x^{1/3}$ are inverses of each other.

Solution   Note that the domain and range of both f and g is $\mathbb{R}$. We have

$$g(f(x)) = g(x^3) = (x^3)^{1/3} = x$$
$$f(g(x)) = f(x^{1/3}) = (x^{1/3})^3 = x$$

So, by the Property of Inverse Functions, f and g are inverses of each other. These equations simply say that the cube function and the cube root function, when composed, cancel each other.   ■

Now let's examine how we compute inverse functions. We first observe from the definition of f^{-1} that

$$y = f(x) \quad \Leftrightarrow \quad f^{-1}(y) = x$$

So, if $y = f(x)$ and if we are able to solve this equation for x in terms of y, then we must have $x = f^{-1}(y)$. If we then interchange x and y, we have $y = f^{-1}(x)$, which is the desired equation.

How to Find the Inverse of a One-to-One Function

1. Write $y = f(x)$.

2. Solve this equation for x in terms of y (if possible).

3. Interchange x and y. The resulting equation is $y = f^{-1}(x)$.

In Example 6 note how f^{-1} reverses the effect of f. The function f is the rule "multiply by 3, then subtract 2," whereas f^{-1} is the rule "add 2, then divide by 3."

Note that Steps 2 and 3 can be reversed. In other words, we can interchange x and y first and then solve for y in terms of x.

Example 6   Finding the Inverse of a Function

Find the inverse of the function $f(x) = 3x - 2$.

Check Your Answer

We use the Inverse Function Property.

$$f^{-1}(f(x)) = f^{-1}(3x - 2)$$
$$= \frac{(3x - 2) + 2}{3}$$
$$= \frac{3x}{3} = x$$

$$f(f^{-1}(x)) = f\left(\frac{x + 2}{3}\right)$$
$$= 3\left(\frac{x + 2}{3}\right) - 2$$
$$= x + 2 - 2 = x \quad \checkmark$$

Solution   First we write $y = f(x)$.

$$y = 3x - 2$$

Then we solve this equation for x:

$$3x = y + 2 \qquad \text{Add 2}$$
$$x = \frac{y + 2}{3} \qquad \text{Divide by 3}$$

Finally, we interchange x and y:

$$y = \frac{x + 2}{3}$$

Therefore, the inverse function is $f^{-1}(x) = \dfrac{x + 2}{3}$.   ■

In Example 7 note how f^{-1} reverses the effect of f. The function f is the rule "take the fifth power, subtract 3, then divide by 2," whereas f^{-1} is the rule "multiply by 2, add 3, then take the fifth root."

Check Your Answer

We use the Inverse Function Property.

$$f^{-1}(f(x)) = f^{-1}\left(\frac{x^5 - 3}{2}\right)$$

$$= \left[2\left(\frac{x^5 - 3}{2}\right) + 3\right]^{1/5}$$

$$= (x^5 - 3 + 3)^{1/5}$$

$$= (x^5)^{1/5} = x$$

$$f(f^{-1}(x)) = f((2x + 3)^{1/5})$$

$$= \frac{[(2x + 3)^{1/5}]^5 - 3}{2}$$

$$= \frac{2x + 3 - 3}{2}$$

$$= \frac{2x}{2} = x \qquad \checkmark$$

Example 7 Finding the Inverse of a Function

Find the inverse of the function $f(x) = \dfrac{x^5 - 3}{2}$.

Solution We first write $y = (x^5 - 3)/2$ and solve for x.

$$y = \frac{x^5 - 3}{2} \qquad \text{Equation defining function}$$

$$2y = x^5 - 3 \qquad \text{Multiply by 2}$$

$$x^5 = 2y + 3 \qquad \text{Add 3}$$

$$x = (2y + 3)^{1/5} \qquad \text{Take fifth roots}$$

Then we interchange x and y to get $y = (2x + 3)^{1/5}$. Therefore, the inverse function is $f^{-1}(x) = (2x + 3)^{1/5}$. ∎

The principle of interchanging x and y to find the inverse function also gives us a method for obtaining the graph of f^{-1} from the graph of f. If $f(a) = b$, then $f^{-1}(b) = a$. Thus, the point (a, b) is on the graph of f if and only if the point (b, a) is on the graph of f^{-1}. But we get the point (b, a) from the point (a, b) by reflecting in the line $y = x$ (see Figure 8). Therefore, as Figure 9 illustrates, the following is true.

> The graph of f^{-1} is obtained by reflecting the graph of f in the line $y = x$.

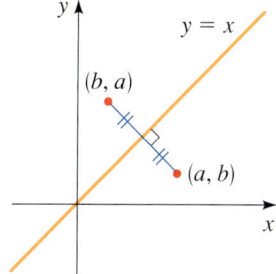

Figure 8

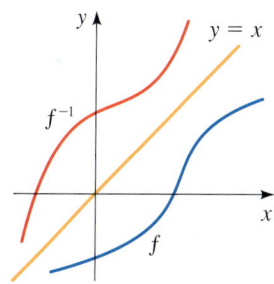

Figure 9

Example 8 Finding the Inverse of a Function

(a) Sketch the graph of $f(x) = \sqrt{x - 2}$.

(b) Use the graph of f to sketch the graph of f^{-1}.

(c) Find an equation for f^{-1}.

Solution

(a) Using the transformations from Section 2.4, we sketch the graph of $y = \sqrt{x - 2}$ by plotting the graph of the function $y = \sqrt{x}$ (Example 1(c) in Section 2.2) and moving it to the right 2 units.

(b) The graph of f^{-1} is obtained from the graph of f in part (a) by reflecting it in the line $y = x$, as shown in Figure 10.

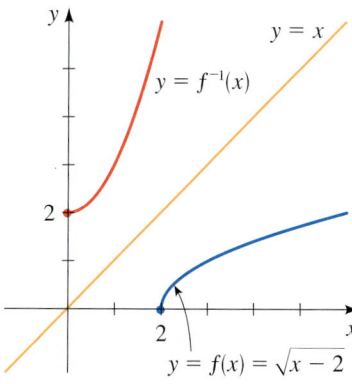

Figure 10

(c) Solve $y = \sqrt{x - 2}$ for x, noting that $y \geq 0$.

$$\sqrt{x - 2} = y$$

$$x - 2 = y^2 \qquad \text{Square each side}$$

$$x = y^2 + 2, \quad y \geq 0 \qquad \text{Add 2}$$

Interchange x and y:

$$y = x^2 + 2, \qquad x \geq 0$$

Thus $$f^{-1}(x) = x^2 + 2, \qquad x \geq 0$$

In Example 8 note how f^{-1} reverses the effect of f. The function f is the rule "subtract 2, then take the square root," whereas f^{-1} is the rule "square, then add 2."

This expression shows that the graph of f^{-1} is the right half of the parabola $y = x^2 + 2$ and, from the graph shown in Figure 10, this seems reasonable.   ■

2.8   Exercises

1–6 ■ The graph of a function f is given. Determine whether f is one-to-one.

1.
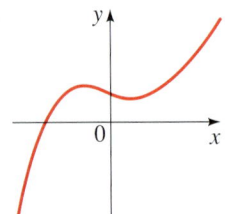

2.
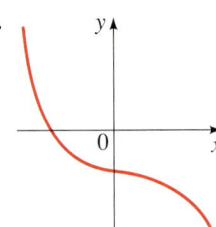

3.
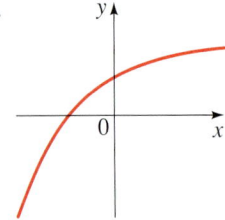

4.
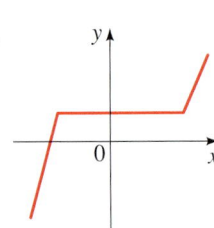

5.
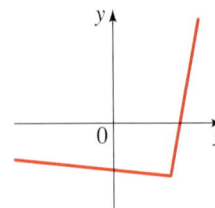

6.
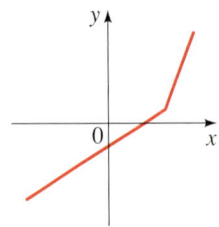

7–16 ■ Determine whether the function is one-to-one.

7. $f(x) = -2x + 4$

8. $f(x) = 3x - 2$

9. $g(x) = \sqrt{x}$

10. $g(x) = |x|$

11. $h(x) = x^2 - 2x$

12. $h(x) = x^3 + 8$

13. $f(x) = x^4 + 5$

14. $f(x) = x^4 + 5, \quad 0 \leq x \leq 2$

15. $f(x) = \dfrac{1}{x^2}$

16. $f(x) = \dfrac{1}{x}$

17–18 ■ Assume f is a one-to-one function.

17. (a) If $f(2) = 7$, find $f^{-1}(7)$.
   (b) If $f^{-1}(3) = -1$, find $f(-1)$.

18. (a) If $f(5) = 18$, find $f^{-1}(18)$.
   (b) If $f^{-1}(4) = 2$, find $f(2)$.

19. If $f(x) = 5 - 2x$, find $f^{-1}(3)$.

20. If $g(x) = x^2 + 4x$ with $x \geq -2$, find $g^{-1}(5)$.

21–30 ■ Use the Inverse Function Property to show that f and g are inverses of each other.

21. $f(x) = x - 6, \quad g(x) = x + 6$

22. $f(x) = 3x, \quad g(x) = \dfrac{x}{3}$

23. $f(x) = 2x - 5; \quad g(x) = \dfrac{x + 5}{2}$

24. $f(x) = \dfrac{3 - x}{4}; \quad g(x) = 3 - 4x$

25. $f(x) = \dfrac{1}{x}, \quad g(x) = \dfrac{1}{x}$

26. $f(x) = x^5, \quad g(x) = \sqrt[5]{x}$

27. $f(x) = x^2 - 4, \quad x \geq 0;$
   $g(x) = \sqrt{x + 4}, \quad x \geq -4$

28. $f(x) = x^3 + 1$; $g(x) = (x - 1)^{1/3}$

29. $f(x) = \dfrac{1}{x - 1}$, $x \neq 1$;

   $g(x) = \dfrac{1}{x} + 1$, $x \neq 0$

30. $f(x) = \sqrt{4 - x^2}$, $0 \leq x \leq 2$;

   $g(x) = \sqrt{4 - x^2}$, $0 \leq x \leq 2$

31–50 ■ Find the inverse function of f.

31. $f(x) = 2x + 1$

32. $f(x) = 6 - x$

33. $f(x) = 4x + 7$

34. $f(x) = 3 - 5x$

35. $f(x) = \dfrac{x}{2}$

36. $f(x) = \dfrac{1}{x^2}$, $x > 0$

37. $f(x) = \dfrac{1}{x + 2}$

38. $f(x) = \dfrac{x - 2}{x + 2}$

39. $f(x) = \dfrac{1 + 3x}{5 - 2x}$

40. $f(x) = 5 - 4x^3$

41. $f(x) = \sqrt{2 + 5x}$

42. $f(x) = x^2 + x$, $x \geq -\frac{1}{2}$

43. $f(x) = 4 - x^2$, $x \geq 0$

44. $f(x) = \sqrt{2x - 1}$

45. $f(x) = 4 + \sqrt[3]{x}$

46. $f(x) = (2 - x^3)^5$

47. $f(x) = 1 + \sqrt{1 + x}$

48. $f(x) = \sqrt{9 - x^2}$, $0 \leq x \leq 3$

49. $f(x) = x^4$, $x \geq 0$

50. $f(x) = 1 - x^3$

51–54 ■ A function f is given.

(a) Sketch the graph of f.

(b) Use the graph of f to sketch the graph of f^{-1}.

(c) Find f^{-1}.

51. $f(x) = 3x - 6$

52. $f(x) = 16 - x^2$, $x \geq 0$

53. $f(x) = \sqrt{x + 1}$

54. $f(x) = x^3 - 1$

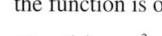 **55–60** ■ Draw the graph of f and use it to determine whether the function is one-to-one.

55. $f(x) = x^3 - x$

56. $f(x) = x^3 + x$

57. $f(x) = \dfrac{x + 12}{x - 6}$

58. $f(x) = \sqrt{x^3 - 4x + 1}$

59. $f(x) = |x| - |x - 6|$

60. $f(x) = x \cdot |x|$

 61–64 ■ A one-to-one function is given.

(a) Find the inverse of the function.

(b) Graph both the function and its inverse on the same screen to verify that the graphs are reflections of each other in the line $y = x$.

61. $f(x) = 2 + x$

62. $f(x) = 2 - \frac{1}{2}x$

63. $g(x) = \sqrt{x + 3}$

64. $g(x) = x^2 + 1$, $x \geq 0$

65–68 ■ The given function is not one-to-one. Restrict its domain so that the resulting function *is* one-to-one. Find the inverse of the function with the restricted domain. (There is more than one correct answer.)

65. $f(x) = 4 - x^2$

66. $g(x) = (x - 1)^2$

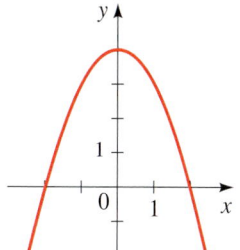
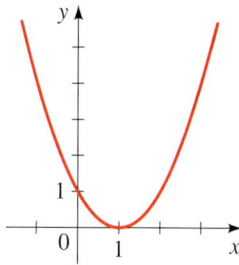

67. $h(x) = (x + 2)^2$

68. $k(x) = |x - 3|$

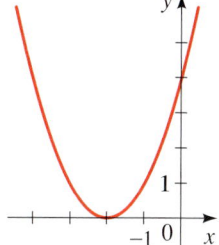
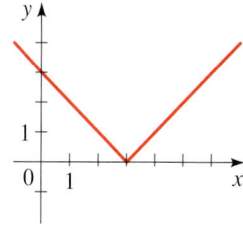

69–70 ■ Use the graph of f to sketch the graph of f^{-1}.

69.

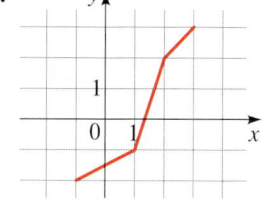

70.

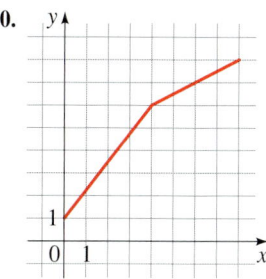

Applications

71. Fee for Service   For his services, a private investigator requires a $500 retention fee plus $80 per hour. Let x represent the number of hours the investigator spends working on a case.

(a) Find a function f that models the investigator's fee as a function of x.

(b) Find f^{-1}. What does f^{-1} represent?

(c) Find $f^{-1}(1220)$. What does your answer represent?

72. Toricelli's Law   A tank holds 100 gallons of water, which drains from a leak at the bottom, causing the tank to empty in 40 minutes. Toricelli's Law gives the volume of water remaining in the tank after t minutes as

$$V(t) = 100\left(1 - \frac{t}{40}\right)^2$$

(a) Find V^{-1}. What does V^{-1} represent?

(b) Find $V^{-1}(15)$. What does your answer represent?

73. Blood Flow   As blood moves through a vein or artery, its velocity v is greatest along the central axis and decreases as the distance r from the central axis increases (see the figure below). For an artery with radius 0.5 cm, v is given as a function of r by

$$v(r) = 18{,}500(0.25 - r^2)$$

(a) Find v^{-1}. What does v^{-1} represent?

(b) Find $v^{-1}(30)$. What does your answer represent?

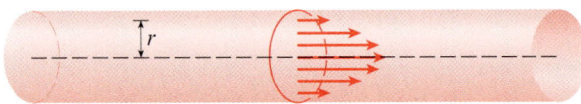

74. Demand Function   The amount of a commodity sold is called the *demand* for the commodity. The demand D for a certain commodity is a function of the price given by

$$D(p) = -3p + 150$$

(a) Find D^{-1}. What does D^{-1} represent?

(b) Find $D^{-1}(30)$. What does your answer represent?

75. Temperature Scales   The relationship between the Fahrenheit (F) and Celsius (C) scales is given by

$$F(C) = \tfrac{9}{5}C + 32$$

(a) Find F^{-1}. What does F^{-1} represent?

(b) Find $F^{-1}(86)$. What does your answer represent?

76. Exchange Rates   The relative value of currencies fluctuates every day. When this problem was written, one Canadian dollar was worth 0.8159 U.S. dollar.

(a) Find a function f that gives the U.S. dollar value $f(x)$ of x Canadian dollars.

(b) Find f^{-1}. What does f^{-1} represent?

(c) How much Canadian money would $12,250 in U.S. currency be worth?

77. Income Tax   In a certain country, the tax on incomes less than or equal to €20,000 is 10%. For incomes

more than €20,000, the tax is €2000 plus 20% of the amount over €20,000.

(a) Find a function f that gives the income tax on an income x. Express f as a piecewise defined function.

(b) Find f^{-1}. What does f^{-1} represent?

(c) How much income would require paying a tax of €10,000?

78. Multiple Discounts   A car dealership advertises a 15% discount on all its new cars. In addition, the manufacturer offers a $1000 rebate on the purchase of a new car. Let x represent the sticker price of the car.

(a) Suppose only the 15% discount applies. Find a function f that models the purchase price of the car as a function of the sticker price x.

(b) Suppose only the $1000 rebate applies. Find a function g that models the purchase price of the car as a function of the sticker price x.

(c) Find a formula for $H = f \circ g$.

(d) Find H^{-1}. What does H^{-1} represent?

(e) Find $H^{-1}(13{,}000)$. What does your answer represent?

79. Pizza Cost   Marcello's Pizza charges a base price of $7 for a large pizza, plus $2 for each topping. Thus, if you order a large pizza with x toppings, the price of your pizza is given by the function $f(x) = 7 + 2x$. Find f^{-1}. What does the function f^{-1} represent?

Discovery • Discussion

80. Determining when a Linear Function Has an Inverse   For the linear function $f(x) = mx + b$ to be one-to-one, what must be true about its slope? If it is one-to-one, find its inverse. Is the inverse linear? If so, what is its slope?

81. Finding an Inverse "In Your Head"   In the margin notes in this section we pointed out that the inverse of a function can be found by simply reversing the operations that make up the function. For instance, in Example 6 we saw that the inverse of

$$f(x) = 3x - 2 \quad \text{is} \quad f^{-1}(x) = \frac{x + 2}{3}$$

because the "reverse" of "multiply by 3 and subtract 2" is "add 2 and divide by 3." Use the same procedure to find the inverse of the following functions.

(a) $f(x) = \dfrac{2x + 1}{5}$ (b) $f(x) = 3 - \dfrac{1}{x}$

(c) $f(x) = \sqrt{x^3 + 2}$ (d) $f(x) = (2x - 5)^3$

Now consider another function:

$$f(x) = x^3 + 2x + 6$$

Is it possible to use the same sort of simple reversal of operations to find the inverse of this function? If so, do it. If not, explain what is different about this function that makes this task difficult.

82. The Identity Function   The function $I(x) = x$ is called the **identity function**. Show that for any function f we have $f \circ I = f$, $I \circ f = f$, and $f \circ f^{-1} = f^{-1} \circ f = I$. (This means that the identity function I behaves for functions and composition just like the number 1 behaves for real numbers and multiplication.)

83. Solving an Equation for an Unknown Function   In Exercise 65 of Section 2.7 you were asked to solve equations in which the unknowns were functions. Now that we know about inverses and the identity function (see Exercise 82), we can use algebra to solve such equations. For

instance, to solve $f \circ g = h$ for the unknown function f, we perform the following steps:

$$f \circ g = h \qquad \textit{Problem: Solve for f}$$
$$f \circ g \circ g^{-1} = h \circ g^{-1} \qquad \textit{Compose with } g^{-1} \textit{ on the right}$$
$$f \circ I = h \circ g^{-1} \qquad g \circ g^{-1} = I$$
$$f = h \circ g^{-1} \qquad f \circ I = f$$

So the solution is $f = h \circ g^{-1}$. Use this technique to solve the equation $f \circ g = h$ for the indicated unknown function.

(a) Solve for f, where $g(x) = 2x + 1$ and $h(x) = 4x^2 + 4x + 7$

(b) Solve for g, where $f(x) = 3x + 5$ and $h(x) = 3x^2 + 3x + 2$

---

2   Review

Concept Check

1. Define each concept in your own words. (Check by referring to the definition in the text.)

   (a) Function

   (b) Domain and range of a function

   (c) Graph of a function

   (d) Independent and dependent variables

2. Give an example of each type of function.

   (a) Constant function

   (b) Linear function

   (c) Quadratic function

3. Sketch by hand, on the same axes, the graphs of the following functions.

   (a) $f(x) = x$ (b) $g(x) = x^2$

   (c) $h(x) = x^3$ (d) $j(x) = x^4$

4. (a) State the Vertical Line Test.

   (b) State the Horizontal Line Test.

5. How is the average rate of change of the function f between two points defined?

6. Define each concept in your own words.

   (a) Increasing function

   (b) Decreasing function

   (c) Constant function

7. Suppose the graph of f is given. Write an equation for each graph that is obtained from the graph of f as follows.

   (a) Shift 3 units upward

   (b) Shift 3 units downward

   (c) Shift 3 units to the right

   (d) Shift 3 units to the left

   (e) Reflect in the x-axis

   (f) Reflect in the y-axis

   (g) Stretch vertically by a factor of 3

   (h) Shrink vertically by a factor of $\frac{1}{3}$

   (i) Stretch horizontally by a factor of 2

   (j) Shrink horizontally by a factor of $\frac{1}{2}$

8. (a) What is an even function? What symmetry does its graph possess? Give an example of an even function.

   (b) What is an odd function? What symmetry does its graph possess? Give an example of an odd function.

9. Write the standard form of a quadratic function.

10. What does it mean to say that $f(3)$ is a local maximum value of f?

11. Suppose that f has domain A and g has domain B.

   (a) What is the domain of $f + g$?

   (b) What is the domain of fg?

   (c) What is the domain of f/g?

12. How is the composite function $f \circ g$ defined?

13. (a) What is a one-to-one function?

(b) How can you tell from the graph of a function whether it is one-to-one?

(c) Suppose f is a one-to-one function with domain A and range B. How is the inverse function f^{-1} defined? What is the domain of f^{-1}? What is the range of f^{-1}?

(d) If you are given a formula for f, how do you find a formula for f^{-1}?

(e) If you are given the graph of f, how do you find the graph of f^{-1}?

Exercises

1. If $f(x) = x^2 - 4x + 6$, find $f(0)$, $f(2)$, $f(-2)$, $f(a)$, $f(-a)$, $f(x + 1)$, $f(2x)$, and $2f(x) - 2$.

2. If $f(x) = 4 - \sqrt{3x - 6}$, find $f(5)$, $f(9)$, $f(a + 2)$, $f(-x)$, $f(x^2)$, and $[f(x)]^2$.

3. The graph of a function f is given.

(a) Find $f(-2)$ and $f(2)$.

(b) Find the domain of f.

(c) Find the range of f.

(d) On what intervals is f increasing? On what intervals is f decreasing?

(e) Is f one-to-one?

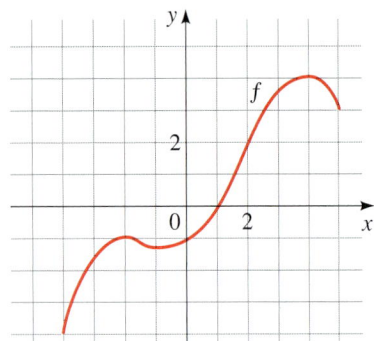

4. Which of the following figures are graphs of functions? Which of the functions are one-to-one?

(a)
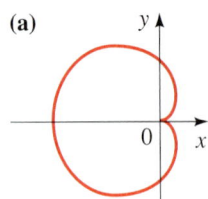

(b)
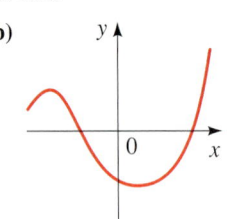

(b)
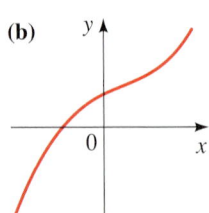

(d)
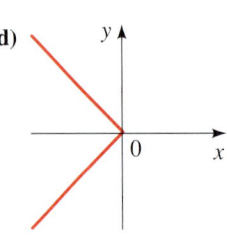

5–6 ■ Find the domain and range of the function.

5. $f(x) = \sqrt{x + 3}$

6. $F(t) = t^2 + 2t + 5$

7–14 ■ Find the domain of the function.

7. $f(x) = 7x + 15$

8. $f(x) = \dfrac{2x + 1}{2x - 1}$

9. $f(x) = \sqrt{x + 4}$

10. $f(x) = 3x - \dfrac{2}{\sqrt{x + 1}}$

11. $f(x) = \dfrac{1}{x} + \dfrac{1}{x + 1} + \dfrac{1}{x + 2}$

12. $g(x) = \dfrac{2x^2 + 5x + 3}{2x^2 - 5x - 3}$

13. $h(x) = \sqrt{4 - x} + \sqrt{x^2 - 1}$

14. $f(x) = \dfrac{\sqrt[3]{2x + 1}}{\sqrt[3]{2x + 2}}$

15–32 ■ Sketch the graph of the function.

15. $f(x) = 1 - 2x$

16. $f(x) = \frac{1}{3}(x - 5)$, $2 \le x \le 8$

17. $f(t) = 1 - \frac{1}{2}t^2$

18. $g(t) = t^2 - 2t$

19. $f(x) = x^2 - 6x + 6$

20. $f(x) = 3 - 8x - 2x^2$

21. $g(x) = 1 - \sqrt{x}$

22. $g(x) = -|x|$

23. $h(x) = \frac{1}{2}x^3$

24. $h(x) = \sqrt{x + 3}$

25. $h(x) = \sqrt[3]{x}$

26. $H(x) = x^3 - 3x^2$

27. $g(x) = \dfrac{1}{x^2}$

28. $G(x) = \dfrac{1}{(x - 3)^2}$

29. $f(x) = \begin{cases} 1 - x & \text{if } x < 0 \\ 1 & \text{if } x \ge 0 \end{cases}$

30. $f(x) = \begin{cases} 1 - 2x & \text{if } x \le 0 \\ 2x - 1 & \text{if } x > 0 \end{cases}$

31. $f(x) = \begin{cases} x + 6 & \text{if } x < -2 \\ x^2 & \text{if } x \ge -2 \end{cases}$

32. $f(x) = \begin{cases} -x & \text{if } x < 0 \\ x^2 & \text{if } 0 \le x < 2 \\ 1 & \text{if } x \ge 2 \end{cases}$

33. Determine which viewing rectangle produces the most appropriate graph of the function $f(x) = 6x^3 - 15x^2 + 4x - 1$.

(i) $[-2, 2]$ by $[-2, 2]$ (ii) $[-8, 8]$ by $[-8, 8]$

(iii) $[-4, 4]$ by $[-12, 12]$ (iv) $[-100, 100]$ by $[-100, 100]$

34. Determine which viewing rectangle produces the most appropriate graph of the function $f(x) = \sqrt{100 - x^3}$.
    (i) $[-4, 4]$ by $[-4, 4]$
    (ii) $[-10, 10]$ by $[-10, 10]$
    (iii) $[-10, 10]$ by $[-10, 40]$
    (iv) $[-100, 100]$ by $[-100, 100]$

35–38 ■ Draw the graph of the function in an appropriate viewing rectangle.

35. $f(x) = x^2 + 25x + 173$

36. $f(x) = 1.1x^3 - 9.6x^2 - 1.4x + 3.2$

37. $f(x) = \dfrac{x}{\sqrt{x^2 + 16}}$

38. $f(x) = |x(x + 2)(x + 4)|$

39. Find, approximately, the domain of the function $f(x) = \sqrt{x^3 - 4x + 1}$.

40. Find, approximately, the range of the function $f(x) = x^4 - x^3 + x^2 + 3x - 6$.

41–44 ■ Find the average rate of change of the function between the given points.

41. $f(x) = x^2 + 3x; \quad x = 0, x = 2$

42. $f(x) = \dfrac{1}{x - 2}; \quad x = 4, x = 8$

43. $f(x) = \dfrac{1}{x}; \quad x = 3, x = 3 + h$

44. $f(x) = (x + 1)^2; \quad x = a, x = a + h$

45–46 ■ Draw a graph of the function f, and determine the intervals on which f is increasing and on which f is decreasing.

45. $f(x) = x^3 - 4x^2$

46. $f(x) = |x^4 - 16|$

47. Suppose the graph of f is given. Describe how the graphs of the following functions can be obtained from the graph of f.
    (a) $y = f(x) + 8$ **(b)** $y = f(x + 8)$
    (c) $y = 1 + 2f(x)$ **(d)** $y = f(x - 2) - 2$
    (e) $y = f(-x)$ **(f)** $y = -f(-x)$
    (g) $y = -f(x)$ **(h)** $y = f^{-1}(x)$

48. The graph of f is given. Draw the graphs of the following functions.
    (a) $y = f(x - 2)$ **(b)** $y = -f(x)$
    (c) $y = 3 - f(x)$ **(d)** $y = \frac{1}{2}f(x) - 1$
    (e) $y = f^{-1}(x)$ **(f)** $y = f(-x)$

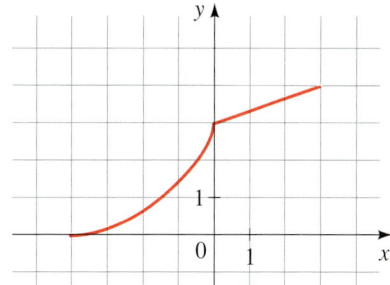

49. Determine whether f is even, odd, or neither.
    (a) $f(x) = 2x^5 - 3x^2 + 2$ **(b)** $f(x) = x^3 - x^7$
    (c) $f(x) = \dfrac{1 - x^2}{1 + x^2}$ **(d)** $f(x) = \dfrac{1}{x + 2}$

50. Determine whether the function in the figure is even, odd, or neither.

(a) 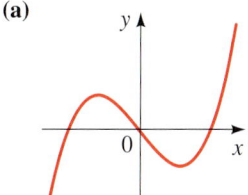   **(b)**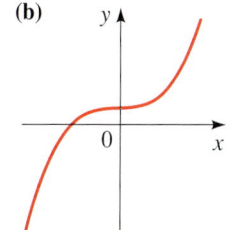

(c) 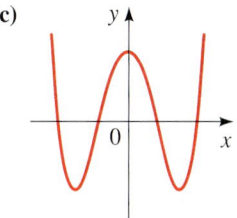   **(d)**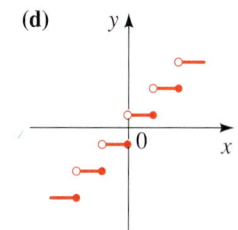

51. Express the quadratic function $f(x) = x^2 + 4x + 1$ in standard form.

52. Express the quadratic function $f(x) = -2x^2 + 12x + 12$ in standard form.

53. Find the minimum value of the function $g(x) = 2x^2 + 4x - 5$.

54. Find the maximum value of the function $f(x) = 1 - x - x^2$.

55. A stone is thrown upward from the top of a building. Its height (in feet) above the ground after t seconds is given by $h(t) = -16t^2 + 48t + 32$. What maximum height does it reach?

56. The profit P (in dollars) generated by selling x units of a certain commodity is given by

$$P(x) = -1500 + 12x - 0.0004x^2$$

What is the maximum profit, and how many units must be sold to generate it?

 57–58 ■ Find the local maximum and minimum values of the function and the values of x at which they occur. State each answer correct to two decimal places.

57. $f(x) = 3.3 + 1.6x - 2.5x^3$

58. $f(x) = x^{2/3}(6 - x)^{1/3}$

59. The number of air conditioners sold by an appliance store depends on the time of year. Sketch a rough graph of the number of A/C units sold as a function of the time of year.

60. An isosceles triangle has a perimeter of 8 cm. Express the area A of the triangle as a function of the length b of the base of the triangle.

61. A rectangle is inscribed in an equilateral triangle with a perimeter of 30 cm as in the figure.

    (a) Express the area A of the rectangle as a function of the length x shown in the figure.

    (b) Find the dimensions of the rectangle with the largest area.

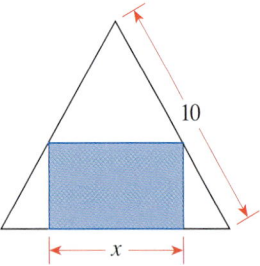

62. A piece of wire 10 m long is cut into two pieces. One piece, of length x, is bent into the shape of a square. The other piece is bent into the shape of an equilateral triangle.

    (a) Express the total area enclosed as a function of x.

    (b) For what value of x is this total area a minimum?

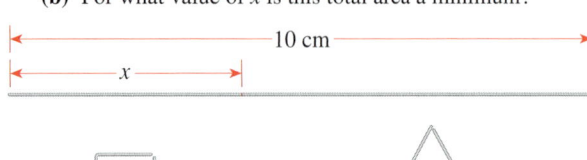

63. If $f(x) = x^2 - 3x + 2$ and $g(x) = 4 - 3x$, find the following functions.

    (a) $f + g$ **(b)** $f - g$ **(c)** fg

    (d) f/g **(e)** $f \circ g$ **(f)** $g \circ f$

64. If $f(x) = 1 + x^2$ and $g(x) = \sqrt{x - 1}$, find the following.

    (a) $f \circ g$ **(b)** $g \circ f$ **(c)** $(f \circ g)(2)$

    (d) $(f \circ f)(2)$ **(e)** $f \circ g \circ f$ **(f)** $g \circ f \circ g$

65–66 ■ Find the functions $f \circ g$, $g \circ f$, $f \circ f$, and $g \circ g$ and their domains.

65. $f(x) = 3x - 1, \quad g(x) = 2x - x^2$

66. $f(x) = \sqrt{x}, \quad g(x) = \dfrac{2}{x - 4}$

67. Find $f \circ g \circ h$, where $f(x) = \sqrt{1 - x}, g(x) = 1 - x^2$, and $h(x) = 1 + \sqrt{x}$.

68. If $T(x) = \dfrac{1}{\sqrt{1 + \sqrt{x}}}$, find functions f, g, and h such that $f \circ g \circ h = T$.

69–74 ■ Determine whether the function is one-to-one.

69. $f(x) = 3 + x^3$

70. $g(x) = 2 - 2x + x^2$

71. $h(x) = \dfrac{1}{x^4}$

72. $r(x) = 2 + \sqrt{x + 3}$

 73. $p(x) = 3.3 + 1.6x - 2.5x^3$

 74. $q(x) = 3.3 + 1.6x + 2.5x^3$

75–78 ■ Find the inverse of the function.

75. $f(x) = 3x - 2$

76. $f(x) = \dfrac{2x + 1}{3}$

77. $f(x) = (x + 1)^3$

78. $f(x) = 1 + \sqrt[5]{x - 2}$

79. (a) Sketch the graph of the function

$$f(x) = x^2 - 4, \quad x \geq 0$$

    (b) Use part (a) to sketch the graph of f^{-1}.

    (c) Find an equation for f^{-1}.

80. (a) Show that the function $f(x) = 1 + \sqrt[4]{x}$ is one-to-one.

    (b) Sketch the graph of f.

    (c) Use part (b) to sketch the graph of f^{-1}.

    (d) Find an equation for f^{-1}.

2   Test

1. Which of the following are graphs of functions? If the graph is that of a function, is it one-to-one?

   (a)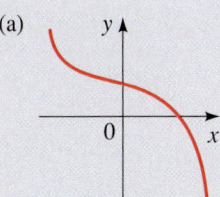

   (b)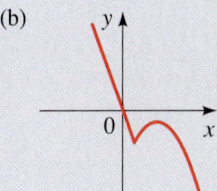

   (c)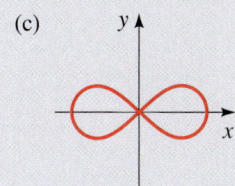

   (d)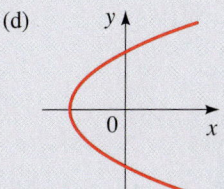

2. Let $f(x) = \dfrac{\sqrt{x+1}}{x}$.

   (a) Evaluate $f(3)$, $f(5)$, and $f(a-1)$.

   (b) Find the domain of f.

3. Determine the average rate of change for the function $f(t) = t^2 - 2t$ between $t = 2$ and $t = 5$.

4. (a) Sketch the graph of the function $f(x) = x^3$.

   (b) Use part (a) to graph the function $g(x) = (x-1)^3 - 2$.

5. (a) How is the graph of $y = f(x-3) + 2$ obtained from the graph of f?

   (b) How is the graph of $y = f(-x)$ obtained from the graph of f?

6. (a) Write the quadratic function $f(x) = 2x^2 - 8x + 13$ in standard form.

   (b) Sketch a graph of f.

   (c) What is the minimum value of f?

7. Let $f(x) = \begin{cases} 1 - x^2 & \text{if } x \le 0 \\ 2x + 1 & \text{if } x > 0 \end{cases}$

   (a) Evaluate $f(-2)$ and $f(1)$.

   (b) Sketch the graph of f.

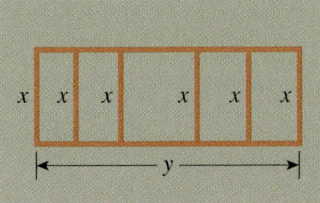

8. (a) If 1800 ft of fencing is available to build five adjacent pens, as shown in the diagram to the left, express the total area of the pens as a function of x.

   (b) What value of x will maximize the total area?

9. If $f(x) = x^2 + 1$ and $g(x) = x - 3$, find the following.

   (a) $f \circ g$ (b) $g \circ f$

   (c) $f(g(2))$ (d) $g(f(2))$

   (e) $g \circ g \circ g$

10. (a) If $f(x) = \sqrt{3 - x}$, find the inverse function f^{-1}.

(b) Sketch the graphs of f and f^{-1} on the same coordinate axes.

11. The graph of a function f is given.

(a) Find the domain and range of f.

(b) Sketch the graph of f^{-1}.

(c) Find the average rate of change of f between $x = 2$ and $x = 6$.

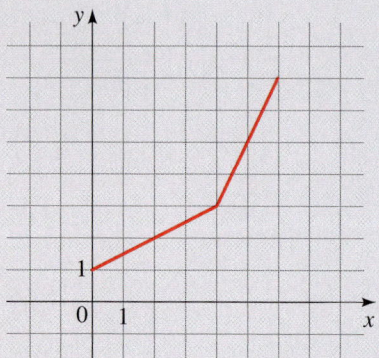

12. Let $f(x) = 3x^4 - 14x^2 + 5x - 3$.

(a) Draw the graph of f in an appropriate viewing rectangle.

(b) Is f one-to-one?

(c) Find the local maximum and minimum values of f and the values of x at which they occur. State each answer correct to two decimal places.

(d) Use the graph to determine the range of f.

(e) Find the intervals on which f is increasing and on which f is decreasing.

A model is a representation of an object or process. For example, a toy Ferrari is a *model* of the actual car; a road map is a model of the streets and highways in a city. A model usually represents just one aspect of the original thing. The toy Ferrari is not an actual car, but it does represent what a real Ferrari looks like; a road map does not contain the actual streets in a city, but it does represent the relationship of the streets to each other.

A **mathematical model** is a mathematical representation of an object or process. Often a mathematical model is a function that describes a certain phenomenon. In Example 12 of Section 1.10 we found that the function $T = -10h + 20$ models the atmospheric temperature T at elevation h. We then used this function to predict the temperature at a certain height. The figure below illustrates the process of mathematical modeling.

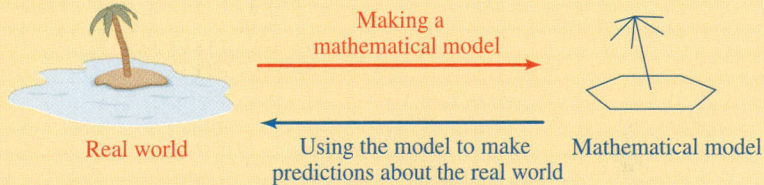

Real world   Using the model to make   Mathematical model
              predictions about the real world

Mathematical models are useful because they enable us to isolate critical aspects of the thing we are studying and then to predict how it will behave. Models are used extensively in engineering, industry, and manufacturing. For example, engineers use computer models of skyscrapers to predict their strength and how they would behave in an earthquake. Aircraft manufacturers use elaborate mathematical models to predict the aerodynamic properties of a new design *before* the aircraft is actually built.

How are mathematical models developed? How are they used to predict the behavior of a process? In the next few pages and in subsequent *Focus on Modeling* sections, we explain how mathematical models can be constructed from real-world data, and we describe some of their applications.

Linear Equations as Models

The data in Table 1 were obtained by measuring pressure at various ocean depths. From the table it appears that pressure increases with depth. To see this trend better, we make a **scatter plot** as in Figure 1. It appears that the data lie more or less along a line. We can try to fit a line visually to approximate the points in the scatter plot (see Figure 2),

Table 1

Depth (ft)	Pressure (lb/in²)
5	15.5
8	20.3
12	20.7
15	20.8
18	23.2
22	23.8
25	24.9
30	29.3

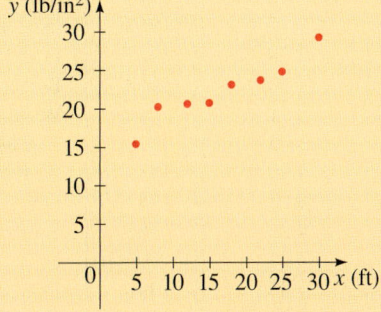

Figure 1
Scatter plot

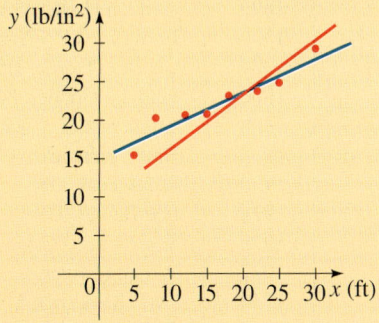

Figure 2
Attempts to fit line to data visually

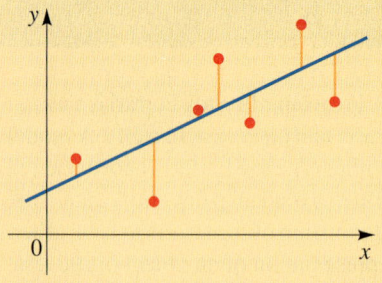

Figure 3

Distances from the points to the line

but this method is not accurate. So how do we find the line that fits the data as best as possible?

It seems reasonable to choose the line that is as close as possible to all the data points. This is the line for which the sum of the distances from the data points to the line is as small as possible (see Figure 3). For technical reasons it is better to find the line where the sum of the squares of these distances is smallest. The resulting line is called the **regression line**. The formula for the regression line is found using calculus. Fortunately, this formula is programmed into most graphing calculators. Using a calculator (see Figure 4(a)), we find that the regression line for the depth-pressure data in Table 1 is

$$P = 0.45d + 14.7 \qquad \text{Model}$$

The regression line and the scatter plot are graphed in Figure 4(b).

```
LinReg
 y=ax+b
 a=.4500365586
 b=14.71813307
```

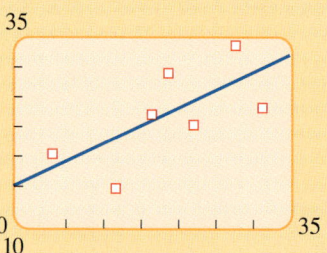

Figure 4

Linear regression on a graphing calculator

(a) Output of the `LinReg` command on a TI-83 calculator

(b) Scatter plot and regression line for depth-pressure data

Example 1 Olympic Pole Vaults

Table 2 gives the men's Olympic pole vault records up to 2004.

(a) Find the regression line for the data.

(b) Make a scatter plot of the data and graph the regression line. Does the regression line appear to be a suitable model for the data?

(c) Use the model to predict the winning pole vault height for the 2008 Olympics.

Table 2

Year	Gold medalist	Height (m)	Year	Gold medalist	Height (m)
1896	William Hoyt, USA	3.30	1956	Robert Richards USA	4.56
1900	Irving Baxter, USA	3.30	1960	Don Bragg, USA	4.70
1904	Charles Dvorak, USA	3.50	1964	Fred Hansen, USA	5.10
1906	Fernand Gonder, France	3.50	1968	Bob Seagren, USA	5.40
1908	A. Gilbert, E. Cook, USA	3.71	1972	W. Nordwig, E. Germany	5.64
1912	Harry Babcock, USA	3.95	1976	Tadeusz Slusarski, Poland	5.64
1920	Frank Foss, USA	4.09	1980	W. Kozakiewicz, Poland	5.78
1924	Lee Barnes, USA	3.95	1984	Pierre Quinon, France	5.75
1928	Sabin Carr, USA	4.20	1988	Sergei Bubka, USSR	5.90
1932	William Miller, USA	4.31	1992	M. Tarassob, Unified Team	5.87
1936	Earle Meadows, USA	4.35	1996	Jean Jalfione, France	5.92
1948	Guinn Smith, USA	4.30	2000	Nick Hysong, USA	5.90
1952	Robert Richards, USA	4.55	2004	Timothy Mack, USA	5.95

```
LinReg
 y=ax+b
 a=.0265652857
 b=3.400989881
```

Output of the `LinReg` function on the TI-83 Plus

Solution

(a) Let x = year − 1900, so that 1896 corresponds to $x = -4$, 1900 to $x = 0$, and so on. Using a calculator, we find the regression line:

$$y = 0.0266x + 3.40$$

(b) The scatter plot and the regression line are shown in Figure 5. The regression line appears to be a good model for the data.

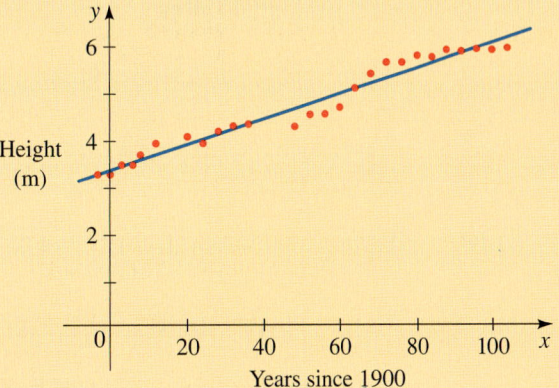

Figure 5

Scatter plot and regression line for pole-vault data

(c) The year 2008 corresponds to $x = 108$ in our model. The model gives

$$y = 0.0266(108) + 3.40 \approx 6.27 \text{ m}$$

If you are reading this after the 2008 Olympics, look up the actual record for 2008 and compare with this prediction. Such predictions are reasonable for points close to our measured data, but we can't predict too far away from the measured data. Is it reasonable to use this model to predict the record 100 years from now?

Example 2   Asbestos Fibers and Cancer

When laboratory rats are exposed to asbestos fibers, some of them develop lung tumors. Table 3 lists the results of several experiments by different scientists.

(a) Find the regression line for the data.

(b) Make a scatter plot of the data and graph the regression line. Does the regression line appear to be a suitable model for the data?

Table 3

Asbestos exposure (fibers/mL)	Percent that develop lung tumors
50	2
400	6
500	5
900	10
1100	26
1600	42
1800	37
2000	28
3000	50

Solution

(a) Using a calculator, we find the regression line (see Figure 6(a)):

$$y = 0.0177x + 0.5405$$

(b) The scatter plot and the regression line are shown in Figure 6(b). The regression line appears to be a reasonable model for the data.

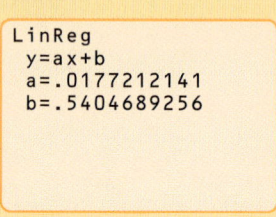

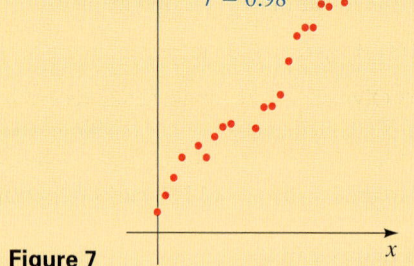

Figure 6

Linear regression for the asbestos-tumor data

(a) Output of the LinReg command on a TI-83 calculator

(b) Scatter plot and regression line ∎

How Good Is the Fit?

For any given set of data it is always possible to find the regression line, even if the data do not tend to lie along a line. Consider the three scatter plots in Figure 7.

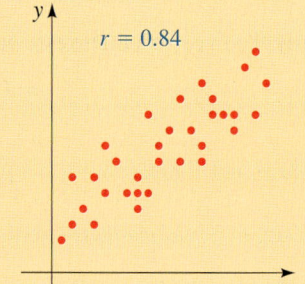

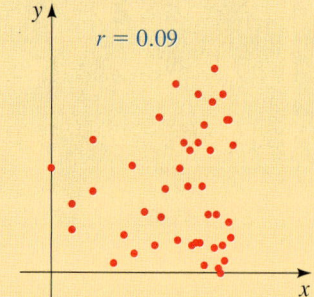

Figure 7

The data in the first scatter plot appear to lie along a line. In the second plot they also appear to display a linear trend, but it seems more scattered. The third does not have a discernible trend. We can easily find the regression lines for each scatter plot using a graphing calculator. But how well do these lines represent the data? The calculator gives a **correlation coefficient** r, which is a statistical measure of how well the data lie along the regression line, or how well the two variables are **correlated**. The correlation coefficient is a number between -1 and 1. A correlation coefficient r close to 1 or -1 indicates strong correlation and a coefficient close to 0 indicates very little correlation; the slope of the line determines whether the correlation coefficient is positive or negative. Also, the more data points we have, the more meaningful the correlation coefficient will be. Using a calculator we find that the correlation coefficient between asbestos fibers and lung tumors in the rats of Example 2 is $r = 0.92$. We can reasonably conclude that the presence of asbestos and the risk of lung tumors in rats are related. Can we conclude that asbestos *causes* lung tumors in rats?

If two variables are correlated, it does not necessarily mean that a change in one variable *causes* a change in the other. For example, the mathematician John Allen Paulos points out that shoe size is strongly correlated to mathematics scores among school children. Does this mean that big feet cause high math scores? Certainly not—both shoe size and math skills increase independently as children get older. So it is important not to jump to conclusions: Correlation and causation are not the same thing. Correlation is a useful tool in bringing important cause-and-effect relationships to light, but to prove causation, we must explain the mechanism by which one variable affects the other. For example, the link between smoking and lung cancer was observed as a correlation long before science found the mechanism through which smoking causes lung cancer.

Problems

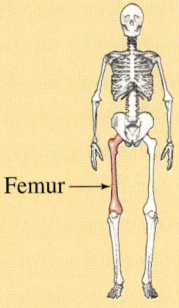

Femur ⟶

1. **Femur Length and Height**   Anthropologists use a linear model that relates femur length to height. The model allows an anthropologist to determine the height of an individual when only a partial skeleton (including the femur) is found. In this problem we find the model by analyzing the data on femur length and height for the eight males given in the table.

   (a) Make a scatter plot of the data.

   (b) Find and graph a linear function that models the data.

   (c) An anthropologist finds a femur of length 58 cm. How tall was the person?

Femur length (cm)	Height (cm)
50.1	178.5
48.3	173.6
45.2	164.8
44.7	163.7
44.5	168.3
42.7	165.0
39.5	155.4
38.0	155.8

2. **Demand for Soft Drinks**   A convenience store manager notices that sales of soft drinks are higher on hotter days, so he assembles the data in the table.

   (a) Make a scatter plot of the data.

   (b) Find and graph a linear function that models the data.

   (c) Use the model to predict soft-drink sales if the temperature is 95 °F.

High temperature (°F)	Number of cans sold
55	340
58	335
64	410
68	460
70	450
75	610
80	735
84	780

3. **Tree Diameter and Age**   To estimate ages of trees, forest rangers use a linear model that relates tree diameter to age. The model is useful because tree diameter is much easier to measure than tree age (which requires special tools for extracting a representative

cross section of the tree and counting the rings). To find the model, use the data in the table collected for a certain variety of oaks.

(a) Make a scatter plot of the data.

(b) Find and graph a linear function that models the data.

(c) Use the model to estimate the age of an oak whose diameter is 18 in.

Diameter (in.)	Age (years)
2.5	15
4.0	24
6.0	32
8.0	56
9.0	49
9.5	76
12.5	90
15.5	89

Year	CO$_2$ level (ppm)
1984	344.3
1986	347.0
1988	351.3
1990	354.0
1992	356.3
1994	358.9
1996	362.7
1998	366.5
2000	369.4

4. **Carbon Dioxide Levels**   The table lists average carbon dioxide (CO$_2$) levels in the atmosphere, measured in parts per million (ppm) at Mauna Loa Observatory from 1984 to 2000.

(a) Make a scatter plot of the data.

(b) Find and graph the regression line.

(c) Use the linear model in part (b) to estimate the CO$_2$ level in the atmosphere in 2001. Compare your answer with the actual CO$_2$ level of 371.1 measured in 2001.

5. **Temperature and Chirping Crickets**   Biologists have observed that the chirping rate of crickets of a certain species appears to be related to temperature. The table shows the chirping rates for various temperatures.

(a) Make a scatter plot of the data.

(b) Find and graph the regression line.

(c) Use the linear model in part (b) to estimate the chirping rate at 100 °F.

Temperature (°F)	Chirping rate (chirps/min)
50	20
55	46
60	79
65	91
70	113
75	140
80	173
85	198
90	211

Income	Ulcer rate
$4,000	14.1
$6,000	13.0
$8,000	13.4
$12,000	12.4
$16,000	12.0
$20,000	12.5
$30,000	10.5
$45,000	9.4
$60,000	8.2

6. **Ulcer Rates**   The table in the margin shows (lifetime) peptic ulcer rates (per 100 population) for various family incomes as reported by the 1989 National Health Interview Survey.

(a) Make a scatter plot of the data.

(b) Find and graph the regression line.

(c) Estimate the peptic ulcer rate for an income level of $25,000 according to the linear model in part (b).

(d) Estimate the peptic ulcer rate for an income level of $80,000 according to the linear model in part (b).

7. **Mosquito Prevalence**   The table lists the relative abundance of mosquitoes (as measured by the mosquito positive rate) versus the flow rate (measured as a percentage of maximum flow) of canal networks in Saga City, Japan.

(a) Make a scatter plot of the data.

(b) Find and graph the regression line.

(c) Use the linear model in part (b) to estimate the mosquito positive rate if the canal flow is 70% of maximum.

Flow rate (%)	Mosquito positive rate (%)
0	22
10	16
40	12
60	11
90	6
100	2

8. **Noise and Intelligibility**   Audiologists study the intelligibility of spoken sentences under different noise levels. Intelligibility, the MRT score, is measured as the percent of a spoken sentence that the listener can decipher at a certain noise level in decibels (dB). The table shows the results of one such test.

(a) Make a scatter plot of the data.

(b) Find and graph the regression line.

(c) Find the correlation coefficient. Is a linear model appropriate?

(d) Use the linear model in part (b) to estimate the intelligibility of a sentence at a 94-dB noise level.

Noise level (dB)	MRT score (%)
80	99
84	91
88	84
92	70
96	47
100	23
104	11

9. **Life Expectancy**   The average life expectancy in the United States has been rising steadily over the past few decades, as shown in the table.

(a) Make a scatter plot of the data.

(b) Find and graph the regression line.

(c) Use the linear model you found in part (b) to predict the life expectancy in the year 2004.

(d) Search the Internet or your campus library to find the actual 2004 average life expectancy. Compare to your answer in part (c).

Year	Life expectancy
1920	54.1
1930	59.7
1940	62.9
1950	68.2
1960	69.7
1970	70.8
1980	73.7
1990	75.4
2000	76.9

Phil Shemeister/Corbis

10. Heights of Tall Buildings   The table gives the heights and number of stories for 11 tall buildings.

(a) Make a scatter plot of the data.

(b) Find and graph the regression line.

(c) What is the slope of your regression line? What does its value indicate?

Building	Height (ft)	Stories
Empire State Building, New York	1250	102
One Liberty Place, Philadelphia	945	61
Canada Trust Tower, Toronto	863	51
Bank of America Tower, Seattle	943	76
Sears Tower, Chicago	1450	110
Petronas Tower I, Malaysia	1483	88
Commerzbank Tower, Germany	850	60
Palace of Culture and Science, Poland	758	42
Republic Plaza, Singapore	919	66
Transamerica Pyramid, San Francisco	853	48
Taipei 101 Building, Taiwan	1679	101

11. Olympic Swimming Records   The tables give the gold medal times in the men's and women's 100-m freestyle Olympic swimming event.

(a) Find the regression lines for the men's data and the women's data.

(b) Sketch both regression lines on the same graph. When do these lines predict that the women will overtake the men in the event? Does this conclusion seem reasonable?

MEN

Year	Gold medalist	Time (s)
1908	C. Daniels, USA	65.6
1912	D. Kahanamoku, USA	63.4
1920	D. Kahanamoku, USA	61.4
1924	J. Weissmuller, USA	59.0
1928	J. Weissmuller, USA	58.6
1932	Y. Miyazaki, Japan	58.2
1936	F. Csik, Hungary	57.6
1948	W. Ris, USA	57.3
1952	C. Scholes, USA	57.4
1956	J. Henricks, Australia	55.4
1960	J. Devitt, Australia	55.2
1964	D. Schollander, USA	53.4
1968	M. Wenden, Australia	52.2
1972	M. Spitz, USA	51.22
1976	J. Montgomery, USA	49.99
1980	J. Woithe, E. Germany	50.40
1984	R. Gaines, USA	49.80
1988	M. Biondi, USA	48.63
1992	A. Popov, Russia	49.02
1996	A. Popov, Russia	48.74
2000	P. van den Hoogenband, Netherlands	48.30
2004	P. van den Hoogenband, Netherlands	48.17

WOMEN

Year	Gold medalist	Time (s)
1912	F. Durack, Australia	82.2
1920	E. Bleibtrey, USA	73.6
1924	E. Lackie, USA	72.4
1928	A. Osipowich, USA	71.0
1932	H. Madison, USA	66.8
1936	H. Mastenbroek, Holland	65.9
1948	G. Andersen, Denmark	66.3
1952	K. Szoke, Hungary	66.8
1956	D. Fraser, Australia	62.0
1960	D. Fraser, Australia	61.2
1964	D. Fraser, Australia	59.5
1968	J. Henne, USA	60.0
1972	S. Nielson, USA	58.59
1976	K. Ender, E. Germany	55.65
1980	B. Krause, E. Germany	54.79
1984	(Tie) C. Steinseifer, USA	55.92
	N. Hogshead, USA	55.92
1988	K. Otto, E. Germany	54.93
1992	Z. Yong, China	54.64
1996	L. Jingyi, China	54.50
2000	I. DeBruijn, Netherlands	53.83
2004	J. Henry, Australia	53.84

12. **Parent Height and Offspring Height**   In 1885 Sir Francis Galton compared the height of children to the height of their parents. His study is considered one of the first uses of regression. The table gives some of Galton's original data. The term "midparent height" means the average of the heights of the father and mother.

(a) Find a linear equation that models the data.

(b) How well does the model predict your own height (based on your parents' heights)?

Midparent height (in.)	Offspring height (in.)
64.5	66.2
65.5	66.2
66.5	67.2
67.5	69.2
68.5	67.2
68.5	69.2
69.5	71.2
69.5	70.2
70.5	69.2
70.5	70.2
72.5	72.2
73.5	73.2

13. **Shoe Size and Height**   Do you think that shoe size and height are correlated? Find out by surveying the shoe sizes and heights of people in your class. (Of course, the data for men and women should be separate.) Find the correlation coefficient.

14. **Demand for Candy Bars**   In this problem you will determine a linear demand equation that describes the demand for candy bars in your class. Survey your classmates to determine what price they would be willing to pay for a candy bar. Your survey form might look like the sample to the left.

(a) Make a table of the number of respondents who answered "yes" at each price level.

(b) Make a scatter plot of your data.

(c) Find and graph the regression line $y = mp + b$, which gives the number of responents y who would buy a candy bar if the price were p cents. This is the *demand equation*. Why is the slope m negative?

(d) What is the p-intercept of the demand equation? What does this intercept tell you about pricing candy bars?

Would you buy a candy bar from the vending machine in the hallway if the price is as indicated?

Price	Yes or No
30¢	
40¢	
50¢	
60¢	
70¢	
80¢	
90¢	
$1.00	
$1.10	
$1.20	

3 Polynomial and Rational Functions

Chapter Overview

Functions defined by polynomial expressions are called polynomial functions. For example,

$$P(x) = 2x^3 - x + 1$$

is a polynomial function. Polynomial functions are easy to evaluate because they are defined using only addition, subtraction, and multiplication. This property makes them the most useful functions in mathematics.

The graphs of polynomial functions can increase and decrease several times. For this reason they are useful in modeling many real-world situations. For example, a factory owner notices that if she increases the number of workers, productivity increases, but if there are too many workers, productivity begins to decrease. This situation is modeled by a polynomial function of degree 2 (a quadratic polynomial). In many animal species the young experience an initial growth spurt, followed by a period of slow growth, followed by another growth spurt. This phenomenon is modeled by a polynomial function of degree 3 (a cubic polynomial).

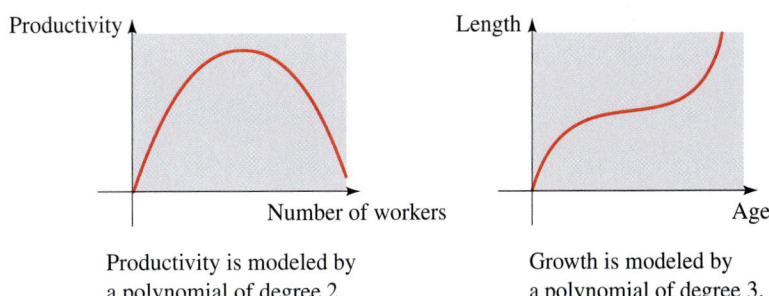

Productivity is modeled by
a polynomial of degree 2.

Growth is modeled by
a polynomial of degree 3.

The graphs of polynomial functions are beautiful, smooth curves that are used in design processes. For example, boat makers put together portions of the graphs of different cubic functions (called cubic splines) to design the natural curves for the hull of a boat.

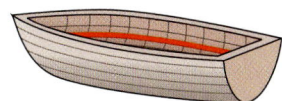

In this chapter we also study rational functions, which are quotients of polynomial functions. We will see that rational functions also have many useful applications.

J. L. Amos/SuperStock

3.1   Polynomial Functions and Their Graphs

Before we work with polynomial functions, we must agree on some terminology.

Polynomial Functions

A **polynomial function of degree n** is a function of the form

$$P(x) = a_n x^n + a_{n-1} x^{n-1} + \cdots + a_1 x + a_0$$

where n is a nonnegative integer and $a_n \neq 0$.

The numbers $a_0, a_1, a_2, \ldots, a_n$ are called the **coefficients** of the polynomial.

The number a_0 is the **constant coefficient** or **constant term**.

The number a_n, the coefficient of the highest power, is the **leading coefficient**, and the term $a_n x^n$ is the **leading term**.

We often refer to polynomial functions simply as *polynomials*. The following polynomial has degree 5, leading coefficient 3, and constant term -6.

Leading coefficient 3   Degree 5   Constant coefficient -6

$$3x^5 + 6x^4 - 2x^3 + x^2 + 7x - 6$$

Leading term $3x^5$

Coefficients 3, 6, -2, 1, 7, and -6

Here are some more examples of polynomials.

$$P(x) = 3 \qquad\qquad \text{Degree 0}$$

$$Q(x) = 4x - 7 \qquad\qquad \text{Degree 1}$$

$$R(x) = x^2 + x \qquad\qquad \text{Degree 2}$$

$$S(x) = 2x^3 - 6x^2 - 10 \qquad \text{Degree 3}$$

If a polynomial consists of just a single term, then it is called a **monomial**. For example, $P(x) = x^3$ and $Q(x) = -6x^5$ are monomials.

Graphs of Polynomials

The graphs of polynomials of degree 0 or 1 are lines (Section 1.10), and the graphs of polynomials of degree 2 are parabolas (Section 2.5). The greater the degree of the polynomial, the more complicated its graph can be. However, the graph of a polynomial function is always a smooth curve; that is, it has no breaks or corners (see Figure 1). The proof of this fact requires calculus.

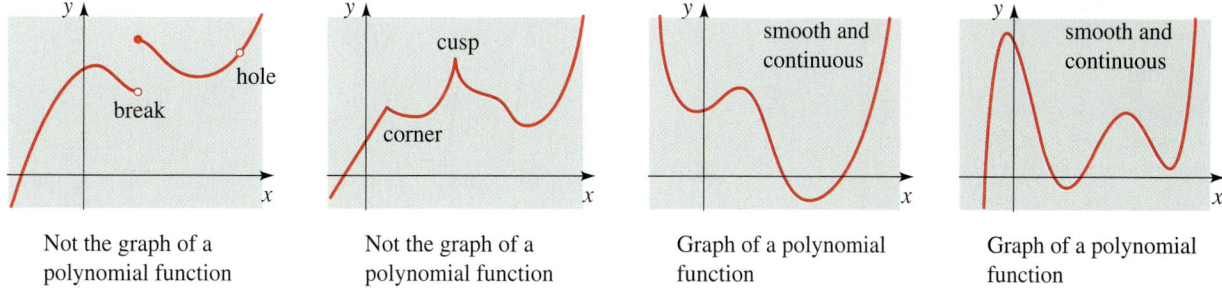

Not the graph of a polynomial function

Not the graph of a polynomial function

Graph of a polynomial function

Graph of a polynomial function

Figure 1

The simplest polynomial functions are the monomials $P(x) = x^n$, whose graphs are shown in Figure 2. As the figure suggests, the graph of $P(x) = x^n$ has the same general shape as $y = x^2$ when n is even, and the same general shape as $y = x^3$ when n is odd. However, as the degree n becomes larger, the graphs become flatter around the origin and steeper elsewhere.

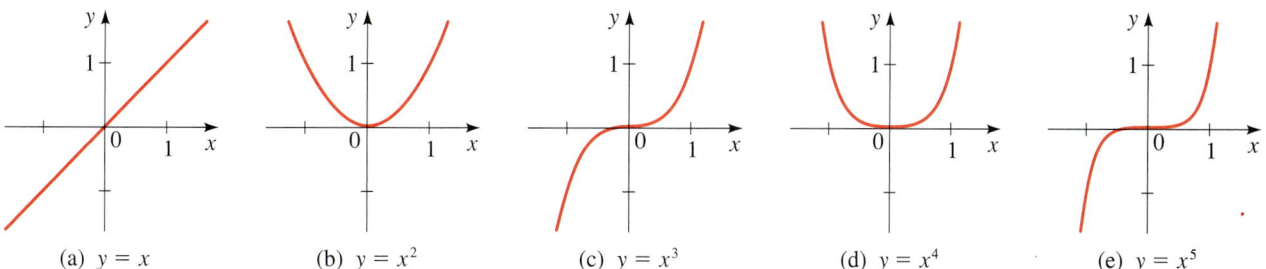

(a) $y = x$ (b) $y = x^2$ (c) $y = x^3$ (d) $y = x^4$ (e) $y = x^5$

Figure 2
Graphs of monomials

Example 1 Transformations of Monomials

Sketch the graphs of the following functions.

(a) $P(x) = -x^3$ (b) $Q(x) = (x - 2)^4$

(c) $R(x) = -2x^5 + 4$

Solution We use the graphs in Figure 2 and transform them using the techniques of Section 2.4.

(a) The graph of $P(x) = -x^3$ is the reflection of the graph of $y = x^3$ in the x-axis, as shown in Figure 3(a) on the following page.

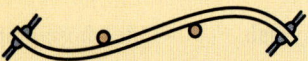
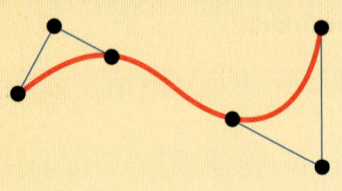
(b) The graph of $Q(x) = (x - 2)^4$ is the graph of $y = x^4$ shifted to the right 2 units, as shown in Figure 3(b).

(c) We begin with the graph of $y = x^5$. The graph of $y = -2x^5$ is obtained by stretching the graph vertically and reflecting it in the x-axis (see the dashed blue graph in Figure 3(c)). Finally, the graph of $R(x) = -2x^5 + 4$ is obtained by shifting upward 4 units (see the red graph in Figure 3(c)).

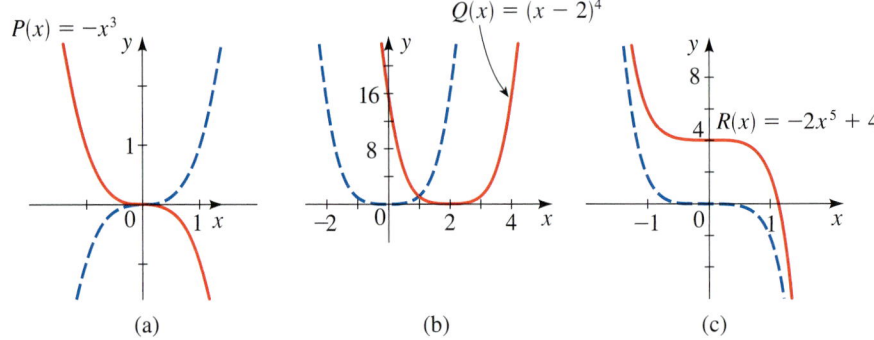

Figure 3

End Behavior and the Leading Term

The **end behavior** of a polynomial is a description of what happens as x becomes large in the positive or negative direction. To describe end behavior, we use the following notation:

$$x \to \infty \qquad \text{means} \qquad \text{"}x \text{ becomes large in the positive direction"}$$

$$x \to -\infty \qquad \text{means} \qquad \text{"}x \text{ becomes large in the negative direction"}$$

For example, the monomial $y = x^2$ in Figure 2(b) has the following end behavior:

$$y \to \infty \quad \text{as} \quad x \to \infty \qquad \text{and} \qquad y \to \infty \quad \text{as} \quad x \to -\infty$$

The monomial $y = x^3$ in Figure 2(c) has the end behavior

$$y \to \infty \quad \text{as} \quad x \to \infty \qquad \text{and} \qquad y \to -\infty \quad \text{as} \quad x \to -\infty$$

For any polynomial, *the end behavior is determined by the term that contains the highest power of x*, because when x is large, the other terms are relatively insignificant in size. The following box shows the four possible types of end behavior, based on the highest power and the sign of its coefficient.

End Behavior of Polynomials

The end behavior of the polynomial $P(x) = a_n x^n + a_{n-1}x^{n-1} + \cdots + a_1 x + a_0$ is determined by the degree n and the sign of the leading coefficient a_n, as indicated in the following graphs.

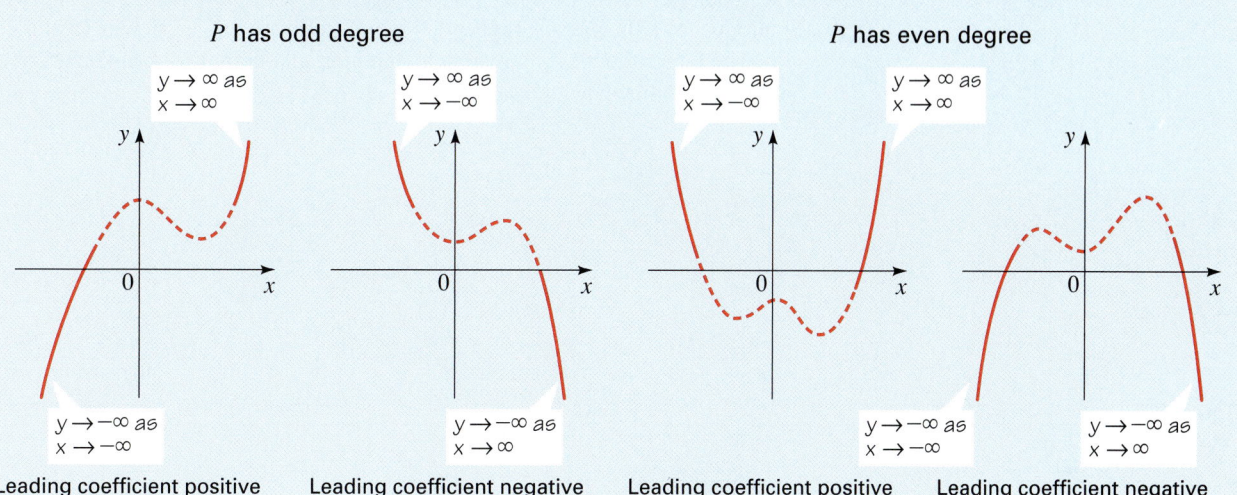

P has odd degree   P has even degree

Leading coefficient positive   Leading coefficient negative   Leading coefficient positive   Leading coefficient negative

Example 2   End Behavior of a Polynomial

Determine the end behavior of the polynomial

$$P(x) = -2x^4 + 5x^3 + 4x - 7$$

Solution   The polynomial P has degree 4 and leading coefficient -2. Thus, P has *even* degree and *negative* leading coefficient, so it has the following end behavior:

$$y \to -\infty \quad \text{as} \quad x \to \infty \qquad \text{and} \qquad y \to -\infty \quad \text{as} \quad x \to -\infty$$

The graph in Figure 4 illustrates the end behavior of P.

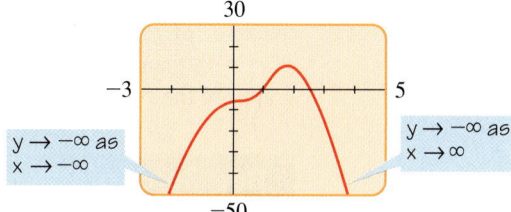

Figure 4

$P(x) = -2x^4 + 5x^3 + 4x - 7$

Example 3   End Behavior of a Polynomial

(a) Determine the end behavior of the polynomial $P(x) = 3x^5 - 5x^3 + 2x$.

(b) Confirm that P and its leading term $Q(x) = 3x^5$ have the same end behavior by graphing them together.

Solution

(a) Since P has odd degree and positive leading coefficient, it has the following end behavior:

$$y \to \infty \quad \text{as} \quad x \to \infty \qquad \text{and} \qquad y \to -\infty \quad \text{as} \quad x \to -\infty$$

(b) Figure 5 shows the graphs of P and Q in progressively larger viewing rectangles. The larger the viewing rectangle, the more the graphs look alike. This confirms that they have the same end behavior.

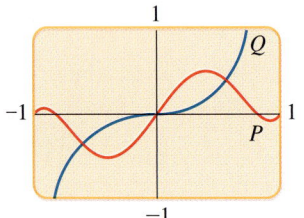 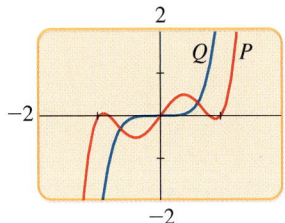 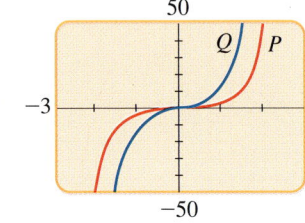 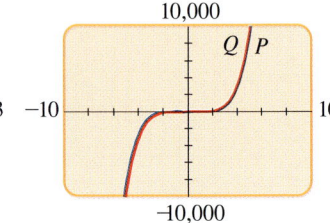

Figure 5

$P(x) = 3x^5 - 5x^3 + 2x$

$Q(x) = 3x^5$

To see algebraically why P and Q in Example 3 have the same end behavior, factor P as follows and compare with Q.

$$P(x) = 3x^5\left(1 - \frac{5}{3x^2} + \frac{2}{3x^4}\right) \qquad\qquad Q(x) = 3x^5$$

When x is large, the terms $5/3x^2$ and $2/3x^4$ are close to 0 (see Exercise 79 on page 12). So for large x, we have

$$P(x) \approx 3x^5(1 - 0 - 0) = 3x^5 = Q(x)$$

So, when x is large, P and Q have approximately the same values. We can also see this numerically by making a table like the one in the margin.

By the same reasoning we can show that the end behavior of *any* polynomial is determined by its leading term.

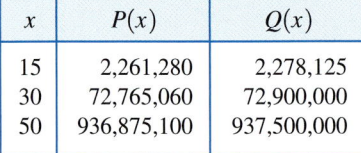

x	$P(x)$	$Q(x)$
15	2,261,280	2,278,125
30	72,765,060	72,900,000
50	936,875,100	937,500,000

Using Zeros to Graph Polynomials

If P is a polynomial function, then c is called a **zero** of P if $P(c) = 0$. In other words, the zeros of P are the solutions of the polynomial equation $P(x) = 0$. Note that if $P(c) = 0$, then the graph of P has an x-intercept at $x = c$, so the x-intercepts of the graph are the zeros of the function.

Real Zeros of Polynomials

If P is a polynomial and c is a real number, then the following are equivalent.

1. c is a zero of P.

2. $x = c$ is a solution of the equation $P(x) = 0$.

3. $x - c$ is a factor of $P(x)$.

4. $x = c$ is an x-intercept of the graph of P.

To find the zeros of a polynomial P, we factor and then use the Zero-Product Property (see page 47). For example, to find the zeros of $P(x) = x^2 + x - 6$, we factor P to get

$$P(x) = (x - 2)(x + 3)$$

From this factored form we easily see that

1. 2 is a zero of P.
2. $x = 2$ is a solution of the equation $x^2 + x - 6 = 0$.
3. $x - 2$ is a factor of $x^2 + x - 6$.
4. $x = 2$ is an x-intercept of the graph of P.

The same facts are true for the other zero, -3.

The following theorem has many important consequences. (See, for instance, the Discovery Project on page 283.) Here we use it to help us graph polynomial functions.

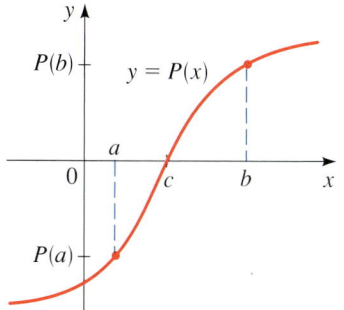

Figure 6

Intermediate Value Theorem for Polynomials

If P is a polynomial function and $P(a)$ and $P(b)$ have opposite signs, then there exists at least one value c between a and b for which $P(c) = 0$.

We will not prove this theorem, but Figure 6 shows why it is intuitively plausible.

One important consequence of this theorem is that between any two successive zeros, the values of a polynomial are either all positive or all negative. That is, between two successive zeros the graph of a polynomial lies *entirely above* or *entirely below* the x-axis. To see why, suppose c_1 and c_2 are successive zeros of P. If P has both positive and negative values between c_1 and c_2, then by the Intermediate Value Theorem P must have another zero between c_1 and c_2. But that's not possible because c_1 and c_2 are successive zeros. This observation allows us to use the following guidelines to graph polynomial functions.

Guidelines for Graphing Polynomial Functions

1. **Zeros.**   Factor the polynomial to find all its real zeros; these are the x-intercepts of the graph.

2. **Test Points.**   Make a table of values for the polynomial. Include test points to determine whether the graph of the polynomial lies above or below the x-axis on the intervals determined by the zeros. Include the y-intercept in the table.

3. **End Behavior.**   Determine the end behavior of the polynomial.

4. **Graph.**   Plot the intercepts and other points you found in the table. Sketch a smooth curve that passes through these points and exhibits the required end behavior.

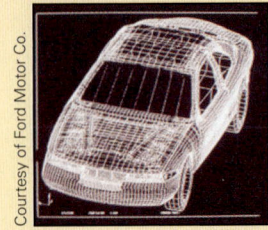
Example 4 **Using Zeros to Graph a Polynomial Function**

Sketch the graph of the polynomial function $P(x) = (x + 2)(x - 1)(x - 3)$.

Solution The zeros are $x = -2$, 1, and 3. These determine the intervals $(-\infty, -2)$, $(-2, 1)$, $(1, 3)$, and $(3, \infty)$. Using test points in these intervals, we get the information in the following sign diagram (see Section 1.7).

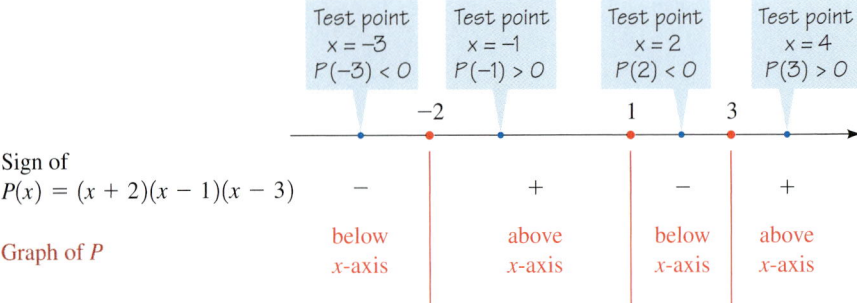

Plotting a few additional points and connecting them with a smooth curve helps us complete the graph in Figure 7.

x	$P(x)$
Test point → −3	−24
−2	0
Test point → −1	8
0	6
1	0
Test point → 2	−4
3	0
Test point → 4	18

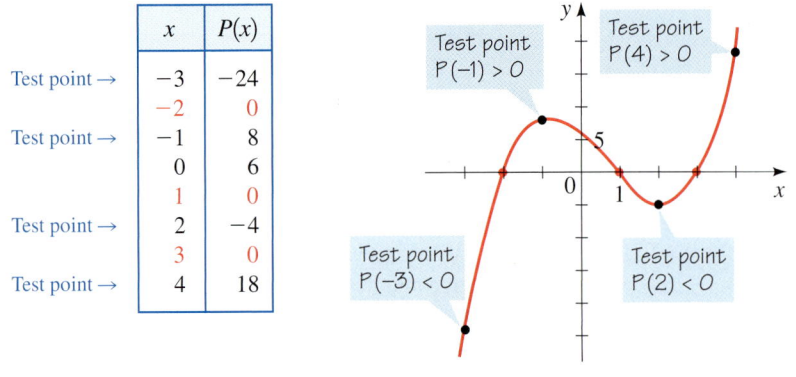

Figure 7
$P(x) = (x + 2)(x - 1)(x - 3)$

Example 5 **Finding Zeros and Graphing a Polynomial Function**

Let $P(x) = x^3 - 2x^2 - 3x$.

(a) Find the zeros of P. (b) Sketch the graph of P.

Solution

(a) To find the zeros, we factor completely.

$$P(x) = x^3 - 2x^2 - 3x$$
$$= x(x^2 - 2x - 3) \qquad \text{Factor } x$$
$$= x(x - 3)(x + 1) \qquad \text{Factor quadratic}$$

Thus, the zeros are $x = 0$, $x = 3$, and $x = -1$.

(b) The x-intercepts are $x = 0$, $x = 3$, and $x = -1$. The y-intercept is $P(0) = 0$. We make a table of values of $P(x)$, making sure we choose test points between (and to the right and left of) successive zeros.

Since P is of odd degree and its leading coefficient is positive, it has the following end behavior:

$$y \to \infty \quad \text{as} \quad x \to \infty \qquad \text{and} \qquad y \to -\infty \quad \text{as} \quad x \to -\infty$$

We plot the points in the table and connect them by a smooth curve to complete the graph, as shown in Figure 8.

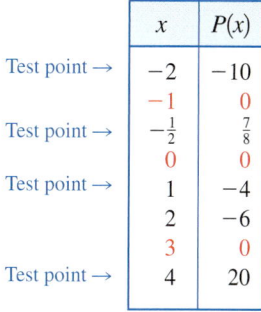

	x	$P(x)$
Test point →	-2	-10
	-1	0
Test point →	$-\frac{1}{2}$	$\frac{7}{8}$
	0	0
Test point →	1	-4
	2	-6
	3	0
Test point →	4	20

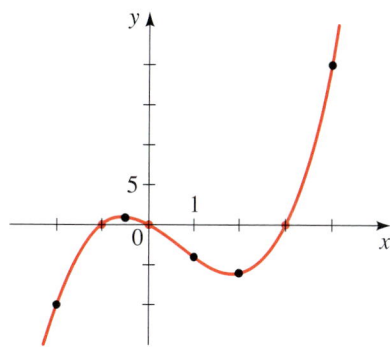

Figure 8

$P(x) = x^3 - 2x^2 - 3x$

Example 6   Finding Zeros and Graphing a Polynomial Function

Let $P(x) = -2x^4 - x^3 + 3x^2$.

(a) Find the zeros of P.   (b) Sketch the graph of P.

Solution

(a) To find the zeros, we factor completely.

$$P(x) = -2x^4 - x^3 + 3x^2$$
$$= -x^2(2x^2 + x - 3) \qquad \text{Factor } -x^2$$
$$= -x^2(2x + 3)(x - 1) \qquad \text{Factor quadratic}$$

Thus, the zeros are $x = 0$, $x = -\frac{3}{2}$, and $x = 1$.

(b) The x-intercepts are $x = 0$, $x = -\frac{3}{2}$, and $x = 1$. The y-intercept is $P(0) = 0$. We make a table of values of $P(x)$, making sure we choose test points between (and to the right and left of) successive zeros.

Since P is of even degree and its leading coefficient is negative, it has the following end behavior:

$$y \to -\infty \quad \text{as} \quad x \to \infty \qquad \text{and} \qquad y \to -\infty \quad \text{as} \quad x \to -\infty$$

We plot the points from the table and connect the points by a smooth curve to complete the graph in Figure 9.

Table of values are most easily calculated using a programmable calculator or a graphing calculator.

x	$P(x)$
-2	-12
-1.5	0
-1	2
-0.5	0.75
0	0
0.5	0.5
1	0
1.5	-6.75

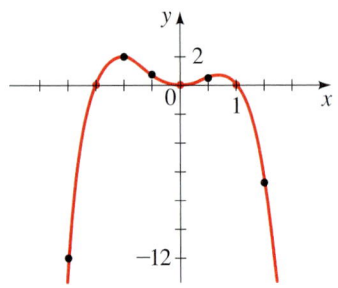

Figure 9

$$P(x) = -2x^4 - x^3 + 3x^2$$

Example 7   Finding Zeros and Graphing a Polynomial Function

Let $P(x) = x^3 - 2x^2 + 4x + 8$.

(a) Find the zeros of P.   (b) Sketch the graph of P.

Solution

(a) To find the zeros, we factor completely.

$$P(x) = x^3 - 2x^2 - 4x + 8$$

$$= x^2(x - 2) - 4(x - 2) \qquad \text{Group and factor}$$

$$= (x^2 - 4)(x - 2) \qquad \text{Factor } x - 2$$

$$= (x + 2)(x - 2)(x - 2) \qquad \text{Difference of squares}$$

$$= (x + 2)(x - 2)^2 \qquad \text{Simplify}$$

Thus, the zeros are $x = -2$ and $x = 2$.

(b) The x-intercepts are $x = -2$ and $x = 2$. The y-intercept is $P(0) = 8$. The table gives additional values of $P(x)$.

Since P is of odd degree and its leading coefficient is positive, it has the following end behavior:

$$y \to \infty \quad \text{as} \quad x \to \infty \qquad \text{and} \qquad y \to -\infty \quad \text{as} \quad x \to -\infty$$

We connect the points by a smooth curve to complete the graph in Figure 10.

x	$P(x)$
-3	-25
-2	0
-1	9
0	8
1	3
2	0
3	5

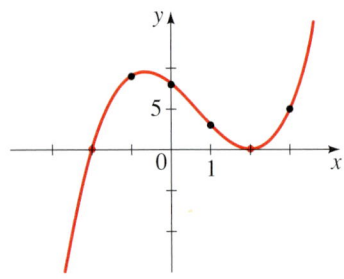

Figure 10

$$P(x) = x^3 - 2x^2 - 4x + 8$$

Shape of the Graph Near a Zero

Although $x = 2$ is a zero of the polynomial in Example 7, the graph does not cross the x-axis at the x-intercept 2. This is because the factor $(x - 2)^2$ corresponding to that zero is raised to an even power, so it doesn't change sign as we test points on either side of 2. In the same way, the graph does not cross the x-axis at $x = 0$ in Example 6.

In general, if c is a zero of P and the corresponding factor $x - c$ occurs exactly m times in the factorization of P then we say that c is a **zero of multiplicity** m. By considering test points on either side of the x-intercept c, we conclude that the graph crosses the x-axis at c if the multiplicity m is odd and does not cross the x-axis if m is even. Moreover, it can be shown using calculus that near $x = c$ the graph has the same general shape as $A(x - c)^m$.

Shape of the Graph Near a Zero of Multiplicity m

Suppose that c is a zero of P of multiplicity m. Then the shape of the graph of P near c is as follows.

Multiplicity of c Shape of the graph of P near the x-intercept c

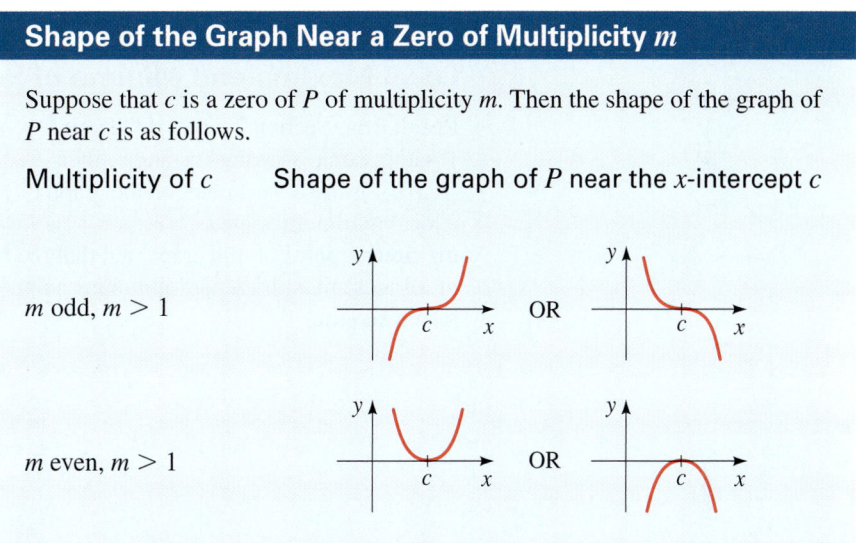

m odd, $m > 1$ OR

m even, $m > 1$ OR

Example 8 Graphing a Polynomial Function Using Its Zeros

Graph the polynomial $P(x) = x^4(x - 2)^3(x + 1)^2$.

Solution The zeros of P are $-1, 0,$ and 2, with multiplicities $2, 4,$ and 3, respectively.

0 is a zero of multiplicity 4.   2 is a zero of multiplicity 3. -1 is a zero of multiplicity 2.

$$P(x) = x^4(x - 2)^3(x + 1)^2$$

The zero 2 has *odd* multiplicity, so the graph crosses the x-axis at the x-intercept 2. But the zeros 0 and -1 have *even* multiplicity, so the graph does not cross the x-axis at the x-intercepts 0 and -1.

Since P is a polynomial of degree 9 and has positive leading coefficient, it has the following end behavior:

$$y \to \infty \quad \text{as} \quad x \to \infty \quad \text{and} \quad y \to -\infty \quad \text{as} \quad x \to -\infty$$

With this information and a table of values, we sketch the graph in Figure 11.

x	$P(x)$
-1.3	-9.2
-1	0
-0.5	-3.9
0	0
1	-4
2	0
2.3	8.2

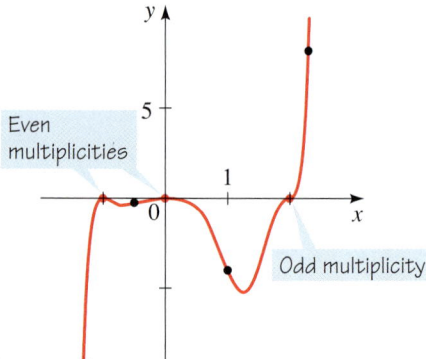

Figure 11

$$P(x) = x^4(x - 2)^3(x + 1)^2$$

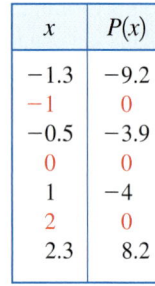 Local Maxima and Minima of Polynomials

Recall from Section 2.5 that if the point $(a, f(a))$ is the highest point on the graph of f within some viewing rectangle, then $f(a)$ is a local maximum value of f, and if $(b, f(b))$ is the lowest point on the graph of f within a viewing rectangle, then $f(b)$ is a local minimum value (see Figure 12). We say that such a point $(a, f(a))$ is a **local maximum point** on the graph and that $(b, f(b))$ is a **local minimum point**. The set of all local maximum and minimum points on the graph of a function is called its **local extrema**.

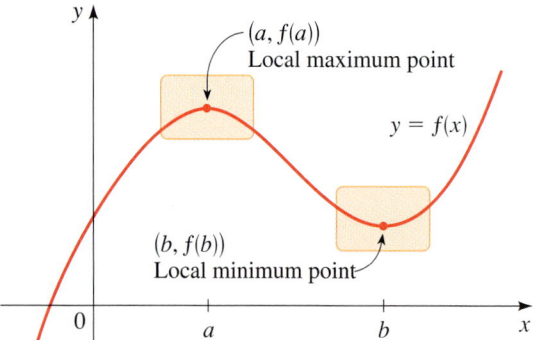

Figure 12

For a polynomial function the number of local extrema must be less than the degree, as the following principle indicates. (A proof of this principle requires calculus.)

Local Extrema of Polynomials

If $P(x) = a_n x^n + a_{n-1} x^{n-1} + \cdots + a_1 x + a_0$ is a polynomial of degree n, then the graph of P has at most $n - 1$ local extrema.

A polynomial of degree n may in fact have less than $n - 1$ local extrema. For example, $P(x) = x^5$ (graphed in Figure 2) has *no* local extrema, even though it is of de-

⊘ gree 5. The preceding principle tells us only that a polynomial of degree n can have no more than $n - 1$ local extrema.

Example 9 The Number of Local Extrema

Determine how many local extrema each polynomial has.

(a) $P_1(x) = x^4 + x^3 - 16x^2 - 4x + 48$

(b) $P_2(x) = x^5 + 3x^4 - 5x^3 - 15x^2 + 4x - 15$ (c) $P_3(x) = 7x^4 + 3x^2 - 10x$

Solution   The graphs are shown in Figure 13.

(a) P_1 has two local minimum points and one local maximum point, for a total of three local extrema.

(b) P_2 has two local minimum points and two local maximum points, for a total of four local extrema.

(c) P_3 has just one local extremum, a local minimum.

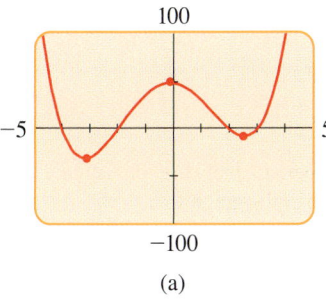

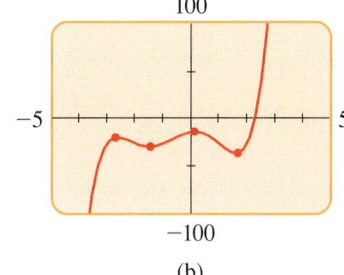

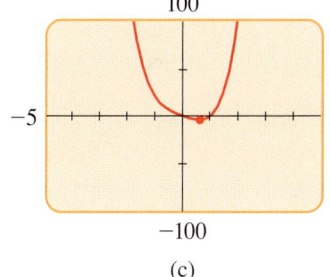

(a)   (b)   (c)

$P_1(x) = x^4 + x^3 - 16x^2 - 4x + 48$ $P_2(x) = x^5 + 3x^4 - 5x^3 - 15x^2 + 4x - 15$ $P_3(x) = 7x^4 + 3x^2 - 10x$

Figure 13

∎

With a graphing calculator we can quickly draw the graphs of many functions at once, on the same viewing screen. This allows us to see how changing a value in the definition of the functions affects the shape of its graph. In the next example we apply this principle to a family of third-degree polynomials.

Example 10 A Family of Polynomials

Sketch the family of polynomials $P(x) = x^3 - cx^2$ for $c = 0, 1, 2,$ and 3. How does changing the value of c affect the graph?

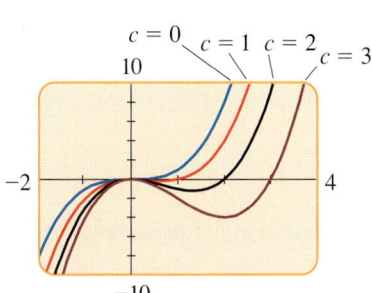

Figure 14

A family of polynomials
$P(x) = x^3 - cx^2$

Solution   The polynomials

$$P_0(x) = x^3 \qquad\qquad P_1(x) = x^3 - x^2$$
$$P_2(x) = x^3 - 2x^2 \qquad\qquad P_3(x) = x^3 - 3x^2$$

are graphed in Figure 14. We see that increasing the value of c causes the graph to develop an increasingly deep "valley" to the right of the y-axis, creating a local maximum at the origin and a local minimum at a point in quadrant IV. This local minimum moves lower and farther to the right as c increases. To see why this happens, factor $P(x) = x^2(x - c)$. The polynomial P has zeros at 0 and c, and the larger c gets, the farther to the right the minimum between 0 and c will be.   ∎

1–4 ■ Sketch the graph of each function by transforming the graph of an appropriate function of the form $y = x^n$ from Figure 2. Indicate all x- and y-intercepts on each graph.

1. (a) $P(x) = x^2 - 4$ (b) $Q(x) = (x - 4)^2$
(c) $R(x) = 2x^2 - 2$ (d) $S(x) = 2(x - 2)^2$

2. (a) $P(x) = x^4 - 16$ (b) $Q(x) = (x + 2)^4$
(c) $R(x) = (x + 2)^4 - 16$ (d) $S(x) = -2(x + 2)^4$

3. (a) $P(x) = x^3 - 8$ (b) $Q(x) = -x^3 + 27$
(c) $R(x) = -(x + 2)^3$ (d) $S(x) = \frac{1}{2}(x - 1)^3 + 4$

4. (a) $P(x) = (x + 3)^5$ (b) $Q(x) = 2(x + 3)^5 - 64$
(c) $R(x) = -\frac{1}{2}(x - 2)^5$ (d) $S(x) = -\frac{1}{2}(x - 2)^5 + 16$

5–10 ■ Match the polynomial function with one of the graphs I–VI. Give reasons for your choice.

5. $P(x) = x(x^2 - 4)$ **6.** $Q(x) = -x^2(x^2 - 4)$

7. $R(x) = -x^5 + 5x^3 - 4x$ **8.** $S(x) = \frac{1}{2}x^6 - 2x^4$

9. $T(x) = x^4 + 2x^3$ **10.** $U(x) = -x^3 + 2x^2$

I
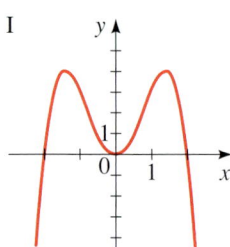

II
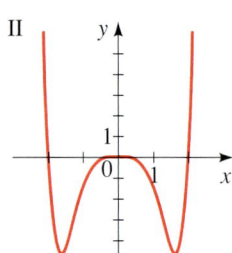

III
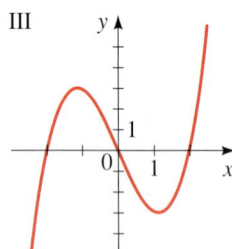

IV
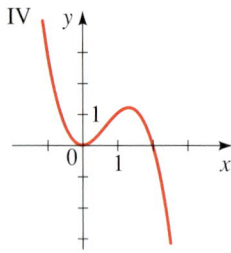

V
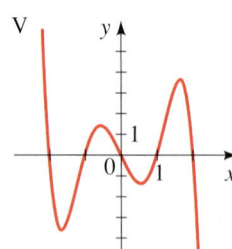

VI
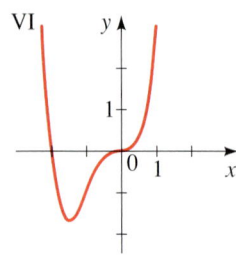

11–22 ■ Sketch the graph of the polynomial function. Make sure your graph shows all intercepts and exhibits the proper end behavior.

11. $P(x) = (x - 1)(x + 2)$

12. $P(x) = (x - 1)(x + 1)(x - 2)$

13. $P(x) = x(x - 3)(x + 2)$

14. $P(x) = (2x - 1)(x + 1)(x + 3)$

15. $P(x) = (x - 3)(x + 2)(3x - 2)$

16. $P(x) = \frac{1}{5}x(x - 5)^2$

17. $P(x) = (x - 1)^2(x - 3)$

18. $P(x) = \frac{1}{4}(x + 1)^3(x - 3)$

19. $P(x) = \frac{1}{12}(x + 2)^2(x - 3)^2$

20. $P(x) = (x - 1)^2(x + 2)^3$

21. $P(x) = x^3(x + 2)(x - 3)^2$

22. $P(x) = (x - 3)^2(x + 1)^2$

23–36 ■ Factor the polynomial and use the factored form to find the zeros. Then sketch the graph.

23. $P(x) = x^3 - x^2 - 6x$

24. $P(x) = x^3 + 2x^2 - 8x$

25. $P(x) = -x^3 + x^2 + 12x$

26. $P(x) = -2x^3 - x^2 + x$

27. $P(x) = x^4 - 3x^3 + 2x^2$

28. $P(x) = x^5 - 9x^3$

29. $P(x) = x^3 + x^2 - x - 1$

30. $P(x) = x^3 + 3x^2 - 4x - 12$

31. $P(x) = 2x^3 - x^2 - 18x + 9$

32. $P(x) = \frac{1}{8}(2x^4 + 3x^3 - 16x - 24)^2$

33. $P(x) = x^4 - 2x^3 - 8x + 16$

34. $P(x) = x^4 - 2x^3 + 8x - 16$

35. $P(x) = x^4 - 3x^2 - 4$

36. $P(x) = x^6 - 2x^3 + 1$

37–42 ■ Determine the end behavior of P. Compare the graphs of P and Q on large and small viewing rectangles, as in Example 3(b).

37. $P(x) = 3x^3 - x^2 + 5x + 1$; $Q(x) = 3x^3$

38. $P(x) = -\frac{1}{8}x^3 + \frac{1}{4}x^2 + 12x$; $Q(x) = -\frac{1}{8}x^3$

39. $P(x) = x^4 - 7x^2 + 5x + 5;$ $Q(x) = x^4$

40. $P(x) = -x^5 + 2x^2 + x;$ $Q(x) = -x^5$

41. $P(x) = x^{11} - 9x^9;$ $Q(x) = x^{11}$

42. $P(x) = 2x^2 - x^{12};$ $Q(x) = -x^{12}$

43–46 ■ The graph of a polynomial function is given. From the graph, find

(a) the x- and y-intercepts

(b) the coordinates of all local extrema

43. $P(x) = -x^2 + 4x$

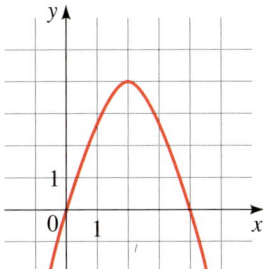

44. $P(x) = \frac{2}{9}x^3 - x^2$

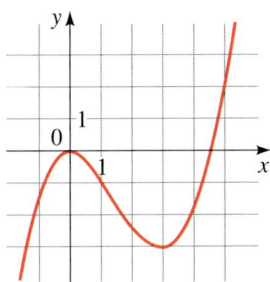

45. $P(x) = -\frac{1}{2}x^3 + \frac{3}{2}x - 1$

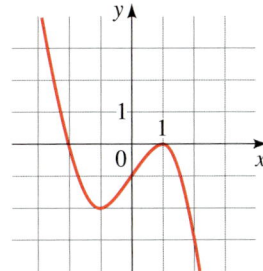

46. $P(x) = \frac{1}{9}x^4 - \frac{4}{9}x^3$

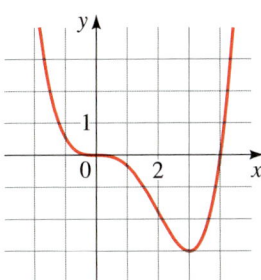

 47–54 ■ Graph the polynomial in the given viewing rectangle. Find the coordinates of all local extrema. State each answer correct to two decimal places.

47. $y = -x^2 + 8x,$ $[-4, 12]$ by $[-50, 30]$

48. $y = x^3 - 3x^2,$ $[-2, 5]$ by $[-10, 10]$

49. $y = x^3 - 12x + 9,$ $[-5, 5]$ by $[-30, 30]$

50. $y = 2x^3 - 3x^2 - 12x - 32,$ $[-5, 5]$ by $[-60, 30]$

51. $y = x^4 + 4x^3,$ $[-5, 5]$ by $[-30, 30]$

52. $y = x^4 - 18x^2 + 32,$ $[-5, 5]$ by $[-100, 100]$

53. $y = 3x^5 - 5x^3 + 3,$ $[-3, 3]$ by $[-5, 10]$

54. $y = x^5 - 5x^2 + 6,$ $[-3, 3]$ by $[-5, 10]$

 55–64 ■ Graph the polynomial and determine how many local maxima and minima it has.

55. $y = -2x^2 + 3x + 5$

56. $y = x^3 + 12x$

57. $y = x^3 - x^2 - x$

58. $y = 6x^3 + 3x + 1$

59. $y = x^4 - 5x^2 + 4$

60. $y = 1.2x^5 + 3.75x^4 - 7x^3 - 15x^2 + 18x$

61. $y = (x - 2)^5 + 32$

62. $y = (x^2 - 2)^3$

63. $y = x^8 - 3x^4 + x$

64. $y = \frac{1}{3}x^7 - 17x^2 + 7$

 65–70 ■ Graph the family of polynomials in the same viewing rectangle, using the given values of c. Explain how changing the value of c affects the graph.

65. $P(x) = cx^3;$ $c = 1, 2, 5, \frac{1}{2}$

66. $P(x) = (x - c)^4;$ $c = -1, 0, 1, 2$

67. $P(x) = x^4 + c;$ $c = -1, 0, 1, 2$

68. $P(x) = x^3 + cx;$ $c = 2, 0, -2, -4$

69. $P(x) = x^4 - cx;$ $c = 0, 1, 8, 27$

70. $P(x) = x^c;$ $c = 1, 3, 5, 7$

71. (a) On the same coordinate axes, sketch graphs (as accurately as possible) of the functions

$$y = x^3 - 2x^2 - x + 2 \quad \text{and} \quad y = -x^2 + 5x + 2$$

(b) Based on your sketch in part (a), at how many points do the two graphs appear to intersect?

(c) Find the coordinates of all intersection points.

72. Portions of the graphs of $y = x^2, y = x^3, y = x^4, y = x^5,$ and $y = x^6$ are plotted in the figures. Determine which function belongs to each graph.

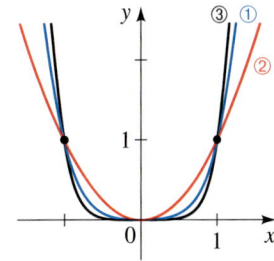

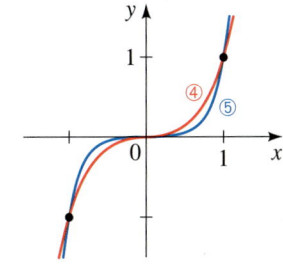

73. Recall that a function f is *odd* if $f(-x) = -f(x)$ or *even* if $f(-x) = f(x)$ for all real x.

   (a) Show that a polynomial $P(x)$ that contains only odd powers of x is an odd function.

   (b) Show that a polynomial $P(x)$ that contains only even powers of x is an even function.

   (c) Show that if a polynomial $P(x)$ contains both odd and even powers of x, then it is neither an odd nor an even function.

   (d) Express the function

$$P(x) = x^5 + 6x^3 - x^2 - 2x + 5$$

   as the sum of an odd function and an even function.

 74. (a) Graph the function $P(x) = (x-1)(x-3)(x-4)$ and find all local extrema, correct to the nearest tenth.

   (b) Graph the function

$$Q(x) = (x-1)(x-3)(x-4) + 5$$

   and use your answers to part (a) to find all local extrema, correct to the nearest tenth.

 75. (a) Graph the function $P(x) = (x-2)(x-4)(x-5)$ and determine how many local extrema it has.

   (b) If $a < b < c$, explain why the function

$$P(x) = (x-a)(x-b)(x-c)$$

   must have two local extrema.

 76. (a) How many x-intercepts and how many local extrema does the polynomial $P(x) = x^3 - 4x$ have?

   (b) How many x-intercepts and how many local extrema does the polynomial $Q(x) = x^3 + 4x$ have?

   (c) If $a > 0$, how many x-intercepts and how many local extrema does each of the polynomials $P(x) = x^3 - ax$ and $Q(x) = x^3 + ax$ have? Explain your answer.

Applications

 77. Market Research A market analyst working for a small-appliance manufacturer finds that if the firm produces and sells x blenders annually, the total profit (in dollars) is

$$P(x) = 8x + 0.3x^2 - 0.0013x^3 - 372$$

Graph the function P in an appropriate viewing rectangle and use the graph to answer the following questions.

   (a) When just a few blenders are manufactured, the firm loses money (profit is negative). (For example, $P(10) = -263.3$, so the firm loses \$263.30 if it produces and sells only 10 blenders.) How many blenders must the firm produce to break even?

   (b) Does profit increase indefinitely as more blenders are produced and sold? If not, what is the largest possible profit the firm could have?

 78. Population Change The rabbit population on a small island is observed to be given by the function

$$P(t) = 120t - 0.4t^4 + 1000$$

where t is the time (in months) since observations of the island began.

   (a) When is the maximum population attained, and what is that maximum population?

   (b) When does the rabbit population disappear from the island?

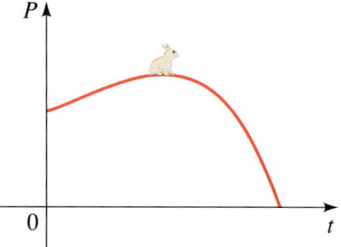

79. Volume of a Box An open box is to be constructed from a piece of cardboard 20 cm by 40 cm by cutting squares of side length x from each corner and folding up the sides, as shown in the figure.

   (a) Express the volume V of the box as a function of x.

   (b) What is the domain of V? (Use the fact that length and volume must be positive.)

    (c) Draw a graph of the function V and use it to estimate the maximum volume for such a box.

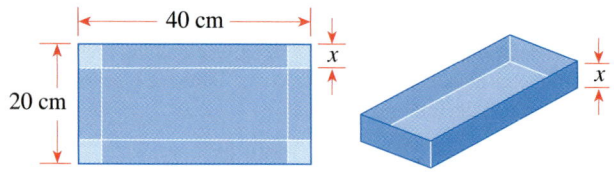

80. Volume of a Box A cardboard box has a square base, with each edge of the base having length x inches, as shown in the figure. The total length of all 12 edges of the box is 144 in.

   (a) Show that the volume of the box is given by the function $V(x) = 2x^2(18 - x)$.

(b) What is the domain of V? (Use the fact that length and volume must be positive.)

 (c) Draw a graph of the function V and use it to estimate the maximum volume for such a box.

Discovery • Discussion

81. Graphs of Large Powers Graph the functions $y = x^2$, $y = x^3$, $y = x^4$, and $y = x^5$, for $-1 \le x \le 1$, on the same coordinate axes. What do you think the graph of $y = x^{100}$ would look like on this same interval? What about $y = x^{101}$? Make a table of values to confirm your answers.

82. Maximum Number of Local Extrema What is the smallest possible degree that the polynomial whose graph is shown can have? Explain.

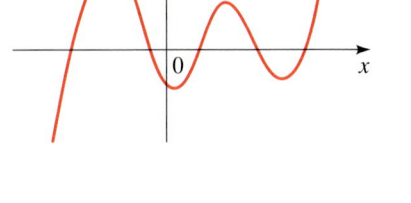

83. Possible Number of Local Extrema Is it possible for a third-degree polynomial to have exactly one local extremum? Can a fourth-degree polynomial have exactly two local extrema? How many local extrema can polynomials of third, fourth, fifth, and sixth degree have? (Think about the end behavior of such polynomials.) Now give an example of a polynomial that has six local extrema.

84. Impossible Situation? Is it possible for a polynomial to have two local maxima and no local minimum? Explain.

3.2 Dividing Polynomials

So far in this chapter we have been studying polynomial functions *graphically*. In this section we begin to study polynomials *algebraically*. Most of our work will be concerned with factoring polynomials, and to factor, we need to know how to divide polynomials.

Long Division of Polynomials

Dividing polynomials is much like the familiar process of dividing numbers. When we divide 38 by 7, the quotient is 5 and the remainder is 3. We write

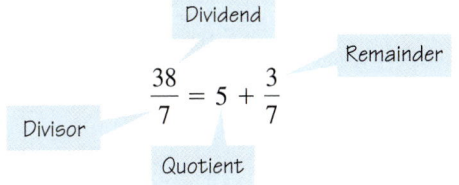

$$\frac{38}{7} = 5 + \frac{3}{7}$$

To divide polynomials, we use long division, as in the next example.

Example 1   **Long Division of Polynomials**

Divide $6x^2 - 26x + 12$ by $x - 4$.

Solution   The *dividend* is $6x^2 - 26x + 12$ and the *divisor* is $x - 4$. We begin by arranging them as follows:

$$x - 4 \overline{)6x^2 - 26x + 12}$$

Next we divide the leading term in the dividend by the leading term in the divisor to get the first term of the quotient: $6x^2/x = 6x$. Then we multiply the divisor by $6x$ and subtract the result from the dividend.

$$
\begin{array}{r}
6x \\
x - 4 \overline{)6x^2 - 26x + 12} \\
\underline{6x^2 - 24x} \\
-2x + 12
\end{array}
$$

Divide leading terms: $\dfrac{6x^2}{x} = 6x$

Multiply: $6x(x - 4) = 6x^2 - 24x$

Subtract and "bring down" 12

We repeat the process using the last line $-2x + 12$ as the dividend.

$$
\begin{array}{r}
6x - 2 \\
x - 4 \overline{)6x^2 - 26x + 12} \\
\underline{6x^2 - 24x} \\
-2x + 12 \\
\underline{-2x + 8} \\
4
\end{array}
$$

Divide leading terms: $\dfrac{-2x}{x} = -2$

Multiply: $-2(x - 4) = -2x + 8$

Subtract

The division process ends when the last line is of lesser degree than the divisor. The last line then contains the *remainder*, and the top line contains the *quotient*. The result of the division can be interpreted in either of two ways.

$$\frac{6x^2 - 26x + 12}{x - 4} = 6x - 2 + \frac{4}{x - 4}$$

or

$$6x^2 - 26x + 12 = (x - 4)(6x - 2) + 4$$

Dividend   Divisor   Quotient   Remainder

We summarize the long division process in the following theorem.

Division Algorithm

If $P(x)$ and $D(x)$ are polynomials, with $D(x) \neq 0$, then there exist unique polynomials $Q(x)$ and $R(x)$, where $R(x)$ is either 0 or of degree less than the degree of $D(x)$, such that

$$P(x) = D(x) \cdot Q(x) + R(x)$$

Dividend   Divisor   Quotient   Remainder

The polynomials $P(x)$ and $D(x)$ are called the **dividend** and **divisor**, respectively, $Q(x)$ is the **quotient**, and $R(x)$ is the **remainder**.

To write the division algorithm another way, divide through by $D(x)$:

$$\frac{P(x)}{D(x)} = Q(x) + \frac{R(x)}{D(x)}$$

Example 2   Long Division of Polynomials

Let $P(x) = 8x^4 + 6x^2 - 3x + 1$ and $D(x) = 2x^2 - x + 2$. Find polynomials $Q(x)$ and $R(x)$ such that $P(x) = D(x) \cdot Q(x) + R(x)$.

Solution   We use long division after first inserting the term $0x^3$ into the dividend to ensure that the columns line up correctly.

$$
\begin{array}{r}
4x^2 + 2x \\
2x^2 - x + 2{\overline{\smash{\big)}\,8x^4 + 0x^3 + 6x^2 - 3x + 1}} \\
\underline{8x^4 - 4x^3 + 8x^2} \quad\quad\quad\quad\quad \\
4x^3 - 2x^2 - 3x \quad\quad \\
\underline{4x^3 - 2x^2 + 4x} \quad\quad \\
-7x + 1
\end{array}
$$

Multiply divisor by $4x^2$
Subtract
Multiply divisor by $2x$
Subtract

The process is complete at this point because $-7x + 1$ is of lesser degree than the divisor $2x^2 - x + 2$. From the above long division we see that $Q(x) = 4x^2 + 2x$ and $R(x) = -7x + 1$, so

$$8x^4 + 6x^2 - 3x + 1 = (2x^2 - x + 2)(4x^2 + 2x) + (-7x + 1) \quad \blacksquare$$

Synthetic Division

Synthetic division is a quick method of dividing polynomials; it can be used when the divisor is of the form $x - c$. In synthetic division we write only the essential parts of the long division. Compare the following long and synthetic divisions, in which we divide $2x^3 - 7x^2 + 5$ by $x - 3$. (We'll explain how to perform the synthetic division in Example 3.)

Long Division

$$
\begin{array}{r}
2x^2 - x - 3 \\
x - 3{\overline{\smash{\big)}\,2x^3 - 7x^2 + 0x + 5}} \\
\underline{2x^3 - 6x^2} \quad\quad\quad\quad\quad \\
-x^2 + 0x \quad\quad \\
\underline{-x^2 + 3x} \quad\quad \\
-3x + 5 \\
\underline{-3x + 9} \\
-4
\end{array}
$$

Quotient

Remainder

Synthetic Division

$$
\begin{array}{c|ccccc}
3 & 2 & -7 & 0 & 5 \\
  &   & 6 & -3 & -9 \\
\hline
  & 2 & -1 & -3 & -4
\end{array}
$$

Quotient   Remainder

Note that in synthetic division we abbreviate $2x^3 - 7x^2 + 5$ by writing only the coefficients: $2 \quad -7 \quad 0 \quad 5$, and instead of $x - 3$, we simply write 3. (Writing 3 instead of -3 allows us to add instead of subtract, but this changes the sign of all the numbers that appear in the gold boxes.)

The next example shows how synthetic division is performed.

Example 3   **Synthetic Division**

Use synthetic division to divide $2x^3 - 7x^2 + 5$ by $x - 3$.

Solution   We begin by writing the appropriate coefficients to represent the divisor and the dividend.

$$\text{Divisor } x - 3 \quad\longrightarrow\quad 3 \;\Big|\; \begin{array}{cccc} 2 & -7 & 0 & 5 \end{array} \quad \begin{array}{l}\text{Dividend}\\ 2x^3 - 7x^2 + 0x + 5\end{array}$$

We bring down the 2, multiply $3 \cdot 2 = 6$, and write the result in the middle row. Then we add:

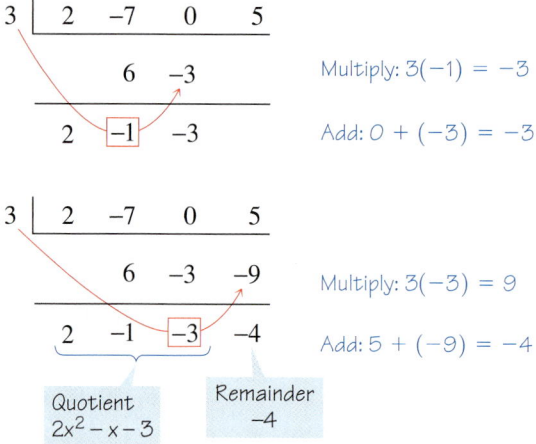

We repeat this process of multiplying and then adding until the table is complete.

From the last line of the synthetic division, we see that the quotient is $2x^2 - x - 3$ and the remainder is -4. Thus

$$2x^3 - 7x^2 + 5 = (x - 3)(2x^2 - x - 3) - 4 \qquad\blacksquare$$

The Remainder and Factor Theorems

The next theorem shows how synthetic division can be used to evaluate polynomials easily.

Remainder Theorem

If the polynomial $P(x)$ is divided by $x - c$, then the remainder is the value $P(c)$.

■ **Proof** If the divisor in the Division Algorithm is of the form $x - c$ for some real number c, then the remainder must be a constant (since the degree of the remainder is less than the degree of the divisor). If we call this constant r, then

$$P(x) = (x - c) \cdot Q(x) + r$$

Setting $x = c$ in this equation, we get $P(c) = (c - c) \cdot Q(x) + r = 0 + r = r$, that is, $P(c)$ is the remainder r. ■

Example 4 Using the Remainder Theorem to Find the Value of a Polynomial

Let $P(x) = 3x^5 + 5x^4 - 4x^3 + 7x + 3$.

(a) Find the quotient and remainder when $P(x)$ is divided by $x + 2$.

(b) Use the Remainder Theorem to find $P(-2)$.

Solution

(a) Since $x + 2 = x - (-2)$, the synthetic division for this problem takes the following form.

$$
\begin{array}{r|rrrrrr}
-2 & 3 & 5 & -4 & 0 & 7 & 3 \\
   &   & -6 & 2 & 4 & -8 & 2 \\
\hline
   & 3 & -1 & -2 & 4 & -1 & 5
\end{array}
$$

Remainder is 5, so $P(-2) = 5$.

The quotient is $3x^4 - x^3 - 2x^2 + 4x - 1$ and the remainder is 5.

(b) By the Remainder Theorem, $P(-2)$ is the remainder when $P(x)$ is divided by $x - (-2) = x + 2$. From part (a) the remainder is 5, so $P(-2) = 5$. ■

The next theorem says that *zeros* of polynomials correspond to *factors*; we used this fact in Section 3.1 to graph polynomials.

Factor Theorem

c is a zero of P if and only if $x - c$ is a factor of $P(x)$.

■ **Proof** If $P(x)$ factors as $P(x) = (x - c) \cdot Q(x)$, then

$$P(c) = (c - c) \cdot Q(c) = 0 \cdot Q(c) = 0$$

Conversely, if $P(c) = 0$, then by the Remainder Theorem

$$P(x) = (x - c) \cdot Q(x) + 0 = (x - c) \cdot Q(x)$$

so $x - c$ is a factor of $P(x)$. ■

Example 5 Factoring a Polynomial Using the Factor Theorem

Let $P(x) = x^3 - 7x + 6$. Show that $P(1) = 0$, and use this fact to factor $P(x)$ completely.

Solution Substituting, we see that $P(1) = 1^3 - 7 \cdot 1 + 6 = 0$. By the Factor Theorem, this means that $x - 1$ is a factor of $P(x)$. Using synthetic or long division

$$
\begin{array}{r|rrrr}
1 & 1 & 0 & -7 & 6 \\
  &   & 1 & 1 & -6 \\
\hline
  & 1 & 1 & -6 & 0
\end{array}
$$

$$\begin{array}{r} x^2 + x - 6 \\ x - 1\overline{)x^3 + 0x^2 - 7x + 6} \\ \underline{x^3 - x^2} \\ x^2 - 7x \\ \underline{x^2 - x} \\ -6x + 6 \\ \underline{-6x + 6} \\ 0 \end{array}$$

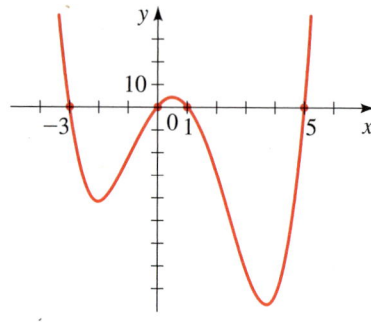

Figure 1
$P(x) = (x + 3)x(x - 1)(x - 5)$
has zeros $-3, 0, 1,$ and 5.

(shown in the margin), we see that

$$P(x) = x^3 - 7x + 6$$

$$= (x - 1)(x^2 + x - 6) \qquad \textit{See margin}$$

$$= (x - 1)(x - 2)(x + 3) \qquad \textit{Factor quadratic } x^2 + x - 6 \qquad \blacksquare$$

Example 6   Finding a Polynomial with Specified Zeros

Find a polynomial of degree 4 that has zeros $-3, 0, 1,$ and 5.

Solution   By the Factor Theorem, $x - (-3), x - 0, x - 1,$ and $x - 5$ must all be factors of the desired polynomial, so let

$$P(x) = (x + 3)(x - 0)(x - 1)(x - 5) = x^4 - 3x^3 - 13x^2 + 15x$$

Since $P(x)$ is of degree 4 it is a solution of the problem. Any other solution of the problem must be a constant multiple of $P(x)$, since only multiplication by a constant does not change the degree. $\blacksquare$

The polynomial P of Example 6 is graphed in Figure 1. Note that the zeros of P correspond to the x-intercepts of the graph.

3.2   Exercises

1–6 ■ Two polynomials P and D are given. Use either synthetic or long division to divide $P(x)$ by $D(x)$, and express P in the form $P(x) = D(x) \cdot Q(x) + R(x)$.

1. $P(x) = 3x^2 + 5x - 4, \quad D(x) = x + 3$

2. $P(x) = x^3 + 4x^2 - 6x + 1, \quad D(x) = x - 1$

3. $P(x) = 2x^3 - 3x^2 - 2x, \quad D(x) = 2x - 3$

4. $P(x) = 4x^3 + 7x + 9, \quad D(x) = 2x + 1$

5. $P(x) = x^4 - x^3 + 4x + 2, \quad D(x) = x^2 + 3$

6. $P(x) = 2x^5 + 4x^4 - 4x^3 - x - 3, \quad D(x) = x^2 - 2$

7–12 ■ Two polynomials P and D are given. Use either synthetic or long division to divide $P(x)$ by $D(x)$, and express the quotient $P(x)/D(x)$ in the form

$$\frac{P(x)}{D(x)} = Q(x) + \frac{R(x)}{D(x)}$$

7. $P(x) = x^2 + 4x - 8, \quad D(x) = x + 3$

8. $P(x) = x^3 + 6x + 5, \quad D(x) = x - 4$

9. $P(x) = 4x^2 - 3x - 7, \quad D(x) = 2x - 1$

10. $P(x) = 6x^3 + x^2 - 12x + 5, \quad D(x) = 3x - 4$

11. $P(x) = 2x^4 - x^3 + 9x^2, \quad D(x) = x^2 + 4$

12. $P(x) = x^5 + x^4 - 2x^3 + x + 1, \quad D(x) = x^2 + x - 1$

13–22 ■ Find the quotient and remainder using long division.

13. $\dfrac{x^2 - 6x - 8}{x - 4}$

14. $\dfrac{x^3 - x^2 - 2x + 6}{x - 2}$

15. $\dfrac{4x^3 + 2x^2 - 2x - 3}{2x + 1}$

16. $\dfrac{x^3 + 3x^2 + 4x + 3}{3x + 6}$

17. $\dfrac{x^3 + 6x + 3}{x^2 - 2x + 2}$

18. $\dfrac{3x^4 - 5x^3 - 20x - 5}{x^2 + x + 3}$

19. $\dfrac{6x^3 + 2x^2 + 22x}{2x^2 + 5}$

20. $\dfrac{9x^2 - x + 5}{3x^2 - 7x}$

21. $\dfrac{x^6 + x^4 + x^2 + 1}{x^2 + 1}$

22. $\dfrac{2x^5 - 7x^4 - 13}{4x^2 - 6x + 8}$

23–36 ■ Find the quotient and remainder using synthetic division.

23. $\dfrac{x^2 - 5x + 4}{x - 3}$

24. $\dfrac{x^2 - 5x + 4}{x - 1}$

25. $\dfrac{3x^2 + 5x}{x - 6}$

26. $\dfrac{4x^2 - 3}{x + 5}$

27. $\dfrac{x^3 + 2x^2 + 2x + 1}{x + 2}$

28. $\dfrac{3x^3 - 12x^2 - 9x + 1}{x - 5}$

29. $\dfrac{x^3 - 8x + 2}{x + 3}$

30. $\dfrac{x^4 - x^3 + x^2 - x + 2}{x - 2}$

31. $\dfrac{x^5 + 3x^3 - 6}{x - 1}$

32. $\dfrac{x^3 - 9x^2 + 27x - 27}{x - 3}$

33. $\dfrac{2x^3 + 3x^2 - 2x + 1}{x - \frac{1}{2}}$

34. $\dfrac{6x^4 + 10x^3 + 5x^2 + x + 1}{x + \frac{2}{3}}$

35. $\dfrac{x^3 - 27}{x - 3}$

36. $\dfrac{x^4 - 16}{x + 2}$

37–49 ■ Use synthetic division and the Remainder Theorem to evaluate $P(c)$.

37. $P(x) = 4x^2 + 12x + 5, \quad c = -1$

38. $P(x) = 2x^2 + 9x + 1, \quad c = \frac{1}{2}$

39. $P(x) = x^3 + 3x^2 - 7x + 6, \quad c = 2$

40. $P(x) = x^3 - x^2 + x + 5, \quad c = -1$

41. $P(x) = x^3 + 2x^2 - 7, \quad c = -2$

42. $P(x) = 2x^3 - 21x^2 + 9x - 200, \quad c = 11$

43. $P(x) = 5x^4 + 30x^3 - 40x^2 + 36x + 14, \quad c = -7$

44. $P(x) = 6x^5 + 10x^3 + x + 1, \quad c = -2$

45. $P(x) = x^7 - 3x^2 - 1, \quad c = 3$

46. $P(x) = -2x^6 + 7x^5 + 40x^4 - 7x^2 + 10x + 112, \quad c = -3$

47. $P(x) = 3x^3 + 4x^2 - 2x + 1, \quad c = \frac{2}{3}$

48. $P(x) = x^3 - x + 1, \quad c = \frac{1}{4}$

49. $P(x) = x^3 + 2x^2 - 3x - 8, \quad c = 0.1$

50. Let

$$P(x) = 6x^7 - 40x^6 + 16x^5 - 200x^4$$
$$- 60x^3 - 69x^2 + 13x - 139$$

Calculate $P(7)$ by **(a)** using synthetic division and **(b)** substituting $x = 7$ into the polynomial and evaluating directly.

51–54 ■ Use the Factor Theorem to show that $x - c$ is a factor of $P(x)$ for the given value(s) of c.

51. $P(x) = x^3 - 3x^2 + 3x - 1, \quad c = 1$

52. $P(x) = x^3 + 2x^2 - 3x - 10, \quad c = 2$

53. $P(x) = 2x^3 + 7x^2 + 6x - 5, \quad c = \frac{1}{2}$

54. $P(x) = x^4 + 3x^3 - 16x^2 - 27x + 63, \quad c = 3, -3$

55–56 ■ Show that the given value(s) of c are zeros of $P(x)$, and find all other zeros of $P(x)$.

55. $P(x) = x^3 - x^2 - 11x + 15, \quad c = 3$

56. $P(x) = 3x^4 - x^3 - 21x^2 - 11x + 6, \quad c = \frac{1}{3}, -2$

57–60 ■ Find a polynomial of the specified degree that has the given zeros.

57. Degree 3; zeros $-1, 1, 3$

58. Degree 4; zeros $-2, 0, 2, 4$

59. Degree 4; zeros $-1, 1, 3, 5$

60. Degree 5; zeros $-2, -1, 0, 1, 2$

61. Find a polynomial of degree 3 that has zeros $1, -2$, and 3, and in which the coefficient of x^2 is 3.

62. Find a polynomial of degree 4 that has integer coefficients and zeros $1, -1, 2$, and $\frac{1}{2}$.

63–66 ■ Find the polynomial of the specified degree whose graph is shown.

63. Degree 3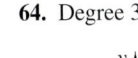

64. Degree 3

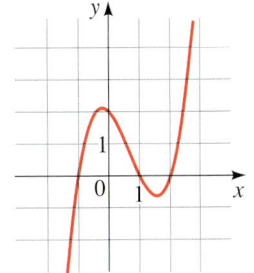
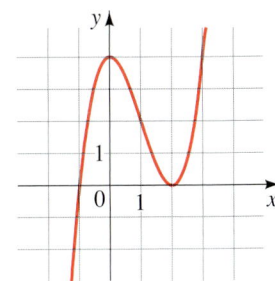

65. Degree 4

66. Degree 4

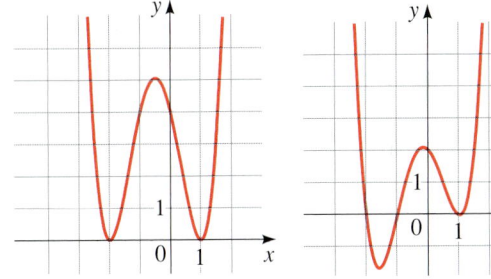

Discovery • Discussion

67. Impossible Division? Suppose you were asked to solve the following two problems on a test:

A. Find the remainder when $6x^{1000} - 17x^{562} + 12x + 26$ is divided by $x + 1$.

B. Is $x - 1$ a factor of $x^{567} - 3x^{400} + x^9 + 2$?

Obviously, it's impossible to solve these problems by dividing, because the polynomials are of such large degree. Use one or more of the theorems in this section to solve these problems *without* actually dividing.

68. Nested Form of a Polynomial   Expand Q to prove that the polynomials P and Q are the same.

$$P(x) = 3x^4 - 5x^3 + x^2 - 3x + 5$$

$$Q(x) = (((3x - 5)x + 1)x - 3)x + 5$$

Try to evaluate $P(2)$ and $Q(2)$ in your head, using the forms given. Which is easier? Now write the polynomial $R(x) = x^5 - 2x^4 + 3x^3 - 2x^2 + 3x + 4$ in "nested" form, like the polynomial Q. Use the nested form to find $R(3)$ in your head.

Do you see how calculating with the nested form follows the same arithmetic steps as calculating the value of a polynomial using synthetic division?

3.3   Real Zeros of Polynomials

The Factor Theorem tells us that finding the zeros of a polynomial is really the same thing as factoring it into linear factors. In this section we study some algebraic methods that help us find the real zeros of a polynomial, and thereby factor the polynomial. We begin with the *rational* zeros of a polynomial.

Rational Zeros of Polynomials

To help us understand the next theorem, let's consider the polynomial

$$P(x) = (x - 2)(x - 3)(x + 4) \qquad \text{Factored form}$$
$$= x^3 - x^2 - 14x + 24 \qquad \text{Expanded form}$$

From the factored form we see that the zeros of P are 2, 3, and -4. When the polynomial is expanded, the constant 24 is obtained by multiplying $(-2) \times (-3) \times 4$. This means that the zeros of the polynomial are all factors of the constant term. The following generalizes this observation.

Rational Zeros Theorem

If the polynomial $P(x) = a_n x^n + a_{n-1} x^{n-1} + \cdots + a_1 x + a_0$ has integer coefficients, then every rational zero of P is of the form

$$\frac{p}{q}$$

where p is a factor of the constant coefficient a_0
and q is a factor of the leading coefficient a_n.

■ **Proof**   If p/q is a rational zero, in lowest terms, of the polynomial P, then we have

$$a_n \left(\frac{p}{q} \right)^n + a_{n-1} \left(\frac{p}{q} \right)^{n-1} + \cdots + a_1 \left(\frac{p}{q} \right) + a_0 = 0$$

$$a_n p^n + a_{n-1} p^{n-1} q + \cdots + a_1 p q^{n-1} + a_0 q^n = 0 \qquad \text{Multiply by } q^n$$

$$p(a_n p^{n-1} + a_{n-1} p^{n-2} q + \cdots + a_1 q^{n-1}) = -a_0 q^n \qquad \begin{array}{l}\text{Subtract } a_0 q^n \\ \text{and factor LHS}\end{array}$$

Now p is a factor of the left side, so it must be a factor of the right as well. Since p/q is in lowest terms, p and q have no factor in common, and so p must be a factor of a_0. A similar proof shows that q is a factor of a_n.   ■

We see from the Rational Zeros Theorem that if the leading coefficient is 1 or -1, then the rational zeros must be factors of the constant term.

Evariste Galois (1811–1832) is one of the very few mathematicians to have an entire theory named in his honor. Not yet 21 when he died, he completely settled the central problem in the theory of equations by describing a criterion that reveals whether a polynomial equation can be solved by algebraic operations. Galois was one of the greatest mathematicians in the world at that time, although no one knew it but him. He repeatedly sent his work to the eminent mathematicians Cauchy and Poisson, who either lost his letters or did not understand his ideas. Galois wrote in a terse style and included few details, which probably played a role in his failure to pass the entrance exams at the Ecole Polytechnique in Paris. A political radical, Galois spent several months in prison for his revolutionary activities. His brief life came to a tragic end when he was killed in a duel over a love affair. The night before his duel, fearing he would die, Galois wrote down the essence of his ideas and entrusted them to his friend Auguste Chevalier. He concluded by writing ". . . there will, I hope, be people who will find it to their advantage to decipher all this mess." The mathematician Camille Jordan did just that, 14 years later.

Example 1   Finding Rational Zeros (Leading Coefficient 1)

Find the rational zeros of $P(x) = x^3 - 3x + 2$.

Solution   Since the leading coefficient is 1, any rational zero must be a divisor of the constant term 2. So the possible rational zeros are ± 1 and ± 2. We test each of these possibilities.

$$P(1) = (1)^3 - 3(1) + 2 = 0$$
$$P(-1) = (-1)^3 - 3(-1) + 2 = 4$$
$$P(2) = (2)^3 - 3(2) + 2 = 4$$
$$P(-2) = (-2)^3 - 3(-2) + 2 = 0$$

The rational zeros of P are 1 and -2.   ∎

Example 2   Using the Rational Zeros Theorem to Factor a Polynomial

Factor the polynomial $P(x) = 2x^3 + x^2 - 13x + 6$.

Solution   By the Rational Zeros Theorem the rational zeros of P are of the form

$$\text{possible rational zero of } P = \frac{\text{factor of constant term}}{\text{factor of leading coefficient}}$$

The constant term is 6 and the leading coefficient is 2, so

$$\text{possible rational zero of } P = \frac{\text{factor of } 6}{\text{factor of } 2}$$

The factors of 6 are ± 1, ± 2, ± 3, ± 6 and the factors of 2 are ± 1, ± 2. Thus, the possible rational zeros of P are

$$\pm\frac{1}{1}, \quad \pm\frac{2}{1}, \quad \pm\frac{3}{1}, \quad \pm\frac{6}{1}, \quad \pm\frac{1}{2}, \quad \pm\frac{2}{2}, \quad \pm\frac{3}{2}, \quad \pm\frac{6}{2}$$

Simplifying the fractions and eliminating duplicates, we get the following list of possible rational zeros:

$$\pm 1, \quad \pm 2, \quad \pm 3, \quad \pm 6, \quad \pm\frac{1}{2}, \quad \pm\frac{3}{2}$$

To check which of these *possible* zeros actually *are* zeros, we need to evaluate P at each of these numbers. An efficient way to do this is to use synthetic division.

	Test if 1 is a zero		**Test if 2 is a zero**	

```
       Test if 1 is a zero                    Test if 2 is a zero

  1  | 2    1   -13    6              2  | 2    1   -13    6
     |      2     3  -10                 |      4    10   -6
     -------------------                 -------------------
       2    3   -10   -4                   2    5    -3    0
```

Remainder is not 0, so 1 is not a zero.

Remainder is 0, so 2 is a zero.

From the last synthetic division we see that 2 is a zero of P and that P factors as

$$P(x) = 2x^3 + x^2 - 13x + 6$$
$$= (x - 2)(2x^2 + 5x - 3)$$
$$= (x - 2)(2x - 1)(x + 3) \qquad \text{Factor } 2x^2 + 5x - 3 \qquad \blacksquare$$

The following box explains how we use the Rational Zeros Theorem with synthetic division to factor a polynomial.

Finding the Rational Zeros of a Polynomial

1. **List Possible Zeros.**   List all possible rational zeros using the Rational Zeros Theorem.

2. **Divide.**   Use synthetic division to evaluate the polynomial at each of the candidates for rational zeros that you found in Step 1. When the remainder is 0, note the quotient you have obtained.

3. **Repeat.**   Repeat Steps 1 and 2 for the quotient. Stop when you reach a quotient that is quadratic or factors easily, and use the quadratic formula or factor to find the remaining zeros.

Example 3   Using the Rational Zeros Theorem and the Quadratic Formula

Let $P(x) = x^4 - 5x^3 - 5x^2 + 23x + 10$.

(a) Find the zeros of P.

(b) Sketch the graph of P.

Solution

(a) The leading coefficient of P is 1, so all the rational zeros are integers: They are divisors of the constant term 10. Thus, the possible candidates are

$$\pm 1, \quad \pm 2, \quad \pm 5, \quad \pm 10$$

Using synthetic division (see the margin) we find that 1 and 2 are not zeros, but that 5 is a zero and that P factors as

$$x^4 - 5x^3 - 5x^2 + 23x + 10 = (x - 5)(x^3 - 5x - 2)$$

We now try to factor the quotient $x^3 - 5x - 2$. Its possible zeros are the divisors of -2, namely,

$$\pm 1, \quad \pm 2$$

Since we already know that 1 and 2 are not zeros of the original polynomial P, we don't need to try them again. Checking the remaining candidates -1 and -2, we see that -2 is a zero (see the margin), and P factors as

$$x^4 - 5x^3 - 5x^2 + 23x + 10 = (x - 5)(x^3 - 5x - 2)$$
$$= (x - 5)(x + 2)(x^2 - 2x - 1)$$

$$
\begin{array}{r|rrrrr}
1 & 1 & -5 & -5 & 23 & 10 \\
  &   & 1 & -4 & -9 & 14 \\
\hline
  & 1 & -4 & -9 & 14 & 24 \\
\end{array}
$$

$$
\begin{array}{r|rrrrr}
2 & 1 & -5 & -5 & 23 & 10 \\
  &   & 2 & -6 & -22 & 2 \\
\hline
  & 1 & -3 & -11 & 1 & 12 \\
\end{array}
$$

$$
\begin{array}{r|rrrrr}
5 & 1 & -5 & -5 & 23 & 10 \\
  &   & 5 & 0 & -25 & -10 \\
\hline
  & 1 & 0 & -5 & -2 & 0 \\
\end{array}
$$

$$
\begin{array}{r|rrrr}
-2 & 1 & 0 & -5 & -2 \\
   &   & -2 & 4 & 2 \\
\hline
   & 1 & -2 & -1 & 0 \\
\end{array}
$$

Now we use the quadratic formula to obtain the two remaining zeros of P:

$$x = \frac{2 \pm \sqrt{(-2)^2 - 4(1)(-1)}}{2} = 1 \pm \sqrt{2}$$

The zeros of P are 5, -2, $1 + \sqrt{2}$, and $1 - \sqrt{2}$.

(b) Now that we know the zeros of P, we can use the methods of Section 3.1 to sketch the graph. If we want to use a graphing calculator instead, knowing the zeros allows us to choose an appropriate viewing rectangle—one that is wide enough to contain all the x-intercepts of P. Numerical approximations to the zeros of P are

$$5, \qquad -2, \qquad 2.4, \qquad \text{and} \qquad -0.4$$

So in this case we choose the rectangle $[-3, 6]$ by $[-50, 50]$ and draw the graph shown in Figure 1. ◼

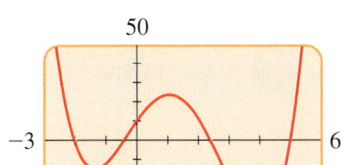

Figure 1
$P(x) = x^4 - 5x^3 - 5x^2 + 23x + 10$

Descartes' Rule of Signs and Upper and Lower Bounds for Roots

In some cases, the following rule—discovered by the French philosopher and mathematician René Descartes around 1637 (see page 112)—is helpful in eliminating candidates from lengthy lists of possible rational roots. To describe this rule, we need the concept of *variation in sign*. If $P(x)$ is a polynomial with real coefficients, written with descending powers of x (and omitting powers with coefficient 0), then a **variation in sign** occurs whenever adjacent coefficients have opposite signs. For example,

$$P(x) = 5x^7 - 3x^5 - x^4 + 2x^2 + x - 3$$

has three variations in sign.

Polynomial	Variations in sign
$x^2 + 4x + 1$	0
$2x^3 + x - 6$	1
$x^4 - 3x^2 - x + 4$	2

Descartes' Rule of Signs

Let P be a polynomial with real coefficients.

1. The number of positive real zeros of $P(x)$ is either equal to the number of variations in sign in $P(x)$ or is less than that by an even whole number.

2. The number of negative real zeros of $P(x)$ is either equal to the number of variations in sign in $P(-x)$ or is less than that by an even whole number.

Example 4 **Using Descartes' Rule**

Use Descartes' Rule of Signs to determine the possible number of positive and negative real zeros of the polynomial

$$P(x) = 3x^6 + 4x^5 + 3x^3 - x - 3$$

Solution The polynomial has one variation in sign and so it has one positive zero. Now

$$P(-x) = 3(-x)^6 + 4(-x)^5 + 3(-x)^3 - (-x) - 3$$
$$= 3x^6 - 4x^5 - 3x^3 + x - 3$$

So, $P(-x)$ has three variations in sign. Thus, $P(x)$ has either three or one negative zero(s), making a total of either two or four real zeros. ◼

We say that a is a **lower bound** and b is an **upper bound** for the zeros of a polynomial if every real zero c of the polynomial satisfies $a \le c \le b$. The next theorem helps us find such bounds for the zeros of a polynomial.

The Upper and Lower Bounds Theorem

Let P be a polynomial with real coefficients.

1. If we divide $P(x)$ by $x - b$ (with $b > 0$) using synthetic division, and if the row that contains the quotient and remainder has no negative entry, then b is an upper bound for the real zeros of P.

2. If we divide $P(x)$ by $x - a$ (with $a < 0$) using synthetic division, and if the row that contains the quotient and remainder has entries that are alternately nonpositive and nonnegative, then a is a lower bound for the real zeros of P.

A proof of this theorem is suggested in Exercise 91. The phrase "alternately nonpositive and nonnegative" simply means that the signs of the numbers alternate, with 0 considered to be positive or negative as required.

Example 5   Upper and Lower Bounds for Zeros of a Polynomial

Show that all the real zeros of the polynomial $P(x) = x^4 - 3x^2 + 2x - 5$ lie between -3 and 2.

Solution   We divide $P(x)$ by $x - 2$ and $x + 3$ using synthetic division.

$$
\begin{array}{r|rrrrr}
2 & 1 & 0 & -3 & 2 & -5 \\
  &   & 2 & 4 & 2 & 8 \\
\hline
  & 1 & 2 & 1 & 4 & 3
\end{array}
$$

All entries positive

$$
\begin{array}{r|rrrrr}
-3 & 1 & 0 & -3 & 2 & -5 \\
   &   & -3 & 9 & -18 & 48 \\
\hline
   & 1 & -3 & 6 & -16 & 43
\end{array}
$$

Entries alternate in sign.

By the Upper and Lower Bounds Theorem, -3 is a lower bound and 2 is an upper bound for the zeros. Since neither -3 nor 2 is a zero (the remainders are not 0 in the division table), all the real zeros lie between these numbers. ∎

Example 6   Factoring a Fifth-Degree Polynomial

Factor completely the polynomial

$$P(x) = 2x^5 + 5x^4 - 8x^3 - 14x^2 + 6x + 9$$

Solution   The possible rational zeros of P are $\pm\frac{1}{2}$, ±1, $\pm\frac{3}{2}$, ±3, $\pm\frac{9}{2}$, and ±9. We check the positive candidates first, beginning with the smallest.

$$
\begin{array}{r|rrrrrr}
\frac{1}{2} & 2 & 5 & -8 & -14 & 6 & 9 \\
  &   & 1 & 3 & -\frac{5}{2} & -\frac{33}{4} & -\frac{9}{8} \\
\hline
  & 2 & 6 & -5 & -\frac{33}{2} & -\frac{9}{4} & \frac{63}{8}
\end{array}
$$

$\frac{1}{2}$ is not a zero

$$
\begin{array}{r|rrrrrr}
1 & 2 & 5 & -8 & -14 & 6 & 9 \\
  &   & 2 & 7 & -1 & -15 & -9 \\
\hline
  & 2 & 7 & -1 & -15 & -9 & 0
\end{array}
$$

$P(1) = 0$

So 1 is a zero, and $P(x) = (x - 1)(2x^4 + 7x^3 - x^2 - 15x - 9)$. We continue by factoring the quotient. We still have the same list of possible zeros except that $\frac{1}{2}$ has been eliminated.

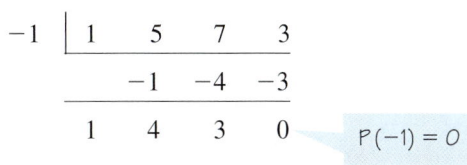

$$
\begin{array}{r|rrrrr}
1 & 2 & 7 & -1 & -15 & -9 \\
  &   & 2 & 9 & 8 & -7 \\
\hline
  & 2 & 9 & 8 & -7 & -16 \\
\end{array}
$$
1 is not a zero.

$$
\begin{array}{r|rrrrr}
\frac{3}{2} & 2 & 7 & -1 & -15 & -9 \\
  &   & 3 & 15 & 21 & 9 \\
\hline
  & 2 & 10 & 14 & 6 & 0 \\
\end{array}
$$
$P(\frac{3}{2}) = 0$, all entries nonnegative

We see that $\frac{3}{2}$ is both a zero and an upper bound for the zeros of $P(x)$, so we don't need to check any further for positive zeros, because all the remaining candidates are greater than $\frac{3}{2}$.

$$P(x) = (x - 1)(x - \tfrac{3}{2})(2x^3 + 10x^2 + 14x + 6)$$
$$= (x - 1)(2x - 3)(x^3 + 5x^2 + 7x + 3) \quad \text{Factor 2 from last factor, multiply into second factor}$$

By Descartes' Rule of Signs, $x^3 + 5x^2 + 7x + 3$ has no positive zero, so its only possible rational zeros are -1 and -3.

$$
\begin{array}{r|rrrr}
-1 & 1 & 5 & 7 & 3 \\
   &   & -1 & -4 & -3 \\
\hline
   & 1 & 4 & 3 & 0 \\
\end{array}
$$
$P(-1) = 0$

Therefore

$$P(x) = (x - 1)(2x - 3)(x + 1)(x^2 + 4x + 3)$$
$$= (x - 1)(2x - 3)(x + 1)^2(x + 3) \quad \text{Factor quadratic}$$

This means that the zeros of P are 1, $\frac{3}{2}$, -1, and -3. The graph of the polynomial is shown in Figure 2. ∎

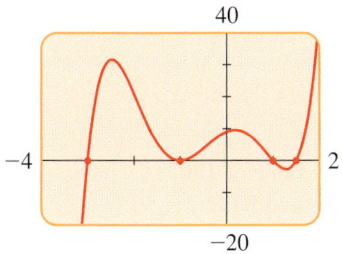

Figure 2
$P(x) = 2x^5 + 5x^4 - 8x^3 - 14x^2 + 6x + 9$
$= (x - 1)(2x - 3)(x + 1)^2(x + 3)$

Using Algebra and Graphing Devices to Solve Polynomial Equations

In Section 1.9 we used graphing devices to solve equations graphically. We can now use the algebraic techniques we've learned to select an appropriate viewing rectangle when solving a polynomial equation graphically.

Example 7   **Solving a Fourth-Degree Equation Graphically**

Find all real solutions of the following equation, correct to the nearest tenth.

$$3x^4 + 4x^3 - 7x^2 - 2x - 3 = 0$$

Solution   To solve the equation graphically, we graph

$$P(x) = 3x^4 + 4x^3 - 7x^2 - 2x - 3$$

We use the Upper and Lower Bounds Theorem to see where the roots can be found.

First we use the Upper and Lower Bounds Theorem to find two numbers between which all the solutions must lie. This allows us to choose a viewing rectangle that is certain to contain all the x-intercepts of P. We use synthetic division and proceed by trial and error.

To find an upper bound, we try the whole numbers, 1, 2, 3, . . . as potential candidates. We see that 2 is an upper bound for the roots.

$$
\begin{array}{r|rrrrr}
2 & 3 & 4 & -7 & -2 & -3 \\
  &   & 6 & 20 & 26 & 48 \\
\hline
  & 3 & 10 & 13 & 24 & 45
\end{array}
$$

All positive

Now we look for a lower bound, trying the numbers -1, -2, and -3 as potential candidates. We see that -3 is a lower bound for the roots.

$$
\begin{array}{r|rrrrr}
-3 & 3 & 4 & -7 & -2 & -3 \\
   &   & -9 & 15 & -24 & 78 \\
\hline
   & 3 & -5 & 8 & -26 & 75
\end{array}
$$

Entries alternate in sign.

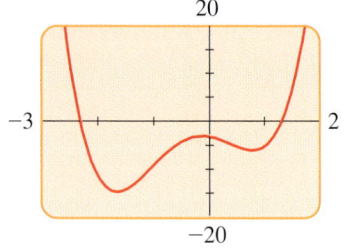

Figure 3
$y = 3x^4 + 4x^3 - 7x^2 - 2x - 3$

Thus, all the roots lie between -3 and 2. So the viewing rectangle $[-3, 2]$ by $[-20, 20]$ contains all the x-intercepts of P. The graph in Figure 3 has two x-intercepts, one between -3 and -2 and the other between 1 and 2. Zooming in, we find that the solutions of the equation, to the nearest tenth, are -2.3 and 1.3. ∎

Example 8   Determining the Size of a Fuel Tank

A fuel tank consists of a cylindrical center section 4 ft long and two hemispherical end sections, as shown in Figure 4. If the tank has a volume of 100 ft^3, what is the radius r shown in the figure, correct to the nearest hundredth of a foot?

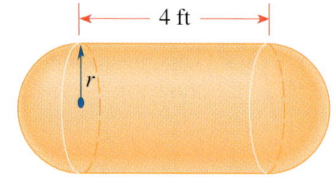

Figure 4

Solution   Using the volume formula listed on the inside front cover of this book, we see that the volume of the cylindrical section of the tank is

Volume of a cylinder: $V = \pi r^2 h$

$$\pi \cdot r^2 \cdot 4$$

The two hemispherical parts together form a complete sphere whose volume is

Volume of a sphere: $V = \frac{4}{3}\pi r^3$

$$\tfrac{4}{3}\pi r^3$$

Because the total volume of the tank is 100 ft^3, we get the following equation:

$$\tfrac{4}{3}\pi r^3 + 4\pi r^2 = 100$$

A negative solution for r would be meaningless in this physical situation, and by substitution we can verify that $r = 3$ leads to a tank that is over 226 ft^3 in volume, much larger than the required 100 ft^3. Thus, we know the correct radius lies somewhere between 0 and 3 ft, and so we use a viewing rectangle of $[0, 3]$ by $[50, 150]$

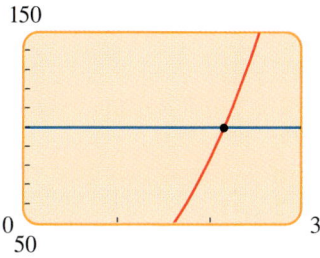

Figure 5
$y = \frac{4}{3}\pi x^3 + 4\pi x^2$ and $y = 100$

to graph the function $y = \frac{4}{3}\pi x^3 + 4\pi x^2$, as shown in Figure 5. Since we want the value of this function to be 100, we also graph the horizontal line $y = 100$ in the same viewing rectangle. The correct radius will be the x-coordinate of the point of intersection of the curve and the line. Using the cursor and zooming in, we see that at the point of intersection $x \approx 2.15$, correct to two decimal places. Thus, the tank has a radius of about 2.15 ft. ■

Note that we also could have solved the equation in Example 8 by first writing it as

$$\tfrac{4}{3}\pi r^3 + 4\pi r^2 - 100 = 0$$

and then finding the x-intercept of the function $y = \frac{4}{3}\pi x^3 + 4\pi x^2 - 100$.

3.3 Exercises

1–6 ■ List all possible rational zeros given by the Rational Zeros Theorem (but don't check to see which actually are zeros).

1. $P(x) = x^3 - 4x^2 + 3$

2. $Q(x) = x^4 - 3x^3 - 6x + 8$

3. $R(x) = 2x^5 + 3x^3 + 4x^2 - 8$

4. $S(x) = 6x^4 - x^2 + 2x + 12$

5. $T(x) = 4x^4 - 2x^2 - 7$

6. $U(x) = 12x^5 + 6x^3 - 2x - 8$

7–10 ■ A polynomial function P and its graph are given.

(a) List all possible rational zeros of P given by the Rational Zeros Theorem.

(b) From the graph, determine which of the possible rational zeros actually turn out to be zeros.

7. $P(x) = 5x^3 - x^2 - 5x + 1$

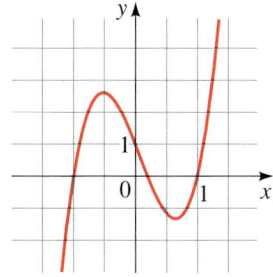

8. $P(x) = 3x^3 + 4x^2 - x - 2$

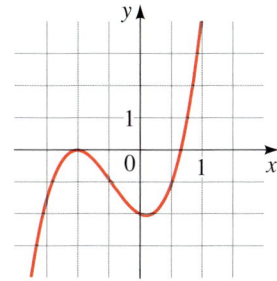

9. $P(x) = 2x^4 - 9x^3 + 9x^2 + x - 3$

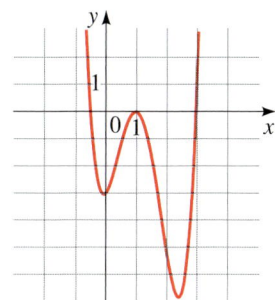

10. $P(x) = 4x^4 - x^3 - 4x + 1$

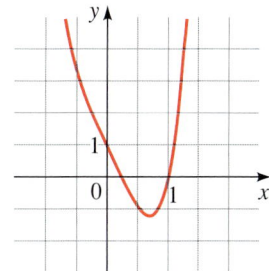

11–40 ■ Find all rational zeros of the polynomial.

11. $P(x) = x^3 + 3x^2 - 4$

12. $P(x) = x^3 - 7x^2 + 14x - 8$

13. $P(x) = x^3 - 3x - 2$

14. $P(x) = x^3 + 4x^2 - 3x - 18$

15. $P(x) = x^3 - 6x^2 + 12x - 8$

16. $P(x) = x^3 - x^2 - 8x + 12$

17. $P(x) = x^3 - 4x^2 + x + 6$

18. $P(x) = x^3 - 4x^2 - 7x + 10$

19. $P(x) = x^3 + 3x^2 + 6x + 4$

20. $P(x) = x^3 - 2x^2 - 2x - 3$

21. $P(x) = x^4 - 5x^2 + 4$

22. $P(x) = x^4 - 2x^3 - 3x^2 + 8x - 4$

23. $P(x) = x^4 + 6x^3 + 7x^2 - 6x - 8$

24. $P(x) = x^4 - x^3 - 23x^2 - 3x + 90$

25. $P(x) = 4x^4 - 25x^2 + 36$

26. $P(x) = x^4 - x^3 - 5x^2 + 3x + 6$

27. $P(x) = x^4 + 8x^3 + 24x^2 + 32x + 16$

28. $P(x) = 2x^3 + 7x^2 + 4x - 4$

29. $P(x) = 4x^3 + 4x^2 - x - 1$

30. $P(x) = 2x^3 - 3x^2 - 2x + 3$

31. $P(x) = 4x^3 - 7x + 3$

32. $P(x) = 8x^3 + 10x^2 - x - 3$

33. $P(x) = 4x^3 + 8x^2 - 11x - 15$

34. $P(x) = 6x^3 + 11x^2 - 3x - 2$

35. $P(x) = 2x^4 - 7x^3 + 3x^2 + 8x - 4$

36. $P(x) = 6x^4 - 7x^3 - 12x^2 + 3x + 2$

37. $P(x) = x^5 + 3x^4 - 9x^3 - 31x^2 + 36$

38. $P(x) = x^5 - 4x^4 - 3x^3 + 22x^2 - 4x - 24$

39. $P(x) = 3x^5 - 14x^4 - 14x^3 + 36x^2 + 43x + 10$

40. $P(x) = 2x^6 - 3x^5 - 13x^4 + 29x^3 - 27x^2 + 32x - 12$

41–50 ■ Find all the real zeros of the polynomial. Use the quadratic formula if necessary, as in Example 3(a).

41. $P(x) = x^3 + 4x^2 + 3x - 2$

42. $P(x) = x^3 - 5x^2 + 2x + 12$

43. $P(x) = x^4 - 6x^3 + 4x^2 + 15x + 4$

44. $P(x) = x^4 + 2x^3 - 2x^2 - 3x + 2$

45. $P(x) = x^4 - 7x^3 + 14x^2 - 3x - 9$

46. $P(x) = x^5 - 4x^4 - x^3 + 10x^2 + 2x - 4$

47. $P(x) = 4x^3 - 6x^2 + 1$

48. $P(x) = 3x^3 - 5x^2 - 8x - 2$

49. $P(x) = 2x^4 + 15x^3 + 17x^2 + 3x - 1$

50. $P(x) = 4x^5 - 18x^4 - 6x^3 + 91x^2 - 60x + 9$

51–58 ■ A polynomial P is given.
(a) Find all the real zeros of P.
(b) Sketch the graph of P.

51. $P(x) = x^3 - 3x^2 - 4x + 12$

52. $P(x) = -x^3 - 2x^2 + 5x + 6$

53. $P(x) = 2x^3 - 7x^2 + 4x + 4$

54. $P(x) = 3x^3 + 17x^2 + 21x - 9$

55. $P(x) = x^4 - 5x^3 + 6x^2 + 4x - 8$

56. $P(x) = -x^4 + 10x^2 + 8x - 8$

57. $P(x) = x^5 - x^4 - 5x^3 + x^2 + 8x + 4$

58. $P(x) = x^5 - x^4 - 6x^3 + 14x^2 - 11x + 3$

59–64 ■ Use Descartes' Rule of Signs to determine how many positive and how many negative real zeros the polynomial can have. Then determine the possible total number of real zeros.

59. $P(x) = x^3 - x^2 - x - 3$

60. $P(x) = 2x^3 - x^2 + 4x - 7$

61. $P(x) = 2x^6 + 5x^4 - x^3 - 5x - 1$

62. $P(x) = x^4 + x^3 + x^2 + x + 12$

63. $P(x) = x^5 + 4x^3 - x^2 + 6x$

64. $P(x) = x^8 - x^5 + x^4 - x^3 + x^2 - x + 1$

65–68 ■ Show that the given values for a and b are lower and upper bounds for the real zeros of the polynomial.

65. $P(x) = 2x^3 + 5x^2 + x - 2$; $a = -3, b = 1$

66. $P(x) = x^4 - 2x^3 - 9x^2 + 2x + 8$; $a = -3, b = 5$

67. $P(x) = 8x^3 + 10x^2 - 39x + 9$; $a = -3, b = 2$

68. $P(x) = 3x^4 - 17x^3 + 24x^2 - 9x + 1$; $a = 0, b = 6$

69–72 ■ Find integers that are upper and lower bounds for the real zeros of the polynomial.

69. $P(x) = x^3 - 3x^2 + 4$

70. $P(x) = 2x^3 - 3x^2 - 8x + 12$

71. $P(x) = x^4 - 2x^3 + x^2 - 9x + 2$

72. $P(x) = x^5 - x^4 + 1$

73–78 ■ Find all rational zeros of the polynomial, and then find the irrational zeros, if any. Whenever appropriate, use the Rational Zeros Theorem, the Upper and Lower Bounds Theorem, Descartes' Rule of Signs, the quadratic formula, or other factoring techniques.

73. $P(x) = 2x^4 + 3x^3 - 4x^2 - 3x + 2$

74. $P(x) = 2x^4 + 15x^3 + 31x^2 + 20x + 4$

75. $P(x) = 4x^4 - 21x^2 + 5$

76. $P(x) = 6x^4 - 7x^3 - 8x^2 + 5x$

77. $P(x) = x^5 - 7x^4 + 9x^3 + 23x^2 - 50x + 24$

78. $P(x) = 8x^5 - 14x^4 - 22x^3 + 57x^2 - 35x + 6$

79–82 ■ Show that the polynomial does not have any rational zeros.

79. $P(x) = x^3 - x - 2$

80. $P(x) = 2x^4 - x^3 + x + 2$

81. $P(x) = 3x^3 - x^2 - 6x + 12$

82. $P(x) = x^{50} - 5x^{25} + x^2 - 1$

 83–86 ■ The real solutions of the given equation are rational. List all possible rational roots using the Rational Zeros Theorem, and then graph the polynomial in the given viewing rectangle to determine which values are actually solutions. (All solutions can be seen in the given viewing rectangle.)

83. $x^3 - 3x^2 - 4x + 12 = 0$; $[-4, 4]$ by $[-15, 15]$

84. $x^4 - 5x^2 + 4 = 0$; $[-4, 4]$ by $[-30, 30]$

85. $2x^4 - 5x^3 - 14x^2 + 5x + 12 = 0$; $[-2, 5]$ by $[-40, 40]$

86. $3x^3 + 8x^2 + 5x + 2 = 0$; $[-3, 3]$ by $[-10, 10]$

 87–90 ■ Use a graphing device to find all real solutions of the equation, correct to two decimal places.

87. $x^4 - x - 4 = 0$

88. $2x^3 - 8x^2 + 9x - 9 = 0$

89. $4.00x^4 + 4.00x^3 - 10.96x^2 - 5.88x + 9.09 = 0$

90. $x^5 + 2.00x^4 + 0.96x^3 + 5.00x^2 + 10.00x + 4.80 = 0$

91. Let $P(x)$ be a polynomial with real coefficients and let $b > 0$. Use the Division Algorithm to write

$$P(x) = (x - b) \cdot Q(x) + r$$

Suppose that $r \geq 0$ and that all the coefficients in $Q(x)$ are nonnegative. Let $z > b$.

(a) Show that $P(z) > 0$.

(b) Prove the first part of the Upper and Lower Bounds Theorem.

(c) Use the first part of the Upper and Lower Bounds Theorem to prove the second part. [*Hint:* Show that if $P(x)$ satisfies the second part of the theorem, then $P(-x)$ satisfies the first part.]

92. Show that the equation

$$x^5 - x^4 - x^3 - 5x^2 - 12x - 6 = 0$$

has exactly one rational root, and then prove that it must have either two or four irrational roots.

Applications

 93. Volume of a Silo A grain silo consists of a cylindrical main section and a hemispherical roof. If the total volume of the silo (including the part inside the roof section) is

15,000 ft^3 and the cylindrical part is 30 ft tall, what is the radius of the silo, correct to the nearest tenth of a foot?

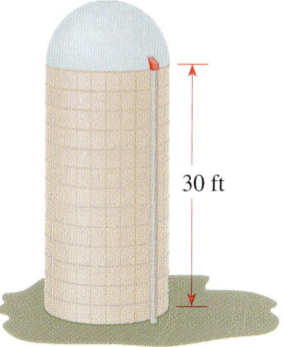

94. Dimensions of a Lot A rectangular parcel of land has an area of 5000 ft^2. A diagonal between opposite corners is measured to be 10 ft longer than one side of the parcel. What are the dimensions of the land, correct to the nearest foot?

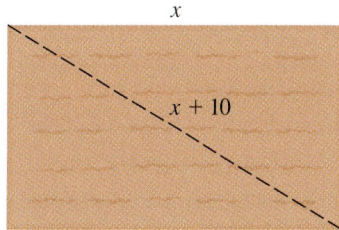

95. Depth of Snowfall Snow began falling at noon on Sunday. The amount of snow on the ground at a certain location at time t was given by the function

$$h(t) = 11.60t - 12.41t^2 + 6.20t^3$$
$$- 1.58t^4 + 0.20t^5 - 0.01t^6$$

where t is measured in days from the start of the snowfall and $h(t)$ is the depth of snow in inches. Draw a graph of this function and use your graph to answer the following questions.

(a) What happened shortly after noon on Tuesday?

(b) Was there ever more than 5 in. of snow on the ground? If so, on what day(s)?

(c) On what day and at what time (to the nearest hour) did the snow disappear completely?

96. Volume of a Box An open box with a volume of 1500 cm^3 is to be constructed by taking a piece of cardboard 20 cm by 40 cm, cutting squares of side length x cm from each corner, and folding up the sides. Show that this can be done in two different ways, and find the exact dimensions of the box in each case.

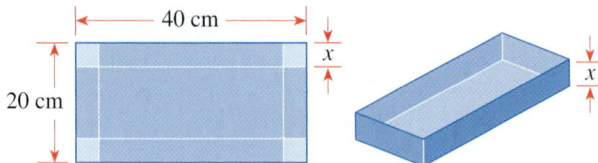

97. Volume of a Rocket   A rocket consists of a right circular cylinder of height 20 m surmounted by a cone whose height and diameter are equal and whose radius is the same as that of the cylindrical section. What should this radius be (correct to two decimal places) if the total volume is to be $500\pi/3$ m^3?

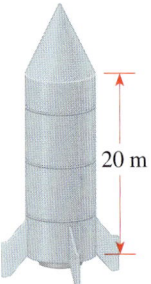

20 m

98. Volume of a Box   A rectangular box with a volume of $2\sqrt{2}$ ft^3 has a square base as shown below. The diagonal of the box (between a pair of opposite corners) is 1 ft longer than each side of the base.

(a) If the base has sides of length x feet, show that

$$x^6 - 2x^5 - x^4 + 8 = 0$$

(b) Show that two different boxes satisfy the given conditions. Find the dimensions in each case, correct to the nearest hundredth of a foot.

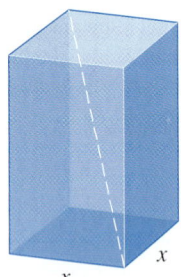

99. Girth of a Box   A box with a square base has length plus girth of 108 in. (Girth is the distance "around" the box.) What is the length of the box if its volume is 2200 in^3?

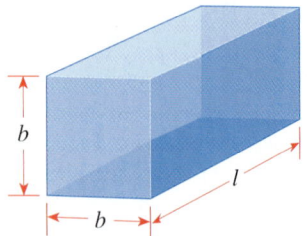

Discovery • Discussion

100. How Many Real Zeros Can a Polynomial Have?   Give examples of polynomials that have the following properties, or explain why it is impossible to find such a polynomial.

(a) A polynomial of degree 3 that has no real zeros

(b) A polynomial of degree 4 that has no real zeros

(c) A polynomial of degree 3 that has three real zeros, only one of which is rational

(d) A polynomial of degree 4 that has four real zeros, none of which is rational

What must be true about the degree of a polynomial with integer coefficients if it has no real zeros?

101. The Depressed Cubic   The most general cubic (third-degree) equation with rational coefficients can be written as

$$x^3 + ax^2 + bx + c = 0$$

(a) Show that if we replace x by $X - a/3$ and simplify, we end up with an equation that doesn't have an X^2 term, that is, an equation of the form

$$X^3 + pX + q = 0$$

This is called a *depressed cubic*, because we have "depressed" the quadratic term.

(b) Use the procedure described in part (a) to depress the equation $x^3 + 6x^2 + 9x + 4 = 0$.

102. The Cubic Formula   The quadratic formula can be used to solve any quadratic (or second-degree) equation. You may have wondered if similar formulas exist for cubic (third-degree), quartic (fourth-degree), and higher-degree equations. For the depressed cubic $x^3 + px + q = 0$, Cardano (page 296) found the following formula for one solution:

$$x = \sqrt[3]{-\frac{q}{2} + \sqrt{\frac{q^2}{4} + \frac{p^3}{27}}} + \sqrt[3]{-\frac{q}{2} - \sqrt{\frac{q^2}{4} + \frac{p^3}{27}}}$$

A formula for quartic equations was discovered by the Italian mathematician Ferrari in 1540. In 1824 the Norwegian mathematician Niels Henrik Abel proved that it is impossible to write a quintic formula, that is, a formula for fifth-degree equations. Finally, Galois (page 273) gave a criterion for determining which equations can be solved by a formula involving radicals.

Use the cubic formula to find a solution for the following equations. Then solve the equations using the methods you learned in this section. Which method is easier?

(a) $x^3 - 3x + 2 = 0$

(b) $x^3 - 27x - 54 = 0$

(c) $x^3 + 3x + 4 = 0$

DISCOVERY PROJECT

Zeroing in on a Zero

We have seen how to find the zeros of a polynomial algebraically and graphically. Let's work through a **numerical method** for finding the zeros. With this method we can find the value of any real zero to as many decimal places as we wish.

The Intermediate Value Theorem states: If P is a polynomial and if $P(a)$ and $P(b)$ are of opposite sign, then P has a zero between a and b. (See page 255.) The Intermediate Value Theorem is an example of an **existence theorem**—it tells us that a zero exists, but doesn't tell us exactly where it is. Nevertheless, we can use the theorem to zero in on the zero.

For example, consider the polynomial $P(x) = x^3 + 8x - 30$. Notice that $P(2) < 0$ and $P(3) > 0$. By the Intermediate Value Theorem P must have a zero between 2 and 3. To "trap" the zero in a smaller interval, we evaluate P at successive tenths between 2 and 3 until we find where P changes sign, as in Table 1. From the table we see that the zero we are looking for lies between 2.2 and 2.3, as shown in Figure 1.

Table 1

x	$P(x)$
2.1	−3.94
2.2	−1.75
2.3	0.57

}change of sign

Table 2

x	$P(x)$
2.26	−0.38
2.27	−0.14
2.28	0.09

}change of sign

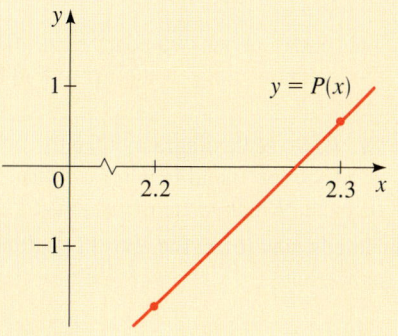

Figure 1

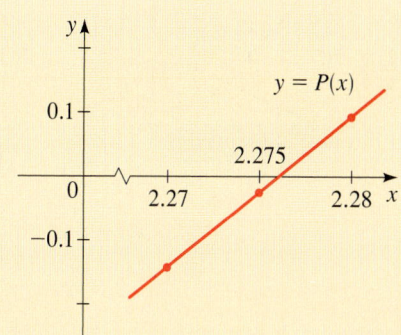

Figure 2

We can repeat this process by evaluating P at successive 100ths between 2.2 and 2.3, as in Table 2. By repeating this process over and over again, we can get a numerical value for the zero as accurately as we want. From Table 2 we see that the zero is between 2.27 and 2.28. To see whether it is closer to 2.27 or 2.28, we check the value of P halfway between these two numbers: $P(2.275) \approx -0.03$. Since this value is negative, the zero we are looking for lies between 2.275 and 2.28, as illustrated in Figure 2. Correct to the nearest 100th, the zero is 2.28.

1. (a) Show that $P(x) = x^2 - 2$ has a zero between 1 and 2.

(b) Find the zero of P to the nearest tenth.

(c) Find the zero of P to the nearest 100th.

(d) Explain why the zero you found is an approximation to $\sqrt{2}$. Repeat the process several times to obtain $\sqrt{2}$ correct to three decimal places. Compare your results to $\sqrt{2}$ obtained by a calculator.

2. Find a polynomial that has $\sqrt[3]{5}$ as a zero. Use the process described here to zero in on $\sqrt[3]{5}$ to four decimal places.

3. Show that the polynomial has a zero between the given integers, and then zero in on that zero, correct to two decimals.

(a) $P(x) = x^3 + x - 7$; between 1 and 2

(b) $P(x) = x^3 - x^2 - 5$; between 2 and 3

(c) $P(x) = 2x^4 - 4x^2 + 1$; between 1 and 2

(d) $P(x) = 2x^4 - 4x^2 + 1$; between -1 and 0

4. Find the indicated irrational zero, correct to two decimals.

(a) The positive zero of $P(x) = x^4 + 2x^3 + x^2 - 1$

(b) The negative zero of $P(x) = x^4 + 2x^3 + x^2 - 1$

5. In a passageway between two buildings, two ladders are propped up from the base of each building to the wall of the other so that they cross, as shown in the figure. If the ladders have lengths $a = 3$ m and $b = 2$ m and the crossing point is at height $c = 1$ m, then it can be shown that the distance x between the buildings is a solution of the equation

$$x^8 - 22x^6 + 163x^4 - 454x^2 + 385 = 0$$

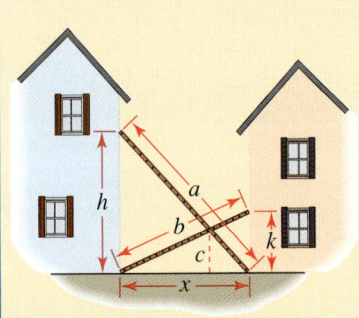

(a) This equation has two positive solutions, which lie between 1 and 2. Use the technique of "zeroing in" to find both of these correct to the nearest tenth.

(b) Draw two scale diagrams, like the figure, one for each of the two values of x that you found in part (a). Measure the height of the crossing point on each. Which value of x seems to be the correct one?

(c) Here is how to get the above equation. First, use similar triangles to show that

$$\frac{1}{c} = \frac{1}{h} + \frac{1}{k}$$

Then use the Pythagorean Theorem to rewrite this as

$$\frac{1}{c} = \frac{1}{\sqrt{a^2 - x^2}} + \frac{1}{\sqrt{b^2 - x^2}}$$

Substitute $a = 3$, $b = 2$, and $c = 1$, then simplify to obtain the desired equation. [Note that you must square twice in this process to eliminate both square roots. This is why you obtain an extraneous solution in part (a). (See the *Warning* on page 53.)]

3.4   ## Complex Numbers

In Section 1.5 we saw that if the discriminant of a quadratic equation is negative, the equation has no real solution. For example, the equation

$$x^2 + 4 = 0$$

has no real solution. If we try to solve this equation, we get $x^2 = -4$, so

$$x = \pm\sqrt{-4}$$

But this is impossible, since the square of any real number is positive. [For example, $(-2)^2 = 4$, a positive number.] Thus, negative numbers don't have real square roots.

To make it possible to solve *all* quadratic equations, mathematicians invented an expanded number system, called the *complex number system*. First they defined the new number

$$i = \sqrt{-1}$$

This means $i^2 = -1$. A complex number is then a number of the form $a + bi$, where a and b are real numbers.

See the note on Cardano, page 296, for an example of how complex numbers are used to find real solutions of polynomial equations.

Definition of Complex Numbers

A **complex number** is an expression of the form

$$a + bi$$

where a and b are real numbers and $i^2 = -1$. The **real part** of this complex number is a and the **imaginary part** is b. Two complex numbers are **equal** if and only if their real parts are equal and their imaginary parts are equal.

Note that both the real and imaginary parts of a complex number are real numbers.

Example 1   Complex Numbers

The following are examples of complex numbers.

$3 + 4i$ Real part 3, imaginary part 4

$\frac{1}{2} - \frac{2}{3}i$ Real part $\frac{1}{2}$, imaginary part $-\frac{2}{3}$

$6i$ Real part 0, imaginary part 6

-7 Real part -7, imaginary part 0   ∎

A number such as $6i$, which has real part 0, is called a **pure imaginary number**. A real number like -7 can be thought of as a complex number with imaginary part 0.

In the complex number system every quadratic equation has solutions. The numbers $2i$ and $-2i$ are solutions of $x^2 = -4$ because

$$(2i)^2 = 2^2i^2 = 4(-1) = -4 \qquad \text{and} \qquad (-2i)^2 = (-2)^2i^2 = 4(-1) = -4$$

Although we use the term *imaginary* in this context, imaginary numbers should not be thought of as any less "real" (in the ordinary rather than the mathematical sense of that word) than negative numbers or irrational numbers. All numbers (except possibly the positive integers) are creations of the human mind—the numbers -1 and $\sqrt{2}$ as well as the number i. We study complex numbers because they complete, in a useful and elegant fashion, our study of the solutions of equations. In fact, imaginary numbers are useful not only in algebra and mathematics, but in the other sciences as well. To give just one example, in electrical theory the *reactance* of a circuit is a quantity whose measure is an imaginary number.

Arithmetic Operations on Complex Numbers

Complex numbers are added, subtracted, multiplied, and divided just as we would any number of the form $a + b\sqrt{c}$. The only difference we need to keep in mind is that $i^2 = -1$. Thus, the following calculations are valid.

$$(a + bi)(c + di) = ac + (ad + bc)i + bdi^2 \qquad \text{Multiply and collect like terms}$$
$$= ac + (ad + bc)i + bd(-1) \qquad i^2 = -1$$
$$= (ac - bd) + (ad + bc)i \qquad \text{Combine real and imaginary parts}$$

We therefore define the sum, difference, and product of complex numbers as follows.

Adding, Subtracting, and Multiplying Complex Numbers

Definition	Description
Addition	
$(a + bi) + (c + di) = (a + c) + (b + d)i$	To add complex numbers, add the real parts and the imaginary parts.
Subtraction	
$(a + bi) - (c + di) = (a - c) + (b - d)i$	To subtract complex numbers, subtract the real parts and the imaginary parts.
Multiplication	
$(a + bi) \cdot (c + di) = (ac - bd) + (ad + bc)i$	Multiply complex numbers like binomials, using $i^2 = -1$.

Example 2 Adding, Subtracting, and Multiplying Complex Numbers

Express the following in the form $a + bi$.

(a) $(3 + 5i) + (4 - 2i)$ (b) $(3 + 5i) - (4 - 2i)$

(c) $(3 + 5i)(4 - 2i)$ (d) i^{23}

Graphing calculators can perform arithmetic operations on complex numbers.

```
(3+5i)+(4-2i)
              7+3i
(3+5i)*(4-2i)
            22+14i
```

Solution

(a) According to the definition, we add the real parts and we add the imaginary parts.

$$(3 + 5i) + (4 - 2i) = (3 + 4) + (5 - 2)i = 7 + 3i$$

(b) $(3 + 5i) - (4 - 2i) = (3 - 4) + [5 - (-2)]i = -1 + 7i$

(c) $(3 + 5i)(4 - 2i) = [3 \cdot 4 - 5(-2)] + [3(-2) + 5 \cdot 4]i = 22 + 14i$

(d) $i^{23} = i^{22+1} = (i^2)^{11}i = (-1)^{11}i = (-1)i = -i$ ∎

Complex Conjugates

Number	Conjugate
$3 + 2i$	$3 - 2i$
$1 - i$	$1 + i$
$4i$	$-4i$
5	5

Division of complex numbers is much like rationalizing the denominator of a radical expression, which we considered in Section 1.2. For the complex number $z = a + bi$ we define its **complex conjugate** to be $\bar{z} = a - bi$. Note that

$$z \cdot \bar{z} = (a + bi)(a - bi) = a^2 + b^2$$

So the product of a complex number and its conjugate is always a nonnegative real number. We use this property to divide complex numbers.

Dividing Complex Numbers

To simplify the quotient $\dfrac{a + bi}{c + di}$, multiply the numerator and the denominator by the complex conjugate of the denominator:

$$\frac{a + bi}{c + di} = \left(\frac{a + bi}{c + di}\right)\left(\frac{c - di}{c - di}\right) = \frac{(ac + bd) + (bc - ad)i}{c^2 + d^2}$$

Rather than memorize this entire formula, it's easier to just remember the first step and then multiply out the numerator and the denominator as usual.

Example 3   Dividing Complex Numbers

Express the following in the form $a + bi$.

(a) $\dfrac{3 + 5i}{1 - 2i}$ (b) $\dfrac{7 + 3i}{4i}$

Solution   We multiply both the numerator and denominator by the complex conjugate of the denominator to make the new denominator a real number.

(a) The complex conjugate of $1 - 2i$ is $\overline{1 - 2i} = 1 + 2i$.

$$\frac{3 + 5i}{1 - 2i} = \left(\frac{3 + 5i}{1 - 2i}\right)\left(\frac{1 + 2i}{1 + 2i}\right) = \frac{-7 + 11i}{5} = -\frac{7}{5} + \frac{11}{5}i$$

(b) The complex conjugate of $4i$ is $-4i$. Therefore

$$\frac{7 + 3i}{4i} = \left(\frac{7 + 3i}{4i}\right)\left(\frac{-4i}{-4i}\right) = \frac{12 - 28i}{16} = \frac{3}{4} - \frac{7}{4}i$$ ∎

Square Roots of Negative Numbers

Just as every positive real number r has two square roots ($\sqrt{r}$ and $-\sqrt{r}$), every negative number has two square roots as well. If $-r$ is a negative number, then its square roots are $\pm i\sqrt{r}$, because $(i\sqrt{r})^2 = i^2r = -r$ and $(-i\sqrt{r})^2 = i^2r = -r$.

Leonhard Euler (1707–1783) was born in Basel, Switzerland, the son of a pastor. At age 13 his father sent him to the University at Basel to study theology, but Euler soon decided to devote himself to the sciences. Besides theology he studied mathematics, medicine, astronomy, physics, and Asian languages. It is said that Euler could calculate as effortlessly as "men breathe or as eagles fly." One hundred years before Euler, Fermat (see page 652) had conjectured that $2^{2^n} + 1$ is a prime number for all n. The first five of these numbers are 5, 17, 257, 65537, and 4,294,967,297. It's easy to show that the first four are prime. The fifth was also thought to be prime until Euler, with his phenomenal calculating ability, showed that it is the product $641 \times 6,700,417$ and so is not prime. Euler published more than any other mathematician in history. His collected works comprise 75 large volumes. Although he was blind for the last 17 years of his life, he continued to work and publish. In his writings he popularized the use of the symbols π, e, and i, which you will find in this textbook. One of Euler's most lasting contributions is his development of complex numbers.

Square Roots of Negative Numbers

If $-r$ is negative, then the **principal square root** of $-r$ is

$$\sqrt{-r} = i\sqrt{r}$$

The two square roots of $-r$ are $i\sqrt{r}$ and $-i\sqrt{r}$.

We usually write $i\sqrt{b}$ instead of $\sqrt{b}\,i$ to avoid confusion with $\sqrt{bi}$.

Example 4 Square Roots of Negative Numbers

(a) $\sqrt{-1} = i\sqrt{1} = i$

(b) $\sqrt{-16} = i\sqrt{16} = 4i$

(c) $\sqrt{-3} = i\sqrt{3}$ ∎

Special care must be taken when performing calculations involving square roots of negative numbers. Although $\sqrt{a} \cdot \sqrt{b} = \sqrt{ab}$ when a and b are positive, this is *not* true when both are negative. For example,

$$\sqrt{-2} \cdot \sqrt{-3} = i\sqrt{2} \cdot i\sqrt{3} = i^2\sqrt{6} = -\sqrt{6}$$

but

$$\sqrt{(-2)(-3)} = \sqrt{6}$$

so

$$\sqrt{-2} \cdot \sqrt{-3} \;\;\neq\;\; \sqrt{(-2)(-3)}$$

 When multiplying radicals of negative numbers, express them first in the form $i\sqrt{r}$ (where $r > 0$) to avoid possible errors of this type.

Example 5 Using Square Roots of Negative Numbers

Evaluate $(\sqrt{12} - \sqrt{-3})(3 + \sqrt{-4})$ and express in the form $a + bi$.

Solution

$$(\sqrt{12} - \sqrt{-3})(3 + \sqrt{-4}) = (\sqrt{12} - i\sqrt{3})(3 + i\sqrt{4})$$
$$= (2\sqrt{3} - i\sqrt{3})(3 + 2i)$$
$$= (6\sqrt{3} + 2\sqrt{3}) + i(2 \cdot 2\sqrt{3} - 3\sqrt{3})$$
$$= 8\sqrt{3} + i\sqrt{3}$$ ∎

Complex Roots of Quadratic Equations

We have already seen that, if $a \neq 0$, then the solutions of the quadratic equation $ax^2 + bx + c = 0$ are

$$x = \frac{-b \pm \sqrt{b^2 - 4ac}}{2a}$$

If $b^2 - 4ac < 0$, then the equation has no real solution. But in the complex number system, this equation will always have solutions, because negative numbers have square roots in this expanded setting.

Example 6 **Quadratic Equations with Complex Solutions**

Solve each equation.

(a) $x^2 + 9 = 0$ (b) $x^2 + 4x + 5 = 0$

Solution

(a) The equation $x^2 + 9 = 0$ means $x^2 = -9$, so

$$x = \pm\sqrt{-9} = \pm i\sqrt{9} = \pm 3i$$

The solutions are therefore $3i$ and $-3i$.

(b) By the quadratic formula we have

$$x = \frac{-4 \pm \sqrt{4^2 - 4\cdot 5}}{2}$$

$$= \frac{-4 \pm \sqrt{-4}}{2}$$

$$= \frac{-4 \pm 2i}{2} = \frac{2(-2 \pm i)}{2} = -2 \pm i$$

So, the solutions are $-2 + i$ and $-2 - i$. ∎

The two solutions of *any* quadratic equation that has real coefficients are complex conjugates of each other. To understand why this is true, think about the ± sign in the quadratic formula.

Example 7 **Complex Conjugates as Solutions of a Quadratic**

Show that the solutions of the equation

$$4x^2 - 24x + 37 = 0$$

are complex conjugates of each other.

Solution   We use the quadratic formula to get

$$x = \frac{24 \pm \sqrt{(24)^2 - 4(4)(37)}}{2(4)}$$

$$= \frac{24 \pm \sqrt{-16}}{8} = \frac{24 \pm 4i}{8} = 3 \pm \frac{1}{2}i$$

So, the solutions are $3 + \frac{1}{2}i$ and $3 - \frac{1}{2}i$, and these are complex conjugates. ∎

3.4 Exercises

1–10 ■ Find the real and imaginary parts of the complex number.

1. $5 - 7i$

2. $-6 + 4i$

3. $\dfrac{-2 - 5i}{3}$

4. $\dfrac{4 + 7i}{2}$

5. 3

6. $-\frac{1}{2}$

7. $-\frac{2}{3}i$

8. $i\sqrt{3}$

9. $\sqrt{3} + \sqrt{-4}$

10. $2 - \sqrt{-5}$

11–22 ■ Perform the addition or subtraction and write the result in the form $a + bi$.

11. $(2 - 5i) + (3 + 4i)$

12. $(2 + 5i) + (4 - 6i)$

13. $(-6 + 6i) + (9 - i)$

14. $(3 - 2i) + \left(-5 - \frac{1}{3}i\right)$

15. $3i + (6 - 4i)$

16. $\left(\frac{1}{2} - \frac{1}{3}i\right) + \left(\frac{1}{2} + \frac{1}{3}i\right)$

17. $\left(7 - \frac{1}{2}i\right) - \left(5 + \frac{3}{2}i\right)$

18. $(-4 + i) - (2 - 5i)$

19. $(-12 + 8i) - (7 + 4i)$

20. $6i - (4 - i)$

21. $\frac{1}{3}i - \left(\frac{1}{4} - \frac{1}{6}i\right)$

22. $(0.1 - 1.1i) - (1.2 - 3.6i)$

23–56 ■ Evaluate the expression and write the result in the form $a + bi$.

23. $4(-1 + 2i)$

24. $2i\left(\frac{1}{2} - i\right)$

25. $(7 - i)(4 + 2i)$

26. $(5 - 3i)(1 + i)$

27. $(3 - 4i)(5 - 12i)$

28. $\left(\frac{2}{3} + 12i\right)\left(\frac{1}{6} + 24i\right)$

29. $(6 + 5i)(2 - 3i)$

30. $(-2 + i)(3 - 7i)$

31. $\dfrac{1}{i}$

32. $\dfrac{1}{1 + i}$

33. $\dfrac{2 - 3i}{1 - 2i}$

34. $\dfrac{5 - i}{3 + 4i}$

35. $\dfrac{26 + 39i}{2 - 3i}$

36. $\dfrac{25}{4 - 3i}$

37. $\dfrac{10i}{1 - 2i}$

38. $(2 - 3i)^{-1}$

39. $\dfrac{4 + 6i}{3i}$

40. $\dfrac{-3 + 5i}{15i}$

41. $\dfrac{1}{1 + i} - \dfrac{1}{1 - i}$

42. $\dfrac{(1 + 2i)(3 - i)}{2 + i}$

43. i^3

44. $(2i)^4$

45. i^{100}

46. i^{1002}

47. $\sqrt{-25}$

48. $\sqrt{\dfrac{-9}{4}}$

49. $\sqrt{-3}\sqrt{-12}$

50. $\sqrt{\frac{1}{3}}\sqrt{-27}$

51. $(3 - \sqrt{-5})(1 + \sqrt{-1})$

52. $\dfrac{1 - \sqrt{-1}}{1 + \sqrt{-1}}$

53. $\dfrac{2 + \sqrt{-8}}{1 + \sqrt{-2}}$

54. $(\sqrt{3} - \sqrt{-4})(\sqrt{6} - \sqrt{-8})$

55. $\dfrac{\sqrt{-36}}{\sqrt{-2}\sqrt{-9}}$

56. $\dfrac{\sqrt{-7}\sqrt{-49}}{\sqrt{28}}$

57–70 ■ Find all solutions of the equation and express them in the form $a + bi$.

57. $x^2 + 9 = 0$

58. $9x^2 + 4 = 0$

59. $x^2 - 4x + 5 = 0$

60. $x^2 + 2x + 2 = 0$

61. $x^2 + x + 1 = 0$

62. $x^2 - 3x + 3 = 0$

63. $2x^2 - 2x + 1 = 0$

64. $2x^2 + 3 = 2x$

65. $t + 3 + \dfrac{3}{t} = 0$

66. $z + 4 + \dfrac{12}{z} = 0$

67. $6x^2 + 12x + 7 = 0$

68. $4x^2 - 16x + 19 = 0$

69. $\frac{1}{2}x^2 - x + 5 = 0$

70. $x^2 + \frac{1}{2}x + 1 = 0$

71–78 ■ Recall that the symbol $\bar{z}$ represents the complex conjugate of z. If $z = a + bi$ and $w = c + di$, prove each statement.

71. $\overline{z + w} = \bar{z} + \bar{w}$

72. $\overline{zw} = \bar{z} \cdot \bar{w}$

73. $(\bar{z})^2 = \overline{z^2}$

74. $\bar{\bar{z}} = z$

75. $z + \bar{z}$ is a real number

76. $z - \bar{z}$ is a pure imaginary number

77. $z \cdot \bar{z}$ is a real number

78. $z = \bar{z}$ if and only if z is real

Discovery • Discussion

79. Complex Conjugate Roots   Suppose that the equation $ax^2 + bx + c = 0$ has real coefficients and complex roots. Why must the roots be complex conjugates of each other? (Think about how you would find the roots using the quadratic formula.)

80. Powers of i   Calculate the first 12 powers of i, that is, $i, i^2, i^3, \ldots, i^{12}$. Do you notice a pattern? Explain how you would calculate any whole number power of i, using the pattern you have discovered. Use this procedure to calculate i^{4446}.

81. Complex Radicals   The number 8 has one real cube root, $\sqrt[3]{8} = 2$. Calculate $(-1 + i\sqrt{3})^3$ and $(-1 - i\sqrt{3})^3$ to verify that 8 has at least two other complex cube roots. Can you find four fourth roots of 16?

3.5 # Complex Zeros and the Fundamental Theorem of Algebra

We have already seen that an nth-degree polynomial can have at most n real zeros. In the complex number system an nth-degree polynomial has exactly n zeros, and so can be factored into exactly n linear factors. This fact is a consequence of the Fundamental Theorem of Algebra, which was proved by the German mathematician C. F. Gauss in 1799 (see page 294).

The Fundamental Theorem of Algebra and Complete Factorization

The following theorem is the basis for much of our work in factoring polynomials and solving polynomial equations.

Fundamental Theorem of Algebra

Every polynomial

$$P(x) = a_n x^n + a_{n-1} x^{n-1} + \cdots + a_1 x + a_0 \qquad (n \geq 1, a_n \neq 0)$$

with complex coefficients has at least one complex zero.

Because any real number is also a complex number, the theorem applies to polynomials with real coefficients as well.

The Fundamental Theorem of Algebra and the Factor Theorem together show that a polynomial can be factored completely into linear factors, as we now prove.

Complete Factorization Theorem

If $P(x)$ is a polynomial of degree $n \geq 1$, then there exist complex numbers a, $c_1, c_2, \ldots, c_n$ (with $a \neq 0$) such that

$$P(x) = a(x - c_1)(x - c_2) \cdots (x - c_n)$$

■ **Proof**   By the Fundamental Theorem of Algebra, P has at least one zero. Let's call it c_1. By the Factor Theorem, $P(x)$ can be factored as

$$P(x) = (x - c_1) \cdot Q_1(x)$$

where $Q_1(x)$ is of degree $n - 1$. Applying the Fundamental Theorem to the quotient $Q_1(x)$ gives us the factorization

$$P(x) = (x - c_1) \cdot (x - c_2) \cdot Q_2(x)$$

where $Q_2(x)$ is of degree $n - 2$ and c_2 is a zero of $Q_1(x)$. Continuing this process for n steps, we get a final quotient $Q_n(x)$ of degree 0, a nonzero constant that we will call a. This means that P has been factored as

$$P(x) = a(x - c_1)(x - c_2) \cdots (x - c_n)$$

■

To actually find the complex zeros of an nth-degree polynomial, we usually first factor as much as possible, then use the quadratic formula on parts that we can't factor further.

Example 1 Factoring a Polynomial Completely

Let $P(x) = x^3 - 3x^2 + x - 3$.

(a) Find all the zeros of P.

(b) Find the complete factorization of P.

Solution

(a) We first factor P as follows.

$$
\begin{aligned}
P(x) &= x^3 - 3x^2 + x - 3 & &\text{Given}\\
&= x^2(x - 3) + (x - 3) & &\text{Group terms}\\
&= (x - 3)(x^2 + 1) & &\text{Factor } x - 3
\end{aligned}
$$

We find the zeros of P by setting each factor equal to 0:

$$P(x) = (x - 3)(x^2 + 1)$$

This factor is 0 when x = 3.   This factor is 0 when x = i or $-i$.

Setting $x - 3 = 0$, we see that $x = 3$ is a zero. Setting $x^2 + 1 = 0$, we get $x^2 = -1$, so $x = \pm i$. So the zeros of P are 3, i, and $-i$.

(b) Since the zeros are 3, i, and $-i$, by the Complete Factorization Theorem P factors as

$$
\begin{aligned}
P(x) &= (x - 3)(x - i)[x - (-i)]\\
&= (x - 3)(x - i)(x + i)
\end{aligned}
$$

Example 2 Factoring a Polynomial Completely

Let $P(x) = x^3 - 2x + 4$.

(a) Find all the zeros of P.

(b) Find the complete factorization of P.

Solution

$$
\begin{array}{r|rrrr}
-2 & 1 & 0 & -2 & 4\\
   &   & -2 & 4 & -4\\
\hline
   & 1 & -2 & 2 & 0
\end{array}
$$

(a) The possible rational zeros are the factors of 4, which are $\pm 1, \pm 2, \pm 4$. Using synthetic division (see the margin) we find that -2 is a zero, and the polynomial factors as

$$P(x) = (x + 2)(x^2 - 2x + 2)$$

This factor is 0 when x = -2.   Use the quadratic formula to find when this factor is 0.

To find the zeros, we set each factor equal to 0. Of course, $x + 2 = 0$ means $x = -2$. We use the quadratic formula to find when the other factor is 0.

$$x^2 - 2x + 2 = 0 \qquad \text{Set factor equal to 0}$$

$$x = \frac{2 \pm \sqrt{4 - 8}}{2} \qquad \text{Quadratic formula}$$

$$x = \frac{2 \pm 2i}{2} \qquad \text{Take square root}$$

$$x = 1 \pm i \qquad \text{Simplify}$$

So the zeros of P are -2, $1 + i$, and $1 - i$.

(b) Since the zeros are -2, $1 + i$, and $1 - i$, by the Complete Factorization Theorem P factors as

$$P(x) = [x - (-2)][x - (1 + i)][x - (1 - i)]$$
$$= (x + 2)(x - 1 - i)(x - 1 + i) \qquad \blacksquare$$

Zeros and Their Multiplicities

In the Complete Factorization Theorem the numbers $c_1, c_2, \ldots, c_n$ are the zeros of P. These zeros need not all be different. If the factor $x - c$ appears k times in the complete factorization of $P(x)$, then we say that c is a zero of **multiplicity k** (see page 259). For example, the polynomial

$$P(x) = (x - 1)^3(x + 2)^2(x + 3)^5$$

has the following zeros:

$$1 \, (\text{multiplicity } 3), \qquad -2 \, (\text{multiplicity } 2), \qquad -3 \, (\text{multiplicity } 5)$$

The polynomial P has the same number of zeros as its degree—it has degree 10 and has 10 zeros, provided we count multiplicities. This is true for all polynomials, as we prove in the following theorem.

Zeros Theorem

Every polynomial of degree $n \geq 1$ has exactly n zeros, provided that a zero of multiplicity k is counted k times.

■ **Proof**   Let P be a polynomial of degree n. By the Complete Factorization Theorem

$$P(x) = a(x - c_1)(x - c_2) \cdots (x - c_n)$$

Now suppose that c is a zero of P other than $c_1, c_2, \ldots, c_n$. Then

$$P(c) = a(c - c_1)(c - c_2) \cdots (c - c_n) = 0$$

Thus, by the Zero-Product Property one of the factors $c - c_i$ must be 0, so $c = c_i$ for some i. It follows that P has exactly the n zeros $c_1, c_2, \ldots, c_n$.   ■

Example 3   Factoring a Polynomial with Complex Zeros

Find the complete factorization and all five zeros of the polynomial

$$P(x) = 3x^5 + 24x^3 + 48x$$

Solution   Since $3x$ is a common factor, we have

$$P(x) = 3x(x^4 + 8x^2 + 16)$$

$$= 3x(x^2 + 4)^2$$

This factor is 0 when $x = 0$.   This factor is 0 when $x = 2i$ or $x = -2i$.

To factor $x^2 + 4$, note that $2i$ and $-2i$ are zeros of this polynomial. Thus $x^2 + 4 = (x - 2i)(x + 2i)$, and so

$$P(x) = 3x[(x - 2i)(x + 2i)]^2$$

$$= 3x(x - 2i)^2(x + 2i)^2$$

0 is a zero of multiplicity 1. $2i$ is a zero of multiplicity 2. $-2i$ is a zero of multiplicity 2.

The zeros of P are 0, $2i$, and $-2i$. Since the factors $x - 2i$ and $x + 2i$ each occur twice in the complete factorization of P, the zeros $2i$ and $-2i$ are of multiplicity 2 (or *double* zeros). Thus, we have found all five zeros. ∎

The following table gives further examples of polynomials with their complete factorizations and zeros.

Corbis

Carl Friedrich Gauss (1777–1855) is considered the greatest mathematician of modern times. His contemporaries called him the "Prince of Mathematics." He was born into a poor family; his father made a living as a mason. As a very small child, Gauss found a calculation error in his father's accounts, the first of many incidents that gave evidence of his mathematical precocity. (See also page 834.) At 19 Gauss demonstrated that the regular 17-sided polygon can be constructed with straight-edge and compass alone. This was remarkable because, since the time of Euclid, it was thought that the only regular polygons constructible in this way were the triangle and pentagon. Because of this discovery Gauss decided to pursue a career in mathematics instead of languages, his other passion. In his doctoral dissertation, written at the age of 22, Gauss proved the Fundamental Theorem of Algebra: A polynomial of degree n with complex coefficients has n roots. His other accomplishments range over every branch of mathematics, as well as physics and astronomy.

Degree	Polynomial	Zero(s)	Number of zeros
1	$P(x) = x - 4$	4	1
2	$P(x) = x^2 - 10x + 25$ $= (x - 5)(x - 5)$	5 (multiplicity 2)	2
3	$P(x) = x^3 + x$ $= x(x - i)(x + i)$	$0, i, -i$	3
4	$P(x) = x^4 + 18x^2 + 81$ $= (x - 3i)^2(x + 3i)^2$	$3i$ (multiplicity 2), $-3i$ (multiplicity 2)	4
5	$P(x) = x^5 - 2x^4 + x^3$ $= x^3(x - 1)^2$	0 (multiplicity 3), 1 (multiplicity 2)	5

Example 4   Finding Polynomials with Specified Zeros

(a) Find a polynomial $P(x)$ of degree 4, with zeros i, $-i$, 2, and -2 and with $P(3) = 25$.

(b) Find a polynomial $Q(x)$ of degree 4, with zeros -2 and 0, where -2 is a zero of multiplicity 3.

Solution

(a) The required polynomial has the form

$$P(x) = a(x - i)(x - (-i))(x - 2)(x - (-2))$$

$$= a(x^2 + 1)(x^2 - 4) \qquad \text{Difference of squares}$$

$$= a(x^4 - 3x^2 - 4) \qquad \text{Multiply}$$

We know that $P(3) = a(3^4 - 3 \cdot 3^2 - 4) = 50a = 25$, so $a = \frac{1}{2}$. Thus

$$P(x) = \tfrac{1}{2}x^4 - \tfrac{3}{2}x^2 - 2$$

(b) We require

$$Q(x) = a[x - (-2)]^3(x - 0)$$

$$= a(x + 2)^3 x$$

$$= a(x^3 + 6x^2 + 12x + 8)x \qquad \text{Special Product Formula 4 (Section 1.3)}$$

$$= a(x^4 + 6x^3 + 12x^2 + 8x)$$

Since we are given no information about Q other than its zeros and their multiplicity, we can choose any number for a. If we use $a = 1$, we get

$$Q(x) = x^4 + 6x^3 + 12x^2 + 8x \qquad \blacksquare$$

Example 5   Finding All the Zeros of a Polynomial

Find all four zeros of $P(x) = 3x^4 - 2x^3 - x^2 - 12x - 4$.

Solution   Using the Rational Zeros Theorem from Section 3.3, we obtain the following list of possible rational zeros: $\pm 1, \pm 2, \pm 4, \pm\frac{1}{3}, \pm\frac{2}{3}, \pm\frac{4}{3}$. Checking these using synthetic division, we find that 2 and $-\frac{1}{3}$ are zeros, and we get the following factorization.

$$P(x) = 3x^4 - 2x^3 - x^2 - 12x - 4$$

$$= (x - 2)(3x^3 + 4x^2 + 7x + 2) \qquad \text{Factor } x - 2$$

$$= (x - 2)(x + \tfrac{1}{3})(3x^2 + 3x + 6) \qquad \text{Factor } x + \tfrac{1}{3}$$

$$= 3(x - 2)(x + \tfrac{1}{3})(x^2 + x + 2) \qquad \text{Factor } 3$$

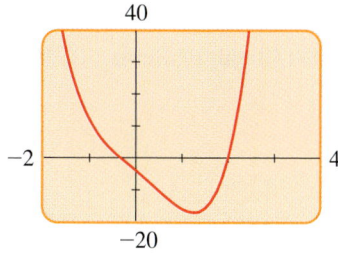

Figure 1

$P(x) = 3x^4 - 2x^3 - x^2 - 12x - 4$

Figure 1 shows the graph of the polynomial P in Example 5. The x-intercepts correspond to the real zeros of P. The imaginary zeros cannot be determined from the graph.

The zeros of the quadratic factor are

$$x = \frac{-1 \pm \sqrt{1 - 8}}{2} = -\frac{1}{2} \pm i\frac{\sqrt{7}}{2} \qquad \text{Quadratic formula}$$

so the zeros of $P(x)$ are

$$2, \quad -\frac{1}{3}, \quad -\frac{1}{2} + i\frac{\sqrt{7}}{2}, \qquad \text{and} \qquad -\frac{1}{2} - i\frac{\sqrt{7}}{2} \qquad \blacksquare$$

Complex Zeros Come in Conjugate Pairs

As you may have noticed from the examples so far, the complex zeros of polynomials with real coefficients come in pairs. Whenever $a + bi$ is a zero, its complex conjugate $a - bi$ is also a zero.

Conjugate Zeros Theorem

If the polynomial P has real coefficients, and if the complex number z is a zero of P, then its complex conjugate $\bar{z}$ is also a zero of P.

■ **Proof**   Let

$$P(x) = a_n x^n + a_{n-1} x^{n-1} + \cdots + a_1 x + a_0$$

where each coefficient is real. Suppose that $P(z) = 0$. We must prove that $P(\bar{z}) = 0$. We use the facts that the complex conjugate of a sum of two complex numbers is the sum of the conjugates and that the conjugate of a product is the product of the conjugates (see Exercises 71 and 72 in Section 3.4).

$$
\begin{aligned}
P(\bar{z}) &= a_n (\bar{z})^n + a_{n-1} (\bar{z})^{n-1} + \cdots + a_1 \bar{z} + a_0 \\
&= \overline{a_n}\, \overline{z^n} + \overline{a_{n-1}}\, \overline{z^{n-1}} + \cdots + \overline{a_1}\, \overline{z} + \overline{a_0} \qquad \text{\textcolor{blue}{Because the coefficients are real}} \\
&= \overline{a_n z^n} + \overline{a_{n-1} z^{n-1}} + \cdots + \overline{a_1 z} + \overline{a_0} \\
&= \overline{a_n z^n + a_{n-1} z^{n-1} + \cdots + a_1 z + a_0} \\
&= \overline{P(z)} = \overline{0} = 0
\end{aligned}
$$

This shows that $\bar{z}$ is also a zero of $P(x)$, which proves the theorem.   ■

Example 6   A Polynomial with a Specified Complex Zero

Find a polynomial $P(x)$ of degree 3 that has integer coefficients and zeros $\frac{1}{2}$ and $3 - i$.

Solution   Since $3 - i$ is a zero, then so is $3 + i$ by the Conjugate Zeros Theorem. This means that $P(x)$ has the form

$$
\begin{aligned}
P(x) &= a\left(x - \tfrac{1}{2}\right)[x - (3 - i)][x - (3 + i)] \\
&= a\left(x - \tfrac{1}{2}\right)[(x - 3) + i][(x - 3) - i] \qquad \text{\textcolor{blue}{Regroup}} \\
&= a\left(x - \tfrac{1}{2}\right)[(x - 3)^2 - i^2] \qquad \text{\textcolor{blue}{Difference of Squares Formula}} \\
&= a\left(x - \tfrac{1}{2}\right)(x^2 - 6x + 10) \qquad \text{\textcolor{blue}{Expand}} \\
&= a\left(x^3 - \tfrac{13}{2}x^2 + 13x - 5\right) \qquad \text{\textcolor{blue}{Expand}}
\end{aligned}
$$

To make all coefficients integers, we set $a = 2$ and get

$$P(x) = 2x^3 - 13x^2 + 26x - 10$$

Any other polynomial that satisfies the given requirements must be an integer multiple of this one.   ■

Example 7 Using Descartes' Rule to Count Real and Imaginary Zeros

Without actually factoring, determine how many positive real zeros, negative real zeros, and imaginary zeros the following polynomial could have:

$$P(x) = x^4 + 6x^3 - 12x^2 - 14x - 24$$

Solution Since there is one change of sign, by Descartes' Rule of Signs, P has one positive real zero. Also, $P(-x) = x^4 - 6x^3 - 12x^2 + 14x - 24$ has three changes of sign, so there are either three or one negative real zero(s). So P has a total of either four or two real zeros. Since P is of degree 4, it has four zeros in all, which gives the following possibilities.

Positive real zeros	Negative real zeros	Imaginary zeros
1	3	0
1	1	2

Linear and Quadratic Factors

We have seen that a polynomial factors completely into linear factors if we use complex numbers. If we don't use complex numbers, then a polynomial with real coefficients can always be factored into linear and quadratic factors. We use this property in Section 9.8 when we study partial fractions. A quadratic polynomial with no real zeros is called **irreducible** over the real numbers. Such a polynomial cannot be factored without using complex numbers.

Linear and Quadratic Factors Theorem

Every polynomial with real coefficients can be factored into a product of linear and irreducible quadratic factors with real coefficients.

■ **Proof** We first observe that if $c = a + bi$ is a complex number, then

$$\begin{aligned}
(x - c)(x - \bar{c}) &= [x - (a + bi)][x - (a - bi)] \\
&= [(x - a) - bi][(x - a) + bi] \\
&= (x - a)^2 - (bi)^2 \\
&= x^2 - 2ax + (a^2 + b^2)
\end{aligned}$$

The last expression is a quadratic with *real* coefficients.

Now, if P is a polynomial with real coefficients, then by the Complete Factorization Theorem

$$P(x) = a(x - c_1)(x - c_2) \cdots (x - c_n)$$

Since the complex roots occur in conjugate pairs, we can multiply the factors corresponding to each such pair to get a quadratic factor with real coefficients. This results in P being factored into linear and irreducible quadratic factors. ■

Example 8 **Factoring a Polynomial into Linear and Quadratic Factors**

Let $P(x) = x^4 + 2x^2 - 8$.

(a) Factor P into linear and irreducible quadratic factors with real coefficients.

(b) Factor P completely into linear factors with complex coefficients.

Solution

(a)
$$P(x) = x^4 + 2x^2 - 8$$
$$= (x^2 - 2)(x^2 + 4)$$
$$= (x - \sqrt{2})(x + \sqrt{2})(x^2 + 4)$$

The factor $x^2 + 4$ is irreducible since it has only the imaginary zeros $\pm 2i$.

(b) To get the complete factorization, we factor the remaining quadratic factor.

$$P(x) = (x - \sqrt{2})(x + \sqrt{2})(x^2 + 4)$$
$$= (x - \sqrt{2})(x + \sqrt{2})(x - 2i)(x + 2i)$$ ∎

3.5 Exercises

1–12 ■ A polynomial P is given.

(a) Find all zeros of P, real and complex.

(b) Factor P completely.

1. $P(x) = x^4 + 4x^2$
2. $P(x) = x^5 + 9x^3$

3. $P(x) = x^3 - 2x^2 + 2x$
4. $P(x) = x^3 + x^2 + x$

5. $P(x) = x^4 + 2x^2 + 1$
6. $P(x) = x^4 - x^2 - 2$

7. $P(x) = x^4 - 16$
8. $P(x) = x^4 + 6x^2 + 9$

9. $P(x) = x^3 + 8$
10. $P(x) = x^3 - 8$

11. $P(x) = x^6 - 1$
12. $P(x) = x^6 - 7x^3 - 8$

13–30 ■ Factor the polynomial completely and find all its zeros. State the multiplicity of each zero.

13. $P(x) = x^2 + 25$
14. $P(x) = 4x^2 + 9$

15. $Q(x) = x^2 + 2x + 2$
16. $Q(x) = x^2 - 8x + 17$

17. $P(x) = x^3 + 4x$
18. $P(x) = x^3 - x^2 + x$

19. $Q(x) = x^4 - 1$
20. $Q(x) = x^4 - 625$

21. $P(x) = 16x^4 - 81$
22. $P(x) = x^3 - 64$

23. $P(x) = x^3 + x^2 + 9x + 9$
24. $P(x) = x^6 - 729$

25. $Q(x) = x^4 + 2x^2 + 1$
26. $Q(x) = x^4 + 10x^2 + 25$

27. $P(x) = x^4 + 3x^2 - 4$
28. $P(x) = x^5 + 7x^3$

29. $P(x) = x^5 + 6x^3 + 9x$
30. $P(x) = x^6 + 16x^3 + 64$

31–40 ■ Find a polynomial with integer coefficients that satisfies the given conditions.

31. P has degree 2, and zeros $1 + i$ and $1 - i$.

32. P has degree 2, and zeros $1 + i\sqrt{2}$ and $1 - i\sqrt{2}$.

33. Q has degree 3, and zeros 3, $2i$, and $-2i$.

34. Q has degree 3, and zeros 0 and i.

35. P has degree 3, and zeros 2 and i.

36. Q has degree 3, and zeros -3 and $1 + i$.

37. R has degree 4, and zeros $1 - 2i$ and 1, with 1 a zero of multiplicity 2.

38. S has degree 4, and zeros $2i$ and $3i$.

39. T has degree 4, zeros i and $1 + i$, and constant term 12.

40. U has degree 5, zeros $\frac{1}{2}$, -1, and $-i$, and leading coefficient 4; the zero -1 has multiplicity 2.

41–58 ■ Find all zeros of the polynomial.

41. $P(x) = x^3 + 2x^2 + 4x + 8$

42. $P(x) = x^3 - 7x^2 + 17x - 15$

43. $P(x) = x^3 - 2x^2 + 2x - 1$

44. $P(x) = x^3 + 7x^2 + 18x + 18$

45. $P(x) = x^3 - 3x^2 + 3x - 2$

46. $P(x) = x^3 - x - 6$

47. $P(x) = 2x^3 + 7x^2 + 12x + 9$

48. $P(x) = 2x^3 - 8x^2 + 9x - 9$

49. $P(x) = x^4 + x^3 + 7x^2 + 9x - 18$

50. $P(x) = x^4 - 2x^3 - 2x^2 - 2x - 3$

51. $P(x) = x^5 - x^4 + 7x^3 - 7x^2 + 12x - 12$

52. $P(x) = x^5 + x^3 + 8x^2 + 8$ [*Hint:* Factor by grouping.]

53. $P(x) = x^4 - 6x^3 + 13x^2 - 24x + 36$

54. $P(x) = x^4 - x^2 + 2x + 2$

55. $P(x) = 4x^4 + 4x^3 + 5x^2 + 4x + 1$

56. $P(x) = 4x^4 + 2x^3 - 2x^2 - 3x - 1$

57. $P(x) = x^5 - 3x^4 + 12x^3 - 28x^2 + 27x - 9$

58. $P(x) = x^5 - 2x^4 + 2x^3 - 4x^2 + x - 2$

59–64 ■ A polynomial P is given.

(a) Factor P into linear and irreducible quadratic factors with real coefficients.

(b) Factor P completely into linear factors with complex coefficients.

59. $P(x) = x^3 - 5x^2 + 4x - 20$

60. $P(x) = x^3 - 2x - 4$

61. $P(x) = x^4 + 8x^2 - 9$

62. $P(x) = x^4 + 8x^2 + 16$

63. $P(x) = x^6 - 64$

64. $P(x) = x^5 - 16x$

 65. By the Zeros Theorem, every nth-degree polynomial equation has exactly n solutions (including possibly some that are repeated). Some of these may be real and some may be imaginary. Use a graphing device to determine how many real and imaginary solutions each equation has.

(a) $x^4 - 2x^3 - 11x^2 + 12x = 0$

(b) $x^4 - 2x^3 - 11x^2 + 12x - 5 = 0$

(c) $x^4 - 2x^3 - 11x^2 + 12x + 40 = 0$

66–68 ■ So far we have worked only with polynomials that have real coefficients. These exercises involve polynomials with real and imaginary coefficients.

66. Find all solutions of the equation.

(a) $2x + 4i = 1$

(b) $x^2 - ix = 0$

(c) $x^2 + 2ix - 1 = 0$

(d) $ix^2 - 2x + i = 0$

67. (a) Show that $2i$ and $1 - i$ are both solutions of the equation
$$x^2 - (1 + i)x + (2 + 2i) = 0$$
but that their complex conjugates $-2i$ and $1 + i$ are not.

(b) Explain why the result of part (a) does not violate the Conjugate Zeros Theorem.

68. (a) Find the polynomial with *real* coefficients of the smallest possible degree for which i and $1 + i$ are zeros and in which the coefficient of the highest power is 1.

(b) Find the polynomial with *complex* coefficients of the smallest possible degree for which i and $1 + i$ are zeros and in which the coefficient of the highest power is 1.

Discovery • Discussion

69. Polynomials of Odd Degree   The Conjugate Zeros Theorem says that the complex zeros of a polynomial with real coefficients occur in complex conjugate pairs. Explain how this fact proves that a polynomial with real coefficients and odd degree has at least one real zero.

70. Roots of Unity   There are two square roots of 1, namely 1 and -1. These are the solutions of $x^2 = 1$. The fourth roots of 1 are the solutions of the equation $x^4 = 1$ or $x^4 - 1 = 0$. How many fourth roots of 1 are there? Find them. The cube roots of 1 are the solutions of the equation $x^3 = 1$ or $x^3 - 1 = 0$. How many cube roots of 1 are there? Find them. How would you find the sixth roots of 1? How many are there? Make a conjecture about the number of nth roots of 1.

3.6   # Rational Functions

A rational function is a function of the form

$$r(x) = \frac{P(x)}{Q(x)}$$

where P and Q are polynomials. We assume that $P(x)$ and $Q(x)$ have no factor in common. Even though rational functions are constructed from polynomials, their graphs look quite different than the graphs of polynomial functions.

Rational Functions and Asymptotes

Domains of rational expressions are discussed in Section 1.4.

The *domain* of a rational function consists of all real numbers x except those for which the denominator is zero. When graphing a rational function, we must pay special attention to the behavior of the graph near those x-values. We begin by graphing a very simple rational function.

Example 1 A Simple Rational Function

Sketch a graph of the rational function $f(x) = \dfrac{1}{x}$.

Solution The function f is not defined for $x = 0$. The following tables show that when x is close to zero, the value of $|f(x)|$ is large, and the closer x gets to zero, the larger $|f(x)|$ gets.

For positive real numbers,

$$\frac{1}{\text{BIG NUMBER}} = \text{small number}$$

$$\frac{1}{\text{small number}} = \text{BIG NUMBER}$$

x	$f(x)$
-0.1	-10
-0.01	-100
-0.00001	$-100{,}000$

Approaching 0^- Approaching $-\infty$

x	$f(x)$
0.1	10
0.01	100
0.00001	$100{,}000$

Approaching 0^+ Approaching ∞

We describe this behavior in words and in symbols as follows. The first table shows that as x approaches 0 from the left, the values of $y = f(x)$ decrease without bound. In symbols,

$$f(x) \to -\infty \quad \text{as} \quad x \to 0^-$$
"y approaches negative infinity as x approaches 0 from the left"

The second table shows that as x approaches 0 from the right, the values of $f(x)$ increase without bound. In symbols,

$$f(x) \to \infty \quad \text{as} \quad x \to 0^+$$
"y approaches infinity as x approaches 0 from the right"

The next two tables show how $f(x)$ changes as $|x|$ becomes large.

x	$f(x)$
-10	-0.1
-100	-0.01
$-100{,}000$	-0.00001

Approaching $-\infty$ Approaching 0

x	$f(x)$
10	0.1
100	0.01
$100{,}000$	0.00001

Approaching ∞ Approaching 0

These tables show that as $|x|$ becomes large, the value of $f(x)$ gets closer and closer to zero. We describe this situation in symbols by writing

$$f(x) \to 0 \quad \text{as} \quad x \to -\infty \qquad \text{and} \qquad f(x) \to 0 \quad \text{as} \quad x \to \infty$$

Using the information in these tables and plotting a few additional points, we obtain the graph shown in Figure 1.

x	$f(x) = \frac{1}{x}$
-2	$-\frac{1}{2}$
-1	-1
$-\frac{1}{2}$	-2
$\frac{1}{2}$	2
1	1
2	$\frac{1}{2}$

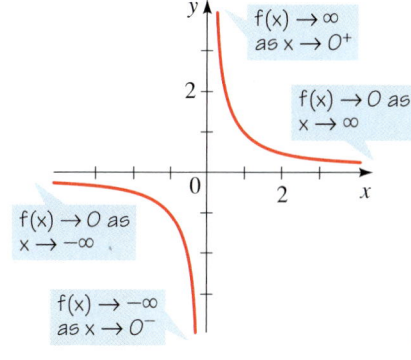

Figure 1

$f(x) = \frac{1}{x}$

In Example 1 we used the following arrow notation.

Symbol	Meaning
$x \to a^-$	x approaches a from the left
$x \to a^+$	x approaches a from the right
$x \to -\infty$	x goes to negative infinity; that is, x decreases without bound
$x \to \infty$	x goes to infinity; that is, x increases without bound

The line $x = 0$ is called a *vertical asymptote* of the graph in Figure 1, and the line $y = 0$ is a *horizontal asymptote*. Informally speaking, an asymptote of a function is a line that the graph of the function gets closer and closer to as one travels along that line.

Definition of Vertical and Horizontal Asymptotes

1. The line $x = a$ is a **vertical asymptote** of the function $y = f(x)$ if y approaches $\pm\infty$ as x approaches a from the right or left.

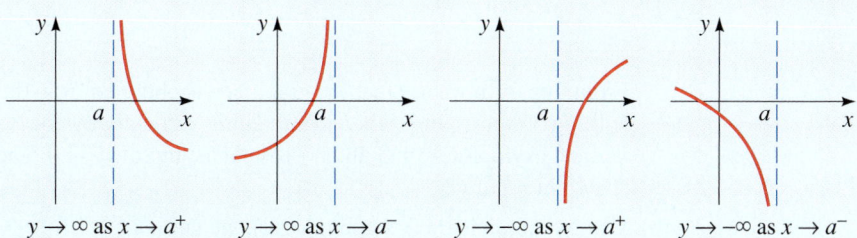

$y \to \infty$ as $x \to a^+$ ⠀⠀⠀ $y \to \infty$ as $x \to a^-$ ⠀⠀⠀ $y \to -\infty$ as $x \to a^+$ ⠀⠀⠀ $y \to -\infty$ as $x \to a^-$

2. The line $y = b$ is a **horizontal asymptote** of the function $y = f(x)$ if y approaches b as x approaches $\pm\infty$.

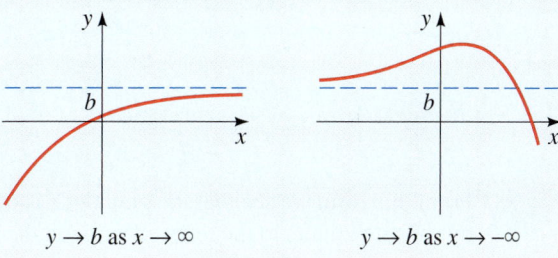

$y \to b$ as $x \to \infty$ ⠀⠀⠀⠀⠀⠀⠀ $y \to b$ as $x \to -\infty$

A rational function has vertical asymptotes where the function is undefined, that is, where the denominator is zero.

Transformations of $y = \dfrac{1}{x}$

A rational function of the form

$$r(x) = \frac{ax + b}{cx + d}$$

can be graphed by shifting, stretching, and/or reflecting the graph of $f(x) = \frac{1}{x}$ shown in Figure 1, using the transformations studied in Section 2.4. (Such functions are called *linear fractional transformations*.)

Example 2   Using Transformations to Graph Rational Functions

Sketch a graph of each rational function.

(a) $r(x) = \dfrac{2}{x - 3}$

(b) $s(x) = \dfrac{3x + 5}{x + 2}$

Solution

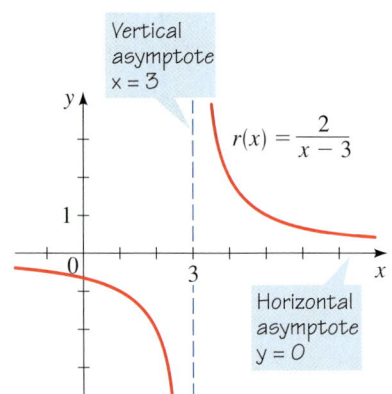

Figure 2

$$
\begin{array}{r}
3 \\
x + 2\overline{)3x + 5} \\
\underline{3x + 6} \\
-1
\end{array}
$$

(a) Let $f(x) = \frac{1}{x}$. Then we can express r in terms of f as follows:

$$r(x) = \frac{2}{x - 3}$$

$$= 2\left(\frac{1}{x - 3}\right) \qquad \text{Factor 2}$$

$$= 2(f(x - 3)) \qquad \text{Since } f(x) = \tfrac{1}{x}$$

From this form we see that the graph of r is obtained from the graph of f by shifting 3 units to the right and stretching vertically by a factor of 2. Thus, r has vertical asymptote $x = 3$ and horizontal asymptote $y = 0$. The graph of r is shown in Figure 2.

(b) Using long division (see the margin), we get $s(x) = 3 - \frac{1}{x + 2}$. Thus, we can express s in terms of f as follows:

$$s(x) = 3 - \frac{1}{x + 2}$$

$$= -\frac{1}{x + 2} + 3 \qquad \text{Rearrange terms}$$

$$= -f(x + 2) + 3 \qquad \text{Since } f(x) = \tfrac{1}{x}$$

From this form we see that the graph of s is obtained from the graph of f by shifting 2 units to the left, reflecting in the x-axis, and shifting upward

3 units. Thus, s has vertical asymptote $x = -2$ and horizontal asymptote $y = 3$. The graph of s is shown in Figure 3.

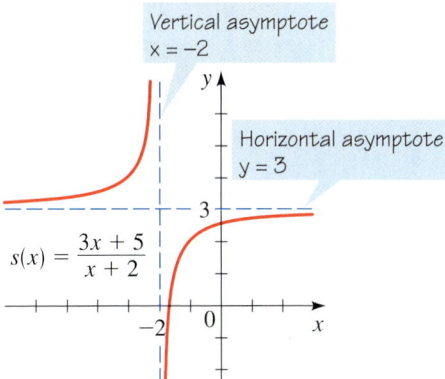

Figure 3

Asymptotes of Rational Functions

The methods of Example 2 work only for simple rational functions. To graph more complicated ones, we need to take a closer look at the behavior of a rational function near its vertical and horizontal asymptotes.

Example 3 Asymptotes of a Rational Function

Graph the rational function $r(x) = \dfrac{2x^2 - 4x + 5}{x^2 - 2x + 1}$.

Solution

VERTICAL ASYMPTOTE: We first factor the denominator

$$r(x) = \frac{2x^2 - 4x + 5}{(x - 1)^2}$$

The line $x = 1$ is a vertical asymptote because the denominator of r is zero when $x = 1$.

To see what the graph of r looks like near the vertical asymptote, we make tables of values for x-values to the left and to the right of 1. From the tables shown below we see that

$$y \to \infty \quad \text{as} \quad x \to 1^- \qquad \text{and} \qquad y \to \infty \quad \text{as} \quad x \to 1^+$$

$x \to 1^-$

x	y
0	5
0.5	14
0.9	302
0.99	30,002

$x \to 1^+$

x	y
2	5
1.5	14
1.1	302
1.01	30,002

Approaching 1^- Approaching ∞ Approaching 1^+ Approaching ∞

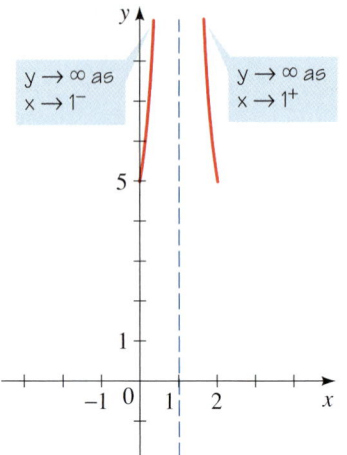

Figure 4

Thus, near the vertical asymptote $x = 1$, the graph of r has the shape shown in Figure 4.

HORIZONTAL ASYMPTOTE: The horizontal asymptote is the value y approaches as $x \to \pm\infty$. To help us find this value, we divide both numerator and denominator by x^2, the highest power of x that appears in the expression:

$$y = \frac{2x^2 - 4x + 5}{x^2 - 2x + 1} \cdot \frac{\dfrac{1}{x^2}}{\dfrac{1}{x^2}} = \frac{2 - \dfrac{4}{x} + \dfrac{5}{x^2}}{1 - \dfrac{2}{x} + \dfrac{1}{x^2}}$$

The fractional expressions $\frac{4}{x}$, $\frac{5}{x^2}$, $\frac{2}{x}$, and $\frac{1}{x^2}$ all approach 0 as $x \to \pm\infty$ (see Exercise 79, Section 1.1). So as $x \to \pm\infty$, we have

These terms approach 0.

$$y = \frac{2 - \dfrac{4}{x} + \dfrac{5}{x^2}}{1 - \dfrac{2}{x} + \dfrac{1}{x^2}} \qquad \longrightarrow \qquad \frac{2 - 0 + 0}{1 - 0 + 0} = 2$$

These terms approach 0.

Thus, the horizontal asymptote is the line $y = 2$.

Since the graph must approach the horizontal asymptote, we can complete it as in Figure 5.

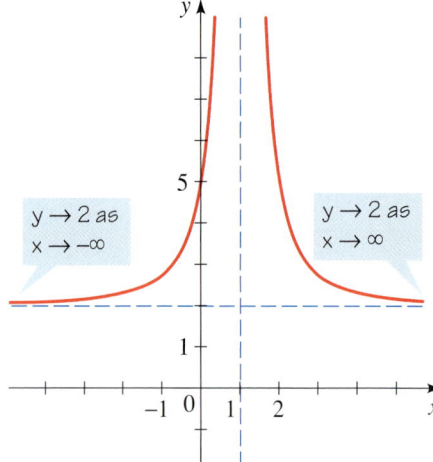

Figure 5

$$r(x) = \frac{2x^2 - 4x + 5}{x^2 - 2x + 1}$$

From Example 3 we see that the horizontal asymptote is determined by the leading coefficients of the numerator and denominator, since after dividing through by x^2 (the highest power of x) all other terms approach zero. In general, if

$r(x) = P(x)/Q(x)$ and the degrees of P and Q are the same (both n, say), then dividing both numerator and denominator by x^n shows that the horizontal asymptote is

$$y = \frac{\text{leading coefficient of } P}{\text{leading coefficient of } Q}$$

The following box summarizes the procedure for finding asymptotes.

Asymptotes of Rational Functions

Let r be the rational function

$$r(x) = \frac{a_n x^n + a_{n-1} x^{n-1} + \cdots + a_1 x + a_0}{b_m x^m + b_{m-1} x^{m-1} + \cdots + b_1 x + b_0}$$

1. The vertical asymptotes of r are the lines $x = a$, where a is a zero of the denominator.

2. (a) If $n < m$, then r has horizontal asymptote $y = 0$.

    (b) If $n = m$, then r has horizontal asymptote $y = \dfrac{a_n}{b_m}$.

    (c) If $n > m$, then r has no horizontal asymptote.

Example 4   Asymptotes of a Rational Function

Find the vertical and horizontal asymptotes of $r(x) = \dfrac{3x^2 - 2x - 1}{2x^2 + 3x - 2}$.

Solution

VERTICAL ASYMPTOTES: We first factor

$$r(x) = \frac{3x^2 - 2x - 1}{(2x - 1)(x + 2)}$$

> This factor is 0 when $x = \frac{1}{2}$.

> This factor is 0 when $x = -2$.

The vertical asymptotes are the lines $x = \frac{1}{2}$ and $x = -2$.

HORIZONTAL ASYMPTOTE: The degrees of the numerator and denominator are the same and

$$\frac{\text{leading coefficient of numerator}}{\text{leading coefficient of denominator}} = \frac{3}{2}$$

Thus, the horizontal asymptote is the line $y = \frac{3}{2}$.

To confirm our results, we graph r using a graphing calculator (see Figure 6).

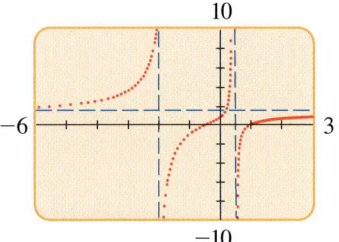

Graph is drawn using dot mode to avoid extraneous lines.

Figure 6

$$r(x) = \frac{3x^2 - 2x - 1}{2x^2 + 3x - 2}$$

Graphing Rational Functions

We have seen that asymptotes are important when graphing rational functions. In general, we use the following guidelines to graph rational functions.

Sketching Graphs of Rational Functions

1. **Factor.**   Factor the numerator and denominator.

A fraction is 0 if and only if its numerator is 0.

2. **Intercepts.**   Find the x-intercepts by determining the zeros of the numerator, and the y-intercept from the value of the function at $x = 0$.

3. **Vertical Asymptotes.**   Find the vertical asymptotes by determining the zeros of the denominator, and then see if $y \to \infty$ or $y \to -\infty$ on each side of each vertical asymptote by using test values.

4. **Horizontal Asymptote.**   Find the horizontal asymptote (if any) by dividing both numerator and denominator by the highest power of x that appears in the denominator, and then letting $x \to \pm\infty$.

5. **Sketch the Graph.**   Graph the information provided by the first four steps. Then plot as many additional points as needed to fill in the rest of the graph of the function.

Example 5   **Graphing a Rational Function**

Graph the rational function $r(x) = \dfrac{2x^2 + 7x - 4}{x^2 + x - 2}$.

Solution   We factor the numerator and denominator, find the intercepts and asymptotes, and sketch the graph.

FACTOR: $y = \dfrac{(2x - 1)(x + 4)}{(x - 1)(x + 2)}$

x-INTERCEPTS: The x-intercepts are the zeros of the numerator, $x = \frac{1}{2}$ and $x = -4$.

***y*-INTERCEPT:**   To find the *y*-intercept, we substitute $x = 0$ into the original form of the function:

$$r(0) = \frac{2(0)^2 + 7(0) - 4}{(0)2 + (0) - 2} = \frac{-4}{-2} = 2$$

The *y*-intercept is 2.

VERTICAL ASYMPTOTES: The vertical asymptotes occur where the denominator is 0, that is, where the function is undefined. From the factored form we see that the vertical asymptotes are the lines $x = 1$ and $x = -2$.

BEHAVIOR NEAR VERTICAL ASYMPTOTES: We need to know whether $y \to \infty$ or $y \to -\infty$ on each side of each vertical asymptote. To determine the sign of *y* for *x*-values near the vertical asymptotes, we use test values. For instance, as $x \to 1^-$, we use a test value close to and to the left of 1 ($x = 0.9$, say) to check whether *y* is positive or negative to the left of $x = 1$:

When choosing test values, we must make sure that there is no x-intercept between the test point and the vertical asymptote.

$$y = \frac{(2(0.9) - 1)((0.9) + 4)}{((0.9) - 1)((0.9) + 2)} \qquad \text{whose sign is} \qquad \frac{(+)(+)}{(-)(+)} \quad \text{(negative)}$$

So $y \to -\infty$ as $x \to 1^-$. On the other hand, as $x \to 1^+$, we use a test value close to and to the right of 1 ($x = 1.1$, say), to get

$$y = \frac{(2(1.1) - 1)((1.1) + 4)}{((1.1) - 1)((1.1) + 2)} \qquad \text{whose sign is} \qquad \frac{(+)(+)}{(+)(+)} \quad \text{(positive)}$$

So $y \to \infty$ as $x \to 1^+$. The other entries in the following table are calculated similarly.

As $x \to$	-2^-	-2^+	1^-	1^+
the sign of $y = \dfrac{(2x - 1)(x + 4)}{(x - 1)(x + 2)}$ is	$\dfrac{(-)(+)}{(-)(-)}$	$\dfrac{(-)(+)}{(-)(+)}$	$\dfrac{(+)(+)}{(-)(+)}$	$\dfrac{(+)(+)}{(+)(+)}$
so $y \to$	$-\infty$	∞	$-\infty$	∞

HORIZONTAL ASYMPTOTE: The degrees of the numerator and denominator are the same and

$$\frac{\text{leading coefficient of numerator}}{\text{leading coefficient of denominator}} = \frac{2}{1} = 2$$

Thus, the horizontal asymptote is the line $y = 2$.

ADDITIONAL VALUES: **GRAPH:**

x	y
-6	0.93
-3	-1.75
-1	4.50
1.5	6.29
2	4.50
3	3.50

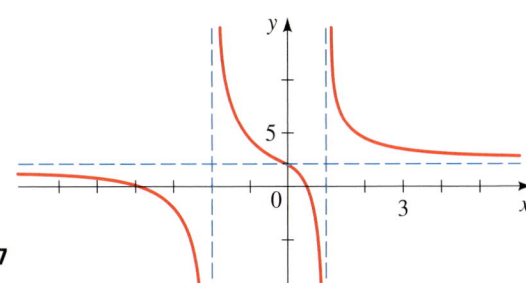

Figure 7

$$r(x) = \frac{2x^2 + 7x - 4}{x^2 + x - 2}$$

Example 6 Graphing a Rational Function

Graph the rational function $r(x) = \dfrac{5x + 21}{x^2 + 10x + 25}$.

Solution

FACTOR: $y = \dfrac{5x + 21}{(x + 5)^2}$

x-INTERCEPT: $-\dfrac{21}{5}$, from $5x + 21 = 0$

y-INTERCEPT: $\dfrac{21}{25}$, because $r(0) = \dfrac{5 \cdot 0 + 21}{0^2 + 10 \cdot 0 + 25}$

$$= \dfrac{21}{25}$$

VERTICAL ASYMPTOTE: $x = -5$, from the zeros of the denominator

BEHAVIOR NEAR VERTICAL ASYMPTOTE:

As $x \to$	-5^-	-5^+
the sign of $y = \dfrac{5x + 21}{(x + 5)^2}$ is	$\dfrac{(-)}{(-)(-)}$	$\dfrac{(-)}{(+)(+)}$
so $y \to$	$-\infty$	$-\infty$

HORIZONTAL ASYMPTOTE: $y = 0$, because degree of numerator is less than degree of denominator

ADDITIONAL VALUES: **GRAPH:**

x	y
-15	-0.5
-10	-1.2
-3	1.5
-1	1.0
3	0.6
5	0.5
10	0.3

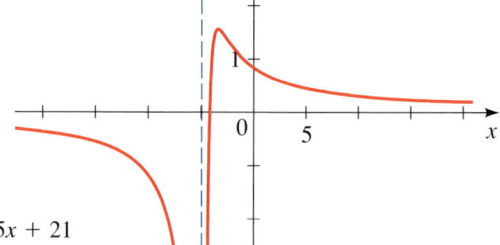

Figure 8

$$r(x) = \dfrac{5x + 21}{x^2 + 10x + 25}$$

    From the graph in Figure 8 we see that, contrary to the common misconception, a graph may cross a horizontal asymptote. The graph in Figure 8 crosses the x-axis (the horizontal asymptote) from below, reaches a maximum value near $x = -3$, and then approaches the x-axis from above as $x \to \infty$.

The RSA code is an example of a "public key encryption" code. In such codes, anyone can code a message using a publicly known procedure based on N, but to decode the message they must know p and q, the factors of N. When the RSA code was developed, it was thought that a carefully selected 80-digit number would provide an unbreakable code. But interestingly, recent advances in the study of factoring have made much larger numbers necessary.

Example 7 Graphing a Rational Function

Graph the rational function $r(x) = \dfrac{x^2 - 3x - 4}{2x^2 + 4x}$.

Solution

FACTOR: $y = \dfrac{(x+1)(x-4)}{2x(x+2)}$

x-INTERCEPTS: -1 and 4, from $x + 1 = 0$ and $x - 4 = 0$

y-INTERCEPT: None, because $r(0)$ is undefined

VERTICAL ASYMPTOTES: $x = 0$ and $x = -2$, from the zeros of the denominator

BEHAVIOR NEAR VERTICAL ASYMPTOTES:

As $x \to$	-2^-	-2^+	0^-	0^+
the sign of $y = \dfrac{(x+1)(x-4)}{2x(x+2)}$ is	$\dfrac{(-)(-)}{(-)(-)}$	$\dfrac{(-)(-)}{(-)(+)}$	$\dfrac{(+)(-)}{(-)(+)}$	$\dfrac{(+)(-)}{(+)(+)}$
so $y \to$	∞	$-\infty$	∞	$-\infty$

HORIZONTAL ASYMPTOTE: $y = \frac{1}{2}$, because degree of numerator and denominator are the same and

$$\frac{\text{leading coefficient of numerator}}{\text{leading coefficient of denominator}} = \frac{1}{2}$$

ADDITIONAL VALUES: **GRAPH:**

x	y
-3	2.33
-2.5	3.90
-0.5	1.50
1	-1.00
3	-0.13
5	0.09

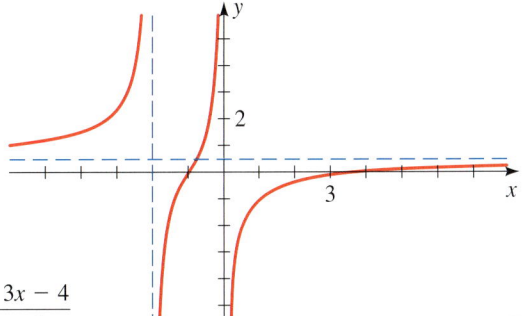

Figure 9

$r(x) = \dfrac{x^2 - 3x - 4}{2x^2 + 4x}$

Slant Asymptotes and End Behavior

If $r(x) = P(x)/Q(x)$ is a rational function in which the degree of the numerator is one more than the degree of the denominator, we can use the Division Algorithm to express the function in the form

$$r(x) = ax + b + \frac{R(x)}{Q(x)}$$

where the degree of R is less than the degree of Q and $a \neq 0$. This means that as

$x \to \pm\infty$, $R(x)/Q(x) \to 0$, so for large values of $|x|$, the graph of $y = r(x)$ approaches the graph of the line $y = ax + b$. In this situation we say that $y = ax + b$ is a **slant asymptote**, or an **oblique asymptote**.

Example 8   A Rational Function with a Slant Asymptote

Graph the rational function $r(x) = \dfrac{x^2 - 4x - 5}{x - 3}$.

Solution

FACTOR: $y = \dfrac{(x + 1)(x - 5)}{x - 3}$

x-INTERCEPTS: -1 and 5, from $x + 1 = 0$ and $x - 5 = 0$

y-INTERCEPTS: $\dfrac{5}{3}$, because $r(0) = \dfrac{0^2 - 4 \cdot 0 - 5}{0 - 3} = \dfrac{5}{3}$

HORIZONTAL ASYMPTOTE: None, because degree of numerator is greater than degree of denominator

VERTICAL ASYMPTOTE: $x = 3$, from the zero of the denominator

BEHAVIOR NEAR VERTICAL ASYMPTOTE: $y \to \infty$ as $x \to 3^-$ and $y \to -\infty$ as $x \to 3^+$

$$
\begin{array}{r}
x - 1 \\
x - 3 \overline{\smash{)}\, x^2 - 4x - 5} \\
\underline{x^2 - 3x} \\
-x - 5 \\
\underline{-x + 3} \\
-8
\end{array}
$$

SLANT ASYMPTOTE: Since the degree of the numerator is one more than the degree of the denominator, the function has a slant asymptote. Dividing (see the margin), we obtain

$$r(x) = x - 1 - \frac{8}{x - 3}$$

Thus, $y = x - 1$ is the slant asymptote.

ADDITIONAL VALUES: **GRAPH:**

x	y
-2	-1.4
1	4
2	9
4	-5
6	2.33

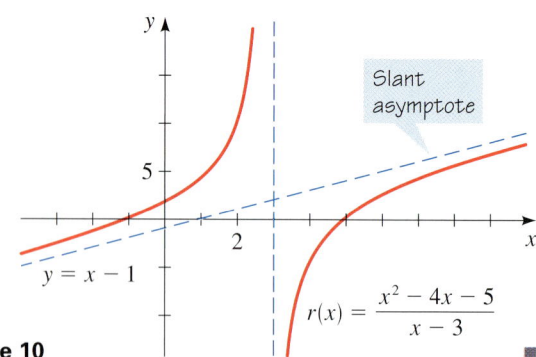

Figure 10

So far we have considered only horizontal and slant asymptotes as end behaviors for rational functions. In the next example we graph a function whose end behavior is like that of a parabola.

Example 9   End Behavior of a Rational Function

Graph the rational function

$$r(x) = \frac{x^3 - 2x^2 + 3}{x - 2}$$

and describe its end behavior.

Solution

FACTOR: $y = \dfrac{(x + 1)(x^2 - 3x + 3)}{x - 2}$

x-INTERCEPTS: -1, from $x + 1 = 0$ (The other factor in the numerator has no real zeros.)

y-INTERCEPTS: $-\dfrac{3}{2}$, because $r(0) = \dfrac{0^3 - 2 \cdot 0^2 + 3}{0 - 2} = -\dfrac{3}{2}$

VERTICAL ASYMPTOTE: $x = 2$, from the zero of the denominator

BEHAVIOR NEAR VERTICAL ASYMPTOTE: $y \to -\infty$ as $x \to 2^-$ and $y \to \infty$ as $x \to 2^+$

HORIZONTAL ASYMPTOTE: None, because degree of numerator is greater than degree of denominator

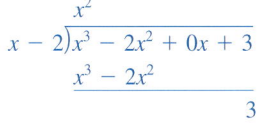

END BEHAVIOR: Dividing (see the margin), we get

$$r(x) = x^2 + \frac{3}{x - 2}$$

This shows that the end behavior of r is like that of the parabola $y = x^2$ because $3/(x - 2)$ is small when $|x|$ is large. That is, $3/(x - 2) \to 0$ as $x \to \pm\infty$. This means that the graph of r will be close to the graph of $y = x^2$ for large $|x|$.

GRAPH: In Figure 11(a) we graph r in a small viewing rectangle; we can see the intercepts, the vertical asymptotes, and the local minimum. In Figure 11(b) we graph r in a larger viewing rectangle; here the graph looks almost like the graph of a parabola. In Figure 11(c) we graph both $y = r(x)$ and $y = x^2$; these graphs are very close to each other except near the vertical asymptote.

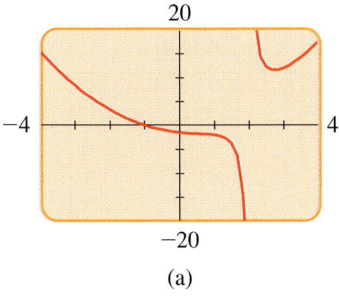

(a)

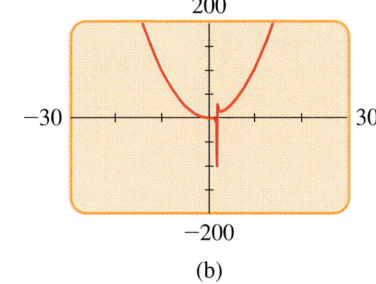

(b)

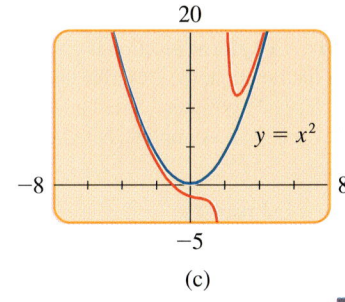

(c)

Figure 11

$r(x) = \dfrac{x^3 - 2x^2 + 3}{x - 2}$

Applications

Rational functions occur frequently in scientific applications of algebra. In the next example we analyze the graph of a function from the theory of electricity.

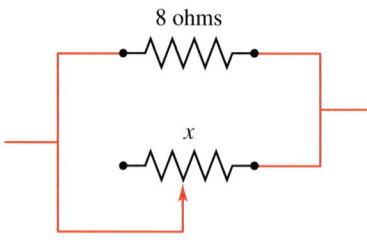

Figure 12

Example 10   Electrical Resistance

When two resistors with resistances R_1 and R_2 are connected in parallel, their combined resistance R is given by the formula

$$R = \frac{R_1 R_2}{R_1 + R_2}$$

Suppose that a fixed 8-ohm resistor is connected in parallel with a variable resistor, as shown in Figure 12. If the resistance of the variable resistor is denoted by x, then the combined resistance R is a function of x. Graph R and give a physical interpretation of the graph.

Solution   Substituting $R_1 = 8$ and $R_2 = x$ into the formula gives the function

$$R(x) = \frac{8x}{8 + x}$$

Since resistance cannot be negative, this function has physical meaning only when $x > 0$. The function is graphed in Figure 13(a) using the viewing rectangle $[0, 20]$ by $[0, 10]$. The function has no vertical asymptote when x is restricted to positive values. The combined resistance R increases as the variable resistance x increases. If we widen the viewing rectangle to $[0, 100]$ by $[0, 10]$, we obtain the graph in Figure 13(b). For large x, the combined resistance R levels off, getting closer and closer to the horizontal asymptote $R = 8$. No matter how large the variable resistance x, the combined resistance is never greater than 8 ohms.

Figure 13

$$R(x) = \frac{8x}{8 + x}$$

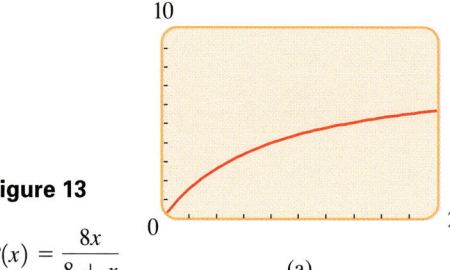

(a)

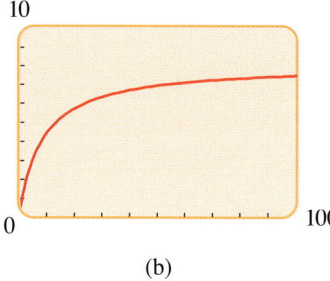

(b)

3.6   Exercises

1–4 ■ A rational function is given. **(a)** Complete each table for the function. **(b)** Describe the behavior of the function near its vertical asymptote, based on Tables 1 and 2. **(c)** Determine the horizontal asymptote, based on Tables 3 and 4.

Table 1

x	$r(x)$
1.5	
1.9	
1.99	
1.999	

Table 2

x	$r(x)$
2.5	
2.1	
2.01	
2.001	

Table 3

x	$r(x)$
10	
50	
100	
1000	

Table 4

x	$r(x)$
-10	
-50	
-100	
-1000	

1. $r(x) = \dfrac{x}{x - 2}$

2. $r(x) = \dfrac{4x + 1}{x - 2}$

3. $r(x) = \dfrac{3x - 10}{(x - 2)^2}$

4. $r(x) = \dfrac{3x^2 + 1}{(x - 2)^2}$

5–10 ■ Find the *x*- and *y*-intercepts of the rational function.

5. $r(x) = \dfrac{x - 1}{x + 4}$ **6.** $s(x) = \dfrac{3x}{x - 5}$

7. $t(x) = \dfrac{x^2 - x - 2}{x - 6}$ **8.** $r(x) = \dfrac{2}{x^2 + 3x - 4}$

9. $r(x) = \dfrac{x^2 - 9}{x^2}$ **10.** $r(x) = \dfrac{x^3 + 8}{x^2 + 4}$

11–14 ■ From the graph, determine the *x*- and *y*-intercepts and the vertical and horizontal asymptotes.

11. 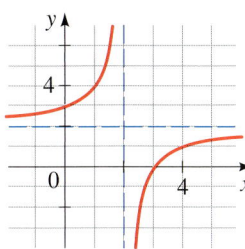   **12.**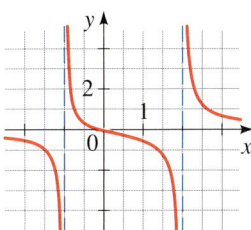

13. 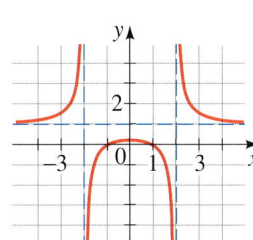   **14.**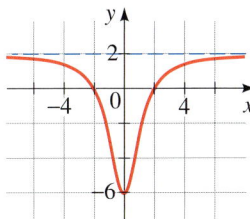

15–24 ■ Find all horizontal and vertical asymptotes (if any).

15. $r(x) = \dfrac{3}{x + 2}$ **16.** $s(x) = \dfrac{2x + 3}{x - 1}$

17. $t(x) = \dfrac{x^2}{x^2 - x - 6}$ **18.** $r(x) = \dfrac{2x - 4}{x^2 + 2x + 1}$

19. $s(x) = \dfrac{6}{x^2 + 2}$ **20.** $t(x) = \dfrac{(x - 1)(x - 2)}{(x - 3)(x - 4)}$

21. $r(x) = \dfrac{6x - 2}{x^2 + 5x - 6}$ **22.** $s(x) = \dfrac{3x^2}{x^2 + 2x + 5}$

23. $t(x) = \dfrac{x^2 + 2}{x - 1}$ **24.** $r(x) = \dfrac{x^3 + 3x^2}{x^2 - 4}$

25–32 ■ Use transformations of the graph of $y = \frac{1}{x}$ to graph the rational function, as in Example 2.

25. $r(x) = \dfrac{1}{x - 1}$ **26.** $r(x) = \dfrac{1}{x + 4}$

27. $s(x) = \dfrac{3}{x + 1}$ **28.** $s(x) = \dfrac{-2}{x - 2}$

29. $t(x) = \dfrac{2x - 3}{x - 2}$ **30.** $t(x) = \dfrac{3x - 3}{x + 2}$

31. $r(x) = \dfrac{x + 2}{x + 3}$ **32.** $r(x) = \dfrac{2x - 9}{x - 4}$

33–56 ■ Find the intercepts and asymptotes, and then sketch a graph of the rational function. Use a graphing device to confirm your answer.

33. $r(x) = \dfrac{4x - 4}{x + 2}$ **34.** $r(x) = \dfrac{2x + 6}{-6x + 3}$

35. $s(x) = \dfrac{4 - 3x}{x + 7}$ **36.** $s(x) = \dfrac{1 - 2x}{2x + 3}$

37. $r(x) = \dfrac{18}{(x - 3)^2}$ **38.** $r(x) = \dfrac{x - 2}{(x + 1)^2}$

39. $s(x) = \dfrac{4x - 8}{(x - 4)(x + 1)}$ **40.** $s(x) = \dfrac{x + 2}{(x + 3)(x - 1)}$

41. $s(x) = \dfrac{6}{x^2 - 5x - 6}$ **42.** $s(x) = \dfrac{2x - 4}{x^2 + x - 2}$

43. $t(x) = \dfrac{3x + 6}{x^2 + 2x - 8}$ **44.** $t(x) = \dfrac{x - 2}{x^2 - 4x}$

45. $r(x) = \dfrac{(x - 1)(x + 2)}{(x + 1)(x - 3)}$ **46.** $r(x) = \dfrac{2x(x + 2)}{(x - 1)(x - 4)}$

47. $r(x) = \dfrac{x^2 - 2x + 1}{x^2 + 2x + 1}$ **48.** $r(x) = \dfrac{4x^2}{x^2 - 2x - 3}$

49. $r(x) = \dfrac{2x^2 + 10x - 12}{x^2 + x - 6}$ **50.** $r(x) = \dfrac{2x^2 + 2x - 4}{x^2 + x}$

51. $r(x) = \dfrac{x^2 - x - 6}{x^2 + 3x}$ **52.** $r(x) = \dfrac{x^2 + 3x}{x^2 - x - 6}$

53. $r(x) = \dfrac{3x^2 + 6}{x^2 - 2x - 3}$ **54.** $r(x) = \dfrac{5x^2 + 5}{x^2 + 4x + 4}$

55. $s(x) = \dfrac{x^2 - 2x + 1}{x^3 - 3x^2}$ **56.** $t(x) = \dfrac{x^3 - x^2}{x^3 - 3x - 2}$

57–64 ■ Find the slant asymptote, the vertical asymptotes, and sketch a graph of the function.

57. $r(x) = \dfrac{x^2}{x - 2}$ **58.** $r(x) = \dfrac{x^2 + 2x}{x - 1}$

59. $r(x) = \dfrac{x^2 - 2x - 8}{x}$ **60.** $r(x) = \dfrac{3x - x^2}{2x - 2}$

61. $r(x) = \dfrac{x^2 + 5x + 4}{x - 3}$ **62.** $r(x) = \dfrac{x^3 + 4}{2x^2 + x - 1}$

63. $r(x) = \dfrac{x^3 + x^2}{x^2 - 4}$ **64.** $r(x) = \dfrac{2x^3 + 2x}{x^2 - 1}$

65–68 ■ Graph the rational function f and determine all vertical asymptotes from your graph. Then graph f and g in a sufficiently large viewing rectangle to show that they have the same end behavior.

65. $f(x) = \dfrac{2x^2 + 6x + 6}{x + 3}, \quad g(x) = 2x$

66. $f(x) = \dfrac{-x^3 + 6x^2 - 5}{x^2 - 2x}, \quad g(x) = -x + 4$

67. $f(x) = \dfrac{x^3 - 2x^2 + 16}{x - 2}, \quad g(x) = x^2$

68. $f(x) = \dfrac{-x^4 + 2x^3 - 2x}{(x - 1)^2}, \quad g(x) = 1 - x^2$

 69–74 ■ Graph the rational function and find all vertical asymptotes, x- and y-intercepts, and local extrema, correct to the nearest decimal. Then use long division to find a polynomial that has the same end behavior as the rational function, and graph both functions in a sufficiently large viewing rectangle to verify that the end behaviors of the polynomial and the rational function are the same.

69. $y = \dfrac{2x^2 - 5x}{2x + 3}$

70. $y = \dfrac{x^4 - 3x^3 + x^2 - 3x + 3}{x^2 - 3x}$

71. $y = \dfrac{x^5}{x^3 - 1}$　　**72.** $y = \dfrac{x^4}{x^2 - 2}$

73. $r(x) = \dfrac{x^4 - 3x^3 + 6}{x - 3}$　　**74.** $r(x) = \dfrac{4 + x^2 - x^4}{x^2 - 1}$

Applications

 75. Population Growth　Suppose that the rabbit population on Mr. Jenkins' farm follows the formula

$$p(t) = \frac{3000t}{t + 1}$$

where $t \geq 0$ is the time (in months) since the beginning of the year.

(a) Draw a graph of the rabbit population.

(b) What eventually happens to the rabbit population?

76. Drug Concentration　After a certain drug is injected into a patient, the concentration c of the drug in the bloodstream is monitored. At time $t \geq 0$ (in minutes since the injection), the concentration (in mg/L) is given by

$$c(t) = \frac{30t}{t^2 + 2}$$

(a) Draw a graph of the drug concentration.

(b) What eventually happens to the concentration of drug in the bloodstream?

 77. Drug Concentration　A drug is administered to a patient and the concentration of the drug in the bloodstream is monitored. At time $t \geq 0$ (in hours since giving the drug), the concentration (in mg/L) is given by

$$c(t) = \frac{5t}{t^2 + 1}$$

Graph the function c with a graphing device.

(a) What is the highest concentration of drug that is reached in the patient's bloodstream?

(b) What happens to the drug concentration after a long period of time?

(c) How long does it take for the concentration to drop below 0.3 mg/L?

 78. Flight of a Rocket　Suppose a rocket is fired upward from the surface of the earth with an initial velocity v (measured in m/s). Then the maximum height h (in meters) reached by the rocket is given by the function

$$h(v) = \frac{Rv^2}{2gR - v^2}$$

where $R = 6.4 \times 10^6$ m is the radius of the earth and $g = 9.8$ m/s^2 is the acceleration due to gravity. Use a graphing device to draw a graph of the function h. (Note that h and v must both be positive, so the viewing rectangle need not contain negative values.) What does the vertical asymptote represent physically?

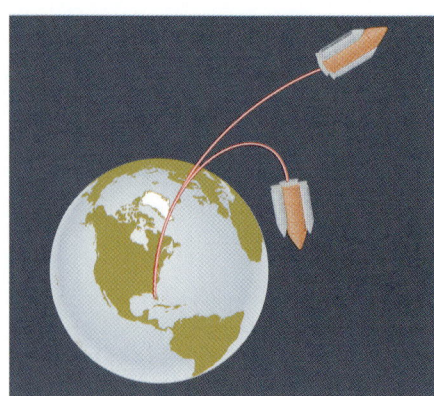

79. The Doppler Effect   As a train moves toward an observer (see the figure), the pitch of its whistle sounds higher to the observer than it would if the train were at rest, because the crests of the sound waves are compressed closer together. This phenomenon is called the *Doppler effect*. The observed pitch P is a function of the speed v of the train and is given by

$$P(v) = P_0\left(\frac{s_0}{s_0 - v}\right)$$

where P_0 is the actual pitch of the whistle at the source and $s_0 = 332$ m/s is the speed of sound in air. Suppose that a train has a whistle pitched at $P_0 = 440$ Hz. Graph the function $y = P(v)$ using a graphing device. How can the vertical asymptote of this function be interpreted physically?

80. Focusing Distance   For a camera with a lens of fixed focal length F to focus on an object located a distance x from the lens, the film must be placed a distance y behind the lens, where F, x, and y are related by

$$\frac{1}{x} + \frac{1}{y} = \frac{1}{F}$$

(See the figure.) Suppose the camera has a 55-mm lens ($F = 55$).

(a) Express y as a function of x and graph the function.

(b) What happens to the focusing distance y as the object moves far away from the lens?

(c) What happens to the focusing distance y as the object moves close to the lens?

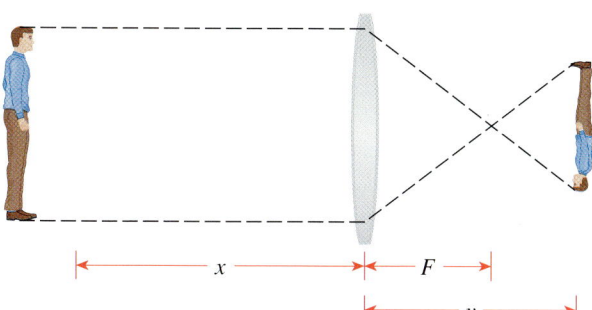

Discovery • Discussion

81. Constructing a Rational Function from Its Asymptotes   Give an example of a rational function that has vertical asymptote $x = 3$. Now give an example of one that has vertical asymptote $x = 3$ *and* horizontal asymptote $y = 2$. Now give an example of a rational function with vertical asymptotes $x = 1$ and $x = -1$, horizontal asymptote $y = 0$, and x-intercept 4.

82. A Rational Function with No Asymptote   Explain how you can tell (without graphing it) that the function

$$r(x) = \frac{x^6 + 10}{x^4 + 8x^2 + 15}$$

has no x-intercept and no horizontal, vertical, or slant asymptote. What is its end behavior?'

83. Graphs with Holes   In this chapter we adopted the convention that in rational functions, the numerator and denominator don't share a common factor. In this exercise we consider the graph of a rational function that doesn't satisfy this rule.

(a) Show that the graph of

$$r(x) = \frac{3x^2 - 3x - 6}{x - 2}$$

is the line $y = 3x + 3$ with the point $(2, 9)$ removed. [*Hint:* Factor. What is the domain of r?]

(b) Graph the rational functions:

$$s(x) = \frac{x^2 + x - 20}{x + 5}$$

$$t(x) = \frac{2x^2 - x - 1}{x - 1}$$

$$u(x) = \frac{x - 2}{x^2 - 2x}$$

84. Transformations of $y = 1/x^2$ In Example 2 we saw that some simple rational functions can be graphed by shifting, stretching, or reflecting the graph of $y = 1/x$. In this exercise we consider rational functions that can be graphed by transforming the graph of $y = 1/x^2$, shown on the following page.

(a) Graph the function

$$r(x) = \frac{1}{(x - 2)^2}$$

by transforming the graph of $y = 1/x^2$.

(b) Use long division and factoring to show that the function

$$s(x) = \frac{2x^2 + 4x + 5}{x^2 + 2x + 1}$$

can be written as

$$s(x) = 2 + \frac{3}{(x + 1)^2}$$

Then graph s by transforming the graph of $y = 1/x^2$.

(c) One of the following functions can be graphed by transforming the graph of $y = 1/x^2$; the other cannot. Use transformations to graph the one that can be,

and explain why this method doesn't work for the other one.

$$p(x) = \frac{2 - 3x^2}{x^2 - 4x + 4} \qquad q(x) = \frac{12x - 3x^2}{x^2 - 4x + 4}$$

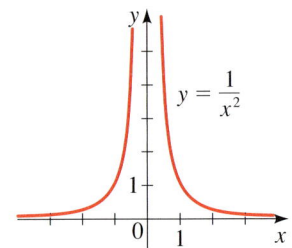

---

3 Review

Concept Check

1. (a) Write the defining equation for a polynomial P of degree n.

  (b) What does it mean to say that c is a zero of P?

2. Sketch graphs showing the possible end behaviors of polynomials of odd degree and of even degree.

3. What steps would you follow to graph a polynomial by hand?

4. (a) What is meant by a local maximum point or local minimum point of a polynomial?

  (b) How many local extrema can a polynomial of degree n have?

5. State the Division Algorithm and identify the dividend, divisor, quotient, and remainder.

6. How does synthetic division work?

7. (a) State the Remainder Theorem.

  (b) State the Factor Theorem.

8. (a) State the Rational Zeros Theorem.

  (b) What steps would you take to find the rational zeros of a polynomial?

9. State Descartes' Rule of Signs.

10. (a) What does it mean to say that a is a lower bound and b is an upper bound for the zeros of a polynomial?

  (b) State the Upper and Lower Bounds Theorem.

11. (a) What is a complex number?

  (b) What are the real and imaginary parts of a complex number?

  (c) What is the complex conjugate of a complex number?

  (d) How do you add, subtract, multiply, and divide complex numbers?

12. (a) State the Fundamental Theorem of Algebra.

  (b) State the Complete Factorization Theorem.

  (c) What does it mean to say that c is a zero of multiplicity k of a polynomial P?

  (d) State the Zeros Theorem.

  (e) State the Conjugate Zeros Theorem.

13. (a) What is a rational function?

  (b) What does it mean to say that $x = a$ is a vertical asymptote of $y = f(x)$?

  (c) How do you locate a vertical asymptote?

  (d) What does it mean to say that $y = b$ is a horizontal asymptote of $y = f(x)$?

  (e) How do you locate a horizontal asymptote?

  (f) What steps do you follow to sketch the graph of a rational function by hand?

  (g) Under what circumstances does a rational function have a slant asymptote? If one exists, how do you find it?

  (h) How do you determine the end behavior of a rational function?

Exercises

1–6 ■ Graph the polynomial by transforming an appropriate graph of the form $y = x^n$. Show clearly all x- and y-intercepts.

1. $P(x) = -x^3 + 64$

2. $P(x) = 2x^3 - 16$

3. $P(x) = 2(x + 1)^4 - 32$

4. $P(x) = 81 - (x - 3)^4$

5. $P(x) = 32 + (x - 1)^5$

6. $P(x) = -3(x + 2)^5 + 96$

 7–10 ■ Use a graphing device to graph the polynomial. Find the x- and y-intercepts and the coordinates of all local extrema, correct to the nearest decimal. Describe the end behavior of the polynomial.

7. $P(x) = x^3 - 4x + 1$

8. $P(x) = -2x^3 + 6x^2 - 2$

9. $P(x) = 3x^4 - 4x^3 - 10x - 1$

10. $P(x) = x^5 + x^4 - 7x^3 - x^2 + 6x + 3$

11. The strength S of a wooden beam of width x and depth y is given by the formula $S = 13.8xy^2$. A beam is to be cut from a log of diameter 10 in., as shown in the figure.

    (a) Express the strength S of this beam as a function of x only.

    (b) What is the domain of the function S?

     (c) Draw a graph of S.

     (d) What width will make the beam the strongest?

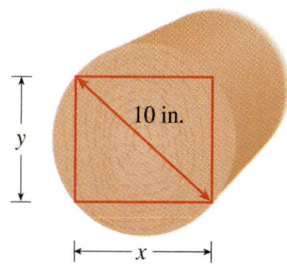

10 in.

12. A small shelter for delicate plants is to be constructed of thin plastic material. It will have square ends and a rectangular top and back, with an open bottom and front, as shown in the figure. The total area of the four plastic sides is to be 1200 in².

    (a) Express the volume V of the shelter as a function of the depth x.

     (b) Draw a graph of V.

     (c) What dimensions will maximize the volume of the shelter?

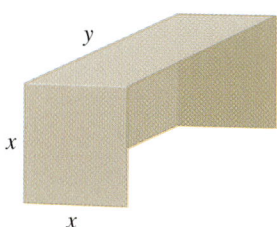

13–20 ■ Find the quotient and remainder.

13. $\dfrac{x^2 - 3x + 5}{x - 2}$

14. $\dfrac{x^2 + x - 12}{x - 3}$

15. $\dfrac{x^3 - x^2 + 11x + 2}{x - 4}$

16. $\dfrac{x^3 + 2x^2 - 10}{x + 3}$

17. $\dfrac{x^4 - 8x^2 + 2x + 7}{x + 5}$

18. $\dfrac{2x^4 + 3x^3 - 12}{x + 4}$

19. $\dfrac{2x^3 + x^2 - 8x + 15}{x^2 + 2x - 1}$

20. $\dfrac{x^4 - 2x^2 + 7x}{x^2 - x + 3}$

21–22 ■ Find the indicated value of the polynomial using the Remainder Theorem.

21. $P(x) = 2x^3 - 9x^2 - 7x + 13$; find $P(5)$

22. $Q(x) = x^4 + 4x^3 + 7x^2 + 10x + 15$; find $Q(-3)$

23. Show that $\frac{1}{2}$ is a zero of the polynomial

$$P(x) = 2x^4 + x^3 - 5x^2 + 10x - 4$$

24. Use the Factor Theorem to show that $x + 4$ is a factor of the polynomial

$$P(x) = x^5 + 4x^4 - 7x^3 - 23x^2 + 23x + 12$$

25. What is the remainder when the polynomial

$$P(x) = x^{500} + 6x^{201} - x^2 - 2x + 4$$

is divided by $x - 1$?

26. What is the remainder when $x^{101} - x^4 + 2$ is divided by $x + 1$?

27–28 ■ A polynomial P is given.

(a) List all possible rational zeros (without testing to see if they actually are zeros).

(b) Determine the possible number of positive and negative real zeros using Descartes' Rule of Signs.

27. $P(x) = x^5 - 6x^3 - x^2 + 2x + 18$

28. $P(x) = 6x^4 + 3x^3 + x^2 + 3x + 4$

29–36 ■ A polynomial P is given.

(a) Find all real zeros of P and state their multiplicities.

(b) Sketch the graph of P.

29. $P(x) = x^3 - 16x$

30. $P(x) = x^3 - 3x^2 - 4x$

31. $P(x) = x^4 + x^3 - 2x^2$

32. $P(x) = x^4 - 5x^2 + 4$

33. $P(x) = x^4 - 2x^3 - 7x^2 + 8x + 12$

34. $P(x) = x^4 - 2x^3 - 2x^2 + 8x - 8$

35. $P(x) = 2x^4 + x^3 + 2x^2 - 3x - 2$

36. $P(x) = 9x^5 - 21x^4 + 10x^3 + 6x^2 - 3x - 1$

37–46 ■ Evaluate the expression and write in the form $a + bi$.

37. $(2 - 3i) + (1 + 4i)$ **38.** $(3 - 6i) - (6 - 4i)$

39. $(2 + i)(3 - 2i)$ **40.** $4i(2 - \frac{1}{2}i)$

41. $\dfrac{4 + 2i}{2 - i}$ **42.** $\dfrac{8 + 3i}{4 + 3i}$

43. i^{25} **44.** $(1 + i)^3$

45. $(1 - \sqrt{-1})(1 + \sqrt{-1})$ **46.** $\sqrt{-10} \cdot \sqrt{-40}$

47. Find a polynomial of degree 3 with constant coefficient 12 and zeros $-\frac{1}{2}$, 2, and 3.

48. Find a polynomial of degree 4 having integer coefficients and zeros $3i$ and 4, with 4 a double zero.

49. Does there exist a polynomial of degree 4 with integer coefficients that has zeros i, $2i$, $3i$, and $4i$? If so, find it. If not, explain why.

50. Prove that the equation $3x^4 + 5x^2 + 2 = 0$ has no real root.

51–60 ■ Find all rational, irrational, and complex zeros (and state their multiplicities). Use Descartes' Rule of Signs, the Upper and Lower Bounds Theorem, the quadratic formula, or other factoring techniques to help you whenever possible.

51. $P(x) = x^3 - 3x^2 - 13x + 15$

52. $P(x) = 2x^3 + 5x^2 - 6x - 9$

53. $P(x) = x^4 + 6x^3 + 17x^2 + 28x + 20$

54. $P(x) = x^4 + 7x^3 + 9x^2 - 17x - 20$

55. $P(x) = x^5 - 3x^4 - x^3 + 11x^2 - 12x + 4$

56. $P(x) = x^4 - 81$

57. $P(x) = x^6 - 64$

58. $P(x) = 18x^3 + 3x^2 - 4x - 1$

59. $P(x) = 6x^4 - 18x^3 + 6x^2 - 30x + 36$

60. $P(x) = x^4 + 15x^2 + 54$

61–64 ■ Use a graphing device to find all real solutions of the equation.

61. $2x^2 = 5x + 3$

62. $x^3 + x^2 - 14x - 24 = 0$

63. $x^4 - 3x^3 - 3x^2 - 9x - 2 = 0$

64. $x^5 = x + 3$

65–70 ■ Graph the rational function. Show clearly all x- and y-intercepts and asymptotes.

65. $r(x) = \dfrac{3x - 12}{x + 1}$ **66.** $r(x) = \dfrac{1}{(x + 2)^2}$

67. $r(x) = \dfrac{x - 2}{x^2 - 2x - 8}$ **68.** $r(x) = \dfrac{2x^2 - 6x - 7}{x - 4}$

69. $r(x) = \dfrac{x^2 - 9}{2x^2 + 1}$ **70.** $r(x) = \dfrac{x^3 + 27}{x + 4}$

71–74 ■ Use a graphing device to analyze the graph of the rational function. Find all x- and y-intercepts; and all vertical, horizontal, and slant asymptotes. If the function has no horizontal or slant asymptote, find a polynomial that has the same end behavior as the rational function.

71. $r(x) = \dfrac{x - 3}{2x + 6}$ **72.** $r(x) = \dfrac{2x - 7}{x^2 + 9}$

73. $r(x) = \dfrac{x^3 + 8}{x^2 - x - 2}$ **74.** $r(x) = \dfrac{2x^3 - x^2}{x + 1}$

75. Find the coordinates of all points of intersection of the graphs of

$$y = x^4 + x^2 + 24x \quad \text{and} \quad y = 6x^3 + 20$$

3   Test

1. Graph the polynomial $P(x) = -(x + 2)^3 + 27$, showing clearly all x- and y-intercepts.

2. (a) Use synthetic division to find the quotient and remainder when $x^4 - 4x^2 + 2x + 5$ is divided by $x - 2$.

   (b) Use long division to find the quotient and remainder when $2x^5 + 4x^4 - x^3 - x^2 + 7$ is divided by $2x^2 - 1$.

3. Let $P(x) = 2x^3 - 5x^2 - 4x + 3$.

   (a) List all possible rational zeros of P.

   (b) Find the complete factorization of P.

   (c) Find the zeros of P.

   (d) Sketch the graph of P.

4. Perform the indicated operation and write the result in the form $a + bi$.

   (a) $(3 - 2i) + (4 + 3i)$

   (b) $(3 - 2i) - (4 + 3i)$

   (c) $(3 - 2i)(4 + 3i)$

   (d) $\dfrac{3 - 2i}{4 + 3i}$

   (e) i^{48}

   (f) $(\sqrt{2} - \sqrt{-2})(\sqrt{8} + \sqrt{-2})$

5. Find all real and complex zeros of $P(x) = x^3 - x^2 - 4x - 6$.

6. Find the complete factorization of $P(x) = x^4 - 2x^3 + 5x^2 - 8x + 4$.

7. Find a fourth-degree polynomial with integer coefficients that has zeros $3i$ and -1, with -1 a zero of multiplicity 2.

8. Let $P(x) = 2x^4 - 7x^3 + x^2 - 18x + 3$.

   (a) Use Descartes' Rule of Signs to determine how many positive and how many negative real zeros P can have.

   (b) Show that 4 is an upper bound and -1 is a lower bound for the real zeros of P.

   (c) Draw a graph of P and use it to estimate the real zeros of P, correct to two decimal places.

   (d) Find the coordinates of all local extrema of P, correct to two decimals.

9. Consider the following rational functions:

   $$r(x) = \frac{2x - 1}{x^2 - x - 2} \qquad s(x) = \frac{x^3 + 27}{x^2 + 4} \qquad t(x) = \frac{x^3 - 9x}{x + 2} \qquad u(x) = \frac{x^2 + x - 6}{x^2 - 25}$$

   (a) Which of these rational functions has a horizontal asymptote?

   (b) Which of these functions has a slant asymptote?

   (c) Which of these functions has no vertical asymptote?

   (d) Graph $y = u(x)$, showing clearly any asymptotes and x- and y-intercepts the function may have.

   (e) Use long division to find a polynomial P that has the same end behavior as t. Graph both P and t on the same screen to verify that they have the same end behavior.

We have learned how to fit a line to data (see *Focus on Modeling*, page 239). The line models the increasing or decreasing trend in the data. If the data exhibits more variability, such as an increase followed by a decrease, then to model the data we need to use a curve rather than a line. Figure 1 shows a scatter plot with three possible models that appear to fit the data. Which model fits the data best?

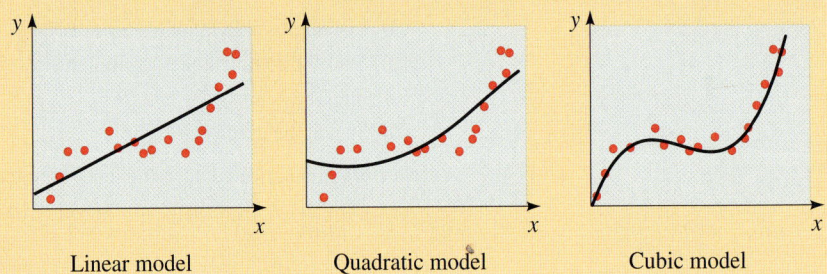

Linear model   Quadratic model   Cubic model

Figure 1

Polynomial Functions as Models

Polynomial functions are ideal for modeling data where the scatter plot has peaks or valleys (that is, local maxima or minima). For example, if the data have a single peak as in Figure 2(a), then it may be appropriate to use a quadratic polynomial to model the data. The more peaks or valleys the data exhibit, the higher the degree of the polynomial needed to model the data (see Figure 2).

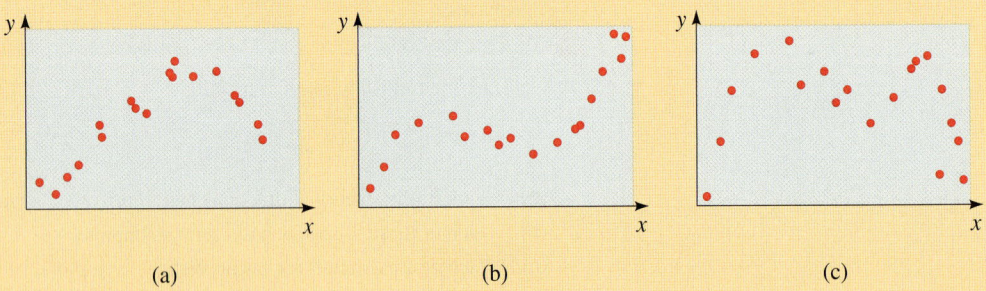

(a)   (b)   (c)

Figure 2

Graphing calculators are programmed to find the **polynomial of best fit** of a specified degree. As is the case for lines (see pages 239–240), a polynomial of a given degree fits the data *best* if the sum of the squares of the distances between the graph of the polynomial and the data points is minimized.

Ted Wood/The Image Bank/Getty Images

Example 1 **Rainfall and Crop Yield**

Rain is essential for crops to grow, but too much rain can diminish crop yields. The data give rainfall and cotton yield per acre for several seasons in a certain county.

(a) Make a scatter plot of the data. What degree polynomial seems appropriate for modeling the data?

(b) Use a graphing calculator to find the polynomial of best fit. Graph the polynomial on the scatter plot.

(c) Use the model you found to estimate the yield if there are 25 in. of rainfall.

Season	Rainfall (in.)	Yield (kg/acre)
1	23.3	5311
2	20.1	4382
3	18.1	3950
4	12.5	3137
5	30.9	5113
6	33.6	4814
7	35.8	3540
8	15.5	3850
9	27.6	5071
10	34.5	3881

Solution

(a) The scatter plot is shown in Figure 3. The data appear to have a peak, so it is appropriate to model the data by a quadratic polynomial (degree 2).

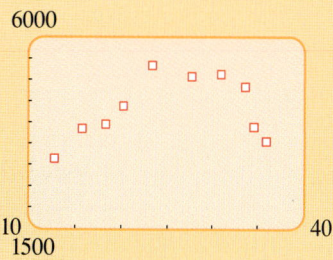

Figure 3

Scatter plot of yield vs. rainfall data

(b) Using a graphing calculator, we find that the quadratic polynomial of best fit is

$$y = -12.6x^2 + 651.5x - 3283.2$$

The calculator output and the scatter plot, together with the graph of the quadratic model, are shown in Figure 4.

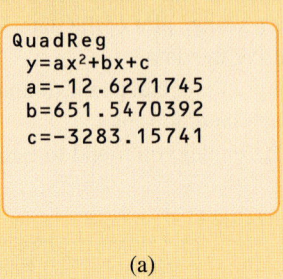

```
QuadReg
y=ax²+bx+c
a=-12.6271745
b=651.5470392
c=-3283.15741
```

(a)

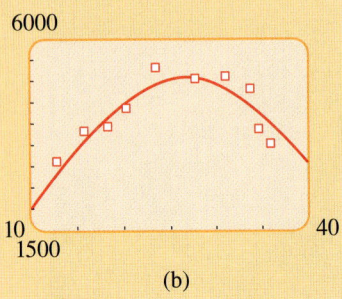

(b)

Figure 4

(c) Using the model with $x = 25$, we get

$$y = -12.6(25)^2 + 651.5(25) - 3283.2 \approx 5129.3$$

We estimate the yield to be about 5130 kg per acre.   ■

Example 2   Length-at-Age Data for Fish

Otoliths ("earstones") are tiny structures found in the heads of fish. Microscopic growth rings on the otoliths, not unlike growth rings on a tree, record the age of a fish. The table gives the lengths of rock bass of different ages, as determined by the otoliths. Scientists have proposed a cubic polynomial to model this data.

(a) Use a graphing calculator to find the cubic polynomial of best fit for the data.

(b) Make a scatter plot of the data and graph the polynomial from part (a).

(c) A fisherman catches a rock bass 20 in. long. Use the model to estimate its age.

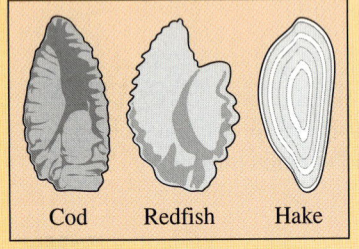

Cod   Redfish   Hake

Otoliths for several fish species.

Age (yr)	Length (in.)	Age (yr)	Length (in.)
1	4.8	9	18.2
2	8.8	9	17.1
2	8.0	10	18.8
3	7.9	10	19.5
4	11.9	11	18.9
5	14.4	12	21.7
6	14.1	12	21.9
6	15.8	13	23.8
7	15.6	14	26.9
8	17.8	14	25.1

Solution

(a) Using a graphing calculator (see Figure 5(a)), we find the cubic polynomial of best fit

$$y = 0.0155x^3 - 0.372x^2 + 3.95x + 1.21$$

(b) The scatter plot of the data and the cubic polynomial are graphed in Figure 5(b).

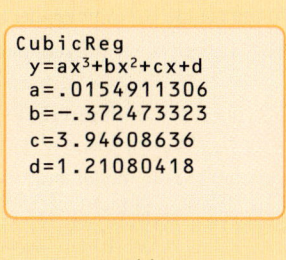

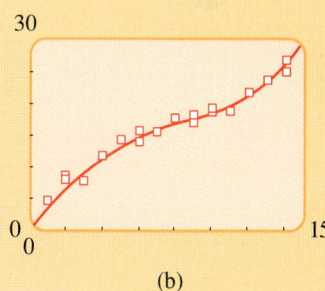

Figure 5   (a)   (b)

(c) Moving the cursor along the graph of the polynomial, we find that $y = 20$ when $x \approx 10.8$. Thus, the fish is about 11 years old. ∎

Problems

Pressure (lb/in²)	Tire life (mi)
26	50,000
28	66,000
31	78,000
35	81,000
38	74,000
42	70,000
45	59,000

1. Tire Inflation and Treadware   Car tires need to be inflated properly. Overinflation or underinflation can cause premature treadwear. The data and scatter plot show tire life for different inflation values for a certain type of tire.

(a) Find the quadratic polynomial that best fits the data.

(b) Draw a graph of the polynomial from part (a) together with a scatter plot of the data.

(c) Use your result from part (b) to estimate the pressure that gives the longest tire life.

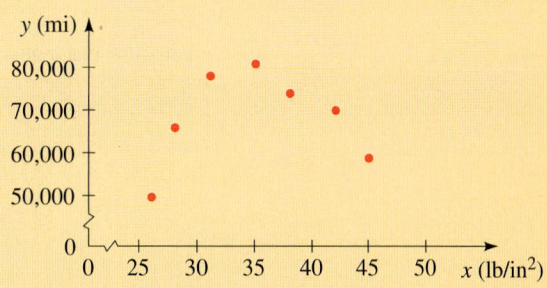

Density (plants/acre)	Crop yield (bushels/acre)
15,000	43
20,000	98
25,000	118
30,000	140
35,000	142
40,000	122
45,000	93
50,000	67

2. Too Many Corn Plants per Acre?   The more corn a farmer plants per acre the greater the yield that he can expect, but only up to a point. Too many plants per acre can cause overcrowding and decrease yields. The data give crop yields per acre for various densities of corn plantings, as found by researchers at a university test farm.

(a) Find the quadratic polynomial that best fits the data.

(b) Draw a graph of the polynomial from part (a) together with a scatter plot of the data.

(c) Use your result from part (b) to estimate the yield for 37,000 plants per acre.

3. How Fast Can You List Your Favorite Things?   If you are asked to make a list of objects in a certain category, how fast you can list them follows a predictable pattern. For example, if you try to name as many vegetables as you can, you'll probably think of several right away—for example, carrots, peas, beans, corn, and so on. Then after a pause you may think of ones you eat less frequently—perhaps zucchini, eggplant, and asparagus. Finally a few more exotic vegetables might come to mind—artichokes, jicama, bok choy, and the like. A psychologist performs this experiment on a number of subjects. The table below gives the average number of vegetables that the subjects named by a given number of seconds.

(a) Find the cubic polynomial that best fits the data.

(b) Draw a graph of the polynomial from part (a) together with a scatter plot of the data.

(c) Use your result from part (b) to estimate the number of vegetables that subjects would be able to name in 40 seconds.

(d) According to the model, how long (to the nearest 0.1 s) would it take a person to name five vegetables?

Seconds	Number of Vegetables
1	2
2	6
5	10
10	12
15	14
20	15
25	18
30	21

4. Clothing Sales Are Seasonal   Clothing sales tend to vary by season with more clothes sold in spring and fall. The table gives sales figures for each month at a certain clothing store.

(a) Find the quartic (fourth-degree) polynomial that best fits the data.

(b) Draw a graph of the polynomial from part (a) together with a scatter plot of the data.

(c) Do you think that a quartic polynomial is a good model for these data? Explain.

Month	Sales ($)
January	8,000
February	18,000
March	22,000
April	31,000
May	29,000
June	21,000
July	22,000
August	26,000
September	38,000
October	40,000
November	27,000
December	15,000

5. Height of a Baseball   A baseball is thrown upward and its height measured at 0.5-second intervals using a strobe light. The resulting data are given in the table.

(a) Draw a scatter plot of the data. What degree polynomial is appropriate for modeling the data?

(b) Find a polynomial model that best fits the data, and graph it on the scatter plot.

(c) Find the times when the ball is 20 ft above the ground.

(d) What is the maximum height attained by the ball?

Time (s)	Height (ft)
0	4.2
0.5	26.1
1.0	40.1
1.5	46.0
2.0	43.9
2.5	33.7
3.0	15.8

6. Torricelli's Law   Water in a tank will flow out of a small hole in the bottom faster when the tank is nearly full than when it is nearly empty. According to Torricelli's Law, the height $h(t)$ of water remaining at time t is a quadratic function of t.

A certain tank is filled with water and allowed to drain. The height of the water is measured at different times as shown in the table.

(a) Find the quadratic polynomial that best fits the data.

(b) Draw a graph of the polynomial from part (a) together with a scatter plot of the data.

(c) Use your graph from part (b) to estimate how long it takes for the tank to drain completely.

Time (min)	Height (ft)
0	5.0
4	3.1
8	1.9
12	0.8
16	0.2

4

Exponential and Logarithmic Functions

Chapter Overview

In this chapter we study a new class of functions called *exponential functions*. For example,

$$f(x) = 2^x$$

is an exponential function (with base 2). Notice how quickly the values of this function increase:

$$f(3) = 2^3 = 8$$

$$f(10) = 2^{10} = 1024$$

$$f(30) = 2^{30} = 1{,}073{,}741{,}824$$

Compare this with the function $g(x) = x^2$, where $g(30) = 30^2 = 900$. The point is, when the variable is in the exponent, even a small change in the variable can cause a dramatic change in the value of the function.

In spite of this incomprehensibly huge growth, exponential functions are appropriate for modeling population growth for all living things, from bacteria to elephants. To understand how a population grows, consider the case of a single bacterium, which divides every hour. After one hour we would have 2 bacteria, after two hours 2^2 or 4 bacteria, after three hours 2^3 or 8 bacteria, and so on. After x hours we would have 2^x bacteria. This leads us to model the bacteria population by the function $f(x) = 2^x$.

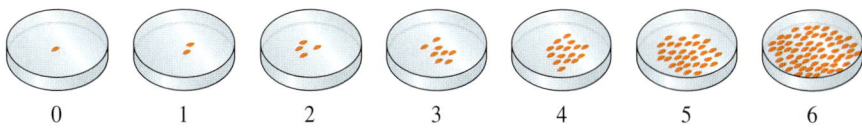

0	1	2	3	4	5	6

The principle governing population growth is the following: The larger the population, the greater the number of offspring. This same principle is present in many other real-life situations. For example, the larger your bank account, the more interest you get. So we also use exponential functions to find compound interest.

We use *logarithmic functions*, which are inverses of exponential functions, to help us answer such questions as, When will my investment grow to $100,000? In *Focus on Modeling* (page 386) we explore how to fit exponential and logarithmic models to data.

Theo Allofs /The Image Bank /Getty Images

4.1	**Exponential Functions**

So far, we have studied polynomial and rational functions. We now study one of the most important functions in mathematics, the *exponential function*. This function is used to model such natural processes as population growth and radioactive decay.

Exponential Functions

In Section 1.2 we defined a^x for $a > 0$ and x a rational number, but we have not yet defined irrational powers. So, what is meant by $5^{\sqrt{3}}$ or 2^{π}? To define a^x when x is irrational, we approximate x by rational numbers. For example, since

$$\sqrt{3} \approx 1.73205\ldots$$

is an irrational number, we successively approximate $a^{\sqrt{3}}$ by the following rational powers:

$$a^{1.7}, a^{1.73}, a^{1.732}, a^{1.7320}, a^{1.73205}, \ldots$$

Intuitively, we can see that these rational powers of a are getting closer and closer to $a^{\sqrt{3}}$. It can be shown using advanced mathematics that there is exactly one number that these powers approach. We define $a^{\sqrt{3}}$ to be this number.

For example, using a calculator we find

$$5^{\sqrt{3}} \approx 5^{1.732}$$

$$\approx 16.2411\ldots$$

The more decimal places of $\sqrt{3}$ we use in our calculation, the better our approximation of $5^{\sqrt{3}}$.

The Laws of Exponents are listed on page 14.

It can be proved that the *Laws of Exponents are still true when the exponents are real numbers*.

Exponential Functions

The **exponential function with base a** is defined for all real numbers x by

$$f(x) = a^x$$

where $a > 0$ and $a \neq 1$.

We assume $a \neq 1$ because the function $f(x) = 1^x = 1$ is just a constant function. Here are some examples of exponential functions:

$$f(x) = 2^x \qquad g(x) = 3^x \qquad h(x) = 10^x$$

Base 2   Base 3   Base 10

Example 1 **Evaluating Exponential Functions**

Let $f(x) = 3^x$ and evaluate the following:

(a) $f(2)$ (b) $f(-\frac{2}{3})$ (c) $f(\pi)$ (d) $f(\sqrt{2})$

Solution   We use a calculator to obtain the values of f.

	Calculator keystrokes	Output
(a) $f(2) = 3^2 = 9$	3 ∧ 2 ENTER	9
(b) $f(-\frac{2}{3}) = 3^{-2/3} \approx 0.4807$	3 ∧ ((−) 2 ÷ 3) ENTER	0.4807498
(c) $f(\pi) = 3^\pi \approx 31.544$	3 ∧ π ENTER	31.5442807
(d) $f(\sqrt{2}) = 3^{\sqrt{2}} \approx 4.7288$	3 ∧ √ 2 ENTER	4.7288043

■

Graphs of Exponential Functions

We first graph exponential functions by plotting points. We will see that the graphs of such functions have an easily recognizable shape.

Example 2 **Graphing Exponential Functions by Plotting Points**

Draw the graph of each function.

(a) $f(x) = 3^x$ (b) $g(x) = \left(\dfrac{1}{3}\right)^x$

Solution   We calculate values of $f(x)$ and $g(x)$ and plot points to sketch the graphs in Figure 1.

x	$f(x) = 3^x$	$g(x) = \left(\frac{1}{3}\right)^x$
-3	$\frac{1}{27}$	27
-2	$\frac{1}{9}$	9
-1	$\frac{1}{3}$	3
0	1	1
1	3	$\frac{1}{3}$
2	9	$\frac{1}{9}$
3	27	$\frac{1}{27}$

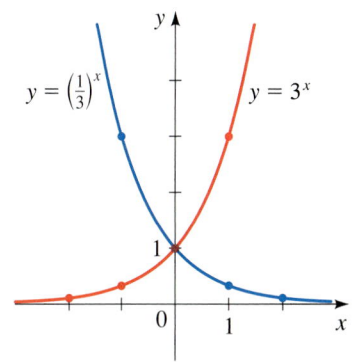

Figure 1

Notice that

$$g(x) = \left(\frac{1}{3}\right)^x = \frac{1}{3^x} = 3^{-x} = f(-x)$$

Reflecting graphs is explained in Section 2.4.

and so we could have obtained the graph of g from the graph of f by reflecting in the y-axis.

■

To see just how quickly $f(x) = 2^x$ increases, let's perform the following thought experiment. Suppose we start with a piece of paper a thousandth of an inch thick, and we fold it in half 50 times. Each time we fold the paper, the thickness of the paper stack doubles, so the thickness of the resulting stack would be $2^{50}/1000$ inches. How thick do you think that is? It works out to be more than 17 million miles!

Figure 2 shows the graphs of the family of exponential functions $f(x) = a^x$ for various values of the base a. All of these graphs pass through the point $(0, 1)$ because $a^0 = 1$ for $a \neq 0$. You can see from Figure 2 that there are two kinds of exponential functions: If $0 < a < 1$, the exponential function decreases rapidly. If $a > 1$, the function increases rapidly (see the margin note).

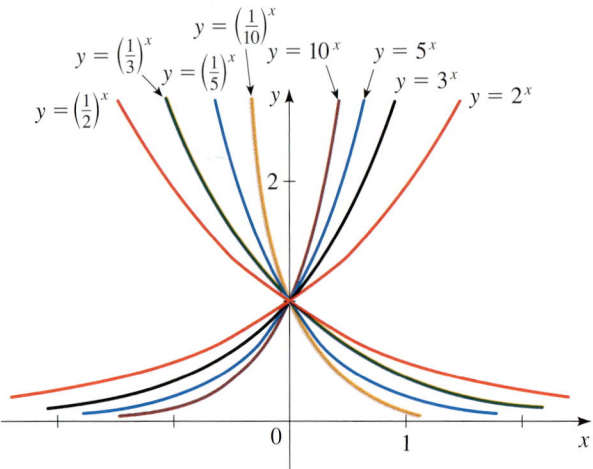

Figure 2

A family of exponential functions

See Section 3.6, page 301, where the "arrow notation" used here is explained.

The x-axis is a horizontal asymptote for the exponential function $f(x) = a^x$. This is because when $a > 1$, we have $a^x \to 0$ as $x \to -\infty$, and when $0 < a < 1$, we have $a^x \to 0$ as $x \to \infty$ (see Figure 2). Also, $a^x > 0$ for all $x \in \mathbb{R}$, so the function $f(x) = a^x$ has domain $\mathbb{R}$ and range $(0, \infty)$. These observations are summarized in the following box.

Graphs of Exponential Functions

The exponential function

$$f(x) = a^x \qquad (a > 0, a \neq 1)$$

has domain $\mathbb{R}$ and range $(0, \infty)$. The line $y = 0$ (the x-axis) is a horizontal asymptote of f. The graph of f has one of the following shapes.

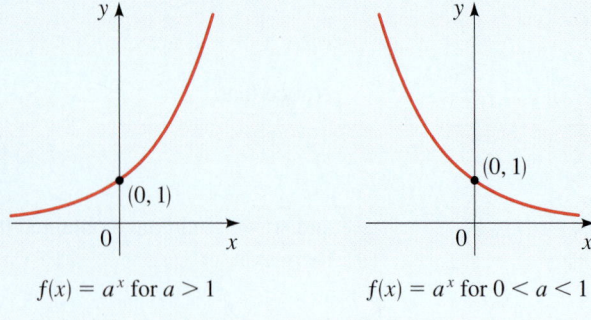

$f(x) = a^x$ for $a > 1$ $\qquad\qquad$ $f(x) = a^x$ for $0 < a < 1$

The **Gateway Arch** in St. Louis, Missouri, is shaped in the form of the graph of a combination of exponential functions (*not* a parabola, as it might first appear). Specifically, it is a **catenary**, which is the graph of an equation of the form

$$y = a(e^{bx} + e^{-bx})$$

(see Exercise 57). This shape was chosen because it is optimal for distributing the internal structural forces of the arch. Chains and cables suspended between two points (for example, the stretches of cable between pairs of telephone poles) hang in the shape of a catenary.

Example 3   Identifying Graphs of Exponential Functions

Find the exponential function $f(x) = a^x$ whose graph is given.

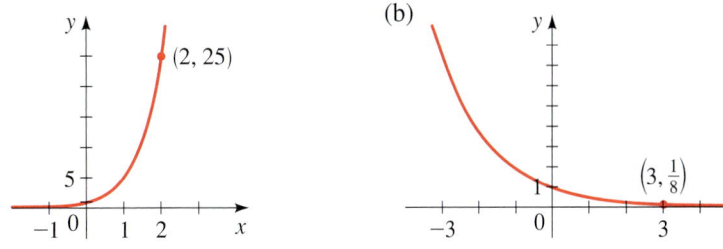

(a)   (b)

Solution

(a) Since $f(2) = a^2 = 25$, we see that the base is $a = 5$. So $f(x) = 5^x$.

(b) Since $f(3) = a^3 = \frac{1}{8}$, we see that the base is $a = \frac{1}{2}$. So $f(x) = \left(\frac{1}{2}\right)^x$. ∎

In the next example we see how to graph certain functions, not by plotting points, but by taking the basic graphs of the exponential functions in Figure 2 and applying the shifting and reflecting transformations of Section 2.4.

Example 4   Transformations of Exponential Functions

Use the graph of $f(x) = 2^x$ to sketch the graph of each function.

(a) $g(x) = 1 + 2^x$ (b) $h(x) = -2^x$ (c) $k(x) = 2^{x-1}$

Solution

(a) To obtain the graph of $g(x) = 1 + 2^x$, we start with the graph of $f(x) = 2^x$ and shift it upward 1 unit. Notice from Figure 3(a) that the line $y = 1$ is now a horizontal asymptote.

(b) Again we start with the graph of $f(x) = 2^x$, but here we reflect in the x-axis to get the graph of $h(x) = -2^x$ shown in Figure 3(b).

(c) This time we start with the graph of $f(x) = 2^x$ and shift it to the right by 1 unit, to get the graph of $k(x) = 2^{x-1}$ shown in Figure 3(c).

Shifting and reflecting of graphs is explained in Section 2.4.

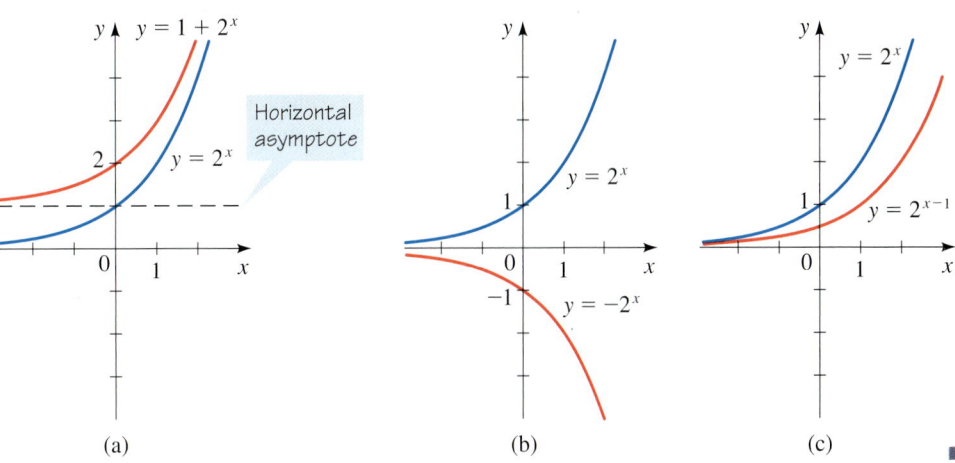

Figure 3

(a)   (b)   (c) ∎

Example 5 Comparing Exponential and Power Functions

Compare the rates of growth of the exponential function $f(x) = 2^x$ and the power function $g(x) = x^2$ by drawing the graphs of both functions in the following viewing rectangles.

(a) $[0, 3]$ by $[0, 8]$ (b) $[0, 6]$ by $[0, 25]$

(c) $[0, 20]$ by $[0, 1000]$

Solution

(a) Figure 4(a) shows that the graph of $g(x) = x^2$ catches up with, and becomes higher than, the graph of $f(x) = 2^x$ at $x = 2$.

(b) The larger viewing rectangle in Figure 4(b) shows that the graph of $f(x) = 2^x$ overtakes that of $g(x) = x^2$ when $x = 4$.

(c) Figure 4(c) gives a more global view and shows that, when x is large, $f(x) = 2^x$ is much larger than $g(x) = x^2$.

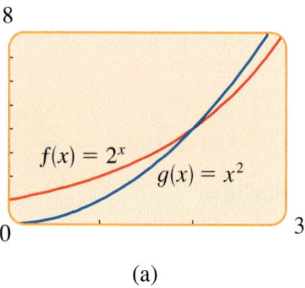

(a)   (b)   (c)

Figure 4
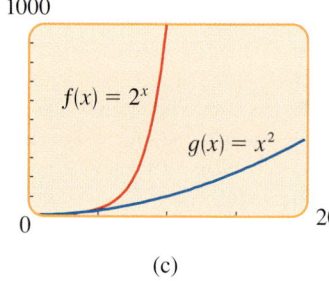
■

The Natural Exponential Function

n	$\left(1 + \dfrac{1}{n}\right)^n$
1	2.00000
5	2.48832
10	2.59374
100	2.70481
1000	2.71692
10,000	2.71815
100,000	2.71827
1,000,000	2.71828

Any positive number can be used as the base for an exponential function, but some bases are used more frequently than others. We will see in the remaining sections of this chapter that the bases 2 and 10 are convenient for certain applications, but the most important base is the number denoted by the letter e.

The number e is defined as the value that $(1 + 1/n)^n$ approaches as n becomes large. (In calculus this idea is made more precise through the concept of a limit. See Exercise 55.) The table in the margin shows the values of the expression $(1 + 1/n)^n$ for increasingly large values of n. It appears that, correct to five decimal places, $e \approx 2.71828$; in fact, the approximate value to 20 decimal places is

$$e \approx 2.71828182845904523536$$

It can be shown that e is an irrational number, so we cannot write its exact value in decimal form.

Why use such a strange base for an exponential function? It may seem at first that a base such as 10 is easier to work with. We will see, however, that in certain applications the number e is the best possible base. In this section we study how e occurs in the description of compound interest.

The notation e was chosen by Leonhard Euler (see page 288), probably because it is the first letter of the word *exponential*.

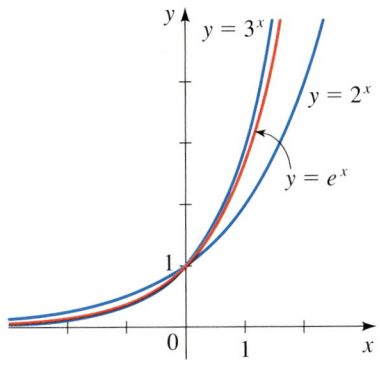

Figure 5

Graph of the natural exponential function

The Natural Exponential Function

The **natural exponential function** is the exponential function

$$f(x) = e^x$$

with base e. It is often referred to as *the* exponential function.

Since $2 < e < 3$, the graph of the natural exponential function lies between the graphs of $y = 2^x$ and $y = 3^x$, as shown in Figure 5.

Scientific calculators have a special key for the function $f(x) = e^x$. We use this key in the next example.

Example 6 Evaluating the Exponential Function

Evaluate each expression correct to five decimal places.

(a) e^3 (b) $2e^{-0.53}$ (c) $e^{4.8}$

Solution We use the $\boxed{e^x}$ key on a calculator to evaluate the exponential function.

(a) $e^3 \approx 20.08554$

(b) $2e^{-0.53} \approx 1.17721$

(c) $e^{4.8} \approx 121.51042$ ■

Example 7 Transformations of the Exponential Function

Sketch the graph of each function.

(a) $f(x) = e^{-x}$ (b) $g(x) = 3e^{0.5x}$

Solution

(a) We start with the graph of $y = e^x$ and reflect in the y-axis to obtain the graph of $y = e^{-x}$ as in Figure 6.

(b) We calculate several values, plot the resulting points, then connect the points with a smooth curve. The graph is shown in Figure 7.

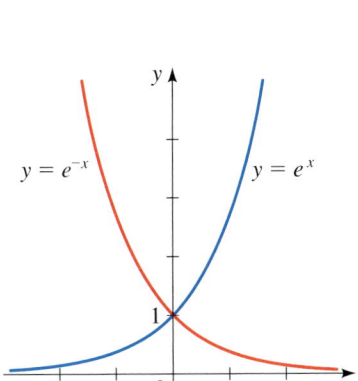

Figure 6

x	$f(x) = 3e^{0.5x}$
-3	0.67
-2	1.10
-1	1.82
0	3.00
1	4.95
2	8.15
3	13.45

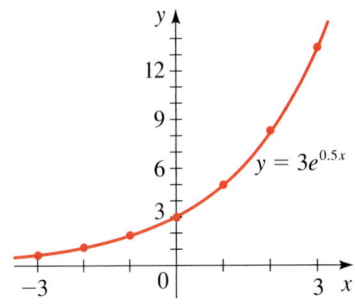

Figure 7   ■

Example 8   An Exponential Model for the Spread of a Virus

An infectious disease begins to spread in a small city of population 10,000. After t days, the number of persons who have succumbed to the virus is modeled by the function

$$v(t) = \frac{10,000}{5 + 1245e^{-0.97t}}$$

(a) How many infected people are there initially (at time $t = 0$)?

(b) Find the number of infected people after one day, two days, and five days.

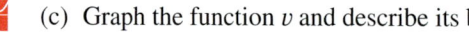 (c) Graph the function v and describe its behavior.

Solution

(a) Since $v(0) = 10,000/(5 + 1245e^0) = 10,000/1250 = 8$, we conclude that 8 people initially have the disease.

(b) Using a calculator, we evaluate $v(1)$, $v(2)$, and $v(5)$, and then round off to obtain the following values.

Days	Infected people
1	21
2	54
5	678

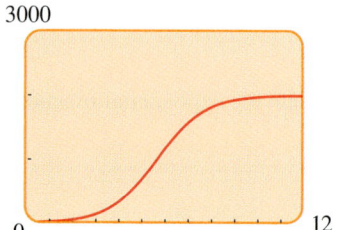

Figure 8

$$v(t) = \frac{10,000}{5 + 1245e^{-0.97t}}$$

(c) From the graph in Figure 8, we see that the number of infected people first rises slowly; then rises quickly between day 3 and day 8, and then levels off when about 2000 people are infected. ∎

The graph in Figure 8 is called a *logistic curve* or a *logistic growth model*. Curves like it occur frequently in the study of population growth. (See Exercises 69–72.)

Compound Interest

Exponential functions occur in calculating compound interest. If an amount of money P, called the **principal**, is invested at an interest rate i per time period, then after one time period the interest is Pi, and the amount A of money is

$$A = P + Pi = P(1 + i)$$

If the interest is reinvested, then the new principal is $P(1 + i)$, and the amount after another time period is $A = P(1 + i)(1 + i) = P(1 + i)^2$. Similarly, after a third time period the amount is $A = P(1 + i)^3$. In general, after k periods the amount is

$$A = P(1 + i)^k$$

Notice that this is an exponential function with base $1 + i$.

If the annual interest rate is r and if interest is compounded n times per year, then in each time period the interest rate is $i = r/n$, and there are nt time periods in t years. This leads to the following formula for the amount after t years.

Compound Interest

Compound interest is calculated by the formula

$$A(t) = P\left(1 + \frac{r}{n}\right)^{nt}$$

where $A(t) =$ amount after t years

$P =$ principal

$r =$ interest rate per year

$n =$ number of times interest is compounded per year

$t =$ number of years

r is often referred to as the nominal annual interest rate.

Example 9   Calculating Compound Interest

A sum of $1000 is invested at an interest rate of 12% per year. Find the amounts in the account after 3 years if interest is compounded annually, semiannually, quarterly, monthly, and daily.

Solution   We use the compound interest formula with $P = \$1000$, $r = 0.12$, and $t = 3$.

Compounding	n	Amount after 3 years
Annual	1	$1000\left(1 + \dfrac{0.12}{1}\right)^{1(3)} = \$1404.93$
Semiannual	2	$1000\left(1 + \dfrac{0.12}{2}\right)^{2(3)} = \$1418.52$
Quarterly	4	$1000\left(1 + \dfrac{0.12}{4}\right)^{4(3)} = \$1425.76$
Monthly	12	$1000\left(1 + \dfrac{0.12}{12}\right)^{12(3)} = \$1430.77$
Daily	365	$1000\left(1 + \dfrac{0.12}{365}\right)^{365(3)} = \$1433.24$

■

We see from Example 9 that the interest paid increases as the number of compounding periods n increases. Let's see what happens as n increases indefinitely. If we let $m = n/r$, then

$$A(t) = P\left(1 + \frac{r}{n}\right)^{nt} = P\left[\left(1 + \frac{r}{n}\right)^{n/r}\right]^{rt} = P\left[\left(1 + \frac{1}{m}\right)^{m}\right]^{rt}$$

Recall that as m becomes large, the quantity $(1 + 1/m)^m$ approaches the number e. Thus, the amount approaches $A = Pe^{rt}$. This expression gives the amount when the interest is compounded at "every instant."

Continuously Compounded Interest

Continuously compounded interest is calculated by the formula

$$A(t) = Pe^{rt}$$

where $A(t) =$ amount after t years

$P =$ principal

$r =$ interest rate per year

$t =$ number of years

Example 10   Calculating Continuously Compounded Interest

Find the amount after 3 years if $1000 is invested at an interest rate of 12% per year, compounded continuously.

Solution   We use the formula for continuously compounded interest with $P = \$1000$, $r = 0.12$, and $t = 3$ to get

$$A(3) = 1000e^{(0.12)3} = 1000e^{0.36} = \$1433.33$$

Compare this amount with the amounts in Example 9. ∎

4.1   Exercises

1–4 ■ Use a calculator to evaluate the function at the indicated values. Round your answers to three decimals.

1. $f(x) = 4^x$; $f(0.5), f(\sqrt{2}), f(\pi), f(\frac{1}{3})$

2. $f(x) = 3^{x+1}$; $f(-1.5), f(\sqrt{3}), f(e), f(-\frac{5}{4})$

3. $g(x) = (\frac{2}{3})^{x-1}$; $g(1.3), g(\sqrt{5}), g(2\pi), g(-\frac{1}{2})$

4. $g(x) = (\frac{3}{4})^{2x}$; $g(0.7), g(\sqrt{7}/2), g(1/\pi), g(\frac{2}{3})$

5–10 ■ Sketch the graph of the function by making a table of values. Use a calculator if necessary.

5. $f(x) = 2^x$ **6.** $g(x) = 8^x$

7. $f(x) = (\frac{1}{3})^x$ **8.** $h(x) = (1.1)^x$

9. $g(x) = 3e^x$ **10.** $h(x) = 2e^{-0.5x}$

11–14 ■ Graph both functions on one set of axes.

11. $f(x) = 2^x$ and $g(x) = 2^{-x}$

12. $f(x) = 3^{-x}$ and $g(x) = (\frac{1}{3})^x$

13. $f(x) = 4^x$ and $g(x) = 7^x$

14. $f(x) = (\frac{2}{3})^x$ and $g(x) = (\frac{4}{3})^x$

15–18 ■ Find the exponential function $f(x) = a^x$ whose graph is given.

15.

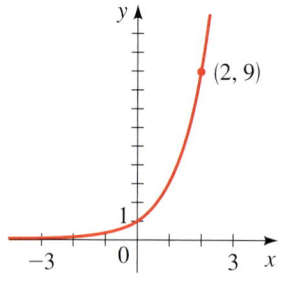

16.

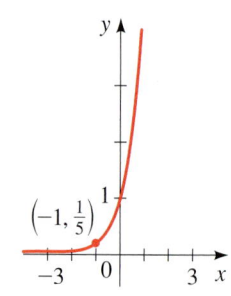

17.

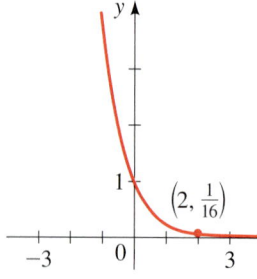

18.

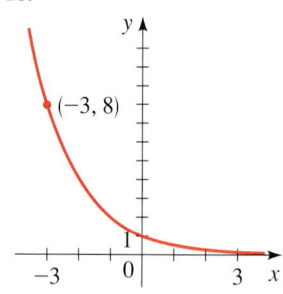

19–24 ■ Match the exponential function with one of the graphs labeled I–VI.

19. $f(x) = 5^x$

20. $f(x) = -5^x$

21. $f(x) = 5^{-x}$

22. $f(x) = 5^x + 3$

23. $f(x) = 5^{x-3}$

24. $f(x) = 5^{x+1} - 4$

I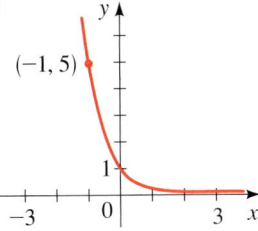

II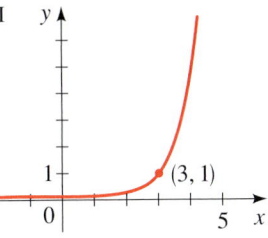

III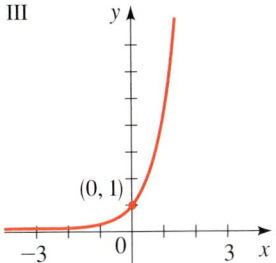

IV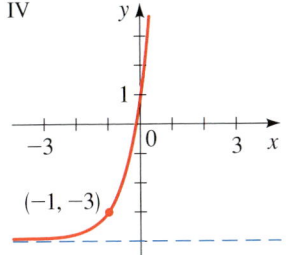

V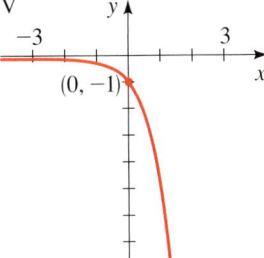

VI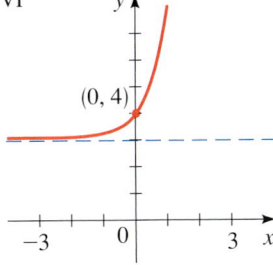

25–38 ■ Graph the function, not by plotting points, but by starting from the graphs in Figures 2 and 5. State the domain, range, and asymptote.

25. $f(x) = -3^x$

26. $f(x) = 10^{-x}$

27. $g(x) = 2^x - 3$

28. $g(x) = 2^{x-3}$

29. $h(x) = 4 + \left(\frac{1}{2}\right)^x$

30. $h(x) = 6 - 3^x$

31. $f(x) = 10^{x+3}$

32. $f(x) = -\left(\frac{1}{5}\right)^x$

33. $f(x) = -e^x$

34. $y = 1 - e^x$

35. $y = e^{-x} - 1$

36. $f(x) = -e^{-x}$

37. $f(x) = e^{x-2}$

38. $y = e^{x-3} + 4$

39–40 ■ Find the function of the form $f(x) = Ca^x$ whose graph is given.

39. **40.**

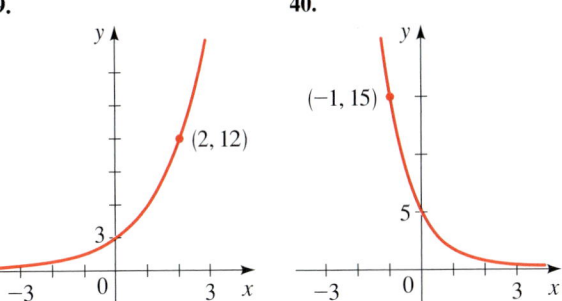

41. (a) Sketch the graphs of $f(x) = 2^x$ and $g(x) = 3(2^x)$.

(b) How are the graphs related?

42. (a) Sketch the graphs of $f(x) = 9^{x/2}$ and $g(x) = 3^x$.

(b) Use the Laws of Exponents to explain the relationship between these graphs.

43. If $f(x) = 10^x$, show that

$$\frac{f(x + h) - f(x)}{h} = 10^x\left(\frac{10^h - 1}{h}\right)$$

44. Compare the functions $f(x) = x^3$ and $g(x) = 3^x$ by evaluating both of them for $x = 0, 1, 2, 3, 4, 5, 6, 7, 8, 9, 10, 15,$ and 20. Then draw the graphs of f and g on the same set of axes.

45. The *hyperbolic cosine function* is defined by

$$\cosh(x) = \frac{e^x + e^{-x}}{2}$$

Sketch the graphs of the functions $y = \frac{1}{2}e^x$ and $y = \frac{1}{2}e^{-x}$ on the same axes and use graphical addition (see Section 2.7) to sketch the graph of $y = \cosh(x)$.

46. The *hyperbolic sine function* is defined by

$$\sinh(x) = \frac{e^x - e^{-x}}{2}$$

Sketch the graph of this function using graphical addition as in Exercise 45.

47–50 ■ Use the definitions in Exercises 45 and 46 to prove the identity.

47. $\cosh(-x) = \cosh(x)$

48. $\sinh(-x) = -\sinh(x)$

49. $[\cosh(x)]^2 - [\sinh(x)]^2 = 1$

50. $\sinh(x + y) = \sinh(x)\cosh(y) + \cosh(x)\sinh(y)$

51. (a) Compare the rates of growth of the functions $f(x) = 2^x$ and $g(x) = x^5$ by drawing the graphs of both functions in the following viewing rectangles.

    (i) $[0, 5]$ by $[0, 20]$

    (ii) $[0, 25]$ by $[0, 10^7]$

    (iii) $[0, 50]$ by $[0, 10^8]$

(b) Find the solutions of the equation $2^x = x^5$, correct to one decimal place.

52. (a) Compare the rates of growth of the functions $f(x) = 3^x$ and $g(x) = x^4$ by drawing the graphs of both functions in the following viewing rectangles:

    (i) $[-4, 4]$ by $[0, 20]$ (ii) $[0, 10]$ by $[0, 5000]$

    (iii) $[0, 20]$ by $[0, 10^5]$

(b) Find the solutions of the equation $3^x = x^4$, correct to two decimal places.

53–54 ■ Draw graphs of the given family of functions for $c = 0.25, 0.5, 1, 2, 4$. How are the graphs related?

53. $f(x) = c2^x$ **54.** $f(x) = 2^{cx}$

55. Illustrate the definition of the number e by graphing the curve $y = (1 + 1/x)^x$ and the line $y = e$ on the same screen using the viewing rectangle $[0, 40]$ by $[0, 4]$.

56. Investigate the behavior of the function

$$f(x) = \left(1 - \frac{1}{x}\right)^x$$

as $x \to \infty$ by graphing f and the line $y = 1/e$ on the same screen using the viewing rectangle $[0, 20]$ by $[0, 1]$.

57. (a) Draw the graphs of the family of functions

$$f(x) = \frac{a}{2}(e^{x/a} + e^{-x/a})$$

    for $a = 0.5, 1, 1.5,$ and 2.

(b) How does a larger value of a affect the graph?

58–59 ■ Graph the function and comment on vertical and horizontal asymptotes.

58. $y = 2^{1/x}$ **59.** $y = \dfrac{e^x}{x}$

60–61 ■ Find the local maximum and minimum values of the function and the value of x at which each occurs. State each answer correct to two decimal places.

60. $g(x) = x^x$ $(x > 0)$ **61.** $g(x) = e^x + e^{-3x}$

62–63 ■ Find, correct to two decimal places, (a) the intervals on which the function is increasing or decreasing, and (b) the range of the function.

62. $y = 10^{x-x^2}$ **63.** $y = xe^{-x}$

Applications

64. Medical Drugs When a certain medical drug is administered to a patient, the number of milligrams remaining in the patient's bloodstream after t hours is modeled by

$$D(t) = 50e^{-0.2t}$$

How many milligrams of the drug remain in the patient's bloodstream after 3 hours?

65. Radioactive Decay A radioactive substance decays in such a way that the amount of mass remaining after t days is given by the function

$$m(t) = 13e^{-0.015t}$$

where $m(t)$ is measured in kilograms.

(a) Find the mass at time $t = 0$.

(b) How much of the mass remains after 45 days?

66. Radioactive Decay Radioactive iodine is used by doctors as a tracer in diagnosing certain thyroid gland disorders. This type of iodine decays in such a way that the mass remaining after t days is given by the function

$$m(t) = 6e^{-0.087t}$$

where $m(t)$ is measured in grams.

(a) Find the mass at time $t = 0$.

(b) How much of the mass remains after 20 days?

67. Sky Diving A sky diver jumps from a reasonable height above the ground. The air resistance she experiences is proportional to her velocity, and the constant of proportionality is 0.2. It can be shown that the downward velocity of the sky diver at time t is given by

$$v(t) = 80(1 - e^{-0.2t})$$

where t is measured in seconds and $v(t)$ is measured in feet per second (ft/s).

(a) Find the initial velocity of the sky diver.

(b) Find the velocity after 5 s and after 10 s.

(c) Draw a graph of the velocity function $v(t)$.

(d) The maximum velocity of a falling object with wind resistance is called its *terminal velocity*. From the graph in part (c) find the terminal velocity of this sky diver.

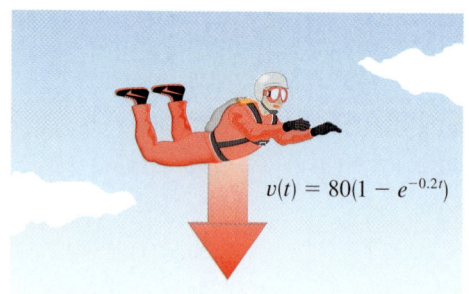

$$v(t) = 80(1 - e^{-0.2t})$$

68. Mixtures and Concentrations A 50-gallon barrel is filled completely with pure water. Salt water with a concentration of 0.3 lb/gal is then pumped into the barrel, and the resulting mixture overflows at the same rate. The amount of salt in the barrel at time t is given by

$$Q(t) = 15(1 - e^{-0.04t})$$

where t is measured in minutes and $Q(t)$ is measured in pounds.

(a) How much salt is in the barrel after 5 min?

(b) How much salt is in the barrel after 10 min?

 (c) Draw a graph of the function $Q(t)$.

 (d) Use the graph in part (c) to determine the value that the amount of salt in the barrel approaches as t becomes large. Is this what you would expect?

$$Q(t) = 15(1 - e^{-0.04t})$$

69. Logistic Growth Animal populations are not capable of unrestricted growth because of limited habitat and food supplies. Under such conditions the population follows a *logistic growth model*

$$P(t) = \frac{d}{1 + ke^{-ct}}$$

where c, d, and k are positive constants. For a certain fish population in a small pond $d = 1200$, $k = 11$, $c = 0.2$, and t is measured in years. The fish were introduced into the pond at time $t = 0$.

(a) How many fish were originally put in the pond?

(b) Find the population after 10, 20, and 30 years.

(c) Evaluate $P(t)$ for large values of t. What value does the population approach as $t \to \infty$? Does the graph shown confirm your calculations?

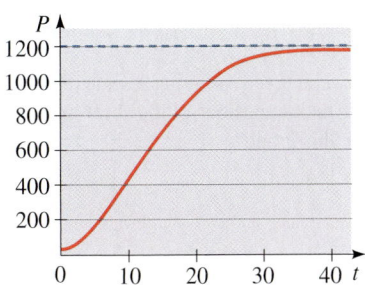

70. Bird Population The population of a certain species of bird is limited by the type of habitat required for nesting. The population behaves according to the logistic growth model

$$n(t) = \frac{5600}{0.5 + 27.5e^{-0.044t}}$$

where t is measured in years.

(a) Find the initial bird population.

(b) Draw a graph of the function $n(t)$.

(c) What size does the population approach as time goes on?

71. Tree Diameter For a certain type of tree the diameter D (in feet) depends on the tree's age t (in years) according to the logistic growth model

$$D(t) = \frac{5.4}{1 + 2.9e^{-0.01t}}$$

Find the diameter of a 20-year-old tree.

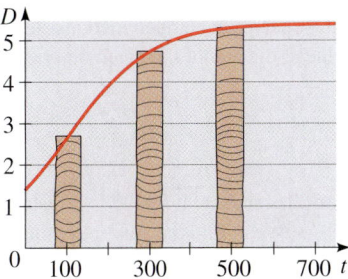

72. Rabbit Population Assume that a population of rabbits behaves according to the logistic growth model

$$n(t) = \frac{300}{0.05 + \left(\dfrac{300}{n_0} - 0.05\right)e^{-0.55t}}$$

where n_0 is the initial rabbit population.

(a) If the initial population is 50 rabbits, what will the population be after 12 years?

(b) Draw graphs of the function $n(t)$ for $n_0 = 50, 500, 2000, 8000,$ and $12{,}000$ in the viewing rectangle $[0, 15]$ by $[0, 12{,}000]$.

(c) From the graphs in part (b), observe that, regardless of the initial population, the rabbit population seems to approach a certain number as time goes on. What is that number? (This is the number of rabbits that the island can support.)

73–74 ■ **Compound Interest**   An investment of $5000 is deposited into an account in which interest is compounded monthly. Complete the table by filling in the amounts to which the investment grows at the indicated times or interest rates.

73. $r = 4\%$

Time (years)	Amount
1	
2	
3	
4	
5	
6	

74. $t = 5$ years

Rate per year	Amount
1%	
2%	
3%	
4%	
5%	
6%	

75. Compound Interest   If $10,000 is invested at an interest rate of 10% per year, compounded semiannually, find the value of the investment after the given number of years.

(a) 5 years

(b) 10 years

(c) 15 years

76. Compound Interest   If $4000 is borrowed at a rate of 16% interest per year, compounded quarterly, find the amount due at the end of the given number of years.

(a) 4 years

(b) 6 years

(c) 8 years

77. Compound Interest   If $3000 is invested at an interest rate of 9% per year, find the amount of the investment at the end of 5 years for the following compounding methods.

(a) Annual

(b) Semiannual

(c) Monthly

(d) Weekly

(e) Daily

(f) Hourly

(g) Continuously

78. Compound Interest   If $4000 is invested in an account for which interest is compounded quarterly, find the amount of the investment at the end of 5 years for the following interest rates.

(a) 6%

(b) $6\frac{1}{2}\%$

(c) 7%

(d) 8%

79. Compound Interest   Which of the given interest rates and compounding periods would provide the best investment?

(i) $8\frac{1}{2}\%$ per year, compounded semiannually

(ii) $8\frac{1}{4}\%$ per year, compounded quarterly

(iii) 8% per year, compounded continuously

80. Compound Interest   Which of the given interest rates and compounding periods would provide the better investment?

(i) $9\frac{1}{4}\%$ per year, compounded semiannually

(ii) 9% per year, compounded continuously

81. Present Value   The **present value** of a sum of money is the amount that must be invested now, at a given rate of interest, to produce the desired sum at a later date.

(a) Find the present value of $10,000 if interest is paid at a rate of 9% per year, compounded semiannually, for 3 years.

(b) Find the present value of $100,000 if interest is paid at a rate of 8% per year, compounded monthly, for 5 years.

82. Investment   A sum of $5000 is invested at an interest rate of 9% per year, compounded semiannually.

(a) Find the value $A(t)$ of the investment after t years.

(b) Draw a graph of $A(t)$.

(c) Use the graph of $A(t)$ to determine when this investment will amount to $25,000.

Discovery • Discussion

83. Growth of an Exponential Function   Suppose you are offered a job that lasts one month, and you are to be very well paid. Which of the following methods of payment is more profitable for you?

(a) One million dollars at the end of the month

(b) Two cents on the first day of the month, 4 cents on the second day, 8 cents on the third day, and, in general, 2^n cents on the nth day

84. The Height of the Graph of an Exponential Function   Your mathematics instructor asks you to sketch a graph of the exponential function

$$f(x) = 2^x$$

for x between 0 and 40, using a scale of 10 units to one inch. What are the dimensions of the sheet of paper you will need to sketch this graph?

DISCOVERY PROJECT

Exponential Explosion

To help us grasp just how explosive exponential growth is, let's try a thought experiment.

Suppose you put a penny in your piggy bank today, two pennies tomorrow, four pennies the next day, and so on, doubling the number of pennies you add to the bank each day (see the table). How many pennies will you put in your piggy bank on day 30? The answer is 2^{30} pennies. That's simple, but can you guess how many dollars that is? 2^{30} pennies is more than 10 million dollars!

Day	Pennies
0	1
1	2
2	4
3	8
4	16
⋮	⋮
n	2^n
⋮	⋮

As you can see, the exponential function $f(x) = 2^x$ grows extremely fast. This is the principle behind atomic explosions. An atom splits releasing two neutrons, which cause two atoms to split, each releasing two neutrons, causing four atoms to split, and so on. At the nth stage 2^n atoms split—an exponential explosion!

Populations also grow exponentially. Let's see what this means for a type of bacteria that splits every minute. Suppose that at 12:00 noon a single bacterium colonizes a discarded food can. The bacterium and his descendants are all happy, but they fear the time when the can is completely full of bacteria—doomsday.

1. How many bacteria are in the can at 12:05? At 12:10?

2. The can is completely full of bacteria at 1:00 P.M. At what time was the can only half full of bacteria?

3. When the can is exactly half full, the president of the bacteria colony reassures his constituents that doomsday is far away—after all, there is as much room left in the can as has been used in the entire previous history of the colony. Is the president correct? How much time is left before doomsday?

4. When the can is one-quarter full, how much time remains till doomsday?

5. A wise bacterium decides to start a new colony in another can and slow down splitting time to 2 minutes. How much time does this new colony have?

4.2   Logarithmic Functions

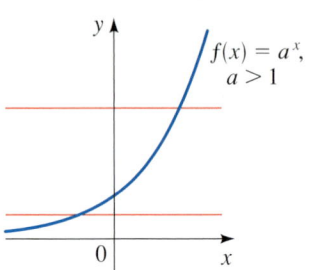

Figure 1
$f(x) = a^x$ is one-to-one

In this section we study the inverse of exponential functions.

Logarithmic Functions

Every exponential function $f(x) = a^x$, with $a > 0$ and $a \neq 1$, is a one-to-one function by the Horizontal Line Test (see Figure 1 for the case $a > 1$) and therefore has an inverse function. The inverse function f^{-1} is called the *logarithmic function with base a* and is denoted by $\log_a$. Recall from Section 2.8 that f^{-1} is defined by

$$f^{-1}(x) = y \quad \Leftrightarrow \quad f(y) = x$$

This leads to the following definition of the logarithmic function.

> **Definition of the Logarithmic Function**
>
> Let a be a positive number with $a \neq 1$. The **logarithmic function with base a**, denoted by $\log_a$, is defined by
>
> $$\log_a x = y \quad \Leftrightarrow \quad a^y = x$$
>
> So, $\log_a x$ is the *exponent* to which the base a must be raised to give x.

We read $\log_a x = y$ as "log base a of x is y."

By tradition, the name of the logarithmic function is $\log_a$, not just a single letter. Also, we usually omit the parentheses in the function notation and write
$$\log_a(x) = \log_a x$$

When we use the definition of logarithms to switch back and forth between the **logarithmic form** $\log_a x = y$ and the **exponential form** $a^y = x$, it's helpful to notice that, in both forms, the base is the same:

Logarithmic form

Exponent

$$\log_a x = y$$

Base

Exponential form

Exponent

$$a^y = x$$

Base

Example 1   Logarithmic and Exponential Forms

The logarithmic and exponential forms are equivalent equations—if one is true, then so is the other. So, we can switch from one form to the other as in the following illustrations.

Logarithmic form	Exponential form
$\log_{10} 100{,}000 = 5$	$10^5 = 100{,}000$
$\log_2 8 = 3$	$2^3 = 8$
$\log_2\left(\frac{1}{8}\right) = -3$	$2^{-3} = \frac{1}{8}$
$\log_5 s = r$	$5^r = s$

■

It's important to understand that $\log_a x$ is an *exponent*. For example, the numbers in the right column of the table in the margin are the logarithms (base 10) of the

x	$\log_{10} x$
10^4	4
10^3	3
10^2	2
10	1
1	0
10^{-1}	-1
10^{-2}	-2
10^{-3}	-3
10^{-4}	-4

Inverse Function Property:

$$f^{-1}(f(x)) = x$$
$$f(f^{-1}(x)) = x$$

numbers in the left column. This is the case for all bases, as the following example illustrates.

Example 2   Evaluating Logarithms

(a) $\log_{10} 1000 = 3$ because $10^3 = 1000$

(b) $\log_2 32 = 5$ because $2^5 = 32$

(c) $\log_{10} 0.1 = -1$ because $10^{-1} = 0.1$

(d) $\log_{16} 4 = \frac{1}{2}$ because $16^{1/2} = 4$ ∎

When we apply the Inverse Function Property described on page 227 to $f(x) = a^x$ and $f^{-1}(x) = \log_a x$, we get

$$\log_a(a^x) = x \qquad x \in \mathbb{R}$$
$$a^{\log_a x} = x \qquad x > 0$$

We list these and other properties of logarithms discussed in this section.

Properties of Logarithms

Property	Reason
1. $\log_a 1 = 0$	We must raise a to the power 0 to get 1.
2. $\log_a a = 1$	We must raise a to the power 1 to get a.
3. $\log_a a^x = x$	We must raise a to the power x to get a^x.
4. $a^{\log_a x} = x$	$\log_a x$ is the power to which a must be raised to get x.

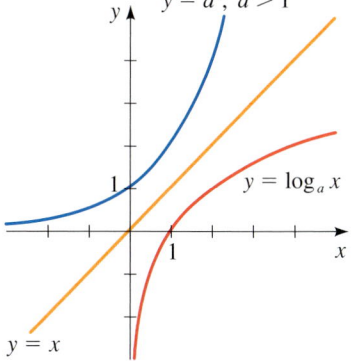

$y = a^x,\ a > 1$

$y = \log_a x$

$y = x$

Figure 2

Graph of the logarithmic function $f(x) = \log_a x$

Arrow notation is explained on page 301.

Example 3   Applying Properties of Logarithms

We illustrate the properties of logarithms when the base is 5.

$\log_5 1 = 0$ *Property 1* $\log_5 5 = 1$ *Property 2*

$\log_5 5^8 = 8$ *Property 3* $5^{\log_5 12} = 12$ *Property 4*   ∎

Graphs of Logarithmic Functions

Recall that if a one-to-one function f has domain A and range B, then its inverse function f^{-1} has domain B and range A. Since the exponential function $f(x) = a^x$ with $a \neq 1$ has domain $\mathbb{R}$ and range $(0, \infty)$, we conclude that its inverse function, $f^{-1}(x) = \log_a x$, has domain $(0, \infty)$ and range $\mathbb{R}$.

The graph of $f^{-1}(x) = \log_a x$ is obtained by reflecting the graph of $f(x) = a^x$ in the line $y = x$. Figure 2 shows the case $a > 1$. The fact that $y = a^x$ (for $a > 1$) is a very rapidly increasing function for $x > 0$ implies that $y = \log_a x$ is a very slowly increasing function for $x > 1$ (see Exercise 84).

Since $\log_a 1 = 0$, the x-intercept of the function $y = \log_a x$ is 1. The y-axis is a vertical asymptote of $y = \log_a x$ because $\log_a x \to -\infty$ as $x \to 0^+$.

Example 4   Graphing a Logarithmic Function by Plotting Points

Sketch the graph of $f(x) = \log_2 x$.

Solution   To make a table of values, we choose the x-values to be powers of 2 so that we can easily find their logarithms. We plot these points and connect them with a smooth curve as in Figure 3.

x	$\log_2 x$
2^3	3
2^2	2
2	1
1	0
2^{-1}	-1
2^{-2}	-2
2^{-3}	-3
2^{-4}	-4

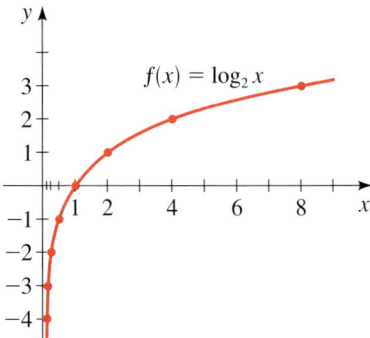

Figure 3

Figure 4 shows the graphs of the family of logarithmic functions with bases 2, 3, 5, and 10. These graphs are drawn by reflecting the graphs of $y = 2^x$, $y = 3^x$, $y = 5^x$, and $y = 10^x$ (see Figure 2 in Section 4.1) in the line $y = x$. We can also plot points as an aid to sketching these graphs, as illustrated in Example 4.

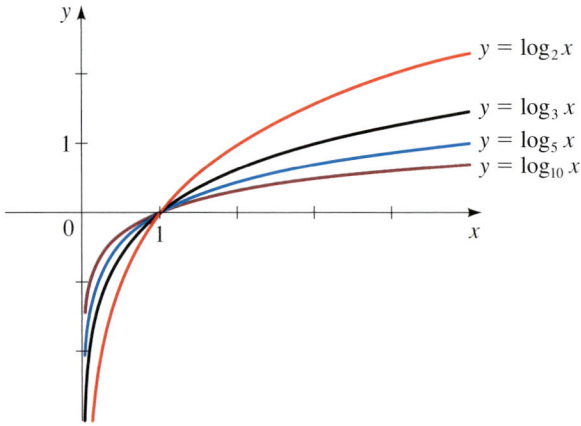

Figure 4

A family of logarithmic functions

In the next two examples we graph logarithmic functions by starting with the basic graphs in Figure 4 and using the transformations of Section 2.4.

be programmed into a computer to give a picture of how a person's appearance changes over time. These pictures aid law enforcement agencies in locating missing persons.

Example 5 Reflecting Graphs of Logarithmic Functions

Sketch the graph of each function.

(a) $g(x) = -\log_2 x$ (b) $h(x) = \log_2(-x)$

Solution

(a) We start with the graph of $f(x) = \log_2 x$ and reflect in the x-axis to get the graph of $g(x) = -\log_2 x$ in Figure 5(a).

(b) We start with the graph of $f(x) = \log_2 x$ and reflect in the y-axis to get the graph of $h(x) = \log_2(-x)$ in Figure 5(b).

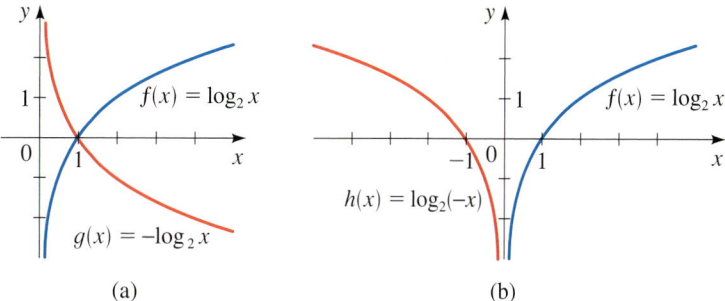

Figure 5   (a)   (b)   ∎

Example 6 Shifting Graphs of Logarithmic Functions

Find the domain of each function, and sketch the graph.

(a) $g(x) = 2 + \log_5 x$ (b) $h(x) = \log_{10}(x - 3)$

Solution

(a) The graph of g is obtained from the graph of $f(x) = \log_5 x$ (Figure 4) by shifting upward 2 units (see Figure 6). The domain of f is $(0, \infty)$.

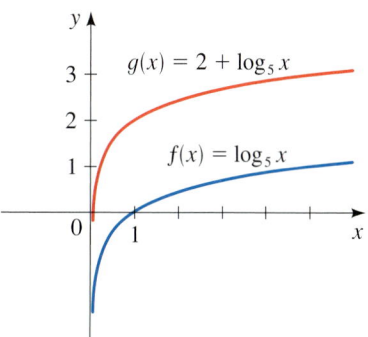

Figure 6

(b) The graph of h is obtained from the graph of $f(x) = \log_{10} x$ (Figure 4) by shifting to the right 3 units (see Figure 7 on the next page). The line $x = 3$ is a vertical asymptote. Since $\log_{10} x$ is defined only when $x > 0$, the domain

John Napier (1550–1617) was a Scottish landowner for whom mathematics was a hobby. We know him today because of his key invention—logarithms, which he published in 1614 under the title *A Description of the Marvelous Rule of Logarithms*. In Napier's time, logarithms were used exclusively for simplifying complicated calculations. For example, to multiply two large numbers we would write them as powers of 10. The exponents are simply the logarithms of the numbers. For instance,

4532×57783

$\approx 10^{3.65629} \times 10^{4.76180}$

$= 10^{8.41809}$

$\approx 261{,}872{,}564$

The idea is that multiplying powers of 10 is easy (we simply add their exponents). Napier produced extensive tables giving the logarithms (or exponents) of numbers. Since the advent of calculators and computers, logarithms are no longer used for this purpose. The logarithmic functions, however, have found many applications, some of which are described in this chapter.

　　Napier wrote on many topics. One of his most colorful works is a book entitled *A Plaine Discovery of the Whole Revelation of Saint John*, in which he predicted that the world would end in the year 1700.

of $h(x) = \log_{10}(x - 3)$ is

$$\{x \mid x - 3 > 0\} = \{x \mid x > 3\} = (3, \infty)$$

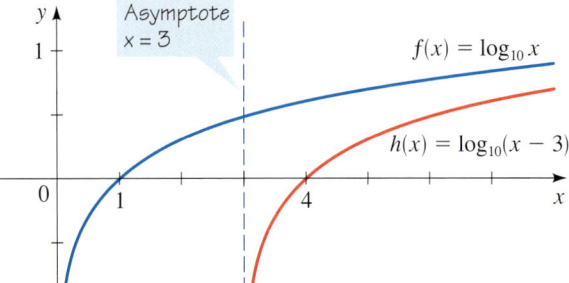

Figure 7

Common Logarithms

We now study logarithms with base 10.

Common Logarithm

The logarithm with base 10 is called the **common logarithm** and is denoted by omitting the base:

$$\log x = \log_{10} x$$

From the definition of logarithms we can easily find that

$$\log 10 = 1 \quad \text{and} \quad \log 100 = 2$$

But how do we find log 50? We need to find the exponent y such that $10^y = 50$. Clearly, 1 is too small and 2 is too large. So

$$1 < \log 50 < 2$$

To get a better approximation, we can experiment to find a power of 10 closer to 50. Fortunately, scientific calculators are equipped with a $\boxed{\text{LOG}}$ key that directly gives values of common logarithms.

Example 7   Evaluating Common Logarithms

Use a calculator to find appropriate values of $f(x) = \log x$ and use the values to sketch the graph.

Solution   We make a table of values, using a calculator to evaluate the function at those values of x that are not powers of 10. We plot those points and connect them by a smooth curve as in Figure 8.

x	$\log x$
0.01	-2
0.1	-1
0.5	-0.301
1	0
4	0.602
5	0.699
10	1

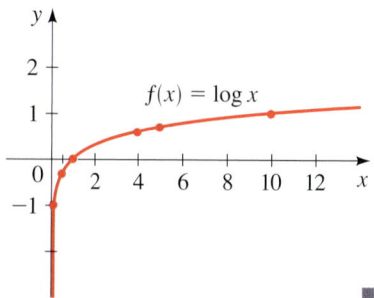

Figure 8

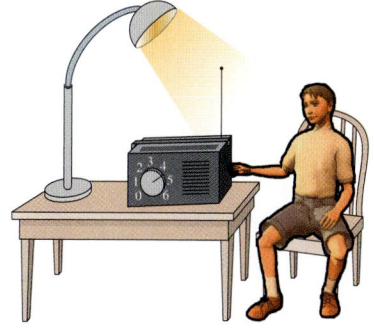

Human response to sound and
light intensity is logarithmic.

Scientists model human response to stimuli (such as sound, light, or pressure) using logarithmic functions. For example, the intensity of a sound must be increased manyfold before we "feel" that the loudness has simply doubled. The psychologist Gustav Fechner formulated the law as

$$S = k \log\left(\frac{I}{I_0}\right)$$

where S is the subjective intensity of the stimulus, I is the physical intensity of the stimulus, I_0 stands for the threshold physical intensity, and k is a constant that is different for each sensory stimulus.

Example 8   Common Logarithms and Sound

We study the decibel scale in more detail in Section 4.5.

The perception of the loudness B (in decibels, dB) of a sound with physical intensity I (in W/m^2) is given by

$$B = 10 \log\left(\frac{I}{I_0}\right)$$

where I_0 is the physical intensity of a barely audible sound. Find the decibel level (loudness) of a sound whose physical intensity I is 100 times that of I_0.

Solution   We find the decibel level B by using the fact that $I = 100I_0$.

$$B = 10 \log\left(\frac{I}{I_0}\right) \qquad \text{Definition of } B$$

$$= 10 \log\left(\frac{100I_0}{I_0}\right) \qquad I = 100I_0$$

$$= 10 \log 100 \qquad \text{Cancel } I_0$$

$$= 10 \cdot 2 = 20 \qquad \text{Definition of log}$$

The loudness of the sound is 20 dB.   ■

Natural Logarithms

Of all possible bases a for logarithms, it turns out that the most convenient choice for the purposes of calculus is the number e, which we defined in Section 4.1.

Natural Logarithm

The notation ln is an abbreviation for the Latin name *logarithmus naturalis*.

The logarithm with base e is called the **natural logarithm** and is denoted by **ln**:

$$\ln x = \log_e x$$

The natural logarithmic function $y = \ln x$ is the inverse function of the exponential function $y = e^x$. Both functions are graphed in Figure 9. By the definition of inverse functions we have

$$\ln x = y \iff e^y = x$$

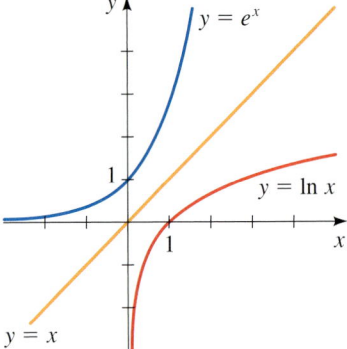

Figure 9

Graph of the natural logarithmic function

If we substitute $a = e$ and write "ln" for "$\log_e$" in the properties of logarithms mentioned earlier, we obtain the following properties of natural logarithms.

Properties of Natural Logarithms

Property	Reason
1. $\ln 1 = 0$	We must raise e to the power 0 to get 1.
2. $\ln e = 1$	We must raise e to the power 1 to get e.
3. $\ln e^x = x$	We must raise e to the power x to get e^x.
4. $e^{\ln x} = x$	$\ln x$ is the power to which e must be raised to get x.

Calculators are equipped with an $\boxed{\text{LN}}$ key that directly gives the values of natural logarithms.

Example 9   Evaluating the Natural Logarithm Function

(a) $\ln e^8 = 8$ Definition of natural logarithm

(b) $\ln\left(\dfrac{1}{e^2}\right) = \ln e^{-2} = -2$ Definition of natural logarithm

(c) $\ln 5 \approx 1.609$ Use $\boxed{\text{LN}}$ key on calculator   ∎

Example 10 Finding the Domain of a Logarithmic Function

Find the domain of the function $f(x) = \ln(4 - x^2)$.

Solution As with any logarithmic function, $\ln x$ is defined when $x > 0$. Thus, the domain of f is

$$\{x \mid 4 - x^2 > 0\} = \{x \mid x^2 < 4\} = \{x \mid |x| < 2\}$$
$$= \{x \mid -2 < x < 2\} = (-2, 2) \qquad \blacksquare$$

Example 11 Drawing the Graph of a Logarithmic Function

Draw the graph of the function $y = x \ln(4 - x^2)$ and use it to find the asymptotes and local maximum and minimum values.

Solution As in Example 10 the domain of this function is the interval $(-2, 2)$, so we choose the viewing rectangle $[-3, 3]$ by $[-3, 3]$. The graph is shown in Figure 10, and from it we see that the lines $x = -2$ and $x = 2$ are vertical asymptotes.

The function has a local maximum point to the right of $x = 1$ and a local minimum point to the left of $x = -1$. By zooming in and tracing along the graph with the cursor, we find that the local maximum value is approximately 1.13 and this occurs when $x \approx 1.15$. Similarly (or by noticing that the function is odd), we find that the local minimum value is about -1.13, and it occurs when $x \approx -1.15$. $\blacksquare$

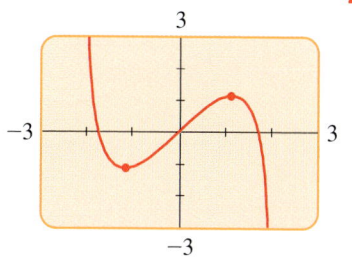

Figure 10

$y = x \ln(4 - x^2)$

4.2 Exercises

1–2 ■ Complete the table by finding the appropriate logarithmic or exponential form of the equation, as in Example 1.

1.

Logarithmic form	Exponential form
$\log_8 8 = 1$	
$\log_8 64 = 2$	
	$8^{2/3} = 4$
	$8^3 = 512$
$\log_8\left(\tfrac{1}{8}\right) = -1$	
	$8^{-2} = \tfrac{1}{64}$

2.

Logarithmic form	Exponential form
	$4^3 = 64$
$\log_4 2 = \tfrac{1}{2}$	
	$4^{3/2} = 8$
$\log_4\left(\tfrac{1}{16}\right) = -2$	
$\log_4\left(\tfrac{1}{2}\right) = -\tfrac{1}{2}$	
	$4^{-5/2} = \tfrac{1}{32}$

3–8 ■ Express the equation in exponential form.

3. (a) $\log_5 25 = 2$ (b) $\log_5 1 = 0$

4. (a) $\log_{10} 0.1 = -1$ (b) $\log_8 512 = 3$

5. (a) $\log_8 2 = \tfrac{1}{3}$ (b) $\log_2\left(\tfrac{1}{8}\right) = -3$

6. (a) $\log_3 81 = 4$ (b) $\log_8 4 = \tfrac{2}{3}$

7. (a) $\ln 5 = x$ (b) $\ln y = 5$

8. (a) $\ln(x + 1) = 2$ (b) $\ln(x - 1) = 4$

9–14 ■ Express the equation in logarithmic form.

9. (a) $5^3 = 125$ (b) $10^{-4} = 0.0001$

10. (a) $10^3 = 1000$ (b) $81^{1/2} = 9$

11. (a) $8^{-1} = \tfrac{1}{8}$ (b) $2^{-3} = \tfrac{1}{8}$

12. (a) $4^{-3/2} = 0.125$ (b) $7^3 = 343$

13. (a) $e^x = 2$ (b) $e^3 = y$

14. (a) $e^{x+1} = 0.5$ (b) $e^{0.5x} = t$

15–24 ■ Evaluate the expression.

15. (a) $\log_3 3$ (b) $\log_3 1$ (c) $\log_3 3^2$

16. (a) $\log_5 5^4$ (b) $\log_4 64$ (c) $\log_9 9$

17. (a) $\log_6 36$　　**(b)** $\log_9 81$　　**(c)** $\log_7 7^{10}$

18. (a) $\log_2 32$　　**(b)** $\log_8 8^{17}$　　**(c)** $\log_6 1$

19. (a) $\log_3\left(\frac{1}{27}\right)$　　**(b)** $\log_{10} \sqrt{10}$　　**(c)** $\log_5 0.2$

20. (a) $\log_5 125$　　**(b)** $\log_{49} 7$　　**(c)** $\log_9 \sqrt{3}$

21. (a) $2^{\log_2 37}$　　**(b)** $3^{\log_3 8}$　　**(c)** $e^{\ln \sqrt{5}}$

22. (a) $e^{\ln \pi}$　　**(b)** $10^{\log 5}$　　**(c)** $10^{\log 87}$

23. (a) $\log_8 0.25$　　**(b)** $\ln e^4$　　**(c)** $\ln(1/e)$

24. (a) $\log_4 \sqrt{2}$　　**(b)** $\log_4\left(\frac{1}{2}\right)$　　**(c)** $\log_4 8$

25–32 ■ Use the definition of the logarithmic function to find x.

25. (a) $\log_2 x = 5$　　　　**(b)** $\log_2 16 = x$

26. (a) $\log_5 x = 4$　　　　**(b)** $\log_{10} 0.1 = x$

27. (a) $\log_3 243 = x$　　　**(b)** $\log_3 x = 3$

28. (a) $\log_4 2 = x$　　　　**(b)** $\log_4 x = 2$

29. (a) $\log_{10} x = 2$　　　**(b)** $\log_5 x = 2$

30. (a) $\log_x 1000 = 3$　　　**(b)** $\log_x 25 = 2$

31. (a) $\log_x 16 = 4$　　　　**(b)** $\log_x 8 = \frac{3}{2}$

32. (a) $\log_x 6 = \frac{1}{2}$　　　**(b)** $\log_x 3 = \frac{1}{3}$

33–36 ■ Use a calculator to evaluate the expression, correct to four decimal places.

33. (a) $\log 2$　　　**(b)** $\log 35.2$　　　**(c)** $\log\left(\frac{2}{3}\right)$

34. (a) $\log 50$　　　**(b)** $\log \sqrt{2}$　　　**(c)** $\log(3\sqrt{2})$

35. (a) $\ln 5$　　　**(b)** $\ln 25.3$　　　**(c)** $\ln(1 + \sqrt{3})$

36. (a) $\ln 27$　　　**(b)** $\ln 7.39$　　　**(c)** $\ln 54.6$

37–40 ■ Find the function of the form $y = \log_a x$ whose graph is given.

37.
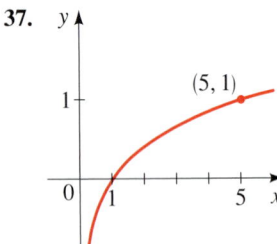

38.
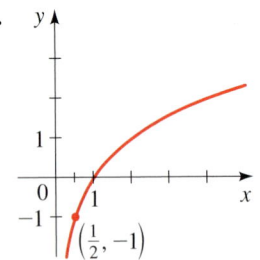

39.
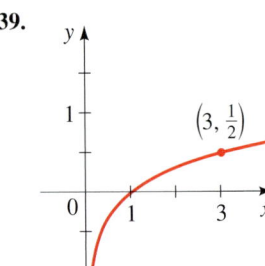

40.
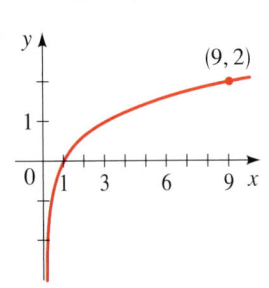

41–46 ■ Match the logarithmic function with one of the graphs labeled I–VI.

41. $f(x) = -\ln x$　　　　**42.** $f(x) = \ln(x - 2)$

43. $f(x) = 2 + \ln x$　　　**44.** $f(x) = \ln(-x)$

45. $f(x) = \ln(2 - x)$　　　**46.** $f(x) = -\ln(-x)$

I
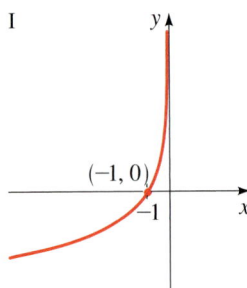

II
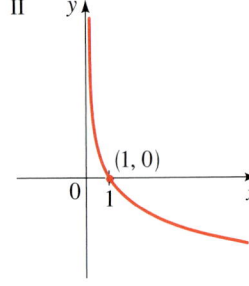

III
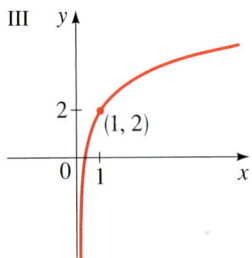

IV
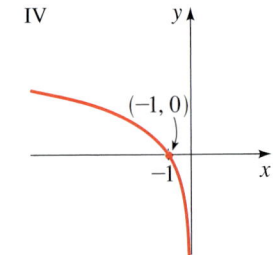

V
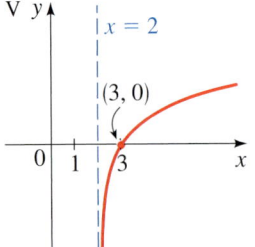

VI
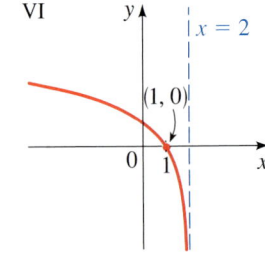

47. Draw the graph of $y = 4^x$, then use it to draw the graph of $y = \log_4 x$.

48. Draw the graph of $y = 3^x$, then use it to draw the graph of $y = \log_3 x$.

49–58 ■ Graph the function, not by plotting points, but by starting from the graphs in Figures 4 and 9. State the domain, range, and asymptote.

49. $f(x) = \log_2(x - 4)$　　**50.** $f(x) = -\log_{10} x$

51. $g(x) = \log_5(-x)$　　　**52.** $g(x) = \ln(x + 2)$

53. $y = 2 + \log_3 x$　　　　**54.** $y = \log_3(x - 1) - 2$

55. $y = 1 - \log_{10} x$　　　**56.** $y = 1 + \ln(-x)$

57. $y = |\ln x|$　　　　　　**58.** $y = \ln |x|$

59–64 ■ Find the domain of the function.

59. $f(x) = \log_{10}(x + 3)$

60. $f(x) = \log_5(8 - 2x)$

61. $g(x) = \log_3(x^2 - 1)$

62. $g(x) = \ln(x - x^2)$

63. $h(x) = \ln x + \ln(2 - x)$

64. $h(x) = \sqrt{x - 2} - \log_5(10 - x)$

 65–70 ■ Draw the graph of the function in a suitable viewing rectangle and use it to find the domain, the asymptotes, and the local maximum and minimum values.

65. $y = \log_{10}(1 - x^2)$

66. $y = \ln(x^2 - x)$

67. $y = x + \ln x$

68. $y = x(\ln x)^2$

69. $y = \dfrac{\ln x}{x}$

70. $y = x \log_{10}(x + 10)$

 71. Compare the rates of growth of the functions $f(x) = \ln x$ and $g(x) = \sqrt{x}$ by drawing their graphs on a common screen using the viewing rectangle $[-1, 30]$ by $[-1, 6]$.

 72. (a) By drawing the graphs of the functions

$$f(x) = 1 + \ln(1 + x) \quad \text{and} \quad g(x) = \sqrt{x}$$

in a suitable viewing rectangle, show that even when a logarithmic function starts out higher than a root function, it is ultimately overtaken by the root function.

(b) Find, correct to two decimal places, the solutions of the equation $\sqrt{x} = 1 + \ln(1 + x)$.

 73–74 ■ A family of functions is given.

(a) Draw graphs of the family for $c = 1, 2, 3,$ and 4.

(b) How are the graphs in part (a) related?

73. $f(x) = \log(cx)$

74. $f(x) = c \log x$

75–76 ■ A function $f(x)$ is given.

(a) Find the domain of the function f.

(b) Find the inverse function of f.

75. $f(x) = \log_2(\log_{10} x)$

76. $f(x) = \ln(\ln(\ln x))$

77. (a) Find the inverse of the function $f(x) = \dfrac{2^x}{1 + 2^x}$.

(b) What is the domain of the inverse function?

Applications

78. Absorption of Light A spectrophotometer measures the concentration of a sample dissolved in water by shining a light through it and recording the amount of light that emerges. In other words, if we know the amount of light absorbed, we can calculate the concentration of the sample.

For a certain substance, the concentration (in moles/liter) is found using the formula

$$C = -2500 \ln\left(\frac{I}{I_0}\right)$$

where I_0 is the intensity of the incident light and I is the intensity of light that emerges. Find the concentration of the substance if the intensity I is 70% of I_0.

79. Carbon Dating The age of an ancient artifact can be determined by the amount of radioactive carbon-14 remaining in it. If D_0 is the original amount of carbon-14 and D is the amount remaining, then the artifact's age A (in years) is given by

$$A = -8267 \ln\left(\frac{D}{D_0}\right)$$

Find the age of an object if the amount D of carbon-14 that remains in the object is 73% of the original amount D_0.

80. Bacteria Colony A certain strain of bacteria divides every three hours. If a colony is started with 50 bacteria, then the time t (in hours) required for the colony to grow to N bacteria is given by

$$t = 3\frac{\log(N/50)}{\log 2}$$

Find the time required for the colony to grow to a million bacteria.

81. Investment The time required to double the amount of an investment at an interest rate r compounded continuously is given by

$$t = \frac{\ln 2}{r}$$

Find the time required to double an investment at 6%, 7%, and 8%.

82. Charging a Battery The rate at which a battery charges is slower the closer the battery is to its maximum charge C_0. The time (in hours) required to charge a fully discharged battery to a charge C is given by

$$t = -k \ln\left(1 - \frac{C}{C_0}\right)$$

where k is a positive constant that depends on the battery. For a certain battery, $k = 0.25$. If this battery is fully discharged, how long will it take to charge to 90% of its maximum charge C_0?

83. Difficulty of a Task The difficulty in "acquiring a target" (such as using your mouse to click on an icon on your computer screen) depends on the distance to the target and the size of the target. According to Fitts's Law, the index of difficulty (ID) is given by

$$ID = \frac{\log(2A/W)}{\log 2}$$

where W is the width of the target and A is the distance to the center of the target. Compare the difficulty of clicking on an icon that is 5 mm wide to one that is 10 mm wide. In each case, assume the mouse is 100 mm from the icon.

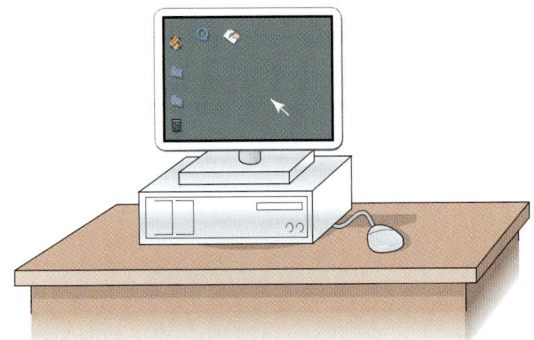

Discovery • Discussion

84. The Height of the Graph of a Logarithmic Function Suppose that the graph of $y = 2^x$ is drawn on a coordinate plane where the unit of measurement is an inch.

(a) Show that at a distance 2 ft to the right of the origin the height of the graph is about 265 mi.

(b) If the graph of $y = \log_2 x$ is drawn on the same set of axes, how far to the right of the origin do we have to go before the height of the curve reaches 2 ft?

85. The Googolplex A **googol** is 10^{100}, and a **googolplex** is 10^{googol}. Find

$$\log(\log(\text{googol})) \quad \text{and} \quad \log(\log(\log(\text{googolplex})))$$

86. Comparing Logarithms Which is larger, $\log_4 17$ or $\log_5 24$? Explain your reasoning.

87. The Number of Digits in an Integer Compare log 1000 to the number of digits in 1000. Do the same for 10,000. How many digits does any number between 1000 and 10,000 have? Between what two values must the common logarithm of such a number lie? Use your observations to explain why the number of digits in any positive integer x is $[\![\log x]\!] + 1$. (The symbol $[\![n]\!]$ is the greatest integer function defined in Section 2.2.) How many digits does the number 2^{100} have?

4.3	**Laws of Logarithms**

In this section we study properties of logarithms. These properties give logarithmic functions a wide range of applications, as we will see in Section 4.5.

Laws of Logarithms

Since logarithms are exponents, the Laws of Exponents give rise to the Laws of Logarithms.

Laws of Logarithms

Let a be a positive number, with $a \neq 1$. Let A, B, and C be any real numbers with $A > 0$ and $B > 0$.

Law	Description
1. $\log_a(AB) = \log_a A + \log_a B$	The logarithm of a product of numbers is the sum of the logarithms of the numbers.
2. $\log_a\left(\dfrac{A}{B}\right) = \log_a A - \log_a B$	The logarithm of a quotient of numbers is the difference of the logarithms of the numbers.
3. $\log_a(A^C) = C \log_a A$	The logarithm of a power of a number is the exponent times the logarithm of the number.

■ **Proof** We make use of the property $\log_a a^x = x$ from Section 4.2.

Law 1. Let $\log_a A = u$ and $\log_a B = v$. When written in exponential form, these equations become

$$a^u = A \quad \text{and} \quad a^v = B$$

Thus
$$\log_a(AB) = \log_a(a^u a^v) = \log_a(a^{u+v})$$
$$= u + v = \log_a A + \log_a B$$

Law 2. Using Law 1, we have

$$\log_a A = \log_a\left[\left(\frac{A}{B}\right)B\right] = \log_a\left(\frac{A}{B}\right) + \log_a B$$

so
$$\log_a\left(\frac{A}{B}\right) = \log_a A - \log_a B$$

Law 3. Let $\log_a A = u$. Then $a^u = A$, so

$$\log_a(A^C) = \log_a(a^u)^C = \log_a(a^{uC}) = uC = C\log_a A \qquad ■$$

Example 1 **Using the Laws of Logarithms to Evaluate Expressions**

Evaluate each expression.

(a) $\log_4 2 + \log_4 32$ (b) $\log_2 80 - \log_2 5$ (c) $-\frac{1}{3}\log 8$

Solution

(a) $\log_4 2 + \log_4 32 = \log_4(2 \cdot 32)$ Law 1
$$= \log_4 64 = 3 \qquad \text{Because } 64 = 4^3$$

(b) $\log_2 80 - \log_2 5 = \log_2\left(\frac{80}{5}\right)$ Law 2
$$= \log_2 16 = 4 \qquad \text{Because } 16 = 2^4$$

(c) $-\frac{1}{3}\log 8 = \log 8^{-1/3}$ Law 3
$$= \log\left(\tfrac{1}{2}\right) \qquad \text{Property of negative exponents}$$
$$\approx -0.301 \qquad \text{Calculator} \qquad ■$$

Expanding and Combining Logarithmic Expressions

The laws of logarithms allow us to write the logarithm of a product or a quotient as the sum or difference of logarithms. This process, called *expanding* a logarithmic expression, is illustrated in the next example.

Example 2 **Expanding Logarithmic Expressions**

Use the Laws of Logarithms to expand each expression.

(a) $\log_2(6x)$ (b) $\log_5(x^3 y^6)$ (c) $\ln\left(\dfrac{ab}{\sqrt[3]{c}}\right)$

Solution

(a) $\log_2(6x) = \log_2 6 + \log_2 x$ Law 1

(b) $\log_5(x^3y^6) = \log_5 x^3 + \log_5 y^6$ Law 1

$= 3\log_5 x + 6\log_5 y$ Law 3

(c) $\ln\left(\dfrac{ab}{\sqrt[3]{c}}\right) = \ln(ab) - \ln\sqrt[3]{c}$ Law 2

$= \ln a + \ln b - \ln c^{1/3}$ Law 1

$= \ln a + \ln b - \tfrac{1}{3}\ln c$ Law 3   ■

The laws of logarithms also allow us to reverse the process of expanding done in Example 2. That is, we can write sums and differences of logarithms as a single logarithm. This process, called *combining* logarithmic expressions, is illustrated in the next example.

Example 3   Combining Logarithmic Expressions

Combine $3\log x + \tfrac{1}{2}\log(x+1)$ into a single logarithm.

Solution

$$3\log x + \tfrac{1}{2}\log(x+1) = \log x^3 + \log(x+1)^{1/2}\quad \text{Law 3}$$

$$= \log(x^3(x+1)^{1/2})\quad \text{Law 1}\quad ■$$

Example 4   Combining Logarithmic Expressions

Combine $3\ln s + \tfrac{1}{2}\ln t - 4\ln(t^2+1)$ into a single logarithm.

Solution

$$3\ln s + \tfrac{1}{2}\ln t - 4\ln(t^2+1) = \ln s^3 + \ln t^{1/2} - \ln(t^2+1)^4\quad \text{Law 3}$$

$$= \ln(s^3 t^{1/2}) - \ln(t^2+1)^4\quad \text{Law 1}$$

$$= \ln\left(\frac{s^3\sqrt{t}}{(t^2+1)^4}\right)\quad \text{Law 2}\quad ■$$

WARNING   Although the Laws of Logarithms tell us how to compute the logarithm of a product or a quotient, *there is no corresponding rule for the logarithm of a sum or a difference.* For instance,

$$\log_a(x+y) \;\cancel{\neq}\; \log_a x + \log_a y$$

In fact, we know that the right side is equal to $\log_a(xy)$. Also, don't improperly simplify quotients or powers of logarithms. For instance,

$$\frac{\log 6}{\log 2} \;\cancel{\neq}\; \log\left(\frac{6}{2}\right) \qquad \text{and} \qquad (\log_2 x)^3 \;\cancel{\neq}\; 3\log_2 x$$

Logarithmic functions are used to model a variety of situations involving human behavior. One such behavior is how quickly we forget things we have learned. For example, if you learn algebra at a certain performance level (say 90% on a test) and then don't use algebra for a while, how much will you retain after a week, a month, or a year? Hermann Ebbinghaus (1850–1909) studied this phenomenon and formulated the law described in the next example.

Forgetting what we've learned depends logarithmically on how long ago we learned it.

Example 5 The Law of Forgetting

Ebbinghaus' Law of Forgetting states that if a task is learned at a performance level P_0, then after a time interval t the performance level P satisfies

$$\log P = \log P_0 - c \log(t + 1)$$

where c is a constant that depends on the type of task and t is measured in months.

(a) Solve for P.

(b) If your score on a history test is 90, what score would you expect to get on a similar test after two months? After a year? (Assume $c = 0.2$.)

Solution

(a) We first combine the right-hand side.

$$\log P = \log P_0 - c \log(t + 1) \qquad \text{Given equation}$$

$$\log P = \log P_0 - \log(t + 1)^c \qquad \text{Law 3}$$

$$\log P = \log \frac{P_0}{(t + 1)^c} \qquad \text{Law 2}$$

$$P = \frac{P_0}{(t + 1)^c} \qquad \text{Because log is one-to-one}$$

(b) Here $P_0 = 90$, $c = 0.2$, and t is measured in months.

In two months: $t = 2$ and $P = \dfrac{90}{(2 + 1)^{0.2}} \approx 72$

In one year: $t = 12$ and $P = \dfrac{90}{(12 + 1)^{0.2}} \approx 54$

Your expected scores after two months and one year are 72 and 54, respectively. ∎

Change of Base

For some purposes, we find it useful to change from logarithms in one base to logarithms in another base. Suppose we are given $\log_a x$ and want to find $\log_b x$. Let

$$y = \log_b x$$

We write this in exponential form and take the logarithm, with base a, of each side.

$$b^y = x \qquad \text{Exponential form}$$

$$\log_a(b^y) = \log_a x \qquad \text{Take } \log_a \text{ of each side}$$

$$y \log_a b = \log_a x \qquad \text{Law 3}$$

$$y = \frac{\log_a x}{\log_a b} \qquad \text{Divide by } \log_a b$$

This proves the following formula.

We may write the Change of Base Formula as

$$\log_b x = \left(\frac{1}{\log_a b}\right)\log_a x$$

So, $\log_b x$ is just a constant multiple of $\log_a x$; the constant is $\dfrac{1}{\log_a b}$.

Change of Base Formula

$$\log_b x = \frac{\log_a x}{\log_a b}$$

In particular, if we put $x = a$, then $\log_a a = 1$ and this formula becomes

$$\log_b a = \frac{1}{\log_a b}$$

We can now evaluate a logarithm to *any* base by using the Change of Base Formula to express the logarithm in terms of common logarithms or natural logarithms and then using a calculator.

Example 6   Evaluating Logarithms with the Change of Base Formula

Use the Change of Base Formula and common or natural logarithms to evaluate each logarithm, correct to five decimal places.

(a) $\log_8 5$ (b) $\log_9 20$

Solution

We get the same answer whether we use $\log_{10}$ or ln:

$$\log_8 5 = \frac{\ln 5}{\ln 8} \approx 0.77398$$

(a) We use the Change of Base Formula with $b = 8$ and $a = 10$:

$$\log_8 5 = \frac{\log_{10} 5}{\log_{10} 8} \approx 0.77398$$

(b) We use the Change of Base Formula with $b = 9$ and $a = e$:

$$\log_9 20 = \frac{\ln 20}{\ln 9} \approx 1.36342$$ ∎

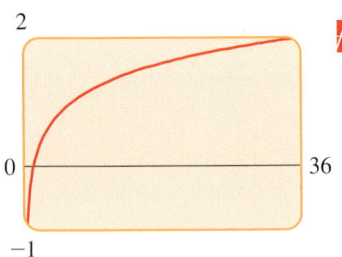

Figure 1
$f(x) = \log_6 x = \dfrac{\ln x}{\ln 6}$

Example 7   Using the Change of Base Formula to Graph a Logarithmic Function

Use a graphing calculator to graph $f(x) = \log_6 x$.

Solution   Calculators don't have a key for $\log_6$, so we use the Change of Base Formula to write

$$f(x) = \log_6 x = \frac{\ln x}{\ln 6}$$

Since calculators do have an ⎡LN⎤ key, we can enter this new form of the function and graph it. The graph is shown in Figure 1. ∎

4.3   Exercises

1–12 ■ Evaluate the expression.

1. $\log_3 \sqrt{27}$

2. $\log_2 160 - \log_2 5$

3. $\log 4 + \log 25$

4. $\log \dfrac{1}{\sqrt{1000}}$

5. $\log_4 192 - \log_4 3$

6. $\log_{12} 9 + \log_{12} 16$

7. $\log_2 6 - \log_2 15 + \log_2 20$

8. $\log_3 100 - \log_3 18 - \log_3 50$

9. $\log_4 16^{100}$ **10.** $\log_2 8^{33}$

11. $\log(\log 10^{10,000})$ **12.** $\ln(\ln e^{e^{200}})$

13–38 ■ Use the Laws of Logarithms to expand the expression.

13. $\log_2(2x)$ **14.** $\log_3(5y)$

15. $\log_2(x(x - 1))$ **16.** $\log_5 \dfrac{x}{2}$

17. $\log 6^{10}$ **18.** $\ln \sqrt{z}$

19. $\log_2(AB^2)$ **20.** $\log_6 \sqrt[4]{17}$

21. $\log_3(x\sqrt{y})$ **22.** $\log_2(xy)^{10}$

23. $\log_5 \sqrt[3]{x^2 + 1}$ **24.** $\log_a\left(\dfrac{x^2}{yz^3}\right)$

25. $\ln \sqrt{ab}$ **26.** $\ln \sqrt[3]{3r^2 s}$

27. $\log\left(\dfrac{x^3 y^4}{z^6}\right)$ **28.** $\log\left(\dfrac{a^2}{b^4 \sqrt{c}}\right)$

29. $\log_2\left(\dfrac{x(x^2 + 1)}{\sqrt{x^2 - 1}}\right)$ **30.** $\log_5 \sqrt{\dfrac{x - 1}{x + 1}}$

31. $\ln\left(x\sqrt{\dfrac{y}{z}}\right)$ **32.** $\ln \dfrac{3x^2}{(x + 1)^{10}}$

33. $\log \sqrt[4]{x^2 + y^2}$ **34.** $\log\left(\dfrac{x}{\sqrt[3]{1 - x}}\right)$

35. $\log\sqrt{\dfrac{x^2 + 4}{(x^2 + 1)(x^3 - 7)^2}}$ **36.** $\log \sqrt{x\sqrt{y\sqrt{z}}}$

37. $\ln\left(\dfrac{x^3 \sqrt{x - 1}}{3x + 4}\right)$ **38.** $\log\left(\dfrac{10^x}{x(x^2 + 1)(x^4 + 2)}\right)$

39–48 ■ Use the Laws of Logarithms to combine the expression.

39. $\log_3 5 + 5 \log_3 2$ **40.** $\log 12 + \frac{1}{2} \log 7 - \log 2$

41. $\log_2 A + \log_2 B - 2 \log_2 C$

42. $\log_5(x^2 - 1) - \log_5(x - 1)$

43. $4 \log x - \frac{1}{3} \log(x^2 + 1) + 2 \log(x - 1)$

44. $\ln(a + b) + \ln(a - b) - 2 \ln c$

45. $\ln 5 + 2 \ln x + 3 \ln(x^2 + 5)$

46. $2(\log_5 x + 2 \log_5 y - 3 \log_5 z)$

47. $\frac{1}{3} \log(2x + 1) + \frac{1}{2}[\log(x - 4) - \log(x^4 - x^2 - 1)]$

48. $\log_a b + c \log_a d - r \log_a s$

49–56 ■ Use the Change of Base Formula and a calculator to evaluate the logarithm, correct to six decimal places. Use either natural or common logarithms.

49. $\log_2 5$ **50.** $\log_5 2$

51. $\log_3 16$ **52.** $\log_6 92$

53. $\log_7 2.61$ **54.** $\log_6 532$

55. $\log_4 125$ **56.** $\log_{12} 2.5$

 57. Use the Change of Base Formula to show that

$$\log_3 x = \frac{\ln x}{\ln 3}$$

Then use this fact to draw the graph of the function $f(x) = \log_3 x$.

58. Draw graphs of the family of functions $y = \log_a x$ for $a = 2, e, 5,$ and 10 on the same screen, using the viewing rectangle $[0, 5]$ by $[-3, 3]$. How are these graphs related?

59. Use the Change of Base Formula to show that

$$\log e = \frac{1}{\ln 10}$$

60. Simplify: $(\log_2 5)(\log_5 7)$

61. Show that $-\ln(x - \sqrt{x^2 - 1}) = \ln(x + \sqrt{x^2 - 1})$.

Applications

62. Forgetting   Use the Ebbinghaus Forgetting Law (Example 5) to estimate a student's score on a biology test two years after he got a score of 80 on a test covering the same material. Assume $c = 0.3$ and t is measured in months.

63. Wealth Distribution   Vilfredo Pareto (1848–1923) observed that most of the wealth of a country is owned by a few members of the population. **Pareto's Principle** is

$$\log P = \log c - k \log W$$

where W is the wealth level (how much money a person has) and P is the number of people in the population having that much money.

(a) Solve the equation for P.

(b) Assume $k = 2.1$, $c = 8000$, and W is measured in millions of dollars. Use part (a) to find the number of people who have \$2 million or more. How many people have \$10 million or more?

64. Biodiversity   Some biologists model the number of species S in a fixed area A (such as an island) by the Species-Area relationship

$$\log S = \log c + k \log A$$

where c and k are positive constants that depend on the type of species and habitat.

(a) Solve the equation for S.

(b) Use part (a) to show that if $k = 3$ then doubling the area increases the number of species eightfold.

65. Magnitude of Stars   The magnitude M of a star is a measure of how bright a star appears to the human eye. It is defined by

$$M = -2.5 \log\left(\frac{B}{B_0}\right)$$

where B is the actual brightness of the star and B_0 is a constant.

(a) Expand the right-hand side of the equation.

(b) Use part (a) to show that the brighter a star, the less its magnitude.

(c) Betelgeuse is about 100 times brighter than Albiero. Use part (a) to show that Betelgeuse is 5 magnitudes less than Albiero.

Discovery • Discussion

66. True or False?   Discuss each equation and determine whether it is true for all possible values of the variables. (Ignore values of the variables for which any term is undefined.)

(a) $\log\left(\dfrac{x}{y}\right) = \dfrac{\log x}{\log y}$

(b) $\log_2(x - y) = \log_2 x - \log_2 y$

(c) $\log_5\left(\dfrac{a}{b^2}\right) = \log_5 a - 2\log_5 b$

(d) $\log 2^z = z \log 2$

(e) $(\log P)(\log Q) = \log P + \log Q$

(f) $\dfrac{\log a}{\log b} = \log a - \log b$

(g) $(\log_2 7)^x = x \log_2 7$

(h) $\log_a a^a = a$

(i) $\log(x - y) = \dfrac{\log x}{\log y}$

(j) $-\ln\left(\dfrac{1}{A}\right) = \ln A$

67. Find the Error   What is wrong with the following argument?

$$\log 0.1 < 2 \log 0.1$$
$$= \log(0.1)^2$$
$$= \log 0.01$$
$$\log 0.1 < \log 0.01$$
$$0.1 < 0.01$$

68. Shifting, Shrinking, and Stretching Graphs of Functions   Let $f(x) = x^2$. Show that $f(2x) = 4f(x)$, and explain how this shows that shrinking the graph of f horizontally has the same effect as stretching it vertically. Then use the identities $e^{2+x} = e^2 e^x$ and $\ln(2x) = \ln 2 + \ln x$ to show that for $g(x) = e^x$, a horizontal shift is the same as a vertical stretch and for $h(x) = \ln x$, a horizontal shrinking is the same as a vertical shift.

4.4   Exponential and Logarithmic Equations

In this section we solve equations that involve exponential or logarithmic functions. The techniques we develop here will be used in the next section for solving applied problems.

Exponential Equations

An *exponential equation* is one in which the variable occurs in the exponent. For example,

$$2^x = 7$$

The variable x presents a difficulty because it is in the exponent. To deal with this difficulty, we take the logarithm of each side and then use the Laws of Logarithms to "bring down x" from the exponent.

$$2^x = 7 \qquad \text{\textit{Given equation}}$$

$$\ln 2^x = \ln 7 \qquad \text{\textit{Take ln of each side}}$$

$$x \ln 2 = \ln 7 \qquad \text{\textit{Law 3 (bring down exponent)}}$$

$$x = \frac{\ln 7}{\ln 2} \qquad \text{\textit{Solve for x}}$$

$$\approx 2.807 \qquad \text{\textit{Calculator}}$$

Recall that Law 3 of the Laws of Logarithms says that $\log_a A^C = C \log_a A$.

The method we used to solve $2^x = 7$ is typical of how we solve exponential equations in general.

Guidelines for Solving Exponential Equations

1. Isolate the exponential expression on one side of the equation.

2. Take the logarithm of each side, then use the Laws of Logarithms to "bring down the exponent."

3. Solve for the variable.

Example 1 Solving an Exponential Equation

Find the solution of the equation $3^{x+2} = 7$, correct to six decimal places.

Solution We take the common logarithm of each side and use Law 3.

$$3^{x+2} = 7 \qquad \text{\textit{Given equation}}$$

$$\log(3^{x+2}) = \log 7 \qquad \text{\textit{Take log of each side}}$$

$$(x + 2)\log 3 = \log 7 \qquad \text{\textit{Law 3 (bring down exponent)}}$$

$$x + 2 = \frac{\log 7}{\log 3} \qquad \text{\textit{Divide by log 3}}$$

$$x = \frac{\log 7}{\log 3} - 2 \qquad \text{\textit{Subtract 2}}$$

$$\approx -0.228756 \qquad \text{\textit{Calculator}} \qquad \blacksquare$$

We could have used natural logarithms instead of common logarithms. In fact, using the same steps, we get

$$x = \frac{\ln 7}{\ln 3} - 2 \approx -0.228756$$

Check Your Answer Substituting $x = -0.228756$ into the original equation and using a calculator, we get

$$3^{(-0.228756)+2} \approx 7 \quad \checkmark$$

Radiocarbon dating is a method archeologists use to determine the age of ancient objects. The carbon dioxide in the atmosphere always contains a fixed fraction of radioactive carbon, carbon-14 (^{14}C), with a half-life of about 5730 years. Plants absorb carbon dioxide from the atmosphere, which then makes its way to animals through the food chain. Thus, all living creatures contain the same fixed proportions of ^{14}C to nonradioactive ^{12}C as the atmosphere.

After an organism dies, it stops assimilating ^{14}C, and the amount of ^{14}C in it begins to decay exponentially. We can then determine the time elapsed since the death of the organism by measuring the amount of ^{14}C left in it.

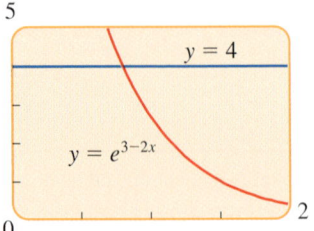

For example, if a donkey bone contains 73% as much ^{14}C as a living donkey and it died t years ago, then by the formula for radioactive decay (Section 4.5),

$$0.73 = (1.00)e^{-(t \ln 2)/5730}$$

We solve this exponential equation to find $t \approx 2600$, so the bone is about 2600 years old.

Example 2   Solving an Exponential Equation

Solve the equation $8e^{2x} = 20$.

Solution   We first divide by 8 in order to isolate the exponential term on one side of the equation.

$$8e^{2x} = 20 \qquad \text{Given equation}$$
$$e^{2x} = \frac{20}{8} \qquad \text{Divide by 8}$$
$$\ln e^{2x} = \ln 2.5 \qquad \text{Take ln of each side}$$
$$2x = \ln 2.5 \qquad \text{Property of ln}$$
$$x = \frac{\ln 2.5}{2} \qquad \text{Divide by 2}$$
$$\approx 0.458 \qquad \text{Calculator}$$ ∎

Check Your Answer   Substituting $x = 0.458$ into the original equation and using a calculator, we get

$$8e^{2(0.458)} \approx 20 \quad \checkmark$$

Example 3   Solving an Exponential Equation Algebraically and Graphically

Solve the equation $e^{3-2x} = 4$ algebraically and graphically.

Solution 1: Algebraic

Since the base of the exponential term is e, we use natural logarithms to solve this equation.

$$e^{3-2x} = 4 \qquad \text{Given equation}$$
$$\ln(e^{3-2x}) = \ln 4 \qquad \text{Take ln of each side}$$
$$3 - 2x = \ln 4 \qquad \text{Property of ln}$$
$$2x = 3 - \ln 4$$
$$x = \tfrac{1}{2}(3 - \ln 4) \approx 0.807$$

You should check that this answer satisfies the original equation.

Solution 2: Graphical

We graph the equations $y = e^{3-2x}$ and $y = 4$ in the same viewing rectangle as in Figure 1. The solutions occur where the graphs intersect. Zooming in on the point of intersection of the two graphs, we see that $x \approx 0.81$.   ∎

Figure 1

Example 4 **An Exponential Equation of Quadratic Type**

Solve the equation $e^{2x} - e^x - 6 = 0$.

Solution   To isolate the exponential term, we factor.

$$e^{2x} - e^x - 6 = 0 \qquad \text{Given equation}$$

$$(e^x)^2 - e^x - 6 = 0 \qquad \text{Law of Exponents}$$

$$(e^x - 3)(e^x + 2) = 0 \qquad \text{Factor (a quadratic in } e^x)$$

$$e^x - 3 = 0 \quad \text{or} \quad e^x + 2 = 0 \qquad \text{Zero-Product Property}$$

$$e^x = 3 \qquad\qquad e^x = -2$$

If we let $w = e^x$, we get the quadratic equation

$$w^2 - w - 6 = 0$$

which factors as

$$(w - 3)(w + 2) = 0$$

The equation $e^x = 3$ leads to $x = \ln 3$. But the equation $e^x = -2$ has no solution because $e^x > 0$ for all x. Thus, $x = \ln 3 \approx 1.0986$ is the only solution. You should check that this answer satisfies the original equation. ∎

Example 5 **Solving an Exponential Equation**

Solve the equation $3xe^x + x^2e^x = 0$.

Solution   First we factor the left side of the equation.

$$3xe^x + x^2e^x = 0 \qquad \text{Given equation}$$

$$x(3 + x)e^x = 0 \qquad \text{Factor out common factors}$$

$$x(3 + x) = 0 \qquad \text{Divide by } e^x \text{ (because } e^x \neq 0)$$

$$x = 0 \quad \text{or} \quad 3 + x = 0 \qquad \text{Zero-Product Property}$$

Thus, the solutions are $x = 0$ and $x = -3$. ∎

Check Your Answers

$x = 0$:

$$3(0)e^0 + 0^2e^0 = 0 \qquad ✓$$

$x = -3$:

$$3(-3)e^{-3} + (-3)^2e^{-3}$$

$$= -9e^{-3} + 9e^{-3} = 0 \qquad ✓$$

Logarithmic Equations

A *logarithmic equation* is one in which a logarithm of the variable occurs. For example,

$$\log_2(x + 2) = 5$$

To solve for x, we write the equation in exponential form.

$$x + 2 = 2^5 \qquad \text{Exponential form}$$

$$x = 32 - 2 = 30 \qquad \text{Solve for } x$$

Another way of looking at the first step is to raise the base, 2, to each side of the equation.

$$2^{\log_2(x+2)} = 2^5 \qquad \text{Raise 2 to each side}$$

$$x + 2 = 2^5 \qquad \text{Property of logarithms}$$

$$x = 32 - 2 = 30 \qquad \text{Solve for } x$$

The method used to solve this simple problem is typical. We summarize the steps as follows.

Guidelines for Solving Logarithmic Equations

1. Isolate the logarithmic term on one side of the equation; you may first need to combine the logarithmic terms.

2. Write the equation in exponential form (or raise the base to each side of the equation).

3. Solve for the variable.

Example 6 Solving Logarithmic Equations

Solve each equation for x.

(a) $\ln x = 8$ (b) $\log_2(25 - x) = 3$

Solution

(a)

$$\ln x = 8 \qquad \text{Given equation}$$

$$x = e^8 \qquad \text{Exponential form}$$

Therefore, $x = e^8 \approx 2981$.

We can also solve this problem another way:

$$\ln x = 8 \qquad \text{Given equation}$$

$$e^{\ln x} = e^8 \qquad \text{Raise } e \text{ to each side}$$

$$x = e^8 \qquad \text{Property of ln}$$

(b) The first step is to rewrite the equation in exponential form.

$$\log_2(25 - x) = 3 \qquad \text{Given equation}$$

$$25 - x = 2^3 \qquad \text{Exponential form (or raise 2 to each side)}$$

$$25 - x = 8$$

$$x = 25 - 8 = 17$$

Check Your Answer

If $x = 17$, we get

$\log_2(25 - 17) = \log_2 8 = 3$ ✓

Example 7 Solving a Logarithmic Equation

Solve the equation $4 + 3 \log(2x) = 16$.

Solution We first isolate the logarithmic term. This allows us to write the equation in exponential form.

$$4 + 3 \log(2x) = 16 \qquad \text{Given equation}$$

$$3 \log(2x) = 12 \qquad \text{Subtract 4}$$

$$\log(2x) = 4 \qquad \text{Divide by 3}$$

$$2x = 10^4 \qquad \text{Exponential form (or raise 10 to each side)}$$

$$x = 5000 \qquad \text{Divide by 2}$$

Check Your Answer

If $x = 5000$, we get

$4 + 3 \log 2(5000) = 4 + 3 \log 10{,}000$

$= 4 + 3(4)$

$= 16$ ✓

Example 8 Solving a Logarithmic Equation Algebraically and Graphically

Solve the equation $\log(x + 2) + \log(x - 1) = 1$ algebraically and graphically.

Solution 1: Algebraic

We first combine the logarithmic terms using the Laws of Logarithms.

$$\log[(x + 2)(x - 1)] = 1 \qquad \text{Law 1}$$
$$(x + 2)(x - 1) = 10 \qquad \text{Exponential form (or raise 10 to each side)}$$
$$x^2 + x - 2 = 10 \qquad \text{Expand left side}$$
$$x^2 + x - 12 = 0 \qquad \text{Subtract 10}$$
$$(x + 4)(x - 3) = 0 \qquad \text{Factor}$$
$$x = -4 \qquad \text{or} \qquad x = 3$$

We check these potential solutions in the original equation and find that $x = -4$ is not a solution (because logarithms of negative numbers are undefined), but $x = 3$ is a solution. (See *Check Your Answers*.)

Check Your Answers

$x = -4$:

$\log(-4 + 2) + \log(-4 - 1)$
  $= \log(-2) + \log(-5)$
    undefined   ✗

$x = 3$:

$\log(3 + 2) + \log(3 - 1)$
  $= \log 5 + \log 2 = \log(5 \cdot 2)$
  $= \log 10 = 1$ ✓

Solution 2: Graphical

We first move all terms to one side of the equation:

$$\log(x + 2) + \log(x - 1) - 1 = 0$$

Then we graph

$$y = \log(x + 2) + \log(x - 1) - 1$$

as in Figure 2. The solutions are the x-intercepts of the graph. Thus, the only solution is $x \approx 3$.

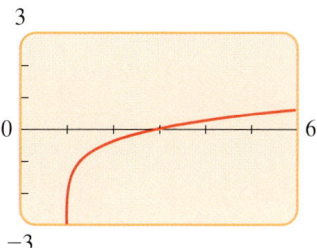

Figure 2

Example 9 Solving a Logarithmic Equation Graphically

Solve the equation $x^2 = 2 \ln(x + 2)$.

In Example 9, it's not possible to isolate x algebraically, so we must solve the equation graphically.

Solution We first move all terms to one side of the equation

$$x^2 - 2 \ln(x + 2) = 0$$

Then we graph

$$y = x^2 - 2 \ln(x + 2)$$

as in Figure 3. The solutions are the x-intercepts of the graph. Zooming in on the x-intercepts, we see that there are two solutions:

$$x \approx -0.71 \qquad \text{and} \qquad x \approx 1.60$$

Figure 3

Logarithmic equations are used in determining the amount of light that reaches various depths in a lake. (This information helps biologists determine the types of life a lake can support.) As light passes through water (or other transparent materials such as glass or plastic), some of the light is absorbed. It's easy to see that the murkier the water the more light is absorbed. The exact relationship between light absorption and the distance light travels in a material is described in the next example.

The intensity of light in a lake diminishes with depth.

Example 10 Transparency of a Lake

If I_0 and I denote the intensity of light before and after going through a material and x is the distance (in feet) the light travels in the material, then according to the **Beer-Lambert Law**

$$-\frac{1}{k}\ln\left(\frac{I}{I_0}\right) = x$$

where k is a constant depending on the type of material.

(a) Solve the equation for I.

(b) For a certain lake $k = 0.025$ and the light intensity is $I_0 = 14$ lumens (lm). Find the light intensity at a depth of 20 ft.

Solution

(a) We first isolate the logarithmic term.

$$-\frac{1}{k}\ln\left(\frac{I}{I_0}\right) = x \qquad \text{Given equation}$$

$$\ln\left(\frac{I}{I_0}\right) = -kx \qquad \text{Multiply by } -k$$

$$\frac{I}{I_0} = e^{-kx} \qquad \text{Exponential form}$$

$$I = I_0 e^{-kx} \qquad \text{Multiply by } I_0$$

(b) We find I using the formula from part (a).

$$I = I_0 e^{-kx} \qquad \text{From part (a)}$$

$$= 14 e^{(-0.025)(20)} \qquad I_0 = 14, k = 0.025, x = 20$$

$$\approx 8.49 \qquad \text{Calculator}$$

The light intensity at a depth of 20 ft is about 8.5 lm.

Compound Interest

Recall the formulas for interest that we found in Section 4.1. If a principal P is invested at an interest rate r for a period of t years, then the amount A of the investment is given by

$$A = P(1 + r) \qquad \text{Simple interest (for one year)}$$

$$A(t) = P\left(1 + \frac{r}{n}\right)^{nt} \qquad \text{Interest compounded } n \text{ times per year}$$

$$A(t) = Pe^{rt} \qquad \text{Interest compounded continuously}$$

We can use logarithms to determine the time it takes for the principal to increase to a given amount.

Example 11   Finding the Term for an Investment to Double

A sum of $5000 is invested at an interest rate of 5% per year. Find the time required for the money to double if the interest is compounded according to the following method.

(a) Semiannual   (b) Continuous

Solution

(a) We use the formula for compound interest with $P = \$5000$, $A(t) = \$10{,}000$, $r = 0.05$, $n = 2$, and solve the resulting exponential equation for t.

$$5000\left(1 + \frac{0.05}{2}\right)^{2t} = 10{,}000 \qquad P\left(1 + \frac{r}{n}\right)^{nt} = A$$

$$(1.025)^{2t} = 2 \qquad \text{Divide by 5000}$$

$$\log 1.025^{2t} = \log 2 \qquad \text{Take log of each side}$$

$$2t \log 1.025 = \log 2 \qquad \text{Law 3 (bring down the exponent)}$$

$$t = \frac{\log 2}{2 \log 1.025} \qquad \text{Divide by 2 log 1.025}$$

$$t \approx 14.04 \qquad \text{Calculator}$$

The money will double in 14.04 years.

(b) We use the formula for continuously compounded interest with $P = \$5000$, $A(t) = \$10{,}000$, $r = 0.05$, and solve the resulting exponential equation for t.

$$5000e^{0.05t} = 10{,}000 \qquad Pe^{rt} = A$$

$$e^{0.05t} = 2 \qquad \text{Divide by 5000}$$

$$\ln e^{0.05t} = \ln 2 \qquad \text{Take ln of each side}$$

$$0.05t = \ln 2 \qquad \text{Property of ln}$$

$$t = \frac{\ln 2}{0.05} \qquad \text{Divide by 0.05}$$

$$t \approx 13.86 \qquad \text{Calculator}$$

The money will double in 13.86 years.

Example 12 Time Required to Grow an Investment

A sum of $1000 is invested at an interest rate of 4% per year. Find the time required for the amount to grow to $4000 if interest is compounded continuously.

Solution We use the formula for continuously compounded interest with $P = \$1000$, $A(t) = \$4000$, $r = 0.04$, and solve the resulting exponential equation for t.

$$1000e^{0.04t} = 4000 \qquad Pe^{rt} = A$$

$$e^{0.04t} = 4 \qquad \text{Divide by 1000}$$

$$0.04t = \ln 4 \qquad \text{Take ln of each side}$$

$$t = \frac{\ln 4}{0.04} \qquad \text{Divide by 0.04}$$

$$t \approx 34.66 \qquad \text{Calculator}$$

The amount will be $4000 in about 34 years and 8 months. ■

If an investment earns compound interest, then the **annual percentage yield** (APY) is the *simple* interest rate that yields the same amount at the end of one year.

Example 13 Calculating the Annual Percentage Yield

Find the annual percentage yield for an investment that earns interest at a rate of 6% per year, compounded daily.

Solution After one year, a principal P will grow to the amount

$$A = P\left(1 + \frac{0.06}{365}\right)^{365} = P(1.06183)$$

The formula for simple interest is

$$A = P(1 + r)$$

Comparing, we see that $1 + r = 1.06183$, so $r = 0.06183$. Thus the annual percentage yield is 6.183%. ■

4.4 Exercises

1–26 ■ Find the solution of the exponential equation, correct to four decimal places.

1. $10^x = 25$

2. $10^{-x} = 4$

3. $e^{-2x} = 7$

4. $e^{3x} = 12$

5. $2^{1-x} = 3$

6. $3^{2x-1} = 5$

7. $3e^x = 10$

8. $2e^{12x} = 17$

9. $e^{1-4x} = 2$

10. $4(1 + 10^{5x}) = 9$

11. $4 + 3^{5x} = 8$

12. $2^{3x} = 34$

13. $8^{0.4x} = 5$

14. $3^{x/14} = 0.1$

15. $5^{-x/100} = 2$

16. $e^{3-5x} = 16$

17. $e^{2x+1} = 200$

18. $\left(\frac{1}{4}\right)^x = 75$

19. $5^x = 4^{x+1}$

20. $10^{1-x} = 6^x$

21. $2^{3x+1} = 3^{x-2}$

22. $7^{x/2} = 5^{1-x}$

23. $\dfrac{50}{1 + e^{-x}} = 4$

24. $\dfrac{10}{1 + e^{-x}} = 2$

25. $100(1.04)^{2t} = 300$

26. $(1.00625)^{12t} = 2$

27–34 ■ Solve the equation.

27. $x^2 2^x - 2^x = 0$

28. $x^2 10^x - x 10^x = 2(10^x)$

29. $4x^3 e^{-3x} - 3x^4 e^{-3x} = 0$

30. $x^2 e^x + x e^x - e^x = 0$

31. $e^{2x} - 3e^x + 2 = 0$

32. $e^{2x} - e^x - 6 = 0$

33. $e^{4x} + 4e^{2x} - 21 = 0$

34. $e^x - 12e^{-x} - 1 = 0$

35–50 ■ Solve the logarithmic equation for x.

35. $\ln x = 10$

36. $\ln(2 + x) = 1$

37. $\log x = -2$

38. $\log(x - 4) = 3$

39. $\log(3x + 5) = 2$

40. $\log_3(2 - x) = 3$

41. $2 - \ln(3 - x) = 0$

42. $\log_2(x^2 - x - 2) = 2$

43. $\log_2 3 + \log_2 x = \log_2 5 + \log_2(x - 2)$

44. $2 \log x = \log 2 + \log(3x - 4)$

45. $\log x + \log(x - 1) = \log(4x)$

46. $\log_5 x + \log_5(x + 1) = \log_5 20$

47. $\log_5(x + 1) - \log_5(x - 1) = 2$

48. $\log x + \log(x - 3) = 1$

49. $\log_9(x - 5) + \log_9(x + 3) = 1$

50. $\ln(x - 1) + \ln(x + 2) = 1$

51. For what value of x is the following true?
$$\log(x + 3) = \log x + \log 3$$

52. For what value of x is it true that $(\log x)^3 = 3 \log x$?

53. Solve for x: $2^{2/\log_5 x} = \frac{1}{16}$

54. Solve for x: $\log_2(\log_3 x) = 4$

55–62 ■ Use a graphing device to find all solutions of the equation, correct to two decimal places.

55. $\ln x = 3 - x$

56. $\log x = x^2 - 2$

57. $x^3 - x = \log(x + 1)$

58. $x = \ln(4 - x^2)$

59. $e^x = -x$

60. $2^{-x} = x - 1$

61. $4^{-x} = \sqrt{x}$

62. $e^{x^2} - 2 = x^3 - x$

63–66 ■ Solve the inequality.

63. $\log(x - 2) + \log(9 - x) < 1$

64. $3 \le \log_2 x \le 4$

65. $2 < 10^x < 5$

66. $x^2 e^x - 2e^x < 0$

Applications

67. Compound Interest   A man invests $5000 in an account that pays 8.5% interest per year, compounded quarterly.
   (a) Find the amount after 3 years.
   (b) How long will it take for the investment to double?

68. Compound Interest   A man invests $6500 in an account that pays 6% interest per year, compounded continuously.
   (a) What is the amount after 2 years?
   (b) How long will it take for the amount to be $8000?

69. Compound Interest   Find the time required for an investment of $5000 to grow to $8000 at an interest rate of 7.5% per year, compounded quarterly.

70. Compound Interest   Nancy wants to invest $4000 in saving certificates that bear an interest rate of 9.75% per year, compounded semiannually. How long a time period should she choose in order to save an amount of $5000?

71. Doubling an Investment   How long will it take for an investment of $1000 to double in value if the interest rate is 8.5% per year, compounded continuously?

72. Interest Rate   A sum of $1000 was invested for 4 years, and the interest was compounded semiannually. If this sum amounted to $1435.77 in the given time, what was the interest rate?

73. Annual Percentage Yield   Find the annual percentage yield for an investment that earns 8% per year, compounded monthly.

74. Annual Percentage Yield   Find the annual percentage yield for an investment that earns $5\frac{1}{2}\%$ per year, compounded continuously.

75. Radioactive Decay   A 15-g sample of radioactive iodine decays in such a way that the mass remaining after t days is given by $m(t) = 15e^{-0.087t}$ where $m(t)$ is measured in grams. After how many days is there only 5 g remaining?

76. Skydiving   The velocity of a sky diver t seconds after jumping is given by $v(t) = 80(1 - e^{-0.2t})$. After how many seconds is the velocity 70 ft/s?

77. Fish Population   A small lake is stocked with a certain species of fish. The fish population is modeled by the function
$$P = \frac{10}{1 + 4e^{-0.8t}}$$
where P is the number of fish in thousands and t is measured in years since the lake was stocked.
   (a) Find the fish population after 3 years.

(b) After how many years will the fish population reach 5000 fish?

78. Transparency of a Lake   Environmental scientists measure the intensity of light at various depths in a lake to find the "transparency" of the water. Certain levels of transparency are required for the biodiversity of the submerged macrophyte population. In a certain lake the intensity of light at depth x is given by

$$I = 10e^{-0.008x}$$

where I is measured in lumens and x in feet.

(a) Find the intensity I at a depth of 30 ft.

(b) At what depth has the light intensity dropped to $I = 5$?

79. Atmospheric Pressure   Atmospheric pressure P (in kilopascals, kPa) at altitude h (in kilometers, km) is governed by the formula

$$\ln\left(\frac{P}{P_0}\right) = -\frac{h}{k}$$

where $k = 7$ and $P_0 = 100$ kPa are constants.

(a) Solve the equation for P.

(b) Use part (a) to find the pressure P at an altitude of 4 km.

80. Cooling an Engine   Suppose you're driving your car on a cold winter day (20 °F outside) and the engine overheats (at about 220 °F). When you park, the engine begins to cool down. The temperature T of the engine t minutes after you park satisfies the equation

$$\ln\left(\frac{T - 20}{200}\right) = -0.11t$$

(a) Solve the equation for T.

(b) Use part (a) to find the temperature of the engine after 20 min ($t = 20$).

81. Electric Circuits   An electric circuit contains a battery that produces a voltage of 60 volts (V), a resistor with a resistance of 13 ohms (Ω), and an inductor with an inductance of 5 henrys (H), as shown in the figure. Using calculus, it can be shown that the current $I = I(t)$ (in amperes, A) t seconds after the switch is closed is $I = \frac{60}{13}(1 - e^{-13t/5})$.

(a) Use this equation to express the time t as a function of the current I.

(b) After how many seconds is the current 2 A?

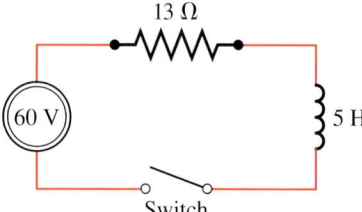

82. Learning Curve   A *learning curve* is a graph of a function $P(t)$ that measures the performance of someone learning a skill as a function of the training time t. At first, the rate of learning is rapid. Then, as performance increases and approaches a maximal value M, the rate of learning decreases. It has been found that the function

$$P(t) = M - Ce^{-kt}$$

where k and C are positive constants and $C < M$ is a reasonable model for learning.

(a) Express the learning time t as a function of the performance level P.

(b) For a pole-vaulter in training, the learning curve is given by

$$P(t) = 20 - 14e^{-0.024t}$$

where $P(t)$ is the height he is able to pole-vault after t months. After how many months of training is he able to vault 12 ft?

(c) Draw a graph of the learning curve in part (b).

Discovery • Discussion

83. Estimating a Solution   Without actually solving the equation, find two whole numbers between which the solution of $9^x = 20$ must lie. Do the same for $9^x = 100$. Explain how you reached your conclusions.

84. A Surprising Equation   Take logarithms to show that the equation

$$x^{1/\log x} = 5$$

has no solution. For what values of k does the equation

$$x^{1/\log x} = k$$

have a solution? What does this tell us about the graph of the function $f(x) = x^{1/\log x}$? Confirm your answer using a graphing device.

85. Disguised Equations   Each of these equations can be transformed into an equation of linear or quadratic type by applying the hint. Solve each equation.

(a) $(x - 1)^{\log(x-1)} = 100(x - 1)$ [Take log of each side.]

(b) $\log_2 x + \log_4 x + \log_8 x = 11$ [Change all logs to base 2.]

(c) $4^x - 2^{x+1} = 3$ [Write as a quadratic in 2^x.]

4.5   Modeling with Exponential and Logarithmic Functions

Many processes that occur in nature, such as population growth, radioactive decay, heat diffusion, and numerous others, can be modeled using exponential functions. Logarithmic functions are used in models for the loudness of sounds, the intensity of earthquakes, and many other phenomena. In this section we study exponential and logarithmic models.

Exponential Models of Population Growth

Biologists have observed that the population of a species doubles its size in a fixed period of time. For example, under ideal conditions a certain population of bacteria doubles in size every 3 hours. If the culture is started with 1000 bacteria, then after 3 hours there will be 2000 bacteria, after another 3 hours there will be 4000, and so on. If we let $n = n(t)$ be the number of bacteria after t hours, then

$$n(0) = 1000$$

$$n(3) = 1000 \cdot 2$$

$$n(6) = (1000 \cdot 2) \cdot 2 = 1000 \cdot 2^2$$

$$n(9) = (1000 \cdot 2^2) \cdot 2 = 1000 \cdot 2^3$$

$$n(12) = (1000 \cdot 2^3) \cdot 2 = 1000 \cdot 2^4$$

From this pattern it appears that the number of bacteria after t hours is modeled by the function

$$n(t) = 1000 \cdot 2^{t/3}$$

In general, suppose that the initial size of a population is n_0 and the doubling period is a. Then the size of the population at time t is modeled by

$$n(t) = n_0 2^{ct}$$

where $c = 1/a$. If we knew the tripling time b, then the formula would be $n(t) = n_0 3^{ct}$ where $c = 1/b$. These formulas indicate that the growth of the bacteria is modeled by an exponential function. But what base should we use? The answer is e, because then it can be shown (using calculus) that the population is modeled by

$$n(t) = n_0 e^{rt}$$

where r is the *relative rate of growth of population, expressed as a proportion of the population at any time*. For instance, if $r = 0.02$, then at any time t the growth rate is 2% of the population at time t.

Notice that the formula for population growth is the same as that for continuously compounded interest. In fact, the same principle is at work in both cases: The growth of a population (or an investment) per time period is proportional to the size of the population (or the amount of the investment). A population of 1,000,000 will increase more in one year than a population of 1000; in exactly the same way, an investment of $1,000,000 will increase more in one year than an investment of $1000.

Exponential Growth Model

A population that experiences **exponential growth** increases according to the model

$$n(t) = n_0 e^{rt}$$

where $n(t)$ = population at time t

n_0 = initial size of the population

r = relative rate of growth (expressed as a proportion of the population)

t = time

In the following examples we assume that the populations grow exponentially.

Example 1   Predicting the Size of a Population

The initial bacterium count in a culture is 500. A biologist later makes a sample count of bacteria in the culture and finds that the relative rate of growth is 40% per hour.

(a) Find a function that models the number of bacteria after t hours.

(b) What is the estimated count after 10 hours?

(c) Sketch the graph of the function $n(t)$.

Solution

(a) We use the exponential growth model with $n_0 = 500$ and $r = 0.4$ to get

$$n(t) = 500 e^{0.4t}$$

where t is measured in hours.

(b) Using the function in part (a), we find that the bacterium count after 10 hours is

$$n(10) = 500 e^{0.4(10)} = 500 e^4 \approx 27{,}300$$

(c) The graph is shown in Figure 1.   ∎

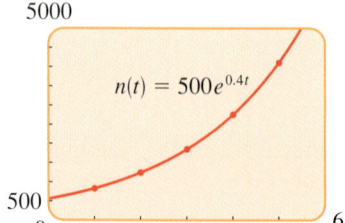

Figure 1

Example 2 Comparing Different Rates of Population Growth

In 2000 the population of the world was 6.1 billion and the relative rate of growth was 1.4% per year. It is claimed that a rate of 1.0% per year would make a significant difference in the total population in just a few decades. Test this claim by estimating the population of the world in the year 2050 using a relative rate of growth of (a) 1.4% per year and (b) 1.0% per year.

Graph the population functions for the next 100 years for the two relative growth rates in the same viewing rectangle.

Solution

(a) By the exponential growth model, we have

$$n(t) = 6.1e^{0.014t}$$

where $n(t)$ is measured in billions and t is measured in years since 2000. Because the year 2050 is 50 years after 2000, we find

$$n(50) = 6.1e^{0.014(50)} = 6.1e^{0.7} \approx 12.3$$

The estimated population in the year 2050 is about 12.3 billion.

(b) We use the function

$$n(t) = 6.1e^{0.010t}$$

and find

$$n(50) = 6.1e^{0.010(50)} = 6.1e^{0.50} \approx 10.1$$

The estimated population in the year 2050 is about 10.1 billion.

The graphs in Figure 2 show that a small change in the relative rate of growth will, over time, make a large difference in population size. ∎

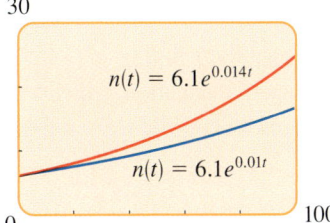

Figure 2

Example 3 Finding the Initial Population

A certain breed of rabbit was introduced onto a small island about 8 years ago. The current rabbit population on the island is estimated to be 4100, with a relative growth rate of 55% per year.

(a) What was the initial size of the rabbit population?

(b) Estimate the population 12 years from now.

Solution

(a) From the exponential growth model, we have

$$n(t) = n_0 e^{0.55t}$$

and we know that the population at time $t = 8$ is $n(8) = 4100$. We substitute what we know into the equation and solve for n_0:

$$4100 = n_0 e^{0.55(8)}$$

$$n_0 = \frac{4100}{e^{0.55(8)}} \approx \frac{4100}{81.45} \approx 50$$

Thus, we estimate that 50 rabbits were introduced onto the island.

Another way to solve part (b) is to let t be the number of years from now. In this case, $n_0 = 4100$ (the current population), and the population 12 years from now will be

$$n(12) = 4100e^{0.55(12)} \approx 3 \text{ million}$$

(b) Now that we know n_0, we can write a formula for population growth:

$$n(t) = 50e^{0.55t}$$

Twelve years from now, $t = 20$ and

$$n(20) = 50e^{0.55(20)} \approx 2{,}993{,}707$$

We estimate that the rabbit population on the island 12 years from now will be about 3 million.   ∎

Can the rabbit population in Example 3(b) actually reach such a high number? In reality, as the island becomes overpopulated with rabbits, the rabbit population growth will be slowed due to food shortage and other factors. One model that takes into account such factors is the *logistic growth model* described in the *Focus on Modeling*, page 392.

Standing Room Only

The population of the world was about 6.1 billion in 2000, and was increasing at 1.4% per year. Assuming that each person occupies an average of 4 ft² of the surface of the earth, the exponential model for population growth projects that by the year 2801 there will be standing room only! (The total land surface area of the world is about 1.8×10^{15} ft².)

Example 4   World Population Projections

The population of the world in 2000 was 6.1 billion, and the estimated relative growth rate was 1.4% per year. If the population continues to grow at this rate, when will it reach 122 billion?

Solution   We use the population growth function with $n_0 = 6.1$ billion, $r = 0.014$, and $n(t) = 122$ billion. This leads to an exponential equation, which we solve for t.

$$6.1e^{0.014t} = 122 \qquad n_0e^{rt} = n(t)$$

$$e^{0.014t} = 20 \qquad \text{Divide by 6.1}$$

$$\ln e^{0.014t} = \ln 20 \qquad \text{Take ln of each side}$$

$$0.014t = \ln 20 \qquad \text{Property of ln}$$

$$t = \frac{\ln 20}{0.014} \qquad \text{Divide by 0.014}$$

$$t \approx 213.98 \qquad \text{Calculator}$$

Thus, the population will reach 122 billion in approximately 214 years, that is, in the year $2000 + 214 = 2214$.   ∎

Example 5   The Number of Bacteria in a Culture

A culture starts with 10,000 bacteria, and the number doubles every 40 min.

(a) Find a function that models the number of bacteria at time t.

(b) Find the number of bacteria after one hour.

(c) After how many minutes will there be 50,000 bacteria?

 (d) Sketch a graph of the number of bacteria at time t.

Solution

(a) To find the function that models this population growth, we need to find the rate r. To do this, we use the formula for population growth with $n_0 = 10{,}000$, $t = 40$, and $n(t) = 20{,}000$, and then solve for r.

$$10{,}000e^{r(40)} = 20{,}000 \qquad n_0e^{rt} = n(t)$$

$$e^{40r} = 2 \qquad \text{Divide by 10,000}$$

$$\ln e^{40r} = \ln 2 \qquad \text{Take ln of each side}$$

$$40r = \ln 2 \qquad \text{Property of ln}$$

$$r = \frac{\ln 2}{40} \qquad \text{Divide by 40}$$

$$r \approx 0.01733 \qquad \text{Calculator}$$

Now that we know $r \approx 0.01733$, we can write the function for the population growth:

$$n(t) = 10{,}000e^{0.01733t}$$

(b) Using the function we found in part (a) with $t = 60$ min (one hour), we get

$$n(60) = 10{,}000e^{0.01733(60)} \approx 28{,}287$$

Thus, the number of bacteria after one hour is approximately 28,000.

(c) We use the function we found in part (a) with $n(t) = 50{,}000$ and solve the resulting exponential equation for t.

$$10{,}000e^{0.01733t} = 50{,}000 \qquad n_0e^{rt} = n(t)$$

$$e^{0.01733t} = 5 \qquad \text{Divide by 10,000}$$

$$\ln e^{0.01733t} = \ln 5 \qquad \text{Take ln of each side}$$

$$0.01733t = \ln 5 \qquad \text{Property of ln}$$

$$t = \frac{\ln 5}{0.01733} \qquad \text{Divide by 0.01733}$$

$$t \approx 92.9 \qquad \text{Calculator}$$

The bacterium count will reach 50,000 in approximately 93 min.

(d) The graph of the function $n(t) = 10{,}000e^{0.01733t}$ is shown in Figure 3.   ■

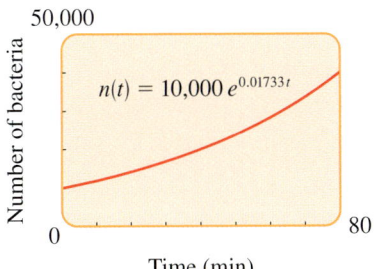

Figure 3

The half-lives of **radioactive elements** vary from very long to very short. Here are some examples.	
Element	**Half-life**
Thorium-232	14.5 billion years
Uranium-235	4.5 billion years
Thorium-230	80,000 years
Plutonium-239	24,360 years
Carbon-14	5,730 years
Radium-226	1,600 years
Cesium-137	30 years
Strontium-90	28 years
Polonium-210	140 days
Thorium-234	25 days
Iodine-135	8 days
Radon-222	3.8 days
Lead-211	3.6 minutes
Krypton-91	10 seconds

Radioactive Decay

Radioactive substances decay by spontaneously emitting radiation. The rate of decay is directly proportional to the mass of the substance. This is analogous to population growth, except that the mass of radioactive material *decreases*. It can be shown that the mass $m(t)$ remaining at time t is modeled by the function

$$m(t) = m_0e^{-rt}$$

where r is the rate of decay expressed as a proportion of the mass and m_0 is the initial mass. Physicists express the rate of decay in terms of **half-life**, the time required for half the mass to decay. We can obtain the rate r from this as follows. If h is the

Radioactive Waste

Harmful radioactive isotopes are produced whenever a nuclear reaction occurs, whether as the result of an atomic bomb test, a nuclear accident such as the one at Chernobyl in 1986, or the uneventful production of electricity at a nuclear power plant.

One radioactive material produced in atomic bombs is the isotope strontium-90 (^{90}Sr), with a half-life of 28 years. This is deposited like calcium in human bone tissue, where it can cause leukemia and other cancers. However, in the decades since atmospheric testing of nuclear weapons was halted, ^{90}Sr levels in the environment have fallen to a level that no longer poses a threat to health.

Nuclear power plants produce radioactive plutonium-239 (^{239}Pu), which has a half-life of 24,360 years. Because of its long half-life, ^{239}Pu could pose a threat to the environment for thousands of years. So, great care must be taken to dispose of it properly. The difficulty of ensuring the safety of the disposed radioactive waste is one reason that nuclear power plants remain controversial.

Joel W. Rogers/Corbis

half-life, then a mass of 1 unit becomes $\frac{1}{2}$ unit when $t = h$. Substituting this into the model, we get

$$\frac{1}{2} = 1 \cdot e^{-rh} \qquad \textcolor{blue}{m(t) = m_o e^{-rt}}$$

$$\ln\left(\tfrac{1}{2}\right) = -rh \qquad \textcolor{blue}{\text{Take ln of each side}}$$

$$r = -\frac{1}{h}\ln(2^{-1}) \qquad \textcolor{blue}{\text{Solve for } r}$$

$$r = \frac{\ln 2}{h} \qquad \textcolor{blue}{\ln 2^{-1} = -\ln 2 \text{ by Law 3}}$$

This last equation allows us to find the rate r from the half-life h.

Radioactive Decay Model

If m_0 is the initial mass of a radioactive substance with half-life h, then the mass remaining at time t is modeled by the function

$$m(t) = m_0 e^{-rt}$$

where $r = \dfrac{\ln 2}{h}$.

Example 6   Radioactive Decay

Polonium-210 (^{210}Po) has a half-life of 140 days. Suppose a sample of this substance has a mass of 300 mg.

(a) Find a function that models the amount of the sample remaining at time t.

(b) Find the mass remaining after one year.

(c) How long will it take for the sample to decay to a mass of 200 mg?

(d) Draw a graph of the sample mass as a function of time.

Solution

(a) Using the model for radioactive decay with $m_0 = 300$ and $r = (\ln 2/140) \approx 0.00495$, we have

$$m(t) = 300 e^{-0.00495t}$$

(b) We use the function we found in part (a) with $t = 365$ (one year).

$$m(365) = 300 e^{-0.00495(365)} \approx 49.256$$

Thus, approximately 49 mg of ^{210}Po remains after one year.

(c) We use the function we found in part (a) with $m(t) = 200$ and solve the resulting exponential equation for t.

$$300 e^{-0.00495t} = 200 \qquad \textcolor{blue}{m(t) = m_o e^{-rt}}$$

$$e^{-0.00495t} = \tfrac{2}{3} \qquad \textcolor{blue}{\text{Divided by 300}}$$

$$\ln e^{-0.00495t} = \ln \tfrac{2}{3} \qquad \textcolor{blue}{\text{Take ln of each side}}$$

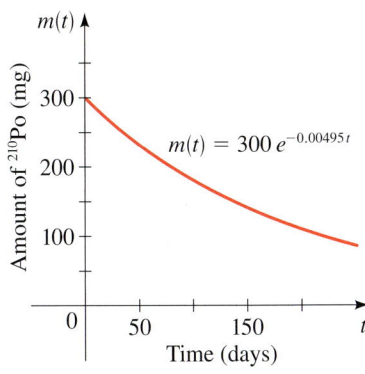

$$m(t) = 300\,e^{-0.00495t}$$

Figure 4

$$-0.00495t = \ln \tfrac{2}{3} \qquad \text{Property of ln}$$

$$t = -\frac{\ln \tfrac{2}{3}}{0.00495} \qquad \text{Divide by } -0.00495$$

$$t \approx 81.9 \qquad \text{Calculator}$$

The time required for the sample to decay to 200 mg is about 82 days.

(d) A graph of the function $m(t) = 300e^{-0.00495t}$ is shown in Figure 4. ∎

Newton's Law of Cooling

Newton's Law of Cooling states that the rate of cooling of an object is proportional to the temperature difference between the object and its surroundings, provided that the temperature difference is not too large. Using calculus, the following model can be deduced from this law.

Newton's Law of Cooling

If D_0 is the initial temperature difference between an object and its surroundings, and if its surroundings have temperature T_s, then the temperature of the object at time t is modeled by the function

$$T(t) = T_s + D_0 e^{-kt}$$

where k is a positive constant that depends on the type of object.

Example 7   Newton's Law of Cooling

A cup of coffee has a temperature of 200 °F and is placed in a room that has a temperature of 70 °F. After 10 min the temperature of the coffee is 150 °F.

(a) Find a function that models the temperature of the coffee at time t.

(b) Find the temperature of the coffee after 15 min.

(c) When will the coffee have cooled to 100 °F?

(d) Illustrate by drawing a graph of the temperature function.

Solution

(a) The temperature of the room is $T_s = 70\,°\text{F}$, and the initial temperature difference is

$$D_0 = 200 - 70 = 130\,°\text{F}$$

So, by Newton's Law of Cooling, the temperature after t minutes is modeled by the function

$$T(t) = 70 + 130e^{-kt}$$

We need to find the constant k associated with this cup of coffee. To do this, we use the fact that when $t = 10$, the temperature is $T(10) = 150$.

So we have

$$70 + 130e^{-10k} = 150 \qquad \textcolor{blue}{T_s + D_0 e^{-kt} = T(t)}$$

$$130e^{-10k} = 80 \qquad \textcolor{blue}{\text{Subtract } 70}$$

$$e^{-10k} = \tfrac{8}{13} \qquad \textcolor{blue}{\text{Divide by } 130}$$

$$-10k = \ln\tfrac{8}{13} \qquad \textcolor{blue}{\text{Take } \ln \text{ of each side}}$$

$$k = -\tfrac{1}{10}\ln\tfrac{8}{13} \qquad \textcolor{blue}{\text{Divide by } -10}$$

$$k \approx 0.04855 \qquad \textcolor{blue}{\text{Caculator}}$$

Substituting this value of k into the expression for $T(t)$, we get

$$T(t) = 70 + 130e^{-0.04855t}$$

(b) We use the function we found in part (a) with $t = 15$.

$$T(15) = 70 + 130e^{-0.04855(15)} \approx 133\,°\text{F}$$

(c) We use the function we found in part (a) with $T(t) = 100$ and solve the resulting exponential equation for t.

$$70 + 130e^{-0.04855t} = 100 \qquad \textcolor{blue}{T_s + D_0 e^{-kt} = T(t)}$$

$$130e^{-0.04855t} = 30 \qquad \textcolor{blue}{\text{Subtract } 70}$$

$$e^{-0.04855t} = \tfrac{3}{13} \qquad \textcolor{blue}{\text{Divide by } 130}$$

$$-0.04855t = \ln\tfrac{3}{13} \qquad \textcolor{blue}{\text{Take } \ln \text{ of each side}}$$

$$t = \frac{\ln\tfrac{3}{13}}{-0.04855} \qquad \textcolor{blue}{\text{Divide by } -0.04855}$$

$$t \approx 30.2 \qquad \textcolor{blue}{\text{Calculator}}$$

The coffee will have cooled to 100°F after about half an hour.

(d) The graph of the temperature function is sketched in Figure 5. Notice that the line $t = 70$ is a horizontal asymptote. (Why?)

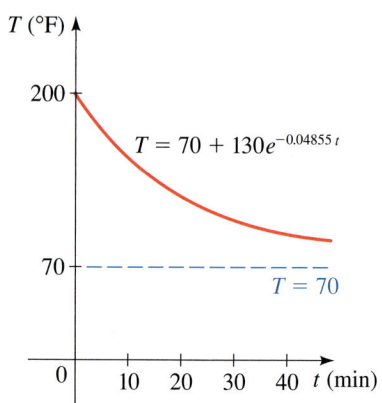

Figure 5

Temperature of coffee after t minutes

Logarithmic Scales

When a physical quantity varies over a very large range, it is often convenient to take its logarithm in order to have a more manageable set of numbers. We discuss three such situations: the pH scale, which measures acidity; the Richter scale, which measures the intensity of earthquakes; and the decibel scale, which measures the loudness of sounds. Other quantities that are measured on logarithmic scales are light intensity, information capacity, and radiation.

THE pH SCALE   Chemists measured the acidity of a solution by giving its hydrogen ion concentration until Sorensen, in 1909, proposed a more convenient measure. He defined

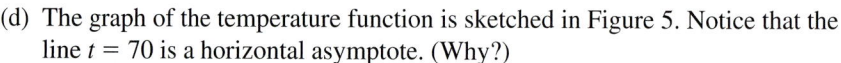

$$\text{pH} = -\log[\text{H}^+]$$

Substance	pH
Milk of Magnesia	10.5
Seawater	8.0–8.4
Human blood	7.3–7.5
Crackers	7.0–8.5
Hominy (lye)	6.9–7.9
Cow's milk	6.4–6.8
Spinach	5.1–5.7
Tomatoes	4.1–4.4
Oranges	3.0–4.0
Apples	2.9–3.3
Limes	1.3–2.0
Battery acid	1.0

where $[H^+]$ is the concentration of hydrogen ions measured in moles per liter (M). He did this to avoid very small numbers and negative exponents. For instance,

$$\text{if} \qquad [H^+] = 10^{-4}\,M, \qquad \text{then} \qquad pH = -\log_{10}(10^{-4}) = -(-4) = 4$$

Solutions with a pH of 7 are defined as *neutral*, those with pH $<$ 7 are *acidic*, and those with pH $>$ 7 are *basic*. Notice that when the pH increases by one unit, $[H^+]$ decreases by a factor of 10.

Example 8   pH Scale and Hydrogen Ion Concentration

(a) The hydrogen ion concentration of a sample of human blood was measured to be $[H^+] = 3.16 \times 10^{-8}$ M. Find the pH and classify the blood as acidic or basic.

(b) The most acidic rainfall ever measured occurred in Scotland in 1974; its pH was 2.4. Find the hydrogen ion concentration.

Solution

(a) A calculator gives

$$pH = -\log[H^+] = -\log(3.16 \times 10^{-8}) \approx 7.5$$

Since this is greater than 7, the blood is basic.

(b) To find the hydrogen ion concentration, we need to solve for $[H^+]$ in the logarithmic equation

$$\log[H^+] = -pH$$

So, we write it in exponential form.

$$[H^+] = 10^{-pH}$$

In this case, pH $= 2.4$, so

$$[H^+] = 10^{-2.4} \approx 4.0 \times 10^{-3}\,M \qquad \blacksquare$$

THE RICHTER SCALE   In 1935 the American geologist Charles Richter (1900–1984) defined the magnitude M of an earthquake to be

$$M = \log \frac{I}{S}$$

where I is the intensity of the earthquake (measured by the amplitude of a seismograph reading taken 100 km from the epicenter of the earthquake) and S is the intensity of a "standard" earthquake (whose amplitude is 1 micron $= 10^{-4}$ cm). The magnitude of a standard earthquake is

$$M = \log \frac{S}{S} = \log 1 = 0$$

Richter studied many earthquakes that occurred between 1900 and 1950. The largest had magnitude 8.9 on the Richter scale, and the smallest had magnitude 0. This corresponds to a ratio of intensities of 800,000,000, so the Richter scale provides more

Largest Earthquakes

Location	Date	Magnitude
Chile	1960	9.5
Alaska	1964	9.2
Alaska	1957	9.1
Kamchatka	1952	9.0
Sumatra	2004	9.0
Ecuador	1906	8.8
Alaska	1965	8.7
Tibet	1950	8.6
Kamchatka	1923	8.5
Indonesia	1938	8.5
Kuril Islands	1963	8.5

manageable numbers to work with. For instance, an earthquake of magnitude 6 is ten times stronger than an earthquake of magnitude 5.

Example 9   Magnitude of Earthquakes

The 1906 earthquake in San Francisco had an estimated magnitude of 8.3 on the Richter scale. In the same year a powerful earthquake occurred on the Colombia-Ecuador border and was four times as intense. What was the magnitude of the Colombia-Ecuador earthquake on the Richter scale?

Solution   If I is the intensity of the San Francisco earthquake, then from the definition of magnitude we have

$$M = \log \frac{I}{S} = 8.3$$

The intensity of the Colombia-Ecuador earthquake was $4I$, so its magnitude was

$$M = \log \frac{4I}{S} = \log 4 + \log \frac{I}{S} = \log 4 + 8.3 \approx 8.9$$ ∎

Example 10   Intensity of Earthquakes

The 1989 Loma Prieta earthquake that shook San Francisco had a magnitude of 7.1 on the Richter scale. How many times more intense was the 1906 earthquake (see Example 9) than the 1989 event?

Solution   If I_1 and I_2 are the intensities of the 1906 and 1989 earthquakes, then we are required to find I_1/I_2. To relate this to the definition of magnitude, we divide numerator and denominator by S.

$$\log \frac{I_1}{I_2} = \log \frac{I_1/S}{I_2/S} \qquad \textit{Divide numerator and denominator by S}$$

$$= \log \frac{I_1}{S} - \log \frac{I_2}{S} \qquad \textit{Law 2 of logarithms}$$

$$= 8.3 - 7.1 = 1.2 \qquad \textit{Definition of earthquake magnitude}$$

Therefore

$$\frac{I_1}{I_2} = 10^{\log(I_1/I_2)} = 10^{1.2} \approx 16$$

The 1906 earthquake was about 16 times as intense as the 1989 earthquake.   ∎

Roger Ressmeyer/Corbis

THE DECIBEL SCALE   The ear is sensitive to an extremely wide range of sound intensities. We take as a reference intensity $I_0 = 10^{-12}$ W/m^2 (watts per square meter) at a frequency of 1000 hertz, which measures a sound that is just barely audible (the threshold of hearing). The psychological sensation of loudness varies with the logarithm of the intensity (the Weber-Fechner Law) and so the **intensity level** B, measured in decibels (dB), is defined as

$$B = 10 \log \frac{I}{I_0}$$

The intensity level of the barely audible reference sound is

$$B = 10 \log \frac{I_0}{I_0} = 10 \log 1 = 0 \text{ dB}$$

Example 11   Sound Intensity of a Jet Takeoff

Find the decibel intensity level of a jet engine during takeoff if the intensity was measured at 100 W/m².

Solution   From the definition of intensity level we see that

$$B = 10 \log \frac{I}{I_0} = 10 \log \frac{10^2}{10^{-12}} = 10 \log 10^{14} = 140 \text{ dB}$$

Thus, the intensity level is 140 dB.   ■

The table in the margin lists decibel intensity levels for some common sounds ranging from the threshold of human hearing to the jet takeoff of Example 11. The threshold of pain is about 120 dB.

4.5   Exercises

1–13 ■ These exercises use the population growth model.

1. Bacteria Culture   The number of bacteria in a culture is modeled by the function

$$n(t) = 500e^{0.45t}$$

where t is measured in hours.

(a) What is the initial number of bacteria?

(b) What is the relative rate of growth of this bacterium population? Express your answer as a percentage.

(c) How many bacteria are in the culture after 3 hours?

(d) After how many hours will the number of bacteria reach 10,000?

2. Fish Population   The number of a certain species of fish is modeled by the function

$$n(t) = 12e^{0.012t}$$

where t is measured in years and $n(t)$ is measured in millions.

(a) What is the relative rate of growth of the fish population? Express your answer as a percentage.

(b) What will the fish population be after 5 years?

(c) After how many years will the number of fish reach 30 million?

(d) Sketch a graph of the fish population function $n(t)$.

3. Fox Population   The fox population in a certain region has a relative growth rate of 8% per year. It is estimated that the population in 2000 was 18,000.

(a) Find a function that models the population t years after 2000.

(b) Use the function from part (a) to estimate the fox population in the year 2008.

(c) Sketch a graph of the fox population function for the years 2000–2008.

4. Population of a Country   The population of a country has a relative growth rate of 3% per year. The government is trying to reduce the growth rate to 2%. The population in 1995 was approximately 110 million. Find the projected population for the year 2020 for the following conditions.

(a) The relative growth rate remains at 3% per year.

(b) The relative growth rate is reduced to 2% per year.

5. Population of a City   The population of a certain city was 112,000 in 1998, and the observed relative growth rate is 4% per year.

(a) Find a function that models the population after t years.

(b) Find the projected population in the year 2004.

(c) In what year will the population reach 200,000?

6. Frog Population   The frog population in a small pond grows exponentially. The current population is 85 frogs, and the relative growth rate is 18% per year.

(a) Find a function that models the population after *t* years.

(b) Find the projected population after 3 years.

(c) Find the number of years required for the frog population to reach 600.

7. Deer Population   The graph shows the deer population in a Pennsylvania county between 1996 and 2000. Assume that the population grows exponentially.

(a) What was the deer population in 1996?

(b) Find a function that models the deer population *t* years after 1996.

(c) What is the projected deer population in 2004?

(d) In what year will the deer population reach 100,000?

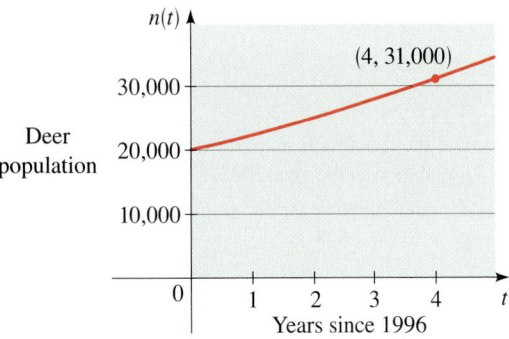

8. Bacteria Culture   A culture contains 1500 bacteria initially and doubles every 30 min.

(a) Find a function that models the number of bacteria *n(t)* after *t* minutes.

(b) Find the number of bacteria after 2 hours.

(c) After how many minutes will the culture contain 4000 bacteria?

9. Bacteria Culture   A culture starts with 8600 bacteria. After one hour the count is 10,000.

(a) Find a function that models the number of bacteria *n(t)* after *t* hours.

(b) Find the number of bacteria after 2 hours.

(c) After how many hours will the number of bacteria double?

10. Bacteria Culture   The count in a culture of bacteria was 400 after 2 hours and 25,600 after 6 hours.

(a) What is the relative rate of growth of the bacteria population? Express your answer as a percentage.

(b) What was the initial size of the culture?

(c) Find a function that models the number of bacteria *n(t)* after *t* hours.

(d) Find the number of bacteria after 4.5 hours.

(e) When will the number of bacteria be 50,000?

11. World Population   The population of the world was 5.7 billion in 1995 and the observed relative growth rate was 2% per year.

(a) By what year will the population have doubled?

(b) By what year will the population have tripled?

12. Population of California   The population of California was 10,586,223 in 1950 and 23,668,562 in 1980. Assume the population grows exponentially.

(a) Find a function that models the population *t* years after 1950.

(b) Find the time required for the population to double.

(c) Use the function from part (a) to predict the population of California in the year 2000. Look up California's actual population in 2000, and compare.

13. Infectious Bacteria   An infectious strain of bacteria increases in number at a relative growth rate of 200% per hour. When a certain critical number of bacteria are present in the bloodstream, a person becomes ill. If a single bacterium infects a person, the critical level is reached in 24 hours. How long will it take for the critical level to be reached if the same person is infected with 10 bacteria?

14–22 ■ These exercises use the radioactive decay model.

14. Radioactive Radium   The half-life of radium-226 is 1600 years. Suppose we have a 22-mg sample.

(a) Find a function that models the mass remaining after *t* years.

(b) How much of the sample will remain after 4000 years?

(c) After how long will only 18 mg of the sample remain?

15. Radioactive Cesium   The half-life of cesium-137 is 30 years. Suppose we have a 10-g sample.

(a) Find a function that models the mass remaining after *t* years.

(b) How much of the sample will remain after 80 years?

(c) After how long will only 2 g of the sample remain?

16. Radioactive Thorium   The mass *m(t)* remaining after *t* days from a 40-g sample of thorium-234 is given by

$$m(t) = 40e^{-0.0277t}$$

(a) How much of the sample will remain after 60 days?

(b) After how long will only 10 g of the sample remain?

(c) Find the half-life of thorium-234.

17. Radioactive Strontium   The half-life of strontium-90 is 28 years. How long will it take a 50-mg sample to decay to a mass of 32 mg?

18. Radioactive Radium   Radium-221 has a half-life of 30 s. How long will it take for 95% of a sample to decay?

19. Finding Half-life   If 250 mg of a radioactive element decays to 200 mg in 48 hours, find the half-life of the element.

20. Radioactive Radon   After 3 days a sample of radon-222 has decayed to 58% of its original amount.

(a) What is the half-life of radon-222?

(b) How long will it take the sample to decay to 20% of its original amount?

21. Carbon-14 Dating   A wooden artifact from an ancient tomb contains 65% of the carbon-14 that is present in living trees. How long ago was the artifact made? (The half-life of carbon-14 is 5730 years.)

22. Carbon-14 Dating   The burial cloth of an Egyptian mummy is estimated to contain 59% of the carbon-14 it contained originally. How long ago was the mummy buried? (The half-life of carbon-14 is 5730 years.)

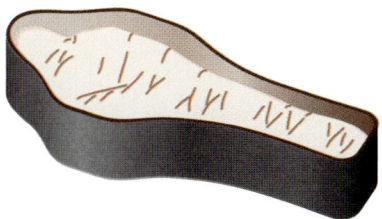

23–26 ■ These exercises use Newton's Law of Cooling.

23. Cooling Soup   A hot bowl of soup is served at a dinner party. It starts to cool according to Newton's Law of Cooling so that its temperature at time t is given by

$$T(t) = 65 + 145e^{-0.05t}$$

where t is measured in minutes and T is measured in °F.

(a) What is the initial temperature of the soup?

(b) What is the temperature after 10 min?

(c) After how long will the temperature be 100°F?

24. Time of Death   Newton's Law of Cooling is used in homicide investigations to determine the time of death. The normal body temperature is 98.6°F. Immediately following death, the body begins to cool. It has been determined experimentally that the constant in Newton's Law of Cooling is approximately $k = 0.1947$, assuming time is measured in hours. Suppose that the temperature of the surroundings is 60°F.

(a) Find a function $T(t)$ that models the temperature t hours after death.

(b) If the temperature of the body is now 72°F, how long ago was the time of death?

25. Cooling Turkey   A roasted turkey is taken from an oven when its temperature has reached 185°F and is placed on a table in a room where the temperature is 75°F.

(a) If the temperature of the turkey is 150°F after half an hour, what is its temperature after 45 min?

(b) When will the turkey cool to 100°F?

 **26. Boiling Water**   A kettle full of water is brought to a boil in a room with temperature 20°C. After 15 min the temperature of the water has decreased from 100°C to 75°C. Find the temperature after another 10 min. Illustrate by graphing the temperature function.

27–41 ■ These exercises deal with logarithmic scales.

27. Finding pH   The hydrogen ion concentration of a sample of each substance is given. Calculate the pH of the substance.

(a) Lemon juice: $[H^+] = 5.0 \times 10^{-3}$ M

(b) Tomato juice: $[H^+] = 3.2 \times 10^{-4}$ M

(c) Seawater: $[H^+] = 5.0 \times 10^{-9}$ M

28. Finding pH   An unknown substance has a hydrogen ion concentration of $[H^+] = 3.1 \times 10^{-8}$ M. Find the pH and classify the substance as acidic or basic.

29. Ion Concentration   The pH reading of a sample of each substance is given. Calculate the hydrogen ion concentration of the substance.

(a) Vinegar: pH = 3.0

(b) Milk: pH = 6.5

30. Ion Concentration   The pH reading of a glass of liquid is given. Find the hydrogen ion concentration of the liquid.

(a) Beer: pH = 4.6

(b) Water: pH = 7.3

31. Finding pH   The hydrogen ion concentrations in cheeses range from 4.0×10^{-7} M to 1.6×10^{-5} M. Find the corresponding range of pH readings.

32. Ion Concentration in Wine   The pH readings for wines vary from 2.8 to 3.8. Find the corresponding range of hydrogen ion concentrations.

33. Earthquake Magnitudes   If one earthquake is 20 times as intense as another, how much larger is its magnitude on the Richter scale?

34. Earthquake Magnitudes   The 1906 earthquake in San Francisco had a magnitude of 8.3 on the Richter scale. At the same time in Japan an earthquake with magnitude 4.9

caused only minor damage. How many times more intense was the San Francisco earthquake than the Japanese earthquake?

35. Earthquake Magnitudes   The Alaska earthquake of 1964 had a magnitude of 8.6 on the Richter scale. How many times more intense was this than the 1906 San Francisco earthquake? (See Exercise 34.)

36. Earthquake Magnitudes   The Northridge, California, earthquake of 1994 had a magnitude of 6.8 on the Richter scale. A year later, a 7.2-magnitude earthquake struck Kobe, Japan. How many times more intense was the Kobe earthquake than the Northridge earthquake?

37. Earthquake Magnitudes   The 1985 Mexico City earthquake had a magnitude of 8.1 on the Richter scale. The 1976 earthquake in Tangshan, China, was 1.26 times as intense. What was the magnitude of the Tangshan earthquake?

38. Traffic Noise   The intensity of the sound of traffic at a busy intersection was measured at 2.0×10^{-5} W/m². Find the intensity level in decibels.

39. Subway Noise   The intensity of the sound of a subway train was measured at 98 dB. Find the intensity in W/m².

40. Comparing Decibel Levels   The noise from a power mower was measured at 106 dB. The noise level at a rock concert was measured at 120 dB. Find the ratio of the intensity of the rock music to that of the power mower.

41. Inverse Square Law for Sound   A law of physics states that the intensity of sound is inversely proportional to the square of the distance d from the source: $I = k/d^2$.

(a) Use this model and the equation

$$B = 10 \log \frac{I}{I_0}$$

(described in this section) to show that the decibel levels B_1 and B_2 at distances d_1 and d_2 from a sound source are related by the equation

$$B_2 = B_1 + 20 \log \frac{d_1}{d_2}$$

(b) The intensity level at a rock concert is 120 dB at a distance 2 m from the speakers. Find the intensity level at a distance of 10 m.

4   Review

Concept Check

1. (a) Write an equation that defines the exponential function with base a.
   (b) What is the domain of this function?
   (c) What is the range of this function?
   (d) Sketch the general shape of the graph of the exponential function for each case.
      (i) $a > 1$ (ii) $0 < a < 1$

2. If x is large, which function grows faster, $y = 2^x$ or $y = x^2$?

3. (a) How is the number e defined?
   (b) What is the natural exponential function?

4. (a) How is the logarithmic function $y = \log_a x$ defined?
   (b) What is the domain of this function?
   (c) What is the range of this function?
   (d) Sketch the general shape of the graph of the function $y = \log_a x$ if $a > 1$.
   (e) What is the natural logarithm?
   (f) What is the common logarithm?

5. State the three Laws of Logarithms.

6. State the Change of Base Formula.

7. (a) How do you solve an exponential equation?
   (b) How do you solve a logarithmic equation?

8. Suppose an amount P is invested at an interest rate r and A is the amount after t years.
   (a) Write an expression for A if the interest is compounded n times per year.
   (b) Write an expression for A if the interest is compounded continuously.

9. If the initial size of a population is n_0 and the population grows exponentially with relative growth rate r, write an expression for the population $n(t)$ at time t.

10. (a) What is the half-life of a radioactive substance?
   (b) If a radioactive substance has initial mass m_0 and half-life h, write an expression for the mass $m(t)$ remaining at time t.

11. What does Newton's Law of Cooling say?

12. What do the pH scale, the Richter scale, and the decibel scale have in common? What do they measure?

Exercises

1–12 ■ Sketch the graph of the function. State the domain, range, and asymptote.

1. $f(x) = 2^{-x+1}$

2. $f(x) = 3^{x-2}$

3. $g(x) = 3 + 2^x$

4. $g(x) = 5^{-x} - 5$

5. $f(x) = \log_3(x - 1)$

6. $g(x) = \log(-x)$

7. $f(x) = 2 - \log_2 x$

8. $f(x) = 3 + \log_5(x + 4)$

9. $F(x) = e^x - 1$

10. $G(x) = \frac{1}{2}e^{x-1}$

11. $g(x) = 2 \ln x$

12. $g(x) = \ln(x^2)$

13–16 ■ Find the domain of the function.

13. $f(x) = 10^{x^2} + \log(1 - 2x)$ **14.** $g(x) = \ln(2 + x - x^2)$

15. $h(x) = \ln(x^2 - 4)$

16. $k(x) = \ln|x|$

17–20 ■ Write the equation in exponential form.

17. $\log_2 1024 = 10$

18. $\log_6 37 = x$

19. $\log x = y$

20. $\ln c = 17$

21–24 ■ Write the equation in logarithmic form.

21. $2^6 = 64$

22. $49^{-1/2} = \frac{1}{7}$

23. $10^x = 74$

24. $e^k = m$

25–40 ■ Evaluate the expression without using a calculator.

25. $\log_2 128$

26. $\log_8 1$

27. $10^{\log 45}$

28. $\log 0.000001$

29. $\ln(e^6)$

30. $\log_4 8$

31. $\log_3\left(\frac{1}{27}\right)$

32. $2^{\log_2 13}$

33. $\log_5 \sqrt{5}$

34. $e^{2\ln 7}$

35. $\log 25 + \log 4$

36. $\log_3 \sqrt{243}$

37. $\log_2 16^{23}$

38. $\log_5 250 - \log_5 2$

39. $\log_8 6 - \log_8 3 + \log_8 2$ **40.** $\log \log 10^{100}$

41–46 ■ Expand the logarithmic expression.

41. $\log(AB^2C^3)$

42. $\log_2(x\sqrt{x^2 + 1})$

43. $\ln\sqrt{\dfrac{x^2 - 1}{x^2 + 1}}$

44. $\log\left(\dfrac{4x^3}{y^2(x - 1)^5}\right)$

45. $\log_5\left(\dfrac{x^2(1 - 5x)^{3/2}}{\sqrt{x^3 - x}}\right)$

46. $\ln\left(\dfrac{\sqrt[3]{x^4 + 12}}{(x + 16)\sqrt{x - 3}}\right)$

47–52 ■ Combine into a single logarithm.

47. $\log 6 + 4 \log 2$

48. $\log x + \log(x^2 y) + 3 \log y$

49. $\frac{3}{2}\log_2(x - y) - 2\log_2(x^2 + y^2)$

50. $\log_5 2 + \log_5(x + 1) - \frac{1}{3}\log_5(3x + 7)$

51. $\log(x - 2) + \log(x + 2) - \frac{1}{2}\log(x^2 + 4)$

52. $\frac{1}{2}[\ln(x - 4) + 5\ln(x^2 + 4x)]$

53–62 ■ Solve the equation. Find the exact solution if possible; otherwise approximate to two decimals.

53. $\log_2(1 - x) = 4$

54. $2^{3x-5} = 7$

55. $5^{5-3x} = 26$

56. $\ln(2x - 3) = 14$

57. $e^{3x/4} = 10$

58. $2^{1-x} = 3^{2x+5}$

59. $\log x + \log(x + 1) = \log 12$

60. $\log_8(x + 5) - \log_8(x - 2) = 1$

61. $x^2 e^{2x} + 2xe^{2x} = 8e^{2x}$

62. $2^{3^x} = 5$

63–66 ■ Use a calculator to find the solution of the equation, correct to six decimal places.

63. $5^{-2x/3} = 0.63$

64. $2^{3x-5} = 7$

65. $5^{2x+1} = 3^{4x-1}$

66. $e^{-15k} = 10,000$

67–70 ■ Draw a graph of the function and use it to determine the asymptotes and the local maximum and minimum values.

67. $y = e^{x/(x+2)}$

68. $y = 2x^2 - \ln x$

69. $y = \log(x^3 - x)$

70. $y = 10^x - 5^x$

71–72 ■ Find the solutions of the equation, correct to two decimal places.

71. $3 \log x = 6 - 2x$

72. $4 - x^2 = e^{-2x}$

73–74 ■ Solve the inequality graphically.

73. $\ln x > x - 2$

74. $e^x < 4x^2$

75. Use a graph of $f(x) = e^x - 3e^{-x} - 4x$ to find, approximately, the intervals on which f is increasing and on which f is decreasing.

76. Find an equation of the line shown in the figure.

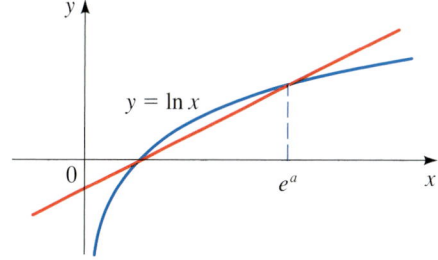

77. Evaluate $\log_4 15$, correct to six decimal places.

78. Solve the inequality: $0.2 \le \log x < 2$

79. Which is larger, $\log_4 258$ or $\log_5 620$?

80. Find the inverse of the function $f(x) = 2^{3^x}$ and state its domain and range.

81. If $12,000 is invested at an interest rate of 10% per year, find the amount of the investment at the end of 3 years for each compounding method.

    (a) Semiannual   **(b)** Monthly

    (c) Daily   **(d)** Continuous

82. A sum of $5000 is invested at an interest rate of $8\frac{1}{2}\%$ per year, compounded semiannually.

    (a) Find the amount of the investment after $1\frac{1}{2}$ years.

    (b) After what period of time will the investment amount to $7000?

83. The stray-cat population in a small town grows exponentially. In 1999, the town had 30 stray cats and the relative growth rate was 15% per year.

    (a) Find a function that models the stray-cat population $n(t)$ after t years.

    (b) Find the projected population after 4 years.

    (c) Find the number of years required for the stray-cat population to reach 500.

84. A culture contains 10,000 bacteria initially. After an hour the bacteria count is 25,000.

    (a) Find the doubling period.

    (b) Find the number of bacteria after 3 hours.

85. Uranium-234 has a half-life of 2.7×10^5 years.

    (a) Find the amount remaining from a 10-mg sample after a thousand years.

    (b) How long will it take this sample to decompose until its mass is 7 mg?

86. A sample of bismuth-210 decayed to 33% of its original mass after 8 days.

    (a) Find the half-life of this element.

    (b) Find the mass remaining after 12 days.

87. The half-life of radium-226 is 1590 years.

    (a) If a sample has a mass of 150 mg, find a function that models the mass that remains after t years.

    (b) Find the mass that will remain after 1000 years.

    (c) After how many years will only 50 mg remain?

88. The half-life of palladium-100 is 4 days. After 20 days a sample has been reduced to a mass of 0.375 g.

    (a) What was the initial mass of the sample?

    (b) Find a function that models the mass remaining after t days.

    (c) What is the mass after 3 days?

    (d) After how many days will only 0.15 g remain?

89. The graph shows the population of a rare species of bird, where t represents years since 1999 and $n(t)$ is measured in thousands.

    (a) Find a function that models the bird population at time t in the form $n(t) = n_0 e^{rt}$.

    (b) What is the bird population expected to be in the year 2010?

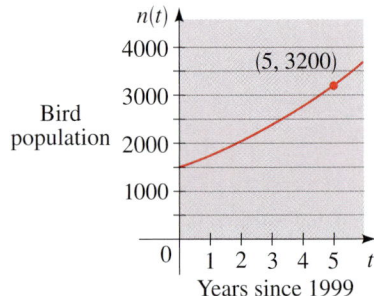

90. A car engine runs at a temperature of $190\,°F$. When the engine is turned off, it cools according to Newton's Law of Cooling with constant $k = 0.0341$, where the time is measured in minutes. Find the time needed for the engine to cool to $90\,°F$ if the surrounding temperature is $60\,°F$.

91. The hydrogen ion concentration of fresh egg whites was measured as

$$[\text{H}^+] = 1.3 \times 10^{-8}\ \text{M}$$

Find the pH, and classify the substance as acidic or basic.

92. The pH of lime juice is 1.9. Find the hydrogen ion concentration.

93. If one earthquake has magnitude 6.5 on the Richter scale, what is the magnitude of another quake that is 35 times as intense?

94. The drilling of a jackhammer was measured at 132 dB. The sound of whispering was measured at 28 dB. Find the ratio of the intensity of the drilling to that of the whispering.

4 Test

1. Graph the functions $y = 2^x$ and $y = \log_2 x$ on the same axes.

2. Sketch the graph of the function $f(x) = \log(x + 1)$ and state the domain, range, and asymptote.

3. Evaluate each logarithmic expression.

    (a) $\log_3 \sqrt{27}$

    (b) $\log_2 80 - \log_2 10$

    (c) $\log_8 4$

    (d) $\log_6 4 + \log_6 9$

4. Use the Laws of Logarithms to expand the expression.

$$\log \sqrt[3]{\frac{x + 2}{x^4(x^2 + 4)}}$$

5. Combine into a single logarithm: $\ln x - 2 \ln(x^2 + 1) + \frac{1}{2}\ln(3 - x^4)$

6. Find the solution of the equation, correct to two decimal places.

    (a) $2^{x-1} = 10$

    (b) $5 \ln(3 - x) = 4$

    (c) $10^{x+3} = 6^{2x}$

    (d) $\log_2(x + 2) + \log_2(x - 1) = 2$

7. The initial size of a culture of bacteria is 1000. After one hour the bacteria count is 8000.

    (a) Find a function that models the population after t hours.

    (b) Find the population after 1.5 hours.

    (c) When will the population reach 15,000?

    (d) Sketch the graph of the population function.

8. Suppose that \$12,000 is invested in a savings account paying 5.6% interest per year.

    (a) Write the formula for the amount in the account after t years if interest is compounded monthly.

    (b) Find the amount in the account after 3 years if interest is compounded daily.

    (c) How long will it take for the amount in the account to grow to \$20,000 if interest is compounded semiannually?

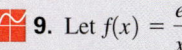 9. Let $f(x) = \dfrac{e^x}{x^3}$.

    (a) Graph f in an appropriate viewing rectangle.

    (b) State the asymptotes of f.

    (c) Find, correct to two decimal places, the local minimum value of f and the value of x at which it occurs.

    (d) Find the range of f.

    (e) Solve the equation $\dfrac{e^x}{x^3} = 2x + 1$. State each solution correct to two decimal places.

In *Focus on Modeling* (page 320) we learned that the shape of a scatter plot helps us choose the type of curve to use in modeling data. The first plot in Figure 1 fairly screams for a line to be fitted through it, and the second one points to a cubic polynomial. For the third plot it is tempting to fit a second-degree polynomial. But what if an exponential curve fits better? How do we decide this? In this section we learn how to fit exponential and power curves to data and how to decide which type of curve fits the data better. We also learn that for scatter plots like those in the last two plots in Figure 1, the data can be modeled by logarithmic or logistic functions.

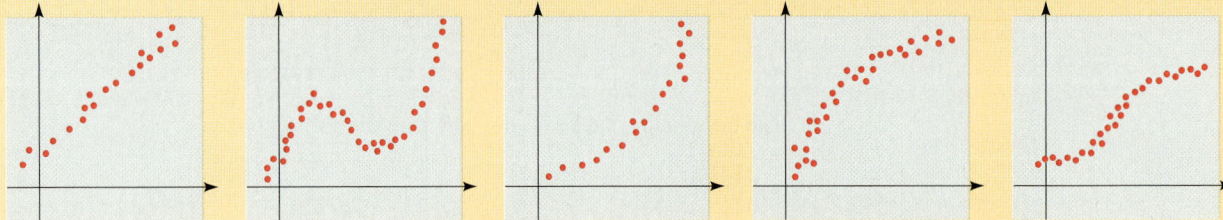

Figure 1

Modeling with Exponential Functions

If a scatter plot shows that the data increases rapidly, we might want to model the data using an *exponential model*, that is, a function of the form

$$f(x) = Ce^{kx}$$

where C and k are constants. In the first example we model world population by an exponential model. Recall from Section 4.5 that population tends to increase exponentially.

Table 1 World population

Year (t)	World population (P in millions)
1900	1650
1910	1750
1920	1860
1930	2070
1940	2300
1950	2520
1960	3020
1970	3700
1980	4450
1990	5300
2000	6060

Example 1 An Exponential Model for World Population

Table 1 gives the population of the world in the 20th century.

(a) Draw a scatter plot and note that a linear model is not appropriate.

(b) Find an exponential function that models population growth.

(c) Draw a graph of the function you found together with the scatter plot. How well does the model fit the data?

(d) Use the model you found to predict world population in the year 2020.

Solution

(a) The scatter plot is shown in Figure 2. The plotted points do not appear to lie

Chabruken / The Image Bank / Getty Images

The population of the world increases exponentially

along a straight line, so a linear model is not appropriate.

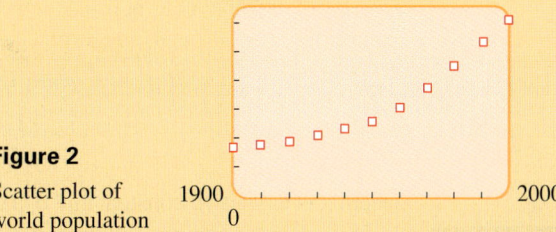

Figure 2

Scatter plot of world population

(b) Using a graphing calculator and the `ExpReg` command (see Figure 3(a)), we get the exponential model

$$P(t) = (0.0082543) \cdot (1.0137186)^t$$

This is a model of the form $y = Cb^t$. To convert this to the form $y = Ce^{kt}$, we use the properties of exponentials and logarithms as follows:

$$1.0137186^t = e^{\ln 1.0137186^t} \qquad A = e^{\ln A}$$
$$= e^{t \ln 1.0137186} \qquad \ln A^B = B \ln A$$
$$= e^{0.013625t} \qquad \ln 1.0137186 \approx 0.013625$$

Thus, we can write the model as

$$P(t) = 0.0082543e^{0.013625t}$$

(c) From the graph in Figure 3(b), we see that the model appears to fit the data fairly well. The period of relatively slow population growth is explained by the depression of the 1930s and the two world wars.

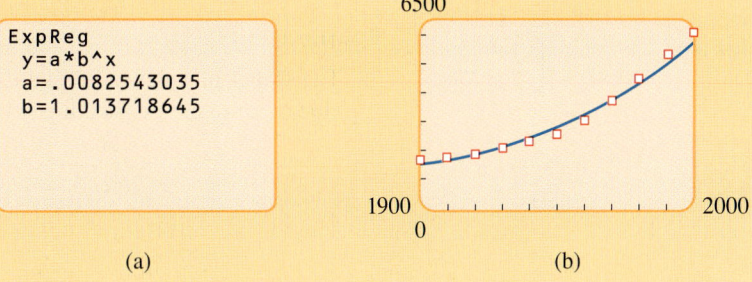

Figure 3

Exponential model for world population　　　　　(a)　　　　　　　　　　　(b)

(d) The model predicts that the world population in 2020 will be

$$P(2020) = 0.0082543e^{(0.013625)(2020)}$$
$$\approx 7,405,400,000$$

∎

Modeling with Power Functions

If the scatter plot of the data we are studying resembles the graph of $y = ax^2$, $y = ax^{1.32}$, or some other power function, then we seek a *power model*, that is, a function of the form

$$f(x) = ax^n$$

where a is a positive constant and n is any real number.

In the next example we seek a power model for some astronomical data. In astronomy, distance in the solar system is often measured in astronomical units. An *astronomical unit* (AU) is the mean distance from the earth to the sun. The *period* of a planet is the time it takes the planet to make a complete revolution around the sun (measured in earth years). In this example we derive the remarkable relationship, first discovered by Johannes Kepler (see page 780), between the mean distance of a planet from the sun and its period.

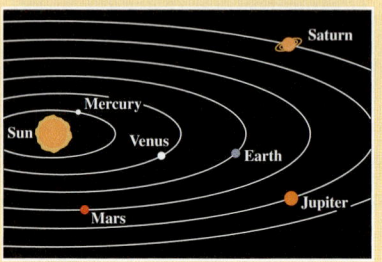

Example 2   A Power Model for Planetary Periods

Table 2   Distances and periods of the planets

Planet	d	T
Mercury	0.387	0.241
Venus	0.723	0.615
Earth	1.000	1.000
Mars	1.523	1.881
Jupiter	5.203	11.861
Saturn	9.541	29.457
Uranus	19.190	84.008
Neptune	30.086	164.784
Pluto	39.507	248.350

Table 2 gives the mean distance d of each planet from the sun in astronomical units and its period T in years.

(a) Sketch a scatter plot. Is a linear model appropriate?

(b) Find a power function that models the data.

(c) Draw a graph of the function you found and the scatter plot on the same graph. How well does the model fit the data?

(d) Use the model you found to find the period of an asteroid whose mean distance from the sun is 5 AU.

Solution

(a) The scatter plot shown in Figure 4 indicates that the plotted points do not lie along a straight line, so a linear model is not appropriate.

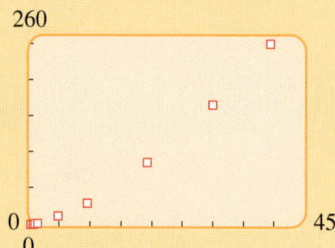

Figure 4
Scatter plot of planetary data

(b) Using a graphing calculator and the `PwrReg` command (see Figure 5(a)), we get the power model

$$T = 1.000396d^{1.49966}$$

If we round both the coefficient and the exponent to three significant figures, we can write the model as

$$T = d^{1.5}$$

This is the relationship discovered by Kepler (see page 780). Sir Isaac Newton later used his Law of Gravity to derive this relationship theoretically, thereby providing strong scientific evidence that the Law of Gravity must be true.

(c) The graph is shown in Figure 5(b). The model appears to fit the data very well.

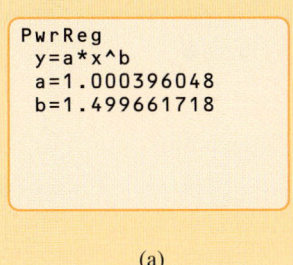

(a)

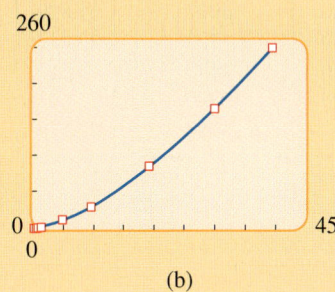

(b)

Figure 5

Power model for planetary data

(d) In this case, $d = 5$ AU and so our model gives

$$T = 1.00039 \cdot 5^{1.49966} \approx 11.22$$

The period of the asteroid is about 11.2 years. ∎

Linearizing Data

We have used the shape of a scatter plot to decide which type of model to use—linear, exponential, or power. This works well if the data points lie on a straight line. But it's difficult to distinguish a scatter plot that is exponential from one that requires a power model. So, to help decide which model to use, we can *linearize* the data, that is, apply a function that "straightens" the scatter plot. The inverse of the linearizing function is then an appropriate model. We now describe how to linearize data that can be modeled by exponential or power functions.

■ **Linearizing exponential data**

If we suspect that the data points (x, y) lie on an exponential curve $y = Ce^{kx}$, then the points

$$(x, \ln y)$$

should lie on a straight line. We can see this from the following calculations:

$$\ln y = \ln Ce^{kx} \qquad \text{Assume } y = Ce^{kx} \text{ and take } \ln$$

$$= \ln e^{kx} + \ln C \qquad \text{Property of ln}$$

$$= kx + \ln C \qquad \text{Property of ln}$$

To see that $\ln y$ is a linear function of x, let $Y = \ln y$ and $A = \ln C$; then

$$Y = kx + A$$

Table 3 World population data

	Population P	
t	(in millions)	$\ln P$
1900	1650	21.224
1910	1750	21.283
1920	1860	21.344
1930	2070	21.451
1940	2300	21.556
1950	2520	21.648
1960	3020	21.829
1970	3700	22.032
1980	4450	22.216
1990	5300	22.391
2000	6060	22.525

We apply this technique to the world population data (t, P) to obtain the points $(t, \ln P)$ in Table 3. The scatter plot in Figure 6 shows that the linearized data lie approximately on a straight line, so an exponential model should be appropriate.

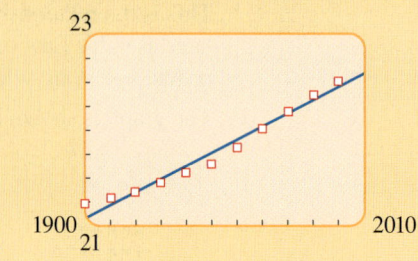

Figure 6

▪ Linearizing power data

If we suspect that the data points (x, y) lie on a power curve $y = ax^n$, then the points

$$(\ln x, \ln y)$$

should be on a straight line. We can see this from the following calculations:

$$\ln y = \ln ax^n \qquad \text{Assume } y = ax^n \text{ and take ln}$$

$$= \ln a + \ln x^n \qquad \text{Property of ln}$$

$$= \ln a + n \ln x \qquad \text{Property of ln}$$

To see that $\ln y$ is a linear function of $\ln x$, let $Y = \ln y$, $X = \ln x$, and $A = \ln a$; then

$$Y = nX + A$$

We apply this technique to the planetary data (d, T) in Table 2, to obtain the points $(\ln d, \ln T)$ in Table 4. The scatter plot in Figure 7 shows that the data lie on a straight line, so a power model seems appropriate.

Table 4 Log-log table

$\ln d$	$\ln T$
-0.94933	-1.4230
-0.32435	-0.48613
0	0
0.42068	0.6318
1.6492	2.4733
2.2556	3.3829
2.9544	4.4309
3.4041	5.1046
3.6765	5.5148

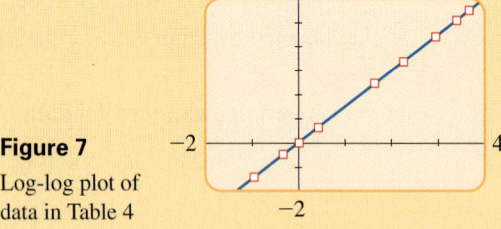

Figure 7

Log-log plot of
data in Table 4

An Exponential or Power Model?

Suppose that a scatter plot of the data points (x, y) shows a rapid increase. Should we use an exponential function or a power function to model the data? To help us decide, we draw two scatter plots—one for the points $(x, \ln y)$ and the other for the points $(\ln x, \ln y)$. If the first scatter plot appears to lie along a line, then an exponential model is appropriate. If the second plot appears to lie along a line, then a power model is appropriate.

Example 3 An Exponential or Power Model?

Data points (x, y) are shown in Table 5.

(a) Draw a scatter plot of the data.

(b) Draw scatter plots of $(x, \ln y)$ and $(\ln x, \ln y)$.

(c) Is an exponential function or a power function appropriate for modeling this data?

(d) Find an appropriate function to model the data.

Table 5

x	y
1	2
2	6
3	14
4	22
5	34
6	46
7	64
8	80
9	102
10	130

Solution

(a) The scatter plot of the data is shown in Figure 8.

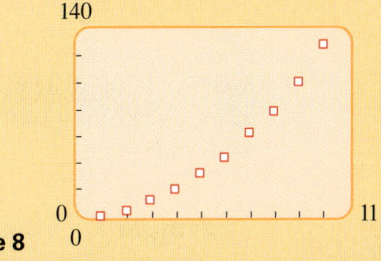

Figure 8

(b) We use the values from Table 6 to graph the scatter plots in Figures 9 and 10.

Table 6

x	$\ln x$	$\ln y$
1	0	0.7
2	0.7	1.8
3	1.1	2.6
4	1.4	3.1
5	1.6	3.5
6	1.8	3.8
7	1.9	4.2
8	2.1	4.4
9	2.2	4.6
10	2.3	4.9

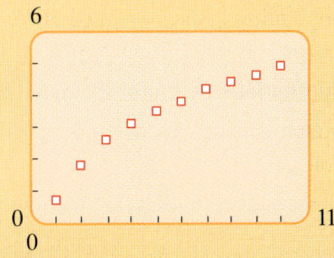

Figure 9

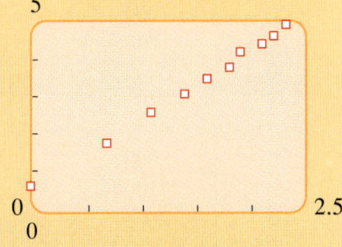

Figure 10

(c) The scatter plot of $(x, \ln y)$ in Figure 9 does not appear to be linear, so an exponential model is not appropriate. On the other hand, the scatter plot of $(\ln x, \ln y)$ in Figure 10 is very nearly linear, so a power model is appropriate.

(d) Using the PwrReg command on a graphing calculator, we find that the power function that best fits the data point is

$$y = 1.85x^{1.82}$$

The graph of this function and the original data points are shown in Figure 11.

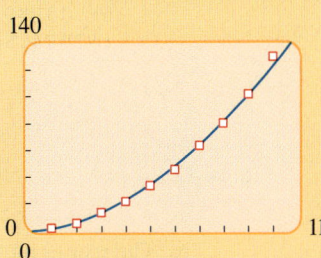

Figure 11

Before graphing calculators and statistical software became common, exponential and power models for data were often constructed by first finding a linear model for the linearized data. Then the model for the actual data was found by taking exponentials. For instance, if we find that $\ln y = A \ln x + B$, then by taking exponentials we get the model $y = e^B \cdot e^{A \ln x}$, or $y = Cx^A$ (where $C = e^B$). Special graphing paper called "log paper" or "log-log paper" was used to facilitate this process.

Modeling with Logistic Functions

A logistic growth model is a function of the form

$$f(t) = \frac{c}{1 + ae^{-bt}}$$

where a, b, and c are positive constants. Logistic functions are used to model populations where the growth is constrained by available resources. (See Exercises 69–72 of Section 4.1.)

Example 4   Stocking a Pond with Catfish

Table 7

Week	Catfish
0	1000
15	1500
30	3300
45	4400
60	6100
75	6900
90	7100
105	7800
120	7900

Much of the fish sold in supermarkets today is raised on commercial fish farms, not caught in the wild. A pond on one such farm is initially stocked with 1000 catfish, and the fish population is then sampled at 15-week intervals to estimate its size. The population data are given in Table 7.

(a) Find an appropriate model for the data.

(b) Make a scatter plot of the data and graph the model you found in part (a) on the scatter plot.

(c) How does the model predict that the fish population will change with time?

Solution

(a) Since the catfish population is restricted by its habitat (the pond), a logistic model is appropriate. Using the Logistic command on a calculator (see Figure 12(a)), we find the following model for the catfish population $P(t)$:

$$P(t) = \frac{7925}{1 + 7.7e^{-0.052t}}$$

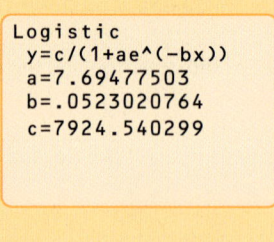

```
Logistic
y=c/(1+ae^(-bx))
a=7.69477503
b=.0523020764
c=7924.540299
```

Figure 12   (a)   (b) Catfish population $y = P(t)$

(b) The scatter plot and the logistic curve are shown in Figure 12(b).

(c) From the graph of P in Figure 12(b), we see that the catfish population increases rapidly until about $t = 80$ weeks. Then growth slows down, and at about $t = 120$ weeks the population levels off and remains more or less constant at slightly over 7900. ∎

The behavior exhibited by the catfish population in Example 4 is typical of logistic growth. After a rapid growth phase, the population approaches a constant level called the **carrying capacity** of the environment. This occurs because as $t \to \infty$, we have $e^{-bt} \to 0$ (see Section 4.1), and so

$$P(t) = \frac{c}{1 + ae^{-bt}} \quad \rightarrow \quad \frac{c}{1 + 0} = c$$

Thus, the carrying capacity is c.

Problems

1. **U.S. Population**   The U.S. Constitution requires a census every 10 years. The census data for 1790–2000 is given in the table.

   (a) Make a scatter plot of the data.

   (b) Use a calculator to find an exponential model for the data.

   (c) Use your model to predict the population at the 2010 census.

   (d) Use your model to estimate the population in 1965.

   (e) Compare your answers from parts (c) and (d) to the values in the table. Do you think an exponential model is appropriate for these data?

Year	Population (in millions)	Year	Population (in millions)	Year	Population (in millions)
1790	3.9	1870	38.6	1950	151.3
1800	5.3	1880	50.2	1960	179.3
1810	7.2	1890	63.0	1970	203.3
1820	9.6	1900	76.2	1980	226.5
1830	12.9	1910	92.2	1990	248.7
1840	17.1	1920	106.0	2000	281.4
1850	23.2	1930	123.2		
1860	31.4	1940	132.2		

Time (s)	Distance (m)
0.1	0.048
0.2	0.197
0.3	0.441
0.4	0.882
0.5	1.227
0.6	1.765
0.7	2.401
0.8	3.136
0.9	3.969
1.0	4.902

2. **A Falling Ball**   In a physics experiment a lead ball is dropped from a height of 5 m. The students record the distance the ball has fallen every one-tenth of a second. (This can be done using a camera and a strobe light.)

   (a) Make a scatter plot of the data.

   (b) Use a calculator to find a power model.

   (c) Use your model to predict how far a dropped ball would fall in 3 s.

3. **Health-care Expenditures**   The U.S. health-care expenditures for 1970–2001 are given in the table on the next page, and a scatter plot of the data is shown in the figure.

   (a) Does the scatter plot shown suggest an exponential model?

   (b) Make a table of the values $(t, \ln E)$ and a scatter plot. Does the scatter plot appear to be linear?

Year	Health expenditures (in billions of dollars)
1970	74.3
1980	251.1
1985	434.5
1987	506.2
1990	696.6
1992	820.3
1994	937.2
1996	1039.4
1998	1150.0
2000	1310.0
2001	1424.5

(c) Find the regression line for the data in part (b).

(d) Use the results of part (c) to find an exponential model for the growth of health-care expenditures.

(e) Use your model to predict the total health-care expenditures in 2009.

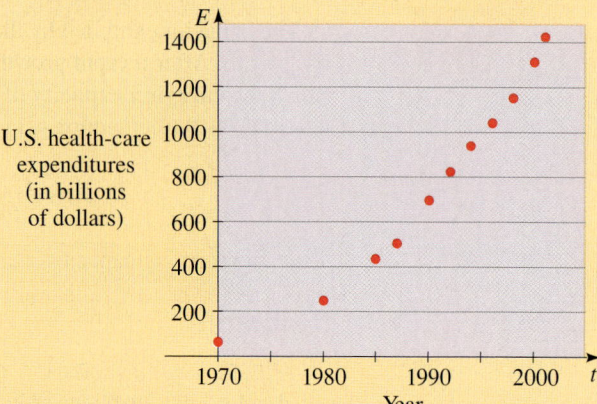

U.S. health-care expenditures (in billions of dollars)

Year

Time (h)	Amount of ^{131}I (g)
0	4.80
8	4.66
16	4.51
24	4.39
32	4.29
40	4.14
48	4.04

4. **Half-life of Radioactive Iodine**   A student is trying to determine the half-life of radioactive iodine-131. He measures the amount of iodine-131 in a sample solution every 8 hours. His data are shown in the table in the margin.

(a) Make a scatter plot of the data.

(b) Use a calculator to find an exponential model.

(c) Use your model to find the half-life of iodine-131.

5. **The Beer-Lambert Law**   As sunlight passes through the waters of lakes and oceans, the light is absorbed and the deeper it penetrates, the more its intensity diminishes. The light intensity I at depth x is given by the Beer-Lambert Law:

$$I = I_0 e^{-kx}$$

where I_0 is the light intensity at the surface and k is a constant that depends on the murkiness of the water (see page 364). A biologist uses a photometer to investigate light penetration in a northern lake, obtaining the data in the table.

(a) Use a graphing calculator to find an exponential function of the form given by the Beer-Lambert Law to model these data. What is the light intensity I_0 at the surface on this day, and what is the "murkiness" constant k for this lake? [*Hint:* If your calculator gives you a function of the form $I = ab^x$, convert this to the form you want using the identities $b^x = e^{\ln(b^x)} = e^{x \ln b}$. See Example 1(b).]

(b) Make a scatter plot of the data and graph the function that you found in part (a) on your scatter plot.

(c) If the light intensity drops below 0.15 lumens (lm), a certain species of algae can't survive because photosynthesis is impossible. Use your model from part (a) to determine the depth below which there is insufficient light to support this algae.

Depth (ft)	Light intensity (lm)	Depth (ft)	Light intensity (lm)
5	13.0	25	1.8
10	7.6	30	1.1
15	4.5	35	0.5
20	2.7	40	0.3

Light intensity decreases exponentially with depth.

6. Experimenting with "Forgetting" Curves Every one of us is all too familiar with the phenomenon of forgetting. Facts that we clearly understood at the time we first learned them sometimes fade from our memory by the time the final exam rolls around. Psychologists have proposed several ways to model this process. One such model is Ebbinghaus' Forgetting Curve, described on page 355. Other models use exponential or logarithmic functions. To develop her own model, a psychologist performs an experiment on a group of volunteers by asking them to memorize a list of 100 related words. She then tests how many of these words they can recall after various periods of time. The average results for the group are shown in the table.

(a) Use a graphing calculator to find a *power* function of the form $y = at^b$ that models the average number of words y that the volunteers remember after t hours. Then find an *exponential* function of the form $y = ab^t$ to model the data.

(b) Make a scatter plot of the data and graph both the functions that you found in part (a) on your scatter plot.

(c) Which of the two functions seems to provide the better model?

Time	Words recalled
15 min	64.3
1 h	45.1
8 h	37.3
1 day	32.8
2 days	26.9
3 days	25.6
5 days	22.9

7. Lead Emissions The table below gives U.S. lead emissions into the environment in millions of metric tons for 1970–1992.

(a) Find an exponential model for these data.

(b) Find a fourth-degree polynomial model for these data.

(c) Which of these curves gives a better model for the data? Use graphs of the two models to decide.

(d) Use each model to estimate the lead emissions in 1972 and 1982.

Year	Lead emissions
1970	199.1
1975	143.8
1980	68.0
1985	18.3
1988	5.9
1989	5.5
1990	5.1
1991	4.5
1992	4.7

8. Auto Exhaust Emissions   A study by the U.S. Office of Science and Technology in 1972 estimated the cost of reducing automobile emissions by certain percentages. Find an exponential model that captures the "diminishing returns" trend of these data shown in the table below.

Reduction in emissions (%)	Cost per car ($)
50	45
55	55
60	62
65	70
70	80
75	90
80	100
85	200
90	375
95	600

9. Exponential or Power Model?   Data points (x, y) are shown in the table.

(a) Draw a scatter plot of the data.

(b) Draw scatter plots of $(x, \ln y)$ and $(\ln x, \ln y)$.

(c) Which is more appropriate for modeling this data—an exponential function or a power function?

(d) Find an appropriate function to model the data.

x	y
2	0.08
4	0.12
6	0.18
8	0.25
10	0.36
12	0.52
14	0.73
16	1.06

x	y
10	29
20	82
30	151
40	235
50	330
60	430
70	546
80	669
90	797

10. Exponential or Power Model?   Data points (x, y) are shown in the table in the margin.

(a) Draw a scatter plot of the data.

(b) Draw scatter plots of $(x, \ln y)$ and $(\ln x, \ln y)$.

(c) Which is more appropriate for modeling this data—an exponential function or a power function?

(d) Find an appropriate function to model the data.

11. Logistic Population Growth   The table and scatter plot give the population of black flies in a closed laboratory container over an 18-day period.

(a) Use the `Logistic` command on your calculator to find a logistic model for these data.

(b) Use the model to estimate the time when there were 400 flies in the container.

Time (days)	Number of flies
0	10
2	25
4	66
6	144
8	262
10	374
12	446
16	492
18	498

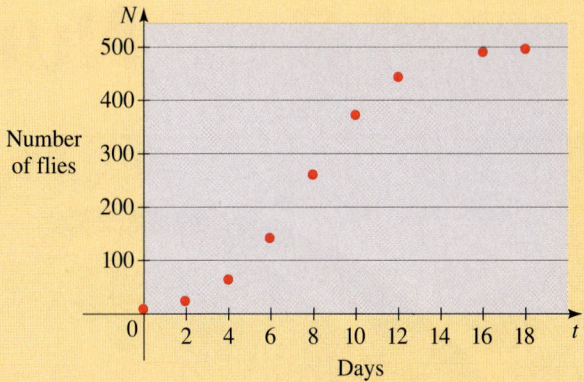

12. Logarithmic Models   A **logarithmic model** is a function of the form

$$y = a + b \ln x$$

Many relationships between variables in the real world can be modeled by this type of function. The table and the scatter plot show the coal production (in metric tons) from a small mine in northern British Columbia.

(a) Use the `LnReg` command on your calculator to find a logarithmic model for these production figures.

(b) Use the model to predict coal production from this mine in 2010.

Year	Metric tons of coal
1950	882
1960	889
1970	894
1980	899
1990	905
2000	909

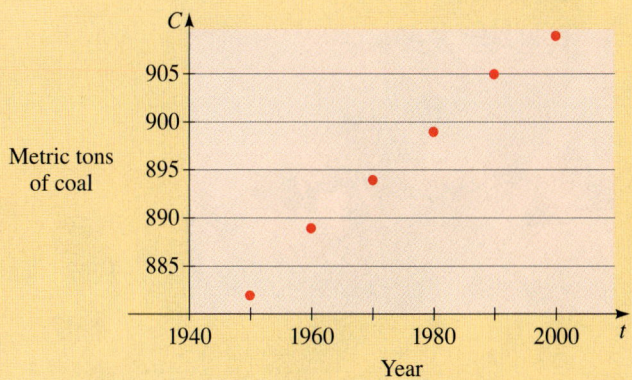

5 Trigonometric Functions of Real Numbers

Chapter Overview

In this chapter and the next we introduce new functions called the trigonometric functions. The trigonometric functions can be defined in two different but equivalent ways—as functions of angles (Chapter 6) or functions of real numbers (Chapter 5). The two approaches to trigonometry are independent of each other, so either Chapter 5 or Chapter 6 may be studied first. We study both approaches because different applications require that we view these functions differently. The approach in this chapter lends itself to modeling periodic motion.

If you've ever taken a ferris wheel ride, then you know about periodic motion—that is, motion that repeats over and over. This type of motion is common in nature. Think about the daily rising and setting of the sun (day, night, day, night, ...), the daily variation in tide levels (high, low, high, low, ...), the vibrations of a leaf in the wind (left, right, left, right, ...), or the pressure in the cylinders of a car engine (high, low, high, low, ...). To describe such motion mathematically we need a function whose values increase, then decrease, then increase, ..., repeating this pattern indefinitely. To understand how to define such a function, let's look at the ferris wheel again. A person riding on the wheel goes up and down, up and down, ... The graph shows how high the person is above the center of the ferris wheel at time t. Notice that as the wheel turns the graph goes up and down repeatedly.

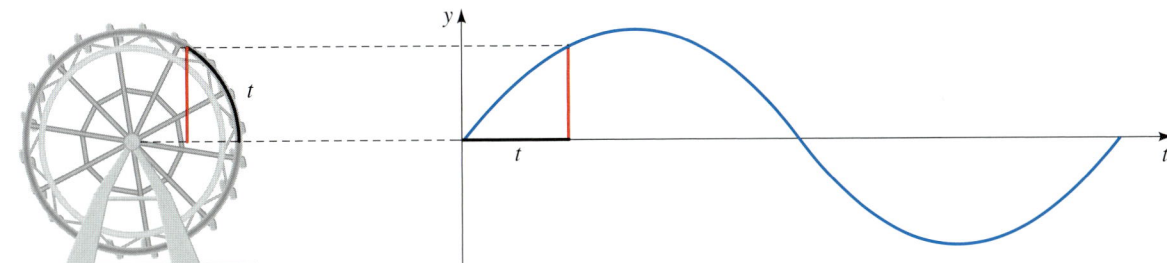

We define the trigonometric function called *sine* in a similar way. We start with a circle of radius 1, and for each distance t along the arc of the circle ending at (x, y) we define the value of the function $\sin t$ to be the height (or y-coordinate) of that point. To apply this function to real-world situations we use the transformations we learned in Chapter 2 to stretch, shrink, or shift the function to fit the variation we are modeling.

There are six trigonometric functions, each with its special properties. In this chapter we study their definitions, graphs, and applications. In Section 5.5 we see how trigonometric functions can be used to model harmonic motion.

5.1	**The Unit Circle**

In this section we explore some properties of the circle of radius 1 centered at the origin. These properties are used in the next section to define the trigonometric functions.

The Unit Circle

The set of points at a distance 1 from the origin is a circle of radius 1 (see Figure 1). In Section 1.8 we learned that the equation of this circle is $x^2 + y^2 = 1$.

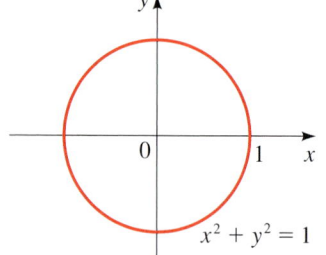

Figure 1
The unit circle

> **The Unit Circle**
>
> The **unit circle** is the circle of radius 1 centered at the origin in the xy-plane. Its equation is
>
> $$x^2 + y^2 = 1$$

Example 1   A Point on the Unit Circle

Show that the point $P\left(\dfrac{\sqrt{3}}{3}, \dfrac{\sqrt{6}}{3}\right)$ is on the unit circle.

Solution   We need to show that this point satisfies the equation of the unit circle, that is, $x^2 + y^2 = 1$. Since

$$\left(\frac{\sqrt{3}}{3}\right)^2 + \left(\frac{\sqrt{6}}{3}\right)^2 = \frac{3}{9} + \frac{6}{9} = 1$$

P is on the unit circle. ∎

Example 2   Locating a Point on the Unit Circle

The point $P(\sqrt{3}/2, y)$ is on the unit circle in quadrant IV. Find its y-coordinate.

Solution   Since the point is on the unit circle, we have

$$\left(\frac{\sqrt{3}}{2}\right)^2 + y^2 = 1$$

$$y^2 = 1 - \frac{3}{4} = \frac{1}{4}$$

$$y = \pm\frac{1}{2}$$

Since the point is in quadrant IV, its y-coordinate must be negative, so $y = -\frac{1}{2}$. ∎

Terminal Points on the Unit Circle

Suppose t is a real number. Let's mark off a distance t along the unit circle, starting at the point $(1, 0)$ and moving in a counterclockwise direction if t is positive or in a clockwise direction if t is negative (Figure 2). In this way we arrive at a point $P(x, y)$ on the unit circle. The point $P(x, y)$ obtained in this way is called the **terminal point** determined by the real number t.

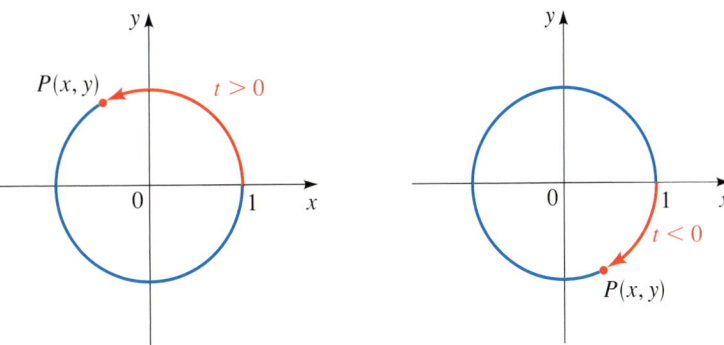

Figure 2

(a) Terminal point $P(x, y)$ determined by $t > 0$

(b) Terminal point $P(x, y)$ determined by $t < 0$

The circumference of the unit circle is $C = 2\pi(1) = 2\pi$. So, if a point starts at $(1, 0)$ and moves counterclockwise all the way around the unit circle and returns to $(1, 0)$, it travels a distance of 2π. To move halfway around the circle, it travels a distance of $\frac{1}{2}(2\pi) = \pi$. To move a quarter of the distance around the circle, it travels a distance of $\frac{1}{4}(2\pi) = \pi/2$. Where does the point end up when it travels these distances along the circle? From Figure 3 we see, for example, that when it travels a distance of π starting at $(1, 0)$, its terminal point is $(-1, 0)$.

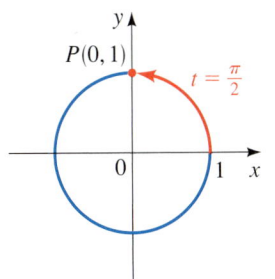 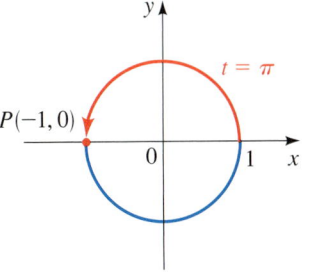 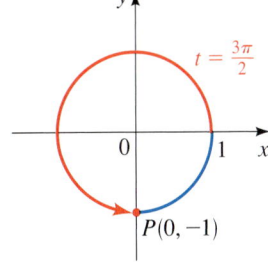 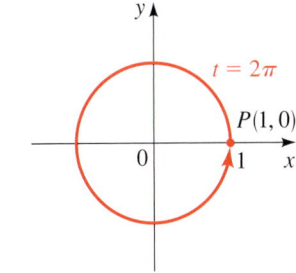

Figure 3

Terminal points determined by $t = \frac{\pi}{2}, \pi, \frac{3\pi}{2},$ and 2π

Example 3   Finding Terminal Points

Find the terminal point on the unit circle determined by each real number t.

(a) $t = 3\pi$ (b) $t = -\pi$ (c) $t = -\dfrac{\pi}{2}$

Solution   From Figure 4 we get the following.

(a) The terminal point determined by 3π is $(-1, 0)$.

(b) The terminal point determined by $-\pi$ is $(-1, 0)$.

(c) The terminal point determined by $-\pi/2$ is $(0, -1)$.

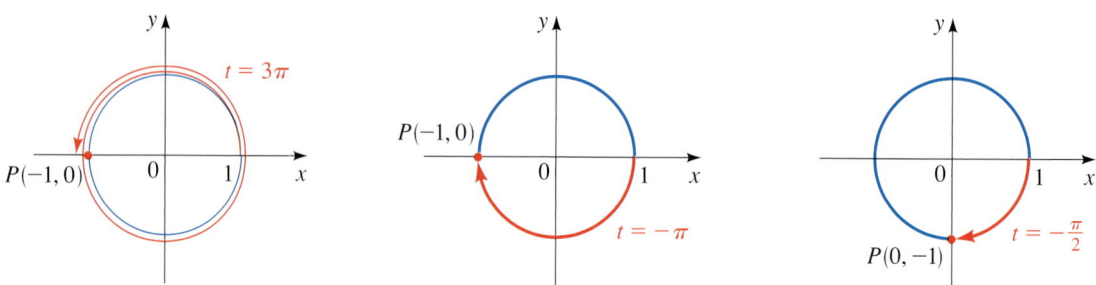

Figure 4

Notice that different values of t can determine the same terminal point. ■

The terminal point $P(x, y)$ determined by $t = \pi/4$ is the same distance from $(1, 0)$ as from $(0, 1)$ along the unit circle (see Figure 5).

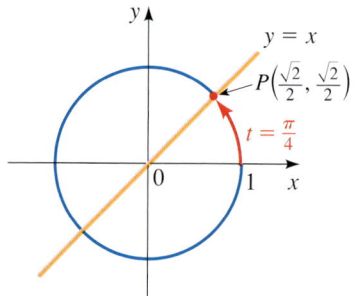

Figure 5

Since the unit circle is symmetric with respect to the line $y = x$, it follows that P lies on the line $y = x$. So P is the point of intersection (in the first quadrant) of the circle $x^2 + y^2 = 1$ and the line $y = x$. Substituting x for y in the equation of the circle, we get

$$x^2 + x^2 = 1$$

$$2x^2 = 1 \qquad \text{Combine like terms}$$

$$x^2 = \frac{1}{2} \qquad \text{Divide by 2}$$

$$x = \pm\frac{1}{\sqrt{2}} \qquad \text{Take square roots}$$

Since P is in the first quadrant, $x = 1/\sqrt{2}$ and since $y = x$, we have $y = 1/\sqrt{2}$ also. Thus, the terminal point determined by $\pi/4$ is

$$P\left(\frac{1}{\sqrt{2}}, \frac{1}{\sqrt{2}}\right) = P\left(\frac{\sqrt{2}}{2}, \frac{\sqrt{2}}{2}\right)$$

Similar methods can be used to find the terminal points determined by $t = \pi/6$ and $t = \pi/3$ (see Exercises 55 and 56). Table 1 and Figure 6 give the terminal points for some special values of t.

Table 1

t	Terminal point determined by t
0	$(1,0)$
$\dfrac{\pi}{6}$	$\left(\dfrac{\sqrt{3}}{2}, \dfrac{1}{2}\right)$
$\dfrac{\pi}{4}$	$\left(\dfrac{\sqrt{2}}{2}, \dfrac{\sqrt{2}}{2}\right)$
$\dfrac{\pi}{3}$	$\left(\dfrac{1}{2}, \dfrac{\sqrt{3}}{2}\right)$
$\dfrac{\pi}{2}$	$(0,1)$

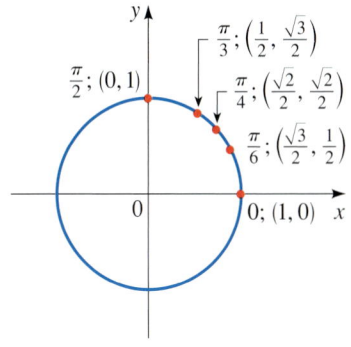

Figure 6

Example 4   Finding Terminal Points

Find the terminal point determined by each given real number t.

(a) $t = -\dfrac{\pi}{4}$ (b) $t = \dfrac{3\pi}{4}$ (c) $t = -\dfrac{5\pi}{6}$

Solution

(a) Let P be the terminal point determined by $-\pi/4$, and let Q be the terminal point determined by $\pi/4$. From Figure 7(a) we see that the point P has the same coordinates as Q except for sign. Since P is in quadrant IV, its x-coordinate is positive and its y-coordinate is negative. Thus, the terminal point is $P\left(\sqrt{2}/2, -\sqrt{2}/2\right)$.

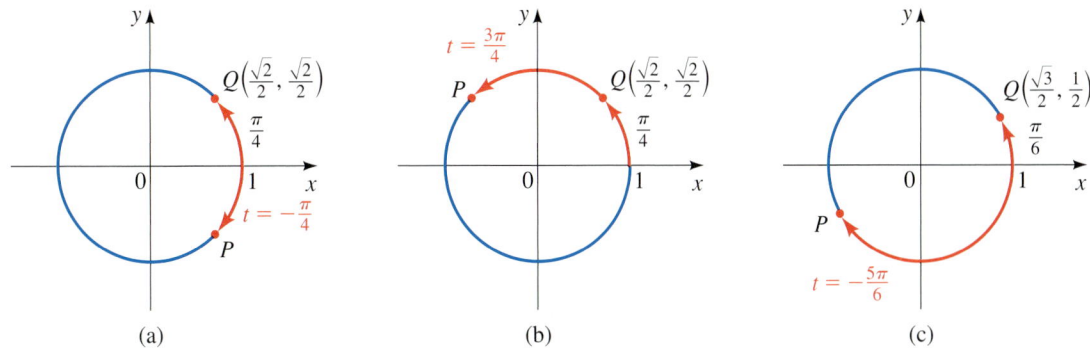

Figure 7   (a)   (b)   (c)

(b) Let P be the terminal point determined by $3\pi/4$, and let Q be the terminal point determined by $\pi/4$. From Figure 7(b) we see that the point P has the same coordinates as Q except for sign. Since P is in quadrant II, its x-coordinate is negative and its y-coordinate is positive. Thus, the terminal point is $P\left(-\sqrt{2}/2, \sqrt{2}/2\right)$.

(c) Let P be the terminal point determined by $-5\pi/6$, and let Q be the terminal point determined by $\pi/6$. From Figure 7(c) we see that the point P has the same coordinates as Q except for sign. Since P is in quadrant III, its coordinates are both negative. Thus, the terminal point is $P\left(-\sqrt{3}/2, -\frac{1}{2}\right)$.   ∎

The Reference Number

From Examples 3 and 4, we see that to find a terminal point in any quadrant we need only know the "corresponding" terminal point in the first quadrant. We use the idea of the *reference number* to help us find terminal points.

Reference Number

Let t be a real number. The **reference number** $\bar{t}$ associated with t is the shortest distance along the unit circle between the terminal point determined by t and the x-axis.

Figure 8 shows that to find the reference number $\bar{t}$ it's helpful to know the quadrant in which the terminal point determined by t lies. If the terminal point lies in quadrants I or IV, where x is positive, we find $\bar{t}$ by moving along the circle to the *positive* x-axis. If it lies in quadrants II or III, where x is negative, we find $\bar{t}$ by moving along the circle to the *negative* x-axis.

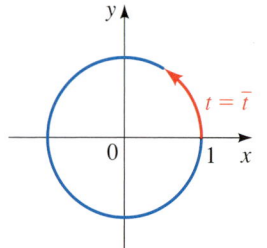 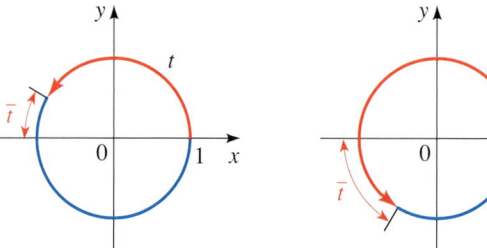 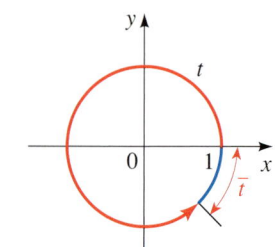

Figure 8

The reference number $\bar{t}$ for t

Example 5   Finding Reference Numbers

Find the reference number for each value of t.

(a) $t = \dfrac{5\pi}{6}$ (b) $t = \dfrac{7\pi}{4}$ (c) $t = -\dfrac{2\pi}{3}$ (d) $t = 5.80$

Solution   From Figure 9 we find the reference numbers as follows.

(a) $\bar{t} = \pi - \dfrac{5\pi}{6} = \dfrac{\pi}{6}$

(b) $\bar{t} = 2\pi - \dfrac{7\pi}{4} = \dfrac{\pi}{4}$

(c) $\bar{t} = \pi - \dfrac{2\pi}{3} = \dfrac{\pi}{3}$

(d) $\bar{t} = 2\pi - 5.80 \approx 0.48$

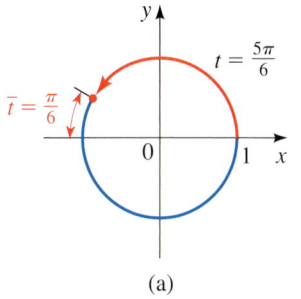
(a)

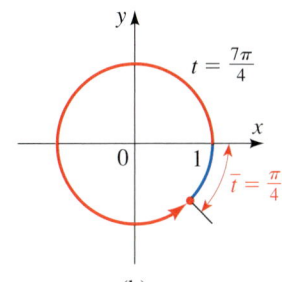
(b)

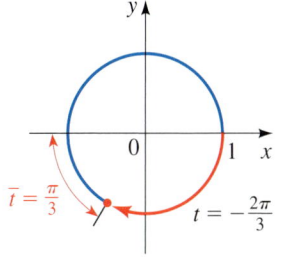
(c)

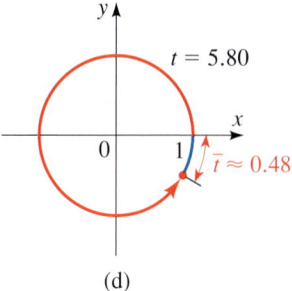
(d)

Figure 9

Using Reference Numbers to Find Terminal Points

To find the terminal point P determined by any value of t, we use the following steps:

1. Find the reference number $\bar{t}$.

2. Find the terminal point $Q(a, b)$ determined by $\bar{t}$.

3. The terminal point determined by t is $P(\pm a, \pm b)$, where the signs are chosen according to the quadrant in which this terminal point lies.

Example 6   Using Reference Numbers to Find Terminal Points

Find the terminal point determined by each given real number t.

(a) $t = \dfrac{5\pi}{6}$ (b) $t = \dfrac{7\pi}{4}$ (c) $t = -\dfrac{2\pi}{3}$

Solution   The reference numbers associated with these values of t were found in Example 5.

(a) The reference number is $\bar{t} = \pi/6$, which determines the terminal point $\left(\sqrt{3}/2, \tfrac{1}{2}\right)$ from Table 1. Since the terminal point determined by t is in quadrant II, its x-coordinate is negative and its y-coordinate is positive. Thus, the desired terminal point is

$$\left(-\dfrac{\sqrt{3}}{2}, \dfrac{1}{2}\right)$$

(b) The reference number is $\bar{t} = \pi/4$, which determines the terminal point $\left(\sqrt{2}/2, \sqrt{2}/2\right)$ from Table 1. Since the terminal point is in quadrant IV, its

x-coordinate is positive and its y-coordinate is negative. Thus, the desired terminal point is

$$\left(\frac{\sqrt{2}}{2}, -\frac{\sqrt{2}}{2} \right)$$

(c) The reference number is $\bar{t} = \pi/3$, which determines the terminal point $\left(\frac{1}{2}, \sqrt{3}/2 \right)$ from Table 1. Since the terminal point determined by t is in quadrant III, its coordinates are both negative. Thus, the desired terminal point is

$$\left(-\frac{1}{2}, -\frac{\sqrt{3}}{2} \right)$$ ∎

Since the circumference of the unit circle is 2π, the terminal point determined by t is the same as that determined by $t + 2\pi$ or $t - 2\pi$. In general, we can add or subtract 2π any number of times without changing the terminal point determined by t. We use this observation in the next example to find terminal points for large t.

Example 7   Finding the Terminal Point for Large t

Find the terminal point determined by $t = \dfrac{29\pi}{6}$.

Solution   Since

$$t = \frac{29\pi}{6} = 4\pi + \frac{5\pi}{6}$$

we see that the terminal point of t is the same as that of $5\pi/6$ (that is, we subtract 4π). So by Example 6(a) the terminal point is $\left(-\sqrt{3}/2, \frac{1}{2} \right)$. (See Figure 10.) ∎

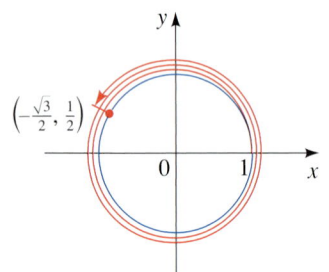

Figure 10

1–6 ■ Show that the point is on the unit circle.

1. $\left(\dfrac{4}{5}, -\dfrac{3}{5} \right)$ **2.** $\left(-\dfrac{5}{13}, \dfrac{12}{13} \right)$ **3.** $\left(\dfrac{7}{25}, \dfrac{24}{25} \right)$

4. $\left(-\dfrac{5}{7}, -\dfrac{2\sqrt{6}}{7} \right)$ **5.** $\left(-\dfrac{\sqrt{5}}{3}, \dfrac{2}{3} \right)$ **6.** $\left(\dfrac{\sqrt{11}}{6}, \dfrac{5}{6} \right)$

7–12 ■ Find the missing coordinate of P, using the fact that P lies on the unit circle in the given quadrant.

Coordinates	Quadrant
7. $P\left(-\frac{3}{5}, \quad \right)$	III
8. $P\left(\quad, -\frac{7}{25} \right)$	IV
9. $P\left(\quad, \frac{1}{3} \right)$	II
10. $P\left(\frac{2}{5}, \quad \right)$	I
11. $P\left(\quad, -\frac{2}{7} \right)$	IV
12. $P\left(-\frac{2}{3}, \quad \right)$	II

13–18 ■ The point P is on the unit circle. Find $P(x, y)$ from the given information.

13. The x-coordinate of P is $\frac{4}{5}$ and the y-coordinate is positive.

14. The y-coordinate of P is $-\frac{1}{3}$ and the x-coordinate is positive.

15. The y-coordinate of P is $\frac{2}{3}$ and the x-coordinate is negative.

16. The x-coordinate of P is positive and the y-coordinate of P is $-\sqrt{5}/5$.

17. The x-coordinate of P is $-\sqrt{2}/3$ and P lies below the x-axis.

18. The x-coordinate of P is $-\frac{2}{5}$ and P lies above the x-axis.

19–20 ■ Find t and the terminal point determined by t for each point in the figure. In Exercise 19, t increases in increments of $\pi/4$; in Exercise 20, t increases in increments of $\pi/6$.

19.

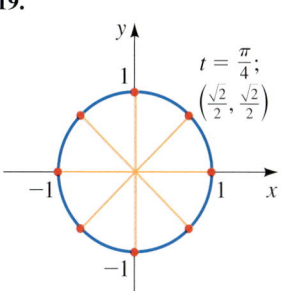

20.

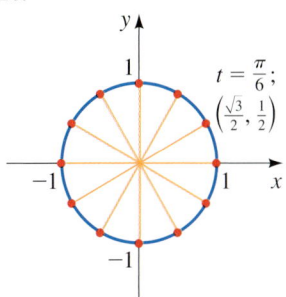

21–30 ■ Find the terminal point $P(x, y)$ on the unit circle determined by the given value of t.

21. $t = \dfrac{\pi}{2}$

22. $t = \dfrac{3\pi}{2}$

23. $t = \dfrac{5\pi}{6}$

24. $t = \dfrac{7\pi}{6}$

25. $t = -\dfrac{\pi}{3}$

26. $t = \dfrac{5\pi}{3}$

27. $t = \dfrac{2\pi}{3}$

28. $t = -\dfrac{\pi}{2}$

29. $t = -\dfrac{3\pi}{4}$

30. $t = \dfrac{11\pi}{6}$

31. Suppose that the terminal point determined by t is the point $\left(\frac{3}{5}, \frac{4}{5}\right)$ on the unit circle. Find the terminal point determined by each of the following.
  (a) $\pi - t$ (b) $-t$
  (c) $\pi + t$ (d) $2\pi + t$

32. Suppose that the terminal point determined by t is the point $\left(\frac{3}{4}, \sqrt{7}/4\right)$ on the unit circle. Find the terminal point determined by each of the following.
  (a) $-t$ (b) $4\pi + t$
  (c) $\pi - t$ (d) $t - \pi$

33–36 ■ Find the reference number for each value of t.

33. (a) $t = \dfrac{5\pi}{4}$ (b) $t = \dfrac{7\pi}{3}$

  (c) $t = -\dfrac{4\pi}{3}$ (d) $t = \dfrac{\pi}{6}$

34. (a) $t = \dfrac{5\pi}{6}$ (b) $t = \dfrac{7\pi}{6}$

  (c) $t = \dfrac{11\pi}{3}$ (d) $t = -\dfrac{7\pi}{4}$

35. (a) $t = \dfrac{5\pi}{7}$ (b) $t = -\dfrac{7\pi}{9}$

  (c) $t = -3$ (d) $t = 5$

36. (a) $t = \dfrac{11\pi}{5}$ (b) $t = -\dfrac{9\pi}{7}$

  (c) $t = 6$ (d) $t = -7$

37–50 ■ Find (a) the reference number for each value of t, and (b) the terminal point determined by t.

37. $t = \dfrac{2\pi}{3}$

38. $t = \dfrac{4\pi}{3}$

39. $t = \dfrac{3\pi}{4}$

40. $t = \dfrac{7\pi}{3}$

41. $t = -\dfrac{2\pi}{3}$

42. $t = -\dfrac{7\pi}{6}$

43. $t = \dfrac{13\pi}{4}$

44. $t = \dfrac{13\pi}{6}$

45. $t = \dfrac{7\pi}{6}$

46. $t = \dfrac{17\pi}{4}$

47. $t = -\dfrac{11\pi}{3}$

48. $t = \dfrac{31\pi}{6}$

49. $t = \dfrac{16\pi}{3}$

50. $t = -\dfrac{41\pi}{4}$

51–54 ■ Use the figure to find the terminal point determined by the real number t, with coordinates correct to one decimal place.

51. $t = 1$

52. $t = 2.5$

53. $t = -1.1$

54. $t = 4.2$

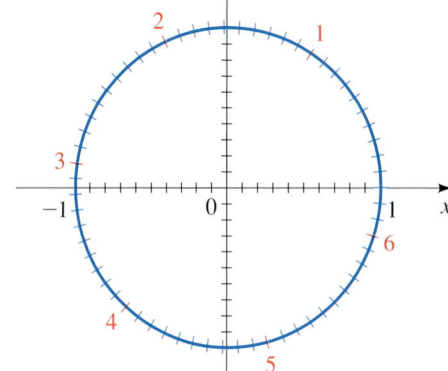

Discovery • Discussion

55. Finding the Terminal Point for $\pi/6$ Suppose the terminal point determined by $t = \pi/6$ is $P(x, y)$ and the points Q and R are as shown in the figure on the next page. Why are the distances PQ and PR the same? Use this fact, together with the Distance Formula, to show that the coordinates of

P satisfy the equation $2y = \sqrt{x^2 + (y-1)^2}$. Simplify this equation using the fact that $x^2 + y^2 = 1$. Solve the simplified equation to find $P(x, y)$.

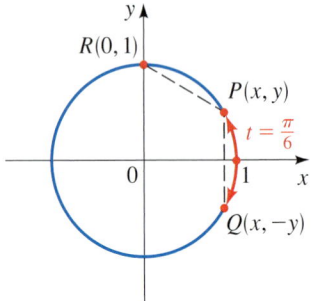

56. Finding the Terminal Point for $\pi/3$ Now that you know the terminal point determined by $t = \pi/6$, use symmetry to find the terminal point determined by $t = \pi/3$ (see the figure). Explain your reasoning.

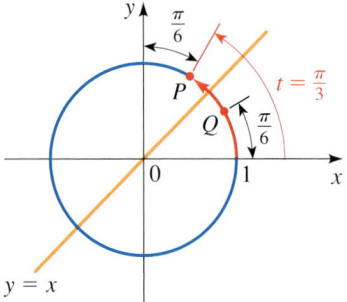

$y = x$

| 5.2 | **Trigonometric Functions of Real Numbers** |

A function is a rule that assigns to each real number another real number. In this section we use properties of the unit circle from the preceding section to define the trigonometric functions.

The Trigonometric Functions

Recall that to find the terminal point $P(x, y)$ for a given real number t, we move a distance t along the unit circle, starting at the point $(1, 0)$. We move in a counterclockwise direction if t is positive and in a clockwise direction if t is negative (see Figure 1). We now use the x- and y-coordinates of the point $P(x, y)$ to define several functions. For instance, we define the function called *sine* by assigning to each real number t the y-coordinate of the terminal point $P(x, y)$ determined by t. The functions *cosine, tangent, cosecant, secant,* and *cotangent* are also defined using the coordinates of $P(x, y)$.

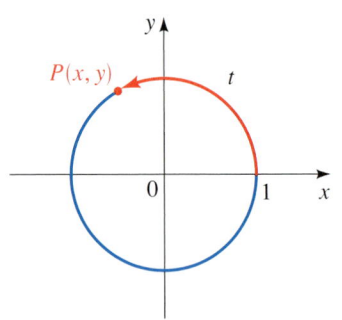

Figure 1

Definition of the Trigonometric Functions

Let t be any real number and let $P(x, y)$ be the terminal point on the unit circle determined by t. We define

$$\sin t = y \qquad\qquad \cos t = x \qquad\qquad \tan t = \frac{y}{x} \quad (x \neq 0)$$

$$\csc t = \frac{1}{y} \quad (y \neq 0) \qquad \sec t = \frac{1}{x} \quad (x \neq 0) \qquad \cot t = \frac{x}{y} \quad (y \neq 0)$$

Because the trigonometric functions can be defined in terms of the unit circle, they are sometimes called the **circular functions**.

Relationship to the Trigonometric Functions of Angles

If you have previously studied trigonometry of right triangles (Chapter 6), you are probably wondering how the sine and cosine of an *angle* relate to those of this section. To see how, let's start with a right triangle, ΔOPQ.

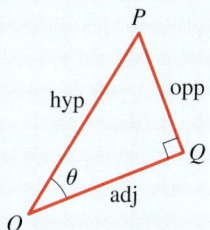

Right triangle *OPQ*

Place the triangle in the coordinate plane as shown, with angle θ in standard position.

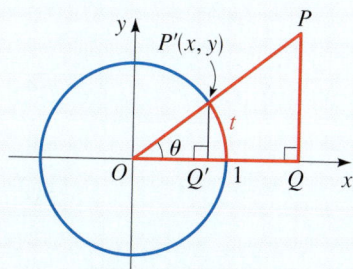

$P'(x, y)$ is the terminal point determined by *t*.

The point $P'(x, y)$ in the figure is the terminal point determined by the arc t. Note that triangle OPQ is similar to the small triangle $OP'Q'$ whose legs have lengths x and y.

Now, by the definition of the trigonometric functions of the *angle* θ we have

$$\sin \theta = \frac{\text{opp}}{\text{hyp}} = \frac{PQ}{OP} = \frac{P'Q'}{OP'}$$

$$= \frac{y}{1} = y$$

$$\cos \theta = \frac{\text{adj}}{\text{hyp}} = \frac{OQ}{OP} = \frac{OQ'}{OP'}$$

$$= \frac{x}{1} = x$$

By the definition of the trigonometric functions of the *real number t*, we have

$$\sin t = y \qquad \cos t = x$$

Now, if θ is measured in radians, then $\theta = t$ (see the figure). So the trigonometric functions of the angle with radian measure θ are exactly the same as the trigonometric functions defined in terms of the terminal point determined by the real number t.

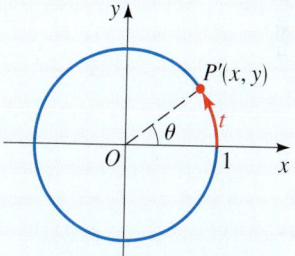

The radian measure of angle θ is t.

Why then study trigonometry in two different ways? Because different applications require that we view the trigonometric functions differently. (Compare Section 5.5 with Sections 6.2, 6.4, and 6.5.)

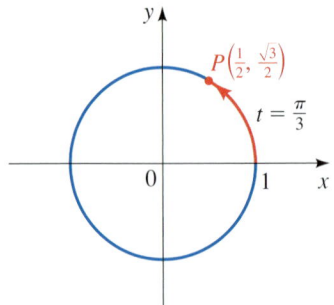

Figure 2

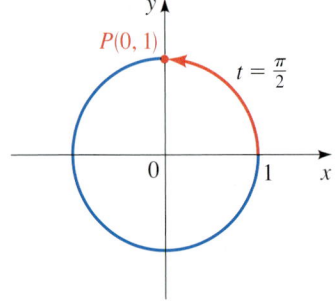

Figure 3

Example 1   Evaluating Trigonometric Functions

Find the six trigonometric functions of each given real number t.

(a) $t = \dfrac{\pi}{3}$ (b) $t = \dfrac{\pi}{2}$

Solution

(a) From Table 1 on page 403, we see that the terminal point determined by $t = \pi/3$ is $P\left(\frac{1}{2}, \sqrt{3}/2\right)$. (See Figure 2.) Since the coordinates are $x = \frac{1}{2}$ and $y = \sqrt{3}/2$, we have

$$\sin \frac{\pi}{3} = \frac{\sqrt{3}}{2} \qquad \cos \frac{\pi}{3} = \frac{1}{2} \qquad \tan \frac{\pi}{3} = \frac{\sqrt{3}/2}{1/2} = \sqrt{3}$$

$$\csc \frac{\pi}{3} = \frac{2\sqrt{3}}{3} \qquad \sec \frac{\pi}{3} = 2 \qquad \cot \frac{\pi}{4} = \frac{1/2}{\sqrt{3}/2} = \frac{\sqrt{3}}{3}$$

(b) The terminal point determined by $\pi/2$ is $P(0, 1)$. (See Figure 3.) So

$$\sin \frac{\pi}{2} = 1 \qquad \cos \frac{\pi}{2} = 0 \qquad \csc \frac{\pi}{2} = \frac{1}{1} = 1 \qquad \cot \frac{\pi}{2} = \frac{0}{1} = 0$$

But $\tan \pi/2$ and $\sec \pi/2$ are undefined because $x = 0$ appears in the denominator in each of their definitions. ∎

Some special values of the trigonometric functions are listed in Table 1. This table is easily obtained from Table 1 of Section 5.1, together with the definitions of the trigonometric functions.

Table 1   Special values of the trigonometric functions

t	$\sin t$	$\cos t$	$\tan t$	$\csc t$	$\sec t$	$\cot t$
0	0	1	0	—	1	—
$\dfrac{\pi}{6}$	$\dfrac{1}{2}$	$\dfrac{\sqrt{3}}{2}$	$\dfrac{\sqrt{3}}{3}$	2	$\dfrac{2\sqrt{3}}{3}$	$\sqrt{3}$
$\dfrac{\pi}{4}$	$\dfrac{\sqrt{2}}{2}$	$\dfrac{\sqrt{2}}{2}$	1	$\sqrt{2}$	$\sqrt{2}$	1
$\dfrac{\pi}{3}$	$\dfrac{\sqrt{3}}{2}$	$\dfrac{1}{2}$	$\sqrt{3}$	$\dfrac{2\sqrt{3}}{3}$	2	$\dfrac{\sqrt{3}}{3}$
$\dfrac{\pi}{2}$	1	0	—	1	—	0

We can easily remember the sines and cosines of the basic angles by writing them in the form $\sqrt{\blacksquare}/2$:

t	$\sin t$	$\cos t$
0	$\sqrt{0}/2$	$\sqrt{4}/2$
$\pi/6$	$\sqrt{1}/2$	$\sqrt{3}/2$
$\pi/4$	$\sqrt{2}/2$	$\sqrt{2}/2$
$\pi/3$	$\sqrt{3}/2$	$\sqrt{1}/2$
$\pi/2$	$\sqrt{4}/2$	$\sqrt{0}/2$

Example 1 shows that some of the trigonometric functions fail to be defined for certain real numbers. So we need to determine their domains. The functions sine and cosine are defined for all values of t. Since the functions cotangent and cosecant have y in the denominator of their definitions, they are not defined whenever the y-coordinate of the terminal point $P(x, y)$ determined by t is 0. This happens when $t = n\pi$ for any integer n, so their domains do not include these points. The functions tangent and secant have x in the denominator in their definitions, so they are not defined whenever $x = 0$. This happens when $t = (\pi/2) + n\pi$ for any integer n.

Domains of the Trigonometric Functions	
Function	**Domain**
sin, cos	All real numbers
tan, sec	All real numbers other than $\dfrac{\pi}{2} + n\pi$ for any integer n
cot, csc	All real numbers other than $n\pi$ for any integer, n

Values of the Trigonometric Functions

To compute other values of the trigonometric functions, we first determine their signs. The signs of the trigonometric functions depend on the quadrant in which the terminal point of t lies. For example, if the terminal point $P(x, y)$ determined by t lies in quadrant III, then its coordinates are both negative. So sin t, cos t, csc t, and sec t are all negative, whereas tan t and cot t are positive. You can check the other entries in the following box.

The following mnemonic device will help you remember which trigonometric functions are positive in each quadrant: **A**ll of them, **S**ine, **T**angent, or **C**osine.

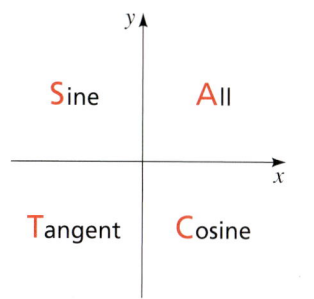

You can remember this as "**A**ll **S**tudents **T**ake **C**alculus."

Signs of the Trigonometric Functions		
Quadrant	**Positive Functions**	**Negative functions**
I	all	none
II	sin, csc	cos, sec, tan, cot
III	tan, cot	sin, csc, cos, sec
IV	cos, sec	sin, csc, tan, cot

Example 2   Determining the Sign of a Trigonometric Function

(a) $\cos \dfrac{\pi}{3} > 0$, because the terminal point of $t = \dfrac{\pi}{3}$ is in quadrant I.

(b) $\tan 4 > 0$, because the terminal point of $t = 4$ is in quadrant III.

(c) If $\cos t < 0$ and $\sin t > 0$, then the terminal point of t must be in quadrant II.

■

In Section 5.1 we used the reference number to find the terminal point determined by a real number t. Since the trigonometric functions are defined in terms of the coordinates of terminal points, we can use the reference number to find values of the trigonometric functions. Suppose that $\bar{t}$ is the reference number for t. Then the terminal point of $\bar{t}$ has the same coordinates, except possibly for sign, as the terminal point of t. So the values of the trigonometric functions at t are the same, except possibly for sign, as their values at $\bar{t}$. We illustrate this procedure in the next example.

Example 3   Evaluating Trigonometric Functions

Find each value.

(a) $\cos \dfrac{2\pi}{3}$ (b) $\tan\left(-\dfrac{\pi}{3}\right)$ (c) $\sin \dfrac{19\pi}{4}$

Solution

(a) The reference number for $2\pi/3$ is $\pi/3$ (see Figure 4(a)). Since the terminal point of $2\pi/3$ is in quadrant II, $\cos(2\pi/3)$ is negative. Thus

$$\cos \frac{2\pi}{3} = -\cos \frac{\pi}{3} = -\frac{1}{2}$$

Sign   Reference   From
number   Table 1

(b) The reference number for $-\pi/3$ is $\pi/3$ (see Figure 4(b)). Since the terminal point of $-\pi/3$ is in quadrant IV, $\tan(-\pi/3)$ is negative. Thus

$$\tan\left(-\frac{\pi}{3}\right) = -\tan \frac{\pi}{3} = -\sqrt{3}$$

Sign   Reference   From
number   Table 1

(c) Since $(19\pi/4) - 4\pi = 3\pi/4$, the terminal points determined by $19\pi/4$ and $3\pi/4$ are the same. The reference number for $3\pi/4$ is $\pi/4$ (see Figure 4(c)). Since the terminal point of $3\pi/4$ is in quadrant II, $\sin(3\pi/4)$ is positive. Thus

$$\sin \frac{19\pi}{4} = \sin \frac{3\pi}{4} = +\sin \frac{\pi}{4} = \frac{\sqrt{2}}{2}$$

Subtract 4π Sign   Reference   From
number   Table 1   ∎

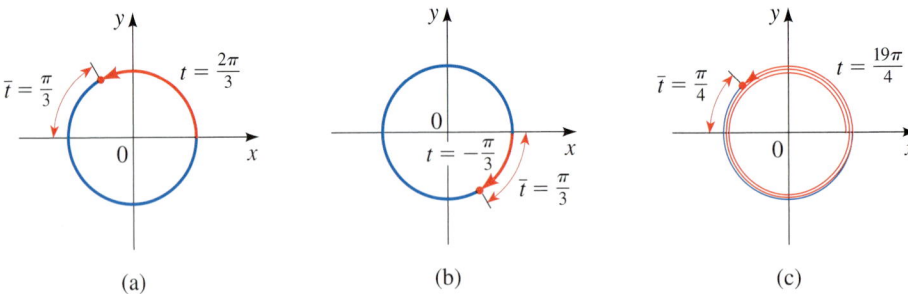

Figure 4   (a)   (b)   (c)

So far we have been able to compute the values of the trigonometric functions only for certain values of t. In fact, we can compute the values of the trigonometric functions whenever t is a multiple of $\pi/6$, $\pi/4$, $\pi/3$, and $\pi/2$. How can we compute the trigonometric functions for other values of t? For example, how can we find $\sin 1.5$? One way is to carefully sketch a diagram and read the value (see Exercises 37–44); however, this method is not very accurate. Fortunately, programmed directly into scientific calculators are mathematical procedures (see the margin note on page 436) that find the values of *sine*, *cosine*, and *tangent* correct to the number of digits in the display. The calculator must be put in *radian mode* to evaluate these functions. To find values of cosecant, secant, and cotangent using a calculator, we need to use the following *reciprocal relations*:

$$\csc t = \frac{1}{\sin t} \qquad \sec t = \frac{1}{\cos t} \qquad \cot t = \frac{1}{\tan t}$$

These identities follow from the definitions of the trigonometric functions. For instance, since $\sin t = y$ and $\csc t = 1/y$, we have $\csc t = 1/y = 1/(\sin t)$. The others follow similarly.

Example 4 Using a Calculator to Evaluate Trigonometric Functions

Making sure our calculator is set to radian mode and rounding the results to six decimal places, we get

(a) $\sin 2.2 \approx 0.808496$

(b) $\cos 1.1 \approx 0.453596$

(c) $\cot 28 = \dfrac{1}{\tan 28} \approx -3.553286$

(d) $\csc 0.98 = \dfrac{1}{\sin 0.98} \approx 1.204098$ ∎

Let's consider the relationship between the trigonometric functions of t and those of $-t$. From Figure 5 we see that

$$\sin(-t) = -y = -\sin t$$

$$\cos(-t) = x = \cos t$$

$$\tan(-t) = \frac{-y}{x} = -\frac{y}{x} = -\tan t$$

These equations show that sine and tangent are odd functions, whereas cosine is an even function. It's easy to see that the reciprocal of an even function is even and the reciprocal of an odd function is odd. This fact, together with the reciprocal relations, completes our knowledge of the even-odd properties for all the trigonometric functions.

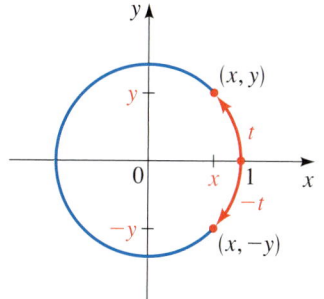

Figure 5

Even and odd functions are defined in Section 2.4.

Even-Odd Properties

Sine, cosecant, tangent, and cotangent are odd functions; cosine and secant are even functions.

$$\sin(-t) = -\sin t \qquad \cos(-t) = \cos t \qquad \tan(-t) = -\tan t$$

$$\csc(-t) = -\csc t \qquad \sec(-t) = \sec t \qquad \cot(-t) = -\cot t$$

Example 5 Even and Odd Trigonometric Functions

Use the even-odd properties of the trigonometric functions to determine each value.

(a) $\sin\left(-\dfrac{\pi}{6}\right)$

(b) $\cos\left(-\dfrac{\pi}{4}\right)$

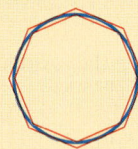
Solution   By the even-odd properties and Table 1, we have

(a) $\sin\left(-\dfrac{\pi}{6}\right) = -\sin\dfrac{\pi}{6} = -\dfrac{1}{2}$ *Sine is odd*

(b) $\cos\left(-\dfrac{\pi}{4}\right) = \cos\dfrac{\pi}{4} = \dfrac{\sqrt{2}}{2}$ *Cosine is even*   ■

Fundamental Identities

The trigonometric functions are related to each other through equations called **trigonometric identities**. We give the most important ones in the following box.*

Fundamental Identities

Reciprocal Identities

$$\csc t = \frac{1}{\sin t} \qquad \sec t = \frac{1}{\cos t} \qquad \cot t = \frac{1}{\tan t}$$

$$\tan t = \frac{\sin t}{\cos t} \qquad \cot t = \frac{\cos t}{\sin t}$$

Pythagorean Identities

$$\sin^2 t + \cos^2 t = 1 \qquad \tan^2 t + 1 = \sec^2 t \qquad 1 + \cot^2 t = \csc^2 t$$

■ **Proof**   The reciprocal identities follow immediately from the definition on page 408. We now prove the Pythagorean identities. By definition, $\cos t = x$ and $\sin t = y$, where x and y are the coordinates of a point $P(x, y)$ on the unit circle. Since $P(x, y)$ is on the unit circle, we have $x^2 + y^2 = 1$. Thus

$$\sin^2 t + \cos^2 t = 1$$

Dividing both sides by $\cos^2 t$ (provided $\cos t \neq 0$), we get

$$\frac{\sin^2 t}{\cos^2 t} + \frac{\cos^2 t}{\cos^2 t} = \frac{1}{\cos^2 t}$$

$$\left(\frac{\sin t}{\cos t}\right)^2 + 1 = \left(\frac{1}{\cos t}\right)^2$$

$$\tan^2 t + 1 = \sec^2 t$$

We have used the reciprocal identities $\sin t/\cos t = \tan t$ and $1/\cos t = \sec t$. Similarly, dividing both sides of the first Pythagorean identity by $\sin^2 t$ (provided $\sin t \neq 0$) gives us $1 + \cot^2 t = \csc^2 t$.   ■

---

*We follow the usual convention of writing $\sin^2 t$ for $(\sin t)^2$. In general, we write $\sin^n t$ for $(\sin t)^n$ for all integers n except $n = -1$. The exponent $n = -1$ will be assigned another meaning in Section 7.4. Of course, the same convention applies to the other five trigonometric functions.

As their name indicates, the fundamental identities play a central role in trigonometry because we can use them to relate any trigonometric function to any other. So, if we know the value of any one of the trigonometric functions at t, then we can find the values of all the others at t.

Example 6 Finding All Trigonometric Functions from the Value of One

If $\cos t = \frac{3}{5}$ and t is in quadrant IV, find the values of all the trigonometric functions at t.

Solution From the Pythagorean identities we have

$$\sin^2 t + \cos^2 t = 1$$

$$\sin^2 t + \left(\tfrac{3}{5}\right)^2 = 1 \qquad \text{\textit{Substitute cos t} = $\tfrac{3}{5}$}$$

$$\sin^2 t = 1 - \tfrac{9}{25} = \tfrac{16}{25} \qquad \text{\textit{Solve for sin}^{2}\textit{t}}$$

$$\sin t = \pm\tfrac{4}{5} \qquad \text{\textit{Take square roots}}$$

Since this point is in quadrant IV, $\sin t$ is negative, so $\sin t = -\frac{4}{5}$. Now that we know both $\sin t$ and $\cos t$, we can find the values of the other trigonometric functions using the reciprocal identities:

$$\sin t = -\frac{4}{5} \qquad \cos t = \frac{3}{5} \qquad \tan t = \frac{\sin t}{\cos t} = \frac{-\frac{4}{5}}{\frac{3}{5}} = -\frac{4}{3}$$

$$\csc t = \frac{1}{\sin t} = -\frac{5}{4} \qquad \sec t = \frac{1}{\cos t} = \frac{5}{3} \qquad \cot t = \frac{1}{\tan t} = -\frac{3}{4} \qquad \blacksquare$$

Example 7 Writing One Trigonometric Function in Terms of Another

Write $\tan t$ in terms of $\cos t$, where t is in quadrant III.

Solution Since $\tan t = \sin t/\cos t$, we need to write $\sin t$ in terms of $\cos t$. By the Pythagorean identities we have

$$\sin^2 t + \cos^2 t = 1$$

$$\sin^2 t = 1 - \cos^2 t \qquad \text{\textit{Solve for sin}^{2}\textit{t}}$$

$$\sin t = \pm\sqrt{1 - \cos^2 t} \qquad \text{\textit{Take square roots}}$$

Since $\sin t$ is negative in quadrant III, the negative sign applies here. Thus

$$\tan t = \frac{\sin t}{\cos t} = \frac{-\sqrt{1 - \cos^2 t}}{\cos t} \qquad \blacksquare$$

5.2   Exercises

1–2 ■ Find sin *t* and cos *t* for the values of *t* whose terminal points are shown on the unit circle in the figure. In Exercise 1, *t* increases in increments of $\pi/4$; in Exercise 2, *t* increases in increments of $\pi/6$. (See Exercises 19 and 20 in Section 5.1.)

1.

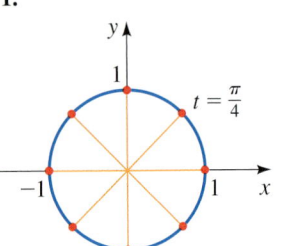

2.

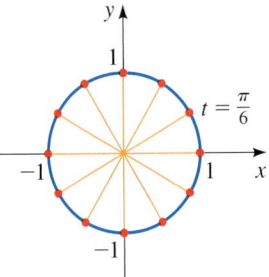

3–22 ■ Find the exact value of the trigonometric function at the given real number.

3. (a) $\sin \dfrac{2\pi}{3}$ (b) $\cos \dfrac{2\pi}{3}$ (c) $\tan \dfrac{2\pi}{3}$

4. (a) $\sin \dfrac{5\pi}{6}$ (b) $\cos \dfrac{5\pi}{6}$ (c) $\tan \dfrac{5\pi}{6}$

5. (a) $\sin \dfrac{7\pi}{6}$ (b) $\sin\left(-\dfrac{\pi}{6}\right)$ (c) $\sin \dfrac{11\pi}{6}$

6. (a) $\cos \dfrac{5\pi}{3}$ (b) $\cos\left(-\dfrac{5\pi}{3}\right)$ (c) $\cos \dfrac{7\pi}{3}$

7. (a) $\cos \dfrac{3\pi}{4}$ (b) $\cos \dfrac{5\pi}{4}$ (c) $\cos \dfrac{7\pi}{4}$

8. (a) $\sin \dfrac{3\pi}{4}$ (b) $\sin \dfrac{5\pi}{4}$ (c) $\sin \dfrac{7\pi}{4}$

9. (a) $\sin \dfrac{7\pi}{3}$ (b) $\csc \dfrac{7\pi}{3}$ (c) $\cot \dfrac{7\pi}{3}$

10. (a) $\cos\left(-\dfrac{\pi}{3}\right)$ (b) $\sec\left(-\dfrac{\pi}{3}\right)$ (c) $\tan\left(-\dfrac{\pi}{3}\right)$

11. (a) $\sin\left(-\dfrac{\pi}{2}\right)$ (b) $\cos\left(-\dfrac{\pi}{2}\right)$ (c) $\cot\left(-\dfrac{\pi}{2}\right)$

12. (a) $\sin\left(-\dfrac{3\pi}{2}\right)$ (b) $\cos\left(-\dfrac{3\pi}{2}\right)$ (c) $\cot\left(-\dfrac{3\pi}{2}\right)$

13. (a) $\sec \dfrac{11\pi}{3}$ (b) $\csc \dfrac{11\pi}{3}$ (c) $\sec\left(-\dfrac{\pi}{3}\right)$

14. (a) $\cos \dfrac{7\pi}{6}$ (b) $\sec \dfrac{7\pi}{6}$ (c) $\csc \dfrac{7\pi}{6}$

15. (a) $\tan \dfrac{5\pi}{6}$ (b) $\tan \dfrac{7\pi}{6}$ (c) $\tan \dfrac{11\pi}{6}$

16. (a) $\cot\left(-\dfrac{\pi}{3}\right)$ (b) $\cot \dfrac{2\pi}{3}$ (c) $\cot \dfrac{5\pi}{3}$

17. (a) $\cos\left(-\dfrac{\pi}{4}\right)$ (b) $\csc\left(-\dfrac{\pi}{4}\right)$ (c) $\cot\left(-\dfrac{\pi}{4}\right)$

18. (a) $\sin \dfrac{5\pi}{4}$ (b) $\sec \dfrac{5\pi}{4}$ (c) $\tan \dfrac{5\pi}{4}$

19. (a) $\csc\left(-\dfrac{\pi}{2}\right)$ (b) $\csc \dfrac{\pi}{2}$ (c) $\csc \dfrac{3\pi}{2}$

20. (a) $\sec(-\pi)$ (b) $\sec \pi$ (c) $\sec 4\pi$

21. (a) $\sin 13\pi$ (b) $\cos 14\pi$ (c) $\tan 15\pi$

22. (a) $\sin \dfrac{25\pi}{2}$ (b) $\cos \dfrac{25\pi}{2}$ (c) $\cot \dfrac{25\pi}{2}$

23–26 ■ Find the value of each of the six trigonometric functions (if it is defined) at the given real number *t*. Use your answers to complete the table.

23. $t = 0$ **24.** $t = \dfrac{\pi}{2}$ **25.** $t = \pi$ **26.** $t = \dfrac{3\pi}{2}$

t	$\sin t$	$\cos t$	$\tan t$	$\csc t$	$\sec t$	$\cot t$
0	0	1		undefined		
$\dfrac{\pi}{2}$						
π			0			undefined
$\dfrac{3\pi}{2}$						

27–36 ■ The terminal point $P(x, y)$ determined by a real number *t* is given. Find sin *t*, cos *t*, and tan *t*.

27. $\left(\dfrac{3}{5}, \dfrac{4}{5}\right)$ **28.** $\left(-\dfrac{3}{5}, \dfrac{4}{5}\right)$

29. $\left(\dfrac{\sqrt{5}}{4}, -\dfrac{\sqrt{11}}{4}\right)$ **30.** $\left(-\dfrac{1}{3}, -\dfrac{2\sqrt{2}}{3}\right)$

31. $\left(-\dfrac{6}{7}, \dfrac{\sqrt{13}}{7}\right)$ **32.** $\left(\dfrac{40}{41}, \dfrac{9}{41}\right)$

33. $\left(-\dfrac{5}{13}, -\dfrac{12}{13}\right)$ **34.** $\left(\dfrac{\sqrt{5}}{5}, \dfrac{2\sqrt{5}}{5}\right)$

35. $\left(-\dfrac{20}{29}, \dfrac{21}{29}\right)$ **36.** $\left(\dfrac{24}{25}, -\dfrac{7}{25}\right)$

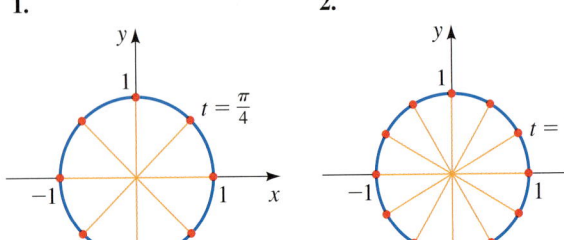

37–44 ■ Find the approximate value of the given trigonometric function by using **(a)** the figure and **(b)** a calculator. Compare the two values.

37. sin 1

38. cos 0.8

39. sin 1.2

40. cos 5

41. tan 0.8

42. tan(−1.3)

43. cos 4.1

44. sin(−5.2)

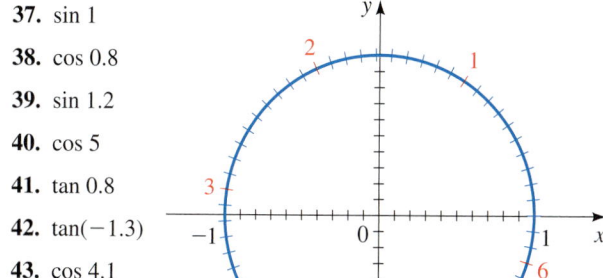

45–48 ■ Find the sign of the expression if the terminal point determined by t is in the given quadrant.

45. $\sin t \cos t$, quadrant II   **46.** $\tan t \sec t$, quadrant IV

47. $\dfrac{\tan t \sin t}{\cot t}$, quadrant III   **48.** $\cos t \sec t$, any quadrant

49–52 ■ From the information given, find the quadrant in which the terminal point determined by t lies.

49. $\sin t > 0$ and $\cos t < 0$ **50.** $\tan t > 0$ and $\sin t < 0$

51. $\csc t > 0$ and $\sec t < 0$ **52.** $\cos t < 0$ and $\cot t < 0$

53–62 ■ Write the first expression in terms of the second if the terminal point determined by t is in the given quadrant.

53. $\sin t, \cos t$;   quadrant II   **54.** $\cos t, \sin t$;   quadrant IV

55. $\tan t, \sin t$;   quadrant IV   **56.** $\tan t, \cos t$;   quadrant III

57. $\sec t, \tan t$;   quadrant II   **58.** $\csc t, \cot t$;   quadrant III

59. $\tan t, \sec t$;   quadrant III   **60.** $\sin t, \sec t$;   quadrant IV

61. $\tan^2 t, \sin t$;   any quadrant

62. $\sec^2 t \sin^2 t, \cos t$;   any quadrant

63–70 ■ Find the values of the trigonometric functions of t from the given information.

63. $\sin t = \frac{3}{5}$,   terminal point of t is in quadrant II

64. $\cos t = -\frac{4}{5}$,   terminal point of t is in quadrant III

65. $\sec t = 3$,   terminal point of t is in quadrant IV

66. $\tan t = \frac{1}{4}$,   terminal point of t is in quadrant III

67. $\tan t = -\frac{3}{4}$, $\cos t > 0$ **68.** $\sec t = 2$, $\sin t < 0$

69. $\sin t = -\frac{1}{4}$, $\sec t < 0$ **70.** $\tan t = -4$, $\csc t > 0$

71–78 ■ Determine whether the function is even, odd, or neither.

71. $f(x) = x^2 \sin x$ **72.** $f(x) = x^2 \cos 2x$

73. $f(x) = \sin x \cos x$ **74.** $f(x) = \sin x + \cos x$

75. $f(x) = |x| \cos x$ **76.** $f(x) = x \sin^3 x$

77. $f(x) = x^3 + \cos x$ **78.** $f(x) = \cos(\sin x)$

Applications

79. Harmonic Motion   The displacement from equilibrium of an oscillating mass attached to a spring is given by $y(t) = 4 \cos 3\pi t$ where y is measured in inches and t in seconds. Find the displacement at the times indicated in the table.

t	$y(t)$
0	
0.25	
0.50	
0.75	
1.00	
1.25	

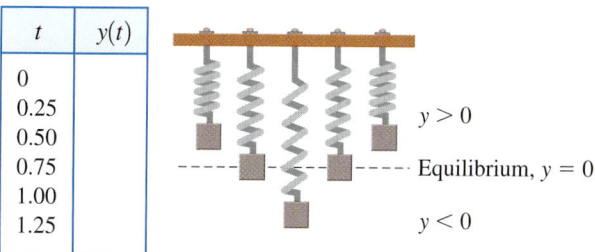

80. Circadian Rhythms   Everybody's blood pressure varies over the course of the day. In a certain individual the resting diastolic blood pressure at time t is given by $B(t) = 80 + 7 \sin(\pi t/12)$, where t is measured in hours since midnight and $B(t)$ in mmHg (millimeters of mercury). Find this person's diastolic blood pressure at

(a) 6:00 A.M.   **(b)** 10:30 A.M.   **(c)** Noon   **(d)** 8:00 P.M.

81. Electric Circuit   After the switch is closed in the circuit shown, the current t seconds later is $I(t) = 0.8e^{-3t}\sin 10t$. Find the current at the times

(a) $t = 0.1$ s and **(b)** $t = 0.5$ s.

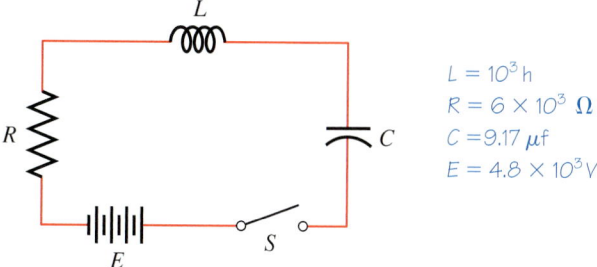

$L = 10^3$ h
$R = 6 \times 10^3\ \Omega$
$C = 9.17\ \mu$f
$E = 4.8 \times 10^3$ V

82. Bungee Jumping   A bungee jumper plummets from a high bridge to the river below and then bounces back over and over again. At time t seconds after her jump, her height H (in meters) above the river is given by

$H(t) = 100 + 75e^{-t/20}\cos(\frac{\pi}{4}t)$. Find her height at the times indicated in the table.

t	$H(t)$
0	
1	
2	
4	
6	
8	
12	

Discovery • Discussion

83. Reduction Formulas   A *reduction formula* is one that can be used to "reduce" the number of terms in the input for a trigonometric function. Explain how the figure shows that the following reduction formulas are valid:

$$\sin(t + \pi) = -\sin t \qquad \cos(t + \pi) = -\cos t$$

$$\tan(t + \pi) = \tan t$$

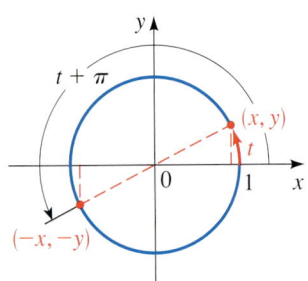

84. More Reduction Formulas   By the "Angle-Side-Angle" theorem from elementary geometry, triangles *CDO* and *AOB* in the figure are congruent. Explain how this proves that if *B* has coordinates (x, y), then *D* has coordinates $(-y, x)$. Then explain how the figure shows that the following reduction formulas are valid:

$$\sin\left(t + \frac{\pi}{2}\right) = \cos t$$

$$\cos\left(t + \frac{\pi}{2}\right) = -\sin t$$

$$\tan\left(t + \frac{\pi}{2}\right) = -\cot t$$

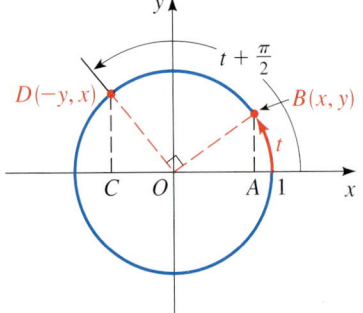

5.3   Trigonometric Graphs

The graph of a function gives us a better idea of its behavior. So, in this section we graph the sine and cosine functions and certain transformations of these functions. The other trigonometric functions are graphed in the next section.

Graphs of the Sine and Cosine Functions

To help us graph the sine and cosine functions, we first observe that these functions repeat their values in a regular fashion. To see exactly how this happens, recall that the circumference of the unit circle is 2π. It follows that the terminal point $P(x, y)$ determined by the real number t is the same as that determined by $t + 2\pi$. Since the sine and cosine functions are defined in terms of the coordinates of $P(x, y)$, it follows that their values are unchanged by the addition of any integer multiple of 2π. In other words,

$$\sin(t + 2n\pi) = \sin t \qquad \text{for any integer } n$$

$$\cos(t + 2n\pi) = \cos t \qquad \text{for any integer } n$$

Thus, the sine and cosine functions are *periodic* according to the following definition: A function f is **periodic** if there is a positive number p such that $f(t + p) = f(t)$ for every t. The least such positive number (if it exists) is the **period** of f. If f has period p, then the graph of f on any interval of length p is called **one complete period** of f.

Periodic Properties of Sine and Cosine

The functions sine and cosine have period 2π:

$$\sin(t + 2\pi) = \sin t \qquad \cos(t + 2\pi) = \cos t$$

Table 1

t	$\sin t$	$\cos t$
$0 \to \dfrac{\pi}{2}$	$0 \to 1$	$1 \to 0$
$\dfrac{\pi}{2} \to \pi$	$1 \to 0$	$0 \to -1$
$\pi \to \dfrac{3\pi}{2}$	$0 \to -1$	$-1 \to 0$
$\dfrac{3\pi}{2} \to 2\pi$	$-1 \to 0$	$0 \to 1$

So the sine and cosine functions repeat their values in any interval of length 2π. To sketch their graphs, we first graph one period. To sketch the graphs on the interval $0 \le t \le 2\pi$, we could try to make a table of values and use those points to draw the graph. Since no such table can be complete, let's look more closely at the definitions of these functions.

Recall that $\sin t$ is the y-coordinate of the terminal point $P(x, y)$ on the unit circle determined by the real number t. How does the y-coordinate of this point vary as t increases? It's easy to see that the y-coordinate of $P(x, y)$ increases to 1, then decreases to -1 repeatedly as the point $P(x, y)$ travels around the unit circle. (See Figure 1.) In fact, as t increases from 0 to $\pi/2$, $y = \sin t$ increases from 0 to 1. As t increases from $\pi/2$ to π, the value of $y = \sin t$ decreases from 1 to 0. Table 1 shows the variation of the sine and cosine functions for t between 0 and 2π.

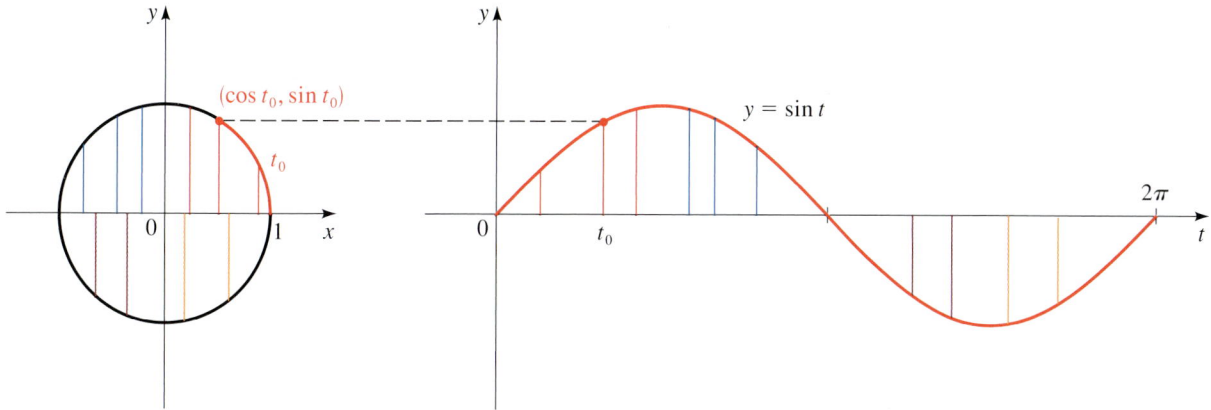

Figure 1

To draw the graphs more accurately, we find a few other values of $\sin t$ and $\cos t$ in Table 2. We could find still other values with the aid of a calculator.

Table 2

t	0	$\dfrac{\pi}{6}$	$\dfrac{\pi}{3}$	$\dfrac{\pi}{2}$	$\dfrac{2\pi}{3}$	$\dfrac{5\pi}{6}$	π	$\dfrac{7\pi}{6}$	$\dfrac{4\pi}{3}$	$\dfrac{3\pi}{2}$	$\dfrac{5\pi}{3}$	$\dfrac{11\pi}{6}$	2π
$\sin t$	0	$\dfrac{1}{2}$	$\dfrac{\sqrt{3}}{2}$	1	$\dfrac{\sqrt{3}}{2}$	$\dfrac{1}{2}$	0	$-\dfrac{1}{2}$	$-\dfrac{\sqrt{3}}{2}$	-1	$-\dfrac{\sqrt{3}}{2}$	$-\dfrac{1}{2}$	0
$\cos t$	1	$\dfrac{\sqrt{3}}{2}$	$\dfrac{1}{2}$	0	$-\dfrac{1}{2}$	$-\dfrac{\sqrt{3}}{2}$	-1	$-\dfrac{\sqrt{3}}{2}$	$-\dfrac{1}{2}$	0	$\dfrac{1}{2}$	$\dfrac{\sqrt{3}}{2}$	1

Now we use this information to graph the functions sin t and cos t for t between 0 and 2π in Figures 2 and 3. These are the graphs of one period. Using the fact that these functions are periodic with period 2π, we get their complete graphs by continuing the same pattern to the left and to the right in every successive interval of length 2π.

The graph of the sine function is symmetric with respect to the origin. This is as expected, since sine is an odd function. Since the cosine function is an even function, its graph is symmetric with respect to the y-axis.

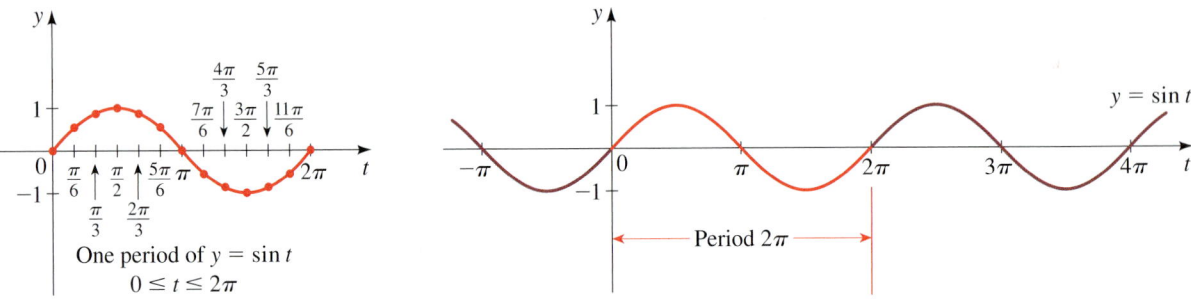

Figure 2

Graph of sin t

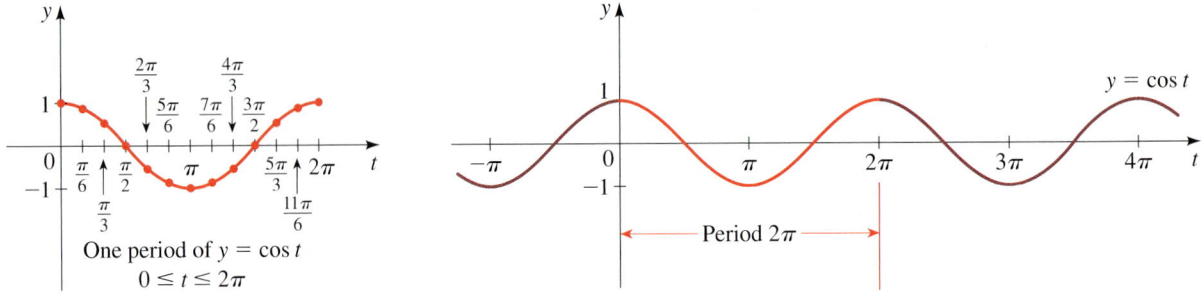

Figure 3

Graph of cos t

Graphs of Transformations of Sine and Cosine

We now consider graphs of functions that are transformations of the sine and cosine functions. Thus, the graphing techniques of Section 2.4 are very useful here. The graphs we obtain are important for understanding applications to physical situations such as harmonic motion (see Section 5.5), but some of them are beautiful graphs that are interesting in their own right.

It's traditional to use the letter x to denote the variable in the domain of a function. So, from here on we use the letter x and write $y = \sin x$, $y = \cos x$, $y = \tan x$, and so on to denote these functions.

Example 1   Cosine Curves

Sketch the graph of each function.

(a) $f(x) = 2 + \cos x$ (b) $g(x) = -\cos x$

Solution

(a) The graph of $y = 2 + \cos x$ is the same as the graph of $y = \cos x$, but shifted up 2 units (see Figure 4(a)).

(b) The graph of $y = -\cos x$ in Figure 4(b) is the reflection of the graph of $y = \cos x$ in the x-axis.

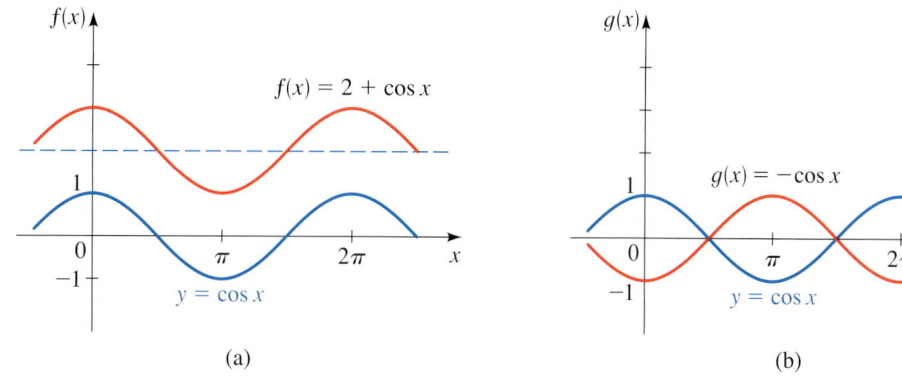

Figure 4   (a)   (b)

Let's graph $y = 2 \sin x$. We start with the graph of $y = \sin x$ and multiply the y-coordinate of each point by 2. This has the effect of stretching the graph vertically by a factor of 2. To graph $y = \frac{1}{2} \sin x$, we start with the graph of $y = \sin x$ and multiply the y-coordinate of each point by $\frac{1}{2}$. This has the effect of shrinking the graph vertically by a factor of $\frac{1}{2}$ (see Figure 5).

Vertical stretching and shrinking of graphs is discussed in Section 2.4.

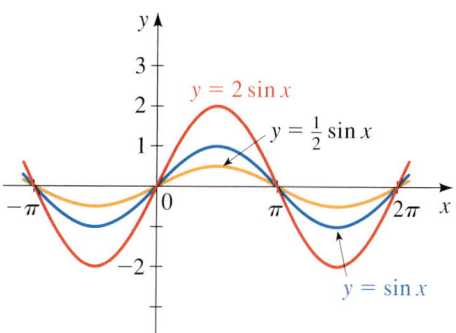

Figure 5

In general, for the functions

$$y = a \sin x \qquad \text{and} \qquad y = a \cos x$$

the number $|a|$ is called the **amplitude** and is the largest value these functions attain. Graphs of $y = a \sin x$ for several values of a are shown in Figure 6.

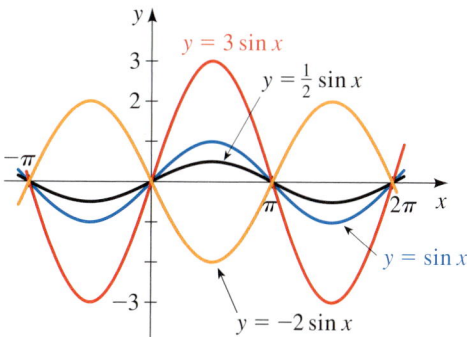

Figure 6

Example 2 Stretching a Cosine Curve

Find the amplitude of $y = -3 \cos x$ and sketch its graph.

Solution The amplitude is $|-3| = 3$, so the largest value the graph attains is 3 and the smallest value is -3. To sketch the graph, we begin with the graph of $y = \cos x$, stretch the graph vertically by a factor of 3, and reflect in the x-axis, arriving at the graph in Figure 7.

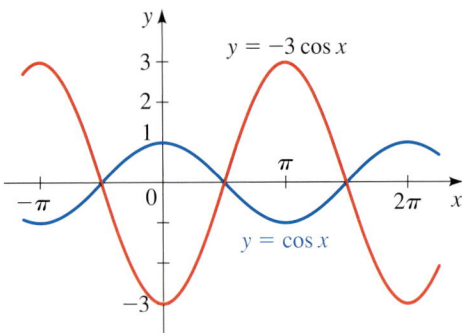

Figure 7

Since the sine and cosine functions have period 2π, the functions

$$y = a \sin kx \qquad \text{and} \qquad y = a \cos kx \qquad (k > 0)$$

complete one period as kx varies from 0 to 2π, that is, for $0 \le kx \le 2\pi$ or for $0 \le x \le 2\pi/k$. So these functions complete one period as x varies between 0 and $2\pi/k$ and thus have period $2\pi/k$. The graphs of these functions are called **sine curves** and **cosine curves**, respectively. (Collectively, sine and cosine curves are often referred to as **sinusoidal** curves.)

Sine and Cosine Curves

The sine and cosine curves

$$y = a \sin kx \qquad \text{and} \qquad y = a \cos kx \qquad (k > 0)$$

have amplitude $|a|$ and period $2\pi/k$.

An appropriate interval on which to graph one complete period is $[0, 2\pi/k]$.

To see how the value of k affects the graph of $y = \sin kx$, let's graph the sine curve $y = \sin 2x$. Since the period is $2\pi/2 = \pi$, the graph completes one period in the interval $0 \le x \le \pi$ (see Figure 8(a)). For the sine curve $y = \sin \frac{1}{2}x$, the period is $2\pi \div \frac{1}{2} = 4\pi$, and so the graph completes one period in the interval $0 \le x \le 4\pi$ (see Figure 8(b)). We see that the effect is to *shrink* the graph horizontally if $k > 1$ or to *stretch* the graph horizontally if $k < 1$.

Horizontal stretching and shrinking of graphs is discussed in Section 2.4.

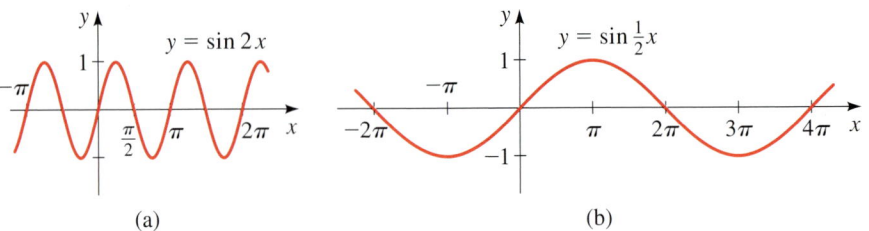

Figure 8   (a)   (b)

For comparison, in Figure 9 we show the graphs of one period of the sine curve $y = a \sin kx$ for several values of k.

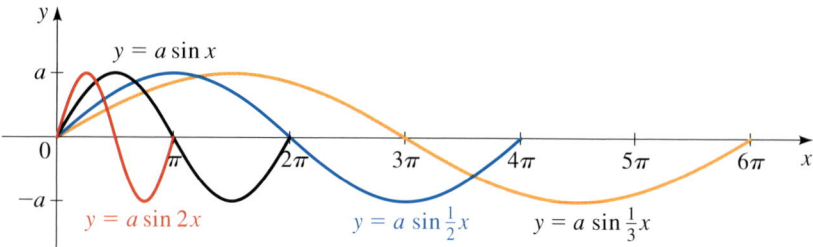

Figure 9

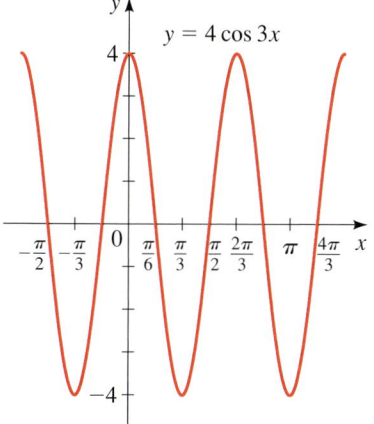

Figure 10

Example 3 Amplitude and Period

Find the amplitude and period of each function, and sketch its graph.

(a) $y = 4 \cos 3x$ (b) $y = -2 \sin \frac{1}{2}x$

Solution

(a) We get the amplitude and period from the form of the function as follows:

$$\text{amplitude} = |a| = 4$$

$$y = 4 \cos 3x$$

$$\text{period} = \frac{2\pi}{k} = \frac{2\pi}{3}$$

The amplitude is 4 and the period is $2\pi/3$. The graph is shown in Figure 10.

(b) For $y = -2 \sin \frac{1}{2}x$,

$$\text{amplitude} = |a| = |-2| = 2$$

$$\text{period} = \frac{2\pi}{\frac{1}{2}} = 4\pi$$

The graph is shown in Figure 11. ■

The graphs of functions of the form $y = a \sin k(x - b)$ and $y = a \cos k(x - b)$ are simply sine and cosine curves shifted horizontally by an amount $|b|$. They are shifted to the right if $b > 0$ or to the left if $b < 0$. The number b is the **phase shift**. We summarize the properties of these functions in the following box.

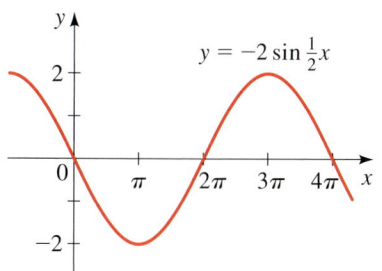

Figure 11

Shifted Sine and Cosine Curves

The sine and cosine curves

$$y = a \sin k(x - b) \quad \text{and} \quad y = a \cos k(x - b) \quad (k > 0)$$

have amplitude $|a|$, period $2\pi/k$, and phase shift b.

An appropriate interval on which to graph one complete period is $[b, b + (2\pi/k)]$.

The graphs of $y = \sin\left(x - \dfrac{\pi}{3}\right)$ and $y = \sin\left(x + \dfrac{\pi}{6}\right)$ are shown in Figure 12.

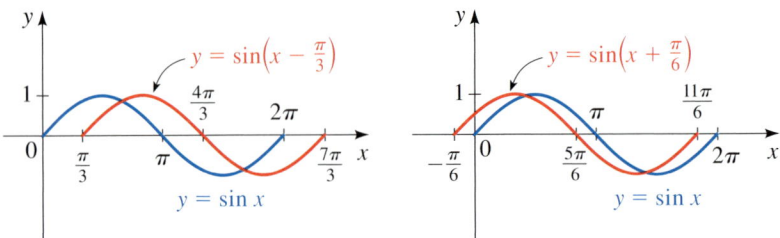

Figure 12

Example 4   A Shifted Sine Curve

Find the amplitude, period, and phase shift of $y = 3\sin 2\left(x - \dfrac{\pi}{4}\right)$, and graph one complete period.

Solution   We get the amplitude, period, and phase shift from the form of the function as follows:

$$\text{amplitude} = |a| = 3 \qquad \text{period} = \frac{2\pi}{k} = \frac{2\pi}{2} = \pi$$

$$y = 3\sin 2\left(x - \frac{\pi}{4}\right)$$

$$\text{phase shift} = \frac{\pi}{4}\ (\text{to the right})$$

Here is another way to find an appropriate interval on which to graph one complete period. Since the period of $y = \sin x$ is 2π, the function $y = 3\sin 2(x - \frac{\pi}{4})$ will go through one complete period as $2(x - \frac{\pi}{4})$ varies from 0 to 2π.

Start of period: End of period:

$2(x - \frac{\pi}{4}) = 0$ $2(x - \frac{\pi}{4}) = 2\pi$

$x - \frac{\pi}{4} = 0$ $x - \frac{\pi}{4} = \pi$

$x = \frac{\pi}{4}$ $x = \frac{5\pi}{4}$

So we graph one period on the interval $\left[\frac{\pi}{4}, \frac{5\pi}{4}\right]$.

Since the phase shift is $\pi/4$ and the period is π, one complete period occurs on the interval

$$\left[\frac{\pi}{4}, \frac{\pi}{4} + \pi\right] = \left[\frac{\pi}{4}, \frac{5\pi}{4}\right]$$

As an aid in sketching the graph, we divide this interval into four equal parts, then graph a sine curve with amplitude 3 as in Figure 13.

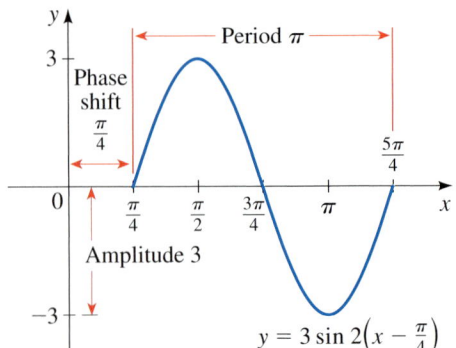

Figure 13

Example 5 A Shifted Cosine Curve

Find the amplitude, period, and phase shift of

$$y = \frac{3}{4} \cos\left(2x + \frac{2\pi}{3}\right)$$

and graph one complete period.

Solution We first write this function in the form $y = a \cos k(x - b)$. To do this, we factor 2 from the expression $2x + \frac{2\pi}{3}$ to get

$$y = \frac{3}{4} \cos 2\left[x - \left(-\frac{\pi}{3}\right)\right]$$

We can also find one complete period as follows:

Start of period: End of period:

$2x + \frac{2\pi}{3} = 0$ $2x + \frac{2\pi}{3} = 2\pi$

$\quad 2x = -\frac{2\pi}{3}$ $\quad 2x = \frac{4\pi}{3}$

$\quad\quad x = -\frac{\pi}{3}$ $\quad\quad x = \frac{2\pi}{3}$

So we graph one period on the interval $\left[-\frac{\pi}{3}, \frac{2\pi}{3}\right]$.

Thus, we have

$$\text{amplitude} = |a| = \frac{3}{4}$$

$$\text{period} = \frac{2\pi}{k} = \frac{2\pi}{2} = \pi$$

$$\text{phase shift} = b = -\frac{\pi}{3} \qquad \textit{Shift } \frac{\pi}{3} \textit{ to the left}$$

From this information it follows that one period of this cosine curve begins at $-\pi/3$ and ends at $(-\pi/3) + \pi = 2\pi/3$. To sketch the graph over the interval $[-\pi/3, 2\pi/3]$, we divide this interval into four equal parts and graph a cosine curve with amplitude $\frac{3}{4}$ as shown in Figure 14.

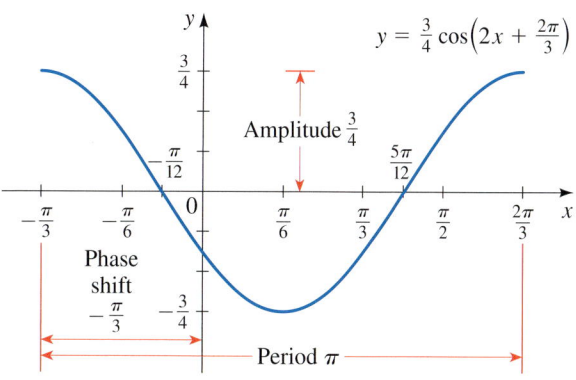

Figure 14

Using Graphing Devices to Graph Trigonometric Functions

When using a graphing calculator or a computer to graph a function, it is important to choose the viewing rectangle carefully in order to produce a reasonable graph of the function. This is especially true for trigonometric functions; the next example shows that, if care is not taken, it's easy to produce a very misleading graph of a trigonometric function.

See Section 1.9 for guidelines on choosing an appropriate viewing rectangle.

Example 6 Choosing the Viewing Rectangle

Graph the function $f(x) = \sin 50x$ in an appropriate viewing rectangle.

Solution Figure 15(a) shows the graph of f produced by a graphing calculator using the viewing rectangle $[-12, 12]$ by $[-1.5, 1.5]$. At first glance the graph appears to be reasonable. But if we change the viewing rectangle to the ones shown in Figure 15, the graphs look very different. Something strange is happening.

The appearance of the graphs in Figure 15 depends on the machine used. The graphs you get with your own graphing device might not look like these figures, but they will also be quite inaccurate.

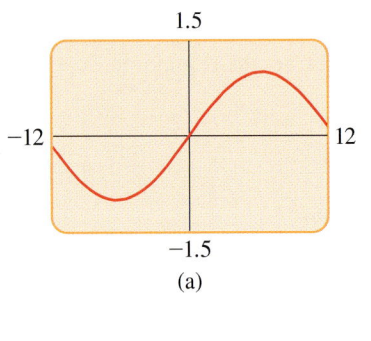

(a)

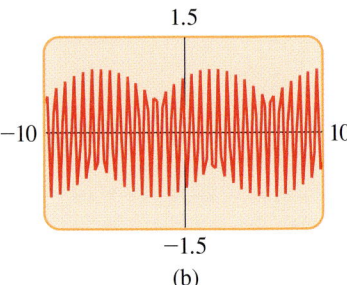

(b)

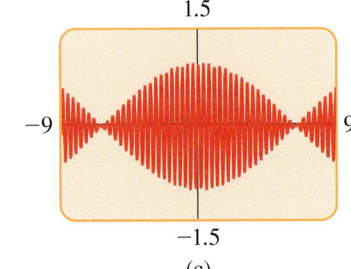

(c)

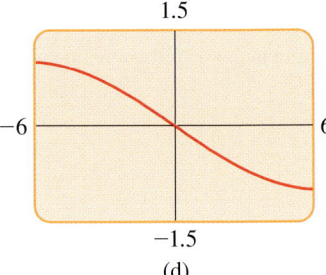

(d)

Figure 15

Graphs of $f(x) = \sin 50x$ in different viewing rectangles

To explain the big differences in appearance of these graphs and to find an appropriate viewing rectangle, we need to find the period of the function $y = \sin 50x$:

$$\text{period} = \frac{2\pi}{50} = \frac{\pi}{25} \approx 0.126$$

This suggests that we should deal only with small values of x in order to show just a few oscillations of the graph. If we choose the viewing rectangle $[-0.25, 0.25]$ by $[-1.5, 1.5]$, we get the graph shown in Figure 16.

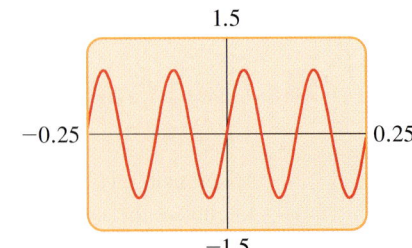

Figure 16

$f(x) = \sin 50x$

Now we see what went wrong in Figure 15. The oscillations of $y = \sin 50x$ are so rapid that when the calculator plots points and joins them, it misses most of the

The function h in Example 7 is **periodic** with period 2π. In general, functions that are sums of functions from the following list

$$1, \cos kx, \cos 2kx, \cos 3kx, \ldots$$

$$\sin kx, \sin 2kx, \sin 3kx, \ldots$$

are periodic. Although these functions appear to be special, they are actually fundamental to describing all periodic functions that arise in practice. The French mathematician J. B. J. Fourier (see page 536) discovered that nearly every periodic function can be written as a sum (usually an infinite sum) of these functions. This is remarkable because it means that any situation in which periodic variation occurs can be described mathematically using the functions sine and cosine. A modern application of Fourier's discovery is the digital encoding of sound on compact discs.

maximum and minimum points and therefore gives a very misleading impression of the graph. ∎

Example 7   A Sum of Sine and Cosine Curves

Graph $f(x) = 2 \cos x$, $g(x) = \sin 2x$, and $h(x) = 2 \cos x + \sin 2x$ on a common screen to illustrate the method of graphical addition.

Solution   Notice that $h = f + g$, so its graph is obtained by adding the corresponding y-coordinates of the graphs of f and g. The graphs of f, g, and h are shown in Figure 17.

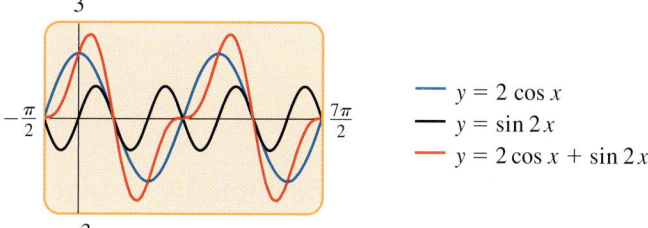

$$y = 2 \cos x$$
$$y = \sin 2x$$
$$y = 2 \cos x + \sin 2x$$

Figure 17 ∎

Example 8   A Cosine Curve with Variable Amplitude

Graph the functions $y = x^2$, $y = -x^2$, and $y = x^2 \cos 6\pi x$ on a common screen. Comment on and explain the relationship among the graphs.

Solution   Figure 18 shows all three graphs in the viewing rectangle $[-1.5, 1.5]$ by $[-2, 2]$. It appears that the graph of $y = x^2 \cos 6\pi x$ lies between the graphs of the functions $y = x^2$ and $y = -x^2$.

To understand this, recall that the values of $\cos 6\pi x$ lie between -1 and 1, that is,

$$-1 \le \cos 6\pi x \le 1$$

for all values of x. Multiplying the inequalities by x^2, and noting that $x^2 \ge 0$, we get

$$-x^2 \le x^2 \cos 6\pi x \le x^2$$

This explains why the functions $y = x^2$ and $y = -x^2$ form a boundary for the graph of $y = x^2 \cos 6\pi x$. (Note that the graphs touch when $\cos 6\pi x = \pm 1$.) ∎

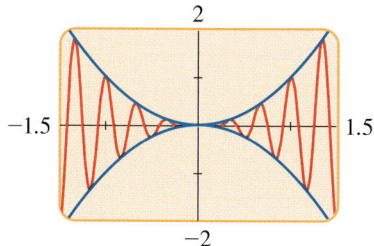

Figure 18
$y = x^2 \cos 6\pi x$

Example 8 shows that the function $y = x^2$ controls the amplitude of the graph of $y = x^2 \cos 6\pi x$. In general, if $f(x) = a(x) \sin kx$ or $f(x) = a(x) \cos kx$, the function a determines how the amplitude of f varies, and the graph of f lies between the graphs of $y = -a(x)$ and $y = a(x)$. Here is another example.

Example 9   A Cosine Curve with Variable Amplitude

Graph the function $f(x) = \cos 2\pi x \cos 16\pi x$.

Solution   The graph is shown in Figure 19 on the next page. Although it was drawn by a computer, we could have drawn it by hand, by first sketching the bound-

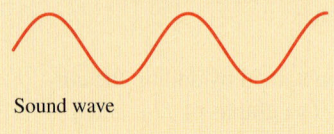
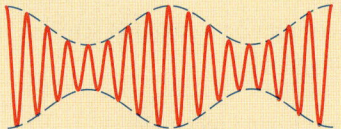
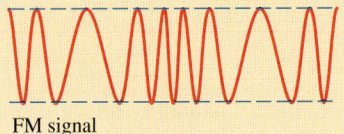
ary curves $y = \cos 2\pi x$ and $y = -\cos 2\pi x$. The graph of f is a cosine curve that lies between the graphs of these two functions.

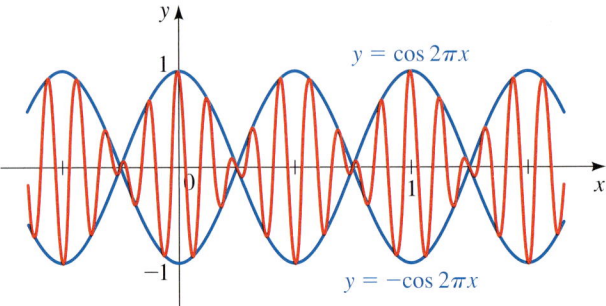

Figure 19

$$f(x) = \cos 2\pi x \cos 16\pi x$$

Example 10   A Sine Curve with Decaying Amplitude

The function $f(x) = \dfrac{\sin x}{x}$ is important in calculus. Graph this function and comment on its behavior when x is close to 0.

Solution   The viewing rectangle $[-15, 15]$ by $[-0.5, 1.5]$ shown in Figure 20(a) gives a good global view of the graph of f. The viewing rectangle $[-1, 1]$ by $[-0.5, 1.5]$ in Figure 20(b) focuses on the behavior of f when $x \approx 0$. Notice that although $f(x)$ is not defined when $x = 0$ (in other words, 0 is not in the domain of f), the values of f seem to approach 1 when x gets close to 0. This fact is crucial in calculus.

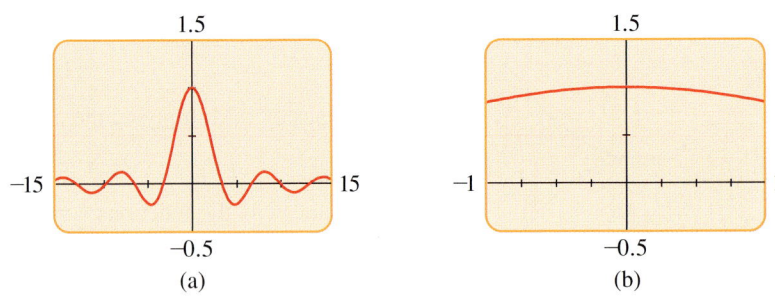

(a)   (b)

Figure 20

$$f(x) = \frac{\sin x}{x}$$

The function in Example 10 can be written as

$$f(x) = \frac{1}{x} \sin x$$

and may thus be viewed as a sine function whose amplitude is controlled by the function $a(x) = 1/x$.

5.3   Exercises

1–14 ■ Graph the function.

1. $f(x) = 1 + \cos x$

2. $f(x) = 3 + \sin x$

3. $f(x) = -\sin x$

4. $f(x) = 2 - \cos x$

5. $f(x) = -2 + \sin x$

6. $f(x) = -1 + \cos x$

7. $g(x) = 3 \cos x$

8. $g(x) = 2 \sin x$

9. $g(x) = -\frac{1}{2} \sin x$

10. $g(x) = -\frac{2}{3} \cos x$

11. $g(x) = 3 + 3 \cos x$

12. $g(x) = 4 - 2 \sin x$

13. $h(x) = |\cos x|$

14. $h(x) = |\sin x|$

15–26 ■ Find the amplitude and period of the function, and sketch its graph.

15. $y = \cos 2x$

16. $y = -\sin 2x$

17. $y = -3 \sin 3x$

18. $y = \frac{1}{2} \cos 4x$

19. $y = 10 \sin \frac{1}{2}x$

20. $y = 5 \cos \frac{1}{4}x$

21. $y = -\frac{1}{3} \cos \frac{1}{3}x$

22. $y = 4 \sin(-2x)$

23. $y = -2 \sin 2\pi x$

24. $y = -3 \sin \pi x$

25. $y = 1 + \frac{1}{2} \cos \pi x$

26. $y = -2 + \cos 4\pi x$

27–40 ■ Find the amplitude, period, and phase shift of the function, and graph one complete period.

27. $y = \cos\left(x - \frac{\pi}{2}\right)$

28. $y = 2 \sin\left(x - \frac{\pi}{3}\right)$

29. $y = -2 \sin\left(x - \frac{\pi}{6}\right)$

30. $y = 3 \cos\left(x + \frac{\pi}{4}\right)$

31. $y = -4 \sin 2\left(x + \frac{\pi}{2}\right)$

32. $y = \sin \frac{1}{2}\left(x + \frac{\pi}{4}\right)$

33. $y = 5 \cos\left(3x - \frac{\pi}{4}\right)$

34. $y = 2 \sin\left(\frac{2}{3}x - \frac{\pi}{6}\right)$

35. $y = \frac{1}{2} - \frac{1}{2} \cos\left(2x - \frac{\pi}{3}\right)$

36. $y = 1 + \cos\left(3x + \frac{\pi}{2}\right)$

37. $y = 3 \cos \pi(x + \frac{1}{2})$

38. $y = 3 + 2 \sin 3(x + 1)$

39. $y = \sin(\pi + 3x)$

40. $y = \cos\left(\frac{\pi}{2} - x\right)$

41–48 ■ The graph of one complete period of a sine or cosine curve is given.

(a) Find the amplitude, period, and phase shift.

(b) Write an equation that represents the curve in the form

$$y = a \sin k(x - b) \quad\text{or}\quad y = a \cos k(x - b)$$

41.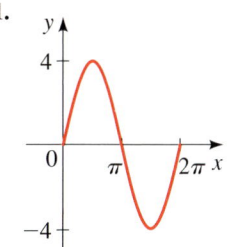

42.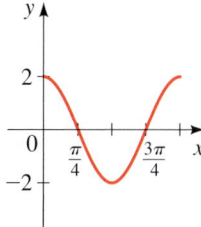

43.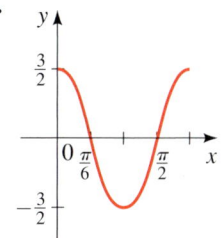

44.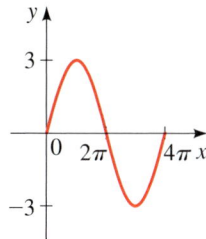

45.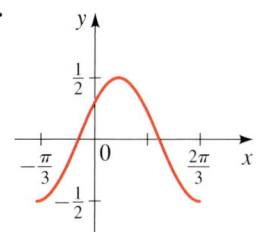

46.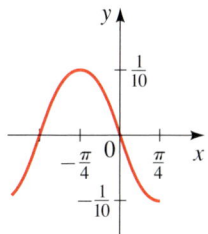

47.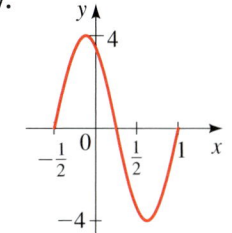

48.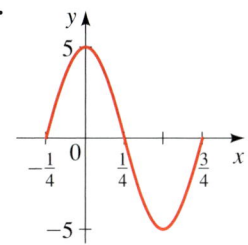

49–56 ■ Determine an appropriate viewing rectangle for each function, and use it to draw the graph.

49. $f(x) = \cos 100x$

50. $f(x) = 3 \sin 120x$

51. $f(x) = \sin(x/40)$

52. $f(x) = \cos(x/80)$

53. $y = \tan 25x$

54. $y = \csc 40x$

55. $y = \sin^2 20x$

56. $y = \sqrt{\tan 10\pi x}$

 57–58 ■ Graph f, g, and $f + g$ on a common screen to illustrate graphical addition.

57. $f(x) = x$, $g(x) = \sin x$

58. $f(x) = \sin x$, $g(x) = \sin 2x$

 59–64 ■ Graph the three functions on a common screen. How are the graphs related?

59. $y = x^2$, $y = -x^2$, $y = x^2 \sin x$

60. $y = x$, $y = -x$, $y = x \cos x$

61. $y = \sqrt{x}$, $y = -\sqrt{x}$, $y = \sqrt{x} \sin 5\pi x$

62. $y = \dfrac{1}{1 + x^2}$, $y = -\dfrac{1}{1 + x^2}$, $y = \dfrac{\cos 2\pi x}{1 + x^2}$

63. $y = \cos 3\pi x$, $y = -\cos 3\pi x$, $y = \cos 3\pi x \cos 21\pi x$

64. $y = \sin 2\pi x$, $y = -\sin 2\pi x$, $y = \sin 2\pi x \sin 10\pi x$

 65–68 ■ Find the maximum and minimum values of the function.

65. $y = \sin x + \sin 2x$

66. $y = x - 2 \sin x, 0 \le x \le 2\pi$

67. $y = 2 \sin x + \sin^2 x$

68. $y = \dfrac{\cos x}{2 + \sin x}$

 69–72 ■ Find all solutions of the equation that lie in the interval $[0, \pi]$. State each answer correct to two decimal places.

69. $\cos x = 0.4$

70. $\tan x = 2$

71. $\csc x = 3$

72. $\cos x = x$

 73–74 ■ A function f is given.

(a) Is f even, odd, or neither?

(b) Find the x-intercepts of the graph of f.

(c) Graph f in an appropriate viewing rectangle.

(d) Describe the behavior of the function as $x \to \pm\infty$.

(e) Notice that $f(x)$ is not defined when $x = 0$. What happens as x approaches 0?

73. $f(x) = \dfrac{1 - \cos x}{x}$

74. $f(x) = \dfrac{\sin 4x}{2x}$

Applications

75. Height of a Wave   As a wave passes by an offshore piling, the height of the water is modeled by the function

$$h(t) = 3 \cos\left(\frac{\pi}{10} t\right)$$

where $h(t)$ is the height in feet above mean sea level at time t seconds.

(a) Find the period of the wave.

(b) Find the wave height, that is, the vertical distance between the trough and the crest of the wave.

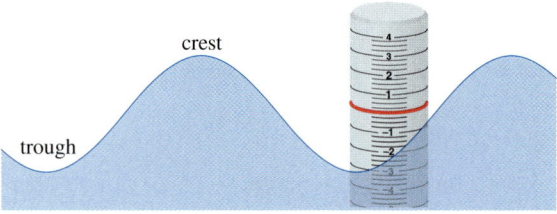

76. Sound Vibrations   A tuning fork is struck, producing a pure tone as its tines vibrate. The vibrations are modeled by the function

$$v(t) = 0.7 \sin(880\pi t)$$

where $v(t)$ is the displacement of the tines in millimeters at time t seconds.

(a) Find the period of the vibration.

(b) Find the frequency of the vibration, that is, the number of times the fork vibrates per second.

(c) Graph the function v.

77. Blood Pressure   Each time your heart beats, your blood pressure first increases and then decreases as the heart rests between beats. The maximum and minimum blood pressures are called the *systolic* and *diastolic* pressures, respectively. Your *blood pressure reading* is written as systolic/diastolic. A reading of 120/80 is considered normal.

A certain person's blood pressure is modeled by the function

$$p(t) = 115 + 25 \sin(160\pi t)$$

where $p(t)$ is the pressure in mmHg, at time t measured in minutes.

(a) Find the period of p.

(b) Find the number of heartbeats per minute.

(c) Graph the function p.

(d) Find the blood pressure reading. How does this compare to normal blood pressure?

78. Variable Stars   Variable stars are ones whose brightness
varies periodically. One of the most visible is R Leonis; its
brightness is modeled by the function

$$b(t) = 7.9 - 2.1 \cos\left(\frac{\pi}{156}t\right)$$

where t is measured in days.

(a) Find the period of R Leonis.

(b) Find the maximum and minimum brightness.

(c) Graph the function b.

Discovery • Discussion

79. Compositions Involving Trigonometric Functions
This exercise explores the effect of the inner function g
on a composite function $y = f(g(x))$.

(a) Graph the function $y = \sin \sqrt{x}$ using the viewing
rectangle $[0, 400]$ by $[-1.5, 1.5]$. In what ways does
this graph differ from the graph of the sine function?

(b) Graph the function $y = \sin(x^2)$ using the viewing
rectangle $[-5, 5]$ by $[-1.5, 1.5]$. In what ways does this
graph differ from the graph of the sine function?

80. Periodic Functions I   Recall that a function f is *periodic*
if there is a positive number p such that $f(t + p) = f(t)$
for every t, and the least such p (if it exists) is the *period*
of f. The graph of a function of period p looks the same on
each interval of length p, so we can easily determine the
period from the graph. Determine whether the function
whose graph is shown is periodic; if it is periodic, find
the period.

(a)

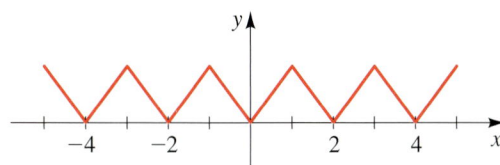

(b)

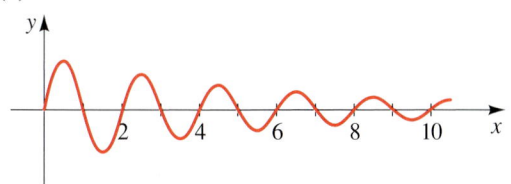

(c)

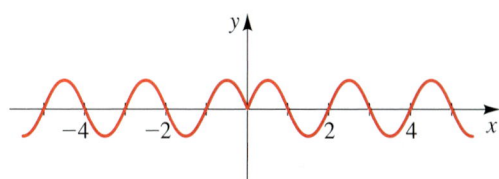

(d)

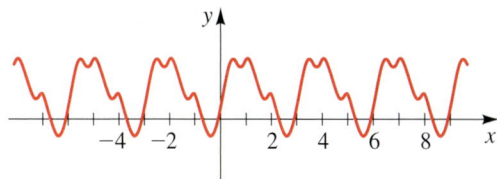

81. Periodic Functions II   Use a graphing device to graph the
following functions. From the graph, determine whether the
function is periodic; if it is periodic find the period. (See
page 162 for the definition of $[\![x]\!]$.)

(a) $y = |\sin x|$

(b) $y = \sin|x|$

(c) $y = 2^{\cos x}$

(d) $y = x - [\![x]\!]$

(e) $y = \cos(\sin x)$

(f) $y = \cos(x^2)$

82. Sinusoidal Curves   The graph of $y = \sin x$ is the same as
the graph of $y = \cos x$ shifted to the right $\pi/2$ units. So the
sine curve $y = \sin x$ is also at the same time a cosine curve:
$y = \cos(x - \pi/2)$. In fact, any sine curve is also a cosine
curve with a different phase shift, and any cosine curve is
also a sine curve. Sine and cosine curves are collectively
referred to as *sinusoidal*. For the curve whose graph is
shown, find all possible ways of expressing it as a sine curve
$y = a \sin(x - b)$ or as a cosine curve $y = a \cos(x - b)$.
Explain why you think you have found all possible choices
for a and b in each case.

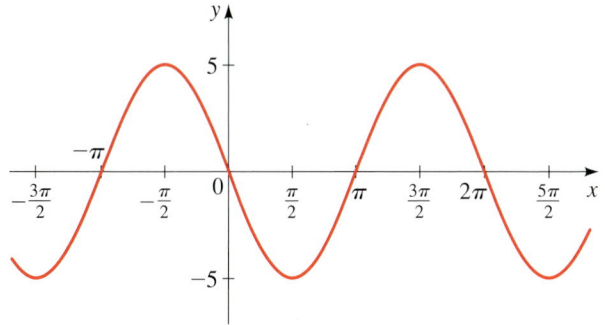

DISCOVERY PROJECT

Predator/Prey Models

Sine and cosine functions are used primarily in physics and engineering to model oscillatory behavior, such as the motion of a pendulum or the current in an AC electrical circuit. (See Section 5.5.) But these functions also arise in the other sciences. In this project, we consider an application to biology—we use sine functions to model the population of a predator and its prey.

An isolated island is inhabited by two species of mammals: lynx and hares. The lynx are *predators* who feed on the hares, their *prey*. The lynx and hare populations change cyclically, as graphed in Figure 1. In part A of the graph, hares are abundant, so the lynx have plenty to eat and their population increases. By the time portrayed in part B, so many lynx are feeding on the hares that the hare population declines. In part C, the hare population has declined so much that there is not enough food for the lynx, so the lynx population starts to decrease. In part D, so many lynx have died that the hares have few enemies, and their population increases again. This takes us back to where we started, and the cycle repeats over and over again.

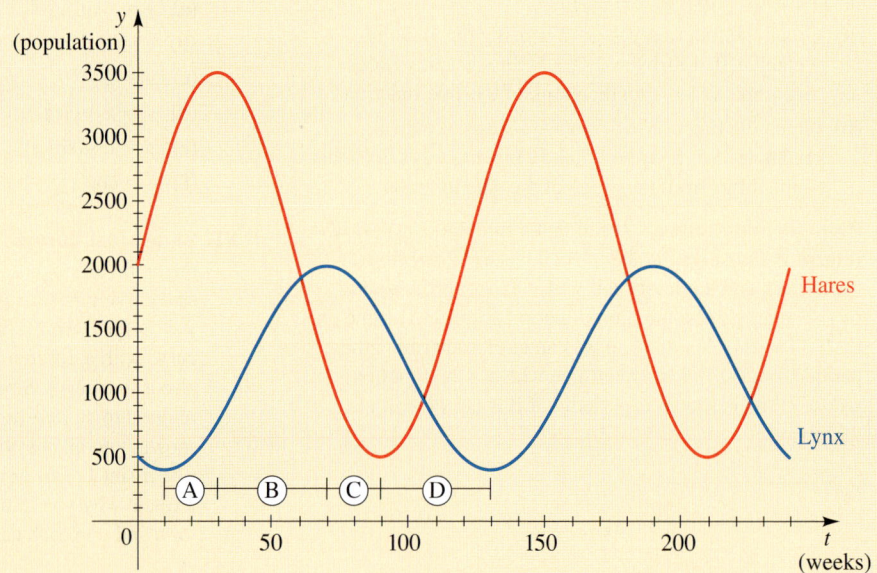

Figure 1

The graphs in Figure 1 are sine curves that have been shifted upward, so they are graphs of functions of the form

$$y = a \sin k(t - b) + c$$

Here c is the amount by which the sine curve has been shifted vertically (see Section 2.4). Note that c is the average value of the function, halfway between the highest and lowest values on the graph. The amplitude $|a|$ is the amount by which the graph varies above and below the average value (see Figure 2).

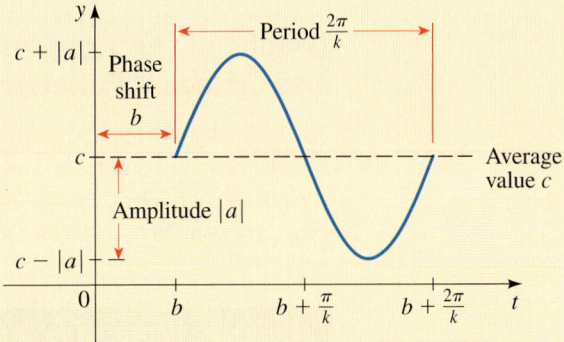

Figure 2
$y = a \sin k(t - b) + c$

1. Find functions of the form $y = a \sin k(t - b) + c$ that model the lynx and hare populations graphed in Figure 1. Graph both functions on your calculator and compare to Figure 1 to verify that your functions are the right ones.

2. Add the lynx and hare population functions to get a new function that models the total *mammal* population on this island. Graph this function on your calculator, and find its average value, amplitude, period, and phase shift. How are the average value and period of the mammal population function related to the average value and period of the lynx and hare population functions?

3. A small lake on the island contains two species of fish: hake and redfish. The hake are predators that eat the redfish. The fish population in the lake varies periodically with period 180 days. The number of hake varies between 500 and 1500, and the number of redfish varies between 1000 and 3000. The hake reach their maximum population 30 days after the redfish have reached *their* maximum population in the cycle.

    (a) Sketch a graph (like the one in Figure 1) that shows two complete periods of the population cycle for these species of fish. Assume that $t = 0$ corresponds to a time when the redfish population is at a maximum.

    (b) Find cosine functions of the form $y = a \cos k(t - b) + c$ that model the hake and redfish populations in the lake.

4. In real life, most predator/prey populations do not behave as simply as the examples we have described here. In most cases, the populations of predator and prey oscillate, but the amplitude of the oscillations gets smaller and smaller, so that eventually both populations stabilize near a constant value. Sketch a rough graph that illustrates how the populations of predator and prey might behave in this case.

| 5.4 | **More Trigonometric Graphs** |

In this section we graph the tangent, cotangent, secant, and cosecant functions, and transformations of these functions.

Graphs of the Tangent, Cotangent, Secant, and Cosecant Function

We begin by stating the periodic properties of these functions. Recall that sine and cosine have period 2π. Since cosecant and secant are the reciprocals of sine and cosine, respectively, they also have period 2π (see Exercise 53). Tangent and cotangent, however, have period π (see Exercise 83 of Section 5.2).

Periodic Properties

The functions tangent and cotangent have period π:

$$\tan(x + \pi) = \tan x \qquad \cot(x + \pi) = \cot x$$

The functions cosecant and secant have period 2π:

$$\csc(x + 2\pi) = \csc x \qquad \sec(x + 2\pi) = \sec x$$

x	$\tan x$
0	0
$\dfrac{\pi}{6}$	0.58
$\dfrac{\pi}{4}$	1.00
$\dfrac{\pi}{3}$	1.73
1.4	5.80
1.5	14.10
1.55	48.08
1.57	$1,255.77$
1.5707	$10,381.33$

We first sketch the graph of tangent. Since it has period π, we need only sketch the graph on any interval of length π and then repeat the pattern to the left and to the right. We sketch the graph on the interval $(-\pi/2, \pi/2)$. Since $\tan \pi/2$ and $\tan(-\pi/2)$ aren't defined, we need to be careful in sketching the graph at points near $\pi/2$ and $-\pi/2$. As x gets near $\pi/2$ through values less than $\pi/2$, the value of $\tan x$ becomes large. To see this, notice that as x gets close to $\pi/2$, $\cos x$ approaches 0 and $\sin x$ approaches 1 and so $\tan x = \sin x/\cos x$ is large. A table of values of $\tan x$ for x close to $\pi/2$ (≈ 1.570796) is shown in the margin.

Thus, by choosing x close enough to $\pi/2$ through values less than $\pi/2$, we can make the value of $\tan x$ larger than any given positive number. We express this by writing

$$\tan x \to \infty \qquad \text{as} \qquad x \to \frac{\pi}{2}^-$$

This is read "$\tan x$ approaches infinity as x approaches $\pi/2$ from the left."

In a similar way, by choosing x close to $-\pi/2$ through values greater than $-\pi/2$, we can make $\tan x$ smaller than any given negative number. We write this as

$$\tan x \to -\infty \qquad \text{as} \qquad x \to -\frac{\pi}{2}^+$$

Arrow notation is discussed in Section 3.6.

This is read "$\tan x$ approaches negative infinity as x approaches $-\pi/2$ from the right."

Thus, the graph of $y = \tan x$ approaches the vertical lines $x = \pi/2$ and $x = -\pi/2$. So these lines are **vertical asymptotes**. With the information we have so far, we sketch the graph of $y = \tan x$ for $-\pi/2 < x < \pi/2$ in Figure 1. The complete graph

Asymptotes are discussed in Section 3.6.

of tangent (see Figure 5(a) on page 436) is now obtained using the fact that tangent is periodic with period π.

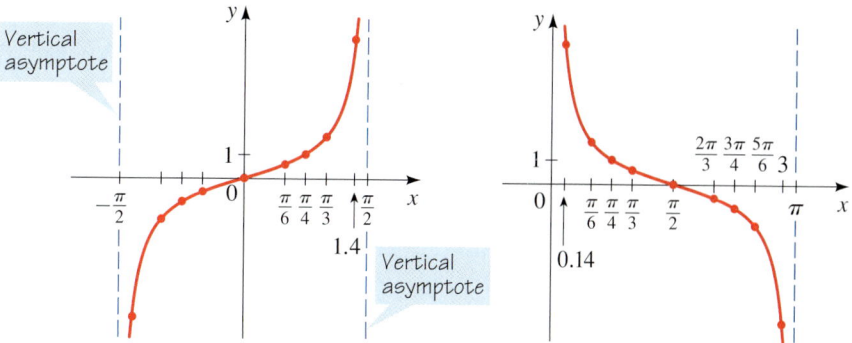

Figure 1

One period of $y = \tan x$

Figure 2

One period of $y = \cot x$

The function $y = \cot x$ is graphed on the interval $(0, \pi)$ by a similar analysis (see Figure 2). Since $\cot x$ is undefined for $x = n\pi$ with n an integer, its complete graph (in Figure 5(b) on page 436) has vertical asymptotes at these values.

To graph the cosecant and secant functions, we use the reciprocal identities

$$\csc x = \frac{1}{\sin x} \quad \text{and} \quad \sec x = \frac{1}{\cos x}$$

So, to graph $y = \csc x$, we take the reciprocals of the y-coordinates of the points of the graph of $y = \sin x$. (See Figure 3.) Similarly, to graph $y = \sec x$, we take the reciprocals of the y-coordinates of the points of the graph of $y = \cos x$. (See Figure 4.)

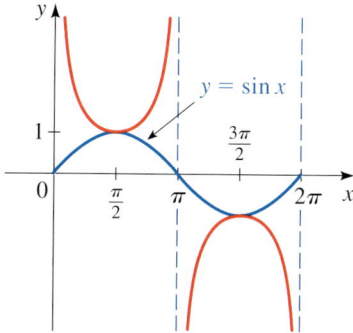

Figure 3

One period of $y = \csc x$

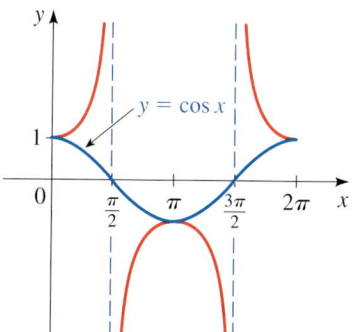

Figure 4

One period of $y = \sec x$

Let's consider more closely the graph of the function $y = \csc x$ on the interval $0 < x < \pi$. We need to examine the values of the function near 0 and π since at these values $\sin x = 0$, and $\csc x$ is thus undefined. We see that

$$\csc x \to \infty \quad \text{as} \quad x \to 0^+$$

$$\csc x \to \infty \quad \text{as} \quad x \to \pi^-$$

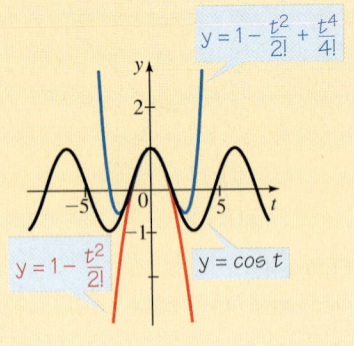

Thus, the lines $x = 0$ and $x = \pi$ are vertical asymptotes. In the interval $\pi < x < 2\pi$ the graph is sketched in the same way. The values of $\csc x$ in that interval are the same as those in the interval $0 < x < \pi$ except for sign (see Figure 3). The complete graph in Figure 5(c) is now obtained from the fact that the function cosecant is periodic with period 2π. Note that the graph has vertical asymptotes at the points where $\sin x = 0$, that is, at $x = n\pi$, for n an integer.

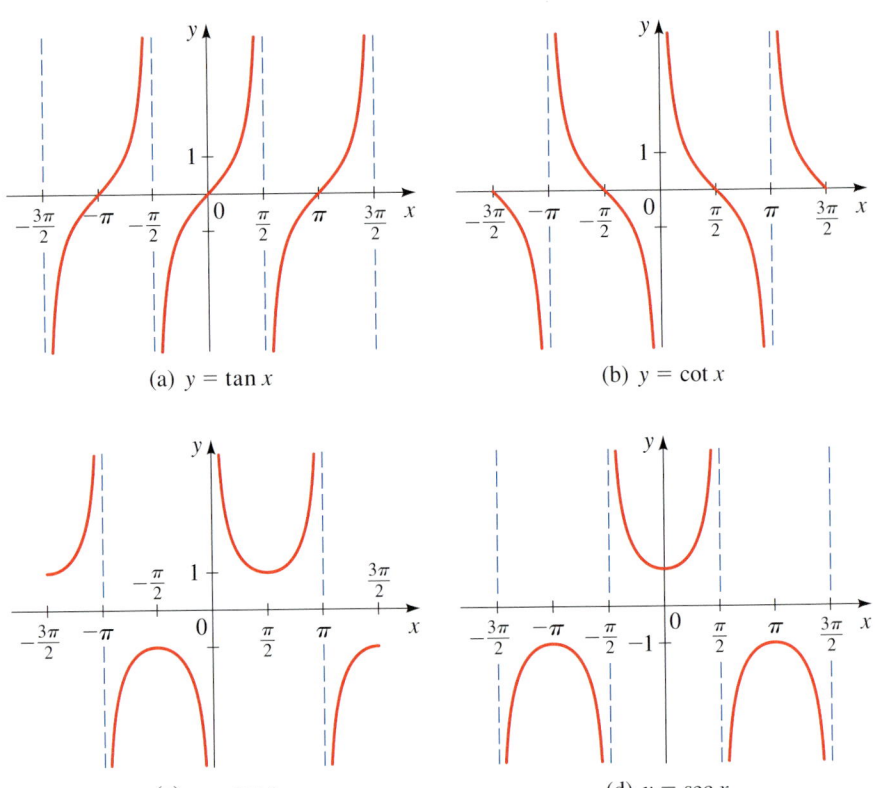

Figure 5

The graph of $y = \sec x$ is sketched in a similar manner. Observe that the domain of $\sec x$ is the set of all real numbers other than $x = (\pi/2) + n\pi$, for n an integer, so the graph has vertical asymptotes at those points. The complete graph is shown in Figure 5(d).

It is apparent that the graphs of $y = \tan x$, $y = \cot x$, and $y = \csc x$ are symmetric about the origin, whereas that of $y = \sec x$ is symmetric about the y-axis. This is because tangent, cotangent, and cosecant are odd functions, whereas secant is an even function.

Graphs Involving Tangent and Cotangent Functions

We now consider graphs of transformations of the tangent and cotangent functions.

Example 1 Graphing Tangent Curves

Graph each function.

(a) $y = 2 \tan x$ (b) $y = -\tan x$

Solution   We first graph $y = \tan x$ and then transform it as required.

(a) To graph $y = 2 \tan x$, we multiply the y-coordinate of each point on the graph of $y = \tan x$ by 2. The resulting graph is shown in Figure 6(a).

(b) The graph of $y = -\tan x$ in Figure 6(b) is obtained from that of $y = \tan x$ by reflecting in the x-axis.

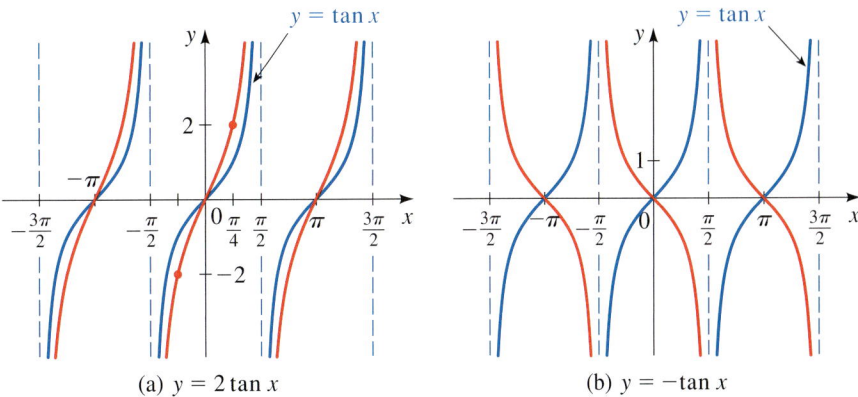

(a) $y = 2 \tan x$ (b) $y = -\tan x$

Figure 6

Since the tangent and cotangent functions have period π, the functions

$$y = a \tan kx \quad \text{and} \quad y = a \cot kx \quad (k > 0)$$

complete one period as kx varies from 0 to π, that is, for $0 \le kx \le \pi$. Solving this inequality, we get $0 \le x \le \pi/k$. So they each have period π/k.

Tangent and Cotangent Curves

The functions

$$y = a \tan kx \quad \text{and} \quad y = a \cot kx \quad (k > 0)$$

have period π/k.

Thus, one complete period of the graphs of these functions occurs on any interval of length π/k. To sketch a complete period of these graphs, it's convenient to select an interval between vertical asymptotes:

To graph one period of $y = a \tan kx$, an appropriate interval is $\left(-\dfrac{\pi}{2k}, \dfrac{\pi}{2k} \right)$.

To graph one period of $y = a \cot kx$, an appropriate interval is $\left(0, \dfrac{\pi}{k} \right)$.

Example 2   **Graphing Tangent Curves**

Graph each function.

(a) $y = \tan 2x$ (b) $y = \tan 2\left(x - \dfrac{\pi}{4}\right)$

Solution

(a) The period is $\pi/2$ and an appropriate interval is $(-\pi/4, \pi/4)$. The endpoints $x = -\pi/4$ and $x = \pi/4$ are vertical asymptotes. Thus, we graph one complete period of the function on $(-\pi/4, \pi/4)$. The graph has the same shape as that of the tangent function, but is shrunk horizontally by a factor of $\frac{1}{2}$. We then repeat that portion of the graph to the left and to the right. See Figure 7(a).

(b) The graph is the same as that in part (a), but it is shifted to the right $\pi/4$, as shown in Figure 7(b).

Since $y = \tan x$ completes one period between $x = -\frac{\pi}{2}$ and $x = \frac{\pi}{2}$, the function $y = \tan 2(x - \frac{\pi}{4})$ completes one period as $2(x - \frac{\pi}{4})$ varies from $-\frac{\pi}{2}$ to $\frac{\pi}{2}$.

Start of period:	End of period:
$2\left(x - \frac{\pi}{4}\right) = -\frac{\pi}{2}$	$2\left(x - \frac{\pi}{4}\right) = \frac{\pi}{2}$
$x - \frac{\pi}{4} = -\frac{\pi}{4}$	$x - \frac{\pi}{4} = \frac{\pi}{4}$
$x = 0$	$x = \frac{\pi}{2}$

So we graph one period on the interval $(0, \frac{\pi}{2})$.

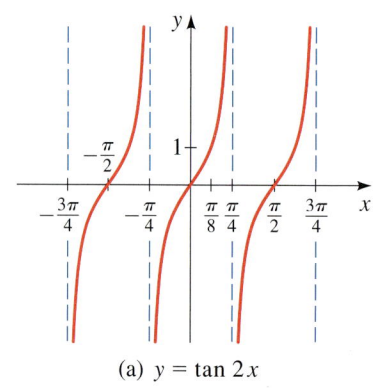

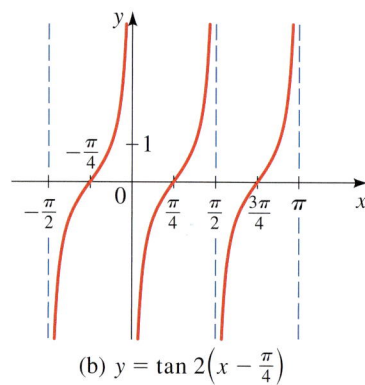

Figure 7   (a) $y = \tan 2x$ (b) $y = \tan 2\left(x - \frac{\pi}{4}\right)$

Example 3   **A Shifted Cotangent Curve**

Graph $y = 2 \cot\left(3x - \dfrac{\pi}{2}\right)$.

Solution   We first put this in the form $y = a \cot k(x - b)$ by factoring 3 from the expression $3x - \dfrac{\pi}{2}$:

$$y = 2 \cot\left(3x - \frac{\pi}{2}\right) = 2 \cot 3\left(x - \frac{\pi}{6}\right)$$

Since $y = \cot x$ completes one period between $x = 0$ and $x = \pi$, the function $y = 2 \cot(3x - \frac{\pi}{2})$ completes one period as $3x - \frac{\pi}{2}$ varies from 0 to π.

Start of period:	End of period:
$3x - \frac{\pi}{2} = 0$	$3x - \frac{\pi}{2} = \pi$
$3x = \frac{\pi}{2}$	$3x = \frac{3\pi}{2}$
$x = \frac{\pi}{6}$	$x = \frac{\pi}{2}$

So we graph one period on the interval $(\frac{\pi}{6}, \frac{\pi}{2})$.

Thus, the graph is the same as that of $y = 2 \cot 3x$, but is shifted to the right $\pi/6$. The period of $y = 2 \cot 3x$ is $\pi/3$, and an appropriate interval is $(0, \pi/3)$. To get the corresponding interval for the desired graph, we shift this interval to the right $\pi/6$. This gives

$$\left(0 + \frac{\pi}{6}, \frac{\pi}{3} + \frac{\pi}{6}\right) = \left(\frac{\pi}{6}, \frac{\pi}{2}\right)$$

Finally, we graph one period in the shape of cotangent on the interval $(\pi/6, \pi/2)$ and repeat that portion of the graph to the left and to the right. (See Figure 8.)

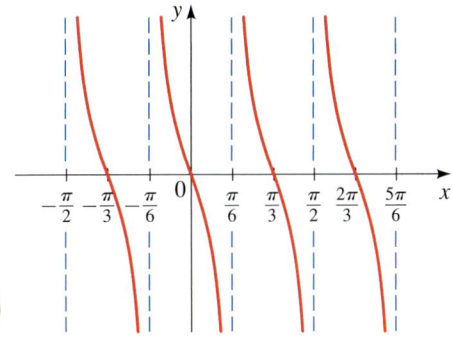

Figure 8

$$y = 2 \cot\left(3x - \frac{\pi}{2} \right)$$

Graphs Involving the Cosecant and Secant Functions

We have already observed that the cosecant and secant functions are the reciprocals of the sine and cosine functions. Thus, the following result is the counterpart of the result for sine and cosine curves in Section 5.3.

Cosecant and Secant Curves

The functions

$$y = a \csc kx \quad \text{and} \quad y = a \sec kx \quad (k > 0)$$

have period $2\pi/k$.

An appropriate interval on which to graph one complete period is $[0, 2\pi/k]$.

Example 4   Graphing Cosecant Curves

Graph each function.

(a) $y = \dfrac{1}{2} \csc 2x$ 　　　(b) $y = \dfrac{1}{2} \csc\left(2x + \dfrac{\pi}{2} \right)$

Solution

(a) The period is $2\pi/2 = \pi$. An appropriate interval is $[0, \pi]$, and the asymptotes occur in this interval whenever $\sin 2x = 0$. So the asymptotes in this interval are $x = 0$, $x = \pi/2$, and $x = \pi$. With this information we sketch on the interval $[0, \pi]$ a graph with the same general shape as that of one period of the cosecant function. The complete graph in Figure 9(a) is obtained by repeating this portion of the graph to the left and to the right.

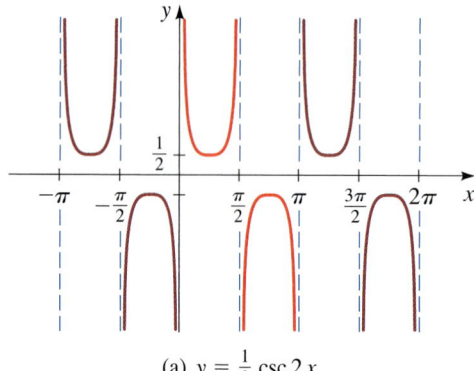

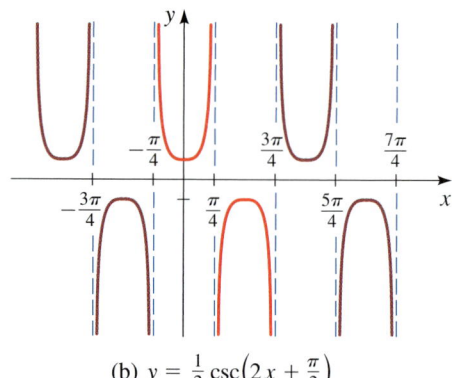

Figure 9   (a) $y = \frac{1}{2} \csc 2x$ (b) $y = \frac{1}{2} \csc\left(2x + \frac{\pi}{2}\right)$

Since $y = \csc x$ completes one period between $x = 0$ and $x = 2\pi$, the function $y = \frac{1}{2} \csc(2x + \frac{\pi}{2})$ completes one period as $2x + \frac{\pi}{2}$ varies from 0 to 2π.

Start of period: End of period:

$2x + \frac{\pi}{2} = 0$ $2x + \frac{\pi}{2} = 2\pi$

$2x = -\frac{\pi}{2}$ $2x = \frac{3\pi}{2}$

$x = -\frac{\pi}{4}$ $x = \frac{3\pi}{4}$

So we graph one period on the interval $\left(-\frac{\pi}{4}, \frac{3\pi}{4}\right)$.

(b) We first write

$$y = \frac{1}{2} \csc\left(2x + \frac{\pi}{2}\right) = \frac{1}{2} \csc 2\left(x + \frac{\pi}{4}\right)$$

From this we see that the graph is the same as that in part (a), but shifted to the left $\pi/4$. The graph is shown in Figure 9(b). ∎

Example 5   Graphing a Secant Curve

Graph $y = 3 \sec \frac{1}{2}x$.

Solution   The period is $2\pi \div \frac{1}{2} = 4\pi$. An appropriate interval is $[0, 4\pi]$, and the asymptotes occur in this interval wherever $\cos \frac{1}{2}x = 0$. Thus, the asymptotes in this interval are $x = \pi, x = 3\pi$. With this information we sketch on the interval $[0, 4\pi]$ a graph with the same general shape as that of one period of the secant function. The complete graph in Figure 10 is obtained by repeating this portion of the graph to the left and to the right.

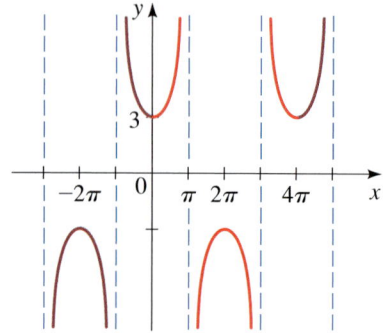

Figure 10
$y = 3 \sec \frac{1}{2}x$

∎

5.4   Exercises

1–6 ■ Match the trigonometric function with one of the graphs I–VI.

1. $f(x) = \tan\left(x + \dfrac{\pi}{4}\right)$

2. $f(x) = \sec 2x$

3. $f(x) = \cot 2x$

4. $f(x) = -\tan x$

5. $f(x) = 2 \sec x$

6. $f(x) = 1 + \csc x$

I
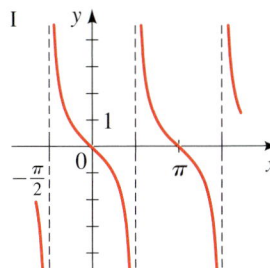

II
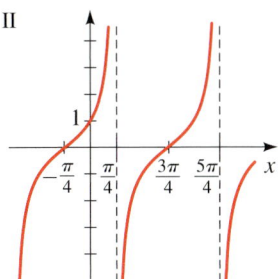

III
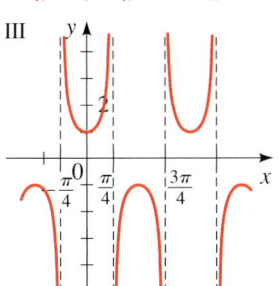

IV
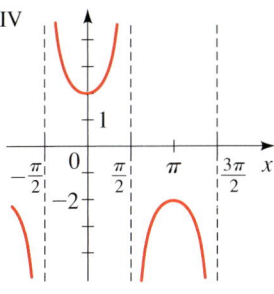

V
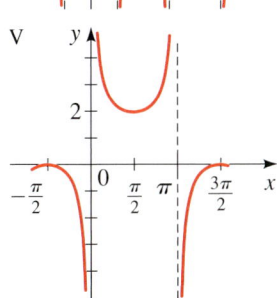

VI
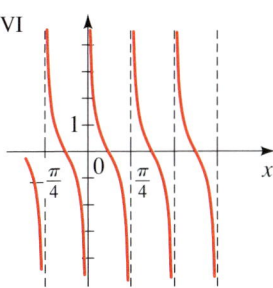

7–52 ■ Find the period and graph the function.

7. $y = 4 \tan x$

8. $y = -4 \tan x$

9. $y = -\frac{1}{2} \tan x$

10. $y = \frac{1}{2} \tan x$

11. $y = -\cot x$

12. $y = 2 \cot x$

13. $y = 2 \csc x$

14. $y = \frac{1}{2} \csc x$

15. $y = 3 \sec x$

16. $y = -3 \sec x$

17. $y = \tan\left(x + \dfrac{\pi}{2}\right)$

18. $y = \tan\left(x - \dfrac{\pi}{4}\right)$

19. $y = \csc\left(x - \dfrac{\pi}{2}\right)$

20. $y = \sec\left(x + \dfrac{\pi}{4}\right)$

21. $y = \cot\left(x + \dfrac{\pi}{4}\right)$

22. $y = 2 \csc\left(x - \dfrac{\pi}{3}\right)$

23. $y = \dfrac{1}{2} \sec\left(x - \dfrac{\pi}{6}\right)$

24. $y = 3 \csc\left(x + \dfrac{\pi}{2}\right)$

25. $y = \tan 2x$

26. $y = \tan \frac{1}{2}x$

27. $y = \tan \dfrac{\pi}{4}x$

28. $y = \cot \dfrac{\pi}{2}x$

29. $y = \sec 2x$

30. $y = 5 \csc 3x$

31. $y = \csc 2x$

32. $y = \csc \frac{1}{2}x$

33. $y = 2 \tan 3\pi x$

34. $y = 2 \tan \dfrac{\pi}{2}x$

35. $y = 5 \csc \dfrac{3\pi}{2}x$

36. $y = 5 \sec 2\pi x$

37. $y = \tan 2\left(x + \dfrac{\pi}{2}\right)$

38. $y = \csc 2\left(x + \dfrac{\pi}{2}\right)$

39. $y = \tan 2(x - \pi)$

40. $y = \sec 2\left(x - \dfrac{\pi}{2}\right)$

41. $y = \cot\left(2x - \dfrac{\pi}{2}\right)$

42. $y = \frac{1}{2} \tan(\pi x - \pi)$

43. $y = 2 \csc\left(\pi x - \dfrac{\pi}{3}\right)$

44. $y = 2 \sec\left(\dfrac{1}{2}x - \dfrac{\pi}{3}\right)$

45. $y = 5 \sec\left(3x - \dfrac{\pi}{2}\right)$

46. $y = \frac{1}{2} \sec(2\pi x - \pi)$

47. $y = \tan\left(\dfrac{2}{3}x - \dfrac{\pi}{6}\right)$

48. $y = \tan \dfrac{1}{2}\left(x + \dfrac{\pi}{4}\right)$

49. $y = 3 \sec \pi\left(x + \dfrac{1}{2}\right)$

50. $y = \sec\left(3x + \dfrac{\pi}{2}\right)$

51. $y = -2 \tan\left(2x - \dfrac{\pi}{3}\right)$

52. $y = 2 \csc(3x + 3)$

53. (a) Prove that if f is periodic with period p, then $1/f$ is also periodic with period p.

(b) Prove that cosecant and secant each have period 2π.

54. Prove that if f and g are periodic with period p, then f/g is also periodic, but its period could be smaller than p.

Applications

55. Lighthouse   The beam from a lighthouse completes one rotation every two minutes. At time t, the distance d shown in the figure on the next page is

$$d(t) = 3 \tan \pi t$$

where t is measured in minutes and d in miles.

(a) Find $d(0.15)$, $d(0.25)$, and $d(0.45)$.

(b) Sketch a graph of the function d for $0 \le t < \frac{1}{2}$.

(c) What happens to the distance d as t approaches $\frac{1}{2}$?

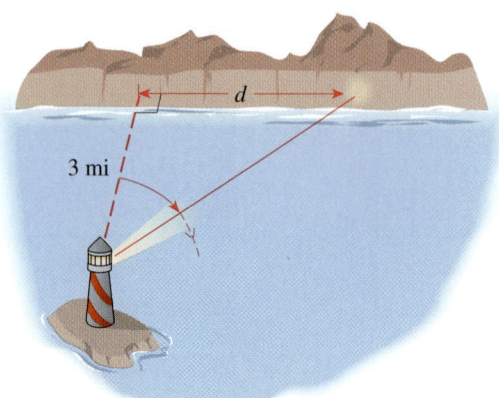

3 mi

56. Length of a Shadow On a day when the sun passes directly overhead at noon, a six-foot-tall man casts a shadow of length

$$S(t) = 6 \left| \cot \frac{\pi}{12} t \right|$$

where S is measured in feet and t is the number of hours since 6 A.M.

(a) Find the length of the shadow at 8:00 A.M., noon, 2:00 P.M., and 5:45 P.M.

(b) Sketch a graph of the function S for $0 < t < 12$.

(c) From the graph determine the values of t at which the length of the shadow equals the man's height. To what time of day does each of these values correspond?

(d) Explain what happens to the shadow as the time approaches 6 P.M. (that is, as $t \to 12^-$).

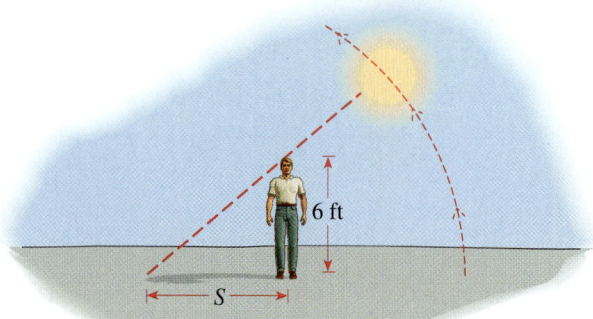

6 ft

S

Discovery • Discussion

57. Reduction Formulas Use the graphs in Figure 5 to explain why the following formulas are true.

$$\tan\left(x - \frac{\pi}{2}\right) = -\cot x$$

$$\sec\left(x - \frac{\pi}{2}\right) = \csc x$$

5.5   Modeling Harmonic Motion

Periodic behavior—behavior that repeats over and over again—is common in nature. Perhaps the most familiar example is the daily rising and setting of the sun, which results in the repetitive pattern of day, night, day, night, ... Another example is the daily variation of tide levels at the beach, which results in the repetitive pattern of high tide, low tide, high tide, low tide, ... Certain animal populations increase and decrease in a predictable periodic pattern: A large population exhausts the food supply, which causes the population to dwindle; this in turn results in a more plentiful food supply, which makes it possible for the population to increase; and the pattern then repeats over and over (see pages 432–433).

Other common examples of periodic behavior involve motion that is caused by vibration or oscillation. A mass suspended from a spring that has been compressed and then allowed to vibrate vertically is a simple example. This same "back and forth" motion also occurs in such diverse phenomena as sound waves, light waves, alternating electrical current, and pulsating stars, to name a few. In this section we consider the problem of modeling periodic behavior.

Modeling Periodic Behavior

The trigonometric functions are ideally suited for modeling periodic behavior. A glance at the graphs of the sine and cosine functions, for instance, tells us that these functions themselves exhibit periodic behavior. Figure 1 shows the graph of $y = \sin t$. If we think of t as time, we see that as time goes on, $y = \sin t$ increases and decreases over and over again. Figure 2 shows that the motion of a vibrating mass on a spring is modeled very accurately by $y = \sin t$.

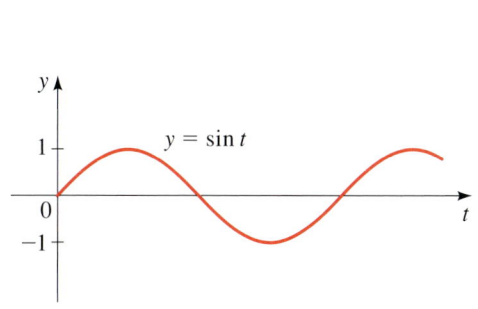

Figure 1

$y = \sin t$

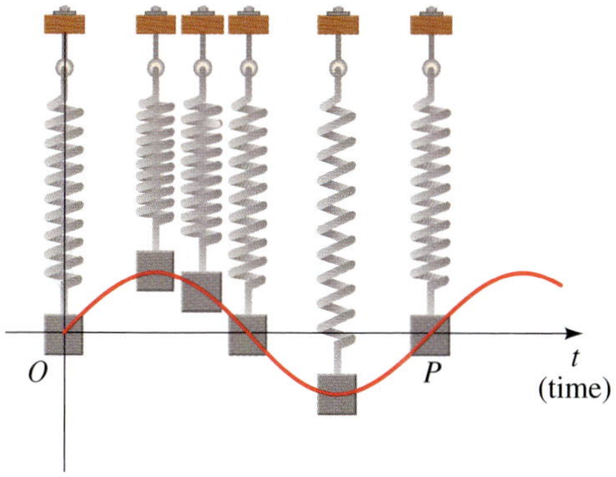

Figure 2

Motion of a vibrating spring is modeled by $y = \sin t$.

Notice that the mass returns to its original position over and over again. A **cycle** is one complete vibration of an object, so the mass in Figure 2 completes one cycle of its motion between O and P. Our observations about how the sine and cosine functions model periodic behavior are summarized in the following box.

The main difference between the two equations describing simple harmonic motion is the starting point. At $t = 0$, we get

$$y = a \sin \omega \cdot 0 = 0$$

$$y = a \cos \omega \cdot 0 = a$$

In the first case the motion "starts" with zero displacement, whereas in the second case the motion "starts" with the displacement at maximum (at the amplitude a).

Simple Harmonic Motion

If the equation describing the displacement y of an object at time t is

$$y = a \sin \omega t \qquad \text{or} \qquad y = a \cos \omega t$$

then the object is in **simple harmonic motion**. In this case,

amplitude $= |a|$ Maximum displacement of the object

period $= \dfrac{2\pi}{\omega}$ Time required to complete one cycle

frequency $= \dfrac{\omega}{2\pi}$ Number of cycles per unit of time

The symbol ω is the lowercase Greek letter "omega," and ν is the letter "nu."

Notice that the functions

$$y = a \sin 2\pi\nu t \qquad \text{and} \qquad y = a \cos 2\pi\nu t$$

have frequency ν, because $2\pi\nu/(2\pi) = \nu$. Since we can immediately read the frequency from these equations, we often write equations of simple harmonic motion in this form.

Example 1   A Vibrating Spring

The displacement of a mass suspended by a spring is modeled by the function

$$y = 10 \sin 4\pi t$$

where y is measured in inches and t in seconds (see Figure 3).

(a) Find the amplitude, period, and frequency of the motion of the mass.

(b) Sketch the graph of the displacement of the mass.

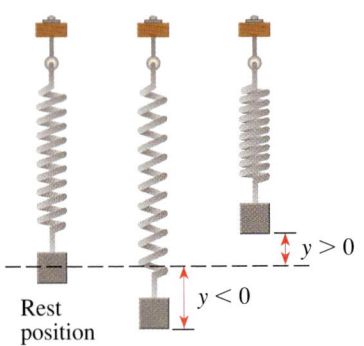

Rest position

Figure 3

Solution

(a) From the formulas for amplitude, period, and frequency, we get

$$\text{amplitude} = |a| = 10 \text{ in.}$$

$$\text{period} = \frac{2\pi}{\omega} = \frac{2\pi}{4\pi} = \frac{1}{2}\text{ s}$$

$$\text{frequency} = \frac{\omega}{2\pi} = \frac{4\pi}{2\pi} = 2 \text{ Hz}$$

(b) The graph of the displacement of the mass at time t is shown in Figure 4.   ∎

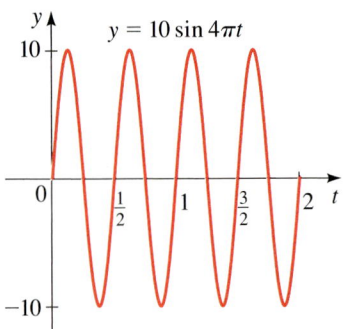

$y = 10 \sin 4\pi t$

Figure 4

   An important situation where simple harmonic motion occurs is in the production of sound. Sound is produced by a regular variation in air pressure from the normal pressure. If the pressure varies in simple harmonic motion, then a pure sound is produced. The tone of the sound depends on the frequency and the loudness depends on the amplitude.

Example 2   Vibrations of a Musical Note

A tuba player plays the note E and sustains the sound for some time. For a pure E the variation in pressure from normal air pressure is given by

$$V(t) = 0.2 \sin 80\pi t$$

where V is measured in pounds per square inch and t in seconds.

(a) Find the amplitude, period, and frequency of V.

(b) Sketch a graph of V.

(c) If the tuba player increases the loudness of the note, how does the equation for V change?

(d) If the player is playing the note incorrectly and it is a little flat, how does the equation for V change?

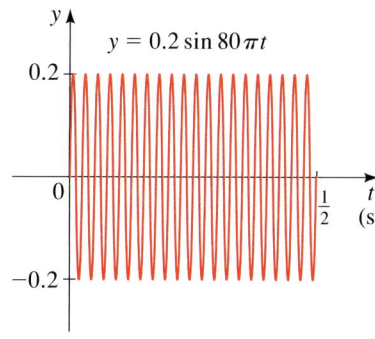

$y = 0.2 \sin 80\pi t$

Figure 5

Solution

(a) From the formulas for amplitude, period, and frequency, we get

$$\text{amplitude} = |\,0.2\,| = 0.2$$

$$\text{period} = \frac{2\pi}{80\pi} = \frac{1}{40}$$

$$\text{frequency} = \frac{80\pi}{2\pi} = 40$$

(b) The graph of V is shown in Figure 5.

(c) If the player increases the loudness the amplitude increases. So the number 0.2 is replaced by a larger number.

(d) If the note is flat, then the frequency is decreased. Thus, the coefficient of t is less than 80π. ∎

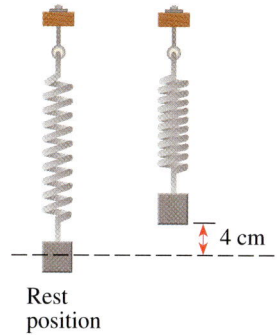

Rest
position

4 cm

Example 3   Modeling a Vibrating Spring

A mass is suspended from a spring. The spring is compressed a distance of 4 cm and then released. It is observed that the mass returns to the compressed position after $\frac{1}{3}$ s.

(a) Find a function that models the displacement of the mass.

(b) Sketch the graph of the displacement of the mass.

Solution

(a) The motion of the mass is given by one of the equations for simple harmonic motion. The amplitude of the motion is 4 cm. Since this amplitude is reached at time $t = 0$, an appropriate function that models the displacement is of the form

$$y = a \cos \omega t$$

Since the period is $p = \frac{1}{3}$, we can find ω from the following equation:

$$\text{period} = \frac{2\pi}{\omega}$$

$$\frac{1}{3} = \frac{2\pi}{\omega} \qquad \textcolor{blue}{\textit{Period} = \frac{1}{3}}$$

$$\omega = 6\pi \qquad \textcolor{blue}{\textit{Solve for } \omega}$$

So, the motion of the mass is modeled by the function

$$y = 4 \cos 6\pi t$$

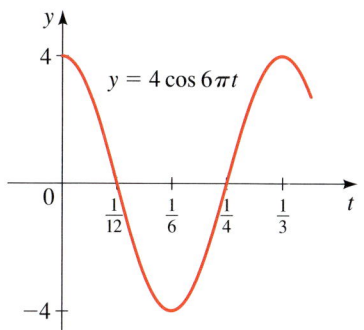

$y = 4 \cos 6\pi t$

where y is the displacement from the rest position at time t. Notice that when $t = 0$, the displacement is $y = 4$, as we expect.

(b) The graph of the displacement of the mass at time t is shown in Figure 6. ∎

Figure 6

In general, the sine or cosine functions representing harmonic motion may be shifted horizontally or vertically. In this case, the equations take the form

$$y = a\sin(\omega(t - c)) + b \qquad \text{or} \qquad y = a\cos(\omega(t - c)) + b$$

The vertical shift b indicates that the variation occurs around an average value b. The horizontal shift c indicates the position of the object at $t = 0$. (See Figure 7.)

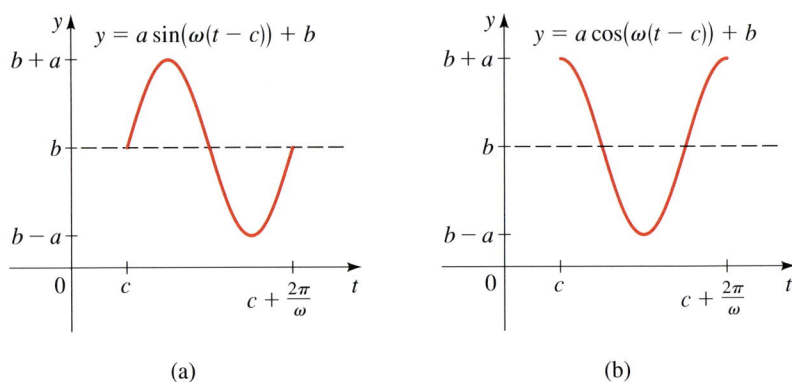

Figure 7   (a)   (b)

Example 4   Modeling the Brightness of a Variable Star

A variable star is one whose brightness alternately increases and decreases. For the variable star Delta Cephei, the time between periods of maximum brightness is 5.4 days. The average brightness (or magnitude) of the star is 4.0, and its brightness varies by ±0.35 magnitude.

(a) Find a function that models the brightness of Delta Cephei as a function of time.

(b) Sketch a graph of the brightness of Delta Cephei as a function of time.

Solution

(a) Let's find a function in the form

$$y = a\cos(\omega(t - c)) + b$$

The amplitude is the maximum variation from average brightness, so the amplitude is $a = 0.35$ magnitude. We are given that the period is 5.4 days, so

$$\omega = \frac{2\pi}{5.4} \approx 1.164$$

Since the brightness varies from an average value of 4.0 magnitudes, the graph is shifted upward by $b = 4.0$. If we take $t = 0$ to be a time when the star is at maximum brightness, there is no horizontal shift, so $c = 0$ (because a cosine curve achieves its maximum at $t = 0$). Thus, the function we want is

$$y = 0.35\cos(1.16t) + 4.0$$

where t is the number of days from a time when the star is at maximum brightness.

(b) The graph is sketched in Figure 8.   ∎

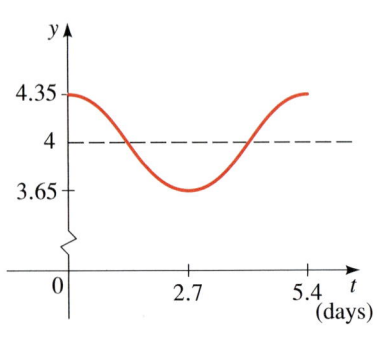

Figure 8

∎

The number of hours of daylight varies throughout the course of a year. In the Northern Hemisphere, the longest day is June 21, and the shortest is December 21. The average length of daylight is 12 h, and the variation from this average depends on the latitude. (For example, Fairbanks, Alaska, experiences more than 20 h of daylight on the longest day and less than 4 h on the shortest day!) The graph in Figure 9 shows the number of hours of daylight at different times of the year for various latitudes. It's apparent from the graph that the variation in hours of daylight is simple harmonic.

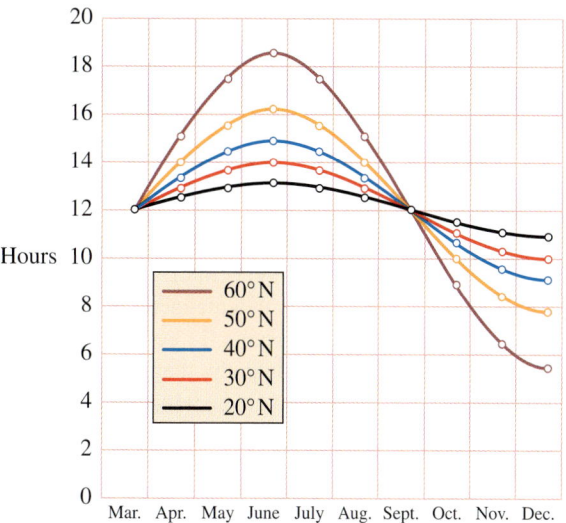

Figure 9

Graph of the length of daylight from March 21 through December 21 at various latitudes

Source: Lucia C. Harrison, *Daylight, Twilight, Darkness and Time* (New York: Silver, Burdett, 1935), page 40.

Example 5 Modeling the Number of Hours of Daylight

In Philadelphia (40° N latitude), the longest day of the year has 14 h 50 min of daylight and the shortest day has 9 h 10 min of daylight.

(a) Find a function L that models the length of daylight as a function of t, the number of days from January 1.

(b) An astronomer needs at least 11 hours of darkness for a long exposure astronomical photograph. On what days of the year are such long exposures possible?

Solution

(a) We need to find a function in the form

$$y = a \sin(\omega(t - c)) + b$$

whose graph is the 40° N latitude curve in Figure 9. From the information given, we see that the amplitude is

$$a = \tfrac{1}{2}\left(14\tfrac{5}{6} - 9\tfrac{1}{6}\right) \approx 2.83 \text{ h}$$

Since there are 365 days in a year, the period is 365, so

$$\omega = \frac{2\pi}{365} \approx 0.0172$$

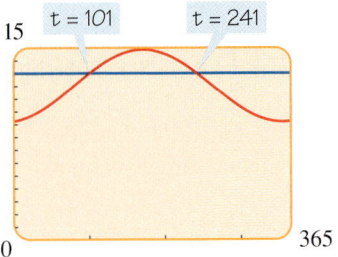

Figure 10

Since the average length of daylight is 12 h, the graph is shifted upward by 12, so $b = 12$. Since the curve attains the average value (12) on March 21, the 80th day of the year, the curve is shifted 80 units to the right. Thus, $c = 80$. So a function that models the number of hours of daylight is

$$y = 2.83 \sin(0.0172(t - 80)) + 12$$

where t is the number of days from January 1.

(b) A day has 24 h, so 11 h of night correspond to 13 h of daylight. So we need to solve the inequality $y \leq 13$. To solve this inequality graphically, we graph $y = 2.83 \sin 0.0172(t - 80) + 12$ and $y = 13$ on the same graph. From the graph in Figure 10 we see that there are fewer than 13 h of daylight between day 1 (January 1) and day 101 (April 11) and from day 241 (August 29) to day 365 (December 31). ∎

Another situation where simple harmonic motion occurs is in alternating current (AC) generators. Alternating current is produced when an armature rotates about its axis in a magnetic field.

Figure 11 represents a simple version of such a generator. As the wire passes through the magnetic field, a voltage E is generated in the wire. It can be shown that the voltage generated is given by

$$E(t) = E_0 \cos \omega t$$

where E_0 is the maximum voltage produced (which depends on the strength of the magnetic field) and $\omega/(2\pi)$ is the number of revolutions per second of the armature (the frequency).

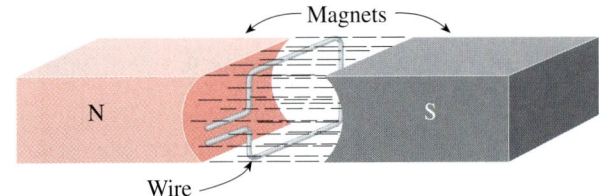

Figure 11

Magnets

N   S

Wire

Why do we say that household current is 110 V when the maximum voltage produced is 155 V? From the symmetry of the cosine function, we see that the average voltage produced is zero. This average value would be the same for all AC generators and so gives no information about the voltage generated. To obtain a more informative measure of voltage, engineers use the **root-mean-square** (rms) method. It can be shown that the rms voltage is $1/\sqrt{2}$ times the maximum voltage. So, for household current the rms voltage is

$$155 \times \frac{1}{\sqrt{2}} \approx 110 \text{ V}$$

Example 6   Modeling Alternating Current

Ordinary 110-V household alternating current varies from $+155$ V to -155 V with a frequency of 60 Hz (cycles per second). Find an equation that describes this variation in voltage.

Solution   The variation in voltage is simple harmonic. Since the frequency is 60 cycles per second, we have

$$\frac{\omega}{2\pi} = 60 \qquad \text{or} \qquad \omega = 120\pi$$

Let's take $t = 0$ to be a time when the voltage is $+155$ V. Then

$$E(t) = a \cos \omega t = 155 \cos 120\pi t$$ ∎

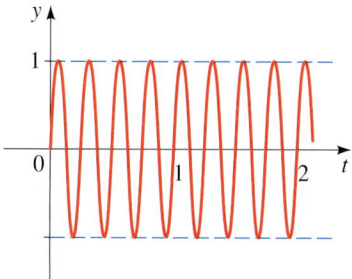

(a) Harmonic motion: $y = \sin 8\pi t$

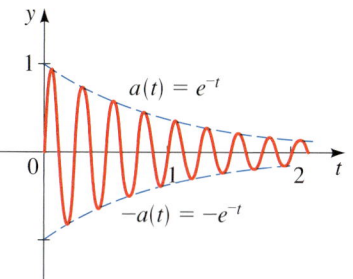

(b) Damped harmonic motion:
$y = e^{-t} \sin 8\pi t$

Figure 12

Hz is the abbreviation for hertz. One
hertz is one cycle per second.

Damped Harmonic Motion

The spring in Figure 2 on page 443 is assumed to oscillate in a frictionless environment. In this hypothetical case, the amplitude of the oscillation will not change. In the presence of friction, however, the motion of the spring eventually "dies down"; that is, the amplitude of the motion decreases with time. Motion of this type is called *damped harmonic motion*.

Damped Harmonic Motion

If the equation describing the displacement y of an object at time t is

$$y = ke^{-ct} \sin \omega t \quad \text{or} \quad y = ke^{-ct} \cos \omega t \quad (c > 0)$$

then the object is in **damped harmonic motion**. The constant c is the **damping constant**, k is the initial amplitude, and $2\pi/\omega$ is the period.*

Damped harmonic motion is simply harmonic motion for which the amplitude is governed by the function $a(t) = ke^{-ct}$. Figure 12 shows the difference between harmonic motion and damped harmonic motion.

Example 7 Modeling Damped Harmonic Motion

Two mass-spring systems are experiencing damped harmonic motion, both at 0.5 cycles per second, and both with an initial maximum displacement of 10 cm. The first has a damping constant of 0.5 and the second has a damping constant of 0.1.

(a) Find functions of the form $g(t) = ke^{-ct} \cos \omega t$ to model the motion in each case.

(b) Graph the two functions you found in part (a). How do they differ?

Solution

(a) At time $t = 0$, the displacement is 10 cm. Thus $g(0) = ke^{-c\cdot 0} \cos(\omega \cdot 0) = k$, and so $k = 10$. Also, the frequency is $f = 0.5$ Hz, and since $\omega = 2\pi f$ (see page 443), we get $\omega = 2\pi(0.5) = \pi$. Using the given damping constants, we find that the motions of the two springs are given by the functions

$$g_1(t) = 10e^{-0.5t} \cos \pi t \quad \text{and} \quad g_2(t) = 10e^{-0.1t} \cos \pi t$$

(b) The functions g_1 and g_2 are graphed in Figure 13. From the graphs we see that in the first case (where the damping constant is larger) the motion dies down quickly, whereas in the second case, perceptible motion continues much longer.

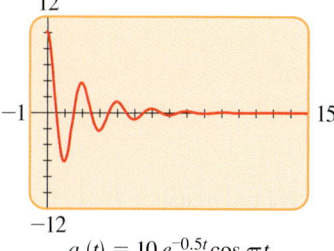

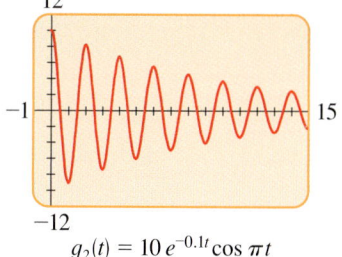

Figure 13 $g_1(t) = 10\,e^{-0.5t} \cos \pi t$ $g_2(t) = 10\,e^{-0.1t} \cos \pi t$ ∎

---

*In the case of damped harmonic motion, the term *quasi-period* is often used instead of *period* because the motion is not actually periodic—it diminishes with time. However, we will continue to use the term *period* to avoid confusion.

As the preceding example indicates, the larger the damping constant c, the quicker the oscillation dies down. When a guitar string is plucked and then allowed to vibrate freely, a point on that string undergoes damped harmonic motion. We hear the damping of the motion as the sound produced by the vibration of the string fades. How fast the damping of the string occurs (as measured by the size of the constant c) is a property of the size of the string and the material it is made of. Another example of damped harmonic motion is the motion that a shock absorber on a car undergoes when the car hits a bump in the road. In this case, the shock absorber is engineered to damp the motion as quickly as possible (large c) and to have the frequency as small as possible (small ω). On the other hand, the sound produced by a tuba player playing a note is undamped as long as the player can maintain the loudness of the note. The electromagnetic waves that produce light move in simple harmonic motion that is not damped.

Example 8   A Vibrating Violin String

The G-string on a violin is pulled a distance of 0.5 cm above its rest position, then released and allowed to vibrate. The damping constant c for this string is determined to be 1.4. Suppose that the note produced is a pure G (frequency = 200 Hz). Find an equation that describes the motion of the point at which the string was plucked.

Solution   Let P be the point at which the string was plucked. We will find a function $f(t)$ that gives the distance at time t of the point P from its original rest position. Since the maximum displacement occurs at $t = 0$, we find an equation in the form

$$y = ke^{-ct} \cos \omega t$$

From this equation, we see that $f(0) = k$. But we know that the original displacement of the string is 0.5 cm. Thus, $k = 0.5$. Since the frequency of the vibration is 200, we have $\omega = 2\pi f = 2\pi(200) = 400\pi$. Finally, since we know that the damping constant is 1.4, we get

$$f(t) = 0.5e^{-1.4t} \cos 400\pi t \qquad \blacksquare$$

Example 9   Ripples on a Pond

A stone is dropped in a calm lake, causing waves to form. The up-and-down motion of a point on the surface of the water is modeled by damped harmonic motion. At some time the amplitude of the wave is measured, and 20 s later it is found that the amplitude has dropped to $\frac{1}{10}$ of this value. Find the damping constant c.

Solution   The amplitude is governed by the coefficient ke^{-ct} in the equations for damped harmonic motion. Thus, the amplitude at time t is ke^{-ct}, and 20 s later, it is $ke^{-c(t+20)}$. So, because the later value is $\frac{1}{10}$ the earlier value, we have

$$ke^{-c(t+20)} = \tfrac{1}{10}ke^{-ct}$$

We now solve this equation for c. Canceling k and using the Laws of Exponents, we get

$$e^{-ct} \cdot e^{-20c} = \tfrac{1}{10}e^{-ct}$$

$$e^{-20c} = \tfrac{1}{10} \qquad \text{\textcolor{teal}{\textit{Cancel }} } e^{-ct}$$

$$e^{20c} = 10 \qquad \text{\textcolor{teal}{\textit{Take reciprocals}}}$$

Taking the natural logarithm of each side gives

$$20c = \ln(10)$$

$$c = \tfrac{1}{20} \ln(10) \approx \tfrac{1}{20}(2.30) \approx 0.12$$

Thus, the damping constant is $c \approx 0.12$. ■

5.5 Exercises

1–8 ■ The given function models the displacement of an object moving in simple harmonic motion.

(a) Find the amplitude, period, and frequency of the motion.

(b) Sketch a graph of the displacement of the object over one complete period.

1. $y = 2 \sin 3t$

2. $y = 3 \cos \tfrac{1}{2}t$

3. $y = -\cos 0.3t$

4. $y = 2.4 \sin 3.6t$

5. $y = -0.25 \cos\left(1.5t - \dfrac{\pi}{3}\right)$ **6.** $y = -\tfrac{3}{2} \sin(0.2t + 1.4)$

7. $y = 5 \cos(\tfrac{2}{3}t + \tfrac{3}{4})$ **8.** $y = 1.6 \sin(t - 1.8)$

9–12 ■ Find a function that models the simple harmonic motion having the given properties. Assume that the displacement is zero at time $t = 0$.

9. amplitude 10 cm, period 3 s

10. amplitude 24 ft, period 2 min

11. amplitude 6 in., frequency $5/\pi$ Hz

12. amplitude 1.2 m, frequency 0.5 Hz

13–16 ■ Find a function that models the simple harmonic motion having the given properties. Assume that the displacement is at its maximum at time $t = 0$.

13. amplitude 60 ft, period 0.5 min

14. amplitude 35 cm, period 8 s

15. amplitude 2.4 m, frequency 750 Hz

16. amplitude 6.25 in., frequency 60 Hz

17–24 ■ An initial amplitude k, damping constant c, and frequency f or period p are given. (Recall that frequency and period are related by the equation $f = 1/p$.)

(a) Find a function that models the damped harmonic motion. Use a function of the form $y = ke^{-ct} \cos \omega t$ in Exercises 17–20, and of the form $y = ke^{-ct} \sin \omega t$ in Exercises 21–24.

(b) Graph the function.

17. $k = 2$, $c = 1.5$, $f = 3$

18. $k = 15$, $c = 0.25$, $f = 0.6$

19. $k = 100$, $c = 0.05$, $p = 4$

20. $k = 0.75$, $c = 3$, $p = 3\pi$

21. $k = 7$, $c = 10$, $p = \pi/6$

22. $k = 1$, $c = 1$, $p = 1$

23. $k = 0.3$, $c = 0.2$, $f = 20$

24. $k = 12$, $c = 0.01$, $f = 8$

Applications

25. A Bobbing Cork A cork floating in a lake is bobbing in simple harmonic motion. Its displacement above the bottom of the lake is modeled by

$$y = 0.2 \cos 20\pi t + 8$$

where y is measured in meters and t is measured in minutes.

(a) Find the frequency of the motion of the cork.

(b) Sketch a graph of y.

(c) Find the maximum displacement of the cork above the lake bottom.

26. FM Radio Signals The carrier wave for an FM radio signal is modeled by the function

$$y = a \sin(2\pi(9.15 \times 10^7)t)$$

where t is measured in seconds. Find the period and frequency of the carrier wave.

27. Predator Population Model In a predator/prey model (see page 432), the predator population is modeled by the function

$$y = 900 \cos 2t + 8000$$

where t is measured in years.

(a) What is the maximum population?

(b) Find the length of time between successive periods of maximum population.

28. Blood Pressure Each time your heart beats, your blood pressure increases, then decreases as the heart rests between beats. A certain person's blood pressure is modeled by the function

$$p(t) = 115 + 25 \sin(160\pi t)$$

where $p(t)$ is the pressure in mmHg at time t, measured in minutes.

(a) Find the amplitude, period, and frequency of p.

(b) Sketch a graph of p.

(c) If a person is exercising, his heart beats faster. How does this affect the period and frequency of p?

29. Spring–Mass System   A mass attached to a spring is moving up and down in simple harmonic motion. The graph gives its displacement $d(t)$ from equilibrium at time t. Express the function d in the form $d(t) = a \sin \omega t$.

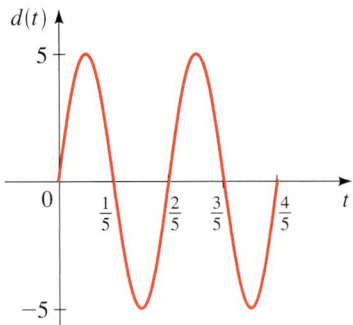

30. Tides   The graph shows the variation of the water level relative to mean sea level in Commencement Bay at Tacoma, Washington, for a particular 24-hour period. Assuming that this variation is modeled by simple harmonic motion, find an equation of the form $y = a \sin \omega t$ that describes the variation in water level as a function of the number of hours after midnight.

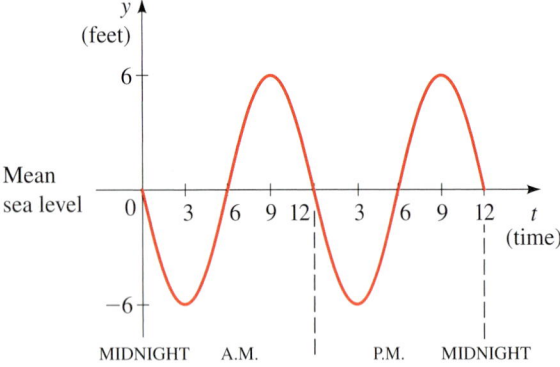

31. Tides   The Bay of Fundy in Nova Scotia has the highest tides in the world. In one 12-hour period the water starts at mean sea level, rises to 21 ft above, drops to 21 ft below, then returns to mean sea level. Assuming that the motion of the tides is simple harmonic, find an equation that describes the height of the tide in the Bay of Fundy above

mean sea level. Sketch a graph that shows the level of the tides over a 12-hour period.

32. Spring–Mass System   A mass suspended from a spring is pulled down a distance of 2 ft from its rest position, as shown in the figure. The mass is released at time $t = 0$ and allowed to oscillate. If the mass returns to this position after 1 s, find an equation that describes its motion.

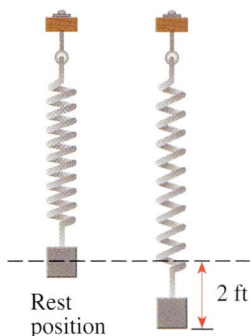

33. Spring–Mass System   A mass is suspended on a spring. The spring is compressed so that the mass is located 5 cm above its rest position. The mass is released at time $t = 0$ and allowed to oscillate. It is observed that the mass reaches its lowest point $\frac{1}{2}$ s after it is released. Find an equation that describes the motion of the mass.

34. Spring–Mass System   The frequency of oscillation of an object suspended on a spring depends on the stiffness k of the spring (called the *spring constant*) and the mass m of the object. If the spring is compressed a distance a and then allowed to oscillate, its displacement is given by

$$f(t) = a \cos \sqrt{k/m}\, t$$

(a) A 10-g mass is suspended from a spring with stiffness $k = 3$. If the spring is compressed a distance 5 cm and then released, find the equation that describes the oscillation of the spring.

(b) Find a general formula for the frequency (in terms of k and m).

(c) How is the frequency affected if the mass is increased? Is the oscillation faster or slower?

(d) How is the frequency affected if a stiffer spring is used (larger k)? Is the oscillation faster or slower?

35. Ferris Wheel   A ferris wheel has a radius of 10 m, and the bottom of the wheel passes 1 m above the ground. If the ferris wheel makes one complete revolution every 20 s, find an equation that gives the height above the ground of a person on the ferris wheel as a function of time.

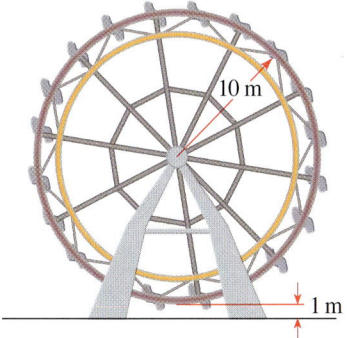

36. Clock Pendulum The pendulum in a grandfather clock makes one complete swing every 2 s. The maximum angle that the pendulum makes with respect to its rest position is 10°. We know from physical principles that the angle θ between the pendulum and its rest position changes in simple harmonic fashion. Find an equation that describes the size of the angle θ as a function of time. (Take $t = 0$ to be a time when the pendulum is vertical.)

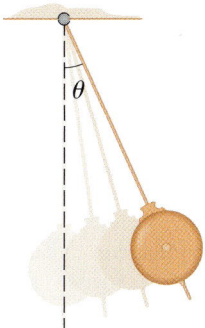

37. Variable Stars The variable star Zeta Gemini has a period of 10 days. The average brightness of the star is 3.8 magnitudes, and the maximum variation from the average is 0.2 magnitude. Assuming that the variation in brightness is simple harmonic, find an equation that gives the brightness of the star as a function of time.

38. Variable Stars Astronomers believe that the radius of a variable star increases and decreases with the brightness of the star. The variable star Delta Cephei (Example 4) has an average radius of 20 million miles and changes by a maximum of 1.5 million miles from this average during a single pulsation. Find an equation that describes the radius of this star as a function of time.

39. Electric Generator The armature in an electric generator is rotating at the rate of 100 revolutions per second (rps). If the maximum voltage produced is 310 V, find an equation that describes this variation in voltage. What is the rms voltage? (See Example 6 and the margin note adjacent to it.)

40. Biological Clocks *Circadian rhythms* are biological processes that oscillate with a period of approximately 24 hours. That is, a circadian rhythm is an internal daily biological clock. Blood pressure appears to follow such a rhythm. For a certain individual the average resting blood pressure varies from a maximum of 100 mmHg at 2:00 P.M. to a minimum of 80 mmHg at 2:00 A.M. Find a sine function of the form

$$f(t) = a \sin(\omega(t - c)) + b$$

that models the blood pressure at time t, measured in hours from midnight.

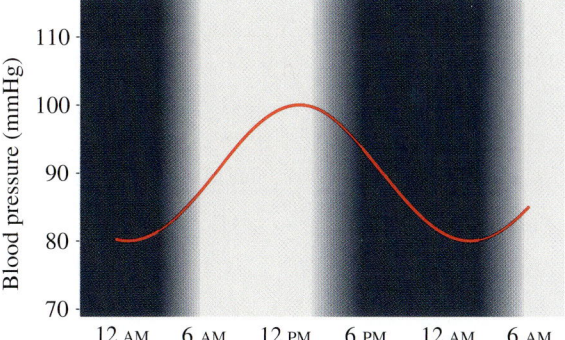

41. Electric Generator The graph shows an oscilloscope reading of the variation in voltage of an AC current produced by a simple generator.

(a) Find the maximum voltage produced.

(b) Find the frequency (cycles per second) of the generator.

(c) How many revolutions per second does the armature in the generator make?

(d) Find a formula that describes the variation in voltage as a function of time.

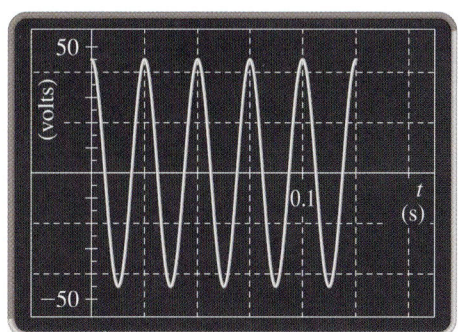

42. Doppler Effect   When a car with its horn blowing drives by an observer, the pitch of the horn seems higher as it approaches and lower as it recedes (see the figure). This phenomenon is called the **Doppler effect**. If the sound source is moving at speed v relative to the observer and if the speed of sound is v_0, then the perceived frequency f is related to the actual frequency f_0 as follows:

$$f = f_0 \left(\frac{v_0}{v_0 \pm v} \right)$$

We choose the minus sign if the source is moving toward the observer and the plus sign if it is moving away.

Suppose that a car drives at 110 ft/s past a woman standing on the shoulder of a highway, blowing its horn, which has a frequency of 500 Hz. Assume that the speed of sound is 1130 ft/s. (This is the speed in dry air at $70°$ F.)

(a) What are the frequencies of the sounds that the woman hears as the car approaches her and as it moves away from her?

(b) Let A be the amplitude of the sound. Find functions of the form

$$y = A \sin \omega t$$

that model the perceived sound as the car approaches the woman and as it recedes.

43. Motion of a Building   A strong gust of wind strikes a tall building, causing it to sway back and forth in damped harmonic motion. The frequency of the oscillation is 0.5 cycle per second and the damping constant is $c = 0.9$. Find an equation that describes the motion of the building. (Assume

$k = 1$ and take $t = 0$ to be the instant when the gust of wind strikes the building.)

44. Shock Absorber   When a car hits a certain bump on the road, a shock absorber on the car is compressed a distance of 6 in., then released (see the figure). The shock absorber vibrates in damped harmonic motion with a frequency of 2 cycles per second. The damping constant for this particular shock absorber is 2.8.

(a) Find an equation that describes the displacement of the shock absorber from its rest position as a function of time. Take $t = 0$ to be the instant that the shock absorber is released.

(b) How long does it take for the amplitude of the vibration to decrease to 0.5 in?

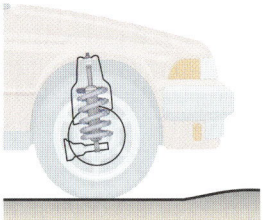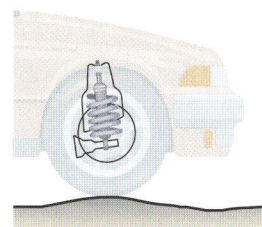

45. Tuning Fork   A tuning fork is struck and oscillates in damped harmonic motion. The amplitude of the motion is measured, and 3 s later it is found that the amplitude has dropped to $\frac{1}{4}$ of this value. Find the damping constant c for this tuning fork.

46. Guitar String   A guitar string is pulled at point P a distance of 3 cm above its rest position. It is then released and vibrates in damped harmonic motion with a frequency of 165 cycles per second. After 2 s, it is observed that the amplitude of the vibration at point P is 0.6 cm.

(a) Find the damping constant c.

(b) Find an equation that describes the position of point P above its rest position as a function of time. Take $t = 0$ to be the instant that the string is released.

5	**Review**

Concept Check

1. (a) What is the unit circle?

   (b) Use a diagram to explain what is meant by the terminal point determined by a real number t.

   (c) What is the reference number $\bar{t}$ associated with t?

   (d) If t is a real number and $P(x, y)$ is the terminal point determined by t, write equations that define $\sin t$, $\cos t$, $\tan t$, $\cot t$, $\sec t$, and $\csc t$.

(e) What are the domains of the six functions that you defined in part (d)?

(f) Which trigonometric functions are positive in quadrants I, II, III, and IV?

2. (a) What is an even function?

(b) Which trigonometric functions are even?

(c) What is an odd function?

(d) Which trigonometric functions are odd?

3. (a) State the reciprocal identities.

(b) State the Pythagorean identities.

4. (a) What is a periodic function?

(b) What are the periods of the six trigonometric functions?

5. Graph the sine and cosine functions. How is the graph of cosine related to the graph of sine?

6. Write expressions for the amplitude, period, and phase shift of the sine curve $y = a \sin k(x - b)$ and the cosine curve $y = a \cos k(x - b)$.

7. (a) Graph the tangent and cotangent functions.

(b) State the periods of the tangent curve $y = a \tan kx$ and the cotangent curve $y = a \cot kx$.

8. (a) Graph the secant and cosecant functions.

(b) State the periods of the secant curve $y = a \sec kx$ and the cosecant curve $y = a \csc kx$.

9. (a) What is simple harmonic motion?

(b) What is damped harmonic motion?

(c) Give three real-life examples of simple harmonic motion and of damped harmonic motion.

Exercises

1–2 ■ A point $P(x, y)$ is given.

(a) Show that P is on the unit circle.

(b) Suppose that P is the terminal point determined by t. Find $\sin t$, $\cos t$, and $\tan t$.

1. $P\left(-\dfrac{\sqrt{3}}{2}, \dfrac{1}{2}\right)$

2. $P\left(\dfrac{3}{5}, -\dfrac{4}{5}\right)$

3–6 ■ A real number t is given.

(a) Find the reference number for t.

(b) Find the terminal point $P(x, y)$ on the unit circle determined by t.

(c) Find the six trigonometric functions of t.

3. $t = \dfrac{2\pi}{3}$

4. $t = \dfrac{5\pi}{3}$

5. $t = -\dfrac{11\pi}{4}$

6. $t = -\dfrac{7\pi}{6}$

7–16 ■ Find the value of the trigonometric function. If possible, give the exact value; otherwise, use a calculator to find an approximate value correct to five decimal places.

7. (a) $\sin \dfrac{3\pi}{4}$

(b) $\cos \dfrac{3\pi}{4}$

8. (a) $\tan \dfrac{\pi}{3}$

(b) $\tan\left(-\dfrac{\pi}{3}\right)$

9. (a) $\sin 1.1$

(b) $\cos 1.1$

10. (a) $\cos \dfrac{\pi}{5}$

(b) $\cos\left(-\dfrac{\pi}{5}\right)$

11. (a) $\cos \dfrac{9\pi}{2}$

(b) $\sec \dfrac{9\pi}{2}$

12. (a) $\sin \dfrac{\pi}{7}$

(b) $\csc \dfrac{\pi}{7}$

13. (a) $\tan \dfrac{5\pi}{2}$

(b) $\cot \dfrac{5\pi}{2}$

14. (a) $\sin 2\pi$

(b) $\csc 2\pi$

15. (a) $\tan \dfrac{5\pi}{6}$

(b) $\cot \dfrac{5\pi}{6}$

16. (a) $\cos \dfrac{\pi}{3}$

(b) $\sin \dfrac{\pi}{6}$

17–20 ■ Use the fundamental identities to write the first expression in terms of the second.

17. $\dfrac{\tan t}{\cos t}$, $\sin t$

18. $\tan^2 t \sec t$, $\cos t$

19. $\tan t$, $\sin t$; t in quadrant IV

20. $\sec t$, $\sin t$; t in quadrant II

21–24 ■ Find the values of the remaining trigonometric functions at t from the given information.

21. $\sin t = \frac{5}{13}$, $\cos t = -\frac{12}{13}$

22. $\sin t = -\frac{1}{2}$, $\cos t > 0$

23. $\cot t = -\frac{1}{2}$, $\csc t = \sqrt{5}/2$

24. $\cos t = -\frac{3}{5}$, $\tan t < 0$

25. If $\tan t = \frac{1}{4}$ and the terminal point for t is in quadrant III, find $\sec t + \cot t$.

26. If $\sin t = -\frac{8}{17}$ and the terminal point for t is in quadrant IV, find $\csc t + \sec t$.

27. If $\cos t = \frac{3}{5}$ and the terminal point for t is in quadrant I, find $\tan t + \sec t$.

28. If $\sec t = -5$ and the terminal point for t is in quadrant II, find $\sin^2 t + \cos^2 t$.

29–36 ■ A trigonometric function is given.

(a) Find the amplitude, period, and phase shift of the function.

(b) Sketch the graph.

29. $y = 10 \cos \frac{1}{2}x$ **30.** $y = 4 \sin 2\pi x$

31. $y = -\sin \frac{1}{2}x$ **32.** $y = 2 \sin\left(x - \frac{\pi}{4}\right)$

33. $y = 3 \sin(2x - 2)$ **34.** $y = \cos 2\left(x - \frac{\pi}{2}\right)$

35. $y = -\cos\left(\frac{\pi}{2}x + \frac{\pi}{6}\right)$ **36.** $y = 10 \sin\left(2x - \frac{\pi}{2}\right)$

37–40 ■ The graph of one period of a function of the form $y = a \sin k(x - b)$ or $y = a \cos k(x - b)$ is shown. Determine the function.

37.
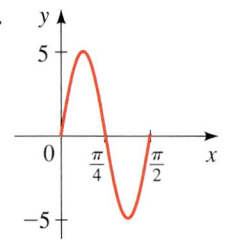

38.
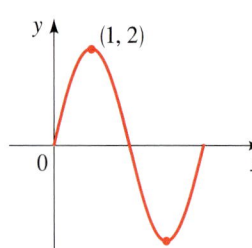

39.   40.
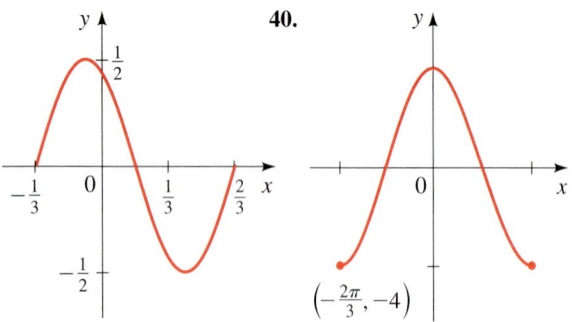

41–48 ■ Find the period, and sketch the graph.

41. $y = 3 \tan x$ **42.** $y = \tan \pi x$

43. $y = 2 \cot\left(x - \frac{\pi}{2}\right)$ **44.** $y = \sec\left(\frac{1}{2}x - \frac{\pi}{2}\right)$

45. $y = 4 \csc(2x + \pi)$ **46.** $y = \tan\left(x + \frac{\pi}{6}\right)$

47. $y = \tan\left(\frac{1}{2}x - \frac{\pi}{8}\right)$ **48.** $y = -4 \sec 4\pi x$

49–54 ■ A function is given.

(a) Use a graphing device to graph the function.

(b) Determine from the graph whether the function is periodic and, if so, determine the period.

(c) Determine from the graph whether the function is odd, even, or neither.

49. $y = |\cos x|$ **50.** $y = \sin(\cos x)$

51. $y = \cos(2^{0.1x})$ **52.** $y = 1 + 2^{\cos x}$

53. $y = |x| \cos 3x$ **54.** $y = \sqrt{x} \sin 3x$ $(x > 0)$

55–58 ■ Graph the three functions on a common screen. How are the graphs related?

55. $y = x$, $y = -x$, $y = x \sin x$

56. $y = 2^{-x}$, $y = -2^{-x}$, $y = 2^{-x} \cos 4\pi x$

57. $y = x$, $y = \sin 4x$, $y = x + \sin 4x$

58. $y = \sin^2 x$, $y = \cos^2 x$, $y = \sin^2 x + \cos^2 x$

59–60 ■ Find the maximum and minimum values of the function.

59. $y = \cos x + \sin 2x$ **60.** $y = \cos x + \sin^2 x$

61. Find the solutions of $\sin x = 0.3$ in the interval $[0, 2\pi]$.

62. Find the solutions of $\cos 3x = x$ in the interval $[0, \pi]$.

63. Let $f(x) = \dfrac{\sin^2 x}{x}$.

(a) Is the function f even, odd, or neither?

(b) Find the x-intercepts of the graph of f.

(c) Graph f in an appropriate viewing rectangle.

(d) Describe the behavior of the function as x becomes large.

(e) Notice that $f(x)$ is not defined when $x = 0$. What happens as x approaches 0?

64. Let $y_1 = \cos(\sin x)$ and $y_2 = \sin(\cos x)$.

(a) Graph y_1 and y_2 in the same viewing rectangle.

(b) Determine the period of each of these functions from its graph.

(c) Find an inequality between $\sin(\cos x)$ and $\cos(\sin x)$ that is valid for all x.

65. A point P moving in simple harmonic motion completes 8 cycles every second. If the amplitude of the motion is 50 cm, find an equation that describes the motion of P as a function of time. Assume the point P is at its maximum displacement when $t = 0$.

66. A mass suspended from a spring oscillates in simple harmonic motion at a frequency of 4 cycles per second. The distance from the highest to the lowest point of the oscillation is 100 cm. Find an equation that describes the distance of the mass from its rest position as a function of time. Assume the mass is at its lowest point when $t = 0$.

67. The graph shows the variation of the water level relative to mean sea level in the Long Beach harbor for a particular 24-hour period. Assuming that this variation is simple harmonic, find an equation of the form $y = a \cos \omega t$ that describes the variation in water level as a function of the number of hours after midnight.

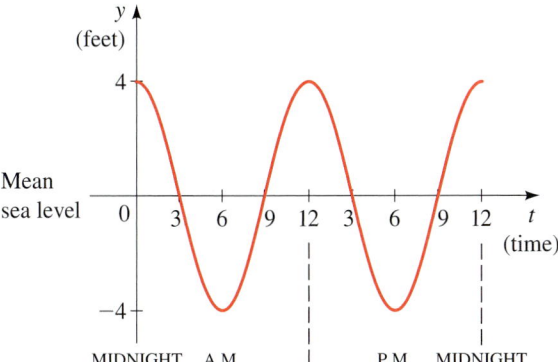

68. The top floor of a building undergoes damped harmonic motion after a sudden brief earthquake. At time $t = 0$ the displacement is at a maximum, 16 cm from the normal position. The damping constant is $c = 0.72$ and the building vibrates at 1.4 cycles per second.

   (a) Find a function of the form $y = ke^{-ct} \cos \omega t$ to model the motion.

     (b) Graph the function you found in part (a).

   (c) What is the displacement at time $t = 10$ s?

5   Test

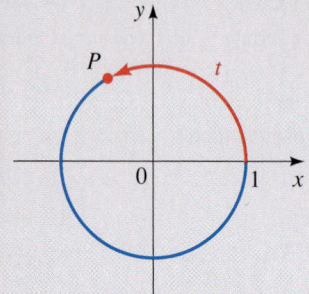

1. The point $P(x, y)$ is on the unit circle in quadrant IV. If $x = \sqrt{11}/6$, find y.

2. The point P in the figure at the left has y-coordinate $\frac{4}{5}$. Find:

(a) $\sin t$ (b) $\cos t$

(c) $\tan t$ (d) $\sec t$

3. Find the exact value.

(a) $\sin \dfrac{7\pi}{6}$ (b) $\cos \dfrac{13\pi}{4}$

(c) $\tan\left(-\dfrac{5\pi}{3}\right)$ (d) $\csc \dfrac{3\pi}{2}$

4. Express $\tan t$ in terms of $\sin t$, if the terminal point determined by t is in quadrant II.

5. If $\cos t = -\frac{8}{17}$ and if the terminal point determined by t is in quadrant III, find $\tan t \cot t + \csc t$.

6–7 ■ A trigonometric function is given.

(a) Find the amplitude, period, and phase shift of the function.

(b) Sketch the graph.

6. $y = -5 \cos 4x$ **7.** $y = 2 \sin\left(\dfrac{1}{2}x - \dfrac{\pi}{6}\right)$

8–9 ■ Find the period, and graph the function.

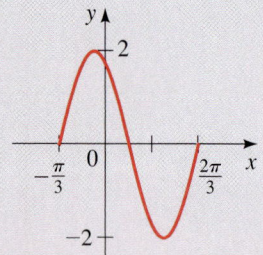

8. $y = -\csc 2x$ **9.** $y = \tan\left(2x - \dfrac{\pi}{2}\right)$

10. The graph shown at left is one period of a function of the form $y = a \sin k(x - b)$. Determine the function.

11. Let $f(x) = \dfrac{\cos x}{1 + x^2}$.

(a) Use a graphing device to graph f in an appropriate viewing rectangle.

(b) Determine from the graph if f is even, odd, or neither.

(c) Find the minimum and maximum values of f.

12. A mass suspended from a spring oscillates in simple harmonic motion. The mass completes 2 cycles every second and the distance between the highest point and the lowest point of the oscillation is 10 cm. Find an equation of the form $y = a \sin \omega t$ that gives the distance of the mass from its rest position as a function of time.

13. An object is moving up and down in damped harmonic motion. Its displacement at time $t = 0$ is 16 in; this is its maximum displacement. The damping constant is $c = 0.1$ and the frequency is 12 Hz.

(a) Find a function that models this motion.

(b) Graph the function.

In the *Focus on Modeling* that follows Chapter 2 (page 239), we learned how to construct linear models from data. Figure 1 shows some scatter plots of data; the first plot appears to be linear but the others are not. What do we do when the data we are studying are not linear? In this case, our model would be some other type of function that best fits the data. If the scatter plot indicates simple harmonic motion, then we might try to model the data with a sine or cosine function. The next example illustrates this process.

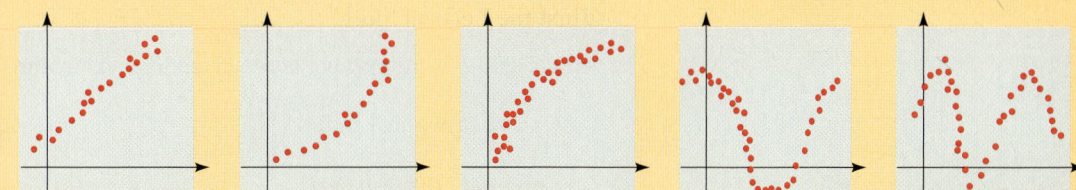

Figure 1

Example 1 Modeling the Height of a Tide

The water depth in a narrow channel varies with the tides. Table 1 shows the water depth over a 12-hour period.

(a) Make a scatter plot of the water depth data.

(b) Find a function that models the water depth with respect to time.

(c) If a boat needs at least 11 ft of water to cross the channel, during which times can it safely do so?

Solution

(a) A scatter plot of the data is shown in Figure 2.

Table 1

Time	Depth (ft)
12:00 A.M.	9.8
1:00 A.M.	11.4
2:00 A.M.	11.6
3:00 A.M.	11.2
4:00 A.M.	9.6
5:00 A.M.	8.5
6:00 A.M.	6.5
7:00 A.M.	5.7
8:00 A.M.	5.4
9:00 A.M.	6.0
10:00 A.M.	7.0
11:00 A.M.	8.6
12:00 P.M.	10.0

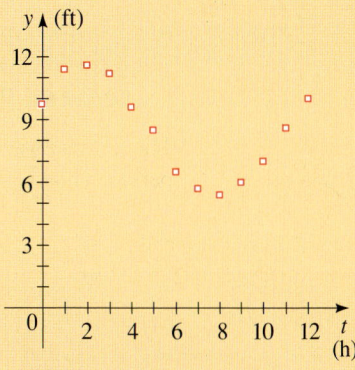

Figure 2

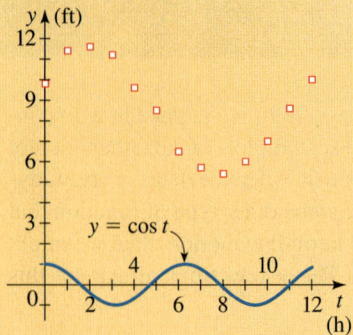

Figure 3

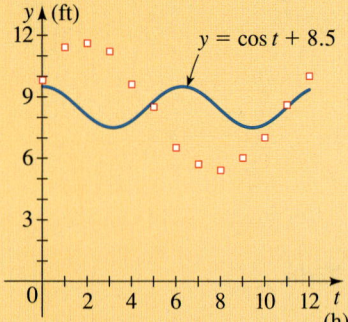

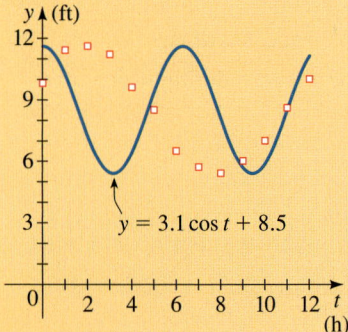

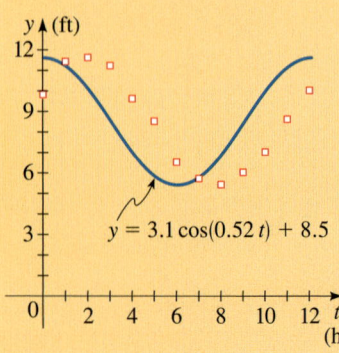

(b) The data appear to lie on a cosine (or sine) curve. But if we graph $y = \cos t$ on the same graph as the scatter plot, the result in Figure 3 is not even close to the data—to fit the data we need to adjust the vertical shift, amplitude, period, and phase shift of the cosine curve. In other words, we need to find a function of the form

$$y = a\cos(\omega(t - c)) + b$$

We use the following steps, which are illustrated by the graphs in the margin.

- **Adjust the Vertical Shift**

The vertical shift b is the average of the maximum and minimum values:

$$b = \text{vertical shift}$$

$$= \frac{1}{2} \cdot (\text{maximum value} + \text{minimum value})$$

$$= \frac{1}{2}(11.6 + 5.4) = 8.5$$

- **Adjust the Amplitude**

The amplitude a is half of the difference between the maximum and minimum values:

$$a = \text{amplitude}$$

$$= \frac{1}{2} \cdot (\text{maximum value} - \text{minimum value})$$

$$= \frac{1}{2}(11.6 - 5.4) = 3.1$$

- **Adjust the Period**

The time between consecutive maximum and minimum values is half of one period. Thus

$$\frac{2\pi}{\omega} = \text{period}$$

$$= 2 \cdot (\text{time of maximum value} - \text{time of minimum value})$$

$$= 2(8 - 2) = 12$$

Thus, $\omega = 2\pi/12 = 0.52$.

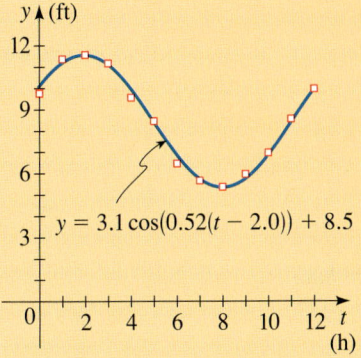

Figure 4

- **Adjust the Horizontal Shift**

Since the maximum value of the data occurs at approximately $t = 2.0$, it represents a cosine curve shifted 2 h to the right. So

$$c = \text{phase shift}$$
$$= \text{time of maximum value}$$
$$= 2.0$$

- **The Model**

We have shown that a function that models the tides over the given time period is given by

$$y = 3.1 \cos(0.52(t - 2.0)) + 8.5$$

A graph of the function and the scatter plot are shown in Figure 4. It appears that the model we found is a good approximation to the data.

(c) We need to solve the inequality $y \geq 11$. We solve this inequality graphically by graphing $y = 3.1 \cos 0.52(t - 2.0) + 8.5$ and $y = 11$ on the same graph. From the graph in Figure 5 we see the water depth is higher than 11 ft between $t \approx 0.8$ and $t \approx 3.2$. This corresponds to the times 12:48 A.M. to 3:12 A.M. ∎

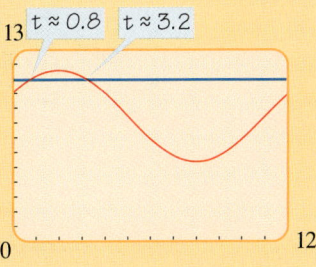

Figure 5

In Example 1 we used the scatter plot to guide us in finding a cosine curve that gives an approximate model of the data. Some graphing calculators are capable of finding a sine or cosine curve that best fits a given set of data points. The method these calculators use is similar to the method of finding a line of best fit, as explained on pages 239–240.

For the TI-83 and TI-86 the command S i n R e g (for sine regression) finds the sine curve that best fits the given data.

Example 2   Fitting a Sine Curve to Data

(a) Use a graphing device to find the sine curve that best fits the depth of water data in Table 1 on page 459.

(b) Compare your result to the model found in Example 1.

```
SinReg
y=a*sin(bx+c)+d
a=3.097877596
b=.5268322697
c=.5493035195
d=8.424021899
```

Output of the SinReg function on the TI-83.

Solution

(a) Using the data in Table 1 and the SinReg command on the TI-83 calculator, we get a function of the form

$$y = a \sin(bt + c) + d$$

where

$$a = 3.1 \qquad b = 0.53$$
$$c = 0.55 \qquad d = 8.42$$

So, the sine function that best fits the data is

$$y = 3.1 \sin(0.53t + 0.55) + 8.42$$

(b) To compare this with the function in Example 1, we change the sine function to a cosine function by using the reduction formula $\sin u = \cos(u - \pi/2)$.

$$y = 3.1 \sin(0.53t + 0.55) + 8.42$$

$$= 3.1 \cos\left(0.53t + 0.55 - \frac{\pi}{2}\right) + 8.42 \qquad \text{Reduction formula}$$

$$= 3.1 \cos(0.53t - 1.02) + 8.42$$

$$= 3.1 \cos(0.53(t - 1.92)) + 8.42 \qquad \text{Factor 0.53}$$

Comparing this with the function we obtained in Example 1, we see that there are small differences in the coefficients. In Figure 6 we graph a scatter plot of the data together with the sine function of best fit.

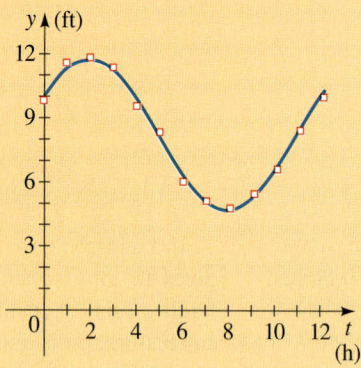

Figure 6

In Example 1 we estimated the values of the amplitude, period, and shifts from the data. In Example 2 the calculator computed the sine curve that best fits the data (that is, the curve that deviates least from the data as explained on page 240). The different ways of obtaining the model account for the differences in the functions.

Problems

1–4 ■ Modeling Periodic Data   A set of data is given.

(a) Make a scatter plot of the data.

(b) Find a cosine function of the form $y = a \cos(\omega(t - c)) + b$ that models the data, as in Example 1.

(c) Graph the function you found in part (b) together with the scatter plot. How well does the curve fit the data?

 (d) Use a graphing calculator to find the sine function that best fits the data, as in Example 2.

(e) Compare the functions you found in parts (b) and (d). [Use the reduction formula $\sin u = \cos(u - \pi/2)$.]

1.

t	y
0	2.1
2	1.1
4	−0.8
6	−2.1
8	−1.3
10	0.6
12	1.9
14	1.5

2.

t	y
0	190
25	175
50	155
75	125
100	110
125	95
150	105
175	120
200	140
225	165
250	185
275	200
300	195
325	185
350	165

3.

t	y
0.1	21.1
0.2	23.6
0.3	24.5
0.4	21.7
0.5	17.5
0.6	12.0
0.7	5.6
0.8	2.2
0.9	1.0
1.0	3.5
1.1	7.6
1.2	13.2
1.3	18.4
1.4	23.0
1.5	25.1

4.

t	y
0.0	0.56
0.5	0.45
1.0	0.29
1.5	0.13
2.0	0.05
2.5	−0.10
3.0	0.02
3.5	0.12
4.0	0.26
4.5	0.43
5.0	0.54
5.5	0.63
6.0	0.59

5. Annual Temperature Change   The table gives the average monthly temperature in Montgomery County, Maryland.

(a) Make a scatter plot of the data.

(b) Find a cosine curve that models the data (as in Example 1).

(c) Graph the function you found in part (b) together with the scatter plot.

 (d) Use a graphing calculator to find the sine curve that best fits the data (as in Example 2).

Month	Average temperature (°F)	Month	Average temperature (°F)
January	40.0	July	85.8
February	43.1	August	83.9
March	54.6	September	76.9
April	64.2	October	66.8
May	73.8	November	55.5
June	81.8	December	44.5

6. Circadian Rhythms   Circadian rhythm (from the Latin *circa*—about, and *diem*—day) is the daily biological pattern by which body temperature, blood pressure, and other physiological variables change. The data in the table below show typical changes in human body temperature over a 24-hour period ($t = 0$ corresponds to midnight).

(a) Make a scatter plot of the data.

(b) Find a cosine curve that models the data (as in Example 1).

(c) Graph the function you found in part (b) together with the scatter plot.

 (d) Use a graphing calculator to find the sine curve that best fits the data (as in Example 2).

Time	Body temperature (°C)	Time	Body temperature (°C)
0	36.8	14	37.3
2	36.7	16	37.4
4	36.6	18	37.3
6	36.7	20	37.2
8	36.8	22	37.0
10	37.0	24	36.8
12	37.2		

7. Predator Population   When two species interact in a predator/prey relationship (see page 432), the populations of both species tend to vary in a sinusoidal fashion. In a certain midwestern county, the main food source for barn owls consists of field mice and other small mammals. The table gives the population of barn owls in this county every July 1 over a 12-year period.

(a) Make a scatter plot of the data.

(b) Find a sine curve that models the data (as in Example 1).

(c) Graph the function you found in part (b) together with the scatter plot.

 (d) Use a graphing calculator to find the sine curve that best fits the data (as in Example 2). Compare to your answer from part (b).

Year	Owl population
0	50
1	62
2	73
3	80
4	71
5	60
6	51
7	43
8	29
9	20
10	28
11	41
12	49

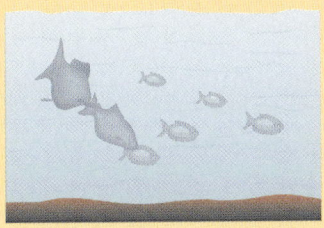

8. Salmon Survival   For reasons not yet fully understood, the number of fingerling salmon that survive the trip from their riverbed spawning grounds to the open ocean varies approximately sinusoidally from year to year. The table shows the number of salmon that hatch in a certain British Columbia creek and then make their way to the Strait of Georgia. The data is given in thousands of fingerlings, over a period of 16 years.

(a) Make a scatter plot of the data.

(b) Find a sine curve that models the data (as in Example 1).

(c) Graph the function you found in part (b) together with the scatter plot.

 (d) Use a graphing calculator to find the sine curve that best fits the data (as in Example 2). Compare to your answer from part (b).

Year	Salmon (× 1000)	Year	Salmon (× 1000)
1985	43	1993	56
1986	36	1994	63
1987	27	1995	57
1988	23	1996	50
1989	26	1997	44
1990	33	1998	38
1991	43	1999	30
1992	50	2000	22

9. Sunspot Activity   Sunspots are relatively "cool" regions on the sun that appear as dark spots when observed through special solar filters. The number of sunspots varies in an 11-year cycle. The table gives the average daily sunspot count for the years 1975–2004.

(a) Make a scatter plot of the data.

(b) Find a cosine curve that models the data (as in Example 1).

(c) Graph the function you found in part (b) together with the scatter plot.

 (d) Use a graphing calculator to find the sine curve that best fits the data (as in Example 2). Compare to your answer in part (b).

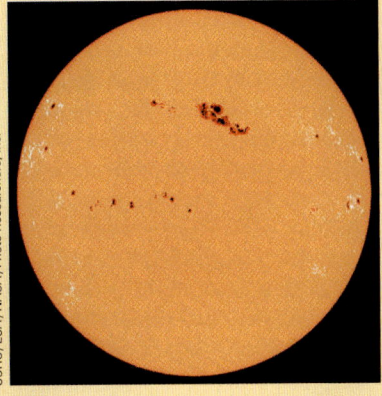

Year	Sunspots	Year	Sunspots	Year	Sunspots
1975	16	1985	18	1995	18
1976	13	1986	13	1996	9
1977	28	1987	29	1997	21
1978	93	1988	100	1998	64
1979	155	1989	158	1999	93
1980	155	1990	143	2000	119
1981	140	1991	146	2001	111
1982	116	1992	94	2002	104
1983	67	1993	55	2003	64
1984	46	1994	30	2004	40

6

Trigonometric Functions of Angles

Chapter Overview

The trigonometric functions can be defined in two different but equivalent ways—as functions of real numbers (Chapter 5) or as functions of angles (Chapter 6). The two approaches to trigonometry are independent of each other, so either Chapter 5 or Chapter 6 may be studied first. We study both approaches because different applications require that we view these functions differently. The approach in this chapter lends itself to geometric problems involving finding angles and distances.

Suppose we want to find the distance to the sun. Using a tape measure is of course impractical, so we need something besides simple measurement to tackle this problem. Angles are easy to measure—for example, we can find the angle formed by the sun, earth, and moon by simply pointing to the sun with one arm and the moon with the other and estimating the angle between them. The key idea then is to find a relationship between angles and distances. So if we had a way to determine distances from angles, we'd be able to find the distance to the sun without going there. The trigonometric functions provide us with just the tools we need.

If ABC is a right triangle with acute angle θ as in the figure, then we define $\sin \theta$ to be the ratio y/r. Triangle $A'B'C'$ is similar to triangle ABC, so

$$\frac{y}{r} = \frac{y'}{r'}$$

Although the distances y' and r' are different from y and r, the given ratio is the same. Thus, in *any* right triangle with acute angle θ, the ratio of the side opposite angle θ to the hypotenuse is the same and is called $\sin \theta$. The other trigonometric ratios are defined in a similar fashion.

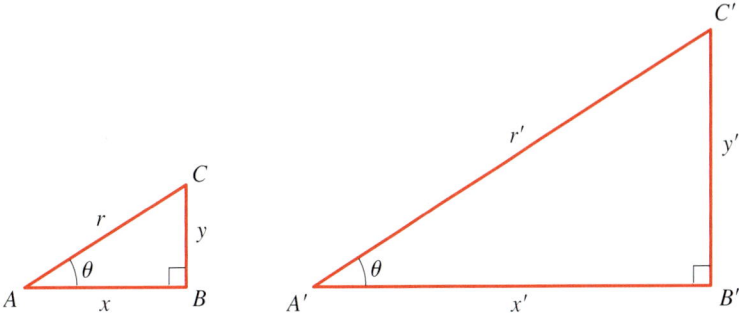

In this chapter we learn how trigonometric functions can be used to measure distances on the earth and in space. In Exercises 61 and 62 on page 487, we actually determine the distance to the sun using trigonometry. Right triangle trigonometry has many other applications, from determining the optimal cell structure in a beehive (Exercise 67, page 497) to explaining the shape of a rainbow (Exercise 69, page 498). In the *Focus on Modeling*, pages 522–523, we see how a surveyor uses trigonometry to map a town.

6.1 Angle Measure

An **angle** *AOB* consists of two rays R_1 and R_2 with a common vertex *O* (see Figure 1). We often interpret an angle as a rotation of the ray R_1 onto R_2. In this case, R_1 is called the **initial side**, and R_2 is called the **terminal side** of the angle. If the rotation is counterclockwise, the angle is considered **positive**, and if the rotation is clockwise, the angle is considered **negative**.

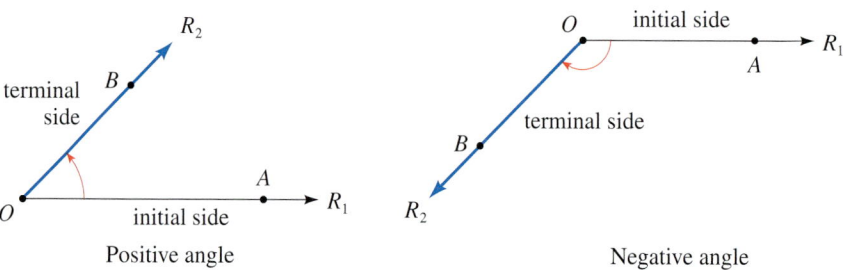

Figure 1

Angle Measure

The **measure** of an angle is the amount of rotation about the vertex required to move R_1 onto R_2. Intuitively, this is how much the angle "opens." One unit of measurement for angles is the **degree**. An angle of measure 1 degree is formed by rotating the initial side $\frac{1}{360}$ of a complete revolution. In calculus and other branches of mathematics, a more natural method of measuring angles is used—*radian measure*. The amount an angle opens is measured along the arc of a circle of radius 1 with its center at the vertex of the angle.

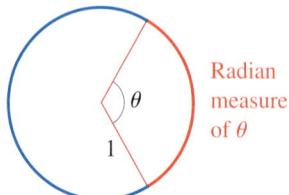

Figure 2

Definition of Radian Measure

If a circle of radius 1 is drawn with the vertex of an angle at its center, then the measure of this angle in **radians** (abbreviated **rad**) is the length of the arc that subtends the angle (see Figure 2).

The circumference of the circle of radius 1 is 2π and so a complete revolution has measure 2π rad, a straight angle has measure π rad, and a right angle has measure

$\pi/2$ rad. An angle that is subtended by an arc of length 2 along the unit circle has radian measure 2 (see Figure 3).

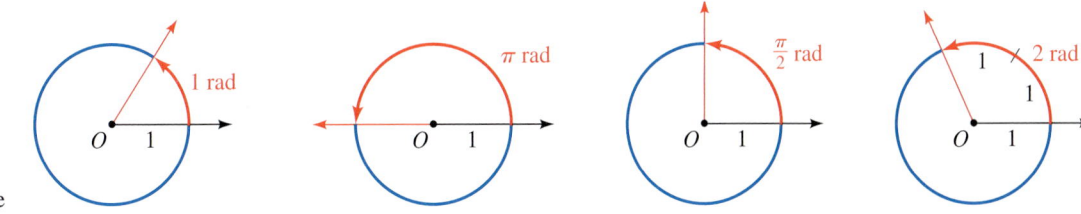

Figure 3
Radian measure

Since a complete revolution measured in degrees is 360° and measured in radians is 2π rad, we get the following simple relationship between these two methods of angle measurement.

Relationship between Degrees and Radians

$$180° = \pi \text{ rad} \qquad 1 \text{ rad} = \left(\frac{180}{\pi}\right)° \qquad 1° = \frac{\pi}{180} \text{ rad}$$

1. To convert degrees to radians, multiply by $\dfrac{\pi}{180}$.

2. To convert radians to degrees, multiply by $\dfrac{180}{\pi}$.

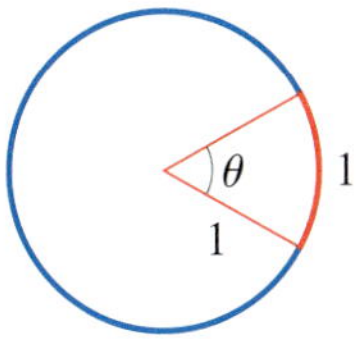

Measure of $\theta = 1$ rad
Measure of $\theta \approx 57.296°$

Figure 4

To get some idea of the size of a radian, notice that

$$1 \text{ rad} \approx 57.296° \qquad \text{and} \qquad 1° \approx 0.01745 \text{ rad}$$

An angle θ of measure 1 rad is shown in Figure 4.

Example 1   Converting between Radians and Degrees

(a) Express 60° in radians.   (b) Express $\dfrac{\pi}{6}$ rad in degrees.

Solution   The relationship between degrees and radians gives

(a) $60° = 60\left(\dfrac{\pi}{180}\right) \text{ rad} = \dfrac{\pi}{3} \text{ rad}$ (b) $\dfrac{\pi}{6} \text{ rad} = \left(\dfrac{\pi}{6}\right)\left(\dfrac{180}{\pi}\right) = 30°$ ∎

A note on terminology: We often use a phrase such as "a 30° angle" to mean *an angle whose measure is 30°*. Also, for an angle θ, we write $\theta = 30°$ or $\theta = \pi/6$ to mean *the measure of θ is 30° or $\pi/6$ rad*. When no unit is given, the angle is assumed to be measured in radians.

Angles in Standard Position

An angle is in **standard position** if it is drawn in the xy-plane with its vertex at the origin and its initial side on the positive x-axis. Figure 5 gives examples of angles in standard position.

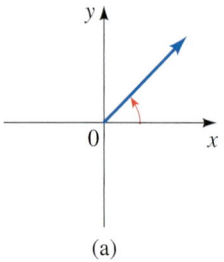

(a)

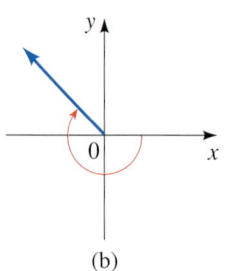

(b)

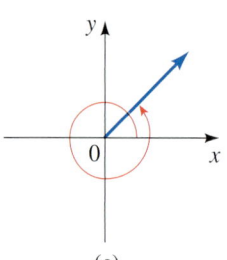

(c)

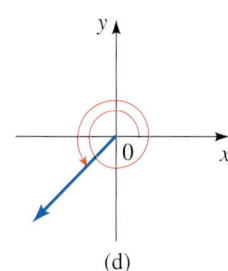

(d)

Figure 5

Angles in standard position

Two angles in standard position are **coterminal** if their sides coincide. In Figure 5 the angles in (a) and (c) are coterminal.

Example 2 Coterminal Angles

(a) Find angles that are coterminal with the angle $\theta = 30°$ in standard position.

(b) Find angles that are coterminal with the angle $\theta = \dfrac{\pi}{3}$ in standard position.

Solution

(a) To find positive angles that are coterminal with θ, we add any multiple of $360°$. Thus

$$30° + 360° = 390° \qquad \text{and} \qquad 30° + 720° = 750°$$

are coterminal with $\theta = 30°$. To find negative angles that are coterminal with θ, we subtract any multiple of $360°$. Thus

$$30° - 360° = -330° \qquad \text{and} \qquad 30° - 720° = -690°$$

are coterminal with θ. (See Figure 6.)

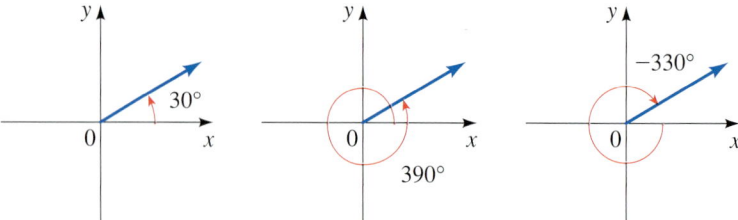

Figure 6

(b) To find positive angles that are coterminal with θ, we add any multiple of 2π. Thus

$$\frac{\pi}{3} + 2\pi = \frac{7\pi}{3} \qquad \text{and} \qquad \frac{\pi}{3} + 4\pi = \frac{13\pi}{3}$$

are coterminal with $\theta = \pi/3$. To find negative angles that are coterminal with θ, we subtract any multiple of 2π. Thus

$$\frac{\pi}{3} - 2\pi = -\frac{5\pi}{3} \qquad \text{and} \qquad \frac{\pi}{3} - 4\pi = -\frac{11\pi}{3}$$

are coterminal with θ. (See Figure 7.)

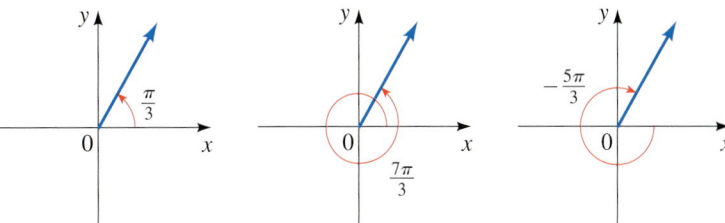

Figure 7

Example 3   Coterminal Angles

Find an angle with measure between $0°$ and $360°$ that is coterminal with the angle of measure $1290°$ in standard position.

Solution   We can subtract $360°$ as many times as we wish from $1290°$, and the resulting angle will be coterminal with $1290°$. Thus, $1290° - 360° = 930°$ is coterminal with $1290°$, and so is the angle $1290° - 2(360)° = 570°$.

To find the angle we want between $0°$ and $360°$, we subtract $360°$ from $1290°$ as many times as necessary. An efficient way to do this is to determine how many times $360°$ goes into $1290°$, that is, divide 1290 by 360, and the remainder will be the angle we are looking for. We see that 360 goes into 1290 three times with a remainder of 210. Thus, $210°$ is the desired angle (see Figure 8).

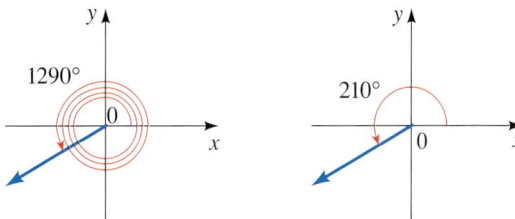

Figure 8

Length of a Circular Arc

An angle whose radian measure is θ is subtended by an arc that is the fraction $\theta/(2\pi)$ of the circumference of a circle. Thus, in a circle of radius r, the length s of an arc that subtends the angle θ (see Figure 9) is

$$s = \frac{\theta}{2\pi} \times \text{circumference of circle}$$

$$= \frac{\theta}{2\pi}(2\pi r) = \theta r$$

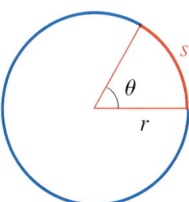

Figure 9

$s = \theta r$

Length of a Circular Arc

In a circle of radius r, the length s of an arc that subtends a central angle of θ radians is

$$s = r\theta$$

Solving for θ, we get the important formula

$$\theta = \frac{s}{r}$$

This formula allows us to define radian measure using a circle of any radius r: The radian measure of an angle θ is s/r, where s is the length of the circular arc that subtends θ in a circle of radius r (see Figure 10).

Figure 10

The radian measure of θ is the number of "radiuses" that can fit in the arc that subtends θ; hence the term *radian*.

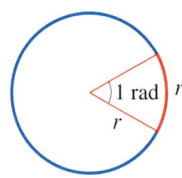
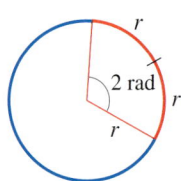

Example 4   Arc Length and Angle Measure

(a) Find the length of an arc of a circle with radius 10 m that subtends a central angle of 30°.

(b) A central angle θ in a circle of radius 4 m is subtended by an arc of length 6 m. Find the measure of θ in radians.

Solution

(a) From Example 1(b) we see that $30° = \pi/6$ rad. So the length of the arc is

⊘ The formula $s = r\theta$ is true only when θ is measured in radians.

$$s = r\theta = (10)\frac{\pi}{6} = \frac{5\pi}{3} \text{ m}$$

(b) By the formula $\theta = s/r$, we have

$$\theta = \frac{s}{r} = \frac{6}{4} = \frac{3}{2} \text{ rad}$$ ∎

Area of a Circular Sector

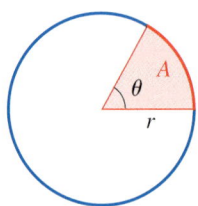

The area of a circle of radius r is $A = \pi r^2$. A sector of this circle with central angle θ has an area that is the fraction $\theta/(2\pi)$ of the area of the entire circle (see Figure 11). So the area of this sector is

$$A = \frac{\theta}{2\pi} \times \text{area of circle}$$

Figure 11
$A = \frac{1}{2}r^2\theta$

$$= \frac{\theta}{2\pi}(\pi r^2) = \frac{1}{2}r^2\theta$$

Area of a Circular Sector

In a circle of radius r, the area A of a sector with a central angle of θ radians is

$$A = \frac{1}{2}r^2\theta$$

Example 5   Area of a Sector

Find the area of a sector of a circle with central angle $60°$ if the radius of the circle is 3 m.

Solution   To use the formula for the area of a circular sector, we must find the central angle of the sector in radians: $60° = 60(\pi/180)$ rad $= \pi/3$ rad. Thus, the area of the sector is

$$A = \frac{1}{2}r^2\theta = \frac{1}{2}(3)^2\left(\frac{\pi}{3}\right) = \frac{3\pi}{2}\,\text{m}^2 \qquad \blacksquare$$

> ⊘ The formula $A = \frac{1}{2}r^2\theta$ is true only when θ is measured in radians.

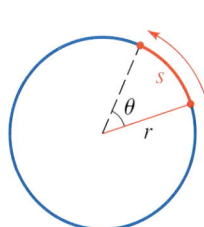

Figure 12

Circular Motion

Suppose a point moves along a circle as shown in Figure 12. There are two ways to describe the motion of the point—linear speed and angular speed. **Linear speed** is the rate at which the distance traveled is changing, so linear speed is the distance traveled divided by the time elapsed. **Angular speed** is the rate at which the central angle θ is changing, so angular speed is the number of radians this angle changes divided by the time elapsed.

Linear Speed and Angular Speed

Suppose a point moves along a circle of radius r and the ray from the center of the circle to the point traverses θ radians in time t. Let $s = r\theta$ be the distance the point travels in time t. Then the speed of the object is given by

Angular speed $\omega = \dfrac{\theta}{t}$

Linear speed $v = \dfrac{s}{t}$

> The symbol ω is the Greek letter "omega."

Example 6   Finding Linear and Angular Speed

A boy rotates a stone in a 3-ft-long sling at the rate of 15 revolutions every 10 seconds. Find the angular and linear velocities of the stone.

Solution   In 10 s, the angle θ changes by $15 \cdot 2\pi = 30\pi$ radians. So the *angular speed* of the stone is

$$\omega = \frac{\theta}{t} = \frac{30\pi \text{ rad}}{10 \text{ s}} = 3\pi \text{ rad/s}$$

The distance traveled by the stone in 10 s is $s = 15 \cdot 2\pi r = 15 \cdot 2\pi \cdot 3 = 90\pi$ ft. So the *linear speed* of the stone is

$$v = \frac{s}{t} = \frac{90\pi \text{ ft}}{10 \text{ s}} = 9\pi \text{ ft/s} \qquad \blacksquare$$

Notice that angular speed does *not* depend on the radius of the circle, but only on the angle θ. However, if we know the angular speed ω and the radius r, we can find linear speed as follows: $v = s/t = r\theta/t = r(\theta/t) = r\omega$.

Relationship between Linear and Angular Speed

If a point moves along a circle of radius r with angular speed ω, then its linear speed v is given by

$$v = r\omega$$

Example 7 Finding Linear Speed from Angular Speed

A woman is riding a bicycle whose wheels are 26 inches in diameter. If the wheels rotate at 125 revolutions per minute (rpm), find the speed at which she is traveling, in mi/h.

Solution The angular speed of the wheels is $2\pi \cdot 125 = 250\pi$ rad/min. Since the wheels have radius 13 in. (half the diameter), the linear speed is

$$v = r\omega = 13 \cdot 250\pi \approx 10{,}210.2 \text{ in./min}$$

Since there are 12 inches per foot, 5280 feet per mile, and 60 minutes per hour, her speed in miles per hour is

$$\frac{10{,}210.2 \text{ in./min} \times 60 \text{ min/h}}{12 \text{ in./ft} \times 5280 \text{ ft/mi}} = \frac{612{,}612 \text{ in./h}}{63{,}360 \text{ in./mi}}$$

$$\approx 9.7 \text{ mi/h} \qquad \blacksquare$$

6.1 Exercises

1–12 ■ Find the radian measure of the angle with the given degree measure.

1. $72°$ **2.** $54°$ **3.** $-45°$

4. $-60°$ **5.** $-75°$ **6.** $-300°$

7. $1080°$ **8.** $3960°$ **9.** $96°$

10. $15°$ **11.** $7.5°$ **12.** $202.5°$

13–24 ■ Find the degree measure of the angle with the given radian measure.

13. $\dfrac{7\pi}{6}$ **14.** $\dfrac{11\pi}{3}$ **15.** $-\dfrac{5\pi}{4}$

16. $-\dfrac{3\pi}{2}$ **17.** 3 **18.** -2

19. -1.2 **20.** 3.4 **21.** $\dfrac{\pi}{10}$

22. $\dfrac{5\pi}{18}$ **23.** $-\dfrac{2\pi}{15}$ **24.** $-\dfrac{13\pi}{12}$

25–30 ■ The measure of an angle in standard position is given. Find two positive angles and two negative angles that are coterminal with the given angle.

25. $50°$ **26.** $135°$ **27.** $\dfrac{3\pi}{4}$

28. $\dfrac{11\pi}{6}$

29. $-\dfrac{\pi}{4}$

30. $-45°$

31–36 ■ The measures of two angles in standard position are given. Determine whether the angles are coterminal.

31. $70°, \quad 430°$

32. $-30°, \quad 330°$

33. $\dfrac{5\pi}{6}, \quad \dfrac{17\pi}{6}$

34. $\dfrac{32\pi}{3}, \quad \dfrac{11\pi}{3}$

35. $155°, \quad 875°$

36. $50°, \quad 340°$

37–42 ■ Find an angle between 0° and 360° that is coterminal with the given angle.

37. $733°$

38. $361°$

39. $1110°$

40. $-100°$

41. $-800°$

42. $1270°$

43–48 ■ Find an angle between 0 and 2π that is coterminal with the given angle.

43. $\dfrac{17\pi}{6}$

44. $-\dfrac{7\pi}{3}$

45. 87π

46. 10

47. $\dfrac{17\pi}{4}$

48. $\dfrac{51\pi}{2}$

49. Find the length of the arc s in the figure.

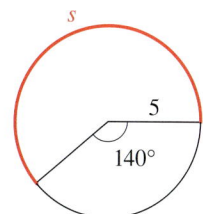

50. Find the angle θ in the figure.

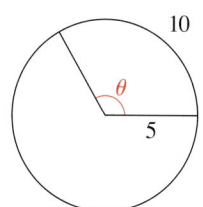

51. Find the radius r of the circle in the figure.

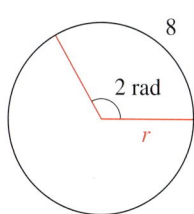

52. Find the length of an arc that subtends a central angle of 45° in a circle of radius 10 m.

53. Find the length of an arc that subtends a central angle of 2 rad in a circle of radius 2 mi.

54. A central angle θ in a circle of radius 5 m is subtended by an arc of length 6 m. Find the measure of θ in degrees and in radians.

55. An arc of length 100 m subtends a central angle θ in a circle of radius 50 m. Find the measure of θ in degrees and in radians.

56. A circular arc of length 3 ft subtends a central angle of 25°. Find the radius of the circle.

57. Find the radius of the circle if an arc of length 6 m on the circle subtends a central angle of $\pi/6$ rad.

58. Find the radius of the circle if an arc of length 4 ft on the circle subtends a central angle of 135°.

59. Find the area of the sector shown in each figure.

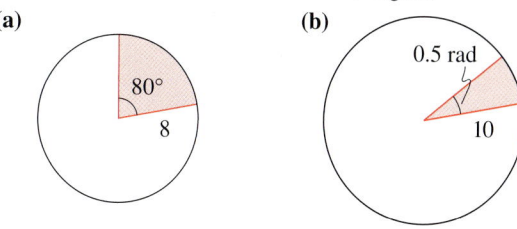

(a)   **(b)**

60. Find the radius of each circle if the area of the sector is 12.

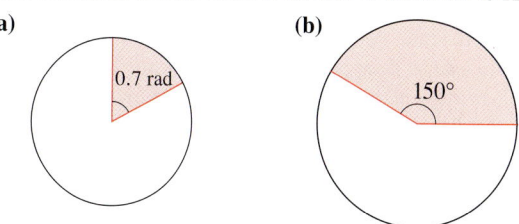

(a)   **(b)**

61. Find the area of a sector with central angle 1 rad in a circle of radius 10 m.

62. A sector of a circle has a central angle of 60°. Find the area of the sector if the radius of the circle is 3 mi.

63. The area of a sector of a circle with a central angle of 2 rad is 16 m². Find the radius of the circle.

64. A sector of a circle of radius 24 mi has an area of 288 mi². Find the central angle of the sector.

65. The area of a circle is 72 cm². Find the area of a sector of this circle that subtends a central angle of $\pi/6$ rad.

66. Three circles with radii 1, 2, and 3 ft are externally tangent to one another, as shown in the figure on the next page. Find the area of the sector of the circle of radius 1 that is cut off by the line segments joining the center of that circle to the centers of the other two circles.

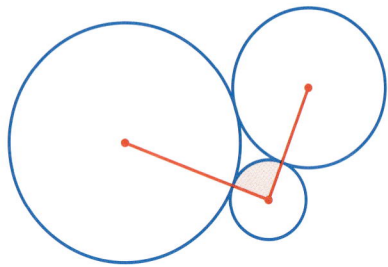

Applications

67. Travel Distance   A car's wheels are 28 in. in diameter. How far (in miles) will the car travel if its wheels revolve 10,000 times without slipping?

68. Wheel Revolutions   How many revolutions will a car wheel of diameter 30 in. make as the car travels a distance of one mile?

69. Latitudes   Pittsburgh, Pennsylvania, and Miami, Florida, lie approximately on the same meridian. Pittsburgh has a latitude of 40.5° N and Miami, 25.5° N. Find the distance between these two cities. (The radius of the earth is 3960 mi.)

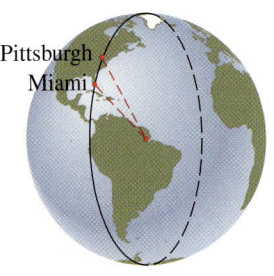

70. Latitudes   Memphis, Tennessee, and New Orleans, Louisiana, lie approximately on the same meridian. Memphis has latitude 35° N and New Orleans, 30° N. Find the distance between these two cities. (The radius of the earth is 3960 mi.)

71. Orbit of the Earth   Find the distance that the earth travels in one day in its path around the sun. Assume that a year has 365 days and that the path of the earth around the sun is a circle of radius 93 million miles. [The path of the earth around the sun is actually an *ellipse* with the sun at one focus (see Section 10.2). This ellipse, however, has very small eccentricity, so it is nearly circular.]

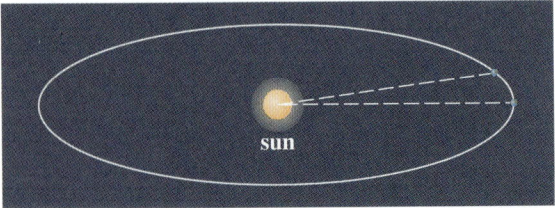

72. Circumference of the Earth   The Greek mathematician Eratosthenes (ca. 276–195 B.C.) measured the circumference of the earth from the following observations. He noticed that on a certain day the sun shone directly down a deep well in Syene (modern Aswan). At the same time in Alexandria, 500 miles north (on the same meridian), the rays of the sun shone at an angle of 7.2° to the zenith. Use this information and the figure to find the radius and circumference of the earth.

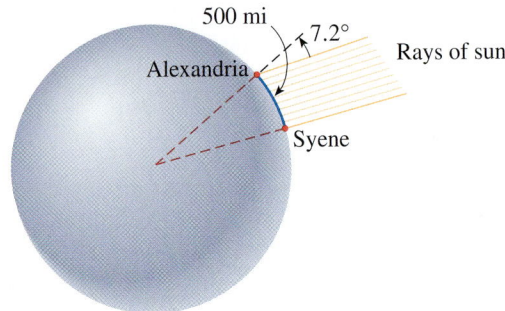

73. Nautical Miles   Find the distance along an arc on the surface of the earth that subtends a central angle of 1 minute (1 minute $= \frac{1}{60}$ degree). This distance is called a *nautical mile*. (The radius of the earth is 3960 mi.)

74. Irrigation   An irrigation system uses a straight sprinkler pipe 300 ft long that pivots around a central point as shown. Due to an obstacle the pipe is allowed to pivot through 280° only. Find the area irrigated by this system.

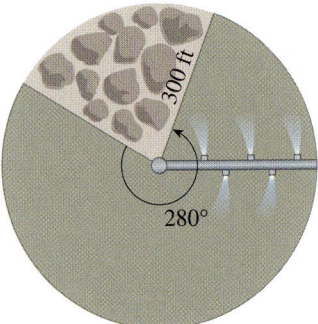

75. Windshield Wipers   The top and bottom ends of a windshield wiper blade are 34 in. and 14 in. from the pivot point, respectively. While in operation the wiper sweeps through 135°. Find the area swept by the blade.

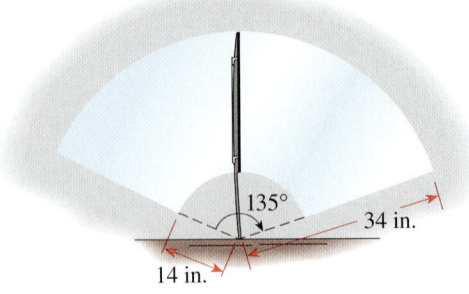

76. The Tethered Cow   A cow is tethered by a 100-ft rope to the inside corner of an L-shaped building, as shown in the figure. Find the area that the cow can graze.

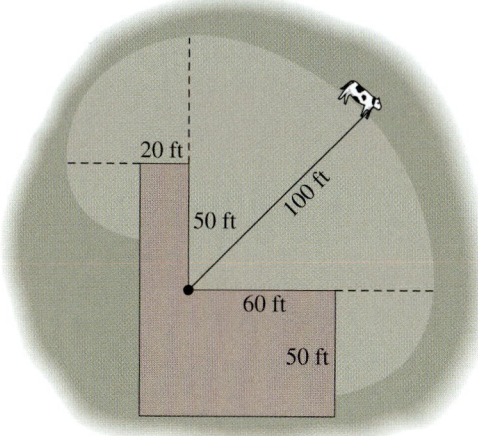

77. Winch   A winch of radius 2 ft is used to lift heavy loads. If the winch makes 8 revolutions every 15 s, find the speed at which the load is rising.

78. Fan   A ceiling fan with 16-in. blades rotates at 45 rpm.

(a) Find the angular speed of the fan in rad/min.

(b) Find the linear speed of the tips of the blades in in./min.

79. Radial Saw   A radial saw has a blade with a 6-in. radius. Suppose that the blade spins at 1000 rpm.

(a) Find the angular speed of the blade in rad/min.

(b) Find the linear speed of the sawteeth in ft/s.

80. Speed at Equator   The earth rotates about its axis once every 23 h 56 min 4 s, and the radius of the earth is 3960 mi. Find the linear speed of a point on the equator in mi/h.

81. Speed of a Car   The wheels of a car have radius 11 in. and are rotating at 600 rpm. Find the speed of the car in mi/h.

82. Truck Wheels   A truck with 48-in.-diameter wheels is traveling at 50 mi/h.

(a) Find the angular speed of the wheels in rad/min.

(b) How many revolutions per minute do the wheels make?

83. Speed of a Current   To measure the speed of a current, scientists place a paddle wheel in the stream and observe the rate at which it rotates. If the paddle wheel has radius 0.20 m and rotates at 100 rpm, find the speed of the current in m/s.

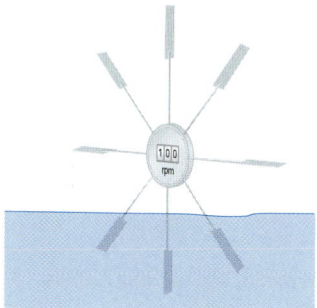

84. Bicycle Wheel   The sprockets and chain of a bicycle are shown in the figure. The pedal sprocket has a radius of 4 in., the wheel sprocket a radius of 2 in., and the wheel a radius of 13 in. The cyclist pedals at 40 rpm.

(a) Find the angular speed of the wheel sprocket.

(b) Find the speed of the bicycle. (Assume that the wheel turns at the same rate as the wheel sprocket.)

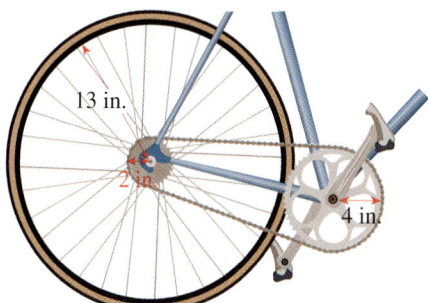

85. Conical Cup   A conical cup is made from a circular piece of paper with radius 6 cm by cutting out a sector and joining the edges as shown. Suppose $\theta = 5\pi/3$.

(a) Find the circumference C of the opening of the cup.

(b) Find the radius r of the opening of the cup. [*Hint*: Use $C = 2\pi r$.]

(c) Find the height h of the cup. [*Hint*: Use the Pythagorean Theorem.]

(d) Find the volume of the cup.

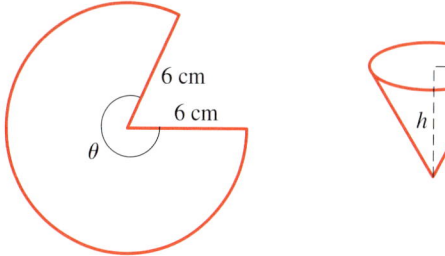

86. Conical Cup In this exercise we find the volume of the conical cup in Exercise 85 for any angle θ.

(a) Follow the steps in Exercise 85 to show that the volume of the cup as a function of θ is

$$V(\theta) = \frac{9}{\pi^2}\theta^2\sqrt{4\pi^2 - \theta^2}, \qquad 0 < \theta < 2\pi$$

(b) Graph the function V.

(c) For what angle θ is the volume of the cup a maximum?

Discovery • Discussion

87. Different Ways of Measuring Angles The custom of measuring angles using degrees, with $360°$ in a circle, dates back to the ancient Babylonians, who used a number system based on groups of 60. Another system of measuring angles divides the circle into 400 units, called *grads*. In this system

a right angle is 100 grad, so this fits in with our base 10 number system.

Write a short essay comparing the advantages and disadvantages of these two systems and the radian system of measuring angles. Which system do you prefer?

88. Clocks and Angles In one hour, the minute hand on a clock moves through a complete circle, and the hour hand moves through $\frac{1}{12}$ of a circle. Through how many radians do the minute and the hour hand move between 1:00 P.M. and 6:45 P.M. (on the same day)?

6.2   Trigonometry of Right Triangles

In this section we study certain ratios of the sides of right triangles, called trigonometric ratios, and give several applications.

Trigonometric Ratios

Consider a right triangle with θ as one of its acute angles. The trigonometric ratios are defined as follows (see Figure 1).

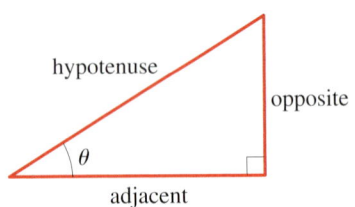

The Trigonometric Ratios

$$\sin\theta = \frac{\text{opposite}}{\text{hypotenuse}} \qquad \cos\theta = \frac{\text{adjacent}}{\text{hypotenuse}} \qquad \tan\theta = \frac{\text{opposite}}{\text{adjacent}}$$

$$\csc\theta = \frac{\text{hypotenuse}}{\text{opposite}} \qquad \sec\theta = \frac{\text{hypotenuse}}{\text{adjacent}} \qquad \cot\theta = \frac{\text{adjacent}}{\text{opposite}}$$

Figure 1

The symbols we use for these ratios are abbreviations for their full names: **sine**, **cosine**, **tangent**, **cosecant**, **secant**, **cotangent**. Since any two right triangles with angle θ are similar, these ratios are the same, regardless of the size of the triangle; the trigonometric ratios depend only on the angle θ (see Figure 2).

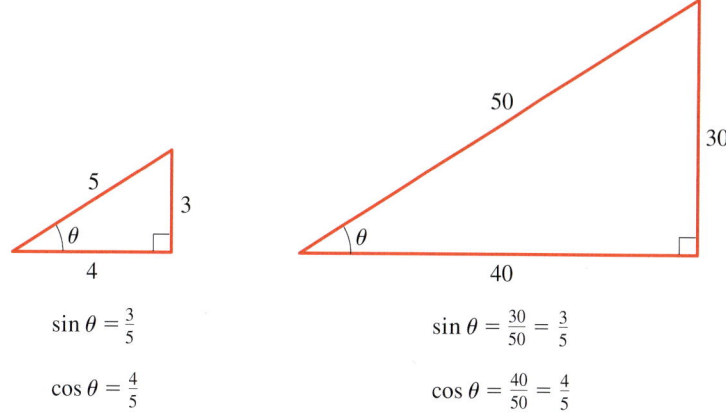

$$\sin \theta = \frac{3}{5}$$

$$\cos \theta = \frac{4}{5}$$

$$\sin \theta = \frac{30}{50} = \frac{3}{5}$$

$$\cos \theta = \frac{40}{50} = \frac{4}{5}$$

Figure 2

Example 1   Finding Trigonometric Ratios

Find the six trigonometric ratios of the angle θ in Figure 3.

Solution

$$\sin \theta = \frac{2}{3} \qquad \cos \theta = \frac{\sqrt{5}}{3} \qquad \tan \theta = \frac{2}{\sqrt{5}}$$

$$\csc \theta = \frac{3}{2} \qquad \sec \theta = \frac{3}{\sqrt{5}} \qquad \cot \theta = \frac{\sqrt{5}}{2}$$

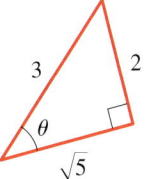

Figure 3

Example 2   Finding Trigonometric Ratios

If $\cos \alpha = \frac{3}{4}$, sketch a right triangle with acute angle α, and find the other five trigonometric ratios of α.

Solution   Since $\cos \alpha$ is defined as the ratio of the adjacent side to the hypotenuse, we sketch a triangle with hypotenuse of length 4 and a side of length 3 adjacent to α. If the opposite side is x, then by the Pythagorean Theorem, $3^2 + x^2 = 4^2$ or $x^2 = 7$, so $x = \sqrt{7}$. We then use the triangle in Figure 4 to find the ratios.

$$\sin \alpha = \frac{\sqrt{7}}{4} \qquad \cos \alpha = \frac{3}{4} \qquad \tan \alpha = \frac{\sqrt{7}}{3}$$

$$\csc \alpha = \frac{4}{\sqrt{7}} \qquad \sec \alpha = \frac{4}{3} \qquad \cot \alpha = \frac{3}{\sqrt{7}}$$

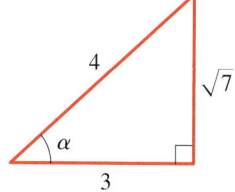

Figure 4

Special Triangles

Certain right triangles have ratios that can be calculated easily from the Pythagorean Theorem. Since they are used frequently, we mention them here.

The first triangle is obtained by drawing a diagonal in a square of side 1 (see Figure 5 on page 480). By the Pythagorean Theorem this diagonal has length $\sqrt{2}$. The resulting triangle has angles 45°, 45°, and 90° (or $\pi/4$, $\pi/4$, and $\pi/2$). To get the second triangle, we start with an equilateral triangle ABC of side 2 and draw the perpendicular bisector DB of the base, as in Figure 6. By the Pythagorean Theorem the length of DB is $\sqrt{3}$. Since DB bisects angle ABC, we obtain a triangle with angles 30°, 60°, and 90° (or $\pi/6$, $\pi/3$, and $\pi/2$).

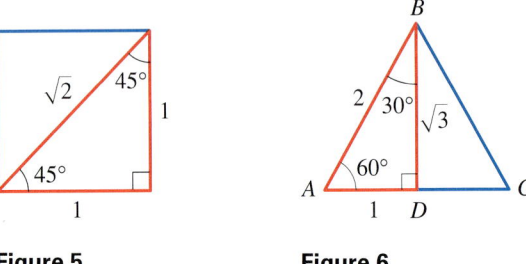

Figure 5 **Figure 6**

We can now use the special triangles in Figures 5 and 6 to calculate the trigonometric ratios for angles with measures 30°, 45°, and 60° (or $\pi/6$, $\pi/4$, and $\pi/3$). These are listed in Table 1.

Table 1 Values of the trigonometric ratios for special angles

θ in degrees	θ in radians	$\sin\theta$	$\cos\theta$	$\tan\theta$	$\csc\theta$	$\sec\theta$	$\cot\theta$
30°	$\frac{\pi}{6}$	$\frac{1}{2}$	$\frac{\sqrt{3}}{2}$	$\frac{\sqrt{3}}{3}$	2	$\frac{2\sqrt{3}}{3}$	$\sqrt{3}$
45°	$\frac{\pi}{4}$	$\frac{\sqrt{2}}{2}$	$\frac{\sqrt{2}}{2}$	1	$\sqrt{2}$	$\sqrt{2}$	1
60°	$\frac{\pi}{3}$	$\frac{\sqrt{3}}{2}$	$\frac{1}{2}$	$\sqrt{3}$	$\frac{2\sqrt{3}}{3}$	2	$\frac{\sqrt{3}}{3}$

It's useful to remember these special trigonometric ratios because they occur often. Of course, they can be recalled easily if we remember the triangles from which they are obtained.

To find the values of the trigonometric ratios for other angles, we use a calculator. Mathematical methods (called *numerical methods*) used in finding the trigonometric ratios are programmed directly into scientific calculators. For instance, when the ⌈SIN⌉ key is pressed, the calculator computes an approximation to the value of the sine of the given angle. Calculators give the values of sine, cosine, and tangent; the other ratios can be easily calculated from these using the following *reciprocal relations*:

For an explanation of numerical methods, see the margin note on page 436.

$$\csc t = \frac{1}{\sin t} \qquad \sec t = \frac{1}{\cos t} \qquad \cot t = \frac{1}{\tan t}$$

You should check that these relations follow immediately from the definitions of the trigonometric ratios.

We follow the convention that when we write sin *t, we mean the sine of the angle whose radian measure is t.* For instance, sin 1 means the sine of the angle whose radian measure is 1. When using a calculator to find an approximate value for this number, set your calculator to radian mode; you will find that

$$\sin 1 \approx 0.841471$$

If you want to find the sine of the angle whose measure is 1°, set your calculator to degree mode; you will find that

$$\sin 1° \approx 0.0174524$$

Example 3   Using a Calculator to Find Trigonometric Ratios

With our calculator in degree mode, and writing the results correct to five decimal places, we find

$$\sin 17° \approx 0.29237 \qquad \sec 88° = \frac{1}{\cos 88°} \approx 28.65371$$

With our calculator in radian mode, and writing the results correct to five decimal places, we find

$$\cos 1.2 \approx 0.36236 \qquad \cot 1.54 = \frac{1}{\tan 1.54} \approx 0.03081 \qquad ■$$

Applications of Trigonometry of Right Triangles

A triangle has six parts: three angles and three sides. To **solve a triangle** means to determine all of its parts from the information known about the triangle, that is, to determine the lengths of the three sides and the measures of the three angles.

Example 4   Solving a Right Triangle

Solve triangle ABC, shown in Figure 7.

Solution   It's clear that $\angle B = 60°$. To find a, we look for an equation that relates a to the lengths and angles we already know. In this case, we have $\sin 30° = a/12$, so

$$a = 12 \sin 30° = 12\left(\tfrac{1}{2}\right) = 6$$

Similarly, $\cos 30° = b/12$, so

$$b = 12 \cos 30° = 12\left(\frac{\sqrt{3}}{2}\right) = 6\sqrt{3} \qquad ■$$

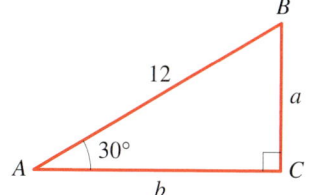

Figure 7

It's very useful to know that, using the information given in Figure 8, the lengths of the legs of a right triangle are

$$a = r \sin \theta \qquad \text{and} \qquad b = r \cos \theta$$

The ability to solve right triangles using the trigonometric ratios is fundamental to many problems in navigation, surveying, astronomy, and the measurement of distances. The applications we consider in this section always involve right triangles but, as we will see in the next three sections, trigonometry is also useful in solving triangles that are not right triangles.

To discuss the next examples, we need some terminology. If an observer is looking at an object, then the line from the eye of the observer to the object is called

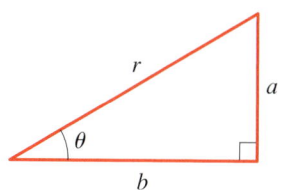

Figure 8
$a = r \sin \theta$
$b = r \cos \theta$

Thales of Miletus (circa 625–547 B.C.) is the legendary founder of Greek geometry. It is said that he calculated the height of a Greek column by comparing the length of the shadow of his staff with that of the column. Using properties of similar triangles, he argued that the ratio of the height h of the column to the height h' of his staff was equal to the ratio of the length s of the column's shadow to the length s' of the staff's shadow:

$$\frac{h}{h'} = \frac{s}{s'}$$

Since three of these quantities are known, Thales was able to calculate the height of the column.

According to legend, Thales used a similar method to find the height of the Great Pyramid in Egypt, a feat that impressed Egypt's king. Plutarch wrote that "although he [the king of Egypt] admired you [Thales] for other things, yet he particularly liked the manner by which you measured the height of the pyramid without any trouble or instrument." The principle Thales used, the fact that ratios of corresponding sides of similar triangles are equal, is the foundation of the subject of trigonometry.

the **line of sight** (Figure 9). If the object being observed is above the horizontal, then the angle between the line of sight and the horizontal is called the **angle of elevation**. If the object is below the horizontal, then the angle between the line of sight and the horizontal is called the **angle of depression**. In many of the examples and exercises in this chapter, angles of elevation and depression will be given for a hypothetical observer at ground level. If the line of sight follows a physical object, such as an inclined plane or a hillside, we use the term **angle of inclination**.

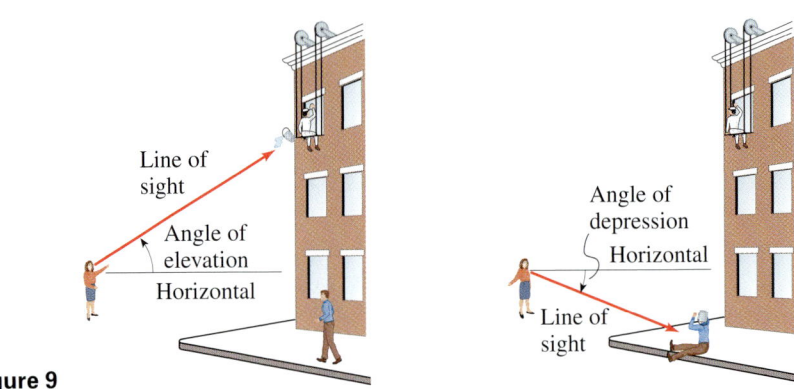

Figure 9

The next example gives an important application of trigonometry to the problem of measurement: We measure the height of a tall tree without having to climb it! Although the example is simple, the result is fundamental to understanding how the trigonometric ratios are applied to such problems.

Example 5   Finding the Height of a Tree

A giant redwood tree casts a shadow 532 ft long. Find the height of the tree if the angle of elevation of the sun is 25.7°.

Solution   Let the height of the tree be h. From Figure 10 we see that

$$\frac{h}{532} = \tan 25.7° \qquad \text{Definition of tangent}$$
$$h = 532 \tan 25.7° \qquad \text{Multiply by 532}$$
$$\approx 532(0.48127) \approx 256 \qquad \text{Use a calculator}$$

Therefore, the height of the tree is about 256 ft.

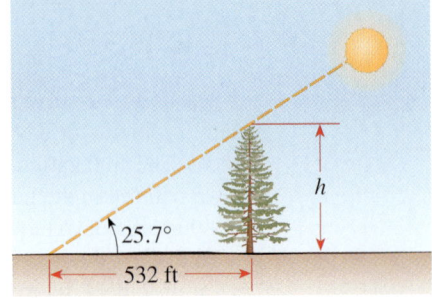

Figure 10

Example 6   A Problem Involving Right Triangles

From a point on the ground 500 ft from the base of a building, an observer finds that the angle of elevation to the top of the building is 24° and that the angle of elevation to the top of a flagpole atop the building is 27°. Find the height of the building and the length of the flagpole.

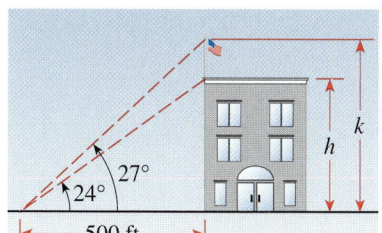

Figure 11

Solution   Figure 11 illustrates the situation. The height of the building is found in the same way that we found the height of the tree in Example 5.

$$\frac{h}{500} = \tan 24° \qquad \textit{Definition of tangent}$$

$$h = 500 \tan 24° \qquad \textit{Multiply by 500}$$

$$\approx 500(0.4452) \approx 223 \qquad \textit{Use a calculator}$$

The height of the building is approximately 223 ft.

To find the length of the flagpole, let's first find the height from the ground to the top of the pole:

$$\frac{k}{500} = \tan 27°$$

$$k = 500 \tan 27°$$

$$\approx 500(0.5095)$$

$$\approx 255$$

To find the length of the flagpole, we subtract h from k. So the length of the pole is approximately $255 - 223 = 32$ ft.   ∎

The key labels $\boxed{\text{SIN}^{-1}}$ or $\boxed{\text{INV}}$ $\boxed{\text{SIN}}$ stand for "inverse sine." We study the inverse trigonometric functions in Section 7.4.

In some problems we need to find an angle in a right triangle whose sides are given. To do this, we use Table 1 (page 480) "backward"; that is, we find the *angle* with the specified trigonometric ratio. For example, if $\sin \theta = \frac{1}{2}$, what is the angle θ? From Table 1 we can tell that $\theta = 30°$. To find an angle whose sine is not given in the table, we use the $\boxed{\text{SIN}^{-1}}$ or $\boxed{\text{INV}}$ $\boxed{\text{SIN}}$ or $\boxed{\text{ARCSIN}}$ keys on a calculator. For example, if $\sin \theta = 0.8$, we apply the $\boxed{\text{SIN}^{-1}}$ key to 0.8 to get $\theta = 53.13°$ or 0.927 rad. The calculator also gives angles whose cosine or tangent are known, using the $\boxed{\text{COS}^{-1}}$ or $\boxed{\text{TAN}^{-1}}$ key.

Example 7   Solving for an Angle in a Right Triangle

A 40-ft ladder leans against a building. If the base of the ladder is 6 ft from the base of the building, what is the angle formed by the ladder and the building?

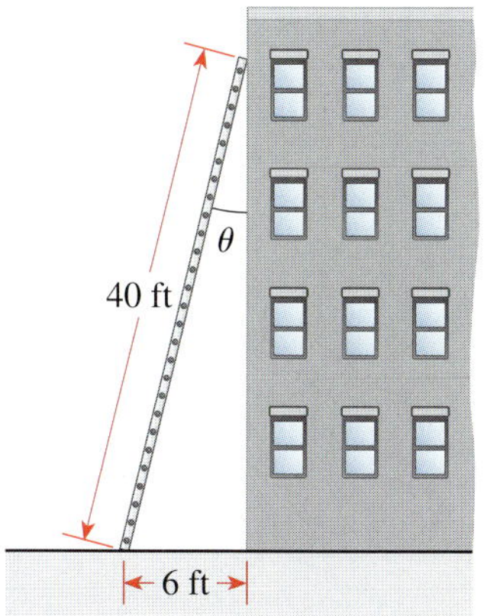

Solution   First we sketch a diagram as in Figure 12. If θ is the angle between the ladder and the building, then

$$\sin \theta = \frac{6}{40} = 0.15$$

So θ is the angle whose sine is 0.15. To find the angle θ, we use the $\boxed{\text{SIN}^{-1}}$ key on a calculator. With our calculator in degree mode, we get

$$\theta \approx 8.6°$$ ∎

Figure 12

6.2 Exercises

1–6 ■ Find the exact values of the six trigonometric ratios of the angle θ in the triangle.

1.
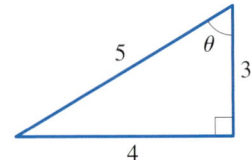

2.
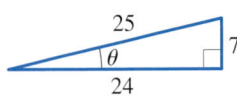

3.
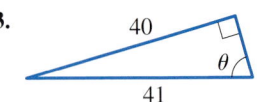

4.
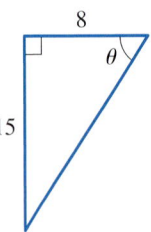

5.
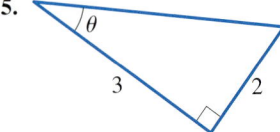

6.
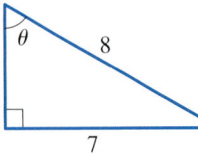

7–8 ■ Find (a) $\sin \alpha$ and $\cos \beta$, (b) $\tan \alpha$ and $\cot \beta$, and (c) $\sec \alpha$ and $\csc \beta$.

7.
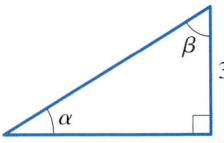

8.
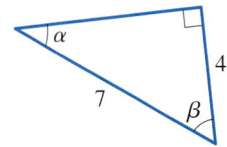

9–14 ■ Find the side labeled x. In Exercises 13 and 14 state your answer correct to five decimal places.

9.
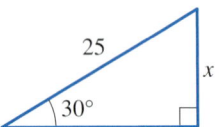

10.
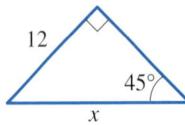

11.
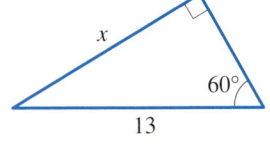

12.
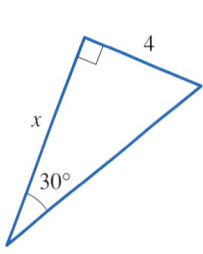

13.
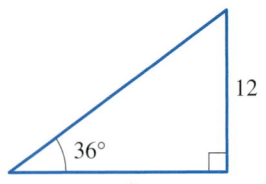

14.
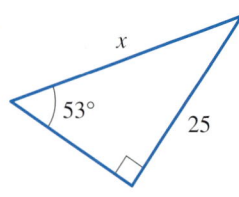

15–16 ■ Express x and y in terms of trigonometric ratios of θ.

15.
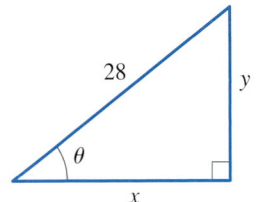

16.
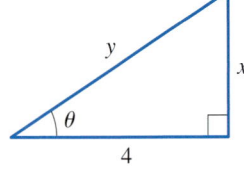

17–22 ■ Sketch a triangle that has acute angle θ, and find the other five trigonometric ratios of θ.

17. $\sin \theta = \frac{3}{5}$

18. $\cos \theta = \frac{9}{40}$

19. $\cot \theta = 1$

20. $\tan \theta = \sqrt{3}$

21. $\sec \theta = \frac{7}{2}$

22. $\csc \theta = \frac{13}{12}$

23–28 ■ Evaluate the expression without using a calculator.

23. $\sin \dfrac{\pi}{6} + \cos \dfrac{\pi}{6}$

24. $\sin 30° \csc 30°$

25. $\sin 30° \cos 60° + \sin 60° \cos 30°$

26. $(\sin 60°)^2 + (\cos 60°)^2$

27. $(\cos 30°)^2 - (\sin 30°)^2$

28. $\left(\sin \dfrac{\pi}{3} \cos \dfrac{\pi}{4} - \sin \dfrac{\pi}{4} \cos \dfrac{\pi}{3} \right)^2$

29–36 ■ Solve the right triangle.

29.
(triangle with 45°, side 16)

30.
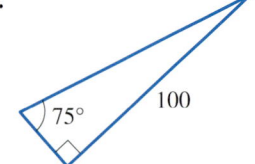

31.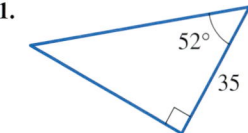

32.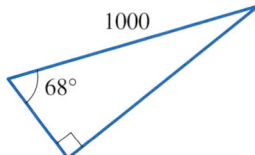

33.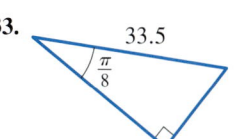

34.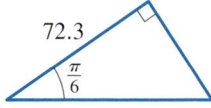

35.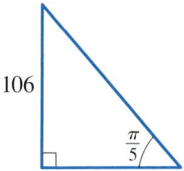

36.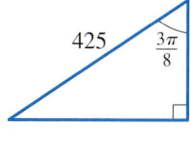

37. Use a ruler to carefully measure the sides of the triangle, and then use your measurements to estimate the six trigonometric ratios of θ.

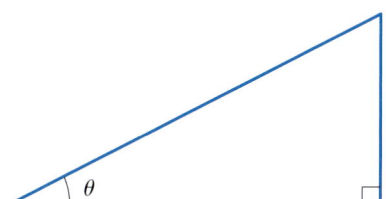

38. Using a protractor, sketch a right triangle that has the acute angle 40°. Measure the sides carefully, and use your results to estimate the six trigonometric ratios of 40°.

39–42 ■ Find x correct to one decimal place.

39.

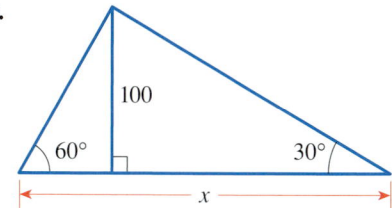

40.

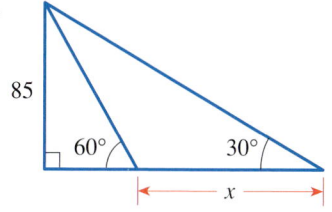

41.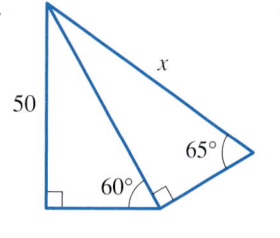

42.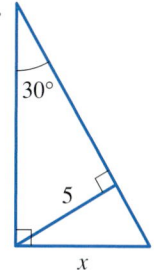

43. Express the length x in terms of the trigonometric ratios of θ.

44. Express the length a, b, c, and d in the figure in terms of the trigonometric ratios of θ.

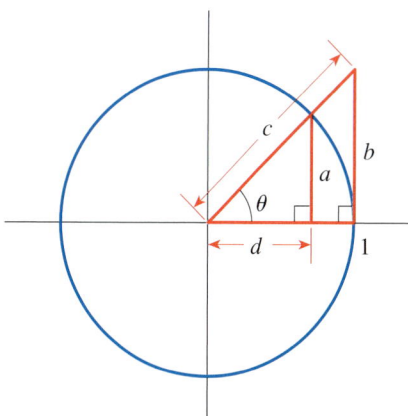

Applications

45. Height of a Building The angle of elevation to the top of the Empire State Building in New York is found to be 11° from the ground at a distance of 1 mi from the base of the building. Using this information, find the height of the Empire State Building.

46. Gateway Arch A plane is flying within sight of the Gateway Arch in St. Louis, Missouri, at an elevation of 35,000 ft. The pilot would like to estimate her distance from the Gateway Arch. She finds that the angle of depression to a point on the ground below the arch is 22°.

(a) What is the distance between the plane and the arch?

(b) What is the distance between a point on the ground directly below the plane and the arch?

47. Deviation of a Laser Beam A laser beam is to be directed toward the center of the moon, but the beam strays 0.5° from its intended path.

(a) How far has the beam diverged from its assigned target when it reaches the moon? (The distance from the earth to the moon is 240,000 mi.)

(b) The radius of the moon is about 1000 mi. Will the beam strike the moon?

48. Distance at Sea From the top of a 200-ft lighthouse, the angle of depression to a ship in the ocean is 23°. How far is the ship from the base of the lighthouse?

49. Leaning Ladder A 20-ft ladder leans against a building so that the angle between the ground and the ladder is 72°. How high does the ladder reach on the building?

50. Leaning Ladder A 20-ft ladder is leaning against a building. If the base of the ladder is 6 ft from the base of the building, what is the angle of elevation of the ladder? How high does the ladder reach on the building?

51. Angle of the Sun A 96-ft tree casts a shadow that is 120 ft long. What is the angle of elevation of the sun?

52. Height of a Tower A 600-ft guy wire is attached to the top of a communications tower. If the wire makes an angle of 65° with the ground, how tall is the communications tower?

53. Elevation of a Kite A man is lying on the beach, flying a kite. He holds the end of the kite string at ground level, and estimates the angle of elevation of the kite to be 50°. If the string is 450 ft long, how high is the kite above the ground?

54. Determining a Distance A woman standing on a hill sees a flagpole that she knows is 60 ft tall. The angle of depression to the bottom of the pole is 14°, and the angle of elevation to the top of the pole is 18°. Find her distance x from the pole.

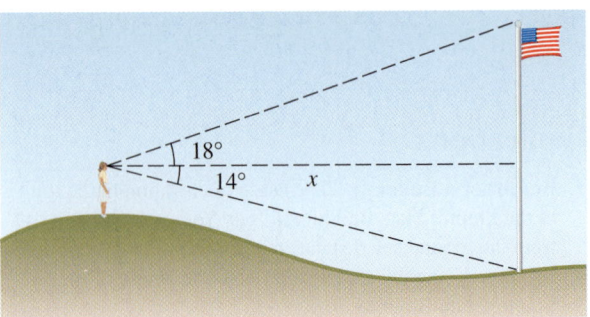

55. Height of a Tower A water tower is located 325 ft from a building (see the figure). From a window in the building, an observer notes that the angle of elevation to the top of the tower is 39° and that the angle of depression to the bottom of the tower is 25°. How tall is the tower? How high is the window?

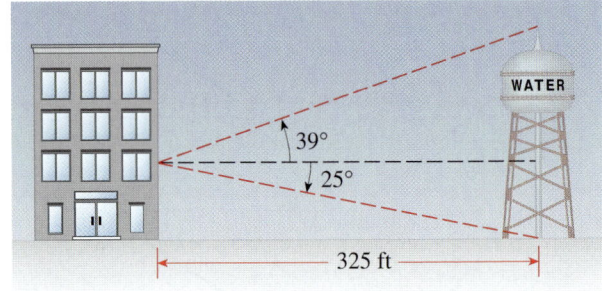

56. Determining a Distance An airplane is flying at an elevation of 5150 ft, directly above a straight highway. Two motorists are driving cars on the highway on opposite sides of the plane, and the angle of depression to one car is 35° and to the other is 52°. How far apart are the cars?

57. Determining a Distance If both cars in Exercise 56 are on one side of the plane and if the angle of depression to one car is 38° and to the other car is 52°, how far apart are the cars?

58. Height of a Balloon A hot-air balloon is floating above a straight road. To estimate their height above the ground, the balloonists simultaneously measure the angle of depression to two consecutive mileposts on the road on the same side of the balloon. The angles of depression are found to be 20° and 22°. How high is the balloon?

59. Height of a Mountain To estimate the height of a mountain above a level plain, the angle of elevation to the top of the mountain is measured to be 32°. One thousand feet closer to the mountain along the plain, it is found that the angle of elevation is 35°. Estimate the height of the mountain.

60. Height of Cloud Cover To measure the height of the cloud cover at an airport, a worker shines a spotlight upward at an angle 75° from the horizontal. An observer 600 m away measures the angle of elevation to the spot of light to be 45°. Find the height h of the cloud cover.

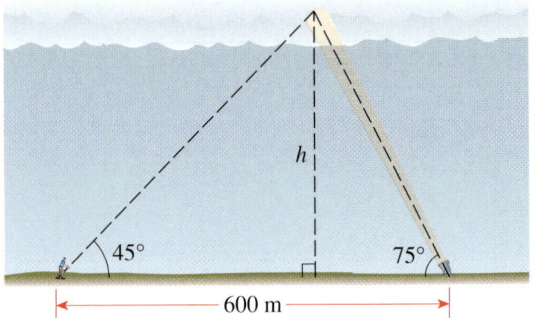

61. Distance to the Sun When the moon is exactly half full, the earth, moon, and sun form a right angle (see the figure). At that time the angle formed by the sun, earth, and moon is measured to be 89.85°. If the distance from the earth to the moon is 240,000 mi, estimate the distance from the earth to the sun.

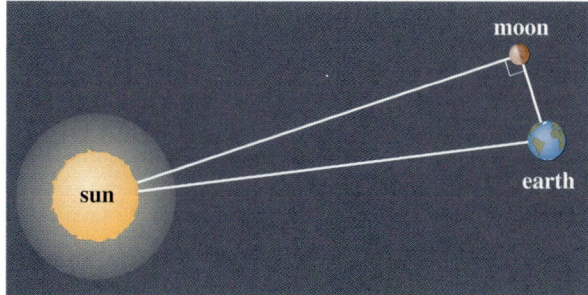

62. Distance to the Moon To find the distance to the sun as in Exercise 61, we needed to know the distance to the moon. Here is a way to estimate that distance: When the moon is seen at its zenith at a point A on the earth, it is observed to be at the horizon from point B (see the figure). Points A and B are 6155 mi apart, and the radius of the earth is 3960 mi.

(a) Find the angle θ in degrees.

(b) Estimate the distance from point A to the moon.

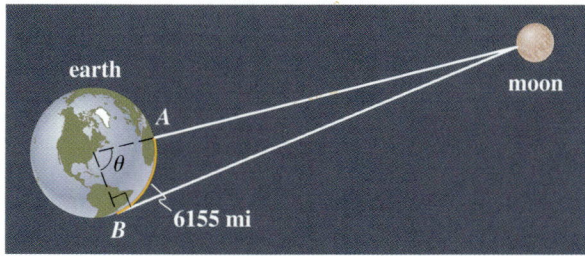

63. Radius of the Earth In Exercise 72 of Section 6.1 a method was given for finding the radius of the earth. Here is a more modern method: From a satellite 600 mi above the earth, it is observed that the angle formed by the vertical and the line of sight to the horizon is 60.276°. Use this information to find the radius of the earth.

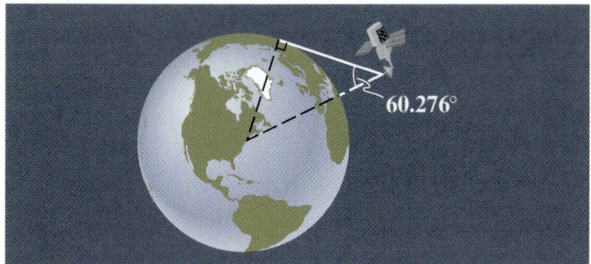

64. Parallax To find the distance to nearby stars, the method of parallax is used. The idea is to find a triangle with the star at one vertex and with a base as large as possible. To do this, the star is observed at two different times exactly 6 months apart, and its apparent change in position is recorded. From these two observations, $\angle E_1 S E_2$ can be calculated. (The times are chosen so that $\angle E_1 S E_2$ is as large as possible, which guarantees that $\angle E_1 O S$ is 90°.) The angle $E_1 S O$ is called the *parallax* of the star. Alpha Centauri, the star nearest the earth, has a parallax of 0.000211°. Estimate the distance to this star. (Take the distance from the earth to the sun to be 9.3×10^7 mi.)

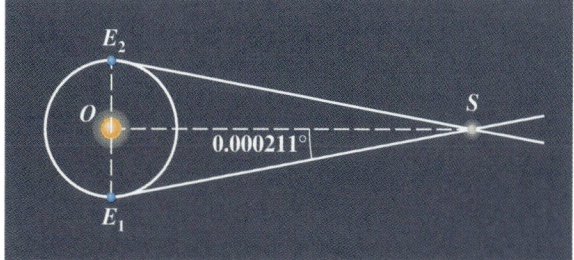

65. Distance from Venus to the Sun The **elongation** α of a planet is the angle formed by the planet, earth, and sun (see the figure). When Venus achieves its maximum elongation of 46.3°, the earth, Venus, and the sun form a triangle with a right angle at Venus. Find the distance between Venus and the sun in Astronomical Units (AU). (By definition, the distance between the earth and the sun is 1 AU.)

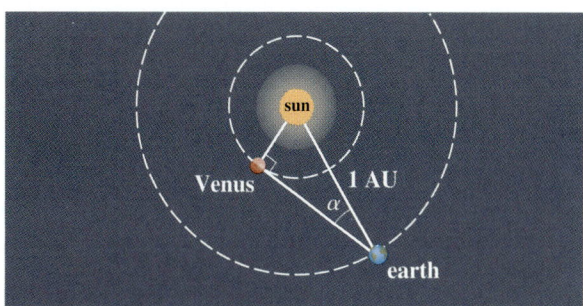

Discovery • Discussion

66. Similar Triangles If two triangles are similar, what properties do they share? Explain how these properties make it possible to define the trigonometric ratios without regard to the size of the triangle.

| 6.3 | **Trigonometric Functions of Angles** |

In the preceding section we defined the trigonometric ratios for acute angles. Here we extend the trigonometric ratios to all angles by defining the trigonometric functions of angles. With these functions we can solve practical problems that involve angles which are not necessarily acute.

Trigonometric Functions of Angles

Let POQ be a right triangle with acute angle θ as shown in Figure 1(a). Place θ in standard position as shown in Figure 1(b).

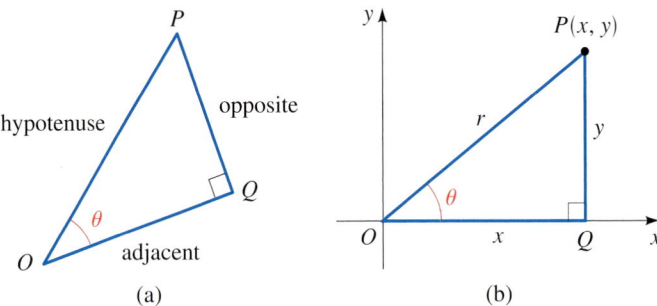

Figure 1   (a)   (b)

Then $P = P(x, y)$ is a point on the terminal side of θ. In triangle POQ, the opposite side has length y and the adjacent side has length x. Using the Pythagorean Theorem, we see that the hypotenuse has length $r = \sqrt{x^2 + y^2}$. So

$$\sin \theta = \frac{y}{r} \qquad \cos \theta = \frac{x}{r} \qquad \tan \theta = \frac{y}{x}$$

The other trigonometric ratios can be found in the same way.

These observations allow us to extend the trigonometric ratios to any angle. We define the trigonometric functions of angles as follows (see Figure 2).

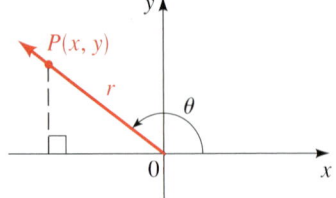

Figure 2

Definition of the Trigonometric Functions

Let θ be an angle in standard position and let $P(x, y)$ be a point on the terminal side. If $r = \sqrt{x^2 + y^2}$ is the distance from the origin to the point $P(x, y)$, then

$$\sin \theta = \frac{y}{r} \qquad\qquad \cos \theta = \frac{x}{r} \qquad\qquad \tan \theta = \frac{y}{x} \quad (x \neq 0)$$

$$\csc \theta = \frac{r}{y} \quad (y \neq 0) \qquad \sec \theta = \frac{r}{x} \quad (x \neq 0) \qquad \cot \theta = \frac{x}{y} \quad (y \neq 0)$$

Relationship to the Trigonometric Functions of Real Numbers

You may have already studied the trigonometric functions defined using the unit circle (Chapter 5). To see how they relate to the trigonometric functions of an *angle*, let's start with the unit circle in the coordinate plan.

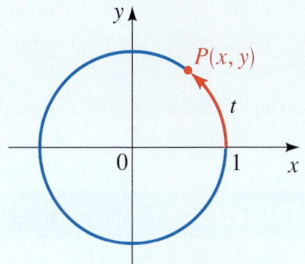

$P(x, y)$ is the terminal point determined by t.

Let $P(x, y)$ be the terminal point determined by an arc of length t on the unit circle. Then t subtends an angle θ at the center of the circle. If we drop a perpendicular from P onto the point Q on the x-axis, then triangle $\triangle OPQ$ is a right triangle with legs of length x and y, as shown in the figure.

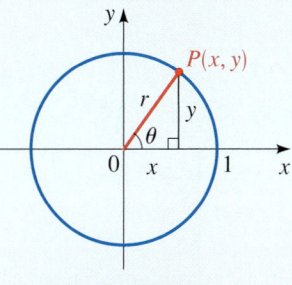

Triangle OPQ is a right triangle.

Now, by the definition of the trigonometric functions of the *real number t*, we have

$$\sin t = y$$
$$\cos t = x$$

By the definition of the trigonometric functions of the *angle* θ, we have

$$\sin \theta = \frac{\text{opp}}{\text{hyp}} = \frac{y}{1} = y$$

$$\cos \theta = \frac{\text{adj}}{\text{hyp}} = \frac{x}{1} = x$$

If θ is measured in radians, then $\theta = t$. (See the figure below.) Comparing the two ways of defining the trigonometric functions, we see that they are identical. In other words, as functions, they assign identical values to a given real number (the real number is the radian measure of θ in one case or the length t of an arc in the other).

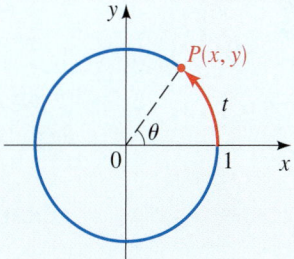

The radian measure of angle θ is t.

Why then do we study trigonometry in two different ways? Because different applications require that we view the trigonometric functions differently. (See *Focus on Modeling*, pages 459, 522, and 575, and Sections 6.2, 6.4, and 6.5.)

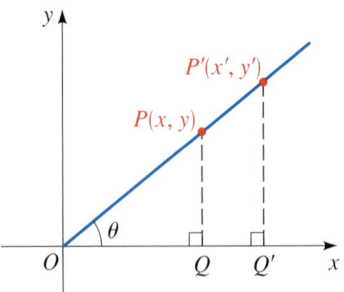

Figure 3

The following mnemonic device can be used to remember which trigonometric functions are positive in each quadrant: **A**ll of them, **S**ine, **T**angent, or **C**osine.

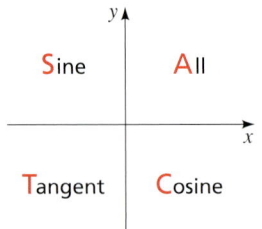

You can remember this as "All Students Take Calculus."

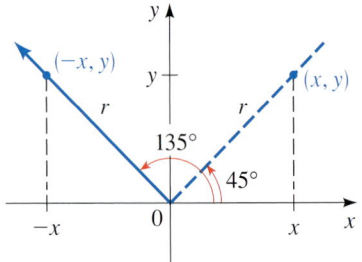

Figure 4

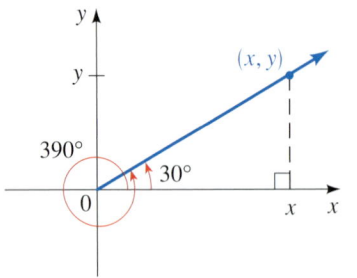

Figure 5

Since division by 0 is an undefined operation, certain trigonometric functions are not defined for certain angles. For example, $\tan 90° = y/x$ is undefined because $x = 0$. The angles for which the trigonometric functions may be undefined are the angles for which either the x- or y-coordinate of a point on the terminal side of the angle is 0. These are **quadrantal angles**—angles that are coterminal with the coordinate axes.

It is a crucial fact that the values of the trigonometric functions do *not* depend on the choice of the point $P(x, y)$. This is because if $P'(x', y')$ is any other point on the terminal side, as in Figure 3, then triangles POQ and $P'OQ'$ are similar.

Evaluating Trigonometric Functions at Any Angle

From the definition we see that the values of the trigonometric functions are all positive if the angle θ has its terminal side in quadrant I. This is because x and y are positive in this quadrant. [Of course, r is always positive, since it is simply the distance from the origin to the point $P(x, y)$.] If the terminal side of θ is in quadrant II, however, then x is negative and y is positive. Thus, in quadrant II the functions $\sin \theta$ and $\csc \theta$ are positive, and all the other trigonometric functions have negative values. You can check the other entries in the following table.

Signs of the Trigonometric Functions

Quadrant	Positive functions	Negative functions
I	all	none
II	sin, csc	cos, sec, tan, cot
III	tan, cot	sin, csc, cos, sec
IV	cos, sec	sin, csc, tan, cot

We now turn our attention to finding the values of the trigonometric functions for angles that are not acute.

Example 1   Finding Trigonometric Functions of Angles

Find (a) $\cos 135°$ and (b) $\tan 390°$.

Solution

(a) From Figure 4 we see that $\cos 135° = -x/r$. But $\cos 45° = x/r$, and since $\cos 45° = \sqrt{2}/2$, we have

$$\cos 135° = -\frac{\sqrt{2}}{2}$$

(b) The angles $390°$ and $30°$ are coterminal. From Figure 5 it's clear that $\tan 390° = \tan 30°$ and, since $\tan 30° = \sqrt{3}/3$, we have

$$\tan 390° = \frac{\sqrt{3}}{3}$$

∎

From Example 1 we see that the trigonometric functions for angles that aren't acute have the same value, except possibly for sign, as the corresponding trigonometric functions of an acute angle. That acute angle will be called the *reference angle*.

Reference Angle

Let θ be an angle in standard position. The **reference angle** $\bar{\theta}$ associated with θ is the acute angle formed by the terminal side of θ and the x-axis.

Figure 6 shows that to find a reference angle it's useful to know the quadrant in which the terminal side of the angle lies.

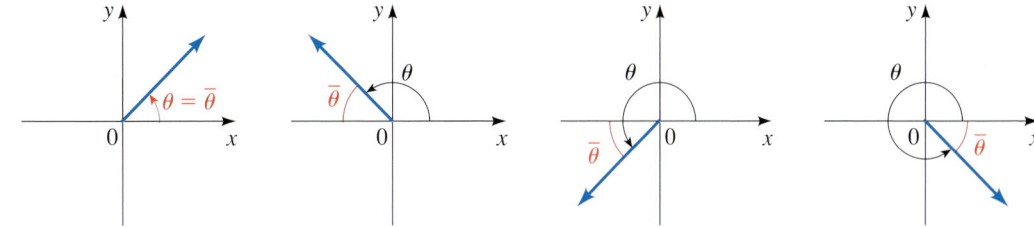

Figure 6

The reference angle $\bar{\theta}$ for an angle θ

Example 2   Finding Reference Angles

Find the reference angle for (a) $\theta = \dfrac{5\pi}{3}$ and (b) $\theta = 870°$.

Solution

(a) The reference angle is the acute angle formed by the terminal side of the angle $5\pi/3$ and the x-axis (see Figure 7). Since the terminal side of this angle is in quadrant IV, the reference angle is

$$\bar{\theta} = 2\pi - \frac{5\pi}{3} = \frac{\pi}{3}$$

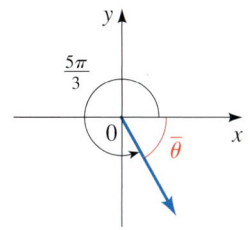

Figure 7

(b) The angles 870° and 150° are coterminal [because $870 - 2(360) = 150$]. Thus, the terminal side of this angle is in quadrant II (see Figure 8). So the reference angle is

$$\bar{\theta} = 180° - 150° = 30°$$

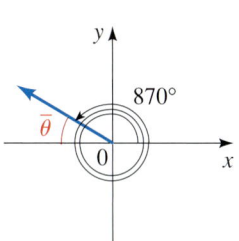

Figure 8

Evaluating Trigonometric Functions for Any Angle

To find the values of the trigonometric functions for any angle θ, we carry out the following steps.

1. Find the reference angle $\bar{\theta}$ associated with the angle θ.

2. Determine the sign of the trigonometric function of θ by noting the quadrant in which θ lies.

3. The value of the trigonometric function of θ is the same, except possibly for sign, as the value of the trigonometric function of $\bar{\theta}$.

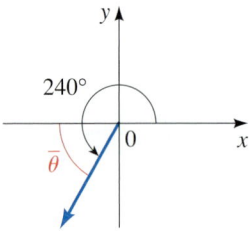

Figure 9

$\frac{S \mid A}{T \mid C}$ sin 240° is *negative*.

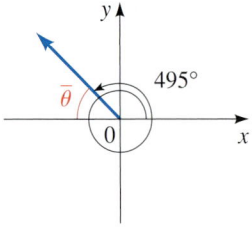

Figure 10

$\frac{S \mid A}{T \mid C}$ tan 495° *is negative,*
so cot 495° *is negative.*

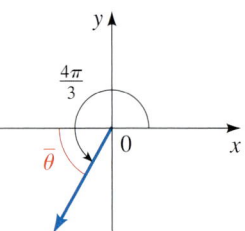

Figure 11

$\frac{S \mid A}{T \mid C}$ sin $\frac{16\pi}{3}$ *is negative.*

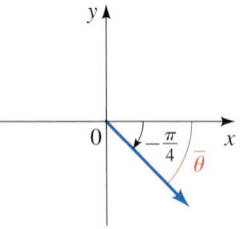

Figure 12

$\frac{S \mid A}{T \mid C}$ $\cos(-\frac{\pi}{4})$ *is positive,*
so $\sec(-\frac{\pi}{4})$ *is positive.*

Example 3 Using the Reference Angle to Evaluate Trigonometric Functions

Find (a) sin 240° and (b) cot 495°.

Solution

(a) This angle has its terminal side in quadrant III, as shown in Figure 9. The reference angle is therefore $240° - 180° = 60°$, and the value of sin 240° is negative. Thus

$$\sin 240° = -\sin 60° = -\frac{\sqrt{3}}{2}$$

Sign   Reference angle

(b) The angle 495° is coterminal with the angle 135°, and the terminal side of this angle is in quadrant II, as shown in Figure 10. So the reference angle is $180° - 135° = 45°$, and the value of cot 495° is negative. We have

$$\cot 495° = \cot 135° = -\cot 45° = -1$$

Coterminal angles   Sign   Reference angle

Example 4 Using the Reference Angle to Evaluate Trigonometric Functions

Find (a) $\sin \dfrac{16\pi}{3}$ and (b) $\sec\left(-\dfrac{\pi}{4}\right)$.

Solution

(a) The angle $16\pi/3$ is coterminal with $4\pi/3$, and these angles are in quadrant III (see Figure 11). Thus, the reference angle is $(4\pi/3) - \pi = \pi/3$. Since the value of sine is negative in quadrant III, we have

$$\sin \frac{16\pi}{3} = \sin \frac{4\pi}{3} = -\sin \frac{\pi}{3} = -\frac{\sqrt{3}}{2}$$

Coterminal angles   Sign   Reference angle

(b) The angle $-\pi/4$ is in quadrant IV, and its reference angle is $\pi/4$ (see Figure 12). Since secant is positive in this quadrant, we get

$$\sec\left(-\frac{\pi}{4}\right) = +\sec \frac{\pi}{4} = \frac{\sqrt{2}}{2}$$

Sign   Reference angle

Trigonometric Identities

The trigonometric functions of angles are related to each other through several important equations called **trigonometric identities**. We've already encountered the reciprocal identities. These identities continue to hold for any angle θ, provided both sides of the equation are defined. The Pythagorean identities are a consequence of the Pythagorean Theorem.*

Fundamental Identities

Reciprocal Identities

$$\csc \theta = \frac{1}{\sin \theta} \qquad \sec \theta = \frac{1}{\cos \theta} \qquad \cot \theta = \frac{1}{\tan \theta}$$

$$\tan \theta = \frac{\sin \theta}{\cos \theta} \qquad \cot \theta = \frac{\cos \theta}{\sin \theta}$$

Pythagorean Identities

$$\sin^2\theta + \cos^2\theta = 1 \qquad \tan^2\theta + 1 = \sec^2\theta \qquad 1 + \cot^2\theta = \csc^2\theta$$

■ **Proof** Let's prove the first Pythagorean identity. Using $x^2 + y^2 = r^2$ (the Pythagorean Theorem) in Figure 13, we have

$$\sin^2\theta + \cos^2\theta = \left(\frac{y}{r}\right)^2 + \left(\frac{x}{r}\right)^2 = \frac{x^2 + y^2}{r^2} = \frac{r^2}{r^2} = 1$$

Thus, $\sin^2\theta + \cos^2\theta = 1$. (Although the figure indicates an acute angle, you should check that the proof holds for all angles θ.) ■

See Exercises 59 and 60 for the proofs of the other two Pythagorean identities.

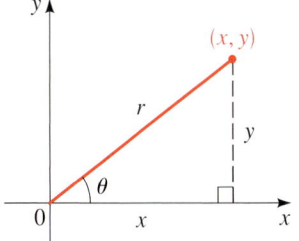

Figure 13

Example 5   Expressing One Trigonometric Function in Terms of Another

(a) Express $\sin \theta$ in terms of $\cos \theta$.

(b) Express $\tan \theta$ in terms of $\sin \theta$, where θ is in quadrant II.

Solution

(a) From the first Pythagorean identity we get

$$\sin \theta = \pm\sqrt{1 - \cos^2\theta}$$

where the sign depends on the quadrant. If θ is in quadrant I or II, then $\sin \theta$ is positive, and hence

$$\sin \theta = \sqrt{1 - \cos^2\theta}$$

whereas if θ is in quadrant III or IV, $\sin \theta$ is negative and so

$$\sin \theta = -\sqrt{1 - \cos^2\theta}$$

---

* We follow the usual convention of writing $\sin^2\theta$ for $(\sin \theta)^2$. In general, we write $\sin^n\theta$ for $(\sin \theta)^n$ for all integers n except $n = -1$. The exponent $n = -1$ will be assigned another meaning in Section 7.4. Of course, the same convention applies to the other five trigonometric functions.

(b) Since $\tan \theta = \sin \theta / \cos \theta$, we need to write $\cos \theta$ in terms of $\sin \theta$. By part (a)

$$\cos \theta = \pm \sqrt{1 - \sin^2\theta}$$

and since $\cos \theta$ is negative in quadrant II, the negative sign applies here. Thus

$$\tan \theta = \frac{\sin \theta}{\cos \theta} = \frac{\sin \theta}{-\sqrt{1 - \sin^2\theta}} \qquad \blacksquare$$

Example 6   Evaluating a Trigonometric Function

If $\tan \theta = \frac{2}{3}$ and θ is in quadrant III, find $\cos \theta$.

Solution 1   We need to write $\cos \theta$ in terms of $\tan \theta$. From the identity $\tan^2\theta + 1 = \sec^2\theta$, we get $\sec \theta = \pm\sqrt{\tan^2\theta + 1}$. In quadrant III, $\sec \theta$ is negative, so

$$\sec \theta = -\sqrt{\tan^2\theta + 1}$$

Thus
$$\cos \theta = \frac{1}{\sec \theta} = \frac{1}{-\sqrt{\tan^2\theta + 1}}$$

If you wish to rationalize the denominator, you can express $\cos \theta$ as

$$-\frac{3}{\sqrt{13}} \cdot \frac{\sqrt{13}}{\sqrt{13}} = -\frac{3\sqrt{3}}{13}$$

$$= \frac{1}{-\sqrt{\left(\frac{2}{3}\right)^2 + 1}} = \frac{1}{-\sqrt{\frac{13}{9}}} = -\frac{3}{\sqrt{13}}$$

Solution 2   This problem can be solved more easily using the method of Example 2 of Section 6.2. Recall that, except for sign, the values of the trigonometric functions of any angle are the same as those of an acute angle (the reference angle). So, ignoring the sign for the moment, let's sketch a right triangle with an acute angle $\bar{\theta}$ satisfying $\tan \bar{\theta} = \frac{2}{3}$ (see Figure 14). By the Pythagorean Theorem the hypotenuse of this triangle has length $\sqrt{13}$. From the triangle in Figure 14 we immediately see that $\cos \bar{\theta} = 3/\sqrt{13}$. Since θ is in quadrant III, $\cos \theta$ is negative and so

$$\cos \theta = -\frac{3}{\sqrt{13}} \qquad \blacksquare$$

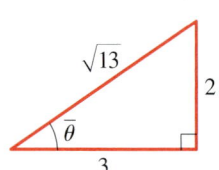

Figure 14

Example 7   Evaluating Trigonometric Functions

If $\sec \theta = 2$ and θ is in quadrant IV, find the other five trigonometric functions of θ.

Solution   We sketch a triangle as in Figure 15 so that $\sec \bar{\theta} = 2$. Taking into account the fact that θ is in quadrant IV, we get

$$\sin \theta = -\frac{\sqrt{3}}{2} \qquad \cos \theta = \frac{1}{2} \qquad \tan \theta = -\sqrt{3}$$

$$\csc \theta = -\frac{2}{\sqrt{3}} \qquad \sec \theta = 2 \qquad \cot \theta = -\frac{1}{\sqrt{3}} \qquad \blacksquare$$

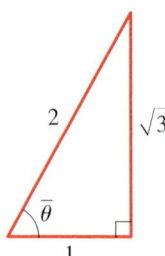

Figure 15

Areas of Triangles

We conclude this section with an application of the trigonometric functions that involves angles that are not necessarily acute. More extensive applications appear in the next two sections.

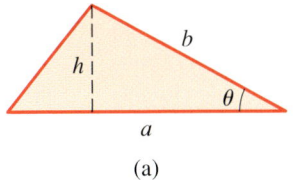

(a)

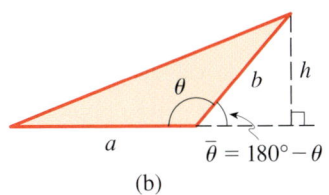

(b)

Figure 16

The area of a triangle is $\mathcal{A} = \frac{1}{2} \times$ base $\times$ height. If we know two sides and the included angle of a triangle, then we can find the height using the trigonometric functions, and from this we can find the area.

If θ is an acute angle, then the height of the triangle in Figure 16(a) is given by $h = b \sin \theta$. Thus, the area is

$$\mathcal{A} = \tfrac{1}{2} \times \text{base} \times \text{height} = \tfrac{1}{2}ab \sin \theta$$

If the angle θ is not acute, then from Figure 16(b) we see that the height of the triangle is

$$h = b \sin(180° - \theta) = b \sin \theta$$

This is so because the reference angle of θ is the angle $180° - \theta$. Thus, in this case also, the area of the triangle is

$$\mathcal{A} = \tfrac{1}{2} \times \text{base} \times \text{height} = \tfrac{1}{2}ab \sin \theta$$

Area of a Triangle

The area $\mathcal{A}$ of a triangle with sides of lengths a and b and with included angle θ is

$$\mathcal{A} = \tfrac{1}{2}ab \sin \theta$$

Example 8 Finding the Area of a Triangle

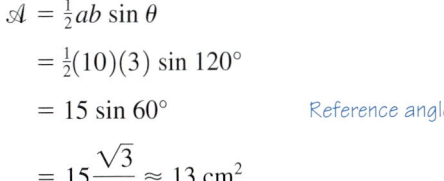

Figure 17

Find the area of triangle ABC shown in Figure 17.

Solution   The triangle has sides of length 10 cm and 3 cm, with included angle 120°. Therefore

$$\mathcal{A} = \tfrac{1}{2}ab \sin \theta$$
$$= \tfrac{1}{2}(10)(3) \sin 120°$$
$$= 15 \sin 60° \qquad \textit{Reference angle}$$
$$= 15\frac{\sqrt{3}}{2} \approx 13 \text{ cm}^2 \qquad ■$$

6.3   Exercises

1–8 ■ Find the reference angle for the given angle.

1. (a) 150°   **(b)** 330°   **(c)** −30°

2. (a) 120°   **(b)** −210°   **(c)** 780°

3. (a) 225°   **(b)** 810°   **(c)** −105°

4. (a) 99°   **(b)** −199°   **(c)** 359°

5. (a) $\dfrac{11\pi}{4}$ **(b)** $-\dfrac{11\pi}{6}$ **(c)** $\dfrac{11\pi}{3}$

6. (a) $\dfrac{4\pi}{3}$ **(b)** $\dfrac{33\pi}{4}$ **(c)** $-\dfrac{23\pi}{6}$

7. (a) $\dfrac{5\pi}{7}$ **(b)** -1.4π **(c)** 1.4

8. (a) 2.3π **(b)** 2.3   **(c)** -10π

9–32 ■ Find the exact value of the trigonometric function.

9. $\sin 150°$ **10.** $\sin 225°$ **11.** $\cos 135°$

12. $\cos(-60°)$ **13.** $\tan(-60°)$ **14.** $\sec 300°$

15. $\csc(-630°)$ **16.** $\cot 210°$ **17.** $\cos 570°$

18. $\sec 120°$ **19.** $\tan 750°$ **20.** $\cos 660°$

21. $\sin \dfrac{2\pi}{3}$ **22.** $\sin \dfrac{5\pi}{3}$ **23.** $\sin \dfrac{3\pi}{2}$

24. $\cos \dfrac{7\pi}{3}$ **25.** $\cos\left(-\dfrac{7\pi}{3}\right)$ **26.** $\tan \dfrac{5\pi}{6}$

27. $\sec \dfrac{17\pi}{3}$ **28.** $\csc \dfrac{5\pi}{4}$ **29.** $\cot\left(-\dfrac{\pi}{4}\right)$

30. $\cos \dfrac{7\pi}{4}$ **31.** $\tan \dfrac{5\pi}{2}$ **32.** $\sin \dfrac{11\pi}{6}$

33–36 ■ Find the quadrant in which θ lies from the information given.

33. $\sin \theta < 0$ and $\cos \theta < 0$

34. $\tan \theta < 0$ and $\sin \theta < 0$

35. $\sec \theta > 0$ and $\tan \theta < 0$

36. $\csc \theta > 0$ and $\cos \theta < 0$

37–42 ■ Write the first trigonometric function in terms of the second for θ in the given quadrant.

37. $\tan \theta$, $\cos \theta$; θ in quadrant III

38. $\cot \theta$, $\sin \theta$; θ in quadrant II

39. $\cos \theta$, $\sin \theta$; θ in quadrant IV

40. $\sec \theta$, $\sin \theta$; θ in quadrant I

41. $\sec \theta$, $\tan \theta$; θ in quadrant II

42. $\csc \theta$, $\cot \theta$; θ in quadrant III

43–50 ■ Find the values of the trigonometric functions of θ from the information given.

43. $\sin \theta = \frac{3}{5}$, θ in quadrant II

44. $\cos \theta = -\frac{7}{12}$, θ in quadrant III

45. $\tan \theta = -\frac{3}{4}$, $\cos \theta > 0$

46. $\sec \theta = 5$, $\sin \theta < 0$

47. $\csc \theta = 2$, θ in quadrant I

48. $\cot \theta = \frac{1}{4}$, $\sin \theta < 0$

49. $\cos \theta = -\frac{2}{7}$, $\tan \theta < 0$

50. $\tan \theta = -4$, $\sin \theta > 0$

51. If $\theta = \pi/3$, find the value of each expression.
(a) $\sin 2\theta$, $2 \sin \theta$ (b) $\sin \frac{1}{2}\theta$, $\frac{1}{2}\sin \theta$
(c) $\sin^2\theta$, $\sin(\theta^2)$

52. Find the area of a triangle with sides of length 7 and 9 and included angle $72°$.

53. Find the area of a triangle with sides of length 10 and 22 and included angle $10°$.

54. Find the area of an equilateral triangle with side of length 10.

55. A triangle has an area of 16 in^2, and two of the sides of the triangle have lengths 5 in. and 7 in. Find the angle included by these two sides.

56. An isosceles triangle has an area of 24 cm^2, and the angle between the two equal sides is $5\pi/6$. What is the length of the two equal sides?

57–58 ■ Find the area of the shaded region in the figure.

57.   **58.**

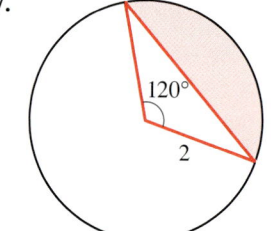

59. Use the first Pythagorean identity to prove the second. [*Hint:* Divide by $\cos^2\theta$.]

60. Use the first Pythagorean identity to prove the third.

Applications

61. Height of a Rocket   A rocket fired straight up is tracked by an observer on the ground a mile away.

(a) Show that when the angle of elevation is θ, the height of the rocket in feet is $h = 5280 \tan \theta$.

(b) Complete the table to find the height of the rocket at the given angles of elevation.

θ	20°	60°	80°	85°
h				

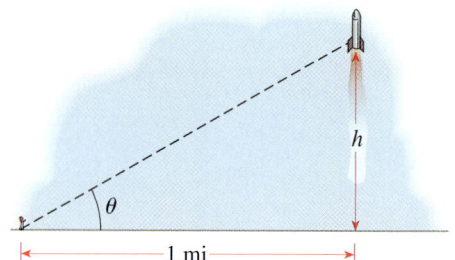

62. Rain Gutter A rain gutter is to be constructed from a metal sheet of width 30 cm by bending up one-third of the sheet on each side through an angle θ.

(a) Show that the cross-sectional area of the gutter is modeled by the function

$$A(\theta) = 100 \sin \theta + 100 \sin \theta \cos \theta$$

(b) Graph the function A for $0 \le \theta \le \pi/2$.

(c) For what angle θ is the largest cross-sectional area achieved?

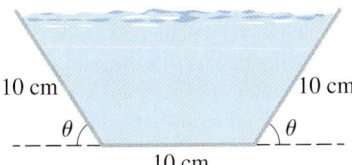

63. Wooden Beam A rectangular beam is to be cut from a cylindrical log of diameter 20 cm. The figures show different ways this can be done.

(a) Express the cross-sectional area of the beam as a function of the angle θ in the figures.

(b) Graph the function you found in part (a).

(c) Find the dimensions of the beam with largest cross-sectional area.

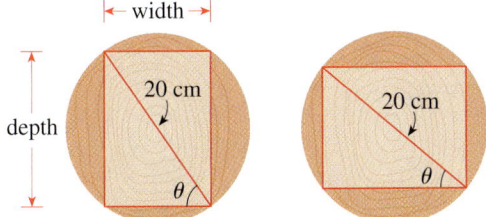

64. Strength of a Beam The strength of a beam is proportional to the width and the square of the depth. A beam is cut from a log as in Exercise 63. Express the strength of the beam as a function of the angle θ in the figures.

65. Throwing a Shot Put The range R and height H of a shot put thrown with an initial velocity of v_0 ft/s at an angle θ are given by

$$R = \frac{v_0^2 \sin(2\theta)}{g}$$

$$H = \frac{v_0^2 \sin^2\theta}{2g}$$

On the earth $g = 32$ ft/s^2 and on the moon $g = 5.2$ ft/s^2.

Find the range and height of a shot put thrown under the given conditions.

(a) On the earth with $v_0 = 12$ ft/s and $\theta = \pi/6$

(b) On the moon with $v_0 = 12$ ft/s and $\theta = \pi/6$

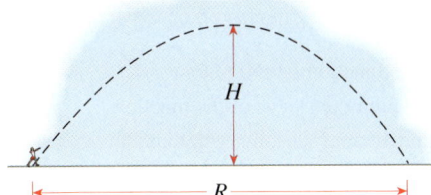

66. Sledding The time in seconds that it takes for a sled to slide down a hillside inclined at an angle θ is

$$t = \sqrt{\frac{d}{16 \sin \theta}}$$

where d is the length of the slope in feet. Find the time it takes to slide down a 2000-ft slope inclined at 30°.

67. Beehives In a beehive each cell is a regular hexagonal prism, as shown in the figure. The amount of wax W in the cell depends on the apex angle θ and is given by

$$W = 3.02 - 0.38 \cot \theta + 0.65 \csc \theta$$

Bees instinctively choose θ so as to use the least amount of wax possible.

(a) Use a graphing device to graph W as a function of θ for $0 < \theta < \pi$.

(b) For what value of θ does W have its minimum value? [*Note*: Biologists have discovered that bees rarely deviate from this value by more than a degree or two.]

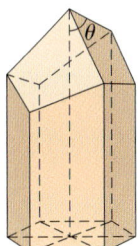

68. Turning a Corner A steel pipe is being carried down a hallway 9 ft wide. At the end of the hall there is a right-angled turn into a narrower hallway 6 ft wide.

(a) Show that the length of the pipe in the figure is modeled by the function

$$L(\theta) = 9 \csc \theta + 6 \sec \theta$$

(b) Graph the function L for $0 < \theta < \pi/2$.

(c) Find the minimum value of the function L.

(d) Explain why the value of L you found in part (c) is the length of the longest pipe that can be carried around the corner.

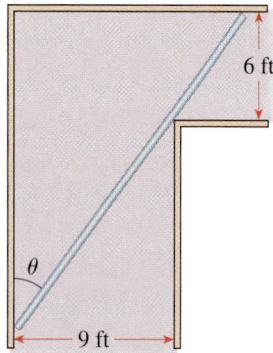

69. Rainbows Rainbows are created when sunlight of different wavelengths (colors) is refracted and reflected in raindrops. The angle of elevation θ of a rainbow is always the same. It can be shown that $\theta = 4\beta - 2\alpha$ where

$$\sin \alpha = k \sin \beta$$

and $\alpha = 59.4°$ and $k = 1.33$ is the index of refraction of water. Use the given information to find the angle of elevation θ of a rainbow. (For a mathematical explanation of rainbows see *Calculus*, 5th Edition, by James Stewart, pages 288–289.)

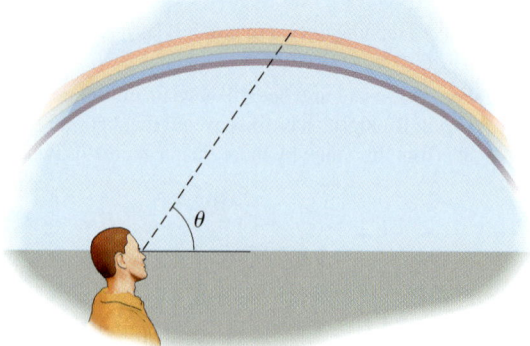

Discovery • Discussion

70. Using a Calculator To solve a certain problem, you need to find the sine of 4 rad. Your study partner uses his calculator and tells you that

$$\sin 4 = 0.0697564737$$

On your calculator you get

$$\sin 4 = -0.7568024953$$

What is wrong? What mistake did your partner make?

71. Viète's Trigonometric Diagram In the 16th century, the French mathematician François Viète (see page 49) published the following remarkable diagram. Each of the six trigonometric functions of θ is equal to the length of a line segment in the figure. For instance, $\sin \theta = |PR|$, since from $\triangle OPR$ we see that

$$\sin \theta = \frac{\text{opp}}{\text{hyp}}$$

$$= \frac{|PR|}{|OR|}$$

$$= \frac{|PR|}{1}$$

$$= |PR|$$

For each of the five other trigonometric functions, find a line segment in the figure whose length equals the value of the function at θ. (*Note:* The radius of the circle is 1, the center is O, segment QS is tangent to the circle at R, and $\angle SOQ$ is a right angle.)

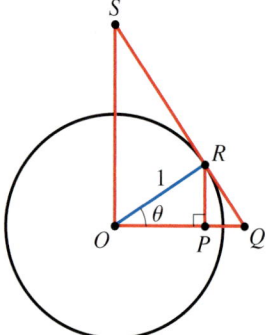

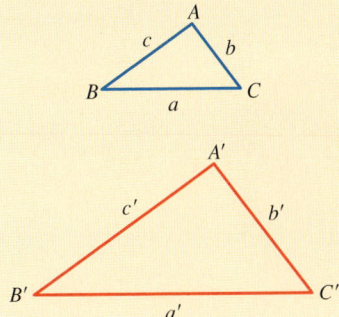

Thales used similar triangles to find the height of a tall column. (See page 482.)

Similarity

In geometry you learned that two triangles are similar if they have the same angles. In this case, the ratios of corresponding sides are equal. Triangles ABC and $A'B'C'$ in the margin are similar, so

$$\frac{a'}{a} = \frac{b'}{b} = \frac{c'}{c}$$

Similarity is the crucial idea underlying trigonometry. We can define $\sin\theta$ as the ratio of the opposite side to the hypotenuse in *any* right triangle with an angle θ, because all such right triangles are similar. So the ratio represented by $\sin\theta$ does not depend on the size of the right triangle but only on the angle θ. This is a powerful idea because angles are often easier to measure than distances. For example, the angle formed by the sun, earth, and moon can be measured from the earth. The secret to finding the distance to the sun is that the trigonometric ratios are the same for the huge triangle formed by the sun, earth, and moon as for any other similar triangle (see Exercise 61 in Section 6.2).

In general, two objects are **similar** if they have the same shape even though they may not be the same size.* For example, we recognize the following as representations of the letter A because they are all similar.

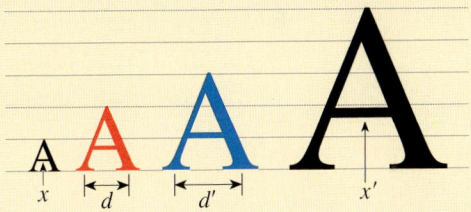

If two figures are similar, then the distances between corresponding points in the figures are proportional. The blue and red A's above are similar—the ratio of distances between corresponding points is $\frac{3}{2}$. We say that the **similarity ratio** is $s = \frac{3}{2}$. To obtain the distance d' between any two points in the blue A, we multiply the corresponding distance d in the red A by $\frac{3}{2}$. So

$$d' = sd \quad \text{or} \quad d' = \tfrac{3}{2}d$$

Likewise, the similarity ratio between the first and last letters is $s = 5$, so $x' = 5x$.

1. Write a short paragraph explaining how the concept of similarity is used to define the trigonometric ratios.

2. How is similarity used in map making? How are distances on a city road map related to actual distances?

3. How is your yearbook photograph similar to you? Compare distances between different points on your face (such as distance between ears, length of

---

* If they have the same shape *and* size, they are congruent, which is a special case of similarity.

nose, distance between eyes, and so on) to the corresponding distances in a photograph. What is the similarity ratio?

4. The figure illustrates a method for drawing an apple twice the size of a given apple. Use the method to draw a tie 3 times the size (similarity ratio 3) of the blue tie.

5. Give conditions under which two rectangles are similar to each other. Do the same for two isosceles triangles.

6. Suppose that two similar triangles have similarity ratio s.
   (a) How are the perimeters of the triangles related?
   (b) How are the areas of the triangles related?

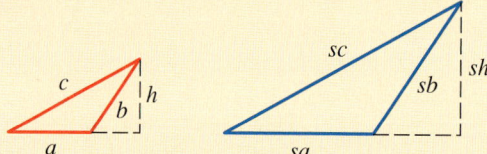

7. (a) If two squares have similarity ratio s, show that their areas A_1 and A_2 have the property that $A_2 = s^2 A_1$.
   (b) If the side of a square is tripled, its area is multiplied by what factor?
   (c) A plane figure can be approximated by squares (as shown). Explain how we can conclude that for any two plane figures with similarity ratio s, their areas satisfy $A_2 = s^2 A_1$. (Use part (a).)

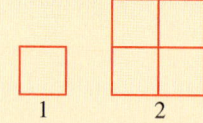

If the side of a square is doubled, its area is multiplied by 2^2.

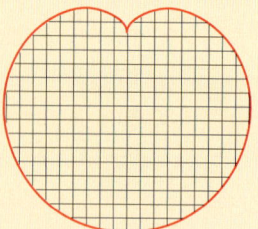

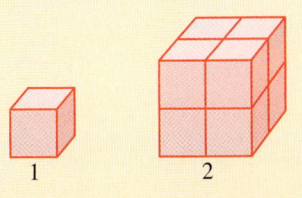

If the side of a cube is doubled, its volume is multiplied by 2^3.

8. (a) If two cubes have similarity ratio s, show that their volumes V_1 and V_2 have the property that $V_2 = s^3 V_1$.
(b) If the side of a cube is multiplied by 10, by what factor is the volume multiplied?
(c) How can we use the fact that a solid object can be "filled" by little cubes to show that for any two solids with similarity ratio s, the volumes satisfy $V_2 = s^3 V_1$?

9. King Kong is 10 times as tall as Joe, a normal-sized 300-lb gorilla. Assuming that King Kong and Joe are similar, use the results from Problems 7 and 8 to answer the following questions.
(a) How much does King Kong weigh?
(b) If Joe's hand is 13 in. long, how long is King Kong's hand?
(c) If it takes 2 square yards of material to make a shirt for Joe, how much material would a shirt for King Kong require?

6.4 The Law of Sines

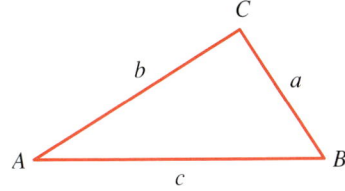

Figure 1

In Section 6.2 we used the trigonometric ratios to solve right triangles. The trigonometric functions can also be used to solve *oblique triangles*, that is, triangles with no right angles. To do this, we first study the Law of Sines here and then the Law of Cosines in the next section. To state these laws (or formulas) more easily, we follow the convention of labeling the angles of a triangle as A, B, C, and the lengths of the corresponding opposite sides as a, b, c, as in Figure 1.

To solve a triangle, we need to know certain information about its sides and angles. To decide whether we have enough information, it's often helpful to make a sketch. For instance, if we are given two angles and the included side, then it's clear that one and only one triangle can be formed (see Figure 2(a)). Similarly, if two sides and the included angle are known, then a unique triangle is determined (Figure 2(c)). But if we know all three angles and no sides, we cannot uniquely determine the triangle because many triangles can have the same three angles. (All these triangles would be similar, of course.) So we won't consider this last case.

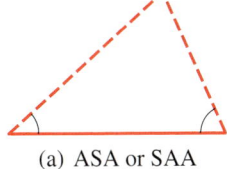
(a) ASA or SAA

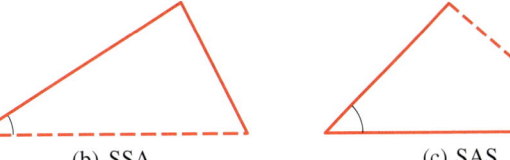
(b) SSA

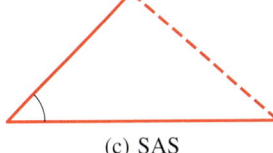
(c) SAS

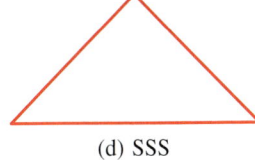
(d) SSS

Figure 2

In general, a triangle is determined by three of its six parts (angles and sides) as long as at least one of these three parts is a side. So, the possibilities, illustrated in Figure 2, are as follows.

Case 1 One side and two angles (ASA or SAA)

Case 2 Two sides and the angle opposite one of those sides (SSA)

Case 3 Two sides and the included angle (SAS)

Case 4 Three sides (SSS)

Cases 1 and 2 are solved using the Law of Sines; Cases 3 and 4 require the Law of Cosines.

The Law of Sines

The **Law of Sines** says that in any triangle the lengths of the sides are proportional to the sines of the corresponding opposite angles.

The Law of Sines

In triangle ABC we have

$$\frac{\sin A}{a} = \frac{\sin B}{b} = \frac{\sin C}{c}$$

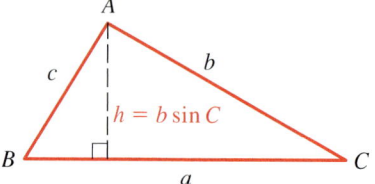

Figure 3

■ **Proof** To see why the Law of Sines is true, refer to Figure 3. By the formula in Section 6.3 the area of triangle ABC is $\frac{1}{2}ab \sin C$. By the same formula the area of this triangle is also $\frac{1}{2}ac \sin B$ and $\frac{1}{2}bc \sin A$. Thus

$$\tfrac{1}{2}bc \sin A = \tfrac{1}{2}ac \sin B = \tfrac{1}{2}ab \sin C$$

Multiplying by $2/(abc)$ gives the Law of Sines. ■

Example 1 Tracking a Satellite (ASA)

A satellite orbiting the earth passes directly overhead at observation stations in Phoenix and Los Angeles, 340 mi apart. At an instant when the satellite is between these two stations, its angle of elevation is simultaneously observed to be 60° at Phoenix and 75° at Los Angeles. How far is the satellite from Los Angeles? In other words, find the distance AC in Figure 4.

Solution Whenever two angles in a triangle are known, the third angle can be determined immediately because the sum of the angles of a triangle is 180°. In this case, $\angle C = 180° - (75° + 60°) = 45°$ (see Figure 4), so we have

$$\frac{\sin B}{b} = \frac{\sin C}{c} \qquad \textit{Law of Sines}$$

$$\frac{\sin 60°}{b} = \frac{\sin 45°}{340} \qquad \textit{Substitute}$$

$$b = \frac{340 \sin 60°}{\sin 45°} \approx 416 \qquad \textit{Solve for } b$$

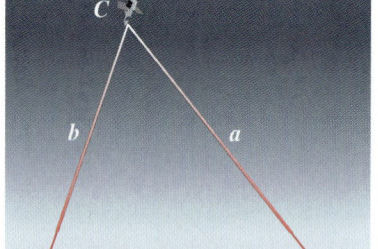

Figure 4

The distance of the satellite from Los Angeles is approximately 416 mi. ■

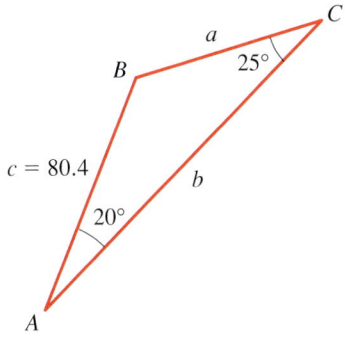

Figure 5

Example 2   Solving a Triangle (SAA)

Solve the triangle in Figure 5.

Solution   First, $\angle B = 180° - (20° + 25°) = 135°$. Since side c is known, to find side a we use the relation

$$\frac{\sin A}{a} = \frac{\sin C}{c} \qquad \text{Law of Sines}$$

$$a = \frac{c \sin A}{\sin C} = \frac{80.4 \sin 20°}{\sin 25°} \approx 65.1 \qquad \text{Solve for } a$$

Similarly, to find b we use

$$\frac{\sin B}{b} = \frac{\sin C}{c} \qquad \text{Law of Sines}$$

$$b = \frac{c \sin B}{\sin C} = \frac{80.4 \sin 135°}{\sin 25°} \approx 134.5 \qquad \text{Solve for } b \qquad ■$$

The Ambiguous Case

In Examples 1 and 2 a unique triangle was determined by the information given. This is always true of Case 1 (ASA or SAA). But in Case 2 (SSA) there may be two triangles, one triangle, or no triangle with the given properties. For this reason, Case 2 is sometimes called the **ambiguous case**. To see why this is so, we show in Figure 6 the possibilities when angle A and sides a and b are given. In part (a) no solution is possible, since side a is too short to complete the triangle. In part (b) the solution is a right triangle. In part (c) two solutions are possible, and in part (d) there is a unique triangle with the given properties. We illustrate the possibilities of Case 2 in the following examples.

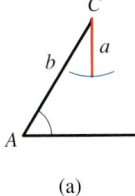 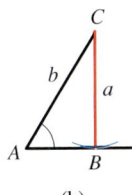 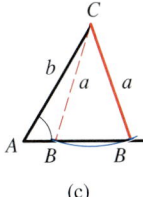 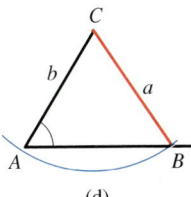

Figure 6
The ambiguous case   (a)   (b)   (c)   (d)

Example 3   SSA, the One-Solution Case

Solve triangle ABC, where $\angle A = 45°$, $a = 7\sqrt{2}$, and $b = 7$.

Solution   We first sketch the triangle with the information we have (see Figure 7). Our sketch is necessarily tentative, since we don't yet know the other angles. Nevertheless, we can now see the possibilities.
    We first find $\angle B$.

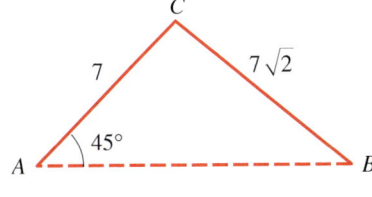

Figure 7

$$\frac{\sin A}{a} = \frac{\sin B}{b} \qquad \text{Law of Sines}$$

$$\sin B = \frac{b \sin A}{a} = \frac{7}{7\sqrt{2}} \sin 45° = \left(\frac{1}{\sqrt{2}}\right)\left(\frac{\sqrt{2}}{2}\right) = \frac{1}{2} \qquad \text{Solve for } \sin B$$

We consider only angles smaller than 180°, since no triangle can contain an angle of 180° or larger.

Which angles B have $\sin B = \frac{1}{2}$? From the preceding section we know that there are two such angles smaller than 180° (they are 30° and 150°). Which of these angles is compatible with what we know about triangle ABC? Since $\angle A = 45°$, we cannot have $\angle B = 150°$, because $45° + 150° > 180°$. So $\angle B = 30°$, and the remaining angle is $\angle C = 180° - (30° + 45°) = 105°$.

Now we can find side c.

$$\frac{\sin B}{b} = \frac{\sin C}{c} \qquad \text{\textit{Law of Sines}}$$

$$c = \frac{b \sin C}{\sin B} = \frac{7 \sin 105°}{\sin 30°} = \frac{7 \sin 105°}{\frac{1}{2}} \approx 13.5 \qquad \text{\textit{Solve for } c} \qquad \blacksquare$$

 In Example 3 there were two possibilities for angle B, and one of these was not compatible with the rest of the information. In general, if $\sin A < 1$, we must check the angle and its supplement as possibilities, because any angle smaller than 180° can be in the triangle. To decide whether either possibility works, we check to see whether the resulting sum of the angles exceeds 180°. It can happen, as in Figure 6(c), that both possibilities are compatible with the given information. In that case, two different triangles are solutions to the problem.

The *supplement* of an angle θ (where $0 \le \theta \le 180°$) is the angle $180° - \theta$.

Example 4 SSA, the Two-Solution Case

Solve triangle ABC if $\angle A = 43.1°$, $a = 186.2$, and $b = 248.6$.

Solution   From the given information we sketch the triangle shown in Figure 8. Note that side a may be drawn in two possible positions to complete the triangle. From the Law of Sines

$$\sin B = \frac{b \sin A}{a} = \frac{248.6 \sin 43.1°}{186.2} \approx 0.91225$$

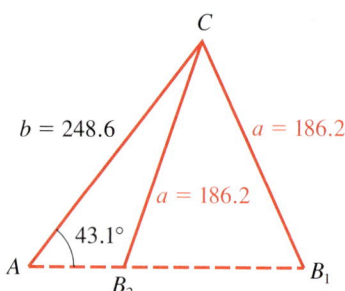

Figure 8

There are two possible angles B between 0° and 180° such that $\sin B = 0.91225$. Using the $\boxed{\text{SIN}^{-1}}$ key on a calculator (or $\boxed{\text{INV}}$ $\boxed{\text{SIN}}$ or $\boxed{\text{ARCSIN}}$), we find that one of these angles is approximately 65.8°. The other is approximately $180° - 65.8° = 114.2°$. We denote these two angles by B_1 and B_2 so that

$$\angle B_1 \approx 65.8° \qquad \text{and} \qquad \angle B_2 \approx 114.2°$$

Surveying is a method of land measurement used for mapmaking. Surveyors use a process called *triangulation* in which a network of thousands of interlocking triangles is created on the area to be mapped. The process is started by measuring the length of a *baseline* between two surveying stations. Then, using an instrument called a *theodolite*, the angles between these two stations and a third station are measured. The Law of Sines is then used to calculate the two other sides of the triangle formed by the three stations. The calculated sides are used as baselines, and the process is repeated over and over to create a network of triangles. In this method, the only distance measured is the initial baseline; all *(continued)*

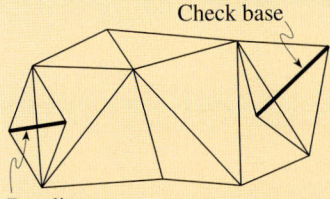
Thus, two triangles satisfy the given conditions: triangle $A_1B_1C_1$ and triangle $A_2B_2C_2$.

Solve triangle $A_1B_1C_1$:

$$\angle C_1 \approx 180° - (43.1° + 65.8°) = 71.1° \qquad \text{Find } \angle C_1$$

Thus $\qquad c_1 = \dfrac{a_1 \sin C_1}{\sin A_1} \approx \dfrac{186.2 \sin 71.1°}{\sin 43.1°} \approx 257.8 \qquad$ Law of Sines

Solve triangle $A_2B_2C_2$:

$$\angle C_2 \approx 180° - (43.1° + 114.2°) = 22.7° \qquad \text{Find } \angle C_2$$

Thus $\qquad c_2 = \dfrac{a_2 \sin C_2}{\sin A_2} \approx \dfrac{186.2 \sin 22.7°}{\sin 43.1°} \approx 105.2 \qquad$ Law of Sines

Triangles $A_1B_1C_1$ and $A_2B_2C_2$ are shown in Figure 9.

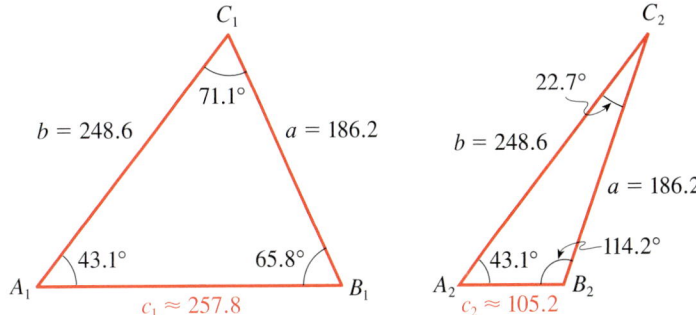

Figure 9

The next example presents a situation for which no triangle is compatible with the given data.

Example 5   SSA, the No-Solution Case

Solve triangle ABC, where $\angle A = 42°$, $a = 70$, and $b = 122$.

Solution   To organize the given information, we sketch the diagram in Figure 10. Let's try to find $\angle B$. We have

$$\frac{\sin A}{a} = \frac{\sin B}{b} \qquad \text{Law of Sines}$$

$$\sin B = \frac{b \sin A}{a} = \frac{122 \sin 42°}{70} \approx 1.17 \qquad \text{Solve for } \sin B$$

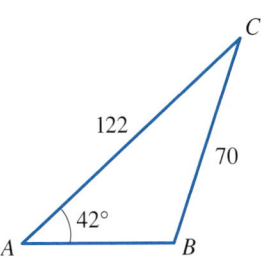

Figure 10

Since the sine of an angle is never greater than 1, we conclude that no triangle satisfies the conditions given in this problem.

6.4   Exercises

1–6 ■ Use the Law of Sines to find the indicated side x or angle θ.

1.
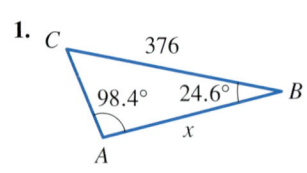

2.
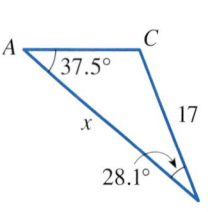

3.
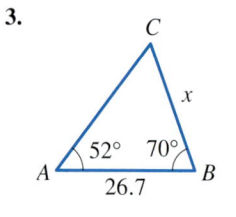

4.
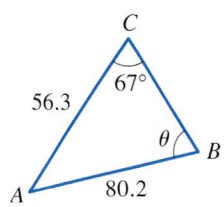

5.
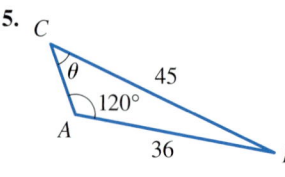

6.
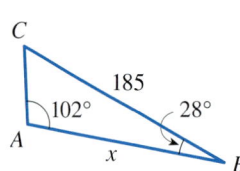

7–10 ■ Solve the triangle using the Law of Sines.

7.
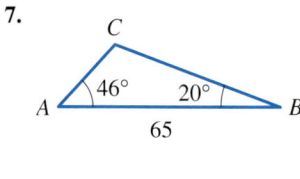

8.
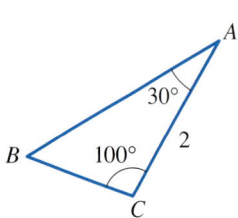

9.
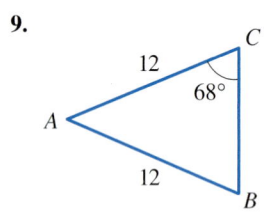

10.
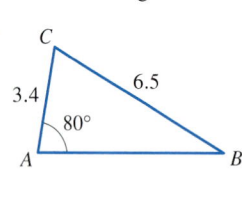

11–16 ■ Sketch each triangle and then solve the triangle using the Law of Sines.

11. $\angle A = 50°$, $\angle B = 68°$, $c = 230$

12. $\angle A = 23°$, $\angle B = 110°$, $c = 50$

13. $\angle A = 30°$, $\angle C = 65°$, $b = 10$

14. $\angle A = 22°$, $\angle B = 95°$, $a = 420$

15. $\angle B = 29°$, $\angle C = 51°$, $b = 44$

16. $\angle B = 10°$, $\angle C = 100°$, $c = 115$

17–26 ■ Use the Law of Sines to solve for all possible triangles that satisfy the given conditions.

17. $a = 28$, $b = 15$, $\angle A = 110°$

18. $a = 30$, $c = 40$, $\angle A = 37°$

19. $a = 20$, $c = 45$, $\angle A = 125°$

20. $b = 45$, $c = 42$, $\angle C = 38°$

21. $b = 25$, $c = 30$, $\angle B = 25°$

22. $a = 75$, $b = 100$, $\angle A = 30°$

23. $a = 50$, $b = 100$, $\angle A = 50°$

24. $a = 100$, $b = 80$, $\angle A = 135°$

25. $a = 26$, $c = 15$, $\angle C = 29°$

26. $b = 73$, $c = 82$, $\angle B = 58°$

27. For the triangle shown, find
   (a) $\angle BCD$ and
   (b) $\angle DCA$.

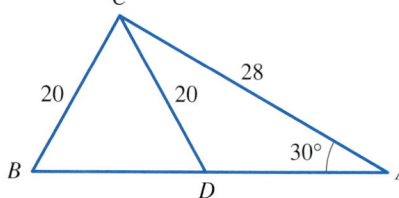

28. For the triangle shown, find the length AD.

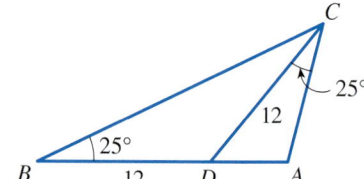

29. In triangle ABC, $\angle A = 40°$, $a = 15$, and $b = 20$.
   (a) Show that there are two triangles, ABC and $A'B'C'$, that satisfy these conditions.
   (b) Show that the areas of the triangles in part (a) are proportional to the sines of the angles C and C', that is,
   $$\frac{\text{area of } \triangle ABC}{\text{area of } \triangle A'B'C'} = \frac{\sin C}{\sin C'}$$

30. Show that, given the three angles A, B, C of a triangle and one side, say a, the area of the triangle is
   $$\text{area} = \frac{a^2 \sin B \sin C}{2 \sin A}$$

Applications

31. Tracking a Satellite   The path of a satellite orbiting the earth causes it to pass directly over two tracking stations A and B, which are 50 mi apart. When the satellite is on one side of the two stations, the angles of elevation at A and B are measured to be 87.0° and 84.2°, respectively.

(a) How far is the satellite from station A?

(b) How high is the satellite above the ground?

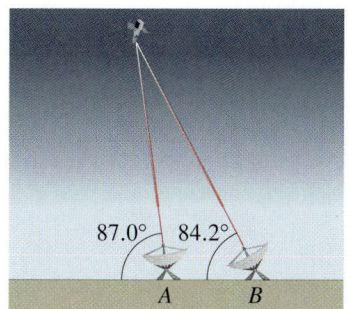

32. Flight of a Plane   A pilot is flying over a straight highway. He determines the angles of depression to two mileposts, 5 mi apart, to be 32° and 48°, as shown in the figure.

(a) Find the distance of the plane from point A.

(b) Find the elevation of the plane.

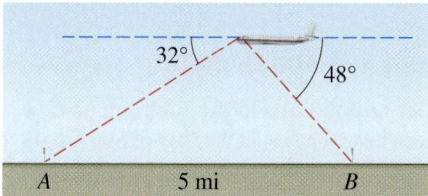

33. Distance Across a River   To find the distance across a river, a surveyor chooses points A and B, which are 200 ft apart on one side of the river (see the figure). She then chooses a reference point C on the opposite side of the river and finds that $\angle BAC \approx 82°$ and $\angle ABC \approx 52°$. Approximate the distance from A to C.

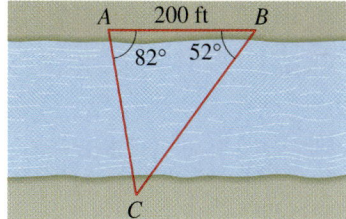

34. Distance Across a Lake   Points A and B are separated by a lake. To find the distance between them, a surveyor locates a point C on land such that $\angle CAB = 48.6°$. He also measures CA as 312 ft and CB as 527 ft. Find the distance between A and B.

35. The Leaning Tower of Pisa   The bell tower of the cathedral in Pisa, Italy, leans 5.6° from the vertical. A tourist stands 105 m from its base, with the tower leaning directly toward her. She measures the angle of elevation to the top of the tower to be 29.2°. Find the length of the tower to the nearest meter.

36. Radio Antenna   A short-wave radio antenna is supported by two guy wires, 165 ft and 180 ft long. Each wire is attached to the top of the antenna and anchored to the ground, at two anchor points on opposite sides of the antenna. The shorter wire makes an angle of 67° with the ground. How far apart are the anchor points?

37. Height of a Tree   A tree on a hillside casts a shadow 215 ft down the hill. If the angle of inclination of the hillside is 22° to the horizontal and the angle of elevation of the sun is 52°, find the height of the tree.

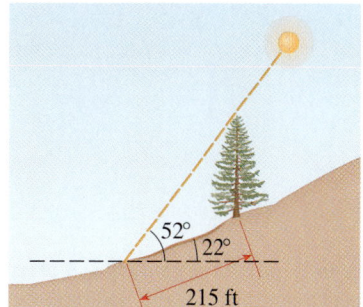

38. Length of a Guy Wire   A communications tower is located at the top of a steep hill, as shown. The angle of inclination of the hill is 58°. A guy wire is to be attached to the top of the tower and to the ground, 100 m downhill from the base of the tower. The angle α in the figure is determined to be 12°. Find the length of cable required for the guy wire.

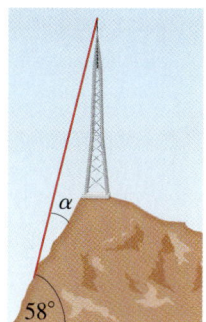

39. Calculating a Distance   Observers at P and Q are located on the side of a hill that is inclined 32° to the horizontal, as shown. The observer at P determines the angle of elevation to a hot-air balloon to be 62°. At the same instant, the observer at Q measures the angle of elevation to the balloon to be 71°. If P is 60 m down the hill from Q, find the distance from Q to the balloon.

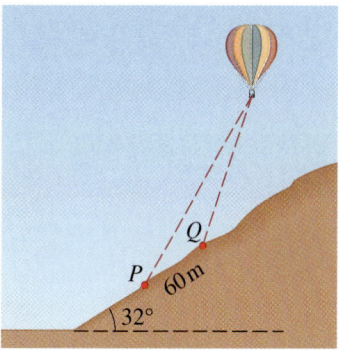

40. Calculating an Angle A water tower 30 m tall is located at the top of a hill. From a distance of 120 m down the hill, it is observed that the angle formed between the top and base of the tower is 8°. Find the angle of inclination of the hill.

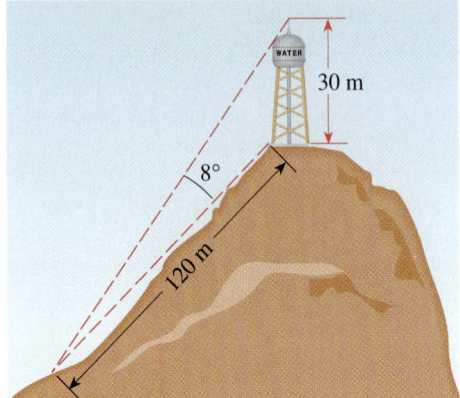

41. Distances to Venus The *elongation* α of a planet is the angle formed by the planet, earth, and sun (see the figure). It is known that the distance from the sun to Venus is 0.723 AU (see Exercise 65 in Section 6.2). At a certain time the elongation of Venus is found to be 39.4°. Find the possible distances from the earth to Venus at that time in Astronomical Units (AU).

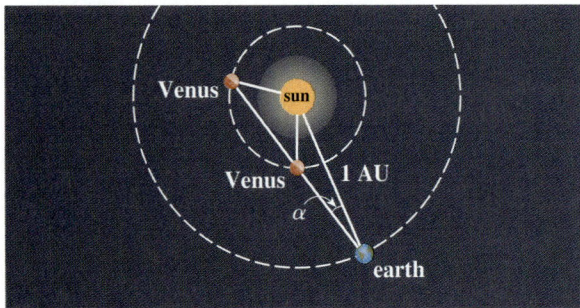

42. Soap Bubbles When two bubbles cling together in midair, their common surface is part of a sphere whose center *D* lies on the line passing throught the centers of the bubbles (see the figure). Also, angles *ACB* and *ACD* each have measure 60°.

(a) Show that the radius *r* of the common face is given by

$$r = \frac{ab}{a - b}$$

[*Hint:* Use the Law of Sines together with the fact that an angle θ and its supplement 180° − θ have the same sine.]

(b) Find the radius of the common face if the radii of the bubbles are 4 cm and 3 cm.

(c) What shape does the common face take if the two bubbles have equal radii?

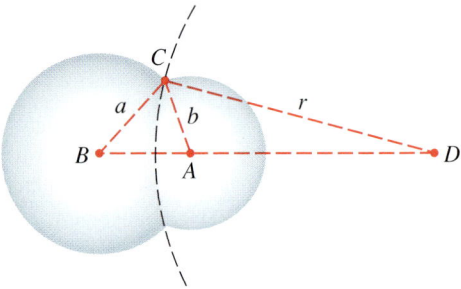

Discovery • Discussion

43. Number of Solutions in the Ambiguous Case We have seen that when using the Law of Sines to solve a triangle in the SSA case, there may be two, one, or no solution(s). Sketch triangles like those in Figure 6 to verify the criteria in the table for the number of solutions if you are given ∠*A* and sides *a* and *b*.

Criterion	Number of Solutions
$a \geq b$	1
$b > a > b \sin A$	2
$a = b \sin A$	1
$a < b \sin A$	0

If ∠*A* = 30° and *b* = 100, use these criteria to find the range of values of *a* for which the triangle *ABC* has two solutions, one solution, or no solution.

6.5   The Law of Cosines

The Law of Sines cannot be used directly to solve triangles if we know two sides and the angle between them or if we know all three sides (these are Cases 3 and 4 of the preceding section). In these two cases, the **Law of Cosines** applies.

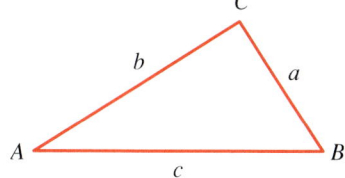

Figure 1

The Law of Cosines

In any triangle ABC (see Figure 1), we have

$$a^2 = b^2 + c^2 - 2bc \cos A$$
$$b^2 = a^2 + c^2 - 2ac \cos B$$
$$c^2 = a^2 + b^2 - 2ab \cos C$$

■ **Proof** To prove the Law of Cosines, place triangle ABC so that $\angle A$ is at the origin, as shown in Figure 2. The coordinates of the vertices B and C are $(c, 0)$ and $(b \cos A, b \sin A)$, respectively. (You should check that the coordinates of these points will be the same if we draw angle A as an acute angle.) Using the Distance Formula, we get

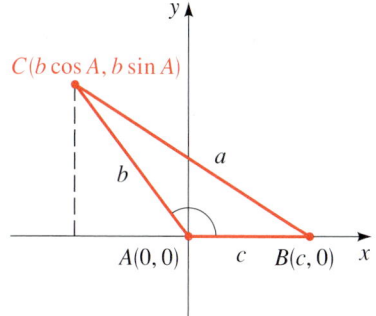

Figure 2

$$a^2 = (b \cos A - c)^2 + (b \sin A - 0)^2$$
$$= b^2 \cos^2 A - 2bc \cos A + c^2 + b^2 \sin^2 A$$
$$= b^2(\cos^2 A + \sin^2 A) - 2bc \cos A + c^2$$
$$= b^2 + c^2 - 2bc \cos A \qquad \text{Because } \sin^2 A + \cos^2 A = 1$$

This proves the first formula. The other two formulas are obtained in the same way by placing each of the other vertices of the triangle at the origin and repeating the preceding argument. ■

In words, the Law of Cosines says that the square of any side of a triangle is equal to the sum of the squares of the other two sides, minus twice the product of those two sides times the cosine of the included angle.

If one of the angles of a triangle, say $\angle C$, is a right angle, then $\cos C = 0$ and the Law of Cosines reduces to the Pythagorean Theorem, $c^2 = a^2 + b^2$. Thus, the Pythagorean Theorem is a special case of the Law of Cosines.

Example 1 Length of a Tunnel

A tunnel is to be built through a mountain. To estimate the length of the tunnel, a surveyor makes the measurements shown in Figure 3. Use the surveyor's data to approximate the length of the tunnel.

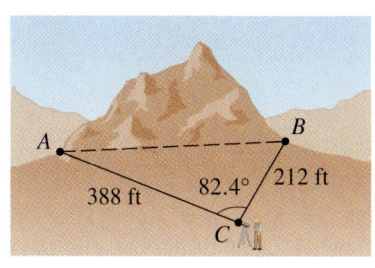

Figure 3

Solution To approximate the length c of the tunnel, we use the Law of Cosines:

$$c^2 = a^2 + b^2 - 2ab \cos C \qquad \text{Law of Cosines}$$
$$= 388^2 + 212^2 - 2(388)(212) \cos 82.4° \qquad \text{Substitute}$$
$$\approx 173730.2367 \qquad \text{Use a calculator}$$
$$c \approx \sqrt{173730.2367} \approx 416.8 \qquad \text{Take square roots}$$

Thus, the tunnel will be approximately 417 ft long. ■

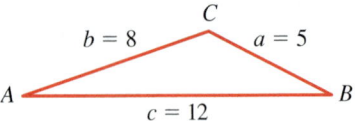

Figure 4

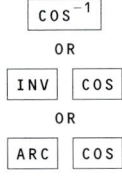

Example 2 **SSS, the Law of Cosines**

The sides of a triangle are $a = 5$, $b = 8$, and $c = 12$ (see Figure 4). Find the angles of the triangle.

Solution   We first find $\angle A$. From the Law of Cosines, we have $a^2 = b^2 + c^2 - 2bc \cos A$. Solving for $\cos A$, we get

$$\cos A = \frac{b^2 + c^2 - a^2}{2bc} = \frac{8^2 + 12^2 - 5^2}{2(8)(12)} = \frac{183}{192} = 0.953125$$

Using a calculator, we find that $\angle A \approx 18°$. In the same way the equations

$$\cos B = \frac{a^2 + c^2 - b^2}{2ac} = \frac{5^2 + 12^2 - 8^2}{2(5)(12)} = 0.875$$

$$\cos C = \frac{a^2 + b^2 - c^2}{2ab} = \frac{5^2 + 8^2 - 12^2}{2(5)(8)} = -0.6875$$

give $\angle B \approx 29°$ and $\angle C \approx 133°$. Of course, once two angles are calculated, the third can more easily be found from the fact that the sum of the angles of a triangle is 180°. However, it's a good idea to calculate all three angles using the Law of Cosines and add the three angles as a check on your computations. ■

Example 3 **SAS, the Law of Cosines**

Solve triangle ABC, where $\angle A = 46.5°$, $b = 10.5$, and $c = 18.0$.

Solution   We can find a using the Law of Cosines.

$$a^2 = b^2 + c^2 - 2bc \cos A$$

$$= (10.5)^2 + (18.0)^2 - 2(10.5)(18.0)(\cos 46.5°) \approx 174.05$$

Thus, $a \approx \sqrt{174.05} \approx 13.2$. We also use the Law of Cosines to find $\angle B$ and $\angle C$, as in Example 2.

$$\cos B = \frac{a^2 + c^2 - b^2}{2ac} = \frac{13.2^2 + 18.0^2 - 10.5^2}{2(13.2)(18.0)} \approx 0.816477$$

$$\cos C = \frac{a^2 + b^2 - c^2}{2ab} = \frac{13.2^2 + 10.5^2 - 18.0^2}{2(13.2)(10.5)} \approx -0.142532$$

Using a calculator, we find that $\angle B \approx 35.3°$ and $\angle C \approx 98.2°$.

To summarize: $\angle B \approx 35.3°$, $\angle C \approx 98.2°$, and $a \approx 13.2$. (See Figure 5.) ■

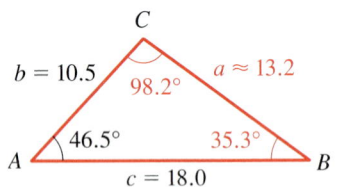

Figure 5

We could have used the Law of Sines to find $\angle B$ and $\angle C$ in Example 3, since we knew all three sides and an angle in the triangle. But knowing the sine of an angle does not uniquely specify the angle, since an angle θ and its supplement $180° - \theta$ both have the same sine. Thus we would need to decide which of the two angles is the correct choice. This ambiguity does not arise when we use the Law of Cosines, because every angle between 0° and 180° has a unique cosine. So using only the Law of Cosines is preferable in problems like Example 3.

Navigation: Heading and Bearing

In navigation a direction is often given as a **bearing**, that is, as an acute angle measured from due north or due south. The bearing N 30° E, for example, indicates a direction that points 30° to the east of due north (see Figure 6).

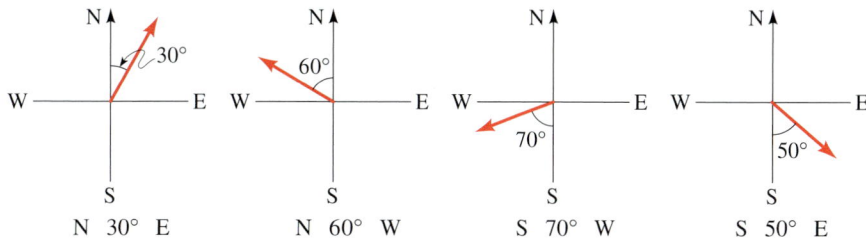

Figure 6

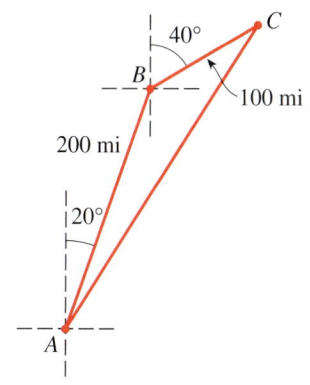

Figure 7

Example 4 Navigation

A pilot sets out from an airport and heads in the direction N 20° E, flying at 200 mi/h. After one hour, he makes a course correction and heads in the direction N 40° E. Half an hour after that, engine trouble forces him to make an emergency landing.

(a) Find the distance between the airport and his final landing point.

(b) Find the bearing from the airport to his final landing point.

Solution

(a) In one hour the plane travels 200 mi, and in half an hour it travels 100 mi, so we can plot the pilot's course as in Figure 7. When he makes his course correction, he turns 20° to the right, so the angle between the two legs of his trip is $180° - 20° = 160°$. So by the Law of Cosines we have

$$b^2 = 200^2 + 100^2 - 2 \cdot 200 \cdot 100 \cos 160°$$

$$\approx 87{,}587.70$$

Thus, $b \approx 295.95$. The pilot lands about 296 mi from his starting point.

(b) We first use the Law of Sines to find $\angle A$.

$$\frac{\sin A}{100} = \frac{\sin 160°}{295.95}$$

$$\sin A = 100 \cdot \frac{\sin 160°}{295.95}$$

$$\approx 0.11557$$

Another angle with sine 0.11557 is $180° - 6.636° = 173.364°$. But this is clearly too large to be $\angle A$ in $\triangle ABC$.

Using the $\boxed{\text{SIN}^{-1}}$ key on a calculator, we find that $\angle A \approx 6.636°$. From Figure 7 we see that the line from the airport to the final landing site points in the direction $20° + 6.636° = 26.636°$ east of due north. Thus, the bearing is about N 26.6° E. ∎

The Area of a Triangle

An interesting application of the Law of Cosines involves a formula for finding the area of a triangle from the lengths of its three sides (see Figure 8).

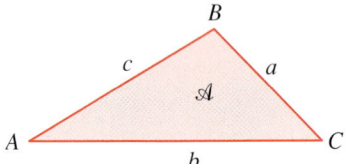

Figure 8

> ### Heron's Formula
>
> The area $\mathcal{A}$ of triangle ABC is given by
> $$\mathcal{A} = \sqrt{s(s-a)(s-b)(s-c)}$$
> where $s = \frac{1}{2}(a+b+c)$ is the **semiperimeter** of the triangle; that is, s is half the perimeter.

■ **Proof** We start with the formula $\mathcal{A} = \frac{1}{2}ab \sin C$ from Section 6.3. Thus

$$\mathcal{A}^2 = \tfrac{1}{4}a^2b^2 \sin^2 C$$

$$= \tfrac{1}{4}a^2b^2(1 - \cos^2 C) \qquad \text{Pythagorean identity}$$

$$= \tfrac{1}{4}a^2b^2(1 - \cos C)(1 + \cos C) \qquad \text{Factor}$$

Next, we write the expressions $1 - \cos C$ and $1 + \cos C$ in terms of a, b and c. By the Law of Cosines we have

$$\cos C = \frac{a^2 + b^2 - c^2}{2ab} \qquad \text{Law of Cosines}$$

$$1 + \cos C = 1 + \frac{a^2 + b^2 - c^2}{2ab} \qquad \text{Add 1}$$

$$= \frac{2ab + a^2 + b^2 - c^2}{2ab} \qquad \text{Common denominator}$$

$$= \frac{(a+b)^2 - c^2}{2ab} \qquad \text{Factor}$$

$$= \frac{(a+b+c)(a+b-c)}{2ab} \qquad \text{Difference of squares}$$

Similarly

$$1 - \cos C = \frac{(c+a-b)(c-a+b)}{2ab}$$

Substituting these expressions in the formula we obtained for $\mathcal{A}^2$ gives

To see that the factors in the last two products are equal, note for example that

$$\frac{a+b-c}{2} = \frac{a+b+c}{2} - c$$

$$= s - c$$

$$\mathcal{A}^2 = \tfrac{1}{4}a^2b^2 \frac{(a+b+c)(a+b-c)}{2ab} \frac{(c+a-b)(c-a+b)}{2ab}$$

$$= \frac{(a+b+c)}{2} \frac{(a+b-c)}{2} \frac{(c+a-b)}{2} \frac{(c-a+b)}{2}$$

$$= s(s-c)(s-b)(s-a)$$

Heron's Formula now follows by taking the square root of each side. ■

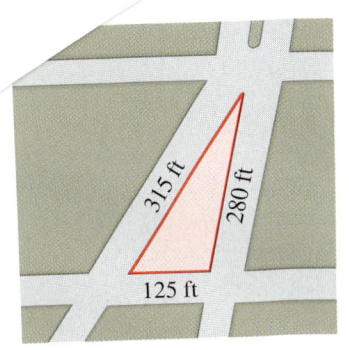

Figure 9

Example 5 Area of a Lot

A businessman wishes to buy a triangular lot in a busy downtown location (see Figure 9). The lot frontages on the three adjacent streets are 125, 280, and 315 ft. Find the area of the lot.

Solution The semiperimeter of the lot is

$$s = \frac{125 + 280 + 315}{2} = 360$$

By Heron's Formula the area is

$$\mathcal{A} = \sqrt{360(360 - 125)(360 - 280)(360 - 315)} \approx 17{,}451.6$$

Thus, the area is approximately 17,452 ft^2. ∎

6.5 Exercises

1–8 ■ Use the Law of Cosines to determine the indicated side x or angle θ.

1.
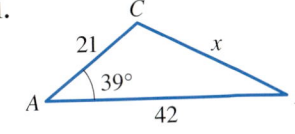

2.
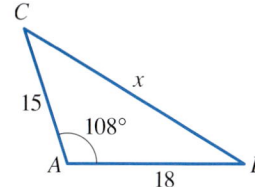

3.
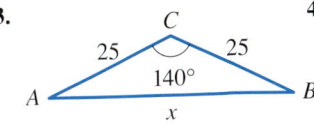

4.
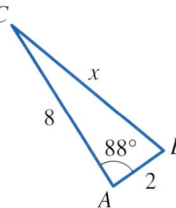

5.
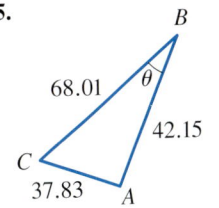

6.
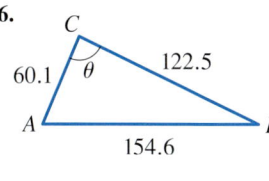

7.
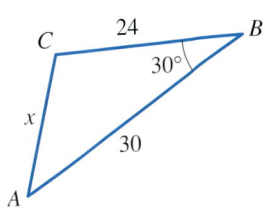

8.
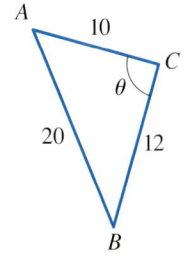

9–18 ■ Solve triangle ABC.

9.
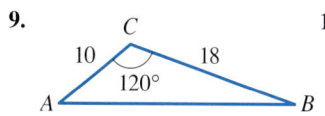

10.
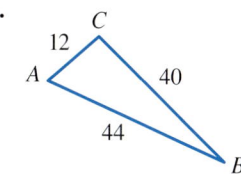

11. $a = 3.0,\quad b = 4.0,\quad \angle C = 53°$

12. $b = 60,\quad c = 30,\quad \angle A = 70°$

13. $a = 20,\quad b = 25,\quad c = 22$

14. $a = 10,\quad b = 12,\quad c = 16$

15. $b = 125,\quad c = 162,\quad \angle B = 40°$

16. $a = 65,\quad c = 50,\quad \angle C = 52°$

17. $a = 50,\quad b = 65,\quad \angle A = 55°$

18. $a = 73.5,\quad \angle B = 61°,\quad \angle C = 83°$

19–26 ■ Find the indicated side x or angle θ. (Use either the Law of Sines or the Law of Cosines, as appropriate.)

19.
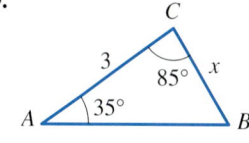

20.
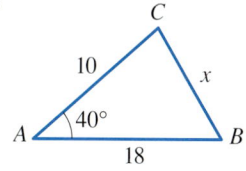

21.
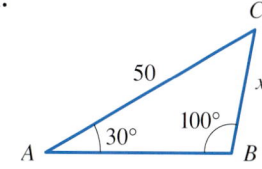

22.
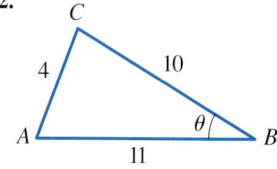

23.

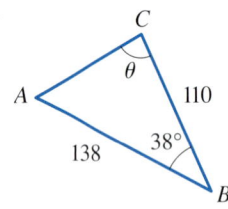

24.

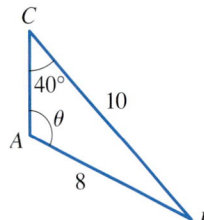

25.

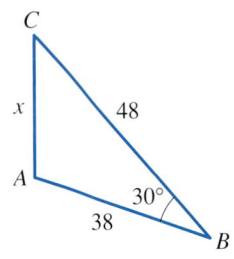

26.

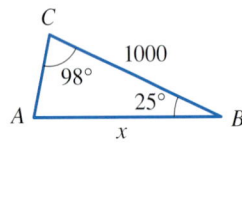

27–30 ■ Find the area of the triangle whose sides have the given lengths.

27. $a = 9$, $b = 12$, $c = 15$ **28.** $a = 1$, $b = 2$, $c = 2$

29. $a = 7$, $b = 8$, $c = 9$

30. $a = 11$, $b = 100$, $c = 101$

31–34 ■ Find the area of the shaded figure, correct to two decimals.

31.

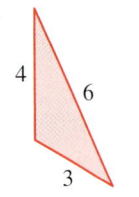

32.

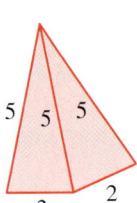

33.

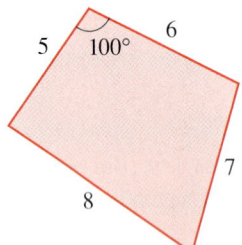

34.

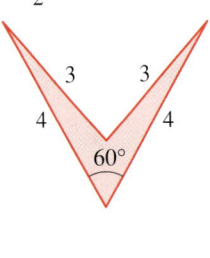

35. Three circles of radii 4, 5, and 6 cm are mutually tangent. Find the shaded area enclosed between the circles.

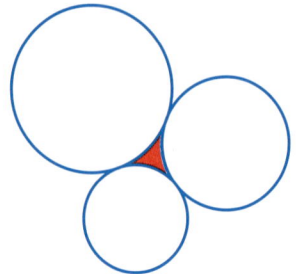

36. Prove that in triangle ABC

$$a = b \cos C + c \cos B$$

$$b = c \cos A + a \cos C$$

$$c = a \cos B + b \cos A$$

These are called the *Projection Laws*. [*Hint:* To get the first equation, add the second and third equations in the Law of Cosines and solve for a.]

Applications

37. Surveying   To find the distance across a small lake, a surveyor has taken the measurements shown. Find the distance across the lake using this information.

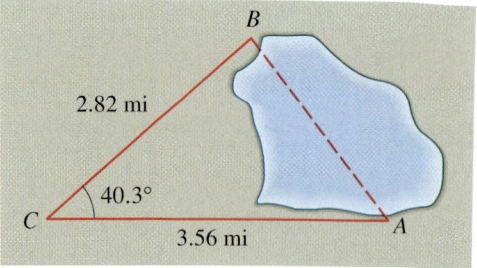

38. Geometry   A parallelogram has sides of lengths 3 and 5, and one angle is 50°. Find the lengths of the diagonals.

39. Calculating Distance   Two straight roads diverge at an angle of 65°. Two cars leave the intersection at 2:00 P.M., one traveling at 50 mi/h and the other at 30 mi/h. How far apart are the cars at 2:30 P.M.?

40. Calculating Distance   A car travels along a straight road, heading east for 1 h, then traveling for 30 min on another road that leads northeast. If the car has maintained a constant speed of 40 mi/h, how far is it from its starting position?

41. Dead Reckoning   A pilot flies in a straight path for 1 h 30 min. She then makes a course correction, heading 10° to the right of her original course, and flies 2 h in the new direction. If she maintains a constant speed of 625 mi/h, how far is she from her starting position?

42. Navigation   Two boats leave the same port at the same time. One travels at a speed of 30 mi/h in the direction N 50° E and the other travels at a speed of 26 mi/h in a direction S 70° E (see the figure). How far apart are the two boats after one hour?

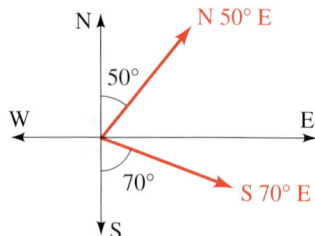

43. Navigation   A fisherman leaves his home port and heads in the direction N 70° W. He travels 30 mi and reaches Egg Island. The next day he sails N 10° E for 50 mi, reaching Forrest Island.

(a) Find the distance between the fisherman's home port and Forrest Island.

(b) Find the bearing from Forrest Island back to his home port.

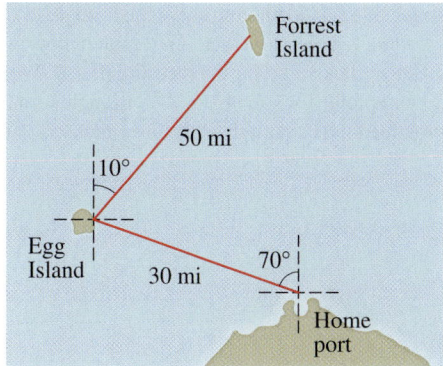

44. Navigation   Airport B is 300 mi from airport A at a bearing N 50° E (see the figure). A pilot wishing to fly from A to B mistakenly flies due east at 200 mi/h for 30 minutes, when he notices his error.

(a) How far is the pilot from his destination at the time he notices the error?

(b) What bearing should he head his plane in order to arrive at airport B?

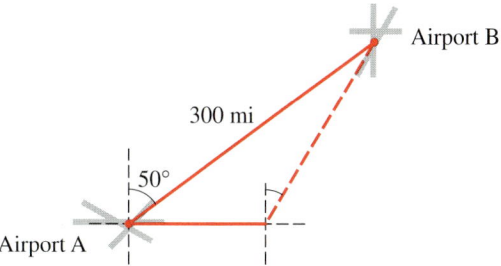

45. Triangular Field   A triangular field has sides of lengths 22, 36, and 44 yd. Find the largest angle.

46. Towing a Barge   Two tugboats that are 120 ft apart pull a barge, as shown. If the length of one cable is 212 ft and the length of the other is 230 ft, find the angle formed by the two cables.

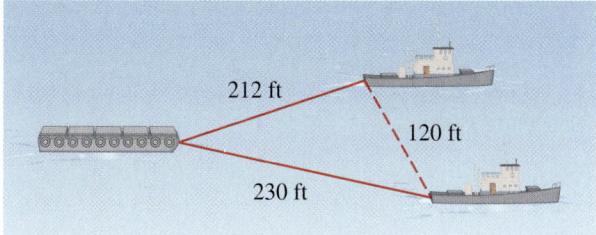

47. Flying Kites   A boy is flying two kites at the same time. He has 380 ft of line out to one kite and 420 ft to the other. He estimates the angle between the two lines to be 30°. Approximate the distance between the kites.

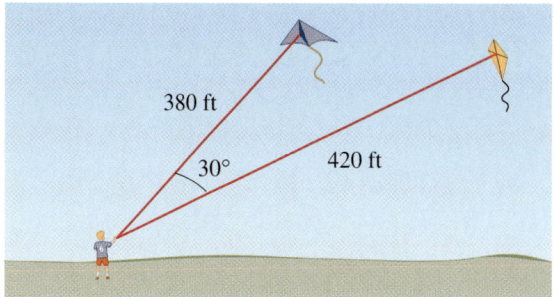

48. Securing a Tower   A 125-ft tower is located on the side of a mountain that is inclined 32° to the horizontal. A guy wire is to be attached to the top of the tower and anchored at a point 55 ft downhill from the base of the tower. Find the shortest length of wire needed.

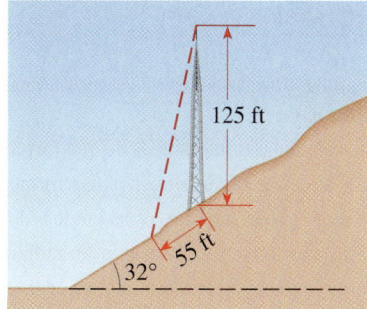

49. Cable Car   A steep mountain is inclined 74° to the horizontal and rises 3400 ft above the surrounding plain. A cable car is to be installed from a point 800 ft from the base to the top of the mountain, as shown. Find the shortest length of cable needed.

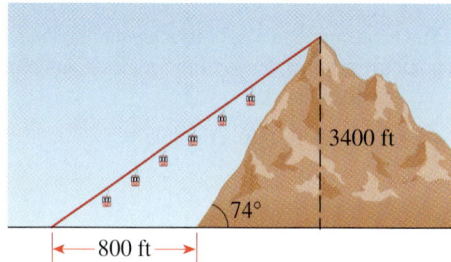

50. CN Tower   The CN Tower in Toronto, Canada, is the tallest free-standing structure in the world. A woman on the observation deck, 1150 ft above the ground, wants to determine the distance between two landmarks on the ground below. She observes that the angle formed by the lines of sight to these two landmarks is 43°. She also observes that the angle between the vertical and the line of sight to one of the landmarks is 62° and to the other landmark is 54°. Find the distance between the two landmarks.

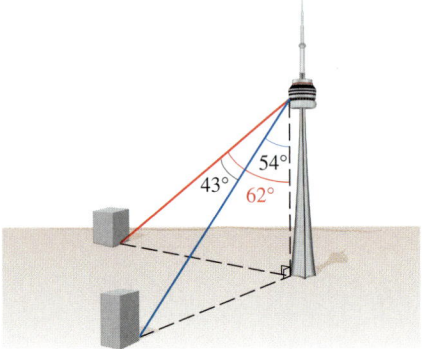

51. Land Value Land in downtown Columbia is valued at $20 a square foot. What is the value of a triangular lot with sides of lengths 112, 148, and 190 ft?

Discovery • Discussion

52. Solving for the Angles in a Triangle The paragraph that follows the solution of Example 3 on page 510 explains an alternative method for finding $\angle B$ and $\angle C$, using the Law of Sines. Use this method to solve the triangle in the example, finding $\angle B$ first and then $\angle C$. Explain how you chose the appropriate value for the measure of $\angle B$. Which method do you prefer for solving an SAS triangle problem, the one explained in Example 3 or the one you used in this exercise?

---

6 Review

Concept Check

1. (a) Explain the difference between a positive angle and a negative angle.

   (b) How is an angle of measure 1 degree formed?

   (c) How is an angle of measure 1 radian formed?

   (d) How is the radian measure of an angle θ defined?

   (e) How do you convert from degrees to radians?

   (f) How do you convert from radians to degrees?

2. (a) When is an angle in standard position?

   (b) When are two angles coterminal?

3. (a) What is the length s of an arc of a circle with radius r that subtends a central angle of θ radians?

   (b) What is the area A of a sector of a circle with radius r and central angle θ radians?

4. If θ is an acute angle in a right triangle, define the six trigonometric ratios in terms of the adjacent and opposite sides and the hypotenuse.

5. What does it mean to solve a triangle?

6. If θ is an angle in standard position, $P(x, y)$ is a point on the terminal side, and r is the distance from the origin to P, write expressions for the six trigonometric functions of θ.

7. Which trigonometric functions are positive in quadrants I, II, III, and IV?

8. If θ is an angle in standard position, what is its reference angle $\bar{\theta}$?

9. (a) State the reciprocal identities.

   (b) State the Pythagorean identities.

10. (a) What is the area of a triangle with sides of length a and b and with included angle θ?

   (b) What is the area of a triangle with sides of length a, b, and c?

11. (a) State the Law of Sines.

   (b) State the Law of Cosines.

12. Explain the ambiguous case in the Law of Sines.

Exercises

1–2 ■ Find the radian measure that corresponds to the given degree measure.

1. (a) 60°   **(b)** 330°   **(c)** −135°   **(d)** −90°

2. (a) 24°   **(b)** −330°   **(c)** 750°   **(d)** 5°

3–4 ■ Find the degree measure that corresponds to the given radian measure.

3. (a) $\dfrac{5\pi}{2}$ **(b)** $-\dfrac{\pi}{6}$ **(c)** $\dfrac{9\pi}{4}$ **(d)** 3.1

4. (a) 8   **(b)** $-\dfrac{5}{2}$ **(c)** $\dfrac{11\pi}{6}$ **(d)** $\dfrac{3\pi}{5}$

5. Find the length of an arc of a circle of radius 8 m if the arc subtends a central angle of 1 rad.

6. Find the measure of a central angle θ in a circle of radius 5 ft if the angle is subtended by an arc of length 7 ft.

7. A circular arc of length 100 ft subtends a central angle of $70°$. Find the radius of the circle.

8. How many revolutions will a car wheel of diameter 28 in. make over a period of half an hour if the car is traveling at 60 mi/h?

9. New York and Los Angeles are 2450 mi apart. Find the angle that the arc between these two cities subtends at the center of the earth. (The radius of the earth is 3960 mi.)

10. Find the area of a sector with central angle 2 rad in a circle of radius 5 m.

11. Find the area of a sector with central angle $52°$ in a circle of radius 200 ft.

12. A sector in a circle of radius 25 ft has an area of 125 ft². Find the central angle of the sector.

13. A potter's wheel with radius 8 in. spins at 150 rpm. Find the angular and linear speeds of a point on the rim of the wheel.

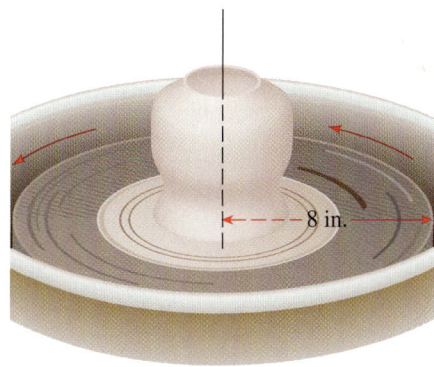

14. In an automobile transmission a *gear ratio* g is the ratio

$$g = \frac{\text{angular speed of engine}}{\text{angular speed of wheels}}$$

The angular speed of the engine is shown on the tachometer (in rpm).

A certain sports car has wheels with radius 11 in. Its gear ratios are shown in the following table. Suppose the car is in fourth gear and the tachometer reads 3500 rpm.

(a) Find the angular speed of the engine.

(b) Find the angular speed of the wheels.

(c) How fast (in mi/h) is the car traveling?

Gear	Ratio
1st	4.1
2nd	3.0
3rd	1.6
4th	0.9
5th	0.7

15–16 ■ Find the values of the six trigonometric ratios of θ.

15.   **16.**

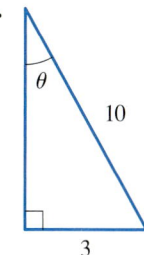

17–20 ■ Find the sides labeled x and y, correct to two decimal places.

17.   **18.**

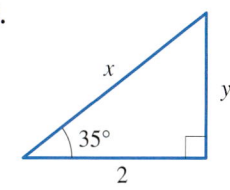

19.   **20.**

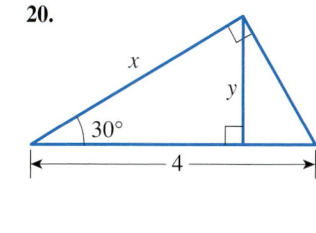

21–22 ■ Solve the triangle.

21.   **22.**

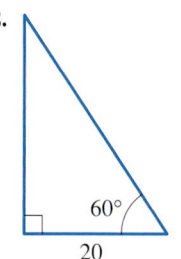

23. Express the lengths a and b in the figure in terms of the trigonometric ratios of θ.

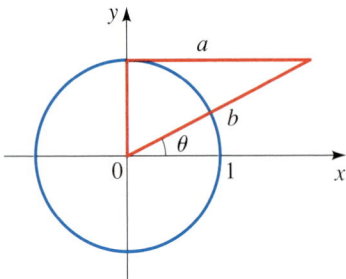

24. The highest free-standing tower in the world is the CN Tower in Toronto, Canada. From a distance of 1 km from its base, the angle of elevation to the top of the tower is 28.81°. Find the height of the tower.

25. Find the perimeter of a regular hexagon that is inscribed in a circle of radius 8 m.

26. The pistons in a car engine move up and down repeatedly to turn the crankshaft, as shown. Find the height of the point P above the center O of the crankshaft in terms of the angle θ.

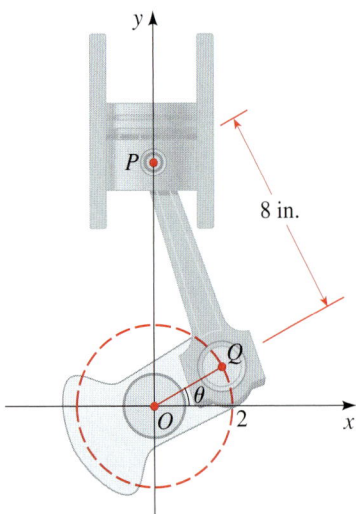

27. As viewed from the earth, the angle subtended by the full moon is 0.518°. Use this information and the fact that the distance AB from the earth to the moon is 236,900 mi to find the radius of the moon.

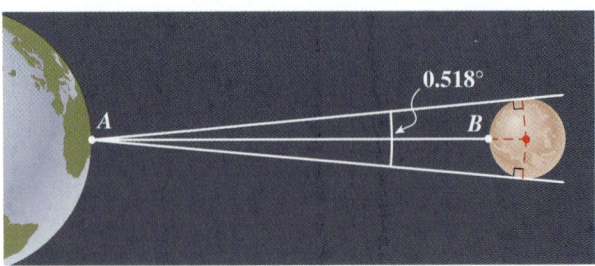

28. A pilot measures the angles of depression to two ships to be 40° and 52° (see the figure). If the pilot is flying at an elevation of 35,000 ft, find the distance between the two ships.

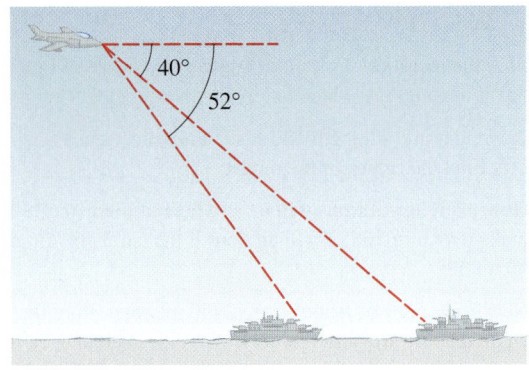

29–40 ■ Find the exact value.

29. $\sin 315°$

30. $\csc \dfrac{9\pi}{4}$

31. $\tan(-135°)$

32. $\cos \dfrac{5\pi}{6}$

33. $\cot\left(-\dfrac{22\pi}{3}\right)$

34. $\sin 405°$

35. $\cos 585°$

36. $\sec \dfrac{22\pi}{3}$

37. $\csc \dfrac{8\pi}{3}$

38. $\sec \dfrac{13\pi}{6}$

39. $\cot(-390°)$

40. $\tan \dfrac{23\pi}{4}$

41. Find the values of the six trigonometric ratios of the angle θ in standard position if the point $(-5, 12)$ is on the terminal side of θ.

42. Find $\sin\theta$ if θ is in standard position and its terminal side intersects the circle of radius 1 centered at the origin at the point $(-\sqrt{3}/2, \frac{1}{2})$.

43. Find the acute angle that is formed by the line $y - \sqrt{3}x + 1 = 0$ and the x-axis.

44. Find the six trigonometric ratios of the angle θ in standard position if its terminal side is in quadrant III and is parallel to the line $4y - 2x - 1 = 0$.

45–48 ■ Write the first expression in terms of the second, for θ in the given quadrant.

45. $\tan\theta$, $\cos\theta$; θ in quadrant II

46. $\sec\theta$, $\sin\theta$; θ in quadrant III

47. $\tan^2\theta$, $\sin\theta$; θ in any quadrant

48. $\csc^2\theta \ \cos^2\theta$, $\sin\theta$; θ in any quadrant

49–52 ■ Find the values of the six trigonometric functions of θ from the information given.

49. $\tan \theta = \sqrt{7}/3, \quad \sec \theta = \frac{4}{3}$ **50.** $\sec \theta = \frac{41}{40}, \quad \csc \theta = -\frac{41}{9}$

51. $\sin \theta = \frac{3}{5}, \quad \cos \theta < 0$ **52.** $\sec \theta = -\frac{13}{5}, \quad \tan \theta > 0$

53. If $\tan \theta = -\frac{1}{2}$ for θ in quadrant II, find $\sin \theta + \cos \theta$.

54. If $\sin \theta = \frac{1}{2}$ for θ in quadrant I, find $\tan \theta + \sec \theta$.

55. If $\tan \theta = -1$, find $\sin^2\theta + \cos^2\theta$.

56. If $\cos \theta = -\sqrt{3}/2$ and $\pi/2 < \theta < \pi$, find $\sin 2\theta$.

57–62 ■ Find the side labeled x.

57.

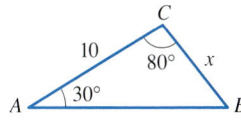

58.

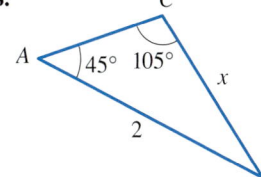

59.

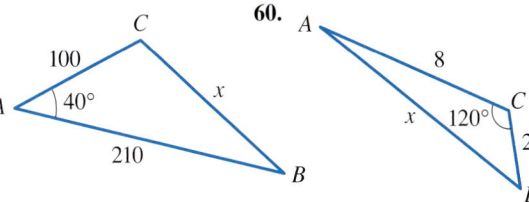

60.

61.

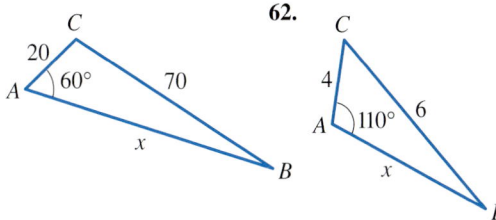

62.

63. Two ships leave a port at the same time. One travels at 20 mi/h in a direction N 32° E, and the other travels at 28 mi/h in a direction S 42° E (see the figure). How far apart are the two ships after 2 h?

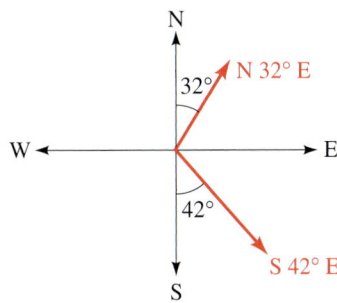

64. From a point A on the ground, the angle of elevation to the top of a tall building is 24.1°. From a point B, which is 600 ft closer to the building, the angle of elevation is measured to be 30.2°. Find the height of the building.

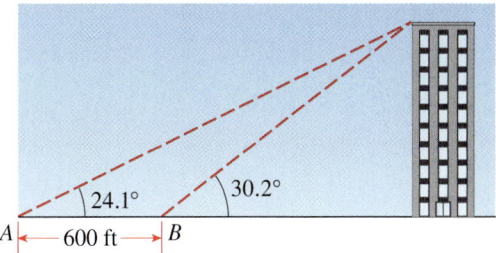

65. Find the distance between points A and B on opposite sides of a lake from the information shown.

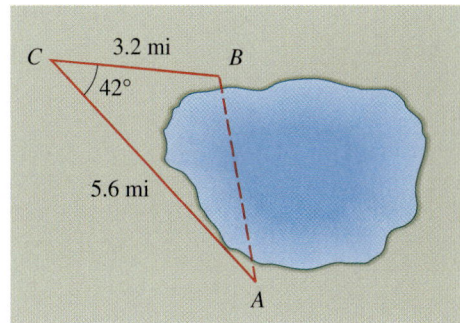

66. A boat is cruising the ocean off a straight shoreline. Points A and B are 120 mi apart on the shore, as shown. It is found that $\angle A = 42.3°$ and $\angle B = 68.9°$. Find the shortest distance from the boat to the shore.

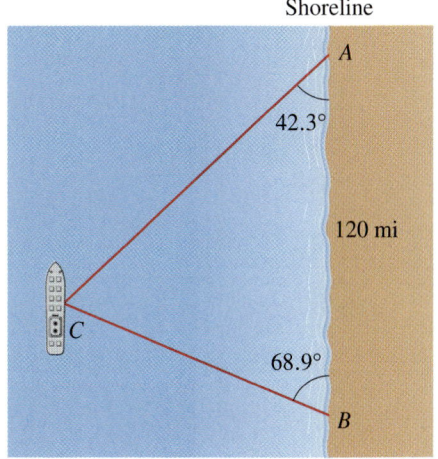

67. Find the area of a triangle with sides of length 8 and 14 and included angle 35°.

68. Find the area of a triangle with sides of length 5, 6, and 8.

6   Test

1. Find the radian measures that correspond to the degree measures $330°$ and $-135°$.

2. Find the degree measures that correspond to the radian measures $\dfrac{4\pi}{3}$ and -1.3.

3. The rotor blades of a helicopter are 16 ft long and are rotating at 120 rpm.
    (a) Find the angular speed of the rotor.
    (b) Find the linear speed of a point on the tip of a blade.

4. Find the exact value of each of the following.
    (a) $\sin 405°$ (b) $\tan(-150°)$
    (c) $\sec \dfrac{5\pi}{3}$ (d) $\csc \dfrac{5\pi}{2}$

5. Find $\tan\theta + \sin\theta$ for the angle θ shown.

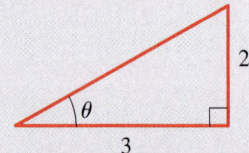

6. Express the lengths a and b shown in the figure in terms of θ.

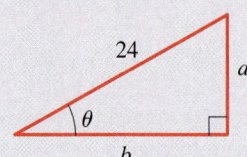

7. If $\cos\theta = -\frac{1}{3}$ and θ is in quadrant III, find $\tan\theta \cot\theta + \csc\theta$.

8. If $\sin\theta = \frac{5}{13}$ and $\tan\theta = -\frac{5}{12}$, find $\sec\theta$.

9. Express $\tan\theta$ in terms of $\sec\theta$ for θ in quadrant II.

10. The base of the ladder in the figure is 6 ft from the building, and the angle formed by the ladder and the ground is $73°$. How high up the building does the ladder touch?

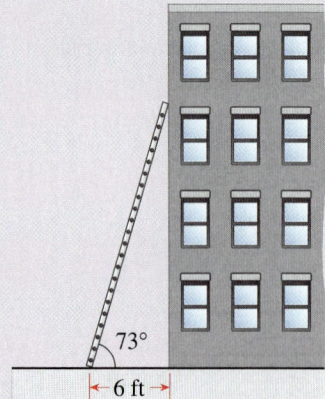

11–14 ■ Find the side labeled x.

11.
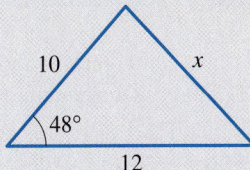

12.
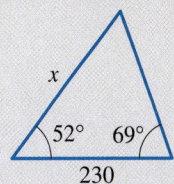

13.
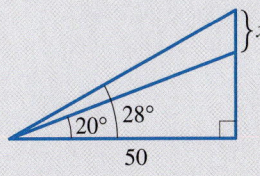

14.
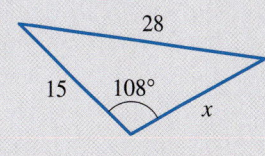

15. Refer to the figure below.

   (a) Find the area of the shaded region.

   (b) Find the perimeter of the shaded region.

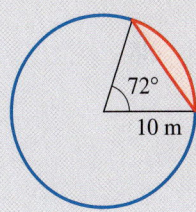

16. Refer to the figure below.

   (a) Find the angle opposite the longest side.

   (b) Find the area of the triangle.

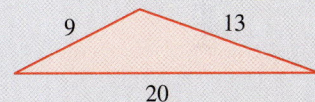

17. Two wires tether a balloon to the ground, as shown. How high is the balloon above the ground?

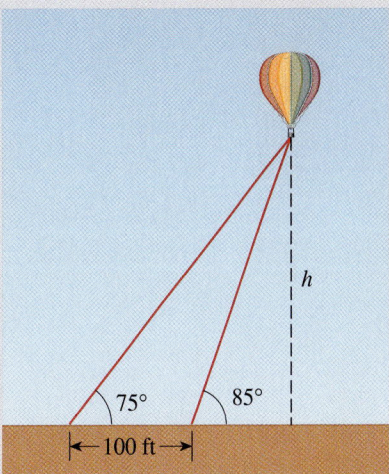

Focus on Modeling
Surveying

How can we measure the height of a mountain, or the distance across a lake? Obviously it may be difficult, inconvenient, or impossible to measure these distances directly (that is, using a tape measure or a yard stick). On the other hand, it is easy to measure *angles* to distant objects. That's where trigonometry comes in—the trigonometric ratios relate angles to distances, so they can be used to *calculate* distances from the *measured* angles. In this *Focus* we examine how trigonometry is used to map a town. Modern map making methods use satellites and the Global Positioning System, but mathematics remains at the core of the process.

Mapping a Town

A student wants to draw a map of his hometown. To construct an accurate map (or scale model), he needs to find distances between various landmarks in the town. The student makes the measurements shown in Figure 1. Note that only one distance is measured, between City Hall and the first bridge. All other measurements are angles.

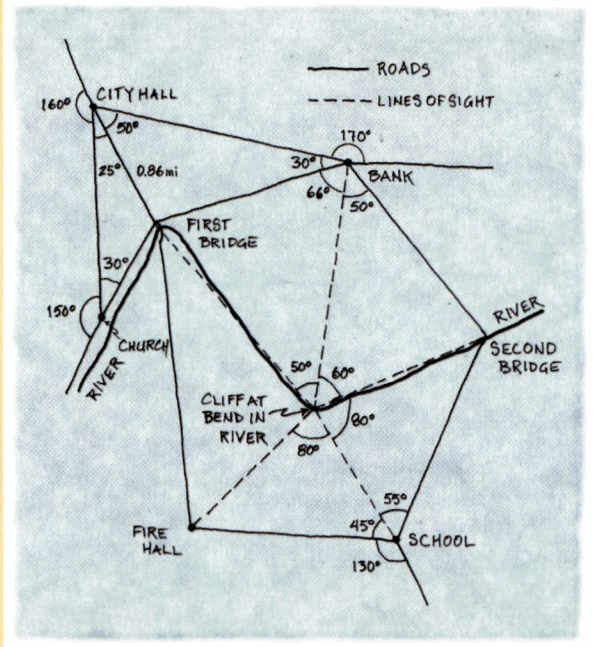

Figure 1

The distances between other landmarks can now be found using the Law of Sines. For example, the distance x from the bank to the first bridge is calculated by applying the Law of Sines to the triangle with vertices at City Hall, the bank, and the first bridge:

$$\frac{x}{\sin 50°} = \frac{0.86}{\sin 30°} \qquad \text{Law of Sines}$$

$$x = \frac{0.86 \sin 50°}{\sin 30°} \qquad \text{Solve for } x$$

$$\approx 1.32 \text{ mi} \qquad \text{Calculator}$$

522

So the distance between the bank and the first bridge is 1.32 mi.

The distance we just found can now be used to find other distances. For instance, we find the distance y between the bank and the cliff as follows:

$$\frac{y}{\sin 64°} = \frac{1.32}{\sin 50°} \qquad \text{Law of Sines}$$

$$y = \frac{1.32 \sin 64°}{\sin 50°} \qquad \text{Solve for } y$$

$$\approx 1.55 \text{ mi} \qquad \text{Calculator}$$

Continuing in this fashion, we can calculate all the distances between the landmarks shown in the rough sketch in Figure 1. We can use this information to draw the map shown in Figure 2.

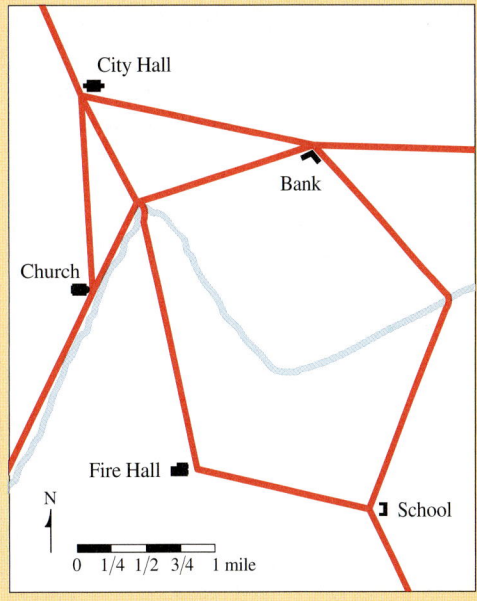

Figure 2

To make a topographic map, we need to measure elevation. This concept is explored in Problems 4–6.

Problems

1. **Completing the Map**   Find the distance between the church and City Hall.

2. **Completing the Map**   Find the distance between the fire hall and the school. (You will need to find other distances first.)

3. Determining a Distance   A surveyor on one side of a river wishes to find the distance between points A and B on the opposite side of the river. On her side, she chooses points C and D, which are 20 m apart, and measures the angles shown in the figure below. Find the distance between A and B.

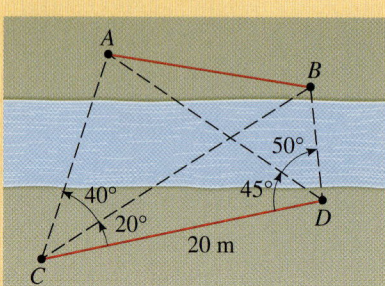

4. Height of a Cliff   To measure the height of an inaccessible cliff on the opposite side of a river, a surveyor makes the measurements shown in the figure at the left. Find the height of the cliff.

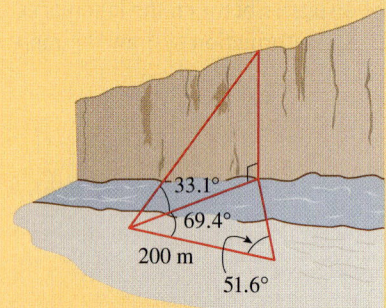

5. Height of a Mountain   To calculate the height h of a mountain, angle α, β, and distance d are measured, as shown in the figure below.

(a) Show that

$$h = \frac{d}{\cot \alpha - \cot \beta}$$

(b) Show that

$$h = d \frac{\sin \alpha \sin \beta}{\sin(\beta - \alpha)}$$

(c) Use the formulas from parts (a) and (b) to find the height of a mountain if $\alpha = 25°$, $\beta = 29°$, and $d = 800$ ft. Do you get the same answer from each formula?

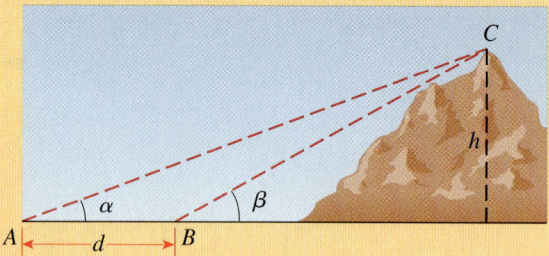

6. Determining a Distance   A surveyor has determined that a mountain is 2430 ft high. From the top of the mountain he measures the angles of depression to two landmarks at the base of the mountain, and finds them to be 42° and 39°. (Observe that these are the same as the angles of elevation from the landmarks as shown in the figure at the left.) The angle between the lines of sight to the landmarks is 68°. Calculate the distance between the two landmarks.

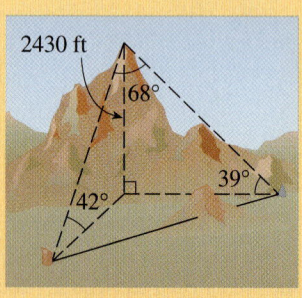

7. Surveying Building Lots   A surveyor surveys two adjacent lots and makes the following rough sketch showing his measurements. Calculate all the distances shown in the figure and use your result to draw an accurate map of the two lots.

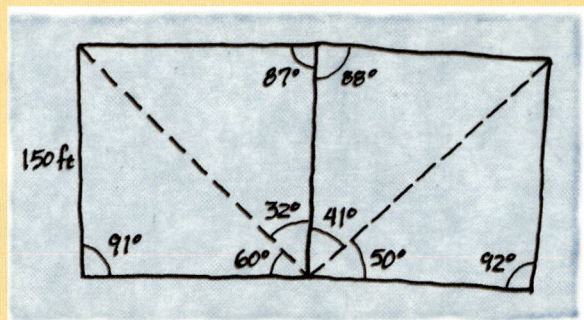

8. Great Survey of India   The Great Trigonometric Survey of India was one of the most massive mapping projects ever undertaken (see the margin note on page 504). Do some research at your library or on the Internet to learn more about the Survey, and write a report on your findings.

7 Analytic Trigonometry

Chapter Overview

In Chapters 5 and 6 we studied the graphical and geometric properties of trigonometric functions. In this chapter we study the algebraic aspects of trigonometry, that is, simplifying and factoring expressions and solving equations that involve trigonometric functions. The basic tools in the algebra of trigonometry are trigonometric identities.

A *trigonometric identity* is an equation involving the trigonometric functions that holds for all values of the variable. For example, from the definitions of sine and cosine it follows that for any θ we have

$$\sin^2\theta + \cos^2\theta = 1$$

Here are some other identities that we will study in this chapter:

$$\sin 2\theta = 2 \sin\theta \cos\theta \qquad \sin A \cos B = \tfrac{1}{2}[\sin(A + B) + \sin(A - B)]$$

Using identities we can simplify a complicated expression involving the trigonometric functions into a much simpler expression, thereby allowing us to better understand what the expression means. For example, the area of the rectangle in the figure at the left is $A = 2 \sin\theta \cos\theta$; then using one of the above identities we see that $A = \sin 2\theta$.

A *trigonometric equation* is an equation involving the trigonometric functions. For example, the equation

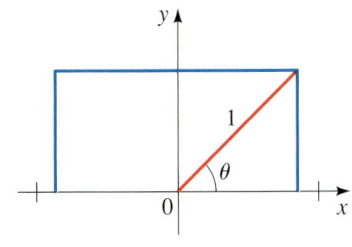

$$\sin\theta - \frac{1}{2} = 0$$

is a trigonometric equation. To solve this equation we need to find all the values of θ that satisfy the equation. A graph of $y = \sin\theta$ shows that $\sin\theta = \frac{1}{2}$ infinitely many times, so the equation has infinitely many solutions. Two of these solutions are $\theta = \frac{\pi}{6}$ and $\frac{5\pi}{6}$; we can get the others by adding multiples of 2π to these solutions.

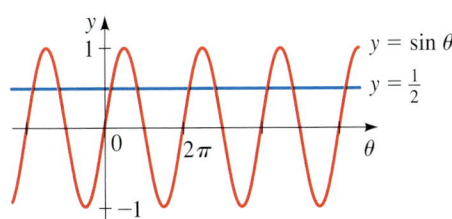

We also study the *inverse trigonometric functions*. In order to define the inverse of a trigonometric function, we first restrict its domain to an interval on which the function is one-to-one. For example, we restrict the domain of the sine function to $[-\pi/2, \pi/2]$. On this interval $\sin\frac{\pi}{6} = \frac{1}{2}$, so $\sin^{-1}\frac{1}{2} = \frac{\pi}{6}$. We will see that these inverse functions are useful in solving trigonometric equations.

In *Focus on Modeling* (page 575) we study some applications of the concepts of this chapter to the motion of waves.

7.1   Trigonometric Identities

We begin by listing some of the basic trigonometric identities. We studied most of these in Chapters 5 and 6; you are asked to prove the cofunction identities in Exercise 100.

Fundamental Trigonometric Identities

Reciprocal Identities

$$\csc x = \frac{1}{\sin x} \qquad \sec x = \frac{1}{\cos x} \qquad \cot x = \frac{1}{\tan x}$$

$$\tan x = \frac{\sin x}{\cos x} \qquad \cot x = \frac{\cos x}{\sin x}$$

Pythagorean Identities

$$\sin^2 x + \cos^2 x = 1 \qquad \tan^2 x + 1 = \sec^2 x \qquad 1 + \cot^2 x = \csc^2 x$$

Even-Odd Identities

$$\sin(-x) = -\sin x \qquad \cos(-x) = \cos x \qquad \tan(-x) = -\tan x$$

Cofunction Identities

$$\sin\left(\frac{\pi}{2} - u\right) = \cos u \qquad \tan\left(\frac{\pi}{2} - u\right) = \cot u \qquad \sec\left(\frac{\pi}{2} - u\right) = \csc u$$

$$\cos\left(\frac{\pi}{2} - u\right) = \sin u \qquad \cot\left(\frac{\pi}{2} - u\right) = \tan u \qquad \csc\left(\frac{\pi}{2} - u\right) = \sec u$$

Simplifying Trigonometric Expressions

Identities enable us to write the same expression in different ways. It is often possible to rewrite a complicated looking expression as a much simpler one. To simplify algebraic expressions, we used factoring, common denominators, and the Special Product Formulas. To simplify trigonometric expressions, we use these same techniques together with the fundamental trigonometric identities.

Example 1 **Simplifying a Trigonometric Expression**

Simplify the expression $\cos t + \tan t \sin t$.

Solution We start by rewriting the expression in terms of sine and cosine.

$$\cos t + \tan t \sin t = \cos t + \left(\frac{\sin t}{\cos t}\right)\sin t \qquad \text{Reciprocal identity}$$

$$= \frac{\cos^2 t + \sin^2 t}{\cos t} \qquad \text{Common denominator}$$

$$= \frac{1}{\cos t} \qquad \text{Pythagorean identity}$$

$$= \sec t \qquad \text{Reciprocal identity} \qquad \blacksquare$$

Example 2 **Simplifying by Combining Fractions**

Simplify the expression $\dfrac{\sin \theta}{\cos \theta} + \dfrac{\cos \theta}{1 + \sin \theta}$.

Solution We combine the fractions by using a common denominator.

$$\frac{\sin \theta}{\cos \theta} + \frac{\cos \theta}{1 + \sin \theta} = \frac{\sin \theta\,(1 + \sin \theta) + \cos^2 \theta}{\cos \theta\,(1 + \sin \theta)} \qquad \text{Common denominator}$$

$$= \frac{\sin \theta + \sin^2 \theta + \cos^2 \theta}{\cos \theta\,(1 + \sin \theta)} \qquad \text{Distribute } \sin \theta$$

$$= \frac{\sin \theta + 1}{\cos \theta\,(1 + \sin \theta)} \qquad \text{Pythagorean identity}$$

$$= \frac{1}{\cos \theta} = \sec \theta \qquad \text{Cancel and use reciprocal identity} \qquad \blacksquare$$

Proving Trigonometric Identities

Many identities follow from the fundamental identities. In the examples that follow, we learn how to prove that a given trigonometric equation is an identity, and in the process we will see how to discover new identities.

First, it's easy to decide when a given equation is *not* an identity. All we need to do is show that the equation does not hold for some value of the variable (or variables). Thus, the equation

$$\sin x + \cos x = 1$$

is not an identity, because when $x = \pi/4$, we have

$$\sin \frac{\pi}{4} + \cos \frac{\pi}{4} = \frac{\sqrt{2}}{2} + \frac{\sqrt{2}}{2} = \sqrt{2} \neq 1$$

To verify that a trigonometric equation is an identity, we transform one side of the equation into the other side by a series of steps, each of which is itself an identity.

Guidelines for Proving Trigonometric Identities

1. **Start with one side.**   Pick one side of the equation and write it down. Your goal is to transform it into the other side. It's usually easier to start with the more complicated side.

2. **Use known identities.**   Use algebra and the identities you know to change the side you started with. Bring fractional expressions to a common denominator, factor, and use the fundamental identities to simplify expressions.

3. **Convert to sines and cosines.**   If you are stuck, you may find it helpful to rewrite all functions in terms of sines and cosines.

    Warning: To prove an identity, we do *not* just perform the same operations on both sides of the equation. For example, if we start with an equation that is not an identity, such as

(1) $$\sin x = -\sin x$$

and square both sides, we get the equation

(2) $$\sin^2 x = \sin^2 x$$

which is clearly an identity. Does this mean that the original equation is an identity? Of course not. The problem here is that the operation of squaring is not **reversible** in the sense that we cannot arrive back at (1) from (2) by taking square roots (reversing the procedure). Only operations that are reversible will necessarily transform an identity into an identity.

Example 3   Proving an Identity by Rewriting in Terms of Sine and Cosine

Verify the identity $\cos\theta\,(\sec\theta - \cos\theta) = \sin^2\theta$.

Solution   The left-hand side looks more complicated, so we start with it and try to transform it into the right-hand side.

$$\text{LHS} = \cos\theta\,(\sec\theta - \cos\theta)$$

$$= \cos\theta\left(\frac{1}{\cos\theta} - \cos\theta\right) \qquad \text{Reciprocal identity}$$

$$= 1 - \cos^2\theta \qquad\qquad\qquad \text{Expand}$$

$$= \sin^2\theta = \text{RHS} \qquad\qquad \text{Pythagorean identity} \blacksquare$$

In Example 3 it isn't easy to see how to change the right-hand side into the left-hand side, but it's definitely possible. Simply notice that each step is reversible. In other words, if we start with the last expression in the proof and work backward through the steps, the right side is transformed into the left side. You will probably agree, however, that it's more difficult to prove the identity this way. That's why it's often better to change the more complicated side of the identity into the simpler side.

Example 4 Proving an Identity by Combining Fractions

Verify the identity

$$2 \tan x \sec x = \frac{1}{1 - \sin x} - \frac{1}{1 + \sin x}$$

Solution   Finding a common denominator and combining the fractions on the right-hand side of this equation, we get

$$
\begin{aligned}
\text{RHS} &= \frac{1}{1 - \sin x} - \frac{1}{1 + \sin x} \\[2mm]
&= \frac{(1 + \sin x) - (1 - \sin x)}{(1 - \sin x)(1 + \sin x)} && \textit{Common denominator} \\[2mm]
&= \frac{2 \sin x}{1 - \sin^2 x} && \textit{Simplify} \\[2mm]
&= \frac{2 \sin x}{\cos^2 x} && \textit{Pythagorean identity} \\[2mm]
&= 2\, \frac{\sin x}{\cos x}\left(\frac{1}{\cos x}\right) && \textit{Factor} \\[2mm]
&= 2 \tan x \sec x = \text{LHS} && \textit{Reciprocal identities} \quad ■
\end{aligned}
$$

See *Focus on Problem Solving*, pages 138–145.

In Example 5 we introduce "something extra" to the problem by multiplying the numerator and the denominator by a trigonometric expression, chosen so that we can simplify the result.

Example 5 Proving an Identity by Introducing Something Extra

Verify the identity $\dfrac{\cos u}{1 - \sin u} = \sec u + \tan u$.

Solution   We start with the left-hand side and multiply numerator and denominator by $1 + \sin u$.

We multiply by $1 + \sin u$ because we know by the difference of squares formula that $(1 - \sin u)(1 + \sin u) = 1 - \sin^2 u$, and this is just $\cos^2 u$, a simpler expression.

$$
\begin{aligned}
\text{LHS} &= \frac{\cos u}{1 - \sin u} \\[2mm]
&= \frac{\cos u}{1 - \sin u} \cdot \frac{1 + \sin u}{1 + \sin u} && \textit{Multiply numerator and denominator by } 1 + \sin u \\[2mm]
&= \frac{\cos u\,(1 + \sin u)}{1 - \sin^2 u} && \textit{Expand denominator}
\end{aligned}
$$

$$= \frac{\cos u\,(1 + \sin u)}{\cos^2 u} \qquad \textit{Pythagorean identity}$$

$$= \frac{1 + \sin u}{\cos u} \qquad \textit{Cancel common factor}$$

$$= \frac{1}{\cos u} + \frac{\sin u}{\cos u} \qquad \textit{Separate into two fractions}$$

$$= \sec u + \tan u \qquad \textit{Reciprocal identities} \qquad \blacksquare$$

Here is another method for proving that an equation is an identity. If we can transform each side of the equation *separately*, by way of identities, to arrive at the same result, then the equation is an identity. Example 6 illustrates this procedure.

Example 6   Proving an Identity by Working with Both Sides Separately

Verify the identity $\dfrac{1 + \cos \theta}{\cos \theta} = \dfrac{\tan^2 \theta}{\sec \theta - 1}$.

Solution   We prove the identity by changing each side separately into the same expression. Supply the reasons for each step.

$$\text{LHS} = \frac{1 + \cos \theta}{\cos \theta} = \frac{1}{\cos \theta} + \frac{\cos \theta}{\cos \theta} = \sec \theta + 1$$

$$\text{RHS} = \frac{\tan^2 \theta}{\sec \theta - 1} = \frac{\sec^2 \theta - 1}{\sec \theta - 1} = \frac{(\sec \theta - 1)(\sec \theta + 1)}{\sec \theta - 1} = \sec \theta + 1$$

It follows that LHS = RHS, so the equation is an identity. $\qquad \blacksquare$

We conclude this section by describing the technique of *trigonometric substitution*, which we use to convert algebraic expressions to trigonometric ones. This is often useful in calculus, for instance, in finding the area of a circle or an ellipse.

Example 7   Trigonometric Substitution

Substitute $\sin \theta$ for x in the expression $\sqrt{1 - x^2}$ and simplify. Assume that $0 \le \theta \le \pi/2$.

Solution   Setting $x = \sin \theta$, we have

$$\sqrt{1 - x^2} = \sqrt{1 - \sin^2 \theta} \qquad \textit{Substitute } x = \sin \theta$$

$$= \sqrt{\cos^2 \theta} \qquad \textit{Pythagorean identity}$$

$$= \cos \theta \qquad \textit{Take square root}$$

The last equality is true because $\cos \theta \ge 0$ for the values of θ in question. $\qquad \blacksquare$

7.1 Exercises

1–10 ■ Write the trigonometric expression in terms of sine and cosine, and then simplify.

1. $\cos t \tan t$

2. $\cos t \csc t$

3. $\sin \theta \sec \theta$

4. $\tan \theta \csc \theta$

5. $\tan^2 x - \sec^2 x$

6. $\dfrac{\sec x}{\csc x}$

7. $\sin u + \cot u \cos u$

8. $\cos^2 \theta (1 + \tan^2 \theta)$

9. $\dfrac{\sec \theta - \cos \theta}{\sin \theta}$

10. $\dfrac{\cot \theta}{\csc \theta - \sin \theta}$

11–24 ■ Simplify the trigonometric expression.

11. $\dfrac{\sin x \sec x}{\tan x}$

12. $\cos^3 x + \sin^2 x \cos x$

13. $\dfrac{1 + \cos y}{1 + \sec y}$

14. $\dfrac{\tan x}{\sec(-x)}$

15. $\dfrac{\sec^2 x - 1}{\sec^2 x}$

16. $\dfrac{\sec x - \cos x}{\tan x}$

17. $\dfrac{1 + \csc x}{\cos x + \cot x}$

18. $\dfrac{\sin x}{\csc x} + \dfrac{\cos x}{\sec x}$

19. $\dfrac{1 + \sin u}{\cos u} + \dfrac{\cos u}{1 + \sin u}$

20. $\tan x \cos x \csc x$

21. $\dfrac{2 + \tan^2 x}{\sec^2 x} - 1$

22. $\dfrac{1 + \cot A}{\csc A}$

23. $\tan \theta + \cos(-\theta) + \tan(-\theta)$

24. $\dfrac{\cos x}{\sec x + \tan x}$

25–88 ■ Verify the identity.

25. $\dfrac{\sin \theta}{\tan \theta} = \cos \theta$

26. $\dfrac{\tan x}{\sec x} = \sin x$

27. $\dfrac{\cos u \sec u}{\tan u} = \cot u$

28. $\dfrac{\cot x \sec x}{\csc x} = 1$

29. $\dfrac{\tan y}{\csc y} = \sec y - \cos y$

30. $\dfrac{\cos v}{\sec v \sin v} = \csc v - \sin v$

31. $\sin B + \cos B \cot B = \csc B$

32. $\cos(-x) - \sin(-x) = \cos x + \sin x$

33. $\cot(-\alpha) \cos(-\alpha) + \sin(-\alpha) = -\csc \alpha$

34. $\csc x [\csc x + \sin(-x)] = \cot^2 x$

35. $\tan \theta + \cot \theta = \sec \theta \csc \theta$

36. $(\sin x + \cos x)^2 = 1 + 2 \sin x \cos x$

37. $(1 - \cos \beta)(1 + \cos \beta) = \dfrac{1}{\csc^2 \beta}$

38. $\dfrac{\cos x}{\sec x} + \dfrac{\sin x}{\csc x} = 1$

39. $\dfrac{(\sin x + \cos x)^2}{\sin^2 x - \cos^2 x} = \dfrac{\sin^2 x - \cos^2 x}{(\sin x - \cos x)^2}$

40. $(\sin x + \cos x)^4 = (1 + 2 \sin x \cos x)^2$

41. $\dfrac{\sec t - \cos t}{\sec t} = \sin^2 t$

42. $\dfrac{1 - \sin x}{1 + \sin x} = (\sec x - \tan x)^2$

43. $\dfrac{1}{1 - \sin^2 y} = 1 + \tan^2 y$ **44.** $\csc x - \sin x = \cos x \cot x$

45. $(\cot x - \csc x)(\cos x + 1) = -\sin x$

46. $\sin^4 \theta - \cos^4 \theta = \sin^2 \theta - \cos^2 \theta$

47. $(1 - \cos^2 x)(1 + \cot^2 x) = 1$

48. $\cos^2 x - \sin^2 x = 2 \cos^2 x - 1$

49. $2 \cos^2 x - 1 = 1 - 2 \sin^2 x$

50. $(\tan y + \cot y) \sin y \cos y = 1$

51. $\dfrac{1 - \cos \alpha}{\sin \alpha} = \dfrac{\sin \alpha}{1 + \cos \alpha}$

52. $\sin^2 \alpha + \cos^2 \alpha + \tan^2 \alpha = \sec^2 \alpha$

53. $\tan^2 \theta - \sin^2 \theta = \tan^2 \theta \sin^2 \theta$

54. $\cot^2 \theta \cos^2 \theta = \cot^2 \theta - \cos^2 \theta$

55. $\dfrac{\sin x - 1}{\sin x + 1} = \dfrac{-\cos^2 x}{(\sin x + 1)^2}$ **56.** $\dfrac{\sin w}{\sin w + \cos w} = \dfrac{\tan w}{1 + \tan w}$

57. $\dfrac{(\sin t + \cos t)^2}{\sin t \cos t} = 2 + \sec t \csc t$

58. $\sec t \csc t (\tan t + \cot t) = \sec^2 t + \csc^2 t$

59. $\dfrac{1 + \tan^2 u}{1 - \tan^2 u} = \dfrac{1}{\cos^2 u - \sin^2 u}$

60. $\dfrac{1 + \sec^2 x}{1 + \tan^2 x} = 1 + \cos^2 x$

61. $\dfrac{\sec x}{\sec x - \tan x} = \sec x (\sec x + \tan x)$

62. $\dfrac{\sec x + \csc x}{\tan x + \cot x} = \sin x + \cos x$

63. $\sec v - \tan v = \dfrac{1}{\sec v + \tan v}$

64. $\dfrac{\sin A}{1 - \cos A} - \cot A = \csc A$

65. $\dfrac{\sin x + \cos x}{\sec x + \csc x} = \sin x \cos x$

66. $\dfrac{1 - \cos x}{\sin x} + \dfrac{\sin x}{1 - \cos x} = 2 \csc x$

67. $\dfrac{\csc x - \cot x}{\sec x - 1} = \cot x$ **68.** $\dfrac{\csc^2 x - \cot^2 x}{\sec^2 x} = \cos^2 x$

69. $\tan^2 u - \sin^2 u = \tan^2 u \sin^2 u$

70. $\dfrac{\tan v \sin v}{\tan v + \sin v} = \dfrac{\tan v - \sin v}{\tan v \sin v}$

71. $\sec^4 x - \tan^4 x = \sec^2 x + \tan^2 x$

72. $\dfrac{\cos \theta}{1 - \sin \theta} = \sec \theta + \tan \theta$

73. $\dfrac{\cos \theta}{1 - \sin \theta} = \dfrac{\sin \theta - \csc \theta}{\cos \theta - \cot \theta}$

74. $\dfrac{1 + \tan x}{1 - \tan x} = \dfrac{\cos x + \sin x}{\cos x - \sin x}$

75. $\dfrac{\cos^2 t + \tan^2 t - 1}{\sin^2 t} = \tan^2 t$

76. $\dfrac{1}{1 - \sin x} - \dfrac{1}{1 + \sin x} = 2 \sec x \tan x$

77. $\dfrac{1}{\sec x + \tan x} + \dfrac{1}{\sec x - \tan x} = 2 \sec x$

78. $\dfrac{1 + \sin x}{1 - \sin x} - \dfrac{1 - \sin x}{1 + \sin x} = 4 \tan x \sec x$

79. $(\tan x + \cot x)^2 = \sec^2 x + \csc^2 x$

80. $\tan^2 x - \cot^2 x = \sec^2 x - \csc^2 x$

81. $\dfrac{\sec u - 1}{\sec u + 1} = \dfrac{1 - \cos u}{1 + \cos u}$ **82.** $\dfrac{\cot x + 1}{\cot x - 1} = \dfrac{1 + \tan x}{1 - \tan x}$

83. $\dfrac{\sin^3 x + \cos^3 x}{\sin x + \cos x} = 1 - \sin x \cos x$

84. $\dfrac{\tan v - \cot v}{\tan^2 v - \cot^2 v} = \sin v \cos v$

85. $\dfrac{1 + \sin x}{1 - \sin x} = (\tan x + \sec x)^2$

86. $\dfrac{\tan x + \tan y}{\cot x + \cot y} = \tan x \tan y$

87. $(\tan x + \cot x)^4 = \csc^4 x \sec^4 x$

88. $(\sin \alpha - \tan \alpha)(\cos \alpha - \cot \alpha) = (\cos \alpha - 1)(\sin \alpha - 1)$

89–94 ■ Make the indicated trigonometric substitution in the given algebraic expression and simplify (see Example 7). Assume $0 \le \theta < \pi/2$.

89. $\dfrac{x}{\sqrt{1 - x^2}}$, $x = \sin \theta$ **90.** $\sqrt{1 + x^2}$, $x = \tan \theta$

91. $\sqrt{x^2 - 1}$, $x = \sec \theta$ **92.** $\dfrac{1}{x^2 \sqrt{4 + x^2}}$, $x = 2 \tan \theta$

93. $\sqrt{9 - x^2}$, $x = 3 \sin \theta$ **94.** $\dfrac{\sqrt{x^2 - 25}}{x}$, $x = 5 \sec \theta$

95–98 ■ Graph f and g in the same viewing rectangle. Do the graphs suggest that the equation $f(x) = g(x)$ is an identity? Prove your answer.

95. $f(x) = \cos^2 x - \sin^2 x$, $g(x) = 1 - 2\sin^2 x$

96. $f(x) = \tan x\,(1 + \sin x)$, $g(x) = \dfrac{\sin x \cos x}{1 + \sin x}$

97. $f(x) = (\sin x + \cos x)^2$, $g(x) = 1$

98. $f(x) = \cos^4 x - \sin^4 x$, $g(x) = 2\cos^2 x - 1$

99. Show that the equation is not an identity.

    (a) $\sin 2x = 2 \sin x$ **(b)** $\sin(x + y) = \sin x + \sin y$

    (c) $\sec^2 x + \csc^2 x = 1$

    (d) $\dfrac{1}{\sin x + \cos x} = \csc x + \sec x$

Discovery • Discussion

100. Cofunction Identities   In the right triangle shown, explain why $v = (\pi/2) - u$. Explain how you can obtain all six cofunction identities from this triangle, for $0 < u < \pi/2$.

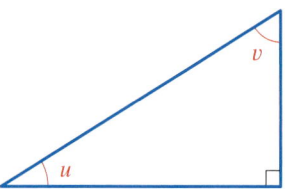

101. Graphs and Identities   Suppose you graph two functions, f and g, on a graphing device, and their graphs appear identical in the viewing rectangle. Does this prove that the equation $f(x) = g(x)$ is an identity? Explain.

102. Making Up Your Own Identity   If you start with a trigonometric expression and rewrite it or simplify it, then setting the original expression equal to the rewritten expression yields a trigonometric identity. For instance, from Example 1 we get the identity

$$\cos t + \tan t \sin t = \sec t$$

Use this technique to make up your own identity, then give it to a classmate to verify.

7.2 Addition and Subtraction Formulas

We now derive identities for trigonometric functions of sums and differences.

Addition and Subtraction Formulas

Formulas for sine:

$$\sin(s + t) = \sin s \cos t + \cos s \sin t$$
$$\sin(s - t) = \sin s \cos t - \cos s \sin t$$

Formulas for cosine:

$$\cos(s + t) = \cos s \cos t - \sin s \sin t$$
$$\cos(s - t) = \cos s \cos t + \sin s \sin t$$

Formulas for tangent:

$$\tan(s + t) = \frac{\tan s + \tan t}{1 - \tan s \tan t}$$

$$\tan(s - t) = \frac{\tan s - \tan t}{1 + \tan s \tan t}$$

■ **Proof of Addition Formula for Cosine** To prove the formula $\cos(s + t) = \cos s \cos t - \sin s \sin t$, we use Figure 1. In the figure, the distances t, $s + t$, and $-s$ have been marked on the unit circle, starting at $P_0(1, 0)$ and terminating at Q_1, P_1, and Q_0, respectively. The coordinates of these points are

$$P_0(1, 0) \qquad\qquad Q_0(\cos(-s), \sin(-s))$$
$$P_1(\cos(s + t), \sin(s + t)) \qquad Q_1(\cos t, \sin t)$$

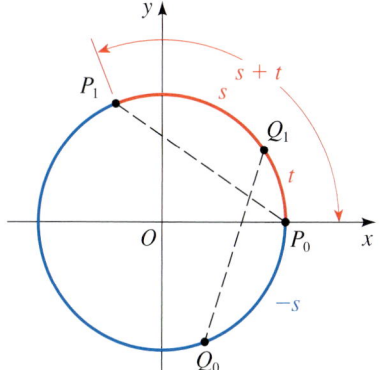

Figure 1

Since $\cos(-s) = \cos s$ and $\sin(-s) = -\sin s$, it follows that the point Q_0 has the coordinates $Q_0(\cos s, -\sin s)$. Notice that the distances between P_0 and P_1 and between Q_0 and Q_1 measured along the arc of the circle are equal. Since equal arcs are subtended by equal chords, it follows that $d(P_0, P_1) = d(Q_0, Q_1)$. Using the Distance Formula, we get

$$\sqrt{[\cos(s + t) - 1]^2 + [\sin(s + t) - 0]^2} = \sqrt{(\cos t - \cos s)^2 + (\sin t + \sin s)^2}$$

Squaring both sides and expanding, we have

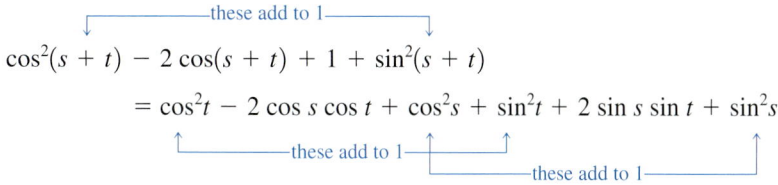

Using the Pythagorean identity $\sin^2\theta + \cos^2\theta = 1$ three times gives

$$2 - 2\cos(s + t) = 2 - 2\cos s \cos t + 2\sin s \sin t$$

Finally, subtracting 2 from each side and dividing both sides by -2, we get

$$\cos(s + t) = \cos s \cos t - \sin s \sin t$$

which proves the addition formula for cosine.

■

■ **Proof of Subtraction Formula for Cosine**   Replacing t with $-t$ in the addition formula for cosine, we get

$$\cos(s - t) = \cos(s + (-t))$$

$$= \cos s \cos(-t) - \sin s \sin(-t) \quad \text{Addition formula for cosine}$$

$$= \cos s \cos t + \sin s \sin t \quad \text{Even-odd identities}$$

This proves the subtraction formula for cosine.   ■

See Exercises 56 and 57 for proofs of the other addition formulas.

Example 1   Using the Addition and Subtraction Formulas

Find the exact value of each expression.

(a) $\cos 75°$ (b) $\cos \dfrac{\pi}{12}$

Solution

(a) Notice that $75° = 45° + 30°$. Since we know the exact values of sine and cosine at $45°$ and $30°$, we use the addition formula for cosine to get

$$\cos 75° = \cos(45° + 30°)$$

$$= \cos 45° \cos 30° - \sin 45° \sin 30°$$

$$= \frac{\sqrt{2}}{2}\frac{\sqrt{3}}{2} - \frac{\sqrt{2}}{2}\frac{1}{2} = \frac{\sqrt{2}\sqrt{3} - \sqrt{2}}{4} = \frac{\sqrt{6} - \sqrt{2}}{4}$$

(b) Since $\dfrac{\pi}{12} = \dfrac{\pi}{4} - \dfrac{\pi}{6}$, the subtraction formula for cosine gives

$$\cos \frac{\pi}{12} = \cos\left(\frac{\pi}{4} - \frac{\pi}{6}\right)$$

$$= \cos \frac{\pi}{4} \cos \frac{\pi}{6} + \sin \frac{\pi}{4} \sin \frac{\pi}{6}$$

$$= \frac{\sqrt{2}}{2}\frac{\sqrt{3}}{2} + \frac{\sqrt{2}}{2}\frac{1}{2} = \frac{\sqrt{6} + \sqrt{2}}{4} \qquad ■$$

Example 2   Using the Addition Formula for Sine

Find the exact value of the expression $\sin 20° \cos 40° + \cos 20° \sin 40°$.

Solution   We recognize the expression as the right-hand side of the addition formula for sine with $s = 20°$ and $t = 40°$. So we have

$$\sin 20° \cos 40° + \cos 20° \sin 40° = \sin(20° + 40°) = \sin 60° = \frac{\sqrt{3}}{2} \qquad ■$$

Example 3   Proving a Cofunction Identity

Prove the cofunction identity $\cos\left(\dfrac{\pi}{2} - u\right) = \sin u$.

Solution   By the subtraction formula for cosine,

$$\cos\left(\frac{\pi}{2} - u\right) = \cos\frac{\pi}{2}\cos u + \sin\frac{\pi}{2}\sin u$$

$$= 0 \cdot \cos u + 1 \cdot \sin u = \sin u \qquad \blacksquare$$

Example 4   Proving an Identity

Verify the identity $\dfrac{1 + \tan x}{1 - \tan x} = \tan\left(\dfrac{\pi}{4} + x\right)$.

Solution   Starting with the right-hand side and using the addition formula for tangent, we get

$$\text{RHS} = \tan\left(\frac{\pi}{4} + x\right) = \frac{\tan\dfrac{\pi}{4} + \tan x}{1 - \tan\dfrac{\pi}{4}\tan x}$$

$$= \frac{1 + \tan x}{1 - \tan x} = \text{LHS} \qquad \blacksquare$$

The next example is a typical use of the addition and subtraction formulas in calculus.

Example 5   An Identity from Calculus

If $f(x) = \sin x$, show that

$$\frac{f(x + h) - f(x)}{h} = \sin x\left(\frac{\cos h - 1}{h}\right) + \cos x\left(\frac{\sin h}{h}\right)$$

Solution

$$\frac{f(x + h) - f(x)}{h} = \frac{\sin(x + h) - \sin x}{h} \qquad \textcolor{blue}{\text{Definition of } f}$$

$$= \frac{\sin x \cos h + \cos x \sin h - \sin x}{h} \qquad \textcolor{blue}{\text{Addition formula for sine}}$$

$$= \frac{\sin x\,(\cos h - 1) + \cos x \sin h}{h} \qquad \textcolor{blue}{\text{Factor}}$$

$$= \sin x\left(\frac{\cos h - 1}{h}\right) + \cos x\left(\frac{\sin h}{h}\right) \qquad \textcolor{blue}{\text{Separate the fraction}} \quad \blacksquare$$

Expressions of the Form $A \sin x + B \cos x$

We can write expressions of the form $A \sin x + B \cos x$ in terms of a single trigonometric function using the addition formula for sine. For example, consider the expression

$$\frac{1}{2} \sin x + \frac{\sqrt{3}}{2} \cos x$$

If we set $\phi = \pi/3$, then $\cos \phi = \frac{1}{2}$ and $\sin \phi = \sqrt{3}/2$, and we can write

$$\frac{1}{2} \sin x + \frac{\sqrt{3}}{2} \cos x = \cos \phi \sin x + \sin \phi \cos x$$

$$= \sin(x + \phi) = \sin\left(x + \frac{\pi}{3}\right)$$

We are able to do this because the coefficients $\frac{1}{2}$ and $\sqrt{3}/2$ are precisely the cosine and sine of a particular number, in this case, $\pi/3$. We can use this same idea in general to write $A \sin x + B \cos x$ in the form $k \sin(x + \phi)$. We start by multiplying the numerator and denominator by $\sqrt{A^2 + B^2}$ to get

$$A \sin x + B \cos x = \sqrt{A^2 + B^2}\left(\frac{A}{\sqrt{A^2 + B^2}} \sin x + \frac{B}{\sqrt{A^2 + B^2}} \cos x\right)$$

We need a number ϕ with the property that

$$\cos \phi = \frac{A}{\sqrt{A^2 + B^2}} \qquad \text{and} \qquad \sin \phi = \frac{B}{\sqrt{A^2 + B^2}}$$

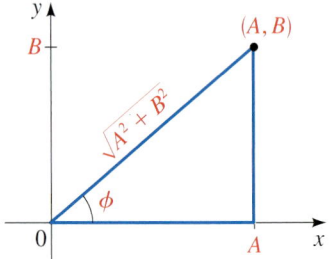

Figure 2

Figure 2 shows that the point (A, B) in the plane determines a number ϕ with precisely this property. With this ϕ, we have

$$A \sin x + B \cos x = \sqrt{A^2 + B^2}(\cos \phi \sin x + \sin \phi \cos x)$$

$$= \sqrt{A^2 + B^2} \sin(x + \phi)$$

We have proved the following theorem.

Sums of Sines and Cosines

If A and B are real numbers, then

$$A \sin x + B \cos x = k \sin(x + \phi)$$

where $k = \sqrt{A^2 + B^2}$ and ϕ satisfies

$$\cos \phi = \frac{A}{\sqrt{A^2 + B^2}} \qquad \text{and} \qquad \sin \phi = \frac{B}{\sqrt{A^2 + B^2}}$$

Example 6   A Sum of Sine and Cosine Terms

Express $3 \sin x + 4 \cos x$ in the form $k \sin(x + \phi)$.

Solution   By the preceding theorem, $k = \sqrt{A^2 + B^2} = \sqrt{3^2 + 4^2} = 5$. The angle ϕ has the property that $\sin \phi = \frac{4}{5}$ and $\cos \phi = \frac{3}{5}$. Using a calculator, we find $\phi \approx 53.1°$. Thus

$$3 \sin x + 4 \cos x \approx 5 \sin(x + 53.1°)$$
∎

Example 7   Graphing a Trigonometric Function

Write the function $f(x) = -\sin 2x + \sqrt{3} \cos 2x$ in the form $k \sin(2x + \phi)$ and use the new form to graph the function.

Solution   Since $A = -1$ and $B = \sqrt{3}$, we have $k = \sqrt{A^2 + B^2} = \sqrt{1 + 3} = 2$. The angle ϕ satisfies $\cos \phi = -\frac{1}{2}$ and $\sin \phi = \sqrt{3}/2$. From the signs of these quantities we conclude that ϕ is in quadrant II. Thus, $\phi = 2\pi/3$. By the preceding theorem we can write

$$f(x) = -\sin 2x + \sqrt{3} \cos 2x = 2 \sin\left(2x + \frac{2\pi}{3}\right)$$

Using the form

$$f(x) = 2 \sin 2\left(x + \frac{\pi}{3}\right)$$

we see that the graph is a sine curve with amplitude 2, period $2\pi/2 = \pi$, and phase shift $-\pi/3$. The graph is shown in Figure 3.
∎

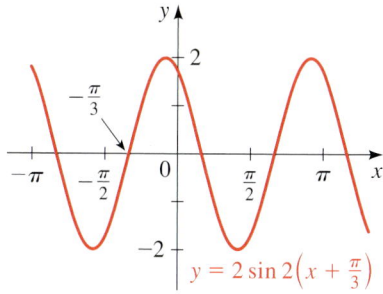

$$y = 2 \sin 2\left(x + \tfrac{\pi}{3}\right)$$

Figure 3

7.2   Exercises

1–12 ■ Use an addition or subtraction formula to find the exact value of the expression, as demonstrated in Example 1.

1. $\sin 75°$

2. $\sin 15°$

3. $\cos 105°$

4. $\cos 195°$

5. $\tan 15°$

6. $\tan 165°$

7. $\sin \dfrac{19\pi}{12}$

8. $\cos \dfrac{17\pi}{12}$

9. $\tan\left(-\dfrac{\pi}{12}\right)$

10. $\sin\left(-\dfrac{5\pi}{12}\right)$

11. $\cos \dfrac{11\pi}{12}$

12. $\tan \dfrac{7\pi}{12}$

13–18 ■ Use an addition or subtraction formula to write the expression as a trigonometric function of one number, and then find its exact value.

13. $\sin 18° \cos 27° + \cos 18° \sin 27°$

14. $\cos 10° \cos 80° - \sin 10° \sin 80°$

15. $\cos \dfrac{3\pi}{7} \cos \dfrac{2\pi}{21} + \sin \dfrac{3\pi}{7} \sin \dfrac{2\pi}{21}$

16. $\dfrac{\tan \dfrac{\pi}{18} + \tan \dfrac{\pi}{9}}{1 - \tan \dfrac{\pi}{18} \tan \dfrac{\pi}{9}}$

17. $\dfrac{\tan 73° - \tan 13°}{1 + \tan 73° \tan 13°}$

18. $\cos \dfrac{13\pi}{15} \cos\left(-\dfrac{\pi}{5}\right) - \sin \dfrac{13\pi}{15} \sin\left(-\dfrac{\pi}{5}\right)$

19–22 ■ Prove the cofunction identity using the addition and subtraction formulas.

19. $\tan\left(\dfrac{\pi}{2} - u\right) = \cot u$

20. $\cot\left(\dfrac{\pi}{2} - u\right) = \tan u$

21. $\sec\left(\dfrac{\pi}{2} - u\right) = \csc u$

22. $\csc\left(\dfrac{\pi}{2} - u\right) = \sec u$

23–40 ■ Prove the identity.

23. $\sin\left(x - \dfrac{\pi}{2}\right) = -\cos x$

24. $\cos\left(x - \dfrac{\pi}{2}\right) = \sin x$

25. $\sin(x - \pi) = -\sin x$ **26.** $\cos(x - \pi) = -\cos x$

27. $\tan(x - \pi) = \tan x$

28. $\sin\left(\dfrac{\pi}{2} - x\right) = \sin\left(\dfrac{\pi}{2} + x\right)$

29. $\cos\left(x + \dfrac{\pi}{6}\right) + \sin\left(x - \dfrac{\pi}{3}\right) = 0$

30. $\tan\left(x - \dfrac{\pi}{4}\right) = \dfrac{\tan x - 1}{\tan x + 1}$

31. $\sin(x + y) - \sin(x - y) = 2 \cos x \sin y$

32. $\cos(x + y) + \cos(x - y) = 2 \cos x \cos y$

33. $\cot(x - y) = \dfrac{\cot x \cot y + 1}{\cot y - \cot x}$

34. $\cot(x + y) = \dfrac{\cot x \cot y - 1}{\cot x + \cot y}$

35. $\tan x - \tan y = \dfrac{\sin(x - y)}{\cos x \cos y}$

36. $1 - \tan x \tan y = \dfrac{\cos(x + y)}{\cos x \cos y}$

37. $\dfrac{\sin(x + y) - \sin(x - y)}{\cos(x + y) + \cos(x - y)} = \tan y$

38. $\cos(x + y) \cos(x - y) = \cos^2 x - \sin^2 y$

39. $\sin(x + y + z) = \sin x \cos y \cos z + \cos x \sin y \cos z$
$\qquad\qquad + \cos x \cos y \sin z - \sin x \sin y \sin z$

40. $\tan(x - y) + \tan(y - z) + \tan(z - x)$
$\qquad\qquad = \tan(x - y) \tan(y - z) \tan(z - x)$

41–44 ■ Write the expression in terms of sine only.

41. $-\sqrt{3} \sin x + \cos x$ **42.** $\sin x + \cos x$

43. $5(\sin 2x - \cos 2x)$ **44.** $3 \sin \pi x + 3\sqrt{3} \cos \pi x$

45–46 ■ **(a)** Express the function in terms of sine only.
(b) Graph the function.

45. $f(x) = \sin x + \cos x$ **46.** $g(x) = \cos 2x + \sqrt{3} \sin 2x$

47. Show that if $\beta - \alpha = \pi/2$, then

$$\sin(x + \alpha) + \cos(x + \beta) = 0$$

48. Let $g(x) = \cos x$. Show that

$$\frac{g(x + h) - g(x)}{h} = -\cos x \left(\frac{1 - \cos h}{h}\right) - \sin x \left(\frac{\sin h}{h}\right)$$

49. Refer to the figure. Show that $\alpha + \beta = \gamma$, and find $\tan \gamma$.

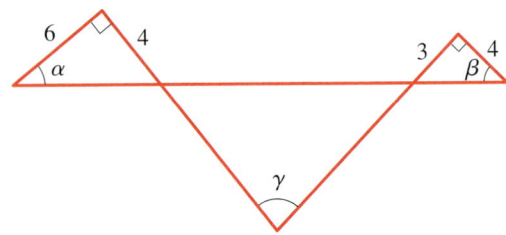

50. **(a)** If L is a line in the plane and θ is the angle formed by the line and the x-axis as shown in the figure, show that the slope m of the line is given by

$$m = \tan \theta$$

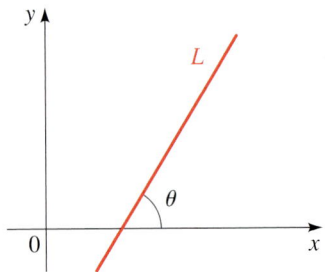

(b) Let L_1 and L_2 be two nonparallel lines in the plane with slopes m_1 and m_2, respectively. Let ψ be the acute angle formed by the two lines (see the figure). Show that

$$\tan \psi = \frac{m_2 - m_1}{1 + m_1 m_2}$$

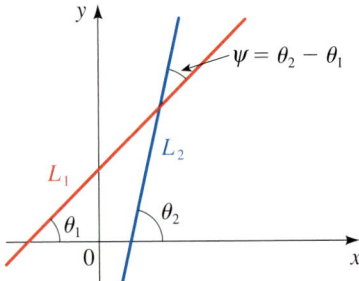

(c) Find the acute angle formed by the two lines

$$y = \tfrac{1}{3}x + 1 \qquad \text{and} \qquad y = -\tfrac{1}{2}x - 3$$

(d) Show that if two lines are perpendicular, then the slope of one is the negative reciprocal of the slope of the other. [*Hint:* First find an expression for $\cot \psi$.]

51–52 ■ **(a)** Graph the function and make a conjecture, then **(b)** prove that your conjecture is true.

51. $y = \sin^2\left(x + \dfrac{\pi}{4}\right) + \sin^2\left(x - \dfrac{\pi}{4}\right)$

52. $y = -\frac{1}{2}[\cos(x + \pi) + \cos(x - \pi)]$

53. Find $\angle A + \angle B + \angle C$ in the figure. [*Hint:* First use an addition formula to find $\tan(A + B)$.]

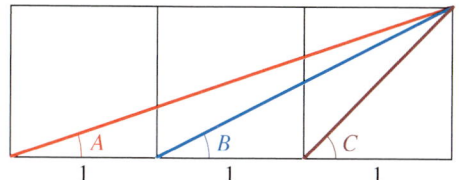

Applications

 54. Adding an Echo A digital delay-device echoes an input signal by repeating it a fixed length of time after it is received. If such a device receives the pure note $f_1(t) = 5 \sin t$ and echoes the pure note $f_2(t) = 5 \cos t$, then the combined sound is $f(t) = f_1(t) + f_2(t)$.

(a) Graph $y = f(t)$ and observe that the graph has the form of a sine curve $y = k \sin(t + \phi)$.

(b) Find k and ϕ.

55. Interference Two identical tuning forks are struck, one a fraction of a second after the other. The sounds produced are modeled by $f_1(t) = C \sin \omega t$ and $f_2(t) = C \sin(\omega t + \alpha)$. The two sound waves interfere to produce a single sound modeled by the sum of these functions

$$f(t) = C \sin \omega t + C \sin(\omega t + \alpha)$$

(a) Use the addition formula for sine to show that f can be written in the form $f(t) = A \sin \omega t + B \cos \omega t$, where A and B are constants that depend on α.

(b) Suppose that $C = 10$ and $\alpha = \pi/3$. Find constants k and ϕ so that $f(t) = k \sin(\omega t + \phi)$.

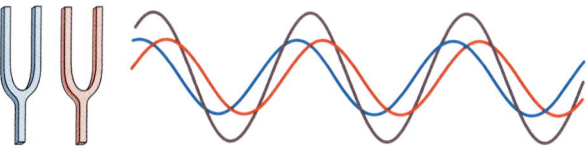

Discovery • Discussion

56. Addition Formula for Sine In the text we proved only the addition and subtraction formulas for cosine. Use these formulas and the cofunction identities

$$\sin x = \cos\left(\frac{\pi}{2} - x\right)$$

$$\cos x = \sin\left(\frac{\pi}{2} - x\right)$$

to prove the addition formula for sine. [*Hint:* To get started, use the first cofunction identity to write

$$\sin(s + t) = \cos\left(\frac{\pi}{2} - (s + t)\right)$$

$$= \cos\left(\left(\frac{\pi}{2} - s\right) - t\right)$$

and use the subtraction formula for cosine.]

57. Addition Formula for Tangent Use the addition formulas for cosine and sine to prove the addition formula for tangent. [*Hint:* Use

$$\tan(s + t) = \frac{\sin(s + t)}{\cos(s + t)}$$

and divide the numerator and denominator by $\cos s \cos t$.]

7.3 # Double-Angle, Half-Angle, and Product-Sum Formulas

The identities we consider in this section are consequences of the addition formulas. The **double-angle formulas** allow us to find the values of the trigonometric functions at $2x$ from their values at x. The **half-angle formulas** relate the values of the trigonometric functions at $\frac{1}{2}x$ to their values at x. The **product-sum formulas** relate products of sines and cosines to sums of sines and cosines.

Double-Angle Formulas

The formulas in the following box are immediate consequences of the addition formulas, which we proved in the preceding section.

Double-Angle Formulas

Formula for sine: $\quad \sin 2x = 2 \sin x \cos x$

Formulas for cosine: $\quad \cos 2x = \cos^2 x - \sin^2 x$

$$= 1 - 2 \sin^2 x$$

$$= 2 \cos^2 x - 1$$

Formula for tangent: $\quad \tan 2x = \dfrac{2 \tan x}{1 - \tan^2 x}$

The proofs for the formulas for cosine are given here. You are asked to prove the remaining formulas in Exercises 33 and 34.

■ **Proof of Double-Angle Formulas for Cosine**

$$\cos 2x = \cos(x + x)$$

$$= \cos x \cos x - \sin x \sin x$$

$$= \cos^2 x - \sin^2 x$$

The second and third formulas for $\cos 2x$ are obtained from the formula we just proved and the Pythagorean identity. Substituting $\cos^2 x = 1 - \sin^2 x$ gives

$$\cos 2x = \cos^2 x - \sin^2 x$$

$$= (1 - \sin^2 x) - \sin^2 x$$

$$= 1 - 2 \sin^2 x$$

The third formula is obtained in the same way, by substituting $\sin^2 x = 1 - \cos^2 x$. ■

Example 1 Using the Double-Angle Formulas

If $\cos x = -\frac{2}{3}$ and x is in quadrant II, find $\cos 2x$ and $\sin 2x$.

Solution Using one of the double-angle formulas for cosine, we get

$$\cos 2x = 2 \cos^2 x - 1$$

$$= 2\left(-\frac{2}{3}\right)^2 - 1 = \frac{8}{9} - 1 = -\frac{1}{9}$$

To use the formula $\sin 2x = 2 \sin x \cos x$, we need to find $\sin x$ first. We have

$$\sin x = \sqrt{1 - \cos^2 x} = \sqrt{1 - \left(-\frac{2}{3}\right)^2} = \frac{\sqrt{5}}{3}$$

where we have used the positive square root because $\sin x$ is positive in quadrant II. Thus

$$\sin 2x = 2 \sin x \cos x$$

$$= 2\left(\frac{\sqrt{5}}{3}\right)\left(-\frac{2}{3}\right) = -\frac{4\sqrt{5}}{9}$$ ■

Example 2 A Triple-Angle Formula

Write $\cos 3x$ in terms of $\cos x$.

Solution

$$\cos 3x = \cos(2x + x)$$

$$= \cos 2x \cos x - \sin 2x \sin x \qquad \text{Addition formula}$$

$$= (2\cos^2 x - 1)\cos x - (2\sin x \cos x)\sin x \qquad \text{Double-angle formulas}$$

$$= 2\cos^3 x - \cos x - 2\sin^2 x \cos x \qquad \text{Expand}$$

$$= 2\cos^3 x - \cos x - 2\cos x(1 - \cos^2 x) \qquad \text{Pythagorean identity}$$

$$= 2\cos^3 x - \cos x - 2\cos x + 2\cos^3 x \qquad \text{Expand}$$

$$= 4\cos^3 x - 3\cos x \qquad \text{Simplify} \qquad \blacksquare$$

Example 2 shows that $\cos 3x$ can be written as a polynomial of degree 3 in $\cos x$. The identity $\cos 2x = 2\cos^2 x - 1$ shows that $\cos 2x$ is a polynomial of degree 2 in $\cos x$. In fact, for any natural number n, we can write $\cos nx$ as a polynomial in $\cos x$ of degree n (see Exercise 87). The analogous result for $\sin nx$ is not true in general.

Example 3 Proving an Identity

Prove the identity $\dfrac{\sin 3x}{\sin x \cos x} = 4\cos x - \sec x$.

Solution We start with the left-hand side.

$$\frac{\sin 3x}{\sin x \cos x} = \frac{\sin(x + 2x)}{\sin x \cos x}$$

$$= \frac{\sin x \cos 2x + \cos x \sin 2x}{\sin x \cos x} \qquad \text{Addition formula}$$

$$= \frac{\sin x(2\cos^2 x - 1) + \cos x(2\sin x \cos x)}{\sin x \cos x} \qquad \text{Double-angle formulas}$$

$$= \frac{\sin x(2\cos^2 x - 1)}{\sin x \cos x} + \frac{\cos x(2\sin x \cos x)}{\sin x \cos x} \qquad \text{Separate fraction}$$

$$= \frac{2\cos^2 x - 1}{\cos x} + 2\cos x \qquad \text{Cancel}$$

$$= 2\cos x - \frac{1}{\cos x} + 2\cos x \qquad \text{Separate fraction}$$

$$= 4\cos x - \sec x \qquad \text{Reciprocal identity} \qquad \blacksquare$$

Half-Angle Formulas

The following formulas allow us to write any trigonometric expression involving even powers of sine and cosine in terms of the first power of cosine only. This technique is important in calculus. The half-angle formulas are immediate consequences of these formulas.

Formulas for Lowering Powers

$$\sin^2 x = \frac{1 - \cos 2x}{2} \qquad \cos^2 x = \frac{1 + \cos 2x}{2}$$

$$\tan^2 x = \frac{1 - \cos 2x}{1 + \cos 2x}$$

■ **Proof** The first formula is obtained by solving for $\sin^2 x$ in the double-angle formula $\cos 2x = 1 - 2\sin^2 x$. Similarly, the second formula is obtained by solving for $\cos^2 x$ in the double-angle formula $\cos 2x = 2\cos^2 x - 1$.

The last formula follows from the first two and the reciprocal identities:

$$\tan^2 x = \frac{\sin^2 x}{\cos^2 x} = \frac{\dfrac{1 - \cos 2x}{2}}{\dfrac{1 + \cos 2x}{2}} = \frac{1 - \cos 2x}{1 + \cos 2x} \qquad ■$$

Example 4 Lowering Powers in a Trigonometric Expression

Express $\sin^2 x \cos^2 x$ in terms of the first power of cosine.

Solution We use the formulas for lowering powers repeatedly.

$$\sin^2 x \cos^2 x = \left(\frac{1 - \cos 2x}{2}\right)\left(\frac{1 + \cos 2x}{2}\right)$$

$$= \frac{1 - \cos^2 2x}{4} = \frac{1}{4} - \frac{1}{4}\cos^2 2x$$

$$= \frac{1}{4} - \frac{1}{4}\left(\frac{1 + \cos 4x}{2}\right) = \frac{1}{4} - \frac{1}{8} - \frac{\cos 4x}{8}$$

$$= \frac{1}{8} - \frac{1}{8}\cos 4x = \frac{1}{8}(1 - \cos 4x)$$

Another way to obtain this identity is to use the double-angle formula for sine in the form $\sin x \cos x = \frac{1}{2}\sin 2x$. Thus

$$\sin^2 x \cos^2 x = \frac{1}{4}\sin^2 2x = \frac{1}{4}\left(\frac{1 - \cos 4x}{2}\right) = \frac{1}{8}(1 - \cos 4x) \qquad ■$$

Half-Angle Formulas

$$\sin\frac{u}{2} = \pm\sqrt{\frac{1 - \cos u}{2}} \qquad \cos\frac{u}{2} = \pm\sqrt{\frac{1 + \cos u}{2}}$$

$$\tan\frac{u}{2} = \frac{1 - \cos u}{\sin u} = \frac{\sin u}{1 + \cos u}$$

The choice of the $+$ or $-$ sign depends on the quadrant in which $u/2$ lies.

■ **Proof**   We substitute $x = u/2$ in the formulas for lowering powers and take the square root of each side. This yields the first two half-angle formulas. In the case of the half-angle formula for tangent, we get

$$\tan \frac{u}{2} = \pm \sqrt{\frac{1 - \cos u}{1 + \cos u}}$$

$$= \pm \sqrt{\left(\frac{1 - \cos u}{1 + \cos u}\right)\left(\frac{1 - \cos u}{1 - \cos u}\right)} \qquad \text{Multiply numerator and denominator by } 1 - \cos u$$

$$= \pm \sqrt{\frac{(1 - \cos u)^2}{1 - \cos^2 u}} \qquad \text{Simplify}$$

$$= \pm \frac{|1 - \cos u|}{|\sin u|} \qquad \sqrt{A^2} = |A| \\ \text{and } 1 - \cos^2 u = \sin^2 u$$

Now, $1 - \cos u$ is nonnegative for all values of u. It is also true that $\sin u$ and $\tan(u/2)$ always have the same sign. (Verify this.) It follows that

$$\tan \frac{u}{2} = \frac{1 - \cos u}{\sin u}$$

The other half-angle formula for tangent is derived from this by multiplying the numerator and denominator by $1 + \cos u$.   ■

Example 5   Using a Half-Angle Formula

Find the exact value of $\sin 22.5°$.

Solution   Since $22.5°$ is half of $45°$, we use the half-angle formula for sine with $u = 45°$. We choose the $+$ sign because $22.5°$ is in the first quadrant.

$$\sin \frac{45°}{2} = \sqrt{\frac{1 - \cos 45°}{2}} \qquad \text{Half-angle formula}$$

$$= \sqrt{\frac{1 - \sqrt{2}/2}{2}} \qquad \cos 45° = \sqrt{2}/2$$

$$= \sqrt{\frac{2 - \sqrt{2}}{4}} \qquad \text{Common denominator}$$

$$= \tfrac{1}{2}\sqrt{2 - \sqrt{2}} \qquad \text{Simplify}$$ ■

Example 6   Using a Half-Angle Formula

Find $\tan(u/2)$ if $\sin u = \tfrac{2}{5}$ and u is in quadrant II.

Solution   To use the half-angle formulas for tangent, we first need to find $\cos u$. Since cosine is negative in quadrant II, we have

$$\cos u = -\sqrt{1 - \sin^2 u}$$

$$= -\sqrt{1 - \left(\tfrac{2}{5}\right)^2} = -\frac{\sqrt{21}}{5}$$

Thus
$$\tan \frac{u}{2} = \frac{1 - \cos u}{\sin u}$$

$$= \frac{1 + \sqrt{21}/5}{\frac{2}{5}} = \frac{5 + \sqrt{21}}{2}$$ ∎

Product-Sum Formulas

It is possible to write the product $\sin u \cos v$ as a sum of trigonometric functions. To see this, consider the addition and subtraction formulas for the sine function:

$$\sin(u + v) = \sin u \cos v + \cos u \sin v$$

$$\sin(u - v) = \sin u \cos v - \cos u \sin v$$

Adding the left- and right-hand sides of these formulas gives

$$\sin(u + v) + \sin(u - v) = 2 \sin u \cos v$$

Dividing by 2 yields the formula

$$\sin u \cos v = \tfrac{1}{2}[\sin(u + v) + \sin(u - v)]$$

The other three **product-to-sum formulas** follow from the addition formulas in a similar way.

Product-to-Sum Formulas

$$\sin u \cos v = \tfrac{1}{2}[\sin(u + v) + \sin(u - v)]$$
$$\cos u \sin v = \tfrac{1}{2}[\sin(u + v) - \sin(u - v)]$$
$$\cos u \cos v = \tfrac{1}{2}[\cos(u + v) + \cos(u - v)]$$
$$\sin u \sin v = \tfrac{1}{2}[\cos(u - v) - \cos(u + v)]$$

Example 7   **Expressing a Trigonometric Product as a Sum**

Express $\sin 3x \sin 5x$ as a sum of trigonometric functions.

Solution   Using the fourth product-to-sum formula with $u = 3x$ and $v = 5x$ and the fact that cosine is an even function, we get

$$\sin 3x \sin 5x = \tfrac{1}{2}[\cos(3x - 5x) - \cos(3x + 5x)]$$

$$= \tfrac{1}{2}\cos(-2x) - \tfrac{1}{2}\cos 8x$$

$$= \tfrac{1}{2}\cos 2x - \tfrac{1}{2}\cos 8x$$ ∎

The product-to-sum formulas can also be used as sum-to-product formulas. This is possible because the right-hand side of each product-to-sum formula is a sum and the left side is a product. For example, if we let

$$u = \frac{x + y}{2} \qquad \text{and} \qquad v = \frac{x - y}{2}$$

in the first product-to-sum formula, we get

$$\sin \frac{x+y}{2} \cos \frac{x-y}{2} = \tfrac{1}{2}(\sin x + \sin y)$$

so

$$\sin x + \sin y = 2 \sin \frac{x+y}{2} \cos \frac{x-y}{2}$$

The remaining three of the following **sum-to-product formulas** are obtained in a similar manner.

Sum-to-Product Formulas

$$\sin x + \sin y = 2 \sin \frac{x+y}{2} \cos \frac{x-y}{2}$$

$$\sin x - \sin y = 2 \cos \frac{x+y}{2} \sin \frac{x-y}{2}$$

$$\cos x + \cos y = 2 \cos \frac{x+y}{2} \cos \frac{x-y}{2}$$

$$\cos x - \cos y = -2 \sin \frac{x+y}{2} \sin \frac{x-y}{2}$$

Example 8 Expressing a Trigonometric Sum as a Product

Write $\sin 7x + \sin 3x$ as a product.

Solution The first sum-to-product formula gives

$$\sin 7x + \sin 3x = 2 \sin \frac{7x+3x}{2} \cos \frac{7x-3x}{2}$$

$$= 2 \sin 5x \cos 2x \qquad \blacksquare$$

Example 9 Proving an Identity

Verify the identity $\dfrac{\sin 3x - \sin x}{\cos 3x + \cos x} = \tan x$.

Solution We apply the second sum-to-product formula to the numerator and the third formula to the denominator.

$$\text{LHS} = \frac{\sin 3x - \sin x}{\cos 3x + \cos x} = \frac{2 \cos \dfrac{3x+x}{2} \sin \dfrac{3x-x}{2}}{2 \cos \dfrac{3x+x}{2} \cos \dfrac{3x-x}{2}} \qquad \text{Sum-to-product formulas}$$

$$= \frac{2 \cos 2x \sin x}{2 \cos 2x \cos x} \qquad \text{Simplify}$$

$$= \frac{\sin x}{\cos x} = \tan x = \text{RHS} \qquad \text{Cancel} \qquad \blacksquare$$

7.3 Exercises

1–8 ■ Find $\sin 2x$, $\cos 2x$, and $\tan 2x$ from the given information.

1. $\sin x = \frac{5}{13}$, x in quadrant I

2. $\tan x = -\frac{4}{3}$, x in quadrant II

3. $\cos x = \frac{4}{5}$, $\csc x < 0$ 4. $\csc x = 4$, $\tan x < 0$

5. $\sin x = -\frac{3}{5}$, x in quadrant III

6. $\sec x = 2$, x in quadrant IV

7. $\tan x = -\frac{1}{3}$, $\cos x > 0$

8. $\cot x = \frac{2}{3}$, $\sin x > 0$

9–14 ■ Use the formulas for lowering powers to rewrite the expression in terms of the first power of cosine, as in Example 4.

9. $\sin^4 x$ 10. $\cos^4 x$

11. $\cos^2 x \sin^4 x$ 12. $\cos^4 x \sin^2 x$

13. $\cos^4 x \sin^4 x$ 14. $\cos^6 x$

15–26 ■ Use an appropriate half-angle formula to find the exact value of the expression.

15. $\sin 15°$ 16. $\tan 15°$

17. $\tan 22.5°$ 18. $\sin 75°$

19. $\cos 165°$ 20. $\cos 112.5°$

21. $\tan \dfrac{\pi}{8}$ 22. $\cos \dfrac{3\pi}{8}$

23. $\cos \dfrac{\pi}{12}$ 24. $\tan \dfrac{5\pi}{12}$

25. $\sin \dfrac{9\pi}{8}$ 26. $\sin \dfrac{11\pi}{12}$

27–32 ■ Simplify the expression by using a double-angle formula or a half-angle formula.

27. (a) $2 \sin 18° \cos 18°$ (b) $2 \sin 3\theta \cos 3\theta$

28. (a) $\dfrac{2 \tan 7°}{1 - \tan^2 7°}$ (b) $\dfrac{2 \tan 7\theta}{1 - \tan^2 7\theta}$

29. (a) $\cos^2 34° - \sin^2 34°$ (b) $\cos^2 5\theta - \sin^2 5\theta$

30. (a) $\cos^2 \dfrac{\theta}{2} - \sin^2 \dfrac{\theta}{2}$ (b) $2 \sin \dfrac{\theta}{2} \cos \dfrac{\theta}{2}$

31. (a) $\dfrac{\sin 8°}{1 + \cos 8°}$ (b) $\dfrac{1 - \cos 4\theta}{\sin 4\theta}$

32. (a) $\sqrt{\dfrac{1 - \cos 30°}{2}}$ (b) $\sqrt{\dfrac{1 - \cos 8\theta}{2}}$

33. Use the addition formula for sine to prove the double-angle formula for sine.

34. Use the addition formula for tangent to prove the double-angle formula for tangent.

35–40 ■ Find $\sin \dfrac{x}{2}$, $\cos \dfrac{x}{2}$, and $\tan \dfrac{x}{2}$ from the given information.

35. $\sin x = \frac{3}{5}$, $0° < x < 90°$

36. $\cos x = -\frac{4}{5}$, $180° < x < 270°$

37. $\csc x = 3$, $90° < x < 180°$

38. $\tan x = 1$, $0° < x < 90°$

39. $\sec x = \frac{3}{2}$, $270° < x < 360°$

40. $\cot x = 5$, $180° < x < 270°$

41–46 ■ Write the product as a sum.

41. $\sin 2x \cos 3x$ 42. $\sin x \sin 5x$

43. $\cos x \sin 4x$ 44. $\cos 5x \cos 3x$

45. $3 \cos 4x \cos 7x$ 46. $11 \sin \dfrac{x}{2} \cos \dfrac{x}{4}$

47–52 ■ Write the sum as a product.

47. $\sin 5x + \sin 3x$ 48. $\sin x - \sin 4x$

49. $\cos 4x - \cos 6x$ 50. $\cos 9x + \cos 2x$

51. $\sin 2x - \sin 7x$ 52. $\sin 3x + \sin 4x$

53–58 ■ Find the value of the product or sum.

53. $2 \sin 52.5° \sin 97.5°$ 54. $3 \cos 37.5° \cos 7.5°$

55. $\cos 37.5° \sin 7.5°$ 56. $\sin 75° + \sin 15°$

57. $\cos 255° - \cos 195°$ 58. $\cos \dfrac{\pi}{12} + \cos \dfrac{5\pi}{12}$

59–76 ■ Prove the identity.

59. $\cos^2 5x - \sin^2 5x = \cos 10x$

60. $\sin 8x = 2 \sin 4x \cos 4x$

61. $(\sin x + \cos x)^2 = 1 + \sin 2x$

62. $\dfrac{2 \tan x}{1 + \tan^2 x} = \sin 2x$ 63. $\dfrac{\sin 4x}{\sin x} = 4 \cos x \cos 2x$

64. $\dfrac{1 + \sin 2x}{\sin 2x} = 1 + \frac{1}{2} \sec x \csc x$

65. $\dfrac{2(\tan x - \cot x)}{\tan^2 x - \cot^2 x} = \sin 2x$ 66. $\cot 2x = \dfrac{1 - \tan^2 x}{2 \tan x}$

67. $\tan 3x = \dfrac{3 \tan x - \tan^3 x}{1 - 3 \tan^2 x}$

68. $4(\sin^6 x + \cos^6 x) = 4 - 3 \sin^2 2x$

69. $\cos^4 x - \sin^4 x = \cos 2x$

70. $\tan^2\left(\dfrac{x}{2} + \dfrac{\pi}{4}\right) = \dfrac{1 + \sin x}{1 - \sin x}$

71. $\dfrac{\sin x + \sin 5x}{\cos x + \cos 5x} = \tan 3x$ **72.** $\dfrac{\sin 3x + \sin 7x}{\cos 3x - \cos 7x} = \cot 2x$

73. $\dfrac{\sin 10x}{\sin 9x + \sin x} = \dfrac{\cos 5x}{\cos 4x}$

74. $\dfrac{\sin x + \sin 3x + \sin 5x}{\cos x + \cos 3x + \cos 5x} = \tan 3x$

75. $\dfrac{\sin x + \sin y}{\cos x + \cos y} = \tan\left(\dfrac{x + y}{2}\right)$

76. $\tan y = \dfrac{\sin(x + y) - \sin(x - y)}{\cos(x + y) + \cos(x - y)}$

77. Show that $\sin 130° - \sin 110° = -\sin 10°$.

78. Show that $\cos 100° - \cos 200° = \sin 50°$.

79. Show that $\sin 45° + \sin 15° = \sin 75°$.

80. Show that $\cos 87° + \cos 33° = \sin 63°$.

81. Prove the identity

$$\dfrac{\sin x + \sin 2x + \sin 3x + \sin 4x + \sin 5x}{\cos x + \cos 2x + \cos 3x + \cos 4x + \cos 5x} = \tan 3x$$

82. Use the identity

$$\sin 2x = 2 \sin x \cos x$$

n times to show that

$$\sin(2^n x) = 2^n \sin x \cos x \cos 2x \cos 4x \cdots \cos 2^{n-1} x$$

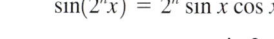 **83. (a)** Graph $f(x) = \dfrac{\sin 3x}{\sin x} - \dfrac{\cos 3x}{\cos x}$ and make a conjecture.

  (b) Prove the conjecture you made in part (a).

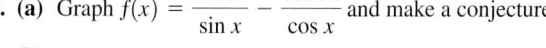 **84. (a)** Graph $f(x) = \cos 2x + 2 \sin^2 x$ and make a conjecture.

  (b) Prove the conjecture you made in part (a).

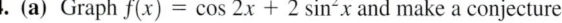 **85.** Let $f(x) = \sin 6x + \sin 7x$.

  (a) Graph $y = f(x)$.

  (b) Verify that $f(x) = 2 \cos \frac{1}{2}x \sin \frac{13}{2}x$.

  (c) Graph $y = 2 \cos \frac{1}{2}x$ and $y = -2 \cos \frac{1}{2}x$, together with the graph in part (a), in the same viewing rectangle. How are these graphs related to the graph of f?

86. Let $3x = \pi/3$ and let $y = \cos x$. Use the result of Example 2 to show that y satisfies the equation

$$8y^3 - 6y - 1 = 0$$

NOTE This equation has roots of a certain kind that are used to show that the angle $\pi/3$ cannot be trisected using a ruler and compass only.

87. (a) Show that there is a polynomial $P(t)$ of degree 4 such that $\cos 4x = P(\cos x)$ (see Example 2).

  (b) Show that there is a polynomial $Q(t)$ of degree 5 such that $\cos 5x = Q(\cos x)$.

NOTE In general, there is a polynomial $P_n(t)$ of degree n such that $\cos nx = P_n(\cos x)$. These polynomials are called *Tchebycheff polynomials*, after the Russian mathematician P. L. Tchebycheff (1821–1894).

88. In triangle ABC (see the figure) the line segment s bisects angle C. Show that the length of s is given by

$$s = \dfrac{2ab \cos x}{a + b}$$

[*Hint:* Use the Law of Sines.]

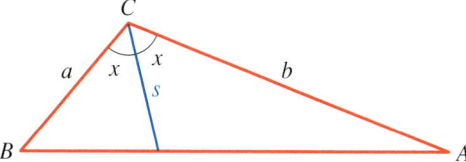

89. If A, B, and C are the angles in a triangle, show that

$$\sin 2A + \sin 2B + \sin 2C = 4 \sin A \sin B \sin C$$

90. A rectangle is to be inscribed in a semicircle of radius 5 cm as shown in the figure.

  (a) Show that the area of the rectangle is modeled by the function

$$A(\theta) = 25 \sin 2\theta$$

  (b) Find the largest possible area for such an inscribed rectangle.

  (c) Find the dimensions of the inscribed rectangle with the largest possible area.

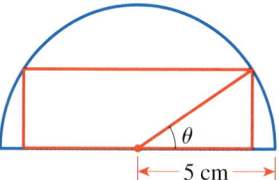

Applications

91. Sawing a Wooden Beam   A rectangular beam is to be cut from a cylindrical log of diameter 20 in.

  (a) Show that the cross-sectional area of the beam is modeled by the function

$$A(\theta) = 200 \sin 2\theta$$

where θ is as shown in the figure on the next page.

(b) Show that the maximum cross-sectional area of such a beam is 200 in^2. [*Hint:* Use the fact that $\sin u$ achieves its maximum value at $u = \pi/2$.]

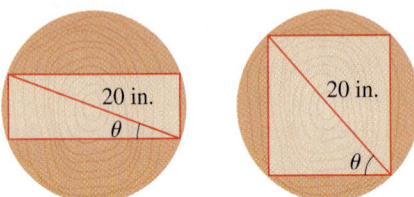

92. Length of a Fold The lower right-hand corner of a long piece of paper 6 in. wide is folded over to the left-hand edge as shown. The length L of the fold depends on the angle θ. Show that

$$L = \frac{3}{\sin\theta \cos^2\theta}$$

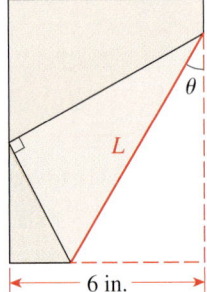

93. Sound Beats When two pure notes that are close in frequency are played together, their sounds interfere to produce *beats*; that is, the loudness (or amplitude) of the sound alternately increases and decreases. If the two notes are given by

$$f_1(t) = \cos 11t \quad \text{and} \quad f_2(t) = \cos 13t$$

the resulting sound is $f(t) = f_1(t) + f_2(t)$.
(a) Graph the function $y = f(t)$.
(b) Verify that $f(t) = 2 \cos t \cos 12t$.
(c) Graph $y = 2 \cos t$ and $y = -2 \cos t$, together with the graph in part (a), in the same viewing rectangle. How do these graphs describe the variation in the loudness of the sound?

94. Touch-Tone Telephones When a key is pressed on a touch-tone telephone, the keypad generates two pure tones, which combine to produce a sound that uniquely identifies the key. The figure shows the low frequency f_1 and the high frequency f_2 associated with each key. Pressing a key produces the sound wave $y = \sin(2\pi f_1 t) + \sin(2\pi f_2 t)$.
(a) Find the function that models the sound produced when the 4 key is pressed.
(b) Use a sum-to-product formula to express the sound generated by the 4 key as a product of a sine and a cosine function.
(c) Graph the sound wave generated by the 4 key, from $t = 0$ to $t = 0.006$ s.

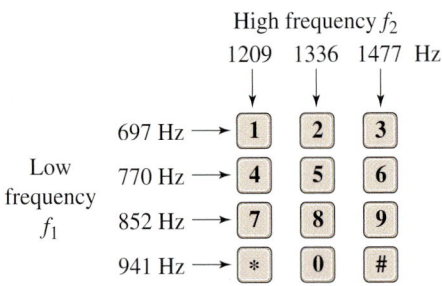

Discovery • Discussion

95. Geometric Proof of a Double-Angle Formula Use the figure to prove that $\sin 2\theta = 2 \sin\theta \cos\theta$.

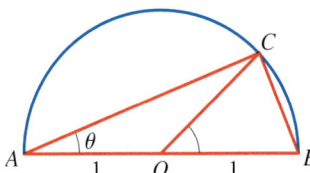

Hint: Find the area of triangle ABC in two different ways. You will need the following facts from geometry:

An angle inscribed in a semicircle is a right angle, so $\angle ACB$ is a right angle.

The central angle subtended by the chord of a circle is twice the angle subtended by the chord on the circle, so $\angle BOC$ is 2θ.

7.4	**Inverse Trigonometric Functions**

If f is a one-to-one function with domain A and range B, then its inverse f^{-1} is the function with domain B and range A defined by

$$f^{-1}(x) = y \quad \Leftrightarrow \quad f(y) = x$$

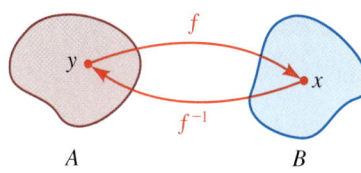

Figure 1

$f^{-1}(x) = y \iff f(y) = x$

(See Section 2.8.) In other words, f^{-1} is the rule that reverses the action of f. Figure 1 represents the actions of f and f^{-1} graphically.

For a function to have an inverse, it must be one-to-one. Since the trigonometric functions are not one-to-one, they do not have inverses. It is possible, however, to restrict the domains of the trigonometric functions in such a way that the resulting functions are one-to-one.

The Inverse Sine Function

Let's first consider the sine function. There are many ways to restrict the domain of sine so that the new function is one-to-one. A natural way to do this is to restrict the domain to the interval $[-\pi/2, \pi/2]$. The reason for this choice is that sine attains each of its values exactly once on this interval. As we see in Figure 2, on this restricted domain the sine function is one-to-one (by the Horizontal Line Test), and so has an inverse.

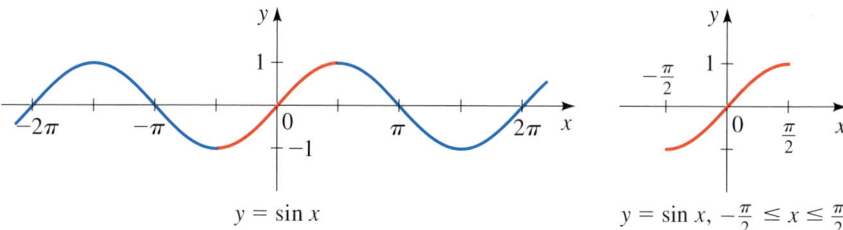

Figure 2

$y = \sin x$ $y = \sin x,\ -\dfrac{\pi}{2} \le x \le \dfrac{\pi}{2}$

The inverse of the function sin is the function $\sin^{-1}$ defined by

$$\sin^{-1}x = y \quad \iff \quad \sin y = x$$

for $-1 \le x \le 1$ and $-\pi/2 \le y \le \pi/2$. The graph of $y = \sin^{-1}x$ is shown in Figure 3; it is obtained by reflecting the graph of $y = \sin x$, $-\pi/2 \le x \le \pi/2$, in the line $y = x$.

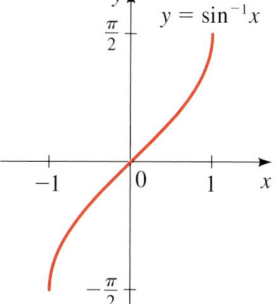

Figure 3

Definition of the Inverse Sine Function

The **inverse sine function** is the function $\sin^{-1}$ with domain $[-1, 1]$ and range $[-\pi/2, \pi/2]$ defined by

$$\sin^{-1}x = y \quad \iff \quad \sin y = x$$

The inverse sine function is also called **arcsine**, denoted by **arcsin**.

Thus, $\sin^{-1}x$ *is the number in the interval* $[-\pi/2, \pi/2]$ *whose sine is x.* In other words, $\sin(\sin^{-1}x) = x$. In fact, from the general properties of inverse functions studied in Section 2.8, we have the following relations.

$$\sin(\sin^{-1}x) = x \qquad \text{for } -1 \le x \le 1$$

$$\sin^{-1}(\sin x) = x \qquad \text{for } -\frac{\pi}{2} \le x \le \frac{\pi}{2}$$

Example 1 **Evaluating the Inverse Sine Function**

Find: (a) $\sin^{-1}\frac{1}{2}$, (b) $\sin^{-1}\left(-\frac{1}{2}\right)$, and (c) $\sin^{-1}\frac{3}{2}$.

Solution

(a) The number in the interval $[-\pi/2, \pi/2]$ whose sine is $\frac{1}{2}$ is $\pi/6$. Thus, $\sin^{-1}\frac{1}{2} = \pi/6$.

(b) The number in the interval $[-\pi/2, \pi/2]$ whose sine is $-\frac{1}{2}$ is $-\pi/6$. Thus, $\sin^{-1}\left(-\frac{1}{2}\right) = -\pi/6$.

(c) Since $\frac{3}{2} > 1$, it is not in the domain of $\sin^{-1}x$, so $\sin^{-1}\frac{3}{2}$ is not defined. ∎

Example 2 **Using a Calculator to Evaluate Inverse Sine**

Find approximate values for (a) $\sin^{-1}(0.82)$ and (b) $\sin^{-1}\frac{1}{3}$.

Solution Since no rational multiple of π has a sine of 0.82 or $\frac{1}{3}$, we use a calculator to approximate these values. Using the INV SIN, or SIN⁻¹, or ARC SIN key(s) on the calculator (with the calculator in radian mode), we get

(a) $\sin^{-1}(0.82) \approx 0.96141$ (b) $\sin^{-1}\frac{1}{3} \approx 0.33984$ ∎

Example 3 **Composing Trigonometric Functions and Their Inverses**

Find $\cos\left(\sin^{-1}\frac{3}{5}\right)$.

Solution 1 It's easy to find $\sin\left(\sin^{-1}\frac{3}{5}\right)$. In fact, by the properties of inverse functions, this value is exactly $\frac{3}{5}$. To find $\cos\left(\sin^{-1}\frac{3}{5}\right)$, we reduce this to the easier problem by writing the cosine function in terms of the sine function. Let $u = \sin^{-1}\frac{3}{5}$. Since $-\pi/2 \le u \le \pi/2$, $\cos u$ is positive and we can write

$$\cos u = +\sqrt{1 - \sin^2 u}$$

Thus
$$\cos\left(\sin^{-1}\frac{3}{5}\right) = \sqrt{1 - \sin^2\left(\sin^{-1}\frac{3}{5}\right)}$$
$$= \sqrt{1 - \left(\frac{3}{5}\right)^2} = \sqrt{1 - \frac{9}{25}} = \sqrt{\frac{16}{25}} = \frac{4}{5}$$

Solution 2 Let $\theta = \sin^{-1}\frac{3}{5}$. Then θ is the number in the interval $[-\pi/2, \pi/2]$ whose sine is $\frac{3}{5}$. Let's interpret θ as an angle and draw a right triangle with θ as one of its acute angles, with opposite side 3 and hypotenuse 5 (see Figure 4). The remaining leg of the triangle is found by the Pythagorean Theorem to be 4. From the figure we get

$$\cos\left(\sin^{-1}\frac{3}{5}\right) = \cos\theta = \frac{4}{5}$$ ∎

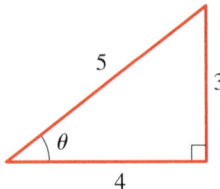

Figure 4

From Solution 2 of Example 3 we can immediately find the values of the other trigonometric functions of $\theta = \sin^{-1}\frac{3}{5}$ from the triangle. Thus

$$\tan\left(\sin^{-1}\frac{3}{5}\right) = \frac{3}{4} \qquad \sec\left(\sin^{-1}\frac{3}{5}\right) = \frac{5}{4} \qquad \csc\left(\sin^{-1}\frac{3}{5}\right) = \frac{5}{3}$$

The Inverse Cosine Function

If the domain of the cosine function is restricted to the interval $[0, \pi]$, the resulting function is one-to-one and so has an inverse. We choose this interval because on it, cosine attains each of its values exactly once (see Figure 5).

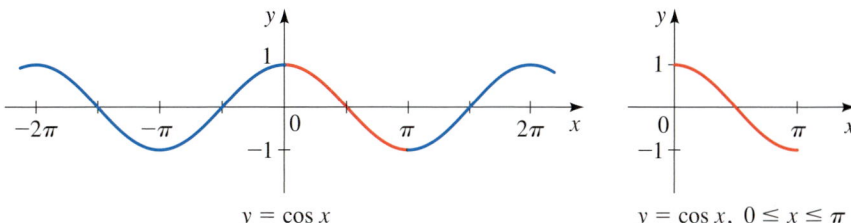

Figure 5 $y = \cos x$ $y = \cos x, \ 0 \le x \le \pi$

Definition of the Inverse Cosine Function

The **inverse cosine function** is the function $\cos^{-1}$ with domain $[-1, 1]$ and range $[0, \pi]$ defined by

$$\cos^{-1} x = y \quad \Leftrightarrow \quad \cos y = x$$

The inverse cosine function is also called **arccosine**, denoted by **arccos**.

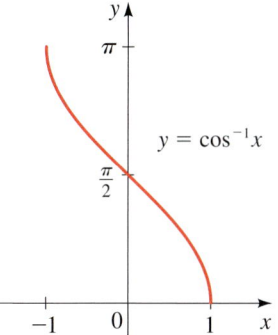

$y = \cos^{-1} x$

Figure 6

Thus, $y = \cos^{-1} x$ *is the number in the interval* $[0, \pi]$ *whose cosine is x.* The following relations follow from the inverse function properties.

$$\cos(\cos^{-1} x) = x \qquad \text{for } -1 \le x \le 1$$
$$\cos^{-1}(\cos x) = x \qquad \text{for } 0 \le x \le \pi$$

The graph of $y = \cos^{-1} x$ is shown in Figure 6; it is obtained by reflecting the graph of $y = \cos x, \ 0 \le x \le \pi$, in the line $y = x$.

Example 4   Evaluating the Inverse Cosine Function

Find: (a) $\cos^{-1}(\sqrt{3}/2)$,   (b) $\cos^{-1} 0$,   and   (c) $\cos^{-1} \frac{5}{7}$.

Solution

(a) The number in the interval $[0, \pi]$ whose cosine is $\sqrt{3}/2$ is $\pi/6$. Thus,
$\cos^{-1}(\sqrt{3}/2) = \pi/6$.

(b) The number in the interval $[0, \pi]$ whose cosine is 0 is $\pi/2$. Thus,
$\cos^{-1} 0 = \pi/2$.

(c) Since no rational multiple of π has cosine $\frac{5}{7}$, we use a calculator (in radian mode) to find this value approximately: $\cos^{-1} \frac{5}{7} \approx 0.77519$. ∎

Example 5   Composing Trigonometric Functions and Their Inverses

Write $\sin(\cos^{-1}x)$ and $\tan(\cos^{-1}x)$ as algebraic expressions in x for $-1 \le x \le 1$.

Solution 1   Let $u = \cos^{-1}x$. We need to find $\sin u$ and $\tan u$ in terms of x. As in Example 3 the idea here is to write sine and tangent in terms of cosine. We have

$$\sin u = \pm\sqrt{1 - \cos^2 u} \qquad \text{and} \qquad \tan u = \frac{\sin u}{\cos u} = \frac{\pm\sqrt{1 - \cos^2 u}}{\cos u}$$

To choose the proper signs, note that u lies in the interval $[0, \pi]$ because $u = \cos^{-1}x$. Since $\sin u$ is positive on this interval, the $+$ sign is the correct choice. Substituting $u = \cos^{-1}x$ in the displayed equations and using the relation $\cos(\cos^{-1}x) = x$ gives

$$\sin(\cos^{-1}x) = \sqrt{1 - x^2} \qquad \text{and} \qquad \tan(\cos^{-1}x) = \frac{\sqrt{1 - x^2}}{x}$$

Solution 2   Let $\theta = \cos^{-1}x$, so $\cos\theta = x$. In Figure 7 we draw a right triangle with an acute angle θ, adjacent side x, and hypotenuse 1. By the Pythagorean Theorem, the remaining leg is $\sqrt{1 - x^2}$. From the figure,

$$\sin(\cos^{-1}x) = \sin\theta = \sqrt{1 - x^2} \qquad \text{and} \qquad \tan(\cos^{-1}x) = \tan\theta = \frac{\sqrt{1 - x^2}}{x} \quad \blacksquare$$

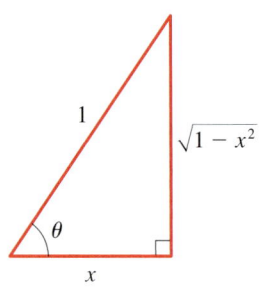

Figure 7

$\cos\theta = \dfrac{x}{1} = x$

NOTE In Solution 2 of Example 5 it may seem that because we are sketching a triangle, the angle $\theta = \cos^{-1}x$ must be acute. But it turns out that the triangle method works for any θ and for any x. The domains and ranges of all six inverse trigonometric functions have been chosen in such a way that we can always use a triangle to find $S(T^{-1}(x))$, where S and T are any trigonometric functions.

Example 6   Composing a Trigonometric Function and an Inverse

Write $\sin(2\cos^{-1}x)$ as an algebraic expression in x for $-1 \le x \le 1$.

Solution   Let $\theta = \cos^{-1}x$ and sketch a triangle as shown in Figure 8. We need to find $\sin 2\theta$, but from the triangle we can find trigonometric functions only of θ, not of 2θ. The double-angle identity for sine is useful here. We have

$$\sin(2\cos^{-1}x) = \sin 2\theta$$

$$= 2\sin\theta\cos\theta \qquad \text{\color{blue}Double-angle formula}$$

$$= 2\left(\sqrt{1 - x^2}\right)x \qquad \text{\color{blue}From triangle}$$

$$= 2x\sqrt{1 - x^2} \qquad\qquad \blacksquare$$

Figure 8

$\cos\theta = \dfrac{x}{1} = x$

The Inverse Tangent Function

We restrict the domain of the tangent function to the interval $(-\pi/2, \pi/2)$ in order to obtain a one-to-one function.

Definition of the Inverse Tangent Function

The **inverse tangent function** is the function $\tan^{-1}$ with domain $\mathbb{R}$ and range $(-\pi/2, \pi/2)$ defined by

$$\tan^{-1}x = y \quad \Leftrightarrow \quad \tan y = x$$

The inverse tangent function is also called **arctangent**, denoted by **arctan**.

Thus, $\tan^{-1}x$ *is the number in the interval* $(-\pi/2, \pi/2)$ *whose tangent is x*. The following relations follow from the inverse function properties.

$$\tan(\tan^{-1}x) = x \qquad \text{for } x \in \mathbb{R}$$
$$\tan^{-1}(\tan x) = x \qquad \text{for } -\frac{\pi}{2} < x < \frac{\pi}{2}$$

Figure 9 shows the graph of $y = \tan x$ on the interval $(-\pi/2, \pi/2)$ and the graph of its inverse function, $y = \tan^{-1}x$.

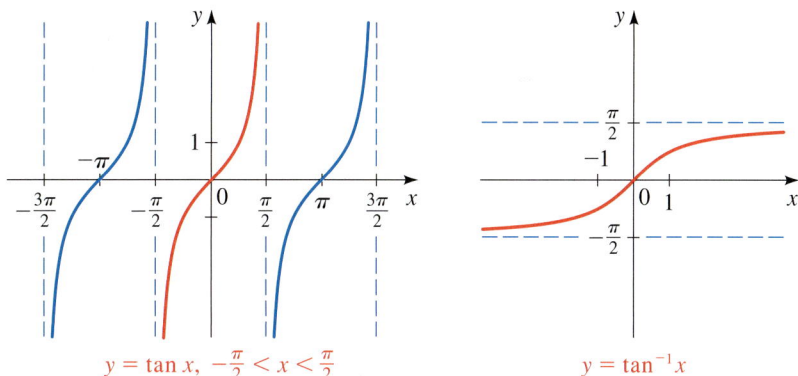

Figure 9

$$y = \tan x, \ -\frac{\pi}{2} < x < \frac{\pi}{2} \qquad\qquad y = \tan^{-1}x$$

Example 7   Evaluating the Inverse Tangent Function

Find: (a) $\tan^{-1}1$,   (b) $\tan^{-1}\sqrt{3}$,   and   (c) $\tan^{-1}(-20)$.

Solution

(a) The number in the interval $(-\pi/2, \pi/2)$ with tangent 1 is $\pi/4$. Thus, $\tan^{-1}1 = \pi/4$.

(b) The number in the interval $(-\pi/2, \pi/2)$ with tangent $\sqrt{3}$ is $\pi/3$. Thus, $\tan^{-1}\sqrt{3} = \pi/3$.

(c) We use a calculator to find that $\tan^{-1}(-20) \approx -1.52084$.

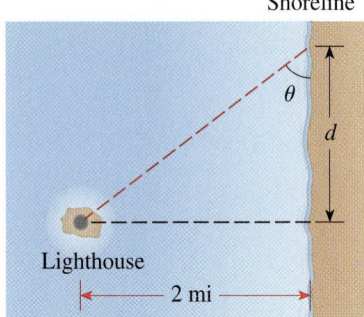

Shoreline

θ

d

Lighthouse

2 mi

Figure 10

See Exercise 59 for a way of finding the values of these inverse trigonometric functions on a calculator.

Example 8   The Angle of a Beam of Light

A lighthouse is located on an island that is 2 mi off a straight shoreline (see Figure 10). Express the angle formed by the beam of light and the shoreline in terms of the distance d in the figure.

Solution   From the figure we see that $\tan \theta = 2/d$. Taking the inverse tangent of both sides, we get

$$\tan^{-1}(\tan \theta) = \tan^{-1}\left(\frac{2}{d}\right)$$

$$\theta = \tan^{-1}\left(\frac{2}{d}\right) \qquad \textit{Cancellation property} \qquad ■$$

The Inverse Secant, Cosecant, and Cotangent Functions

To define the inverse functions of the secant, cosecant, and cotangent functions, we restrict the domain of each function to a set on which it is one-to-one and on which it attains all its values. Although any interval satisfying these criteria is appropriate, we choose to restrict the domains in a way that simplifies the choice of sign in computations involving inverse trigonometric functions. The choices we make are also appropriate for calculus. This explains the seemingly strange restriction for the domains of the secant and cosecant functions. We end this section by displaying the graphs of the secant, cosecant, and cotangent functions with their restricted domains and the graphs of their inverse functions (Figures 11–13).

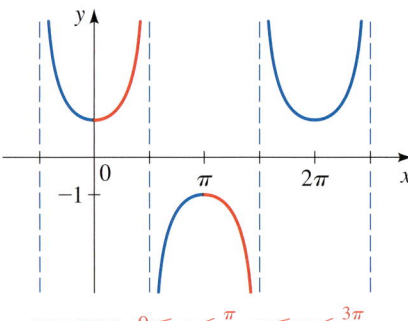

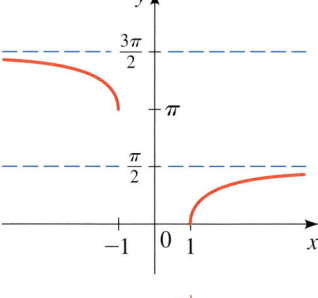

Figure 11

The inverse secant function

$y = \sec x,\ 0 \le x < \frac{\pi}{2},\ \pi \le x < \frac{3\pi}{2}$

$y = \sec^{-1}x$

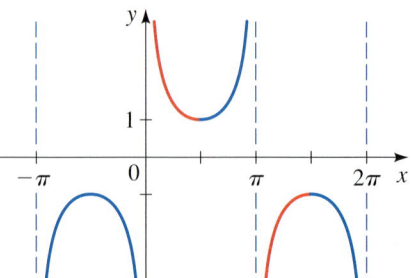

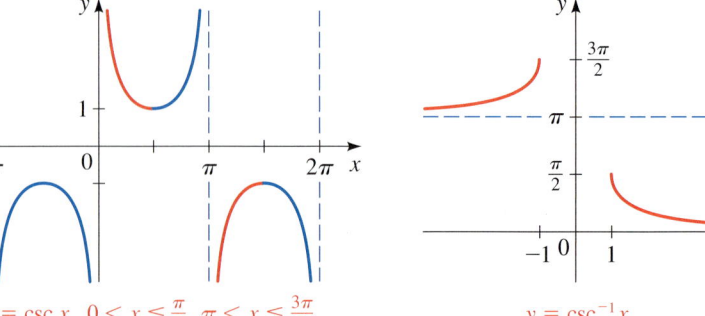

Figure 12

The inverse cosecant function

$y = \csc x,\ 0 < x \le \frac{\pi}{2},\ \pi < x \le \frac{3\pi}{2}$

$y = \csc^{-1}x$

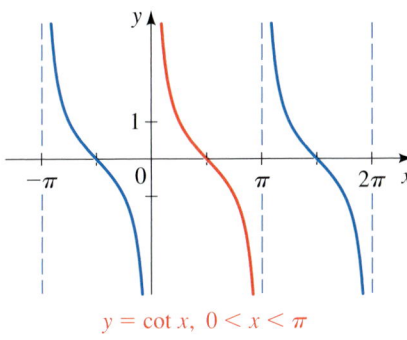

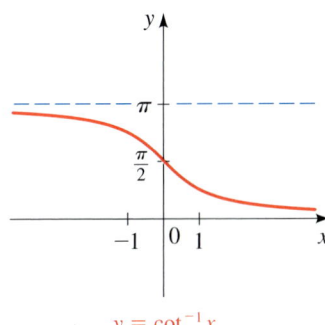

Figure 13

The inverse cotangent function

$y = \cot x, \ 0 < x < \pi$

$y = \cot^{-1} x$

7.4   Exercises

1–8 ■ Find the exact value of each expression, if it is defined.

1. (a) $\sin^{-1} \frac{1}{2}$ (b) $\cos^{-1} \frac{1}{2}$ (c) $\cos^{-1} 2$

2. (a) $\sin^{-1} \dfrac{\sqrt{3}}{2}$ (b) $\cos^{-1} \dfrac{\sqrt{3}}{2}$ (c) $\cos^{-1}\left(-\dfrac{\sqrt{3}}{2}\right)$

3. (a) $\sin^{-1} \dfrac{\sqrt{2}}{2}$ (b) $\cos^{-1} \dfrac{\sqrt{2}}{2}$ (c) $\sin^{-1}\left(-\dfrac{\sqrt{2}}{2}\right)$

4. (a) $\tan^{-1} \sqrt{3}$ (b) $\tan^{-1}(-\sqrt{3})$ (c) $\sin^{-1} \sqrt{3}$

5. (a) $\sin^{-1} 1$ (b) $\cos^{-1} 1$ (c) $\cos^{-1}(-1)$

6. (a) $\tan^{-1} 1$ (b) $\tan^{-1}(-1)$ (c) $\tan^{-1} 0$

7. (a) $\tan^{-1} \dfrac{\sqrt{3}}{3}$ (b) $\tan^{-1}\left(-\dfrac{\sqrt{3}}{3}\right)$ (c) $\sin^{-1}(-2)$

8. (a) $\sin^{-1} 0$ (b) $\cos^{-1} 0$ (c) $\cos^{-1}\left(-\frac{1}{2}\right)$

9–12 ■ Use a calculator to find an approximate value of each expression correct to five decimal places, if it is defined.

9. (a) $\sin^{-1}(0.13844)$

   (b) $\cos^{-1}(-0.92761)$

10. (a) $\cos^{-1}(0.31187)$

   (b) $\tan^{-1}(26.23110)$

11. (a) $\tan^{-1}(1.23456)$

   (b) $\sin^{-1}(1.23456)$

12. (a) $\cos^{-1}(-0.25713)$

   (b) $\tan^{-1}(-0.25713)$

13–28 ■ Find the exact value of the expression, if it is defined.

13. $\sin\left(\sin^{-1} \frac{1}{4}\right)$

14. $\cos\left(\cos^{-1} \frac{2}{3}\right)$

15. $\tan(\tan^{-1} 5)$

16. $\sin(\sin^{-1} 5)$

17. $\cos^{-1}\left(\cos \dfrac{\pi}{3}\right)$

18. $\tan^{-1}\left(\tan \dfrac{\pi}{6}\right)$

19. $\sin^{-1}\left(\sin\left(-\dfrac{\pi}{6}\right)\right)$

20. $\sin^{-1}\left(\sin \dfrac{5\pi}{6}\right)$

21. $\tan^{-1}\left(\tan \dfrac{2\pi}{3}\right)$

22. $\cos^{-1}\left(\cos\left(-\dfrac{\pi}{4}\right)\right)$

23. $\tan\left(\sin^{-1} \frac{1}{2}\right)$

24. $\sin(\sin^{-1} 0)$

25. $\cos\left(\sin^{-1} \dfrac{\sqrt{3}}{2}\right)$

26. $\tan\left(\sin^{-1} \dfrac{\sqrt{2}}{2}\right)$

27. $\tan^{-1}\left(2 \sin \dfrac{\pi}{3}\right)$

28. $\cos^{-1}\left(\sqrt{3} \sin \dfrac{\pi}{6}\right)$

29–40 ■ Evaluate the expression by sketching a triangle, as in Solution 2 of Example 3.

29. $\sin\left(\cos^{-1}\frac{3}{5}\right)$

30. $\tan\left(\sin^{-1}\frac{4}{5}\right)$

31. $\sin\left(\tan^{-1}\frac{12}{5}\right)$

32. $\cos\left(\tan^{-1}5\right)$

33. $\sec\left(\sin^{-1}\frac{12}{13}\right)$

34. $\csc\left(\cos^{-1}\frac{7}{25}\right)$

35. $\cos\left(\tan^{-1}2\right)$

36. $\cot\left(\sin^{-1}\frac{2}{3}\right)$

37. $\sin\left(2 \cos^{-1}\frac{3}{5}\right)$

38. $\tan\left(2 \tan^{-1}\frac{5}{13}\right)$

39. $\sin\left(\sin^{-1}\frac{1}{2} + \cos^{-1}\frac{1}{2}\right)$

40. $\cos\left(\sin^{-1}\frac{3}{5} - \cos^{-1}\frac{3}{5}\right)$

41–48 ■ Rewrite the expression as an algebraic expression in x.

41. $\cos\left(\sin^{-1}x\right)$

42. $\sin\left(\tan^{-1}x\right)$

43. $\tan\left(\sin^{-1}x\right)$

44. $\cos\left(\tan^{-1}x\right)$

45. $\cos\left(2 \tan^{-1}x\right)$

46. $\sin\left(2 \sin^{-1}x\right)$

47. $\cos\left(\cos^{-1}x + \sin^{-1}x\right)$

48. $\sin\left(\tan^{-1}x - \sin^{-1}x\right)$

 49–50 ■ (a) Graph the function and make a conjecture, and (b) prove that your conjecture is true.

49. $y = \sin^{-1}x + \cos^{-1}x$

50. $y = \tan^{-1}x + \tan^{-1}\dfrac{1}{x}$

 51–52 ■ (a) Use a graphing device to find all solutions of the equation, correct to two decimal places, and (b) find the exact solution.

51. $\tan^{-1}x + \tan^{-1}2x = \dfrac{\pi}{4}$

52. $\sin^{-1}x - \cos^{-1}x = 0$

Applications

53. Height of the Space Shuttle An observer views the space shuttle from a distance of 2 miles from the launch pad.

(a) Express the height of the space shuttle as a function of the angle of elevation θ.

(b) Express the angle of elevation θ as a function of the height h of the space shuttle.

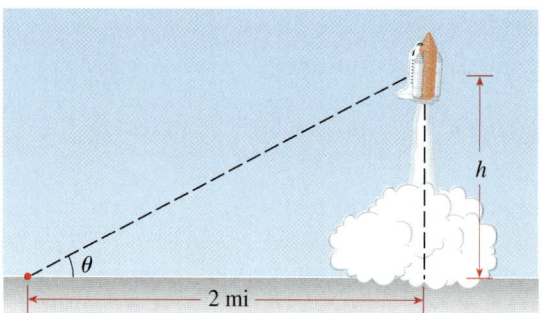

54. Height of a Pole A 50-ft pole casts a shadow as shown in the figure.

(a) Express the angle of elevation θ of the sun as a function of the length s of the shadow.

(b) Find the angle θ of elevation of the sun when the shadow is 20 ft long.

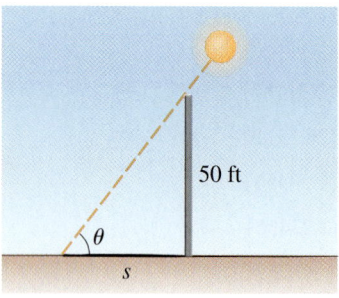

55. Height of a Balloon A 680-ft rope anchors a hot-air balloon as shown in the figure.

(a) Express the angle θ as a function of the height h of the balloon.

(b) Find the angle θ if the balloon is 500 ft high.

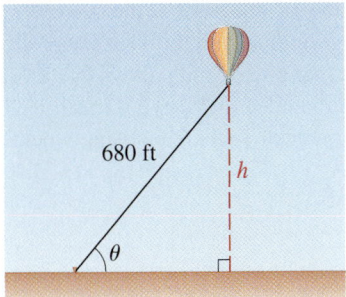

56. View from a Satellite The figures indicate that the higher the orbit of a satellite, the more of the earth the satellite can "see." Let θ, s, and h be as in the figure, and assume the earth is a sphere of radius 3960 mi.

(a) Express the angle θ as a function of h.

(b) Express the distance s as a function of θ.

(c) Express the distance s as a function of h.
[*Hint:* Find the composition of the functions in parts (a) and (b).]

(d) If the satellite is 100 mi above the earth, what is the distance s that it can see?

(e) How high does the satellite have to be in order to see both Los Angeles and New York, 2450 mi apart?

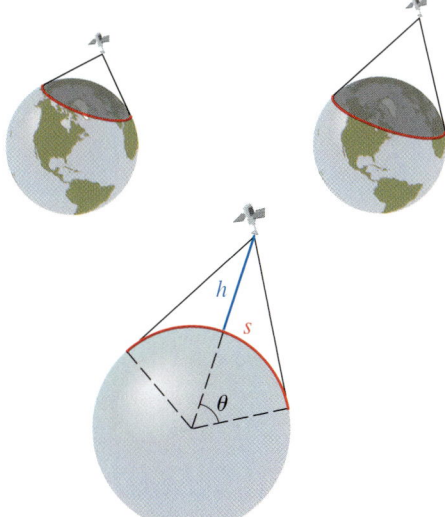

57. Surfing the Perfect Wave For a wave to be surfable it can't break all at once. Robert Guza and Tony Bowen have shown that a wave has a surfable shoulder if it hits the shoreline at an angle θ given by

$$\theta = \sin^{-1}\left(\frac{1}{(2n+1)\tan\beta}\right)$$

where β is the angle at which the beach slopes down and where $n = 0, 1, 2, \ldots$.

(a) For $\beta = 10°$, find θ when $n = 3$.

(b) For $\beta = 15°$, find θ when $n = 2$, 3, and 4. Explain why the formula does not give a value for θ when $n = 0$ or 1.

Discovery • Discussion

58. Two Different Compositions The functions

$$f(x) = \sin(\sin^{-1}x) \qquad \text{and} \qquad g(x) = \sin^{-1}(\sin x)$$

both simplify to just x for suitable values of x. But these functions are not the same for all x. Graph both f and g to show how the functions differ. (Think carefully about the domain and range of $\sin^{-1}$.)

59. Inverse Trigonometric Functions on a Calculator Most calculators do not have keys for $\sec^{-1}$, $\csc^{-1}$, or $\cot^{-1}$. Prove the following identities, then use these identities and a calculator to find $\sec^{-1}2$, $\csc^{-1}3$, and $\cot^{-1}4$.

$$\sec^{-1}x = \cos^{-1}\left(\frac{1}{x}\right), \quad x \geq 1$$

$$\csc^{-1}x = \sin^{-1}\left(\frac{1}{x}\right), \quad x \geq 1$$

$$\cot^{-1}x = \tan^{-1}\left(\frac{1}{x}\right), \quad x > 0$$

Where to Sit at the Movies

Everyone knows that the apparent size of an object depends on its distance from the viewer. The farther away an object, the smaller its apparent size. The apparent size is determined by the angle the object subtends at the eye of the viewer.

If you are looking at a painting hanging on a wall, how far away should you stand to get the maximum view? If the painting is hung above eye level, then the following figures show that the angle subtended at the eye is small if you are too close or too far away. The same situation occurs when choosing where to sit in a movie theatre.

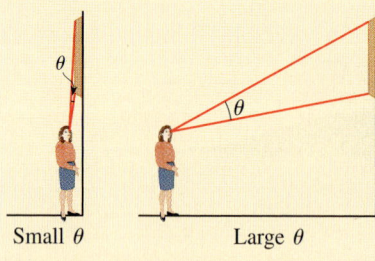

Small θ Large θ

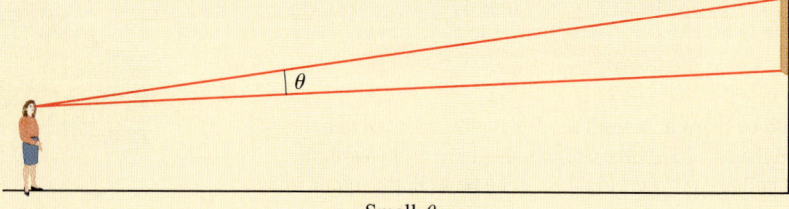

Small θ

1. The screen in a theatre is 22 ft high and is positioned 10 ft above the floor, which is flat. The first row of seats is 7 ft from the screen and the rows are 3 ft apart. You decide to sit in the row where you get the maximum view, that is, where the angle θ subtended by the screen at your eyes is a maximum. Suppose your eyes are 4 ft above the floor, as in the figure, and you sit at a distance x from the screen.

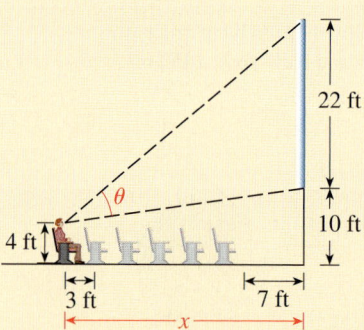

(a) Show that $\theta = \tan^{-1}\left(\dfrac{28}{x}\right) - \tan^{-1}\left(\dfrac{6}{x}\right)$.

(b) Use the subtraction formula for tangent to show that

$$\theta = \tan^{-1}\left(\frac{22x}{x^2 + 168}\right)$$

(c) Use a graphing device to graph θ as a function of x. What value of x maximizes θ? In which row should you sit? What is the viewing angle in this row?

2. Now suppose that, starting with the first row of seats, the floor of the seating area is inclined at an angle of $\alpha = 25°$ above the horizontal, and the distance that you sit up the incline is x, as shown in the figure.

(a) Use the Law of Cosines to show that

$$\theta = \cos^{-1}\left(\frac{a^2 + b^2 - 484}{2ab}\right)$$

where

$$a^2 = (7 + x\cos\alpha)^2 + (28 - x\sin\alpha)^2$$

and

$$b^2 = (7 + x\cos\alpha)^2 + (x\sin\alpha - 6)^2$$

(b) Use a graphing device to graph θ as a function of x, and estimate the value of x that maximizes θ. In which row should you sit? What is the viewing angle θ in this row?

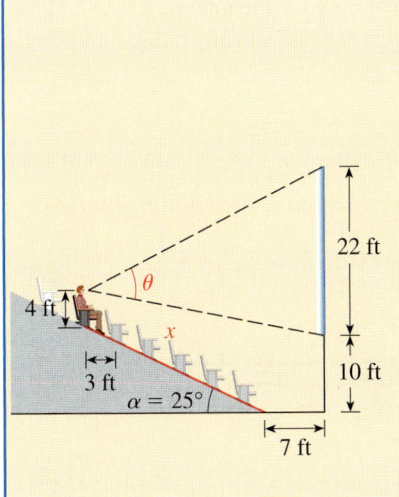

7.5 Trigonometric Equations

An equation that contains trigonometric functions is called a **trigonometric equation**. For example, the following are trigonometric equations:

$$\sin^2 x + \cos^2 x = 1 \qquad 2\sin x - 1 = 0 \qquad \tan^2 2x - 1 = 0$$

The first equation is an *identity*—that is, it is true for every value of the variable x. The other two equations are true only for certain values of x. To solve a trigonometric equation, we find all the values of the variable that make the equation true. (Except in some applied problems, we will always use radian measure for the variable.)

Solving Trigonometric Equations

To solve a trigonometric equation, we use the rules of algebra to isolate the trigonometric function on one side of the equal sign. Then we use our knowledge of the values of the trigonometric functions to solve for the variable.

Example 1 Solving a Trigonometric Equation

Solve the equation $2\sin x - 1 = 0$.

Solution We start by isolating $\sin x$.

$$2\sin x - 1 = 0 \qquad \text{Given equation}$$
$$2\sin x = 1 \qquad \text{Add 1}$$
$$\sin x = \frac{1}{2} \qquad \text{Divide by 2}$$

Because sine has period 2π, we first find the solutions in the interval $[0, 2\pi)$. These are $x = \pi/6$ and $x = 5\pi/6$. To get all other solutions, we add any integer multiple of 2π to these solutions. Thus, the solutions are

$$x = \frac{\pi}{6} + 2k\pi, \qquad x = \frac{5\pi}{6} + 2k\pi$$

where k is any integer. Figure 1 gives a graphical representation of the solutions.

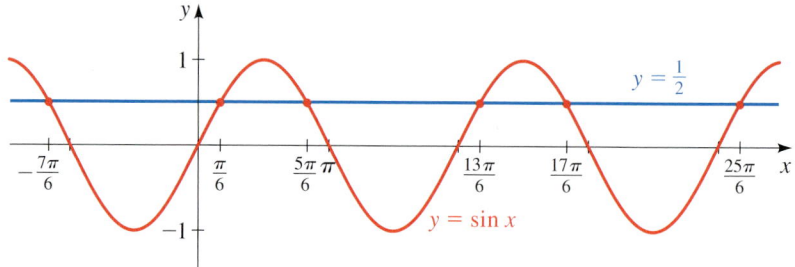

Figure 1

Example 2   Solving a Trigonometric Equation

Solve the equation $\tan^2 x - 3 = 0$.

Solution   We start by isolating $\tan x$.

$$\tan^2 x - 3 = 0 \qquad \text{Given equation}$$
$$\tan^2 x = 3 \qquad \text{Add 3}$$
$$\tan x = \pm\sqrt{3} \qquad \text{Take square roots}$$

Because tangent has period π, we first find the solutions in the interval $(-\pi/2, \pi/2)$. These are $x = -\pi/3$ and $x = \pi/3$. To get all other solutions, we add any integer multiple of π to these solutions. Thus, the solutions are

$$x = -\frac{\pi}{3} + k\pi, \qquad x = \frac{\pi}{3} + k\pi$$

where k is any integer.

Example 3   Finding Intersection Points

Find the values of x for which the graphs of $f(x) = \sin x$ and $g(x) = \cos x$ intersect.

Solution 1: Graphical

The graphs intersect where $f(x) = g(x)$. In Figure 2 we graph $y_1 = \sin x$ and $y_2 = \cos x$ on the same screen, for x between 0 and 2π. Using TRACE or the Intersect command on the graphing calculator, we see that the two points of intersection in this interval occur where $x \approx 0.785$ and $x \approx 3.927$. Since sine and cosine are periodic with period 2π, the intersection points occur where

$$x \approx 0.785 + 2k\pi \qquad \text{and} \qquad x \approx 3.927 + 2k\pi$$

where k is any integer.

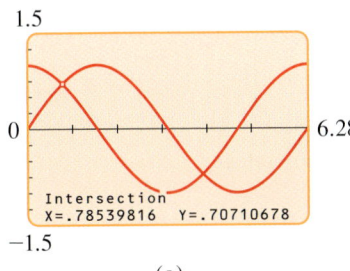

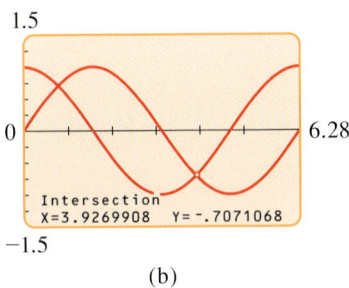

Figure 2   (a)   (b)

Solution 2: Algebraic

To find the exact solution, we set $f(x) = g(x)$ and solve the resulting equation algebraically.

$$\sin x = \cos x \qquad \textcolor{blue}{\text{Equate functions}}$$

Since the numbers x for which $\cos x = 0$ are not solutions of the equation, we can divide both sides by $\cos x$.

$$\frac{\sin x}{\cos x} = 1 \qquad \textcolor{blue}{\text{Divide by cos x}}$$

$$\tan x = 1 \qquad \textcolor{blue}{\text{Reciprocal identity}}$$

Because tangent has period π, we first find the solutions in the interval $(-\pi/2, \pi/2)$. The only solution in this interval is $x = \pi/4$. To get all solutions, we add any integer multiple of π to this solution. Thus, the solutions are

$$x = \frac{\pi}{4} + k\pi$$

where k is any integer. The graphs intersect for these values of x. You should use your calculator to check that, correct to three decimals, these are the same values as we obtained in Solution 1.   ■

Solving Trigonometric Equations by Factoring

Zero-Product Property

If $AB = 0$, then $A = 0$ or $B = 0$.

Factoring is one of the most useful techniques for solving equations, including trigonometric equations. The idea is to move all terms to one side of the equation, factor, then use the Zero-Product Property (see Section 1.5).

Example 4   An Equation of Quadratic Type

Solve the equation $2\cos^2 x - 7\cos x + 3 = 0$.

Solution   We factor the left-hand side of the equation.

Equation of Quadratic Type

$2C^2 - 7C + 3 = 0$

$(2C - 1)(C - 3) = 0$

$$2\cos^2 x - 7\cos x + 3 = 0 \qquad \textcolor{blue}{\text{Given equation}}$$

$$(2\cos x - 1)(\cos x - 3) = 0 \qquad \textcolor{blue}{\text{Factor}}$$

$$2\cos x - 1 = 0 \quad \text{or} \quad \cos x - 3 = 0 \qquad \textcolor{blue}{\text{Set each factor equal to 0}}$$

$$\cos x = \tfrac{1}{2} \quad \text{or} \quad \cos x = 3 \qquad \textcolor{blue}{\text{Solve for cos x}}$$

Because cosine has period 2π, we first find the solutions in the interval $[0, 2\pi)$. For the first equation these are $x = \pi/3$ and $x = 5\pi/3$. The second equation has no solutions because $\cos x$ is never greater than 1. Thus, the solutions are

$$x = \frac{\pi}{3} + 2k\pi, \qquad x = \frac{5\pi}{3} + 2k\pi$$

where k is any integer.   ■

Example 5   Using a Trigonometric Identity

Solve the equation $1 + \sin x = 2\cos^2 x$.

Solution   We use a trigonometric identity to rewrite the equation in terms of a single trigonometric function.

Equation of Quadratic Type

$2S^2 + S - 1 = 0$

$(2S - 1)(S + 1) = 0$

$1 + \sin x = 2\cos^2 x$	Given equation
$1 + \sin x = 2(1 - \sin^2 x)$	Pythagorean identity
$2\sin^2 x + \sin x - 1 = 0$	Put all terms on one side of the equation
$(2\sin x - 1)(\sin x + 1) = 0$	Factor
$2\sin x - 1 = 0$ or $\sin x + 1 = 0$	Set each factor equal to 0
$\sin x = \dfrac{1}{2}$ or $\sin x = -1$	Solve for $\sin x$
$x = \dfrac{\pi}{6}, \dfrac{5\pi}{6}$ or $x = \dfrac{3\pi}{2}$	Solve for x in the interval $[0, 2\pi)$

Because sine has period 2π, we get all the solutions of the equation by adding any integer multiple of 2π to these solutions. Thus, the solutions are

$$x = \frac{\pi}{6} + 2k\pi, \qquad x = \frac{5\pi}{6} + 2k\pi, \qquad x = \frac{3\pi}{2} + 2k\pi$$

where k is any integer.   ■

Example 6   Using a Trigonometric Identity

Solve the equation $\sin 2x - \cos x = 0$.

Solution   The first term is a function of $2x$ and the second is a function of x, so we begin by using a trigonometric identity to rewrite the first term as a function of x only.

$\sin 2x - \cos x = 0$	Given equation
$2\sin x \cos x - \cos x = 0$	Double-angle formula
$\cos x\,(2\sin x - 1) = 0$	Factor

$$\cos x = 0 \qquad \text{or} \qquad 2 \sin x - 1 = 0 \qquad \text{Set each factor equal to 0}$$

$$\sin x = \frac{1}{2} \qquad \text{Solve for sin x}$$

$$x = \frac{\pi}{2}, \frac{3\pi}{2} \qquad \text{or} \qquad x = \frac{\pi}{6}, \frac{5\pi}{6} \qquad \text{Solve for x in the interval } [0, 2\pi)$$

Both sine and cosine have period 2π, so we get all the solutions of the equation by adding any integer multiple of 2π to these solutions. Thus, the solutions are

$$x = \frac{\pi}{2} + 2k\pi, \qquad x = \frac{3\pi}{2} + 2k\pi, \qquad x = \frac{\pi}{6} + 2k\pi, \qquad x = \frac{5\pi}{6} + 2k\pi$$

where k is any integer. ∎

Example 7   Squaring and Using an Identify

Solve the equation $\cos x + 1 = \sin x$ in the interval $[0, 2\pi)$.

Solution   To get an equation that involves either sine only or cosine only, we square both sides and use a Pythagorean identity.

$$\cos x + 1 = \sin x \qquad \text{Given equation}$$

$$\cos^2 x + 2 \cos x + 1 = \sin^2 x \qquad \text{Square both sides}$$

$$\cos^2 x + 2 \cos x + 1 = 1 - \cos^2 x \qquad \text{Pythagorean identity}$$

$$2 \cos^2 x + 2 \cos x = 0 \qquad \text{Simplify}$$

$$2 \cos x \, (\cos x + 1) = 0 \qquad \text{Factor}$$

$$2 \cos x = 0 \qquad \text{or} \qquad \cos x + 1 = 0 \qquad \text{Set each factor equal to 0}$$

$$\cos x = 0 \qquad \text{or} \qquad \cos x = -1 \qquad \text{Solve for cos x}$$

$$x = \frac{\pi}{2}, \frac{3\pi}{2} \qquad \text{or} \qquad x = \pi \qquad \begin{array}{l}\text{Solve for x in the} \\ \text{interval } [0, 2\pi)\end{array}$$

Because we squared both sides, we need to check for extraneous solutions. From *Check Your Answers*, we see that the solutions of the given equation are $\pi/2$ and π. ∎

Check Your Answers

$$x = \frac{\pi}{2}: \qquad\qquad\qquad x = \frac{3\pi}{2}: \qquad\qquad\qquad x = \pi:$$

$$\cos \frac{\pi}{2} + 1 \stackrel{?}{=} \sin \frac{\pi}{2} \qquad \cos \frac{3\pi}{2} + 1 \stackrel{?}{=} \sin \frac{3\pi}{2} \qquad \cos \pi + 1 \stackrel{?}{=} \sin \pi$$

$$0 + 1 = 1 \quad \checkmark \qquad\qquad 0 + 1 \stackrel{?}{=} -1 \quad \times \qquad\qquad -1 + 1 = 0 \quad \checkmark$$

   If we perform an operation on an equation that may introduce new roots, such as squaring both sides, then we must check that the solutions obtained are not extraneous; that is, we must verify that they satisfy the original equation, as in Example 7.

Equations with Trigonometric Functions of Multiple Angles

When solving trigonometric equations that involve functions of multiples of angles, we first solve for the multiple of the angle, then divide to solve for the angle.

Example 8 Trigonometric Functions of Multiple Angles

Consider the equation $2 \sin 3x - 1 = 0$.

(a) Find all solutions of the equation.

(b) Find the solutions in the interval $[0, 2\pi)$.

Solution

(a) We start by isolating $\sin 3x$, and then solve for the multiple angle $3x$.

$$2 \sin 3x - 1 = 0 \qquad \text{Given equation}$$

$$2 \sin 3x = 1 \qquad \text{Add 1}$$

$$\sin 3x = \frac{1}{2} \qquad \text{Divide by 2}$$

$$3x = \frac{\pi}{6}, \frac{5\pi}{6} \qquad \text{Solve for } 3x \text{ in the interval } [0, 2\pi)$$

To get all solutions, we add any integer multiple of 2π to these solutions. Thus, the solutions are of the form

$$3x = \frac{\pi}{6} + 2k\pi, \qquad 3x = \frac{5\pi}{6} + 2k\pi$$

To solve for x, we divide by 3 to get the solutions

$$x = \frac{\pi}{18} + \frac{2k\pi}{3}, \qquad x = \frac{5\pi}{18} + \frac{2k\pi}{3}$$

where k is any integer.

(b) The solutions from part (a) that are in the interval $[0, 2\pi)$ correspond to $k = 0$, 1, and 2. For all other values of k, the corresponding values of x lie outside this interval. Thus, the solutions in the interval $[0, 2\pi)$ are

$$x = \frac{\pi}{18}, \frac{5\pi}{18}, \frac{13\pi}{18}, \frac{17\pi}{18}, \frac{25\pi}{18}, \frac{29\pi}{18} \qquad \blacksquare$$

Example 9 Trigonometric Functions of Multiple Angles

Consider the equation $\sqrt{3} \tan \frac{x}{2} - 1 = 0$.

(a) Find all solutions of the equation.

(b) Find the solutions in the interval $[0, 4\pi)$.

Solution

(a) We start by isolating $\tan(x/2)$.

$$\sqrt{3}\tan\frac{x}{2} - 1 = 0 \qquad \textit{Given equation}$$

$$\sqrt{3}\tan\frac{x}{2} = 1 \qquad \textit{Add 1}$$

$$\tan\frac{x}{2} = \frac{1}{\sqrt{3}} \qquad \textit{Divide by }\sqrt{3}$$

$$\frac{x}{2} = \frac{\pi}{6} \qquad \textit{Solve for }\frac{x}{2}\textit{ in the interval }\left(-\frac{\pi}{2},\frac{\pi}{2}\right)$$

Since tangent has period π, to get all solutions we add any integer multiple of π to this solution. Thus, the solutions are of the form

$$\frac{x}{2} = \frac{\pi}{6} + k\pi$$

Multiplying by 2, we get the solutions

$$x = \frac{\pi}{3} + 2k\pi$$

where k is any integer.

(b) The solutions from part (a) that are in the interval $[0, 4\pi)$ correspond to $k = 0$ and $k = 1$. For all other values of k, the corresponding values of x lie outside this interval. Thus, the solutions in the interval $[0, 4\pi)$ are

$$x = \frac{\pi}{3}, \frac{7\pi}{3} \qquad \blacksquare$$

Using Inverse Trigonometric Functions to Solve Trigonometric Equations

So far, all the equations we've solved have had solutions like $\pi/4$, $\pi/3$, $5\pi/6$, and so on. We were able to find these solutions from the special values of the trigonometric functions that we've memorized. We now consider equations whose solution requires us to use the inverse trigonometric functions.

Example 10   Using Inverse Trigonometric Functions

Solve the equation $\tan^2 x - \tan x - 2 = 0$.

Solution   We start by factoring the left-hand side.

Equation of Quadratic Type

$$T^2 - T - 2 = 0$$
$$(T - 2)(T + 1) = 0$$

$$\tan^2 x - \tan x - 2 = 0 \qquad \textit{Given equation}$$

$$(\tan x - 2)(\tan x + 1) = 0 \qquad \textit{Factor}$$

$$\tan x - 2 = 0 \quad \text{or} \quad \tan x + 1 = 0 \qquad \textit{Set each factor equal to 0}$$

$$\tan x = 2 \quad \text{or} \quad \tan x = -1 \qquad \textit{Solve for tan x}$$

$$x = \tan^{-1}2 \quad \text{or} \quad x = -\frac{\pi}{4} \qquad \textit{Solve for x in the interval } \left(-\frac{\pi}{2},\frac{\pi}{2}\right)$$

Because tangent has period π, we get all solutions by adding integer multiples of π to these solutions. Thus, all the solutions are

$$x = \tan^{-1}2 + k\pi, \qquad x = -\frac{\pi}{4} + k\pi$$

where k is any integer. ∎

If we are using inverse trigonometric functions to solve an equation, we must keep in mind that $\sin^{-1}$ and $\tan^{-1}$ give values in quadrants I and IV, and $\cos^{-1}$ gives values in quadrants I and II. To find other solutions, we must look at the quadrant where the trigonometric function in the equation can take on the value we need.

Example 11   Using Inverse Trigonometric Functions

(a) Solve the equation $3 \sin \theta - 2 = 0$.

(b) Use a calculator to approximate the solutions in the interval $[0, 2\pi)$, correct to five decimals.

Solution

(a) We start by isolating $\sin \theta$.

$$3 \sin \theta - 2 = 0 \qquad \text{Given equation}$$
$$3 \sin \theta = 2 \qquad \text{Add 2}$$
$$\sin \theta = \frac{2}{3} \qquad \text{Divide by 3}$$

From Figure 3 we see that $\sin \theta$ equals $\frac{2}{3}$ in quadrants I and II. The solution in quadrant I is $\theta = \sin^{-1}\frac{2}{3}$. The solution in quadrant II is $\theta = \pi - \sin^{-1}\frac{2}{3}$. Since these are the solutions in the interval $[0, 2\pi)$, we get all other solutions by adding integer multiples of 2π to these. Thus, all the solutions of the equation are

$$\theta = \left(\sin^{-1}\tfrac{2}{3}\right) + 2k\pi, \qquad \theta = \left(\pi - \sin^{-1}\tfrac{2}{3}\right) + 2k\pi$$

where k is any integer.

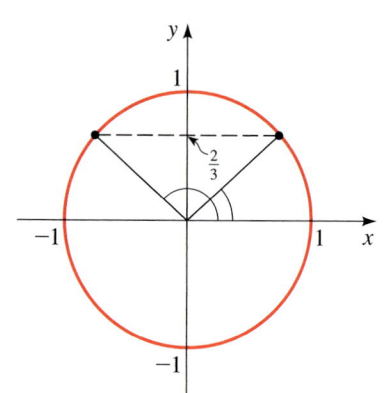

Figure 3

(b) Using a calculator set in radian mode, we see that $\sin^{-1}\frac{2}{3} \approx 0.72973$ and $\pi - \sin^{-1}\frac{2}{3} \approx 2.41186$, so the solutions in the interval $[0, 2\pi)$ are

$$\theta \approx 0.72973, \qquad \theta \approx 2.41186$$
∎

7.5   Exercises

1–40 ■ Find all solutions of the equation.

1. $\cos x + 1 = 0$
2. $\sin x + 1 = 0$
3. $2 \sin x - 1 = 0$
4. $\sqrt{2} \cos x - 1 = 0$
5. $\sqrt{3} \tan x + 1 = 0$
6. $\cot x + 1 = 0$
7. $4 \cos^2 x - 1 = 0$
8. $2 \cos^2 x - 1 = 0$
9. $\sec^2 x - 2 = 0$
10. $\csc^2 x - 4 = 0$
11. $3 \csc^2 x - 4 = 0$
12. $1 - \tan^2 x = 0$
13. $\cos x (2 \sin x + 1) = 0$
14. $\sec x (2 \cos x - \sqrt{2}) = 0$
15. $(\tan x + \sqrt{3})(\cos x + 2) = 0$
16. $(2 \cos x + \sqrt{3})(2 \sin x - 1) = 0$
17. $\cos x \sin x - 2 \cos x = 0$
18. $\tan x \sin x + \sin x = 0$
19. $4 \cos^2 x - 4 \cos x + 1 = 0$
20. $2 \sin^2 x - \sin x - 1 = 0$

21. $\sin^2 x = 2 \sin x + 3$

22. $3 \tan^3 x = \tan x$

23. $\sin^2 x = 4 - 2 \cos^2 x$

24. $2 \cos^2 x + \sin x = 1$

25. $2 \sin 3x + 1 = 0$

26. $2 \cos 2x + 1 = 0$

27. $\sec 4x - 2 = 0$

28. $\sqrt{3} \tan 3x + 1 = 0$

29. $\sqrt{3} \sin 2x = \cos 2x$

30. $\cos 3x = \sin 3x$

31. $\cos \dfrac{x}{2} - 1 = 0$

32. $2 \sin \dfrac{x}{3} + \sqrt{3} = 0$

33. $\tan \dfrac{x}{4} + \sqrt{3} = 0$

34. $\sec \dfrac{x}{2} = \cos \dfrac{x}{2}$

35. $\tan^5 x - 9 \tan x = 0$

36. $3 \tan^3 x - 3 \tan^2 x - \tan x + 1 = 0$

37. $4 \sin x \cos x + 2 \sin x - 2 \cos x - 1 = 0$

38. $\sin 2x = 2 \tan 2x$

39. $\cos^2 2x - \sin^2 2x = 0$

40. $\sec x - \tan x = \cos x$

41–48 ■ Find all solutions of the equation in the interval $[0, 2\pi)$.

41. $2 \cos 3x = 1$

42. $3 \csc^2 x = 4$

43. $2 \sin x \tan x - \tan x = 1 - 2 \sin x$

44. $\sec x \tan x - \cos x \cot x = \sin x$

45. $\tan x - 3 \cot x = 0$

46. $2 \sin^2 x - \cos x = 1$

47. $\tan 3x + 1 = \sec 3x$

48. $3 \sec^2 x + 4 \cos^2 x = 7$

49–56 ■ **(a)** Find all solutions of the equation. **(b)** Use a calculator to solve the equation in the interval $[0, 2\pi)$, correct to five decimal places.

49. $\cos x = 0.4$

50. $2 \tan x = 13$

51. $\sec x - 5 = 0$

52. $3 \sin x = 7 \cos x$

53. $5 \sin^2 x - 1 = 0$

54. $2 \sin 2x - \cos x = 0$

55. $3 \sin^2 x - 7 \sin x + 2 = 0$

56. $\tan^4 x - 13 \tan^2 x + 36 = 0$

57–60 ■ Graph f and g on the same axes, and find their points of intersection.

57. $f(x) = 3 \cos x + 1, \quad g(x) = \cos x - 1$

58. $f(x) = \sin 2x, \quad g(x) = 2 \sin 2x + 1$

59. $f(x) = \tan x, \quad g(x) = \sqrt{3}$

60. $f(x) = \sin x - 1, \quad g(x) = \cos x$

61–64 ■ Use an addition or subtraction formula to simplify the equation. Then find all solutions in the interval $[0, 2\pi)$.

61. $\cos x \cos 3x - \sin x \sin 3x = 0$

62. $\cos x \cos 2x + \sin x \sin 2x = \frac{1}{2}$

63. $\sin 2x \cos x + \cos 2x \sin x = \sqrt{3}/2$

64. $\sin 3x \cos x - \cos 3x \sin x = 0$

65–68 ■ Use a double- or half-angle formula to solve the equation in the interval $[0, 2\pi)$.

65. $\sin 2x + \cos x = 0$

66. $\tan \dfrac{x}{2} - \sin x = 0$

67. $\cos 2x + \cos x = 2$

68. $\tan x + \cot x = 4 \sin 2x$

69–72 ■ Solve the equation by first using a sum-to-product formula.

69. $\sin x + \sin 3x = 0$

70. $\cos 5x - \cos 7x = 0$

71. $\cos 4x + \cos 2x = \cos x$

72. $\sin 5x - \sin 3x = \cos 4x$

73–78 ■ Use a graphing device to find the solutions of the equation, correct to two decimal places.

73. $\sin 2x = x$

74. $\cos x = \dfrac{x}{3}$

75. $2^{\sin x} = x$

76. $\sin x = x^3$

77. $\dfrac{\cos x}{1 + x^2} = x^2$

78. $\cos x = \frac{1}{2}(e^x + e^{-x})$

Applications

79. Range of a Projectile If a projectile is fired with velocity v_0 at an angle θ, then its *range*, the horizontal distance it travels (in feet), is modeled by the function

$$R(\theta) = \frac{v_0^2 \sin 2\theta}{32}$$

(See page 818.) If $v_0 = 2200$ ft/s, what angle (in degrees) should be chosen for the projectile to hit a target on the ground 5000 ft away?

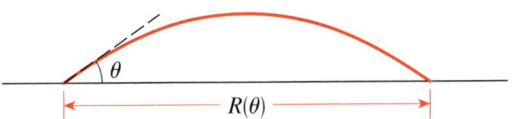

80. Damped Vibrations The displacement of a spring vibrating in damped harmonic motion is given by

$$y = 4e^{-3t} \sin 2\pi t$$

Find the times when the spring is at its equilibrium position $(y = 0)$.

81. Refraction of Light It has been observed since ancient times that light refracts or "bends" as it travels from one medium to another (from air to water, for example). If v_1 is the speed of light in one medium and v_2 its speed in another medium, then according to **Snell's Law,**

$$\frac{\sin \theta_1}{\sin \theta_2} = \frac{v_1}{v_2}$$

where θ_1 is the *angle of incidence* and θ_2 is the *angle of refraction* (see the figure). The number v_1/v_2 is called the *index of refraction*. The index of refraction for several substances is given in the table. If a ray of light passes through the surface of a lake at an angle of incidence of 70°, find the angle of refraction.

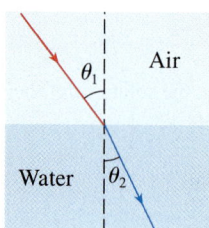

Substance	Refraction from air to substance
Water	1.33
Alcohol	1.36
Glass	1.52
Diamond	2.41

82. Total Internal Reflection   When light passes from a more-dense to a less-dense medium—from glass to air, for example—the angle of refraction predicted by Snell's Law (see Exercise 81) can be 90° or larger. In this case, the light beam is actually reflected back into the denser medium. This phenomenon, called *total internal reflection*, is the principle behind fiber optics.

Set $\theta_2 = 90°$ in Snell's Law and solve for θ_1 to determine the critical angle of incidence at which total internal reflection begins to occur when light passes from glass to air. (Note that the index of refraction from glass to air is the reciprocal of the index from air to glass.)

83. Hours of Daylight   In Philadelphia the number of hours of daylight on day t (where t is the number of days after January 1) is modeled by the function

$$L(t) = 12 + 2.83 \sin\left(\frac{2\pi}{365}(t - 80)\right)$$

(a) Which days of the year have about 10 hours of daylight?

(b) How many days of the year have more than 10 hours of daylight?

84. Phases of the Moon   As the moon revolves around the earth, the side that faces the earth is usually just partially illuminated by the sun. The phases of the moon describe how much of the surface appears to be in sunlight. An astronomical measure of phase is given by the fraction F of the lunar disc that is lit. When the angle between the sun, earth,

and moon is θ ($0 \le \theta \le 360°$), then

$$F = \tfrac{1}{2}(1 - \cos \theta)$$

Determine the angles θ that correspond to the following phases.

(a) $F = 0$ (new moon)

(b) $F = 0.25$ (a crescent moon)

(c) $F = 0.5$ (first or last quarter)

(d) $F = 1$ (full moon)

85. Belts and Pulleys   A thin belt of length L surrounds two pulleys of radii R and r, as shown in the figure.

(a) Show that the angle θ (in radians) where the belt crosses itself satisfies the equation

$$\theta + 2 \cot \frac{\theta}{2} = \frac{L}{R + r} - \pi$$

[*Hint:* Express L in terms of R, r, and θ by adding up the lengths of the curved and straight parts of the belt.]

 (b) Suppose that $R = 2.42$ ft, $r = 1.21$ ft, and $L = 27.78$ ft. Find θ by solving the equation in part (a) graphically. Express your answer both in radians and in degrees.

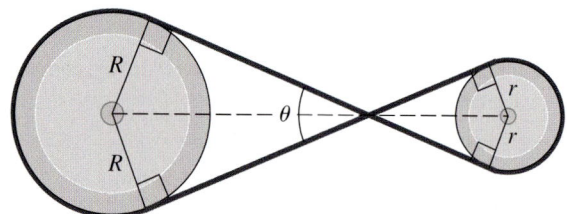

Discovery • Discussion

86. Equations and Identities   Which of the following statements is true?

   A. Every identity is an equation.

   B. Every equation is an identity.

Give examples to illustrate your answer. Write a short paragraph to explain the difference between an equation and an identity.

87. A Special Trigonometric Equation   What makes the equation $\sin(\cos x) = 0$ different from all the other equations we've looked at in this section? Find all solutions of this equation.

7   Review

Concept Check

1. (a) State the reciprocal identities.

(b) State the Pythagorean identities.

(c) State the even-odd identities.

(d) State the cofunction identities.

2. Explain the difference between an equation and an identity.

3. How do you prove a trigonometric identity?

4. (a) State the addition formulas for sine, cosine, and tangent.

(b) State the subtraction formulas for sine, cosine, and tangent.

5. (a) State the double-angle formulas for sine, cosine, and tangent.

(b) State the formulas for lowering powers.

(c) State the half-angle formulas.

6. (a) State the product-to-sum formulas.

(b) State the sum-to-product formulas.

7. (a) Define the inverse sine function $\sin^{-1}$. What are its domain and range?

(b) For what values of x is the equation $\sin(\sin^{-1}x) = x$ true?

(c) For what values of x is the equation $\sin^{-1}(\sin x) = x$ true?

8. (a) Define the inverse cosine function $\cos^{-1}$. What are its domain and range?

(b) For what values of x is the equation $\cos(\cos^{-1}x) = x$ true?

(c) For what values of x is the equation $\cos^{-1}(\cos x) = x$ true?

9. (a) Define the inverse tangent function $\tan^{-1}$. What are its domain and range?

(b) For what values of x is the equation $\tan(\tan^{-1}x) = x$ true?

(c) For what values of x is the equation $\tan^{-1}(\tan x) = x$ true?

10. Explain how you solve a trigonometric equation by factoring.

Exercises

1–24 ■ Verify the identity.

1. $\sin \theta \, (\cot \theta + \tan \theta) = \sec \theta$

2. $(\sec \theta - 1)(\sec \theta + 1) = \tan^2\theta$

3. $\cos^2x \csc x - \csc x = -\sin x$

4. $\dfrac{1}{1 - \sin^2x} = 1 + \tan^2x$

5. $\dfrac{\cos^2x - \tan^2x}{\sin^2x} = \cot^2x - \sec^2x$

6. $\dfrac{1 + \sec x}{\sec x} = \dfrac{\sin^2x}{1 - \cos x}$ **7.** $\dfrac{\cos^2x}{1 - \sin x} = \dfrac{\cos x}{\sec x - \tan x}$

8. $(1 - \tan x)(1 - \cot x) = 2 - \sec x \csc x$

9. $\sin^2x \cot^2x + \cos^2x \tan^2x = 1$

10. $(\tan x + \cot x)^2 = \csc^2x \sec^2x$

11. $\dfrac{\sin 2x}{1 + \cos 2x} = \tan x$

12. $\dfrac{\cos(x + y)}{\cos x \sin y} = \cot y - \tan x$

13. $\tan \dfrac{x}{2} = \csc x - \cot x$

14. $\dfrac{\sin(x + y) + \sin(x - y)}{\cos(x + y) + \cos(x - y)} = \tan x$

15. $\sin(x + y) \sin(x - y) = \sin^2x - \sin^2y$

16. $\csc x - \tan \dfrac{x}{2} = \cot x$ **17.** $1 + \tan x \tan \dfrac{x}{2} = \sec x$

18. $\dfrac{\sin 3x + \cos 3x}{\cos x - \sin x} = 1 + 2 \sin 2x$

19. $\left(\cos \dfrac{x}{2} - \sin \dfrac{x}{2} \right)^2 = 1 - \sin x$

20. $\dfrac{\cos 3x - \cos 7x}{\sin 3x + \sin 7x} = \tan 2x$

21. $\dfrac{\sin 2x}{\sin x} - \dfrac{\cos 2x}{\cos x} = \sec x$

22. $(\cos x + \cos y)^2 + (\sin x - \sin y)^2 = 2 + 2 \cos(x + y)$

23. $\tan\left(x + \dfrac{\pi}{4} \right) = \dfrac{1 + \tan x}{1 - \tan x}$ **24.** $\dfrac{\sec x - 1}{\sin x \sec x} = \tan \dfrac{x}{2}$

25–28 ■ **(a)** Graph f and g. **(b)** Do the graphs suggest that the equation $f(x) = g(x)$ is an identity? Prove your answer.

25. $f(x) = 1 - \left(\cos \dfrac{x}{2} - \sin \dfrac{x}{2} \right)^2$, $g(x) = \sin x$

26. $f(x) = \sin x + \cos x$, $g(x) = \sqrt{\sin^2 x + \cos^2 x}$

27. $f(x) = \tan x \tan \dfrac{x}{2}$, $g(x) = \dfrac{1}{\cos x}$

28. $f(x) = 1 - 8 \sin^2 x + 8 \sin^4 x$, $g(x) = \cos 4x$

 29–30 ■ **(a)** Graph the function(s) and make a conjecture, and **(b)** prove your conjecture.

29. $f(x) = 2 \sin^2 3x + \cos 6x$

30. $f(x) = \sin x \cot \dfrac{x}{2}$, $g(x) = \cos x$

31–46 ■ Solve the equation in the interval $[0, 2\pi)$.

31. $\cos x \sin x - \sin x = 0$ **32.** $\sin x - 2 \sin^2 x = 0$

33. $2 \sin^2 x - 5 \sin x + 2 = 0$

34. $\sin x - \cos x - \tan x = -1$

35. $2 \cos^2 x - 7 \cos x + 3 = 0$ **36.** $4 \sin^2 x + 2 \cos^2 x = 3$

37. $\dfrac{1 - \cos x}{1 + \cos x} = 3$ **38.** $\sin x = \cos 2x$

39. $\tan^3 x + \tan^2 x - 3 \tan x - 3 = 0$

40. $\cos 2x \csc^2 x = 2 \cos 2x$ **41.** $\tan \tfrac{1}{2} x + 2 \sin 2x = \csc x$

42. $\cos 3x + \cos 2x + \cos x = 0$

43. $\tan x + \sec x = \sqrt{3}$ **44.** $2 \cos x - 3 \tan x = 0$

 45. $\cos x = x^2 - 1$ **46.** $e^{\sin x} = x$

47. If a projectile is fired with velocity v_0 at an angle θ, then the maximum height it reaches (in feet) is modeled by the function

$$M(\theta) = \dfrac{v_0^2 \sin^2 \theta}{64}$$

Suppose $v_0 = 400$ ft/s.

(a) At what angle θ should the projectile be fired so that the maximum height it reaches is 2000 ft?

(b) Is it possible for the projectile to reach a height of 3000 ft?

(c) Find the angle θ for which the projectile will travel highest.

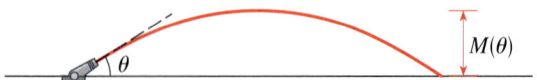

48. The displacement of an automobile shock absorber is modeled by the function

$$f(t) = 2^{-0.2t} \sin 4\pi t$$

Find the times when the shock absorber is at its equilibrium position (that is, when $f(t) = 0$). [*Hint:* $2^x > 0$ for all real x.]

49–58 ■ Find the exact value of the expression.

49. $\cos 15°$

50. $\sin \dfrac{5\pi}{12}$

51. $\tan \dfrac{\pi}{8}$

52. $2 \sin \dfrac{\pi}{12} \cos \dfrac{\pi}{12}$

53. $\sin 5° \cos 40° + \cos 5° \sin 40°$

54. $\dfrac{\tan 66° - \tan 6°}{1 + \tan 66° \tan 6°}$

55. $\cos^2 \dfrac{\pi}{8} - \sin^2 \dfrac{\pi}{8}$

56. $\dfrac{1}{2} \cos \dfrac{\pi}{12} + \dfrac{\sqrt{3}}{2} \sin \dfrac{\pi}{12}$

57. $\cos 37.5° \cos 7.5°$

58. $\cos 67.5° + \cos 22.5°$

59–64 ■ Find the exact value of the expression given that $\sec x = \tfrac{3}{2}$, $\csc y = 3$, and x and y are in quadrant I.

59. $\sin(x + y)$

60. $\cos(x - y)$

61. $\tan(x + y)$

62. $\sin 2x$

63. $\cos \dfrac{y}{2}$

64. $\tan \dfrac{y}{2}$

65–72 ■ Find the exact value of the expression.

65. $\sin^{-1}(\sqrt{3}/2)$

66. $\tan^{-1}(\sqrt{3}/3)$

67. $\cos(\tan^{-1} \sqrt{3})$

68. $\sin(\cos^{-1}(\sqrt{3}/2))$

69. $\tan(\sin^{-1} \tfrac{2}{5})$

70. $\sin(\cos^{-1} \tfrac{3}{8})$

71. $\cos(2 \sin^{-1} \tfrac{1}{3})$

72. $\cos(\sin^{-1} \tfrac{5}{13} - \cos^{-1} \tfrac{4}{5})$

73–74 ■ Rewrite the expression as an algebraic function of x.

73. $\sin(\tan^{-1} x)$

74. $\sec(\sin^{-1} x)$

75–76 ■ Express θ in terms of x.

75.   **76.**

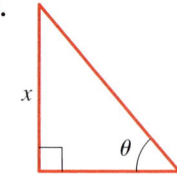

77. A 10-ft-wide highway sign is adjacent to a roadway, as shown in the figure. As a driver approaches the sign, the viewing angle θ changes.

   (a) Express viewing angle θ as a function of the distance x between the driver and the sign.

   (b) The sign is legible when the viewing angle is $2°$ or greater. At what distance x does the sign first become legible?

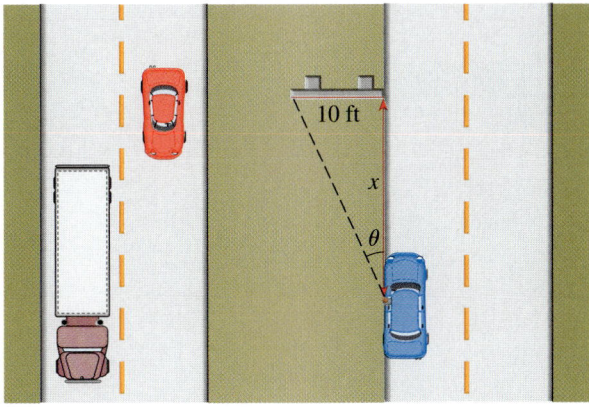

78. A 380-ft-tall building supports a 40-ft communications tower (see the figure). As a driver approaches the building, the viewing angle θ of the tower changes.

   (a) Express the viewing angle θ as a function of the distance x between the driver and the building.

 (b) At what distance from the building is the viewing angle θ as large as possible?

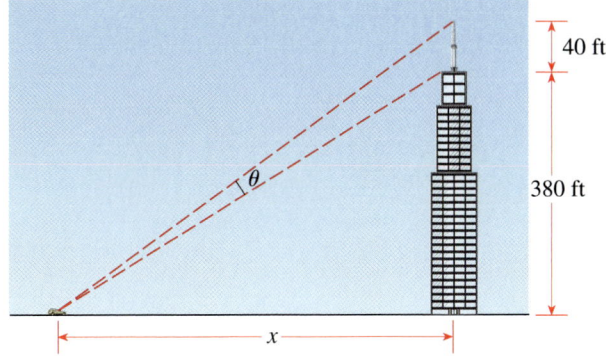

7 | Test

1. Verify each identity.

   (a) $\tan \theta \sin \theta + \cos \theta = \sec \theta$

   (b) $\dfrac{\tan x}{1 - \cos x} = \csc x \, (1 + \sec x)$

   (c) $\dfrac{2 \tan x}{1 + \tan^2 x} = \sin 2x$

2. Let $x = 2 \sin \theta$, $-\pi/2 < \theta < \pi/2$. Simplify the expression

$$\frac{x}{\sqrt{4 - x^2}}$$

3. Find the exact value of each expression.

   (a) $\sin 8° \cos 22° + \cos 8° \sin 22°$ (b) $\sin 75°$ (c) $\sin \dfrac{\pi}{12}$

4. For the angles α and β in the figures, find $\cos(\alpha + \beta)$.

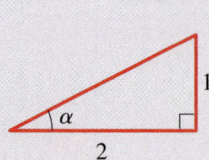
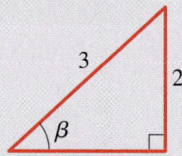

5. (a) Write $\sin 3x \cos 5x$ as a sum of trigonometric functions.

   (b) Write $\sin 2x - \sin 5x$ as a product of trigonometric functions.

6. If $\sin \theta = -\frac{4}{5}$ and θ is in quadrant III, find $\tan(\theta/2)$.

7. Graph $y = \sin x$ and $y = \sin^{-1} x$, and specify the domain of each function.

8. Express θ in each figure in terms of x.

   (a)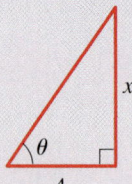
        (b)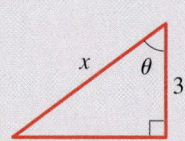

9. Solve each trigonometric equation in the interval $[0, 2\pi)$.

   (a) $2 \cos^2 x + 5 \cos x + 2 = 0$ (b) $\sin 2x - \cos x = 0$

10. Find all solutions in the interval $[0, 2\pi)$, correct to five decimal places:

$$5 \cos 2x = 2$$

11. Find the exact value of $\cos\left(\tan^{-1} \frac{9}{40}\right)$.

We've learned that the position of a particle in simple harmonic motion is described by a function of the form $y = A \sin \omega t$ (see Section 5.5). For example, if a string is moved up and down as in Figure 1, then the red dot on the string moves up and down in simple harmonic motion. Of course, the same holds true for each point on the string.

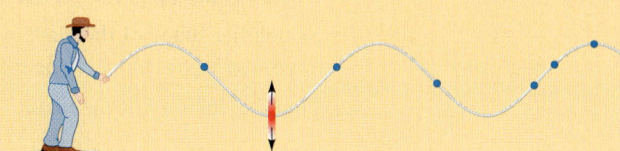

Figure 1

What function describes the shape of the whole string? If we fix an instant in time $(t = 0)$ and snap a photograph of the string, we get the shape in Figure 2, which is modeled by

$$y = A \sin kx$$

where y is the height of the string above the x-axis at the point x.

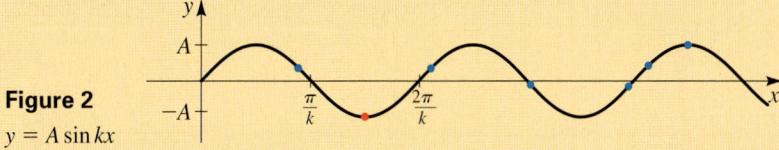

Figure 2

$y = A \sin kx$

Traveling Waves

If we snap photographs of the string at other instants, as in Figure 3, it appears that the waves in the string "travel" or shift to the right.

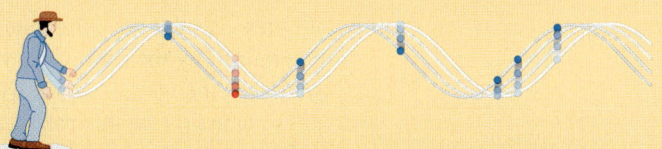

Figure 3

The **velocity** of the wave is the rate at which it moves to the right. If the wave has velocity v, then it moves to the right a distance vt in time t. So the graph of the shifted wave at time t is

$$y(x, t) = A \sin k(x - vt)$$

This function models the position of any point x on the string at any time t. We use the notation $y(x, t)$ to indicate that the function depends on the *two* variables x and t. Here is how this function models the motion of the string.

- **If we fix x**, then $y(x, t)$ is a function of t only, which gives the position of the fixed point x at time t.

- **If we fix t**, then $y(x, t)$ is a function of x only, whose graph is the shape of the string at the fixed time t.

Example 1   A Traveling Wave

A traveling wave is described by the function

$$y(x, t) = 3 \sin\left(2x - \frac{\pi}{2}t\right), \qquad x \geq 0$$

(a) Find the function that models the position of the point $x = \pi/6$ at any time t. Observe that the point moves in simple harmonic motion.

(b) Sketch the shape of the wave when $t = 0, 0.5, 1.0, 1.5,$ and 2.0. Does the wave appear to be traveling to the right?

(c) Find the velocity of the wave.

Solution

(a) Substituting $x = \pi/6$ we get

$$y\left(\frac{\pi}{6}, t\right) = 3 \sin\left(2 \cdot \frac{\pi}{6} - \frac{\pi}{2}t\right) = 3 \sin\left(\frac{\pi}{3} - \frac{\pi}{2}t\right)$$

The function $y = 3 \sin(\frac{\pi}{3} - \frac{\pi}{2}t)$ describes simple harmonic motion with amplitude 3 and period $2\pi/(\pi/2) = 4$.

(b) The graphs are shown in Figure 4. As t increases, the wave moves to the right.

(c) We express the given function in the standard form $y(x, t) = A \sin k(x - vt)$:

$$y(x, t) = 3 \sin\left(2x - \frac{\pi}{2}t\right) \qquad \textit{Given}$$

$$= 3 \sin 2\left(x - \frac{\pi}{4}t\right) \qquad \textit{Factor 2}$$

Comparing this to the standard form, we see that the wave is moving with velocity $v = \pi/4$. ∎

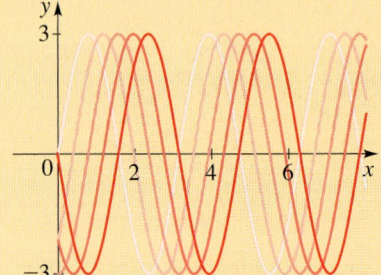

Figure 4

Traveling wave

Standing Waves

If two waves are traveling along the same string, then the movement of the string is determined by the sum of the two waves. For example, if the string is attached to a wall, then the waves bounce back with the same amplitude and speed but in the opposite direction. In this case, one wave is described by $y = A \sin k(x - vt)$ and the reflected wave by $y = A \sin k(x + vt)$. The resulting wave is

$$y(x, t) = A \sin k(x - vt) + A \sin k(x + vt) \qquad \textit{Add the two waves}$$

$$= 2A \sin kx \cos kvt \qquad \textit{Sum-to-product formula}$$

The points where kx is a multiple of 2π are special, because at these points $y = 0$ for any time t. In other words, these points never move. Such points are called **nodes**. Figure 5 shows the graph of the wave for several values of t. We see that the wave does not travel, but simply vibrates up and down. Such a wave is called a **standing wave**.

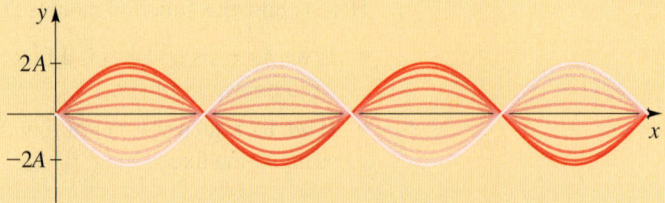

Figure 5

A standing wave

Example 2   A Standing Wave

Traveling waves are generated at each end of a wave tank 30 ft long, with equations

$$y = 1.5 \sin\left(\frac{\pi}{5}x - 3t\right) \quad \text{and} \quad y = 1.5 \sin\left(\frac{\pi}{5}x + 3t\right)$$

(a) Find the equation of the combined wave, and find the nodes.

(b) Sketch the graph for $t = 0, 0.17, 0.34, 0.51, 0.68, 0.85,$ and 1.02. Is this a standing wave?

Solution

(a) The combined wave is obtained by adding the two equations:

$$y = 1.5 \sin\left(\frac{\pi}{5}x - 3t\right) + 1.5 \sin\left(\frac{\pi}{5}x + 3t\right) \qquad \textit{Add the two waves}$$

$$= 3 \sin\frac{\pi}{5}x \cos 3t \qquad \textit{Sum-to-product formula}$$

The nodes occur at the values of x for which $\sin\frac{\pi}{5}x = 0$, that is, where $\frac{\pi}{5}x = k\pi$ (k an integer). Solving for x we get $x = 5k$. So the nodes occur at

$$x = 0, 5, 10, 15, 20, 25, 30$$

(b) The graphs are shown in Figure 6. From the graphs we see that this is a standing wave.

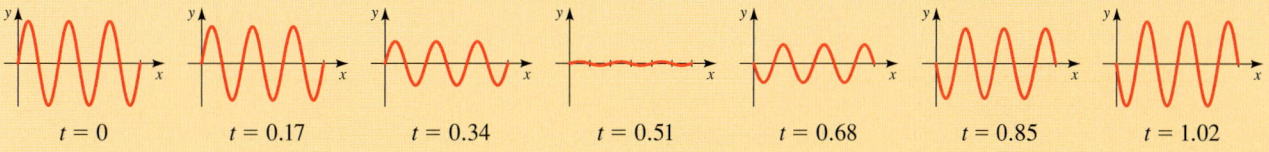

| $t = 0$ | $t = 0.17$ | $t = 0.34$ | $t = 0.51$ | $t = 0.68$ | $t = 0.85$ | $t = 1.02$ |

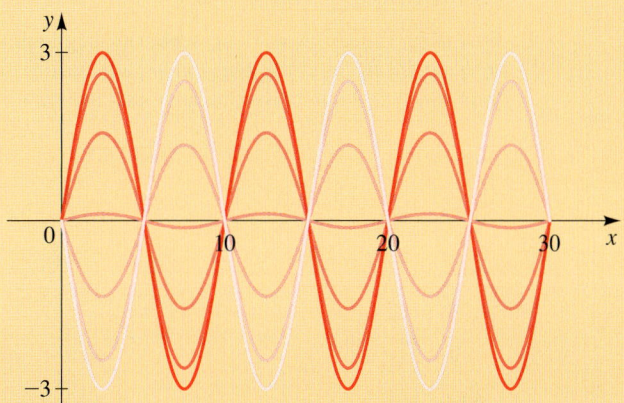

Figure 6

$$y(x, t) = 3 \sin\frac{\pi}{5}x \cos 3t$$

Problems

1. **Wave on a Canal**   A wave on the surface of a long canal is described by the function

$$y(x, t) = 5 \sin\left(2x - \frac{\pi}{2}t\right), \quad x \geq 0$$

   (a) Find the function that models the position of the point $x = 0$ at any time t.
   (b) Sketch the shape of the wave when $t = 0, 0.4, 0.8, 1.2$, and 1.6. Is this a traveling wave?
   (c) Find the velocity of the wave.

2. **Wave in a Rope**   Traveling waves are generated at each end of a tightly stretched rope 24 ft long, with equations

$$y = 0.2 \sin(1.047x - 0.524t) \qquad \text{and} \qquad y = 0.2 \sin(1.047x + 0.524t)$$

   (a) Find the equation of the combined wave, and find the nodes.
   (b) Sketch the graph for $t = 0, 1, 2, 3, 4, 5$, and 6. Is this a standing wave?

3. **Traveling Wave**   A traveling wave is graphed at the instant $t = 0$. If it is moving to the right with velocity 6, find an equation of the form $y(x, t) = A \sin(kx - kvt)$ for this wave.

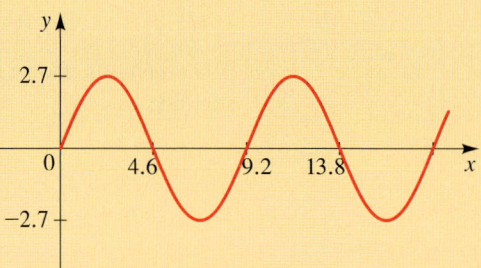

4. **Traveling Wave**   A traveling wave has period $2\pi/3$, amplitude 5, and velocity 0.5.
   (a) Find the equation of the wave.
   (b) Sketch the graph for $t = 0, 0.5, 1, 1.5$, and 2.

5. **Standing Wave**   A standing wave with amplitude 0.6 is graphed at several times t as shown in the figure. If the vibration has a frequency of 20 Hz, find an equation of the form $y(x, t) = A \sin \alpha x \cos \beta t$ that models this wave.

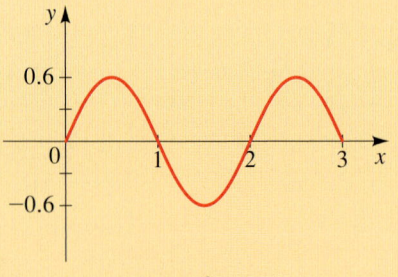

$t = 0$ s

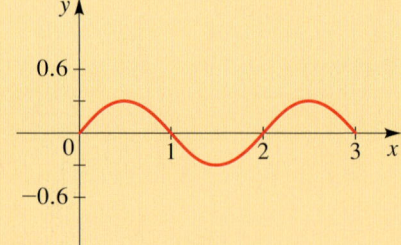

$t = 0.010$ s

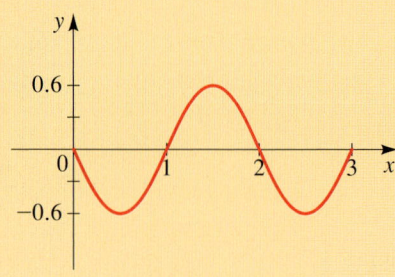

$t = 0.025$ s

6. Standing Wave   A standing wave has maximum amplitude 7 and nodes at 0, $\pi/2$, π, $3\pi/2$, 2π, as shown in the figure. Each point that is not a node moves up and down with period 4π. Find a function of the form $y(x, t) = A \sin \alpha x \cos \beta t$ that models this wave.

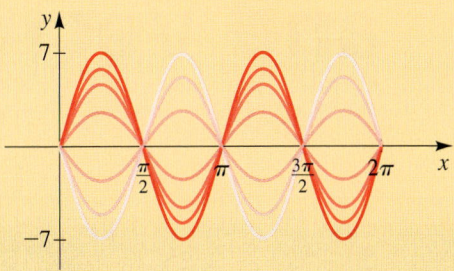

7. Vibrating String   When a violin string vibrates, the sound produced results from a combination of standing waves that have evenly placed nodes. The figure illustrates some of the possible standing waves. Let's assume that the string has length π.

(a) For fixed t, the string has the shape of a sine curve $y = A \sin \alpha x$. Find the appropriate value of α for each of the illustrated standing waves.

(b) Do you notice a pattern in the values of α that you found in part (a)? What would the next two values of α be? Sketch rough graphs of the standing waves associated with these new values of α.

(c) Suppose that for fixed t, each point on the string that is not a node vibrates with frequency 440 Hz. Find the value of β for which an equation of the form $y = A \cos \beta t$ would model this motion.

(d) Combine your answers for parts (a) and (c) to find functions of the form $y(x, t) = A \sin \alpha x \cos \beta t$ that model each of the standing waves in the figure. (Assume $A = 1$.)

8. Waves in a Tube   Standing waves in a violin string must have nodes at the ends of the string because the string is fixed at its endpoints. But this need not be the case with sound waves in a tube (such as a flute or an organ pipe). The figure shows some possible standing waves in a tube.

Suppose that a standing wave in a tube 37.7 ft long is modeled by the function

$$y(x, t) = 0.3 \cos \tfrac{1}{2}x \cos 50\pi t$$

Here $y(x, t)$ represents the variation from normal air pressure at the point x feet from the end of the tube, at time t seconds.

(a) At what points x are the nodes located? Are the endpoints of the tube nodes?

(b) At what frequency does the air vibrate at points that are not nodes?

8 Polar Coordinates and Vectors

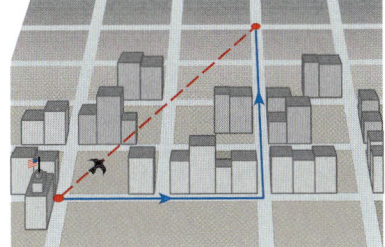

Chapter Overview

In this chapter we study polar coordinates, a new way of describing the location of points in a plane.

A coordinate system is a method for specifying the location of a point in the plane. We are familiar with rectangular (or Cartesian) coordinates. In rectangular coordinates the location of a point is given by an ordered pair (x, y), which gives the distance of the point to two perpendicular axes. Using rectangular coordinates is like describing a location in a city by saying that it's at the corner of 2nd Street and 4th Avenue. But we might also describe this same location by saying that it's $1\frac{1}{2}$ miles northeast of City Hall. So instead of specifying the location with respect to a grid of streets and avenues, we specify it by giving its distance and direction from a fixed reference point. That is what we do in the polar coordinate system. In polar coordinates the location of a point is given by an ordered pair (r, θ) where r is the distance from the origin (or pole) and θ is the angle from the positive x-axis (see the figure below).

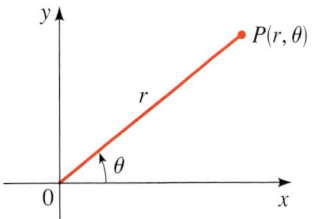

Why do we study different coordinate systems? Because certain curves are more naturally described in one coordinate system rather than the other. In rectangular coordinates we can give simple equations for lines, parabolas, or cubic curves, but the equation of a circle is rather complicated (and it is not a function). In polar coordinates we can give simple equations for circles, ellipses, roses, and figure 8's—curves that are difficult to describe in rectangular coordinates. So, for example, it is more natural to describe a planet's path around the sun in terms of distance from the sun and angle of travel—in other words, in polar coordinates. We will also give polar representations of complex numbers. As you will see, it is easy to multiply complex numbers if they are written in polar form.

In this chapter we also use coordinates to describe directed quantities, or *vectors*. When we talk about temperature, mass, or area, we need only one number. For example, we say the temperature is 70°F. But quantities such as velocity or force are *directed quantities*, because they involve direction as well as magnitude. Thus we say

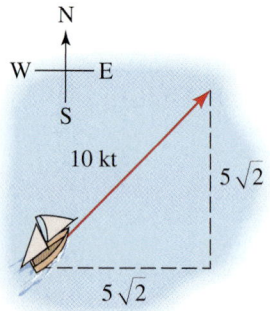

that a boat is sailing at 10 knots to the northeast. We can also express this graphically by drawing an arrow of length 10 in the direction of travel. The velocity can be completely described by the displacement of the arrow from tail to head, which we express as the vector $\langle 5\sqrt{2}, 5\sqrt{2} \rangle$ (see the figure).

In the *Focus on Modeling* (page 630) we will see how polar coordinates are used to draw a (flat) map of a (spherical) world. In the *Discovery Project* on page 626 we explore how an analysis of the vector forces of wind and current can be used to navigate a sailboat.

8.1   Polar Coordinates

In this section we define polar coordinates, and we learn how polar coordinates are related to rectangular coordinates.

Definition of Polar Coordinates

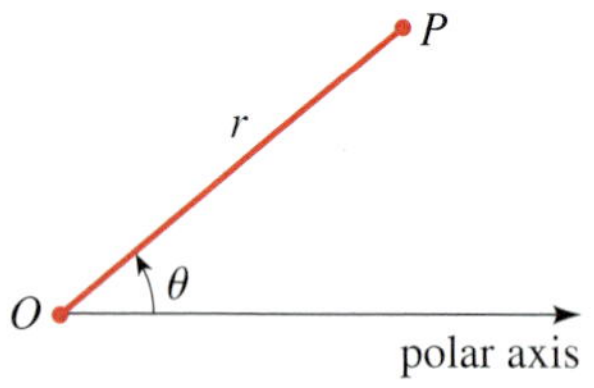

The **polar coordinate system** uses distances and directions to specify the location of a point in the plane. To set up this system, we choose a fixed point O in the plane called the **pole** (or **origin**) and draw from O a ray (half-line) called the **polar axis** as in Figure 1. Then each point P can be assigned polar coordinates $P(r, \theta)$ where

r is the *distance* from O to P

θ is the *angle* between the polar axis and the segment $\overline{OP}$

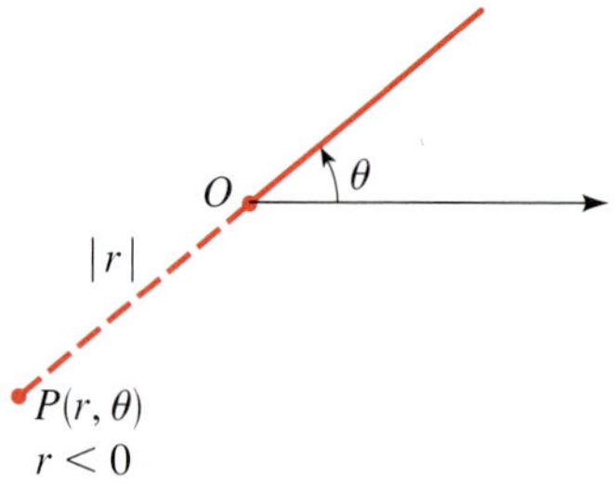

We use the convention that θ is positive if measured in a counterclockwise direction from the polar axis or negative if measured in a clockwise direction. If r is negative, then $P(r, \theta)$ is defined to be the point that lies $|r|$ units from the pole in the direction opposite to that given by θ (see Figure 2).

Figure 1

Figure 2

Example 1   Plotting Points in Polar Coordinates

Plot the points whose polar coordinates are given.

(a) $(1, 3\pi/4)$ (b) $(3, -\pi/6)$ (c) $(3, 3\pi)$ (d) $(-4, \pi/4)$

Solution   The points are plotted in Figure 3. Note that the point in part (d) lies 4 units from the origin along the angle $5\pi/4$, because the given value of r is negative.

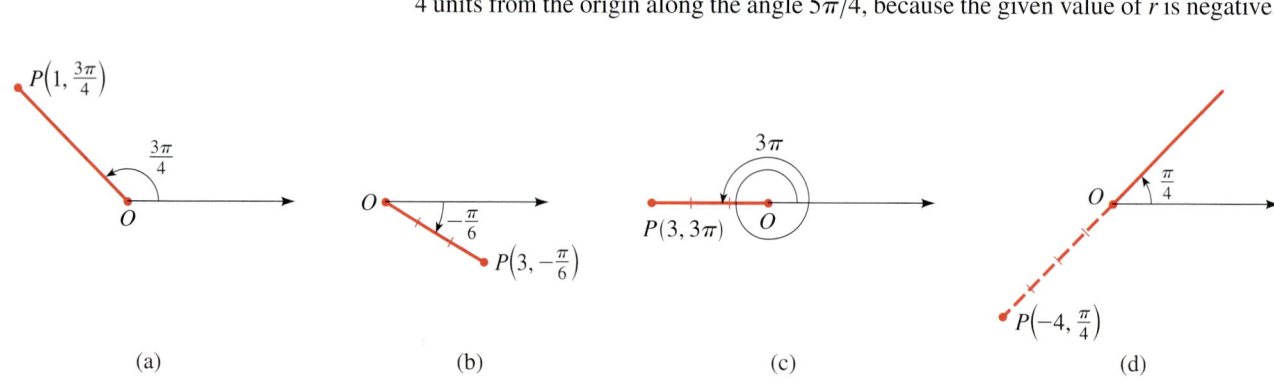

(a)   (b)   (c)   (d)

Figure 3

Note that the coordinates (r, θ) and $(-r, \theta + \pi)$ represent the same point, as shown in Figure 4. Moreover, because the angles $\theta + 2n\pi$ (where n is any integer) all have the same terminal side as the angle θ, each point in the plane has infinitely many representations in polar coordinates. In fact, any point $P(r, \theta)$ can also be represented by

$$P(r, \theta + 2n\pi) \qquad \text{and} \qquad P(-r, \theta + (2n + 1)\pi)$$

for any integer n.

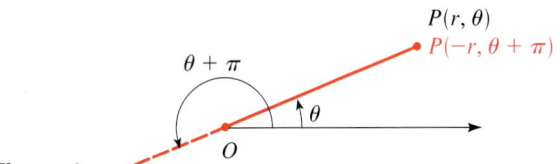

Figure 4

Example 2   Different Polar Coordinates for the Same Point

(a) Graph the point with polar coordinates $P(2, \pi/3)$.

(b) Find two other polar coordinate representations of P with $r > 0$, and two with $r < 0$.

Solution

(a) The graph is shown in Figure 5(a).

(b) Other representations with $r > 0$ are

$$\left(2, \frac{\pi}{3} + 2\pi\right) = \left(2, \frac{7\pi}{3}\right) \qquad \text{Add } 2\pi \text{ to } \theta$$

$$\left(2, \frac{\pi}{3} - 2\pi\right) = \left(2, -\frac{5\pi}{3}\right) \qquad \text{Add } -2\pi \text{ to } \theta$$

Other representations with $r < 0$ are

$$\left(-2, \frac{\pi}{3} + \pi\right) = \left(-2, \frac{4\pi}{3}\right) \qquad \text{Replace } r \text{ by } -r \text{ and add } \pi \text{ to } \theta$$

$$\left(-2, \frac{\pi}{3} - \pi\right) = \left(-2, -\frac{2\pi}{3}\right) \qquad \text{Replace } r \text{ by } -r \text{ and add } -\pi \text{ to } \theta$$

The graphs in Figure 5 explain why these coordinates represent the same point.

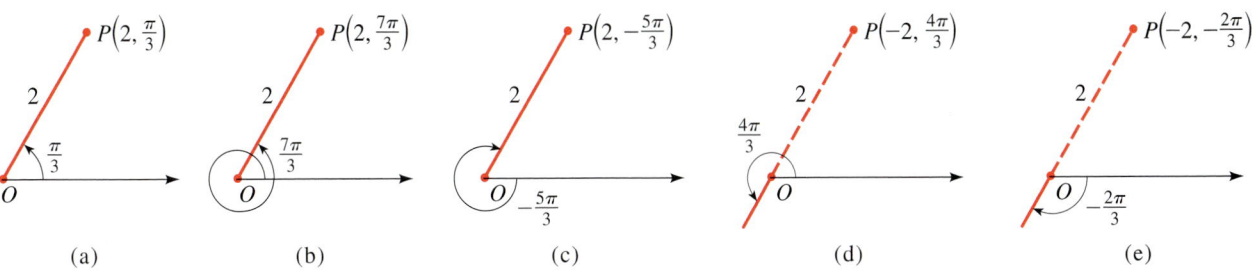

Figure 5

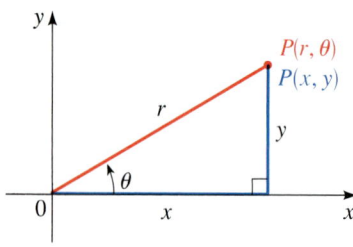

Figure 6

Relationship between Polar and Rectangular Coordinates

Situations often arise in which we need to consider polar and rectangular coordinates simultaneously. The connection between the two systems is illustrated in Figure 6, where the polar axis coincides with the positive x-axis. The formulas in the following box are obtained from the figure using the definitions of the trigonometric functions and the Pythagorean Theorem. (Although we have pictured the case where $r > 0$ and θ is acute, the formulas hold for any angle θ and for any value of r.)

Relationship between Polar and Rectangular Coordinates

1. To change from polar to rectangular coordinates, use the formulas

$$x = r \cos \theta \qquad \text{and} \qquad y = r \sin \theta$$

2. To change from rectangular to polar coordinates, use the formulas

$$r^2 = x^2 + y^2 \qquad \text{and} \qquad \tan \theta = \frac{y}{x} \quad (x \neq 0)$$

Example 3 Converting Polar Coordinates to Rectangular Coordinates

Find rectangular coordinates for the point that has polar coordinates $(4, 2\pi/3)$.

Solution Since $r = 4$ and $\theta = 2\pi/3$, we have

$$x = r \cos \theta = 4 \cos \frac{2\pi}{3} = 4 \cdot \left(-\frac{1}{2} \right) = -2$$

$$y = r \sin \theta = 4 \sin \frac{2\pi}{3} = 4 \cdot \frac{\sqrt{3}}{2} = 2\sqrt{3}$$

Thus, the point has rectangular coordinates $(-2, 2\sqrt{3})$. ■

Example 4 Converting Rectangular Coordinates to Polar Coordinates

Find polar coordinates for the point that has rectangular coordinates $(2, -2)$.

Solution Using $x = 2$, $y = -2$, we get

$$r^2 = x^2 + y^2 = 2^2 + (-2)^2 = 8$$

so $r = 2\sqrt{2}$ or $-2\sqrt{2}$. Also

$$\tan \theta = \frac{y}{x} = \frac{-2}{2} = -1$$

so $\theta = 3\pi/4$ or $-\pi/4$. Since the point $(2, -2)$ lies in quadrant IV (see Figure 7), we can represent it in polar coordinates as $(2\sqrt{2}, -\pi/4)$ or $(-2\sqrt{2}, 3\pi/4)$. ■

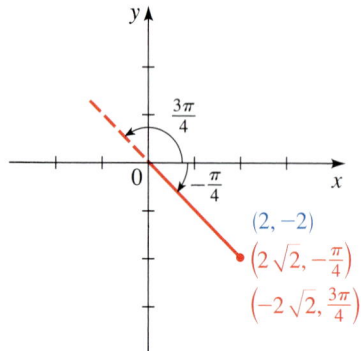

Figure 7

Note that the equations relating polar and rectangular coordinates do not uniquely determine r or θ. When we use these equations to find the polar coordinates of a point, we must be careful that the values we choose for r and θ give us a point in the correct quadrant, as we saw in Example 4.

Polar Equations

In Examples 3 and 4 we converted points from one coordinate system to the other. Now we consider the same problem for equations.

Example 5 Converting an Equation from Rectangular to Polar Coordinates

Express the equation $x^2 = 4y$ in polar coordinates.

Solution   We use the formulas $x = r \cos \theta$ and $y = r \sin \theta$.

$$x^2 = 4y \qquad \text{\textit{Rectangular equation}}$$

$$(r \cos \theta)^2 = 4(r \sin \theta) \qquad \text{\textit{Substitute}}\ x = r\cos\theta,\ y = r\sin\theta$$

$$r^2 \cos^2\theta = 4r \sin \theta \qquad \text{\textit{Expand}}$$

$$r = 4\,\frac{\sin \theta}{\cos^2\theta} \qquad \text{\textit{Divide by}}\ r\cos^2\theta$$

$$r = 4 \sec \theta \tan \theta \qquad \text{\textit{Simplify}} \qquad\blacksquare$$

As Example 5 shows, converting from rectangular to polar coordinates is straightforward—just replace x by $r \cos \theta$ and y by $r \sin \theta$, and then simplify. But converting polar equations to rectangular form often requires more thought.

Example 6 Converting Equations from Polar to Rectangular Coordinates

Express the polar equation in rectangular coordinates. If possible, determine the graph of the equation from its rectangular form.

(a) $r = 5 \sec \theta$ (b) $r = 2 \sin \theta$ (c) $r = 2 + 2 \cos \theta$

Solution

(a) Since $\sec \theta = 1/\cos \theta$, we multiply both sides by $\cos \theta$.

$$r = 5 \sec \theta$$

$$r \cos \theta = 5 \qquad \text{\textit{Multiply by}}\ \cos\theta$$

$$x = 5 \qquad \text{\textit{Substitute}}\ x = r\cos\theta$$

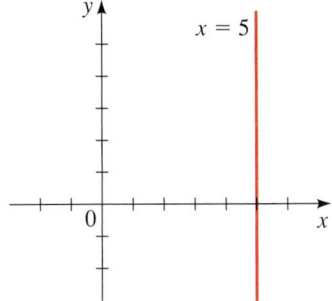

Figure 8

The graph of $x = 5$ is the vertical line in Figure 8.

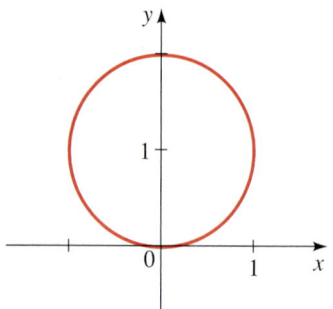

Figure 9

(b) We multiply both sides of the equation by r, because then we can use the formulas $r^2 = x^2 + y^2$ and $r \sin \theta = y$.

$$r^2 = 2r \sin \theta \qquad \text{Multiply by } r$$
$$x^2 + y^2 = 2y \qquad r^2 = x^2 + y^2 \text{ and } r\sin \theta = y$$
$$x^2 + y^2 - 2y = 0 \qquad \text{Subtract } 2y$$
$$x^2 + (y - 1)^2 = 1 \qquad \text{Complete the square in } y$$

This is the equation of a circle of radius 1 centered at the point $(0, 1)$. It is graphed in Figure 9.

(c) We first multiply both sides of the equation by r:

$$r^2 = 2r + 2r \cos \theta$$

Using $r^2 = x^2 + y^2$ and $x = r \cos \theta$, we can convert two of the three terms in the equation into rectangular coordinates, but eliminating the remaining r requires more work:

$$x^2 + y^2 = 2r + 2x \qquad r^2 = x^2 + y^2 \text{ and } r\cos \theta = x$$
$$x^2 + y^2 - 2x = 2r \qquad \text{Subtract } 2x$$
$$(x^2 + y^2 - 2x)^2 = 4r^2 \qquad \text{Square both sides}$$
$$(x^2 + y^2 - 2x)^2 = 4(x^2 + y^2) \qquad r^2 = x^2 + y^2$$

In this case, the rectangular equation looks more complicated than the polar equation. Although we cannot easily determine the graph of the equation from its rectangular form, we will see in the next section how to graph it using the polar equation. ∎

8.1 Exercises

1–6 ■ Plot the point that has the given polar coordinates.

1. $(4, \pi/4)$ **2.** $(1, 0)$ **3.** $(6, -7\pi/6)$

4. $(3, -2\pi/3)$ **5.** $(-2, 4\pi/3)$ **6.** $(-5, -17\pi/6)$

7–12 ■ Plot the point that has the given polar coordinates. Then give two other polar coordinate representations of the point, one with $r < 0$ and the other with $r > 0$.

7. $(3, \pi/2)$ **8.** $(2, 3\pi/4)$ **9.** $(-1, 7\pi/6)$

10. $(-2, -\pi/3)$ **11.** $(-5, 0)$ **12.** $(3, 1)$

13–20 ■ Determine which point in the figure, P, Q, R, or S, has the given polar coordinates.

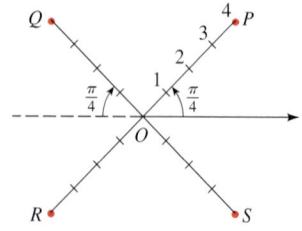

13. $(4, 3\pi/4)$ **14.** $(4, -3\pi/4)$

15. $(-4, -\pi/4)$ **16.** $(-4, 13\pi/4)$

17. $(4, -23\pi/4)$ **18.** $(-4, 23\pi/4)$

19. $(-4, 101\pi/4)$ **20.** $(4, 103\pi/4)$

21–22 ■ A point is graphed in rectangular form. Find polar coordinates for the point, with $r > 0$ and $0 < \theta < 2\pi$.

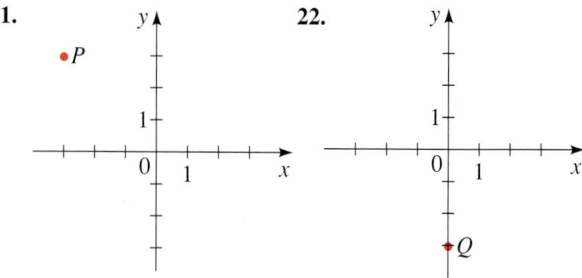

23–24 ■ A point is graphed in polar form. Find its rectangular coordinates.

23.

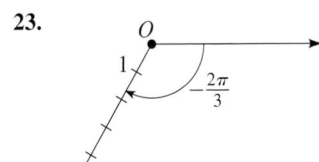

24.

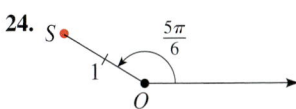

25–32 ■ Find the rectangular coordinates for the point whose polar coordinates are given.

25. $(4, \pi/6)$ **26.** $(6, 2\pi/3)$

27. $(\sqrt{2}, -\pi/4)$ **28.** $(-1, 5\pi/2)$

29. $(5, 5\pi)$ **30.** $(0, 13\pi)$

31. $(6\sqrt{2}, 11\pi/6)$ **32.** $(\sqrt{3}, -5\pi/3)$

33–40 ■ Convert the rectangular coordinates to polar coordinates with $r > 0$ and $0 \le \theta < 2\pi$.

33. $(-1, 1)$ **34.** $(3\sqrt{3}, -3)$

35. $(\sqrt{8}, \sqrt{8})$ **36.** $(-\sqrt{6}, -\sqrt{2})$

37. $(3, 4)$ **38.** $(1, -2)$

39. $(-6, 0)$ **40.** $(0, -\sqrt{3})$

41–46 ■ Convert the equation to polar form.

41. $x = y$ **42.** $x^2 + y^2 = 9$

43. $y = x^2$ **44.** $y = 5$

45. $x = 4$ **46.** $x^2 - y^2 = 1$

47–60 ■ Convert the polar equation to rectangular coordinates.

47. $r = 7$ **48.** $\theta = \pi$

49. $r \cos \theta = 6$ **50.** $r = 6 \cos \theta$

51. $r^2 = \tan \theta$ **52.** $r^2 = \sin 2\theta$

53. $r = \dfrac{1}{\sin \theta - \cos \theta}$ **54.** $r = \dfrac{1}{1 + \sin \theta}$

55. $r = 1 + \cos \theta$ **56.** $r = \dfrac{4}{1 + 2 \sin \theta}$

57. $r = 2 \sec \theta$ **58.** $r = 2 - \cos \theta$

59. $\sec \theta = 2$ **60.** $\cos 2\theta = 1$

Discovery • Discussion

61. The Distance Formula in Polar Coordinates

   (a) Use the Law of Cosines to prove that the distance between the polar points (r_1, θ_1) and (r_2, θ_2) is

$$d = \sqrt{r_1^2 + r_2^2 - 2r_1 r_2 \cos(\theta_2 - \theta_1)}$$

   (b) Find the distance between the points whose polar coordinates are $(3, 3\pi/4)$ and $(1, 7\pi/6)$, using the formula from part (a).

   (c) Now convert the points in part (b) to rectangular coordinates. Find the distance between them using the usual Distance Formula. Do you get the same answer?

8.2 **Graphs of Polar Equations**

The **graph of a polar equation** $r = f(\theta)$ consists of all points P that have at least one polar representation (r, θ) whose coordinates satisfy the equation. Many curves that arise in mathematics and its applications are more easily and naturally represented by polar equations rather than rectangular equations.

A rectangular grid is helpful for plotting points in rectangular coordinates (see Figure 1(a) on the next page). To plot points in polar coordinates, it is convenient to use a grid consisting of circles centered at the pole and rays emanating from the pole, as in Figure 1(b). We will use such grids to help us sketch polar graphs.

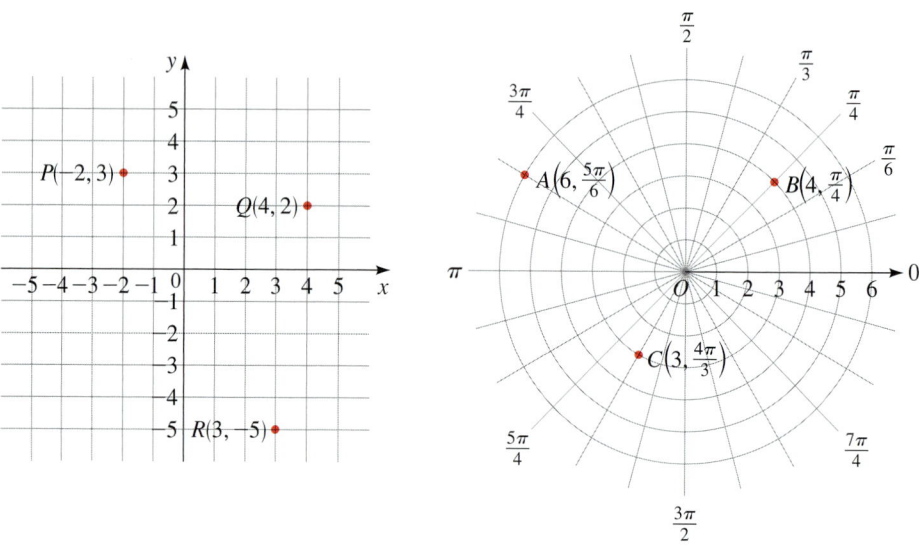

Figure 1   (a) Grid for rectangular coordinates   (b) Grid for polar coordinates

In Examples 1 and 2 we see that circles centered at the origin and lines that pass through the origin have particularly simple equations in polar coordinates.

Example 1   Sketching the Graph of a Polar Equation

Sketch the graph of the equation $r = 3$ and express the equation in rectangular coordinates.

Solution   The graph consists of all points whose r-coordinate is 3, that is, all points that are 3 units away from the origin. So the graph is a circle of radius 3 centered at the origin, as shown in Figure 2.

Squaring both sides of the equation, we get

$$r^2 = 3^2 \qquad \textcolor{blue}{\text{Square both sides}}$$

$$x^2 + y^2 = 9 \qquad \textcolor{blue}{\text{Substitute } r^2 = x^2 + y^2}$$

So the equivalent equation in rectangular coordinates is $x^2 + y^2 = 9$.   ∎

In general, the graph of the equation $r = a$ is a circle of radius $|a|$ centered at the origin. Squaring both sides of this equation, we see that the equivalent equation in rectangular coordinates is $x^2 + y^2 = a^2$.

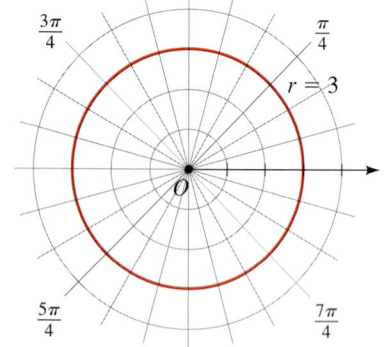

Figure 2

Example 2   Sketching the Graph of a Polar Equation

Sketch the graph of the equation $\theta = \pi/3$ and express the equation in rectangular coordinates.

Solution   The graph consists of all points whose θ-coordinate is $\pi/3$. This is the straight line that passes through the origin and makes an angle of $\pi/3$ with the polar

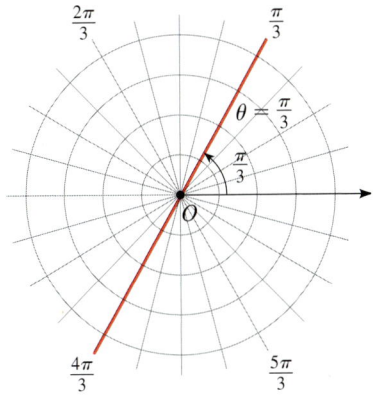

Figure 3

axis (see Figure 3). Note that the points $(r, \pi/3)$ on the line with $r > 0$ lie in quadrant I, whereas those with $r < 0$ lie in quadrant III. If the point (x, y) lies on this line, then

$$\frac{y}{x} = \tan \theta = \tan \frac{\pi}{3} = \sqrt{3}$$

Thus, the rectangular equation of this line is $y = \sqrt{3}\,x$. ∎

To sketch a polar curve whose graph isn't as obvious as the ones in the preceding examples, we plot points calculated for sufficiently many values of θ and then join them in a continuous curve. (This is what we did when we first learned to graph functions in rectangular coordinates.)

Example 3 Sketching the Graph of a Polar Equation

Sketch the graph of the polar equation $r = 2 \sin \theta$.

Solution We first use the equation to determine the polar coordinates of several points on the curve. The results are shown in the following table.

θ	0	$\pi/6$	$\pi/4$	$\pi/3$	$\pi/2$	$2\pi/3$	$3\pi/4$	$5\pi/6$	π
$r = 2 \sin \theta$	0	1	$\sqrt{2}$	$\sqrt{3}$	2	$\sqrt{3}$	$\sqrt{2}$	1	0

We plot these points in Figure 4 and then join them to sketch the curve. The graph appears to be a circle. We have used values of θ only between 0 and π, since the same points (this time expressed with negative r-coordinates) would be obtained if we allowed θ to range from π to 2π.

The polar equation $r = 2 \sin \theta$ in rectangular coordinates is

$$x^2 + (y - 1)^2 = 1$$

(See Section 8.1, Example 6(b)). From the rectangular form of the equation we see that the graph is a circle of radius 1 centered at $(0, 1)$.

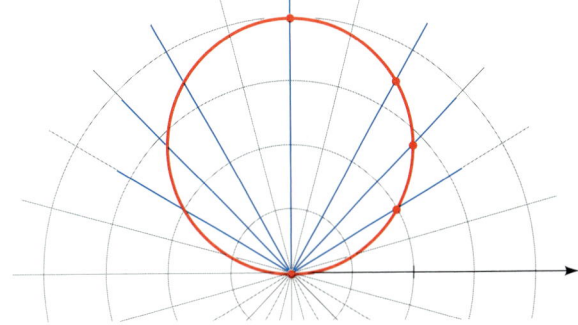

Figure 4
$r = 2 \sin \theta$

∎

In general, the graphs of equations of the form

$$r = 2a \sin \theta \qquad \text{and} \qquad r = 2a \cos \theta$$

are circles with radius $|a|$ centered at the points with polar coordinates $(a, \pi/2)$ and $(a, 0)$, respectively.

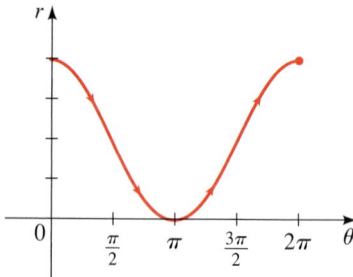

Figure 5

$r = 2 + 2 \cos \theta$

Example 4   Sketching the Graph of a Polar Equation

Sketch the graph of $r = 2 + 2 \cos \theta$.

Solution   Instead of plotting points as in Example 3, we first sketch the graph of $r = 2 + 2 \cos \theta$ in *rectangular* coordinates in Figure 5. We can think of this graph as a table of values that enables us to read at a glance the values of r that correspond to increasing values of θ. For instance, we see that as θ increases from 0 to $\pi/2$, r (the distance from O) decreases from 4 to 2, so we sketch the corresponding part of the polar graph in Figure 6(a). As θ increases from $\pi/2$ to π, Figure 5 shows that r decreases from 2 to 0, so we sketch the next part of the graph as in Figure 6(b). As θ increases from π to $3\pi/2$, r increases from 0 to 2, as shown in part (c). Finally, as θ increases from $3\pi/2$ to 2π, r increases from 2 to 4, as shown in part (d). If we let θ increase beyond 2π or decrease beyond 0, we would simply retrace our path. Combining the portions of the graph from parts (a) through (d) of Figure 6, we sketch the complete graph in part (e).

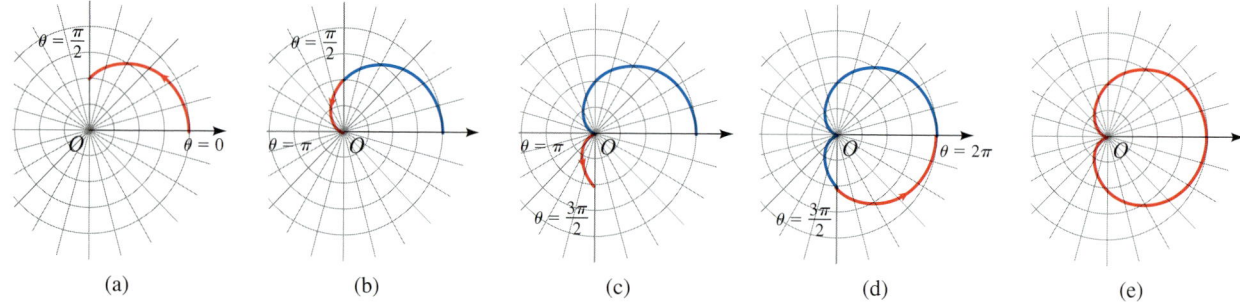

| (a) | (b) | (c) | (d) | (e) |

Figure 6   Steps in sketching $r = 2 + 2 \cos \theta$ ∎

The polar equation $r = 2 + 2 \cos \theta$ in rectangular coordinates is

$$(x^2 + y^2 - 2x)^2 = 4(x^2 + y^2)$$

(See Section 8.1, Example 6(c)). The simpler form of the polar equation shows that it is more natural to describe cardioids using polar coordinates.

The curve in Figure 6 is called a **cardioid** because it is heart-shaped. In general, the graph of any equation of the form

$$r = a(1 \pm \cos \theta) \qquad \text{or} \qquad r = a(1 \pm \sin \theta)$$

is a cardioid.

Example 5   Sketching the Graph of a Polar Equation

Sketch the curve $r = \cos 2\theta$.

Solution   As in Example 4, we first sketch the graph of $r = \cos 2\theta$ in *rectangular* coordinates, as shown in Figure 7. As θ increases from 0 to $\pi/4$, Figure 7 shows that r decreases from 1 to 0, and so we draw the corresponding portion of the polar curve in Figure 8 (indicated by ①). As θ increases from $\pi/4$ to $\pi/2$, the value of r goes from 0 to -1. This means that the distance from the origin increases from 0 to 1, but instead of being in quadrant I, this portion of the polar curve (indicated by ②) lies on the opposite side of the origin in quadrant III. The remainder of the curve is drawn in a similar fashion, with the arrows and numbers indicating the order in which the portions are traced out. The resulting curve has four petals and is called a **four-leaved rose**.

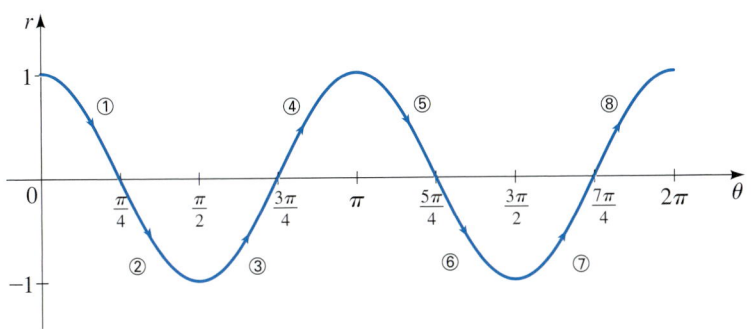

Figure 7

Graph of $r = \cos 2\theta$ sketched in rectangular coordinates

Figure 8

Four-leaved rose $r = \cos 2\theta$ sketched in polar coordinates

In general, the graph of an equation of the form

$$r = a \cos n\theta \qquad \text{or} \qquad r = a \sin n\theta$$

is an **n-leaved rose** if n is odd or a $2n$-leaved rose if n is even (as in Example 5).

Symmetry

When graphing a polar equation, it's often helpful to take advantage of symmetry. We list three tests for symmetry; Figure 9 shows why these tests work.

Tests for Symmetry

1. If a polar equation is unchanged when we replace θ by $-\theta$, then the graph is symmetric about the polar axis (Figure 9(a)).

2. If the equation is unchanged when we replace r by $-r$, then the graph is symmetric about the pole (Figure 9(b)).

3. If the equation is unchanged when we replace θ by $\pi - \theta$, the graph is symmetric about the vertical line $\theta = \pi/2$ (the y-axis) (Figure 9(c)).

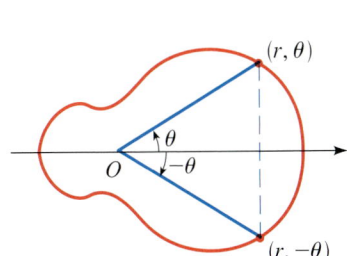

(a) Symmetry about the polar axis

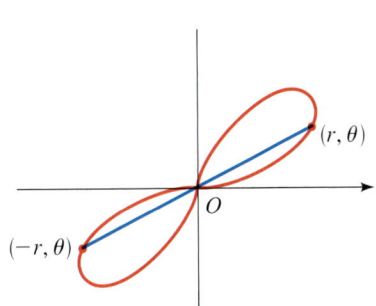

(b) Symmetry about the pole

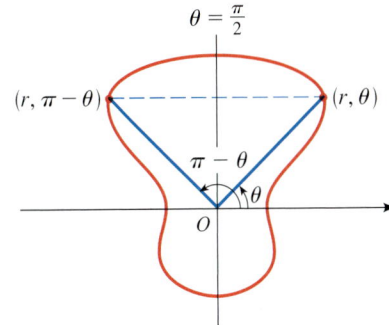

(c) Symmetry about the line $\theta = \frac{\pi}{2}$

Figure 9

The graphs in Figures 2, 6(e), and 8 are symmetric about the polar axis. The graph in Figure 8 is also symmetric about the pole. Figures 4 and 8 show graphs that are symmetric about $\theta = \pi/2$. Note that the four-leaved rose in Figure 8 meets all three tests for symmetry.

In rectangular coordinates, the zeros of the function $y = f(x)$ correspond to the x-intercepts of the graph. In polar coordinates, the zeros of the function $r = f(\theta)$ are the angles θ at which the curve crosses the pole. The zeros help us sketch the graph, as illustrated in the next example.

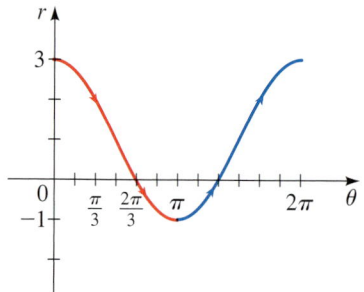

Figure 10

Example 6 Using Symmetry to Sketch a Polar Graph

Sketch the graph of the equation $r = 1 + 2\cos\theta$.

Solution We use the following as aids in sketching the graph.

■ **Symmetry** Since the equation is unchanged when θ is replaced by $-\theta$, the graph is symmetric about the polar axis.

■ **Zeros** To find the zeros, we solve

$$0 = 1 + 2\cos\theta$$

$$\cos\theta = -\frac{1}{2}$$

$$\theta = \frac{2\pi}{3}, \frac{4\pi}{3}$$

■ **Table of values** As in Example 4, we sketch the graph of $r = 1 + 2\cos\theta$ in *rectangular* coordinates to serve as a table of values (Figure 10).

Now we sketch the polar graph of $r = 1 + 2\cos\theta$ from $\theta = 0$ to $\theta = \pi$, and then use symmetry to complete the graph in Figure 11. ■

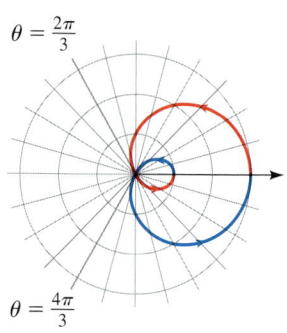

Figure 11
$r = 1 + 2\cos\theta$

The curve in Figure 11 is called a **limaçon**, after the Middle French word for snail. In general, the graph of an equation of the form

$$r = a \pm b\cos\theta \qquad \text{or} \qquad r = a \pm b\sin\theta$$

is a limaçon. The shape of the limaçon depends on the relative size of a and b (see the table on page 594).

Figure 12
$r = \sin\theta + \sin^3(5\theta/2)$

Graphing Polar Equations with Graphing Devices

Although it's useful to be able to sketch simple polar graphs by hand, we need a graphing calculator or computer when the graph is as complicated as the one in Figure 12. Fortunately, most graphing calculators are capable of graphing polar equations directly.

Example 7   Drawing the Graph of a Polar Equation

Graph the equation $r = \cos(2\theta/3)$.

Solution   We need to determine the domain for θ. So we ask ourselves: How many complete rotations are required before the graph starts to repeat itself? The graph repeats itself when the same value of r is obtained at θ and $\theta + 2n\pi$. Thus, we need to find an integer n, so that

$$\cos \frac{2(\theta + 2n\pi)}{3} = \cos \frac{2\theta}{3}$$

For this equality to hold, $4n\pi/3$ must be a multiple of 2π, and this first happens when $n = 3$. Therefore, we obtain the entire graph if we choose values of θ between $\theta = 0$ and $\theta = 0 + 2(3)\pi = 6\pi$. The graph is shown in Figure 13.

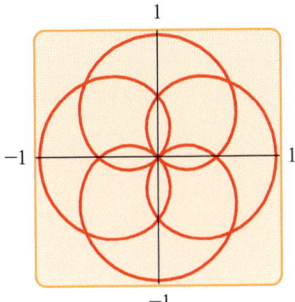

Figure 13
$r = \cos(2\theta/3)$

Example 8   A Family of Polar Equations

Graph the family of polar equations $r = 1 + c \sin \theta$ for $c = 3, 2.5, 2, 1.5, 1$. How does the shape of the graph change as c changes?

Solution   Figure 14 shows computer-drawn graphs for the given values of c. For $c > 1$, the graph has an inner loop; the loop decreases in size as c decreases. When $c = 1$, the loop disappears and the graph becomes a cardioid (see Example 4).

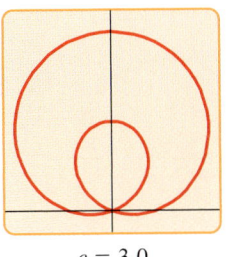
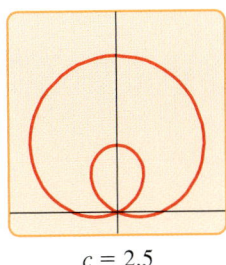
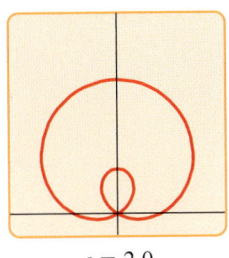
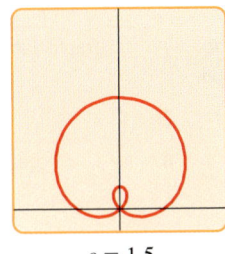
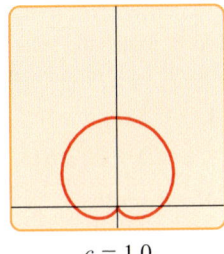

$c = 3.0$ $c = 2.5$ $c = 2.0$ $c = 1.5$ $c = 1.0$

Figure 14   A family of limaçons $r = 1 + c \sin \theta$ in the viewing rectangle $[-2.5, 2.5]$ by $[-0.5, 4.5]$

The following box gives a summary of some of the basic polar graphs used in calculus.

Some Common Polar Curves

Circles and Spiral

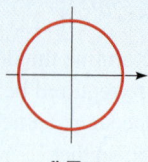 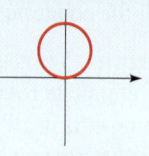 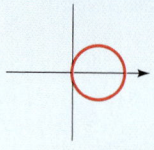 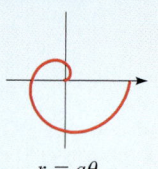

$r = a$	$r = a \sin \theta$	$r = a \cos \theta$	$r = a\theta$
circle	circle	circle	spiral

Limaçons

$r = a \pm b \sin \theta$

$r = a \pm b \cos \theta$

$(a > 0, b > 0)$

Orientation depends on
the trigonometric function
(sine or cosine) and the sign of b.

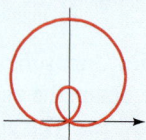 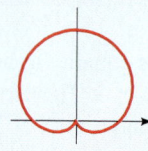 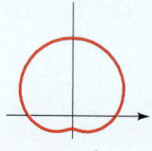 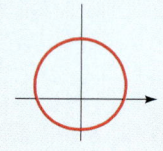

$a < b$	$a = b$	$a > b$	$a \geq 2b$
limaçon with inner loop	cardioid	dimpled limaçon	convex limaçon

Roses

$r = a \sin n\theta$

$r = a \cos n\theta$

n-leaved if n is odd

$2n$-leaved if n is even

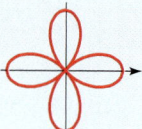 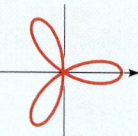  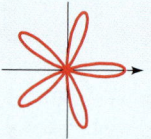

$r = a \cos 2\theta$	$r = a \cos 3\theta$	$r = a \cos 4\theta$	$r = a \cos 5\theta$
4-leaved rose	3-leaved rose	8-leaved rose	5-leaved rose

Lemniscates
Figure-eight-shaped
curves

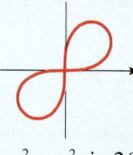

$r^2 = a^2 \sin 2\theta$	$r^2 = a^2 \cos 2\theta$
lemniscate	lemniscate

8.2   Exercises

1–6 ■ Match the polar equation with the graphs labeled I–VI.
Use the table above to help you.

1. $r = 3 \cos \theta$

2. $r = 3$

3. $r = 2 + 2 \sin \theta$

4. $r = 1 + 2 \cos \theta$

5. $r = \sin 3\theta$

6. $r = \sin 4\theta$

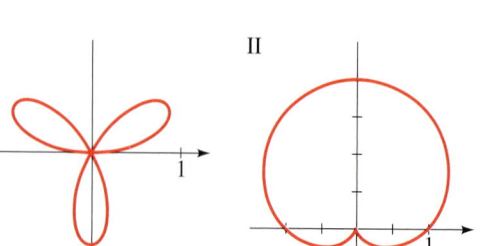

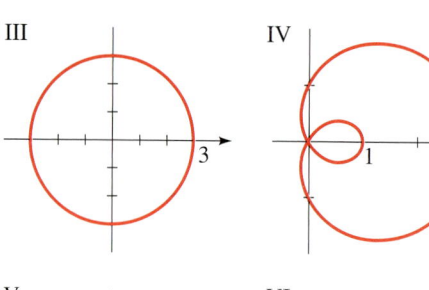

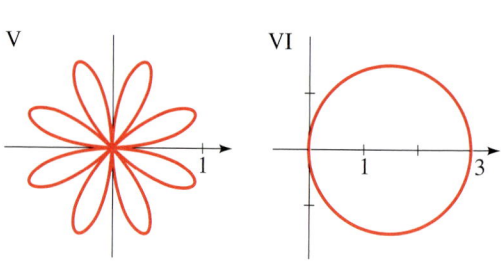

7–14 ■ Test the polar equation for symmetry with respect to the polar axis, the pole, and the line $\theta = \pi/2$.

7. $r = 2 - \sin\theta$

8. $r = 4 + 8\cos\theta$

9. $r = 3\sec\theta$

10. $r = 5\cos\theta\ \csc\theta$

11. $r = \dfrac{4}{3 - 2\sin\theta}$

12. $r = \dfrac{5}{1 + 3\cos\theta}$

13. $r^2 = 4\cos 2\theta$

14. $r^2 = 9\sin\theta$

15–36 ■ Sketch the graph of the polar equation.

15. $r = 2$

16. $r = -1$

17. $\theta = -\pi/2$

18. $\theta = 5\pi/6$

19. $r = 6\sin\theta$

20. $r = \cos\theta$

21. $r = -2\cos\theta$

22. $r = 2\sin\theta + 2\cos\theta$

23. $r = 2 - 2\cos\theta$

24. $r = 1 + \sin\theta$

25. $r = -3(1 + \sin\theta)$

26. $r = \cos\theta - 1$

27. $r = \theta, \quad \theta \geq 0$ (spiral)

28. $r\theta = 1, \quad \theta > 0$ (reciprocal spiral)

29. $r = \sin 2\theta$ (four-leaved rose)

30. $r = 2\cos 3\theta$ (three-leaved rose)

31. $r^2 = \cos 2\theta$ (lemniscate)

32. $r^2 = 4\sin 2\theta$ (lemniscate)

33. $r = 2 + \sin\theta$ (limaçon)

34. $r = 1 - 2\cos\theta$ (limaçon)

35. $r = 2 + \sec\theta$ (conchoid)

36. $r = \sin\theta\ \tan\theta$ (cissoid)

 37–40 ■ Use a graphing device to graph the polar equation. Choose the domain of θ to make sure you produce the entire graph.

37. $r = \cos(\theta/2)$

38. $r = \sin(8\theta/5)$

39. $r = 1 + 2\sin(\theta/2)$ (nephroid)

40. $r = \sqrt{1 - 0.8\sin^2\theta}$ (hippopede)

 41. Graph the family of polar equations $r = 1 + \sin n\theta$ for $n = 1, 2, 3, 4,$ and 5. How is the number of loops related to n?

 42. Graph the family of polar equations $r = 1 + c\sin 2\theta$ for $c = 0.3, 0.6, 1, 1.5,$ and 2. How does the graph change as c increases?

43–46 ■ Match the polar equation with the graphs labeled I–IV. Give reasons for your answers.

43. $r = \sin(\theta/2)$

44. $r = 1/\sqrt{\theta}$

45. $r = \theta\sin\theta$

46. $r = 1 + 3\cos(3\theta)$

I
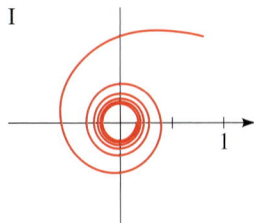

II
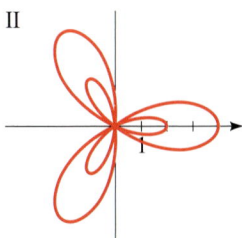

III
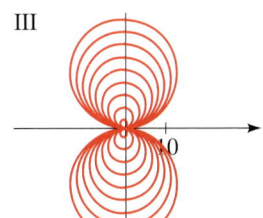

IV
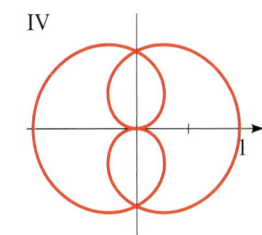

47–50 ■ Sketch a graph of the rectangular equation. [*Hint:* First convert the equation to polar coordinates.]

47. $(x^2 + y^2)^3 = 4x^2y^2$

48. $(x^2 + y^2)^3 = (x^2 - y^2)^2$

49. $(x^2 + y^2)^2 = x^2 - y^2$

50. $x^2 + y^2 = (x^2 + y^2 - x)^2$

51. Show that the graph of $r = a\cos\theta + b\sin\theta$ is a circle, and find its center and radius.

 52. (a) Graph the polar equation $r = \tan\theta\ \sec\theta$ in the viewing rectangle $[-3, 3]$ by $[-1, 9]$.

  (b) Note that your graph in part (a) looks like a parabola (see Section 2.5). Confirm this by converting the equation to rectangular coordinates.

Applications

 53. Orbit of a Satellite   Scientists and engineers often use polar equations to model the motion of satellites in earth orbit. Let's consider a satellite whose orbit is modeled by the equation $r = 22500/(4 - \cos\theta)$, where r is the distance in miles between the satellite and the center of the earth and θ is the angle shown in the figure on the next page.

  (a) On the same viewing screen, graph the circle $r = 3960$ (to represent the earth, which we will assume to be a sphere of radius 3960 mi) and the polar equation of the satellite's orbit. Describe the motion of the satellite as θ increases from 0 to 2π.

(b) For what angle θ is the satellite closest to the earth? Find the height of the satellite above the earth's surface for this value of θ.

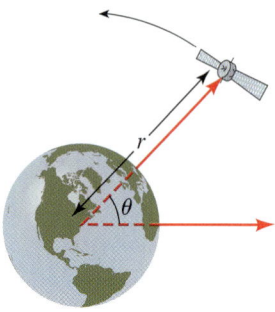

 54. An Unstable Orbit   The orbit described in Exercise 53 is stable because the satellite traverses the same path over and over as θ increases. Suppose that a meteor strikes the satellite and changes its orbit to

$$r = \frac{22500\left(1 - \dfrac{\theta}{40}\right)}{4 - \cos\theta}$$

(a) On the same viewing screen, graph the circle $r = 3960$ and the new orbit equation, with θ increasing from 0 to 3π. Describe the new motion of the satellite.

(b) Use the $\boxed{\text{TRACE}}$ feature on your graphing calculator to find the value of θ at the moment the satellite crashes into the earth.

Discovery • Discussion

55. A Transformation of Polar Graphs   How are the graphs of $r = 1 + \sin(\theta - \pi/6)$ and $r = 1 + \sin(\theta - \pi/3)$ related to the graph of $r = 1 + \sin\theta$? In general, how is the graph of $r = f(\theta - \alpha)$ related to the graph of $r = f(\theta)$?

56. Choosing a Convenient Coordinate System   Compare the polar equation of the circle $r = 2$ with its equation in rectangular coordinates. In which coordinate system is the equation simpler? Do the same for the equation of the four-leaved rose $r = \sin 2\theta$. Which coordinate system would you choose to study these curves?

57. Choosing a Convenient Coordinate System   Compare the rectangular equation of the line $y = 2$ with its polar equation. In which coordinate system is the equation simpler? Which coordinate system would you choose to study lines?

8.3	**Polar Form of Complex Numbers; DeMoivre's Theorem**

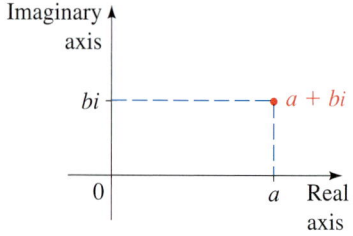

Figure 1

In this section we represent complex numbers in polar (or trigonometric) form. This enables us to find the nth roots of complex numbers. To describe the polar form of complex numbers, we must first learn to work with complex numbers graphically.

Graphing Complex Numbers

To graph real numbers or sets of real numbers, we have been using the number line, which has just one dimension. Complex numbers, however, have two components: a real part and an imaginary part. This suggests that we need two axes to graph complex numbers: one for the real part and one for the imaginary part. We call these the **real axis** and the **imaginary axis**, respectively. The plane determined by these two axes is called the **complex plane**. To graph the complex number $a + bi$, we plot the ordered pair of numbers (a, b) in this plane, as indicated in Figure 1.

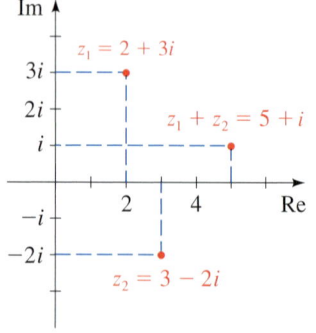

Figure 2

Example 1   Graphing Complex Numbers

Graph the complex numbers $z_1 = 2 + 3i$, $z_2 = 3 - 2i$, and $z_1 + z_2$.

Solution   We have $z_1 + z_2 = (2 + 3i) + (3 - 2i) = 5 + i$. The graph is shown in Figure 2.   ∎

Example 2   **Graphing Sets of Complex Numbers**

Graph each set of complex numbers.

(a) $S = \{a + bi \mid a \geq 0\}$ (b) $T = \{a + bi \mid a < 1, b \geq 0\}$

Solution

(a) S is the set of complex numbers whose real part is nonnegative. The graph is shown in Figure 3(a).

(b) T is the set of complex numbers for which the real part is less than 1 and the imaginary part is nonnegative. The graph is shown in Figure 3(b).

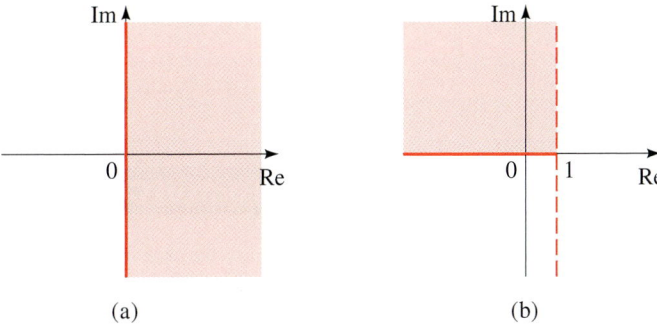

Figure 3   (a)   (b)   ■

Recall that the absolute value of a real number can be thought of as its distance from the origin on the real number line (see Section 1.1). We define absolute value for complex numbers in a similar fashion. Using the Pythagorean Theorem, we can see from Figure 4 that the distance between $a + bi$ and the origin in the complex plane is $\sqrt{a^2 + b^2}$. This leads to the following definition.

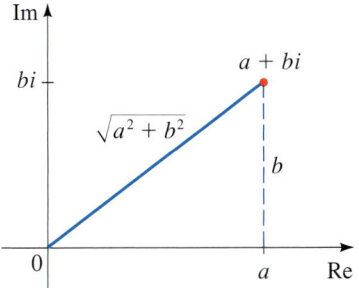

Figure 4

> The **modulus** (or **absolute value**) of the complex number $z = a + bi$ is
> $$|z| = \sqrt{a^2 + b^2}$$

The plural of *modulus* is *moduli*.

Example 3   **Calculating the Modulus**

Find the moduli of the complex numbers $3 + 4i$ and $8 - 5i$.

Solution

$$|3 + 4i| = \sqrt{3^2 + 4^2} = \sqrt{25} = 5$$
$$|8 - 5i| = \sqrt{8^2 + (-5)^2} = \sqrt{89}$$ ■

Example 4   **Absolute Value of Complex Numbers**

Graph each set of complex numbers.

(a) $C = \{z \mid |z| = 1\}$ (b) $D = \{z \mid |z| \leq 1\}$

Solution

(a) C is the set of complex numbers whose distance from the origin is 1. Thus, C is a circle of radius 1 with center at the origin, as shown in Figure 5.

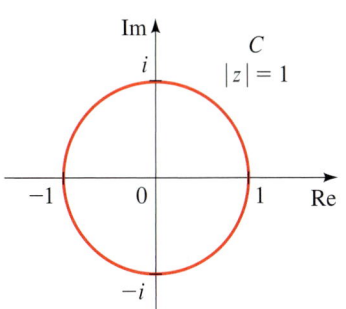

Figure 5

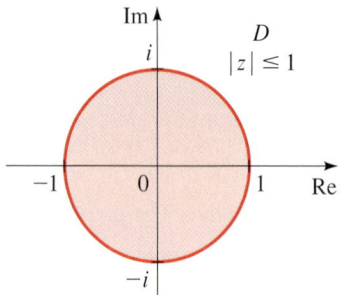

Figure 6

(b) D is the set of complex numbers whose distance from the origin is less than or equal to 1. Thus, D is the disk that consists of all complex numbers on and inside the circle C of part (a), as shown in Figure 6.   ∎

Polar Form of Complex Numbers

Let $z = a + bi$ be a complex number, and in the complex plane let's draw the line segment joining the origin to the point $a + bi$ (see Figure 7). The length of this line segment is $r = |z| = \sqrt{a^2 + b^2}$. If θ is an angle in standard position whose terminal side coincides with this line segment, then by the definitions of sine and cosine (see Section 6.2)

$$a = r \cos \theta \quad \text{and} \quad b = r \sin \theta$$

so $z = r \cos \theta + ir \sin \theta = r(\cos \theta + i \sin \theta)$. We have shown the following.

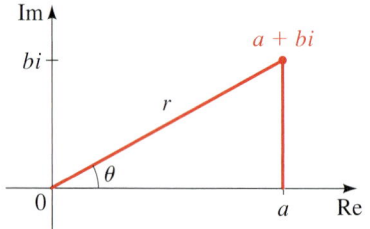

Figure 7

Polar Form of Complex Numbers

A complex number $z = a + bi$ has the **polar form** (or **trigonometric form**)

$$z = r(\cos \theta + i \sin \theta)$$

where $r = |z| = \sqrt{a^2 + b^2}$ and $\tan \theta = b/a$. The number r is the **modulus** of z, and θ is an **argument** of z.

The argument of z is not unique, but any two arguments of z differ by a multiple of 2π.

Example 5   Writing Complex Numbers in Polar Form

Write each complex number in trigonometric form.

(a) $1 + i$ (b) $-1 + \sqrt{3}i$ (c) $-4\sqrt{3} - 4i$ (d) $3 + 4i$

Solution   These complex numbers are graphed in Figure 8, which helps us find their arguments.

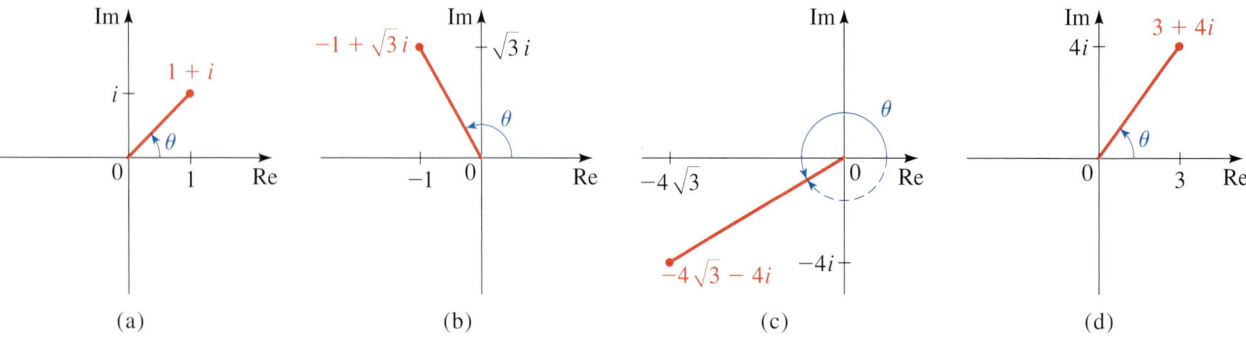

(a)   (b)   (c)   (d)

Figure 8

$\tan \theta = \frac{1}{1} = 1$

$\theta = \frac{\pi}{4}$

(a) An argument is $\theta = \pi/4$ and $r = \sqrt{1 + 1} = \sqrt{2}$. Thus

$$1 + i = \sqrt{2}\left(\cos \frac{\pi}{4} + i \sin \frac{\pi}{4}\right)$$

$\tan \theta = \frac{\sqrt{3}}{-1} = -\sqrt{3}$

$\theta = \frac{2\pi}{3}$

(b) An argument is $\theta = 2\pi/3$ and $r = \sqrt{1 + 3} = 2$. Thus

$$-1 + \sqrt{3}i = 2\left(\cos \frac{2\pi}{3} + i \sin \frac{2\pi}{3}\right)$$

$\tan \theta = \frac{-4}{-4\sqrt{3}} = \frac{1}{\sqrt{3}}$

$\theta = \frac{7\pi}{6}$

(c) An argument is $\theta = 7\pi/6$ (or we could use $\theta = -5\pi/6$), and $r = \sqrt{48 + 16} = 8$. Thus

$$-4\sqrt{3} - 4i = 8\left(\cos \frac{7\pi}{6} + i \sin \frac{7\pi}{6}\right)$$

$\tan \theta = \frac{4}{3}$

$\theta = \tan^{-1}\frac{4}{3}$

(d) An argument is $\theta = \tan^{-1}\frac{4}{3}$ and $r = \sqrt{3^2 + 4^2} = 5$. So

$$3 + 4i = 5\left[\cos\left(\tan^{-1}\frac{4}{3}\right) + i \sin\left(\tan^{-1}\frac{4}{3}\right)\right]$$ ■

The addition formulas for sine and cosine that we discussed in Section 7.2 greatly simplify the multiplication and division of complex numbers in polar form. The following theorem shows how.

Multiplication and Division of Complex Numbers

If the two complex numbers z_1 and z_2 have the polar forms

$$z_1 = r_1(\cos \theta_1 + i \sin \theta_1) \qquad \text{and} \qquad z_2 = r_2(\cos \theta_2 + i \sin \theta_2)$$

then

$$z_1 z_2 = r_1 r_2[\cos(\theta_1 + \theta_2) + i \sin(\theta_1 + \theta_2)] \qquad \text{Multiplication}$$

$$\frac{z_1}{z_2} = \frac{r_1}{r_2}[\cos(\theta_1 - \theta_2) + i \sin(\theta_1 - \theta_2)] \qquad (z_2 \neq 0) \qquad \text{Division}$$

This theorem says:

   To multiply two complex numbers, multiply the moduli and add the arguments.

   To divide two complex numbers, divide the moduli and subtract the arguments.

■ **Proof**   To prove the multiplication formula, we simply multiply the two complex numbers.

$$z_1 z_2 = r_1 r_2(\cos \theta_1 + i \sin \theta_1)(\cos \theta_2 + i \sin \theta_2)$$

$$= r_1 r_2[\cos \theta_1 \cos \theta_2 - \sin \theta_1 \sin \theta_2 + i(\sin \theta_1 \cos \theta_2 + \cos \theta_1 \sin \theta_2)]$$

$$= r_1 r_2[\cos(\theta_1 + \theta_2) + i \sin(\theta_1 + \theta_2)]$$

In the last step we used the addition formulas for sine and cosine.   ■

The proof of the division formula is left as an exercise.

Mathematics in the Modern World

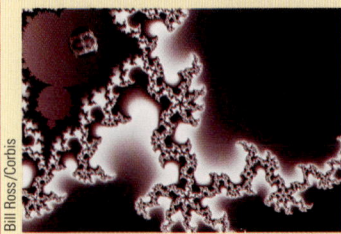

Bill Ross/Corbis

Fractals

Many of the things we model in this book have regular predictable shapes. But recent advances in mathematics have made it possible to model such seemingly random or even chaotic shapes as those of a cloud, a flickering flame, a mountain, or a jagged coastline. The basic tools in this type of modeling are the fractals invented by the mathematician Benoit Mandelbrot. A *fractal* is a geometric shape built up from a simple basic shape by scaling and repeating the shape indefinitely according to a given rule. Fractals have infinite detail; this means the closer you look, the more you see. They are also *self-similar*; that is, zooming in on a portion of the fractal yields the same detail as the original shape. Because of their beautiful shapes, fractals are used by movie makers to create fictional landscapes and exotic backgrounds.

Although a fractal is a complex shape, it is produced according to very simple rules (see page 605). This property of fractals is exploited in a process of storing pictures on a computer called *fractal image compression*. In this process a picture is stored as a simple basic shape and a rule; repeating the shape according to the rule produces the original picture. This is an extremely efficient method of storage; that's how thousands of color pictures can be put on a single compact disc.

Example 6 Multiplying and Dividing Complex Numbers

Let

$$z_1 = 2\left(\cos\frac{\pi}{4} + i\sin\frac{\pi}{4}\right) \qquad \text{and} \qquad z_2 = 5\left(\cos\frac{\pi}{3} + i\sin\frac{\pi}{3}\right)$$

Find (a) $z_1 z_2$ and (b) z_1/z_2.

Solution

(a) By the multiplication formula

$$z_1 z_2 = (2)(5)\left[\cos\left(\frac{\pi}{4} + \frac{\pi}{3}\right) + i\sin\left(\frac{\pi}{4} + \frac{\pi}{3}\right)\right]$$

$$= 10\left(\cos\frac{7\pi}{12} + i\sin\frac{7\pi}{12}\right)$$

To approximate the answer, we use a calculator in radian mode and get

$$z_1 z_2 \approx 10(-0.2588 + 0.9659i) = -2.588 + 9.659i$$

(b) By the division formula

$$\frac{z_1}{z_2} = \frac{2}{5}\left[\cos\left(\frac{\pi}{4} - \frac{\pi}{3}\right) + i\sin\left(\frac{\pi}{4} - \frac{\pi}{3}\right)\right]$$

$$= \frac{2}{5}\left[\cos\left(-\frac{\pi}{12}\right) + i\sin\left(-\frac{\pi}{12}\right)\right]$$

$$= \frac{2}{5}\left(\cos\frac{\pi}{12} - i\sin\frac{\pi}{12}\right)$$

Using a calculator in radian mode, we get the approximate answer:

$$\frac{z_1}{z_2} \approx \tfrac{2}{5}(0.9659 - 0.2588i) = 0.3864 - 0.1035i$$ ∎

DeMoivre's Theorem

Repeated use of the multiplication formula gives the following useful formula for raising a complex number to a power n for any positive integer n.

DeMoivre's Theorem

If $z = r(\cos\theta + i\sin\theta)$, then for any integer n

$$z^n = r^n(\cos n\theta + i\sin n\theta)$$

This theorem says: *To take the nth power of a complex number, we take the nth power of the modulus and multiply the argument by n.*

■ **Proof** By the multiplication formula

$$z^2 = zz = r^2[\cos(\theta + \theta) + i\sin(\theta + \theta)]$$

$$= r^2(\cos 2\theta + i\sin 2\theta)$$

Now we multiply z^2 by z to get

$$z^3 = z^2 z = r^3[\cos(2\theta + \theta) + i\sin(2\theta + \theta)]$$

$$= r^3(\cos 3\theta + i\sin 3\theta)$$

Repeating this argument, we see that for any positive integer n

$$z^n = r^n(\cos n\theta + i\sin n\theta)$$

A similar argument using the division formula shows that this also holds for negative integers. ∎

Example 7   Finding a Power Using DeMoivre's Theorem

Find $\left(\frac{1}{2} + \frac{1}{2}i\right)^{10}$.

Solution   Since $\frac{1}{2} + \frac{1}{2}i = \frac{1}{2}(1 + i)$, it follows from Example 5(a) that

$$\frac{1}{2} + \frac{1}{2}i = \frac{\sqrt{2}}{2}\left(\cos\frac{\pi}{4} + i\sin\frac{\pi}{4}\right)$$

So by DeMoivre's Theorem,

$$\left(\frac{1}{2} + \frac{1}{2}i\right)^{10} = \left(\frac{\sqrt{2}}{2}\right)^{10}\left(\cos\frac{10\pi}{4} + i\sin\frac{10\pi}{4}\right)$$

$$= \frac{2^5}{2^{10}}\left(\cos\frac{5\pi}{2} + i\sin\frac{5\pi}{2}\right) = \frac{1}{32}i$$ ∎

nth Roots of Complex Numbers

An **nth root** of a complex number z is any complex number w such that $w^n = z$. DeMoivre's Theorem gives us a method for calculating the nth roots of any complex number.

nth Roots of Complex Numbers

If $z = r(\cos\theta + i\sin\theta)$ and n is a positive integer, then z has the n distinct nth roots

$$w_k = r^{1/n}\left[\cos\left(\frac{\theta + 2k\pi}{n}\right) + i\sin\left(\frac{\theta + 2k\pi}{n}\right)\right]$$

for $k = 0, 1, 2, \ldots, n - 1$.

■ **Proof**   To find the nth roots of z, we need to find a complex number w such that

$$w^n = z$$

Let's write z in polar form:

$$z = r(\cos\theta + i\sin\theta)$$

One nth root of z is

$$w = r^{1/n}\left(\cos\frac{\theta}{n} + i\sin\frac{\theta}{n}\right)$$

since by DeMoivre's Theorem, $w^n = z$. But the argument θ of z can be replaced by $\theta + 2k\pi$ for any integer k. Since this expression gives a different value of w for $k = 0$, $1, 2, \ldots, n - 1$, we have proved the formula in the theorem. ∎

The following observations help us use the preceding formula.

1. The modulus of each nth root is $r^{1/n}$.

2. The argument of the first root is θ/n.

3. We repeatedly add $2\pi/n$ to get the argument of each successive root.

These observations show that, when graphed, the nth roots of z are spaced equally on the circle of radius $r^{1/n}$.

Example 8   Finding Roots of a Complex Number

Find the six sixth roots of $z = -64$, and graph these roots in the complex plane.

Solution   In polar form, $z = 64(\cos \pi + i \sin \pi)$. Applying the formula for nth roots with $n = 6$, we get

$$w_k = 64^{1/6}\left[\cos\left(\frac{\pi + 2k\pi}{6}\right) + i \sin\left(\frac{\pi + 2k\pi}{6}\right)\right]$$

for $k = 0, 1, 2, 3, 4, 5$. Using $64^{1/6} = 2$, we find that the six sixth roots of -64 are

We add $2\pi/6 = \pi/3$ to each argument to get the argument of the next root.

$$w_0 = 64^{1/6}\left(\cos\frac{\pi}{6} + i \sin\frac{\pi}{6}\right) = \sqrt{3} + i$$

$$w_1 = 64^{1/6}\left(\cos\frac{\pi}{2} + i \sin\frac{\pi}{2}\right) = 2i$$

$$w_2 = 64^{1/6}\left(\cos\frac{5\pi}{6} + i \sin\frac{5\pi}{6}\right) = -\sqrt{3} + i$$

$$w_3 = 64^{1/6}\left(\cos\frac{7\pi}{6} + i \sin\frac{7\pi}{6}\right) = -\sqrt{3} - i$$

$$w_4 = 64^{1/6}\left(\cos\frac{3\pi}{2} + i \sin\frac{3\pi}{2}\right) = -2i$$

$$w_5 = 64^{1/6}\left(\cos\frac{11\pi}{6} + i \sin\frac{11\pi}{6}\right) = \sqrt{3} - i$$

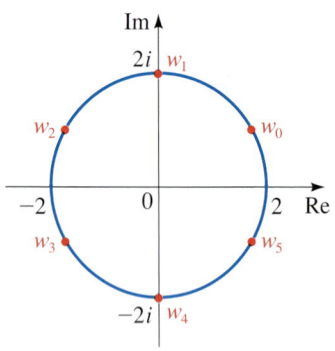

Figure 9
The six sixth roots of $z = -64$

All these points lie on a circle of radius 2, as shown in Figure 9. ∎

When finding roots of complex numbers, we sometimes write the argument θ of the complex number in degrees. In this case, the nth roots are obtained from the formula

$$w_k = r^{1/n}\left[\cos\left(\frac{\theta + 360°k}{n}\right) + i \sin\left(\frac{\theta + 360°k}{n}\right)\right]$$

for $k = 0, 1, 2, \ldots, n - 1$.

Example 9 Finding Cube Roots of a Complex Number

Find the three cube roots of $z = 2 + 2i$, and graph these roots in the complex plane.

Solution First we write z in polar form using degrees. We have
$r = \sqrt{2^2 + 2^2} = 2\sqrt{2}$ and $\theta = 45°$. Thus

$$z = 2\sqrt{2}(\cos 45° + i \sin 45°)$$

$(2\sqrt{2})^{1/3} = (2^{3/2})^{1/3} = 2^{1/2} = \sqrt{2}$

We add $360°/3 = 120°$ to each argument to get the argument of the next root.

Applying the formula for nth roots (in degrees) with $n = 3$, we find the cube roots of z are of the form

$$w_k = (2\sqrt{2})^{1/3}\left[\cos\left(\frac{45° + 360°k}{3}\right) + i \sin\left(\frac{45° + 360°k}{3}\right)\right]$$

where $k = 0, 1, 2$. Thus, the three cube roots are

$$w_0 = \sqrt{2}(\cos 15° + i \sin 15°) \approx 1.366 + 0.366i$$
$$w_1 = \sqrt{2}(\cos 135° + i \sin 135°) = -1 + i$$
$$w_2 = \sqrt{2}(\cos 255° + i \sin 255°) \approx -0.366 - 1.366i$$

The three cube roots of z are graphed in Figure 10. These roots are spaced equally on a circle of radius $\sqrt{2}$. ∎

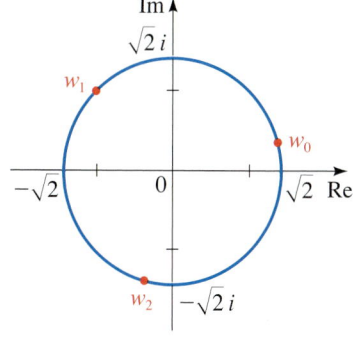

Figure 10
The three cube roots of $z = 2 + 2i$

Example 10 Solving an Equation Using the nth Roots Formula

Solve the equation $z^6 + 64 = 0$.

Solution This equation can be written as $z^6 = -64$. Thus, the solutions are the sixth roots of -64, which we found in Example 8. ∎

8.3　**Exercises**

1–8 ■ Graph the complex number and find its modulus.

1. $4i$　　　　　　**2.** $-3i$

3. -2　　　　　　**4.** 6

5. $5 + 2i$　　　　**6.** $7 - 3i$

7. $\sqrt{3} + i$　　　**8.** $-1 - \dfrac{\sqrt{3}}{3}i$

9. $\dfrac{3 + 4i}{5}$　　　**10.** $\dfrac{-\sqrt{2} + i\sqrt{2}}{2}$

11–12 ■ Sketch the complex number z, and also sketch $2z$, $-z$, and $\frac{1}{2}z$ on the same complex plane.

11. $z = 1 + i$　　　**12.** $z = -1 + i\sqrt{3}$

13–14 ■ Sketch the complex number z and its complex conjugate $\bar{z}$ on the same complex plane.

13. $z = 8 + 2i$　　**14.** $z = -5 + 6i$

15–16 ■ Sketch z_1, z_2, $z_1 + z_2$, and z_1z_2 on the same complex plane.

15. $z_1 = 2 - i$,　$z_2 = 2 + i$

16. $z_1 = -1 + i$,　$z_2 = 2 - 3i$

17–24 ■ Sketch the set in the complex plane.

17. $\{z = a + bi \mid a \le 0, b \ge 0\}$

18. $\{z = a + bi \mid a > 1, b > 1\}$

19. $\{z \mid |z| = 3\}$　　　　**20.** $\{z \mid |z| \ge 1\}$

21. $\{z \mid |z| < 2\}$　　　　**22.** $\{z \mid 2 \le |z| \le 5\}$

23. $\{z = a + bi \mid a + b < 2\}$

24. $\{z = a + bi \mid a \ge b\}$

25–48 ■ Write the complex number in polar form with argument θ between 0 and 2π.

25. $1 + i$

26. $1 + \sqrt{3}\,i$

27. $\sqrt{2} - \sqrt{2}\,i$

28. $1 - i$

29. $2\sqrt{3} - 2i$

30. $-1 + i$

31. $-3i$

32. $-3 - 3\sqrt{3}\,i$

33. $5 + 5i$

34. 4

35. $4\sqrt{3} - 4i$

36. $8i$

37. -20

38. $\sqrt{3} + i$

39. $3 + 4i$

40. $i(2 - 2i)$

41. $3i(1 + i)$

42. $2(1 - i)$

43. $4(\sqrt{3} + i)$

44. $-3 - 3i$

45. $2 + i$

46. $3 + \sqrt{3}\,i$

47. $\sqrt{2} + \sqrt{2}\,i$

48. $-\pi i$

49–56 ■ Find the product $z_1 z_2$ and the quotient z_1/z_2. Express your answer in polar form.

49. $z_1 = \cos \pi + i \sin \pi$, $\quad z_2 = \cos \dfrac{\pi}{3} + i \sin \dfrac{\pi}{3}$

50. $z_1 = \cos \dfrac{\pi}{4} + i \sin \dfrac{\pi}{4}$, $\quad z_2 = \cos \dfrac{3\pi}{4} + i \sin \dfrac{3\pi}{4}$

51. $z_1 = 3\left(\cos \dfrac{\pi}{6} + i \sin \dfrac{\pi}{6}\right)$, $\quad z_2 = 5\left(\cos \dfrac{4\pi}{3} + i \sin \dfrac{4\pi}{3}\right)$

52. $z_1 = 7\left(\cos \dfrac{9\pi}{8} + i \sin \dfrac{9\pi}{8}\right)$, $\quad z_2 = 2\left(\cos \dfrac{\pi}{8} + i \sin \dfrac{\pi}{8}\right)$

53. $z_1 = 4(\cos 120° + i \sin 120°)$,

$\quad z_2 = 2(\cos 30° + i \sin 30°)$

54. $z_1 = \sqrt{2}(\cos 75° + i \sin 75°)$,

$\quad z_2 = 3\sqrt{2}(\cos 60° + i \sin 60°)$

55. $z_1 = 4(\cos 200° + i \sin 200°)$,

$\quad z_2 = 25(\cos 150° + i \sin 150°)$

56. $z_1 = \frac{4}{5}(\cos 25° + i \sin 25°)$,

$\quad z_2 = \frac{1}{5}(\cos 155° + i \sin 155°)$

57–64 ■ Write z_1 and z_2 in polar form, and then find the product $z_1 z_2$ and the quotients z_1/z_2 and $1/z_1$.

57. $z_1 = \sqrt{3} + i$, $\quad z_2 = 1 + \sqrt{3}\,i$

58. $z_1 = \sqrt{2} - \sqrt{2}\,i$, $\quad z_2 = 1 - i$

59. $z_1 = 2\sqrt{3} - 2i$, $\quad z_2 = -1 + i$

60. $z_1 = -\sqrt{2}\,i$, $\quad z_2 = -3 - 3\sqrt{3}\,i$

61. $z_1 = 5 + 5i$, $\quad z_2 = 4$ $\qquad$ **62.** $z_1 = 4\sqrt{3} - 4i$, $\quad z_2 = 8i$

63. $z_1 = -20$, $\quad z_2 = \sqrt{3} + i$ $\quad$ **64.** $z_1 = 3 + 4i$, $\quad z_2 = 2 - 2i$

65–76 ■ Find the indicated power using DeMoivre's Theorem.

65. $(1 + i)^{20}$

66. $(1 - \sqrt{3}\,i)^5$

67. $(2\sqrt{3} + 2i)^5$

68. $(1 - i)^8$

69. $\left(\dfrac{\sqrt{2}}{2} + \dfrac{\sqrt{2}}{2}i\right)^{12}$

70. $(\sqrt{3} - i)^{-10}$

71. $(2 - 2i)^8$

72. $\left(-\dfrac{1}{2} - \dfrac{\sqrt{3}}{2}i\right)^{15}$

73. $(-1 - i)^7$

74. $(3 + \sqrt{3}\,i)^4$

75. $(2\sqrt{3} + 2i)^{-5}$

76. $(1 - i)^{-8}$

77–86 ■ Find the indicated roots, and graph the roots in the complex plane.

77. The square roots of $4\sqrt{3} + 4i$

78. The cube roots of $4\sqrt{3} + 4i$

79. The fourth roots of $-81i$ $\qquad$ **80.** The fifth roots of 32

81. The eighth roots of 1 $\qquad$ **82.** The cube roots of $1 + i$

83. The cube roots of i $\qquad$ **84.** The fifth roots of i

85. The fourth roots of -1

86. The fifth roots of $-16 - 16\sqrt{3}i$

87–92 ■ Solve the equation.

87. $z^4 + 1 = 0$ $\qquad\qquad$ **88.** $z^8 - i = 0$

89. $z^3 - 4\sqrt{3} - 4i = 0$ $\qquad$ **90.** $z^6 - 1 = 0$

91. $z^3 + 1 = -i$ $\qquad\qquad$ **92.** $z^3 - 1 = 0$

93. (a) Let $w = \cos \dfrac{2\pi}{n} + i \sin \dfrac{2\pi}{n}$ where n is a positive integer. Show that $1, w, w^2, w^3, \ldots, w^{n-1}$ are the n distinct nth roots of 1.

(b) If $z \neq 0$ is any complex number and $s^n = z$, show that the n distinct nth roots of z are

$$s, sw, sw^2, sw^3, \ldots, sw^{n-1}$$

Discovery • Discussion

94. Sums of Roots of Unity Find the exact values of all three cube roots of 1 (see Exercise 93) and then add them. Do the same for the fourth, fifth, sixth, and eighth roots of 1. What do you think is the sum of the nth roots of 1, for any n?

95. Products of Roots of Unity Find the product of the three cube roots of 1 (see Exercise 93). Do the same for the fourth, fifth, sixth, and eighth roots of 1. What do you think is the product of the nth roots of 1, for any n?

96. Complex Coefficients and the Quadratic Formula The quadratic formula works whether the coefficients of the equation are real or complex. Solve these equations using the quadratic formula, and, if necessary, DeMoivre's Theorem.

(a) $z^2 + (1 + i)z + i = 0$

(b) $z^2 - iz + 1 = 0$

(c) $z^2 - (2 - i)z - \frac{1}{4}i = 0$

DISCOVERY PROJECT

Fractals

Fractals are geometric objects that exhibit more and more detail the more we magnify them (see *Mathematics in the Modern World* on page 600). Many fractals can be described by iterating functions of complex numbers. The most famous such fractal is illustrated in Figures 1 and 2. It is called the *Mandelbrot set*, named after Benoit Mandelbrot, the mathematician who discovered it in the 1950s.

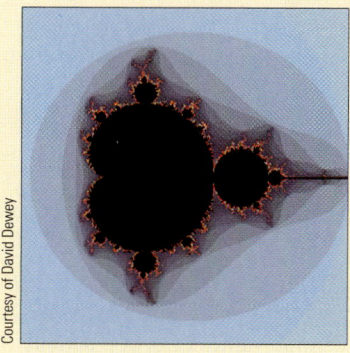

Courtesy of David Dewey

Figure 1

The Mandelbrot set

Stephen Gerard/Photo Researchers, Inc.

Figure 2

Detail from the Mandelbrot set

Here is how the Mandelbrot set is defined. Choose a complex number c, and define the complex quadratic function

$$f(z) = z^2 + c$$

Starting with $z_0 = 0$, we form the iterates of f as follows:

$$z_1 = f(0) = c$$
$$z_2 = f(f(0)) = f(c) = c^2 + c$$
$$z_3 = f(f(f(0))) = f(c^2 + c) = (c^2 + c)^2 + c$$
$$\vdots \qquad \vdots \qquad \vdots \qquad \vdots$$

See page 597 for the definition of *modulus* (plural *moduli*).

As we continue calculating the iterates, one of two things will happen, depending on the value of c. Either the iterates $z_0, z_1, z_2, z_3, \ldots$ form a bounded set (that is, the moduli of the iterates are all less than some fixed number K), or else they eventually grow larger and larger without bound. The calculations in the table on page 606 show that for $c = 0.1 + 0.2i$, the iterates eventually stabilize at about $0.05 + 0.22i$, whereas for $c = 1 - i$, the iterates quickly become so large that a calculator can't handle them.

You can use your calculator to find the iterates, just like in the Discovery Project on page 233. With the TI-83, first put the calculator into $a+bi$ mode. Then press the $\boxed{Y=}$ key and enter the function $Y_1 = X^2 + C$. Now if $c = 1 + i$, for instance, enter the following commands:

$$1 - i \to C$$
$$0 \to X$$
$$Y_1 \to X$$

Press the $\boxed{ENTER}$ key repeatedly to get the list of iterates. (With this value of c, you should end up with the values in the right-hand column of the table.)

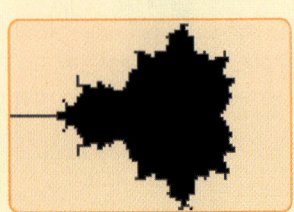

Figure 3

$f(z) = z^2 + 0.1 + 0.2i$	$f(z) = z^2 + 1 - i$
$z_1 = f(z_0) = .1 + .2i$	$z_1 = f(z_0) = 1 - i$
$z_2 = f(z_1) = .07 + .24i$	$z_2 = f(z_1) = 1 - 3i$
$z_3 = f(z_2) = .047 + .234i$	$z_3 = f(z_2) = -7 - 7i$
$z_4 = f(z_3) = .048 + .222i$	$z_4 = f(z_3) = 1 + 97i$
$z_5 = f(z_4) = .053 + .221i$	$z_5 = f(z_4) = -9407 + 193i$
$z_6 = f(z_5) = .054 + .223i$	$z_6 = f(z_5) = 88454401 - 3631103i$
$z_7 = f(z_6) = .053 + .224i$	$z_7 = f(z_6) = 7.8 \times 10^{15} - 6.4 \times 10^{14}i$

The **Mandelbrot set** consists of those complex numbers c for which the iterates of $f(z) = z^2 + c$ are bounded. (In fact, for this function it turns out that if the iterates are bounded, the moduli of all the iterates will be less than $K = 2$.) The numbers c that belong to the Mandelbrot set can be graphed in the complex plane. The result is the black part in Figure 1. The points not in the Mandelbrot set are assigned colors depending on how quickly the iterates become unbounded.

The TI-83 program below draws a rough graph of the Mandelbrot set. The program takes a long time to finish, even though it performs only 10 iterations for each c. For some values of c, you actually have to do many more iterations to tell whether the iterates are unbounded. (See, for instance, Problem 1(f) below.) That's why the program produces only a rough graph. But the calculator output in Figure 3 is actually a good approximation.

```
PROGRAM:MANDLBRT
:ClrDraw
:AxesOff
:(Xmax-Xmin)/94→H
:(Ymax-Ymin)/62→V
:For(I,0,93)
:For(J,0,61)
:Xmin+I*H→X
:Ymin+J*V→Y
:X+Yi→C
:0→Z
:For(N,1,10)
:If abs(Z)≤2
:Z²+C→Z
:End
:If abs(Z)≤2
:Pt-On(real(C),imag(C))
:DispGraph
:End
:End
:StorePic 1
```

Use the viewing rectangle $[-2, 1]$ by $[-1, 1]$ and make sure the calculator is in "$a+bi$" mode

H is the horizontal width of one pixel

V is the vertical height of one pixel

These two "For" loops find the complex number associated with each pixel on the screen

This "For" loop calculates 10 iterates, but stops iterating if Z has modulus larger than 2

If the iterates have modulus less than or equal to 2, the point C is plotted

This stores the final image under "1" so that it can be recalled later

1. Use your calculator as described in the margin on page 606 to decide whether the complex number c is in the Mandelbrot set. (For part (f), calculate at least 60 iterates.)

   (a) $c = 1$ (b) $c = -1$

   (c) $c = -0.7 + 0.15i$ (d) $c = 0.5 + 0.5i$

   (e) $c = i$ (f) $c = -1.0404 + 0.2509i$

2. Use the MANDLBRT program with a smaller viewing rectangle to zoom in on a portion of the Mandelbrot set near its edge. (Store the final image in a different location if you want to keep the complete Mandelbrot picture in "1.") Do you see more detail?

3. (a) Write a calculator program that takes as input a complex number c, iterates the function $f(z) = z^2 + c$ a hundred times, and then gives the following output:

   ■ "UNBOUNDED AT N", if z_N is the first iterate whose modulus is greater than 2

   ■ "BOUNDED" if each iterate from z_1 to z_{100} has modulus less than or equal to 2

   In the first case, the number c is not in the Mandelbrot set, and the index N tells us how "quickly" the iterates become unbounded. In the second case, it is likely that c is in the Mandelbrot set.

   (b) Use your program to test each of the numbers in Problem 1.

   (c) Choose other complex numbers and use your program to test them.

8.4 Vectors

In applications of mathematics, certain quantities are determined completely by their magnitude—for example, length, mass, area, temperature, and energy. We speak of a length of 5 m or a mass of 3 kg; only one number is needed to describe each of these quantities. Such a quantity is called a **scalar**.

On the other hand, to describe the displacement of an object, two numbers are required: the *magnitude* and the *direction* of the displacement. To describe the velocity of a moving object, we must specify both the *speed* and the *direction* of travel. Quantities such as displacement, velocity, acceleration, and force that involve magnitude as well as direction are called *directed quantities*. One way to represent such quantities mathematically is through the use of *vectors*.

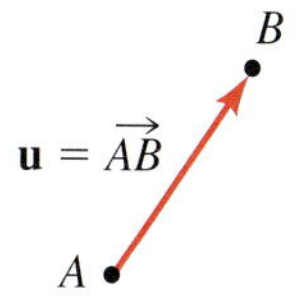

Geometric Description of Vectors

A **vector** in the plane is a line segment with an assigned direction. We sketch a vector as shown in Figure 1 with an arrow to specify the direction. We denote this vector by $\overrightarrow{AB}$. Point A is the **initial point**, and B is the **terminal point** of the vector

$\mathbf{u} = \overrightarrow{AB}$

Figure 1

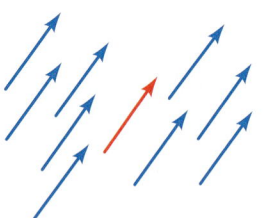

Figure 2

$\overrightarrow{AB}$. The length of the line segment AB is called the **magnitude** or **length** of the vector and is denoted by $|\overrightarrow{AB}|$. We use boldface letters to denote vectors. Thus, we write $\mathbf{u} = \overrightarrow{AB}$.

Two vectors are considered **equal** if they have equal magnitude and the same direction. Thus, all the vectors in Figure 2 are equal. This definition of equality makes sense if we think of a vector as representing a displacement. Two such displacements are the same if they have equal magnitudes and the same direction. So the vectors in Figure 2 can be thought of as the *same* displacement applied to objects in different locations in the plane.

If the displacement $\mathbf{u} = \overrightarrow{AB}$ is followed by the displacement $\mathbf{v} = \overrightarrow{BC}$, then the resulting displacement is $\overrightarrow{AC}$ as shown in Figure 3. In other words, the single displacement represented by the vector $\overrightarrow{AC}$ has the same effect as the other two displacements together. We call the vector $\overrightarrow{AC}$ the **sum** of the vectors $\overrightarrow{AB}$ and $\overrightarrow{BC}$ and we write $\overrightarrow{AC} = \overrightarrow{AB} + \overrightarrow{BC}$. (The **zero vector**, denoted by $\mathbf{0}$, represents no displacement.) Thus, to find the sum of any two vectors $\mathbf{u}$ and $\mathbf{v}$, we sketch vectors equal to $\mathbf{u}$ and $\mathbf{v}$ with the initial point of one at the terminal point of the other (see Figure 4(a)). If we draw $\mathbf{u}$ and $\mathbf{v}$ starting at the same point, then $\mathbf{u} + \mathbf{v}$ is the vector that is the diagonal of the parallelogram formed by $\mathbf{u}$ and $\mathbf{v}$, as shown in Figure 4(b).

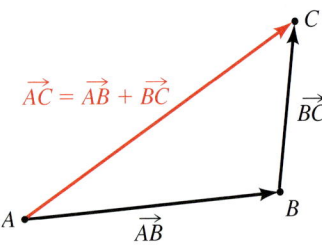

Figure 3

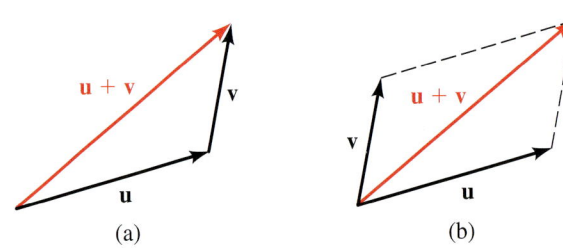

Figure 4

Addition of vectors

(a)   (b)

If a is a real number and $\mathbf{v}$ is a vector, we define a new vector $a\mathbf{v}$ as follows: The vector $a\mathbf{v}$ has magnitude $|a||\mathbf{v}|$ and has the same direction as $\mathbf{v}$ if $a > 0$, or the opposite direction if $a < 0$. If $a = 0$, then $a\mathbf{v} = \mathbf{0}$, the zero vector. This process is called **multiplication of a vector by a scalar**. Multiplying a vector by a scalar has the effect of stretching or shrinking the vector. Figure 5 shows graphs of the vector $a\mathbf{v}$ for different values of a. We write the vector $(-1)\mathbf{v}$ as $-\mathbf{v}$. Thus, $-\mathbf{v}$ is the vector with the same length as $\mathbf{v}$ but with the opposite direction.

The **difference** of two vectors $\mathbf{u}$ and $\mathbf{v}$ is defined by $\mathbf{u} - \mathbf{v} = \mathbf{u} + (-\mathbf{v})$. Figure 6 shows that the vector $\mathbf{u} - \mathbf{v}$ is the other diagonal of the parallelogram formed by $\mathbf{u}$ and $\mathbf{v}$.

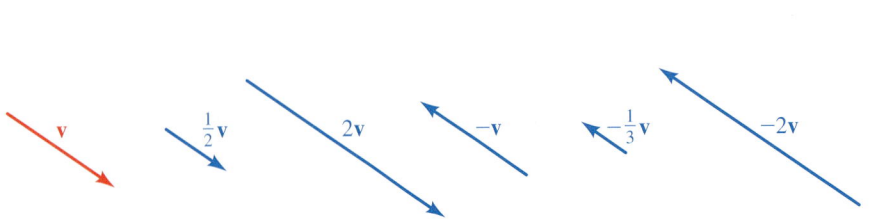

Figure 5

Multiplication of a vector by a scalar

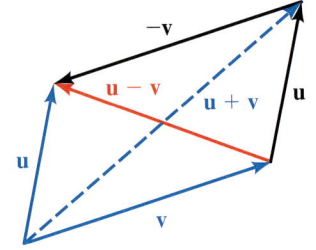

Figure 6

Subtraction of vectors

Vectors in the Coordinate Plane

So far we've discussed vectors geometrically. By placing a vector in a coordinate plane, we can describe it analytically (that is, by using components). In Figure 7(a), to go from the initial point of the vector **v** to the terminal point, we move a units to the right and b units upward. We represent **v** as an ordered pair of real numbers.

Note the distinction between the *vector* $\langle a, b \rangle$ and the *point* (a, b).

$$\mathbf{v} = \langle a, b \rangle$$

where a is the **horizontal component** of **v** and b is the **vertical component** of **v**. Remember that a vector represents a magnitude and a direction, not a particular arrow in the plane. Thus, the vector $\langle a, b \rangle$ has many different representations, depending on its initial point (see Figure 7(b)).

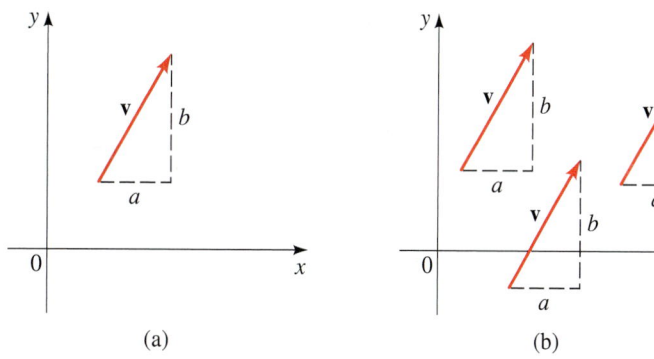

(a)   (b)

Figure 7

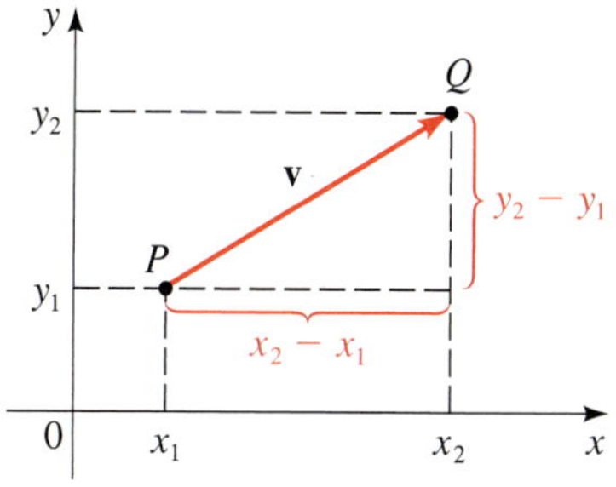

Using Figure 8, the relationship between a geometric representation of a vector and the analytic one can be stated as follows.

Figure 8

Component Form of a Vector

If a vector **v** is represented in the plane with initial point $P(x_1, y_1)$ and terminal point $Q(x_2, y_2)$, then

$$\mathbf{v} = \langle x_2 - x_1, y_2 - y_1 \rangle$$

Example 1   Describing Vectors in Component Form

(a) Find the component form of the vector **u** with initial point $(-2, 5)$ and terminal point $(3, 7)$.

(b) If the vector $\mathbf{v} = \langle 3, 7 \rangle$ is sketched with initial point $(2, 4)$, what is its terminal point?

(c) Sketch representations of the vector $\mathbf{w} = \langle 2, 3 \rangle$ with initial points at $(0, 0)$, $(2, 2)$, $(-2, -1)$, and $(1, 4)$.

Solution

(a) The desired vector is

$$\mathbf{u} = \langle 3 - (-2), 7 - 5 \rangle = \langle 5, 2 \rangle$$

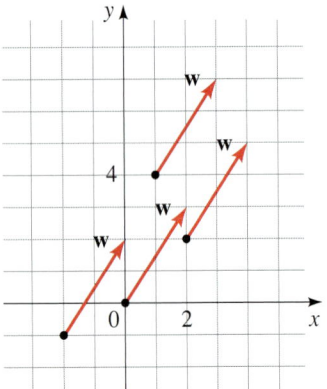

Figure 9

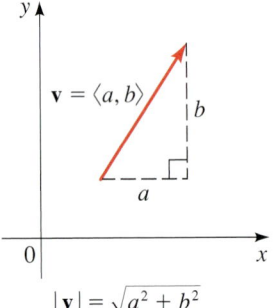

$$|\mathbf{v}| = \sqrt{a^2 + b^2}$$

Figure 10

(b) Let the terminal point of **v** be (x, y). Then

$$\langle x - 2, y - 4 \rangle = \langle 3, 7 \rangle$$

So $x - 2 = 3$ and $y - 4 = 7$, or $x = 5$ and $y = 11$. The terminal point is $(5, 11)$.

(c) Representations of the vector **w** are sketched in Figure 9. ■

We now give analytic definitions of the various operations on vectors that we have described geometrically. Let's start with equality of vectors. We've said that two vectors are equal if they have equal magnitude and the same direction. For the vectors $\mathbf{u} = \langle a_1, b_1 \rangle$ and $\mathbf{v} = \langle a_2, b_2 \rangle$, this means that $a_1 = a_2$ and $b_1 = b_2$. In other words, two vectors are **equal** if and only if their corresponding components are equal. Thus, all the arrows in Figure 7(b) represent the same vector, as do all the arrows in Figure 9.

Applying the Pythagorean Theorem to the triangle in Figure 10, we obtain the following formula for the magnitude of a vector.

Magnitude of a Vector

The **magnitude** or **length** of a vector $\mathbf{v} = \langle a, b \rangle$ is

$$|\mathbf{v}| = \sqrt{a^2 + b^2}$$

Example 2   Magnitudes of Vectors

Find the magnitude of each vector.

(a) $\mathbf{u} = \langle 2, -3 \rangle$ (b) $\mathbf{v} = \langle 5, 0 \rangle$ (c) $\mathbf{w} = \left\langle \frac{3}{5}, \frac{4}{5} \right\rangle$

Solution

(a) $|\mathbf{u}| = \sqrt{2^2 + (-3)^2} = \sqrt{13}$

(b) $|\mathbf{v}| = \sqrt{5^2 + 0^2} = \sqrt{25} = 5$

(c) $|\mathbf{w}| = \sqrt{\left(\frac{3}{5}\right)^2 + \left(\frac{4}{5}\right)^2} = \sqrt{\frac{9}{25} + \frac{16}{25}} = 1$ ■

The following definitions of addition, subtraction, and scalar multiplication of vectors correspond to the geometric descriptions given earlier. Figure 11 shows how the analytic definition of addition corresponds to the geometric one.

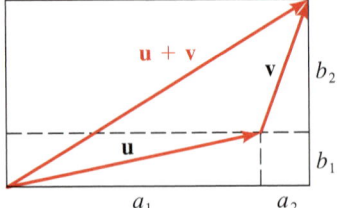

Figure 11

Algebraic Operations on Vectors

If $\mathbf{u} = \langle a_1, b_1 \rangle$ and $\mathbf{v} = \langle a_2, b_2 \rangle$, then

$$\mathbf{u} + \mathbf{v} = \langle a_1 + a_2, b_1 + b_2 \rangle$$

$$\mathbf{u} - \mathbf{v} = \langle a_1 - a_2, b_1 - b_2 \rangle$$

$$c\mathbf{u} = \langle ca_1, cb_1 \rangle, \qquad c \in \mathbb{R}$$

Example 3 Operations with Vectors

If $\mathbf{u} = \langle 2, -3 \rangle$ and $\mathbf{v} = \langle -1, 2 \rangle$, find $\mathbf{u} + \mathbf{v}$, $\mathbf{u} - \mathbf{v}$, $2\mathbf{u}$, $-3\mathbf{v}$, and $2\mathbf{u} + 3\mathbf{v}$.

Solution By the definitions of the vector operations, we have

$$\mathbf{u} + \mathbf{v} = \langle 2, -3 \rangle + \langle -1, 2 \rangle = \langle 1, -1 \rangle$$

$$\mathbf{u} - \mathbf{v} = \langle 2, -3 \rangle - \langle -1, 2 \rangle = \langle 3, -5 \rangle$$

$$2\mathbf{u} = 2\langle 2, -3 \rangle = \langle 4, -6 \rangle$$

$$-3\mathbf{v} = -3\langle -1, 2 \rangle = \langle 3, -6 \rangle$$

$$2\mathbf{u} + 3\mathbf{v} = 2\langle 2, -3 \rangle + 3\langle -1, 2 \rangle = \langle 4, -6 \rangle + \langle -3, 6 \rangle = \langle 1, 0 \rangle \qquad \blacksquare$$

The following properties for vector operations can be easily proved from the definitions. The **zero vector** is the vector $\mathbf{0} = \langle 0, 0 \rangle$. It plays the same role for addition of vectors as the number 0 does for addition of real numbers.

Properties of Vectors

Vector addition	Multiplication by a scalar
$\mathbf{u} + \mathbf{v} = \mathbf{v} + \mathbf{u}$	$c(\mathbf{u} + \mathbf{v}) = c\mathbf{u} + c\mathbf{v}$
$\mathbf{u} + (\mathbf{v} + \mathbf{w}) = (\mathbf{u} + \mathbf{v}) + \mathbf{w}$	$(c + d)\mathbf{u} = c\mathbf{u} + d\mathbf{u}$
$\mathbf{u} + \mathbf{0} = \mathbf{u}$	$(cd)\mathbf{u} = c(d\mathbf{u}) = d(c\mathbf{u})$
$\mathbf{u} + (-\mathbf{u}) = \mathbf{0}$	$1\mathbf{u} = \mathbf{u}$
Length of a vector	$0\mathbf{u} = \mathbf{0}$
$|c\mathbf{u}| = |c|\,|\mathbf{u}|$	$c\mathbf{0} = \mathbf{0}$

A vector of length 1 is called a **unit vector**. For instance, in Example 2(c), the vector $\mathbf{w} = \langle \frac{3}{5}, \frac{4}{5} \rangle$ is a unit vector. Two useful unit vectors are $\mathbf{i}$ and $\mathbf{j}$, defined by

$$\mathbf{i} = \langle 1, 0 \rangle \qquad \mathbf{j} = \langle 0, 1 \rangle$$

These vectors are special because any vector can be expressed in terms of them.

Vectors in Terms of i and j

The vector $\mathbf{v} = \langle a, b \rangle$ can be expressed in terms of $\mathbf{i}$ and $\mathbf{j}$ by

$$\mathbf{v} = \langle a, b \rangle = a\mathbf{i} + b\mathbf{j}$$

Example 4 Vectors in Terms of i and j

(a) Write the vector $\mathbf{u} = \langle 5, -8 \rangle$ in terms of $\mathbf{i}$ and $\mathbf{j}$.

(b) If $\mathbf{u} = 3\mathbf{i} + 2\mathbf{j}$ and $\mathbf{v} = -\mathbf{i} + 6\mathbf{j}$, write $2\mathbf{u} + 5\mathbf{v}$ in terms of $\mathbf{i}$ and $\mathbf{j}$.

Solution

(a) $\mathbf{u} = 5\mathbf{i} + (-8)\mathbf{j} = 5\mathbf{i} - 8\mathbf{j}$

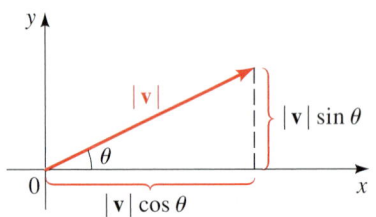

Figure 12

(b) The properties of addition and scalar multiplication of vectors show that we can manipulate vectors in the same way as algebraic expressions. Thus

$$2\mathbf{u} + 5\mathbf{v} = 2(3\mathbf{i} + 2\mathbf{j}) + 5(-\mathbf{i} + 6\mathbf{j})$$
$$= (6\mathbf{i} + 4\mathbf{j}) + (-5\mathbf{i} + 30\mathbf{j})$$
$$= \mathbf{i} + 34\mathbf{j}$$ ∎

Let $\mathbf{v}$ be a vector in the plane with its initial point at the origin. The **direction** of $\mathbf{v}$ is θ, the smallest positive angle in standard position formed by the positive x-axis and $\mathbf{v}$ (see Figure 12). If we know the magnitude and direction of a vector, then Figure 12 shows that we can find the horizontal and vertical components of the vector.

Horizontal and Vertical Components of a Vector

Let $\mathbf{v}$ be a vector with magnitude $|\mathbf{v}|$ and direction θ.
Then $\mathbf{v} = \langle a, b \rangle = a\mathbf{i} + b\mathbf{j}$, where

$$a = |\mathbf{v}| \cos \theta \qquad \text{and} \qquad b = |\mathbf{v}| \sin \theta$$

Thus, we can express $\mathbf{v}$ as

$$\mathbf{v} = |\mathbf{v}| \cos \theta \, \mathbf{i} + |\mathbf{v}| \sin \theta \, \mathbf{j}$$

Example 5 Components and Direction of a Vector

(a) A vector $\mathbf{v}$ has length 8 and direction $\pi/3$. Find the horizontal and vertical components, and write $\mathbf{v}$ in terms of $\mathbf{i}$ and $\mathbf{j}$.

(b) Find the direction of the vector $\mathbf{u} = -\sqrt{3}\,\mathbf{i} + \mathbf{j}$.

Solution

(a) We have $\mathbf{v} = \langle a, b \rangle$, where the components are given by

$$a = 8 \cos \frac{\pi}{3} = 4 \qquad \text{and} \qquad b = 8 \sin \frac{\pi}{3} = 4\sqrt{3}$$

Thus, $\mathbf{v} = \langle 4, 4\sqrt{3} \rangle = 4\mathbf{i} + 4\sqrt{3}\,\mathbf{j}$.

(b) From Figure 13 we see that the direction θ has the property that

$$\tan \theta = \frac{1}{-\sqrt{3}} = -\frac{\sqrt{3}}{3}$$

Thus, the reference angle for θ is $\pi/6$. Since the terminal point of the vector $\mathbf{u}$ is in quadrant II, it follows that $\theta = 5\pi/6$. ∎

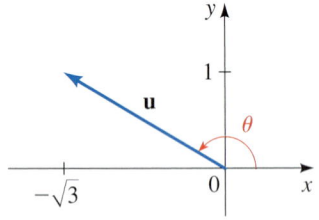

Figure 13

Using Vectors to Model Velocity and Force

The **velocity** of a moving object is modeled by a vector whose direction is the direction of motion and whose magnitude is the speed. Figure 14 shows some vectors $\mathbf{u}$, representing the velocity of wind flowing in the direction N 30° E, and a vector $\mathbf{v}$, representing the velocity of an airplane flying through this wind at the point P. It's obvious from our experience that wind affects both the speed and the direction of an

The use of bearings (such as N 30° E) to describe directions is explained on page 511 in Section 6.5.

airplane. Figure 15 indicates that the true velocity of the plane (relative to the ground) is given by the vector $\mathbf{w} = \mathbf{u} + \mathbf{v}$.

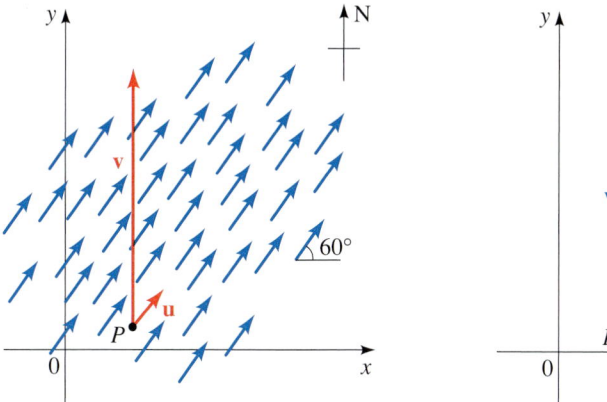

Figure 14

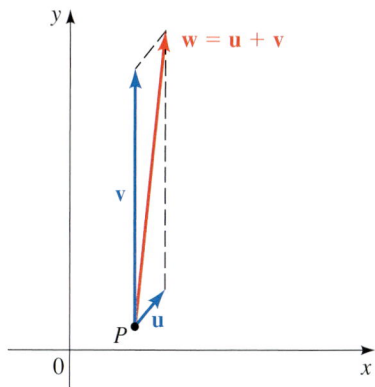

Figure 15

Example 6   The True Speed and Direction of an Airplane

An airplane heads due north at 300 mi/h. It experiences a 40 mi/h crosswind flowing in the direction N 30° E, as shown in Figure 14.

(a) Express the velocity $\mathbf{v}$ of the airplane relative to the air, and the velocity $\mathbf{u}$ of the wind, in component form.

(b) Find the true velocity of the airplane as a vector.

(c) Find the true speed and direction of the airplane.

Solution

(a) The velocity of the airplane relative to the air is $\mathbf{v} = 0\mathbf{i} + 300\mathbf{j} = 300\mathbf{j}$.
    By the formulas for the components of a vector, we find that the velocity of the wind is

$$\mathbf{u} = (40 \cos 60°)\mathbf{i} + (40 \sin 60°)\mathbf{j}$$
$$= 20\mathbf{i} + 20\sqrt{3}\mathbf{j}$$
$$\approx 20\mathbf{i} + 34.64\mathbf{j}$$

(b) The true velocity of the airplane is given by the vector $\mathbf{w} = \mathbf{u} + \mathbf{v}$.

$$\mathbf{w} = \mathbf{u} + \mathbf{v} = (20\mathbf{i} + 20\sqrt{3}\mathbf{j}) + (300\mathbf{j})$$
$$= 20\mathbf{i} + (20\sqrt{3} + 300)\mathbf{j}$$
$$\approx 20\mathbf{i} + 334.64\mathbf{j}$$

(c) The true speed of the airplane is given by the magnitude of $\mathbf{w}$.

$$|\mathbf{w}| \approx \sqrt{(20)^2 + (334.64)^2} \approx 335.2 \text{ mi/h}$$

The direction of the airplane is the direction θ of the vector $\mathbf{w}$. The angle θ has the property that $\tan \theta \approx 334.64/20 = 16.732$ and so $\theta \approx 86.6°$. Thus, the airplane is heading in the direction N 3.4° E.   ∎

Example 7   Calculating a Heading

A woman launches a boat from one shore of a straight river and wants to land at the point directly on the opposite shore. If the speed of the boat (relative to the water) is 10 mi/h and the river is flowing east at the rate of 5 mi/h, in what direction should she head the boat in order to arrive at the desired landing point?

Solution   We choose a coordinate system with the origin at the initial position of the boat as shown in Figure 16. Let **u** and **v** represent the velocities of the river and the boat, respectively. Clearly, **u** = 5**i** and, since the speed of the boat is 10 mi/h, we have $|\mathbf{v}| = 10$, so

$$\mathbf{v} = (10 \cos \theta)\mathbf{i} + (10 \sin \theta)\mathbf{j}$$

where the angle θ is as shown in Figure 16. The true course of the boat is given by the vector **w** = **u** + **v**. We have

$$\mathbf{w} = \mathbf{u} + \mathbf{v} = 5\mathbf{i} + (10 \cos \theta)\mathbf{i} + (10 \sin \theta)\mathbf{j}$$
$$= (5 + 10 \cos \theta)\mathbf{i} + (10 \sin \theta)\mathbf{j}$$

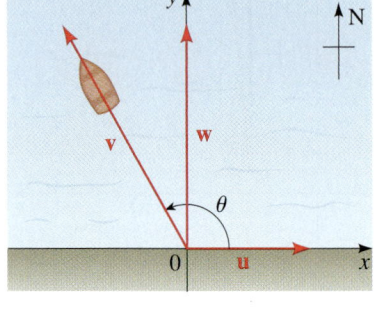

Figure 16

Since the woman wants to land at a point directly across the river, her direction should have horizontal component 0. In other words, she should choose θ in such a way that

$$5 + 10 \cos \theta = 0$$
$$\cos \theta = -\tfrac{1}{2}$$
$$\theta = 120°$$

Thus, she should head the boat in the direction $\theta = 120°$ (or N 30° W).   ∎

Force is also represented by a vector. Intuitively, we can think of force as describing a push or a pull on an object, for example, a horizontal push of a book across a table or the downward pull of the earth's gravity on a ball. Force is measured in pounds (or in newtons, in the metric system). For instance, a man weighing 200 lb exerts a force of 200 lb downward on the ground. If several forces are acting on an object, the **resultant force** experienced by the object is the vector sum of these forces.

Example 8   Resultant Force

Two forces $\mathbf{F}_1$ and $\mathbf{F}_2$ with magnitudes 10 and 20 lb, respectively, act on an object at a point P as shown in Figure 17. Find the resultant force acting at P.

Solution   We write $\mathbf{F}_1$ and $\mathbf{F}_2$ in component form:

$$\mathbf{F}_1 = (10 \cos 45°)\mathbf{i} + (10 \sin 45°)\mathbf{j} = 10\frac{\sqrt{2}}{2}\mathbf{i} + 10\frac{\sqrt{2}}{2}\mathbf{j} = 5\sqrt{2}\mathbf{i} + 5\sqrt{2}\mathbf{j}$$

$$\mathbf{F}_2 = (20 \cos 150°)\mathbf{i} + (20 \sin 150°)\mathbf{j} = -20\frac{\sqrt{3}}{2}\mathbf{i} + 20\left(\frac{1}{2}\right)\mathbf{j}$$

$$= -10\sqrt{3}\mathbf{i} + 10\mathbf{j}$$

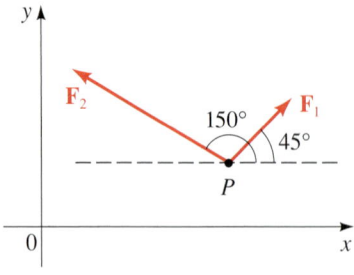

Figure 17

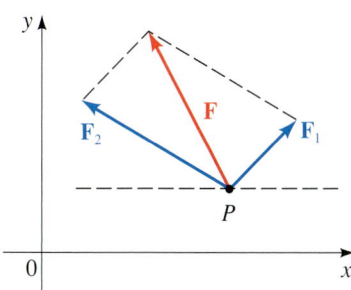

So the resultant force **F** is

$$\mathbf{F} = \mathbf{F}_1 + \mathbf{F}_2$$
$$= (5\sqrt{2}\mathbf{i} + 5\sqrt{2}\mathbf{j}) + (-10\sqrt{3}\mathbf{i} + 10\mathbf{j})$$
$$= (5\sqrt{2} - 10\sqrt{3})\mathbf{i} + (5\sqrt{2} + 10)\mathbf{j}$$
$$\approx -10\mathbf{i} + 17\mathbf{j}$$

The resultant force **F** is shown in Figure 18.

Figure 18

8.4 Exercises

1–6 ■ Sketch the vector indicated. (The vectors **u** and **v** are shown in the figure.)

1. 2**u**

2. −**v**

3. **u** + **v**

4. **u** − **v**

5. **v** − 2**u**

6. 2**u** + **v**

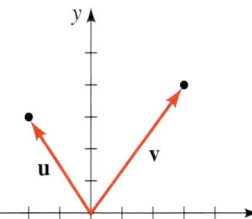

7–16 ■ Express the vector with initial point P and terminal point Q in component form.

7.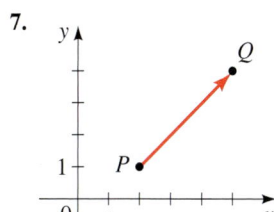

8.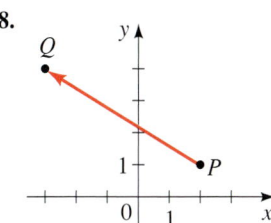

9.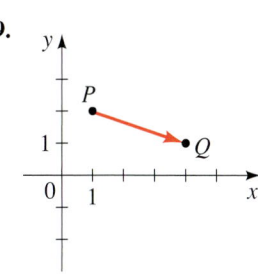

10.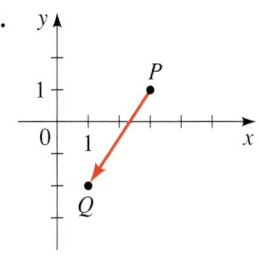

11. $P(3,2)$, $Q(8,9)$

12. $P(1,1)$, $Q(9,9)$

13. $P(5,3)$, $Q(1,0)$

14. $P(-1,3)$, $Q(-6,-1)$

15. $P(-1,-1)$, $Q(-1,1)$

16. $P(-8,-6)$, $Q(-1,-1)$

17–22 ■ Find $2\mathbf{u}$, $-3\mathbf{v}$, $\mathbf{u} + \mathbf{v}$, and $3\mathbf{u} - 4\mathbf{v}$ for the given vectors **u** and **v**.

17. $\mathbf{u} = \langle 2,7 \rangle$, $\mathbf{v} = \langle 3,1 \rangle$

18. $\mathbf{u} = \langle -2,5 \rangle$, $\mathbf{v} = \langle 2,-8 \rangle$

19. $\mathbf{u} = \langle 0,-1 \rangle$, $\mathbf{v} = \langle -2,0 \rangle$

20. $\mathbf{u} = \mathbf{i}$, $\mathbf{v} = -2\mathbf{j}$

21. $\mathbf{u} = 2\mathbf{i}$, $\mathbf{v} = 3\mathbf{i} - 2\mathbf{j}$ **22.** $\mathbf{u} = \mathbf{i} + \mathbf{j}$, $\mathbf{v} = \mathbf{i} - \mathbf{j}$

23–26 ■ Find $|\mathbf{u}|$, $|\mathbf{v}|$, $|2\mathbf{u}|$, $|\frac{1}{2}\mathbf{v}|$, $|\mathbf{u} + \mathbf{v}|$, $|\mathbf{u} - \mathbf{v}|$, and $|\mathbf{u}| - |\mathbf{v}|$.

23. $\mathbf{u} = 2\mathbf{i} + \mathbf{j}$, $\mathbf{v} = 3\mathbf{i} - 2\mathbf{j}$

24. $\mathbf{u} = -2\mathbf{i} + 3\mathbf{j}$, $\mathbf{v} = \mathbf{i} - 2\mathbf{j}$

25. $\mathbf{u} = \langle 10,-1 \rangle$, $\mathbf{v} = \langle -2,-2 \rangle$

26. $\mathbf{u} = \langle -6,6 \rangle$, $\mathbf{v} = \langle -2,-1 \rangle$

27–32 ■ Find the horizontal and vertical components of the vector with given length and direction, and write the vector in terms of the vectors **i** and **j**.

27. $|\mathbf{v}| = 40$, $\theta = 30°$ **28.** $|\mathbf{v}| = 50$, $\theta = 120°$

29. $|\mathbf{v}| = 1$, $\theta = 225°$ **30.** $|\mathbf{v}| = 800$, $\theta = 125°$

31. $|\mathbf{v}| = 4$, $\theta = 10°$ **32.** $|\mathbf{v}| = \sqrt{3}$, $\theta = 300°$

33–38 ■ Find the magnitude and direction (in degrees) of the vector.

33. $\mathbf{v} = \langle 3,4 \rangle$ **34.** $\mathbf{v} = \left\langle -\dfrac{\sqrt{2}}{2}, -\dfrac{\sqrt{2}}{2} \right\rangle$

35. $\mathbf{v} = \langle -12,5 \rangle$ **36.** $\mathbf{v} = \langle 40,9 \rangle$

37. $\mathbf{v} = \mathbf{i} + \sqrt{3}\mathbf{j}$ **38.** $\mathbf{v} = \mathbf{i} + \mathbf{j}$

Applications

39. Components of a Force A man pushes a lawn mower with a force of 30 lb exerted at an angle of 30° to the ground. Find the horizontal and vertical components of the force.

40. Components of a Velocity   A jet is flying in a direction N 20° E with a speed of 500 mi/h. Find the north and east components of the velocity.

41. Velocity   A river flows due south at 3 mi/h. A swimmer attempting to cross the river heads due east swimming at 2 mi/h relative to the water. Find the true velocity of the swimmer as a vector.

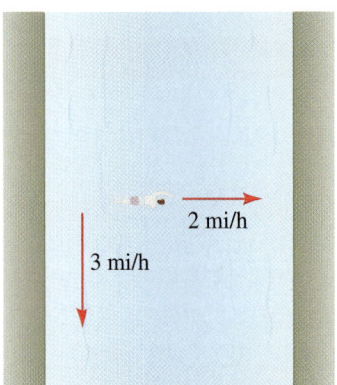

42. Velocity   A migrating salmon heads in the direction N 45° E, swimming at 5 mi/h relative to the water. The prevailing ocean currents flow due east at 3 mi/h. Find the true velocity of the fish as a vector.

43. True Velocity of a Jet   A pilot heads his jet due east. The jet has a speed of 425 mi/h relative to the air. The wind is blowing due north with a speed of 40 mi/h.
 (a) Express the velocity of the wind as a vector in component form.
 (b) Express the velocity of the jet relative to the air as a vector in component form.
 (c) Find the true velocity of the jet as a vector.
 (d) Find the true speed and direction of the jet.

44. True Velocity of a Jet   A jet is flying through a wind that is blowing with a speed of 55 mi/h in the direction N 30° E (see the figure). The jet has a speed of 765 mi/h relative to the air, and the pilot heads the jet in the direction N 45° E.
 (a) Express the velocity of the wind as a vector in component form.
 (b) Express the velocity of the jet relative to the air as a vector in component form.
 (c) Find the true velocity of the jet as a vector.

(d) Find the true speed and direction of the jet.

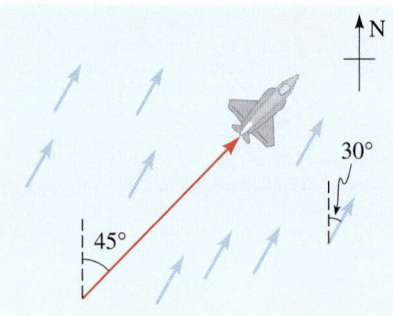

45. True Velocity of a Jet   Find the true speed and direction of the jet in Exercise 44 if the pilot heads the plane in the direction N 30° W.

46. True Velocity of a Jet   In what direction should the pilot in Exercise 44 head the plane for the true course to be due north?

47. Velocity of a Boat   A straight river flows east at a speed of 10 mi/h. A boater starts at the south shore of the river and heads in a direction 60° from the shore (see the figure). The motorboat has a speed of 20 mi/h relative to the water.
 (a) Express the velocity of the river as a vector in component form.
 (b) Express the velocity of the motorboat relative to the water as a vector in component form.
 (c) Find the true velocity of the motorboat.
 (d) Find the true speed and direction of the motorboat.

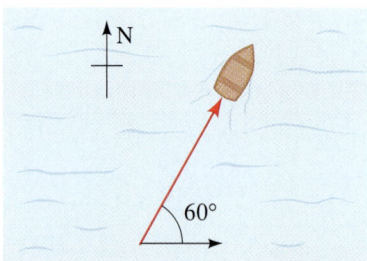

48. Velocity of a Boat   The boater in Exercise 47 wants to arrive at a point on the north shore of the river directly opposite the starting point. In what direction should the boat be headed?

49. Velocity of a Boat   A boat heads in the direction N 72° E. The speed of the boat relative to the water is 24 mi/h. The water is flowing directly south. It is observed that the true direction of the boat is directly east.
 (a) Express the velocity of the boat relative to the water as a vector in component form.

(b) Find the speed of the water and the true speed of the boat.

50. Velocity A woman walks due west on the deck of an ocean liner at 2 mi/h. The ocean liner is moving due north at a speed of 25 mi/h. Find the speed and direction of the woman relative to the surface of the water.

51–56 ■ Equilibrium of Forces The forces $\mathbf{F}_1, \mathbf{F}_2, \ldots, \mathbf{F}_n$ acting at the same point P are said to be in equilibrium if the resultant force is zero, that is, if $\mathbf{F}_1 + \mathbf{F}_2 + \cdots + \mathbf{F}_n = 0$. Find **(a)** the resultant forces acting at P, and **(b)** the additional force required (if any) for the forces to be in equilibrium.

51. $\mathbf{F}_1 = \langle 2, 5 \rangle,\quad \mathbf{F}_2 = \langle 3, -8 \rangle$

52. $\mathbf{F}_1 = \langle 3, -7 \rangle,\quad \mathbf{F}_2 = \langle 4, -2 \rangle,\quad \mathbf{F}_3 = \langle -7, 9 \rangle$

53. $\mathbf{F}_1 = 4\mathbf{i} - \mathbf{j},\quad \mathbf{F}_2 = 3\mathbf{i} - 7\mathbf{j},\quad \mathbf{F}_3 = -8\mathbf{i} + 3\mathbf{j},$
$\mathbf{F}_4 = \mathbf{i} + \mathbf{j}$

54. $\mathbf{F}_1 = \mathbf{i} - \mathbf{j},\quad \mathbf{F}_2 = \mathbf{i} + \mathbf{j},\quad \mathbf{F}_3 = -2\mathbf{i} + \mathbf{j}$

55.

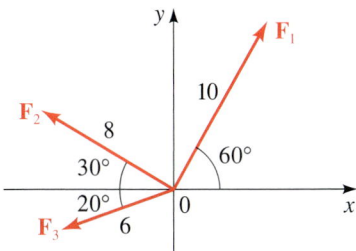

56.

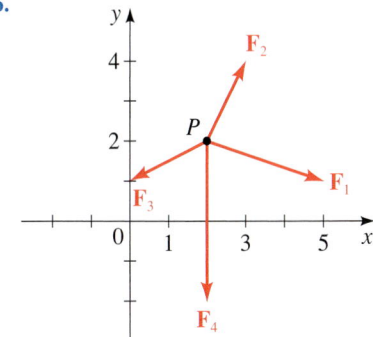

57. Equilibrium of Tensions A 100-lb weight hangs from a string as shown in the figure. Find the tensions $\mathbf{T}_1$ and $\mathbf{T}_2$ in the string.

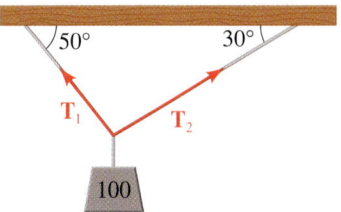

58. Equilibrium of Tensions The cranes in the figure are lifting an object that weighs 18,278 lb. Find the tensions $\mathbf{T}_1$ and $\mathbf{T}_2$.

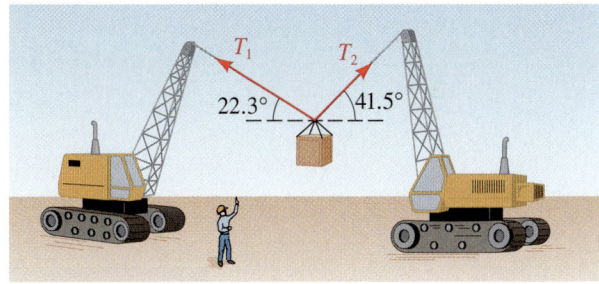

Discovery • Discussion

59. Vectors That Form a Polygon Suppose that n vectors can be placed head to tail in the plane so that they form a polygon. (The figure shows the case of a hexagon.) Explain why the sum of these vectors is **0**.

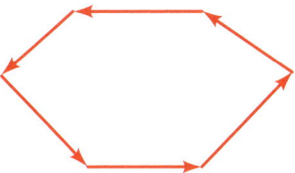

8.5 The Dot Product

In this section we define an operation on vectors called the dot product. This concept is especially useful in calculus and in applications of vectors to physics and engineering.

The Dot Product of Vectors

We begin by defining the dot product of two vectors.

Definition of the Dot Product

If $\mathbf{u} = \langle a_1, b_1 \rangle$ and $\mathbf{v} = \langle a_2, b_2 \rangle$ are vectors, then their **dot product,** denoted by $\mathbf{u} \cdot \mathbf{v}$, is defined by

$$\mathbf{u} \cdot \mathbf{v} = a_1 a_2 + b_1 b_2$$

Thus, to find the dot product of $\mathbf{u}$ and $\mathbf{v}$ we multiply corresponding components and add. The dot product is *not* a vector; it is a real number, or scalar.

Example 1   Calculating Dot Products

(a) If $\mathbf{u} = \langle 3, -2 \rangle$ and $\mathbf{v} = \langle 4, 5 \rangle$ then

$$\mathbf{u} \cdot \mathbf{v} = (3)(4) + (-2)(5) = 2$$

(b) If $\mathbf{u} = 2\mathbf{i} + \mathbf{j}$ and $\mathbf{v} = 5\mathbf{i} - 6\mathbf{j}$, then

$$\mathbf{u} \cdot \mathbf{v} = (2)(5) + (1)(-6) = 4 \qquad ■$$

The proofs of the following properties of the dot product follow easily from the definition.

Properties of the Dot Product

1. $\mathbf{u} \cdot \mathbf{v} = \mathbf{v} \cdot \mathbf{u}$

2. $(a\mathbf{u}) \cdot \mathbf{v} = a(\mathbf{u} \cdot \mathbf{v}) = \mathbf{u} \cdot (a\mathbf{v})$

3. $(\mathbf{u} + \mathbf{v}) \cdot \mathbf{w} = \mathbf{u} \cdot \mathbf{w} + \mathbf{v} \cdot \mathbf{w}$

4. $|\mathbf{u}|^2 = \mathbf{u} \cdot \mathbf{u}$

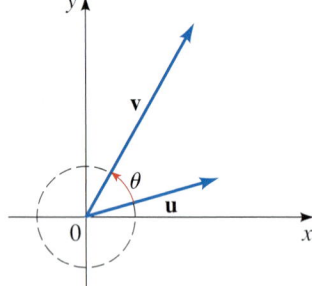

Figure 1

■ **Proof**   We prove only the last property. The proofs of the others are left as exercises. Let $\mathbf{u} = \langle a, b \rangle$. Then

$$\mathbf{u} \cdot \mathbf{u} = \langle a, b \rangle \cdot \langle a, b \rangle = a^2 + b^2 = |\mathbf{u}|^2 \qquad ■$$

Let $\mathbf{u}$ and $\mathbf{v}$ be vectors and sketch them with initial points at the origin. We define the **angle θ between $\mathbf{u}$ and $\mathbf{v}$** to be the smaller of the angles formed by these representations of $\mathbf{u}$ and $\mathbf{v}$ (see Figure 1). Thus, $0 \le \theta \le \pi$. The next theorem relates the angle between two vectors to their dot product.

The Dot Product Theorem

If θ is the angle between two nonzero vectors $\mathbf{u}$ and $\mathbf{v}$, then

$$\mathbf{u} \cdot \mathbf{v} = |\mathbf{u}| |\mathbf{v}| \cos \theta$$

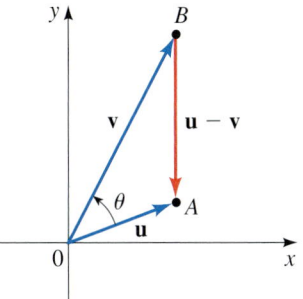

Figure 2

■ **Proof** The proof is a nice application of the Law of Cosines. Applying the Law of Cosines to triangle AOB in Figure 2 gives

$$|\mathbf{u} - \mathbf{v}|^2 = |\mathbf{u}|^2 + |\mathbf{v}|^2 - 2|\mathbf{u}||\mathbf{v}|\cos\theta$$

Using the properties of the dot product, we write the left-hand side as follows:

$$|\mathbf{u} - \mathbf{v}|^2 = (\mathbf{u} - \mathbf{v}) \cdot (\mathbf{u} - \mathbf{v})$$
$$= \mathbf{u} \cdot \mathbf{u} - \mathbf{u} \cdot \mathbf{v} - \mathbf{v} \cdot \mathbf{u} + \mathbf{v} \cdot \mathbf{v}$$
$$= |\mathbf{u}|^2 - 2(\mathbf{u} \cdot \mathbf{v}) + |\mathbf{v}|^2$$

Equating the right-hand sides of the displayed equations, we get

$$|\mathbf{u}|^2 - 2(\mathbf{u} \cdot \mathbf{v}) + |\mathbf{v}|^2 = |\mathbf{u}|^2 + |\mathbf{v}|^2 - 2|\mathbf{u}||\mathbf{v}|\cos\theta$$
$$-2(\mathbf{u} \cdot \mathbf{v}) = -2|\mathbf{u}||\mathbf{v}|\cos\theta$$
$$\mathbf{u} \cdot \mathbf{v} = |\mathbf{u}||\mathbf{v}|\cos\theta$$

This proves the theorem. ■

The Dot Product Theorem is useful because it allows us to find the angle between two vectors if we know the components of the vectors. The angle is obtained simply by solving the equation in the Dot Product Theorem for $\cos\theta$. We state this important result explicitly.

Angle between Two Vectors

If θ is the angle between two nonzero vectors $\mathbf{u}$ and $\mathbf{v}$, then

$$\cos\theta = \frac{\mathbf{u} \cdot \mathbf{v}}{|\mathbf{u}||\mathbf{v}|}$$

Example 2 Finding the Angle between Two Vectors

Find the angle between the vectors $\mathbf{u} = \langle 2, 5 \rangle$ and $\mathbf{v} = \langle 4, -3 \rangle$.

Solution By the formula for the angle between two vectors, we have

$$\cos\theta = \frac{\mathbf{u} \cdot \mathbf{v}}{|\mathbf{u}||\mathbf{v}|} = \frac{(2)(4) + (5)(-3)}{\sqrt{4 + 25}\sqrt{16 + 9}} = \frac{-7}{5\sqrt{29}}$$

Thus, the angle between $\mathbf{u}$ and $\mathbf{v}$ is

$$\theta = \cos^{-1}\left(\frac{-7}{5\sqrt{29}}\right) \approx 105.1°$$

■

Two nonzero vectors $\mathbf{u}$ and $\mathbf{v}$ are called **perpendicular**, or **orthogonal**, if the angle between them is $\pi/2$. The following theorem shows that we can determine if two vectors are perpendicular by finding their dot product.

Orthogonal Vectors

Two nonzero vectors $\mathbf{u}$ and $\mathbf{v}$ are perpendicular if and only if $\mathbf{u} \cdot \mathbf{v} = 0$.

■ **Proof** If **u** and **v** are perpendicular, then the angle between them is $\pi/2$ and so

$$\mathbf{u} \cdot \mathbf{v} = |\mathbf{u}||\mathbf{v}| \cos \frac{\pi}{2} = 0$$

Conversely, if $\mathbf{u} \cdot \mathbf{v} = 0$, then

$$|\mathbf{u}||\mathbf{v}| \cos \theta = 0$$

Since **u** and **v** are nonzero vectors, we conclude that $\cos \theta = 0$, and so $\theta = \pi/2$. Thus, **u** and **v** are orthogonal. ■

Example 3 Checking Vectors for Perpendicularity

Determine whether the vectors in each pair are perpendicular.

(a) $\mathbf{u} = \langle 3, 5 \rangle$ and $\mathbf{v} = \langle 2, -8 \rangle$ (b) $\mathbf{u} = \langle 2, 1 \rangle$ and $\mathbf{v} = \langle -1, 2 \rangle$

Solution

(a) $\mathbf{u} \cdot \mathbf{v} = (3)(2) + (5)(-8) = -34 \neq 0$, so **u** and **v** are not perpendicular.

(b) $\mathbf{u} \cdot \mathbf{v} = (2)(-1) + (1)(2) = 0$, so **u** and **v** are perpendicular. ■

The Component of u Along v

The **component of u along v** (or the **component of u in the direction of v**) is defined to be

*Note that the component of **u** along **v** is a scalar, not a vector.*

$$|\mathbf{u}| \cos \theta$$

where θ is the angle between **u** and **v**. Figure 3 gives a geometric interpretation of this concept. Intuitively, the component of **u** along **v** is the magnitude of the portion of **u** that points in the direction of **v**. Notice that the component of **u** along **v** is negative if $\pi/2 < \theta \leq \pi$.

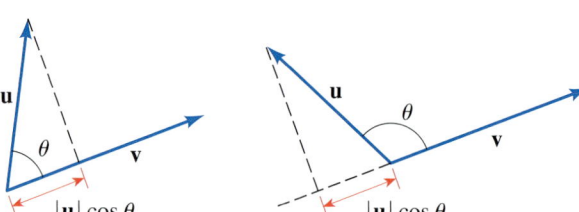

Figure 3

When analyzing forces in physics and engineering, it's often helpful to express a vector as a sum of two vectors lying in perpendicular directions. For example, suppose a car is parked on an inclined driveway as in Figure 4. The weight of the car is a vector **w** that points directly downward. We can write

$$\mathbf{w} = \mathbf{u} + \mathbf{v}$$

where **u** is parallel to the driveway and **v** is perpendicular to the driveway. The vector **u** is the force that tends to roll the car down the driveway, and **v** is the force experienced by the surface of the driveway. The magnitudes of these forces are the components of **w** along **u** and **v**, respectively.

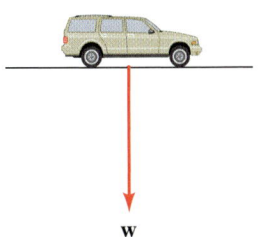
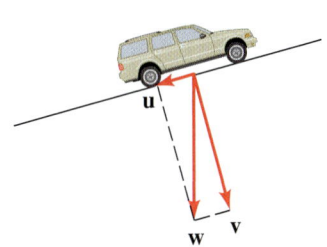
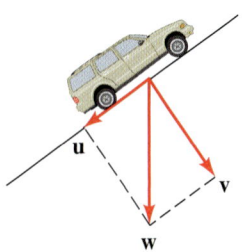
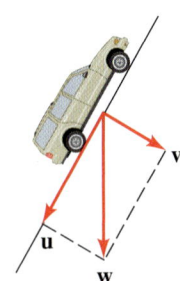

Figure 4

Example 4   Resolving a Force into Components

A car weighing 3000 lb is parked on a driveway that is inclined $15°$ to the horizontal, as shown in Figure 5.

(a) Find the magnitude of the force required to prevent the car from rolling down the driveway.

(b) Find the magnitude of the force experienced by the driveway due to the weight of the car.

Solution   The car exerts a force **w** of 3000 lb directly downward. We resolve **w** into the sum of two vectors **u** and **v**, one parallel to the surface of the driveway and the other perpendicular to it, as shown in Figure 5.

(a) The magnitude of the part of the force **w** that causes the car to roll down the driveway is

$$|\mathbf{u}| = \text{component of } \mathbf{w} \text{ along } \mathbf{u} = 3000 \cos 75° \approx 776$$

Thus, the force needed to prevent the car from rolling down the driveway is about 776 lb.

(b) The magnitude of the force exerted by the car on the driveway is

$$|\mathbf{v}| = \text{component of } \mathbf{w} \text{ along } \mathbf{v} = 3000 \cos 15° \approx 2898$$

The force experienced by the driveway is about 2898 lb. ∎

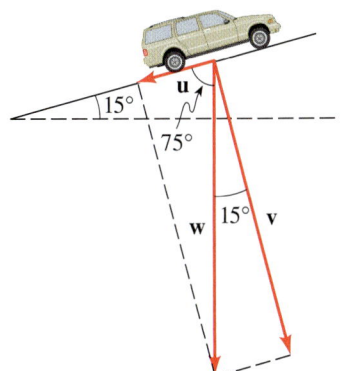

Figure 5

The component of **u** along **v** can be computed using dot products:

$$|\mathbf{u}| \cos \theta = \frac{|\mathbf{v}||\mathbf{u}| \cos \theta}{|\mathbf{v}|} = \frac{\mathbf{u} \cdot \mathbf{v}}{|\mathbf{v}|}$$

We have shown the following.

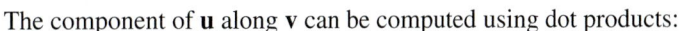

Calculating Components

The component of **u** along **v** is $\dfrac{\mathbf{u} \cdot \mathbf{v}}{|\mathbf{v}|}$.

Example 5 Finding Components

Let $\mathbf{u} = \langle 1, 4 \rangle$ and $\mathbf{v} = \langle -2, 1 \rangle$. Find the component of $\mathbf{u}$ along $\mathbf{v}$.

Solution We have

$$\text{component of } \mathbf{u} \text{ along } \mathbf{v} = \frac{\mathbf{u} \cdot \mathbf{v}}{|\mathbf{v}|} = \frac{(1)(-2) + (4)(1)}{\sqrt{4 + 1}} = \frac{2}{\sqrt{5}}$$ ∎

The Projection of u onto v

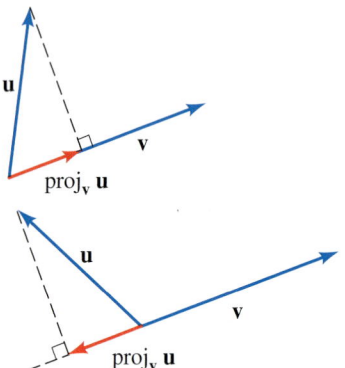

Figure 6

Figure 6 shows representations of the vectors $\mathbf{u}$ and $\mathbf{v}$. The projection of $\mathbf{u}$ onto $\mathbf{v}$, denoted by $\text{proj}_\mathbf{v}\, \mathbf{u}$, is the vector whose *direction* is the same as $\mathbf{v}$ and whose *length* is the component of $\mathbf{u}$ along $\mathbf{v}$. To find an expression for $\text{proj}_\mathbf{v}\, \mathbf{u}$, we first find a unit vector in the direction of $\mathbf{v}$ and then multiply it by the component of $\mathbf{u}$ along $\mathbf{v}$.

$$\text{proj}_\mathbf{v}\, \mathbf{u} = (\text{component of } \mathbf{u} \text{ along } \mathbf{v})(\text{unit vector in direction of } \mathbf{v})$$

$$= \left(\frac{\mathbf{u} \cdot \mathbf{v}}{|\mathbf{v}|} \right) \frac{\mathbf{v}}{|\mathbf{v}|} = \left(\frac{\mathbf{u} \cdot \mathbf{v}}{|\mathbf{v}|^2} \right) \mathbf{v}$$

We often need to **resolve** a vector $\mathbf{u}$ into the sum of two vectors, one parallel to $\mathbf{v}$ and one orthogonal to $\mathbf{v}$. That is, we want to write $\mathbf{u} = \mathbf{u}_1 + \mathbf{u}_2$ where $\mathbf{u}_1$ is parallel to $\mathbf{v}$ and $\mathbf{u}_2$ is orthogonal to $\mathbf{v}$. In this case, $\mathbf{u}_1 = \text{proj}_\mathbf{v}\, \mathbf{u}$ and $\mathbf{u}_2 = \mathbf{u} - \text{proj}_\mathbf{v}\, \mathbf{u}$ (see Exercise 37).

Calculating Projections

The **projection of $\mathbf{u}$ onto $\mathbf{v}$** is the vector $\text{proj}_\mathbf{v}\, \mathbf{u}$ given by

$$\text{proj}_\mathbf{v}\, \mathbf{u} = \left(\frac{\mathbf{u} \cdot \mathbf{v}}{|\mathbf{v}|^2} \right) \mathbf{v}$$

If the vector $\mathbf{u}$ is **resolved** into $\mathbf{u}_1$ and $\mathbf{u}_2$, where $\mathbf{u}_1$ is parallel to $\mathbf{v}$ and $\mathbf{u}_2$ is orthogonal to $\mathbf{v}$, then

$$\mathbf{u}_1 = \text{proj}_\mathbf{v}\, \mathbf{u} \qquad \text{and} \qquad \mathbf{u}_2 = \mathbf{u} - \text{proj}_\mathbf{v}\, \mathbf{u}$$

Example 6 Resolving a Vector into Orthogonal Vectors

Let $\mathbf{u} = \langle -2, 9 \rangle$ and $\mathbf{v} = \langle -1, 2 \rangle$.

(a) Find $\text{proj}_\mathbf{v}\, \mathbf{u}$.

(b) Resolve $\mathbf{u}$ into $\mathbf{u}_1$ and $\mathbf{u}_2$, where $\mathbf{u}_1$ is parallel to $\mathbf{v}$ and $\mathbf{u}_2$ is orthogonal to $\mathbf{v}$.

Solution

(a) By the formula for the projection of one vector onto another we have

$$\text{proj}_\mathbf{v}\, \mathbf{u} = \left(\frac{\mathbf{u} \cdot \mathbf{v}}{|\mathbf{v}|^2} \right) \mathbf{v} \qquad \text{\color{blue}Formula for projection}$$

$$= \left(\frac{\langle -2, 9 \rangle \cdot \langle -1, 2 \rangle}{(-1)^2 + 2^2} \right) \langle -1, 2 \rangle \qquad \text{\color{blue}Definition of } \mathbf{u} \text{ and } \mathbf{v}$$

$$= 4 \langle -1, 2 \rangle = \langle -4, 8 \rangle$$

(b) By the formula in the preceding box we have $\mathbf{u} = \mathbf{u}_1 + \mathbf{u}_2$, where

$$\mathbf{u}_1 = \text{proj}_\mathbf{v}\, \mathbf{u} = \langle -4, 8 \rangle \qquad \textcolor{blue}{\textit{From part (a)}}$$

$$\mathbf{u}_2 = \mathbf{u} - \text{proj}_\mathbf{v}\, \mathbf{u} = \langle -2, 9 \rangle - \langle -4, 8 \rangle = \langle 2, 1 \rangle \qquad \blacksquare$$

Work

One use of the dot product occurs in calculating work. In everyday use, the term *work* means the total amount of effort required to perform a task. In physics, *work* has a technical meaning that conforms to this intuitive meaning. If a constant force of magnitude F moves an object through a distance d along a straight line, then the **work** done is

$$W = Fd \qquad \text{or} \qquad \text{work} = \text{force} \times \text{distance}$$

If F is measured in pounds and d in feet, then the unit of work is a foot-pound (ft-lb). For example, how much work is done in lifting a 20-lb weight 6 ft off the ground? Since a force of 20 lb is required to lift this weight and since the weight moves through a distance of 6 ft, the amount of work done is

$$W = Fd = (20)(6) = 120 \text{ ft-lb}$$

This formula applies only when the force is directed along the direction of motion. In the general case, if the force $\mathbf{F}$ moves an object from P to Q, as in Figure 7, then only the component of the force in the direction of $\mathbf{D} = \overrightarrow{PQ}$ affects the object. Thus, the effective magnitude of the force on the object is

$$\text{component of } \mathbf{F} \text{ along } \mathbf{D} = |\,\mathbf{F}\,|\cos\theta$$

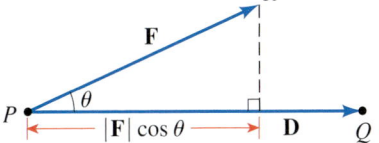

Figure 7

So, the work done is

$$W = \text{force} \times \text{distance} = (|\,\mathbf{F}\,|\cos\theta)|\,\mathbf{D}\,| = |\,\mathbf{F}\,||\,\mathbf{D}\,|\cos\theta = \mathbf{F}\cdot\mathbf{D}$$

We have derived the following simple formula for calculating work.

Work

The **work** W done by a force $\mathbf{F}$ in moving along a vector $\mathbf{D}$ is

$$W = \mathbf{F}\cdot\mathbf{D}$$

Example 7 Calculating Work

A force is given by the vector $\mathbf{F} = \langle 2, 3 \rangle$ and moves an object from the point $(1, 3)$ to the point $(5, 9)$. Find the work done.

Solution The displacement vector is

$$\mathbf{D} = \langle 5 - 1, 9 - 3 \rangle = \langle 4, 6 \rangle$$

So the work done is

$$W = \mathbf{F}\cdot\mathbf{D} = \langle 2, 3 \rangle \cdot \langle 4, 6 \rangle = 26$$

If the unit of force is pounds and the distance is measured in feet, then the work done is 26 ft-lb. $\blacksquare$

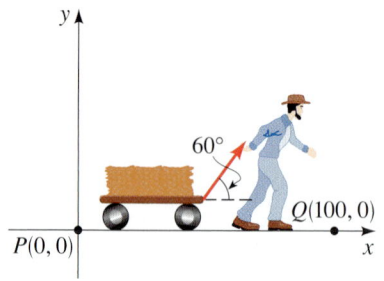

Figure 8

Example 8 Calculating Work

A man pulls a wagon horizontally by exerting a force of 20 lb on the handle. If the handle makes an angle of 60° with the horizontal, find the work done in moving the wagon 100 ft.

Solution   We choose a coordinate system with the origin at the initial position of the wagon (see Figure 8). That is, the wagon moves from the point $P(0, 0)$ to the point $Q(100, 0)$. The vector that represents this displacement is

$$\mathbf{D} = 100\mathbf{i}$$

The force on the handle can be written in terms of components (see Section 8.4) as

$$\mathbf{F} = (20 \cos 60°)\mathbf{i} + (20 \sin 60°)\mathbf{j} = 10\mathbf{i} + 10\sqrt{3}\mathbf{j}$$

Thus, the work done is

$$W = \mathbf{F} \cdot \mathbf{D} = (10\mathbf{i} + 10\sqrt{3}\mathbf{j}) \cdot (100\mathbf{i}) = 1000 \text{ ft-lb} \qquad \blacksquare$$

8.5   Exercises

1–8 ■ Find **(a)** $\mathbf{u} \cdot \mathbf{v}$ and **(b)** the angle between **u** and **v** to the nearest degree.

1. $\mathbf{u} = \langle 2, 0 \rangle, \quad \mathbf{v} = \langle 1, 1 \rangle$

2. $\mathbf{u} = \mathbf{i} + \sqrt{3}\mathbf{j}, \quad \mathbf{v} = -\sqrt{3}\mathbf{i} + \mathbf{j}$

3. $\mathbf{u} = \langle 2, 7 \rangle, \quad \mathbf{v} = \langle 3, 1 \rangle$

4. $\mathbf{u} = \langle -6, 6 \rangle, \quad \mathbf{v} = \langle 1, -1 \rangle$

5. $\mathbf{u} = \langle 3, -2 \rangle, \quad \mathbf{v} = \langle 1, 2 \rangle$

6. $\mathbf{u} = 2\mathbf{i} + \mathbf{j}, \quad \mathbf{v} = 3\mathbf{i} - 2\mathbf{j}$

7. $\mathbf{u} = -5\mathbf{j}, \quad \mathbf{v} = -\mathbf{i} - \sqrt{3}\mathbf{j}$

8. $\mathbf{u} = \mathbf{i} + \mathbf{j}, \quad \mathbf{v} = \mathbf{i} - \mathbf{j}$

9–14 ■ Determine whether the given vectors are orthogonal.

9. $\mathbf{u} = \langle 6, 4 \rangle, \quad \mathbf{v} = \langle -2, 3 \rangle$ **10.** $\mathbf{u} = \langle 0, -5 \rangle, \quad \mathbf{v} = \langle 4, 0 \rangle$

11. $\mathbf{u} = \langle -2, 6 \rangle, \quad \mathbf{v} = \langle 4, 2 \rangle$ **12.** $\mathbf{u} = 2\mathbf{i}, \quad \mathbf{v} = -7\mathbf{j}$

13. $\mathbf{u} = 2\mathbf{i} - 8\mathbf{j}, \quad \mathbf{v} = -12\mathbf{i} - 3\mathbf{j}$

14. $\mathbf{u} = 4\mathbf{i}, \quad \mathbf{v} = -\mathbf{i} + 3\mathbf{j}$

15–18 ■ Find the indicated quantity, assuming $\mathbf{u} = 2\mathbf{i} + \mathbf{j}, \mathbf{v} = \mathbf{i} - 3\mathbf{j}$, and $\mathbf{w} = 3\mathbf{i} + 4\mathbf{j}$.

15. $\mathbf{u} \cdot \mathbf{v} + \mathbf{u} \cdot \mathbf{w}$ **16.** $\mathbf{u} \cdot (\mathbf{v} + \mathbf{w})$

17. $(\mathbf{u} + \mathbf{v}) \cdot (\mathbf{u} - \mathbf{v})$ **18.** $(\mathbf{u} \cdot \mathbf{v})(\mathbf{u} \cdot \mathbf{w})$

19–22 ■ Find the component of **u** along **v**.

19. $\mathbf{u} = \langle 4, 6 \rangle, \quad \mathbf{v} = \langle 3, -4 \rangle$

20. $\mathbf{u} = \langle -3, 5 \rangle, \quad \mathbf{v} = \langle 1/\sqrt{2}, 1/\sqrt{2} \rangle$

21. $\mathbf{u} = 7\mathbf{i} - 24\mathbf{j}, \quad \mathbf{v} = \mathbf{j}$

22. $\mathbf{u} = 7\mathbf{i}, \quad \mathbf{v} = 8\mathbf{i} + 6\mathbf{j}$

23–28 ■ **(a)** Calculate $\text{proj}_\mathbf{v}\, \mathbf{u}$. **(b)** Resolve **u** into $\mathbf{u}_1$ and $\mathbf{u}_2$, where $\mathbf{u}_1$ is parallel to **v** and $\mathbf{u}_2$ is orthogonal to **v**.

23. $\mathbf{u} = \langle -2, 4 \rangle, \quad \mathbf{v} = \langle 1, 1 \rangle$

24. $\mathbf{u} = \langle 7, -4 \rangle, \quad \mathbf{v} = \langle 2, 1 \rangle$

25. $\mathbf{u} = \langle 1, 2 \rangle, \quad \mathbf{v} = \langle 1, -3 \rangle$

26. $\mathbf{u} = \langle 11, 3 \rangle, \quad \mathbf{v} = \langle -3, -2 \rangle$

27. $\mathbf{u} = \langle 2, 9 \rangle, \quad \mathbf{v} = \langle -3, 4 \rangle$

28. $\mathbf{u} = \langle 1, 1 \rangle, \quad \mathbf{v} = \langle 2, -1 \rangle$

29–32 ■ Find the work done by the force **F** in moving an object from P to Q.

29. $\mathbf{F} = 4\mathbf{i} - 5\mathbf{j}; \quad P(0, 0), Q(3, 8)$

30. $\mathbf{F} = 400\mathbf{i} + 50\mathbf{j}; \quad P(-1, 1), Q(200, 1)$

31. $\mathbf{F} = 10\mathbf{i} + 3\mathbf{j}; \quad P(2, 3), Q(6, -2)$

32. $\mathbf{F} = -4\mathbf{i} + 20\mathbf{j}; \quad P(0, 10), Q(5, 25)$

33–36 ■ Let **u**, **v**, and **w** be vectors and let a be a scalar. Prove the given property.

33. $\mathbf{u} \cdot \mathbf{v} = \mathbf{v} \cdot \mathbf{u}$

34. $(a\mathbf{u}) \cdot \mathbf{v} = a(\mathbf{u} \cdot \mathbf{v}) = \mathbf{u} \cdot (a\mathbf{v})$

35. $(\mathbf{u} + \mathbf{v}) \cdot \mathbf{w} = \mathbf{u} \cdot \mathbf{w} + \mathbf{v} \cdot \mathbf{w}$

36. $(\mathbf{u} - \mathbf{v}) \cdot (\mathbf{u} + \mathbf{v}) = |\mathbf{u}|^2 - |\mathbf{v}|^2$

37. Show that the vectors $\text{proj}_\mathbf{v}\, \mathbf{u}$ and $\mathbf{u} - \text{proj}_\mathbf{v}\, \mathbf{u}$ are orthogonal.

38. Evaluate $\mathbf{v} \cdot \text{proj}_\mathbf{v}\, \mathbf{u}$.

Applications

39. Work   The force $\mathbf{F} = 4\mathbf{i} - 7\mathbf{j}$ moves an object 4 ft along the x-axis in the positive direction. Find the work done if the unit of force is the pound.

40. Work   A constant force $\mathbf{F} = \langle 2, 8 \rangle$ moves an object along a straight line from the point $(2, 5)$ to the point $(11, 13)$. Find the work done if the distance is measured in feet and the force is measured in pounds.

41. Work   A lawn mower is pushed a distance of 200 ft along a horizontal path by a constant force of 50 lb. The handle of the lawn mower is held at an angle of 30° from the horizontal (see the figure). Find the work done.

42. Work   A car drives 500 ft on a road that is inclined 12° to the horizontal, as shown in the figure. The car weighs 2500 lb. Thus, gravity acts straight down on the car with a constant force $\mathbf{F} = -2500\mathbf{j}$. Find the work done by the car in overcoming gravity.

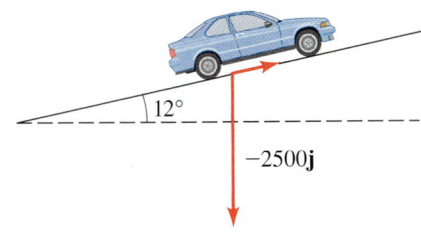

43. Force   A car is on a driveway that is inclined 25° to the horizontal. If the car weighs 2755 lb, find the force required to keep it from rolling down the driveway.

44. Force   A car is on a driveway that is inclined 10° to the horizontal. A force of 490 lb is required to keep the car from rolling down the driveway.
   (a) Find the weight of the car.
   (b) Find the force the car exerts against the driveway.

45. Force   A package that weighs 200 lb is placed on an inclined plane. If a force of 80 lb is just sufficient to keep the package from sliding, find the angle of inclination of the plane. (Ignore the effects of friction.)

46. Force   A cart weighing 40 lb is placed on a ramp inclined at 15° to the horizontal. The cart is held in place by a rope inclined at 60° to the horizontal, as shown in the figure. Find the force that the rope must exert on the cart to keep it from rolling down the ramp.

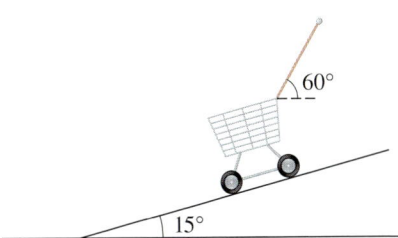

Discovery • Discussion

47. Distance from a Point to a Line   Let L be the line $2x + 4y = 8$ and let P be the point $(3, 4)$.
   (a) Show that the points $Q(0, 2)$ and $R(2, 1)$ lie on L.
   (b) Let $\mathbf{u} = \overrightarrow{QP}$ and $\mathbf{v} = \overrightarrow{QR}$, as shown in the figure. Find $\mathbf{w} = \text{proj}_\mathbf{v}\, \mathbf{u}$.
   (c) Sketch a graph that explains why $|\mathbf{u} - \mathbf{w}|$ is the distance from P to L. Find this distance.
   (d) Write a short paragraph describing the steps you would take to find the distance from a given point to a given line.

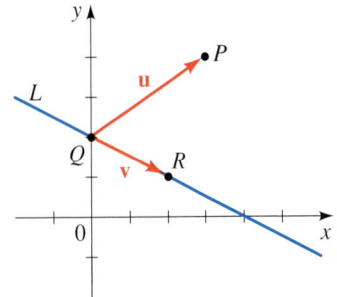

**DISCOVERY
PROJECT**

J. L. Amos/SuperStock

Sailing Against the Wind

Sailors depend on the wind to propel their boats. But what if the wind is blowing in a direction opposite to that in which they want to travel? Although it is obviously impossible to sail directly against the wind, it *is* possible to sail at an angle *into* the wind. Then by *tacking*, that is, zig-zagging on alternate sides of the wind direction, a sailor can make headway against the wind (see Figure 1).

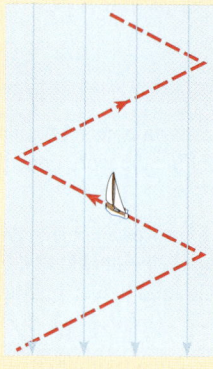

Figure 1
Tacking

How should the sail be aligned to propel the boat in the desired direction into the wind? This question can be answered by modeling the wind as a vector and studying its components along the keel and the sail.

For example, suppose a sailboat headed due north has its sail inclined in the direction N 20° E. The wind is blowing into the sail in the direction S 45° W with a force of magnitude F (see Figure 2).

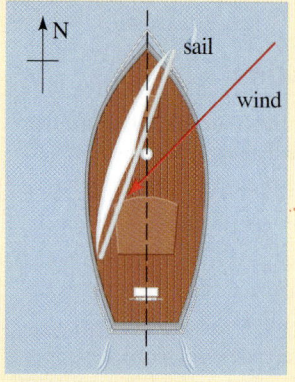

Figure 2

1. Show that the effective force of the wind on the sail is $F \sin 25°$. You can do this by finding the components of the wind parallel to the sail and perpendicular to the sail. The component parallel to the sail slips by and does not propel the boat. Only the perpendicular component pushes against the sail.

2. If the keel of the boat is aligned due north, what fraction of the force F actually drives the boat forward? Only the component of the force found in Problem 1 that is parallel to the keel drives the boat forward.

    (In real life, other factors, including the aerodynamic properties of the sail, influence the speed of the sailboat.)

3. If a boat heading due north has its sail inclined in the direction N $\alpha°$ E, and the wind is blowing with force F in the direction S $\beta°$ W where $0 < \alpha < \beta < 180$, find a formula for the magnitude of the force that actually drives the boat forward.

8   Review

Concept Check

1. Describe how polar coordinates represent the position of a point in the plane.

2. (a) What equations do you use to change from polar to rectangular coordinates?

   (b) What equations do you use to change from rectangular to polar coordinates?

3. How do you sketch the graph of a polar equation $r = f(\theta)$?

4. What type of curve has a polar equation of the given form?

   (a) $r = a \cos \theta$ or $r = a \sin \theta$

   (b) $r = a(1 \pm \cos \theta)$ or $r = a(1 \pm \sin \theta)$

   (c) $r = a \pm b \cos \theta$ or $r = a \pm b \sin \theta$

   (d) $r = a \cos n\theta$ or $r = a \sin n\theta$

5. How do you graph a complex number z? What is the polar form of a complex number z? What is the modulus of z? What is the argument of z?

6. (a) How do you multiply two complex numbers if they are given in polar form?

   (b) How do you divide two such numbers?

7. (a) State DeMoivre's Theorem.

   (b) How do you find the nth roots of a complex number?

8. (a) What is the difference between a scalar and a vector?

   (b) Draw a diagram to show how to add two vectors.

   (c) Draw a diagram to show how to subtract two vectors.

   (d) Draw a diagram to show how to multiply a vector by the scalars $2, \frac{1}{2}, -2$, and $-\frac{1}{2}$.

9. If $\mathbf{u} = \langle a_1, b_1 \rangle$, $\mathbf{v} = \langle a_2, b_2 \rangle$ and c is a scalar, write expressions for $\mathbf{u} + \mathbf{v}, \mathbf{u} - \mathbf{v}, c\mathbf{u}$, and $|\mathbf{u}|$.

10. (a) If $\mathbf{v} = \langle a, b \rangle$, write $\mathbf{v}$ in terms of $\mathbf{i}$ and $\mathbf{j}$.

    (b) Write the components of $\mathbf{v}$ in terms of the magnitude and direction of $\mathbf{v}$.

11. If $\mathbf{u} = \langle a_1, b_1 \rangle$ and $\mathbf{v} = \langle a_2, b_2 \rangle$, what is the dot product $\mathbf{u} \cdot \mathbf{v}$?

12. (a) How do you use the dot product to find the angle between two vectors?

    (b) How do you use the dot product to determine whether two vectors are perpendicular?

13. What is the component of $\mathbf{u}$ along $\mathbf{v}$, and how do you calculate it?

14. What is the projection of $\mathbf{u}$ onto $\mathbf{v}$, and how do you calculate it?

15. How much work is done by the force $\mathbf{F}$ in moving an object along a displacement $\mathbf{D}$?

Exercises

1–6 ■ A point $P(r, \theta)$ is given in polar coordinates.
(a) Plot the point P. (b) Find rectangular coordinates for P.

1. $\left(12, \frac{\pi}{6}\right)$

2. $\left(8, -\frac{3\pi}{4}\right)$

3. $\left(-3, \frac{7\pi}{4}\right)$

4. $\left(-\sqrt{3}, \frac{2\pi}{3}\right)$

5. $\left(4\sqrt{3}, -\frac{5\pi}{3}\right)$

6. $\left(-6\sqrt{2}, -\frac{\pi}{4}\right)$

7–12 ■ A point $P(x, y)$ is given in rectangular coordinates.
(a) Plot the point P.
(b) Find polar coordinates for P with $r \geq 0$.
(c) Find polar coordinates for P with $r \leq 0$.

7. $(8, 8)$

8. $(-\sqrt{2}, \sqrt{6})$

9. $(-6\sqrt{2}, -6\sqrt{2})$

10. $(3\sqrt{3}, 3)$

11. $(-3, \sqrt{3})$

12. $(4, -4)$

13–16 ■ (a) Convert the equation to polar coordinates and simplify. (b) Graph the equation. [*Hint:* Use the form of the equation that you find easier to graph.]

13. $x + y = 4$

14. $xy = 1$

15. $x^2 + y^2 = 4x + 4y$

16. $(x^2 + y^2)^2 = 2xy$

17–24 ■ (a) Sketch the graph of the polar equation.
(b) Express the equation in rectangular coordinates.

17. $r = 3 + 3 \cos \theta$

18. $r = 3 \sin \theta$

19. $r = 2 \sin 2\theta$

20. $r = 4 \cos 3\theta$

21. $r^2 = \sec 2\theta$

22. $r^2 = 4 \sin 2\theta$

23. $r = \sin \theta + \cos \theta$

24. $r = \dfrac{4}{2 + \cos \theta}$

25–28 ■ Use a graphing device to graph the polar equation. Choose the domain of θ to make sure you produce the entire graph.

25. $r = \cos(\theta/3)$ **26.** $r = \sin(9\theta/4)$

27. $r = 1 + 4\cos(\theta/3)$

28. $r = \theta \sin\theta$, $-6\pi \le \theta \le 6\pi$

29–34 ■ A complex number is given.

(a) Graph the complex number in the complex plane.

(b) Find the modulus and argument.

(c) Write the number in polar form.

29. $4 + 4i$ **30.** $-10i$

31. $5 + 3i$ **32.** $1 + \sqrt{3}i$

33. $-1 + i$ **34.** -20

35–38 ■ Use DeMoivre's Theorem to find the indicated power.

35. $(1 - \sqrt{3}i)^4$ **36.** $(1 + i)^8$

37. $(\sqrt{3} + i)^{-4}$ **38.** $\left(\dfrac{1}{2} + \dfrac{\sqrt{3}}{2}i\right)^{20}$

39–42 ■ Find the indicated roots.

39. The square roots of $-16i$

40. The cube roots of $4 + 4\sqrt{3}i$

41. The sixth roots of 1   **42.** The eighth roots of i

43–44 ■ Find $|\mathbf{u}|$, $\mathbf{u} + \mathbf{v}$, $\mathbf{u} - \mathbf{v}$, $2\mathbf{u}$, and $3\mathbf{u} - 2\mathbf{v}$.

43. $\mathbf{u} = \langle -2, 3\rangle$, $\mathbf{v} = \langle 8, 1\rangle$ **44.** $\mathbf{u} = 2\mathbf{i} + \mathbf{j}$, $\mathbf{v} = \mathbf{i} - 2\mathbf{j}$

45. Find the vector $\mathbf{u}$ with initial point $P(0, 3)$ and terminal point $Q(3, -1)$.

46. Find the vector $\mathbf{u}$ having length $|\mathbf{u}| = 20$ and direction $\theta = 60°$.

47. If the vector $5\mathbf{i} - 8\mathbf{j}$ is placed in the plane with its initial point at $P(5, 6)$, find its terminal point.

48. Find the direction of the vector $2\mathbf{i} - 5\mathbf{j}$.

49. Two tugboats are pulling a barge, as shown. One pulls with a force of 2.0×10^4 lb in the direction N 50° E and the other with a force of 3.4×10^4 lb in the direction S 75° E.

(a) Find the resultant force on the barge as a vector.

(b) Find the magnitude and direction of the resultant force.

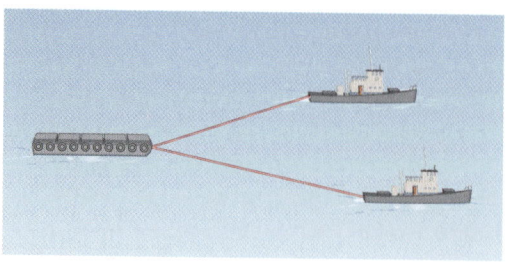

50. An airplane heads N 60° E at a speed of 600 mi/h relative to the air. A wind begins to blow in the direction N 30° W at 50 mi/h.

(a) Find the velocity of the airplane as a vector.

(b) Find the true speed and direction of the airplane.

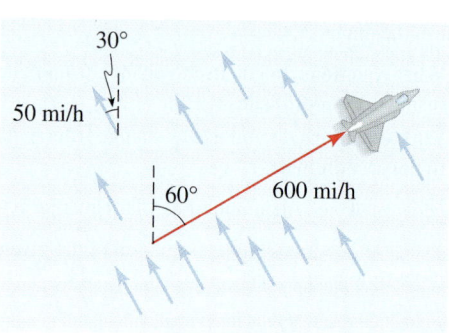

51–54 ■ Find $|\mathbf{u}|$, $\mathbf{u} \cdot \mathbf{u}$, and $\mathbf{u} \cdot \mathbf{v}$.

51. $\mathbf{u} = \langle 4, -3\rangle$, $\mathbf{v} = \langle 9, -8\rangle$

52. $\mathbf{u} = \langle 5, 12\rangle$, $\mathbf{v} = \langle 10, -4\rangle$

53. $\mathbf{u} = -2\mathbf{i} + 2\mathbf{j}$, $\mathbf{v} = \mathbf{i} + \mathbf{j}$

54. $\mathbf{u} = 10\mathbf{j}$, $\mathbf{v} = 5\mathbf{i} - 3\mathbf{j}$

55–58 ■ Are $\mathbf{u}$ and $\mathbf{v}$ orthogonal? If not, find the angle between them.

55. $\mathbf{u} = \langle -4, 2\rangle$, $\mathbf{v} = \langle 3, 6\rangle$

56. $\mathbf{u} = \langle 5, 3\rangle$, $\mathbf{v} = \langle -2, 6\rangle$

57. $\mathbf{u} = 2\mathbf{i} + \mathbf{j}$, $\mathbf{v} = \mathbf{i} + 3\mathbf{j}$

58. $\mathbf{u} = \mathbf{i} - \mathbf{j}$, $\mathbf{v} = \mathbf{i} + \mathbf{j}$

59–60 ■ The vectors $\mathbf{u}$ and $\mathbf{v}$ are given.

(a) Find the component of $\mathbf{u}$ along $\mathbf{v}$.

(b) Find $\text{proj}_\mathbf{v}\,\mathbf{u}$.

(c) Resolve $\mathbf{u}$ into the vectors $\mathbf{u}_1$ and $\mathbf{u}_2$, where $\mathbf{u}_1$ is parallel to $\mathbf{v}$ and $\mathbf{u}_2$ is perpendicular to $\mathbf{v}$.

59. $\mathbf{u} = \langle 3, 1\rangle$, $\mathbf{v} = \langle 6, -1\rangle$

60. $\mathbf{u} = \langle -8, 6\rangle$, $\mathbf{v} = \langle 20, 20\rangle$

61. Find the work done by the force $\mathbf{F} = 2\mathbf{i} + 9\mathbf{j}$ in moving an object from the point $(1, 1)$ to the point $(7, -1)$.

62. A force $\mathbf{F}$ with magnitude 250 lb moves an object in the direction of the vector $\mathbf{D}$ a distance of 20 ft. If the work done is 3800 ft-lb, find the angle between $\mathbf{F}$ and $\mathbf{D}$.

8 Test

1. (a) Convert the point whose polar coordinates are $(8, 5\pi/4)$ to rectangular coordinates.
   (b) Find two polar coordinate representations for the rectangular coordinate point $(-6, 2\sqrt{3})$, one with $r > 0$ and one with $r < 0$, and both with $0 \le \theta < 2\pi$.

2. (a) Graph the polar equation $r = 8 \cos \theta$. What type of curve is this?
   (b) Convert the equation to rectangular coordinates.

3. Let $z = 1 + \sqrt{3}i$.
   (a) Graph z in the complex plane.
   (b) Write z in polar form.
   (c) Find the complex number z^9.

4. Let $z_1 = 4\left(\cos \dfrac{7\pi}{12} + i \sin \dfrac{7\pi}{12} \right)$ and $z_2 = 2\left(\cos \dfrac{5\pi}{12} + i \sin \dfrac{5\pi}{12} \right)$.

   Find $z_1 z_2$ and $\dfrac{z_1}{z_2}$.

5. Find the cube roots of $27i$, and sketch these roots in the complex plane.

6. Let **u** be the vector with initial point $P(3, -1)$ and terminal point $Q(-3, 9)$.
   (a) Express **u** in terms of **i** and **j**.
   (b) Find the length of **u**.

7. Let $\mathbf{u} = \langle 1, 3 \rangle$ and $\mathbf{v} = \langle -6, 2 \rangle$.
   (a) Find $\mathbf{u} - 3\mathbf{v}$. (b) Find $|\mathbf{u} + \mathbf{v}|$.
   (c) Find $\mathbf{u} \cdot \mathbf{v}$. (d) Are **u** and **v** perpendicular?

8. Let $\mathbf{u} = \langle -4\sqrt{3}, 4 \rangle$.
   (a) Graph **u** with initial point $(0, 0)$.
   (b) Find the length and direction of **u**.

9. A river is flowing due east at 8 mi/h. A man heads his motorboat in a direction N 30° E in the river. The speed of the motorboat relative to the water is 12 mi/h.
   (a) Express the true velocity of the motorboat as a vector.
   (b) Find the true speed and direction of the motorboat.

10. Let $\mathbf{u} = 3\mathbf{i} + 2\mathbf{j}$ and $\mathbf{v} = 5\mathbf{i} - \mathbf{j}$.
    (a) Find the angle between **u** and **v**.
    (b) Find the component of **u** along **v**.
    (c) Find $\text{proj}_{\mathbf{v}}\, \mathbf{u}$.

11. Find the work done by the force $\mathbf{F} = 3\mathbf{i} - 5\mathbf{j}$ in moving an object from the point $(2, 2)$ to the point $(7, -13)$.

The method used to survey and map a town (page 522) works well for small areas. But mapping the whole world would introduce a new difficulty: How do we represent the *spherical* world by a *flat* map? Several ingenious methods have been developed.

Cylindrical Projection

One method is the **cylindrical projection**. In this method we imagine a cylinder "wrapped" around the earth at the equator as in Figure 1. Each point on the earth is projected onto the cylinder by a ray emanating from the center of the earth. The "unwrapped" cylinder is the desired flat map of the world. The process is illustrated in Figure 2.

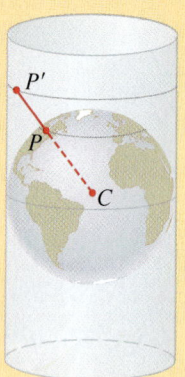

Figure 1

Point P on the earth is projected onto point P' on the cylinder by a ray from the center of the earth C.

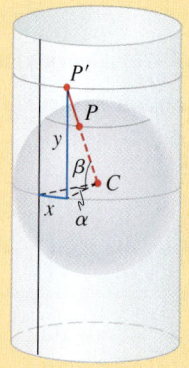

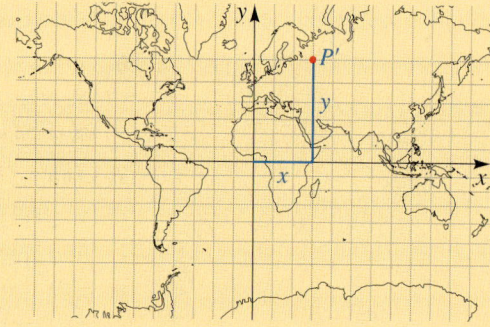

Figure 2   (a) Cylindrical projection   (b) Cylindrical projection map

Of course, we cannot actually wrap a large piece of paper around the world, so this whole process must be done mathematically, and the tool we need is trigonometry. On the unwrapped cylinder we take the x-axis to correspond to the equator and the y-axis to the meridian through Greenwich, England (0° longitude). Let R be the radius of the earth and let P be the point on the earth at $\alpha°$ E longitude and $\beta°$ N latitude. The point P is projected to the point $P'(x, y)$ on the cylinder (viewed as part of the coordinate plane) where

$$x = \left(\frac{\pi}{180}\right)\alpha R \qquad \text{Formula for length of a circular arc}$$

$$y = R \tan \beta \qquad \text{Definition of tangent}$$

See Figure 2(a). These formulas can then be used to draw the map. (Note that West longitude and South latitude correspond to negative values of α and β, respectively.) Of course, using R as the radius of the earth would produce a huge

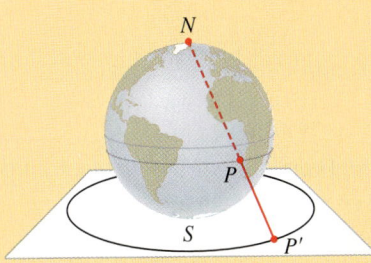

Figure 3

Point P on the earth's surface is projected onto point P' on the plane by a ray from the north pole.

map, so we replace R by a smaller value to get a map at an appropriate scale as in Figure 2(b).

Stereographic Projection

In the **stereographic projection** we imagine the earth placed on the coordinate plane with the south pole at the origin. Points on the earth are projected onto the plane by rays emanating from the north pole (see Figure 3). The earth is placed so that the prime meridian (0° longitude) corresponds to the polar axis. As shown in Figure 4(a), a point P on the earth at $\alpha°$ E longitude and $\beta°$ N latitude is projected onto the point $P'(r, \theta)$ whose polar coordinates are

$$r = 2R \tan\left(\frac{\beta}{2} + 45°\right)$$

$$\theta = \alpha$$

Figure 4(b) shows how the first of these formulas is obtained

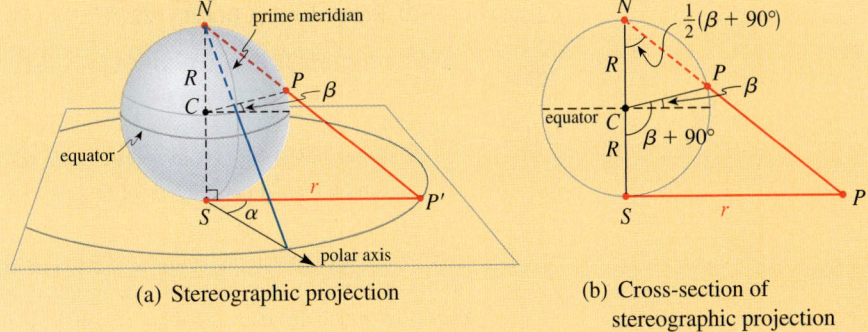

(a) Stereographic projection

(b) Cross-section of stereographic projection

Figure 4

Figure 5 shows a stereographic map of the southern hemisphere.

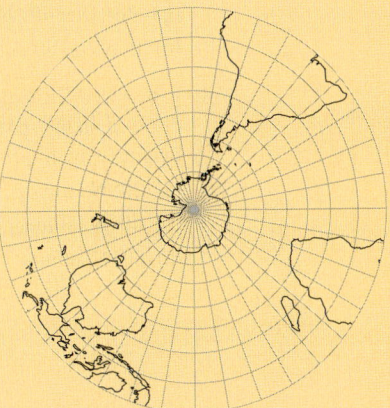

Figure 5

Stereographic projection of the southern hemisphere

Problems

1. **Cylindrical Projection**   A map maker wishes to map the earth using a cylindrical projection. The map is to be 36 inches wide. Thus, the equator is mapped onto a horizontal 36-inch line segment. The radius of the earth is 3960 miles.

   (a) What value of R should he use in the cylindrical projection formulas?

   (b) How many miles does one inch on the map represent at the equator?

2. **Cylindrical Projection**   To map the entire world using the cylindrical projection, the cylinder must extend infinitely far in the vertical direction. So a practical cylindrical map cannot extend all the way to the poles. The map maker in Problem 1 decides that his map should show the earth between 70° N and 70° S latitudes. How tall should his map be?

3. **Cylindrical Projection**   The map maker in Problem 1 places the y-axis (0° longitude) at the center of the map as shown in Figure 2(b). Find the x- and y-coordinates of the following cities on the map.

   (a) Seattle, Washington;   47.6° N, 122.3° W

   (b) Moscow, Russia;   55.8° N, 37.6° E

   (c) Sydney, Australia;   33.9° S, 151.2° E

   (d) Rio de Janeiro, Brazil;   22.9° S, 43.1° W

4. **Stereographic Projection**   A map maker makes a stereographic projection of the southern hemisphere, from the south pole to the equator. The map is to have a radius of 20 in.

   (a) What value of R should he use in the stereographic projection formulas?

   (b) Find the polar coordinates of Sydney, Australia (33.9° S, 151.2° E) on his map.

5–6 ■ The cylindrical projection stretches distances between points not on the equator— the farther from the equator, the more the distances are stretched. In these problems we find the factors by which distances are distorted on the cylindrical projection at various locations.

5. **Projected Distances**   Find the ratio of the projected distance on the cylinder to the actual distance on the sphere between the given latitudes along a meridian (see the figure at the left).

   (a) Between 20° and 21° N latitude

   (b) Between 40° and 41° N latitude

   (c) Between 80° and 81° N latitude

6. **Projected Distances**   Find the ratio of the projected distance on the cylinder to the distance on the sphere along the given parallel of latitude between two points that are 1° longitude apart (see the figure below).

   (a) 20° N latitude

   (b) 40° N latitude

   (c) 80° N latitude

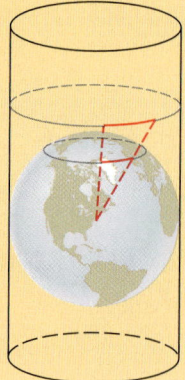

7–8 ■ The stereographic projection also stretches distances—the farther from the south pole, the more distances are stretched. In these problems we find the factors by which distances are distorted on the stereographic projection at various locations.

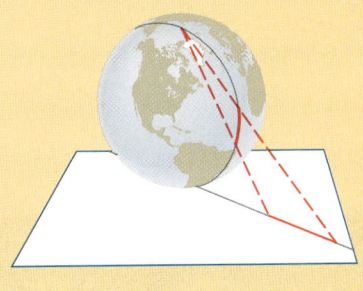

7. Projected Distances   Find the ratio of the projected distance on the plane to the actual distance on the sphere between the given latitudes along a meridian (see the figure at the left).

(a) Between 20° and 21° S latitude

(b) Between 40° and 41° S latitude

(c) Between 80° and 81° S latitude

8. Projected Distances   Find the ratio of the projected distance on the plane to the distance on the sphere along the given parallel of latitude between two points that are 1° longitude apart (see the figure).

(a) 20° S latitude

(b) 40° S latitude

(c) 80° S latitude

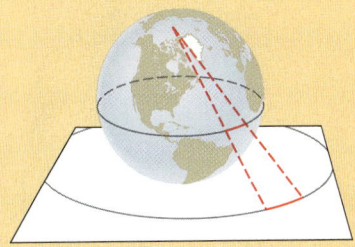

9. Lines of Latitude and Longitude   In this project we see how projection transfers lines of latitude and longitude from a sphere to a flat surface. You will need a round glass bowl, tracing paper, and a light source (a small transparent light bulb). Use a black marker to draw equally spaced lines of latitude and longitude on the outside of the bowl.

(a) To model the stereographic projection, place the bowl on a sheet of tracing paper and use the light source as shown in the figure at the left.

(b) To model the cylindrical projection, wrap the tracing paper around the bowl and use the light source as shown in the figure below.

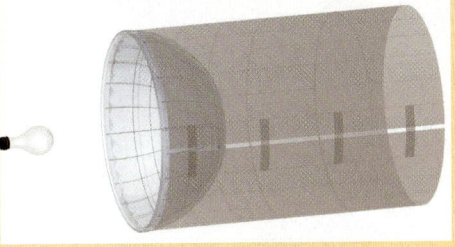

10. Other Projections   There are many other map projections, such as the Albers Conic Projection, the Azimuthal Projection, the Behrmann Cylindrical Equal-Area Projection, the Gall Isographic and Orthographic Projections, the Gnomonic Projection, the Lambert Equal-Area Projection, the Mercator Projection, the Mollweide Projection, the Rectangular Projection, and the Sinusoidal Projection. Research one of these projections in your library or on the Internet and write a report explaining how the map is constructed, and describing its advantages and disadvantages.

9 Systems of Equations and Inequalities

Chapter Overview

Many real-world situations have too many variables to be modeled by a *single* equation. For example, weather depends on many variables, including temperature, wind speed, air pressure, humidity, and so on. So to model (and forecast) the weather, scientists use many equations, each having many variables. Such systems of equations *work together* to describe the weather. Systems of equations with hundreds or even thousands of variables are also used extensively in the air travel and telecommunications industries to establish consistent airline schedules and to find efficient routing for telephone calls. To understand how such systems arise, let's consider the following simple example.

A gas station sells regular gas for $2.20 per gallon and premium for $3.00 per gallon. At the end of a business day 280 gallons of gas were sold and receipts totaled $680. How many gallons of each type of gas were sold? If we let x and y be the number of gallons of regular and premium gasoline sold, respectively, we get the following system of two equations:

$$\begin{cases} x + \quad\;\; y = 280 & \textit{Gallons equation} \\ 2.20x + 3.00y = 680 & \textit{Dollars equation} \end{cases}$$

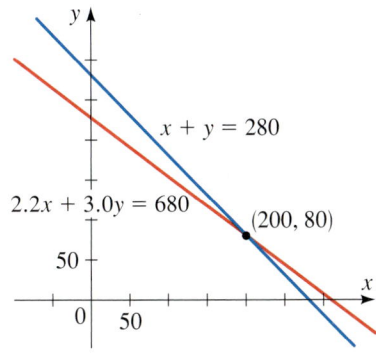

We can solve this system graphically. The point (200, 80) lies on the graph of each equation, so it satisfies both equations.

These equations *work together* to help us find x and y; neither equation alone can tell us the value of x or y. The values $x = 200$ and $y = 80$ satisfy *both* equations, so they form a solution of the system. Thus, the station sold 200 gallons of regular and 80 gallons of premium.

We can also represent a linear system by a rectangular array of numbers called a matrix. The *augmented matrix* of the above system is:

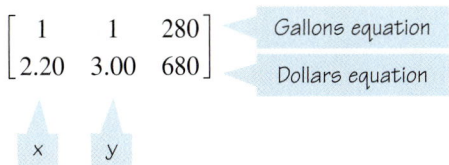

The augmented matrix contains the same information as the system, but in a simpler form. One of the important ideas in this chapter is to think of a matrix as a single object, so we denote a matrix by a single letter, such as A, B, C, and so on. We can add, subtract, and multiply matrices, just as we do ordinary numbers. We will pay special attention to matrix multiplication—it's defined in a way (which may seem complicated at first) that makes it possible to write a linear system as a single *matrix equation*

$$AX = B$$

where X is the unknown matrix. As you will see, solving this matrix equation for the matrix X is analogous to solving the algebraic equation $ax = b$ for the number x.

In this chapter we consider many uses of matrices, including applications to population growth (*Will the Species Survive?* page 688) and to computer graphics (*Computer Graphics I*, page 700).

9.1   Systems of Equations

In this section we study how to solve systems of two equations in two unknowns. We learn three different methods of solving such systems: by substitution, by elimination, and graphically.

Systems of Equations and Their Solutions

A **system of equations** is a set of equations that involve the same variables. A **solution** of a system is an assignment of values for the variables that makes *each* equation in the system true. To **solve** a system means to find all solutions of the system.

Here is an example of a system of two equations in two variables:

$$\begin{cases} 2x - y = 5 & \text{Equation 1} \\ x + 4y = 7 & \text{Equation 2} \end{cases}$$

We can check that $x = 3$ and $y = 1$ is a solution of this system.

Equation 1	**Equation 2**
$2x - y = 5$	$x + 4y = 7$
$2(3) - 1 = 5$ ✓	$3 + 4(1) = 7$ ✓

The solution can also be written as the ordered pair $(3, 1)$.

Note that the graphs of Equations 1 and 2 are lines (see Figure 1). Since the solution $(3, 1)$ satisfies each equation, the point $(3, 1)$ lies on each line. So it is the point of intersection of the two lines.

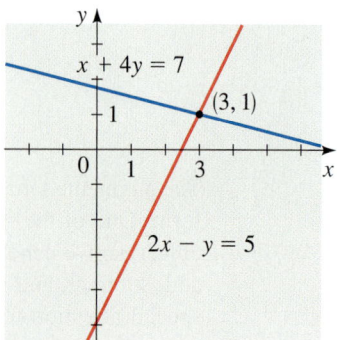

Figure 1

Substitution Method

In the **substitution method** we start with one equation in the system and solve for one variable in terms of the other variable. The following box describes the procedure.

Substitution Method

1. **Solve for One Variable.** Choose one equation and solve for one variable in terms of the other variable.

2. **Substitute.** Substitute the expression you found in Step 1 into the other equation to get an equation in one variable, then solve for that variable.

3. **Back-Substitute.** Substitute the value you found in Step 2 back into the expression found in Step 1 to solve for the remaining variable.

Example 1 Substitution Method

Find all solutions of the system.

$$\begin{cases} 2x + y = 1 & \text{Equation 1} \\ 3x + 4y = 14 & \text{Equation 2} \end{cases}$$

Solution We solve for y in the first equation.

Solve for one variable

$$y = 1 - 2x \qquad \text{Solve for } y \text{ in Equation 1}$$

Now we substitute for y in the second equation and solve for x:

Substitute

$$3x + 4(1 - 2x) = 14 \qquad \text{Substitute } y = 1 - 2x \text{ into Equation 2}$$
$$3x + 4 - 8x = 14 \qquad \text{Expand}$$
$$-5x + 4 = 14 \qquad \text{Simplify}$$
$$-5x = 10 \qquad \text{Subtract 4}$$
$$x = -2 \qquad \text{Solve for } x$$

Next we back-substitute $x = -2$ into the equation $y = 1 - 2x$:

Back-substitute

$$y = 1 - 2(-2) = 5 \qquad \text{Back-substitute}$$

Thus, $x = -2$ and $y = 5$, so the solution is the ordered pair $(-2, 5)$. Figure 2 shows that the graphs of the two equations intersect at the point $(-2, 5)$.

Check Your Answer

$x = -2, y = 5$:
$$\begin{cases} 2(-2) + 5 = 1 \\ 3(-2) + 4(5) = 14 \end{cases} \checkmark$$

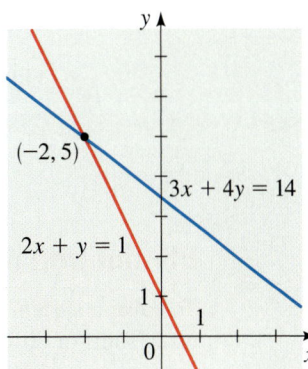

Figure 2

Example 2 Substitution Method

Find all solutions of the system.

$$\begin{cases} x^2 + y^2 = 100 & \text{Equation 1} \\ 3x - y = 10 & \text{Equation 2} \end{cases}$$

Solution We start by solving for y in the second equation.

| Solve for one variable |

$$y = 3x - 10 \qquad \text{Solve for } y \text{ in Equation 2}$$

Next we substitute for y in the first equation and solve for x:

| Substitute |

$$x^2 + (3x - 10)^2 = 100 \qquad \begin{array}{l}\text{Substitute } y = 3x - 10 \\ \text{into Equation 1}\end{array}$$

$$x^2 + (9x^2 - 60x + 100) = 100 \qquad \text{Expand}$$

$$10x^2 - 60x = 0 \qquad \text{Simplify}$$

$$10x(x - 6) = 0 \qquad \text{Factor}$$

$$x = 0 \quad \text{or} \quad x = 6 \qquad \text{Solve for } x$$

Now we back-substitute these values of x into the equation $y = 3x - 10$.

| Back-substitute |

For $x = 0$: $\quad y = 3(0) - 10 = -10 \qquad$ Back-substitute

For $x = 6$: $\quad y = 3(6) - 10 = 8 \qquad$ Back-substitute

So we have two solutions: $(0, -10)$ and $(6, 8)$.

The graph of the first equation is a circle, and the graph of the second equation is a line; Figure 3 shows that the graphs intersect at the two points $(0, -10)$ and $(6, 8)$.

Check Your Answers

$x = 0, y = -10$:

$$\begin{cases} (0)^2 + (-10)^2 = 100 \\ 3(0) - (-10) = 10 \end{cases} \quad ✓$$

$x = 6, y = 8$:

$$\begin{cases} (6)^2 + (8)^2 = 36 + 64 = 100 \\ 3(6) - (8) = 18 - 8 = 10 \end{cases} \quad ✓$$

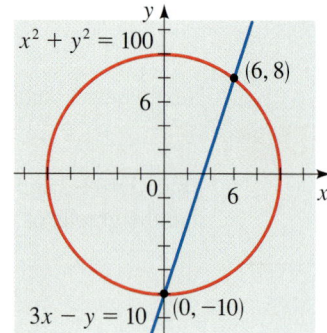

Figure 3

Elimination Method

To solve a system using the **elimination method**, we try to combine the equations using sums or differences so as to eliminate one of the variables.

Elimination Method

1. **Adjust the Coefficients.**   Multiply one or more of the equations by appropriate numbers so that the coefficient of one variable in one equation is the negative of its coefficient in the other equation.

2. **Add the Equations.**   Add the two equations to eliminate one variable, then solve for the remaining variable.

3. **Back-Substitute.**   Substitute the value you found in Step 2 back into one of the original equations, and solve for the remaining variable.

Example 3   Elimination Method

Find all solutions of the system.

$$\begin{cases} 3x + 2y = 14 & \text{Equation 1} \\ x - 2y = 2 & \text{Equation 2} \end{cases}$$

Solution   Since the coefficients of the y-terms are negatives of each other, we can add the equations to eliminate y.

$$\begin{cases} 3x + 2y = 14 \\ \underline{x - 2y = 2} \end{cases} \quad \text{System}$$

$$4x = 16 \quad \text{Add}$$

$$x = 4 \quad \text{Solve for } x$$

Now we back-substitute $x = 4$ into one of the original equations and solve for y. Let's choose the second equation because it looks simpler.

$$x - 2y = 2 \quad \text{Equation 2}$$

$$4 - 2y = 2 \quad \text{Back-substitute } x = 4 \text{ into Equation 2}$$

$$-2y = -2 \quad \text{Subtract 4}$$

$$y = 1 \quad \text{Solve for } y$$

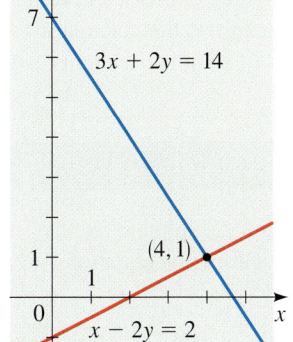

Figure 4

The solution is $(4, 1)$. Figure 4 shows that the graphs of the equations in the system intersect at the point $(4, 1)$. ∎

Example 4   Elimination Method

Find all solutions of the system.

$$\begin{cases} 3x^2 + 2y = 26 & \text{Equation 1} \\ 5x^2 + 7y = 3 & \text{Equation 2} \end{cases}$$

Solution   We choose to eliminate the x-term, so we multiply the first equation by 5 and the second equation by -3. Then we add the two equations and solve for y.

$$\begin{cases} 15x^2 + 10y = 130 & 5 \times \text{Equation 1} \\ \underline{-15x^2 - 21y = {-9}} & (-3) \times \text{Equation 2} \end{cases}$$

$$-11y = 121 \quad \text{Add}$$

$$y = -11 \quad \text{Solve for } y$$

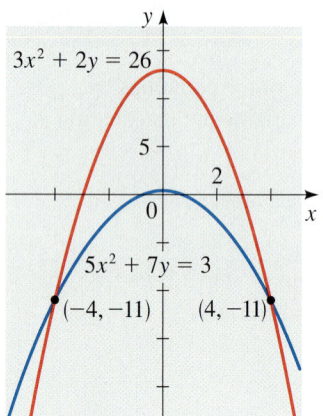

Figure 5

The graphs of quadratic functions $y = ax^2 + bx + c$ are called *parabolas*; see Section 2.5.

Now we back-substitute $y = -11$ into one of the original equations, say $3x^2 + 2y = 26$, and solve for x:

$$3x^2 + 2(-11) = 26 \qquad \text{Back-substitute } y = -11 \text{ into Equation 1}$$

$$3x^2 = 48 \qquad \text{Add 22}$$

$$x^2 = 16 \qquad \text{Divide by 3}$$

$$x = -4 \quad \text{or} \quad x = 4 \qquad \text{Solve for } x$$

So we have two solutions: $(-4, -11)$ and $(4, -11)$.

The graphs of both equations are parabolas; Figure 5 shows that the graphs intersect at the two points $(-4, -11)$ and $(4, -11)$. ∎

Check Your Answers

$x = -4, y = -11:$

$$\begin{cases} 3(-4)^2 + 2(-11) = 26 \\ 5(-4)^2 + 7(-11) = 3 \end{cases} \quad ✓$$

$x = 4, y = -11:$

$$\begin{cases} 3(4)^2 + 2(-11) = 26 \\ 5(4)^2 + 7(-11) = 3 \end{cases} \quad ✓$$

Graphical Method

In the **graphical method** we use a graphing device to solve the system of equations. Note that with many graphing devices, any equation must first be expressed in terms of one or more functions of the form $y = f(x)$ before we can use the calculator to graph it. Not all equations can be readily expressed in this way, so not all systems can be solved by this method.

Graphical Method

1. **Graph Each Equation.**   Express each equation in a form suitable for the graphing calculator by solving for y as a function of x. Graph the equations on the same screen.

2. **Find the Intersection Points.**   The solutions are the x- and y-coordinates of the points of intersection.

It may be more convenient to solve for x in terms of y in the equations. In that case, in Step 1 graph x as a function of y instead.

Example 5   **Graphical Method**

Find all solutions of the system.

$$\begin{cases} x^2 - y = 2 \\ 2x - y = -1 \end{cases}$$

Solution   Solving for y in terms of x, we get the equivalent system

$$\begin{cases} y = x^2 - 2 \\ y = 2x + 1 \end{cases}$$

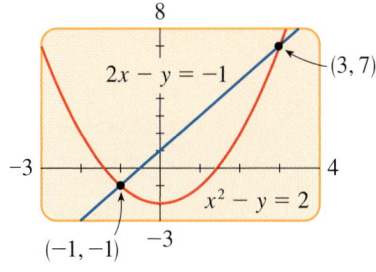

Figure 6

Figure 6 shows that the graphs of these equations intersect at two points. Zooming in, we see that the solutions are

$$(-1, -1) \quad \text{and} \quad (3, 7)$$

Check Your Answers

$x = -1, y = -1$:
$$\begin{cases} (-1)^2 - (-1) = 2 \\ 2(-1) - (-1) = -1 \end{cases} \checkmark$$

$x = 3, y = 7$:
$$\begin{cases} 3^2 - 7 = 2 \\ 2(3) - 7 = -1 \end{cases} \checkmark$$

Example 6   Solving a System of Equations Graphically

Find all solutions of the system, correct to one decimal place.

$$\begin{cases} x^2 + y^2 = 12 & \text{Equation 1} \\ y = 2x^2 - 5x & \text{Equation 2} \end{cases}$$

Solution The graph of the first equation is a circle and the second a parabola. To graph the circle on a graphing calculator, we must first solve for y in terms of x (see Section 2.3).

$$x^2 + y^2 = 12$$
$$y^2 = 12 - x^2 \qquad \text{Isolate } y^2 \text{ on LHS}$$
$$y = \pm\sqrt{12 - x^2} \qquad \text{Take square roots}$$

To graph the circle, we must graph both functions:

$$y = \sqrt{12 - x^2} \quad \text{and} \quad y = -\sqrt{12 - x^2}$$

In Figure 7 the graph of the circle is shown in red and the parabola in blue. The graphs intersect in quadrants I and II. Zooming in, or using the Intersect command, we see that the intersection points are $(-0.559, 3.419)$ and $(2.847, 1.974)$. There also appears to be an intersection point in quadrant IV. However, when we zoom in, we see that the curves come close to each other but don't intersect (see Figure 8). Thus, the system has two solutions; correct to the nearest tenth, they are

$$(-0.6, 3.4) \quad \text{and} \quad (2.8, 2.0)$$

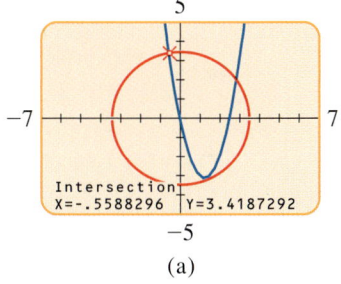

(a)

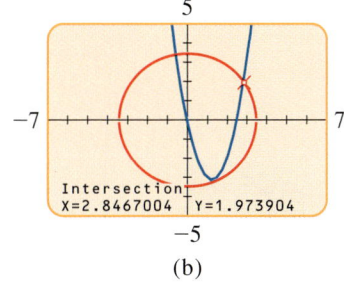

(b)

Figure 7

$x^2 + y^2 = 12, y = 2x^2 - 5x$

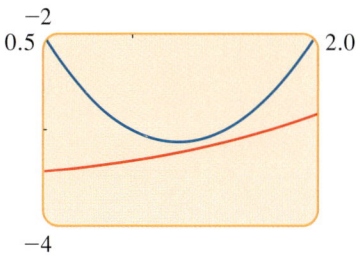

Figure 8

Zooming in

9.1   Exercises

1–8 ■ Use the substitution method to find all solutions of the system of equations.

1. $\begin{cases} x - y = 2 \\ 2x + 3y = 9 \end{cases}$ **2.** $\begin{cases} 2x + y = 7 \\ x + 2y = 2 \end{cases}$

3. $\begin{cases} y = x^2 \\ y = x + 12 \end{cases}$ **4.** $\begin{cases} x^2 + y^2 = 25 \\ y = 2x \end{cases}$

5. $\begin{cases} x^2 + y^2 = 8 \\ x + y = 0 \end{cases}$ **6.** $\begin{cases} x^2 + y = 9 \\ x - y + 3 = 0 \end{cases}$

7. $\begin{cases} x + y^2 = 0 \\ 2x + 5y^2 = 75 \end{cases}$ **8.** $\begin{cases} x^2 - y = 1 \\ 2x^2 + 3y = 17 \end{cases}$

9–16 ■ Use the elimination method to find all solutions of the system of equations.

9. $\begin{cases} x + 2y = 5 \\ 2x + 3y = 8 \end{cases}$ **10.** $\begin{cases} 4x - 3y = 11 \\ 8x + 4y = 12 \end{cases}$

11. $\begin{cases} x^2 - 2y = 1 \\ x^2 + 5y = 29 \end{cases}$ **12.** $\begin{cases} 3x^2 + 4y = 17 \\ 2x^2 + 5y = 2 \end{cases}$

13. $\begin{cases} 3x^2 - y^2 = 11 \\ x^2 + 4y^2 = 8 \end{cases}$ **14.** $\begin{cases} 2x^2 + 4y = 13 \\ x^2 - y^2 = \frac{7}{2} \end{cases}$

15. $\begin{cases} x - y^2 + 3 = 0 \\ 2x^2 + y^2 - 4 = 0 \end{cases}$ **16.** $\begin{cases} x^2 - y^2 = 1 \\ 2x^2 - y^2 = x + 3 \end{cases}$

17–22 ■ Two equations and their graphs are given. Find the intersection point(s) of the graphs by solving the system.

17. $\begin{cases} 2x + y = -1 \\ x - 2y = -8 \end{cases}$ **18.** $\begin{cases} x + y = 2 \\ 2x + y = 5 \end{cases}$

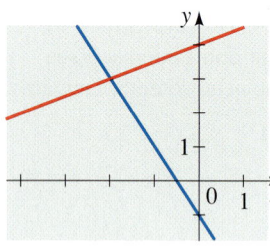   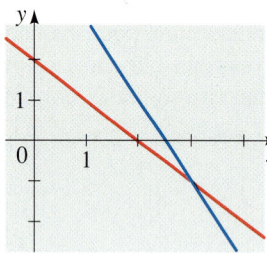

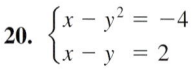

19. $\begin{cases} x^2 + y = 8 \\ x - 2y = -6 \end{cases}$ **20.** $\begin{cases} x - y^2 = -4 \\ x - y = 2 \end{cases}$

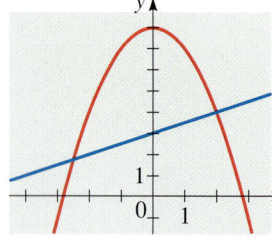   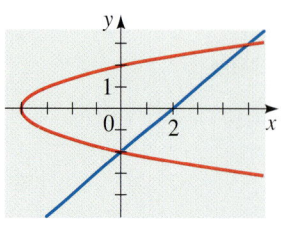

21. $\begin{cases} x^2 + y = 0 \\ x^3 - 2x - y = 0 \end{cases}$ **22.** $\begin{cases} x^2 + y^2 = 4x \\ x = y^2 \end{cases}$

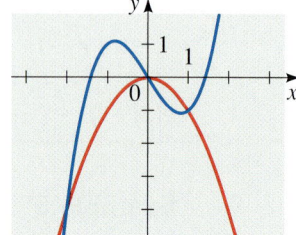   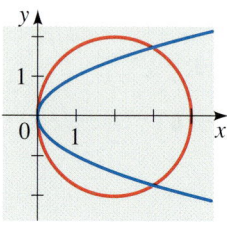

23–36 ■ Find all solutions of the system of equations.

23. $\begin{cases} y + x^2 = 4x \\ y + 4x = 16 \end{cases}$ **24.** $\begin{cases} x - y^2 = 0 \\ y - x^2 = 0 \end{cases}$

25. $\begin{cases} x - 2y = 2 \\ y^2 - x^2 = 2x + 4 \end{cases}$ **26.** $\begin{cases} y = 4 - x^2 \\ y = x^2 - 4 \end{cases}$

27. $\begin{cases} x - y = 4 \\ xy = 12 \end{cases}$ **28.** $\begin{cases} xy = 24 \\ 2x^2 - y^2 + 4 = 0 \end{cases}$

29. $\begin{cases} x^2y = 16 \\ x^2 + 4y + 16 = 0 \end{cases}$ **30.** $\begin{cases} x + \sqrt{y} = 0 \\ y^2 - 4x^2 = 12 \end{cases}$

31. $\begin{cases} x^2 + y^2 = 9 \\ x^2 - y^2 = 1 \end{cases}$ **32.** $\begin{cases} x^2 + 2y^2 = 2 \\ 2x^2 - 3y = 15 \end{cases}$

33. $\begin{cases} 2x^2 - 8y^3 = 19 \\ 4x^2 + 16y^3 = 34 \end{cases}$ **34.** $\begin{cases} x^4 - y^3 = 17 \\ 3x^4 + 5y^3 = 53 \end{cases}$

35. $\begin{cases} \dfrac{2}{x} - \dfrac{3}{y} = 1 \\ -\dfrac{4}{x} + \dfrac{7}{y} = 1 \end{cases}$ **36.** $\begin{cases} \dfrac{4}{x^2} + \dfrac{6}{y^4} = \dfrac{7}{2} \\ \dfrac{1}{x^2} - \dfrac{2}{y^4} = 0 \end{cases}$

37–46 ■ Use the graphical method to find all solutions of the system of equations, correct to two decimal places.

37. $\begin{cases} y = 2x + 6 \\ y = -x + 5 \end{cases}$ **38.** $\begin{cases} y = -2x + 12 \\ y = x + 3 \end{cases}$

39. $\begin{cases} y = x^2 + 8x \\ y = 2x + 16 \end{cases}$ **40.** $\begin{cases} y = x^2 - 4x \\ 2x - y = 2 \end{cases}$

41. $\begin{cases} x^2 + y^2 = 25 \\ x + 3y = 2 \end{cases}$ **42.** $\begin{cases} x^2 + y^2 = 17 \\ x^2 - 2x + y^2 = 13 \end{cases}$

43. $\begin{cases} \dfrac{x^2}{9} + \dfrac{y^2}{18} = 1 \\ y = -x^2 + 6x - 2 \end{cases}$ **44.** $\begin{cases} x^2 - y^2 = 3 \\ y = x^2 - 2x - 8 \end{cases}$

45. $\begin{cases} x^4 + 16y^4 = 32 \\ x^2 + 2x + y = 0 \end{cases}$ **46.** $\begin{cases} y = e^x + e^{-x} \\ y = 5 - x^2 \end{cases}$

Applications

47. Dimensions of a Rectangle A rectangle has an area of 180 cm^2 and a perimeter of 54 cm. What are its dimensions?

48. Legs of a Right Triangle A right triangle has an area of 84 ft^2 and a hypotenuse 25 ft long. What are the lengths of its other two sides?

49. Dimensions of a Rectangle The perimeter of a rectangle is 70 and its diagonal is 25. Find its length and width.

50. Dimensions of a Rectangle A circular piece of sheet metal has a diameter of 20 in. The edges are to be cut off to form a rectangle of area 160 in^2 (see the figure). What are the dimensions of the rectangle?

51. Flight of a Rocket A hill is inclined so that its "slope" is $\frac{1}{2}$, as shown in the figure. We introduce a coordinate system with the origin at the base of the hill and with the scales on the axes measured in meters. A rocket is fired from the base of the hill in such a way that its trajectory is the parabola $y = -x^2 + 401x$. At what point does the rocket strike the hillside? How far is this point from the base of the hill (to the nearest cm)?

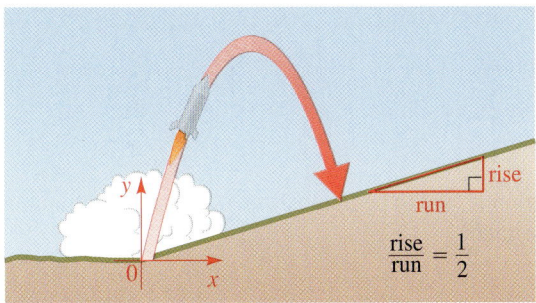

$$\frac{\text{rise}}{\text{run}} = \frac{1}{2}$$

52. Making a Stovepipe A rectangular piece of sheet metal with an area of 1200 in^2 is to be bent into a cylindrical length of stovepipe having a volume of 600 in^3. What are the dimensions of the sheet metal?

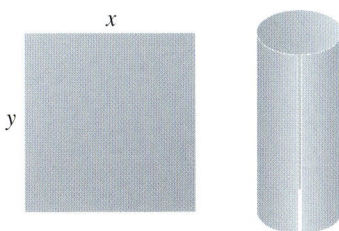

53. Global Positioning System (GPS) The Global Positioning System determines the location of an object from its distances to satellites in orbit around the earth. In the simplified, two-dimensional situation shown in the figure, determine the coordinates of P from the fact that P is 26 units from satellite A and 20 units from satellite B.

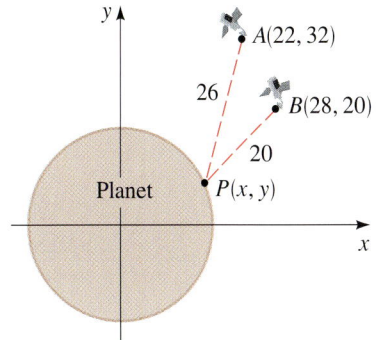

Discovery • Discussion

54. Intersection of a Parabola and a Line On a sheet of graph paper, or using a graphing calculator, draw the parabola $y = x^2$. Then draw the graphs of the linear equation $y = x + k$ on the same coordinate plane for various values of k. Try to choose values of k so that the line and the parabola intersect at two points for some of your k's, and not for others. For what value of k is there exactly one intersection point? Use the results of your experiment to make a conjecture about the values of k for which the following system has two solutions, one solution, and no solution. Prove your conjecture.

$$\begin{cases} y = x^2 \\ y = x + k \end{cases}$$

55. Some Trickier Systems Follow the hints and solve the systems.

(a) $\begin{cases} \log x + \log y = \frac{3}{2} \\ 2 \log x - \log y = 0 \end{cases}$ [*Hint:* Add the equations.]

(b) $\begin{cases} 2^x + 2^y = 10 \\ 4^x + 4^y = 68 \end{cases}$ [*Hint:* Note that $4^x = 2^{2x} = (2^x)^2$.]

(c) $\begin{cases} x - y = 3 \\ x^3 - y^3 = 387 \end{cases}$ [*Hint:* Factor the left side of the second equation.]

(d) $\begin{cases} x^2 + xy = 1 \\ xy + y^2 = 3 \end{cases}$ [*Hint:* Add the equations and factor the result.]

9.2 Systems of Linear Equations in Two Variables

Recall that an equation of the form $Ax + By = C$ is called linear because its graph is a line (see Section 1.10). In this section we study systems of two linear equations in two variables.

Systems of Linear Equations in Two Variables

A **system of two linear equations in two variables** has the form

$$\begin{cases} a_1 x + b_1 y = c_1 \\ a_2 x + b_2 y = c_2 \end{cases}$$

We can use either the substitution method or the elimination method to solve such systems algebraically. But since the elimination method is usually easier for linear systems, we use elimination rather than substitution in our examples.

The graph of a linear system in two variables is a pair of lines, so to solve the system graphically, we must find the intersection point(s) of the lines. Two lines may intersect in a single point, they may be parallel, or they may coincide, as shown in Figure 1. So there are three possible outcomes when solving such a system.

Number of Solutions of a Linear System in Two Variables

For a system of linear equations in two variables, exactly one of the following is true. (See Figure 1.)

1. The system has exactly one solution.

2. The system has no solution.

3. The system has infinitely many solutions.

A system that has no solution is said to be **inconsistent**. A system with infinitely many solutions is called **dependent**.

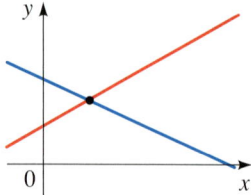

(a) Linear system with one solution. Lines intersect at a single point.

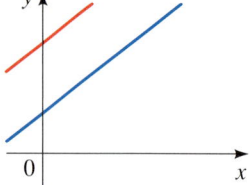

(b) Linear system with no solution. Lines are parallel—they do not intersect.

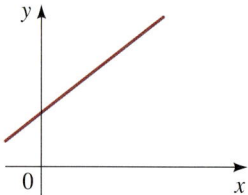

(c) Linear system with infinitely many solutions. Lines coincide—equations are for the same line.

Figure 1

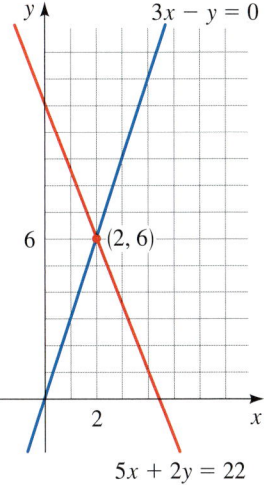

Figure 2

Check Your Answer

$x = 2, y = 6$:

$$\begin{cases} 3(2) - (6) = 0 \\ 5(2) + 2(6) = 22 \end{cases} \checkmark$$

Example 1   **A Linear System with One Solution**

Solve the system and graph the lines.

$$\begin{cases} 3x - y = 0 & \text{Equation 1} \\ 5x + 2y = 22 & \text{Equation 2} \end{cases}$$

Solution   We eliminate y from the equations and solve for x.

$$\begin{cases} 6x - 2y = 0 & \text{2 × Equation 1} \\ 5x + 2y = 22 \end{cases}$$
$$\begin{aligned} 11x &= 22 & \text{Add} \\ x &= 2 & \text{Solve for x} \end{aligned}$$

Now we back-substitute into the first equation and solve for y:

$$\begin{aligned} 6(2) - 2y &= 0 & \text{Back-substitute x = 2} \\ -2y &= -12 & \text{Subtract 6 × 2 = 12} \\ y &= 6 & \text{Solve for y} \end{aligned}$$

The solution of the system is the ordered pair $(2, 6)$, that is,

$$x = 2, \qquad y = 6$$

The graph in Figure 2 shows that the lines in the system intersect at the point $(2, 6)$.

Example 2   **A Linear System with No Solution**

Solve the system.

$$\begin{cases} 8x - 2y = 5 & \text{Equation 1} \\ -12x + 3y = 7 & \text{Equation 2} \end{cases}$$

Solution   This time we try to find a suitable combination of the two equations to eliminate the variable y. Multiplying the first equation by 3 and the second by 2 gives

$$\begin{cases} 24x - 6y = 15 & \text{3 × Equation 1} \\ -24x + 6y = 14 & \text{2 × Equation 2} \end{cases}$$
$$0 = 29 \qquad \text{Add}$$

Adding the two equations eliminates *both x and y* in this case, and we end up with $0 = 29$, which is obviously false. No matter what values we assign to x and y, we cannot make this statement true, so the system has *no solution*. Figure 3 shows that the lines in the system are parallel and do not intersect. The system is inconsistent.

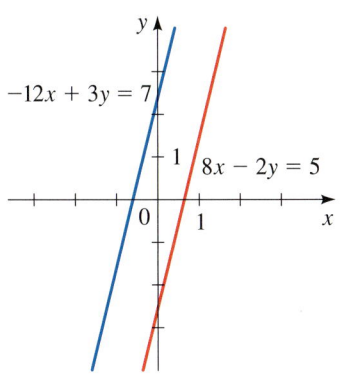

Figure 3

Example 3   **A Linear System with Infinitely Many Solutions**

Solve the system.

$$\begin{cases} 3x - 6y = 12 & \text{Equation 1} \\ 4x - 8y = 16 & \text{Equation 2} \end{cases}$$

Solution   We multiply the first equation by 4 and the second by 3 to prepare for subtracting the equations to eliminate x. The new equations are

$$\begin{cases} 12x - 24y = 48 & \text{4 × Equation 1} \\ 12x - 24y = 48 & \text{3 × Equation 2} \end{cases}$$

We see that the two equations in the original system are simply different ways of expressing the equation of one single line. The coordinates of any point on this line give a solution of the system. Writing the equation in slope-intercept form, we have $y = \frac{1}{2}x - 2$. So if we let t represent any real number, we can write the solution as

$$x = t$$
$$y = \tfrac{1}{2}t - 2$$

We can also write the solution in ordered-pair form as

$$\left(t, \tfrac{1}{2}t - 2\right)$$

where t is any real number. The system has infinitely many solutions (see Figure 4).

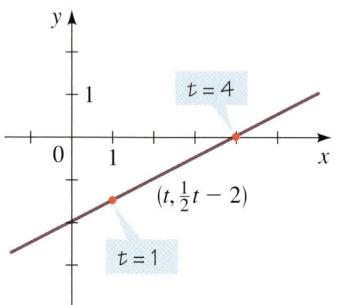

Figure 4

In Example 3, to get specific solutions we have to assign values to t. For instance, if $t = 1$, we get the solution $\left(1, -\frac{3}{2}\right)$. If $t = 4$, we get the solution $(4, 0)$. For every value of t we get a different solution. (See Figure 4.)

Modeling with Linear Systems

Frequently, when we use equations to solve problems in the sciences or in other areas, we obtain systems like the ones we've been considering. When modeling with systems of equations, we use the following guidelines, similar to those in Section 1.6.

Guidelines for Modeling with Systems of Equations

1. **Identify the Variables.**   Identify the quantities the problem asks you to find. These are usually determined by a careful reading of the question posed at the end of the problem. Introduce notation for the variables (call them x and y or some other letters).

2. **Express All Unknown Quantities in Terms of the Variables.**   Read the problem again and express all the quantities mentioned in the problem in terms of the variables you defined in Step 1.

3. **Set Up a System of Equations.**   Find the crucial facts in the problem that give the relationships between the expressions you found in Step 2. Set up a system of equations (or a model) that expresses these relationships.

4. **Solve the System and Interpret the Results.**   Solve the system you found in Step 3, check your solutions, and state your final answer as a sentence that answers the question posed in the problem.

The next two examples illustrate how to model with systems of equations.

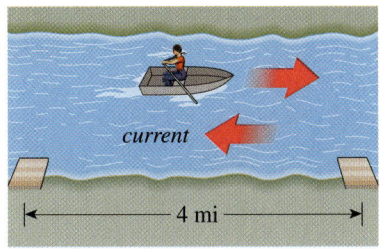

current

|← 4 mi →|

Identify the variables

Example 4   A Distance-Speed-Time Problem

A woman rows a boat upstream from one point on a river to another point 4 mi away in $1\frac{1}{2}$ hours. The return trip, traveling with the current, takes only 45 min. How fast does she row relative to the water, and at what speed is the current flowing?

Solution   We are asked to find the rowing speed and the speed of the current, so we let

$$x = \text{rowing speed (mi/h)}$$

$$y = \text{current speed (mi/h)}$$

The woman's speed when she rows upstream is her rowing speed minus the speed of the current; her speed downstream is her rowing speed plus the speed of the current. Now we translate this information into the language of algebra.

Express unknown quantities in terms of the variable

In Words	In Algebra
Rowing speed	x
Current speed	y
Speed upstream	$x - y$
Speed downstream	$x + y$

The distance upstream and downstream is 4 mi, so using the fact that speed × time = distance for both legs of the trip, we get

$$\text{speed upstream} \times \text{time upstream} = \text{distance traveled}$$

$$\text{speed downstream} \times \text{time downstream} = \text{distance traveled}$$

In algebraic notation this translates into the following equations.

$$(x - y)\tfrac{3}{2} = 4 \qquad \text{Equation 1}$$

$$(x + y)\tfrac{3}{4} = 4 \qquad \text{Equation 2}$$

Set up a system of equations

Solve the system

(The times have been converted to hours, since we are expressing the speeds in miles per *hour*.) We multiply the equations by 2 and 4, respectively, to clear the denominators.

$$\begin{cases} 3x - 3y = 8 & \text{2 × Equation 1} \\ 3x + 3y = 16 & \text{4 × Equation 2} \end{cases}$$

$$6x \qquad = 24 \qquad \text{Add}$$

$$x \qquad = 4 \qquad \text{Solve for x}$$

Back-substituting this value of x into the first equation (the second works just as well) and solving for y gives

$$3(4) - 3y = 8 \qquad \text{Back-substitute } x = 4$$

$$-3y = 8 - 12 \qquad \text{Subtract 12}$$

$$y = \tfrac{4}{3} \qquad \text{Solve for y}$$

The woman rows at 4 mi/h and the current flows at $1\frac{1}{3}$ mi/h. ∎

Check Your Answer

Speed upstream is

$$\frac{\text{distance}}{\text{time}} = \frac{4 \text{ mi}}{1\frac{1}{2}\text{ h}} = 2\tfrac{2}{3}\text{ mi/h}$$

and this should equal

rowing speed − current flow
$$= 4 \text{ mi/h} - \tfrac{4}{3}\text{ mi/h} = 2\tfrac{2}{3}\text{ mi/h}$$

Speed downstream is

$$\frac{\text{distance}}{\text{time}} = \frac{4 \text{ mi}}{\frac{3}{4}\text{ h}} = 5\tfrac{1}{3}\text{ mi/h}$$

and this should equal

rowing speed + current flow
$$= 4 \text{ mi/h} + \tfrac{4}{3}\text{ mi/h} = 5\tfrac{1}{3}\text{ mi/h} \quad ✓$$

Example 5 A Mixture Problem

A vintner fortifies wine that contains 10% alcohol by adding 70% alcohol solution to it. The resulting mixture has an alcoholic strength of 16% and fills 1000 one-liter bottles. How many liters (L) of the wine and of the alcohol solution does he use?

Solution Since we are asked for the amounts of wine and alcohol, we let

$$x = \text{amount of wine used (L)}$$
$$y = \text{amount of alcohol solution used (L)}$$

Identify the variables

From the fact that the wine contains 10% alcohol and the solution 70% alcohol, we get the following.

In Words	In Algebra
Amount of wine used (L)	x
Amount of alcohol solution used (L)	y
Amount of alcohol in wine (L)	$0.10x$
Amount of alcohol in solution (L)	$0.70y$

Express all unknown quantities in terms of the variable

The volume of the mixture must be the total of the two volumes the vintner is adding together, so

$$x + y = 1000$$

Also, the amount of alcohol in the mixture must be the total of the alcohol contributed by the wine and by the alcohol solution, that is

$$0.10x + 0.70y = (0.16)1000$$
$$0.10x + 0.70y = 160 \qquad \text{\color{blue}Simplify}$$
$$x + 7y = 1600 \qquad \text{\color{blue}Multiply by 10 to clear decimals}$$

Thus, we get the system

Set up a system of equations

$$\begin{cases} x + y = 1000 & \text{\color{blue}Equation 1} \\ x + 7y = 1600 & \text{\color{blue}Equation 2} \end{cases}$$

Subtracting the first equation from the second eliminates the variable x, and we get

Solve the system

$$6y = 600 \qquad \text{\color{blue}Subtract Equation 1 from Equation 2}$$
$$y = 100 \qquad \text{\color{blue}Solve for y}$$

We now back-substitute $y = 100$ into the first equation and solve for x:

$$x + 100 = 1000 \qquad \text{\color{blue}Back-substitute } y = 100$$
$$x = 900 \qquad \text{\color{blue}Solve for x}$$

The vintner uses 900 L of wine and 100 L of the alcohol solution. ∎

9.2 Exercises

1–6 ■ Graph each linear system, either by hand or using a graphing device. Use the graph to determine if the system has one solution, no solution, or infinitely many solutions. If there is exactly one solution, use the graph to find it.

1. $\begin{cases} x + y = 4 \\ 2x - y = 2 \end{cases}$
2. $\begin{cases} 2x + y = 11 \\ x - 2y = 4 \end{cases}$

3. $\begin{cases} 2x - 3y = 12 \\ -x + \frac{3}{2}y = 4 \end{cases}$
4. $\begin{cases} 2x + 6y = 0 \\ -3x - 9y = 18 \end{cases}$

5. $\begin{cases} -x + \frac{1}{2}y = -5 \\ 2x - y = 10 \end{cases}$
6. $\begin{cases} 12x + 15y = -18 \\ 2x + \frac{5}{2}y = -3 \end{cases}$

7–34 ■ Solve the system, or show that it has no solution. If the system has infinitely many solutions, express them in the ordered-pair form given in Example 3.

7. $\begin{cases} x + y = 4 \\ -x + y = 0 \end{cases}$
8. $\begin{cases} x - y = 3 \\ x + 3y = 7 \end{cases}$

9. $\begin{cases} 2x - 3y = 9 \\ 4x + 3y = 9 \end{cases}$
10. $\begin{cases} 3x + 2y = 0 \\ -x - 2y = 8 \end{cases}$

11. $\begin{cases} x + 3y = 5 \\ 2x - y = 3 \end{cases}$
12. $\begin{cases} x + y = 7 \\ 2x - 3y = -1 \end{cases}$

13. $\begin{cases} -x + y = 2 \\ 4x - 3y = -3 \end{cases}$
14. $\begin{cases} 4x - 3y = 28 \\ 9x - y = -6 \end{cases}$

15. $\begin{cases} x + 2y = 7 \\ 5x - y = 2 \end{cases}$
16. $\begin{cases} -4x + 12y = 0 \\ 12x + 4y = 160 \end{cases}$

17. $\begin{cases} \frac{1}{2}x + \frac{1}{3}y = 2 \\ \frac{1}{5}x - \frac{2}{3}y = 8 \end{cases}$
18. $\begin{cases} 0.2x - 0.2y = -1.8 \\ -0.3x + 0.5y = 3.3 \end{cases}$

19. $\begin{cases} 3x - 2y = 8 \\ -6x + 4y = 16 \end{cases}$
20. $\begin{cases} 4x + 2y = 16 \\ x - 5y = 70 \end{cases}$

21. $\begin{cases} x + 4y = 8 \\ 3x + 12y = 2 \end{cases}$
22. $\begin{cases} -3x + 5y = 2 \\ 9x - 15y = 6 \end{cases}$

23. $\begin{cases} 2x - 6y = 10 \\ -3x + 9y = -15 \end{cases}$
24. $\begin{cases} 2x - 3y = -8 \\ 14x - 21y = 3 \end{cases}$

25. $\begin{cases} 6x + 4y = 12 \\ 9x + 6y = 18 \end{cases}$
26. $\begin{cases} 25x - 75y = 100 \\ -10x + 30y = -40 \end{cases}$

27. $\begin{cases} 8s - 3t = -3 \\ 5s - 2t = -1 \end{cases}$
28. $\begin{cases} u - 30v = -5 \\ -3u + 80v = 5 \end{cases}$

29. $\begin{cases} \frac{1}{2}x + \frac{3}{5}y = 3 \\ \frac{5}{3}x + 2y = 10 \end{cases}$
30. $\begin{cases} \frac{3}{2}x - \frac{1}{3}y = \frac{1}{2} \\ 2x - \frac{1}{2}y = -\frac{1}{2} \end{cases}$

31. $\begin{cases} 0.4x + 1.2y = 14 \\ 12x - 5y = 10 \end{cases}$
32. $\begin{cases} 26x - 10y = -4 \\ -0.6x + 1.2y = 3 \end{cases}$

33. $\begin{cases} \frac{1}{3}x - \frac{1}{4}y = 2 \\ -8x + 6y = 10 \end{cases}$
34. $\begin{cases} -\frac{1}{10}x + \frac{1}{2}y = 4 \\ 2x - 10y = -80 \end{cases}$

35–38 ■ Use a graphing device to graph both lines in the same viewing rectangle. (Note that you must solve for y in terms of x before graphing if you are using a graphing calculator.) Solve the system correct to two decimal places, either by zooming in and using [TRACE] or by using [Intersect].

35. $\begin{cases} 0.21x + 3.17y = 9.51 \\ 2.35x - 1.17y = 5.89 \end{cases}$

36. $\begin{cases} 18.72x - 14.91y = 12.33 \\ 6.21x - 12.92y = 17.82 \end{cases}$

37. $\begin{cases} 2371x - 6552y = 13,591 \\ 9815x + 992y = 618,555 \end{cases}$

38. $\begin{cases} -435x + 912y = 0 \\ 132x + 455y = 994 \end{cases}$

39–42 ■ Find x and y in terms of a and b.

39. $\begin{cases} x + y = 0 \\ x + ay = 1 \end{cases}$ $(a \neq 1)$

40. $\begin{cases} ax + by = 0 \\ x + y = 1 \end{cases}$ $(a \neq b)$

41. $\begin{cases} ax + by = 1 \\ bx + ay = 1 \end{cases}$ $(a^2 - b^2 \neq 0)$

42. $\begin{cases} ax + by = 0 \\ a^2x + b^2y = 1 \end{cases}$ $(a \neq 0, b \neq 0, a \neq b)$

Applications

43. Number Problem Find two numbers whose sum is 34 and whose difference is 10.

44. Number Problem The sum of two numbers is twice their difference. The larger number is 6 more than twice the smaller. Find the numbers.

45. Value of Coins A man has 14 coins in his pocket, all of which are dimes and quarters. If the total value of his change is $2.75, how many dimes and how many quarters does he have?

46. Admission Fees The admission fee at an amusement park is $1.50 for children and $4.00 for adults. On a certain day, 2200 people entered the park, and the admission fees collected totaled $5050. How many children and how many adults were admitted?

47. Airplane Speed A man flies a small airplane from Fargo to Bismarck, North Dakota—a distance of 180 mi. Because he is flying into a head wind, the trip takes him 2 hours. On the way back, the wind is still blowing at the same speed, so the return trip takes only 1 h 12 min. What is his speed in still air, and how fast is the wind blowing?

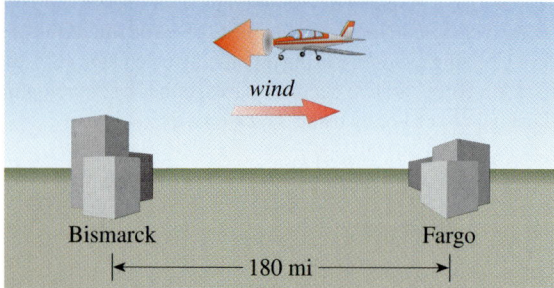

48. Boat Speed A boat on a river travels downstream between two points, 20 mi apart, in one hour. The return trip against the current takes $2\frac{1}{2}$ hours. What is the boat's speed, and how fast does the current in the river flow?

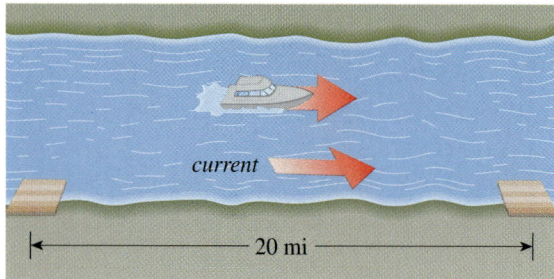

49. Aerobic Exercise A woman keeps fit by bicycling and running every day. On Monday she spends $\frac{1}{2}$ hour at each activity, covering a total of $12\frac{1}{2}$ mi. On Tuesday, she runs for 12 min and cycles for 45 min, covering a total of 16 mi. Assuming her running and cycling speeds don't change from day to day, find these speeds.

50. Mixture Problem A biologist has two brine solutions, one containing 5% salt and another containing 20% salt. How many milliliters of each solution should he mix to obtain 1 L of a solution that contains 14% salt?

51. Nutrition A researcher performs an experiment to test a hypothesis that involves the nutrients niacin and retinol. She feeds one group of laboratory rats a daily diet of precisely 32 units of niacin and 22,000 units of retinol. She uses two types of commercial pellet foods. Food A contains 0.12 unit of niacin and 100 units of retinol per gram. Food B contains 0.20 unit of niacin and 50 units of retinol per gram. How many grams of each food does she feed this group of rats each day?

52. Coffee Mixtures A customer in a coffee shop purchases a blend of two coffees: Kenyan, costing $3.50 a pound, and Sri Lankan, costing $5.60 a pound. He buys 3 lb of the blend, which costs him $11.55. How many pounds of each kind went into the mixture?

53. Mixture Problem A chemist has two large containers of sulfuric acid solution, with different concentrations of acid in each container. Blending 300 mL of the first solution and 600 mL of the second gives a mixture that is 15% acid, whereas 100 mL of the first mixed with 500 mL of the second gives a $12\frac{1}{2}$% acid mixture. What are the concentrations of sulfuric acid in the original containers?

54. Investments A woman invests a total of $20,000 in two accounts, one paying 5% and the other paying 8% simple interest per year. Her annual interest is $1180. How much did she invest at each rate?

55. Investments A man invests his savings in two accounts, one paying 6% and the other paying 10% simple interest per year. He puts twice as much in the lower-yielding account because it is less risky. His annual interest is $3520. How much did he invest at each rate?

56. Distance, Speed, and Time John and Mary leave their house at the same time and drive in opposite directions. John drives at 60 mi/h and travels 35 mi farther than Mary, who drives at 40 mi/h. Mary's trip takes 15 min longer than John's. For what length of time does each of them drive?

57. Number Problem The sum of the digits of a two-digit number is 7. When the digits are reversed, the number is increased by 27. Find the number.

58. Area of a Triangle Find the area of the triangle that lies in the first quadrant (with its base on the x-axis) and that is bounded by the lines $y = 2x - 4$ and $y = -4x + 20$.

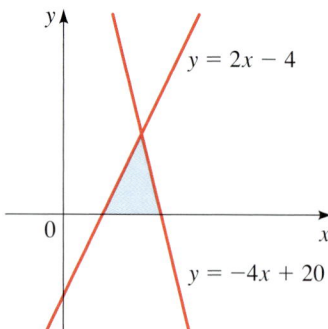

Discovery • Discussion

59. The Least Squares Line The *least squares* line or *regression* line is the line that best fits a set of points in the plane. We studied this line in *Focus on Modeling* (see page 240). Using calculus, it can be shown that the line that best fits the n data points (x_1, y_1), (x_2, y_2), . . . ,(x_n, y_n) is the line $y = ax + b$, where the coefficients a and b satisfy the following pair of linear equations. [The notation $\sum_{k=1}^{n} x_k$ stands for the sum of all the x's. See Section 11.1 for a complete description of sigma ($\sum$) notation.]

$$\left(\sum_{k=1}^{n} x_k \right) a + nb = \sum_{k=1}^{n} y_k$$

$$\left(\sum_{k=1}^{n} x_k^2 \right) a + \left(\sum_{k=1}^{n} x_k \right) b = \sum_{k=1}^{n} x_k y_k$$

Use these equations to find the least squares line for the following data points.

$$(1, 3), \quad (2, 5), \quad (3, 6), \quad (5, 6), \quad (7, 9)$$

Sketch the points and your line to confirm that the line fits these points well. If your calculator computes regression lines, see whether it gives you the same line as the formulas.

9.3 Systems of Linear Equations in Several Variables

A **linear equation in n variables** is an equation that can be put in the form

$$a_1 x_1 + a_2 x_2 + \cdots + a_n x_n = c$$

where $a_1, a_2, \ldots, a_n$ and c are real numbers, and $x_1, x_2, \ldots, x_n$ are the variables. If we have only three or four variables, we generally use x, y, z, and w instead of x_1, x_2, x_3, and x_4. Such equations are called *linear* because if we have just two variables the equation is $a_1 x + a_2 y = c$, which is the equation of a line. Here are some examples of equations in three variables that illustrate the difference between linear and nonlinear equations.

Linear equations	**Nonlinear equations**	
$6x_1 - 3x_2 + \sqrt{5} x_3 = 10$	$x^2 + 3y - \sqrt{z} = 5$	Not linear because it contains the square and the square root of a variable.
$x + y + z = 2w - \frac{1}{2}$	$x_1 x_2 + 6x_3 = -6$	Not linear because it contains a product of variables.

In this section we study systems of linear equations in three or more variables.

Solving a Linear System

The following are two examples of systems of linear equations in three variables. The second system is in **triangular form**; that is, the variable x doesn't appear in the second equation and the variables x and y do not appear in the third equation.

A system of linear equations

$$\begin{cases} x - 2y - z = 1 \\ -x + 3y + 3z = 4 \\ 2x - 3y + z = 10 \end{cases}$$

A system in triangular form

$$\begin{cases} x - 2y - z = 1 \\ y + 2z = 5 \\ z = 3 \end{cases}$$

It's easy to solve a system that is in triangular form using back-substitution. So, our goal in this section is to start with a system of linear equations and change it to a system in triangular form that has the same solutions as the original system. We begin by showing how to use back-substitution to solve a system that is already in triangular form.

Pierre de Fermat (1601–1665) was a French lawyer who became interested in mathematics at the age of 30. Because of his job as a magistrate, Fermat had little time to write complete proofs of his discoveries and often wrote them in the margin of whatever book he was reading at the time. After his death, his copy of Diophantus' *Arithmetica* (see page 20) was found to contain a particularly tantalizing comment. Where Diophantus discusses the solutions of $x^2 + y^2 = z^2$ (for example, $x = 3$, $y = 4$, $z = 5$), Fermat states in the margin that for $n \geq 3$ there are no natural number solutions to the equation $x^n + y^n = z^n$. In other words, it's impossible for a cube to equal the sum of two cubes, a fourth power to equal the sum of two fourth powers, and so on. Fermat writes "I have discovered a truly wonderful proof for this but the margin is too small to contain it." All the other margin comments in Fermat's copy of *Arithmetica* have been proved. This one, however, remained unproved, and it came to be known as "Fermat's Last Theorem."

In 1994, Andrew Wiles of Princeton University announced a proof of Fermat's Last Theorem, an astounding 350 years after it was conjectured. His proof is one of the most widely reported mathematical results in the popular press.

Example 1   Solving a Triangular System Using Back-Substitution

Solve the system using back-substitution:

$$\begin{cases} x - 2y - z = 1 & \text{Equation 1} \\ y + 2z = 5 & \text{Equation 2} \\ z = 3 & \text{Equation 3} \end{cases}$$

Solution   From the last equation we know that $z = 3$. We back-substitute this into the second equation and solve for y.

$$y + 2(3) = 5 \quad \text{Back-substitute } z = 3 \text{ into Equation 2}$$
$$y = -1 \quad \text{Solve for } y$$

Then we back-substitute $y = -1$ and $z = 3$ into the first equation and solve for x.

$$x - 2(-1) - (3) = 1 \quad \text{Back-substitute } y = -1 \text{ and } z = 3 \text{ into Equation 1}$$
$$x = 2 \quad \text{Solve for } x$$

The solution of the system is $x = 2$, $y = -1$, $z = 3$. We can also write the solution as the ordered triple $(2, -1, 3)$. ∎

To change a system of linear equations to an **equivalent system** (that is, a system with the same solutions as the original system), we use the elimination method. This means we can use the following operations.

Operations That Yield an Equivalent System

1. Add a nonzero multiple of one equation to another.
2. Multiply an equation by a nonzero constant.
3. Interchange the positions of two equations.

To solve a linear system, we use these operations to change the system to an equivalent triangular system. Then we use back-substitution as in Example 1. This process is called **Gaussian elimination**.

Example 2   Solving a System of Three Equations in Three Variables

Solve the system using Gaussian elimination.

$$\begin{cases} x - 2y + 3z = 1 & \text{Equation 1} \\ x + 2y - z = 13 & \text{Equation 2} \\ 3x + 2y - 5z = 3 & \text{Equation 3} \end{cases}$$

Solution We need to change this to a triangular system, so we begin by eliminating the x-term from the second equation.

$$
\begin{array}{llll}
x + 2y - z = 13 & \text{\color{blue}Equation 2} \\
\underline{x - 2y + 3z = 1} & \text{\color{blue}Equation 1} \\
4y - 4z = 12 & \text{\color{blue}Equation 2 + (−1) × Equation 1 = new Equation 2}
\end{array}
$$

This gives us a new, equivalent system that is one step closer to triangular form:

$$
\begin{cases}
x - 2y + 3z = 1 & \text{\color{blue}Equation 1} \\
4y - 4z = 12 & \text{\color{blue}Equation 2} \\
3x + 2y - 5z = 3 & \text{\color{blue}Equation 3}
\end{cases}
$$

Now we eliminate the x-term from the third equation.

$$
\begin{array}{lll}
3x + 2y - 5z = 3 \\
\underline{-3x + 6y - 9z = -3} \\
8y - 14z = 0
\end{array}
$$

$$
\begin{cases}
x - 2y + 3z = 1 \\
4y - 4z = 12 \\
8y - 14z = 0 & \text{\color{blue}Equation 3 + (−3) × Equation 1 = new Equation 3}
\end{cases}
$$

Then we eliminate the y-term from the third equation.

$$
\begin{array}{lll}
8y - 14z = 0 \\
\underline{-8y + 8z = -24} \\
-6z = -24
\end{array}
$$

$$
\begin{cases}
x - 2y + 3z = 1 \\
4y - 4z = 12 \\
-6z = -24 & \text{\color{blue}Equation 3 + (−2) × Equation 1 = new Equation 3}
\end{cases}
$$

The system is now in triangular form, but it will be easier to work with if we divide the second and third equations by the common factors of each term.

$$
\begin{cases}
x - 2y + 3z = 1 \\
y - z = 3 & \text{\color{blue}$\frac{1}{4}$ × Equation 2 = new Equation 2} \\
z = 4 & \text{\color{blue}$-\frac{1}{6}$ × Equation 3 = new Equation 3}
\end{cases}
$$

Now we use back-substitution to solve the system. From the third equation we get $z = 4$. We back-substitute this into the second equation and solve for y.

$$
\begin{array}{ll}
y - (4) = 3 & \text{\color{blue}Back-substitute $z = 4$ into Equation 2} \\
 y = 7 & \text{\color{blue}Solve for y}
\end{array}
$$

Then we back-substitute $y = 7$ and $z = 4$ into the first equation and solve for x.

$$
\begin{array}{ll}
x - 2(7) + 3(4) = 1 & \text{\color{blue}Back-substitute $y = 7$ and $z = 4$ into Equation 1} \\
 x = 3 & \text{\color{blue}Solve for x}
\end{array}
$$

The solution of the system is $x = 3$, $y = 7$, $z = 4$, which we can write as the ordered triple $(3, 7, 4)$. ∎

Check Your Answer

We must check that the answer satisfies *all three* equations, $x = 3$, $y = 7$, $z = 4$:

$$
\begin{array}{l}
(3) - 2(7) + 3(4) = 1 \\
(3) + 2(7) - (4) = 13 \\
3(3) + 2(7) - 5(4) = 3 \quad \checkmark
\end{array}
$$

The Number of Solutions of a Linear System

Intersection of Three Planes

When you study calculus or linear algebra, you will learn that the graph of a linear equation in three variables is a *plane* in a three-dimensional coordinate system. For a system of three equations in three variables, the following situations arise:

1. The three planes intersect in a single point.

The system has a unique solution.

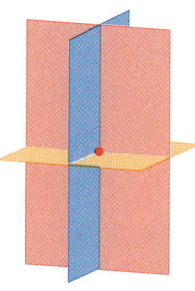

2. The three planes intersect in more than one point.

The system has infinitely many solutions.

3. The three planes have no point in common.

The system has no solution.

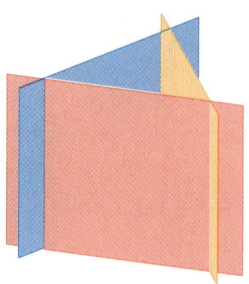

Just as in the case of two variables, a system of equations in several variables may have one solution, no solution, or infinitely many solutions. The graphical interpretation of the solutions of a linear system is analogous to that for systems of equations in two variables (see the margin note).

Number of Solutions of a Linear System

For a system of linear equations, exactly one of the following is true.

1. The system has exactly one solution.

2. The system has no solution.

3. The system has infinitely many solutions.

A system with no solutions is said to be **inconsistent**, and a system with infinitely many solutions is said to be **dependent**. As we see in the next example, a linear system has no solution if we end up with a *false equation* after applying Gaussian elimination to the system.

Example 3 A System with No Solution

Solve the following system.

$$\begin{cases} x + 2y - 2z = 1 & \text{Equation 1} \\ 2x + 2y - z = 6 & \text{Equation 2} \\ 3x + 4y - 3z = 5 & \text{Equation 3} \end{cases}$$

Solution To put this in triangular form, we begin by eliminating the x-terms from the second equation and the third equation.

$$\begin{cases} x + 2y - 2z = 1 \\ \quad\; -2y + 3z = 4 & \text{Equation 2} + (-2) \times \text{Equation 1} = \text{new Equation 2} \\ 3x + 4y - 3z = 5 \end{cases}$$

$$\begin{cases} x + 2y - 2z = 1 \\ \quad\; -2y + 3z = 4 \\ \quad\; -2y + 3z = 2 & \text{Equation 3} + (-3) \times \text{Equation 1} = \text{new Equation 3} \end{cases}$$

Now we eliminate the y-term from the third equation.

$$\begin{cases} x + 2y - 2z = \;\; 1 \\ \quad\; -2y + 3z = \;\; 4 \\ \qquad\qquad\; 0 = -2 & \text{Equation 3} + (-1) \times \text{Equation 2} = \text{new Equation 3} \end{cases}$$

The system is now in triangular form, but the third equation says $0 = -2$, which is false. No matter what values we assign x, y, and z, the third equation will never be true. This means the system has *no solution*. ∎

Example 4 A System with Infinitely Many Solutions

Solve the following system.

$$\begin{cases} x - y + 5z = -2 & \text{Equation 1} \\ 2x + y + 4z = 2 & \text{Equation 2} \\ 2x + 4y - 2z = 8 & \text{Equation 3} \end{cases}$$

Solution To put this in triangular form, we begin by eliminating the x-terms from the second equation and the third equation.

$$\begin{cases} x - y + 5z = -2 \\ 3y - 6z = 6 & \text{Equation 2} + (-2) \times \text{Equation 1} = \text{new Equation 2} \\ 2x + 4y - 2z = 8 \end{cases}$$

$$\begin{cases} x - y + 5z = -2 \\ 3y - 6z = 6 \\ 6y - 12z = 12 & \text{Equation 3} + (-2) \times \text{Equation 1} = \text{new Equation 3} \end{cases}$$

Now we eliminate the y-term from the third equation.

$$\begin{cases} x - y + 5z = -2 \\ 3y - 6z = 6 \\ 0 = 0 & \text{Equation 3} + (-2) \times \text{Equation 2} = \text{new Equation 3} \end{cases}$$

The new third equation is true, but it gives us no new information, so we can drop it from the system. Only two equations are left. We can use them to solve for x and y in terms of z, but z can take on any value, so there are infinitely many solutions.

To find the complete solution of the system we begin by solving for y in terms of z, using the new second equation.

$$3y - 6z = 6 \qquad \text{Equation 2}$$
$$y - 2z = 2 \qquad \text{Multiply by } \tfrac{1}{3}$$
$$y = 2z + 2 \qquad \text{Solve for } y$$

Then we solve for x in terms of z, using the first equation.

$$x - (2z + 2) + 5z = -2 \qquad \text{Substitute } y = 2z + 2 \text{ into Equation 1}$$
$$x + 3z - 2 = -2 \qquad \text{Simplify}$$
$$x = -3z \qquad \text{Solve for } x$$

To describe the complete solution, we let t represent any real number. The solution is

$$x = -3t$$
$$y = 2t + 2$$
$$z = t$$

We can also write this as the ordered triple $(-3t, 2t + 2, t)$. ■

In the solution of Example 4 the variable t is called a **parameter**. To get a specific solution, we give a specific value to the parameter t. For instance, if we set $t = 2$, we get

$$x = -3(2) = -6$$
$$y = 2(2) + 2 = 6$$
$$z = 2$$

Thus, $(-6, 6, 2)$ is a solution of the system. Here are some other solutions of the system obtained by substituting other values for the parameter t.

Parameter t	Solution $(-3t, 2t + 2, t)$
-1	$(3, 0, -1)$
0	$(0, 2, 0)$
3	$(-9, 8, 3)$
10	$(-30, 22, 10)$

You should check that these points satisfy the original equations. There are infinitely many choices for the parameter t, so the system has infinitely many solutions.

Modeling Using Linear Systems

Linear systems are used to model situations that involve several varying quantities. In the next example we consider an application of linear systems to finance.

Example 5   Modeling a Financial Problem Using a Linear System

John receives an inheritance of $50,000. His financial advisor suggests that he invest this in three mutual funds: a money-market fund, a blue-chip stock fund, and a high-tech stock fund. The advisor estimates that the money-market fund will return 5% over the next year, the blue-chip fund 9%, and the high-tech fund 16%. John wants a total first-year return of $4000. To avoid excessive risk, he decides to invest three times as much in the money-market fund as in the high-tech stock fund. How much should he invest in each fund?

Solution   Let

x = amount invested in the money-market fund

y = amount invested in the blue-chip stock fund

z = amount invested in the high-tech stock fund

We convert each fact given in the problem into an equation.

$$x + y + z = 50{,}000 \quad \text{Total amount invested is \$50,000}$$
$$0.05x + 0.09y + 0.16z = 4000 \quad \text{Total investment return is \$4000}$$
$$x = 3z \quad \text{Money-market amount is 3 × high-tech amount}$$

time of radio signals emitted from the satellite. Knowing the distance to three different satellites tells us that we are at the point of intersection of three different spheres. This uniquely determines our position (see Exercise 53, page 643).

Multiplying the second equation by 100 and rewriting the third gives the following system, which we solve using Gaussian elimination.

$$\begin{cases} x + y + z = 50{,}000 \\ 5x + 9y + 16z = 400{,}000 \\ x \quad\quad - 3z = 0 \end{cases}$$
  $100 \times$ Equation 2
  Subtract $3z$

$$\begin{cases} x + y + z = 50{,}000 \\ 4y + 11z = 150{,}000 \\ -y - 4z = -50{,}000 \end{cases}$$
  Equation 2 + $(-5) \times$ Equation 1 = new Equation 2
  Equation 3 + $(-1) \times$ Equation 1 = new Equation 3

$$\begin{cases} x + y + z = 50{,}000 \\ - 5z = -50{,}000 \\ - y - 4z = -50{,}000 \end{cases}$$
  Equation 2 + $4 \times$ Equation 3 = new Equation 2

$$\begin{cases} x + y + z = 50{,}000 \\ z = 10{,}000 \\ y + 4z = 50{,}000 \end{cases}$$
  $\left(-\frac{1}{5}\right) \times$ Equation 2
  $(-1) \times$ Equation 3

$$\begin{cases} x + y + z = 50{,}000 \\ y + 4z = 50{,}000 \\ z = 10{,}000 \end{cases}$$
  Interchange Equations 2 and 3

Now that the system is in triangular form, we use back-substitution to find that $x = 30{,}000$, $y = 10{,}000$, and $z = 10{,}000$. This means that John should invest

$30,000 in the money market fund

$10,000 in the blue-chip stock fund

$10,000 in the high-tech stock fund ■

9.3   Exercises

1–4 ■ State whether the equation or system of equations is linear.

1. $6x - \sqrt{3}y + \frac{1}{2}z = 0$

2. $x^2 + y^2 + z^2 = 4$

3. $\begin{cases} xy - 3y + z = 5 \\ x - y^2 + 5z = 0 \\ 2x + yz = 3 \end{cases}$

4. $\begin{cases} x - 2y + 3z = 10 \\ 2x + 5y = 2 \\ y + 2z = 4 \end{cases}$

5–10 ■ Use back-substitution to solve the triangular system.

5. $\begin{cases} x - 2y + 4z = 3 \\ y + 2z = 7 \\ z = 2 \end{cases}$

6. $\begin{cases} x + y - 3z = 8 \\ y - 3z = 5 \\ z = -1 \end{cases}$

7. $\begin{cases} x + 2y + z = 7 \\ -y + 3z = 9 \\ 2z = 6 \end{cases}$

8. $\begin{cases} x - 2y + 3z = 10 \\ 2y - z = 2 \\ 3z = 12 \end{cases}$

9. $\begin{cases} 2x - y + 6z = 5 \\ y + 4z = 0 \\ -2z = 1 \end{cases}$

10. $\begin{cases} 4x + 3z = 10 \\ 2y - z = -6 \\ \frac{1}{2}z = 4 \end{cases}$

11–14 ■ Perform an operation on the given system that eliminates the indicated variable. Write the new equivalent system.

11. $\begin{cases} x - 2y - z = 4 \\ x - y + 3z = 0 \\ 2x + y + z = 0 \end{cases}$ Eliminate the x-term from the second equation.

12. $\begin{cases} x + y - 3z = 3 \\ -2x + 3y + z = 2 \\ x - y + 2z = 0 \end{cases}$ Eliminate the x-term from the second equation.

13. $\begin{cases} 2x - y + 3z = 2 \\ x + 2y - z = 4 \\ -4x + 5y + z = 10 \end{cases}$ Eliminate the x-term from the third equation.

14. $\begin{cases} x - 4y + z = 3 \\ y - 3z = 10 \\ 3y - 8z = 24 \end{cases}$ Eliminate the y-term from the third equation.

15–32 ■ Find the complete solution of the linear system, or show that it is inconsistent.

15. $\begin{cases} x + y + z = 4 \\ x + 3y + 3z = 10 \\ 2x + y - z = 3 \end{cases}$

16. $\begin{cases} x + y + z = 0 \\ -x + 2y + 5z = 3 \\ 3x - y = 6 \end{cases}$

17. $\begin{cases} x - 4z = 1 \\ 2x - y - 6z = 4 \\ 2x + 3y - 2z = 8 \end{cases}$

18. $\begin{cases} x - y + 2z = 2 \\ 3x + y + 5z = 8 \\ 2x - y - 2z = -7 \end{cases}$

19. $\begin{cases} 2x + 4y - z = 2 \\ x + 2y - 3z = -4 \\ 3x - y + z = 1 \end{cases}$

20. $\begin{cases} 2x + y - z = -8 \\ -x + y + z = 3 \\ -2x + 4z = 18 \end{cases}$

21. $\begin{cases} y - 2z = 0 \\ 2x + 3y = 2 \\ -x - 2y + z = -1 \end{cases}$

22. $\begin{cases} 2y + z = 3 \\ 5x + 4y + 3z = -1 \\ x - 3y = -2 \end{cases}$

23. $\begin{cases} x + 2y - z = 1 \\ 2x + 3y - 4z = -3 \\ 3x + 6y - 3z = 4 \end{cases}$

24. $\begin{cases} -x + 2y + 5z = 4 \\ x - 2z = 0 \\ 4x - 2y - 11z = 2 \end{cases}$

25. $\begin{cases} 2x + 3y - z = 1 \\ x + 2y = 3 \\ x + 3y + z = 4 \end{cases}$

26. $\begin{cases} x - 2y - 3z = 5 \\ 2x + y - z = 5 \\ 4x - 3y - 7z = 5 \end{cases}$

27. $\begin{cases} x + y - z = 0 \\ x + 2y - 3z = -3 \\ 2x + 3y - 4z = -3 \end{cases}$

28. $\begin{cases} x - 2y + z = 3 \\ 2x - 5y + 6z = 7 \\ 2x - 3y - 2z = 5 \end{cases}$

29. $\begin{cases} x + 3y - 2z = 0 \\ 2x + 4z = 4 \\ 4x + 6y = 4 \end{cases}$

30. $\begin{cases} 2x + 4y - z = 3 \\ x + 2y + 4z = 6 \\ x + 2y - 2z = 0 \end{cases}$

31. $\begin{cases} x + z + 2w = 6 \\ y - 2z = -3 \\ x + 2y - z = -2 \\ 2x + y + 3z - 2w = 0 \end{cases}$

32. $\begin{cases} x + y + z + w = 0 \\ x + y + 2z + 2w = 0 \\ 2x + 2y + 3z + 4w = 1 \\ 2x + 3y + 4z + 5w = 2 \end{cases}$

Applications

33–34 ■ **Finance**   An investor has $100,000 to invest in three types of bonds: short-term, intermediate-term, and long-term. How much should she invest in each type to satisfy the given conditions?

33. Short-term bonds pay 4% annually, intermediate-term bonds pay 5%, and long-term bonds pay 6%. The investor wishes to realize a total annual income of 5.1%, with equal amounts invested in short- and intermediate-term bonds.

34. Short-term bonds pay 4% annually, intermediate-term bonds pay 6%, and long-term bonds pay 8%. The investor wishes to have a total annual return of $6700 on her investment, with equal amounts invested in intermediate- and long-term bonds.

35. Nutrition   A biologist is performing an experiment on the effects of various combinations of vitamins. She wishes to feed each of her laboratory rabbits a diet that contains exactly 9 mg of niacin, 14 mg of thiamin, and 32 mg of riboflavin. She has available three different types of commercial rabbit pellets; their vitamin content (per ounce) is given in the table. How many ounces of each type of food should each rabbit be given daily to satisfy the experiment requirements?

	Type A	Type B	Type C
Niacin (mg)	2	3	1
Thiamin (mg)	3	1	3
Riboflavin (mg)	8	5	7

36. Electricity   Using Kirchhoff's Laws, it can be shown that the currents I_1, I_2, and I_3 that pass through the three branches of the circuit in the figure satisfy the given linear system. Solve the system to find I_1, I_2, and I_3.

$$\begin{cases} I_1 + I_2 - I_3 = 0 \\ 16I_1 - 8I_2 \quad\quad = 4 \\ \quad\quad 8I_2 + 4I_3 = 5 \end{cases}$$

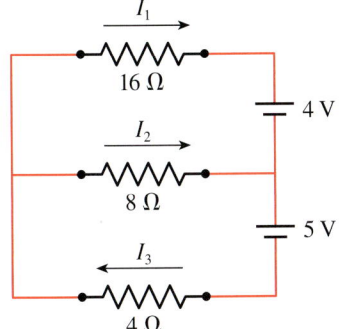

37. Agriculture   A farmer has 1200 acres of land on which he grows corn, wheat, and soybeans. It costs $45 per acre to grow corn, $60 for wheat, and $50 for soybeans. Because of market demand he will grow twice as many acres of wheat as of corn. He has allocated $63,750 for the cost of growing his crops. How many acres of each crop should he plant?

38. Stock Portfolio   An investor owns three stocks: A, B, and C. The closing prices of the stocks on three successive trading days are given in the table.

	Stock A	Stock B	Stock C
Monday	$10	$25	$29
Tuesday	$12	$20	$32
Wednesday	$16	$15	$32

Despite the volatility in the stock prices, the total value of the investor's stocks remained unchanged at $74,000 at the end of each of these three days. How many shares of each stock does the investor own?

Discovery • Discussion

39. Can a Linear System Have Exactly Two Solutions?

(a) Suppose that (x_0, y_0, z_0) and (x_1, y_1, z_1) are solutions of the system

$$\begin{cases} a_1x + b_1y + c_1z = d_1 \\ a_2x + b_2y + c_2z = d_2 \\ a_3x + b_3y + c_3z = d_3 \end{cases}$$

Show that $\left(\dfrac{x_0 + x_1}{2}, \dfrac{y_0 + y_1}{2}, \dfrac{z_0 + z_1}{2} \right)$ is also a solution.

(b) Use the result of part (a) to prove that if the system has two different solutions, then it has infinitely many solutions.

DISCOVERY PROJECT

Best Fit versus Exact Fit

Given several points in the plane, we can find the line that best fits them (see the *Focus on Modeling*, page 239). Of course, not all the points will necessarily lie on the line. We can also find the quadratic polynomial that best fits the points. Again, not every point will necessarily lie on the graph of the polynomial.

However, if we are given just two points, we can find a line of *exact* fit, that is, a line that actually passes through both points. Similarly, given three points (not all on the same line), we can find the quadratic polynomial of *exact* fit. For example, suppose we are given the following three points:

$$(-1, 6), \quad (1, 2), \quad (2, 3)$$

From Figure 1 we see that the points do not lie on a line. Let's find the quadratic polynomial that fits these points exactly. The polynomial must have the form

$$y = ax^2 + bx + c$$

We need to find values for a, b, and c so that the graph of the resulting polynomial contains the given points. Substituting the given points into the equation, we get the following.

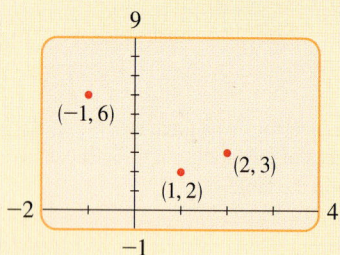

Figure 1

Point	Substitute	Equation
$(-1, 6)$	$x = -1, \quad y = 6$	$6 = a(-1)^2 + b(-1) + c$
$(1, 2)$	$x = 1, \quad y = 2$	$2 = a(1)^2 + b(1) + c$
$(2, 3)$	$x = 2, \quad y = 3$	$3 = a(2)^2 + b(2) + c$

These three equations simplify into the following system.

$$\begin{cases} a - b + c = 6 \\ a + b + c = 2 \\ 4a + 2b + c = 3 \end{cases}$$

Using Gaussian elimination we obtain the solution $a = 1$, $b = -2$, and $c = 3$. So the required quadratic polynomial is

$$y = x^2 - 2x + 3$$

From Figure 2 we see that the graph of the polynomial passes through the given points.

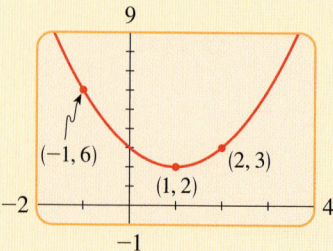

Figure 2

1. Find the quadratic polynomial $y = ax^2 + bx + c$ whose graph passes through the given points.

(a) $(-2, 3)$, $(-1, 1)$, $(1, 9)$
(b) $(-1, -3)$, $(2, 0)$, $(3, -3)$

2. Find the cubic polynomial $y = ax^3 + bx^2 + cx + d$ whose graph passes through the given points.

(a) $(-1, -4)$, $(1, 2)$, $(2, 11)$, $(3, 32)$
(b) $(-2, 10)$, $(-1, 1)$, $(1, -1)$, $(3, 45)$

3. A stone is thrown upward with velocity v from a height h. Its elevation d above the ground at time t is given by

$$d = at^2 + vt + h$$

The elevation is measured at three different times as shown.

Time (s)	1.0	2.0	6.0
Elevation (ft)	144	192	64

(a) Find the constants a, v, and h.
(b) Find the elevation of the stone when $t = 4$ s.

4. (a) Find the quadratic function $y = ax^2 + bx + c$ whose graph passes through the given points. (This is the quadratic curve of *exact* fit.) Graph the points and the quadratic curve that you found.

$$(-2, 10), \quad (1, -5), \quad (2, -6), \quad (4, -2)$$

(b) Now use the QuadReg command on your calculator to find the quadratic curve that *best* fits the points in part (a). How does this compare to the function you found in part (a)?

(c) Show that no quadratic function passes through the points

$$(-2, 11), \quad (1, -6), \quad (2, -5), \quad (4, -1)$$

(d) Use the QuadReg command on your calculator to find the quadratic curve that best fits the points in part (b). Graph the points and the quadratic curve that you found.

(e) Explain how the curve of exact fit differs from the curve of best fit.

9.4 | **Systems of Linear Equations: Matrices**

In this section we express a linear system as a rectangular array of numbers, called a matrix. Matrices* provide us with an efficient way of solving linear systems.

Matrices

We begin by defining the various elements that make up a matrix.

Definition of Matrix

An $m \times n$ **matrix** is a rectangular array of numbers with m **rows** and n **columns**.

$$\left. \begin{bmatrix} a_{11} & a_{12} & a_{13} & \cdots & a_{1n} \\ a_{21} & a_{22} & a_{23} & \cdots & a_{2n} \\ a_{31} & a_{32} & a_{33} & \cdots & a_{3n} \\ \vdots & \vdots & \vdots & \ddots & \vdots \\ a_{m1} & a_{m2} & a_{m3} & \cdots & a_{mn} \end{bmatrix} \begin{array}{c} \leftarrow \\ \leftarrow \\ \leftarrow \\ \\ \leftarrow \end{array} \right\} m \text{ rows}$$

$$\underbrace{\uparrow \qquad \uparrow \qquad \uparrow \qquad\qquad \uparrow}_{n \text{ columns}}$$

We say that the matrix has **dimension** $m \times n$. The numbers a_{ij} are the **entries** of the matrix. The subscript on the entry a_{ij} indicates that it is in the ith row and the jth column.

Here are some examples of matrices.

Matrix	**Dimension**	
$\begin{bmatrix} 1 & 3 & 0 \\ 2 & 4 & -1 \end{bmatrix}$	2×3	2 rows by 3 columns
$\begin{bmatrix} 6 & -5 & 0 & 1 \end{bmatrix}$	1×4	1 row by 4 columns

The Augmented Matrix of a Linear System

We can write a system of linear equations as a matrix, called the **augmented matrix** of the system, by writing only the coefficients and constants that appear in the equations. Here is an example.

Linear system	**Augmented matrix**
$\begin{cases} 3x - 2y + z = 5 \\ x + 3y - z = 0 \\ -x + 4z = 11 \end{cases}$	$\begin{bmatrix} 3 & -2 & 1 & 5 \\ 1 & 3 & -1 & 0 \\ -1 & 0 & 4 & 11 \end{bmatrix}$

---

* The plural of *matrix* is *matrices*.

Notice that a missing variable in an equation corresponds to a 0 entry in the augmented matrix.

Example 1   Finding the Augmented Matrix of a Linear System

Write the augmented matrix of the system of equations.

$$\begin{cases} 6x - 2y - z = 4 \\ x + 3z = 1 \\ 7y + z = 5 \end{cases}$$

Solution   First we write the linear system with the variables lined up in columns.

$$\begin{cases} 6x - 2y - z = 4 \\ x + 3z = 1 \\ 7y + z = 5 \end{cases}$$

The augmented matrix is the matrix whose entries are the coefficients and the constants in this system.

$$\begin{bmatrix} 6 & -2 & -1 & 4 \\ 1 & 0 & 3 & 1 \\ 0 & 7 & 1 & 5 \end{bmatrix}$$
■

Elementary Row Operations

The operations that we used in Section 9.3 to solve linear systems correspond to operations on the rows of the augmented matrix of the system. For example, adding a multiple of one equation to another corresponds to adding a multiple of one row to another.

Elementary Row Operations

1. Add a multiple of one row to another.
2. Multiply a row by a nonzero constant.
3. Interchange two rows.

Note that performing any of these operations on the augmented matrix of a system does not change its solution. We use the following notation to describe the elementary row operations:

Symbol	Description
$R_i + kR_j \rightarrow R_i$	Change the ith row by adding k times row j to it, then put the result back in row i.
kR_i	Multiply the ith row by k.
$R_i \leftrightarrow R_j$	Interchange the ith and jth rows.

In the next example we compare the two ways of writing systems of linear equations.

Example 2   Using Elementary Row Operations to Solve a Linear System

Solve the system of linear equations.

$$\begin{cases} x - y + 3z = 4 \\ x + 2y - 2z = 10 \\ 3x - y + 5z = 14 \end{cases}$$

Solution   Our goal is to eliminate the x-term from the second equation and the x- and y-terms from the third equation. For comparison, we write both the system of equations and its augmented matrix.

<table>
<tr><th colspan="2" style="text-align:center">System</th><th>Augmented matrix</th></tr>
</table>

	System	Augmented matrix
	$\begin{cases} x - y + 3z = 4 \\ x + 2y - 2z = 10 \\ 3x - y + 5z = 14 \end{cases}$	$\begin{bmatrix} 1 & -1 & 3 & 4 \\ 1 & 2 & -2 & 10 \\ 3 & -1 & 5 & 14 \end{bmatrix}$
Add $(-1) \times$ Equation 1 to Equation 2. Add $(-3) \times$ Equation 1 to Equation 3.	$\begin{cases} x - y + 3z = 4 \\ 3y - 5z = 6 \\ 2y - 4z = 2 \end{cases}$	$\xrightarrow[\text{R}_3 - 3\text{R}_1 \rightarrow \text{R}_3]{\text{R}_2 - \text{R}_1 \rightarrow \text{R}_2}$ $\begin{bmatrix} 1 & -1 & 3 & 4 \\ 0 & 3 & -5 & 6 \\ 0 & 2 & -4 & 2 \end{bmatrix}$
Multiply Equation 3 by $\frac{1}{2}$.	$\begin{cases} x - y + 3z = 4 \\ 3y - 5z = 6 \\ y - 2z = 1 \end{cases}$	$\xrightarrow{\frac{1}{2}\text{R}_3}$ $\begin{bmatrix} 1 & -1 & 3 & 4 \\ 0 & 3 & -5 & 6 \\ 0 & 1 & -2 & 1 \end{bmatrix}$
Add $(-3) \times$ Equation 3 to Equation 2 (to eliminate y from Equation 2).	$\begin{cases} x - y + 3z = 4 \\ z = 3 \\ y - 2z = 1 \end{cases}$	$\xrightarrow{\text{R}_2 - 3\text{R}_3 \rightarrow \text{R}_2}$ $\begin{bmatrix} 1 & -1 & 3 & 4 \\ 0 & 0 & 1 & 3 \\ 0 & 1 & -2 & 1 \end{bmatrix}$
Interchange Equations 2 and 3.	$\begin{cases} x - y + 3z = 4 \\ y - 2z = 1 \\ z = 3 \end{cases}$	$\xrightarrow{\text{R}_2 \leftrightarrow \text{R}_3}$ $\begin{bmatrix} 1 & -1 & 3 & 4 \\ 0 & 1 & -2 & 1 \\ 0 & 0 & 1 & 3 \end{bmatrix}$

Now we use back-substitution to find that $x = 2$, $y = 7$, and $z = 3$. The solution is $(2, 7, 3)$. ∎

Gaussian Elimination

In general, to solve a system of linear equations using its augmented matrix, we use elementary row operations to arrive at a matrix in a certain form. This form is described in the following box.

Row-Echelon Form and Reduced Row-Echelon Form of a Matrix

A matrix is in **row-echelon form** if it satisfies the following conditions.

1. The first nonzero number in each row (reading from left to right) is 1. This is called the **leading entry**.

2. The leading entry in each row is to the right of the leading entry in the row immediately above it.

3. All rows consisting entirely of zeros are at the bottom of the matrix.

A matrix is in **reduced row-echelon form** if it is in row-echelon form and also satisfies the following condition.

4. Every number above and below each leading entry is a 0.

In the following matrices the first matrix is in reduced row-echelon form, but the second one is just in row-echelon form. The third matrix is not in row-echelon form. The entries in red are the leading entries.

Reduced row-echelon form	Row-echelon form	Not in row-echelon form
$$\begin{bmatrix} 1 & 3 & 0 & 0 & 0 \\ 0 & 0 & 1 & 0 & -3 \\ 0 & 0 & 0 & 1 & \frac{1}{2} \\ 0 & 0 & 0 & 0 & 0 \end{bmatrix}$$	$$\begin{bmatrix} 1 & 3 & -6 & 10 & 0 \\ 0 & 0 & 1 & 4 & -3 \\ 0 & 0 & 0 & 1 & \frac{1}{2} \\ 0 & 0 & 0 & 0 & 0 \end{bmatrix}$$	$$\begin{bmatrix} 0 & 1 & -\frac{1}{2} & 0 & 7 \\ 1 & 0 & 3 & 4 & -5 \\ 0 & 0 & 0 & 1 & 0.4 \\ 0 & 1 & 1 & 0 & 0 \end{bmatrix}$$
Leading 1's have 0's above and below them.	Leading 1's shift to the right in successive rows.	Leading 1's do not shift to the right in successive rows.

Here is a systematic way to put a matrix in row-echelon form using elementary row operations:

- Start by obtaining 1 in the top left corner. Then obtain zeros below that 1 by adding appropriate multiples of the first row to the rows below it.

- Next, obtain a leading 1 in the next row, and then obtain zeros below that 1.

- At each stage make sure that every leading entry is to the right of the leading entry in the row above it—rearrange the rows if necessary.

- Continue this process until you arrive at a matrix in row-echelon form.

This is how the process might work for a 3×4 matrix:

$$\begin{bmatrix} 1 & \blacksquare & \blacksquare & \blacksquare \\ 0 & \blacksquare & \blacksquare & \blacksquare \\ 0 & \blacksquare & \blacksquare & \blacksquare \end{bmatrix} \rightarrow \begin{bmatrix} 1 & \blacksquare & \blacksquare & \blacksquare \\ 0 & 1 & \blacksquare & \blacksquare \\ 0 & 0 & \blacksquare & \blacksquare \end{bmatrix} \rightarrow \begin{bmatrix} 1 & \blacksquare & \blacksquare & \blacksquare \\ 0 & 1 & \blacksquare & \blacksquare \\ 0 & 0 & 1 & \blacksquare \end{bmatrix}$$

Once an augmented matrix is in row-echelon form, we can solve the corresponding linear system using back-substitution. This technique is called **Gaussian elimination**, in honor of its inventor, the German mathematician C. F. Gauss (see page 294).

> ### Solving a System Using Gaussian Elimination
>
> 1. **Augmented Matrix.**   Write the augmented matrix of the system.
>
> 2. **Row-Echelon Form.**   Use elementary row operations to change the augmented matrix to row-echelon form.
>
> 3. **Back-Substitution.**   Write the new system of equations that corresponds to the row-echelon form of the augmented matrix and solve by back-substitution.

Example 3   **Solving a System Using Row-Echelon Form**

Solve the system of linear equations using Gaussian elimination.

$$\begin{cases} 4x + 8y - 4z = 4 \\ 3x + 8y + 5z = -11 \\ -2x + y + 12z = -17 \end{cases}$$

Solution   We first write the augmented matrix of the system, and then use elementary row operations to put it in row-echelon form.

Need a 1 here.

$$\begin{bmatrix} \boxed{4} & 8 & -4 & 4 \\ 3 & 8 & 5 & -11 \\ -2 & 1 & 12 & -17 \end{bmatrix}$$

$\xrightarrow{\frac{1}{4}R_1}$
$$\begin{bmatrix} 1 & 2 & -1 & 1 \\ \boxed{3} & 8 & 5 & -11 \\ \boxed{-2} & 1 & 12 & -17 \end{bmatrix}$$
Need 0's here.

$\xrightarrow[R_3 + 2R_1 \to R_3]{R_2 - 3R_1 \to R_2}$
$$\begin{bmatrix} 1 & 2 & -1 & 1 \\ 0 & \boxed{2} & 8 & -14 \\ 0 & 5 & 10 & -15 \end{bmatrix}$$
Need a 1 here.

$\xrightarrow{\frac{1}{2}R_2}$
$$\begin{bmatrix} 1 & 2 & -1 & 1 \\ 0 & 1 & 4 & -7 \\ 0 & \boxed{5} & 10 & -15 \end{bmatrix}$$
Need a 0 here.

$\xrightarrow{R_3 - 5R_2 \to R_3}$
$$\begin{bmatrix} 1 & 2 & -1 & 1 \\ 0 & 1 & 4 & -7 \\ 0 & 0 & \boxed{-10} & 20 \end{bmatrix}$$
Need a 1 here.

$\xrightarrow{-\frac{1}{10}R_3}$
$$\begin{bmatrix} 1 & 2 & -1 & 1 \\ 0 & 1 & 4 & -7 \\ 0 & 0 & 1 & -2 \end{bmatrix}$$

We now have an equivalent matrix in row-echelon form, and the corresponding system of equations is

$$\begin{cases} x + 2y - z = 1 \\ y + 4z = -7 \\ z = -2 \end{cases}$$

We use back-substitution to solve the system.

$$y + 4(-2) = -7 \qquad \text{Back-substitute } z = -2 \text{ into Equation 2}$$

$$y = 1 \qquad \text{Solve for } y$$

$$x + 2(1) - (-2) = 1 \qquad \text{Back-substitute } y = 1 \text{ and } z = -2 \text{ into Equation 1}$$

$$x = -3 \qquad \text{Solve for } x$$

So the solution of the system is $(-3, 1, -2)$. ∎

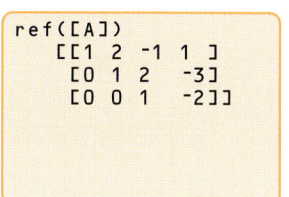

Figure 1

Graphing calculators have a "row-echelon form" command that puts a matrix in row-echelon form. (On the TI-83 this command is `ref`.) For the augmented matrix in Example 3, the `ref` command gives the output shown in Figure 1. Notice that the row-echelon form obtained by the calculator differs from the one we got in Example 3. This is because the calculator used different row operations than we did. You should check that your calculator's row-echelon form leads to the same solution as ours.

Gauss-Jordan Elimination

If we put the augmented matrix of a linear system in *reduced* row-echelon form, then we don't need to back-substitute to solve the system. To put a matrix in reduced row-echelon form, we use the following steps.

■ Use the elementary row operations to put the matrix in row-echelon form.

■ Obtain zeros above each leading entry by adding multiples of the row containing that entry to the rows above it. Begin with the last leading entry and work up.

Here is how the process works for a 3 × 4 matrix:

$$\begin{bmatrix} 1 & \blacksquare & \blacksquare & \blacksquare \\ 0 & 1 & \blacksquare & \blacksquare \\ 0 & 0 & 1 & \blacksquare \end{bmatrix} \rightarrow \begin{bmatrix} 1 & \blacksquare & 0 & \blacksquare \\ 0 & 1 & 0 & \blacksquare \\ 0 & 0 & 1 & \blacksquare \end{bmatrix} \rightarrow \begin{bmatrix} 1 & 0 & 0 & \blacksquare \\ 0 & 1 & 0 & \blacksquare \\ 0 & 0 & 1 & \blacksquare \end{bmatrix}$$

Using the reduced row-echelon form to solve a system is called **Gauss-Jordan elimination**. We illustrate this process in the next example.

Example 4 **Solving a System Using Reduced Row-Echelon Form**

Solve the system of linear equations, using Gauss-Jordan elimination.

$$\begin{cases} 4x + 8y - 4z = 4 \\ 3x + 8y + 5z = -11 \\ -2x + y + 12z = -17 \end{cases}$$

Solution In Example 3 we used Gaussian elimination on the augmented matrix of this system to arrive at an equivalent matrix in row-echelon form. We continue using elementary row operations on the last matrix in Example 3 to arrive at an equivalent matrix in reduced row-echelon form.

$$\begin{bmatrix} 1 & 2 & \boxed{-1} & 1 \\ 0 & 1 & \boxed{4} & -7 \\ 0 & 0 & 1 & -2 \end{bmatrix} \quad \text{Need 0's here.}$$

$$\xrightarrow[\begin{array}{c} R_2 - 4R_3 \to R_2 \\ \hline R_1 + R_3 \to R_1 \end{array}]{} \begin{bmatrix} 1 & \boxed{2} & 0 & -1 \\ 0 & 1 & 0 & 1 \\ 0 & 0 & 1 & -2 \end{bmatrix} \quad \text{Need a 0 here.}$$

$$\xrightarrow[\;R_1 - 2R_2 \to R_1\;]{} \begin{bmatrix} 1 & 0 & 0 & -3 \\ 0 & 1 & 0 & 1 \\ 0 & 0 & 1 & -2 \end{bmatrix}$$

We now have an equivalent matrix in reduced row-echelon form, and the corresponding system of equations is

$$\begin{cases} x = -3 \\ y = 1 \\ z = -2 \end{cases}$$

Since the system is in reduced row-echelon form, back-substitution is not required to get the solution.

Hence we immediately arrive at the solution $(-3, 1, -2)$. ∎

Graphing calculators also have a command that puts a matrix in reduced row-echelon form. (On the TI-83 this command is `rref`.) For the augmented matrix in Example 4, the `rref` command gives the output shown in Figure 2. The calculator gives the same reduced row-echelon form as the one we got in Example 4. This is because every matrix has a *unique* reduced row-echelon form.

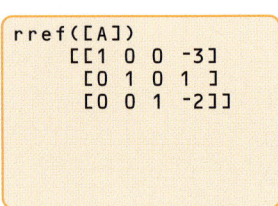

```
rref([A])
    [[1 0 0 -3]
     [0 1 0 1 ]
     [0 0 1 -2]]
```

Figure 2

Inconsistent and Dependent Systems

The systems of linear equations that we considered in Examples 1–4 had exactly one solution. But as we know from Section 9.3 a linear system may have one solution, no solution, or infinitely many solutions. Fortunately, the row-echelon form of a system allows us to determine which of these cases applies, as described in the following box.

First we need some terminology. A **leading variable** in a linear system is one that corresponds to a leading entry in the row-echelon form of the augmented matrix of the system.

The Solutions of a Linear System in Row-Echelon Form

Suppose the augmented matrix of a system of linear equations has been transformed by Gaussian elimination into row-echelon form. Then exactly one of the following is true.

1. **No solution.** If the row-echelon form contains a row that represents the equation $0 = c$ where c is not zero, then the system has no solution. A system with no solution is called **inconsistent**.

2. **One solution.** If each variable in the row-echelon form is a leading variable, then the system has exactly one solution, which we find using back-substitution or Gauss-Jordan elimination.

3. **Infinitely many solutions.** If the variables in the row-echelon form are not all leading variables, and if the system is not inconsistent, then it has infinitely many solutions. In this case, the system is called **dependent**. We solve the system by putting the matrix in reduced row-echelon form and then expressing the leading variables in terms of the nonleading variables. The nonleading variables may take on any real numbers as their values.

The matrices below, all in row-echelon form, illustrate the three cases described in the box.

No solution	One solution	Infinitely many solutions
$\begin{bmatrix} 1 & 2 & 5 & 7 \\ 0 & 1 & 3 & 4 \\ 0 & 0 & 0 & 1 \end{bmatrix}$	$\begin{bmatrix} 1 & 6 & -1 & 3 \\ 0 & 1 & 2 & -2 \\ 0 & 0 & 1 & 8 \end{bmatrix}$	$\begin{bmatrix} 1 & 2 & -3 & 1 \\ 0 & 1 & 5 & -2 \\ 0 & 0 & 0 & 0 \end{bmatrix}$
Last equation says $0 = 1$.	Each variable is a leading variable.	z is not a leading variable.

Example 5   A System with No Solution

Solve the system.

$$\begin{cases} x - 3y + 2z = 12 \\ 2x - 5y + 5z = 14 \\ x - 2y + 3z = 20 \end{cases}$$

Solution   We transform the system into row-echelon form.

$$\begin{bmatrix} 1 & -3 & 2 & 12 \\ 2 & -5 & 5 & 14 \\ 1 & -2 & 3 & 20 \end{bmatrix} \xrightarrow[\text{R}_3 - \text{R}_1 \rightarrow \text{R}_3]{\text{R}_2 - 2\text{R}_1 \rightarrow \text{R}_2} \begin{bmatrix} 1 & -3 & 2 & 12 \\ 0 & 1 & 1 & -10 \\ 0 & 1 & 1 & 8 \end{bmatrix}$$

$$\xrightarrow{\text{R}_3 - \text{R}_2 \rightarrow \text{R}_3} \begin{bmatrix} 1 & -3 & 2 & 12 \\ 0 & 1 & 1 & -10 \\ 0 & 0 & 0 & 18 \end{bmatrix} \xrightarrow{\frac{1}{18}\text{R}_3} \begin{bmatrix} 1 & -3 & 2 & 12 \\ 0 & 1 & 1 & -10 \\ 0 & 0 & 0 & 1 \end{bmatrix}$$

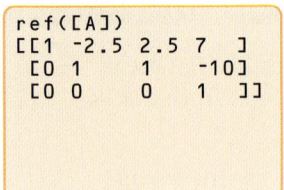

Figure 3

This last matrix is in row-echelon form, so we can stop the Gaussian elimination process. Now if we translate the last row back into equation form, we get $0x + 0y + 0z = 1$, or $0 = 1$, which is false. No matter what values we pick for x, y, and z, the last equation will never be a true statement. This means the system *has no solution*. ∎

Figure 3 shows the row-echelon form produced by a TI-83 calculator for the augmented matrix in Example 5. You should check that this gives the same solution.

Example 6   A System with Infinitely Many Solutions

Find the complete solution of the system.

$$\begin{cases} -3x - 5y + 36z = 10 \\ -x \qquad + 7z = 5 \\ x + y - 10z = -4 \end{cases}$$

Solution   We transform the system into reduced row-echelon form.

$$\begin{bmatrix} -3 & -5 & 36 & 10 \\ -1 & 0 & 7 & 5 \\ 1 & 1 & -10 & -4 \end{bmatrix} \xrightarrow{R_1 \leftrightarrow R_3} \begin{bmatrix} 1 & 1 & -10 & -4 \\ -1 & 0 & 7 & 5 \\ -3 & -5 & 36 & 10 \end{bmatrix}$$

$$\xrightarrow[{R_3 + 3R_1 \to R_3}]{R_2 + R_1 \to R_2} \begin{bmatrix} 1 & 1 & -10 & -4 \\ 0 & 1 & -3 & 1 \\ 0 & -2 & 6 & -2 \end{bmatrix} \xrightarrow{R_3 + 2R_2 \to R_3} \begin{bmatrix} 1 & 1 & -10 & -4 \\ 0 & 1 & -3 & 1 \\ 0 & 0 & 0 & 0 \end{bmatrix}$$

$$\xrightarrow{R_1 - R_2 \to R_1} \begin{bmatrix} 1 & 0 & -7 & -5 \\ 0 & 1 & -3 & 1 \\ 0 & 0 & 0 & 0 \end{bmatrix}$$

The third row corresponds to the equation $0 = 0$. This equation is always true, no matter what values are used for x, y, and z. Since the equation adds no new information about the variables, we can drop it from the system. So the last matrix corresponds to the system

$$\begin{cases} x \qquad - 7z = -5 & \text{Equation 1} \\ \quad y - 3z = 1 & \text{Equation 2} \end{cases}$$

Leading variables

Reduced row-echelon form on the TI-83 calculator:

```
rref([A])
 [[1 0 -7 -5]
  [0 1 -3 1 ]
  [0 0 0  0 ]]
```

Now we solve for the leading variables x and y in terms of the nonleading variable z:

$$x = 7z - 5 \qquad \text{Solve for x in Equation 1}$$

$$y = 3z + 1 \qquad \text{Solve for y in Equation 2}$$

To obtain the complete solution, we let t represent any real number, and we express x, y, and z in terms of t:

$$x = 7t - 5$$

$$y = 3t + 1$$

$$z = t$$

We can also write the solution as the ordered triple $(7t - 5, 3t + 1, t)$, where t is any real number. ∎

In Example 6, to get specific solutions we give a specific value to t. For example, if $t = 1$, then

$$x = 7(1) - 5 = 2$$

$$y = 3(1) + 1 = 4$$

$$z = 1$$

Here are some other solutions of the system obtained by substituting other values for the parameter t.

Parameter t	Solution $(7t - 5, 3t + 1, t)$
-1	$(-12, -2, -1)$
0	$(-5, 1, 0)$
2	$(9, 7, 2)$
5	$(30, 16, 5)$

Example 7   A System with Infinitely Many Solutions

Find the complete solution of the system.

$$\begin{cases} x + 2y - 3z - 4w = 10 \\ x + 3y - 3z - 4w = 15 \\ 2x + 2y - 6z - 8w = 10 \end{cases}$$

Solution   We transform the system into reduced row-echelon form.

$$\begin{bmatrix} 1 & 2 & -3 & -4 & 10 \\ 1 & 3 & -3 & -4 & 15 \\ 2 & 2 & -6 & -8 & 10 \end{bmatrix} \xrightarrow[R_3 - 2R_1 \to R_3]{R_2 - R_1 \to R_2} \begin{bmatrix} 1 & 2 & -3 & -4 & 10 \\ 0 & 1 & 0 & 0 & 5 \\ 0 & -2 & 0 & 0 & -10 \end{bmatrix}$$

$$\xrightarrow{R_3 + 2R_2 \to R_3} \begin{bmatrix} 1 & 2 & -3 & -4 & 10 \\ 0 & 1 & 0 & 0 & 5 \\ 0 & 0 & 0 & 0 & 0 \end{bmatrix} \xrightarrow{R_1 - 2R_2 \to R_1} \begin{bmatrix} 1 & 0 & -3 & -4 & 0 \\ 0 & 1 & 0 & 0 & 5 \\ 0 & 0 & 0 & 0 & 0 \end{bmatrix}$$

This is in reduced row-echelon form. Since the last row represents the equation $0 = 0$, we may discard it. So the last matrix corresponds to the system

$$\begin{cases} x & -3z - 4w = 0 \\ & y = 5 \end{cases}$$

Leading variables

Olga Taussky-Todd (1906–1995) was instrumental in developing applications of Matrix Theory. Described as "in love with anything matrices can do," she successfully applied matrices to aerodynamics, a field used in the design of airplanes and rockets. Taussky-Todd was also famous for her work in Number Theory, which deals with prime numbers and divisibility. Although Number Theory was once considered the least applicable branch of mathematics, it is now used in significant ways throughout the computer industry.

Taussky-Todd studied mathematics at a time when young women rarely aspired to be mathematicians. She said, "When I entered university I had no idea what it meant to study mathematics." One of the most respected mathematicians of her day, she was for many years a professor of mathematics at Caltech in Pasadena.

To obtain the complete solution, we solve for the leading variables x and y in terms of the nonleading variables z and w, and we let z and w be any real numbers. Thus, the complete solution is

$$x = 3s + 4t$$
$$y = 5$$
$$z = s$$
$$w = t$$

where s and t are any real numbers.

We can also express the answer as the ordered quadruple $(3s + 4t, 5, s, t)$. ∎

⊘ Note that s and t do *not* have to be the *same* real number in the solution for Example 7. We can choose arbitrary values for each if we wish to construct a specific solution to the system. For example, if we let $s = 1$ and $t = 2$, then we get the solution $(11, 5, 1, 2)$. You should check that this does indeed satisfy all three of the original equations in Example 7.

Examples 6 and 7 illustrate this general fact: If a system in row-echelon form has n nonzero equations in m variables $(m > n)$, then the complete solution will have $m - n$ nonleading variables. For instance, in Example 6 we arrived at *two* nonzero equations in the *three* variables x, y, and z, which gave us $3 - 2 = 1$ nonleading variable.

Modeling with Linear Systems

Linear equations, often containing hundreds or even thousands of variables, occur frequently in the applications of algebra to the sciences and to other fields. For now, let's consider an example that involves only three variables.

Example 8 Nutritional Analysis Using a System of Linear Equations

A nutritionist is performing an experiment on student volunteers. He wishes to feed one of his subjects a daily diet that consists of a combination of three commercial diet foods: MiniCal, LiquiFast, and SlimQuick. For the experiment it's important that the subject consume exactly 500 mg of potassium, 75 g of protein, and 1150 units of vitamin D every day. The amounts of these nutrients in one ounce of each food are given in the table. How many ounces of each food should the subject eat every day to satisfy the nutrient requirements exactly?

	MiniCal	LiquiFast	SlimQuick
Potassium (mg)	50	75	10
Protein (g)	5	10	3
Vitamin D (units)	90	100	50

Solution Let x, y, and z represent the number of ounces of MiniCal, LiquiFast, and SlimQuick, respectively, that the subject should eat every day. This means that he will get $50x$ mg of potassium from MiniCal, $75y$ mg from LiquiFast, and $10z$ mg from SlimQuick, for a total of $50x + 75y + 10z$ mg potassium in all. Since the

```
rref([A])
  [[1 0 0 5 ]
   [0 1 0 2 ]
   [0 0 1 10]]
```

Figure 4

Check Your Answer

$x = 5, y = 2, z = 10$:

$$\begin{cases} 10(5) + 15(2) + 2(10) = 100 \\ 5(5) + 10(2) + 3(10) = 75 \\ 9(5) + 10(2) + 5(10) = 115 \end{cases} \checkmark$$

potassium requirement is 500 mg, we get the first equation below. Similar reasoning for the protein and vitamin D requirements leads to the system

$$\begin{cases} 50x + 75y + 10z = 500 & \text{Potassium} \\ 5x + 10y + 3z = 75 & \text{Protein} \\ 90x + 100y + 50z = 1150 & \text{Vitamin D} \end{cases}$$

Dividing the first equation by 5 and the third one by 10 gives the system

$$\begin{cases} 10x + 15y + 2z = 100 \\ 5x + 10y + 3z = 75 \\ 9x + 10y + 5z = 115 \end{cases}$$

We can solve this system using Gaussian elimination, or we can use a graphing calculator to find the reduced row-echelon form of the augmented matrix of the system. Using the **rref** command on the TI-83, we get the output in Figure 4. From the reduced row-echelon form we see that $x = 5, y = 2, z = 10$. The subject should be fed 5 oz of MiniCal, 2 oz of LiquiFast, and 10 oz of SlimQuick every day. ■

A more practical application might involve dozens of foods and nutrients rather than just three. Such problems lead to systems with large numbers of variables and equations. Computers or graphing calculators are essential for solving such large systems.

9.4   Exercises

1–6 ■ State the dimension of the matrix.

1. $\begin{bmatrix} 2 & 7 \\ 0 & -1 \\ 5 & -3 \end{bmatrix}$ **2.** $\begin{bmatrix} -1 & 5 & 4 & 0 \\ 0 & 2 & 11 & 3 \end{bmatrix}$ **3.** $\begin{bmatrix} 12 \\ 35 \end{bmatrix}$

4. $\begin{bmatrix} -3 \\ 0 \\ 1 \end{bmatrix}$ **5.** $[1 \ 4 \ 7]$ **6.** $\begin{bmatrix} 1 & 0 \\ 0 & 1 \end{bmatrix}$

7–14 ■ A matrix is given.

(a) Determine whether the matrix is in row-echelon form.

(b) Determine whether the matrix is in reduced row-echelon form.

(c) Write the system of equations for which the given matrix is the augmented matrix.

7. $\begin{bmatrix} 1 & 0 & -3 \\ 0 & 1 & 5 \end{bmatrix}$ **8.** $\begin{bmatrix} 1 & 3 & -3 \\ 0 & 1 & 5 \end{bmatrix}$

9. $\begin{bmatrix} 1 & 2 & 8 & 0 \\ 0 & 1 & 3 & 2 \\ 0 & 0 & 0 & 0 \end{bmatrix}$ **10.** $\begin{bmatrix} 1 & 0 & -7 & 0 \\ 0 & 1 & 3 & 0 \\ 0 & 0 & 0 & 1 \end{bmatrix}$

11. $\begin{bmatrix} 1 & 0 & 0 & 0 \\ 0 & 0 & 0 & 0 \\ 0 & 1 & 5 & 1 \end{bmatrix}$ **12.** $\begin{bmatrix} 1 & 0 & 0 & 1 \\ 0 & 1 & 0 & 2 \\ 0 & 0 & 1 & 3 \end{bmatrix}$

13. $\begin{bmatrix} 1 & 3 & 0 & -1 & 0 \\ 0 & 0 & 1 & 2 & 0 \\ 0 & 0 & 0 & 0 & 1 \\ 0 & 0 & 0 & 0 & 0 \end{bmatrix}$ **14.** $\begin{bmatrix} 1 & 3 & 0 & 1 & 0 & 0 \\ 0 & 1 & 0 & 4 & 0 & 0 \\ 0 & 0 & 0 & 1 & 1 & 2 \\ 0 & 0 & 0 & 1 & 0 & 0 \end{bmatrix}$

15–24 ■ The system of linear equations has a unique solution. Find the solution using Gaussian elimination or Gauss-Jordan elimination.

15. $\begin{cases} x - 2y + z = 1 \\ y + 2z = 5 \\ x + y + 3z = 8 \end{cases}$ **16.** $\begin{cases} x + y + 6z = 3 \\ x + y + 3z = 3 \\ x + 2y + 4z = 7 \end{cases}$

17. $\begin{cases} x + y + z = 2 \\ 2x - 3y + 2z = 4 \\ 4x + y - 3z = 1 \end{cases}$ **18.** $\begin{cases} x + y + z = 4 \\ -x + 2y + 3z = 17 \\ 2x - y = -7 \end{cases}$

19. $\begin{cases} x + 2y - z = -2 \\ x + z = 0 \\ 2x - y - z = -3 \end{cases}$ **20.** $\begin{cases} 2y + z = 4 \\ x + y = 4 \\ 3x + 3y - z = 10 \end{cases}$

21. $\begin{cases} x_1 + 2x_2 - x_3 = 9 \\ 2x_1 - x_3 = -2 \\ 3x_1 + 5x_2 + 2x_3 = 22 \end{cases}$ **22.** $\begin{cases} 2x_1 + x_2 = 7 \\ 2x_1 - x_2 + x_3 = 6 \\ 3x_1 - 2x_2 + 4x_3 = 11 \end{cases}$

23. $\begin{cases} 2x - 3y - z = 13 \\ -x + 2y - 5z = 6 \\ 5x - y - z = 49 \end{cases}$

24. $\begin{cases} 10x + 10y - 20z = 60 \\ 15x + 20y + 30z = -25 \\ -5x + 30y - 10z = 45 \end{cases}$

25–34 ■ Determine whether the system of linear equations is inconsistent or dependent. If it is dependent, find the complete solution.

25. $\begin{cases} x + y + z = 2 \\ y - 3z = 1 \\ 2x + y + 5z = 0 \end{cases}$

26. $\begin{cases} x + 3z = 3 \\ 2x + y - 2z = 5 \\ -y + 8z = 8 \end{cases}$

27. $\begin{cases} 2x - 3y - 9z = -5 \\ x + 3z = 2 \\ -3x + y - 4z = -3 \end{cases}$

28. $\begin{cases} x - 2y + 5z = 3 \\ -2x + 6y - 11z = 1 \\ 3x - 16y + 20z = -26 \end{cases}$

29. $\begin{cases} x - y + 3z = 3 \\ 4x - 8y + 32z = 24 \\ 2x - 3y + 11z = 4 \end{cases}$

30. $\begin{cases} -2x + 6y - 2z = -12 \\ x - 3y + 2z = 10 \\ -x + 3y + 2z = 6 \end{cases}$

31. $\begin{cases} x + 4y - 2z = -3 \\ 2x - y + 5z = 12 \\ 8x + 5y + 11z = 30 \end{cases}$

32. $\begin{cases} 3r + 2s - 3t = 10 \\ r - s - t = -5 \\ r + 4s - t = 20 \end{cases}$

33. $\begin{cases} 2x + y - 2z = 12 \\ -x - \frac{1}{2}y + z = -6 \\ 3x + \frac{3}{2}y - 3z = 18 \end{cases}$

34. $\begin{cases} y - 5z = 7 \\ 3x + 2y = 12 \\ 3x + 10z = 80 \end{cases}$

35–46 ■ Solve the system of linear equations.

35. $\begin{cases} 4x - 3y + z = -8 \\ -2x + y - 3z = -4 \\ x - y + 2z = 3 \end{cases}$

36. $\begin{cases} 2x - 3y + 5z = 14 \\ 4x - y - 2z = -17 \\ -x - y + z = 3 \end{cases}$

37. $\begin{cases} x + 2y - 3z = -5 \\ -2x - 4y - 6z = 10 \\ 3x + 7y - 2z = -13 \end{cases}$

38. $\begin{cases} 3x - y + 2z = -1 \\ 4x - 2y + z = -7 \\ -x + 3y - 2z = -1 \end{cases}$

39. $\begin{cases} -x + 2y + z - 3w = 3 \\ 3x - 4y + z + w = 9 \\ -x - y + z + w = 0 \\ 2x + y + 4z - 2w = 3 \end{cases}$

40. $\begin{cases} x + y - z - w = 6 \\ 2x + z - 3w = 8 \\ x - y + 4w = -10 \\ 3x + 5y - z - w = 20 \end{cases}$

41. $\begin{cases} x + y + 2z - w = -2 \\ 3y + z + 2w = 2 \\ x + y + 3w = 2 \\ -3x + z + 2w = 5 \end{cases}$

42. $\begin{cases} x - 3y + 2z + w = -2 \\ x - 2y - 2w = -10 \\ z + 5w = 15 \\ 3x + 2z + w = -3 \end{cases}$

43. $\begin{cases} x + z + w = 4 \\ y - z = -4 \\ x - 2y + 3z + w = 12 \\ 2x - 2z + 5w = -1 \end{cases}$

44. $\begin{cases} y - z + 2w = 0 \\ 3x + 2y + w = 0 \\ 2x + 4w = 12 \\ -2x - 2z + 5w = 6 \end{cases}$

45. $\begin{cases} x - y + w = 0 \\ 3x - z + 2w = 0 \\ x - 4y + z + 2w = 0 \end{cases}$

46. $\begin{cases} 2x - y + 2z + w = 5 \\ -x + y + 4z - w = 3 \\ 3x - 2y - z = 0 \end{cases}$

Applications

47. Nutrition A doctor recommends that a patient take 50 mg each of niacin, riboflavin, and thiamin daily to alleviate a vitamin deficiency. In his medicine chest at home, the patient finds three brands of vitamin pills. The amounts of the relevant vitamins per pill are given in the table. How many pills of each type should he take every day to get 50 mg of each vitamin?

	VitaMax	Vitron	VitaPlus
Niacin (mg)	5	10	15
Riboflavin (mg)	15	20	0
Thiamin (mg)	10	10	10

48. Mixtures A chemist has three acid solutions at various concentrations. The first is 10% acid, the second is 20%, and the third is 40%. How many milliliters of each should he use to make 100 mL of 18% solution, if he has to use four times as much of the 10% solution as the 40% solution?

49. Distance, Speed, and Time Amanda, Bryce, and Corey enter a race in which they have to run, swim, and cycle over a marked course. Their average speeds are given in the table. Corey finishes first with a total time of 1 h 45 min. Amanda comes in second with a time of 2 h 30 min. Bryce finishes last with a time of 3 h. Find the distance (in miles) for each part of the race.

| | Average speed (mi/h) | | |
	Running	Swimming	Cycling
Amanda	10	4	20
Bryce	$7\frac{1}{2}$	6	15
Corey	15	3	40

50. Classroom Use A small school has 100 students who occupy three classrooms: A, B, and C. After the first period of the school day, half the students in room A move to room B, one-fifth of the students in room B move to room C, and one-third of the students in room C move to room A. Nevertheless, the total number of students in each room is the same for both periods. How many students occupy each room?

51. Manufacturing Furniture A furniture factory makes wooden tables, chairs, and armoires. Each piece of furniture requires three operations: cutting the wood, assembling, and finishing. Each operation requires the number of hours (h) given in the table. The workers in the factory can provide 300 hours of cutting, 400 hours of assembling, and 590 hours of finishing each work week. How many tables, chairs, and armoires should be produced so that all available labor-hours are used? Or is this impossible?

	Table	Chair	Armoire
Cutting (h)	$\frac{1}{2}$	1	1
Assembling (h)	$\frac{1}{2}$	$1\frac{1}{2}$	1
Finishing (h)	1	$1\frac{1}{2}$	2

52. Traffic Flow A section of a city's street network is shown in the figure. The arrows indicate one-way streets, and the numbers show how many cars enter or leave this section of the city via the indicated street in a certain one-hour period. The variables x, y, z, and w represent the number of cars that travel along the portions of First, Second, Avocado, and Birch Streets during this period. Find x, y, z, and w, assuming that none of the cars stop or park on any of the streets shown.

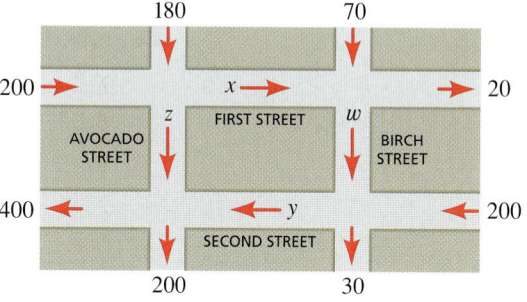

Discovery • Discussion

53. Polynomials Determined by a Set of Points We all know that two points uniquely determine a line $y = ax + b$ in the coordinate plane. Similarly, three points uniquely determine a quadratic (second-degree) polynomial

$$y = ax^2 + bx + c$$

four points uniquely determine a cubic (third-degree) polynomial

$$y = ax^3 + bx^2 + cx + d$$

and so on. (Some exceptions to this rule are if the three points actually lie on a line, or the four points lie on a quadratic or line, and so on.) For the following set of five points, find the line that contains the first two points, the quadratic that contains the first three points, the cubic that contains the first four points, and the fourth-degree polynomial that contains all five points.

$$(0, 0), \quad (1, 12), \quad (2, 40), \quad (3, 6), \quad (-1, -14)$$

Graph the points and functions in the same viewing rectangle using a graphing device.

| 9.5 | **The Algebra of Matrices** |

Thus far we've used matrices simply for notational convenience when solving linear systems. Matrices have many other uses in mathematics and the sciences, and for most of these applications a knowledge of matrix algebra is essential. Like numbers, matrices can be added, subtracted, multiplied, and divided. In this section we learn how to perform these algebraic operations on matrices.

Equality of Matrices

Two matrices are equal if they have the same entries in the same positions.

Equal Matrices

$$\begin{bmatrix} \sqrt{4} & 2^2 & e^0 \\ 0.5 & 1 & 1-1 \end{bmatrix} = \begin{bmatrix} 2 & 4 & 1 \\ \frac{1}{2} & \frac{2}{2} & 0 \end{bmatrix}$$

Unequal Matrices

$$\begin{bmatrix} 1 & 2 \\ 3 & 4 \\ 5 & 6 \end{bmatrix} \neq \begin{bmatrix} 1 & 3 & 5 \\ 2 & 4 & 6 \end{bmatrix}$$

Equality of Matrices

The matrices $A = [a_{ij}]$ and $B = [b_{ij}]$ are **equal** if and only if they have the same dimension $m \times n$, and corresponding entries are equal, that is,

$$a_{ij} = b_{ij}$$

for $i = 1, 2, \ldots, m$ and $j = 1, 2, \ldots, n$.

Example 1   Equal Matrices

Find a, b, c, and d, if

$$\begin{bmatrix} a & b \\ c & d \end{bmatrix} = \begin{bmatrix} 1 & 3 \\ 5 & 2 \end{bmatrix}$$

Solution   Since the two matrices are equal, corresponding entries must be the same. So we must have $a = 1$, $b = 3$, $c = 5$, and $d = 2$. ∎

Addition, Subtraction, and Scalar Multiplication of Matrices

Two matrices can be added or subtracted if they have the same dimension. (Otherwise, their sum or difference is undefined.) We add or subtract the matrices by adding or subtracting corresponding entries. To multiply a matrix by a number, we multiply every element of the matrix by that number. This is called the *scalar product*.

Sum, Difference, and Scalar Product of Matrices

Let $A = [a_{ij}]$ and $B = [b_{ij}]$ be matrices of the same dimension $m \times n$, and let c be any real number.

1. The **sum** $A + B$ is the $m \times n$ matrix obtained by adding corresponding entries of A and B.

$$A + B = [a_{ij} + b_{ij}]$$

2. The **difference** $A - B$ is the $m \times n$ matrix obtained by subtracting corresponding entries of A and B.

$$A - B = [a_{ij} - b_{ij}]$$

3. The **scalar product** cA is the $m \times n$ matrix obtained by multiplying each entry of A by c.

$$cA = [ca_{ij}]$$

Example 2   **Performing Algebraic Operations on Matrices**

Let
$$A = \begin{bmatrix} 2 & -3 \\ 0 & 5 \\ 7 & -\frac{1}{2} \end{bmatrix} \qquad B = \begin{bmatrix} 1 & 0 \\ -3 & 1 \\ 2 & 2 \end{bmatrix}$$

$$C = \begin{bmatrix} 7 & -3 & 0 \\ 0 & 1 & 5 \end{bmatrix} \qquad D = \begin{bmatrix} 6 & 0 & -6 \\ 8 & 1 & 9 \end{bmatrix}$$

Carry out each indicated operation, or explain why it cannot be performed.

(a) $A + B$ (b) $C - D$ (c) $C + A$ (d) $5A$

Solution

(a) $A + B = \begin{bmatrix} 2 & -3 \\ 0 & 5 \\ 7 & -\frac{1}{2} \end{bmatrix} + \begin{bmatrix} 1 & 0 \\ -3 & 1 \\ 2 & 2 \end{bmatrix} = \begin{bmatrix} 3 & -3 \\ -3 & 6 \\ 9 & \frac{3}{2} \end{bmatrix}$

(b) $C - D = \begin{bmatrix} 7 & -3 & 0 \\ 0 & 1 & 5 \end{bmatrix} - \begin{bmatrix} 6 & 0 & -6 \\ 8 & 1 & 9 \end{bmatrix} = \begin{bmatrix} 1 & -3 & 6 \\ -8 & 0 & -4 \end{bmatrix}$

(c) $C + A$ is undefined because we can't add matrices of different dimensions.

(d) $5A = 5 \begin{bmatrix} 2 & -3 \\ 0 & 5 \\ 7 & -\frac{1}{2} \end{bmatrix} = \begin{bmatrix} 10 & -15 \\ 0 & 25 \\ 35 & -\frac{5}{2} \end{bmatrix}$ ■

The properties in the box follow from the definitions of matrix addition and scalar multiplication, and the corresponding properties of real numbers.

Properties of Addition and Scalar Multiplication of Matrices

Let A, B, and C be $m \times n$ matrices and let c and d be scalars.

$A + B = B + A$	Commutative Property of Matrix Addition
$(A + B) + C = A + (B + C)$	Associative Property of Matrix Addition
$c(dA) = (cd)A$	Associative Property of Scalar Multiplication
$(c + d)A = cA + dA$	Distributive Properties of Scalar
$c(A + B) = cA + cB$	Multiplication

Example 3   **Solving a Matrix Equation**

Solve the matrix equation
$$2X - A = B$$

for the unknown matrix X, where

$$A = \begin{bmatrix} 2 & 3 \\ -5 & 1 \end{bmatrix} \qquad B = \begin{bmatrix} 4 & -1 \\ 1 & 3 \end{bmatrix}$$

Julia Robinson (1919–1985) was born in St. Louis, Missouri, and grew up at Point Loma, California. Due to an illness, Robinson missed two years of school but later, with the aid of a tutor, she completed fifth, sixth, seventh, and eighth grades, all in one year. Later at San Diego State University, reading biographies of mathematicians in E. T. Bell's *Men of Mathematics* awakened in her what became a lifelong passion for mathematics. She said, "I cannot overemphasize the importance of such books . . . in the intellectual life of a student." Robinson is famous for her work on Hilbert's tenth problem (page 708), which asks for a general procedure for determining whether an equation has integer solutions. Her ideas led to a complete answer to the problem. Interestingly, the answer involved certain properties of the Fibonacci numbers (page 826) discovered by the then 22-year-old Russian mathematician Yuri Matijasevič. As a result of her brilliant work on Hilbert's tenth problem, Robinson was offered a professorship at the University of California, Berkeley, and became the first woman mathematician elected to the National Academy of Sciences. She also served as president of the American Mathematical Society.

Solution   We use the properties of matrices to solve for X.

$$2X - A = B \qquad \text{\color{blue}{Given equation}}$$

$$2X = B + A \qquad \text{\color{blue}{Add the matrix A to each side}}$$

$$X = \tfrac{1}{2}(B + A) \qquad \text{\color{blue}{Multiply each side by the scalar } \tfrac{1}{2}}$$

So

$$X = \frac{1}{2}\left(\begin{bmatrix} 4 & -1 \\ 1 & 3 \end{bmatrix} + \begin{bmatrix} 2 & 3 \\ -5 & 1 \end{bmatrix} \right) \qquad \text{\color{blue}{Substitute the matrices A and B}}$$

$$= \frac{1}{2}\begin{bmatrix} 6 & 2 \\ -4 & 4 \end{bmatrix} \qquad \text{\color{blue}{Add matrices}}$$

$$= \begin{bmatrix} 3 & 1 \\ -2 & 2 \end{bmatrix} \qquad \text{\color{blue}{Multiply by the scalar } \tfrac{1}{2}} \qquad \blacksquare$$

Multiplication of Matrices

Multiplying two matrices is more difficult to describe than other matrix operations. In later examples we will see why taking the matrix product involves a rather complex procedure, which we now describe.

First, the product AB (or $A \cdot B$) of two matrices A and B is defined only when the number of columns in A is equal to the number of rows in B. This means that if we write their dimensions side by side, the two inner numbers must match:

Matrices	A	B
Dimensions	$m \times n$	$n \times k$
	Columns in A	Rows in B

If the dimensions of A and B match in this fashion, then the product AB is a matrix of dimension $m \times k$. Before describing the procedure for obtaining the elements of AB, we define the *inner product* of a row of A and a column of B.

If $\begin{bmatrix} a_1 & a_2 & \cdots & a_n \end{bmatrix}$ is a row of A, and if $\begin{bmatrix} b_1 \\ b_2 \\ \vdots \\ b_n \end{bmatrix}$ is a column of B, then

their **inner product** is the number $a_1 b_1 + a_2 b_2 + \cdots + a_n b_n$. For example, taking

the inner product of $\begin{bmatrix} 2 & -1 & 0 & 4 \end{bmatrix}$ and $\begin{bmatrix} 5 \\ 4 \\ -3 \\ \tfrac{1}{2} \end{bmatrix}$ gives

$$2 \cdot 5 + (-1) \cdot 4 + 0 \cdot (-3) + 4 \cdot \tfrac{1}{2} = 8$$

We now define the **product** AB of two matrices.

Matrix Multiplication

If $A = [a_{ij}]$ is an $m \times n$ matrix and $B = [b_{ij}]$ an $n \times k$ matrix, then their product is the $m \times k$ matrix

$$C = [c_{ij}]$$

where c_{ij} is the inner product of the ith row of A and the jth column of B. We write the product as

$$C = AB$$

This definition of matrix product says that each entry in the matrix AB is obtained from a *row* of A and a *column* of B as follows: The entry c_{ij} in the ith row and jth column of the matrix AB is obtained by multiplying the entries in the ith row of A with the corresponding entries in the jth column of B and adding the results.

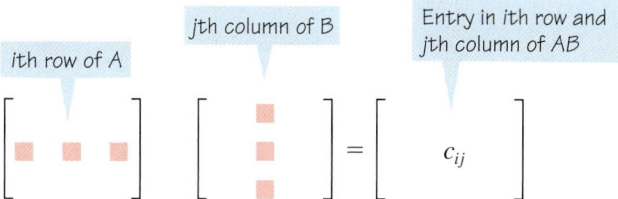

Example 4 Multiplying Matrices

Let

$$A = \begin{bmatrix} 1 & 3 \\ -1 & 0 \end{bmatrix} \quad \text{and} \quad B = \begin{bmatrix} -1 & 5 & 2 \\ 0 & 4 & 7 \end{bmatrix}$$

Calculate, if possible, the products AB and BA.

Solution Since A has dimension 2×2 and B has dimension 2×3, the product AB is defined and has dimension 2×3. We can thus write

$$AB = \begin{bmatrix} 1 & 3 \\ -1 & 0 \end{bmatrix}\begin{bmatrix} -1 & 5 & 2 \\ 0 & 4 & 7 \end{bmatrix} = \begin{bmatrix} ? & ? & ? \\ ? & ? & ? \end{bmatrix}$$

Inner numbers match, so product is defined.

$$\underbrace{2 \times 2}_{} \quad \underbrace{2 \times 3}_{}$$

Outer numbers give dimension of product: 2×3.

where the question marks must be filled in using the rule defining the product of two matrices. If we define $C = AB = [c_{ij}]$, then the entry c_{11} is the inner product of the first row of A and the first column of B:

$$\begin{bmatrix} 1 & 3 \\ -1 & 0 \end{bmatrix}\begin{bmatrix} -1 & 5 & 2 \\ 0 & 4 & 7 \end{bmatrix} \qquad 1 \cdot (-1) + 3 \cdot 0 = -1$$

Similarly, we calculate the remaining entries of the product as follows.

Entry	Inner product of:	Value	Product matrix
c_{12}	$\begin{bmatrix} 1 & 3 \\ -1 & 0 \end{bmatrix}\begin{bmatrix} -1 & 5 & 2 \\ 0 & 4 & 7 \end{bmatrix}$	$1 \cdot 5 + 3 \cdot 4 = 17$	$\begin{bmatrix} -1 & 17 & \\ & & \end{bmatrix}$
c_{13}	$\begin{bmatrix} 1 & 3 \\ -1 & 0 \end{bmatrix}\begin{bmatrix} -1 & 5 & 2 \\ 0 & 4 & 7 \end{bmatrix}$	$1 \cdot 2 + 3 \cdot 7 = 23$	$\begin{bmatrix} -1 & 17 & 23 \\ & & \end{bmatrix}$
c_{21}	$\begin{bmatrix} 1 & 3 \\ -1 & 0 \end{bmatrix}\begin{bmatrix} -1 & 5 & 2 \\ 0 & 4 & 7 \end{bmatrix}$	$(-1) \cdot (-1) + 0 \cdot 0 = 1$	$\begin{bmatrix} -1 & 17 & 23 \\ 1 & & \end{bmatrix}$
c_{22}	$\begin{bmatrix} 1 & 3 \\ -1 & 0 \end{bmatrix}\begin{bmatrix} -1 & 5 & 2 \\ 0 & 4 & 7 \end{bmatrix}$	$(-1) \cdot 5 + 0 \cdot 4 = -5$	$\begin{bmatrix} -1 & 17 & 23 \\ 1 & -5 & \end{bmatrix}$
c_{23}	$\begin{bmatrix} 1 & 3 \\ -1 & 0 \end{bmatrix}\begin{bmatrix} -1 & 5 & 2 \\ 0 & 4 & 7 \end{bmatrix}$	$(-1) \cdot 2 + 0 \cdot 7 = -2$	$\begin{bmatrix} -1 & 17 & 23 \\ 1 & -5 & -2 \end{bmatrix}$

Thus, we have
$$AB = \begin{bmatrix} -1 & 17 & 23 \\ 1 & -5 & -2 \end{bmatrix}$$

> Not equal, so product not defined.
>
> 2×3 2×2

The product BA is not defined, however, because the dimensions of B and A are

$$2 \times 3 \qquad \text{and} \qquad 2 \times 2$$

The inner two numbers are not the same, so the rows and columns won't match up when we try to calculate the product. ■

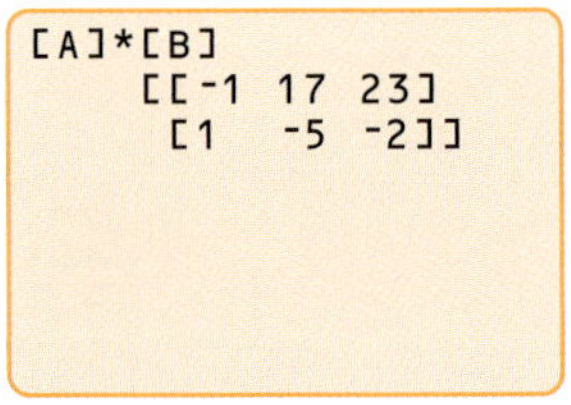

```
[A]*[B]
    [[-1 17 23]
    [1  -5 -2]]
```

Figure 1

Graphing calculators and computers are capable of performing matrix algebra. For instance, if we enter the matrices in Example 4 into the matrix variables `[A]` and `[B]` on a TI-83 calculator, then the calculator finds their product as shown in Figure 1.

Properties of Matrix Multiplication

Although matrix multiplication is not commutative, it does obey the Associative and Distributive Properties.

Properties of Matrix Multiplication

Let A, B, and C be matrices for which the following products are defined. Then

$$A(BC) = (AB)C \qquad \text{Associative Property}$$

$$A(B + C) = AB + AC$$
$$(B + C)A = BA + CA \qquad \text{Distributive Property}$$

The next example shows that even when both *AB* and *BA* are defined, they aren't necessarily equal. This result proves that matrix multiplication is *not* commutative.

Example 5   Matrix Multiplication Is Not Commutative

Let
$$A = \begin{bmatrix} 5 & 7 \\ -3 & 0 \end{bmatrix} \quad \text{and} \quad B = \begin{bmatrix} 1 & 2 \\ 9 & -1 \end{bmatrix}$$

Calculate the products *AB* and *BA*.

Solution   Since both matrices *A* and *B* have dimension 2 × 2, both products *AB* and *BA* are defined, and each product is also a 2 × 2 matrix.

$$AB = \begin{bmatrix} 5 & 7 \\ -3 & 0 \end{bmatrix}\begin{bmatrix} 1 & 2 \\ 9 & -1 \end{bmatrix} = \begin{bmatrix} 5\cdot 1 + 7\cdot 9 & 5\cdot 2 + 7\cdot(-1) \\ (-3)\cdot 1 + 0\cdot 9 & (-3)\cdot 2 + 0\cdot(-1) \end{bmatrix}$$
$$= \begin{bmatrix} 68 & 3 \\ -3 & -6 \end{bmatrix}$$

$$BA = \begin{bmatrix} 1 & 2 \\ 9 & -1 \end{bmatrix}\begin{bmatrix} 5 & 7 \\ -3 & 0 \end{bmatrix} = \begin{bmatrix} 1\cdot 5 + 2\cdot(-3) & 1\cdot 7 + 2\cdot 0 \\ 9\cdot 5 + (-1)\cdot(-3) & 9\cdot 7 + (-1)\cdot 0 \end{bmatrix}$$
$$= \begin{bmatrix} -1 & 7 \\ 48 & 63 \end{bmatrix}$$

This shows that, in general, $AB \neq BA$. In fact, in this example *AB* and *BA* don't even have an entry in common.   ∎

Applications of Matrix Multiplication

We now consider some applied examples that give some indication of why mathematicians chose to define the matrix product in such an apparently bizarre fashion. Example 6 shows how our definition of matrix product allows us to express a system of linear equations as a single matrix equation.

Example 6   Writing a Linear System as a Matrix Equation

Show that the following matrix equation is equivalent to the system of equations in Example 2 of Section 9.4.

$$\begin{bmatrix} 1 & -1 & 3 \\ 1 & 2 & -2 \\ 3 & -1 & 5 \end{bmatrix}\begin{bmatrix} x \\ y \\ z \end{bmatrix} = \begin{bmatrix} 4 \\ 10 \\ 14 \end{bmatrix}$$

Matrix equations like this one are described in more detail on page 694.

Solution   If we perform matrix multiplication on the left side of the equation, we get

$$\begin{bmatrix} x - y + 3z \\ x + 2y - 2z \\ 3x - y + 5z \end{bmatrix} = \begin{bmatrix} 4 \\ 10 \\ 14 \end{bmatrix}$$

Because two matrices are equal only if their corresponding entries are equal, we equate entries to get

$$\begin{cases} x - y + 3z = 4 \\ x + 2y - 2z = 10 \\ 3x - y + 5z = 14 \end{cases}$$

This is exactly the system of equations in Example 2 of Section 9.4.   ∎

Example 7   Representing Demographic Data by Matrices

In a certain city the proportion of voters in each age group who are registered as Democrats, Republicans, or Independents is given by the following matrix.

	Age		
	18–30	31–50	Over 50
Democrat	0.30	0.60	0.50
Republican	0.50	0.35	0.25
Independent	0.20	0.05	0.25

$= A$

The next matrix gives the distribution, by age and sex, of the voting population of this city.

		Male	Female
Age	18–30	5,000	6,000
	31–50	10,000	12,000
	Over 50	12,000	15,000

$= B$

For this problem, let's make the (highly unrealistic) assumption that within each age group, political preference is not related to gender. That is, the percentage of Democrat males in the 18–30 group, for example, is the same as the percentage of Democrat females in this group.

(a) Calculate the product AB.

(b) How many males are registered as Democrats in this city?

(c) How many females are registered as Republicans?

Solution

(a) $AB = \begin{bmatrix} 0.30 & 0.60 & 0.50 \\ 0.50 & 0.35 & 0.25 \\ 0.20 & 0.05 & 0.25 \end{bmatrix} \begin{bmatrix} 5,000 & 6,000 \\ 10,000 & 12,000 \\ 12,000 & 15,000 \end{bmatrix} = \begin{bmatrix} 13,500 & 16,500 \\ 9,000 & 10,950 \\ 4,500 & 5,550 \end{bmatrix}$

(b) When we take the inner product of a row in A with a column in B, we are adding the number of people in each age group who belong to the category in question. For example, the entry c_{21} of AB (the 9000) is obtained by taking the inner product of the Republican row in A with the Male column in B. This

and can give all of his or her votes to one candidate or distribute them among the candidates as he or she sees fit). This last system is often used to select corporate boards of directors. Each system of voting has both advantages and disadvantages.

number is therefore the total number of male Republicans in this city. We can label the rows and columns of AB as follows.

$$
\begin{array}{c}
 & \text{Male} & \text{Female} \\
\begin{array}{c}\text{Democrat}\\\text{Republican}\\\text{Independent}\end{array} & \begin{bmatrix} 13{,}500 & 16{,}500 \\ 9{,}000 & 10{,}950 \\ 4{,}500 & 5{,}550 \end{bmatrix} = AB
\end{array}
$$

Thus, 13,500 males are registered as Democrats in this city.

(c) There are 10,950 females registered as Republicans. ∎

In Example 7, the entries in each column of A add up to 1. (Can you see why this has to be true, given what the matrix describes?) A matrix with this property is called **stochastic**. Stochastic matrices are used extensively in statistics, where they arise frequently in situations like the one described here.

Computer Graphics

One important use of matrices is in the digital representation of images. A digital camera or a scanner converts an image into a matrix by dividing the image into a rectangular array of elements called pixels. Each pixel is assigned a value that represents the color, brightness, or some other feature of that location. For example, in a 256-level gray-scale image each pixel is assigned a value between 0 and 255, where 0 represents white, 255 black, and the numbers in between increasing gradations of gray. The gradations of a much simpler 8-level gray scale are shown in Figure 2. We use this 8-level gray scale to illustrate the process.

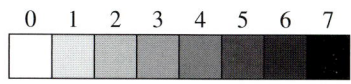

Figure 2

(a) Original image

(b) 10 × 10 grid

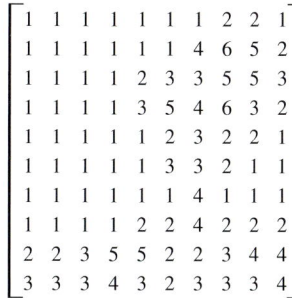
(c) Matrix representation

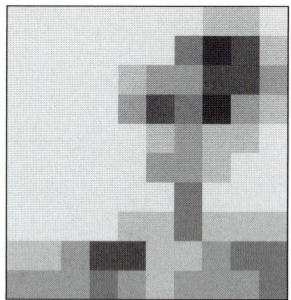
(d) Digital image

Figure 3

To digitize the black and white image in Figure 3(a), we place a grid over the picture as shown in Figure 3(b). Each cell in the grid is compared to the gray scale, and then assigned a value between 0 and 7 depending on which gray square in the scale most closely matches the "darkness" of the cell. (If the cell is not uniformly gray, an average value is assigned.) The values are stored in the matrix shown in Figure 3(c). The digital image corresponding to this matrix is shown in Figure 3(d). Obviously the grid that we have used is far too coarse to provide good image resolution. In practice, currently available high-resolution digital cameras use matrices with dimensions 2048×2048 or larger.

Once the image is stored as a matrix, it can be manipulated using matrix operations. For example, to darken the image, we add a constant to each entry in the matrix; to lighten the image, we subtract. To increase the contrast, we darken the darker areas and lighten the lighter areas, so we could add 1 to each entry that is 4, 5, or 6 and subtract 1 from each entry that is 1, 2, or 3. (Note that we cannot darken an entry of 7 or lighten a 0.) Applying this process to the matrix in Figure 3(c) produces the new matrix in Figure 4(a). This generates the high-contrast image shown in Figure 4(b).

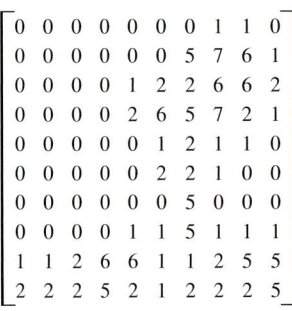

Figure 4

(a) Matrix modified to increase contrast

(b) High-contrast image

Other ways of representing and manipulating images using matrices are discussed in the *Discovery Projects* on pages 700 and 792.

9.5   Exercises

1–2 ■ Determine whether the matrices A and B are equal.

1. $A = \begin{bmatrix} 1 & -2 & 0 \\ \frac{1}{2} & 6 & 0 \end{bmatrix}$, $B = \begin{bmatrix} 1 & -2 \\ \frac{1}{2} & 6 \end{bmatrix}$

2. $A = \begin{bmatrix} \frac{1}{4} & \ln 1 \\ 2 & 3 \end{bmatrix}$, $B = \begin{bmatrix} 0.25 & 0 \\ \sqrt{4} & \frac{6}{2} \end{bmatrix}$

3–10 ■ Perform the matrix operation, or if it is impossible, explain why.

3. $\begin{bmatrix} 2 & 6 \\ -5 & 3 \end{bmatrix} + \begin{bmatrix} -1 & -3 \\ 6 & 2 \end{bmatrix}$

4. $\begin{bmatrix} 0 & 1 & 1 \\ 1 & 1 & 0 \end{bmatrix} - \begin{bmatrix} 2 & 1 & -1 \\ 1 & 3 & -2 \end{bmatrix}$

5. $3 \begin{bmatrix} 1 & 2 \\ 4 & -1 \\ 1 & 0 \end{bmatrix}$

6. $2 \begin{bmatrix} 1 & 1 & 0 \\ 1 & 0 & 1 \\ 0 & 1 & 1 \end{bmatrix} + \begin{bmatrix} 1 & 1 \\ 2 & 1 \\ 3 & 1 \end{bmatrix}$

7. $\begin{bmatrix} 2 & 6 \\ 1 & 3 \\ 2 & 4 \end{bmatrix} \begin{bmatrix} 1 & -2 \\ 3 & 6 \\ -2 & 0 \end{bmatrix}$

8. $\begin{bmatrix} 2 & 1 & 2 \\ 6 & 3 & 4 \end{bmatrix} \begin{bmatrix} 1 & -2 \\ 3 & 6 \\ -2 & 0 \end{bmatrix}$

9. $\begin{bmatrix} 1 & 2 \\ -1 & 4 \end{bmatrix} \begin{bmatrix} 1 & -2 & 3 \\ 2 & 2 & -1 \end{bmatrix}$

10. $\begin{bmatrix} 2 & -3 \\ 0 & 1 \\ 1 & 2 \end{bmatrix} \begin{bmatrix} 5 \\ 1 \end{bmatrix}$

11–16 ■ Solve the matrix equation for the unknown matrix X, or explain why no solution exists.

$$A = \begin{bmatrix} 4 & 6 \\ 1 & 3 \end{bmatrix} \qquad B = \begin{bmatrix} 2 & 5 \\ 3 & 7 \end{bmatrix}$$

$$C = \begin{bmatrix} 2 & 3 \\ 1 & 0 \\ 0 & 2 \end{bmatrix} \qquad D = \begin{bmatrix} 10 & 20 \\ 30 & 20 \\ 10 & 0 \end{bmatrix}$$

11. $2X + A = B$ **12.** $3X - B = C$

13. $2(B - X) = D$ **14.** $5(X - C) = D$

15. $\frac{1}{5}(X + D) = C$ **16.** $2A = B - 3X$

17–38 ■ The matrices A, B, C, D, E, F, and G are defined as follows.

$$A = \begin{bmatrix} 2 & -5 \\ 0 & 7 \end{bmatrix} \quad B = \begin{bmatrix} 3 & \frac{1}{2} & 5 \\ 1 & -1 & 3 \end{bmatrix} \quad C = \begin{bmatrix} 2 & -\frac{5}{2} & 0 \\ 0 & 2 & -3 \end{bmatrix}$$

$$D = \begin{bmatrix} 7 & 3 \end{bmatrix} \qquad E = \begin{bmatrix} 1 \\ 2 \\ 0 \end{bmatrix}$$

$$F = \begin{bmatrix} 1 & 0 & 0 \\ 0 & 1 & 0 \\ 0 & 0 & 1 \end{bmatrix} \qquad G = \begin{bmatrix} 5 & -3 & 10 \\ 6 & 1 & 0 \\ -5 & 2 & 2 \end{bmatrix}$$

Carry out the indicated algebraic operation, or explain why it cannot be performed.

17. $B + C$ **18.** $B + F$

19. $C - B$ **20.** $5A$

21. $3B + 2C$ **22.** $C - 5A$

23. $2C - 6B$ **24.** DA

25. AD **26.** BC

27. BF **28.** GF

29. $(DA)B$ **30.** $D(AB)$

31. GE **32.** A^2

33. A^3 **34.** $DB + DC$

35. B^2 **36.** F^2

37. $BF + FE$ **38.** ABE

39–42 ■ Solve for x and y.

39. $\begin{bmatrix} x & 2y \\ 4 & 6 \end{bmatrix} = \begin{bmatrix} 2 & -2 \\ 2x & -6y \end{bmatrix}$

40. $3\begin{bmatrix} x & y \\ y & x \end{bmatrix} = \begin{bmatrix} 6 & -9 \\ -9 & 6 \end{bmatrix}$

41. $2\begin{bmatrix} x & y \\ x+y & x-y \end{bmatrix} = \begin{bmatrix} 2 & -4 \\ -2 & 6 \end{bmatrix}$

42. $\begin{bmatrix} x & y \\ -y & x \end{bmatrix} - \begin{bmatrix} y & x \\ x & -y \end{bmatrix} = \begin{bmatrix} 4 & -4 \\ -6 & 6 \end{bmatrix}$

43–46 ■ Write the system of equations as a matrix equation (see Example 6).

43. $\begin{cases} 2x - 5y = 7 \\ 3x + 2y = 4 \end{cases}$

44. $\begin{cases} 6x - y + z = 12 \\ 2x \quad\; + z = 7 \\ \quad\; y - 2z = 4 \end{cases}$

45. $\begin{cases} 3x_1 + 2x_2 - x_3 + x_4 = 0 \\ x_1 \quad\quad - x_3 \quad\quad = 5 \\ \quad\; 3x_2 + x_3 - x_4 = 4 \end{cases}$

46. $\begin{cases} x - y + z = 2 \\ 4x - 2y - z = 2 \\ x + y + 5z = 2 \\ -x - y - z = 2 \end{cases}$

47. Let

$$A = \begin{bmatrix} 1 & 0 & 6 & -1 \\ 2 & \frac{1}{2} & 4 & 0 \end{bmatrix}$$

$$B = \begin{bmatrix} 1 & 7 & -9 & 2 \end{bmatrix}$$

$$C = \begin{bmatrix} 1 \\ 0 \\ -1 \\ -2 \end{bmatrix}$$

Determine which of the following products are defined, and calculate the ones that are:

$ABC \qquad ACB \qquad BAC$

$BCA \qquad CAB \qquad CBA$

48. (a) Prove that if A and B are 2×2 matrices, then

$$(A + B)^2 = A^2 + AB + BA + B^2$$

(b) If A and B are 2×2 matrices, is it necessarily true that

$$(A + B)^2 \overset{?}{=} A^2 + 2AB + B^2$$

Applications

49. Fast-Food Sales   A small fast-food chain with restaurants in Santa Monica, Long Beach, and Anaheim sells only hamburgers, hot dogs, and milk shakes. On a certain day, sales were distributed according to the following matrix.

Number of items sold

	Santa Monica	Long Beach	Anaheim	
Hamburgers	4000	1000	3500	
Hot dogs	400	300	200	= A
Milk shakes	700	500	9000	

The price of each item is given by the following matrix.

Hamburger	Hot dog	Milk Shake
[$0.90	$0.80	$1.10] = B

(a) Calculate the product BA.

(b) Interpret the entries in the product matrix BA.

50. Car-Manufacturing Profits   A specialty-car manufacturer has plants in Auburn, Biloxi, and Chattanooga. Three models are produced, with daily production given in the following matrix.

Cars produced each day

	Model K	Model R	Model W	
Auburn	12	10	0	
Biloxi	4	4	20	= A
Chattanooga	8	9	12	

Because of a wage increase, February profits are less than January profits. The profit per car is tabulated by model in the following matrix.

	January	February	
Model K	$1000	$500	
Model R	$2000	$1200	= B
Model W	$1500	$1000	

(a) Calculate AB.

(b) Assuming all cars produced were sold, what was the daily profit in January from the Biloxi plant?

(c) What was the total daily profit (from all three plants) in February?

51. Canning Tomato Products   Jaeger Foods produces tomato sauce and tomato paste, canned in small, medium, large, and giant sized tins. The matrix A gives the size (in ounces) of each container.

	Small	Medium	Large	Giant	
Ounces	[6	10	14	28] = A

The matrix B tabulates one day's production of tomato sauce and tomato paste.

	Cans of sauce	Cans of paste	
Small	2000	2500	
Medium	3000	1500	
Large	2500	1000	= B
Giant	1000	500	

(a) Calculate the product of AB.

(b) Interpret the entries in the product matrix AB.

52. Produce Sales A farmer's three children, Amy, Beth, and Chad, run three roadside produce stands during the summer months. One weekend they all sell watermelons, yellow squash, and tomatoes. The matrices A and B tabulate the number of pounds of each product sold by each sibling on Saturday and Sunday.

	Saturday		
	Melons	Squash	Tomatoes
Amy	120	50	60
Beth	40	25	30
Chad	60	30	20

$$= A$$

	Sunday		
	Melons	Squash	Tomatoes
Amy	100	60	30
Beth	35	20	20
Chad	60	25	30

$$= B$$

The matrix C gives the price per pound (in dollars) for each type of produce that they sell.

Price per pound

Melons	0.10
Squash	0.50
Tomatoes	1.00

$$= C$$

Perform the following matrix operations, and interpret the entries in each result.

(a) AC (b) BC (c) $A + B$ (d) $(A + B)C$

53. Digital Images A four-level gray scale is shown below.

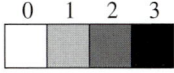

0 1 2 3

(a) Use the gray scale to find a 6×6 matrix that digitally represents the image in the figure.

(b) Find a matrix that represents a darker version of the image in the figure.

(c) The **negative** of an image is obtained by reversing light and dark, as in the negative of a photograph. Find the matrix that represents the negative of the image in the figure. How do you change the elements of the matrix to create the negative?

(d) Increase the contrast of the image by changing each 1 to a 0 and each 2 to a 3 in the matrix you found in part (b). Draw the image represented by the resulting matrix. Does this clarify the image?

(e) Draw the image represented by the matrix I. Can you recognize what this is? If you don't, try increasing the contrast.

$$I = \begin{bmatrix} 1 & 2 & 3 & 3 & 2 & 0 \\ 0 & 3 & 0 & 1 & 0 & 1 \\ 1 & 3 & 2 & 3 & 0 & 0 \\ 0 & 3 & 0 & 1 & 0 & 1 \\ 1 & 3 & 3 & 2 & 3 & 0 \\ 0 & 1 & 0 & 1 & 0 & 1 \end{bmatrix}$$

Discovery • Discussion

54. When Are Both Products Defined? What must be true about the dimensions of the matrices A and B if both products AB and BA are defined?

55. Powers of a Matrix Let

$$A = \begin{bmatrix} 1 & 1 \\ 0 & 1 \end{bmatrix}$$

Calculate $A^2, A^3, A^4, \ldots$ until you detect a pattern. Write a general formula for A^n.

56. Powers of a Matrix Let $A = \begin{bmatrix} 1 & 1 \\ 1 & 1 \end{bmatrix}$. Calculate A^2, A^3, $A^4, \ldots$ until you detect a pattern. Write a general formula for A^n.

57. Square Roots of Matrices A square root of a matrix B is a matrix A with the property that $A^2 = B$. (This is the same definition as for a square root of a number.) Find as many square roots as you can of each matrix:

$$\begin{bmatrix} 4 & 0 \\ 0 & 9 \end{bmatrix} \qquad \begin{bmatrix} 1 & 5 \\ 0 & 9 \end{bmatrix}$$

[*Hint:* If $A = \begin{bmatrix} a & b \\ c & d \end{bmatrix}$, write the equations that a, b, c, and d would have to satisfy if A is the square root of the given matrix.]

DISCOVERY PROJECT

Will the Species Survive?

To study how species survive, mathematicians model their populations by observing the different stages in their life. They consider, for example, the stage at which the animal is fertile, the proportion of the population that reproduces, and the proportion of the young that survive each year. For a certain species, there are three stages: immature, juvenile, and adult. An animal is considered immature for the first year of its life, juvenile for the second year, and an adult from then on. Conservation biologists have collected the following field data for this species:

$$
A = \begin{bmatrix} \text{Immature} & \text{Juvenile} & \text{Adult} \\ 0 & 0 & 0.4 \\ 0.1 & 0 & 0 \\ 0 & 0.3 & 0.8 \end{bmatrix} \begin{matrix} \text{Immature} \\ \text{Juvenile} \\ \text{Adult} \end{matrix}
\qquad
X_0 = \begin{bmatrix} 600 \\ 400 \\ 3500 \end{bmatrix} \begin{matrix} \text{Immature} \\ \text{Juvenile} \\ \text{Adult} \end{matrix}
$$

Art Wolfe/Stone/Getty Images

The entries in the matrix A indicate the proportion of the population that survives *to the next year*. For example, the first column describes what happens to the immature population: None remain immature, 10% survive to become juveniles, and of course none become adults. The second column describes what happens to the juvenile population: None become immature or remain juvenile, and 30% survive to adulthood. The third column describes the adult population: The number of their new offspring is 40% of the adult population, no adults become juveniles, and 80% survive to live another year. The entries in the population matrix X_0 indicate the current population (year 0) of immature, juvenile, and adult animals.

Let $X_1 = AX_0$, $X_2 = AX_1$, $X_3 = AX_2$, and so on.

1. Explain why X_1 gives the population in year 1, X_2 the population in year 2, and so on.

2. Find the population matrix for years 1, 2, 3, and 4. (Round fractional entries to the nearest whole number.) Do you see any trend?

3. Show that $X_2 = A^2 X_0$, $X_3 = A^3 X_0$, and so on.

4. Find the population after 50 years—that is, find X_{50}. (Use your results in Problem 3 and a graphing calculator.) Does it appear that the species will survive?

5. Suppose the environment has improved so that the proportion of immatures that become juveniles each year increases to 0.1 from 0.3, the proportion of juveniles that become adults increases to 0.3 from 0.7, and the proportion of adults that survives to the next year increases from 0.8 to 0.95. Find the population after 50 years with the new matrix A. Does it appear that the species will survive under these new conditions?

6. The survival-rate matrix A given above is called a **transition matrix**. Such matrices occur in many applications of matrix algebra. The following transition matrix T predicts the calculus grades of a class of college students who must take a four-semester sequence of calculus courses. The first column of the matrix, for instance, indicates that of those students who get an A in one course, 70% will get an A in the following course, 15% will get a B, and 10% will get a C. (Students who receive D or F are not permitted to go on to the next course, so are not included in the matrix.) The entries in the matrix Y_0 give the number of incoming students who got A, B, and C, respectively, in their final high school mathematics course.

Let $Y_1 = TY_0$, $Y_2 = TY_1$, $Y_3 = TY_2$, and $Y_4 = TY_3$. Calculate and interpret the entries of Y_1, Y_2, Y_3, and Y_4.

$$T = \begin{array}{c} \\ \end{array}\overset{\begin{array}{ccc}\text{A} & \text{B} & \text{C}\end{array}}{\begin{bmatrix} 0.70 & 0.25 & 0.05 \\ 0.15 & 0.50 & 0.25 \\ 0.05 & 0.15 & 0.45 \end{bmatrix}}\begin{array}{c}\text{A}\\ \text{B}\\ \text{C}\end{array} \qquad Y_0 = \begin{bmatrix} 140 \\ 320 \\ 400 \end{bmatrix}\begin{array}{c}\text{A}\\ \text{B}\\ \text{C}\end{array}$$

9.6 Inverses of Matrices and Matrix Equations

In the preceding section we saw that, when the dimensions are appropriate, matrices can be added, subtracted, and multiplied. In this section we investigate division of matrices. With this operation we can solve equations that involve matrices.

The Inverse of a Matrix

First, we define *identity matrices*, which play the same role for matrix multiplication as the number 1 does for ordinary multiplication of numbers; that is, $1 \cdot a = a \cdot 1 = a$ for all numbers a. In the following definition the term **main diagonal** refers to the entries of a square matrix whose row and column numbers are the same. These entries stretch diagonally down the matrix, from top left to bottom right.

The **identity matrix** I_n is the $n \times n$ matrix for which each main diagonal entry is a 1 and for which all other entries are 0.

Thus, the 2×2, 3×3, and 4×4 identity matrices are

$$I_2 = \begin{bmatrix} 1 & 0 \\ 0 & 1 \end{bmatrix} \qquad I_3 = \begin{bmatrix} 1 & 0 & 0 \\ 0 & 1 & 0 \\ 0 & 0 & 1 \end{bmatrix} \qquad I_4 = \begin{bmatrix} 1 & 0 & 0 & 0 \\ 0 & 1 & 0 & 0 \\ 0 & 0 & 1 & 0 \\ 0 & 0 & 0 & 1 \end{bmatrix}$$

Identity matrices behave like the number 1 in the sense that

$$A \cdot I_n = A \qquad \text{and} \qquad I_n \cdot B = B$$

whenever these products are defined.

Example 1   Identity Matrices

The following matrix products show how multiplying a matrix by an identity matrix of the appropriate dimension leaves the matrix unchanged.

$$\begin{bmatrix} 1 & 0 \\ 0 & 1 \end{bmatrix} \begin{bmatrix} 3 & 5 & 6 \\ -1 & 2 & 7 \end{bmatrix} = \begin{bmatrix} 3 & 5 & 6 \\ -1 & 2 & 7 \end{bmatrix}$$

$$\begin{bmatrix} -1 & 7 & \frac{1}{2} \\ 12 & 1 & 3 \\ -2 & 0 & 7 \end{bmatrix} \begin{bmatrix} 1 & 0 & 0 \\ 0 & 1 & 0 \\ 0 & 0 & 1 \end{bmatrix} = \begin{bmatrix} -1 & 7 & \frac{1}{2} \\ 12 & 1 & 3 \\ -2 & 0 & 7 \end{bmatrix}$$ ∎

If A and B are $n \times n$ matrices, and if $AB = BA = I_n$, then we say that B is the *inverse* of A, and we write $B = A^{-1}$. The concept of the inverse of a matrix is analogous to that of the reciprocal of a real number.

Inverse of a Matrix

Let A be a square $n \times n$ matrix. If there exists an $n \times n$ matrix A^{-1} with the property that

$$AA^{-1} = A^{-1}A = I_n$$

then we say that A^{-1} is the **inverse** of A.

Example 2   Verifying That a Matrix Is an Inverse

Verify that B is the inverse of A, where

$$A = \begin{bmatrix} 2 & 1 \\ 5 & 3 \end{bmatrix} \quad \text{and} \quad B = \begin{bmatrix} 3 & -1 \\ -5 & 2 \end{bmatrix}$$

Solution   We perform the matrix multiplications to show that $AB = I$ and $BA = I$:

$$\begin{bmatrix} 2 & 1 \\ 5 & 3 \end{bmatrix} \begin{bmatrix} 3 & -1 \\ -5 & 2 \end{bmatrix} = \begin{bmatrix} 2 \cdot 3 + 1(-5) & 2(-1) + 1 \cdot 2 \\ 5 \cdot 3 + 3(-5) & 5(-1) + 3 \cdot 2 \end{bmatrix} = \begin{bmatrix} 1 & 0 \\ 0 & 1 \end{bmatrix}$$

$$\begin{bmatrix} 3 & -1 \\ -5 & 2 \end{bmatrix} \begin{bmatrix} 2 & 1 \\ 5 & 3 \end{bmatrix} = \begin{bmatrix} 3 \cdot 2 + (-1)5 & 3 \cdot 1 + (-1)3 \\ (-5)2 + 2 \cdot 5 & (-5)1 + 2 \cdot 3 \end{bmatrix} = \begin{bmatrix} 1 & 0 \\ 0 & 1 \end{bmatrix}$$ ∎

Finding the Inverse of a 2 × 2 Matrix

The following rule provides a simple way for finding the inverse of a 2×2 matrix, when it exists. For larger matrices, there's a more general procedure for finding inverses, which we consider later in this section.

Inverse of a 2 × 2 Matrix

$$\text{If } A = \begin{bmatrix} a & b \\ c & d \end{bmatrix} \quad \text{then} \quad A^{-1} = \frac{1}{ad - bc} \begin{bmatrix} d & -b \\ -c & a \end{bmatrix}$$

If $ad - bc = 0$, then A has no inverse.

Example 3   Finding the Inverse of a 2 × 2 Matrix

Let A be the matrix

$$A = \begin{bmatrix} 4 & 5 \\ 2 & 3 \end{bmatrix}$$

Find A^{-1} and verify that $AA^{-1} = A^{-1}A = I_2$.

Solution   Using the rule for the inverse of a 2 × 2 matrix, we get

$$A^{-1} = \frac{1}{4 \cdot 3 - 5 \cdot 2}\begin{bmatrix} 3 & -5 \\ -2 & 4 \end{bmatrix} = \frac{1}{2}\begin{bmatrix} 3 & -5 \\ -2 & 4 \end{bmatrix} = \begin{bmatrix} \frac{3}{2} & -\frac{5}{2} \\ -1 & 2 \end{bmatrix}$$

To verify that this is indeed the inverse of A, we calculate AA^{-1} and $A^{-1}A$:

$$AA^{-1} = \begin{bmatrix} 4 & 5 \\ 2 & 3 \end{bmatrix}\begin{bmatrix} \frac{3}{2} & -\frac{5}{2} \\ -1 & 2 \end{bmatrix} = \begin{bmatrix} 4 \cdot \frac{3}{2} + 5(-1) & 4(-\frac{5}{2}) + 5 \cdot 2 \\ 2 \cdot \frac{3}{2} + 3(-1) & 2(-\frac{5}{2}) + 3 \cdot 2 \end{bmatrix} = \begin{bmatrix} 1 & 0 \\ 0 & 1 \end{bmatrix}$$

$$A^{-1}A = \begin{bmatrix} \frac{3}{2} & -\frac{5}{2} \\ -1 & 2 \end{bmatrix}\begin{bmatrix} 4 & 5 \\ 2 & 3 \end{bmatrix} = \begin{bmatrix} \frac{3}{2} \cdot 4 + (-\frac{5}{2})2 & \frac{3}{2} \cdot 5 + (-\frac{5}{2})3 \\ (-1)4 + 2 \cdot 2 & (-1)5 + 2 \cdot 3 \end{bmatrix} = \begin{bmatrix} 1 & 0 \\ 0 & 1 \end{bmatrix} \blacksquare$$

The quantity $ad - bc$ that appears in the rule for calculating the inverse of a 2 × 2 matrix is called the **determinant** of the matrix. If the determinant is 0, then the matrix does not have an inverse (since we cannot divide by 0).

Finding the Inverse of an $n \times n$ Matrix

For 3 × 3 and larger square matrices, the following technique provides the most efficient way to calculate their inverses. If A is an $n \times n$ matrix, we first construct the $n \times 2n$ matrix that has the entries of A on the left and of the identity matrix I_n on the right:

$$\begin{bmatrix} a_{11} & a_{12} & \cdots & a_{1n} & | & 1 & 0 & \cdots & 0 \\ a_{21} & a_{22} & \cdots & a_{2n} & | & 0 & 1 & \cdots & 0 \\ \vdots & \vdots & \ddots & \vdots & | & \vdots & \vdots & \ddots & \vdots \\ a_{n1} & a_{n2} & \cdots & a_{nn} & | & 0 & 0 & \cdots & 1 \end{bmatrix}$$

We then use the elementary row operations on this new large matrix to change the left side into the identity matrix. (This means that we are changing the large matrix to reduced row-echelon form.) The right side is transformed automatically into A^{-1}. (We omit the proof of this fact.)

Example 4   Finding the Inverse of a 3 × 3 Matrix

Let A be the matrix

$$A = \begin{bmatrix} 1 & -2 & -4 \\ 2 & -3 & -6 \\ -3 & 6 & 15 \end{bmatrix}$$

(a) Find A^{-1}.

(b) Verify that $AA^{-1} = A^{-1}A = I_3$.

Arthur Cayley (1821–1895) was an English mathematician who was instrumental in developing the theory of matrices. He was the first to use a single symbol such as A to represent a matrix, thereby introducing the idea that a matrix is a single entity rather than just a collection of numbers. Cayley practiced law until the age of 42, but his primary interest from adolescence was mathematics, and he published almost 200 articles on the subject in his spare time. In 1863 he accepted a professorship in mathematics at Cambridge, where he taught until his death. Cayley's work on matrices was of purely theoretical interest in his day, but in the 20th century many of his results found application in physics, the social sciences, business, and other fields. One of the most common uses of matrices today is in computers, where matrices are employed for data storage, error correction, image manipulation, and many other purposes. These applications have made matrix algebra more useful than ever.

Solution

(a) We begin with the 3×6 matrix whose left half is A and whose right half is the identity matrix.

$$\left[\begin{array}{rrr|rrr} 1 & -2 & -4 & 1 & 0 & 0 \\ 2 & -3 & -6 & 0 & 1 & 0 \\ -3 & 6 & 15 & 0 & 0 & 1 \end{array}\right]$$

We then transform the left half of this new matrix into the identity matrix by performing the following sequence of elementary row operations on the *entire* new matrix:

$$\xrightarrow[\begin{array}{c} R_2 - 2R_1 \to R_2 \\ R_3 + 3R_1 \to R_3 \end{array}]{} \left[\begin{array}{rrr|rrr} 1 & -2 & -4 & 1 & 0 & 0 \\ 0 & 1 & 2 & -2 & 1 & 0 \\ 0 & 0 & 3 & 3 & 0 & 1 \end{array}\right]$$

$$\xrightarrow{\frac{1}{3}R_3} \left[\begin{array}{rrr|rrr} 1 & -2 & -4 & 1 & 0 & 0 \\ 0 & 1 & 2 & -2 & 1 & 0 \\ 0 & 0 & 1 & 1 & 0 & \frac{1}{3} \end{array}\right]$$

$$\xrightarrow{R_1 + 2R_2 \to R_1} \left[\begin{array}{rrr|rrr} 1 & 0 & 0 & -3 & 2 & 0 \\ 0 & 1 & 2 & -2 & 1 & 0 \\ 0 & 0 & 1 & 1 & 0 & \frac{1}{3} \end{array}\right]$$

$$\xrightarrow{R_2 - 2R_3 \to R_2} \left[\begin{array}{rrr|rrr} 1 & 0 & 0 & -3 & 2 & 0 \\ 0 & 1 & 0 & -4 & 1 & -\frac{2}{3} \\ 0 & 0 & 1 & 1 & 0 & \frac{1}{3} \end{array}\right]$$

We have now transformed the left half of this matrix into an identity matrix. (This means we've put the entire matrix in reduced row-echelon form.) Note that to do this in as systematic a fashion as possible, we first changed the elements below the main diagonal to zeros, just as we would if we were using Gaussian elimination. We then changed each main diagonal element to a 1 by multiplying by the appropriate constant(s). Finally, we completed the process by changing the remaining entries on the left side to zeros.

The right half is now A^{-1}.

$$A^{-1} = \left[\begin{array}{rrr} -3 & 2 & 0 \\ -4 & 1 & -\frac{2}{3} \\ 1 & 0 & \frac{1}{3} \end{array}\right]$$

(b) We calculate AA^{-1} and $A^{-1}A$, and verify that both products give the identity matrix I_3.

$$AA^{-1} = \left[\begin{array}{rrr} 1 & -2 & -4 \\ 2 & -3 & -6 \\ -3 & 6 & 15 \end{array}\right] \left[\begin{array}{rrr} -3 & 2 & 0 \\ -4 & 1 & -\frac{2}{3} \\ 1 & 0 & \frac{1}{3} \end{array}\right] = \left[\begin{array}{rrr} 1 & 0 & 0 \\ 0 & 1 & 0 \\ 0 & 0 & 1 \end{array}\right]$$

$$A^{-1}A = \left[\begin{array}{rrr} -3 & 2 & 0 \\ -4 & 1 & -\frac{2}{3} \\ 1 & 0 & \frac{1}{3} \end{array}\right] \left[\begin{array}{rrr} 1 & -2 & -4 \\ 2 & -3 & -6 \\ -3 & 6 & 15 \end{array}\right] = \left[\begin{array}{rrr} 1 & 0 & 0 \\ 0 & 1 & 0 \\ 0 & 0 & 1 \end{array}\right]$$

■

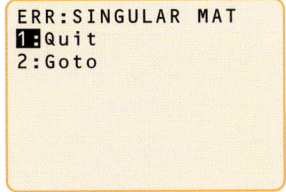

```
[A]⁻¹▶Frac
   [[-3  2  0   ]
    [-4  1 -2/3]
    [1   0  1/3 ]]
```

Figure 1

Graphing calculators are also able to calculate matrix inverses. On the TI-82 and TI-83 calculators, matrices are stored in memory using names such as [A], [B], [C], ... To find the inverse of [A], we key in

$$[A] \quad \boxed{x^{-1}} \quad \boxed{ENTER}$$

For the matrix of Example 4, this results in the output shown in Figure 1 (where we have also used the ▶Frac command to display the output in fraction form rather than in decimal form).

The next example shows that not every square matrix has an inverse.

Example 5   A Matrix That Does Not Have an Inverse

Find the inverse of the matrix.

$$\begin{bmatrix} 2 & -3 & -7 \\ 1 & 2 & 7 \\ 1 & 1 & 4 \end{bmatrix}$$

Solution   We proceed as follows.

$$\begin{bmatrix} 2 & -3 & -7 & | & 1 & 0 & 0 \\ 1 & 2 & 7 & | & 0 & 1 & 0 \\ 1 & 1 & 4 & | & 0 & 0 & 1 \end{bmatrix} \xrightarrow{R_1 \leftrightarrow R_2} \begin{bmatrix} 1 & 2 & 7 & | & 0 & 1 & 0 \\ 2 & -3 & -7 & | & 1 & 0 & 0 \\ 1 & 1 & 4 & | & 0 & 0 & 1 \end{bmatrix}$$

$$\xrightarrow[R_3 - R_1 \to R_3]{R_2 - 2R_1 \to R_2} \begin{bmatrix} 1 & 2 & 7 & | & 0 & 1 & 0 \\ 0 & -7 & -21 & | & 1 & -2 & 0 \\ 0 & -1 & -3 & | & 0 & -1 & 1 \end{bmatrix}$$

$$\xrightarrow{-\frac{1}{7}R_2} \begin{bmatrix} 1 & 2 & 7 & | & 0 & 1 & 0 \\ 0 & 1 & 3 & | & -\frac{1}{7} & \frac{2}{7} & 0 \\ 0 & -1 & -3 & | & 0 & -1 & 1 \end{bmatrix}$$

$$\xrightarrow[R_1 - 2R_2 \to R_1]{R_3 + R_2 \to R_3} \begin{bmatrix} 1 & 0 & 1 & | & \frac{2}{7} & \frac{3}{7} & 0 \\ 0 & 1 & 3 & | & -\frac{1}{7} & \frac{2}{7} & 0 \\ 0 & 0 & 0 & | & -\frac{1}{7} & -\frac{5}{7} & 1 \end{bmatrix}$$

At this point, we would like to change the 0 in the $(3, 3)$ position of this matrix to a 1, without changing the zeros in the $(3, 1)$ and $(3, 2)$ positions. But there is no way to accomplish this, because no matter what multiple of rows 1 and/or 2 we add to row 3, we can't change the third zero in row 3 without changing the first or second zero as well. Thus, we cannot change the left half to the identity matrix, so the original matrix doesn't have an inverse.   ■

```
ERR:SINGULAR MAT
1:Quit
2:Goto
```

Figure 2

If we encounter a row of zeros on the left when trying to find an inverse, as in Example 5, then the original matrix does not have an inverse. If we try to calculate the inverse of the matrix from Example 5 on a TI-83 calculator, we get the error message shown in Figure 2. (A matrix that has no inverse is called *singular*.)

Matrix Equations

We saw in Example 6 in Section 9.5 that a system of linear equations can be written as a single matrix equation. For example, the system

$$\begin{cases} x - 2y - 4z = 7 \\ 2x - 3y - 6z = 5 \\ -3x + 6y + 15z = 0 \end{cases}$$

is equivalent to the matrix equation

$$\underbrace{\begin{bmatrix} 1 & -2 & -4 \\ 2 & -3 & -6 \\ -3 & 6 & 15 \end{bmatrix}}_{A} \underbrace{\begin{bmatrix} x \\ y \\ z \end{bmatrix}}_{X} = \underbrace{\begin{bmatrix} 7 \\ 5 \\ 0 \end{bmatrix}}_{B}$$

If we let

$$A = \begin{bmatrix} 1 & -2 & -4 \\ 2 & -3 & -6 \\ -3 & 6 & 15 \end{bmatrix} \qquad X = \begin{bmatrix} x \\ y \\ z \end{bmatrix} \qquad B = \begin{bmatrix} 7 \\ 5 \\ 0 \end{bmatrix}$$

then this matrix equation can be written as

$$AX = B$$

The matrix A is called the **coefficient matrix**.

We solve this matrix equation by multiplying each side by the inverse of A (provided this inverse exists):

Solving the matrix equation $AX = B$ is very similar to solving the simple real-number equation

$$3x = 12$$

which we do by multiplying each side by the reciprocal (or inverse) of 3:

$$\tfrac{1}{3}(3x) = \tfrac{1}{3}(12)$$

$$x = 4$$

$$AX = B$$

$$A^{-1}(AX) = A^{-1}B \qquad \text{Multiply both sides of equation on the left by } A^{-1}$$

$$(A^{-1}A)X = A^{-1}B \qquad \text{Associative Property}$$

$$I_3 X = A^{-1}B \qquad \text{Property of inverses}$$

$$X = A^{-1}B \qquad \text{Property of identity matrix}$$

In Example 4 we showed that

$$A^{-1} = \begin{bmatrix} -3 & 2 & 0 \\ -4 & 1 & -\tfrac{2}{3} \\ 1 & 0 & \tfrac{1}{3} \end{bmatrix}$$

So, from $X = A^{-1}B$ we have

$$\underbrace{\begin{bmatrix} x \\ y \\ z \end{bmatrix}}_{X} = \underbrace{\begin{bmatrix} -3 & 2 & 0 \\ -4 & 1 & -\tfrac{2}{3} \\ 1 & 0 & \tfrac{1}{3} \end{bmatrix}}_{A^{-1}} \underbrace{\begin{bmatrix} 7 \\ 5 \\ 0 \end{bmatrix}}_{B} = \begin{bmatrix} -11 \\ -23 \\ 7 \end{bmatrix}$$

Thus, $x = -11$, $y = -23$, $z = 7$ is the solution of the original system.

We have proved that the matrix equation $AX = B$ can be solved by the following method.

Solving a Matrix Equation

If A is a square $n \times n$ matrix that has an inverse A^{-1}, and if X is a variable matrix and B a known matrix, both with n rows, then the solution of the matrix equation.

$$AX = B$$

is given by

$$X = A^{-1}B$$

Example 6   Solving a System Using a Matrix Inverse

(a) Write the system of equations as a matrix equation.

(b) Solve the system by solving the matrix equation.

$$\begin{cases} 2x - 5y = 15 \\ 3x - 6y = 36 \end{cases}$$

Solution

(a) We write the system as a matrix equation of the form $AX = B$:

$$\underbrace{\begin{bmatrix} 2 & -5 \\ 3 & -6 \end{bmatrix}}_{A} \underbrace{\begin{bmatrix} x \\ y \end{bmatrix}}_{X} = \underbrace{\begin{bmatrix} 15 \\ 36 \end{bmatrix}}_{B}$$

(b) Using the rule for finding the inverse of a 2×2 matrix, we get

$$A^{-1} = \begin{bmatrix} 2 & -5 \\ 3 & -6 \end{bmatrix}^{-1} = \frac{1}{2(-6) - (-5)3} \begin{bmatrix} -6 & -(-5) \\ -3 & 2 \end{bmatrix} = \frac{1}{3}\begin{bmatrix} -6 & 5 \\ -3 & 2 \end{bmatrix} = \begin{bmatrix} -2 & \frac{5}{3} \\ -1 & \frac{2}{3} \end{bmatrix}$$

Multiplying each side of the matrix equation by this inverse matrix, we get

$$\underbrace{\begin{bmatrix} x \\ y \end{bmatrix}}_{X} = \underbrace{\begin{bmatrix} -2 & \frac{5}{3} \\ -1 & \frac{2}{3} \end{bmatrix}}_{A^{-1}} \underbrace{\begin{bmatrix} 15 \\ 36 \end{bmatrix}}_{B} = \begin{bmatrix} 30 \\ 9 \end{bmatrix}$$

So $x = 30$ and $y = 9$.   ■

Applications

Suppose we need to solve several systems of equations with the same coefficient matrix. Then converting the systems to matrix equations provides an efficient way to obtain the solutions, because we only need to find the inverse of the coefficient matrix once. This procedure is particularly convenient if we use a graphing calculator to perform the matrix operations, as in the next example.

Example 7 Modeling Nutritional Requirements Using Matrix Equations

A pet-store owner feeds his hamsters and gerbils different mixtures of three types of rodent food: KayDee Food, Pet Pellets, and Rodent Chow. He wishes to feed his animals the correct amount of each brand to satisfy their daily requirements for protein, fat, and carbohydrates exactly. Suppose that hamsters require 340 mg of protein, 280 mg of fat, and 440 mg of carbohydrates, and gerbils need 480 mg of protein, 360 mg of fat, and 680 mg of carbohydrates each day. The amount of each nutrient (in mg) in one gram of each brand is given in the following table. How many grams of each food should the storekeeper feed his hamsters and gerbils daily to satisfy their nutrient requirements?

	KayDee Food	Pet Pellets	Rodent Chow
Protein (mg)	10	0	20
Fat (mg)	10	20	10
Carbohydrates (mg)	5	10	30

Solution   We let x_1, x_2, and x_3 be the respective amounts (in grams) of KayDee Food, Pet Pellets, and Rodent Chow that the hamsters should eat and y_1, y_2, and y_3 be the corresponding amounts for the gerbils. Then we want to solve the matrix equations

$$\begin{bmatrix} 10 & 0 & 20 \\ 10 & 20 & 10 \\ 5 & 10 & 30 \end{bmatrix} \begin{bmatrix} x_1 \\ x_2 \\ x_3 \end{bmatrix} = \begin{bmatrix} 340 \\ 280 \\ 440 \end{bmatrix} \qquad \textit{Hamster equation}$$

$$\begin{bmatrix} 10 & 0 & 20 \\ 10 & 20 & 10 \\ 5 & 10 & 30 \end{bmatrix} \begin{bmatrix} y_1 \\ y_2 \\ y_3 \end{bmatrix} = \begin{bmatrix} 480 \\ 360 \\ 680 \end{bmatrix} \qquad \textit{Gerbil equation}$$

Let

$$A = \begin{bmatrix} 10 & 0 & 20 \\ 10 & 20 & 10 \\ 5 & 10 & 30 \end{bmatrix}, \quad B = \begin{bmatrix} 340 \\ 280 \\ 440 \end{bmatrix}, \quad C = \begin{bmatrix} 480 \\ 360 \\ 680 \end{bmatrix}, \quad X = \begin{bmatrix} x_1 \\ x_2 \\ x_3 \end{bmatrix}, \quad Y = \begin{bmatrix} y_1 \\ y_2 \\ y_3 \end{bmatrix}$$

Since Lotka and Volterra's time, more detailed mathematical models of animal populations have been developed. For many species the population is divided into several stages—immature, juvenile, adult, and so on. The proportion of each stage that survives or reproduces in a given time period is entered into a matrix (called a transition matrix); matrix multiplication is then used to predict the population in succeeding time periods. (See the *Discovery Project*, page 688.)

As you can see, the power of mathematics to model and predict is an invaluable tool in the ongoing debate over the environment.

Then we can write these matrix equations as

$$AX = B \qquad \text{Hamster equation}$$

$$AY = C \qquad \text{Gerbil equation}$$

We want to solve for X and Y, so we multiply both sides of each equation by A^{-1}, the inverse of the coefficient matrix. We could find A^{-1} by hand, but it is more convenient to use a graphing calculator as shown in Figure 3.

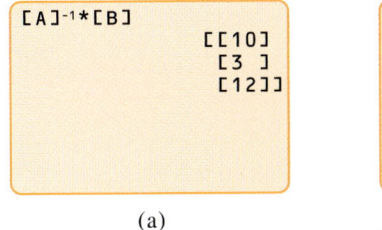

Figure 3   (a)   (b)

From the calculator displays, we see that

$$X = A^{-1}B = \begin{bmatrix} 10 \\ 3 \\ 12 \end{bmatrix}, \qquad Y = A^{-1}C = \begin{bmatrix} 8 \\ 4 \\ 20 \end{bmatrix}$$

Thus, each hamster should be fed 10 g of KayDee Food, 3 g of Pet Pellets, and 12 g of Rodent Chow, and each gerbil should be fed 8 g of KayDee Food, 4 g of Pet Pellets, and 20 g of Rodent Chow daily. ∎

9.6 Exercises

1–4 ■ Calculate the products AB and BA to verify that B is the inverse of A.

1. $A = \begin{bmatrix} 4 & 1 \\ 7 & 2 \end{bmatrix}$, $B = \begin{bmatrix} 2 & -1 \\ -7 & 4 \end{bmatrix}$

2. $A = \begin{bmatrix} 2 & -3 \\ 4 & -7 \end{bmatrix}$, $B = \begin{bmatrix} \frac{7}{2} & -\frac{3}{2} \\ 2 & -1 \end{bmatrix}$

3. $A = \begin{bmatrix} 1 & 3 & -1 \\ 1 & 4 & 0 \\ -1 & -3 & 2 \end{bmatrix}$, $B = \begin{bmatrix} 8 & -3 & 4 \\ -2 & 1 & -1 \\ 1 & 0 & 1 \end{bmatrix}$

4. $A = \begin{bmatrix} 3 & 2 & 4 \\ 1 & 1 & -6 \\ 2 & 1 & 12 \end{bmatrix}$, $B = \begin{bmatrix} 9 & -10 & -8 \\ -12 & 14 & 11 \\ -\frac{1}{2} & \frac{1}{2} & \frac{1}{2} \end{bmatrix}$

5–6 ■ Find the inverse of the matrix and verify that $A^{-1}A = AA^{-1} = I_2$ and $B^{-1}B = BB^{-1} = I_3$.

5. $A = \begin{bmatrix} 7 & 4 \\ 3 & 2 \end{bmatrix}$

6. $B = \begin{bmatrix} 1 & 3 & 2 \\ 0 & 2 & 2 \\ -2 & -1 & 0 \end{bmatrix}$

7–22 ■ Find the inverse of the matrix if it exists.

7. $\begin{bmatrix} 5 & 3 \\ 3 & 2 \end{bmatrix}$

8. $\begin{bmatrix} 3 & 4 \\ 7 & 9 \end{bmatrix}$

9. $\begin{bmatrix} 2 & 5 \\ -5 & -13 \end{bmatrix}$

10. $\begin{bmatrix} -7 & 4 \\ 8 & -5 \end{bmatrix}$

11. $\begin{bmatrix} 6 & -3 \\ -8 & 4 \end{bmatrix}$

12. $\begin{bmatrix} \frac{1}{2} & \frac{1}{3} \\ 5 & 4 \end{bmatrix}$

13. $\begin{bmatrix} 0.4 & -1.2 \\ 0.3 & 0.6 \end{bmatrix}$

14. $\begin{bmatrix} 4 & 2 & 3 \\ 3 & 3 & 2 \\ 1 & 0 & 1 \end{bmatrix}$

15. $\begin{bmatrix} 2 & 4 & 1 \\ -1 & 1 & -1 \\ 1 & 4 & 0 \end{bmatrix}$

16. $\begin{bmatrix} 5 & 7 & 4 \\ 3 & -1 & 3 \\ 6 & 7 & 5 \end{bmatrix}$

17. $\begin{bmatrix} 1 & 2 & 3 \\ 4 & 5 & -1 \\ 1 & -1 & -10 \end{bmatrix}$

18. $\begin{bmatrix} 2 & 1 & 0 \\ 1 & 1 & 4 \\ 2 & 1 & 2 \end{bmatrix}$

19. $\begin{bmatrix} 0 & -2 & 2 \\ 3 & 1 & 3 \\ 1 & -2 & 3 \end{bmatrix}$

20. $\begin{bmatrix} 3 & -2 & 0 \\ 5 & 1 & 1 \\ 2 & -2 & 0 \end{bmatrix}$

21. $\begin{bmatrix} 1 & 2 & 0 & 3 \\ 0 & 1 & 1 & 1 \\ 0 & 1 & 0 & 1 \\ 1 & 2 & 0 & 2 \end{bmatrix}$

22. $\begin{bmatrix} 1 & 0 & 1 & 0 \\ 0 & 1 & 0 & 1 \\ 1 & 1 & 1 & 0 \\ 1 & 1 & 1 & 1 \end{bmatrix}$

23–30 ■ Solve the system of equations by converting to a matrix equation and using the inverse of the coefficient matrix, as in Example 6. Use the inverses from Exercises 7–10, 15, 16, 19, and 21.

23. $\begin{cases} 5x + 3y = 4 \\ 3x + 2y = 0 \end{cases}$

24. $\begin{cases} 3x + 4y = 10 \\ 7x + 9y = 20 \end{cases}$

25. $\begin{cases} 2x + 5y = 2 \\ -5x - 13y = 20 \end{cases}$

26. $\begin{cases} -7x + 4y = 0 \\ 8x - 5y = 100 \end{cases}$

27. $\begin{cases} 2x + 4y + z = 7 \\ -x + y - z = 0 \\ x + 4y = -2 \end{cases}$

28. $\begin{cases} 5x + 7y + 4z = 1 \\ 3x - y + 3z = 1 \\ 6x + 7y + 5z = 1 \end{cases}$

29. $\begin{cases} -2y + 2z = 12 \\ 3x + y + 3z = -2 \\ x - 2y + 3z = 8 \end{cases}$

30. $\begin{cases} x + 2y + 3w = 0 \\ y + z + w = 1 \\ y + w = 2 \\ x + 2y + 2w = 3 \end{cases}$

31–36 ■ Use a calculator that can perform matrix operations to solve the system, as in Example 7.

31. $\begin{cases} x + y - 2z = 3 \\ 2x + 5z = 11 \\ 2x + 3y = 12 \end{cases}$

32. $\begin{cases} 3x + 4y - z = 2 \\ 2x - 3y + z = -5 \\ 5x - 2y + 2z = -3 \end{cases}$

33. $\begin{cases} 12x + \frac{1}{2}y - 7z = 21 \\ 11x - 2y + 3z = 43 \\ 13x + y - 4z = 29 \end{cases}$

34. $\begin{cases} x + \frac{1}{2}y - \frac{1}{3}z = 4 \\ x - \frac{1}{4}y + \frac{1}{6}z = 7 \\ x + y - z = -6 \end{cases}$

35. $\begin{cases} x + y - 3w = 0 \\ x - 2z = 8 \\ 2y - z + w = 5 \\ 2x + 3y - 2w = 13 \end{cases}$

36. $\begin{cases} x + y + z + w = 15 \\ x - y + z - w = 5 \\ x + 2y + 3z + 4w = 26 \\ x - 2y + 3z - 4w = 2 \end{cases}$

37–38 ■ Solve the matrix equation by multiplying each side by the appropriate inverse matrix.

37. $\begin{bmatrix} 3 & -2 \\ -4 & 3 \end{bmatrix} \begin{bmatrix} x & y & z \\ u & v & w \end{bmatrix} = \begin{bmatrix} 1 & 0 & -1 \\ 2 & 1 & 3 \end{bmatrix}$

38. $\begin{bmatrix} 0 & -2 & 2 \\ 3 & 1 & 3 \\ 1 & -2 & 3 \end{bmatrix} \begin{bmatrix} x & u \\ y & v \\ z & w \end{bmatrix} = \begin{bmatrix} 3 & 6 \\ 6 & 12 \\ 0 & 0 \end{bmatrix}$

39–40 ■ Find the inverse of the matrix.

39. $\begin{bmatrix} a & -a \\ a & a \end{bmatrix}$

$(a \neq 0)$

40. $\begin{bmatrix} a & 0 & 0 & 0 \\ 0 & b & 0 & 0 \\ 0 & 0 & c & 0 \\ 0 & 0 & 0 & d \end{bmatrix}$

$(abcd \neq 0)$

41–46 ■ Find the inverse of the matrix. For what value(s) of x, if any, does the matrix have no inverse?

41. $\begin{bmatrix} 2 & x \\ x & x^2 \end{bmatrix}$

42. $\begin{bmatrix} e^x & -e^{2x} \\ e^{2x} & e^{3x} \end{bmatrix}$

43. $\begin{bmatrix} 1 & e^x & 0 \\ e^x & -e^{2x} & 0 \\ 0 & 0 & 2 \end{bmatrix}$

44. $\begin{bmatrix} x & 1 \\ -x & \dfrac{1}{x-1} \end{bmatrix}$

45. $\begin{bmatrix} \cos x & \sin x \\ -\sin x & \cos x \end{bmatrix}$

46. $\begin{bmatrix} \sec x & \tan x \\ \tan x & \sec x \end{bmatrix}$

Applications

47. Nutrition A nutritionist is studying the effects of the nutrients folic acid, choline, and inositol. He has three types of food available, and each type contains the following amounts of these nutrients per ounce:

	Type A	Type B	Type C
Folic acid (mg)	3	1	3
Choline (mg)	4	2	4
Inositol (mg)	3	2	4

(a) Find the inverse of the matrix

$$\begin{bmatrix} 3 & 1 & 3 \\ 4 & 2 & 4 \\ 3 & 2 & 4 \end{bmatrix}$$

and use it to solve the remaining parts of this problem.

(b) How many ounces of each food should the nutritionist feed his laboratory rats if he wants their daily diet to contain 10 mg of folic acid, 14 mg of choline, and 13 mg of inositol?

(c) How much of each food is needed to supply 9 mg of folic acid, 12 mg of choline, and 10 mg of inositol?

(d) Will any combination of these foods supply 2 mg of folic acid, 4 mg of choline, and 11 mg of inositol?

48. Nutrition Refer to Exercise 47. Suppose food type C has been improperly labeled, and it actually contains 4 mg of folic acid, 6 mg of choline, and 5 mg of inositol per ounce. Would it still be possible to use matrix inversion to solve parts (b), (c), and (d) of Exercise 47? Why or why not?

49. Sales Commissions An encyclopedia saleswoman works for a company that offers three different grades of bindings for its encyclopedias: standard, deluxe, and leather. For each set she sells, she earns a commission based on the set's binding grade. One week she sells one standard, one deluxe, and two leather sets and makes $675 in commission. The next week she sells two standard, one deluxe, and one leather set for a $600 commission. The third week she sells one standard, two deluxe, and one leather set, earning $625 in commission.

(a) Let x, y, and z represent the commission she earns on standard, deluxe, and leather sets, respectively. Translate the given information into a system of equations in x, y, and z.

(b) Express the system of equations you found in part (a) as a matrix equation of the form $AX = B$.

(c) Find the inverse of the coefficient matrix A and use it to solve the matrix equation in part (b). How much commission does the saleswoman earn on a set of encyclopedias in each grade of binding?

Discovery • Discussion

50. No Zero-Product Property for Matrices We have used the Zero-Product Property to solve algebraic equations. Matrices do *not* have this property. Let O represent the **2 × 2 zero matrix**:

$$O = \begin{bmatrix} 0 & 0 \\ 0 & 0 \end{bmatrix}$$

Find 2×2 matrices $A \neq O$ and $B \neq O$ such that $AB = O$. Can you find a matrix $A \neq O$ such that $A^2 = O$?

**DISCOVERY
PROJECT**

Computer Graphics I

Matrix algebra is the basic tool used in computer graphics to manipulate images on a computer screen. We will see how matrix multiplication can be used to "move" a point in the plane to a prescribed location. Combining such moves enables us to stretch, compress, rotate, and otherwise transform a figure, as we see in the images below.

Image   Compressed   Rotated   Sheared

Moving Points in the Plane

Let's represent the point (x, y) in the plane by a 2×1 matrix:

$$(x, y) \quad \leftrightarrow \quad \begin{bmatrix} x \\ y \end{bmatrix}$$

For example, the point $(3, 2)$ in the figure is represented by the matrix

$$P = \begin{bmatrix} 3 \\ 2 \end{bmatrix}$$

Multiplying by a 2×2 matrix *moves* the point in the plane. For example, if

$$T = \begin{bmatrix} 1 & 0 \\ 0 & -1 \end{bmatrix}$$

then multiplying P by T we get

$$TP = \begin{bmatrix} 1 & 0 \\ 0 & -1 \end{bmatrix} \begin{bmatrix} 3 \\ 2 \end{bmatrix} = \begin{bmatrix} 3 \\ -2 \end{bmatrix}$$

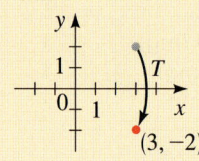

We see that the point $(3, 2)$ has been moved to the point $(3, -2)$. In general, multiplication by this matrix T reflects points in the x-axis. If every point in an image is multiplied by this matrix, then the entire image will be flipped upside down about the x-axis. Matrix multiplication "transforms" a point to a new point in the plane. For this reason, a matrix used in this way is called a **transformation**.

Table 1 gives some standard transformations and their effects on the gray square in the first quadrant.

Table 1

Transformation matrix	Effect
$T = \begin{bmatrix} 1 & 0 \\ 0 & -1 \end{bmatrix}$ Reflection in x-axis	
$T = \begin{bmatrix} c & 0 \\ 0 & 1 \end{bmatrix}$ Expansion (or contraction) in the x-direction	
$T = \begin{bmatrix} 1 & c \\ 0 & 1 \end{bmatrix}$ Shear in x-direction	

Moving Images in the Plane

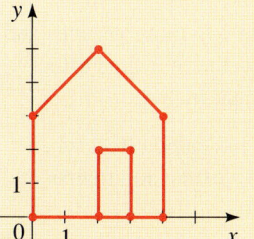

Figure 1

Simple line drawings such as the house in Figure 1 consist of a collection of vertex points and connecting line segments. The entire image in Figure 1 can be represented in a computer by the 2×11 **data matrix**

$$D = \begin{bmatrix} 2 & 0 & 0 & 2 & 4 & 4 & 3 & 3 & 2 & 2 & 3 \\ 0 & 0 & 3 & 5 & 3 & 0 & 0 & 2 & 2 & 0 & 0 \end{bmatrix}$$

The columns of D represent the vertex points of the image. To draw the house, we connect successive points (columns) in D by line segments. Now we can transform the whole house by multiplying D by an appropriate transformation matrix. For example, if we apply the shear transformation $T = \begin{bmatrix} 1 & 0.5 \\ 0 & 1 \end{bmatrix}$, we get the following matrix.

$$TD = \begin{bmatrix} 1 & 0.5 \\ 0 & 1 \end{bmatrix} \begin{bmatrix} 2 & 0 & 0 & 2 & 4 & 4 & 3 & 3 & 2 & 2 & 3 \\ 0 & 0 & 3 & 5 & 3 & 0 & 0 & 2 & 2 & 0 & 0 \end{bmatrix}$$

$$= \begin{bmatrix} 2 & 0 & 1.5 & 4.5 & 5.5 & 4 & 3 & 4 & 3 & 2 & 3 \\ 0 & 0 & 3 & 5 & 3 & 0 & 0 & 2 & 2 & 0 & 0 \end{bmatrix}$$

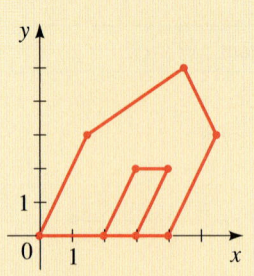

Figure 2

```
PROGRAM:IMAGE
:For(N,1,10)
:Line([A](1,N),
 [A](2,N),[A](1,N+1),
 [A](2,N+1))
:End
```

To draw the image represented by TD, we start with the point $\begin{bmatrix} 2 \\ 0 \end{bmatrix}$, connect it by a line segment to the point $\begin{bmatrix} 0 \\ 0 \end{bmatrix}$, then follow that by a line segment to $\begin{bmatrix} 1.5 \\ 3 \end{bmatrix}$, and so on. The resulting tilted house is shown in Figure 2.

A convenient way to draw an image corresponding to a given data matrix is to use a graphing calculator. The TI-83 program in the margin converts a data matrix stored in [A] into the corresponding image, as shown in Figure 3. (To use this program for a data matrix with m columns, store the matrix in [A] and change the "10" in the For command to $m - 1$.)

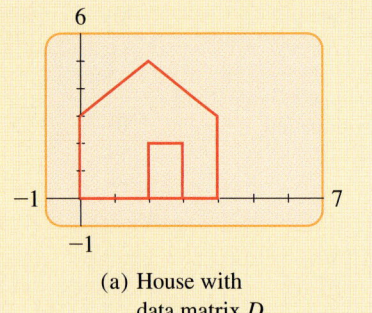

(a) House with
data matrix D

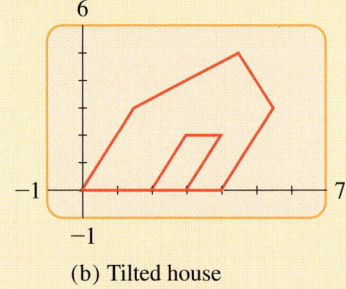

(b) Tilted house
with data matrix TD

Figure 3

We will revisit computer graphics in the *Discovery Project* on page 792, where we will find matrices that rotate an image by any given angle.

1. The gray square in Table 1 has the following vertices.

$$\begin{bmatrix} 0 \\ 0 \end{bmatrix}, \quad \begin{bmatrix} 1 \\ 0 \end{bmatrix}, \quad \begin{bmatrix} 1 \\ 1 \end{bmatrix}, \quad \begin{bmatrix} 0 \\ 1 \end{bmatrix}$$

Apply each of the three transformations given in Table 1 to these vertices and sketch the result, to verify that each transformation has the indicated effect. Use $c = 2$ in the expansion matrix and $c = 1$ in the shear matrix.

2. Verify that multiplication by the given matrix has the indicated effect when applied to the gray square in the table. Use $c = 3$ in the expansion matrix and $c = 1$ in the shear matrix.

$$T_1 = \begin{bmatrix} -1 & 0 \\ 0 & 1 \end{bmatrix} \qquad T_2 = \begin{bmatrix} 1 & 0 \\ 0 & c \end{bmatrix} \qquad T_3 = \begin{bmatrix} 1 & 0 \\ c & 1 \end{bmatrix}$$

Reflection in y-axis   Expansion (or contraction)   Shear in y-direction
in y-direction

3. Let $T = \begin{bmatrix} 1 & 1.5 \\ 0 & 1 \end{bmatrix}$.

(a) What effect does T have on the gray square in the Table 1?

(b) Find T^{-1}.

(c) What effect does T^{-1} have on the gray square?

(d) What happens to the square if we first apply T, then T^{-1}?

4. (a) Let $T = \begin{bmatrix} 3 & 0 \\ 0 & 1 \end{bmatrix}$. What effect does T have on the gray square in Table 1?

(b) Let $S = \begin{bmatrix} 1 & 0 \\ 0 & 2 \end{bmatrix}$. What effect does S have on the gray square in Table 1?

(c) Apply S to the vertices of the square, and then apply T to the result. What is the effect of the combined transformation?

(d) Find the product matrix $W = TS$.

(e) Apply the transformation W to the square. Compare to your final result in part (c). What do you notice?

5. The figure shows three outline versions of the letter **F**. The second one is obtained from the first by shrinking horizontally by a factor of 0.75, and the third is obtained from the first by shearing horizontally by a factor of 0.25.

(a) Find a data matrix D for the first letter **F**.

(b) Find the transformation matrix T that transforms the first **F** into the second. Calculate TD and verify that this is a data matrix for the second **F**.

(c) Find the transformation matrix S that transforms the first **F** into the third. Calculate SD and verify that this is a data matrix for the third **F**.

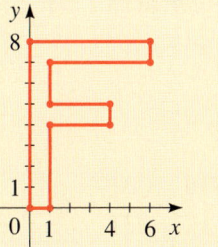 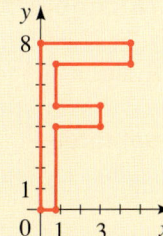 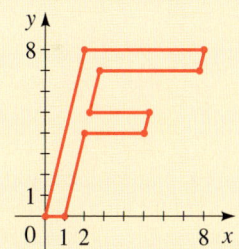

6. Here is a data matrix for a line drawing.

$$D = \begin{bmatrix} 0 & 1 & 2 & 1 & 0 & 0 \\ 0 & 0 & 2 & 4 & 4 & 0 \end{bmatrix}$$

(a) Draw the image represented by D.

(b) Let $T = \begin{bmatrix} 1 & 1 \\ 0 & -1 \end{bmatrix}$. Calculate the matrix product TD and draw the image represented by this product. What is the effect of the transformation T?

(c) Express T as a product of a shear matrix and a reflection matrix. (See problem 2.)

If a matrix is **square** (that is, if it has the same number of rows as columns), then we can assign to it a number called its *determinant*. Determinants can be used to solve systems of linear equations, as we will see later in this section. They are also useful in determining whether a matrix has an inverse.

Determinant of a 2 × 2 Matrix

We denote the determinant of a square matrix A by the symbol $\det(A)$ or $|A|$. We first define $\det(A)$ for the simplest cases. If $A = [a]$ is a 1×1 matrix, then $\det(A) = a$. The following box gives the definition of a 2×2 determinant.

We will use both notations, $\det(A)$ and $|A|$, for the determinant of A. Although the symbol $|A|$ looks like the absolute value symbol, it will be clear from the context which meaning is intended.

Determinant of a 2 × 2 Matrix

The **determinant** of the 2×2 matrix $A = \begin{bmatrix} a & b \\ c & d \end{bmatrix}$ is

$$\det(A) = |A| = \begin{vmatrix} a & b \\ c & d \end{vmatrix} = ad - bc$$

Example 1   Determinant of a 2 × 2 Matrix

Evaluate $|A|$ for $A = \begin{bmatrix} 6 & -3 \\ 2 & 3 \end{bmatrix}$.

Solution

To evaluate a 2×2 determinant, we take the product of the diagonal from top left to bottom right, and subtract the product from top right to bottom left, as indicated by the arrows.

$$\begin{vmatrix} 6 & -3 \\ 2 & 3 \end{vmatrix} = 6 \cdot 3 - (-3)2 = 18 - (-6) = 24 \qquad \blacksquare$$

Determinant of an $n \times n$ Matrix

To define the concept of determinant for an arbitrary $n \times n$ matrix, we need the following terminology.

Let A be an $n \times n$ matrix.

1. The **minor** M_{ij} of the element a_{ij} is the determinant of the matrix obtained by deleting the ith row and jth column of A.

2. The **cofactor** A_{ij} of the element a_{ij} is

$$A_{ij} = (-1)^{i+j} M_{ij}$$

For example, if A is the matrix

$$\begin{bmatrix} 2 & 3 & -1 \\ 0 & 2 & 4 \\ -2 & 5 & 6 \end{bmatrix}$$

then the minor M_{12} is the determinant of the matrix obtained by deleting the first row and second column from A. Thus

$$M_{12} = \begin{vmatrix} 2 & 3 & -1 \\ 0 & 2 & 4 \\ -2 & 5 & 6 \end{vmatrix} = \begin{vmatrix} 0 & 4 \\ -2 & 6 \end{vmatrix} = 0(6) - 4(-2) = 8$$

So, the cofactor $A_{12} = (-1)^{1+2}M_{12} = -8$. Similarly

$$M_{33} = \begin{vmatrix} 2 & 3 & -1 \\ 0 & 2 & 4 \\ -2 & 5 & 6 \end{vmatrix} = \begin{vmatrix} 2 & 3 \\ 0 & 2 \end{vmatrix} = 2 \cdot 2 - 3 \cdot 0 = 4$$

So, $A_{33} = (-1)^{3+3}M_{33} = 4$.

Note that the cofactor of a_{ij} is simply the minor of a_{ij} multiplied by either 1 or -1, depending on whether $i + j$ is even or odd. Thus, in a 3×3 matrix we obtain the cofactor of any element by prefixing its minor with the sign obtained from the following checkerboard pattern:

$$\begin{bmatrix} + & - & + \\ - & + & - \\ + & - & + \end{bmatrix}$$

We are now ready to define the determinant of any square matrix.

The Determinant of a Square Matrix

If A is an $n \times n$ matrix, then the **determinant** of A is obtained by multiplying each element of the first row by its cofactor, and then adding the results. In symbols,

$$\det(A) = |A| = \begin{vmatrix} a_{11} & a_{12} & \cdots & a_{1n} \\ a_{21} & a_{22} & \cdots & a_{2n} \\ \vdots & \vdots & \ddots & \vdots \\ a_{n1} & a_{n2} & \cdots & a_{nn} \end{vmatrix} = a_{11}A_{11} + a_{12}A_{12} + \cdots + a_{1n}A_{1n}$$

Example 2   Determinant of a 3×3 Matrix

Evaluate the determinant of the matrix.

$$A = \begin{bmatrix} 2 & 3 & -1 \\ 0 & 2 & 4 \\ -2 & 5 & 6 \end{bmatrix}$$

Solution

$$\det(A) = \begin{vmatrix} 2 & 3 & -1 \\ 0 & 2 & 4 \\ -2 & 5 & 6 \end{vmatrix} = 2 \begin{vmatrix} 2 & 4 \\ 5 & 6 \end{vmatrix} - 3 \begin{vmatrix} 0 & 4 \\ -2 & 6 \end{vmatrix} + (-1) \begin{vmatrix} 0 & 2 \\ -2 & 5 \end{vmatrix}$$

$$= 2(2 \cdot 6 - 4 \cdot 5) - 3[0 \cdot 6 - 4(-2)] - [0 \cdot 5 - 2(-2)]$$

$$= -16 - 24 - 4$$

$$= -44 \qquad \blacksquare$$

In our definition of the determinant we used the cofactors of elements in the first row only. This is called **expanding the determinant by the first row**. In fact, *we can expand the determinant by any row or column in the same way, and obtain the same result in each case* (although we won't prove this). The next example illustrates this principle.

Example 3 Expanding a Determinant about a Row and a Column

Let A be the matrix of Example 2. Evaluate the determinant of A by expanding

(a) by the second row

(b) by the third column

Verify that each expansion gives the same value.

Solution

(a) Expanding by the second row, we get

$$\det(A) = \begin{vmatrix} 2 & 3 & -1 \\ 0 & 2 & -4 \\ -2 & 5 & 6 \end{vmatrix} = -0 \begin{vmatrix} 3 & -1 \\ 5 & 6 \end{vmatrix} + 2 \begin{vmatrix} 2 & -1 \\ -2 & 6 \end{vmatrix} - 4 \begin{vmatrix} 2 & 3 \\ -2 & 5 \end{vmatrix}$$

$$= 0 + 2[2 \cdot 6 - (-1)(-2)] - 4[2 \cdot 5 - 3(-2)]$$

$$= 0 + 20 - 64 = -44$$

(b) Expanding by the third column gives

$$\det(A) = \begin{vmatrix} 2 & 3 & -1 \\ 0 & 2 & 4 \\ -2 & 5 & 6 \end{vmatrix}$$

$$= -1 \begin{vmatrix} 0 & 2 \\ -2 & 5 \end{vmatrix} - 4 \begin{vmatrix} 2 & 3 \\ -2 & 5 \end{vmatrix} + 6 \begin{vmatrix} 2 & 3 \\ 0 & 2 \end{vmatrix}$$

$$= -[0 \cdot 5 - 2(-2)] - 4[2 \cdot 5 - 3(-2)] + 6(2 \cdot 2 - 3 \cdot 0)$$

$$= -4 - 64 + 24 = -44$$

In both cases, we obtain the same value for the determinant as when we expanded by the first row in Example 2. $\qquad \blacksquare$

Graphing calculators are capable of computing determinants. Here is the output when the TI-83 is used to calculate the determinant in Example 3.

```
[A]
        [[2   3  -1]
         [0   2   4 ]
         [-2  5   6 ]]
det([A])
                    -44
```

The following criterion allows us to determine whether a square matrix has an inverse without actually calculating the inverse. This is one of the most important uses of the determinant in matrix algebra, and it is the reason for the name *determinant*.

Invertibility Criterion

If A is a square matrix, then A has an inverse if and only if $\det(A) \neq 0$.

We will not prove this fact, but from the formula for the inverse of a 2×2 matrix (page 704), you can see why it is true in the 2×2 case.

Example 4   **Using the Determinant to Show That a Matrix Is Not Invertible**

Show that the matrix A has no inverse.

$$A = \begin{bmatrix} 1 & 2 & 0 & 4 \\ 0 & 0 & 0 & 3 \\ 5 & 6 & 2 & 6 \\ 2 & 4 & 0 & 9 \end{bmatrix}$$

Solution   We begin by calculating the determinant of A. Since all but one of the elements of the second row is zero, we expand the determinant by the second row. If we do this, we see from the following equation that only the cofactor A_{24} will have to be calculated.

$$\det(A) = \begin{vmatrix} 1 & 2 & 0 & 4 \\ 0 & 0 & 0 & 3 \\ 5 & 6 & 2 & 6 \\ 2 & 4 & 0 & 9 \end{vmatrix}$$

$$= -0 \cdot A_{21} + 0 \cdot A_{22} - 0 \cdot A_{23} + 3 \cdot A_{24} = 3A_{24}$$

$$= 3 \begin{vmatrix} 1 & 2 & 0 \\ 5 & 6 & 2 \\ 2 & 4 & 0 \end{vmatrix} \qquad \text{Expand this by column 3}$$

$$= 3(-2) \begin{vmatrix} 1 & 2 \\ 2 & 4 \end{vmatrix}$$

$$= 3(-2)(1 \cdot 4 - 2 \cdot 2) = 0$$

Since the determinant of A is zero, A cannot have an inverse, by the Invertibility Criterion.   ◼

Row and Column Transformations

The preceding example shows that if we expand a determinant about a row or column that contains many zeros, our work is reduced considerably because we don't have to evaluate the cofactors of the elements that are zero. The following principle often simplifies the process of finding a determinant by introducing zeros into it without changing its value.

B. H. Ward and K. C. Ward/Corbis

David Hilbert (1862–1943) was born in Königsberg, Germany, and became a professor at Göttingen University. He is considered by many to be the greatest mathematician of the 20th century. At the International Congress of Mathematicians held in Paris in 1900, Hilbert set the direction of mathematics for the about-to-dawn 20th century by posing 23 problems he believed to be of crucial importance. He said that "these are problems whose solutions we expect from the future." Most of Hilbert's problems have now been solved (see Julia Robinson, page 678 and Alan Turing, page 103), and their solutions have led to important new areas of mathematical research. Yet as we enter the new millennium, some of his problems remain unsolved. In his work, Hilbert emphasized structure, logic, and the foundations of mathematics. Part of his genius lay in his ability to see the most general possible statement of a problem. For instance, Euler proved that every whole number is the sum of four squares; Hilbert proved a similar statement for all powers of positive integers.

Row and Column Transformations of a Determinant

If A is a square matrix, and if the matrix B is obtained from A by adding a multiple of one row to another, or a multiple of one column to another, then $\det(A) = \det(B)$.

Example 5 Using Row and Column Transformations to Calculate a Determinant

Find the determinant of the matrix A. Does it have an inverse?

$$A = \begin{bmatrix} 8 & 2 & -1 & -4 \\ 3 & 5 & -3 & 11 \\ 24 & 6 & 1 & -12 \\ 2 & 2 & 7 & -1 \end{bmatrix}$$

Solution If we add -3 times row 1 to row 3, we change all but one element of row 3 to zeros:

$$\begin{bmatrix} 8 & 2 & -1 & -4 \\ 3 & 5 & -3 & 11 \\ 0 & 0 & 4 & 0 \\ 2 & 2 & 7 & -1 \end{bmatrix}$$

This new matrix has the same determinant as A, and if we expand its determinant by the third row, we get

$$\det(A) = 4 \begin{vmatrix} 8 & 2 & -4 \\ 3 & 5 & 11 \\ 2 & 2 & -1 \end{vmatrix}$$

Now, adding 2 times column 3 to column 1 in this determinant gives us

$$\det(A) = 4 \begin{vmatrix} 0 & 2 & -4 \\ 25 & 5 & 11 \\ 0 & 2 & -1 \end{vmatrix} \qquad \text{Expand this by column 1}$$

$$= 4(-25) \begin{vmatrix} 2 & -4 \\ 2 & -1 \end{vmatrix}$$

$$= 4(-25)[2(-1) - (-4)2] = -600$$

Since the determinant of A is not zero, A does have an inverse. ∎

Cramer's Rule

The solutions of linear equations can sometimes be expressed using determinants. To illustrate, let's solve the following pair of linear equations for the variable x.

$$\begin{cases} ax + by = r \\ cx + dy = s \end{cases}$$

To eliminate the variable y, we multiply the first equation by d and the second by b, and subtract.

$$
\begin{aligned}
adx + bdy &= rd \\
bcx + bdy &= bs \\
\hline
adx - bcx &= rd - bs
\end{aligned}
$$

Factoring the left-hand side, we get $(ad - bc)x = rd - bs$. Assuming that $ad - bc \neq 0$, we can now solve this equation for x:

$$
x = \frac{rd - bs}{ad - bc}
$$

Similarly, we find

$$
y = \frac{as - cr}{ad - bc}
$$

The numerator and denominator of the fractions for x and y are determinants of 2×2 matrices. So we can express the solution of the system using determinants as follows.

Cramer's Rule for Systems in Two Variables

The linear system

$$
\begin{cases} ax + by = r \\ cx + dy = s \end{cases}
$$

has the solution

$$
x = \frac{\begin{vmatrix} r & b \\ s & d \end{vmatrix}}{\begin{vmatrix} a & b \\ c & d \end{vmatrix}} \qquad y = \frac{\begin{vmatrix} a & r \\ c & s \end{vmatrix}}{\begin{vmatrix} a & b \\ c & d \end{vmatrix}}
$$

provided $\begin{vmatrix} a & b \\ c & d \end{vmatrix} \neq 0$.

Using the notation

$$
D = \begin{bmatrix} a & b \\ c & d \end{bmatrix} \qquad D_x = \begin{bmatrix} r & b \\ s & d \end{bmatrix} \qquad D_y = \begin{bmatrix} a & r \\ c & s \end{bmatrix}
$$

Coefficient matrix

Replace first column of D by r and s.

Replace second column of D by r and s.

we can write the solution of the system as

$$
x = \frac{|D_x|}{|D|} \quad \text{and} \quad y = \frac{|D_y|}{|D|}
$$

Example 6 Using Cramer's Rule to Solve a System with Two Variables

Use Cramer's Rule to solve the system.

$$\begin{cases} 2x + 6y = -1 \\ x + 8y = 2 \end{cases}$$

Solution For this system we have

$$|D| = \begin{vmatrix} 2 & 6 \\ 1 & 8 \end{vmatrix} = 2 \cdot 8 - 6 \cdot 1 = 10$$

$$|D_x| = \begin{vmatrix} -1 & 6 \\ 2 & 8 \end{vmatrix} = (-1)8 - 6 \cdot 2 = -20$$

$$|D_y| = \begin{vmatrix} 2 & -1 \\ 1 & 2 \end{vmatrix} = 2 \cdot 2 - (-1)1 = 5$$

The solution is

$$x = \frac{|D_x|}{|D|} = \frac{-20}{10} = -2$$

$$y = \frac{|D_y|}{|D|} = \frac{5}{10} = \frac{1}{2} \qquad \blacksquare$$

Cramer's Rule can be extended to apply to any system of n linear equations in n variables in which the determinant of the coefficient matrix is not zero. As we saw in the preceding section, any such system can be written in matrix form as

$$\begin{bmatrix} a_{11} & a_{12} & \cdots & a_{1n} \\ a_{21} & a_{22} & \cdots & a_{2n} \\ \vdots & \vdots & \ddots & \vdots \\ a_{n1} & a_{n2} & \cdots & a_{nn} \end{bmatrix} \begin{bmatrix} x_1 \\ x_2 \\ \vdots \\ x_n \end{bmatrix} = \begin{bmatrix} b_1 \\ b_2 \\ \vdots \\ b_n \end{bmatrix}$$

By analogy with our derivation of Cramer's Rule in the case of two equations in two unknowns, we let D be the coefficient matrix in this system, and D_{x_i} be the matrix obtained by replacing the ith column of D by the numbers $b_1, b_2, \ldots, b_n$ that appear to the right of the equal sign. The solution of the system is then given by the following rule.

Cramer's Rule

If a system of n linear equations in the n variables $x_1, x_2, \ldots, x_n$ is equivalent to the matrix equation $DX = B$, and if $|D| \neq 0$, then its solutions are

$$x_1 = \frac{|D_{x_1}|}{|D|}, \quad x_2 = \frac{|D_{x_2}|}{|D|}, \quad \ldots, \quad x_n = \frac{|D_{x_n}|}{|D|}$$

where D_{x_i} is the matrix obtained by replacing the ith column of D by the $n \times 1$ matrix B.

The Granger Collection

Emmy Noether (1882–1935) was one of the foremost mathematicians of the early 20th century. Her groundbreaking work in abstract algebra provided much of the foundation for this field, and her work in Invariant Theory was essential in the development of Einstein's theory of general relativity. Although women weren't allowed to study at German universities at that time, she audited courses unofficially and went on to receive a doctorate at Erlangen *summa cum laude*, despite the opposition of the academic senate, which declared that women students would "overthrow all academic order." She subsequently taught mathematics at Göttingen, Moscow, and Frankfurt. In 1933 she left Germany to escape Nazi persecution, accepting a position at Bryn Mawr College in suburban Philadelphia. She lectured there and at the Institute for Advanced Study in Princeton, New Jersey, until her untimely death in 1935.

Example 7 Using Cramer's Rule to Solve a System with Three Variables

Use Cramer's Rule to solve the system.

$$\begin{cases} 2x - 3y + 4z = 1 \\ x \quad\quad + 6z = 0 \\ 3x - 2y \quad\quad = 5 \end{cases}$$

Solution First, we evaluate the determinants that appear in Cramer's Rule. Note that D is the coefficient matrix and that D_x, D_y, and D_z are obtained by replacing the first, second, and third columns of D by the constant terms.

$$|D| = \begin{vmatrix} 2 & -3 & 4 \\ 1 & 0 & 6 \\ 3 & -2 & 0 \end{vmatrix} = -38 \qquad |D_x| = \begin{vmatrix} 1 & -3 & 4 \\ 0 & 0 & 6 \\ 5 & -2 & 0 \end{vmatrix} = -78$$

$$|D_y| = \begin{vmatrix} 2 & 1 & 4 \\ 1 & 0 & 6 \\ 3 & 5 & 0 \end{vmatrix} = -22 \qquad |D_z| = \begin{vmatrix} 2 & -3 & 1 \\ 1 & 0 & 0 \\ 3 & -2 & 5 \end{vmatrix} = 13$$

Now we use Cramer's Rule to get the solution:

$$x = \frac{|D_x|}{|D|} = \frac{-78}{-38} = \frac{39}{19}$$

$$y = \frac{|D_y|}{|D|} = \frac{-22}{-38} = \frac{11}{19}$$

$$z = \frac{|D_z|}{|D|} = \frac{13}{-38} = -\frac{13}{38}$$

Solving the system in Example 7 using Gaussian elimination would involve matrices whose elements are fractions with fairly large denominators. Thus, in cases like Examples 6 and 7, Cramer's Rule gives us an efficient way to solve systems of linear equations. But in systems with more than three equations, evaluating the various determinants involved is usually a long and tedious task (unless you are using a graphing calculator). Moreover, the rule doesn't apply if $|D| = 0$ or if D is not a square matrix. So, Cramer's Rule is a useful alternative to Gaussian elimination, but only in some situations.

Areas of Triangles Using Determinants

Determinants provide a simple way to calculate the area of a triangle in the coordinate plane.

Area of a Triangle

If a triangle in the coordinate plane has vertices (a_1, b_1), (a_2, b_2), and (a_3, b_3), then its area is

$$\text{area} = \pm\frac{1}{2}\begin{vmatrix} a_1 & b_1 & 1 \\ a_2 & b_2 & 1 \\ a_3 & b_3 & 1 \end{vmatrix}$$

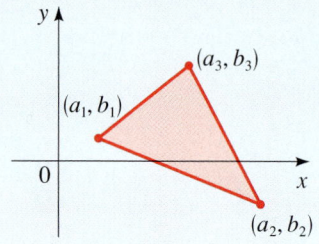

where the sign is chosen to make the area positive.

You are asked to prove this formula in Exercise 59.

Example 8 Area of a Triangle

Find the area of the triangle shown in Figure 1.

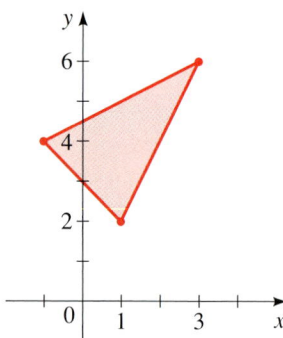

Figure 1

We can calculate the determinant by hand or by using a graphing calculator.

```
[A]
       [[-1 4 1]
        [3  6 1]
        [1  2 1]]
det([A])
              -12
```

Solution The vertices are $(-1, 4)$, $(3, 6)$, and $(1, 2)$. Using the formula in the preceding box, we get:

$$\text{area} = \pm\frac{1}{2}\begin{vmatrix} -1 & 4 & 1 \\ 3 & 6 & 1 \\ 1 & 2 & 1 \end{vmatrix} = \pm\frac{1}{2}(-12)$$

To make the area positive, we choose the negative sign in the formula. Thus, the area of the triangle is

$$\text{area} = -\frac{1}{2}(-12) = 6$$

■

9.7   Exercises

1–8 ■ Find the determinant of the matrix, if it exists.

1. $\begin{bmatrix} 2 & 0 \\ 0 & 3 \end{bmatrix}$

2. $\begin{bmatrix} 0 & -1 \\ 2 & 0 \end{bmatrix}$

3. $\begin{bmatrix} 4 & 5 \\ 0 & -1 \end{bmatrix}$

4. $\begin{bmatrix} -2 & 1 \\ 3 & -2 \end{bmatrix}$

5. $\begin{bmatrix} 2 & 5 \end{bmatrix}$

6. $\begin{bmatrix} 3 \\ 0 \end{bmatrix}$

7. $\begin{bmatrix} \frac{1}{2} & \frac{1}{8} \\ 1 & \frac{1}{2} \end{bmatrix}$

8. $\begin{bmatrix} 2.2 & -1.4 \\ 0.5 & 1.0 \end{bmatrix}$

9–14 ■ Evaluate the minor and cofactor using the matrix A.

$$A = \begin{bmatrix} 1 & 0 & \frac{1}{2} \\ -3 & 5 & 2 \\ 0 & 0 & 4 \end{bmatrix}$$

9. M_{11}, A_{11}

10. M_{33}, A_{33}

11. M_{12}, A_{12}

12. M_{13}, A_{13}

13. M_{23}, A_{23}

14. M_{32}, A_{32}

15–22 ■ Find the determinant of the matrix. Determine whether the matrix has an inverse, but don't calculate the inverse.

15. $\begin{bmatrix} 2 & 1 & 0 \\ 0 & -2 & 4 \\ 0 & 1 & -3 \end{bmatrix}$

16. $\begin{bmatrix} 0 & -1 & 0 \\ 2 & 6 & 4 \\ 1 & 0 & 3 \end{bmatrix}$

17. $\begin{bmatrix} 1 & 3 & 7 \\ 2 & 0 & -1 \\ 0 & 2 & 6 \end{bmatrix}$

18. $\begin{bmatrix} -2 & -\frac{3}{2} & \frac{1}{2} \\ 2 & 4 & 0 \\ \frac{1}{2} & 2 & 1 \end{bmatrix}$

19. $\begin{bmatrix} 30 & 0 & 20 \\ 0 & -10 & -20 \\ 40 & 0 & 10 \end{bmatrix}$

20. $\begin{bmatrix} 1 & 2 & 5 \\ -2 & -3 & 2 \\ 3 & 5 & 3 \end{bmatrix}$

21. $\begin{bmatrix} 1 & 3 & 3 & 0 \\ 0 & 2 & 0 & 1 \\ -1 & 0 & 0 & 2 \\ 1 & 6 & 4 & 1 \end{bmatrix}$

22. $\begin{bmatrix} 1 & 2 & 0 & 2 \\ 3 & -4 & 0 & 4 \\ 0 & 1 & 6 & 0 \\ 1 & 0 & 2 & 0 \end{bmatrix}$

23–26 ■ Evaluate the determinant, using row or column operations whenever possible to simplify your work.

23. $\begin{vmatrix} 0 & 0 & 4 & 6 \\ 2 & 1 & 1 & 3 \\ 2 & 1 & 2 & 3 \\ 3 & 0 & 1 & 7 \end{vmatrix}$

24. $\begin{vmatrix} -2 & 3 & -1 & 7 \\ 4 & 6 & -2 & 3 \\ 7 & 7 & 0 & 5 \\ 3 & -12 & 4 & 0 \end{vmatrix}$

25. $\begin{vmatrix} 1 & 2 & 3 & 4 & 5 \\ 0 & 2 & 4 & 6 & 8 \\ 0 & 0 & 3 & 6 & 9 \\ 0 & 0 & 0 & 4 & 8 \\ 0 & 0 & 0 & 0 & 5 \end{vmatrix}$

26. $\begin{vmatrix} 2 & -1 & 6 & 4 \\ 7 & 2 & -2 & 5 \\ 4 & -2 & 10 & 8 \\ 6 & 1 & 1 & 4 \end{vmatrix}$

27. Let

$$B = \begin{bmatrix} 4 & 1 & 0 \\ -2 & -1 & 1 \\ 4 & 0 & 3 \end{bmatrix}$$

(a) Evaluate $\det(B)$ by expanding by the second row.

(b) Evaluate $\det(B)$ by expanding by the third column.

(c) Do your results in parts (a) and (b) agree?

28. Consider the system

$$\begin{cases} x + 2y + 6z = 5 \\ -3x - 6y + 5z = 8 \\ 2x + 6y + 9z = 7 \end{cases}$$

(a) Verify that $x = -1$, $y = 0$, $z = 1$ is a solution of the system.

(b) Find the determinant of the coefficient matrix.

(c) Without solving the system, determine whether there are any other solutions.

(d) Can Cramer's Rule be used to solve this system? Why or why not?

29–44 ■ Use Cramer's Rule to solve the system.

29. $\begin{cases} 2x - y = -9 \\ x + 2y = 8 \end{cases}$

30. $\begin{cases} 6x + 12y = 33 \\ 4x + 7y = 20 \end{cases}$

31. $\begin{cases} x - 6y = 3 \\ 3x + 2y = 1 \end{cases}$

32. $\begin{cases} \frac{1}{2}x + \frac{1}{3}y = 1 \\ \frac{1}{4}x - \frac{1}{6}y = -\frac{3}{2} \end{cases}$

33. $\begin{cases} 0.4x + 1.2y = 0.4 \\ 1.2x + 1.6y = 3.2 \end{cases}$

34. $\begin{cases} 10x - 17y = 21 \\ 20x - 31y = 39 \end{cases}$

35. $\begin{cases} x - y + 2z = 0 \\ 3x + z = 11 \\ -x + 2y = 0 \end{cases}$

36. $\begin{cases} 5x - 3y + z = 6 \\ 4y - 6z = 22 \\ 7x + 10y = -13 \end{cases}$

37. $\begin{cases} 2x_1 + 3x_2 - 5x_3 = 1 \\ x_1 + x_2 - x_3 = 2 \\ 2x_2 + x_3 = 8 \end{cases}$

38. $\begin{cases} -2a + c = 2 \\ a + 2b - c = 9 \\ 3a + 5b + 2c = 22 \end{cases}$

39. $\begin{cases} \frac{1}{3}x - \frac{1}{5}y + \frac{1}{2}z = \frac{7}{10} \\ -\frac{2}{3}x + \frac{2}{5}y + \frac{3}{2}z = \frac{11}{10} \\ x - \frac{4}{5}y + z = \frac{9}{5} \end{cases}$

40. $\begin{cases} 2x - y = 5 \\ 5x + 3z = 19 \\ 4y + 7z = 17 \end{cases}$

41. $\begin{cases} 3y + 5z = 4 \\ 2x - z = 10 \\ 4x + 7y = 0 \end{cases}$

42. $\begin{cases} 2x - 5y = 4 \\ x + y - z = 8 \\ 3x + 5z = 0 \end{cases}$

43. $\begin{cases} x + y + \; z + w = 0 \\ 2x \qquad\qquad + w = 0 \\ \qquad y - z \qquad\; = 0 \\ x \qquad\; + 2z \qquad = 1 \end{cases}$ **44.** $\begin{cases} x + y = 1 \\ y + z = 2 \\ z + w = 3 \\ w - x = 4 \end{cases}$

45–46 ■ Evaluate the determinants.

45. $\begin{vmatrix} a & 0 & 0 & 0 & 0 \\ 0 & b & 0 & 0 & 0 \\ 0 & 0 & c & 0 & 0 \\ 0 & 0 & 0 & d & 0 \\ 0 & 0 & 0 & 0 & e \end{vmatrix}$ **46.** $\begin{vmatrix} a & a & a & a & a \\ 0 & a & a & a & a \\ 0 & 0 & a & a & a \\ 0 & 0 & 0 & a & a \\ 0 & 0 & 0 & 0 & a \end{vmatrix}$

47–50 ■ Solve for x.

47. $\begin{vmatrix} x & 12 & 13 \\ 0 & x - 1 & 23 \\ 0 & 0 & x - 2 \end{vmatrix} = 0$ **48.** $\begin{vmatrix} x & 1 & 1 \\ 1 & 1 & x \\ x & 1 & x \end{vmatrix} = 0$

49. $\begin{vmatrix} 1 & 0 & x \\ x^2 & 1 & 0 \\ x & 0 & 1 \end{vmatrix} = 0$ **50.** $\begin{vmatrix} a & b & x - a \\ x & x + b & x \\ 0 & 1 & 1 \end{vmatrix} = 0$

51–54 ■ Sketch the triangle with the given vertices and use a determinant to find its area.

51. $(0, 0), (6, 2), (3, 8)$

52. $(1, 0), (3, 5), (-2, 2)$

53. $(-1, 3), (2, 9), (5, -6)$

54. $(-2, 5), (7, 2), (3, -4)$

55. Show that $\begin{vmatrix} 1 & x & x^2 \\ 1 & y & y^2 \\ 1 & z & z^2 \end{vmatrix} = (x - y)(y - z)(z - x)$

Applications

56. Buying Fruit   A roadside fruit stand sells apples at 75¢ a pound, peaches at 90¢ a pound, and pears at 60¢ a pound. Muriel buys 18 pounds of fruit at a total cost of $13.80. Her peaches and pears together cost $1.80 more than her apples.

   (a) Set up a linear system for the number of pounds of apples, peaches, and pears that she bought.

   (b) Solve the system using Cramer's Rule.

57. The Arch of a Bridge   The opening of a railway bridge over a roadway is in the shape of a parabola. A surveyor measures the heights of three points on the bridge, as shown in the figure. He wishes to find an equation of the form

$$y = ax^2 + bx + c$$

to model the shape of the arch.

   (a) Use the surveyed points to set up a system of linear equations for the unknown coefficients a, b, and c.

   (b) Solve the system using Cramer's Rule.

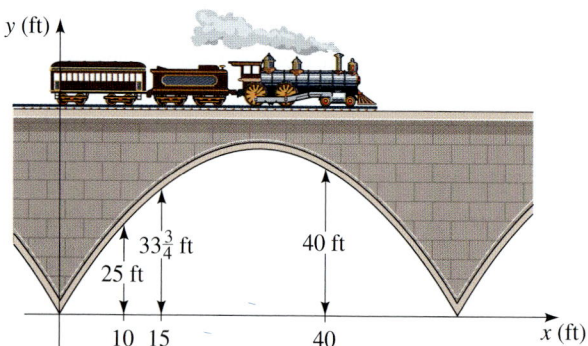

58. A Triangular Plot of Land   An outdoors club is purchasing land to set up a conservation area. The last remaining piece they need to buy is the triangular plot shown in the figure. Use the determinant formula for the area of a triangle to find the area of the plot.

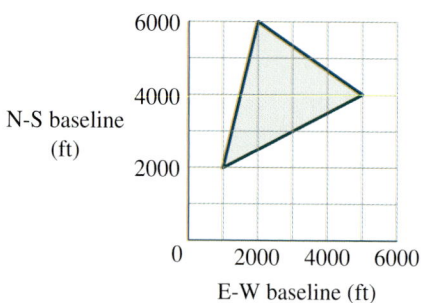

Discovery • Discussion

59. Determinant Formula for the Area of a Triangle   The figure shows a triangle in the plane with vertices (a_1, b_1), (a_2, b_2), and (a_3, b_3).

   (a) Find the coordinates of the vertices of the surrounding rectangle and find its area.

(b) Find the area of the red triangle by subtracting the areas of the three blue triangles from the area of the rectangle.

(c) Use your answer to part (b) to show that the area of the red triangle is given by

$$\text{area} = \pm\frac{1}{2}\begin{vmatrix} a_1 & b_1 & 1 \\ a_2 & b_2 & 1 \\ a_3 & b_3 & 1 \end{vmatrix}$$

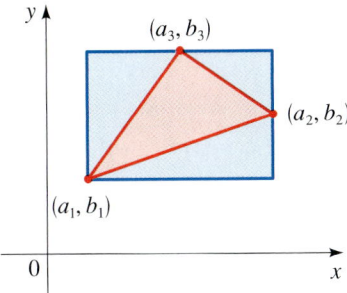

60. Collinear Points and Determinants

(a) If three points lie on a line, what is the area of the "triangle" that they determine? Use the answer to this question, together with the determinant formula for the area of a triangle, to explain why the points (a_1, b_1), (a_2, b_2), and (a_3, b_3) are collinear if and only if

$$\begin{vmatrix} a_1 & b_1 & 1 \\ a_2 & b_2 & 1 \\ a_3 & b_3 & 1 \end{vmatrix} = 0$$

(b) Use a determinant to check whether each set of points is collinear. Graph them to verify your answer.
   (i) $(-6, 4), (2, 10), (6, 13)$
   (ii) $(-5, 10), (2, 6), (15, -2)$

61. Determinant Form for the Equation of a Line

(a) Use the result of Exercise 60(a) to show that the equation of the line containing the points (x_1, y_1) and (x_2, y_2) is

$$\begin{vmatrix} x & y & 1 \\ x_1 & y_1 & 1 \\ x_2 & y_2 & 1 \end{vmatrix} = 0$$

(b) Use the result of part (a) to find an equation for the line containing the points $(20, 50)$ and $(-10, 25)$.

62. Matrices with Determinant Zero Use the definition of determinant and the elementary row and column operations to explain why matrices of the following types have determinant 0.

(a) A matrix with a row or column consisting entirely of zeros

(b) A matrix with two rows the same or two columns the same

(c) A matrix in which one row is a multiple of another row, or one column is a multiple of another column

63. Solving Linear Systems Suppose you have to solve a linear system with five equations and five variables without the assistance of a calculator or computer. Which method would you prefer: Cramer's Rule or Gaussian elimination? Write a short paragraph explaining the reasons for your answer.

9.8 Partial Fractions

To write a sum or difference of fractional expressions as a single fraction, we bring them to a common denominator. For example,

$$\frac{1}{x-1} + \frac{1}{2x+1} = \frac{(2x+1)+(x-1)}{(x-1)(2x+1)} = \frac{3x}{2x^2-x-1}$$

Common denominator ➡

$$\frac{1}{x-1} + \frac{1}{2x+1} = \frac{3x}{2x^2-x-1}$$

⬅ Partial fractions

But for some applications of algebra to calculus, we must reverse this process—that is, we must express a fraction such as $3x/(2x^2 - x - 1)$ as the sum of the simpler fractions $1/(x-1)$ and $1/(2x+1)$. These simpler fractions are called *partial fractions*; we learn how to find them in this section.

The Rhind papyrus is the oldest known mathematical document. It is an Egyptian scroll written in 1650 B.C. by the scribe Ahmes, who explains that it is an exact copy of a scroll written 200 years earlier. Ahmes claims that his papyrus contains "a thorough study of all things, insight into all that exists, knowledge of all obscure secrets." Actually, the document contains rules for doing arithmetic, including multiplication and division of fractions and several exercises with solutions. The exercise shown below reads: A heap and its seventh make nineteen; how large is the heap? In solving problems of this sort, the Egyptians used partial fractions because their number system required all fractions to be written as sums of reciprocals of whole numbers. For example, $\frac{7}{12}$ would be written as $\frac{1}{3} + \frac{1}{4}$.

The papyrus gives a correct formula for the volume of a truncated pyramid (page 143). It also gives the formula $A = \left(\frac{8}{9}d\right)^2$ for the area of a circle with diameter d. How close is this to the actual area?

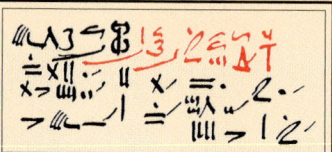

Let r be the rational function

$$r(x) = \frac{P(x)}{Q(x)}$$

where the degree of P is less than the degree of Q. By the Linear and Quadratic Factors Theorem in Section 3.4, every polynomial with real coefficients can be factored completely into linear and irreducible quadratic factors, that is, factors of the form $ax + b$ and $ax^2 + bx + c$, where a, b, and c are real numbers. For instance,

$$x^4 - 1 = (x^2 - 1)(x^2 + 1) = (x - 1)(x + 1)(x^2 + 1)$$

After we have completely factored the denominator Q of r, we can express $r(x)$ as a sum of **partial fractions** of the form

$$\frac{A}{(ax + b)^i} \quad \text{and} \quad \frac{Ax + B}{(ax^2 + bx + c)^j}$$

This sum is called the **partial fraction decomposition** of r. Let's examine the details of the four possible cases.

Case 1: The Denominator Is a Product of Distinct Linear Factors

Suppose that we can factor $Q(x)$ as

$$Q(x) = (a_1x + b_1)(a_2x + b_2) \cdots (a_nx + b_n)$$

with no factor repeated. In this case, the partial fraction decomposition of $P(x)/Q(x)$ takes the form

$$\frac{P(x)}{Q(x)} = \frac{A_1}{a_1x + b_1} + \frac{A_2}{a_2x + b_2} + \cdots + \frac{A_n}{a_nx + b_n}$$

The constants $A_1, A_2, \ldots, A_n$ are determined as in the following example.

Example 1   Distinct Linear Factors

Find the partial fraction decomposition of $\dfrac{5x + 7}{x^3 + 2x^2 - x - 2}$.

Solution   The denominator factors as follows:

$$x^3 + 2x^2 - x - 2 = x^2(x + 2) - (x + 2) = (x^2 - 1)(x + 2)$$
$$= (x - 1)(x + 1)(x + 2)$$

This gives us the partial fraction decomposition

$$\frac{5x + 7}{x^3 + 2x^2 - x - 2} = \frac{A}{x - 1} + \frac{B}{x + 1} + \frac{C}{x + 2}$$

Multiplying each side by the common denominator, $(x - 1)(x + 1)(x + 2)$, we get

$$5x + 7 = A(x + 1)(x + 2) + B(x - 1)(x + 2) + C(x - 1)(x + 1)$$

$$= A(x^2 + 3x + 2) + B(x^2 + x - 2) + C(x^2 - 1) \qquad \text{Expand}$$

$$= (A + B + C)x^2 + (3A + B)x + (2A - 2B - C) \qquad \text{Combine like terms}$$

If two polynomials are equal, then their coefficients are equal. Thus, since $5x + 7$ has no x^2-term, we have $A + B + C = 0$. Similarly, by comparing the coefficients of x, we see that $3A + B = 5$, and by comparing constant terms, we get $2A - 2B - C = 7$. This leads to the following system of linear equations for A, B, and C.

$$\begin{cases} A + B + C = 0 & \text{Equation 1: Coefficients of } x^2 \\ 3A + B = 5 & \text{Equation 2: Coefficients of } x \\ 2A - 2B - C = 7 & \text{Equation 3: Constant coefficients} \end{cases}$$

We use Gaussian elimination to solve this system.

$$\begin{cases} A + B + C = 0 \\ -2B - 3C = 5 & \text{Equation 2} + (-3) \times \text{Equation 1} \\ -4B - 3C = 7 & \text{Equation 3} + (-2) \times \text{Equation 1} \end{cases}$$

$$\begin{cases} A + B + C = 0 \\ -2B - 3C = 5 \\ 3C = -3 & \text{Equation 3} + (-2) \times \text{Equation 2} \end{cases}$$

From the third equation we get $C = -1$. Back-substituting we find that $B = -1$ and $A = 2$. So, the partial fraction decomposition is

$$\frac{5x + 7}{x^3 + 2x^2 - x - 2} = \frac{2}{x - 1} + \frac{-1}{x + 1} + \frac{-1}{x + 2} \qquad \blacksquare$$

The same approach works in the remaining cases. We set up the partial fraction decomposition with the unknown constants, A, B, C, ... Then we multiply each side of the resulting equation by the common denominator, simplify the right-hand side of the equation, and equate coefficients. This gives a set of linear equations that will always have a unique solution (provided that the partial fraction decomposition has been set up correctly).

Case 2: The Denominator Is a Product of Linear Factors, Some of Which Are Repeated

Suppose the complete factorization of $Q(x)$ contains the linear factor $ax + b$ repeated k times; that is, $(ax + b)^k$ is a factor of $Q(x)$. Then, corresponding to each such factor, the partial fraction decomposition for $P(x)/Q(x)$ contains

$$\frac{A_1}{ax + b} + \frac{A_2}{(ax + b)^2} + \cdots + \frac{A_k}{(ax + b)^k}$$

Example 2   Repeated Linear Factors

Find the partial fraction decomposition of $\dfrac{x^2 + 1}{x(x - 1)^3}$.

Solution   Because the factor $x - 1$ is repeated three times in the denominator, the partial fraction decomposition has the form

$$\frac{x^2 + 1}{x(x - 1)^3} = \frac{A}{x} + \frac{B}{x - 1} + \frac{C}{(x - 1)^2} + \frac{D}{(x - 1)^3}$$

Multiplying each side by the common denominator, $x(x - 1)^3$, gives

$$
\begin{aligned}
x^2 + 1 &= A(x - 1)^3 + Bx(x - 1)^2 + Cx(x - 1) + Dx \\
&= A(x^3 - 3x^2 + 3x - 1) + B(x^3 - 2x^2 + x) + C(x^2 - x) + Dx && \text{\small\color{gray}Expand} \\
&= (A + B)x^3 + (-3A - 2B + C)x^2 + (3A + B - C + D)x - A && \text{\small\color{gray}Combine like terms}
\end{aligned}
$$

Equating coefficients, we get the equations

$$
\begin{cases}
A + B &= 0 && \text{\small\color{gray}Coefficients of } x^3 \\
-3A - 2B + C &= 1 && \text{\small\color{gray}Coefficients of } x^2 \\
3A + B - C + D &= 0 && \text{\small\color{gray}Coefficients of } x \\
-A &= 1 && \text{\small\color{gray}Constant coefficients}
\end{cases}
$$

If we rearrange these equations by putting the last one in the first position, we can easily see (using substitution) that the solution to the system is $A = -1$, $B = 1$, $C = 0$, $D = 2$, and so the partial fraction decomposition is

$$\frac{x^2 + 1}{x(x - 1)^3} = \frac{-1}{x} + \frac{1}{x - 1} + \frac{2}{(x - 1)^3}$$ ∎

Case 3: The Denominator Has Irreducible Quadratic Factors, None of Which Is Repeated

Suppose the complete factorization of $Q(x)$ contains the quadratic factor $ax^2 + bx + c$ (which can't be factored further). Then, corresponding to this, the partial fraction decomposition of $P(x)/Q(x)$ will have a term of the form

$$\frac{Ax + B}{ax^2 + bx + c}$$

Example 3   Distinct Quadratic Factors

Find the partial fraction decomposition of $\dfrac{2x^2 - x + 4}{x^3 + 4x}$.

Solution   Since $x^3 + 4x = x(x^2 + 4)$, which can't be factored further, we write

$$\frac{2x^2 - x + 4}{x^3 + 4x} = \frac{A}{x} + \frac{Bx + C}{x^2 + 4}$$

Multiplying by $x(x^2 + 4)$, we get

$$2x^2 - x + 4 = A(x^2 + 4) + (Bx + C)x$$
$$= (A + B)x^2 + Cx + 4A$$

Equating coefficients gives us the equations

$$\begin{cases} A + B = 2 & \text{Coefficients of } x^2 \\ C = -1 & \text{Coefficients of } x \\ 4A = 4 & \text{Constant coefficients} \end{cases}$$

and so $A = 1$, $B = 1$, and $C = -1$. The required partial fraction decomposition is

$$\frac{2x^2 - x + 4}{x^3 + 4x} = \frac{1}{x} + \frac{x - 1}{x^2 + 4}$$

∎

Case 4: The Denominator Has a Repeated Irreducible Quadratic Factor

Suppose the complete factorization of $Q(x)$ contains the factor $(ax^2 + bx + c)^k$, where $ax^2 + bx + c$ can't be factored further. Then the partial fraction decomposition of $P(x)/Q(x)$ will have the terms

$$\frac{A_1x + B_1}{ax^2 + bx + c} + \frac{A_2x + B_2}{(ax^2 + bx + c)^2} + \cdots + \frac{A_kx + B_k}{(ax^2 + bx + c)^k}$$

Example 4   Repeated Quadratic Factors

Write the form of the partial fraction decomposition of

$$\frac{x^5 - 3x^2 + 12x - 1}{x^3(x^2 + x + 1)(x^2 + 2)^3}$$

Solution

$$\frac{x^5 - 3x^2 + 12x - 1}{x^3(x^2 + x + 1)(x^2 + 2)^3}$$

$$= \frac{A}{x} + \frac{B}{x^2} + \frac{C}{x^3} + \frac{Dx + E}{x^2 + x + 1} + \frac{Fx + G}{x^2 + 2} + \frac{Hx + I}{(x^2 + 2)^2} + \frac{Jx + K}{(x^2 + 2)^3}$$

∎

To find the values of A, B, C, D, E, F, G, H, I, J, and K in Example 4, we would have to solve a system of 11 linear equations. Although possible, this would certainly involve a great deal of work!

The techniques we have described in this section apply only to rational functions $P(x)/Q(x)$ in which the degree of P is less than the degree of Q. If this isn't the case, we must first use long division to divide Q into P.

Example 5 **Using Long Division to Prepare for Partial Fractions**

Find the partial fraction decomposition of

$$\frac{2x^4 + 4x^3 - 2x^2 + x + 7}{x^3 + 2x^2 - x - 2}$$

Solution Since the degree of the numerator is larger than the degree of the denominator, we use long division to obtain

$$\require{enclose}
\begin{array}{r}
2x \\
x^3 + 2x^2 - x - 2 \enclose{longdiv}{2x^4 + 4x^3 - 2x^2 + x + 7} \\
\underline{2x^4 + 4x^3 - 2x^2 - 4x} \\
5x + 7
\end{array}$$

$$\frac{2x^4 + 4x^3 - 2x^2 + x + 7}{x^3 + 2x^2 - x - 2} = 2x + \frac{5x + 7}{x^3 + 2x^2 - x - 2}$$

The remainder term now satisfies the requirement that the degree of the numerator is less than the degree of the denominator. At this point we proceed as in Example 1 to obtain the decomposition

$$\frac{2x^4 + 4x^3 - 2x^2 + x + 7}{x^3 + 2x^2 - x - 2} = 2x + \frac{2}{x - 1} + \frac{-1}{x + 1} + \frac{-1}{x + 2}$$

■

9.8 Exercises

1–10 ■ Write the form of the partial fraction decomposition of the function (as in Example 4). Do not determine the numerical values of the coefficients.

1. $\dfrac{1}{(x - 1)(x + 2)}$

2. $\dfrac{x}{x^2 + 3x - 4}$

3. $\dfrac{x^2 - 3x + 5}{(x - 2)^2(x + 4)}$

4. $\dfrac{1}{x^4 - x^3}$

5. $\dfrac{x^2}{(x - 3)(x^2 + 4)}$

6. $\dfrac{1}{x^4 - 1}$

7. $\dfrac{x^3 - 4x^2 + 2}{(x^2 + 1)(x^2 + 2)}$

8. $\dfrac{x^4 + x^2 + 1}{x^2(x^2 + 4)^2}$

9. $\dfrac{x^3 + x + 1}{x(2x - 5)^3(x^2 + 2x + 5)^2}$

10. $\dfrac{1}{(x^3 - 1)(x^2 - 1)}$

11–42 ■ Find the partial fraction decomposition of the rational function.

11. $\dfrac{2}{(x - 1)(x + 1)}$

12. $\dfrac{2x}{(x - 1)(x + 1)}$

13. $\dfrac{5}{(x - 1)(x + 4)}$

14. $\dfrac{x + 6}{x(x + 3)}$

15. $\dfrac{12}{x^2 - 9}$

16. $\dfrac{x - 12}{x^2 - 4x}$

17. $\dfrac{4}{x^2 - 4}$

18. $\dfrac{2x + 1}{x^2 + x - 2}$

19. $\dfrac{x + 14}{x^2 - 2x - 8}$

20. $\dfrac{8x - 3}{2x^2 - x}$

21. $\dfrac{x}{8x^2 - 10x + 3}$

22. $\dfrac{7x - 3}{x^3 + 2x^2 - 3x}$

23. $\dfrac{9x^2 - 9x + 6}{2x^3 - x^2 - 8x + 4}$

24. $\dfrac{-3x^2 - 3x + 27}{(x + 2)(2x^2 + 3x - 9)}$

25. $\dfrac{x^2 + 1}{x^3 + x^2}$

26. $\dfrac{3x^2 + 5x - 13}{(3x + 2)(x^2 - 4x + 4)}$

27. $\dfrac{2x}{4x^2 + 12x + 9}$

28. $\dfrac{x - 4}{(2x - 5)^2}$

29. $\dfrac{4x^2 - x - 2}{x^4 + 2x^3}$

30. $\dfrac{x^3 - 2x^2 - 4x + 3}{x^4}$

31. $\dfrac{-10x^2 + 27x - 14}{(x - 1)^3(x + 2)}$

32. $\dfrac{-2x^2 + 5x - 1}{x^4 - 2x^3 + 2x - 1}$

33. $\dfrac{3x^3 + 22x^2 + 53x + 41}{(x + 2)^2(x + 3)^2}$

34. $\dfrac{3x^2 + 12x - 20}{x^4 - 8x^2 + 16}$

35. $\dfrac{x - 3}{x^3 + 3x}$

36. $\dfrac{3x^2 - 2x + 8}{x^3 - x^2 + 2x - 2}$

37. $\dfrac{2x^3 + 7x + 5}{(x^2 + x + 2)(x^2 + 1)}$

38. $\dfrac{x^2 + x + 1}{2x^4 + 3x^2 + 1}$

39. $\dfrac{x^4 + x^3 + x^2 - x + 1}{x(x^2 + 1)^2}$ **40.** $\dfrac{2x^2 - x + 8}{(x^2 + 4)^2}$

41. $\dfrac{x^5 - 2x^4 + x^3 + x + 5}{x^3 - 2x^2 + x - 2}$

42. $\dfrac{x^5 - 3x^4 + 3x^3 - 4x^2 + 4x + 12}{(x - 2)^2(x^2 + 2)}$

43. Determine A and B in terms of a and b:

$$\frac{ax + b}{x^2 - 1} = \frac{A}{x - 1} + \frac{B}{x + 1}$$

44. Determine A, B, C, and D in terms of a and b:

$$\frac{ax^3 + bx^2}{(x^2 + 1)^2} = \frac{Ax + B}{x^2 + 1} + \frac{Cx + D}{(x^2 + 1)^2}$$

Discovery • Discussion

45. Recognizing Partial Fraction Decompositions For each expression, determine whether it is already a partial fraction decomposition, or whether it can be decomposed further.

(a) $\dfrac{x}{x^2 + 1} + \dfrac{1}{x + 1}$ **(b)** $\dfrac{x}{(x + 1)^2}$

(c) $\dfrac{1}{x + 1} + \dfrac{2}{(x + 1)^2}$ **(d)** $\dfrac{x + 2}{(x^2 + 1)^2}$

46. Assembling and Disassembling Partial Fractions The following expression is a partial fraction decomposition:

$$\frac{2}{x - 1} + \frac{1}{(x - 1)^2} + \frac{1}{x + 1}$$

Use a common denominator to combine the terms into one fraction. Then use the techniques of this section to find its partial fraction decomposition. Did you get back the original expression?

9.9 **Systems of Inequalities**

In this section we study systems of inequalities in two variables from a graphical point of view.

Graphing an Inequality

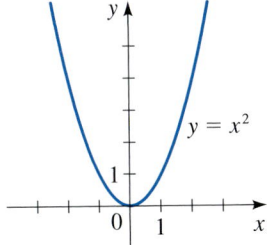

Figure 1

We begin by considering the graph of a single inequality. We already know that the graph of $y = x^2$, for example, is the *parabola* in Figure 1. If we replace the equal sign by the symbol $\geq$, we obtain the *inequality*

$$y \geq x^2$$

Its graph consists of not just the parabola in Figure 1, but also every point whose y-coordinate is *larger* than x^2. We indicate the solution in Figure 2(a) by shading the points *above* the parabola.

Similarly, the graph of $y \leq x^2$ in Figure 2(b) consists of all points on and *below* the parabola. However, the graphs of $y > x^2$ and $y < x^2$ do not include the points on the parabola itself, as indicated by the dashed curves in Figures 2(c) and 2(d).

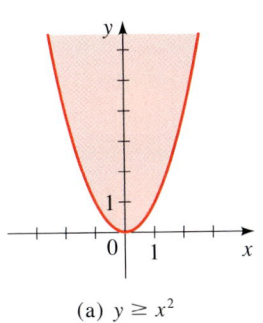

(a) $y \geq x^2$

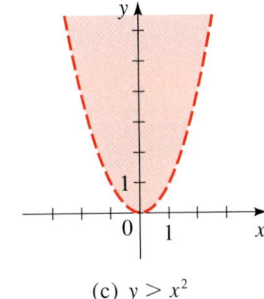

(b) $y \leq x^2$ (c) $y > x^2$

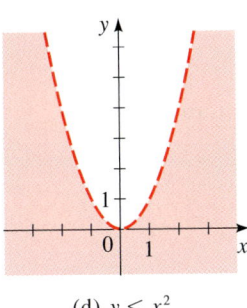

(d) $y < x^2$

Figure 2

The graph of an inequality, in general, consists of a region in the plane whose boundary is the graph of the equation obtained by replacing the inequality sign ($\geq$, $\leq$, $>$, or $<$) with an equal sign. To determine which side of the graph gives the solution set of the inequality, we need only check **test points**.

Graphing Inequalities

To graph an inequality, we carry out the following steps.

1. **Graph Equation.** Graph the equation corresponding to the inequality. Use a dashed curve for $>$ or $<$, and a solid curve for $\leq$ or $\geq$.

2. **Test Points.** Test one point in each region formed by the graph in Step 1. If the point satisfies the inequality, then all the points in that region satisfy the inequality. (In that case, shade the region to indicate it is part of the graph.) If the test point does not satisfy the inequality, then the region isn't part of the graph.

Example 1   Graphs of Inequalities

Graph each inequality.

(a) $x^2 + y^2 < 25$ (b) $x + 2y \geq 5$

Solution

(a) The graph of $x^2 + y^2 = 25$ is a circle of radius 5 centered at the origin. The points on the circle itself do not satisfy the inequality because it is of the form $<$, so we graph the circle with a dashed curve, as shown in Figure 3.

To determine whether the inside or the outside of the circle satisfies the inequality, we use the test points $(0, 0)$ on the inside and $(6, 0)$ on the outside. To do this, we substitute the coordinates of each point into the inequality and check if the result satisfies the inequality. (Note that *any* point inside or outside the circle can serve as a test point. We have chosen these points for simplicity.)

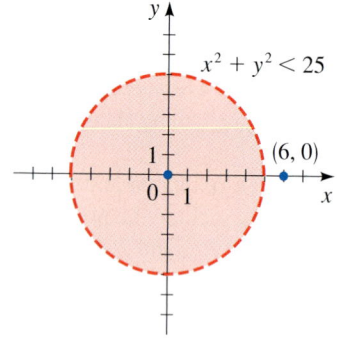

Figure 3

Test point	$x^2 + y^2 < 25$	Conclusion
$(0, 0)$	$0^2 + 0^2 = 0 < 25$	Part of graph
$(6, 0)$	$6^2 + 0^2 = 36 \not< 25$	Not part of graph

Thus, the graph of $x^2 + y^2 < 25$ is the set of all points *inside* the circle (see Figure 3).

(b) The graph of $x + 2y = 5$ is the line shown in Figure 4. We use the test points $(0, 0)$ and $(5, 5)$ on opposite sides of the line.

Test point	$x + 2y \geq 5$	Conclusion
$(0, 0)$	$0 + 2(0) = 0 \not\geq 5$	Not part of graph
$(5, 5)$	$5 + 2(5) = 15 \geq 5$	Part of graph

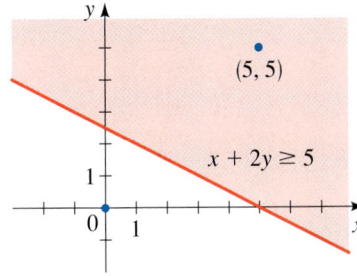

Figure 4

Our check shows that the points *above* the line satisfy the inequality.

Alternatively, we could put the inequality into slope-intercept form and graph it directly:

$$x + 2y \geq 5$$

$$2y \geq -x + 5$$

$$y \geq -\tfrac{1}{2}x + \tfrac{5}{2}$$

From this form we see that the graph includes all points whose y-coordinates are *greater* than those on the line $y = -\tfrac{1}{2}x + \tfrac{5}{2}$; that is, the graph consists of the points *on or above* this line, as shown in Figure 4. ■

Systems of Inequalities

We now consider *systems* of inequalities. The solution of such a system is the set of all points in the coordinate plane that satisfy every inequality in the system.

Example 2   A System of Two Inequalities

Graph the solution of the system of inequalities.

$$\begin{cases} x^2 + y^2 < 25 \\ x + 2y \geq 5 \end{cases}$$

Solution   These are the two inequalities of Example 1. In this example we wish to graph only those points that simultaneously satisfy both inequalities. The solution consists of the intersection of the graphs in Example 1. In Figure 5(a) we show the two regions on the same coordinate plane (in different colors), and in Figure 5(b) we show their intersection.

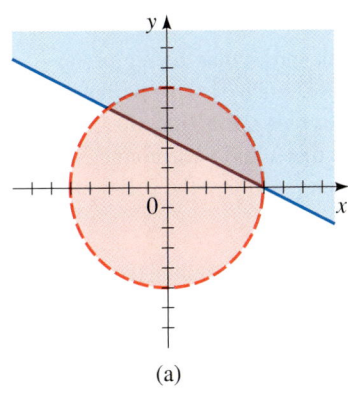

(a)

VERTICES   The points $(-3, 4)$ and $(5, 0)$ in Figure 5(b) are the **vertices** of the solution set. They are obtained by solving the system of *equations*

$$\begin{cases} x^2 + y^2 = 25 \\ x + 2y = 5 \end{cases}$$

We solve this system of equations by substitution. Solving for x in the second equation gives $x = 5 - 2y$, and substituting this into the first equation gives

$$(5 - 2y)^2 + y^2 = 25 \qquad \text{Substitute } x = 5 - 2y$$

$$(25 - 20y + 4y^2) + y^2 = 25 \qquad \text{Expand}$$

$$-20y + 5y^2 = 0 \qquad \text{Simplify}$$

$$-5y(4 - y) = 0 \qquad \text{Factor}$$

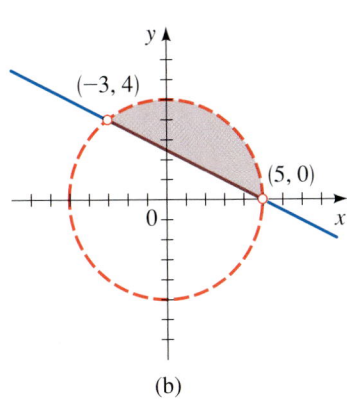

(b)

Figure 5
$$\begin{cases} x^2 + y^2 < 25 \\ x + 2y \geq 5 \end{cases}$$

Thus, $y = 0$ or $y = 4$. When $y = 0$, we have $x = 5 - 2(0) = 5$, and when $y = 4$, we have $x = 5 - 2(4) = -3$. So the points of intersection of these curves are $(5, 0)$ and $(-3, 4)$.

Note that in this case the vertices are not part of the solution set, since they don't satisfy the inequality $x^2 + y^2 < 25$ (and so they are graphed as open circles in the figure). They simply show where the "corners" of the solution set lie. ■

Systems of Linear Inequalities

An inequality is **linear** if it can be put into one of the following forms:

$$ax + by \geq c \qquad ax + by \leq c \qquad ax + by > c \qquad ax + by < c$$

In the next example we graph the solution set of a system of linear inequalities.

Example 3   A System of Four Linear Inequalities

Graph the solution set of the system, and label its vertices.

$$\begin{cases} x + 3y \leq 12 \\ x + y \leq 8 \\ x \geq 0 \\ y \geq 0 \end{cases}$$

Solution   In Figure 6 we first graph the lines given by the equations that correspond to each inequality. To determine the graphs of the linear inequalities, we only need to check one test point. For simplicity let's use the point $(0, 0)$.

Inequality	Test point (0, 0)	Conclusion
$x + 3y \leq 12$	$0 + 3(0) = 0 \leq 12$	Satisfies inequality
$x + y \leq 8$	$0 + 0 = 0 \leq 8$	Satisfies inequality

Since $(0, 0)$ is below the line $x + 3y = 12$, our check shows that the region on or below the line must satisfy the inequality. Likewise, since $(0, 0)$ is below the line $x + y = 8$, our check shows that the region on or below this line must satisfy the inequality. The inequalities $x \geq 0$ and $y \geq 0$ say that x and y are nonnegative. These regions are sketched in Figure 6(a), and the intersection—the solution set—is sketched in Figure 6(b).

VERTICES   The coordinates of each vertex are obtained by simultaneously solving the equations of the lines that intersect at that vertex. From the system

$$\begin{cases} x + 3y = 12 \\ x + y = 8 \end{cases}$$

we get the vertex $(6, 2)$. The other vertices are the x- and y-intercepts of the corresponding lines, $(8, 0)$ and $(0, 4)$, and the origin $(0, 0)$. In this case, all the vertices *are* part of the solution set.

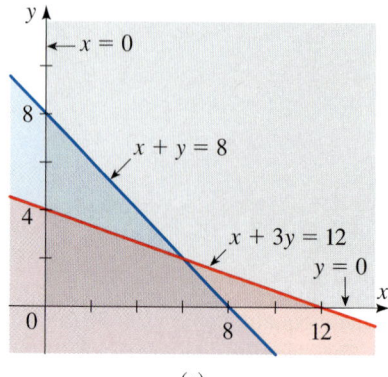

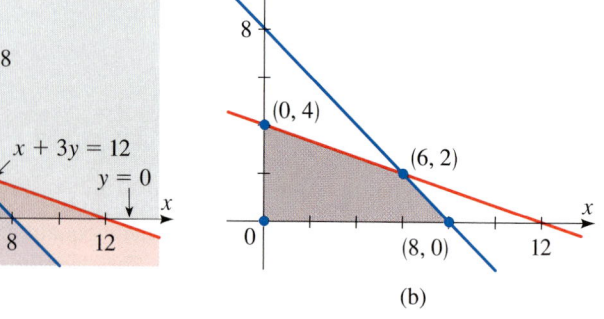

Figure 6   (a)   (b)

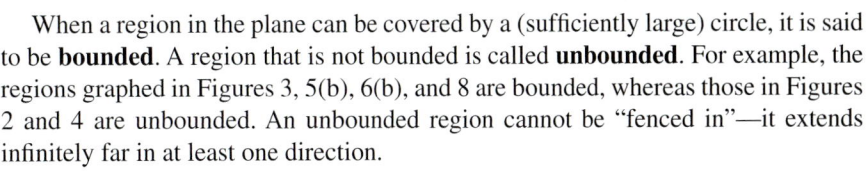

Example 4   A System of Linear Inequalities

Graph the solution set of the system.

$$\begin{cases} x + 2y \geq 8 \\ -x + 2y \leq 4 \\ 3x - 2y \leq 8 \end{cases}$$

Solution   We must graph the lines that correspond to these inequalities and then shade the appropriate regions, as in Example 3. We will use a graphing calculator, so we must first isolate y on the left-hand side of each inequality.

$$\begin{cases} y \geq -\frac{1}{2}x + 4 \\ y \leq \frac{1}{2}x + 2 \\ y \geq \frac{3}{2}x - 4 \end{cases}$$

Using the shading feature of the calculator, we obtain the graph in Figure 7. The solution set is the triangular region that is shaded in all three patterns. We then use $\boxed{\text{TRACE}}$ or the $\texttt{Intersect}$ command to find the vertices of the region. The solution set is graphed in Figure 8. ∎

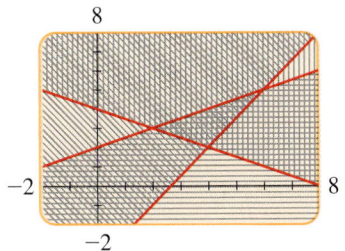

Figure 7

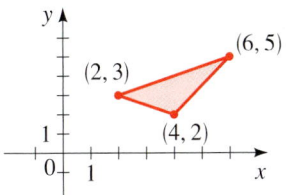

Figure 8

When a region in the plane can be covered by a (sufficiently large) circle, it is said to be **bounded**. A region that is not bounded is called **unbounded**. For example, the regions graphed in Figures 3, 5(b), 6(b), and 8 are bounded, whereas those in Figures 2 and 4 are unbounded. An unbounded region cannot be "fenced in"—it extends infinitely far in at least one direction.

Application: Feasible Regions

Many applied problems involve *constraints* on the variables. For instance, a factory manager has only a certain number of workers that can be assigned to perform jobs on the factory floor. A farmer deciding what crops to cultivate has only a certain amount of land that can be seeded. Such constraints or limitations can usually be expressed as systems of inequalities. When dealing with applied inequalities, we usually refer to the solution set of a system as a *feasible region*, because the points in the solution set represent feasible (or possible) values for the quantities being studied.

Example 5   Restricting Pollutant Outputs

A factory produces two agricultural pesticides, A and B. For every barrel of A, the factory emits 0.25 kg of carbon monoxide (CO) and 0.60 kg of sulfur dioxide (SO_2), and for every barrel of B, it emits 0.50 kg of CO and 0.20 kg of SO_2. Pollution laws restrict the factory's output of CO to a maximum of 75 kg and SO_2 to a maximum of 90 kg per day.

(a) Find a system of inequalities that describes the number of barrels of each pesticide the factory can produce and still satisfy the pollution laws. Graph the feasible region.

(b) Would it be legal for the factory to produce 100 barrels of A and 80 barrels of B per day?

(c) Would it be legal for the factory to produce 60 barrels of A and 160 barrels of B per day?

Solution

(a) To set up the required inequalities, it's helpful to organize the given information into a table.

	A	B	Maximum
CO (kg)	0.25	0.50	75
SO_2 (kg)	0.60	0.20	90

We let

$$x = \text{number of barrels of A produced per day}$$

$$y = \text{number of barrels of B produced per day}$$

From the data in the table and the fact that x and y can't be negative, we obtain the following inequalities.

$$\begin{cases} 0.25x + 0.50y \le 75 & \text{CO inequality} \\ 0.60x + 0.20y \le 90 & \text{SO}_2 \text{ inequality} \\ x \ge 0, \quad y \ge 0 \end{cases}$$

Multiplying the first inequality by 4 and the second by 5 simplifies this to

$$\begin{cases} x + 2y \le 300 \\ 3x + y \le 450 \\ x \ge 0, \quad y \ge 0 \end{cases}$$

The feasible region is the solution of this system of inequalities, shown in Figure 9.

(b) Since the point $(100, 80)$ lies inside the feasible region, this production plan is legal (see Figure 9).

(c) Since the point $(60, 160)$ lies outside the feasible region, this production plan is not legal. It violates the CO restriction, although it does not violate the SO_2 restriction (see Figure 9). ∎

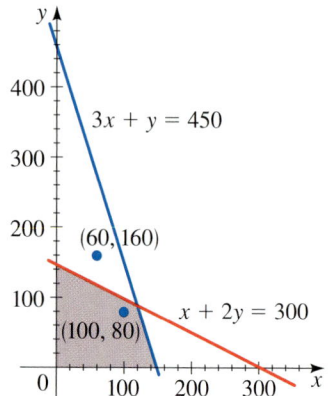

Figure 9

9.9   Exercises

1–14 ■ Graph the inequality.

1. $x < 3$

2. $y \ge -2$

3. $y > x$

4. $y < x + 2$

5. $y \le 2x + 2$

6. $y < -x + 5$

7. $2x - y \le 8$

8. $3x + 4y + 12 > 0$

9. $4x + 5y < 20$

10. $-x^2 + y \ge 10$

11. $y > x^2 + 1$

12. $x^2 + y^2 \ge 9$

13. $x^2 + y^2 \le 25$

14. $x^2 + (y - 1)^2 \le 1$

15–18 ■ An equation and its graph are given. Find an inequality whose solution is the shaded region.

15. $y = \frac{1}{2}x - 1$

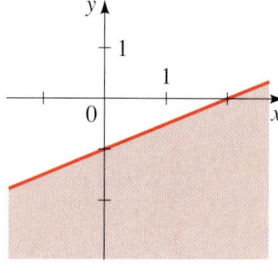

16. $y = x^2 + 2$

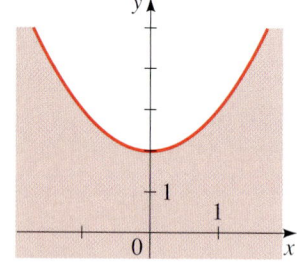

17. $x^2 + y^2 = 4$

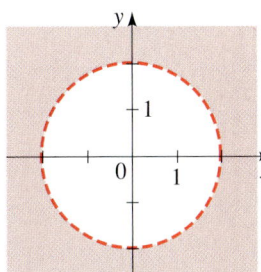

18. $y = x^3 - 4x$

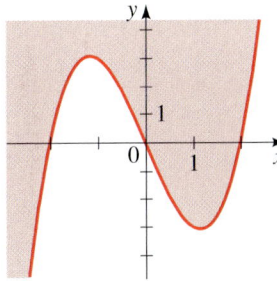

19–40 ■ Graph the solution of the system of inequalities. Find the coordinates of all vertices, and determine whether the solution set is bounded.

19. $\begin{cases} x + y \le 4 \\ \quad\ y \ge x \end{cases}$

20. $\begin{cases} 2x + 3y > 12 \\ 3x - \ y < 21 \end{cases}$

21. $\begin{cases} y < \frac{1}{4}x + 2 \\ y \ge 2x - 5 \end{cases}$

22. $\begin{cases} x - y > 0 \\ 4 + y \le 2x \end{cases}$

23. $\begin{cases} x \ge 0 \\ y \ge 0 \\ 3x + 5y \le 15 \\ 3x + 2y \le 9 \end{cases}$

24. $\begin{cases} x > 2 \\ y < 12 \\ 2x - 4y > 8 \end{cases}$

25. $\begin{cases} y < 9 - x^2 \\ y \ge x + 3 \end{cases}$

26. $\begin{cases} y \ge x^2 \\ x + y \ge 6 \end{cases}$

27. $\begin{cases} x^2 + y^2 \le 4 \\ \quad x - y > 0 \end{cases}$

28. $\begin{cases} x > 0 \\ y > 0 \\ x + y < 10 \\ x^2 + y^2 > 9 \end{cases}$

29. $\begin{cases} x^2 - y \le 0 \\ 2x^2 + y \le 12 \end{cases}$

30. $\begin{cases} x^2 + y^2 < 9 \\ 2x + y^2 \ge 1 \end{cases}$

31. $\begin{cases} x + 2y \le 14 \\ 3x - \ y \ge 0 \\ x - \ y \ge 2 \end{cases}$

32. $\begin{cases} y < x + 6 \\ 3x + 2y \ge 12 \\ x - 2y \le 2 \end{cases}$

33. $\begin{cases} x \ge 0 \\ y \ge 0 \\ x \le 5 \\ x + y \le 7 \end{cases}$

34. $\begin{cases} x \ge 0 \\ y \ge 0 \\ y \le 4 \\ 2x + y \le 8 \end{cases}$

35. $\begin{cases} y > x + 1 \\ x + 2y \le 12 \\ x + 1 > 0 \end{cases}$

36. $\begin{cases} x + y > 12 \\ y < \frac{1}{2}x - 6 \\ 3x + y < 6 \end{cases}$

37. $\begin{cases} x^2 + y^2 \le 8 \\ x \ge 2 \\ y \ge 0 \end{cases}$

38. $\begin{cases} x^2 - y \ge 0 \\ x + y < 6 \\ x - y < 6 \end{cases}$

39. $\begin{cases} x^2 + y^2 < 9 \\ x + y > 0 \\ x \le 0 \end{cases}$

40. $\begin{cases} y \ge x^3 \\ y \le 2x + 4 \\ x + y \ge 0 \end{cases}$

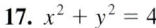 **41–44** ■ Use a graphing calculator to graph the solution of the system of inequalities. Find the coordinates of all vertices, correct to one decimal place.

41. $\begin{cases} y \ge x - 3 \\ y \ge -2x + 6 \\ y \le 8 \end{cases}$

42. $\begin{cases} x + y \ge 12 \\ 2x + y \le 24 \\ x - y \ge -6 \end{cases}$

43. $\begin{cases} y \le 6x - x^2 \\ x + y \ge 4 \end{cases}$

44. $\begin{cases} y \ge x^3 \\ 2x + y \ge 0 \\ y \le 2x + 6 \end{cases}$

Applications

45. Publishing Books A publishing company publishes a total of no more than 100 books every year. At least 20 of these are nonfiction, but the company always publishes at least as much fiction as nonfiction. Find a system of inequalities that describes the possible numbers of fiction and nonfiction books that the company can produce each year consistent with these policies. Graph the solution set.

46. Furniture Manufacturing A man and his daughter manufacture unfinished tables and chairs. Each table requires 3 hours of sawing and 1 hour of assembly. Each chair requires 2 hours of sawing and 2 hours of assembly. The two of them can put in up to 12 hours of sawing and 8 hours of assembly work each day. Find a system of inequalities that describes all possible combinations of tables and chairs that they can make daily. Graph the solution set.

47. Coffee Blends A coffee merchant sells two different coffee blends. The Standard blend uses 4 oz of arabica and 12 oz of robusta beans per package; the Deluxe blend uses 10 oz of arabica and 6 oz of robusta beans per package. The merchant has 80 lb of arabica and 90 lb of robusta beans available. Find a system of inequalities that describes the possible number of Standard and Deluxe packages he can make. Graph the solution set.

48. Nutrition A cat food manufacturer uses fish and beef by-products. The fish contains 12 g of protein and 3 g of fat per ounce. The beef contains 6 g of protein and 9 g of fat per ounce. Each can of cat food must contain at least 60 g of protein and 45 g of fat. Find a system of inequalities that describes the possible number of ounces of fish and beef that can be used in each can to satisfy these minimum requirements. Graph the solution set.

Discovery • Discussion

49. Shading Unwanted Regions   To graph the solution of a system of inequalities, we have shaded the solution of each inequality in a different color; the solution of the system is the region where all the shaded parts overlap. Here is a different method: For each inequality, shade the region that does *not* satisfy the inequality. Explain why the part of the plane that is left unshaded is the solution of the system. Solve the following system by both methods. Which do you prefer?

$$\begin{cases} x + 2y > 4 \\ -x + y < 1 \\ x + 3y < 9 \\ x < 3 \end{cases}$$

---

9   Review

Concept Check

1. Suppose you are asked to solve a system of two equations (not necessarily linear) in two variables. Explain how you would solve the system
   (a) by the substitution method
   (b) by the elimination method
   (c) graphically

2. Suppose you are asked to solve a system of two *linear* equations in two variables.
   (a) Would you prefer to use the substitution method or the elimination method?
   (b) How many solutions are possible? Draw diagrams to illustrate the possibilities.

3. What operations can be performed on a linear system that result in an equivalent system?

4. Explain how Gaussian elimination works. Your explanation should include a discussion of the steps used to obtain a system in triangular form, and back-substitution.

5. What does it mean to say that A is a matrix with dimension $m \times n$?

6. What is the augmented matrix of a system? Describe the role of elementary row operations, row-echelon form, back-substitution, and leading variables when solving a system in matrix form.

7. (a) What is meant by an inconsistent system?
   (b) What is meant by a dependent system?

8. Suppose you have used Gaussian elimination to transform the augmented matrix of a linear system into row-echelon form. How can you tell if the system has
   (a) exactly one solution?
   (b) no solution?
   (c) infinitely many solutions?

9. How can you tell if a matrix is in reduced row-echelon form?

10. How do Gaussian elimination and Gauss-Jordan elimination differ? What advantage does Gauss-Jordan elimination have?

11. If A and B are matrices with the same dimension and k is a real number, how do you find $A + B$, $A - B$, and kA?

12. (a) What must be true of the dimensions of A and B for the product AB to be defined?
   (b) If the product AB is defined, how do you calculate it?

13. (a) What is the identity matrix I_n?
   (b) If A is a square $n \times n$ matrix, what is its inverse matrix?
   (c) Write a formula for the inverse of a 2×2 matrix.
   (d) Explain how you would find the inverse of a 3×3 matrix.

14. (a) Explain how to express a linear system as a matrix equation of the form $AX = B$.
   (b) If A has an inverse, how would you solve the matrix equation $AX = B$?

15. Suppose A is an $n \times n$ matrix.
   (a) What is the minor M_{ij} of the element a_{ij}?
   (b) What is the cofactor A_{ij}?
   (c) How do you find the determinant of A?
   (d) How can you tell if A has an inverse?

16. State Cramer's Rule for solving a system of linear equations in terms of determinants. Do you prefer to use Cramer's Rule or Gaussian elimination? Explain.

17. Explain how to find the partial fraction decomposition of a rational expression. Include in your explanation a discussion of each of the four cases that arise.

18. How do you graph an inequality in two variables?

19. How do you graph the solution set of a system of inequalities?

Exercises

1–4 ■ Two equations and their graphs are given. Find the intersection point(s) of the graphs by solving the system.

1. $\begin{cases} 2x + 3y = 7 \\ x - 2y = 0 \end{cases}$

2. $\begin{cases} 3x + y = 8 \\ y = x^2 - 5x \end{cases}$

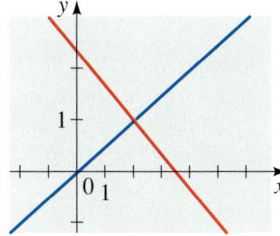
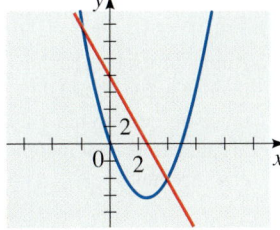

3. $\begin{cases} x^2 + y = 2 \\ x^2 - 3x - y = 0 \end{cases}$

4. $\begin{cases} x - y = -2 \\ x^2 + y^2 - 4y = 4 \end{cases}$

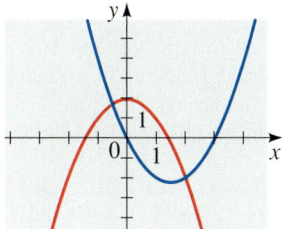
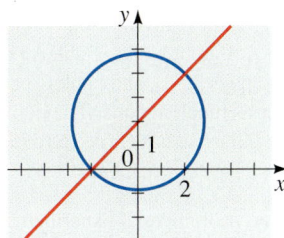

5–10 ■ Solve the system of equations and graph the lines.

5. $\begin{cases} 3x - y = 5 \\ 2x + y = 5 \end{cases}$

6. $\begin{cases} y = 2x + 6 \\ y = -x + 3 \end{cases}$

7. $\begin{cases} 2x - 7y = 28 \\ y = \frac{2}{7}x - 4 \end{cases}$

8. $\begin{cases} 6x - 8y = 15 \\ -\frac{3}{2}x + 2y = -4 \end{cases}$

9. $\begin{cases} 2x - y = 1 \\ x + 3y = 10 \\ 3x + 4y = 15 \end{cases}$

10. $\begin{cases} 2x + 5y = 9 \\ -x + 3y = 1 \\ 7x - 2y = 14 \end{cases}$

11–14 ■ Solve the system of equations.

11. $\begin{cases} y = x^2 + 2x \\ y = 6 + x \end{cases}$

12. $\begin{cases} x^2 + y^2 = 8 \\ y = x + 2 \end{cases}$

13. $\begin{cases} 3x + \dfrac{4}{y} = 6 \\ x - \dfrac{8}{y} = 4 \end{cases}$

14. $\begin{cases} x^2 + y^2 = 10 \\ x^2 + 2y^2 - 7y = 0 \end{cases}$

15–18 ■ Use a graphing device to solve the system, correct to the nearest hundredth.

15. $\begin{cases} 0.32x + 0.43y = 0 \\ 7x - 12y = 341 \end{cases}$

16. $\begin{cases} \sqrt{12}x - 3\sqrt{2}y = 660 \\ 7137x + 3931y = 20{,}000 \end{cases}$

17. $\begin{cases} x - y^2 = 10 \\ x = \frac{1}{22}y + 12 \end{cases}$

18. $\begin{cases} y = 5^x + x \\ y = x^5 + 5 \end{cases}$

19–24 ■ A matrix is given.
(a) State the dimension of the matrix.
(b) Is the matrix in row-echelon form?
(c) Is the matrix in reduced row-echelon form?
(d) Write the system of equations for which the given matrix is the augmented matrix.

19. $\begin{bmatrix} 1 & 2 & -5 \\ 0 & 1 & 3 \end{bmatrix}$

20. $\begin{bmatrix} 1 & 0 & 6 \\ 0 & 1 & 0 \end{bmatrix}$

21. $\begin{bmatrix} 1 & 0 & 8 & 0 \\ 0 & 1 & 5 & -1 \\ 0 & 0 & 0 & 0 \end{bmatrix}$

22. $\begin{bmatrix} 1 & 3 & 6 & 2 \\ 2 & 1 & 0 & 5 \\ 0 & 0 & 1 & 0 \end{bmatrix}$

23. $\begin{bmatrix} 0 & 1 & -3 & 4 \\ 1 & 1 & 0 & 7 \\ 1 & 2 & 1 & 2 \end{bmatrix}$

24. $\begin{bmatrix} 1 & 8 & 6 & -4 \\ 0 & 1 & -3 & 5 \\ 0 & 0 & 2 & -7 \\ 1 & 1 & 1 & 0 \end{bmatrix}$

25–46 ■ Find the complete solution of the system, or show that the system has no solution.

25. $\begin{cases} x + y + 2z = 6 \\ 2x + 5z = 12 \\ x + 2y + 3z = 9 \end{cases}$

26. $\begin{cases} x - 2y + 3z = 1 \\ x - 3y - z = 0 \\ 2x - 6z = 6 \end{cases}$

27. $\begin{cases} x - 2y + 3z = 1 \\ 2x - y + z = 3 \\ 2x - 7y + 11z = 2 \end{cases}$

28. $\begin{cases} x + y + z + w = 2 \\ 2x - 3z = 5 \\ x - 2y + 4w = 9 \\ x + y + 2z + 3w = 5 \end{cases}$

29. $\begin{cases} x + 2y + 2z = 6 \\ x - y = -1 \\ 2x + y + 3z = 7 \end{cases}$

30. $\begin{cases} x - y + z = 2 \\ x + y + 3z = 6 \\ 2y + 3z = 5 \end{cases}$

31. $\begin{cases} x - 2y + 3z = -2 \\ 2x - y + z = 2 \\ 2x - 7y + 11z = -9 \end{cases}$

32. $\begin{cases} x - y + z = 2 \\ x + y + 3z = 6 \\ 3x - y + 5z = 10 \end{cases}$

33. $\begin{cases} x + y + z + w = 0 \\ x - y - 4z - w = -1 \\ x - 2y + 4w = -7 \\ 2x + 2y + 3z + 4w = -3 \end{cases}$

34. $\begin{cases} x + 3z = -1 \\ y - 4w = 5 \\ 2y + z + w = 0 \\ 2x + y + 5z - 4w = 4 \end{cases}$

35. $\begin{cases} x - 3y + z = 4 \\ 4x - y + 15z = 5 \end{cases}$

36. $\begin{cases} 2x - 3y + 4z = 3 \\ 4x - 5y + 9z = 13 \\ 2x + 7z = 0 \end{cases}$

37. $\begin{cases} -x + 4y + z = 8 \\ 2x - 6y + z = -9 \\ x - 6y - 4z = -15 \end{cases}$

38. $\begin{cases} x - z + w = 2 \\ 2x + y - 2w = 12 \\ 3y + z + w = 4 \\ x + y - z = 10 \end{cases}$

39. $\begin{cases} x - y + 3z = 2 \\ 2x + y + z = 2 \\ 3x + 4z = 4 \end{cases}$

40. $\begin{cases} x - y = 1 \\ x + y + 2z = 3 \\ x - 3y - 2z = -1 \end{cases}$

41. $\begin{cases} x - y + z - w = 0 \\ 3x - y - z - w = 2 \end{cases}$

42. $\begin{cases} x - y = 3 \\ 2x + y = 6 \\ x - 2y = 9 \end{cases}$

43. $\begin{cases} x - y + z = 0 \\ 3x + 2y - z = 6 \\ x + 4y - 3z = 3 \end{cases}$

44. $\begin{cases} x + 2y + 3z = 2 \\ 2x - y - 5z = 1 \\ 4x + 3y + z = 6 \end{cases}$

45. $\begin{cases} x + y - z - w = 2 \\ x - y + z - w = 0 \\ 2x + 2w = 2 \\ 2x + 4y - 4z - 2w = 6 \end{cases}$

46. $\begin{cases} x - y - 2z + 3w = 0 \\ y - z + w = 1 \\ 3x - 2y - 7z + 10w = 2 \end{cases}$

47. A man invests his savings in two accounts, one paying 6% interest per year and the other paying 7%. He has twice as much invested in the 7% account as in the 6% account, and his annual interest income is $600. How much is invested in each account?

48. A piggy bank contains 50 coins, all of them nickels, dimes, or quarters. The total value of the coins is $5.60, and the value of the dimes is five times the value of the nickels. How many coins of each type are there?

49. Clarisse invests $60,000 in money-market accounts at three different banks. Bank A pays 2% interest per year, bank B pays 2.5%, and bank C pays 3%. She decides to invest twice as much in bank B as in the other two banks. After one year, Clarisse has earned $1575 in interest. How much did she invest in each bank?

50. A commercial fisherman fishes for haddock, sea bass, and red snapper. He is paid $1.25 a pound for haddock, $0.75 a pound for sea bass, and $2.00 a pound for red snapper. Yesterday he caught 560 lb of fish worth $575. The haddock and red snapper together are worth $320. How many pounds of each fish did he catch?

51–62 ■ Let

$$A = \begin{bmatrix} 2 & 0 & -1 \end{bmatrix} \qquad B = \begin{bmatrix} 1 & 2 & 4 \\ -2 & 1 & 0 \end{bmatrix}$$

$$C = \begin{bmatrix} \frac{1}{2} & 3 \\ 2 & \frac{3}{2} \\ -2 & 1 \end{bmatrix} \qquad D = \begin{bmatrix} 1 & 4 \\ 0 & -1 \\ 2 & 0 \end{bmatrix}$$

$$E = \begin{bmatrix} 2 & -1 \\ -\frac{1}{2} & 1 \end{bmatrix} \qquad F = \begin{bmatrix} 4 & 0 & 2 \\ -1 & 1 & 0 \\ 7 & 5 & 0 \end{bmatrix}$$

$$G = \begin{bmatrix} 5 \end{bmatrix}$$

Carry out the indicated operation, or explain why it cannot be performed.

51. $A + B$ **52.** $C - D$ **53.** $2C + 3D$

54. $5B - 2C$ **55.** GA **56.** AG

57. BC **58.** CB **59.** BF

60. FC **61.** $(C + D)E$ **62.** $F(2C - D)$

63–64 ■ Verify that the matrices A and B are inverses of each other by calculating the products AB and BA.

63. $A = \begin{bmatrix} 2 & -5 \\ -2 & 6 \end{bmatrix}$, $B = \begin{bmatrix} 3 & \frac{5}{2} \\ 1 & 1 \end{bmatrix}$

64. $A = \begin{bmatrix} 2 & -1 & 3 \\ 2 & -2 & 1 \\ 0 & 1 & 1 \end{bmatrix}$, $B = \begin{bmatrix} -\frac{3}{2} & 2 & \frac{5}{2} \\ -1 & 1 & 2 \\ 1 & -1 & -1 \end{bmatrix}$

65–70 ■ Solve the matrix equation for the unknown matrix, X, or show that no solution exists, where

$$A = \begin{bmatrix} 2 & 1 \\ 3 & 2 \end{bmatrix}, \quad B = \begin{bmatrix} 1 & -2 \\ -2 & 4 \end{bmatrix}, \quad C = \begin{bmatrix} 0 & 1 & 3 \\ -2 & 4 & 0 \end{bmatrix}$$

65. $A + 3X = B$ **66.** $\frac{1}{2}(X - 2B) = A$

67. $2(X - A) = 3B$ **68.** $2X + C = 5A$

69. $AX = C$ **70.** $AX = B$

71–78 ■ Find the determinant and, if possible, the inverse of the matrix.

71. $\begin{bmatrix} 1 & 4 \\ 2 & 9 \end{bmatrix}$ **72.** $\begin{bmatrix} 2 & 2 \\ 1 & -3 \end{bmatrix}$

73. $\begin{bmatrix} 4 & -12 \\ -2 & 6 \end{bmatrix}$ **74.** $\begin{bmatrix} 2 & 4 & 0 \\ -1 & 1 & 2 \\ 0 & 3 & 2 \end{bmatrix}$

75. $\begin{bmatrix} 3 & 0 & 1 \\ 2 & -3 & 0 \\ 4 & -2 & 1 \end{bmatrix}$ **76.** $\begin{bmatrix} 1 & 2 & 3 \\ 2 & 4 & 5 \\ 2 & 5 & 6 \end{bmatrix}$

77. $\begin{bmatrix} 1 & 0 & 0 & 1 \\ 0 & 2 & 0 & 2 \\ 0 & 0 & 3 & 3 \\ 0 & 0 & 0 & 4 \end{bmatrix}$ **78.** $\begin{bmatrix} 1 & 0 & 1 & 0 \\ 0 & 1 & 0 & 1 \\ 1 & 1 & 1 & 2 \\ 1 & 2 & 1 & 2 \end{bmatrix}$

79–82 ■ Express the system of linear equations as a matrix equation. Then solve the matrix equation by multiplying each side by the inverse of the coefficient matrix.

79. $\begin{cases} 12x - 5y = 10 \\ 5x - 2y = 17 \end{cases}$ **80.** $\begin{cases} 6x - 5y = 1 \\ 8x - 7y = -1 \end{cases}$

81. $\begin{cases} 2x + y + 5z = \frac{1}{3} \\ x + 2y + 2z = \frac{1}{4} \\ x + 3z = \frac{1}{6} \end{cases}$ **82.** $\begin{cases} 2x + 3z = 5 \\ x + y + 6z = 0 \\ 3x - y + z = 5 \end{cases}$

83–86 ■ Solve the system using Cramer's Rule.

83. $\begin{cases} 2x + 7y = 13 \\ 6x + 16y = 30 \end{cases}$

84. $\begin{cases} 12x - 11y = 140 \\ 7x + 9y = 20 \end{cases}$

85. $\begin{cases} 2x - y + 5z = 0 \\ -x + 7y = 9 \\ 5x + 4y + 3z = -9 \end{cases}$

86. $\begin{cases} 3x + 4y - z = 10 \\ x - 4z = 20 \\ 2x + y + 5z = 30 \end{cases}$

87–88 ■ Use the determinant formula for the area of a triangle to find the area of the triangle in the figure.

87.

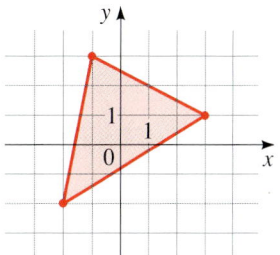

88.

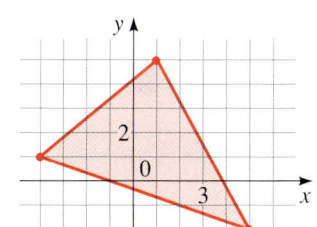

89–94 ■ Find the partial fraction decomposition of the rational function.

89. $\dfrac{3x + 1}{x^2 - 2x - 15}$ **90.** $\dfrac{8}{x^3 - 4x}$

91. $\dfrac{2x - 4}{x(x - 1)^2}$ **92.** $\dfrac{x + 6}{x^3 - 2x^2 + 4x - 8}$

93. $\dfrac{2x - 1}{x^3 + x}$ **94.** $\dfrac{5x^2 - 3x + 10}{x^4 + x^2 - 2}$

95–96 ■ An equation and its graph are given. Find an inequality whose solution is the shaded region.

95. $x + y^2 = 4$ **96.** $x^2 + y^2 = 8$

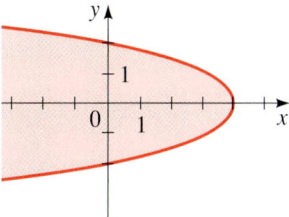

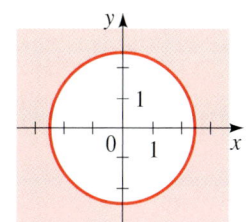

97–100 ■ Graph the inequality.

97. $3x + y \leq 6$ **98.** $y \geq x^2 - 3$

99. $x^2 + y^2 > 9$ **100.** $x - y^2 < 4$

101–104 ■ The figure shows the graphs of the equations corresponding to the given inequalities. Shade the solution set of the system of inequalities.

101. $\begin{cases} y \geq x^2 - 3x \\ y \leq \frac{1}{3}x - 1 \end{cases}$ **102.** $\begin{cases} y \geq x - 1 \\ x^2 + y^2 \leq 1 \end{cases}$

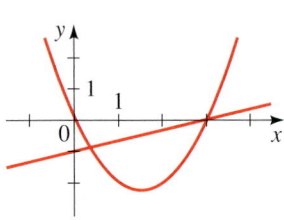   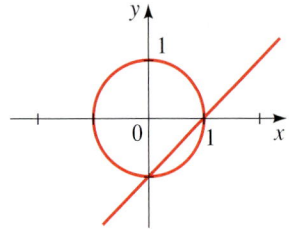

103. $\begin{cases} x + y \geq 2 \\ y - x \leq 2 \\ x \leq 3 \end{cases}$ **104.** $\begin{cases} y \geq -2x \\ y \leq 2x \\ y \leq -\frac{1}{2}x + 2 \end{cases}$

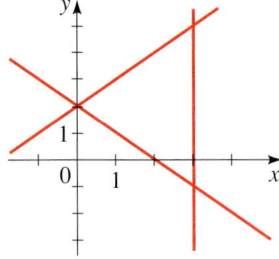   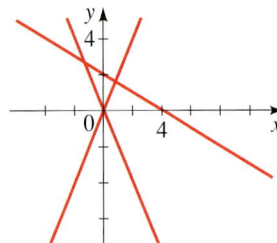

105–108 ■ Graph the solution set of the system of inequalities. Find the coordinates of all vertices, and determine whether the solution set is bounded or unbounded.

105. $\begin{cases} x^2 + y^2 < 9 \\ x + y < 0 \end{cases}$ **106.** $\begin{cases} y - x^2 \geq 4 \\ y < 20 \end{cases}$

107. $\begin{cases} x \geq 0, \quad y \geq 0 \\ x + 2y \leq 12 \\ y \leq x + 4 \end{cases}$ **108.** $\begin{cases} x \geq 4 \\ x + y \geq 24 \\ x \leq 2y + 12 \end{cases}$

109–110 ■ Solve for x, y, and z in terms of a, b, and c.

109. $\begin{cases} -x + y + z = a \\ x - y + z = b \\ x + y - z = c \end{cases}$

110. $\begin{cases} ax + by + cz = a - b + c \\ bx + by + cz = c \\ cx + cy + cz = c \end{cases}$ $(a \neq b, b \neq c, c \neq 0)$

111. For what values of k do the following three lines have a common point of intersection?

$$x + y = 12$$
$$kx - y = 0$$
$$y - x = 2k$$

112. For what value of k does the following system have infinitely many solutions?

$$\begin{cases} kx + y + z = 0 \\ x + 2y + kz = 0 \\ -x + 3z = 0 \end{cases}$$

9   Test

1–2 ■ A system of equations is given.

(a) Determine whether the system is linear or nonlinear.

(b) Find all solutions of the system.

1. $\begin{cases} x + 3y = 7 \\ 5x + 2y = -4 \end{cases}$

2. $\begin{cases} 6x + y^2 = 10 \\ 3x - y = 5 \end{cases}$

3. Use a graphing device to find all solutions of the system correct to two decimal places.

$$\begin{cases} x - 2y = 1 \\ y = x^3 - 2x^2 \end{cases}$$

4. In $2\frac{1}{2}$ h an airplane travels 600 km against the wind. It takes 50 min to travel 300 km with the wind. Find the speed of the wind and the speed of the airplane in still air.

5. Determine whether each matrix is in reduced row-echelon form, row-echelon form, or neither.

(a) $\begin{bmatrix} 1 & 2 & 4 & -6 \\ 0 & 1 & -3 & 0 \end{bmatrix}$

(b) $\begin{bmatrix} 1 & 0 & -1 & 0 & 0 \\ 0 & 1 & 3 & 0 & 0 \\ 0 & 0 & 0 & 1 & 0 \\ 0 & 0 & 0 & 0 & 1 \end{bmatrix}$

(c) $\begin{bmatrix} 1 & 1 & 0 \\ 0 & 0 & 1 \\ 0 & 1 & 3 \end{bmatrix}$

6. Use Gaussian elimination to find the complete solution of the system, or show that no solution exists.

(a) $\begin{cases} x - y + 2z = 0 \\ 2x - 4y + 5z = -5 \\ 2y - 3z = 5 \end{cases}$

(b) $\begin{cases} 2x - 3y + z = 3 \\ x + 2y + 2z = -1 \\ 4x + y + 5z = 4 \end{cases}$

7. Use Gauss-Jordan elimination to find the complete solution of the system.

$$\begin{cases} x + 3y - z = 0 \\ 3x + 4y - 2z = -1 \\ -x + 2y = 1 \end{cases}$$

8. Anne, Barry, and Cathy enter a coffee shop. Anne orders two coffees, one juice, and two donuts, and pays $6.25. Barry orders one coffee and three donuts, and pays $3.75. Cathy orders three coffees, one juice, and four donuts, and pays $9.25. Find the price of coffee, juice, and donuts at this coffee shop.

9. Let

$$A = \begin{bmatrix} 2 & 3 \\ 2 & 4 \end{bmatrix} \qquad B = \begin{bmatrix} 2 & 4 \\ -1 & 1 \\ 3 & 0 \end{bmatrix} \qquad C = \begin{bmatrix} 1 & 0 & 4 \\ -1 & 1 & 2 \\ 0 & 1 & 3 \end{bmatrix}$$

Carry out the indicated operation, or explain why it cannot be performed.

(a) $A + B$　(b) AB　(c) $BA - 3B$　(d) CBA

(e) A^{-1}　(f) B^{-1}　(g) $\det(B)$　(h) $\det(C)$

10. (a) Write a matrix equation equivalent to the following system.

$$\begin{cases} 4x - 3y = 10 \\ 3x - 2y = 30 \end{cases}$$

(b) Find the inverse of the coefficient matrix, and use it to solve the system.

11. Only one of the following matrices has an inverse. Find the determinant of each matrix, and use the determinants to identify the one that has an inverse. Then find the inverse.

$$A = \begin{bmatrix} 1 & 4 & 1 \\ 0 & 2 & 0 \\ 1 & 0 & 1 \end{bmatrix} \qquad B = \begin{bmatrix} 1 & 4 & 0 \\ 0 & 2 & 0 \\ -3 & 0 & 1 \end{bmatrix}$$

12. Solve using Cramer's Rule:

$$\begin{cases} 2x & - & z = 14 \\ 3x - & y + 5z = & 0 \\ 4x + & 2y + 3z = & -2 \end{cases}$$

13. Find the partial fraction decomposition of the rational function.

(a) $\dfrac{4x - 1}{(x - 1)^2(x + 2)}$ (b) $\dfrac{2x - 3}{x^3 + 3x}$

14. Graph the solution set of the system of inequalities. Label the vertices with their coordinates.

(a) $\begin{cases} 2x + & y \leq & 8 \\ x - & y \geq -2 \\ x + & 2y \geq & 4 \end{cases}$ (b) $\begin{cases} x^2 + y \leq 5 \\ y \leq 2x + 5 \end{cases}$

Focus on Modeling
Linear Programming

Linear programming is a modeling technique used to determine the optimal allocation of resources in business, the military, and other areas of human endeavor. For example, a manufacturer who makes several different products from the same raw materials can use linear programming to determine how much of each product should be produced to maximize the profit. This modeling technique is probably the most important practical application of systems of linear inequalities. In 1975 Leonid Kantorovich and T. C. Koopmans won the Nobel Prize in economics for their work in the development of this technique.

Although linear programming can be applied to very complex problems with hundreds or even thousands of variables, we consider only a few simple examples to which the graphical methods of Section 9.9 can be applied. (For large numbers of variables, a linear programming method based on matrices is used.) Let's examine a typical problem.

Example 1 Manufacturing for Maximum Profit

A small shoe manufacturer makes two styles of shoes: oxfords and loafers. Two machines are used in the process: a cutting machine and a sewing machine. Each type of shoe requires 15 min per pair on the cutting machine. Oxfords require 10 min of sewing per pair, and loafers require 20 min of sewing per pair. Because the manufacturer can hire only one operator for each machine, each process is available for just 8 hours per day. If the profit is $15 on each pair of oxfords and $20 on each pair of loafers, how many pairs of each type should be produced per day for maximum profit?

Because loafers produce more profit per pair, it would seem best to manufacture only loafers. Surprisingly, this does not turn out to be the most profitable solution.

Solution First we organize the given information into a table. To be consistent, let's convert all times to hours.

	Oxfords	Loafers	Time available
Time on cutting machine (h)	$\frac{1}{4}$	$\frac{1}{4}$	8
Time on sewing machine (h)	$\frac{1}{6}$	$\frac{1}{3}$	8
Profit	$15	$20	

We describe the model and solve the problem in four steps.

CHOOSING THE VARIABLES   To make a mathematical model, we first give names to the variable quantities. For this problem we let

$$x = \text{number of pairs of oxfords made daily}$$

$$y = \text{number of pairs of loafers made daily}$$

FINDING THE OBJECTIVE FUNCTION   Our goal is to determine which values for x and y give maximum profit. Since each pair of oxfords generates $15 profit and

735

each pair of loafers $20, the total profit is given by

$$P = 15x + 20y$$

This function is called the *objective function*.

GRAPHING THE FEASIBLE REGION   The larger x and y are, the greater the profit. But we cannot choose arbitrarily large values for these variables, because of the restrictions, or *constraints*, in the problem. Each restriction is an inequality in the variables.

In this problem the total number of cutting hours needed is $\frac{1}{4}x + \frac{1}{4}y$. Since only 8 hours are available on the cutting machine, we have

$$\tfrac{1}{4}x + \tfrac{1}{4}y \le 8$$

Similarly, by considering the amount of time needed and available on the sewing machine, we get

$$\tfrac{1}{6}x + \tfrac{1}{3}y \le 8$$

We cannot produce a negative number of shoes, so we also have

$$x \ge 0 \qquad \text{and} \qquad y \ge 0$$

Thus, x and y must satisfy the constraints

$$\begin{cases} \tfrac{1}{4}x + \tfrac{1}{4}y \le 8 \\ \tfrac{1}{6}x + \tfrac{1}{3}y \le 8 \\ \qquad\quad x \ge 0 \\ \qquad\quad y \ge 0 \end{cases}$$

If we multiply the first inequality by 4 and the second by 6, we obtain the simplified system

$$\begin{cases} x + \ y \le 32 \\ x + 2y \le 48 \\ \quad\ x \ge 0 \\ \quad\ y \ge 0 \end{cases}$$

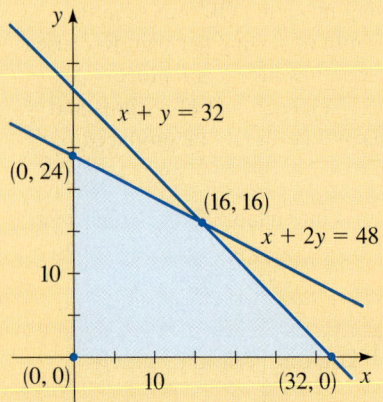

Figure 1

The solution of this system (with vertices labeled) is sketched in Figure 1. The only values that satisfy the restrictions of the problem are the ones that correspond to points of the shaded region in Figure 1. This is called the *feasible region* for the problem.

FINDING MAXIMUM PROFIT   As x or y increases, profit increases as well. Thus, it seems reasonable that the maximum profit will occur at a point on one of the outside edges of the feasible region, where it's impossible to increase x or y without going outside the region. In fact, it can be shown that the maximum value occurs at a vertex. This means that we need to check the profit only at the vertices. The largest value of P occurs at the point $(16, 16)$, where $P = \$560$. Thus, the manufacturer should make 16 pairs of oxfords and 16 pairs of loafers, for a maximum daily profit of $560.

Vertex	$P = 15x + 20y$
$(0, 0)$	0
$(0, 24)$	$15(0)\ \ + 20(24) = \$480$
$(16, 16)$	$15(16) + 20(16) = \$560$
$(32, 0)$	$15(32) + 20(0)\ \ = \$480$

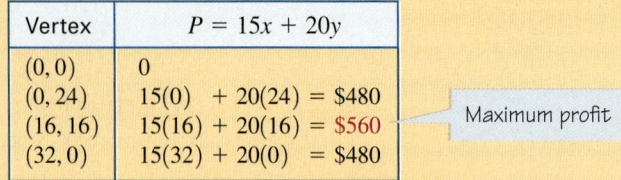

Maximum profit

The linear programming problems that we consider all follow the pattern of Example 1. Each problem involves two variables. The problem describes restrictions, called **constraints**, that lead to a system of linear inequalities whose solution is called the **feasible region**. The function we wish to maximize or minimize is called the **objective function**. This function always attains its largest and smallest values at the **vertices** of the feasible region. This modeling technique involves four steps, summarized in the following box.

Linear programming helps the telephone industry determine the most efficient way to route telephone calls. The computerized routing decisions must be made very rapidly so callers are not kept waiting for connections. Since the database of customers and routes is huge, an extremely fast method for solving linear programming problems is essential. In 1984 the 28-year-old mathematician **Narendra Karmarkar**, working at Bell Labs in Murray Hill, New Jersey, discovered just such a method. His idea is so ingenious and his method so fast that the discovery caused a sensation in the mathematical world. Although mathematical discoveries rarely make the news, this one was reported in *Time*, on December 3, 1984. Today airlines routinely use Karmarkar's technique to minimize costs in scheduling passengers, flight personnel, fuel, baggage, and maintenance workers.

Guidelines for Linear Programming

1. **Choose the Variables.** Decide what variable quantities in the problem should be named x and y.

2. **Find the Objective Function.** Write an expression for the function we want to maximize or minimize.

3. **Graph the Feasible Region.** Express the constraints as a system of inequalities and graph the solution of this system (the feasible region).

4. **Find the Maximum or Minimum.** Evaluate the objective function at the vertices of the feasible region to determine its maximum or minimum value.

Example 2   A Shipping Problem

A car dealer has warehouses in Millville and Trenton and dealerships in Camden and Atlantic City. Every car sold at the dealerships must be delivered from one of the warehouses. On a certain day the Camden dealers sell 10 cars, and the Atlantic City dealers sell 12. The Millville warehouse has 15 cars available, and the Trenton warehouse has 10. The cost of shipping one car is $50 from Millville to Camden, $40 from Millville to Atlantic City, $60 from Trenton to Camden, and $55 from Trenton to Atlantic City. How many cars should be moved from each warehouse to each dealership to fill the orders at minimum cost?

Solution Our first step is to organize the given information. Rather than construct a table, we draw a diagram to show the flow of cars from the warehouses to the dealerships (see Figure 2 on the next page). The diagram shows the number of cars available at each warehouse or required at each dealership and the cost of shipping between these locations.

CHOOSING THE VARIABLES The arrows in Figure 2 indicate four possible routes, so the problem seems to involve four variables. But we let

x = number of cars to be shipped from Millville to Camden

y = number of cars to be shipped from Millville to Atlantic City

To fill the orders, we must have

$10 - x$ = number of cars shipped from Trenton to Camden

$12 - y$ = number of cars shipped from Trenton to Atlantic City

So the only variables in the problem are x and y.

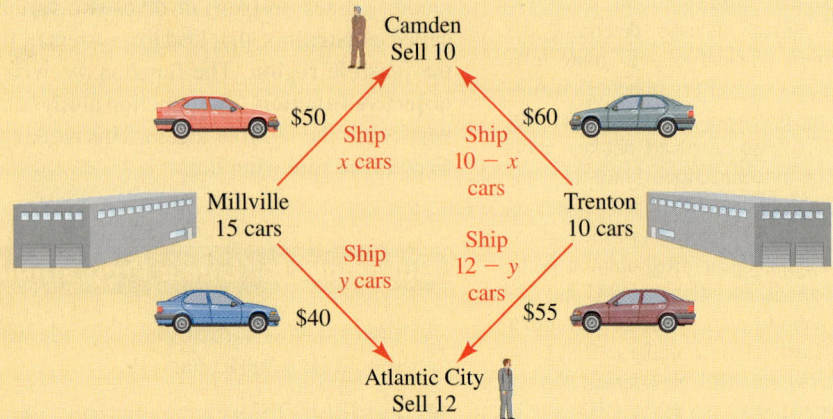

Figure 2

FINDING THE OBJECTIVE FUNCTION   The objective of this problem is to minimize cost. From Figure 2 we see that the total cost C of shipping the cars is

$$C = 50x + 40y + 60(10 - x) + 55(12 - y)$$
$$= 50x + 40y + 600 - 60x + 660 - 55y$$
$$= 1260 - 10x - 15y$$

This is the objective function.

GRAPHING THE FEASIBLE REGION   Now we derive the constraint inequalities that define the feasible region. First, the number of cars shipped on each route can't be negative, so we have

$$x \geq 0 \qquad\qquad y \geq 0$$
$$10 - x \geq 0 \qquad 12 - y \geq 0$$

Second, the total number of cars shipped from each warehouse can't exceed the number of cars available there, so

$$x + y \leq 15$$
$$(10 - x) + (12 - y) \leq 10$$

Simplifying the latter inequality, we get

$$22 - x - y \leq 10$$
$$-x - y \leq -12$$
$$x + y \geq 12$$

The inequalities $10 - x \geq 0$ and $12 - y \geq 0$ can be rewritten as $x \leq 10$ and $y \leq 12$. Thus, the feasible region is described by the constraints

$$\begin{cases} x + y \leq 15 \\ x + y \geq 12 \\ 0 \leq x \leq 10 \\ 0 \leq y \leq 12 \end{cases}$$

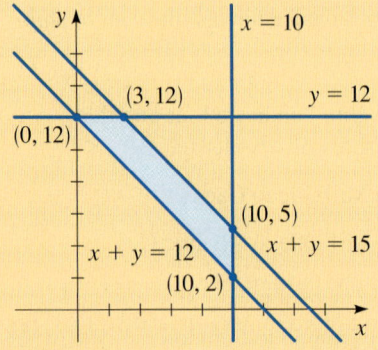

Figure 3

The feasible region is graphed in Figure 3.

FINDING MINIMUM COST   We check the value of the objective function at each vertex of the feasible region.

Vertex	$C = 1260 - 10x - 15y$
$(0, 12)$	$1260 - 10(0)\ - 15(12) = \$1080$
$(3, 12)$	$1260 - 10(3)\ - 15(12) = \$1050$
$(10, 5)$	$1260 - 10(10) - 15(5)\ = \$1085$
$(10, 2)$	$1260 - 10(10) - 15(2)\ = \$1130$

Minimum cost

The lowest cost is incurred at the point $(3, 12)$. Thus, the dealer should ship

3 cars from Millville to Camden
12 cars from Millville to Atlantic City
7 cars from Trenton to Camden
0 cars from Trenton to Atlantic City ∎

In the 1940s mathematicians developed matrix methods for solving linear programming problems that involve more than two variables. These methods were first used by the Allies in World War II to solve supply problems similar to (but, of course, much more complicated than) Example 2. Improving such matrix methods is an active and exciting area of current mathematical research.

Problems

1–4 ∎ Find the maximum and minimum values of the given objective function on the indicated feasible region.

1. $M = 200 - x - y$

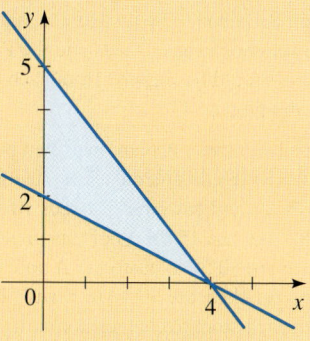

2. $N = \frac{1}{2}x + \frac{1}{4}y + 40$

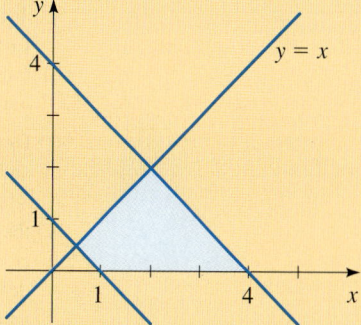

3. $P = 140 - x + 3y$

$$\begin{cases} x \geq 0, \quad y \geq 0 \\ 2x + \ y \leq 10 \\ 2x + 4y \leq 28 \end{cases}$$

4. $Q = 70x + 82y$

$$\begin{cases} x \geq 0, \quad y \geq 0 \\ x \leq 10, \quad y \leq 20 \\ x + y \geq 5 \\ x + 2y \leq 18 \end{cases}$$

5. Making Furniture   A furniture manufacturer makes wooden tables and chairs. The production process involves two basic types of labor: carpentry and finishing. A table requires 2 hours of carpentry and 1 hour of finishing, and a chair requires 3 hours of carpentry and $\frac{1}{2}$ hour of finishing. The profit is $35 per table and $20 per chair. The manufacturer's employees can supply a maximum of 108 hours of carpentry work and 20 hours of finishing work per day. How many tables and chairs should be made each day to maximize profit?

6. A Housing Development A housing contractor has subdivided a farm into 100 building lots. He has designed two types of homes for these lots: colonial and ranch style. A colonial requires $30,000 of capital and produces a profit of $4000 when sold. A ranch-style house requires $40,000 of capital and provides an $8000 profit. If he has $3.6 million of capital on hand, how many houses of each type should he build for maximum profit? Will any of the lots be left vacant?

7. Hauling Fruit A trucker hauls citrus fruit from Florida to Montreal. Each crate of oranges is 4 ft³ in volume and weighs 80 lb. Each crate of grapefruit has a volume of 6 ft³ and weighs 100 lb. Her truck has a maximum capacity of 300 ft³ and can carry no more than 5600 lb. Moreover, she is not permitted to carry more crates of grapefruit than crates of oranges. If her profit is $2.50 on each crate of oranges and $4 on each crate of grapefruit, how many crates of each fruit should she carry for maximum profit?

8. Manufacturing Calculators A manufacturer of calculators produces two models: standard and scientific. Long-term demand for the two models mandates that the company manufacture at least 100 standard and 80 scientific calculators each day. However, because of limitations on production capacity, no more than 200 standard and 170 scientific calculators can be made daily. To satisfy a shipping contract, a total of at least 200 calculators must be shipped every day.

(a) If the production cost is $5 for a standard calculator and $7 for a scientific one, how many of each model should be produced daily to minimize this cost?

(b) If each standard calculator results in a $2 loss but each scientific one produces a $5 profit, how many of each model should be made daily to maximize profit?

9. Shipping Stereos An electronics discount chain has a sale on a certain brand of stereo. The chain has stores in Santa Monica and El Toro and warehouses in Long Beach and Pasadena. To satisfy rush orders, 15 sets must be shipped from the warehouses to the Santa Monica store, and 19 must be shipped to the El Toro store. The cost of shipping a set is $5 from Long Beach to Santa Monica, $6 from Long Beach to El Toro, $4 from Pasadena to Santa Monica, and $5.50 from Pasadena to El Toro. If the Long Beach warehouse has 24 sets and the Pasadena warehouse has 18 sets in stock, how many sets should be shipped from each warehouse to each store to fill the orders at a minimum shipping cost?

10. Delivering Plywood A man owns two building supply stores, one on the east side and one on the west side of a city. Two customers order some $\frac{1}{2}$-inch plywood. Customer A needs 50 sheets and customer B needs 70 sheets. The east-side store has 80 sheets and the west-side store has 45 sheets of this plywood in stock. The east-side store's delivery costs per sheet are $0.50 to customer A and $0.60 to customer B. The west-side store's delivery costs per sheet are $0.40 to A and $0.55 to B. How many sheets should be shipped from each store to each customer to minimize delivery costs?

11. Packaging Nuts A confectioner sells two types of nut mixtures. The standard-mixture package contains 100 g of cashews and 200 g of peanuts and sells for $1.95. The deluxe-mixture package contains 150 g of cashews and 50 g of peanuts and sells for $2.25. The confectioner has 15 kg of cashews and 20 kg of peanuts available. Based on past sales, he needs to have at least as many standard as deluxe packages available. How many bags of each mixture should he package to maximize his revenue?

12. Feeding Lab Rabbits A biologist wishes to feed laboratory rabbits a mixture of two types of foods. Type I contains 8 g of fat, 12 g of carbohydrate, and 2 g of protein per ounce. Type II contains 12 g of fat, 12 g of carbohydrate, and 1 g of protein per ounce.

Type I costs $0.20 per ounce and type II costs $0.30 per ounce. The rabbits each receive a daily minimum of 24 g of fat, 36 g of carbohydrate, and 4 g of protein, but get no more than 5 oz of food per day. How many ounces of each food type should be fed to each rabbit daily to satisfy the dietary requirements at minimum cost?

13. Investing in Bonds   A woman wishes to invest $12,000 in three types of bonds: municipal bonds paying 7% interest per year, bank investment certificates paying 8%, and high-risk bonds paying 12%. For tax reasons, she wants the amount invested in municipal bonds to be at least three times the amount invested in bank certificates. To keep her level of risk manageable, she will invest no more than $2000 in high-risk bonds. How much should she invest in each type of bond to maximize her annual interest yield? [*Hint:* Let x = amount in municipal bonds and y = amount in bank certificates. Then the amount in high-risk bonds will be $12,000 - x - y$.]

14. Annual Interest Yield   Refer to Problem 13. Suppose the investor decides to increase the maximum invested in high-risk bonds to $3000 but leaves the other conditions unchanged. By how much will her maximum possible interest yield increase?

15. Business Strategy   A small software company publishes computer games and educational and utility software. Their business strategy is to market a total of 36 new programs each year, with at least four of these being games. The number of utility programs published is never more than twice the number of educational programs. On average, the company makes an annual profit of $5000 on each computer game, $8000 on each educational program, and $6000 on each utility program. How many of each type of software should they publish annually for maximum profit?

16. Feasible Region   All parts of this problem refer to the following feasible region and objective function.

$$\begin{cases} x \geq 0 \\ x \geq y \\ x + 2y \leq 12 \\ x + y \leq 10 \end{cases}$$

$$P = x + 4y$$

(a) Graph the feasible region.

(b) On your graph from part (a), sketch the graphs of the linear equations obtained by setting P equal to 40, 36, 32, and 28.

(c) If we continue to decrease the value of P, at which vertex of the feasible region will these lines first touch the feasible region?

(d) Verify that the maximum value of P on the feasible region occurs at the vertex you chose in part (c).

10 Analytic Geometry

Chapter Overview

Conic sections are the curves we get when we make a straight cut in a cone, as shown in the figure. For example, if a cone is cut horizontally, the cross section is a circle. So a circle is a conic section. Other ways of cutting a cone produce parabolas, ellipses, and hyperbolas.

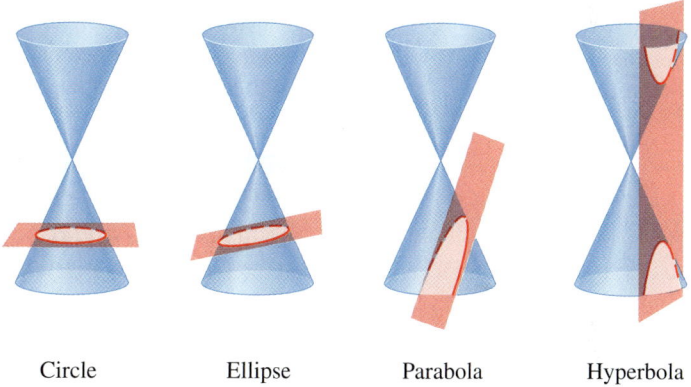

Circle   Ellipse   Parabola   Hyperbola

Our goal in this chapter is to find equations whose graphs are the conic sections. We already know from Section 1.8 that the graph of the equation $x^2 + y^2 = r^2$ is a circle. We will find equations for each of the other conic sections by analyzing their *geometric* properties.

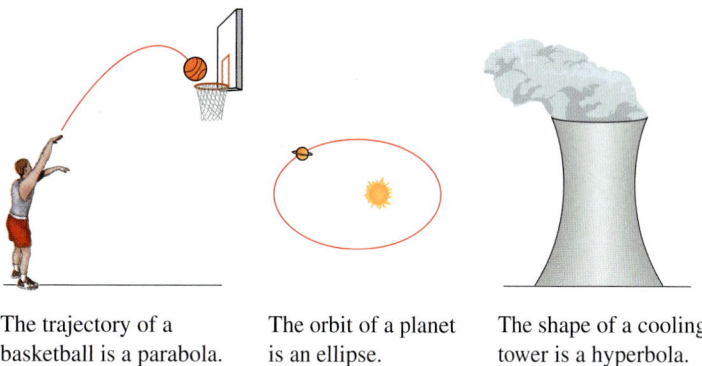

The trajectory of a   The orbit of a planet   The shape of a cooling
basketball is a parabola.   is an ellipse.   tower is a hyperbola.

Conic sections are important because their shapes are hidden in the structure of many things. For example, the path of a planet moving around the sun is an ellipse.

743

The path of a projectile (such as a rocket, a basketball, or water spouting from a fountain) is a parabola—which makes the study of parabolas indispensable in rocket science. The conic sections also occur in many unexpected places. For example, the graph of crop yield as a function of amount of rainfall is a parabola (see page 321). We will examine some uses of the conics in medicine, engineering, navigation, and astronomy.

In Section 10.7 we study parametric equations, which we can use to describe the curve that a moving body traces out over time. In *Focus on Modeling*, page 816, we derive parametric equations for the path of a projectile.

10.1   Parabolas

We saw in Section 2.5 that the graph of the equation $y = ax^2 + bx + c$ is a U-shaped curve called a *parabola* that opens either upward or downward, depending on whether the sign of a is positive or negative.

In this section we study parabolas from a geometric rather than an algebraic point of view. We begin with the geometric definition of a parabola and show how this leads to the algebraic formula that we are already familiar with.

Geometric Definition of a Parabola

A **parabola** is the set of points in the plane equidistant from a fixed point F (called the **focus**) and a fixed line l (called the **directrix**).

This definition is illustrated in Figure 1. The **vertex** V of the parabola lies halfway between the focus and the directrix, and the **axis of symmetry** is the line that runs through the focus perpendicular to the directrix.

In this section we restrict our attention to parabolas that are situated with the vertex at the origin and that have a vertical or horizontal axis of symmetry. (Parabolas in more general positions will be considered in Sections 10.4 and 10.5.) If the focus of such a parabola is the point $F(0, p)$, then the axis of symmetry must be vertical and the directrix has the equation $y = -p$. Figure 2 illustrates the case $p > 0$.

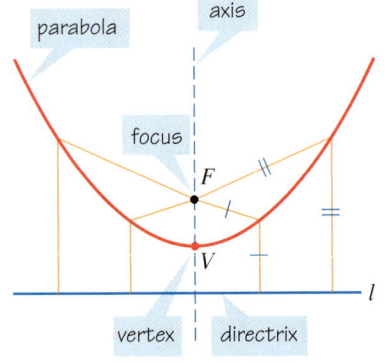

Figure 1

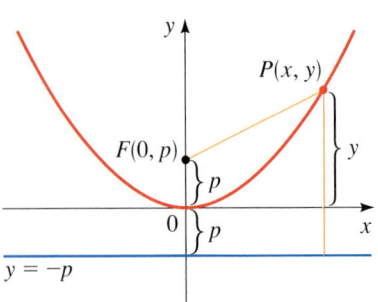

Figure 2

If $P(x, y)$ is any point on the parabola, then the distance from P to the focus F (using the Distance Formula) is

$$\sqrt{x^2 + (y - p)^2}$$

The distance from P to the directrix is

$$|y - (-p)| = |y + p|$$

By the definition of a parabola, these two distances must be equal:

$$\sqrt{x^2 + (y - p)^2} = |y + p|$$

$$x^2 + (y - p)^2 = |y + p|^2 = (y + p)^2 \qquad \textcolor{blue}{\text{Square both sides}}$$

$$x^2 + y^2 - 2py + p^2 = y^2 + 2py + p^2 \qquad \textcolor{blue}{\text{Expand}}$$

$$x^2 - 2py = 2py \qquad \textcolor{blue}{\text{Simplify}}$$

$$x^2 = 4py$$

If $p > 0$, then the parabola opens upward, but if $p < 0$, it opens downward. When x is replaced by $-x$, the equation remains unchanged, so the graph is symmetric about the y-axis.

Equations and Graphs of Parabolas

The following box summarizes what we have just proved about the equation and features of a parabola with a vertical axis.

Parabola with Vertical Axis

The graph of the equation

$$x^2 = 4py$$

is a parabola with the following properties.

VERTEX	$V(0, 0)$
FOCUS	$F(0, p)$
DIRECTRIX	$y = -p$

The parabola opens upward if $p > 0$ or downward if $p < 0$.

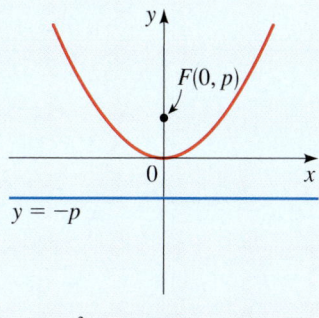

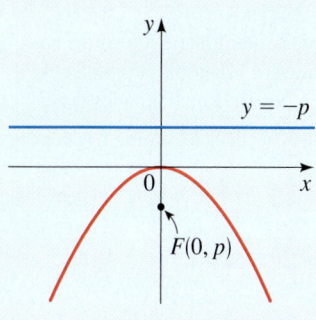

$x^2 = 4py$ with $p > 0$ $x^2 = 4py$ with $p < 0$

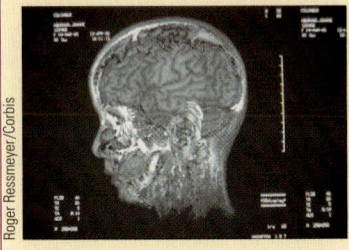
Example 1 Finding the Equation of a Parabola

Find the equation of the parabola with vertex $V(0, 0)$ and focus $F(0, 2)$, and sketch its graph.

Solution Since the focus is $F(0, 2)$, we conclude that $p = 2$ (and so the directrix is $y = -2$). Thus, the equation of the parabola is

$$x^2 = 4(2)y \qquad x^2 = 4py \text{ with } p = 2$$
$$x^2 = 8y$$

Since $p = 2 > 0$, the parabola opens upward. See Figure 3.

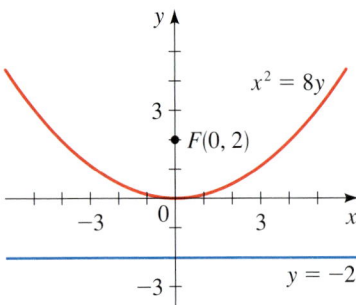

Figure 3

Example 2 Finding the Focus and Directrix of a Parabola from Its Equation

Find the focus and directrix of the parabola $y = -x^2$, and sketch the graph.

Solution To find the focus and directrix, we put the given equation in the standard form $x^2 = -y$. Comparing this to the general equation $x^2 = 4py$, we see that $4p = -1$, so $p = -\frac{1}{4}$. Thus, the focus is $F\left(0, -\frac{1}{4}\right)$ and the directrix is $y = \frac{1}{4}$. The graph of the parabola, together with the focus and the directrix, is shown in Figure 4(a). We can also draw the graph using a graphing calculator as shown in Figure 4(b).

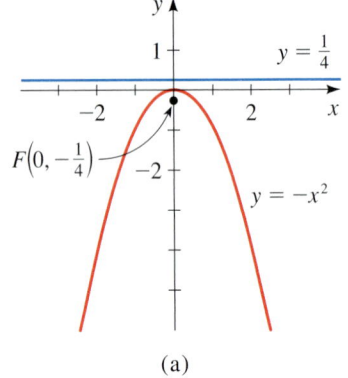

(a)

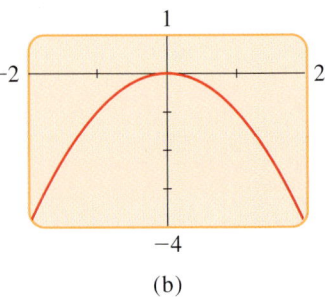

(b)

Figure 4

Reflecting the graph in Figure 2 about the diagonal line $y = x$ has the effect of interchanging the roles of x and y. This results in a parabola with horizontal axis. By the same method as before, we can prove the following properties.

Parabola with Horizontal Axis

The graph of the equation

$$y^2 = 4px$$

is a parabola with the following properties.

VERTEX	$V(0, 0)$
FOCUS	$F(p, 0)$
DIRECTRIX	$x = -p$

The parabola opens to the right if $p > 0$ or to the left if $p < 0$.

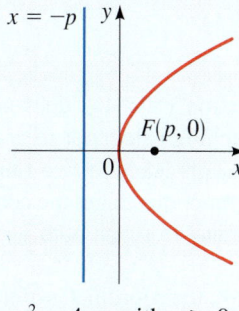
$y^2 = 4px$ with $p > 0$

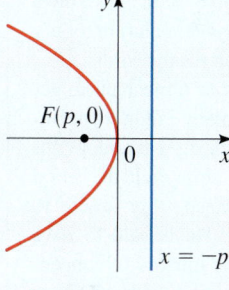
$y^2 = 4px$ with $p < 0$

Example 3 A Parabola with Horizontal Axis

A parabola has the equation $6x + y^2 = 0$.

(a) Find the focus and directrix of the parabola, and sketch the graph.

(b) Use a graphing calculator to draw the graph.

Solution

(a) To find the focus and directrix, we put the given equation in the standard form $y^2 = -6x$. Comparing this to the general equation $y^2 = 4px$, we see that $4p = -6$, so $p = -\frac{3}{2}$. Thus, the focus is $F\left(-\frac{3}{2}, 0\right)$ and the directrix is $x = \frac{3}{2}$. Since $p < 0$, the parabola opens to the left. The graph of the parabola, together with the focus and the directrix, is shown in Figure 5(a) on the next page.

(b) To draw the graph using a graphing calculator, we need to solve for y.

$$6x + y^2 = 0$$

$$y^2 = -6x \qquad \text{Subtract } 6x$$

$$y = \pm\sqrt{-6x} \qquad \text{Take square roots}$$

Archimedes (287–212 B.C.) was the greatest mathematician of the ancient world. He was born in Syracuse, a Greek colony on Sicily, a generation after Euclid (see page 532). One of his many discoveries is the Law of the Lever (see page 69). He famously said, "Give me a place to stand and a fulcrum for my lever, and I can lift the earth."

Renowned as a mechanical genius for his many engineering inventions, he designed pulleys for lifting heavy ships and the spiral screw for transporting water to higher levels. He is said to have used parabolic mirrors to concentrate the rays of the sun to set fire to Roman ships attacking Syracuse.

King Hieron II of Syracuse once suspected a goldsmith of keeping part of the gold intended for the king's crown and replacing it with an equal amount of silver. The king asked Archimedes for advice. While in deep thought at a public bath, Archimedes discovered the solution to the king's problem when he noticed that his body's volume was the same as the volume of water it displaced from the tub. As the story is told, he ran home naked, shouting "Eureka, eureka!" ("I have found it, I have found it!") This incident attests to his enormous powers of concentration.

In spite of his engineering prowess, Archimedes was most proud of his mathematical discov-
(*continued*)

To obtain the graph of the parabola, we graph both functions

$$y = \sqrt{-6x} \qquad \text{and} \qquad y = -\sqrt{-6x}$$

as shown in Figure 5(b).

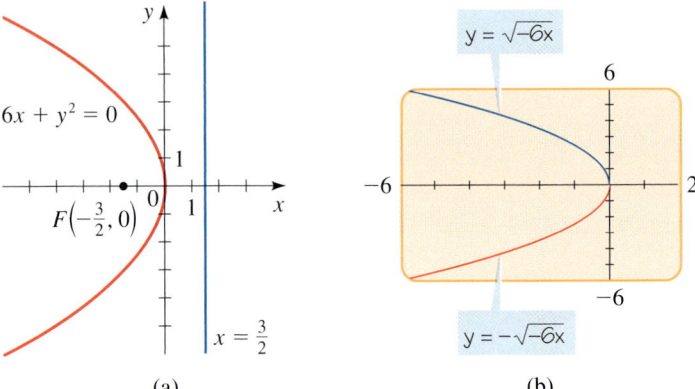

Figure 5   (a)   (b)

The equation $y^2 = 4px$ does not define y as a function of x (see page 164). So, to use a graphing calculator to graph a parabola with horizontal axis, we must first solve for y. This leads to two functions, $y = \sqrt{4px}$ and $y = -\sqrt{4px}$. We need to graph both functions to get the complete graph of the parabola. For example, in Figure 5(b) we had to graph both $y = \sqrt{-6x}$ and $y = -\sqrt{-6x}$ to graph the parabola $y^2 = -6x$.

We can use the coordinates of the focus to estimate the "width" of a parabola when sketching its graph. The line segment that runs through the focus perpendicular to the axis, with endpoints on the parabola, is called the **latus rectum**, and its length is the **focal diameter** of the parabola. From Figure 6 we can see that the distance from an endpoint Q of the latus rectum to the directrix is $|2p|$. Thus, the distance from Q to the focus must be $|2p|$ as well (by the definition of a parabola), and so the focal diameter is $|4p|$. In the next example we use the focal diameter to determine the "width" of a parabola when graphing it.

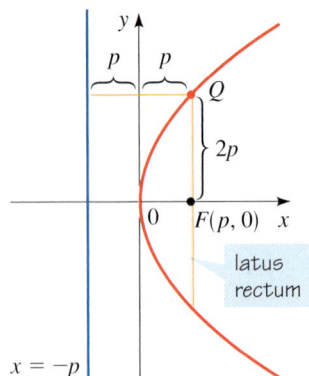

Figure 6 $x = -p$

eries. These include the formulas for the volume of a sphere, $V = \frac{4}{3}\pi r^3$; the surface area of a sphere, $S = 4\pi r^2$; and a careful analysis of the properties of parabolas and other conics.

Example 4 The Focal Diameter of a Parabola

Find the focus, directrix, and focal diameter of the parabola $y = \frac{1}{2}x^2$, and sketch its graph.

Solution   We first put the equation in the form $x^2 = 4py$.

$$y = \frac{1}{2}x^2$$

$$x^2 = 2y \qquad \text{Multiply each side by 2}$$

From this equation we see that $4p = 2$, so the focal diameter is 2. Solving for p gives $p = \frac{1}{2}$, so the focus is $\left(0, \frac{1}{2}\right)$ and the directrix is $y = -\frac{1}{2}$. Since the focal diameter is 2, the latus rectum extends 1 unit to the left and 1 unit to the right of the focus. The graph is sketched in Figure 7.   ■

In the next example we graph a family of parabolas, to show how changing the distance between the focus and the vertex affects the "width" of a parabola.

Example 5 A Family of Parabolas

(a) Find equations for the parabolas with vertex at the origin and foci $F_1\left(0, \frac{1}{8}\right)$, $F_2\left(0, \frac{1}{2}\right)$, $F_3(0, 1)$, and $F_4(0, 4)$.

(b) Draw the graphs of the parabolas in part (a). What do you conclude?

Solution

(a) Since the foci are on the positive y-axis, the parabolas open upward and have equations of the form $x^2 = 4py$. This leads to the following equations.

Focus	p	Equation $x^2 = 4py$	Form of the equation for graphing calculator
$F_1\left(0, \frac{1}{8}\right)$	$p = \frac{1}{8}$	$x^2 = \frac{1}{2}y$	$y = 2x^2$
$F_2\left(0, \frac{1}{2}\right)$	$p = \frac{1}{2}$	$x^2 = 2y$	$y = 0.5x^2$
$F_3(0, 1)$	$p = 1$	$x^2 = 4y$	$y = 0.25x^2$
$F_4(0, 4)$	$p = 4$	$x^2 = 16y$	$y = 0.0625x^2$

(b) The graphs are drawn in Figure 8. We see that the closer the focus to the vertex, the narrower the parabola.

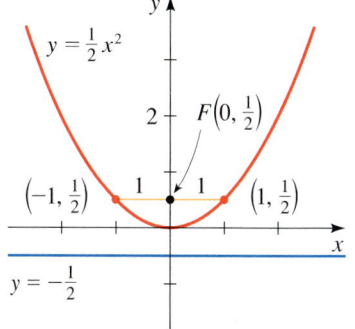

Figure 7

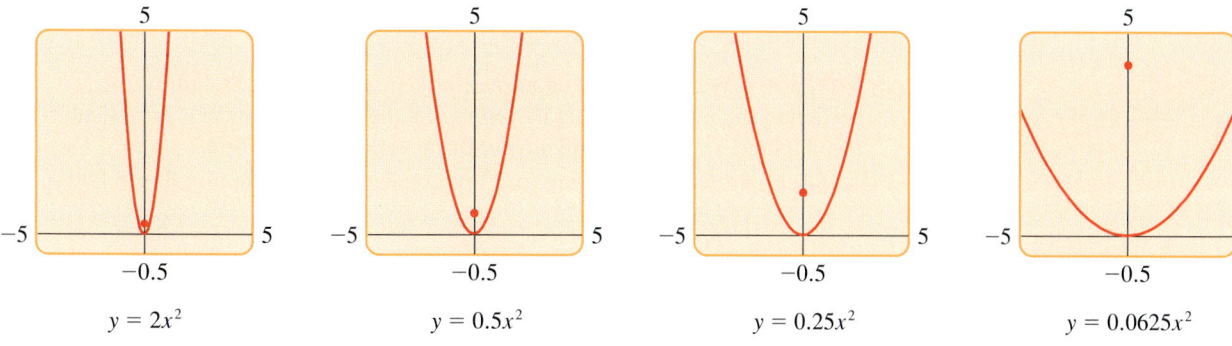

$$y = 2x^2 \qquad\qquad y = 0.5x^2 \qquad\qquad y = 0.25x^2 \qquad\qquad y = 0.0625x^2$$

Figure 8

A family of parabolas   ■

Applications

Parabolas have an important property that makes them useful as reflectors for lamps and telescopes. Light from a source placed at the focus of a surface with parabolic cross section will be reflected in such a way that it travels parallel to the axis of the parabola (see Figure 9). Thus, a parabolic mirror reflects the light into a beam of parallel rays. Conversely, light approaching the reflector in rays parallel to its axis of symmetry is concentrated to the focus. This *reflection property*, which can be proved using calculus, is used in the construction of reflecting telescopes.

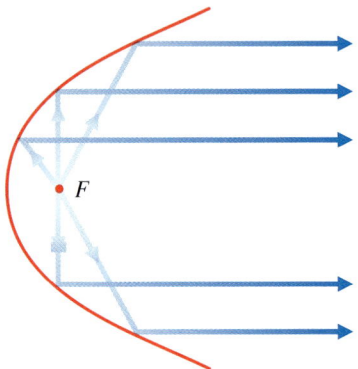

Figure 9
Parabolic reflector

Example 6 **Finding the Focal Point of a Searchlight Reflector**

A searchlight has a parabolic reflector that forms a "bowl," which is 12 in. wide from rim to rim and 8 in. deep, as shown in Figure 10. If the filament of the light bulb is located at the focus, how far from the vertex of the reflector is it?

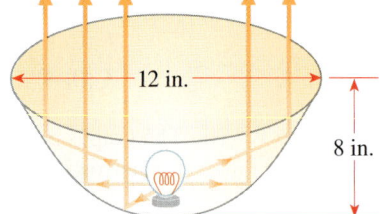

Figure 10
A parabolic reflector

Solution   We introduce a coordinate system and place a parabolic cross section of the reflector so that its vertex is at the origin and its axis is vertical (see Figure 11). Then the equation of this parabola has the form $x^2 = 4py$. From Figure 11 we see that the point $(6, 8)$ lies on the parabola. We use this to find p.

$$6^2 = 4p(8) \qquad \text{The point } (6, 8) \text{ satisfies the equation } x^2 = 4py$$

$$36 = 32p$$

$$p = \tfrac{9}{8}$$

The focus is $F\left(0, \tfrac{9}{8}\right)$, so the distance between the vertex and the focus is $\tfrac{9}{8} = 1\tfrac{1}{8}$ in. Because the filament is positioned at the focus, it is located $1\tfrac{1}{8}$ in. from the vertex of the reflector. ∎

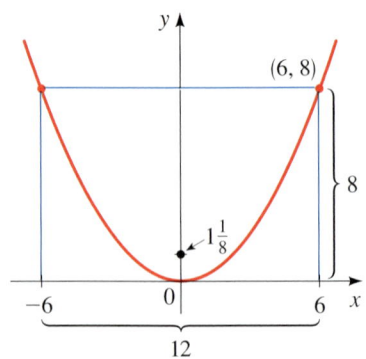

Figure 11

10.1 Exercises

1–6 ■ Match the equation with the graphs labeled I–VI. Give reasons for your answers.

1. $y^2 = 2x$ **2.** $y^2 = -\frac{1}{4}x$ **3.** $x^2 = -6y$

4. $2x^2 = y$ **5.** $y^2 - 8x = 0$ **6.** $12y + x^2 = 0$

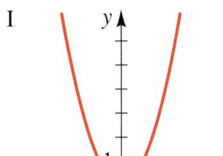
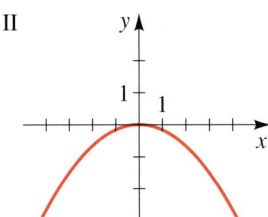
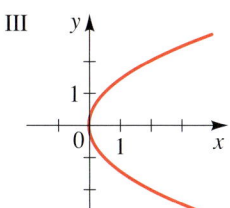
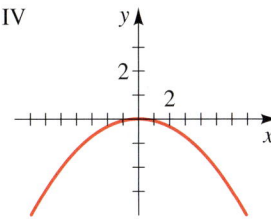
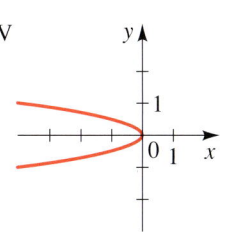
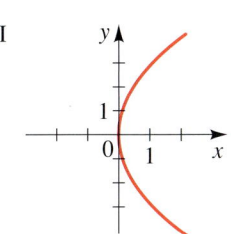

7–18 ■ Find the focus, directrix, and focal diameter of the parabola, and sketch its graph.

7. $y^2 = 4x$ **8.** $x^2 = y$

9. $x^2 = 9y$ **10.** $y^2 = 3x$

11. $y = 5x^2$ **12.** $y = -2x^2$

13. $x = -8y^2$ **14.** $x = \frac{1}{2}y^2$

15. $x^2 + 6y = 0$ **16.** $x - 7y^2 = 0$

17. $5x + 3y^2 = 0$ **18.** $8x^2 + 12y = 0$

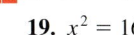 **19–24** ■ Use a graphing device to graph the parabola.

19. $x^2 = 16y$ **20.** $x^2 = -8y$

21. $y^2 = -\frac{1}{3}x$ **22.** $8y^2 = x$

23. $4x + y^2 = 0$ **24.** $x - 2y^2 = 0$

25–36 ■ Find an equation for the parabola that has its vertex at the origin and satisfies the given condition(s).

25. Focus $F(0, 2)$ **26.** Focus $F\left(0, -\frac{1}{2}\right)$

27. Focus $F(-8, 0)$ **28.** Focus $F(5, 0)$

29. Directrix $x = 2$ **30.** Directrix $y = 6$

31. Directrix $y = -10$ **32.** Directrix $x = -\frac{1}{8}$

33. Focus on the positive x-axis, 2 units away from the directrix

34. Directrix has y-intercept 6

35. Opens upward with focus 5 units from the vertex

36. Focal diameter 8 and focus on the negative y-axis

37–46 ■ Find an equation of the parabola whose graph is shown.

37.
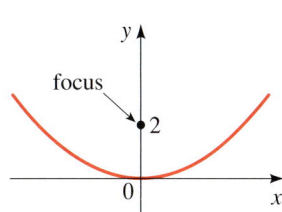

38.
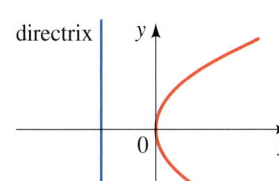

39.
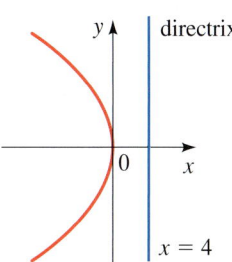

40.
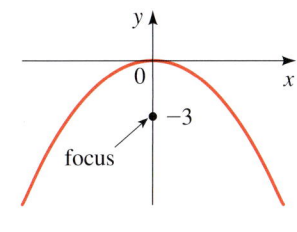

41.
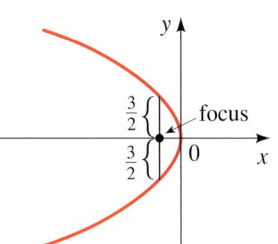

42.
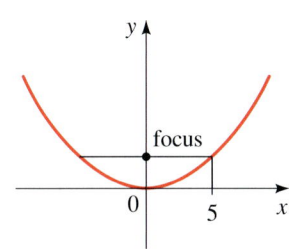

43.
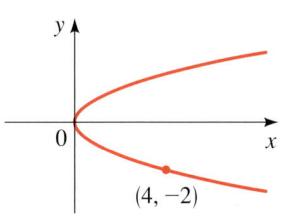

44.
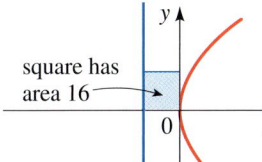

45.

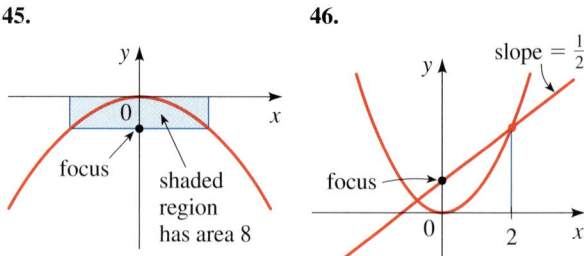

46.

focus

shaded region has area 8

slope = $\frac{1}{2}$

focus

47. (a) Find equations for the family of parabolas with vertex at the origin and with directrixes $y = \frac{1}{2}$, $y = 1$, $y = 4$, and $y = 8$.

(b) Draw the graphs. What do you conclude?

48. (a) Find equations for the family of parabolas with vertex at the origin, focus on the positive y-axis, and with focal diameters 1, 2, 4, and 8.

(b) Draw the graphs. What do you conclude?

Applications

49. Parabolic Reflector   A lamp with a parabolic reflector is shown in the figure. The bulb is placed at the focus and the focal diameter is 12 cm.

(a) Find an equation of the parabola.

(b) Find the diameter $d(C, D)$ of the opening, 20 cm from the vertex.

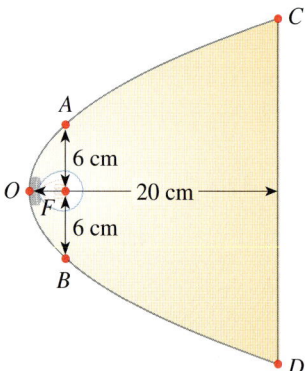

50. Satellite Dish   A reflector for a satellite dish is parabolic in cross section, with the receiver at the focus F. The reflector is 1 ft deep and 20 ft wide from rim to rim (see the figure). How far is the receiver from the vertex of the parabolic reflector?

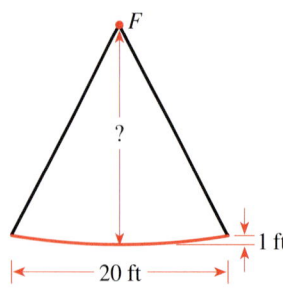

51. Suspension Bridge   In a suspension bridge the shape of the suspension cables is parabolic. The bridge shown in the figure has towers that are 600 m apart, and the lowest

point of the suspension cables is 150 m below the top of the towers. Find the equation of the parabolic part of the cables, placing the origin of the coordinate system at the vertex.

NOTE   This equation is used to find the length of cable needed in the construction of the bridge.

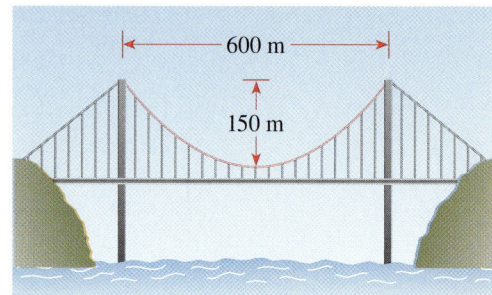

52. Reflecting Telescope   The Hale telescope at the Mount Palomar Observatory has a 200-in. mirror, as shown. The mirror is constructed in a parabolic shape that collects light from the stars and focuses it at the **prime focus**, that is, the focus of the parabola. The mirror is 3.79 in. deep at its center. Find the **focal length** of this parabolic mirror, that is, the distance from the vertex to the focus.

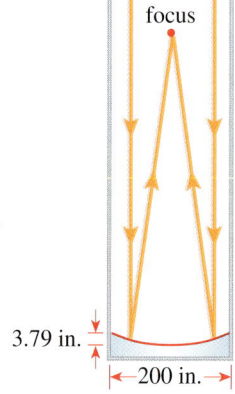

Discovery • Discussion

53. Parabolas in the Real World   Several examples of the uses of parabolas are given in the text. Find other situations in real life where parabolas occur. Consult a scientific encyclopedia in the reference section of your library, or search the Internet.

54. Light Cone from a Flashlight   A flashlight is held to form a lighted area on the ground, as shown in the figure. Is it possible to angle the flashlight in such a way that the boundary of the lighted area is a parabola? Explain your answer.

10.2 Ellipses

An ellipse is an oval curve that looks like an elongated circle. More precisely, we have the following definition.

Geometric Definition of an Ellipse

An **ellipse** is the set of all points in the plane the sum of whose distances from two fixed points F_1 and F_2 is a constant. (See Figure 1.) These two fixed points are the **foci** (plural of **focus**) of the ellipse.

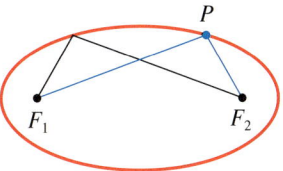

Figure 1

The geometric definition suggests a simple method for drawing an ellipse. Place a sheet of paper on a drawing board and insert thumbtacks at the two points that are to be the foci of the ellipse. Attach the ends of a string to the tacks, as shown in Figure 2(a). With the point of a pencil, hold the string taut. Then carefully move the pencil around the foci, keeping the string taut at all times. The pencil will trace out an ellipse, because the sum of the distances from the point of the pencil to the foci will always equal the length of the string, which is constant.

If the string is only slightly longer than the distance between the foci, then the ellipse traced out will be elongated in shape as in Figure 2(a), but if the foci are close together relative to the length of the string, the ellipse will be almost circular, as shown in Figure 2(b).

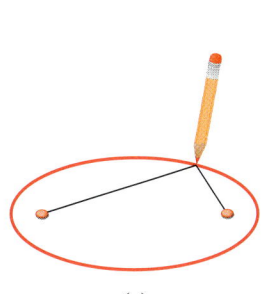

Figure 2   (a)   (b)

To obtain the simplest equation for an ellipse, we place the foci on the x-axis at $F_1(-c, 0)$ and $F_2(c, 0)$, so that the origin is halfway between them (see Figure 3).

For later convenience we let the sum of the distances from a point on the ellipse to the foci be $2a$. Then if $P(x, y)$ is any point on the ellipse, we have

$$d(P, F_1) + d(P, F_2) = 2a$$

So, from the Distance Formula

$$\sqrt{(x + c)^2 + y^2} + \sqrt{(x - c)^2 + y^2} = 2a$$

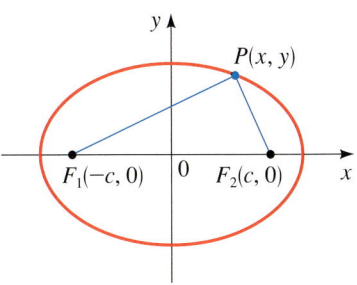

Figure 3   or

$$\sqrt{(x - c)^2 + y^2} = 2a - \sqrt{(x + c)^2 + y^2}$$

Squaring each side and expanding, we get

$$x^2 - 2cx + c^2 + y^2 = 4a^2 - 4a\sqrt{(x + c)^2 + y^2} + (x^2 + 2cx + c^2 + y^2)$$

which simplifies to

$$4a\sqrt{(x + c)^2 + y^2} = 4a^2 + 4cx$$

Dividing each side by 4 and squaring again, we get

$$a^2[(x + c)^2 + y^2] = (a^2 + cx)^2$$

$$a^2x^2 + 2a^2cx + a^2c^2 + a^2y^2 = a^4 + 2a^2cx + c^2x^2$$

$$(a^2 - c^2)x^2 + a^2y^2 = a^2(a^2 - c^2)$$

Since the sum of the distances from P to the foci must be larger than the distance between the foci, we have that $2a > 2c$, or $a > c$. Thus, $a^2 - c^2 > 0$, and we can divide each side of the preceding equation by $a^2(a^2 - c^2)$ to get

$$\frac{x^2}{a^2} + \frac{y^2}{a^2 - c^2} = 1$$

For convenience let $b^2 = a^2 - c^2$ (with $b > 0$). Since $b^2 < a^2$, it follows that $b < a$. The preceding equation then becomes

$$\frac{x^2}{a^2} + \frac{y^2}{b^2} = 1 \qquad \text{with } a > b$$

This is the equation of the ellipse. To graph it, we need to know the x- and y-intercepts. Setting $y = 0$, we get

$$\frac{x^2}{a^2} = 1$$

so $x^2 = a^2$, or $x = \pm a$. Thus, the ellipse crosses the x-axis at $(a, 0)$ and $(-a, 0)$, as in Figure 4. These points are called the **vertices** of the ellipse, and the segment that joins them is called the **major axis**. Its length is $2a$.

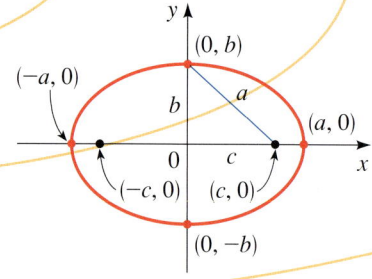

Figure 4
$\dfrac{x^2}{a^2} + \dfrac{y^2}{b^2} = 1$ with $a > b$

Similarly, if we set $x = 0$, we get $y = \pm b$, so the ellipse crosses the y-axis at $(0, b)$ and $(0, -b)$. The segment that joins these points is called the **minor axis**, and it has length $2b$. Note that $2a > 2b$, so the major axis is longer than the minor axis. The origin is the **center** of the ellipse.

If the foci of the ellipse are placed on the y-axis at $(0, \pm c)$ rather than on the x-axis, then the roles of x and y are reversed in the preceding discussion, and we get a vertical ellipse.

Equations and Graphs of Ellipses

The orbits of the planets are ellipses, with the sun at one focus.

The following box summarizes what we have just proved about the equation and features of an ellipse centered at the origin.

Ellipse with Center at the Origin

The graph of each of the following equations is an ellipse with center at the origin and having the given properties.

EQUATION	$\dfrac{x^2}{a^2} + \dfrac{y^2}{b^2} = 1$	$\dfrac{x^2}{b^2} + \dfrac{y^2}{a^2} = 1$
	$a > b > 0$	$a > b > 0$
VERTICES	$(\pm a, 0)$	$(0, \pm a)$
MAJOR AXIS	Horizontal, length $2a$	Vertical, length $2a$
MINOR AXIS	Vertical, length $2b$	Horizontal, length $2b$
FOCI	$(\pm c, 0),\ c^2 = a^2 - b^2$	$(0, \pm c),\ c^2 = a^2 - b^2$
GRAPH		

In the standard equation for an ellipse, a^2 is the *larger* denominator and b^2 is the *smaller*. To find c^2, we subtract: larger denominator minus smaller denominator.

Example 1   Sketching an Ellipse

An ellipse has the equation

$$\frac{x^2}{9} + \frac{y^2}{4} = 1$$

(a) Find the foci, vertices, and the lengths of the major and minor axes, and sketch the graph.

 (b) Draw the graph using a graphing calculator.

Solution

(a) Since the denominator of x^2 is larger, the ellipse has horizontal major axis. This gives $a^2 = 9$ and $b^2 = 4$, so $c^2 = a^2 - b^2 = 9 - 4 = 5$. Thus, $a = 3$, $b = 2$, and $c = \sqrt{5}$.

FOCI	$(\pm\sqrt{5}, 0)$
VERTICES	$(\pm 3, 0)$
LENGTH OF MAJOR AXIS	6
LENGTH OF MINOR AXIS	4

The graph is shown in Figure 5(a) on the next page.

(b) To draw the graph using a graphing calculator, we need to solve for y.

$$\frac{x^2}{9} + \frac{y^2}{4} = 1$$

$$\frac{y^2}{4} = 1 - \frac{x^2}{9} \qquad \text{Subtract } x^2/9$$

$$y^2 = 4\left(1 - \frac{x^2}{9}\right) \qquad \text{Multiply by 4}$$

$$y = \pm 2\sqrt{1 - \frac{x^2}{9}} \qquad \text{Take square roots}$$

Note that the equation of an ellipse does not define y as a function of x (see page 164). That's why we need to graph two functions to graph an ellipse.

To obtain the graph of the ellipse, we graph both functions

$$y = 2\sqrt{1 - x^2/9} \qquad \text{and} \qquad y = -2\sqrt{1 - x^2/9}$$

as shown in Figure 5(b).

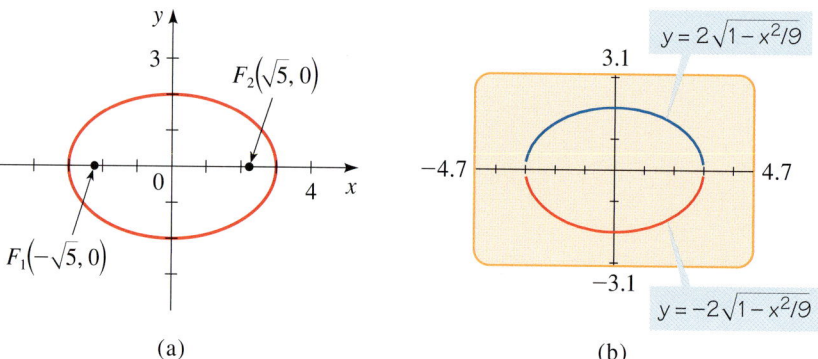

Figure 5
$$\frac{x^2}{9} + \frac{y^2}{4} = 1$$

(a)

(b)

Example 2   Finding the Foci of an Ellipse

Find the foci of the ellipse $16x^2 + 9y^2 = 144$, and sketch its graph.

Solution   First we put the equation in standard form. Dividing by 144, we get

$$\frac{x^2}{9} + \frac{y^2}{16} = 1$$

Since $16 > 9$, this is an ellipse with its foci on the y-axis, and with $a = 4$ and $b = 3$. We have

$$c^2 = a^2 - b^2 = 16 - 9 = 7$$

$$c = \sqrt{7}$$

Thus, the foci are $(0, \pm\sqrt{7})$. The graph is shown in Figure 6(a).

We can also draw the graph using a graphing calculator as shown in Figure 6(b).

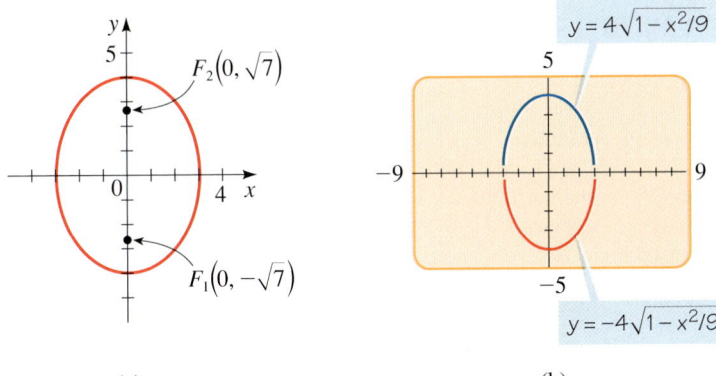

Figure 6

$16x^2 + 9y^2 = 144$

(a)   (b)

Example 3   Finding the Equation of an Ellipse

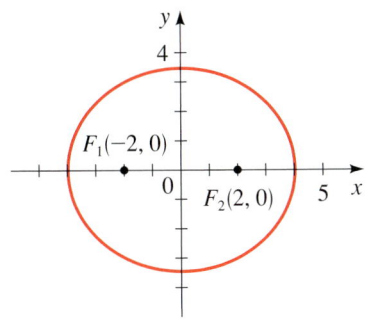

The vertices of an ellipse are $(\pm 4, 0)$ and the foci are $(\pm 2, 0)$. Find its equation and sketch the graph.

Solution   Since the vertices are $(\pm 4, 0)$, we have $a = 4$. The foci are $(\pm 2, 0)$, so $c = 2$. To write the equation, we need to find b. Since $c^2 = a^2 - b^2$, we have

$$2^2 = 4^2 - b^2$$

$$b^2 = 16 - 4 = 12$$

Thus, the equation of the ellipse is

$$\frac{x^2}{16} + \frac{y^2}{12} = 1$$

The graph is shown in Figure 7.

Figure 7

$$\frac{x^2}{16} + \frac{y^2}{12} = 1$$

Eccentricity of an Ellipse

We saw earlier in this section (Figure 2) that if $2a$ is only slightly greater than $2c$, the ellipse is long and thin, whereas if $2a$ is much greater than $2c$, the ellipse is almost circular. We measure the deviation of an ellipse from being circular by the ratio of a and c.

Definition of Eccentricity

For the ellipse $\dfrac{x^2}{a^2} + \dfrac{y^2}{b^2} = 1$ or $\dfrac{x^2}{b^2} + \dfrac{y^2}{a^2} = 1$ (with $a > b > 0$), the **eccentricity** e is the number

$$e = \frac{c}{a}$$

where $c = \sqrt{a^2 - b^2}$. The eccentricity of every ellipse satisfies $0 < e < 1$.

Thus, if e is close to 1, then c is almost equal to a, and the ellipse is elongated in shape, but if e is close to 0, then the ellipse is close to a circle in shape. The eccentricity is a measure of how "stretched" the ellipse is.

In Figure 8 we show a number of ellipses to demonstrate the effect of varying the eccentricity e.

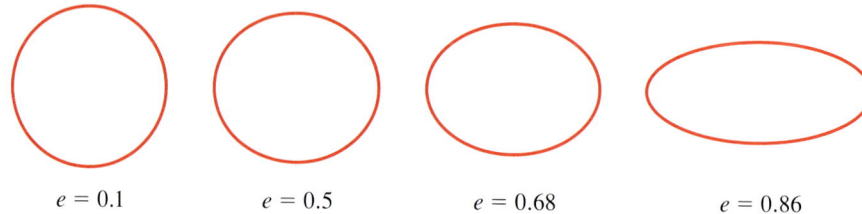

$e = 0.1$ $e = 0.5$ $e = 0.68$ $e = 0.86$

Figure 8
Ellipses with various eccentricities

Example 4   Finding the Equation of an Ellipse from Its Eccentricity and Foci

Find the equation of the ellipse with foci $(0, \pm 8)$ and eccentricity $e = \frac{4}{5}$, and sketch its graph.

Solution   We are given $e = \frac{4}{5}$ and $c = 8$. Thus

$$\frac{4}{5} = \frac{8}{a} \qquad \text{Eccentricity } e = \frac{c}{a}$$

$$4a = 40 \qquad \text{Cross multiply}$$

$$a = 10$$

To find b, we use the fact that $c^2 = a^2 - b^2$.

$$8^2 = 10^2 - b^2$$

$$b^2 = 10^2 - 8^2 = 36$$

$$b = 6$$

Thus, the equation of the ellipse is

$$\frac{x^2}{36} + \frac{y^2}{100} = 1$$

Because the foci are on the y-axis, the ellipse is oriented vertically. To sketch the ellipse, we find the intercepts: The x-intercepts are ± 6 and the y-intercepts are ± 10. The graph is sketched in Figure 9. ∎

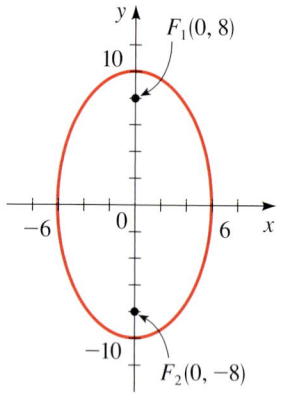

Figure 9
$\dfrac{x^2}{36} + \dfrac{y^2}{100} = 1$

Gravitational attraction causes the planets to move in elliptical orbits around the sun with the sun at one focus. This remarkable property was first observed by Johannes Kepler and was later deduced by Isaac Newton from his inverse square law of gravity, using calculus. The orbits of the planets have different eccentricities, but most are nearly circular (see the margin note above).

Ellipses, like parabolas, have an interesting *reflection property* that leads to a number of practical applications. If a light source is placed at one focus of a reflecting surface with elliptical cross sections, then all the light will be reflected off the surface to the other focus, as shown in Figure 10. This principle, which works for sound waves as well as for light, is used in *lithotripsy*, a treatment for kidney stones. The patient is placed in a tub of water with elliptical cross sections in such a way that the kidney stone is accurately located at one focus. High-intensity sound waves generated at the other focus are reflected to the stone and destroy it with minimal damage to surrounding tissue. The patient is spared the trauma of surgery and recovers within days instead of weeks.

The reflection property of ellipses is also used in the construction of *whispering galleries*. Sound coming from one focus bounces off the walls and ceiling of an elliptical room and passes through the other focus. In these rooms even quiet whispers spoken at one focus can be heard clearly at the other. Famous whispering galleries include the National Statuary Hall of the U.S. Capitol in Washington, D.C. (see page 771), and the Mormon Tabernacle in Salt Lake City, Utah.

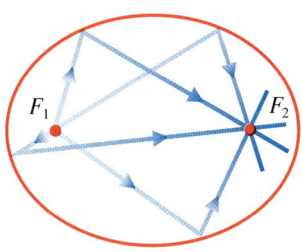

Figure 10

10.2 Exercises

1–4 ■ Match the equation with the graphs labeled I–IV. Give reasons for your answers.

1. $\dfrac{x^2}{16} + \dfrac{y^2}{4} = 1$

2. $x^2 + \dfrac{y^2}{9} = 1$

3. $4x^2 + y^2 = 4$

4. $16x^2 + 25y^2 = 400$

I
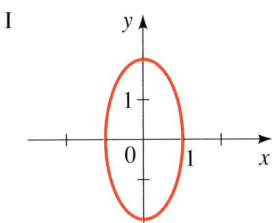

II
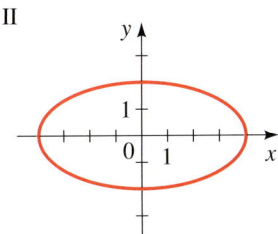

III
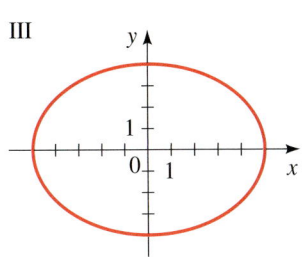

IV
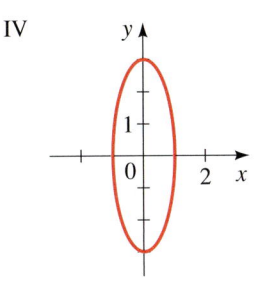

5–18 ■ Find the vertices, foci, and eccentricity of the ellipse. Determine the lengths of the major and minor axes, and sketch the graph.

5. $\dfrac{x^2}{25} + \dfrac{y^2}{9} = 1$

6. $\dfrac{x^2}{16} + \dfrac{y^2}{25} = 1$

7. $9x^2 + 4y^2 = 36$

8. $4x^2 + 25y^2 = 100$

9. $x^2 + 4y^2 = 16$

10. $4x^2 + y^2 = 16$

11. $2x^2 + y^2 = 3$

12. $5x^2 + 6y^2 = 30$

13. $x^2 + 4y^2 = 1$

14. $9x^2 + 4y^2 = 1$

15. $\frac{1}{2}x^2 + \frac{1}{8}y^2 = \frac{1}{4}$

16. $x^2 = 4 - 2y^2$

17. $y^2 = 1 - 2x^2$

18. $20x^2 + 4y^2 = 5$

19–24 ■ Find an equation for the ellipse whose graph is shown.

19.
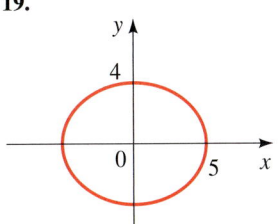

20.
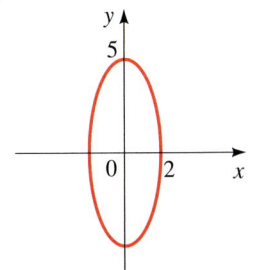

21.

22.

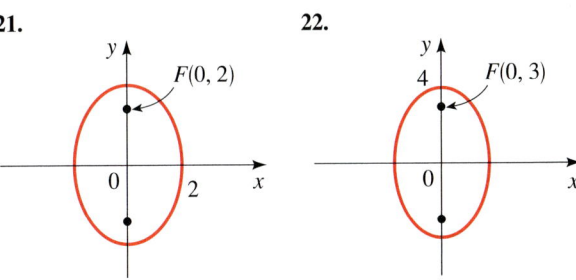

23.

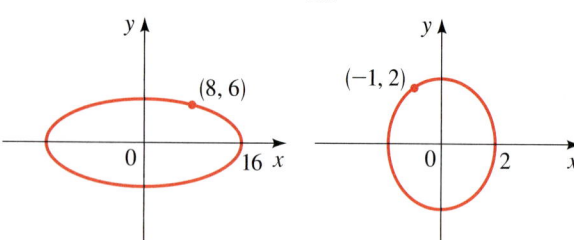

(8, 6)

0   16   *x*

24.

(−1, 2)

0   2   *x*

 25–28 ■ Use a graphing device to graph the ellipse.

25. $\dfrac{x^2}{25} + \dfrac{y^2}{20} = 1$

26. $x^2 + \dfrac{y^2}{12} = 1$

27. $6x^2 + y^2 = 36$

28. $x^2 + 2y^2 = 8$

29–40 ■ Find an equation for the ellipse that satisfies the given conditions.

29. Foci $(\pm 4, 0)$, vertices $(\pm 5, 0)$

30. Foci $(0, \pm 3)$, vertices $(0, \pm 5)$

31. Length of major axis 4, length of minor axis 2, foci on *y*-axis

32. Length of major axis 6, length of minor axis 4, foci on *x*-axis

33. Foci $(0, \pm 2)$, length of minor axis 6

34. Foci $(\pm 5, 0)$, length of major axis 12

35. Endpoints of major axis $(\pm 10, 0)$, distance between foci 6

36. Endpoints of minor axis $(0, \pm 3)$, distance between foci 8

37. Length of major axis 10, foci on *x*-axis, ellipse passes through the point $(\sqrt{5}, 2)$

38. Eccentricity $\frac{1}{9}$, foci $(0, \pm 2)$

39. Eccentricity 0.8, foci $(\pm 1.5, 0)$

40. Eccentricity $\sqrt{3}/2$, foci on *y*-axis, length of major axis 4

41–43 ■ Find the intersection points of the pair of ellipses. Sketch the graphs of each pair of equations on the same coordinate axes and label the points of intersection.

41. $\begin{cases} 4x^2 + y^2 = 4 \\ 4x^2 + 9y^2 = 36 \end{cases}$

42. $\begin{cases} \dfrac{x^2}{16} + \dfrac{y^2}{9} = 1 \\ \dfrac{x^2}{9} + \dfrac{y^2}{16} = 1 \end{cases}$

43. $\begin{cases} 100x^2 + 25y^2 = 100 \\ x^2 + \dfrac{y^2}{9} = 1 \end{cases}$

44. The **ancillary circle** of an ellipse is the circle with radius equal to half the length of the minor axis and center the

same as the ellipse (see the figure). The ancillary circle is thus the largest circle that can fit within an ellipse.

(a) Find an equation for the ancillary circle of the ellipse $x^2 + 4y^2 = 16$.

(b) For the ellipse and ancillary circle of part (a), show that if (s, t) is a point on the ancillary circle, then $(2s, t)$ is a point on the ellipse.

ellipse

ancillary circle

 45. **(a)** Use a graphing device to sketch the top half (the portion in the first and second quadrants) of the family of ellipses $x^2 + ky^2 = 100$ for $k = 4, 10, 25$, and 50.

(b) What do the members of this family of ellipses have in common? How do they differ?

46. If $k > 0$, the following equation represents an ellipse:

$$\frac{x^2}{k} + \frac{y^2}{4 + k} = 1$$

Show that all the ellipses represented by this equation have the same foci, no matter what the value of k.

Applications

47. Perihelion and Aphelion   The planets move around the sun in elliptical orbits with the sun at one focus. The point in the orbit at which the planet is closest to the sun is called **perihelion**, and the point at which it is farthest is called **aphelion**. These points are the vertices of the orbit. The earth's distance from the sun is 147,000,000 km at perihelion and 153,000,000 km at aphelion. Find an equation for the earth's orbit. (Place the origin at the center of the orbit with the sun on the *x*-axis.)

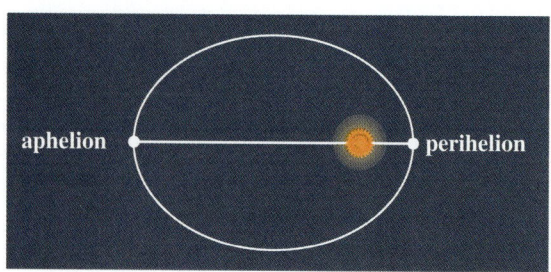

aphelion   perihelion

48. The Orbit of Pluto   With an eccentricity of 0.25, Pluto's orbit is the most eccentric in the solar system. The length of the minor axis of its orbit is approximately 10,000,000,000 km. Find the distance between Pluto and the sun at perihelion and at aphelion. (See Exercise 47.)

49. Lunar Orbit For an object in an elliptical orbit around the moon, the points in the orbit that are closest to and farthest from the center of the moon are called **perilune** and **apolune**, respectively. These are the vertices of the orbit. The center of the moon is at one focus of the orbit. The *Apollo 11* spacecraft was placed in a lunar orbit with perilune at 68 mi and apolune at 195 mi above the surface of the moon. Assuming the moon is a sphere of radius 1075 mi, find an equation for the orbit of *Apollo 11*. (Place the coordinate axes so that the origin is at the center of the orbit and the foci are located on the *x*-axis.)

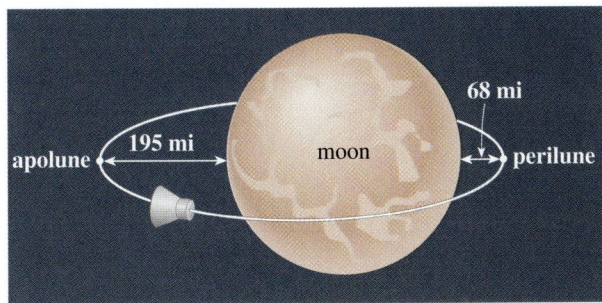

50. Plywood Ellipse A carpenter wishes to construct an elliptical table top from a sheet of plywood, 4 ft by 8 ft. He will trace out the ellipse using the "thumbtack and string" method illustrated in Figures 2 and 3. What length of string should he use, and how far apart should the tacks be located, if the ellipse is to be the largest possible that can be cut out of the plywood sheet?

51. Sunburst Window A "sunburst" window above a doorway is constructed in the shape of the top half of an ellipse, as shown in the figure. The window is 20 in. tall at its highest point and 80 in. wide at the bottom. Find the height of the window 25 in. from the center of the base.

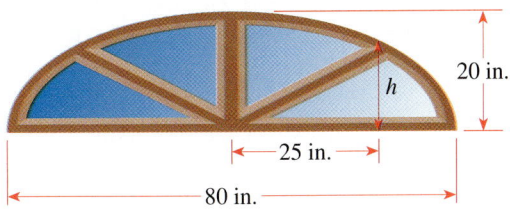

Discovery • Discussion

52. Drawing an Ellipse on a Blackboard Try drawing an ellipse as accurately as possible on a blackboard. How would a piece of string and two friends help this process?

53. Light Cone from a Flashlight A flashlight shines on a wall, as shown in the figure. What is the shape of the boundary of the lighted area? Explain your answer.

54. How Wide Is an Ellipse at Its Foci? A *latus rectum* for an ellipse is a line segment perpendicular to the major axis at a focus, with endpoints on the ellipse, as shown. Show that the length of a latus rectum is $2b^2/a$ for the ellipse

$$\frac{x^2}{a^2} + \frac{y^2}{b^2} = 1 \qquad \text{with } a > b$$

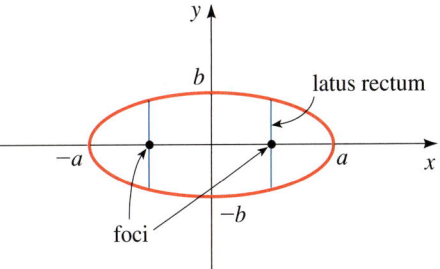

55. Is It an Ellipse? A piece of paper is wrapped around a cylindrical bottle, and then a compass is used to draw a circle on the paper, as shown in the figure. When the paper is laid flat, is the shape drawn on the paper an ellipse? (You don't need to prove your answer, but you may want to do the experiment and see what you get.)

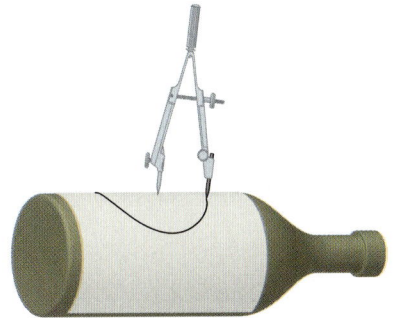

10.3   Hyperbolas

Although ellipses and hyperbolas have completely different shapes, their definitions and equations are similar. Instead of using the *sum* of distances from two fixed foci, as in the case of an ellipse, we use the *difference* to define a hyperbola.

Geometric Definition of a Hyperbola

A **hyperbola** is the set of all points in the plane, the difference of whose distances from two fixed points F_1 and F_2 is a constant. (See Figure 1.) These two fixed points are the **foci** of the hyperbola.

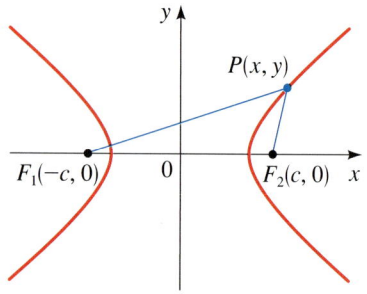

Figure 1

P is on the hyperbola if
$|d(P, F_1) - d(P, F_2)| = 2a$.

As in the case of the ellipse, we get the simplest equation for the hyperbola by placing the foci on the x-axis at $(\pm c, 0)$, as shown in Figure 1. By definition, if $P(x, y)$ lies on the hyperbola, then either $d(P, F_1) - d(P, F_2)$ or $d(P, F_2) - d(P, F_1)$ must equal some positive constant, which we call $2a$. Thus, we have

$$d(P, F_1) - d(P, F_2) = \pm 2a$$

or $$\sqrt{(x + c)^2 + y^2} - \sqrt{(x - c)^2 + y^2} = \pm 2a$$

Proceeding as we did in the case of the ellipse (Section 10.2), we simplify this to

$$(c^2 - a^2)x^2 - a^2 y^2 = a^2(c^2 - a^2)$$

From triangle PF_1F_2 in Figure 1 we see that $|d(P, F_1) - d(P, F_2)| < 2c$. It follows that $2a < 2c$, or $a < c$. Thus, $c^2 - a^2 > 0$, so we can set $b^2 = c^2 - a^2$. We then simplify the last displayed equation to get

$$\frac{x^2}{a^2} - \frac{y^2}{b^2} = 1$$

This is the *equation of the hyperbola*. If we replace x by $-x$ or y by $-y$ in this equation, it remains unchanged, so the hyperbola is symmetric about both the x- and y-axes and about the origin. The x-intercepts are $\pm a$, and the points $(a, 0)$ and $(-a, 0)$ are the **vertices** of the hyperbola. There is no y-intercept, because setting $x = 0$ in the equation of the hyperbola leads to $-y^2 = b^2$, which has no real solution. Furthermore, the equation of the hyperbola implies that

$$\frac{x^2}{a^2} = \frac{y^2}{b^2} + 1 \geq 1$$

so $x^2/a^2 \geq 1$; thus, $x^2 \geq a^2$, and hence $x \geq a$ or $x \leq -a$. This means that the hyperbola consists of two parts, called its **branches**. The segment joining the two vertices on the separate branches is the **transverse axis** of the hyperbola, and the origin is called its **center**.

If we place the foci of the hyperbola on the y-axis rather than on the x-axis, then this has the effect of reversing the roles of x and y in the derivation of the equation of the hyperbola. This leads to a hyperbola with a vertical transverse axis.

Equations and Graphs of Hyperbolas

The main properties of hyperbolas are listed in the following box.

Hyperbola with Center at the Origin

The graph of each of the following equations is a hyperbola with center at the origin and having the given properties.

EQUATION	$\dfrac{x^2}{a^2} - \dfrac{y^2}{b^2} = 1 \quad (a > 0, b > 0)$	$\dfrac{y^2}{a^2} - \dfrac{x^2}{b^2} = 1 \quad (a > 0, b > 0)$
VERTICES	$(\pm a, 0)$	$(0, \pm a)$
TRANSVERSE AXIS	Horizontal, length $2a$	Vertical, length $2a$
ASYMPTOTES	$y = \pm\dfrac{b}{a}x$	$y = \pm\dfrac{a}{b}x$
FOCI	$(\pm c, 0), \quad c^2 = a^2 + b^2$	$(0, \pm c), \quad c^2 = a^2 + b^2$
GRAPH	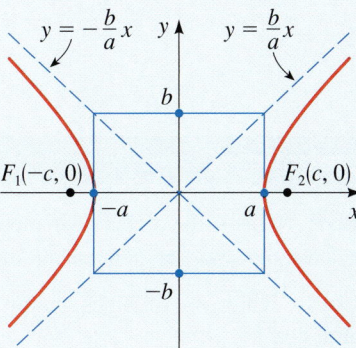	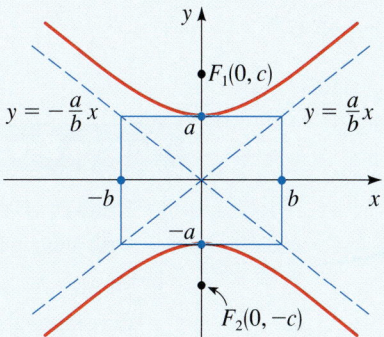

Asymptotes of rational functions are discussed in Section 3.6.

The *asymptotes* mentioned in this box are lines that the hyperbola approaches for large values of x and y. To find the asymptotes in the first case in the box, we solve the equation for y to get

$$y = \pm\frac{b}{a}\sqrt{x^2 - a^2}$$

$$= \pm\frac{b}{a}x\sqrt{1 - \frac{a^2}{x^2}}$$

As x gets large, a^2/x^2 gets closer to zero. In other words, as $x \to \infty$ we have $a^2/x^2 \to 0$. So, for large x the value of y can be approximated as $y = \pm(b/a)x$. This shows that these lines are asymptotes of the hyperbola.

Asymptotes are an essential aid for graphing a hyperbola; they help us determine its shape. A convenient way to find the asymptotes, for a hyperbola with horizontal transverse axis, is to first plot the points $(a, 0)$, $(-a, 0)$, $(0, b)$, and $(0, -b)$. Then sketch horizontal and vertical segments through these points to construct a rectangle, as shown in Figure 2(a) on the next page. We call this rectangle the **central box** of the hyperbola. The slopes of the diagonals of the central box are $\pm b/a$, so by extending them we obtain the asymptotes $y = \pm(b/a)x$, as sketched in part (b) of the figure. Finally, we plot the vertices and use the asymptotes as a guide in sketching the hyperbola shown in part (c). (A similar procedure applies to graphing a hyperbola that has a vertical transverse axis.)

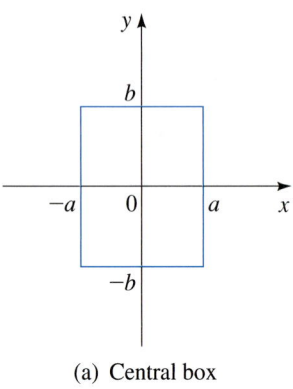

(a) Central box

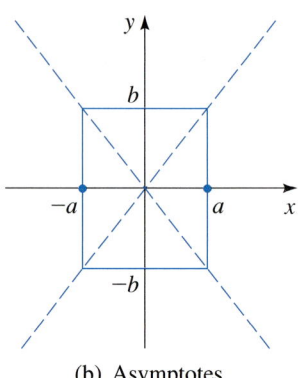

(b) Asymptotes

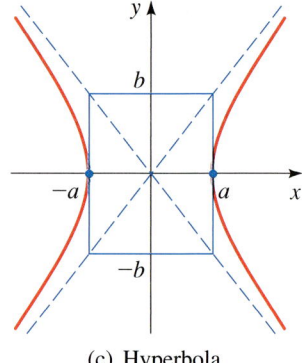

(c) Hyperbola

Figure 2

Steps in graphing the hyperbola $\dfrac{x^2}{a^2} - \dfrac{y^2}{b^2} = 1$

How to Sketch a Hyperbola

1. **Sketch the Central Box.** This is the rectangle centered at the origin, with sides parallel to the axes, that crosses one axis at $\pm a$, the other at $\pm b$.

2. **Sketch the Asymptotes.** These are the lines obtained by extending the diagonals of the central box.

3. **Plot the Vertices.** These are the two x-intercepts or the two y-intercepts.

4. **Sketch the Hyperbola.** Start at a vertex and sketch a branch of the hyperbola, approaching the asymptotes. Sketch the other branch in the same way.

Example 1 **A Hyperbola with Horizontal Transverse Axis**

A hyperbola has the equation

$$9x^2 - 16y^2 = 144$$

(a) Find the vertices, foci, and asymptotes, and sketch the graph.

 (b) Draw the graph using a graphing calculator.

Solution

(a) First we divide both sides of the equation by 144 to put it into standard form:

$$\frac{x^2}{16} - \frac{y^2}{9} = 1$$

Because the x^2-term is positive, the hyperbola has a horizontal transverse axis; its vertices and foci are on the x-axis. Since $a^2 = 16$ and $b^2 = 9$, we get $a = 4$, $b = 3$, and $c = \sqrt{16 + 9} = 5$. Thus, we have

VERTICES	$(\pm 4, 0)$
FOCI	$(\pm 5, 0)$
ASYMPTOTES	$y = \pm \frac{3}{4}x$

After sketching the central box and asymptotes, we complete the sketch of the hyperbola as in Figure 3(a).

(b) To draw the graph using a graphing calculator, we need to solve for y.

$$9x^2 - 16y^2 = 144$$

$$-16y^2 = -9x^2 + 144 \qquad \text{Subtract } 9x^2$$

$$y^2 = 9\left(\frac{x^2}{16} - 1\right) \qquad \text{Divide by } -16 \text{ and factor } 9$$

$$y = \pm 3\sqrt{\frac{x^2}{16} - 1} \qquad \text{Take square roots}$$

To obtain the graph of the hyperbola, we graph the functions

$$y = 3\sqrt{(x^2/16) - 1} \qquad \text{and} \qquad y = -3\sqrt{(x^2/16) - 1}$$

as shown in Figure 3(b).

Note that the equation of a hyperbola does not define y as a function of x (see page 164). That's why we need to graph two functions to graph a hyperbola.

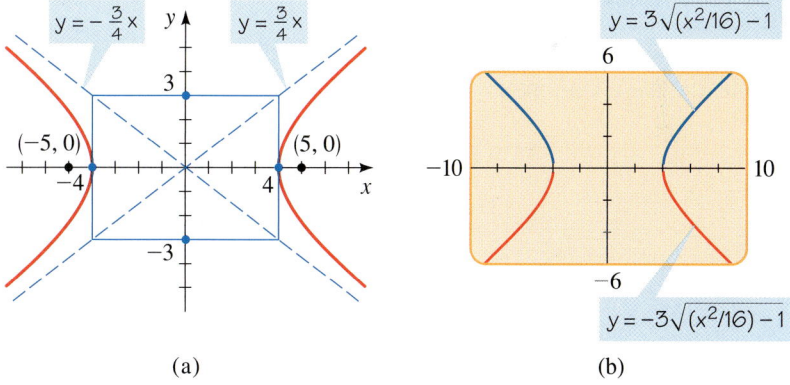

Figure 3
$9x^2 - 16y^2 = 144$

(a)

(b)

Example 2 A Hyperbola with Vertical Transverse Axis

Find the vertices, foci, and asymptotes of the hyperbola, and sketch its graph.

$$x^2 - 9y^2 + 9 = 0$$

Solution We begin by writing the equation in the standard form for a hyperbola.

$$x^2 - 9y^2 = -9$$

$$y^2 - \frac{x^2}{9} = 1 \qquad \text{Divide by } -9$$

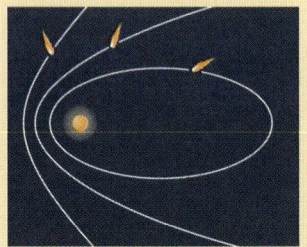
Because the y^2-term is positive, the hyperbola has a vertical transverse axis; its foci and vertices are on the y-axis. Since $a^2 = 1$ and $b^2 = 9$, we get $a = 1$, $b = 3$, and $c = \sqrt{1 + 9} = \sqrt{10}$. Thus, we have

VERTICES	$(0, \pm 1)$
FOCI	$(0, \pm\sqrt{10})$
ASYMPTOTES	$y = \pm\frac{1}{3}x$

We sketch the central box and asymptotes, then complete the graph, as shown in Figure 4(a).

We can also draw the graph using a graphing calculator, as shown in Figure 4(b).

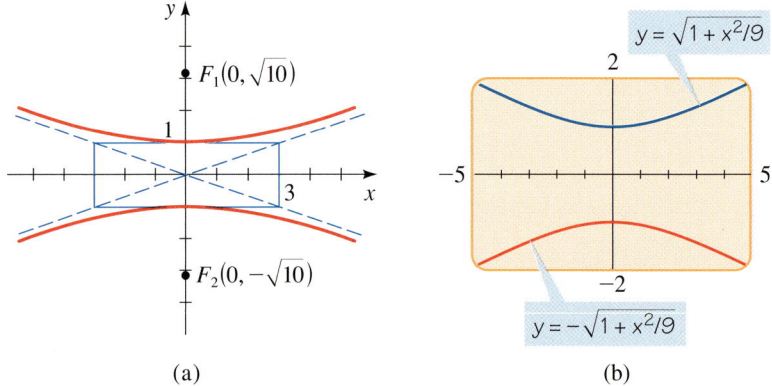

(a)   (b)

Figure 4
$x^2 - 9y^2 + 9 = 0$

Example 3   Finding the Equation of a Hyperbola from Its Vertices and Foci

Find the equation of the hyperbola with vertices $(\pm 3, 0)$ and foci $(\pm 4, 0)$. Sketch the graph.

Solution   Since the vertices are on the x-axis, the hyperbola has a horizontal transverse axis. Its equation is of the form

$$\frac{x^2}{3^2} - \frac{y^2}{b^2} = 1$$

We have $a = 3$ and $c = 4$. To find b, we use the relation $a^2 + b^2 = c^2$:

$$3^2 + b^2 = 4^2$$
$$b^2 = 4^2 - 3^2 = 7$$
$$b = \sqrt{7}$$

Thus, the equation of the hyperbola is

$$\frac{x^2}{9} - \frac{y^2}{7} = 1$$

The graph is shown in Figure 5.

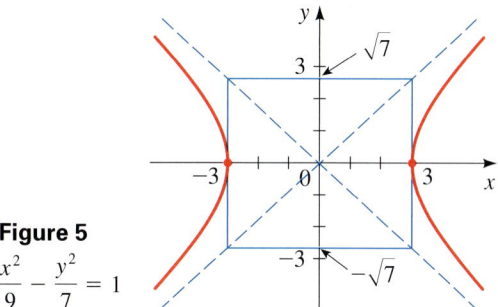

Figure 5

$$\frac{x^2}{9} - \frac{y^2}{7} = 1$$

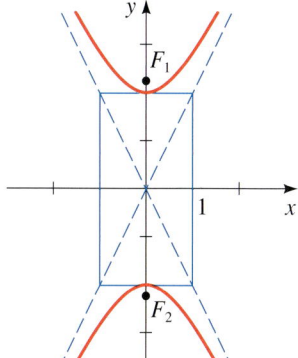

Figure 6

$$\frac{y^2}{4} - x^2 = 1$$

Example 4   Finding the Equation of a Hyperbola from Its Vertices and Asymptotes

Find the equation and the foci of the hyperbola with vertices $(0, \pm 2)$ and asymptotes $y = \pm 2x$. Sketch the graph.

Solution   Since the vertices are on the y-axis, the hyperbola has a vertical transverse axis with $a = 2$. From the asymptote equation we see that $a/b = 2$. Since $a = 2$, we get $2/b = 2$, and so $b = 1$. Thus, the equation of the hyperbola is

$$\frac{y^2}{4} - x^2 = 1$$

To find the foci, we calculate $c^2 = a^2 + b^2 = 2^2 + 1^2 = 5$, so $c = \sqrt{5}$. Thus, the foci are $(0, \pm\sqrt{5})$. The graph is shown in Figure 6. ∎

Like parabolas and ellipses, hyperbolas have an interesting *reflection property*. Light aimed at one focus of a hyperbolic mirror is reflected toward the other focus, as shown in Figure 7. This property is used in the construction of Cassegrain-type telescopes. A hyperbolic mirror is placed in the telescope tube so that light reflected from the primary parabolic reflector is aimed at one focus of the hyperbolic mirror. The light is then refocused at a more accessible point below the primary reflector (Figure 8).

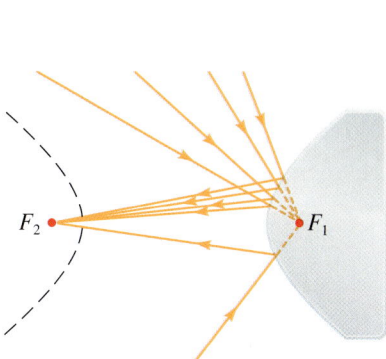

Figure 7

Reflection property of hyperbolas

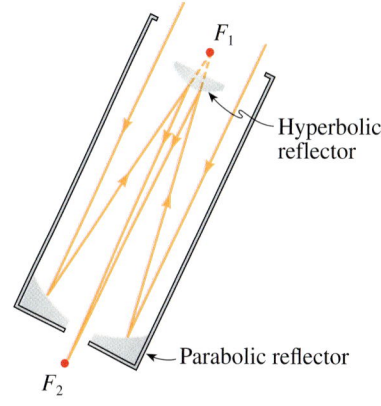

Figure 8

Cassegrain-type telescope

The LORAN (LOng RAnge Navigation) system was used until the early 1990s; it has now been superseded by the GPS system (see page 656). In the LORAN system, hyperbolas are used onboard a ship to determine its location. In Figure 9 radio stations at A and B transmit signals simultaneously for reception by the ship at P. The onboard computer converts the time difference in reception of these signals into a distance difference $d(P, A) - d(P, B)$. From the definition of a hyperbola this locates the ship on one branch of a hyperbola with foci at A and B (sketched in black in the figure). The same procedure is carried out with two other radio stations at C and D, and this locates the ship on a second hyperbola (shown in red in the figure). (In practice, only three stations are needed because one station can be used as a focus for both hyperbolas.) The coordinates of the intersection point of these two hyperbolas, which can be calculated precisely by the computer, give the location of P.

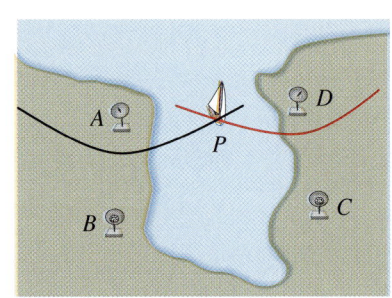

Figure 9

LORAN system for finding the location of a ship

10.3 Exercises

1–4 ■ Match the equation with the graphs labeled I–IV. Give reasons for your answers.

1. $\dfrac{x^2}{4} - y^2 = 1$

2. $y^2 - \dfrac{x^2}{9} = 1$

3. $16y^2 - x^2 = 144$

4. $9x^2 - 25y^2 = 225$

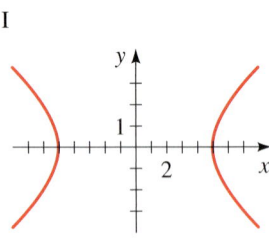

I

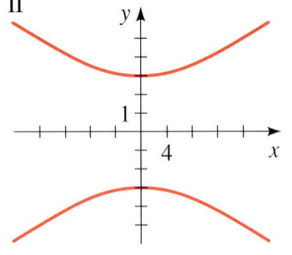

II

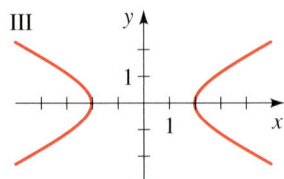

III

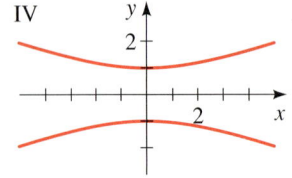

IV

5–16 ■ Find the vertices, foci, and asymptotes of the hyperbola, and sketch its graph.

5. $\dfrac{x^2}{4} - \dfrac{y^2}{16} = 1$

6. $\dfrac{y^2}{9} - \dfrac{x^2}{16} = 1$

7. $y^2 - \dfrac{x^2}{25} = 1$

8. $\dfrac{x^2}{2} - y^2 = 1$

9. $x^2 - y^2 = 1$

10. $9x^2 - 4y^2 = 36$

11. $25y^2 - 9x^2 = 225$

12. $x^2 - y^2 + 4 = 0$

13. $x^2 - 4y^2 - 8 = 0$

14. $x^2 - 2y^2 = 3$

15. $4y^2 - x^2 = 1$

16. $9x^2 - 16y^2 = 1$

17–22 ■ Find the equation for the hyperbola whose graph is shown.

17.

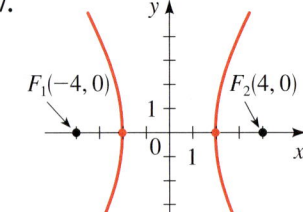

18.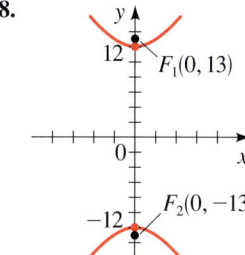

19.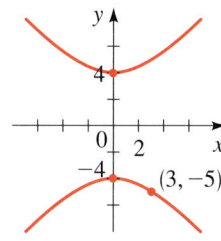

20.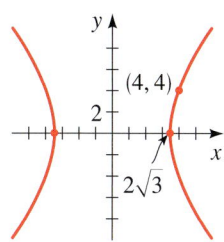

21.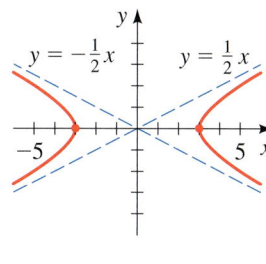

22.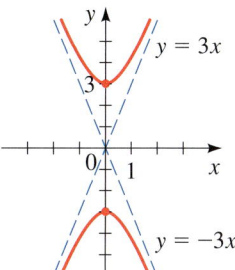

 23–26 ■ Use a graphing device to graph the hyperbola.

23. $x^2 - 2y^2 = 8$

24. $3y^2 - 4x^2 = 24$

25. $\dfrac{y^2}{2} - \dfrac{x^2}{6} = 1$

26. $\dfrac{x^2}{100} - \dfrac{y^2}{64} = 1$

27–38 ■ Find an equation for the hyperbola that satisfies the given conditions.

27. Foci $(\pm 5, 0)$, vertices $(\pm 3, 0)$

28. Foci $(0, \pm 10)$, vertices $(0, \pm 8)$

29. Foci $(0, \pm 2)$, vertices $(0, \pm 1)$

30. Foci $(\pm 6, 0)$, vertices $(\pm 2, 0)$

31. Vertices $(\pm 1, 0)$, asymptotes $y = \pm 5x$

32. Vertices $(0, \pm 6)$, asymptotes $y = \pm \frac{1}{3}x$

33. Foci $(0, \pm 8)$, asymptotes $y = \pm \frac{1}{2}x$

34. Vertices $(0, \pm 6)$, hyperbola passes through $(-5, 9)$

35. Asymptotes $y = \pm x$, hyperbola passes through $(5, 3)$

36. Foci $(\pm 3, 0)$, hyperbola passes through $(4, 1)$

37. Foci $(\pm 5, 0)$, length of transverse axis 6

38. Foci $(0, \pm 1)$, length of transverse axis 1

39. (a) Show that the asymptotes of the hyperbola $x^2 - y^2 = 5$ are perpendicular to each other.

(b) Find an equation for the hyperbola with foci $(\pm c, 0)$ and with asymptotes perpendicular to each other.

40. The hyperbolas

$$\frac{x^2}{a^2} - \frac{y^2}{b^2} = 1 \quad \text{and} \quad \frac{x^2}{a^2} - \frac{y^2}{b^2} = -1$$

are said to be **conjugate** to each other.

(a) Show that the hyperbolas

$$x^2 - 4y^2 + 16 = 0 \quad \text{and} \quad 4y^2 - x^2 + 16 = 0$$

are conjugate to each other, and sketch their graphs on the same coordinate axes.

(b) What do the hyperbolas of part (a) have in common?

(c) Show that any pair of conjugate hyperbolas have the relationship you discovered in part (b).

41. In the derivation of the equation of the hyperbola at the beginning of this section, we said that the equation

$$\sqrt{(x + c)^2 + y^2} - \sqrt{(x - c)^2 + y^2} = \pm 2a$$

simplifies to

$$(c^2 - a^2)x^2 - a^2 y^2 = a^2(c^2 - a^2)$$

Supply the steps needed to show this.

42. (a) For the hyperbola

$$\frac{x^2}{9} - \frac{y^2}{16} = 1$$

determine the values of a, b, and c, and find the coordinates of the foci F_1 and F_2.

(b) Show that the point $P(5, \frac{16}{3})$ lies on this hyperbola.

(c) Find $d(P, F_1)$ and $d(P, F_2)$.

(d) Verify that the difference between $d(P, F_1)$ and $d(P, F_2)$ is $2a$.

43. Hyperbolas are called **confocal** if they have the same foci.

(a) Show that the hyperbolas

$$\frac{y^2}{k} - \frac{x^2}{16 - k} = 1 \quad \text{with } 0 < k < 16$$

are confocal.

(b) Use a graphing device to draw the top branches of the family of hyperbolas in part (a) for $k = 1, 4, 8,$ and 12. How does the shape of the graph change as k increases?

Applications

44. Navigation In the figure, the LORAN stations at A and B are 500 mi apart, and the ship at P receives station A's signal 2640 microseconds (μs) before it receives the signal from B.

(a) Assuming that radio signals travel at 980 ft/μs, find $d(P, A) - d(P, B)$

(b) Find an equation for the branch of the hyperbola indicated in red in the figure. (Use miles as the unit of distance.)

(c) If A is due north of B, and if P is due east of A, how far is P from A?

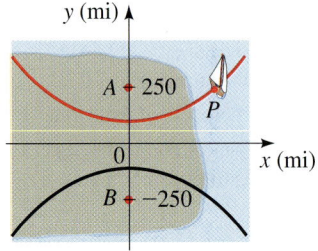

45. Comet Trajectories Some comets, such as Halley's comet, are a permanent part of the solar system, traveling in elliptical orbits around the sun. Others pass through the solar system only once, following a hyperbolic path with the sun at a focus. The figure shows the path of such a comet. Find an equation for the path, assuming that the closest the comet comes to the sun is 2×10^9 mi and that the path the comet was taking before it neared the solar system is at a right angle to the path it continues on after leaving the solar system.

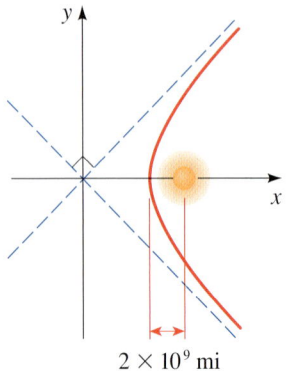

2×10^9 mi

46. Ripples in Pool Two stones are dropped simultaneously in a calm pool of water. The crests of the resulting waves form equally spaced concentric circles, as shown in the figures. The waves interact with each other to create certain interference patterns.

(a) Explain why the red dots lie on an ellipse.

(b) Explain why the blue dots lie on a hyperbola.

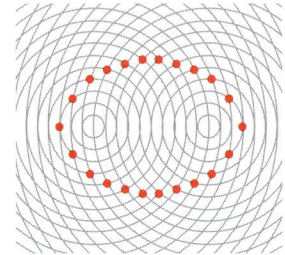
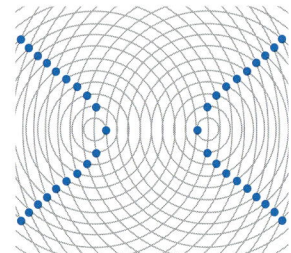

Discovery • Discussion

47. Hyperbolas in the Real World Several examples of the uses of hyperbolas are given in the text. Find other situations in real life where hyperbolas occur. Consult a scientific encyclopedia in the reference section of your library, or search the Internet.

48. Light from a Lamp The light from a lamp forms a lighted area on a wall, as shown in the figure. Why is the boundary of this lighted area a hyperbola? How can one hold a flashlight so that its beam forms a hyperbola on the ground?

DISCOVERY PROJECT

Conics in Architecture

In ancient times architecture was part of mathematics, so architects had to be mathematicians. Many of the structures they built—pyramids, temples, amphitheaters, and irrigation projects—still stand. In modern times architects employ even more sophisticated mathematical principles. The photographs below show some structures that employ conic sections in their design.

Roman Amphitheater in Alexandria, Egypt (circle)

Nik Wheeler/Corbis

Ceiling of Statuary Hall in the U.S. Capitol (ellipse)

Architect of the Capitol

Roof of the Skydome in Toronto, Canada (parabola)

Walter Schmid/Stone/Getty Images

Roof of Washington Dulles Airport (hyperbola and parabola)

Richard T. Nowitz/Corbis

McDonnell Planetarium, St. Louis, MO (hyperbola)

Courtesy of Chamber of Commerce, St. Louis, MO

Attic in La Pedrera, Barcelona, Spain (parabola)

O. Alamany and Vincens/Corbis

Architects have different reasons for using conics in their designs. For example, the Spanish architect Antoni Gaudi used parabolas in the attic of La Pedrera (see photo above). He reasoned that since a rope suspended between two points with an equally distributed load (like in a suspension bridge) has the shape of a parabola, an inverted parabola would provide the best support for a flat roof.

Constructing Conics

The equations of the conics are helpful in manufacturing small objects, because a computer-controlled cutting tool can accurately trace a curve given by an equation. But in a building project, how can we construct a portion of a

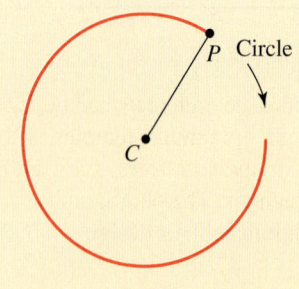

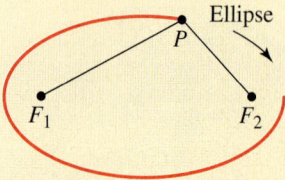

Figure 1

Constructing a circle and an ellipse

parabola, ellipse, or hyperbola that spans the ceiling or walls of a building? The geometric properties of the conics provide practical ways of constructing them. For example, if you were building a circular tower, you would choose a center point, then make sure that the walls of the tower are a fixed distance from that point. Elliptical walls can be constructed using a string anchored at two points, as shown in Figure 1.

To construct a parabola, we can use the apparatus shown in Figure 2. A piece of string of length a is anchored at F and A. The T-square, also of length a, slides along the straight bar L. A pencil at P holds the string taut against the T-square. As the T-square slides to the right the pencil traces out a curve.

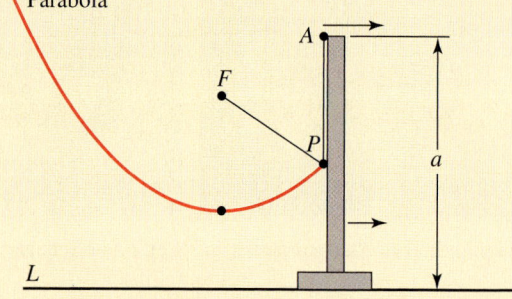

Figure 2

Constructing a parabola

From the figure we see that

$$d(F, P) + d(P, A) = a \qquad \text{The string is of length } a$$

$$d(L, P) + d(P, A) = a \qquad \text{The T-square is of length } a$$

It follows that $d(F, P) + d(P, A) = d(L, P) + d(P, A)$. Subtracting $d(P, A)$ from each side, we get

$$d(F, P) = d(L, P)$$

The last equation says that the distance from F to P is equal to the distance from P to the line L. Thus, the curve is a parabola with focus F and directrix L.

In building projects it's easier to construct a straight line than a curve. So in some buildings, such as in the Kobe Tower (see problem 4), a curved surface is produced by using many straight lines. We can also produce a curve using straight lines, such as the parabola shown in Figure 3.

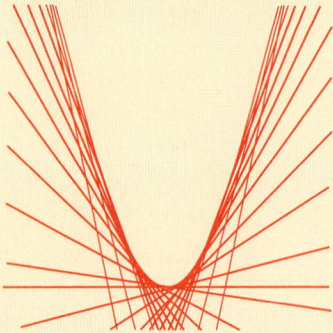

Figure 3

Tangent lines to a parabola

Each line is **tangent** to the parabola; that is, the line meets the parabola at exactly one point and does not cross the parabola. The line tangent to the parabola $y = x^2$ at the point (a, a^2) is

$$y = 2ax - a^2$$

You are asked to show this in problem 6. The parabola is called the **envelope** of all such lines.

1. The photographs on page 771 show six examples of buildings that contain conic sections. Search the Internet to find other examples of structures that employ parabolas, ellipses, or hyperbolas in their design. Find at least one example for each type of conic.

2. In this problem we construct a hyperbola. The wooden bar in the figure can pivot at F_1. A string shorter than the bar is anchored at F_2 and at A, the other end of the bar. A pencil at P holds the string taut against the bar as it moves counterclockwise around F_1.

   (a) Show that the curve traced out by the pencil is one branch of a hyperbola with foci at F_1 and F_2.

   (b) How should the apparatus be reconfigured to draw the other branch of the hyperbola?

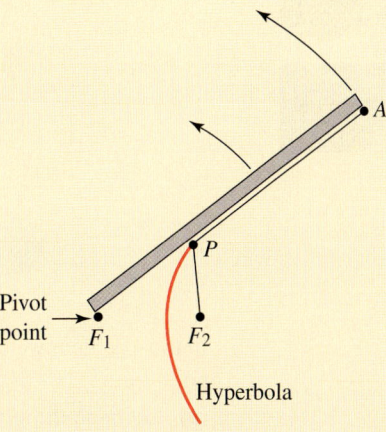

3. The following method can be used to construct a parabola that fits in a given rectangle. The parabola will be approximated by many short line segments.

   First, draw a rectangle. Divide the rectangle in half by a vertical line segment and label the top endpoint V. Next, divide the length and width of each half rectangle into an equal number of parts to form grid lines, as shown in the figure on the next page. Draw lines from V to the endpoints of horizontal grid line 1, and mark the points where these lines cross the vertical grid lines labeled 1. Next, draw lines from V to the endpoints of horizontal grid line 2, and mark the points where these lines cross the vertical grid lines labeled 2. Continue in this way until you have used all the horizontal grid lines.

Now, use line segments to connect the points you have marked to obtain an approximation to the desired parabola. Apply this procedure to draw a parabola that fits into a 6 ft by 10 ft rectangle on a lawn.

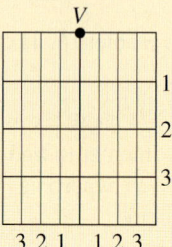 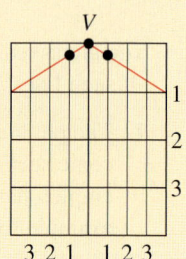 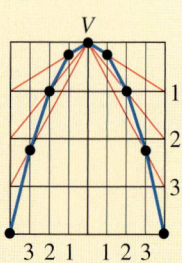

4. In this problem we construct hyperbolic shapes using straight lines. Punch equally spaced holes into the edges of two large plastic lids. Connect corresponding holes with strings of equal lengths as shown in the figure. Holding the strings taut, twist one lid against the other. An imaginary surface passing through the strings has hyperbolic cross sections. (An architectural example of this is the Kobe Tower in Japan shown in the photograph.) What happens to the vertices of the hyperbolic cross sections as the lids are twisted more?

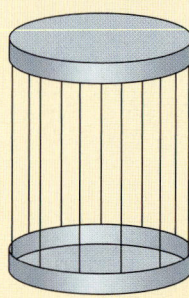 

5. In this problem we show that the line tangent to the parabola $y = x^2$ at the point (a, a^2) has the equation $y = 2ax - a^2$.

(a) Let m be the slope of the tangent line at (a, a^2). Show that the equation of the tangent line is $y - a^2 = m(x - a)$.

(b) Use the fact that the tangent line intersects the parabola at only one point to show that (a, a^2) is the only solution of the system.

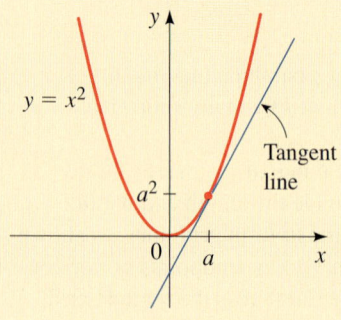

$$\begin{cases} y - a^2 = m(x - a) \\ y = x^2 \end{cases}$$

(c) Eliminate y from the system in part (b) to get a quadratic equation in x. Show that the discriminant of this quadratic is $(m - 2a)^2$. Since the system in (b) has exactly one solution, the discriminant must equal 0. Find m.

(d) Substitute the value for m you found in part (c) into the equation in part (a) and simpify to get the equation of the tangent line.

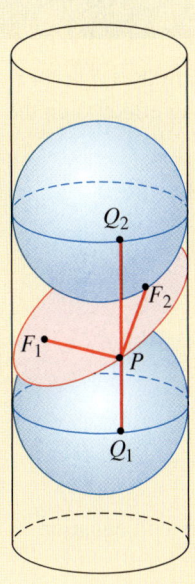

6. In this problem we prove that when a cylinder is cut by a plane an ellipse is formed. An architectural example of this is the Tycho Brahe Planetarium in Copenhagen (see the photograph). In the figure a cylinder is cut by a plane resulting in the red curve. Two spheres with the same radius as the cylinder slide inside the cylinder so that they just touch the plane at F_1 and F_2. Choose an arbitrary point P on the curve and let Q_1 and Q_2 be the two points on the cylinder where a vertical line through P touches the "equator" of each sphere.

(a) Show that $PF_1 = PQ_1$ and $PF_2 = PQ_2$. [*Hint:* Use the fact that all tangents to a sphere from a given point outside the sphere are of the same length.]

(b) Explain why $PQ_1 + PQ_2$ is the same for all points P on the curve.

(c) Show that $PF_1 + PF_2$ is the same for all points P on the curve.

(d) Conclude that the curve is an ellipse with foci F_1 and F_2.

Bob Krist/Corbis

10.4   Shifted Conics

In the preceding sections we studied parabolas with vertices at the origin and ellipses and hyperbolas with centers at the origin. We restricted ourselves to these cases because these equations have the simplest form. In this section we consider conics whose vertices and centers are not necessarily at the origin, and we determine how this affects their equations.

In Section 2.4 we studied transformations of functions that have the effect of shifting their graphs. In general, for any equation in x and y, if we replace x by $x - h$ or by $x + h$, the graph of the new equation is simply the old graph shifted horizontally; if y is replaced by $y - k$ or by $y + k$, the graph is shifted vertically. The following box gives the details.

Shifting Graphs of Equations

If h and k are positive real numbers, then replacing x by $x - h$ or by $x + h$ and replacing y by $y - k$ or by $y + k$ has the following effect(s) on the graph of any equation in x and y.

Replacement	How the graph is shifted
1. x replaced by $x - h$	Right h units
2. x replaced by $x + h$	Left h units
3. y replaced by $y - k$	Upward k units
4. y replaced by $y + k$	Downward k units

Shifted Ellipses

Let's apply horizontal and vertical shifting to the ellipse with equation

$$\frac{x^2}{a^2} + \frac{y^2}{b^2} = 1$$

whose graph is shown in Figure 1. If we shift it so that its center is at the point (h, k) instead of at the origin, then its equation becomes

$$\frac{(x - h)^2}{a^2} + \frac{(y - k)^2}{b^2} = 1$$

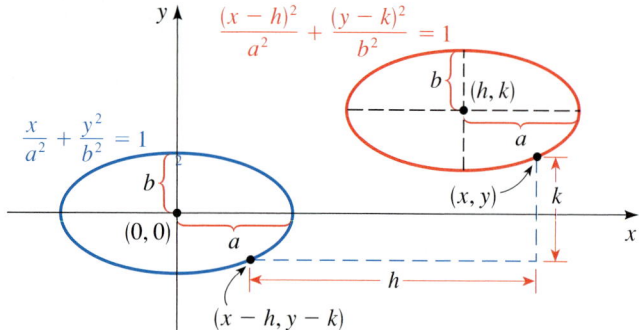

Figure 1
Shifted ellipse

Example 1   Sketching the Graph of a Shifted Ellipse

Sketch the graph of the ellipse

$$\frac{(x + 1)^2}{4} + \frac{(y - 2)^2}{9} = 1$$

and determine the coordinates of the foci.

Solution   The ellipse

$$\frac{(x + 1)^2}{4} + \frac{(y - 2)^2}{9} = 1 \qquad \text{Shifted ellipse}$$

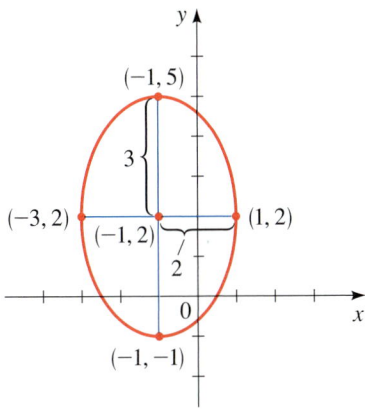

Figure 2

$$\frac{(x+1)^2}{4} + \frac{(y-2)^2}{9} = 1$$

is shifted so that its center is at $(-1, 2)$. It is obtained from the ellipse

$$\frac{x^2}{4} + \frac{y^2}{9} = 1 \qquad \textit{Ellipse with center at origin}$$

by shifting it left 1 unit and upward 2 units. The endpoints of the minor and major axes of the unshifted ellipse are $(2, 0)$, $(-2, 0)$, $(0, 3)$, $(0, -3)$. We apply the required shifts to these points to obtain the corresponding points on the shifted ellipse:

$$(2, 0) \;\rightarrow\; (2 - 1, 0 + 2) = (1, 2)$$

$$(-2, 0) \;\rightarrow\; (-2 - 1, 0 + 2) = (-3, 2)$$

$$(0, 3) \;\rightarrow\; (0 - 1, 3 + 2) = (-1, 5)$$

$$(0, -3) \;\rightarrow\; (0 - 1, -3 + 2) = (-1, -1)$$

This helps us sketch the graph in Figure 2.

To find the foci of the shifted ellipse, we first find the foci of the ellipse with center at the origin. Since $a^2 = 9$ and $b^2 = 4$, we have $c^2 = 9 - 4 = 5$, so $c = \sqrt{5}$. So the foci are $\left(0, \pm\sqrt{5}\right)$. Shifting left 1 unit and upward 2 units, we get

$$\left(0, \sqrt{5}\right) \;\rightarrow\; \left(0 - 1, \sqrt{5} + 2\right) = \left(-1, 2 + \sqrt{5}\right)$$

$$\left(0, -\sqrt{5}\right) \;\rightarrow\; \left(0 - 1, -\sqrt{5} + 2\right) = \left(-1, 2 - \sqrt{5}\right)$$

Thus, the foci of the shifted ellipse are

$$\left(-1, 2 + \sqrt{5}\right) \qquad \text{and} \qquad \left(-1, 2 - \sqrt{5}\right) \qquad\blacksquare$$

Shifted Parabolas

Applying shifts to parabolas leads to the equations and graphs shown in Figure 3.

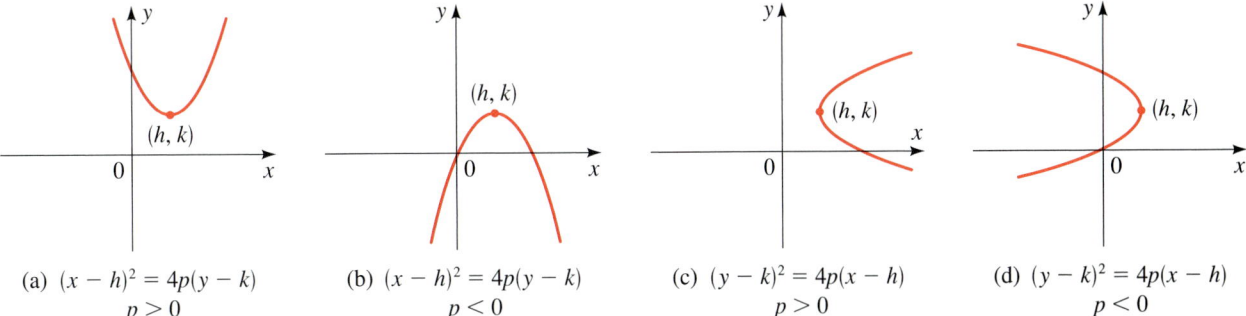

(a) $(x - h)^2 = 4p(y - k)$
$p > 0$

(b) $(x - h)^2 = 4p(y - k)$
$p < 0$

(c) $(y - k)^2 = 4p(x - h)$
$p > 0$

(d) $(y - k)^2 = 4p(x - h)$
$p < 0$

Figure 3

Shifted parabolas

Example 2   **Graphing a Shifted Parabola**

Determine the vertex, focus, and directrix and sketch the graph of the parabola.

$$x^2 - 4x = 8y - 28$$

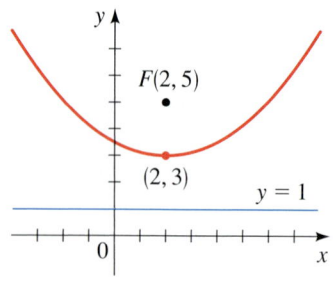

Figure 4

$x^2 - 4x = 8y - 28$

Solution   We complete the square in x to put this equation into one of the forms in Figure 3.

$$x^2 - 4x + 4 = 8y - 28 + 4 \qquad \text{\textit{Add 4 to complete the square}}$$

$$(x - 2)^2 = 8y - 24$$

$$(x - 2)^2 = 8(y - 3) \qquad \text{\textit{Shifted parabola}}$$

This parabola opens upward with vertex at $(2, 3)$. It is obtained from the parabola

$$x^2 = 8y \qquad \text{\textit{Parabola with vertex at origin}}$$

by shifting right 2 units and upward 3 units. Since $4p = 8$, we have $p = 2$, so the focus is 2 units above the vertex and the directrix is 2 units below the vertex. Thus, the focus is $(2, 5)$ and the directrix is $y = 1$. The graph is shown in Figure 4. ∎

Shifted Hyperbolas

Applying shifts to hyperbolas leads to the equations and graphs shown in Figure 5.

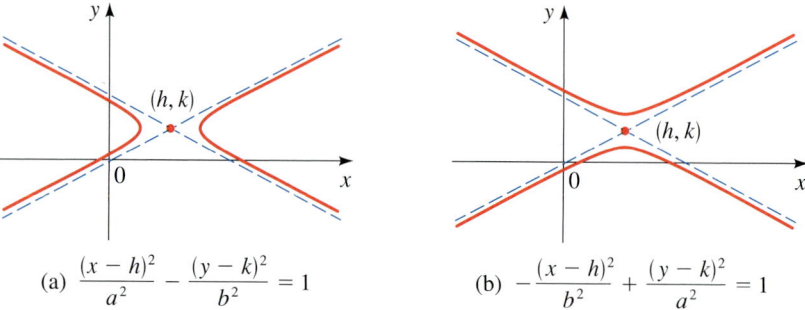

Figure 5
Shifted hyperbolas

(a) $\dfrac{(x - h)^2}{a^2} - \dfrac{(y - k)^2}{b^2} = 1$ \qquad (b) $-\dfrac{(x - h)^2}{b^2} + \dfrac{(y - k)^2}{a^2} = 1$

Example 3   Graphing a Shifted Hyperbola

A shifted conic has the equation

$$9x^2 - 72x - 16y^2 - 32y = 16$$

(a) Complete the square in x and y to show that the equation represents a hyperbola.

(b) Find the center, vertices, foci, and asymptotes of the hyperbola and sketch its graph.

 (c) Draw the graph using a graphing calculator.

Solution

(a) We complete the squares in both x and y:

$$9(x^2 - 8x \quad) - 16(y^2 + 2y \quad) = 16$$

$$9(x^2 - 8x + 16) - 16(y^2 + 2y + 1) = 16 + 9 \cdot 16 - 16 \cdot 1 \quad \text{\textit{Complete the squares}}$$

$$9(x - 4)^2 - 16(y + 1)^2 = 144 \qquad\qquad \text{\textit{Divide this by 144}}$$

$$\frac{(x - 4)^2}{16} - \frac{(y + 1)^2}{9} = 1 \qquad\qquad \text{\textit{Shifted hyperbola}}$$

Comparing this to Figure 5(a), we see that this is the equation of a shifted hyperbola.

(b) The shifted hyperbola has center $(4, -1)$ and a horizontal transverse axis.

$$\text{CENTER} \quad (4, -1)$$

Its graph will have the same shape as the unshifted hyperbola

$$\frac{x^2}{16} - \frac{y^2}{9} = 1 \qquad \textcolor{teal}{\text{Hyperbola with center at origin}}$$

Since $a^2 = 16$ and $b^2 = 9$, we have $a = 4$, $b = 3$, and $c = \sqrt{a^2 + b^2} = \sqrt{16 + 9} = 5$. Thus, the foci lie 5 units to the left and to the right of the center, and the vertices lie 4 units to either side of the center.

$$\text{FOCI} \qquad (-1, -1) \quad \text{and} \quad (9, -1)$$

$$\text{VERTICES} \quad (0, -1) \quad \text{and} \quad (8, -1)$$

The asymptotes of the unshifted hyperbola are $y = \pm\frac{3}{4}x$, so the asymptotes of the shifted hyperbola are found as follows.

$$\text{ASYMPTOTES} \quad y + 1 = \pm\tfrac{3}{4}(x - 4)$$

$$y + 1 = \pm\tfrac{3}{4}x \mp 3$$

$$y = \tfrac{3}{4}x - 4 \qquad \text{and} \qquad y = -\tfrac{3}{4}x + 2$$

To help us sketch the hyperbola, we draw the central box; it extends 4 units left and right from the center and 3 units upward and downward from the center. We then draw the asymptotes and complete the graph of the shifted hyperbola as shown in Figure 6(a).

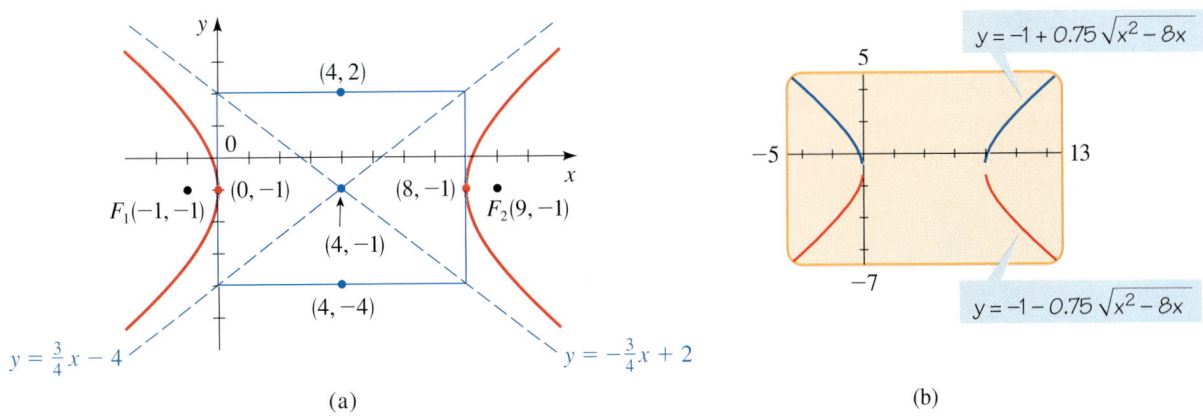

(a)   (b)

Figure 6

$9x^2 - 72x - 16y^2 - 32y = 16$

(c) To draw the graph using a graphing calculator, we need to solve for y. The given equation is a quadratic equation in y, so we use the quadratic formula to solve for y. Writing the equation in the form

$$16y^2 + 32y - 9x^2 + 72x + 16 = 0$$

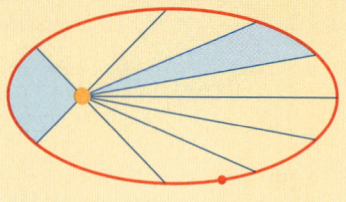

we get

$$y = \frac{-32 \pm \sqrt{32^2 - 4(16)(-9x^2 + 72x + 16)}}{2(16)} \qquad \text{Quadratic formula}$$

$$= \frac{-32 \pm \sqrt{576x^2 - 4608x}}{32} \qquad \text{Expand}$$

$$= \frac{-32 \pm 24\sqrt{x^2 - 8x}}{32} \qquad \text{Factor 576 from under the radical}$$

$$= -1 \pm \tfrac{3}{4}\sqrt{x^2 - 8x} \qquad \text{Simplify}$$

To obtain the graph of the hyperbola, we graph the functions

$$y = -1 + 0.75\sqrt{x^2 - 8x} \qquad \text{and} \qquad y = -1 - 0.75\sqrt{x^2 - 8x}$$

as shown in Figure 6(b). ∎

The General Equation of a Shifted Conic

If we expand and simplify the equations of any of the shifted conics illustrated in Figures 1, 3, and 5, then we will always obtain an equation of the form

$$Ax^2 + Cy^2 + Dx + Ey + F = 0$$

where A and C are not both 0. Conversely, if we begin with an equation of this form, then we can complete the square in x and y to see which type of conic section the equation represents. In some cases, the graph of the equation turns out to be just a pair of lines, a single point, or there may be no graph at all. These cases are called **degenerate conics**. If the equation is not degenerate, then we can tell whether it represents a parabola, an ellipse, or a hyperbola simply by examining the signs of A and C, as described in the following box.

General Equation of a Shifted Conic

The graph of the equation

$$Ax^2 + Cy^2 + Dx + Ey + F = 0$$

where A and C are not both 0, is a conic or a degenerate conic. In the nondegenerate cases, the graph is

1. a parabola if A or C is 0
2. an ellipse if A and C have the same sign (or a circle if $A = C$)
3. a hyperbola if A and C have opposite signs

Example 4   An Equation That Leads to a Degenerate Conic

Sketch the graph of the equation

$$9x^2 - y^2 + 18x + 6y = 0$$

Solution   Because the coefficients of x^2 and y^2 are of opposite sign, this equation looks as if it should represent a hyperbola (like the equation of Example 3). To see whether this is in fact the case, we complete the squares:

$$9(x^2 + 2x \quad\quad) - (y^2 - 6y \quad\quad) = 0 \qquad \text{\color{teal}Group terms and factor 9}$$

$$9(x^2 + 2x + 1) - (y^2 - 6y + 9) = 0 + 9 \cdot 1 - 9 \qquad \text{\color{teal}Complete the square}$$

$$9(x + 1)^2 - (y - 3)^2 = 0 \qquad \text{\color{teal}Factor}$$

$$(x + 1)^2 - \frac{(y - 3)^2}{9} = 0 \qquad \text{\color{teal}Divide by 9}$$

For this to fit the form of the equation of a hyperbola, we would need a nonzero constant to the right of the equal sign. In fact, further analysis shows that this is the equation of a pair of intersecting lines:

$$(y - 3)^2 = 9(x + 1)^2$$

$$y - 3 = \pm 3(x + 1) \qquad \text{\color{teal}Take square roots}$$

$$y = 3(x + 1) + 3 \qquad \text{or} \qquad y = -3(x + 1) + 3$$

$$y = 3x + 6 \qquad\qquad\qquad y = -3x$$

These lines are graphed in Figure 7.   ∎

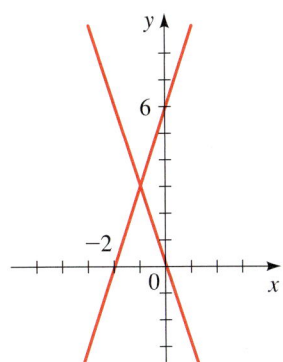

Figure 7
$9x^2 - y^2 + 18x + 6y = 0$

Because the equation in Example 4 looked at first glance like the equation of a hyperbola but, in fact, turned out to represent simply a pair of lines, we refer to its graph as a **degenerate hyperbola**. Degenerate ellipses and parabolas can also arise when we complete the square(s) in an equation that seems to represent a conic. For example, the equation

$$4x^2 + y^2 - 8x + 2y + 6 = 0$$

looks as if it should represent an ellipse, because the coefficients of x^2 and y^2 have the same sign. But completing the squares leads to

$$(x - 1)^2 + \frac{(y + 1)^2}{4} = -\frac{1}{4}$$

which has no solution at all (since the sum of two squares cannot be negative). This equation is therefore degenerate.

10.4   Exercises

1–4 ■ Find the center, foci, and vertices of the ellipse, and determine the lengths of the major and minor axes. Then sketch the graph.

1. $\dfrac{(x - 2)^2}{9} + \dfrac{(y - 1)^2}{4} = 1$ **2.** $\dfrac{(x - 3)^2}{16} + (y + 3)^2 = 1$

3. $\dfrac{x^2}{9} + \dfrac{(y + 5)^2}{25} = 1$ **4.** $\dfrac{(x + 2)^2}{4} + y^2 = 1$

5–8 ■ Find the vertex, focus, and directrix of the parabola, and sketch the graph.

5. $(x - 3)^2 = 8(y + 1)$ **6.** $(y + 5)^2 = -6x + 12$

7. $-4(x + \frac{1}{2})^2 = y$ **8.** $y^2 = 16x - 8$

9–12 ■ Find the center, foci, vertices, and asymptotes of the hyperbola. Then sketch the graph.

9. $\dfrac{(x + 1)^2}{9} - \dfrac{(y - 3)^2}{16} = 1$

10. $(x - 8)^2 - (y + 6)^2 = 1$

11. $y^2 - \dfrac{(x + 1)^2}{4} = 1$

12. $\dfrac{(y - 1)^2}{25} - (x + 3)^2 = 1$

13–18 ■ Find an equation for the conic whose graph is shown.

13.

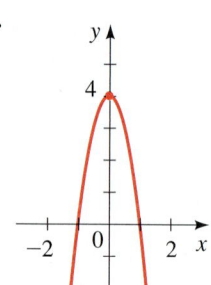

14.

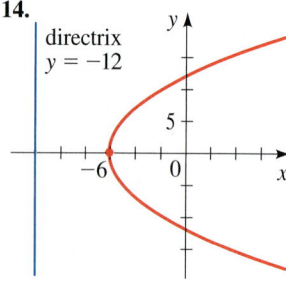

15.

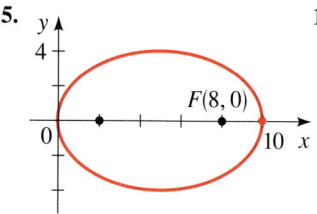

16.

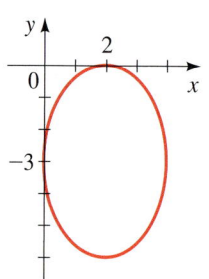

17.

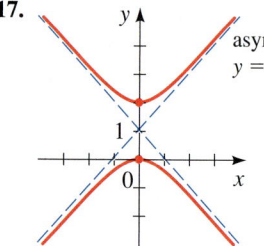

18.

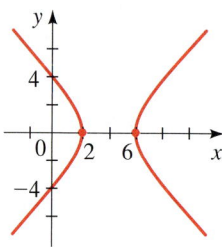

19–30 ■ Complete the square to determine whether the equation represents an ellipse, a parabola, a hyperbola, or a degenerate conic. If the graph is an ellipse, find the center, foci, vertices, and lengths of the major and minor axes. If it is a parabola, find the vertex, focus, and directrix. If it is a hyperbola, find the center, foci, vertices, and asymptotes. Then sketch the graph of the equation. If the equation has no graph, explain why.

19. $9x^2 - 36x + 4y^2 = 0$ **20.** $y^2 = 4(x + 2y)$

21. $x^2 - 4y^2 - 2x + 16y = 20$

22. $x^2 + 6x + 12y + 9 = 0$

23. $4x^2 + 25y^2 - 24x + 250y + 561 = 0$

24. $2x^2 + y^2 = 2y + 1$

25. $16x^2 - 9y^2 - 96x + 288 = 0$

26. $4x^2 - 4x - 8y + 9 = 0$

27. $x^2 + 16 = 4(y^2 + 2x)$ **28.** $x^2 - y^2 = 10(x - y) + 1$

29. $3x^2 + 4y^2 - 6x - 24y + 39 = 0$

30. $x^2 + 4y^2 + 20x - 40y + 300 = 0$

31–34 ■ Use a graphing device to graph the conic.

31. $2x^2 - 4x + y + 5 = 0$

32. $4x^2 + 9y^2 - 36y = 0$

33. $9x^2 + 36 = y^2 + 36x + 6y$

34. $x^2 - 4y^2 + 4x + 8y = 0$

35. Determine what the value of F must be if the graph of the equation

$$4x^2 + y^2 + 4(x - 2y) + F = 0$$

is **(a)** an ellipse, **(b)** a single point, or **(c)** the empty set.

36. Find an equation for the ellipse that shares a vertex and a focus with the parabola $x^2 + y = 100$ and has its other focus at the origin.

37. This exercise deals with **confocal parabolas**, that is, families of parabolas that have the same focus.

(a) Draw graphs of the family of parabolas

$$x^2 = 4p(y + p)$$

for $p = -2, -\frac{3}{2}, -1, -\frac{1}{2}, \frac{1}{2}, 1, \frac{3}{2}, 2$.

(b) Show that each parabola in this family has its focus at the origin.

(c) Describe the effect on the graph of moving the vertex closer to the origin.

Applications

38. Path of a Cannonball   A cannon fires a cannonball as shown in the figure. The path of the cannonball is a parabola with vertex at the highest point of the path. If the cannonball lands 1600 ft from the cannon and the highest point it reaches is 3200 ft above the ground, find an equation for the path of the cannonball. Place the origin at the location of the cannon.

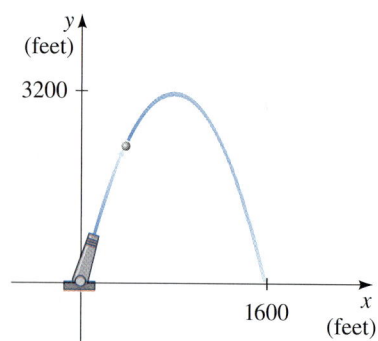

39. Orbit of a Satellite   A satellite is in an elliptical orbit around the earth with the center of the earth at one focus. The height of the satellite above the earth varies between 140 mi and 440 mi. Assume the earth is a sphere with radius

3960 mi. Find an equation for the path of the satellite with the origin at the center of the earth.

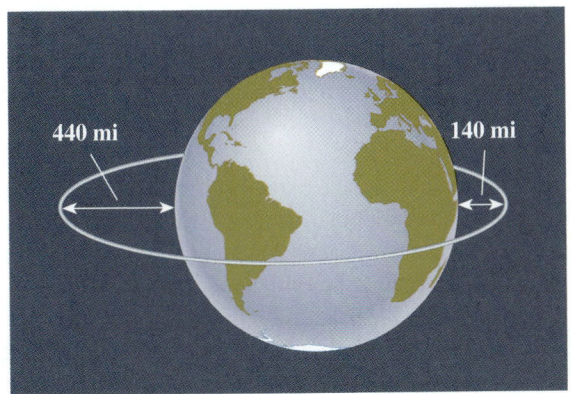

(b) Find equations of two different hyperbolas that have these properties.

(c) Explain why only one parabola satisfies these properties. Find its equation.

(d) Sketch the conics you found in parts (a), (b), and (c) on the same coordinate axes (for the hyperbolas, sketch the top branches only).

(e) How are the ellipses and hyperbolas related to the parabola?

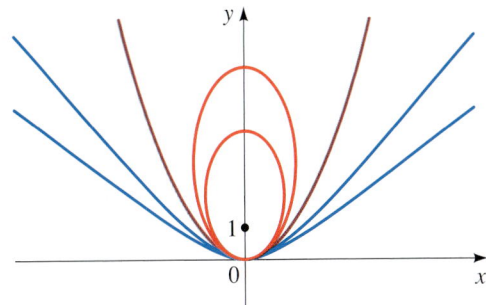

Discovery • Discussion

40. A Family of Confocal Conics   Conics that share a focus are called **confocal**. Consider the family of conics that have a focus at $(0, 1)$ and a vertex at the origin (see the figure).

(a) Find equations of two different ellipses that have these properties.

10.5   **Rotation of Axes**

In Section 10.4 we studied conics with equations of the form

$$Ax^2 + Cy^2 + Dx + Ey + F = 0$$

We saw that the graph is always an ellipse, parabola, or hyperbola with horizontal or vertical axes (except in the degenerate cases). In this section we study the most general second-degree equation

$$Ax^2 + Bxy + Cy^2 + Dx + Ey + F = 0$$

We will see that the graph of an equation of this form is also a conic. In fact, by rotating the coordinate axes through an appropriate angle, we can eliminate the term Bxy and then use our knowledge of conic sections to analyze the graph.

Rotation of Axes

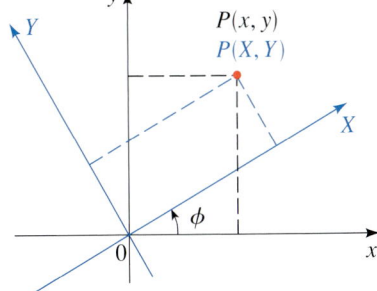

Figure 1

In Figure 1 the x- and y-axes have been rotated through an acute angle ϕ about the origin to produce a new pair of axes, which we call the X- and Y-axes. A point P that has coordinates (x, y) in the old system has coordinates (X, Y) in the new system. If we let r denote the distance of P from the origin and let θ be the angle that the segment OP makes with the new X-axis, then we can see from Figure 2 on the next page (by considering the two right triangles in the figure) that

$$X = r \cos \theta \qquad\qquad Y = r \sin \theta$$

$$x = r \cos(\theta + \phi) \qquad\qquad y = r \sin(\theta + \phi)$$

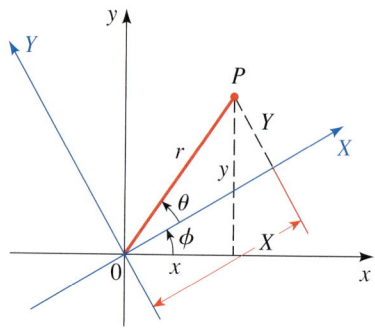

Figure 2

Using the addition formula for cosine, we see that

$$x = r\cos(\theta + \phi)$$
$$= r(\cos\theta\cos\phi - \sin\theta\sin\phi)$$
$$= (r\cos\theta)\cos\phi - (r\sin\theta)\sin\phi$$
$$= X\cos\phi - Y\sin\phi$$

Similarly, we can apply the addition formula for sine to the expression for y to obtain $y = X\sin\phi + Y\cos\phi$. By treating these equations for x and y as a system of linear equations in the variables X and Y (see Exercise 33), we obtain expressions for X and Y in terms of x and y, as detailed in the following box.

Rotation of Axes Formulas

Suppose the x- and y-axes in a coordinate plane are rotated through the acute angle ϕ to produce the X- and Y-axes, as shown in Figure 1. Then the coordinates (x, y) and (X, Y) of a point in the xy- and the XY-planes are related as follows:

$$x = X\cos\phi - Y\sin\phi \qquad X = x\cos\phi + y\sin\phi$$
$$y = X\sin\phi + Y\cos\phi \qquad Y = -x\sin\phi + y\cos\phi$$

Example 1   Rotation of Axes

If the coordinate axes are rotated through $30°$, find the XY-coordinates of the point with xy-coordinates $(2, -4)$.

Solution   Using the Rotation of Axes Formulas with $x = 2$, $y = -4$, and $\phi = 30°$, we get

$$X = 2\cos 30° + (-4)\sin 30° = 2\left(\frac{\sqrt{3}}{2}\right) - 4\left(\frac{1}{2}\right) = \sqrt{3} - 2$$

$$Y = -2\sin 30° + (-4)\cos 30° = -2\left(\frac{1}{2}\right) - 4\left(\frac{\sqrt{3}}{2}\right) = -1 - 2\sqrt{3}$$

The XY-coordinates are $(-2 + \sqrt{3}, -1 - 2\sqrt{3})$.   ∎

Example 2   Rotating a Hyperbola

Rotate the coordinate axes through $45°$ to show that the graph of the equation $xy = 2$ is a hyperbola.

Solution   We use the Rotation of Axes Formulas with $\phi = 45°$ to obtain

$$x = X\cos 45° - Y\sin 45° = \frac{X}{\sqrt{2}} - \frac{Y}{\sqrt{2}}$$

$$y = X\sin 45° + Y\cos 45° = \frac{X}{\sqrt{2}} + \frac{Y}{\sqrt{2}}$$

Substituting these expressions into the original equation gives

$$\left(\frac{X}{\sqrt{2}} - \frac{Y}{\sqrt{2}}\right)\left(\frac{X}{\sqrt{2}} + \frac{Y}{\sqrt{2}}\right) = 2$$

$$\frac{X^2}{2} - \frac{Y^2}{2} = 2$$

$$\frac{X^2}{4} - \frac{Y^2}{4} = 1$$

We recognize this as a hyperbola with vertices $(\pm 2, 0)$ in the XY-coordinate system. Its asymptotes are $Y = \pm X$, which correspond to the coordinate axes in the xy-system (see Figure 3).

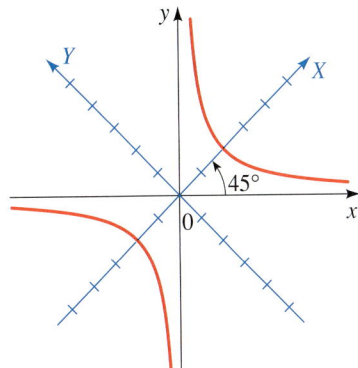

Figure 3

$xy = 2$

General Equation of a Conic

The method of Example 2 can be used to transform any equation of the form

$$Ax^2 + Bxy + Cy^2 + Dx + Ey + F = 0$$

into an equation in X and Y that doesn't contain an XY-term by choosing an appropriate angle of rotation. To find the angle that works, we rotate the axes through an angle ϕ and substitute for x and y using the Rotation of Axes Formulas:

$$A(X \cos \phi - Y \sin \phi)^2 + B(X \cos \phi - Y \sin \phi)(X \sin \phi + Y \cos \phi)$$
$$+ C(X \sin \phi + Y \cos \phi)^2 + D(X \cos \phi - Y \sin \phi)$$
$$+ E(X \sin \phi + Y \cos \phi) + F = 0$$

If we expand this and collect like terms, we obtain an equation of the form

$$A'X^2 + B'XY + C'Y^2 + D'X + E'Y + F' = 0$$

where

$$A' = A \cos^2\phi + B \sin \phi \cos \phi + C \sin^2\phi$$
$$B' = 2(C - A) \sin \phi \cos \phi + B(\cos^2\phi - \sin^2\phi)$$
$$C' = A \sin^2\phi - B \sin \phi \cos \phi + C \cos^2\phi$$

$$D' = D \cos \phi + E \sin \phi$$

$$E' = -D \sin \phi + E \cos \phi$$

$$F' = F$$

To eliminate the XY-term, we would like to choose ϕ so that $B' = 0$, that is,

Double-angle formulas

$\sin 2\phi = 2 \sin \phi \cos \phi$

$\cos 2\phi = \cos^2\phi - \sin^2\phi$

$$2(C - A) \sin \phi \cos \phi + B(\cos^2\phi - \sin^2\phi) = 0 \qquad \textit{Double-angle}$$

$$(C - A) \sin 2\phi + B \cos 2\phi = 0 \qquad \textit{formulas for sine and cosine}$$

$$B \cos 2\phi = (A - C) \sin 2\phi$$

$$\cot 2\phi = \frac{A - C}{B} \qquad \textit{Divide by } B \sin 2\phi$$

The preceding calculation proves the following theorem.

Simplifying the General Conic Equation

To eliminate the xy-term in the general conic equation

$$Ax^2 + Bxy + Cy^2 + Dx + Ey + F = 0$$

rotate the axes through the acute angle ϕ that satisfies

$$\cot 2\phi = \frac{A - C}{B}$$

Example 3   Eliminating the xy-Term

Use a rotation of axes to eliminate the xy-term in the equation

$$6\sqrt{3}x^2 + 6xy + 4\sqrt{3}y^2 = 21\sqrt{3}$$

Identify and sketch the curve.

Solution   To eliminate the xy-term, we rotate the axes through an angle ϕ that satisfies

$$\cot 2\phi = \frac{A - C}{B} = \frac{6\sqrt{3} - 4\sqrt{3}}{6} = \frac{\sqrt{3}}{3}$$

Thus, $2\phi = 60°$ and hence $\phi = 30°$. With this value of ϕ, we get

$$x = X\left(\frac{\sqrt{3}}{2}\right) - Y\left(\frac{1}{2}\right) \qquad \textit{Rotation of Axes Formulas}$$

$$y = X\left(\frac{1}{2}\right) + Y\left(\frac{\sqrt{3}}{2}\right) \qquad \cos\phi = \frac{\sqrt{3}}{2}, \sin\phi = \frac{1}{2}$$

Substituting these values for x and y into the given equation leads to

$$6\sqrt{3}\left(\frac{X\sqrt{3}}{2} - \frac{Y}{2}\right)^2 + 6\left(\frac{X\sqrt{3}}{2} - \frac{Y}{2}\right)\left(\frac{X}{2} + \frac{Y\sqrt{3}}{2}\right) + 4\sqrt{3}\left(\frac{X}{2} + \frac{Y\sqrt{3}}{2}\right)^2 = 21\sqrt{3}$$

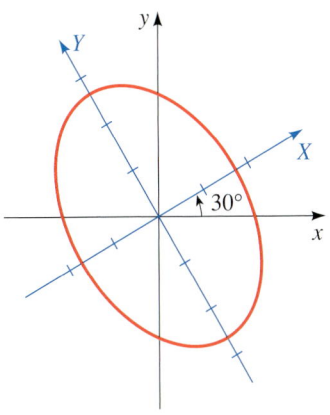

Figure 4
$6\sqrt{3}x^2 + 6xy + 4\sqrt{3}y^2 = 21\sqrt{3}$

Expanding and collecting like terms, we get

$$7\sqrt{3}X^2 + 3\sqrt{3}Y^2 = 21\sqrt{3}$$

$$\frac{X^2}{3} + \frac{Y^2}{7} = 1 \qquad \text{Divide by } 21\sqrt{3}$$

This is the equation of an ellipse in the XY-coordinate system. The foci lie on the Y-axis. Because $a^2 = 7$ and $b^2 = 3$, the length of the major axis is $2\sqrt{7}$, and the length of the minor axis is $2\sqrt{3}$. The ellipse is sketched in Figure 4. ∎

In the preceding example we were able to determine ϕ without difficulty, since we remembered that $\cot 60° = \sqrt{3}/3$. In general, finding ϕ is not quite so easy. The next example illustrates how the following half-angle formulas, which are valid for $0 < \phi < \pi/2$, are useful in determining ϕ (see Section 7.3):

$$\cos\phi = \sqrt{\frac{1 + \cos 2\phi}{2}} \qquad \sin\phi = \sqrt{\frac{1 - \cos 2\phi}{2}}$$

Example 4   Graphing a Rotated Conic

A conic has the equation

$$64x^2 + 96xy + 36y^2 - 15x + 20y - 25 = 0$$

(a) Use a rotation of axes to eliminate the xy-term.

(b) Identify and sketch the graph.

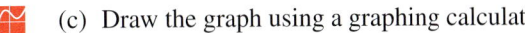

(c) Draw the graph using a graphing calculator.

Solution

(a) To eliminate the xy-term, we rotate the axes through an angle ϕ that satisfies

$$\cot 2\phi = \frac{A - C}{B} = \frac{64 - 36}{96} = \frac{7}{24}$$

In Figure 5 we sketch a triangle with $\cot 2\phi = \frac{7}{24}$. We see that

$$\cos 2\phi = \frac{7}{25}$$

so, using the half-angle formulas, we get

$$\cos\phi = \sqrt{\frac{1 + \frac{7}{25}}{2}} = \sqrt{\frac{16}{25}} = \frac{4}{5}$$

$$\sin\phi = \sqrt{\frac{1 - \frac{7}{25}}{2}} = \sqrt{\frac{9}{25}} = \frac{3}{5}$$

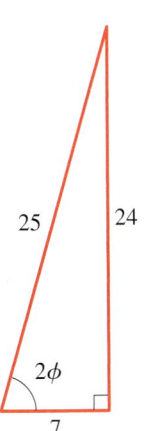

Figure 5

The Rotation of Axes Formulas then give

$$x = \tfrac{4}{5}X - \tfrac{3}{5}Y \qquad \text{and} \qquad y = \tfrac{3}{5}X + \tfrac{4}{5}Y$$

Substituting into the given equation, we have

$$64\left(\tfrac{4}{5}X - \tfrac{3}{5}Y\right)^2 + 96\left(\tfrac{4}{5}X - \tfrac{3}{5}Y\right)\left(\tfrac{3}{5}X + \tfrac{4}{5}Y\right)$$
$$+ 36\left(\tfrac{3}{5}X + \tfrac{4}{5}Y\right)^2 - 15\left(\tfrac{4}{5}X - \tfrac{3}{5}Y\right) + 20\left(\tfrac{3}{5}X + \tfrac{4}{5}Y\right) - 25 = 0$$

Expanding and collecting like terms, we get

$$100X^2 + 25Y - 25 = 0$$

$$-4X^2 = Y - 1 \qquad \text{Simplify}$$

$$X^2 = -\tfrac{1}{4}(Y - 1) \qquad \text{Divide by 4}$$

(b) We recognize this as the equation of a parabola that opens along the negative Y-axis and has vertex $(0, 1)$ in XY-coordinates. Since $4p = -\tfrac{1}{4}$, we have $p = -\tfrac{1}{16}$, so the focus is $\left(0, \tfrac{15}{16}\right)$ and the directrix is $Y = \tfrac{17}{16}$. Using

$$\phi = \cos^{-1}\tfrac{4}{5} \approx 37°$$

we sketch the graph in Figure 6(a).

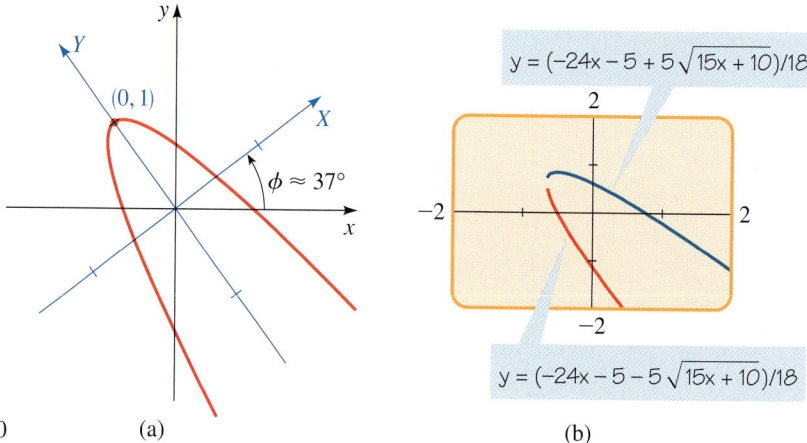

Figure 6

$64x^2 + 96xy + 36y^2 - 15x + 20y - 25 = 0$

(a)

(b)

(c) To draw the graph using a graphing calculator, we need to solve for y. The given equation is a quadratic equation in y, so we can use the quadratic formula to solve for y. Writing the equation in the form

$$36y^2 + (96x + 20)y + (64x^2 - 15x - 25) = 0$$

we get

$$y = \frac{-(96x + 20) \pm \sqrt{(96x + 20)^2 - 4(36)(64x^2 - 15x - 25)}}{2(36)} \qquad \text{Quadratic formula}$$

$$= \frac{-(96x + 20) \pm \sqrt{6000x + 4000}}{72} \qquad \text{Expand}$$

$$= \frac{-96x - 20 \pm 20\sqrt{15x + 10}}{72} \qquad \text{Simplify}$$

$$= \frac{-24x - 5 \pm 5\sqrt{15x + 10}}{18} \qquad \text{Simplify}$$

To obtain the graph of the parabola, we graph the functions

$$y = \left(-24x - 5 + 5\sqrt{15x + 10}\right)/18 \quad \text{and} \quad y = \left(-24x - 5 - 5\sqrt{15x + 10}\right)/18$$

as shown in Figure 6(b). ■

The Discriminant

In Examples 3 and 4 we were able to identify the type of conic by rotating the axes. The next theorem gives rules for identifying the type of conic directly from the equation, without rotating axes.

Identifying Conics by the Discriminant

The graph of the equation

$$Ax^2 + Bxy + Cy^2 + Dx + Ey + F = 0$$

is either a conic or a degenerate conic. In the nondegenerate cases, the graph is

1. a parabola if $B^2 - 4AC = 0$

2. an ellipse if $B^2 - 4AC < 0$

3. a hyperbola if $B^2 - 4AC > 0$

The quantity $B^2 - 4AC$ is called the **discriminant** of the equation.

■ **Proof**   If we rotate the axes through an angle ϕ, we get an equation of the form

$$A'X^2 + B'XY + C'Y^2 + D'X + E'Y + F' = 0$$

where A', B', C', ... are given by the formulas on pages 785–786. A straightforward calculation shows that

$$(B')^2 - 4A'C' = B^2 - 4AC$$

Thus, the expression $B^2 - 4AC$ remains unchanged for any rotation. In particular, if we choose a rotation that eliminates the xy-term $(B' = 0)$, we get

$$A'X^2 + C'Y^2 + D'X + E'Y + F' = 0$$

In this case, $B^2 - 4AC = -4A'C'$. So $B^2 - 4AC = 0$ if either A' or C' is zero; $B^2 - 4AC < 0$ if A' and C' have the same sign; and $B^2 - 4AC > 0$ if A' and C' have opposite signs. According to the box on page 780, these cases correspond to the graph of the last displayed equation being a parabola, an ellipse, or a hyperbola, respectively. ■

In the proof we indicated that the discriminant is unchanged by any rotation; for this reason, the discriminant is said to be **invariant** under rotation.

Example 5   Identifying a Conic by the Discriminant

A conic has the equation

$$3x^2 + 5xy - 2y^2 + x - y + 4 = 0$$

(a) Use the discriminant to identify the conic.

(b) Confirm your answer to part (a) by graphing the conic with a graphing calculator.

Solution

(a) Since $A = 3$, $B = 5$, and $C = -2$, the discriminant is

$$B^2 - 4AC = 5^2 - 4(3)(-2) = 49 > 0$$

So the conic is a hyperbola.

(b) Using the quadratic formula, we solve for y to get

$$y = \frac{5x - 1 \pm \sqrt{49x^2 - 2x + 33}}{4}$$

We graph these functions in Figure 7. The graph confirms that this is a hyperbola.

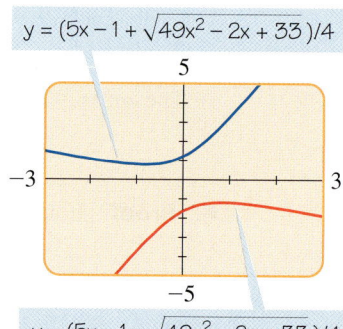

$y = (5x - 1 + \sqrt{49x^2 - 2x + 33})/4$

$y = (5x - 1 - \sqrt{49x^2 - 2x + 33})/4$

Figure 7

10.5   Exercises

1–6 ■ Determine the XY-coordinates of the given point if the coordinate axes are rotated through the indicated angle.

1. $(1, 1)$, $\phi = 45°$

2. $(-2, 1)$, $\phi = 30°$

3. $(3, -\sqrt{3})$, $\phi = 60°$

4. $(2, 0)$, $\phi = 15°$

5. $(0, 2)$, $\phi = 55°$

6. $(\sqrt{2}, 4\sqrt{2})$, $\phi = 45°$

7–12 ■ Determine the equation of the given conic in XY-coordinates when the coordinate axes are rotated through the indicated angle.

7. $x^2 - 3y^2 = 4$, $\phi = 60°$

8. $y = (x - 1)^2$, $\phi = 45°$

9. $x^2 - y^2 = 2y$, $\phi = \cos^{-1}\frac{3}{5}$

10. $x^2 + 2y^2 = 16$, $\phi = \sin^{-1}\frac{3}{5}$

11. $x^2 + 2\sqrt{3}xy - y^2 = 4$, $\phi = 30°$

12. $xy = x + y$, $\phi = \pi/4$

13–26 ■ **(a)** Use the discriminant to determine whether the graph of the equation is a parabola, an ellipse, or a hyperbola. **(b)** Use a rotation of axes to eliminate the xy-term. **(c)** Sketch the graph.

13. $xy = 8$

14. $xy + 4 = 0$

15. $x^2 + 2xy + y^2 + x - y = 0$

16. $13x^2 + 6\sqrt{3}xy + 7y^2 = 16$

17. $x^2 + 2\sqrt{3}xy - y^2 + 2 = 0$

18. $21x^2 + 10\sqrt{3}xy + 31y^2 = 144$

19. $11x^2 - 24xy + 4y^2 + 20 = 0$

20. $25x^2 - 120xy + 144y^2 - 156x - 65y = 0$

21. $\sqrt{3}x^2 + 3xy = 3$

22. $153x^2 + 192xy + 97y^2 = 225$

23. $2\sqrt{3}x^2 - 6xy + \sqrt{3}x + 3y = 0$

24. $9x^2 - 24xy + 16y^2 = 100(x - y - 1)$

25. $52x^2 + 72xy + 73y^2 = 40x - 30y + 75$

26. $(7x + 24y)^2 = 600x - 175y + 25$

27–30 ■ **(a)** Use the discriminant to identify the conic. **(b)** Confirm your answer by graphing the conic using a graphing device.

27. $2x^2 - 4xy + 2y^2 - 5x - 5 = 0$

28. $x^2 - 2xy + 3y^2 = 8$

29. $6x^2 + 10xy + 3y^2 - 6y = 36$

30. $9x^2 - 6xy + y^2 + 6x - 2y = 0$

31. (a) Use rotation of axes to show that the following equation represents a hyperbola:

$$7x^2 + 48xy - 7y^2 - 200x - 150y + 600 = 0$$

(b) Find the XY- and xy-coordinates of the center, vertices, and foci.

(c) Find the equations of the asymptotes in XY- and xy-coordinates.

32. (a) Use rotation of axes to show that the following equation represents a parabola:

$$2\sqrt{2}(x + y)^2 = 7x + 9y$$

(b) Find the XY- and xy-coordinates of the vertex and focus.

(c) Find the equation of the directrix in XY- and xy-coordinates.

33. Solve the equations:

$$x = X \cos \phi - Y \sin \phi$$
$$y = X \sin \phi + Y \cos \phi$$

for X and Y in terms of x and y. [*Hint:* To begin, multiply the first equation by $\cos \phi$ and the second by $\sin \phi$, and then add the two equations to solve for X.]

34. Show that the graph of the equation

$$\sqrt{x} + \sqrt{y} = 1$$

is part of a parabola by rotating the axes through an angle of $45°$. [*Hint:* First convert the equation to one that does not involve radicals.]

Discovery • Discussion

35. Matrix Form of Rotation of Axes Formulas
Let Z, Z', and R be the matrices

$$Z = \begin{bmatrix} x \\ y \end{bmatrix} \qquad Z' = \begin{bmatrix} X \\ Y \end{bmatrix}$$

$$R = \begin{bmatrix} \cos \phi & -\sin \phi \\ \sin \phi & \cos \phi \end{bmatrix}$$

Show that the Rotation of Axes Formulas can be written as

$$Z = RZ' \qquad \text{and} \qquad Z' = R^{-1}Z$$

36. Algebraic Invariants   A quantity is invariant under rotation if it does not change when the axes are rotated. It was stated in the text that for the general equation of a conic, the quantity $B^2 - 4AC$ is invariant under rotation.

(a) Use the formulas for A', B', and C' on page 785 to prove that the quantity $B^2 - 4AC$ is invariant under rotation; that is, show that

$$B^2 - 4AC = B'^2 - 4A'C'$$

(b) Prove that $A + C$ is invariant under rotation.

(c) Is the quantity F invariant under rotation?

37. Geometric Invariants   Do you expect that the distance between two points is invariant under rotation? Prove your answer by comparing the distance $d(P, Q)$ and $d(P', Q')$ where P' and Q' are the images of P and Q under a rotation of axes.

Computer Graphics II

In the *Discovery Project* on page 700 we saw how matrix multiplication is used in computer graphics. We found matrices that reflect, expand, or shear an image. We now consider matrices that rotate an image, as in the graphics shown here.

Rotating Points in the Plane

Recall that a point (x, y) in the plane is represented by the 2×1 matrix $\begin{bmatrix} x \\ y \end{bmatrix}$. The matrix that rotates this point about the origin through an angle ϕ is

$$R = \begin{bmatrix} \cos \phi & -\sin \phi \\ \sin \phi & \cos \phi \end{bmatrix} \qquad \text{Rotation matrix}$$

Compare this matrix with the rotation of axes matrix in Exercise 35, Section 10.5. Note that rotating a point counterclockwise corresponds to rotating the axes clockwise.

When the point $P = \begin{bmatrix} x \\ y \end{bmatrix}$ is rotated clockwise about the origin through an angle ϕ, it moves to a new location $P' = \begin{bmatrix} x' \\ y' \end{bmatrix}$ given by the matrix product $P' = RP$, as shown in Figure 1.

$$P' = RP = \begin{bmatrix} \cos \phi & -\sin \phi \\ \sin \phi & \cos \phi \end{bmatrix} \begin{bmatrix} x \\ y \end{bmatrix} = \begin{bmatrix} x \cos \phi - y \sin \phi \\ x \sin \phi + y \cos \phi \end{bmatrix}$$

For example, if $\phi = 90°$, the rotation matrix is

$$R = \begin{bmatrix} \cos 90° & -\sin 90° \\ \sin 90° & \cos 90° \end{bmatrix} = \begin{bmatrix} 0 & -1 \\ 1 & 0 \end{bmatrix} \qquad \text{Rotation matrix } (\phi = 90°)$$

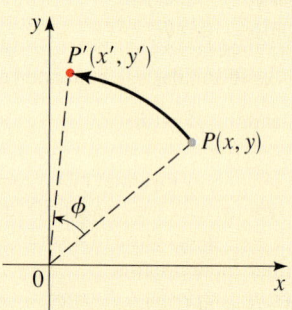

Figure 1

Applying a 90° rotation to the point $P = \begin{bmatrix} 1 \\ 2 \end{bmatrix}$ moves it to the point

$$P' = RP = \begin{bmatrix} 0 & -1 \\ 1 & 0 \end{bmatrix} \begin{bmatrix} 1 \\ 2 \end{bmatrix} = \begin{bmatrix} -2 \\ 1 \end{bmatrix}$$

See Figure 2.

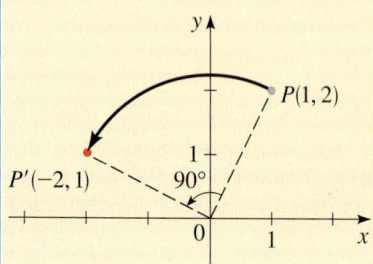

Figure 2

Rotating Images in the Plane

If the rotation matrix is applied to every point in an image, then the entire image is rotated. To rotate the house in Figure 3(a) through a 30° angle about the origin, we multiply its data matrix (described on page 701) by the rotation matrix that has $\phi = 30°$.

$$RD = \begin{bmatrix} \frac{\sqrt{3}}{2} & -\frac{1}{2} \\ \frac{1}{2} & \frac{\sqrt{3}}{2} \end{bmatrix} \begin{bmatrix} 2 & 0 & 0 & 2 & 4 & 4 & 3 & 3 & 2 & 2 & 3 \\ 0 & 0 & 3 & 5 & 3 & 0 & 0 & 2 & 2 & 0 & 0 \end{bmatrix}$$

$$\approx \begin{bmatrix} 1.73 & 0 & -1.50 & -0.77 & 1.96 & 3.46 & 2.60 & 1.60 & 0.73 & 1.73 & 2.60 \\ 1 & 0 & 2.60 & 5.33 & 4.60 & 2 & 1.50 & 3.23 & 2.73 & 1 & 1.50 \end{bmatrix}$$

The new data matrix RD represents the rotated house in Figure 3(b).

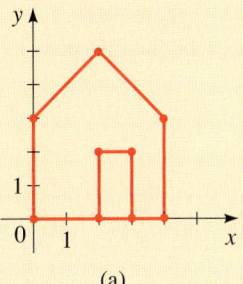

Figure 3   (a)   (b)

The *Discovery Project* on page 702 describes a TI-83 program that draws the image corresponding to a given data matrix. You may find it convenient to use this program in some of the following activities.

1. Use a rotation matrix to find the new coordinates of the given point when it is rotated through the given angle.

(a) $(1, 4)$, $\phi = 90°$ (b) $(-2, 1)$, $\phi = 60°$

(c) $(-2, -2)$, $\phi = 135°$ (d) $(7, 3)$, $\phi = -60°$

2. Find a data matrix for the line drawing in the figure shown in the margin. Multiply the data matrix by a suitable rotation matrix to rotate the image about the origin by $\phi = 120°$. Sketch the rotated image given by the new data matrix.

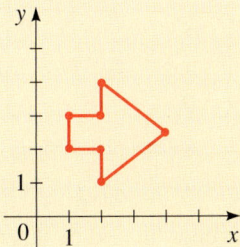

3. Sketch the image represented by the data matrix D.

$$D = \begin{bmatrix} 2 & 3 & 3 & 4 & 4 & 1 & 1 & 2 & 2 \\ 1 & 1 & 3 & 3 & 4 & 4 & 3 & 3 & 1 \end{bmatrix}$$

Find the rotation matrix R that corresponds to a 45° rotation, and the transformation matrix T that corresponds to an expansion by a factor of 2 in the x-direction (see page 701). How does multiplying the data matrix by RT change the image? How about multiplying by TR? Calculate the products RTD and TRD, and sketch the corresponding images to confirm your answers.

4. Let R be the rotation matrix for the angle ϕ. Show that R^{-1} is the rotation matrix for the angle $-\phi$.

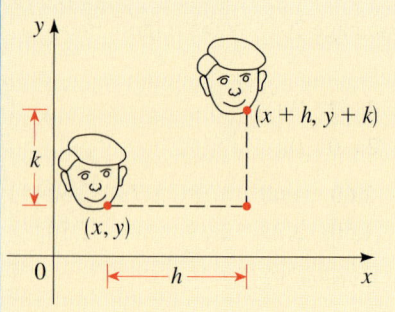

5. To **translate** an image by (h, k), we add h to each x-coordinate and k to each y-coordinate of each point in the image (see the figure in the margin). This can be done by adding an appropriate matrix M to D, but the dimension of M would change depending on the dimension of D. In practice, translation is accomplished by matrix multiplication. To see how this is done, we introduce **homogeneous coordinates**; that is, we represent the point (x, y) by a 3×1 matrix:

$$(x, y) \ \leftrightarrow \ \begin{bmatrix} x \\ y \\ 1 \end{bmatrix}$$

(a) Let T be the matrix

$$T = \begin{bmatrix} 1 & 0 & h \\ 0 & 1 & k \\ 0 & 0 & 1 \end{bmatrix}$$

Show that T translates the point (x, y) to the point $(x + h, y + h)$ by verifying the following matrix multiplication.

$$\begin{bmatrix} 1 & 0 & h \\ 0 & 1 & k \\ 0 & 0 & 1 \end{bmatrix} \begin{bmatrix} x \\ y \\ 1 \end{bmatrix} = \begin{bmatrix} x + h \\ y + k \\ 1 \end{bmatrix}$$

(b) Find T^{-1} and describe how T^{-1} translates points.

(c) Verify that multiplying by the following matrices has the indicated effects on a point (x, y) represented by its homogeneous coordinates $\begin{bmatrix} x \\ y \\ 1 \end{bmatrix}$.

$$\begin{bmatrix} 1 & 0 & 0 \\ 0 & -1 & 0 \\ 0 & 0 & 1 \end{bmatrix} \quad \begin{bmatrix} c & 0 & 0 \\ 0 & 1 & 0 \\ 0 & 0 & 1 \end{bmatrix} \quad \begin{bmatrix} 1 & c & 0 \\ 0 & 1 & 0 \\ 0 & 0 & 1 \end{bmatrix} \quad \begin{bmatrix} \cos \phi & -\sin \phi & 0 \\ \sin \phi & \cos \phi & 0 \\ 0 & 0 & 1 \end{bmatrix}$$

Reflection in x-axis   Expansion (or contraction) in x-direction   Shear in x-direction   Rotation about the origin by the angle ϕ

(d) Sketch the image represented (in homogeneous coordinates) by this data matrix:

$$D = \begin{bmatrix} 3 & 5 & 5 & 7 & 7 & 9 & 9 & 7 & 7 & 5 & 5 & 3 & 3 \\ 7 & 7 & 5 & 5 & 7 & 7 & 9 & 9 & 11 & 11 & 9 & 9 & 7 \\ 1 & 1 & 1 & 1 & 1 & 1 & 1 & 1 & 1 & 1 & 1 & 1 & 1 \end{bmatrix}$$

Find a matrix T that translates the image by $(-6, -8)$ and a matrix R that rotates the image by $45°$. Sketch the images represented by the data matrices TD, RTD, and $T^{-1}RTD$. Describe how an image is changed when its data matrix is multiplied by T, by RT, and by $T^{-1}RT$.

| 10.6 | **Polar Equations of Conics** |

Earlier in this chapter we defined a parabola in terms of a focus and directrix, but we defined the ellipse and hyperbola in terms of two foci. In this section we give a more unified treatment of all three types of conics in terms of a focus and directrix. If we place the focus at the origin, then a conic section has a simple polar equation. Moreover, in polar form, rotation of conics becomes a simple matter. Polar equations of ellipses are crucial in the derivation of Kepler's Laws (see page 780).

Equivalent Description of Conics

Let F be a fixed point (the **focus**), ℓ a fixed line (the **directrix**), and e a fixed positive number (the **eccentricity**). The set of all points P such that the ratio of the distance from P to F to the distance from P to ℓ is the constant e is a conic. That is, the set of all points P such that

$$\frac{d(P, F)}{d(P, \ell)} = e$$

is a conic. The conic is a parabola if $e = 1$, an ellipse if $e < 1$, or a hyperbola if $e > 1$.

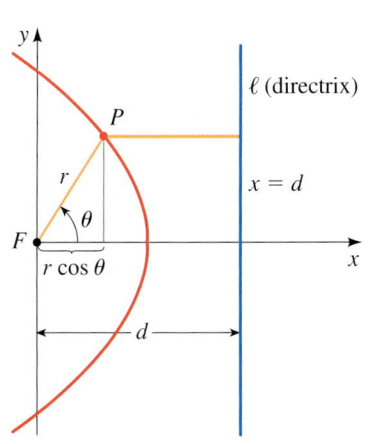

Figure 1

■ **Proof**   If $e = 1$, then $d(P, F) = d(P, \ell)$, and so the given condition becomes the definition of a parabola as given in Section 10.1.

Now, suppose $e \neq 1$. Let's place the focus F at the origin and the directrix parallel to the y-axis and d units to the right. In this case the directrix has equation $x = d$ and is perpendicular to the polar axis. If the point P has polar coordinates (r, θ), we see from Figure 1 that $d(P, F) = r$ and $d(P, \ell) = d - r \cos \theta$. Thus, the condition $d(P, F)/d(P, \ell) = e$, or $d(P, F) = e \cdot d(P, \ell)$, becomes

$$r = e(d - r \cos \theta)$$

If we square both sides of this polar equation and convert to rectangular coordinates, we get

$$x^2 + y^2 = e^2(d - x)^2$$

$$(1 - e^2)x^2 + 2de^2x + y^2 = e^2d^2 \qquad \text{Expand and simplify}$$

$$\left(x + \frac{e^2d}{1 - e^2}\right)^2 + \frac{y^2}{1 - e^2} = \frac{e^2d^2}{(1 - e^2)^2} \qquad \begin{array}{l}\text{Divide by } 1 - e^2 \text{ and complete}\\ \text{the square}\end{array}$$

If $e < 1$, then dividing both sides of this equation by $e^2d^2/(1 - e^2)^2$ gives an equation of the form

$$\frac{(x - h)^2}{a^2} + \frac{y^2}{b^2} = 1$$

where

$$h = \frac{-e^2d}{1 - e^2} \qquad a^2 = \frac{e^2d^2}{(1 - e^2)^2} \qquad b^2 = \frac{e^2d^2}{1 - e^2}$$

This is the equation of an ellipse with center $(h, 0)$. In Section 10.2 we found that the foci of an ellipse are a distance c from the center, where $c^2 = a^2 - b^2$. In our case

$$c^2 = a^2 - b^2 = \frac{e^4 d^2}{(1 - e^2)^2}$$

Thus, $c = e^2 d/(1 - e^2) = -h$, which confirms that the focus defined in the theorem is the same as the focus defined in Section 10.2. It also follows that

$$e = \frac{c}{a}$$

If $e > 1$, a similar proof shows that the conic is a hyperbola with $e = c/a$, where $c^2 = a^2 + b^2$.   ■

In the proof we saw that the polar equation of the conic in Figure 1 is $r = e(d - r \cos \theta)$. Solving for r, we get

$$r = \frac{ed}{1 + e \cos \theta}$$

If the directrix is chosen to be to the *left* of the focus $(x = -d)$, then we get the equation $r = ed/(1 - e \cos \theta)$. If the directrix is *parallel* to the polar axis $(y = d$ or $y = -d)$, then we get $\sin \theta$ instead of $\cos \theta$ in the equation. These observations are summarized in the following box and in Figure 2.

Polar Equations of Conics

A polar equation of the form

$$r = \frac{ed}{1 \pm e \cos \theta} \qquad \text{or} \qquad r = \frac{ed}{1 \pm e \sin \theta}$$

represents a conic with one focus at the origin and with eccentricity e. The conic is

1. a parabola if $e = 1$
2. an ellipse if $0 < e < 1$
3. a hyperbola if $e > 1$

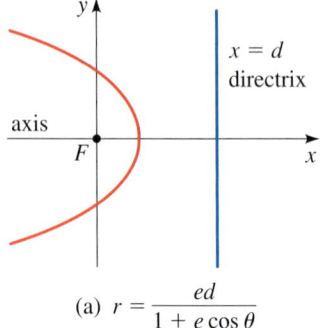

(a) $r = \dfrac{ed}{1 + e \cos \theta}$

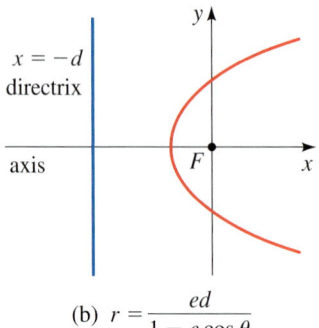

(b) $r = \dfrac{ed}{1 - e \cos \theta}$

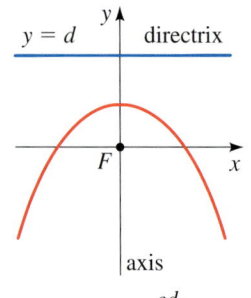

(c) $r = \dfrac{ed}{1 + e \sin \theta}$

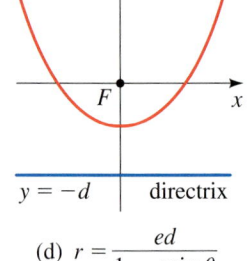

(d) $r = \dfrac{ed}{1 - e \sin \theta}$

Figure 2
The form of the polar equation of a conic indicates the location of the directrix.

To graph the polar equation of a conic, we first determine the location of the directrix from the form of the equation. The four cases that arise are shown in Figure 2. (The figure shows only the parts of the graphs that are close to the focus at the origin. The shape of the rest of the graph depends on whether the equation represents a parabola, an ellipse, or a hyperbola.) The axis of a conic is perpendicular to the directrix—specifically we have the following:

1. For a parabola, the axis of symmetry is perpendicular to the directrix.

2. For an ellipse, the major axis is perpendicular to the directrix.

3. For a hyperbola, the transverse axis is perpendicular to the directrix.

Example 1   Finding a Polar Equation for a Conic

Find a polar equation for the parabola that has its focus at the origin and whose directrix is the line $y = -6$.

Solution   Using $e = 1$ and $d = 6$, and using part (d) of Figure 2, we see that the polar equation of the parabola is

$$r = \frac{6}{1 - \sin \theta}$$ ■

To graph a polar conic, it is helpful to plot the points for which $\theta = 0$, $\pi/2$, π, and $3\pi/2$. Using these points and a knowledge of the type of conic (which we obtain from the eccentricity), we can easily get a rough idea of the shape and location of the graph.

Example 2   Identifying and Sketching a Conic

A conic is given by the polar equation

$$r = \frac{10}{3 - 2 \cos \theta}$$

(a) Show that the conic is an ellipse and sketch the graph.

(b) Find the center of the ellipse, and the lengths of the major and minor axes.

Solution

(a) Dividing the numerator and denominator by 3, we have

$$r = \frac{\frac{10}{3}}{1 - \frac{2}{3} \cos \theta}$$

Since $e = \frac{2}{3} < 1$, the equation represents an ellipse. For a rough graph we plot the points for which $\theta = 0$, $\pi/2$, π, $3\pi/2$ (see Figure 3 on the next page).

(b) Comparing the equation to those in Figure 2, we see that the major axis is horizontal. Thus, the endpoints of the major axis are $V_1(10, 0)$ and $V_2(2, \pi)$.

So the center of the ellipse is at $C(4, 0)$, the midpoint of $V_1 V_2$.

θ	r
0	10
$\pi/2$	$\frac{10}{3}$
π	2
$3\pi/2$	$\frac{10}{3}$

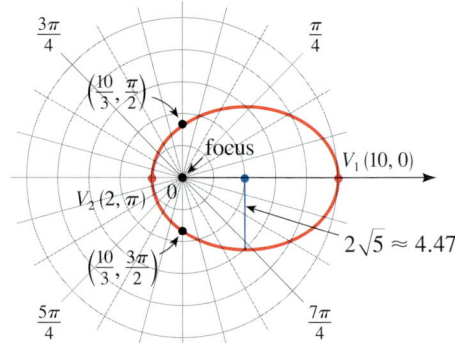

Figure 3

$$r = \frac{10}{3 - 2 \cos \theta}$$

The distance between the vertices V_1 and V_2 is 12; thus, the length of the major axis is $2a = 12$, and so $a = 6$. To determine the length of the minor axis, we need to find b. From page 796 we have $c = ae = 6\left(\frac{2}{3}\right) = 4$, so

$$b^2 = a^2 - c^2 = 6^2 - 4^2 = 20$$

Thus, $b = \sqrt{20} = 2\sqrt{5} \approx 4.47$, and the length of the minor axis is $2b = 4\sqrt{5} \approx 8.94$. ∎

Example 3 Identifying and Sketching a Conic

A conic is given by the polar equation

$$r = \frac{12}{2 + 4 \sin \theta}$$

(a) Show that the conic is a hyperbola and sketch the graph.

(b) Find the center of the hyperbola and sketch the asymptotes.

Solution

(a) Dividing the numerator and denominator by 2, we have

$$r = \frac{6}{1 + 2 \sin \theta}$$

Since $e = 2 > 1$, the equation represents a hyperbola. For a rough graph we plot the points for which $\theta = 0, \pi/2, \pi, 3\pi/2$ (see Figure 4).

(b) Comparing the equation to those in Figure 2, we see that the transverse axis is vertical. Thus, the endpoints of the transverse axis (the vertices of the hyperbola) are $V_1(2, \pi/2)$ and $V_2(-6, 3\pi/2) = V_2(6, \pi/2)$. So the center of the hyperbola is $C(4, \pi/2)$, the midpoint of $V_1 V_2$.

To sketch the asymptotes, we need to find a and b. The distance between V_1 and V_2 is 4; thus, the length of the transverse axis is $2a = 4$, and so $a = 2$. To find b, we first find c. From page 796 we have $c = ae = 2 \cdot 2 = 4$, so

$$b^2 = c^2 - a^2 = 4^2 - 2^2 = 12$$

θ	r
0	6
$\pi/2$	2
π	6
$3\pi/2$	-6

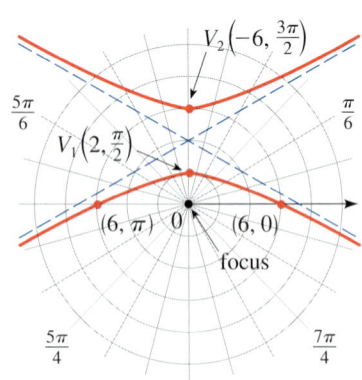

Figure 4

$$r = \frac{12}{2 + 4 \sin \theta}$$

Thus, $b = \sqrt{12} = 2\sqrt{3} \approx 3.46$. Knowing a and b allows us to sketch the central box, from which we obtain the asymptotes shown in Figure 4. ∎

When we rotate conic sections, it is much more convenient to use polar equations than Cartesian equations. We use the fact that the graph of $r = f(\theta - \alpha)$ is the graph of $r = f(\theta)$ rotated counterclockwise about the origin through an angle α (see Exercise 55 in Section 8.2).

Example 4 Rotating an Ellipse

Suppose the ellipse of Example 2 is rotated through an angle $\pi/4$ about the origin. Find a polar equation for the resulting ellipse, and draw its graph.

Solution   We get the equation of the rotated ellipse by replacing θ with $\theta - \pi/4$ in the equation given in Example 2. So the new equation is

$$r = \frac{10}{3 - 2\cos(\theta - \pi/4)}$$

We use this equation to graph the rotated ellipse in Figure 5. Notice that the ellipse has been rotated about the focus at the origin. ∎

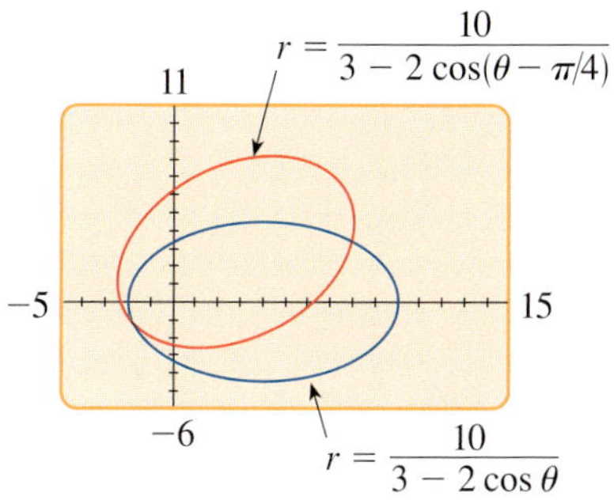

$$r = \frac{10}{3 - 2\cos(\theta - \pi/4)}$$

$$r = \frac{10}{3 - 2\cos\theta}$$

Figure 5

In Figure 6 we use a computer to sketch a number of conics to demonstrate the effect of varying the eccentricity e. Notice that when e is close to 0, the ellipse is nearly circular and becomes more elongated as e increases. When $e = 1$, of course, the conic is a parabola. As e increases beyond 1, the conic is an ever steeper hyperbola.

$e = 0.5$

$e = 0.86$

$e = 1$

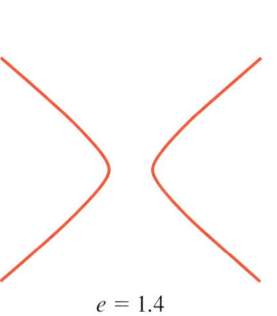

$e = 1.4$

$e = 4$

Figure 6

10.6 Exercises

1–8 ■ Write a polar equation of a conic that has its focus at the origin and satisfies the given conditions.

1. Ellipse, eccentricity $\frac{2}{3}$, directrix $x = 3$

2. Hyperbola, eccentricity $\frac{4}{3}$, directrix $x = -3$

3. Parabola, directrix $y = 2$

4. Ellipse, eccentricity $\frac{1}{2}$, directrix $y = -4$

5. Hyperbola, eccentricity 4, directrix $r = 5\sec\theta$

6. Ellipse, eccentricity 0.6, directrix $r = 2\csc\theta$

7. Parabola, vertex at $(5, \pi/2)$

8. Ellipse, eccentricity 0.4, vertex at $(2, 0)$

9–14 ◼ Match the polar equations with the graphs labeled I–VI. Give reasons for your answer.

9. $r = \dfrac{6}{1 + \cos \theta}$

10. $r = \dfrac{2}{2 - \cos \theta}$

11. $r = \dfrac{3}{1 - 2 \sin \theta}$

12. $r = \dfrac{5}{3 - 3 \sin \theta}$

13. $r = \dfrac{12}{3 + 2 \sin \theta}$

14. $r = \dfrac{12}{2 + 3 \cos \theta}$

I
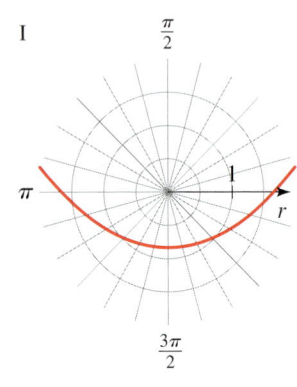

II
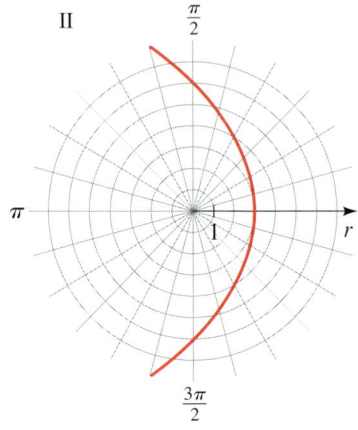

III
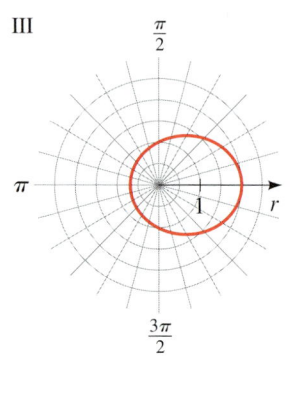

IV
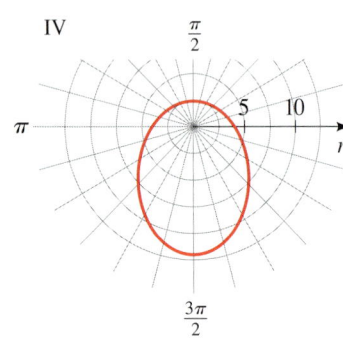

V
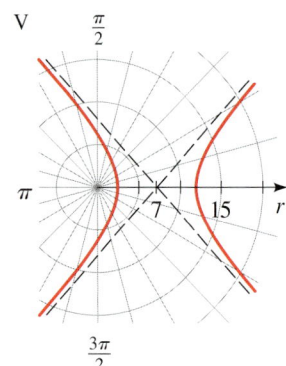

VI
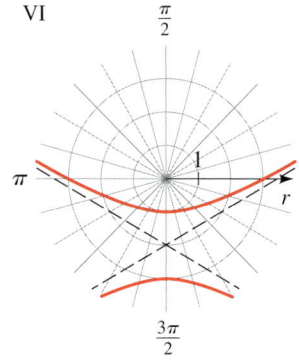

15–22 ◼ **(a)** Find the eccentricity and identify the conic.
(b) Sketch the conic and label the vertices.

15. $r = \dfrac{4}{1 + 3 \cos \theta}$

16. $r = \dfrac{8}{3 + 3 \cos \theta}$

17. $r = \dfrac{2}{1 - \cos \theta}$

18. $r = \dfrac{10}{3 - 2 \sin \theta}$

19. $r = \dfrac{6}{2 + \sin \theta}$

20. $r = \dfrac{5}{2 - 3 \sin \theta}$

21. $r = \dfrac{7}{2 - 5 \sin \theta}$

22. $r = \dfrac{8}{3 + \cos \theta}$

 23. **(a)** Find the eccentricity and directrix of the conic $r = 1/(4 - 3 \cos \theta)$ and graph the conic and its directrix.

(b) If this conic is rotated about the origin through an angle $\pi/3$, write the resulting equation and draw its graph.

 24. Graph the parabola $r = 5/(2 + 2 \sin \theta)$ and its directrix. Also graph the curve obtained by rotating this parabola about its focus through an angle $\pi/6$.

 25. Graph the conics $r = e/(1 - e \cos \theta)$ with $e = 0.4, 0.6, 0.8,$ and 1.0 on a common screen. How does the value of e affect the shape of the curve?

 26. **(a)** Graph the conics

$$r = \frac{ed}{(1 + e \sin \theta)}$$

for $e = 1$ and various values of d. How does the value of d affect the shape of the conic?

(b) Graph these conics for $d = 1$ and various values of e. How does the value of e affect the shape of the conic?

Applications

27. Orbit of the Earth   The polar equation of an ellipse can be expressed in terms of its eccentricity e and the length a of its major axis.

(a) Show that the polar equation of an ellipse with directrix $x = -d$ can be written in the form

$$r = \frac{a(1 - e^2)}{1 - e\cos\theta}$$

[*Hint:* Use the relation $a^2 = e^2d^2/(1 - e^2)^2$ given in the proof on page 795.]

(b) Find an approximate polar equation for the elliptical orbit of the earth around the sun (at one focus) given that the eccentricity is about 0.017 and the length of the major axis is about 2.99×10^8 km.

28. Perihelion and Aphelion   The planets move around the sun in elliptical orbits with the sun at one focus. The positions of a planet that are closest to, and farthest from, the sun are called its **perihelion** and **aphelion**, respectively.

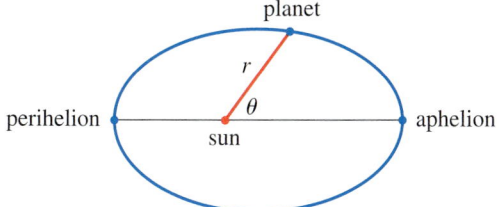

(a) Use Exercise 27(a) to show that the perihelion distance from a planet to the sun is $a(1 - e)$ and the aphelion distance is $a(1 + e)$.

(b) Use the data of Exercise 27(b) to find the distances from the earth to the sun at perihelion and at aphelion.

29. Orbit of Pluto   The distance from the planet Pluto to the sun is 4.43×10^9 km at perihelion and 7.37×10^9 km at aphelion. Use Exercise 28 to find the eccentricity of Pluto's orbit.

Discovery • Discussion

30. Distance to a Focus   When we found polar equations for the conics, we placed one focus at the pole. It's easy to find the distance from that focus to any point on the conic. Explain how the polar equation gives us this distance.

31. Polar Equations of Orbits   When a satellite orbits the earth, its path is an ellipse with one focus at the center of the earth. Why do scientists use polar (rather than rectangular) coordinates to track the position of satellites? [*Hint:* Your answer to Exercise 30 is relevant here.]

| 10.7 | **Plane Curves and Parametric Equations** |

So far we've described a curve by giving an equation (in rectangular or polar coordinates) that the coordinates of all the points on the curve must satisfy. But not all curves in the plane can be described in this way. In this section we study parametric equations, which are a general method for describing any curve.

Plane Curves

We can think of a curve as the path of a point moving in the plane; the x- and y-coordinates of the point are then functions of time. This idea leads to the following definition.

Plane Curves and Parametric Equations

If f and g are functions defined on an interval I, then the set of points $(f(t), g(t))$ is a **plane curve**. The equations

$$x = f(t) \qquad y = g(t)$$

where $t \in I$, are **parametric equations** for the curve, with **parameter** t.

Maria Gaetana Agnesi (1718–1799) is famous for having written *Instituzioni Analitiche*, considered to be the first calculus textbook.

Maria was born into a wealthy family in Milan, Italy, the oldest of 21 children. She was a child prodigy, mastering many languages at an early age, including Latin, Greek, and Hebrew. At the age of 20 she published a series of essays on philosophy and natural science. After Maria's mother died, she took on the task of educating her brothers. In 1748 Agnesi published her famous textbook, which she originally wrote as a text for tutoring her brothers. The book compiled and explained the mathematical knowledge of the day. It contains many carefully chosen examples, one of which is the curve now known as the "witch of Agnesi" (see page 809). One review calls her book an "exposition by examples rather than by theory." The book gained Agnesi immediate recognition. Pope Benedict XIV appointed her to a position at the University of Bologna, writing "we have had the idea that you should be awarded the well-known chair of mathematics, by which it comes of itself that you should not thank us but we you." This appointment was an extremely high honor for a woman, since very few women then were even allowed *(continued)*

Example 1   Sketching a Plane Curve

Sketch the curve defined by the parametric equations

$$x = t^2 - 3t \qquad y = t - 1$$

Solution   For every value of t, we get a point on the curve. For example, if $t = 0$, then $x = 0$ and $y = -1$, so the corresponding point is $(0, -1)$. In Figure 1 we plot the points (x, y) determined by the values of t shown in the following table.

t	x	y
-2	10	-3
-1	4	-2
0	0	-1
1	-2	0
2	-2	1
3	0	2
4	4	3
5	10	4

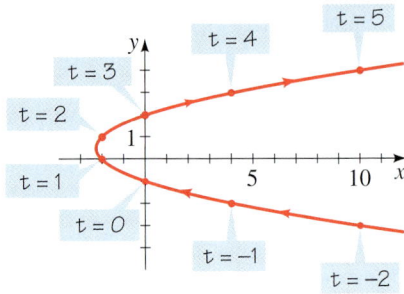

Figure 1

As t increases, a particle whose position is given by the parametric equations moves along the curve in the direction of the arrows.   ∎

If we replace t by $-t$ in Example 1, we obtain the parametric equations

$$x = t^2 + 3t \qquad y = -t - 1$$

The graph of these parametric equations (see Figure 2) is the same as the curve in Figure 1, but traced out in the opposite direction. On the other hand, if we replace t by $2t$ in Example 1, we obtain the parametric equations

$$x = 4t^2 - 6t \qquad y = 2t - 1$$

The graph of these parametric equations (see Figure 3) is again the same, but is traced out "twice as fast." *Thus, a parametrization contains more information than just the shape of the curve; it also indicates* how *the curve is being traced out.*

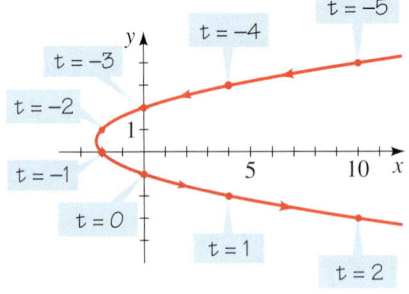

Figure 2
$x = t^2 + 3t,\ y = -t - 1$

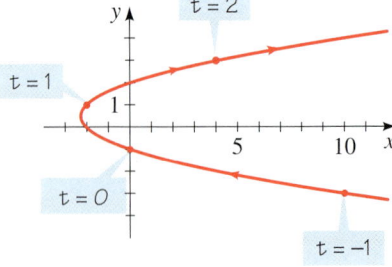

Figure 3
$x = 4t^2 - 6t,\ y = 2t - 1$

to attend university. Just two years later, Agnesi's father died and she left mathematics completely. She became a nun and devoted the rest of her life and her wealth to caring for sick and dying women, herself dying in poverty at a poorhouse of which she had once been director.

Eliminating the Parameter

Often a curve given by parametric equations can also be represented by a single rectangular equation in x and y. The process of finding this equation is called *eliminating the parameter*. One way to do this is to solve for t in one equation, then substitute into the other.

Example 2 Eliminating the Parameter

Eliminate the parameter in the parametric equations of Example 1.

Solution   First we solve for t in the simpler equation, then we substitute into the other equation. From the equation $y = t - 1$, we get $t = y + 1$. Substituting into the equation for x, we get

$$x = t^2 - 3t = (y + 1)^2 - 3(y + 1) = y^2 - y - 2$$

Thus, the curve in Example 1 has the rectangular equation $x = y^2 - y - 2$, so it is a parabola. ■

Eliminating the parameter often helps us identify the shape of a curve, as we see in the next two examples.

Example 3 Eliminating the Parameter

Describe and graph the curve represented by the parametric equations

$$x = \cos t \qquad y = \sin t \qquad 0 \le t \le 2\pi$$

Solution   To identify the curve, we eliminate the parameter. Since $\cos^2 t + \sin^2 t = 1$ and since $x = \cos t$ and $y = \sin t$ for every point (x, y) on the curve, we have

$$x^2 + y^2 = (\cos t)^2 + (\sin t)^2 = 1$$

This means that all points on the curve satisfy the equation $x^2 + y^2 = 1$, so the graph is a circle of radius 1 centered at the origin. As t increases from 0 to 2π, the point given by the parametric equations starts at $(1, 0)$ and moves counterclockwise once around the circle, as shown in Figure 4. Notice that the parameter t can be interpreted as the angle shown in the figure. ■

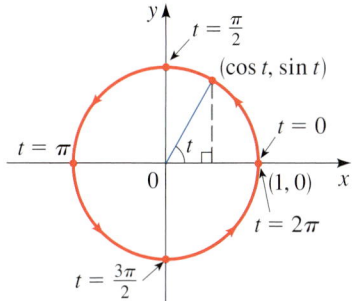

Figure 4

Example 4 Sketching a Parametric Curve

Eliminate the parameter and sketch the graph of the parametric equations

$$x = \sin t \qquad y = 2 - \cos^2 t$$

Solution   To eliminate the parameter, we first use the trigonometric identity $\cos^2 t = 1 - \sin^2 t$ to change the second equation:

$$y = 2 - \cos^2 t = 2 - (1 - \sin^2 t) = 1 + \sin^2 t$$

Now we can substitute $\sin t = x$ from the first equation to get

$$y = 1 + x^2$$

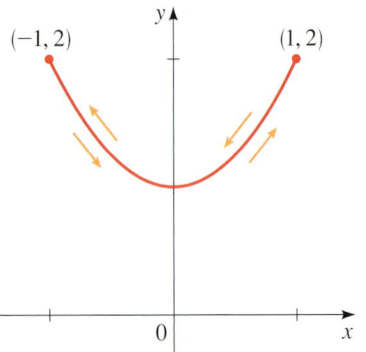

Figure 5

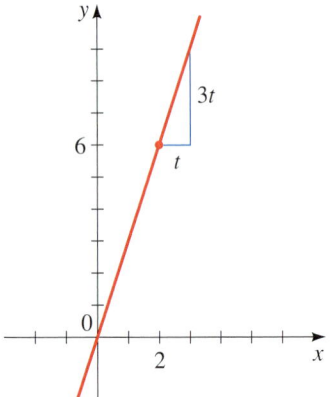

Figure 6

and so the point (x, y) moves along the parabola $y = 1 + x^2$. However, since $-1 \le \sin t \le 1$, we have $-1 \le x \le 1$, so the parametric equations represent only the part of the parabola between $x = -1$ and $x = 1$. Since $\sin t$ is periodic, the point $(x, y) = (\sin t, 2 - \cos^2 t)$ moves back and forth infinitely often along the parabola between the points $(-1, 2)$ and $(1, 2)$ as shown in Figure 5. ■

Finding Parametric Equations for a Curve

It is often possible to find parametric equations for a curve by using some geometric properties that define the curve, as in the next two examples.

Example 5 Finding Parametric Equations for a Graph

Find parametric equations for the line of slope 3 that passes through the point $(2, 6)$.

Solution Let's start at the point $(2, 6)$ and move up and to the right along this line. Because the line has slope 3, for every 1 unit we move to the right, we must move up 3 units. In other words, if we increase the x-coordinate by t units, we must correspondingly increase the y-coordinate by $3t$ units. This leads to the parametric equations

$$x = 2 + t \qquad y = 6 + 3t$$

To confirm that these equations give the desired line, we eliminate the parameter. We solve for t in the first equation and substitute into the second to get

$$y = 6 + 3(x - 2) = 3x$$

Thus, the slope-intercept form of the equation of this line is $y = 3x$, which is a line of slope 3 that does pass through $(2, 6)$ as required. The graph is shown in Figure 6. ■

Example 6 Parametric Equations for the Cycloid

As a circle rolls along a straight line, the curve traced out by a fixed point P on the circumference of the circle is called a **cycloid** (see Figure 7). If the circle has radius a and rolls along the x-axis, with one position of the point P being at the origin, find parametric equations for the cycloid.

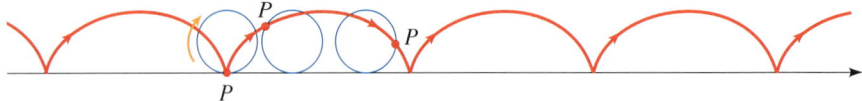

Figure 7

Solution Figure 8 shows the circle and the point P after the circle has rolled through an angle θ (in radians). The distance $d(O, T)$ that the circle has rolled must be the same as the length of the arc PT, which, by the arc length formula, is $a\theta$ (see Section 6.1). This means that the center of the circle is $C(a\theta, a)$.

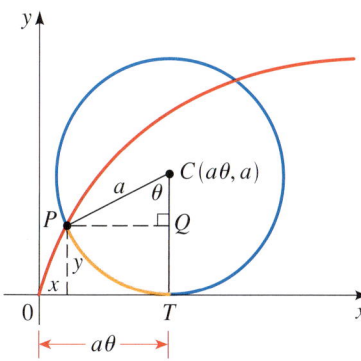

Figure 8

Let the coordinates of P be (x, y). Then from Figure 8 (which illustrates the case $0 < \theta < \pi/2$), we see that

$$x = d(O, T) - d(P, Q) = a\theta - a \sin \theta = a(\theta - \sin \theta)$$

$$y = d(T, C) - d(Q, C) = a - a \cos \theta = a(1 - \cos \theta)$$

so parametric equations for the cycloid are

$$x = a(\theta - \sin \theta) \qquad y = a(1 - \cos \theta)$$ ∎

The cycloid has a number of interesting physical properties. It is the "curve of quickest descent" in the following sense. Let's choose two points P and Q that are not directly above each other, and join them with a wire. Suppose we allow a bead to slide down the wire under the influence of gravity (ignoring friction). Of all possible shapes that the wire can be bent into, the bead will slide from P to Q the fastest when the shape is half of an arch of an inverted cycloid (see Figure 9). The cycloid is also the "curve of equal descent" in the sense that no matter where we place a bead B on a cycloid-shaped wire, it takes the same time to slide to the bottom (see Figure 10). These rather surprising properties of the cycloid were proved (using calculus) in the 17th century by several mathematicians and physicists, including Johann Bernoulli, Blaise Pascal, and Christiaan Huygens.

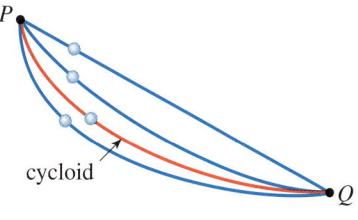

Figure 9

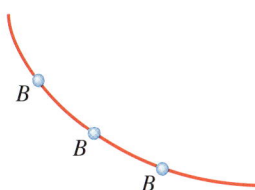

Figure 10

Using Graphing Devices to Graph Parametric Curves

Most graphing calculators and computer graphing programs can be used to graph parametric equations. Such devices are particularly useful when sketching complicated curves like the one shown in Figure 11.

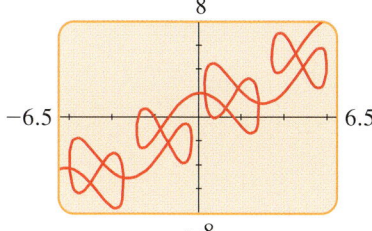

Figure 11

$x = t + 2 \sin 2t, \; y = t + 2 \cos 5t$

Example 7 Graphing Parametric Curves

Use a graphing device to draw the following parametric curves. Discuss their similarities and differences.

(a) $x = \sin 2t$
   $y = 2 \cos t$

(b) $x = \sin 3t$
   $y = 2 \cos t$

Solution In both parts (a) and (b), the graph will lie inside the rectangle given by $-1 \le x \le 1$, $-2 \le y \le 2$, since both the sine and the cosine of any number will be between -1 and 1. Thus, we may use the viewing rectangle $[-1.5, 1.5]$ by $[-2.5, 2.5]$.

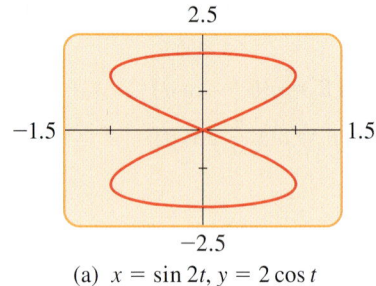

(a) $x = \sin 2t$, $y = 2 \cos t$

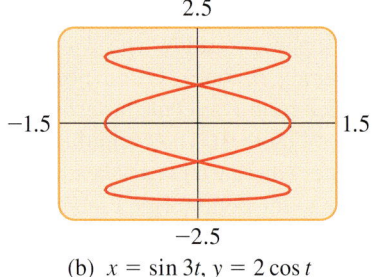

(b) $x = \sin 3t$, $y = 2 \cos t$

Figure 12

(a) Since $2 \cos t$ is periodic with period 2π (see Section 5.3), and since $\sin 2t$ has period π, letting t vary over the interval $0 \le t \le 2\pi$ gives us the complete graph, which is shown in Figure 12(a).

(b) Again, letting t take on values between 0 and 2π gives the complete graph shown in Figure 12(b).

Both graphs are *closed curves*, which means they form loops with the same starting and ending point; also, both graphs cross over themselves. However, the graph in Figure 12(a) has two loops, like a figure eight, whereas the graph in Figure 12(b) has three loops. ∎

The curves graphed in Example 7 are called Lissajous figures. A **Lissajous figure** is the graph of a pair of parametric equations of the form

$$x = A \sin \omega_1 t \qquad y = B \cos \omega_2 t$$

where A, B, ω_1, and ω_2 are real constants. Since $\sin \omega_1 t$ and $\cos \omega_2 t$ are both between -1 and 1, a Lissajous figure will lie inside the rectangle determined by $-A \le x \le A$, $-B \le y \le B$. This fact can be used to choose a viewing rectangle when graphing a Lissajous figure, as in Example 7.

Recall from Section 8.1 that rectangular coordinates (x, y) and polar coordinates (r, θ) are related by the equations $x = r \cos \theta$, $y = r \sin \theta$. Thus, we can graph the polar equation $r = f(\theta)$ by changing it to parametric form as follows:

$$x = r \cos \theta = f(\theta) \cos \theta \qquad \text{Since } r = f(\theta)$$

$$y = r \sin \theta = f(\theta) \sin \theta$$

Replacing θ by the standard parametric variable t, we have the following result.

Polar Equations in Parametric Form

The graph of the polar equation $r = f(\theta)$ is the same as the graph of the parametric equations

$$x = f(t) \cos t \qquad y = f(t) \sin t$$

Example 8 Parametric Form of a Polar Equation

Consider the polar equation $r = \theta$, $1 \le \theta \le 10\pi$.

(a) Express the equation in parametric form.

(b) Draw a graph of the parametric equations from part (a).

Solution

(a) The given polar equation is equivalent to the parametric equations

$$x = t \cos t \qquad y = t \sin t$$

(b) Since $10\pi \approx 31.42$, we use the viewing rectangle $[-32, 32]$ by $[-32, 32]$, and we let t vary from 1 to 10π. The resulting graph shown in Figure 13 is a *spiral*. ∎

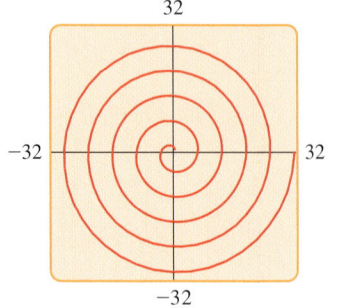

Figure 13

$x = t \cos t$, $y = t \sin t$

10.7 Exercises

1–22 ■ A pair of parametric equations is given.

(a) Sketch the curve represented by the parametric equations.

(b) Find a rectangular-coordinate equation for the curve by eliminating the parameter.

1. $x = 2t,\quad y = t + 6$

2. $x = 6t - 4,\quad y = 3t,\quad t \geq 0$

3. $x = t^2,\quad y = t - 2,\quad 2 \leq t \leq 4$

4. $x = 2t + 1,\quad y = \left(t + \frac{1}{2}\right)^2$

5. $x = \sqrt{t},\quad y = 1 - t$

6. $x = t^2,\quad y = t^4 + 1$

7. $x = \dfrac{1}{t},\quad y = t + 1$

8. $x = t + 1,\quad y = \dfrac{t}{t + 1}$

9. $x = 4t^2,\quad y = 8t^3$

10. $x = |t|,\quad y = |1 - |t||$

11. $x = 2 \sin t,\quad y = 2 \cos t,\quad 0 \leq t \leq \pi$

12. $x = 2 \cos t,\quad y = 3 \sin t,\quad 0 \leq t \leq 2\pi$

13. $x = \sin^2 t,\quad y = \sin^4 t$ **14.** $x = \sin^2 t,\quad y = \cos t$

15. $x = \cos t,\quad y = \cos 2t$

16. $x = \cos 2t,\quad y = \sin 2t$

17 $x = \sec t,\quad y = \tan t,\quad 0 \leq t < \pi/2$

18 $x = \cot t,\quad y = \csc t,\quad 0 < t < \pi$

19 $x = \tan t,\quad y = \cot t,\quad 0 < t < \pi/2$

20. $x = \sec t,\quad y = \tan^2 t,\quad 0 \leq t < \pi/2$

21. $x = \cos^2 t,\quad y = \sin^2 t$

22. $x = \cos^3 t,\quad y = \sin^3 t,\quad 0 \leq t \leq 2\pi$

23–26 ■ Find parametric equations for the line with the given properties.

23. Slope $\frac{1}{2}$, passing through $(4, -1)$

24. Slope -2, passing through $(-10, -20)$

25. Passing through $(6, 7)$ and $(7, 8)$

26. Passing through $(12, 7)$ and the origin

27. Find parametric equations for the circle $x^2 + y^2 = a^2$.

28. Find parametric equations for the ellipse

$$\frac{x^2}{a^2} + \frac{y^2}{b^2} = 1$$

29. Show by eliminating the parameter θ that the following parametric equations represent a hyperbola:

$$x = a \tan \theta \qquad y = b \sec \theta$$

30. Show that the following parametric equations represent a part of the hyperbola of Exercise 29:

$$x = a\sqrt{t} \qquad y = b\sqrt{t + 1}$$

31–34 ■ Sketch the curve given by the parametric equations.

31. $x = t \cos t,\quad y = t \sin t,\quad t \geq 0$

32. $x = \sin t,\quad y = \sin 2t$

33. $x = \dfrac{3t}{1 + t^3},\quad y = \dfrac{3t^2}{1 + t^3}$

34. $x = \cot t,\quad y = 2 \sin^2 t,\quad 0 < t < \pi$

35. If a projectile is fired with an initial speed of v_0 ft/s at an angle α above the horizontal, then its position after t seconds is given by the parametric equations

$$x = (v_0 \cos \alpha)t \qquad y = (v_0 \sin \alpha)t - 16t^2$$

(where x and y are measured in feet). Show that the path of the projectile is a parabola by eliminating the parameter t.

36. Referring to Exercise 35, suppose a gun fires a bullet into the air with an initial speed of 2048 ft/s at an angle of $30°$ to the horizontal.

(a) After how many seconds will the bullet hit the ground?

(b) How far from the gun will the bullet hit the ground?

(c) What is the maximum height attained by the bullet?

37–42 ■ Use a graphing device to draw the curve represented by the parametric equations.

37. $x = \sin t,\quad y = 2 \cos 3t$

38. $x = 2 \sin t,\quad y = \cos 4t$

39. $x = 3 \sin 5t,\quad y = 5 \cos 3t$

40. $x = \sin 4t,\quad y = \cos 3t$

41. $x = \sin(\cos t),\quad y = \cos(t^{3/2}),\quad 0 \leq t \leq 2\pi$

42. $x = 2 \cos t + \cos 2t,\quad y = 2 \sin t - \sin 2t$

43–46 ■ A polar equation is given.

(a) Express the polar equation in parametric form.

(b) Use a graphing device to graph the parametric equations you found in part (a).

43. $r = 2^{\theta/12},\quad 0 \leq \theta \leq 4\pi$ **44.** $r = \sin \theta + 2 \cos \theta$

45. $r = \dfrac{4}{2 - \cos \theta}$ **46.** $r = 2^{\sin \theta}$

47–50 ■ Match the parametric equations with the graphs labeled I–IV. Give reasons for your answers.

47. $x = t^3 - 2t$, $y = t^2 - t$

48. $x = \sin 3t$, $y = \sin 4t$

49. $x = t + \sin 2t$, $y = t + \sin 3t$

50. $x = \sin(t + \sin t)$, $y = \cos(t + \cos t)$

I
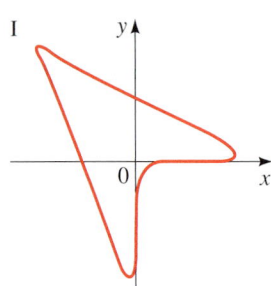

II
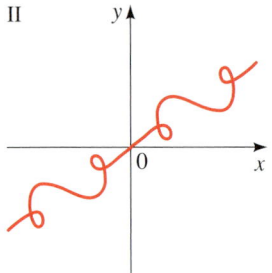

III
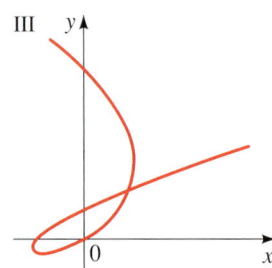

IV
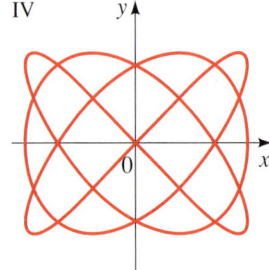

51. (a) In Example 6 suppose the point P that traces out the curve lies not on the edge of the circle, but rather at a fixed point inside the rim, at a distance b from the center (with $b < a$). The curve traced out by P is called a **curtate cycloid** (or **trochoid**). Show that parametric equations for the curtate cycloid are

$$x = a\theta - b \sin \theta \qquad y = a - b \cos \theta$$

(b) Sketch the graph using $a = 3$ and $b = 2$.

52. (a) In Exercise 51 if the point P lies *outside* the circle at a distance b from the center (with $b > a$), then the curve traced out by P is called a **prolate cycloid**. Show that parametric equations for the prolate cycloid are the same as the equations for the curtate cycloid.

(b) Sketch the graph for the case where $a = 1$ and $b = 2$.

53. A circle C of radius b rolls on the inside of a larger circle of radius a centered at the origin. Let P be a fixed point on the smaller circle, with initial position at the point $(a, 0)$ as

shown in the figure. The curve traced out by P is called a **hypocloid**.

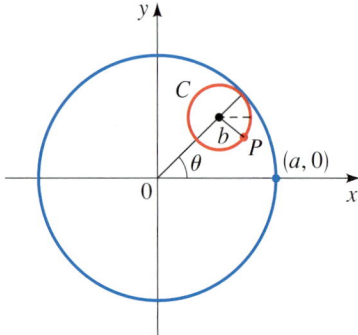

(a) Show that parametric equations for the hypocloid are

$$x = (a - b) \cos \theta + b \cos\left(\frac{a - b}{b}\theta\right)$$

$$y = (a - b) \sin \theta - b \sin\left(\frac{a - b}{b}\theta\right)$$

(b) If $a = 4b$, the hypocloid is called an **astroid**. Show that in this case the parametric equations can be reduced to

$$x = a \cos^3\theta \qquad y = a \sin^3\theta$$

Sketch the curve. Eliminate the parameter to obtain an equation for the astroid in rectangular coordinates.

54. If the circle C of Exercise 53 rolls on the *outside* of the larger circle, the curve traced out by P is called an **epicycloid**. Find parametric equations for the epicycloid.

55. In the figure, the circle of radius a is stationary and, for every θ, the point P is the midpoint of the segment QR. The curve traced out by P for $0 < \theta < \pi$ is called the **longbow curve**. Find parametric equations for this curve.

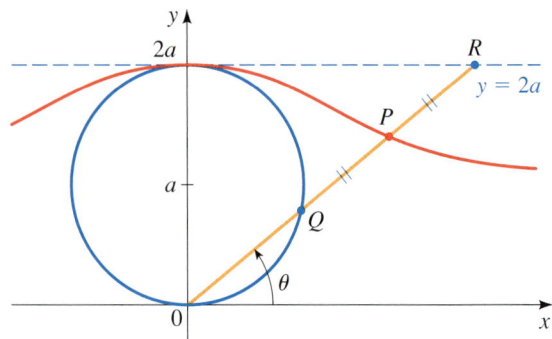

56. Two circles of radius a and b are centered at the origin, as shown in the figure. As the angle θ increases, the point P traces out a curve that lies between the circles.

   (a) Find parametric equations for the curve, using θ as the parameter.

    (b) Graph the curve using a graphing device, with $a = 3$ and $b = 2$.

   (c) Eliminate the parameter and identify the curve.

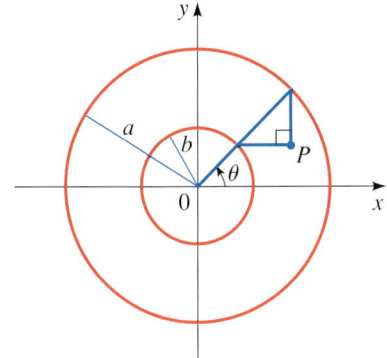

57. Two circles of radius a and b are centered at the origin, as shown in the figure.

   (a) Find parametric equations for the curve traced out by the point P, using the angle θ as the parameter. (Note that the line segment AB is always tangent to the larger circle.)

    (b) Graph the curve using a graphing device, with $a = 3$ and $b = 2$.

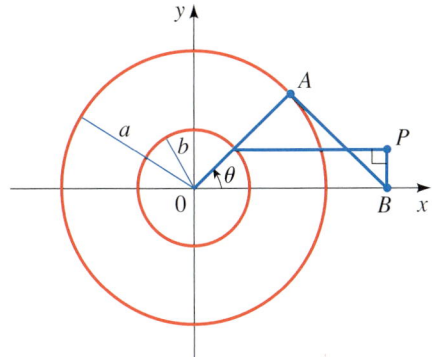

58. A curve, called a **witch of Maria Agnesi**, consists of all points P determined as shown in the figure.

   (a) Show that parametric equations for this curve can be written as

   $$x = 2a \cot \theta \qquad y = 2a \sin^2 \theta$$

 (b) Graph the curve using a graphing device, with $a = 3$.

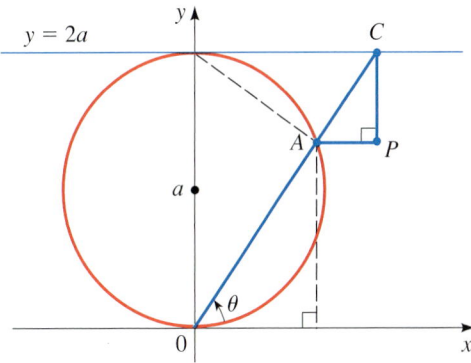

59. Eliminate the parameter θ in the parametric equations for the cycloid (Example 6) to obtain a rectangular coordinate equation for the section of the curve given by $0 \le \theta \le \pi$.

Applications

60. The Rotary Engine   The Mazda RX-8 uses an unconventional engine (invented by Felix Wankel in 1954) in which the pistons are replaced by a triangular rotor that turns in a special housing as shown in the figure. The vertices of the rotor maintain contact with the housing at all times, while the center of the triangle traces out a circle of radius r, turning the drive shaft. The shape of the housing is given by the parametric equations below (where R is the distance between the vertices and center of the rotor).

   $$x = r \cos 3\theta + R \cos \theta \qquad y = r \sin 3\theta + R \sin \theta$$

 (a) Suppose that the drive shaft has radius $r = 1$. Graph the curve given by the parametric equations for the following values of R: 0.5, 1, 3, 5.

   (b) Which of the four values of R given in part (a) seems to best model the engine housing illustrated in the figure?

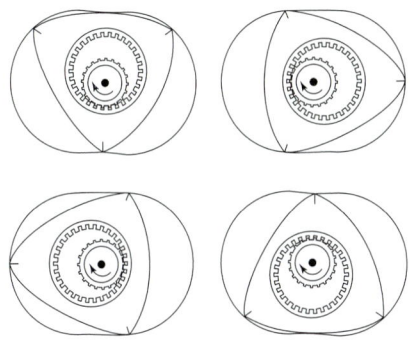

61. Spiral Path of a Dog   A dog is tied to a circular tree trunk of radius 1 ft by a long leash. He has managed to wrap the entire leash around the tree while playing in the yard, and finds himself at the point $(1, 0)$ in the figure. Seeing a squirrel, he runs around the tree counterclockwise, keeping the leash taut while chasing the intruder.

(a) Show that parametric equations for the dog's path (called an **involute of a circle**) are

$$x = \cos \theta + \theta \sin \theta \qquad y = \sin \theta - \theta \cos \theta$$

[*Hint:* Note that the leash is always tangent to the tree, so OT is perpendicular to TD.]

 (b) Graph the path of the dog for $0 \le \theta \le 4\pi$.

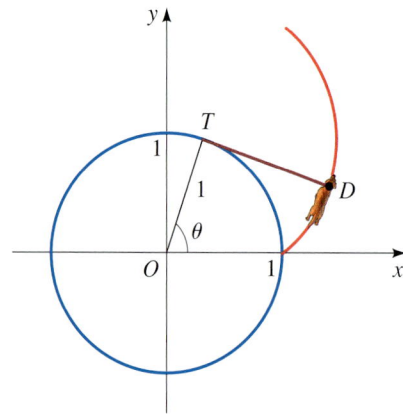

Discovery • Discussion

62. More Information in Parametric Equations   In this section we stated that parametric equations contain more information than just the shape of a curve. Write a short paragraph explaining this statement. Use the following example and your answers to parts (a) and (b) below in your explanation.

The position of a particle is given by the parametric equations

$$x = \sin t \qquad y = \cos t$$

where t represents time. We know that the shape of the path of the particle is a circle.

(a) How long does it take the particle to go once around the circle? Find parametric equations if the particle moves twice as fast around the circle.

(b) Does the particle travel clockwise or counterclockwise around the circle? Find parametric equations if the particle moves in the opposite direction around the circle.

63. Different Ways of Tracing Out a Curve   The curves C, D, E, and F are defined parametrically as follows, where the parameter t takes on all real values unless otherwise stated:

$$C: \quad x = t, \quad y = t^2$$
$$D: \quad x = \sqrt{t}, \quad y = t, \quad t \ge 0$$
$$E: \quad x = \sin t, \quad y = \sin^2 t$$
$$F: \quad x = 3^t, \quad y = 3^{2t}$$

(a) Show that the points on all four of these curves satisfy the same rectangular coordinate equation.

(b) Draw the graph of each curve and explain how the curves differ from one another.

---

10 ## Review

Concept Check

1. (a) Give the geometric definition of a parabola. What are the focus and directrix of the parabola?

(b) Sketch the parabola $x^2 = 4py$ for the case $p > 0$. Identify on your diagram the vertex, focus, and directrix. What happens if $p < 0$?

(c) Sketch the parabola $y^2 = 4px$, together with its vertex, focus, and directrix, for the case $p > 0$. What happens if $p < 0$?

2. (a) Give the geometric definition of an ellipse. What are the foci of the ellipse?

(b) For the ellipse with equation

$$\frac{x^2}{a^2} + \frac{y^2}{b^2} = 1$$

where $a > b > 0$, what are the coordinates of the vertices and the foci? What are the major and minor axes? Illustrate with a graph.

(c) Give an expression for the eccentricity of the ellipse in part (b).

(d) State the equation of an ellipse with foci on the y-axis.

3. (a) Give the geometric definition of a hyperbola. What are the foci of the hyperbola?

(b) For the hyperbola with equation

$$\frac{x^2}{a^2} - \frac{y^2}{b^2} = 1$$

what are the coordinates of the vertices and foci? What are the equations of the asymptotes? What is the transverse axis? Illustrate with a graph.

(c) State the equation of a hyperbola with foci on the y-axis.

(d) What steps would you take to sketch a hyperbola with a given equation?

4. Suppose h and k are positive numbers. What is the effect on the graph of an equation in x and y if

(a) x is replaced by $x - h$? By $x + h$?

(b) y is replaced by $y - k$? By $y + k$?

5. How can you tell whether the following nondegenerate conic is a parabola, an ellipse, or a hyperbola?

$$Ax^2 + Cy^2 + Dx + Ey + F = 0$$

6. Suppose the x- and y-axes are rotated through an acute angle ϕ to produce the X- and Y-axes. Write equations that relate the coordinates (x, y) and (X, Y) of a point in the xy-plane and XY-plane, respectively.

7. (a) How do you eliminate the xy-term in this equation?

$$Ax^2 + Bxy + Cy^2 + Dx + Ey + F = 0$$

(b) What is the discriminant of the conic in part (a)? How can you use the discriminant to determine whether the conic is a parabola, an ellipse, or a hyperbola?

8. (a) Write polar equations that represent a conic with eccentricity e.

(b) For what values of e is the conic an ellipse? A hyperbola? A parabola?

9. A curve is given by the parametric equations $x = f(t), y = g(t)$.

(a) How do you sketch the curve?

(b) How do you eliminate the parameter?

Exercises

1–8 ■ Find the vertex, focus, and directrix of the parabola, and sketch the graph.

1. $y^2 = 4x$

2. $x = \frac{1}{12}y^2$

3. $x^2 + 8y = 0$

4. $2x - y^2 = 0$

5. $x - y^2 + 4y - 2 = 0$

6. $2x^2 + 6x + 5y + 10 = 0$

7. $\frac{1}{2}x^2 + 2x = 2y + 4$

8. $x^2 = 3(x + y)$

9–16 ■ Find the center, vertices, foci, and the lengths of the major and minor axes of the ellipse, and sketch the graph.

9. $\frac{x^2}{9} + \frac{y^2}{25} = 1$

10. $\frac{x^2}{49} + \frac{y^2}{9} = 1$

11. $x^2 + 4y^2 = 16$

12. $9x^2 + 4y^2 = 1$

13. $\frac{(x - 3)^2}{9} + \frac{y^2}{16} = 1$

14. $\frac{(x - 2)^2}{25} + \frac{(y + 3)^2}{16} = 1$

15. $4x^2 + 9y^2 = 36y$

16. $2x^2 + y^2 = 2 + 4(x - y)$

17–24 ■ Find the center, vertices, foci, and asymptotes of the hyperbola, and sketch the graph.

17. $-\frac{x^2}{9} + \frac{y^2}{16} = 1$

18. $\frac{x^2}{49} - \frac{y^2}{32} = 1$

19. $x^2 - 2y^2 = 16$

20. $x^2 - 4y^2 + 16 = 0$

21. $\frac{(x + 4)^2}{16} - \frac{y^2}{16} = 1$

22. $\frac{(x - 2)^2}{8} - \frac{(y + 2)^2}{8} = 1$

23. $9y^2 + 18y = x^2 + 6x + 18$ **24.** $y^2 = x^2 + 6y$

25–30 ■ Find an equation for the conic whose graph is shown.

25.

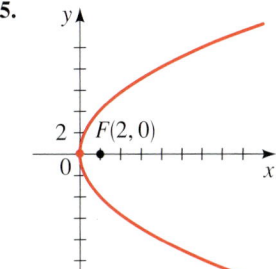

26.

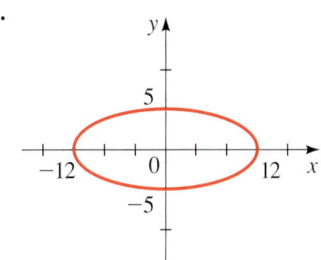

27.

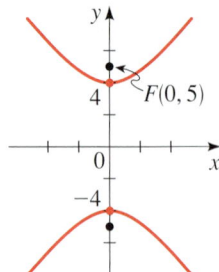

28.

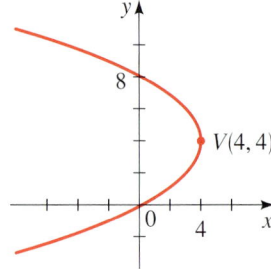

29.

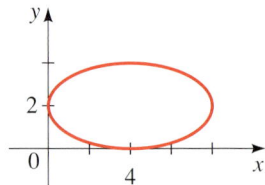

30.

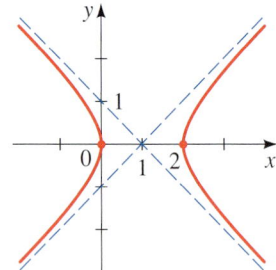

31–42 ■ Determine the type of curve represented by the equation. Find the foci and vertices (if any), and sketch the graph.

31. $\dfrac{x^2}{12} + y = 1$

32. $\dfrac{x^2}{12} + \dfrac{y^2}{144} = \dfrac{y}{12}$

33. $x^2 - y^2 + 144 = 0$

34. $x^2 + 6x = 9y^2$

35. $4x^2 + y^2 = 8(x + y)$

36. $3x^2 - 6(x + y) = 10$

37. $x = y^2 - 16y$

38. $2x^2 + 4 = 4x + y^2$

39. $2x^2 - 12x + y^2 + 6y + 26 = 0$

40. $36x^2 - 4y^2 - 36x - 8y = 31$

41. $9x^2 + 8y^2 - 15x + 8y + 27 = 0$

42. $x^2 + 4y^2 = 4x + 8$

43–50 ■ Find an equation for the conic section with the given properties.

43. The parabola with focus $F(0, 1)$ and directrix $y = -1$

44. The ellipse with center $C(0, 4)$, foci $F_1(0, 0)$ and $F_2(0, 8)$, and major axis of length 10

45. The hyperbola with vertices $V(0, \pm 2)$ and asymptotes $y = \pm \frac{1}{2}x$

46. The hyperbola with center $C(2, 4)$, foci $F_1(2, 1)$ and $F_2(2, 7)$, and vertices $V_1(2, 6)$ and $V_2(2, 2)$

47. The ellipse with foci $F_1(1, 1)$ and $F_2(1, 3)$, and with one vertex on the x-axis

48. The parabola with vertex $V(5, 5)$ and directrix the y-axis

49. The ellipse with vertices $V_1(7, 12)$ and $V_2(7, -8)$, and passing through the point $P(1, 8)$

50. The parabola with vertex $V(-1, 0)$ and horizontal axis of symmetry, and crossing the y-axis at $y = 2$

51. The path of the earth around the sun is an ellipse with the sun at one focus. The ellipse has major axis 186,000,000 mi and eccentricity 0.017. Find the distance between the earth and the sun when the earth is **(a)** closest to the sun and **(b)** farthest from the sun.

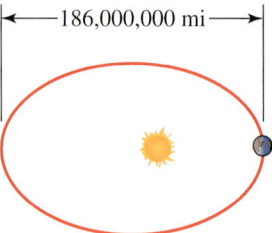

52. A ship is located 40 mi from a straight shoreline. LORAN stations A and B are located on the shoreline, 300 mi apart. From the LORAN signals, the captain determines that his ship is 80 mi closer to A than to B. Find the location of the

ship. (Place A and B on the y-axis with the x-axis halfway between them. Find the x- and y-coordinates of the ship.)

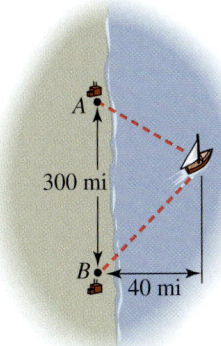

300 mi

40 mi

53. (a) Draw graphs of the following family of ellipses for $k = 1, 2, 4,$ and 8.

$$\frac{x^2}{16 + k^2} + \frac{y^2}{k^2} = 1$$

(b) Prove that all the ellipses in part (a) have the same foci.

 54. (a) Draw graphs of the following family of parabolas for $k = \frac{1}{2}, 1, 2,$ and 4.

$$y = kx^2$$

(b) Find the foci of the parabolas in part (a).

(c) How does the location of the focus change as k increases?

55–58 ■ An equation of a conic is given.

(a) Use the discriminant to determine whether the graph of the equation is a parabola, an ellipse, or a hyperbola.

(b) Use a rotation of axes to eliminate the xy-term.

(c) Sketch the graph.

55. $x^2 + 4xy + y^2 = 1$

56. $5x^2 - 6xy + 5y^2 - 8x + 8y - 8 = 0$

57. $7x^2 - 6\sqrt{3}xy + 13y^2 - 4\sqrt{3}x - 4y = 0$

58. $9x^2 + 24xy + 16y^2 = 25$

 59–62 ■ Use a graphing device to graph the conic. Identify the type of conic from the graph.

59. $5x^2 + 3y^2 = 60$

60. $9x^2 - 12y^2 + 36 = 0$

61. $6x + y^2 - 12y = 30$

62. $52x^2 - 72xy + 73y^2 = 100$

63–66 ■ A polar equation of a conic is given.

(a) Find the eccentricity and identify the conic.

(b) Sketch the conic and label the vertices.

63. $r = \dfrac{1}{1 - \cos\theta}$

64. $r = \dfrac{2}{3 + 2\sin\theta}$

65. $r = \dfrac{4}{1 + 2\sin\theta}$

66. $r = \dfrac{12}{1 - 4\cos\theta}$

67–70 ■ A pair of parametric equations is given.

(a) Sketch the curve represented by the parametric equations.

(b) Find a rectangular-coordinate equation for the curve by eliminating the parameter.

67. $x = 1 - t^2, \quad y = 1 + t$

68. $x = t^2 - 1, \quad y = t^2 + 1$

69. $x = 1 + \cos t, \quad y = 1 - \sin t, \quad 0 \le t \le \pi/2$

70. $x = \dfrac{1}{t} + 2, \quad y = \dfrac{2}{t^2}, \quad 0 < t \le 2$

 71–72 ■ Use a graphing device to draw the parametric curve.

71. $x = \cos 2t, \quad y = \sin 3t$

72. $x = \sin(t + \cos 2t), \quad y = \cos(t + \sin 3t)$

73. In the figure the point P is the midpoint of the segment QR and $0 \le \theta < \pi/2$. Using θ as the parameter, find a parametric representation for the curve traced out by P.

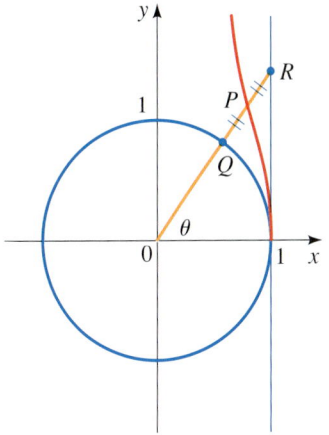

10   Test

1. Find the focus and directrix of the parabola $x^2 = -12y$, and sketch its graph.

2. Find the vertices, foci, and the lengths of the major and minor axes for the ellipse $\dfrac{x^2}{16} + \dfrac{y^2}{4} = 1$. Then sketch its graph.

3. Find the vertices, foci, and asymptotes of the hyperbola $\dfrac{y^2}{9} - \dfrac{x^2}{16} = 1$. Then sketch its graph.

4–6 ■ Find an equation for the conic whose graph is shown.

4.
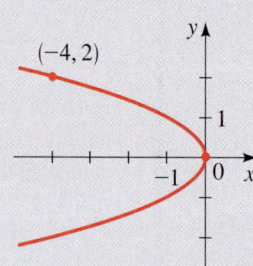

5.
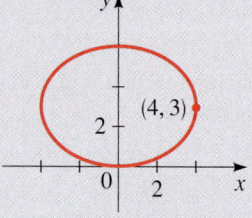

6.
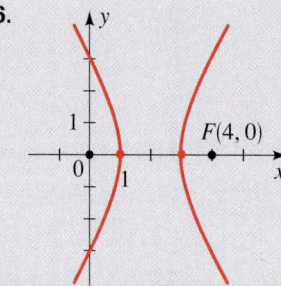

7–9 ■ Sketch the graph of the equation.

7. $16x^2 + 36y^2 - 96x + 36y + 9 = 0$

8. $9x^2 - 8y^2 + 36x + 64y = 164$

9. $2x + y^2 + 8y + 8 = 0$

10. Find an equation for the hyperbola with foci $(0, \pm 5)$ and with asymptotes $y = \pm\frac{3}{4}x$.

11. Find an equation for the parabola with focus $(2, 4)$ and directrix the x-axis.

12. A parabolic reflector for a car headlight forms a bowl shape that is 6 in. wide at its opening and 3 in. deep, as shown in the figure at the left. How far from the vertex should the filament of the bulb be placed if it is to be located at the focus?

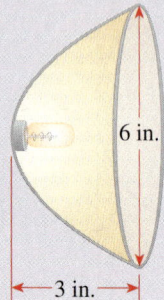

6 in.

3 in.

13. (a) Use the discriminant to determine whether the graph of this equation is a parabola, an ellipse, or a hyperbola:

$$5x^2 + 4xy + 2y^2 = 18$$

(b) Use rotation of axes to eliminate the xy-term in the equation.

(c) Sketch the graph of the equation.

(d) Find the coordinates of the vertices of this conic (in the xy-coordinate system).

14. (a) Find the polar equation of the conic that has a focus at the origin, eccentricity $e = \frac{1}{2}$, and directrix $x = 2$. Sketch the graph.

(b) What type of conic is represented by the following equation? Sketch its graph.

$$r = \frac{3}{2 - \sin \theta}$$

15. (a) Sketch the graph of the parametric curve

$$x = 3 \sin \theta + 3 \qquad y = 2 \cos \theta \qquad 0 \le \theta \le \pi$$

(b) Eliminate the parameter θ in part (a) to obtain an equation for this curve in rectangular coordinates.

Modeling motion is one of the most important ideas in both classical and modern physics. Much of Isaac Newton's work dealt with creating a mathematical model for how objects move and interact—this was the main reason for his invention of calculus. Albert Einstein developed his Special Theory of Relativity in the early 1900s to refine Newton's laws of motion.

In this section we use coordinate geometry to model the motion of a projectile, such as a ball thrown upward into the air, a bullet fired from a gun, or any other sort of missile. A similar model was created by Galileo, but we have the advantage of using our modern mathematical notation to make describing the model much easier than it was for Galileo!

Parametric Equations for the Path of a Projectile

Suppose that we fire a projectile into the air from ground level, with an initial speed v_0 and at an angle θ upward from the ground. The initial *velocity* of the projectile is a vector (see Section 8.4) with horizontal component $v_0 \cos \theta$ and vertical component $v_0 \sin \theta$, as shown in Figure 1.

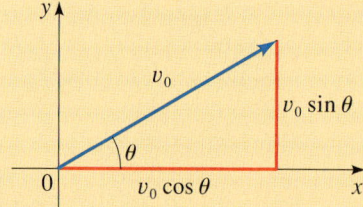

Figure 1

If there were no gravity (and no air resistance), the projectile would just keep moving indefinitely at the same speed and in the same direction. Since distance = speed × time, the projectile's position at time t would therefore be given by the following parametric equations (assuming the origin of our coordinate system is placed at the initial location of the projectile):

$$x = (v_0 \cos \theta)t \qquad y = (v_0 \sin \theta)t \qquad \text{No gravity}$$

But, of course, we know that gravity will pull the projectile back to ground level. Using calculus, it can be shown that the effect of gravity can be accounted for by subtracting $\frac{1}{2}gt^2$ from the vertical position of the projectile. In this expression, g is the gravitational acceleration: $g \approx 32 \text{ ft/s}^2 \approx 9.8 \text{ m/s}^2$. Thus, we have the following parametric equations for the path of the projectile:

$$x = (v_0 \cos \theta)t \qquad y = (v_0 \sin \theta)t - \tfrac{1}{2}gt^2 \qquad \text{With gravity}$$

Example   The Path of a Cannonball

Find parametric equations that model the path of a cannonball fired into the air with an initial speed of 150.0 m/s at a 30° angle of elevation. Sketch the path of the cannonball.

Solution   Substituting the given initial speed and angle into the general parametric equations of the path of a projectile, we get

$$x = (150.0 \cos 30°)t \qquad y = (150.0 \sin 30°)t - \tfrac{1}{2}(9.8)t^2 \qquad \text{Substitute } v_0 = 150.0, \theta = 30°$$

$$x = 129.9t \qquad y = 75.0t - 4.9t^2 \qquad \text{Simplify}$$

This path is graphed in Figure 2.

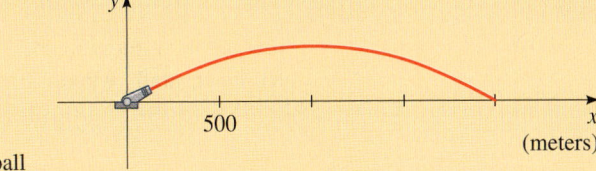

Figure 2
Path of a cannonball

Range of a Projectile

How can we tell where and when the cannonball of the above example hits the ground? Since ground level corresponds to $y = 0$, we substitute this value for y and solve for t.

$$0 = 75.0t - 4.9t^2 \qquad \text{Set } y = 0$$

$$0 = t(75.0 - 4.9t) \qquad \text{Factor}$$

$$t = 0 \quad \text{or} \quad t = \frac{75.0}{4.9} \approx 15.3 \qquad \text{Solve for } t$$

The first solution, $t = 0$, is the time when the cannon was fired; the second solution means that the cannonball hits the ground after 15.3 s of flight. To see *where* this happens, we substitute this value into the equation for x, the horizontal location of the cannonball.

$$x = 129.9(15.3) \approx 1987.5 \text{ m}$$

The cannonball travels almost 2 km before hitting the ground.

Figure 3 shows the paths of several projectiles, all fired with the same initial speed but at different angles. From the graphs we see that if the firing angle is too high or too low, the projectile doesn't travel very far.

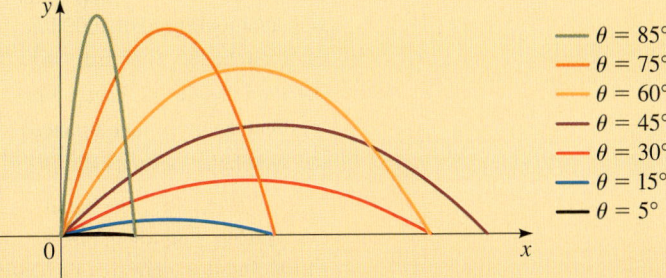

Figure 3
Paths of projectiles

Let's try to find the optimal firing angle—the angle that shoots the projectile as far as possible. We'll go through the same steps as we did in the preceding example, but

The Granger Collection

Galileo Galilei (1564–1642) was born in Pisa, Italy. He studied medicine, but later abandoned this in favor of science and mathematics. At the age of 25 he demonstrated that light objects fall at the same rate as heavier ones, by dropping cannonballs of various sizes from the Leaning Tower of Pisa. This contradicted the then-accepted view of Aristotle that heavier objects fall more quickly. He also showed that the distance an object falls is proportional to the square of the time it has been falling, and from this was able to prove that the path of a projectile is a parabola.

Galileo constructed the first telescope, and using it, discovered the moons of Jupiter. His advocacy of the Copernican view that the earth revolves around the sun (rather than being stationary) led to his being called before the Inquisition. By then an old man, he was forced to recant his views, but he is said to have muttered under his breath "the earth nevertheless does move." Galileo revolutionized science by expressing scientific principles in the language of mathematics. He said, "The great book of nature is written in mathematical symbols."

we'll use the general parametric equations instead. First, we solve for the time when the projectile hits the ground by substituting $y = 0$.

$$0 = (v_0 \sin \theta)t - \tfrac{1}{2}gt^2 \qquad \text{Substitute } y = 0$$

$$0 = t(v_0 \sin \theta - \tfrac{1}{2}gt) \qquad \text{Factor}$$

$$0 = v_0 \sin \theta - \tfrac{1}{2}gt \qquad \text{Set second factor equal to } 0$$

$$t = \frac{2v_0 \sin \theta}{g} \qquad \text{Solve for } t$$

Now we substitute this into the equation for x to see how far the projectile has traveled horizontally when it hits the ground.

$$x = (v_0 \cos \theta)t \qquad \text{Parametric equation for } x$$

$$= (v_0 \cos \theta)\left(\frac{2v_0 \sin \theta}{g}\right) \qquad \text{Substitute } t = (2v_0 \sin \theta)/g$$

$$= \frac{2v_0^2 \sin \theta \cos \theta}{g} \qquad \text{Simplify}$$

$$= \frac{v_0^2 \sin 2\theta}{g} \qquad \text{Use identity } \sin 2\theta = 2 \sin \theta \cos \theta$$

We want to choose θ so that x is as large as possible. The largest value that the sine of any angle can have is 1, the sine of 90°. Thus, we want $2\theta = 90°$, or $\theta = 45°$. So to send the projectile as far as possible, it should be shot up at an angle of 45°. From the last equation in the preceding display, we can see that it will then travel a distance $x = v_0^2/g$.

Problems

1. **Trajectories are Parabolas** From the graphs in Figure 3 the paths of projectiles appear to be parabolas that open downward. Eliminate the parameter t from the general parametric equations to verify that these are indeed parabolas.

2. **Path of a Baseball** Suppose a baseball is thrown at 30 ft/s at a 60° angle to the horizontal, from a height of 4 ft above the ground.
   (a) Find parametric equations for the path of the baseball, and sketch its graph.
   (b) How far does the baseball travel, and when does it hit the ground?

3. **Path of a Rocket** Suppose that a rocket is fired at an angle of 5° from the vertical, with an initial speed of 1000 ft/s.
   (a) Find the length of time the rocket is in the air.
   (b) Find the greatest height it reaches.
   (c) Find the horizontal distance it has traveled when it hits the ground.
   (d) Graph the rocket's path.

4. **Firing a Missile** The initial speed of a missile is 330 m/s.
   (a) At what angle should the missile be fired so that it hits a target 10 km away? (You should find that there are two possible angles.) Graph the missile paths for both angles.
   (b) For which angle is the target hit sooner?

5. **Maximum Height**   Show that the maximum height reached by a projectile as a function of its initial speed v_0 and its firing angle θ is

$$y = \frac{v_0^2 \sin^2 \theta}{2g}$$

6. **Shooting into the Wind**   Suppose that a projectile is fired into a headwind that pushes it back so as to reduce its horizontal speed by a constant amount w. Find parametric equations for the path of the projectile.

7. **Shooting into the Wind**   Using the parametric equations you derived in Problem 6, draw graphs of the path of a projectile with initial speed $v_0 = 32$ ft/s, fired into a headwind of $w = 24$ ft/s, for the angles $\theta = 5°$, $15°$, $30°$, $40°$, $45°$, $55°$, $60°$, and $75°$. Is it still true that the greatest range is attained when firing at $45°$? Draw some more graphs for different angles, and use these graphs to estimate the optimal firing angle.

8. **Simulating the Path of a Projectile**   The path of a projectile can be simulated on a graphing calculator. On the TI-83 use the "Path" graph style to graph the general parametric equations for the path of a projectile and watch as the circular cursor moves, simulating the motion of the projectile. Selecting the size of the **Tstep** determines the speed of the "projectile."

   (a) Simulate the path of a projectile. Experiment with various values of θ. Use $v_0 = 10$ ft/s and **Tstep** $= 0.02$. Part (a) of the figure below shows one such path.

   (b) Simulate the path of two projectiles, fired simultaneously, one at $\theta = 30°$ and the other at $\theta = 60°$. This can be done on the TI-83 using **Simul** mode ("simultaneous" mode). Use $v_0 = 10$ ft/s and **Tstep** $= 0.02$. See part (b) of the figure. Where do the projectiles land? Which lands first?

   (c) Simulate the path of a ball thrown straight up ($\theta = 90°$). Experiment with values of v_0 between 5 and 20 ft/s. Use the "Animate" graph style and **Tstep** $= 0.02$. Simulate the path of two balls thrown simultaneously at different speeds. To better distinguish the two balls, place them at different x-coordinates (for example, $x = 1$ and $x = 2$). See part (c) of the figure. How does doubling v_0 change the maximum height the ball reaches?

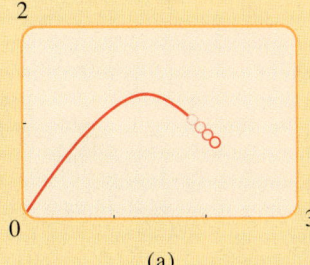
(a)

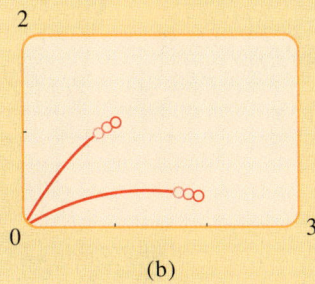
(b)

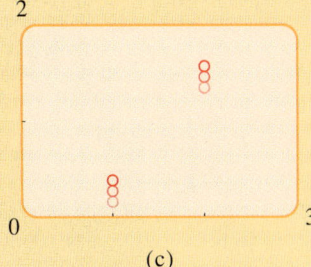
(c)

11 Sequences and Series

Chapter Overview

In this chapter we study sequences and series of numbers. Roughly speaking, a sequence is a list of numbers written in a specific order. The numbers in the sequence are often written as $a_1, a_2, a_3, \ldots$. The dots mean that the list continues forever. A simple example is the sequence

$$
\begin{array}{ccccc}
5, & 10, & 15, & 20, & 25, \ldots \\
\uparrow & \uparrow & \uparrow & \uparrow & \uparrow \\
a_1 & a_2 & a_3 & a_4 & a_5 \ldots
\end{array}
$$

Sequences arise in many real-world situations. For example, if you deposit a sum of money into an interest-bearing account, the interest earned each month forms a sequence. If you drop a ball and let it bounce, the height the ball reaches at each successive bounce is a sequence. An interesting sequence is hidden in the internal structure of a nautilus shell.

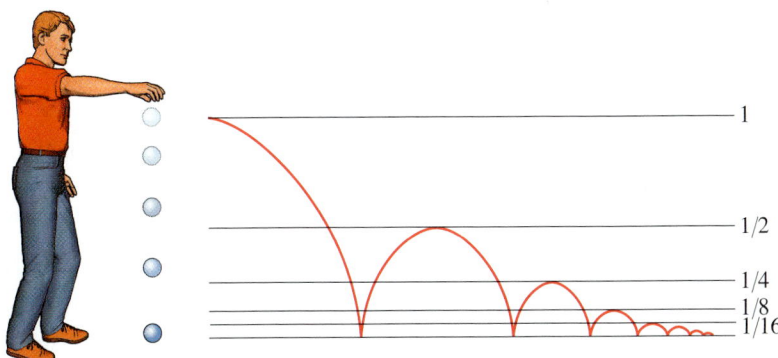

We can describe the pattern of the sequence displayed above by the *formula*:

$$a_n = 5n$$

You may have already thought of a different way to describe the pattern—namely, "you go from one number to the next by adding 5." This natural way of describing the sequence is expressed by the *recursive formula*:

$$a_n = a_{n-1} + 5$$

starting with $a_1 = 5$. Try substituting $n = 1, 2, 3, \ldots$ in each of these formulas to see

Susan Van Etten/PhotoEdit

how they produce the numbers in the sequence.

We often use sequences to model real-world phenomena—for example, the monthly payments on a mortgage form a sequence. We will explore many other applications of sequences in this chapter and in *Focus on Modeling* on page 874.

applications of sequences in this chapter and in *Focus on Modeling* on page 874.

11.1 Sequences and Summation Notation

Many real-world processes generate lists of numbers. For instance, the balance in a bank account at the end of each month forms a list of numbers when tracked over time. Mathematicians call such lists *sequences*. In this section we study sequences and their applications.

Sequences

A *sequence* is a set of numbers written in a specific order:

$$a_1, a_2, a_3, a_4, \ldots, a_n, \ldots$$

The number a_1 is called the *first term*, a_2 is the *second term*, and in general a_n is the *nth term*. Since for every natural number n there is a corresponding number a_n, we can define a sequence as a function.

> ### Definition of a Sequence
>
> A **sequence** is a function f whose domain is the set of natural numbers. The values $f(1), f(2), f(3), \ldots$ are called the **terms** of the sequence.

We usually write a_n instead of the function notation $f(n)$ for the value of the function at the number n.

Here is a simple example of a sequence:

$$2, 4, 6, 8, 10, \ldots$$

The dots indicate that the sequence continues indefinitely. We can write a sequence in this way when it's clear what the subsequent terms of the sequence are. This sequence consists of even numbers. To be more accurate, however, we need to specify a procedure for finding *all* the terms of the sequence. This can be done by giving a formula for the *nth* term a_n of the sequence. In this case,

$$a_n = 2n$$

and the sequence can be written as

2,	4,	6,	8,	$\ldots,$	$2n,$	$\ldots$
1st term	2nd term	3rd term	4th term		nth term	

Another way to write this sequence is to use function notation:

$$a(n) = 2n$$

so $a(1) = 2, a(2) = 4, a(3) = 6, \ldots$

Notice how the formula $a_n = 2n$ gives all the terms of the sequence. For instance, substituting 1, 2, 3, and 4 for n gives the first four terms:

$$a_1 = 2 \cdot 1 = 2 \qquad a_2 = 2 \cdot 2 = 4$$
$$a_3 = 2 \cdot 3 = 6 \qquad a_4 = 2 \cdot 4 = 8$$

To find the 103rd term of this sequence, we use $n = 103$ to get

$$a_{103} = 2 \cdot 103 = 206$$

Example 1   Finding the Terms of a Sequence

Find the first five terms and the 100th term of the sequence defined by each formula.

(a) $a_n = 2n - 1$

(b) $c_n = n^2 - 1$

(c) $t_n = \dfrac{n}{n+1}$

(d) $r_n = \dfrac{(-1)^n}{2^n}$

Solution   To find the first five terms, we substitute $n = 1, 2, 3, 4,$ and 5 in the formula for the nth term. To find the 100th term, we substitute $n = 100$. This gives the following.

	nth term	First five terms	100th term
(a)	$2n - 1$	$1, 3, 5, 7, 9$	199
(b)	$n^2 - 1$	$0, 3, 8, 15, 24$	9999
(c)	$\dfrac{n}{n+1}$	$\dfrac{1}{2}, \dfrac{2}{3}, \dfrac{3}{4}, \dfrac{4}{5}, \dfrac{5}{6}$	$\dfrac{100}{101}$
(d)	$\dfrac{(-1)^n}{2^n}$	$-\dfrac{1}{2}, \dfrac{1}{4}, -\dfrac{1}{8}, \dfrac{1}{16}, -\dfrac{1}{32}$	$\dfrac{1}{2^{100}}$

■

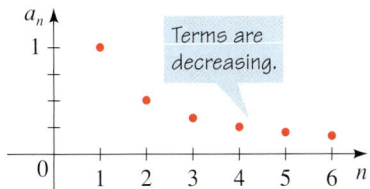

Figure 1

In Example 1(d) the presence of $(-1)^n$ in the sequence has the effect of making successive terms alternately negative and positive.

It is often useful to picture a sequence by sketching its graph. Since a sequence is a function whose domain is the natural numbers, we can draw its graph in the Cartesian plane. For instance, the graph of the sequence

$$1, \frac{1}{2}, \frac{1}{3}, \frac{1}{4}, \frac{1}{5}, \frac{1}{6}, \dots, \frac{1}{n}, \dots$$

is shown in Figure 1. Compare this to the graph of

$$1, -\frac{1}{2}, \frac{1}{3}, -\frac{1}{4}, \frac{1}{5}, -\frac{1}{6}, \dots, \frac{(-1)^{n+1}}{n}, \dots$$

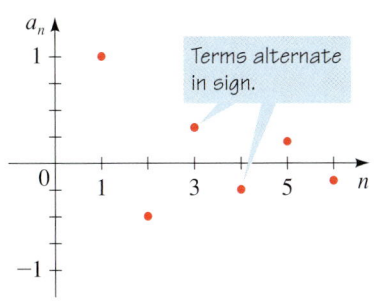

Figure 2

shown in Figure 2. The graph of every sequence consists of isolated points that are *not* connected.

Graphing calculators are useful in analyzing sequences. To work with sequences on a TI-83, we put the calculator in $\mathtt{Seq}$ mode ("sequence" mode) as in

Figure 3(a). If we enter the sequence $u(n) = n/(n + 1)$ of Example 1(c), we can display the terms using the ┌──────┐ TABLE └──────┘ command as shown in Figure 3(b). We can also graph the sequence as shown in Figure 3(c).

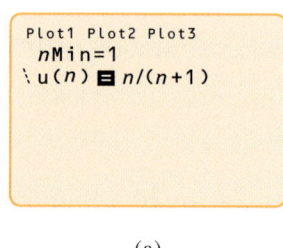

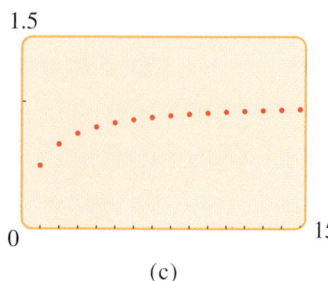

Figure 3
$u(n) = n/(n + 1)$

(a)   (b)   (c)

Finding patterns is an important part of mathematics. Consider a sequence that begins

$$1, 4, 9, 16, \ldots$$

Can you detect a pattern in these numbers? In other words, can you define a sequence whose first four terms are these numbers? The answer to this question seems easy; these numbers are the squares of the numbers 1, 2, 3, 4. Thus, the sequence we are looking for is defined by $a_n = n^2$. However, this is not the *only* sequence whose first four terms are 1, 4, 9, 16. In other words, the answer to our problem is not unique (see Exercise 78). In the next example we are interested in finding an *obvious* sequence whose first few terms agree with the given ones.

Not all sequences can be defined by a formula. For example, there is no known formula for the sequence of prime numbers:

$$2, 3, 5, 7, 11, 13, 17, 19, 23, \ldots$$

Example 2   Finding the nth Term of a Sequence

Find the nth term of a sequence whose first several terms are given.
(a) $\frac{1}{2}, \frac{3}{4}, \frac{5}{6}, \frac{7}{8}, \ldots$ (b) $-2, 4, -8, 16, -32, \ldots$

Solution

(a) We notice that the numerators of these fractions are the odd numbers and the denominators are the even numbers. Even numbers are of the form $2n$, and odd numbers are of the form $2n - 1$ (an odd number differs from an even number by 1). So, a sequence that has these numbers for its first four terms is given by

$$a_n = \frac{2n - 1}{2n}$$

(b) These numbers are powers of 2 and they alternate in sign, so a sequence that agrees with these terms is given by

$$a_n = (-1)^n 2^n$$

You should check that these formulas do indeed generate the given terms.   ∎

Recursively Defined Sequences

Some sequences do not have simple defining formulas like those of the preceding example. The nth term of a sequence may depend on some or all of the terms preceding it. A sequence defined in this way is called **recursive**. Here are two examples.

Example 3 Finding the Terms of a Recursively Defined Sequence

Find the first five terms of the sequence defined recursively by $a_1 = 1$ and

$$a_n = 3(a_{n-1} + 2)$$

Solution   The defining formula for this sequence is recursive. It allows us to find the nth term a_n if we know the preceding term a_{n-1}. Thus, we can find the second term from the first term, the third term from the second term, the fourth term from the third term, and so on. Since we are given the first term $a_1 = 1$, we can proceed as follows.

$$a_2 = 3(a_1 + 2) = 3(1 + 2) = 9$$
$$a_3 = 3(a_2 + 2) = 3(9 + 2) = 33$$
$$a_4 = 3(a_3 + 2) = 3(33 + 2) = 105$$
$$a_5 = 3(a_4 + 2) = 3(105 + 2) = 321$$

Thus, the first five terms of this sequence are

$$1, 9, 33, 105, 321, \ldots \qquad \blacksquare$$

Note that in order to find the 20th term of the sequence in Example 3, we must first find all 19 preceding terms. This is most easily done using a graphing calculator. Figure 4(a) shows how to enter this sequence on the TI-83 calculator. From Figure 4(b) we see that the 20th term of the sequence is $a_{20} = 4{,}649{,}045{,}865$.

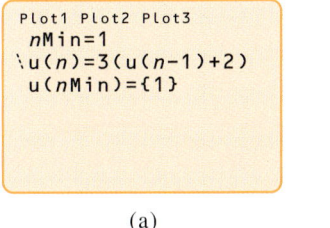

(a)

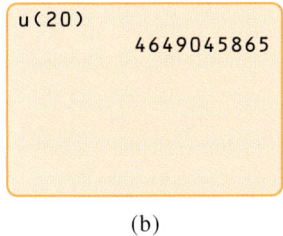

(b)

Figure 4

$u(n) = 3(u(n-1) + 2), u(1) = 1$

Example 4 The Fibonacci Sequence

Find the first 11 terms of the sequence defined recursively by $F_1 = 1$, $F_2 = 1$ and

$$F_n = F_{n-1} + F_{n-2}$$

Solution   To find F_n, we need to find the two preceding terms F_{n-1} and F_{n-2}. Since we are given F_1 and F_2, we proceed as follows.

$$F_3 = F_2 + F_1 = 1 + 1 = 2$$
$$F_4 = F_3 + F_2 = 2 + 1 = 3$$
$$F_5 = F_4 + F_3 = 3 + 2 = 5$$

Fibonacci (1175–1250) was born in Pisa, Italy, and educated in North Africa. He traveled widely in the Mediterranean area and learned the various methods then in use for writing numbers. On returning to Pisa in 1202, Fibonacci advocated the use of the Hindu-Arabic decimal system, the one we use today, over the Roman numeral system used in Europe in his time. His most famous book, *Liber Abaci*, expounds on the advantages of the Hindu-Arabic numerals. In fact, multiplication and division were so complicated using Roman numerals that a college degree was necessary to master these skills. Interestingly, in 1299 the city of Florence outlawed the use of the decimal system for merchants and businesses, requiring numbers to be written in Roman numerals or words. One can only speculate about the reasons for this law.

It's clear what is happening here. Each term is simply the sum of the two terms that precede it, so we can easily write down as many terms as we please. Here are the first 11 terms:

$$1, 1, 2, 3, 5, 8, 13, 21, 34, 55, 89, \ldots$$

The sequence in Example 4 is called the **Fibonacci sequence**, named after the 13th-century Italian mathematician who used it to solve a problem about the breeding of rabbits (see Exercise 77). The sequence also occurs in numerous other applications in nature. (See Figures 5 and 6.) In fact, so many phenomena behave like the Fibonacci sequence that one mathematical journal, the *Fibonacci Quarterly*, is devoted entirely to its properties.

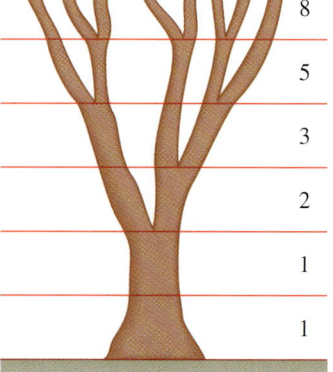

Figure 5

The Fibonacci sequence in the branching of a tree

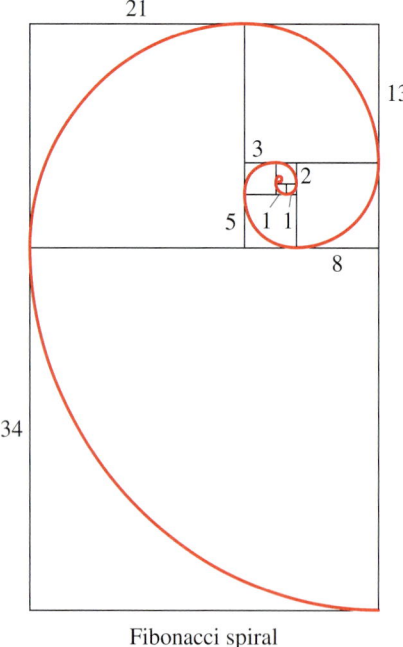

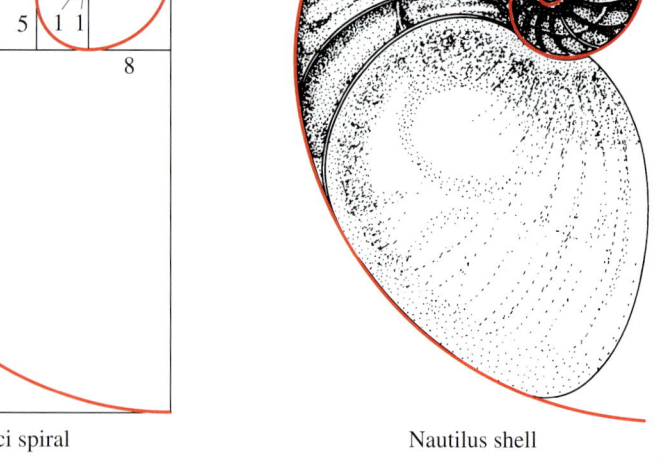

Figure 6

Fibonacci spiral

Nautilus shell

The Partial Sums of a Sequence

In calculus we are often interested in adding the terms of a sequence. This leads to the following definition.

The Partial Sums of a Sequence

For the sequence

$$a_1, a_2, a_3, a_4, \ldots, a_n, \ldots$$

the **partial sums** are

$$S_1 = a_1$$
$$S_2 = a_1 + a_2$$
$$S_3 = a_1 + a_2 + a_3$$
$$S_4 = a_1 + a_2 + a_3 + a_4$$
$$\vdots$$
$$S_n = a_1 + a_2 + a_3 + \cdots + a_n$$
$$\vdots$$

S_1 is called the **first partial sum**, S_2 is the **second partial sum**, and so on. S_n is called the **nth partial sum**. The sequence $S_1, S_2, S_3, \ldots, S_n, \ldots$ is called the **sequence of partial sums**.

Example 5   Finding the Partial Sums of a Sequence

Find the first four partial sums and the nth partial sum of the sequence given by $a_n = 1/2^n$.

Solution   The terms of the sequence are

$$\frac{1}{2}, \frac{1}{4}, \frac{1}{8}, \ldots$$

The first four partial sums are

$$S_1 = \frac{1}{2} \qquad\qquad\qquad = \frac{1}{2}$$

$$S_2 = \frac{1}{2} + \frac{1}{4} \qquad\qquad = \frac{3}{4}$$

$$S_3 = \frac{1}{2} + \frac{1}{4} + \frac{1}{8} \qquad = \frac{7}{8}$$

$$S_4 = \frac{1}{2} + \frac{1}{4} + \frac{1}{8} + \frac{1}{16} = \frac{15}{16}$$

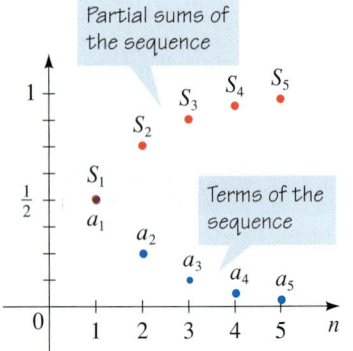

Figure 7

Graph of the sequence a_n and the sequence of partial sums S_n

Notice that in the value of each partial sum the denominator is a power of 2 and the numerator is one less than the denominator. In general, the nth partial sum is

$$S_n = \frac{2^n - 1}{2^n} = 1 - \frac{1}{2^n}$$

The first five terms of a_n and S_n are graphed in Figure 7.

Example 6   **Finding the Partial Sums of a Sequence**

Find the first four partial sums and the nth partial sum of the sequence given by

$$a_n = \frac{1}{n} - \frac{1}{n + 1}$$

Solution   The first four partial sums are

$$S_1 = \left(1 - \frac{1}{2} \right) \qquad\qquad\qquad\qquad\qquad\qquad\qquad = 1 - \frac{1}{2}$$

$$S_2 = \left(1 - \frac{1}{2} \right) + \left(\frac{1}{2} - \frac{1}{3} \right) \qquad\qquad\qquad\qquad = 1 - \frac{1}{3}$$

$$S_3 = \left(1 - \frac{1}{2} \right) + \left(\frac{1}{2} - \frac{1}{3} \right) + \left(\frac{1}{3} - \frac{1}{4} \right) \qquad\quad = 1 - \frac{1}{4}$$

$$S_4 = \left(1 - \frac{1}{2} \right) + \left(\frac{1}{2} - \frac{1}{3} \right) + \left(\frac{1}{3} - \frac{1}{4} \right) + \left(\frac{1}{4} - \frac{1}{5} \right) = 1 - \frac{1}{5}$$

Do you detect a pattern here? Of course. The nth partial sum is

$$S_n = 1 - \frac{1}{n + 1}$$

Sigma Notation

Given a sequence

$$a_1, a_2, a_3, a_4, \ldots$$

we can write the sum of the first n terms using **summation notation**, or **sigma notation**. This notation derives its name from the Greek letter Σ (capital sigma, corresponding to our S for "sum"). Sigma notation is used as follows:

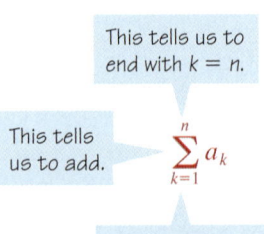

$$\sum_{k=1}^{n} a_k = a_1 + a_2 + a_3 + a_4 + \cdots + a_n$$

The left side of this expression is read "The sum of a_k from $k = 1$ to $k = n$." The letter k is called the **index of summation**, or the **summation variable**, and the idea is to replace k in the expression after the sigma by the integers $1, 2, 3, \ldots, n$, and add the resulting expressions, arriving at the right side of the equation.

The ancient Greeks considered a line segment to be divided into the **golden ratio** if the ratio of the shorter part to the longer part is the same as the ratio of the longer part to the whole segment.

Thus, the segment shown is divided into the golden ratio if

$$\frac{1}{x} = \frac{x}{1+x}$$

This leads to a quadratic equation whose positive solution is

$$x = \frac{1+\sqrt{5}}{2} \approx 1.618$$

This ratio occurs naturally in many places. For instance, psychological experiments show that the most pleasing shape of rectangle is one whose sides are in golden ratio. The ancient Greeks agreed with this and built their temples in this ratio.

The golden ratio is related to the Fibonacci sequence. In fact, it can be shown using calculus* that the ratio of two successive Fibonacci numbers

$$\frac{F_{n+1}}{F_n}$$

gets closer to the golden ratio the larger the value of n. Try finding this ratio for $n = 10$.

*James Stewart, *Calculus*, 5th ed. (Pacific Grove, CA: Brooks/Cole, 2003) p. 748.

Example 7 Sigma Notation

Find each sum.

(a) $\displaystyle\sum_{k=1}^{5} k^2$ (b) $\displaystyle\sum_{j=3}^{5} \frac{1}{j}$ (c) $\displaystyle\sum_{i=5}^{10} i$ (d) $\displaystyle\sum_{i=1}^{6} 2$

Solution

(a) $\displaystyle\sum_{k=1}^{5} k^2 = 1^2 + 2^2 + 3^2 + 4^2 + 5^2 = 55$

(b) $\displaystyle\sum_{j=3}^{5} \frac{1}{j} = \frac{1}{3} + \frac{1}{4} + \frac{1}{5} = \frac{47}{60}$

(c) $\displaystyle\sum_{i=5}^{10} i = 5 + 6 + 7 + 8 + 9 + 10 = 45$

(d) $\displaystyle\sum_{i=1}^{6} 2 = 2 + 2 + 2 + 2 + 2 + 2 = 12$ ∎

We can use a graphing calculator to evaluate sums. For instance, Figure 8 shows how the TI-83 can be used to evaluate the sums in parts (a) and (b) of Example 7.

```
sum(seq(K²,K,1,5,1))
                    55
sum(seq(1/J,J,3,5,
1))▶Frac
                 47/60
```

Figure 8

Example 8 Writing Sums in Sigma Notation

Write each sum using sigma notation.
(a) $1^3 + 2^3 + 3^3 + 4^3 + 5^3 + 6^3 + 7^3$
(b) $\sqrt{3} + \sqrt{4} + \sqrt{5} + \cdots + \sqrt{77}$

Solution

(a) We can write

$$1^3 + 2^3 + 3^3 + 4^3 + 5^3 + 6^3 + 7^3 = \sum_{k=1}^{7} k^3$$

(b) A natural way to write this sum is

$$\sqrt{3} + \sqrt{4} + \sqrt{5} + \cdots + \sqrt{77} = \sum_{k=3}^{77} \sqrt{k}$$

However, there is no unique way of writing a sum in sigma notation. We could also write this sum as

$$\sqrt{3} + \sqrt{4} + \sqrt{5} + \cdots + \sqrt{77} = \sum_{k=0}^{74} \sqrt{k+3}$$

or $$\sqrt{3} + \sqrt{4} + \sqrt{5} + \cdots + \sqrt{77} = \sum_{k=1}^{75} \sqrt{k+2}$$ ∎

The following properties of sums are natural consequences of properties of the real numbers.

Properties of Sums

Let $a_1, a_2, a_3, a_4, \ldots$ and $b_1, b_2, b_3, b_4, \ldots$ be sequences. Then for every positive integer n and any real number c, the following properties hold.

1. $\displaystyle\sum_{k=1}^{n} (a_k + b_k) = \sum_{k=1}^{n} a_k + \sum_{k=1}^{n} b_k$

2. $\displaystyle\sum_{k=1}^{n} (a_k - b_k) = \sum_{k=1}^{n} a_k - \sum_{k=1}^{n} b_k$

3. $\displaystyle\sum_{k=1}^{n} c a_k = c\left(\sum_{k=1}^{n} a_k \right)$

■ **Proof**   To prove Property 1, we write out the left side of the equation to get

$$\sum_{k=1}^{n} (a_k + b_k) = (a_1 + b_1) + (a_2 + b_2) + (a_3 + b_3) + \cdots + (a_n + b_n)$$

Because addition is commutative and associative, we can rearrange the terms on the right side to read

$$\sum_{k=1}^{n} (a_k + b_k) = (a_1 + a_2 + a_3 + \cdots + a_n) + (b_1 + b_2 + b_3 + \cdots + b_n)$$

Rewriting the right side using sigma notation gives Property 1. Property 2 is proved in a similar manner. To prove Property 3, we use the Distributive Property:

$$\sum_{k=1}^{n} c a_k = c a_1 + c a_2 + c a_3 + \cdots + c a_n$$

$$= c(a_1 + a_2 + a_3 + \cdots + a_n) = c\left(\sum_{k=1}^{n} a_k \right) \qquad ■$$

11.1   Exercises

1–10 ■ Find the first four terms and the 100th term of the sequence.

1. $a_n = n + 1$

2. $a_n = 2n + 3$

3. $a_n = \dfrac{1}{n + 1}$

4. $a_n = n^2 + 1$

5. $a_n = \dfrac{(-1)^n}{n^2}$

6. $a_n = \dfrac{1}{n^2}$

7. $a_n = 1 + (-1)^n$

8. $a_n = (-1)^{n+1} \dfrac{n}{n + 1}$

9. $a_n = n^n$

10. $a_n = 3$

11–16 ■ Find the first five terms of the given recursively defined sequence.

11. $a_n = 2(a_{n-1} - 2)$ and $a_1 = 3$

12. $a_n = \dfrac{a_{n-1}}{2}$ and $a_1 = -8$

13. $a_n = 2a_{n-1} + 1$ and $a_1 = 1$

14. $a_n = \dfrac{1}{1 + a_{n-1}}$ and $a_1 = 1$

15. $a_n = a_{n-1} + a_{n-2}$ and $a_1 = 1, a_2 = 2$

16. $a_n = a_{n-1} + a_{n-2} + a_{n-3}$ and $a_1 = a_2 = a_3 = 1$

17–22 ■ Use a graphing calculator to do the following.

(a) Find the first 10 terms of the sequence.

(b) Graph the first 10 terms of the sequence.

17. $a_n = 4n + 3$

18. $a_n = n^2 + n$

19. $a_n = \dfrac{12}{n}$

20. $a_n = 4 - 2(-1)^n$

21. $a_n = \dfrac{1}{a_{n-1}}$ and $a_1 = 2$

22. $a_n = a_{n-1} - a_{n-2}$ and $a_1 = 1, a_2 = 3$

23–30 ■ Find the nth term of a sequence whose first several terms are given.

23. $2, 4, 8, 16, \ldots$

24. $-\frac{1}{3}, \frac{1}{9}, -\frac{1}{27}, \frac{1}{81}, \ldots$

25. $1, 4, 7, 10, \ldots$

26. $5, -25, 125, -625, \ldots$

27. $1, \frac{3}{4}, \frac{5}{9}, \frac{7}{16}, \frac{9}{25}, \ldots$

28. $\frac{3}{4}, \frac{4}{5}, \frac{5}{6}, \frac{6}{7}, \ldots$

29. $0, 2, 0, 2, 0, 2, \ldots$

30. $1, \frac{1}{2}, 3, \frac{1}{4}, 5, \frac{1}{6}, \ldots$

31–34 ■ Find the first six partial sums $S_1, S_2, S_3, S_4, S_5, S_6$ of the sequence.

31. $1, 3, 5, 7, \ldots$

32. $1^2, 2^2, 3^2, 4^2, \ldots$

33. $\dfrac{1}{3}, \dfrac{1}{3^2}, \dfrac{1}{3^3}, \dfrac{1}{3^4}, \ldots$

34. $-1, 1, -1, 1, \ldots$

35–38 ■ Find the first four partial sums and the nth partial sum of the sequence a_n.

35. $a_n = \dfrac{2}{3^n}$

36. $a_n = \dfrac{1}{n+1} - \dfrac{1}{n+2}$

37. $a_n = \sqrt{n} - \sqrt{n+1}$

38. $a_n = \log\left(\dfrac{n}{n+1}\right)$ [*Hint:* Use a property of logarithms to write the nth term as a difference.]

39–46 ■ Find the sum.

39. $\displaystyle\sum_{k=1}^{4} k$

40. $\displaystyle\sum_{k=1}^{4} k^2$

41. $\displaystyle\sum_{k=1}^{3} \dfrac{1}{k}$

42. $\displaystyle\sum_{j=1}^{100} (-1)^j$

43. $\displaystyle\sum_{i=1}^{8} [1 + (-1)^i]$

44. $\displaystyle\sum_{i=4}^{12} 10$

45. $\displaystyle\sum_{k=1}^{5} 2^{k-1}$

46. $\displaystyle\sum_{i=1}^{3} i2^i$

 47–52 ■ Use a graphing calculator to evaluate the sum.

47. $\displaystyle\sum_{k=1}^{10} k^2$

48. $\displaystyle\sum_{k=1}^{100} (3k + 4)$

49. $\displaystyle\sum_{j=7}^{20} j^2(1 + j)$

50. $\displaystyle\sum_{j=5}^{15} \dfrac{1}{j^2 + 1}$

51. $\displaystyle\sum_{n=0}^{22} (-1)^n 2n$

52. $\displaystyle\sum_{n=1}^{100} \dfrac{(-1)^n}{n}$

53–58 ■ Write the sum without using sigma notation.

53. $\displaystyle\sum_{k=1}^{5} \sqrt{k}$

54. $\displaystyle\sum_{i=0}^{4} \dfrac{2i - 1}{2i + 1}$

55. $\displaystyle\sum_{k=0}^{6} \sqrt{k + 4}$

56. $\displaystyle\sum_{k=6}^{9} k(k + 3)$

57. $\displaystyle\sum_{k=3}^{100} x^k$

58. $\displaystyle\sum_{j=1}^{n} (-1)^{j+1} x^j$

59–66 ■ Write the sum using sigma notation.

59. $1 + 2 + 3 + 4 + \cdots + 100$

60. $2 + 4 + 6 + \cdots + 20$

61. $1^2 + 2^2 + 3^2 + \cdots + 10^2$

62. $\dfrac{1}{2 \ln 2} - \dfrac{1}{3 \ln 3} + \dfrac{1}{4 \ln 4} - \dfrac{1}{5 \ln 5} + \cdots + \dfrac{1}{100 \ln 100}$

63. $\dfrac{1}{1 \cdot 2} + \dfrac{1}{2 \cdot 3} + \dfrac{1}{3 \cdot 4} + \cdots + \dfrac{1}{999 \cdot 1000}$

64. $\dfrac{\sqrt{1}}{1^2} + \dfrac{\sqrt{2}}{2^2} + \dfrac{\sqrt{3}}{3^2} + \cdots + \dfrac{\sqrt{n}}{n^2}$

65. $1 + x + x^2 + x^3 + \cdots + x^{100}$

66. $1 - 2x + 3x^2 - 4x^3 + 5x^4 + \cdots - 100x^{99}$

67. Find a formula for the nth term of the sequence

$$\sqrt{2}, \quad \sqrt{2\sqrt{2}}, \quad \sqrt{2\sqrt{2\sqrt{2}}}, \quad \sqrt{2\sqrt{2\sqrt{2\sqrt{2}}}}, \ldots$$

[*Hint:* Write each term as a power of 2.]

 68. Define the sequence

$$G_n = \dfrac{1}{\sqrt{5}}\left(\dfrac{(1 + \sqrt{5})^n - (1 - \sqrt{5})^n}{2^n}\right)$$

Use the $\boxed{\text{TABLE}}$ command on a graphing calculator to find the first 10 terms of this sequence. Compare to the Fibonacci sequence F_n.

Applications

69. Compound Interest   Julio deposits $2000 in a savings account that pays 2.4% interest per year compounded monthly. The amount in the account after n months is given by the sequence

$$A_n = 2000\left(1 + \frac{0.024}{12}\right)^n$$

(a) Find the first six terms of the sequence.

(b) Find the amount in the account after 3 years.

70. Compound Interest Helen deposits $100 at the end of each month into an account that pays 6% interest per year compounded monthly. The amount of interest she has accumulated after n months is given by the sequence

$$I_n = 100\left(\frac{1.005^n - 1}{0.005} - n\right)$$

(a) Find the first six terms of the sequence.

(b) Find the interest she has accumulated after 5 years.

71. Population of a City A city was incorporated in 2004 with a population of 35,000. It is expected that the population will increase at a rate of 2% per year. The population n years after 2004 is given by the sequence

$$P_n = 35,000(1.02)^n$$

(a) Find the first five terms of the sequence.

(b) Find the population in 2014.

72. Paying off a Debt Margarita borrows $10,000 from her uncle and agrees to repay it in monthly installments of $200. Her uncle charges 0.5% interest per month on the balance.

(a) Show that her balance A_n in the nth month is given recursively by $A_0 = 10,000$ and

$$A_n = 1.005A_{n-1} - 200$$

(b) Find her balance after six months.

73. Fish Farming A fish farmer has 5000 catfish in his pond. The number of catfish increases by 8% per month, and the farmer harvests 300 catfish per month.

(a) Show that the catfish population P_n after n months is given recursively by $P_0 = 5000$ and

$$P_n = 1.08P_{n-1} - 300$$

(b) How many fish are in the pond after 12 months?

74. Price of a House The median price of a house in Orange County increases by about 6% per year. In 2002 the median price was $240,000. Let P_n be the median price n years after 2002.

(a) Find a formula for the sequence P_n.

(b) Find the expected median price in 2010.

75. Salary Increases A newly hired salesman is promised a beginning salary of $30,000 a year with a $2000 raise every year. Let S_n be his salary in his nth year of employment.

(a) Find a recursive definition of S_n.

(b) Find his salary in his fifth year of employment.

76. Concentration of a Solution A biologist is trying to find the optimal salt concentration for the growth of a certain species of mollusk. She begins with a brine solution that has 4 g/L of salt and increases the concentration by 10% every day. Let C_0 denote the initial concentration and C_n the concentration after n days.

(a) Find a recursive definition of C_n.

(b) Find the salt concentration after 8 days.

77. Fibonacci's Rabbits Fibonacci posed the following problem: Suppose that rabbits live forever and that every month each pair produces a new pair that becomes productive at age 2 months. If we start with one newborn pair, how many pairs of rabbits will we have in the nth month? Show that the answer is F_n, where F_n is the nth term of the Fibonacci sequence.

Discovery • Discussion

78. Different Sequences That Start the Same

(a) Show that the first four terms of the sequence $a_n = n^2$ are

$$1, 4, 9, 16, \ldots$$

(b) Show that the first four terms of the sequence $a_n = n^2 + (n - 1)(n - 2)(n - 3)(n - 4)$ are also

$$1, 4, 9, 16, \ldots$$

(c) Find a sequence whose first six terms are the same as those of $a_n = n^2$ but whose succeeding terms differ from this sequence.

(d) Find two different sequences that begin

$$2, 4, 8, 16, \ldots$$

79. A Recursively Defined Sequence Find the first 40 terms of the sequence defined by

$$a_{n+1} = \begin{cases} \dfrac{a_n}{2} & \text{if } a_n \text{ is an even number} \\ 3a_n + 1 & \text{if } a_n \text{ is an odd number} \end{cases}$$

and $a_1 = 11$. Do the same if $a_1 = 25$. Make a conjecture about this type of sequence. Try several other values for a_1, to test your conjecture.

80. A Different Type of Recursion Find the first 10 terms of the sequence defined by

$$a_n = a_{n-a_{n-1}} + a_{n-a_{n-2}}$$

with

$$a_1 = 1 \quad \text{and} \quad a_2 = 1$$

How is this recursive sequence different from the others in this section?

11.2   Arithmetic Sequences

In this section we study a special type of sequence, called an arithmetic sequence.

Arithmetic Sequences

Perhaps the simplest way to generate a sequence is to start with a number a and add to it a fixed constant d, over and over again.

Definition of an Arithmetic Sequence

An **arithmetic sequence** is a sequence of the form

$$a, a + d, a + 2d, a + 3d, a + 4d, \ldots$$

The number a is the **first term**, and d is the **common difference** of the sequence. The **nth term** of an arithmetic sequence is given by

$$a_n = a + (n - 1)d$$

The number d is called the common difference because any two consecutive terms of an arithmetic sequence differ by d.

Example 1   Arithmetic Sequences

(a) If $a = 2$ and $d = 3$, then we have the arithmetic sequence

$$2, 2 + 3, 2 + 6, 2 + 9, \ldots$$

or

$$2, 5, 8, 11, \ldots$$

Any two consecutive terms of this sequence differ by $d = 3$. The nth term is $a_n = 2 + 3(n - 1)$.

(b) Consider the arithmetic sequence

$$9, 4, -1, -6, -11, \ldots$$

Here the common difference is $d = -5$. The terms of an arithmetic sequence decrease if the common difference is negative. The nth term is $a_n = 9 - 5(n - 1)$.

(c) The graph of the arithmetic sequence $a_n = 1 + 2(n - 1)$ is shown in Figure 1. Notice that the points in the graph lie on a straight line with slope $d = 2$.

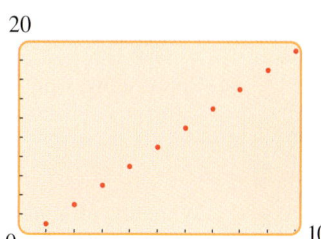

Figure 1

An arithmetic sequence is determined completely by the first term a and the common difference d. Thus, if we know the first two terms of an arithmetic sequence, then we can find a formula for the nth term, as the next example shows.

Example 2   Finding Terms of an Arithmetic Sequence

Find the first six terms and the 300th term of the arithmetic sequence

$$13, 7, \ldots$$

Solution   Since the first term is 13, we have $a = 13$. The common difference is $d = 7 - 13 = -6$. Thus, the nth term of this sequence is

$$a_n = 13 - 6(n - 1)$$

From this we find the first six terms:

$$13, 7, 1, -5, -11, -17, \ldots$$

The 300th term is $a_{300} = 13 - 6(299) = -1781$.   ■

The next example shows that an arithmetic sequence is determined completely by *any* two of its terms.

Example 3   Finding Terms of an Arithmetic Sequence

The 11th term of an arithmetic sequence is 52, and the 19th term is 92. Find the 1000th term.

Solution   To find the nth term of this sequence, we need to find a and d in the formula

$$a_n = a + (n - 1)d$$

From this formula we get

$$a_{11} = a + (11 - 1)d = a + 10d$$

$$a_{19} = a + (19 - 1)d = a + 18d$$

Since $a_{11} = 52$ and $a_{19} = 92$, we get the two equations:

$$\begin{cases} 52 = a + 10d \\ 92 = a + 18d \end{cases}$$

Solving this system for a and d, we get $a = 2$ and $d = 5$. (Verify this.) Thus, the nth term of this sequence is

$$a_n = 2 + 5(n - 1)$$

The 1000th term is $a_{1000} = 2 + 5(999) = 4997$.   ■

Partial Sums of Arithmetic Sequences

Suppose we want to find the sum of the numbers 1, 2, 3, 4, . . . , 100, that is,

$$\sum_{k=1}^{100} k$$

When the famous mathematician C. F. Gauss was a schoolboy, his teacher posed this problem to the class and expected that it would keep the students busy for a long time. But Gauss answered the question almost immediately. His idea was this: Since we are required to divide the portion into two parts, one part to be used by itself, the other by a consortium that will preserve it for later use by a less developed country. The consortium gets first pick.

adding numbers produced according to a fixed pattern, there must also be a pattern (or formula) for finding the sum. He started by writing the numbers from 1 to 100 and below them the same numbers in reverse order. Writing S for the sum and adding corresponding terms gives

$$
\begin{array}{rrrrrrr}
S = & 1 + & 2 + & 3 + \cdots + & 98 + & 99 + & 100 \\
S = & 100 + & 99 + & 98 + \cdots + & 3 + & 2 + & 1 \\
\hline
2S = & 101 + & 101 + & 101 + \cdots + & 101 + & 101 + & 101
\end{array}
$$

It follows that $2S = 100(101) = 10{,}100$ and so $S = 5050$.

Of course, the sequence of natural numbers $1, 2, 3, \ldots$ is an arithmetic sequence (with $a = 1$ and $d = 1$), and the method for summing the first 100 terms of this sequence can be used to find a formula for the nth partial sum of any arithmetic sequence. We want to find the sum of the first n terms of the arithmetic sequence whose terms are $a_k = a + (k - 1)d$; that is, we want to find

$$
S_n = \sum_{k=1}^{n} [a + (k - 1)d]
$$

$$
= a + (a + d) + (a + 2d) + (a + 3d) + \cdots + [a + (n - 1)d]
$$

Using Gauss's method, we write

$$
\begin{array}{rl}
S_n = & a \quad + \quad (a + d) \quad + \cdots + \quad [a + (n - 2)d] + [a + (n - 1)d] \\
S_n = & [a + (n - 1)d] + [a + (n - 2)d] + \cdots + \quad (a + d) \quad + \quad a \\
\hline
2S_n = & [2a + (n - 1)d] + [2a + (n - 1)d] + \cdots + [2a + (n - 1)d] + [2a + (n - 1)d]
\end{array}
$$

There are n identical terms on the right side of this equation, so

$$
2S_n = n[2a + (n - 1)d]
$$

$$
S_n = \frac{n}{2}[2a + (n - 1)d]
$$

Notice that $a_n = a + (n - 1)d$ is the nth term of this sequence. So, we can write

$$
S_n = \frac{n}{2}[a + a + (n - 1)d] = n\left(\frac{a + a_n}{2}\right)
$$

This last formula says that the sum of the first n terms of an arithmetic sequence is the average of the first and nth terms multiplied by n, the number of terms in the sum. We now summarize this result.

Partial Sums of an Arithmetic Sequence

For the arithmetic sequence $a_n = a + (n - 1)d$, the **nth partial sum**

$$
S_n = a + (a + d) + (a + 2d) + (a + 3d) + \cdots + [a + (n - 1)d]
$$

is given by either of the following formulas.

1. $S_n = \dfrac{n}{2}[2a + (n - 1)d]$

2. $S_n = n\left(\dfrac{a + a_n}{2}\right)$

Example 4   **Finding a Partial Sum of an Arithmetic Sequence**

Find the sum of the first 40 terms of the arithmetic sequence

$$3, 7, 11, 15, \ldots$$

Solution   For this arithmetic sequence, $a = 3$ and $d = 4$. Using Formula 1 for the partial sum of an arithmetic sequence, we get

$$S_{40} = \tfrac{40}{2}[2(3) + (40 - 1)4] = 20(6 + 156) = 3240$$ ■

Example 5   **Finding a Partial Sum of an Arithmetic Sequence**

Find the sum of the first 50 odd numbers.

Solution   The odd numbers form an arithmetic sequence with $a = 1$ and $d = 2$. The nth term is $a_n = 1 + 2(n - 1) = 2n - 1$, so the 50th odd number is $a_{50} = 2(50) - 1 = 99$. Substituting in Formula 2 for the partial sum of an arithmetic sequence, we get

$$S_{50} = 50\left(\frac{a + a_{50}}{2}\right) = 50\left(\frac{1 + 99}{2}\right) = 50 \cdot 50 = 2500$$ ■

Example 6   **Finding the Seating Capacity of an Amphitheater**

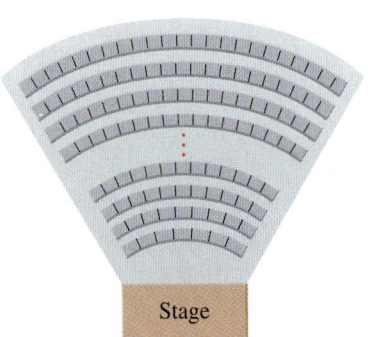

Stage

An amphitheater has 50 rows of seats with 30 seats in the first row, 32 in the second, 34 in the third, and so on. Find the total number of seats.

Solution   The numbers of seats in the rows form an arithmetic sequence with $a = 30$ and $d = 2$. Since there are 50 rows, the total number of seats is the sum

$$S_{50} = \tfrac{50}{2}[2(30) + 49(2)] \qquad S_n = \frac{n}{2}[2a + (n - 1)d]$$

$$= 3950$$

Thus, the amphitheater has 3950 seats. ■

Example 7   **Finding the Number of Terms in a Partial Sum**

How many terms of the arithmetic sequences $5, 7, 9, \ldots$ must be added to get 572?

Solution   We are asked to find n when $S_n = 572$. Substituting $a = 5$, $d = 2$, and $S_n = 572$ in Formula 1 for the partial sum of an arithmetic sequence, we get

$$572 = \frac{n}{2}[2 \cdot 5 + (n - 1)2] \qquad S_n = \frac{n}{2}[2a + (n - 1)d]$$

$$572 = 5n + n(n - 1)$$

$$0 = n^2 + 4n - 572$$

$$0 = (n - 22)(n + 26)$$

This gives $n = 22$ or $n = -26$. But since n is the *number* of terms in this partial sum, we must have $n = 22$. ■

11.2 Exercises

1–4 ■ A sequence is given.
(a) Find the first five terms of the sequence.
(b) What is the common difference d?
(c) Graph the terms you found in (a).

1. $a_n = 5 + 2(n - 1)$ **2.** $a_n = 3 - 4(n - 1)$

3. $a_n = \frac{5}{2} - (n - 1)$ **4.** $a_n = \frac{1}{2}(n - 1)$

5–8 ■ Find the nth term of the arithmetic sequence with given first term a and common difference d. What is the 10th term?

5. $a = 3, d = 5$ **6.** $a = -6, d = 3$

7. $a = \frac{5}{2}, d = -\frac{1}{2}$ **8.** $a = \sqrt{3}, d = \sqrt{3}$

9–16 ■ Determine whether the sequence is arithmetic. If it is arithmetic, find the common difference.

9. $5, 8, 11, 14, \ldots$ **10.** $3, 6, 9, 13, \ldots$

11. $2, 4, 8, 16, \ldots$ **12.** $2, 4, 6, 8, \ldots$

13. $3, \frac{3}{2}, 0, -\frac{3}{2}, \ldots$ **14.** $\ln 2, \ln 4, \ln 8, \ln 16, \ldots$

15. $2.6, 4.3, 6.0, 7.7, \ldots$ **16.** $\frac{1}{2}, \frac{1}{3}, \frac{1}{4}, \frac{1}{5}, \ldots$

17–22 ■ Find the first five terms of the sequence and determine if it is arithmetic. If it is arithmetic, find the common difference and express the nth term of the sequence in the standard form $a_n = a + (n - 1)d$.

17. $a_n = 4 + 7n$ **18.** $a_n = 4 + 2^n$

19. $a_n = \dfrac{1}{1 + 2n}$ **20.** $a_n = 1 + \dfrac{n}{2}$

21. $a_n = 6n - 10$ **22.** $a_n = 3 + (-1)^n n$

23–32 ■ Determine the common difference, the fifth term, the nth term, and the 100th term of the arithmetic sequence.

23. $2, 5, 8, 11, \ldots$ **24.** $1, 5, 9, 13, \ldots$

25. $4, 9, 14, 19, \ldots$ **26.** $11, 8, 5, 2, \ldots$

27. $-12, -8, -4, 0, \ldots$ **28.** $\frac{7}{6}, \frac{5}{3}, \frac{13}{6}, \frac{8}{3}, \ldots$

29. $25, 26.5, 28, 29.5, \ldots$ **30.** $15, 12.3, 9.6, 6.9, \ldots$

31. $2, 2 + s, 2 + 2s, 2 + 3s, \ldots$

32. $-t, -t + 3, -t + 6, -t + 9, \ldots$

33. The tenth term of an arithmetic sequence is $\frac{55}{2}$, and the second term is $\frac{7}{2}$. Find the first term.

34. The 12th term of an arithmetic sequence is 32, and the fifth term is 18. Find the 20th term.

35. The 100th term of an arithmetic sequence is 98, and the common difference is 2. Find the first three terms.

36. The 20th term of an arithmetic sequence is 101, and the common difference is 3. Find a formula for the nth term.

37. Which term of the arithmetic sequence $1, 4, 7, \ldots$ is 88?

38. The first term of an arithmetic sequence is 1, and the common difference is 4. Is 11,937 a term of this sequence? If so, which term is it?

39–44 ■ Find the partial sum S_n of the arithmetic sequence that satisfies the given conditions.

39. $a = 1, d = 2, n = 10$ **40.** $a = 3, d = 2, n = 12$

41. $a = 4, d = 2, n = 20$ **42.** $a = 100, d = -5, n = 8$

43. $a_1 = 55, d = 12, n = 10$ **44.** $a_2 = 8, a_5 = 9.5, n = 15$

45–50 ■ A partial sum of an arithmetic sequence is given. Find the sum.

45. $1 + 5 + 9 + \cdots + 401$

46. $-3 + \left(-\frac{3}{2}\right) + 0 + \frac{3}{2} + 3 + \cdots + 30$

47. $0.7 + 2.7 + 4.7 + \cdots + 56.7$

48. $-10 - 9.9 - 9.8 - \cdots - 0.1$

49. $\displaystyle\sum_{k=0}^{10} (3 + 0.25k)$ **50.** $\displaystyle\sum_{n=0}^{20} (1 - 2n)$

51. Show that a right triangle whose sides are in arithmetic progression is similar to a 3–4–5 triangle.

52. Find the product of the numbers

$$10^{1/10}, 10^{2/10}, 10^{3/10}, 10^{4/10}, \ldots, 10^{19/10}$$

53. A sequence is **harmonic** if the reciprocals of the terms of the sequence form an arithmetic sequence. Determine whether the following sequence is harmonic:

$$1, \frac{3}{5}, \frac{3}{7}, \frac{1}{3}, \ldots$$

54. The **harmonic mean** of two numbers is the reciprocal of the average of the reciprocals of the two numbers. Find the harmonic mean of 3 and 5.

55. An arithmetic sequence has first term $a = 5$ and common difference $d = 2$. How many terms of this sequence must be added to get 2700?

56. An arithmetic sequence has first term $a_1 = 1$ and fourth term $a_4 = 16$. How many terms of this sequence must be added to get 2356?

Applications

57. Depreciation The purchase value of an office computer is $12,500. Its annual depreciation is $1875. Find the value of the computer after 6 years.

58. Poles in a Pile Telephone poles are stored in a pile with 25 poles in the first layer, 24 in the second, and so on. If there are 12 layers, how many telephone poles does the pile contain?

59. Salary Increases A man gets a job with a salary of $30,000 a year. He is promised a $2300 raise each subsequent year. Find his total earnings for a 10-year period.

60. Drive-In Theater A drive-in theater has spaces for 20 cars in the first parking row, 22 in the second, 24 in the third, and so on. If there are 21 rows in the theater, find the number of cars that can be parked.

61. Theater Seating An architect designs a theater with 15 seats in the first row, 18 in the second, 21 in the third, and so on. If the theater is to have a seating capacity of 870, how many rows must the architect use in his design?

62. Falling Ball When an object is allowed to fall freely near the surface of the earth, the gravitational pull is such that the object falls 16 ft in the first second, 48 ft in the next second, 80 ft in the next second, and so on.

(a) Find the total distance a ball falls in 6 s.

(b) Find a formula for the total distance a ball falls in n seconds.

63. The Twelve Days of Christmas In the well-known song "The Twelve Days of Christmas," a person gives his sweetheart k gifts on the kth day for each of the 12 days of Christmas. The person also repeats each gift identically on each subsequent day. Thus, on the 12th day the sweetheart receives a gift for the first day, 2 gifts for the second, 3 gifts for the third, and so on. Show that the number of gifts received on the 12th day is a partial sum of an arithmetic sequence. Find this sum.

Discovery • Discussion

64. Arithmetic Means The **arithmetic mean** (or average) of two numbers a and b is

$$m = \frac{a + b}{2}$$

Note that m is the same distance from a as from b, so a, m, b is an arithmetic sequence. In general, if $m_1, m_2, \ldots, m_k$ are equally spaced between a and b so that

$$a, m_1, m_2, \ldots, m_k, b$$

is an arithmetic sequence, then $m_1, m_2, \ldots, m_k$ are called k arithmetic means between a and b.

(a) Insert two arithmetic means between 10 and 18.

(b) Insert three arithmetic means between 10 and 18.

(c) Suppose a doctor needs to increase a patient's dosage of a certain medicine from 100 mg to 300 mg per day in five equal steps. How many arithmetic means must be inserted between 100 and 300 to give the progression of daily doses, and what are these means?

11.3 Geometric Sequences

In this section we study geometric sequences. This type of sequence occurs frequently in applications to finance, population growth, and other fields.

Geometric Sequences

Recall that an arithmetic sequence is generated when we repeatedly add a number d to an initial term a. A *geometric* sequence is generated when we start with a number a and repeatedly *multiply* by a fixed nonzero constant r.

Definition of a Geometric Sequence

A **geometric sequence** is a sequence of the form

$$a, ar, ar^2, ar^3, ar^4, \ldots$$

The number a is the **first term**, and r is the **common ratio** of the sequence. The **nth term** of a geometric sequence is given by

$$a_n = ar^{n-1}$$

The number r is called the common ratio because the ratio of any two consecutive terms of the sequence is r.

Example 1   Geometric Sequences

(a) If $a = 3$ and $r = 2$, then we have the geometric sequence

$$3, \quad 3 \cdot 2, \quad 3 \cdot 2^2, \quad 3 \cdot 2^3, \quad 3 \cdot 2^4, \quad \ldots$$

or

$$3, 6, 12, 24, 48, \ldots$$

Notice that the ratio of any two consecutive terms is $r = 2$. The nth term is $a_n = 3(2)^{n-1}$.

(b) The sequence

$$2, -10, 50, -250, 1250, \ldots$$

is a geometric sequence with $a = 2$ and $r = -5$. When r is negative, the terms of the sequence alternate in sign. The nth term is $a_n = 2(-5)^{n-1}$.

(c) The sequence

$$1, \ \frac{1}{3}, \ \frac{1}{9}, \ \frac{1}{27}, \ \frac{1}{81}, \ \ldots$$

is a geometric sequence with $a = 1$ and $r = \frac{1}{3}$. The nth term is $a_n = 1\left(\frac{1}{3}\right)^{n-1}$.

(d) The graph of the geometric sequence $a_n = \frac{1}{5} \cdot 2^{n-1}$ is shown in Figure 1. Notice that the points in the graph lie on the graph of the exponential function $y = \frac{1}{5} \cdot 2^{x-1}$.

If $0 < r < 1$, then the terms of the geometric sequence ar^{n-1} decrease, but if $r > 1$, then the terms increase. (What happens if $r = 1$?) ∎

Geometric sequences occur naturally. Here is a simple example. Suppose a ball has elasticity such that when it is dropped it bounces up one-third of the distance it has fallen. If this ball is dropped from a height of 2 m, then it bounces up to a height of $2\left(\frac{1}{3}\right) = \frac{2}{3}$ m. On its second bounce, it returns to a height of $\left(\frac{2}{3}\right)\left(\frac{1}{3}\right) = \frac{2}{9}$ m, and so on (see Figure 2). Thus, the height h_n that the ball reaches on its nth bounce is given by the geometric sequence

$$h_n = \frac{2}{3}\left(\frac{1}{3}\right)^{n-1} = 2\left(\frac{1}{3}\right)^n$$

We can find the nth term of a geometric sequence if we know any two terms, as the following examples show.

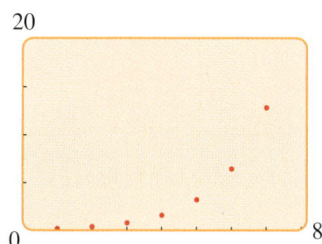

Figure 1

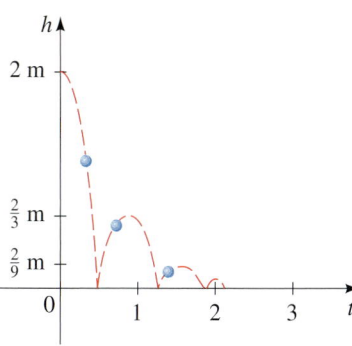

Figure 2

Michael Ng

Srinivasa Ramanujan (1887–1920) was born into a poor family in the small town of Kumbakonam in India. Self-taught in mathematics, he worked in virtual isolation from other mathematicians. At the age of 25 he wrote a letter to G. H. Hardy, the leading British mathematician at the time, listing some of his discoveries. Hardy immediately recognized Ramanujan's genius and for the next six years the two worked together in London until Ramanujan fell ill and returned to his hometown in India, where he died a year later. Ramanujan was a genius with phenomenal ability to see hidden patterns in the properties of numbers. Most of his discoveries were written as complicated infinite series, the importance of which was not recognized until many years after his death. In the last year of his life he wrote 130 pages of mysterious formulas, many of which still defy proof. Hardy tells the story that when he visited Ramanujan in a hospital and arrived in a taxi, he remarked to Ramanujan that the cab's number, 1729, was uninteresting. Ramanujan replied "No, it is a very interesting number. It is the smallest number expressible as the sum of two cubes in two different ways." (See Problem 23 on page 144.)

Example 2   Finding Terms of a Geometric Sequence

Find the eighth term of the geometric sequence 5, 15, 45, ...

Solution   To find a formula for the nth term of this sequence, we need to find a and r. Clearly, $a = 5$. To find r, we find the ratio of any two consecutive terms. For instance, $r = \frac{45}{15} = 3$. Thus

$$a_n = 5(3)^{n-1}$$

The eighth term is $a_8 = 5(3)^{8-1} = 5(3)^7 = 10{,}935$.   ■

Example 3   Finding Terms of a Geometric Sequence

The third term of a geometric sequence is $\frac{63}{4}$, and the sixth term is $\frac{1701}{32}$. Find the fifth term.

Solution   Since this sequence is geometric, its nth term is given by the formula $a_n = ar^{n-1}$. Thus

$$a_3 = ar^{3-1} = ar^2$$
$$a_6 = ar^{6-1} = ar^5$$

From the values we are given for these two terms, we get the following system of equations:

$$\begin{cases} \frac{63}{4} = ar^2 \\ \frac{1701}{32} = ar^5 \end{cases}$$

We solve this system by dividing.

$$\frac{ar^5}{ar^2} = \frac{\frac{1701}{32}}{\frac{63}{4}}$$

$$r^3 = \frac{27}{8} \qquad \textcolor{teal}{\text{Simplify}}$$

$$r = \frac{3}{2} \qquad \textcolor{teal}{\text{Take cube root of each side}}$$

Substituting for r in the first equation, $\frac{63}{4} = ar^2$, gives

$$\frac{63}{4} = a\left(\frac{3}{2}\right)^2$$

$$a = 7 \qquad \textcolor{teal}{\text{Solve for } a}$$

It follows that the nth term of this sequence is

$$a_n = 7\left(\frac{3}{2}\right)^{n-1}$$

Thus, the fifth term is

$$a_5 = 7\left(\frac{3}{2}\right)^{5-1} = 7\left(\frac{3}{2}\right)^4 = \frac{567}{16}$$ ■

Partial Sums of Geometric Sequences

For the geometric sequence $a, ar, ar^2, ar^3, ar^4, \ldots, ar^{n-1}, \ldots$, the nth partial sum is

$$S_n = \sum_{k=1}^{n} ar^{k-1} = a + ar + ar^2 + ar^3 + ar^4 + \cdots + ar^{n-1}$$

To find a formula for S_n, we multiply S_n by r and subtract from S_n:

$$S_n = a + ar + ar^2 + ar^3 + ar^4 + \cdots + ar^{n-1}$$

$$rS_n = \quad\quad ar + ar^2 + ar^3 + ar^4 + \cdots + ar^{n-1} + ar^n$$

$$S_n - rS_n = a - ar^n$$

So,
$$S_n(1 - r) = a(1 - r^n)$$

$$S_n = \frac{a(1 - r^n)}{1 - r} \quad\quad (r \neq 1)$$

We summarize this result.

Partial Sums of a Geometric Sequence

For the geometric sequence $a_n = ar^{n-1}$, the **nth partial sum**

$$S_n = a + ar + ar^2 + ar^3 + ar^4 + \cdots + ar^{n-1} \quad\quad (r \neq 1)$$

is given by

$$S_n = a\frac{1 - r^n}{1 - r}$$

Example 4   Finding a Partial Sum of a Geometric Sequence

Find the sum of the first five terms of the geometric sequence

$$1, 0.7, 0.49, 0.343, \ldots$$

Solution   The required sum is the sum of the first five terms of a geometric sequence with $a = 1$ and $r = 0.7$. Using the formula for S_n with $n = 5$, we get

$$S_5 = 1 \cdot \frac{1 - (0.7)^5}{1 - 0.7} = 2.7731$$

Thus, the sum of the first five terms of this sequence is 2.7731.   ∎

Example 5   Finding a Partial Sum of a Geometric Sequence

Find the sum $\displaystyle\sum_{k=1}^{5} 7\left(-\tfrac{2}{3}\right)^k$.

Solution   The given sum is the fifth partial sum of a geometric sequence with first term $a = 7\left(-\tfrac{2}{3}\right) = -\tfrac{14}{3}$ and common ratio $r = -\tfrac{2}{3}$. Thus, by the formula for S_n, we have

$$S_5 = -\frac{14}{3} \cdot \frac{1 - \left(-\tfrac{2}{3}\right)^5}{1 - \left(-\tfrac{2}{3}\right)} = -\frac{14}{3} \cdot \frac{1 + \tfrac{32}{243}}{\tfrac{5}{3}} = -\frac{770}{243}$$ ∎

What Is an Infinite Series?

An expression of the form

$$a_1 + a_2 + a_3 + a_4 + \cdots$$

is called an **infinite series**. The dots mean that we are to continue the addition indefinitely. What meaning can we attach to the sum of infinitely many numbers? It seems at first that it is not possible to add infinitely many numbers and arrive at a finite number. But consider the following problem. You have a cake and you want to eat it by first eating half the cake, then eating half of what remains, then again eating half of what remains. This process can continue indefinitely because at each stage some of the cake remains. (See Figure 3.)

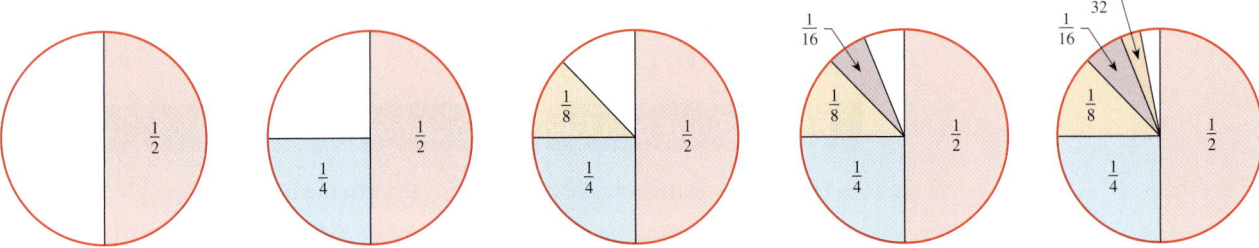

Figure 3

Does this mean that it's impossible to eat all of the cake? Of course not. Let's write down what you have eaten from this cake:

$$\frac{1}{2} + \frac{1}{4} + \frac{1}{8} + \frac{1}{16} + \cdots + \frac{1}{2^n} + \cdots$$

This is an infinite series, and we note two things about it: First, from Figure 3 it's clear that no matter how many terms of this series we add, the total will never exceed 1. Second, the more terms of this series we add, the closer the sum is to 1 (see Figure 3). This suggests that the number 1 can be written as the sum of infinitely many smaller numbers:

$$1 = \frac{1}{2} + \frac{1}{4} + \frac{1}{8} + \frac{1}{16} + \cdots + \frac{1}{2^n} + \cdots$$

To make this more precise, let's look at the partial sums of this series:

$$S_1 = \frac{1}{2} \qquad\qquad = \frac{1}{2}$$

$$S_2 = \frac{1}{2} + \frac{1}{4} \qquad\qquad = \frac{3}{4}$$

$$S_3 = \frac{1}{2} + \frac{1}{4} + \frac{1}{8} \qquad\qquad = \frac{7}{8}$$

$$S_4 = \frac{1}{2} + \frac{1}{4} + \frac{1}{8} + \frac{1}{16} = \frac{15}{16}$$

and, in general (see Example 5 of Section 11.1),

$$S_n = 1 - \frac{1}{2^n}$$

As n gets larger and larger, we are adding more and more of the terms of this series. Intuitively, as n gets larger, S_n gets closer to the sum of the series. Now notice that as n gets large, $1/2^n$ gets closer and closer to 0. Thus, S_n gets close to $1 - 0 = 1$. Using the notation of Section 3.6, we can write

$$S_n \to 1 \qquad \text{as} \qquad n \to \infty$$

In general, if S_n gets close to a finite number S as n gets large, we say that S is the **sum of the infinite series**.

Infinite Geometric Series

An **infinite geometric series** is a series of the form

$$a + ar + ar^2 + ar^3 + ar^4 + \cdots + ar^{n-1} + \cdots$$

We can apply the reasoning used earlier to find the sum of an infinite geometric series. The nth partial sum of such a series is given by the formula

$$S_n = a\,\frac{1 - r^n}{1 - r} \qquad (r \neq 1)$$

It can be shown that if $|r| < 1$, then r^n gets close to 0 as n gets large (you can easily convince yourself of this using a calculator). It follows that S_n gets close to $a/(1 - r)$ as n gets large, or

$$S_n \to \frac{a}{1 - r} \qquad \text{as} \qquad n \to \infty$$

Thus, the sum of this infinite geometric series is $a/(1 - r)$.

> Here is another way to arrive at the formula for the sum of an infinite geometric series:
>
> $$S = a + ar + ar^2 + ar^3 + \cdots$$
> $$= a + r(a + ar + ar^2 + \cdots)$$
> $$= a + rS$$
>
> Solve the equation $S = a + rS$ for S to get
>
> $$S - rS = a$$
> $$(1 - r)S = a$$
> $$S = \frac{a}{1 - r}$$

Sum of an Infinite Geometric Series

If $|r| < 1$, then the infinite geometric series

$$a + ar + ar^2 + ar^3 + ar^4 + \cdots + ar^{n-1} + \cdots$$

has the sum

$$S = \frac{a}{1 - r}$$

Example 6   Finding the Sum of an Infinite Geometric Series

Find the sum of the infinite geometric series

$$2 + \frac{2}{5} + \frac{2}{25} + \frac{2}{125} + \cdots + \frac{2}{5^n} + \cdots$$

Solution   We use the formula for the sum of an infinite geometric series. In this case, $a = 2$ and $r = \frac{1}{5}$. Thus, the sum of this infinite series is

$$S = \frac{2}{1 - \frac{1}{5}} = \frac{5}{2} \qquad \blacksquare$$

Example 7   Writing a Repeated Decimal as a Fraction

Find the fraction that represents the rational number $2.3\overline{51}$.

Solution   This repeating decimal can be written as a series:

$$\frac{23}{10} + \frac{51}{1000} + \frac{51}{100{,}000} + \frac{51}{10{,}000{,}000} + \frac{51}{1{,}000{,}000{,}000} + \cdots$$

After the first term, the terms of this series form an infinite geometric series with

$$a = \frac{51}{1000} \quad \text{and} \quad r = \frac{1}{100}$$

Thus, the sum of this part of the series is

$$S = \frac{\frac{51}{1000}}{1 - \frac{1}{100}} = \frac{\frac{51}{1000}}{\frac{99}{100}} = \frac{51}{1000} \cdot \frac{100}{99} = \frac{51}{990}$$

So,

$$2.3\overline{51} = \frac{23}{10} + \frac{51}{990} = \frac{2328}{990} = \frac{388}{165}$$

11.3 Exercises

1–4 ■ The nth term of a sequence is given.

(a) Find the first five terms of the sequence.

(b) What is the common ratio r?

(c) Graph the terms you found in (a).

1. $a_n = 5(2)^{n-1}$ **2.** $a_n = 3(-4)^{n-1}$

3. $a_n = \frac{5}{2}\left(-\frac{1}{2}\right)^{n-1}$ **4.** $a_n = 3^{n-1}$

5–8 ■ Find the nth term of the geometric sequence with given first term a and common ratio r. What is the fourth term?

5. $a = 3, \quad r = 5$ **6.** $a = -6, \quad r = 3$

7. $a = \frac{5}{2}, \quad r = -\frac{1}{2}$ **8.** $a = \sqrt{3}, \quad r = \sqrt{3}$

9–16 ■ Determine whether the sequence is geometric. If it is geometric, find the common ratio.

9. $2, 4, 8, 16, \ldots$ **10.** $2, 6, 18, 36, \ldots$

11. $3, \frac{3}{2}, \frac{3}{4}, \frac{3}{8}, \ldots$ **12.** $27, -9, 3, -1, \ldots$

13. $\frac{1}{2}, \frac{1}{3}, \frac{1}{4}, \frac{1}{5}, \ldots$ **14.** $e^2, e^4, e^6, e^8, \ldots$

15. $1.0, 1.1, 1.21, 1.331, \ldots$ **16.** $\frac{1}{2}, \frac{1}{4}, \frac{1}{6}, \frac{1}{8}, \ldots$

17–22 ■ Find the first five terms of the sequence and determine if it is geometric. If it is geometric, find the common ratio and express the nth term of the sequence in the standard form $a_n = ar^{n-1}$.

17. $a_n = 2(3)^n$ **18.** $a_n = 4 + 3^n$

19. $a_n = \dfrac{1}{4^n}$ **20.** $a_n = (-1)^n 2^n$

21. $a_n = \ln(5^{n-1})$ **22.** $a_n = n^n$

23–32 ■ Determine the common ratio, the fifth term, and the nth term of the geometric sequence.

23. $2, 6, 18, 54, \ldots$ **24.** $7, \frac{14}{3}, \frac{28}{9}, \frac{56}{27}, \ldots$

25. $0.3, -0.09, 0.027, -0.0081, \ldots$

26. $1, \sqrt{2}, 2, 2\sqrt{2}, \ldots$

27. $144, -12, 1, -\frac{1}{12}, \ldots$ **28.** $-8, -2, -\frac{1}{2}, -\frac{1}{8}, \ldots$

29. $3, 3^{5/3}, 3^{7/3}, 27, \ldots$ **30.** $t, \dfrac{t^2}{2}, \dfrac{t^3}{4}, \dfrac{t^4}{8}, \ldots$

31. $1, s^{2/7}, s^{4/7}, s^{6/7}, \ldots$ **32.** $5, 5^{c+1}, 5^{2c+1}, 5^{3c+1}, \ldots$

33. The first term of a geometric sequence is 8, and the second term is 4. Find the fifth term.

34. The first term of a geometric sequence is 3, and the third term is $\frac{4}{3}$. Find the fifth term.

35. The common ratio in a geometric sequence is $\frac{2}{5}$, and the fourth term is $\frac{5}{2}$. Find the third term.

36. The common ratio in a geometric sequence is $\frac{3}{2}$, and the fifth term is 1. Find the first three terms.

37. Which term of the geometric sequence $2, 6, 18, \ldots$ is 118,098?

38. The second and the fifth terms of a geometric sequence are 10 and 1250, respectively. Is 31,250 a term of this sequence? If so, which term is it?

39–42 ■ Find the partial sum S_n of the geometric sequence that satisfies the given conditions.

39. $a = 5, \quad r = 2, \quad n = 6$ **40.** $a = \frac{2}{3}, \quad r = \frac{1}{3}, \quad n = 4$

41. $a_3 = 28, \quad a_6 = 224, \quad n = 6$

42. $a_2 = 0.12, \quad a_5 = 0.00096, \quad n = 4$

43–46 ■ Find the sum.

43. $1 + 3 + 9 + \cdots + 2187$

44. $1 - \frac{1}{2} + \frac{1}{4} - \frac{1}{8} + \cdots - \frac{1}{512}$

45. $\displaystyle\sum_{k=0}^{10} 3\left(\frac{1}{2}\right)^k$ **46.** $\displaystyle\sum_{j=0}^{5} 7\left(\frac{3}{2}\right)^j$

47–54 ■ Find the sum of the infinite geometric series.

47. $1 + \dfrac{1}{3} + \dfrac{1}{9} + \dfrac{1}{27} + \cdots$ **48.** $1 - \dfrac{1}{2} + \dfrac{1}{4} - \dfrac{1}{8} + \cdots$

49. $1 - \dfrac{1}{3} + \dfrac{1}{9} - \dfrac{1}{27} + \cdots$ **50.** $\dfrac{2}{5} + \dfrac{4}{25} + \dfrac{8}{125} + \cdots$

51. $\dfrac{1}{3^6} + \dfrac{1}{3^8} + \dfrac{1}{3^{10}} + \dfrac{1}{3^{12}} + \cdots$

52. $3 - \dfrac{3}{2} + \dfrac{3}{4} - \dfrac{3}{8} + \cdots$

53. $-\dfrac{100}{9} + \dfrac{10}{3} - 1 + \dfrac{3}{10} - \cdots$

54. $\dfrac{1}{\sqrt{2}} + \dfrac{1}{2} + \dfrac{1}{2\sqrt{2}} + \dfrac{1}{4} + \cdots$

55–60 ■ Express the repeating decimal as a fraction.

55. $0.777\ldots$ **56.** $0.2\overline{53}$

57. $0.030303\ldots$ **58.** $2.11\overline{25}$

59. $0.\overline{112}$ **60.** $0.123123123\ldots$

61. If the numbers $a_1, a_2, \ldots, a_n$ form a geometric sequence, then $a_2, a_3, \ldots, a_{n-1}$ are **geometric means** between a_1 and a_n. Insert three geometric means between 5 and 80.

62. Find the sum of the first ten terms of the sequence

$$a + b, \ a^2 + 2b, \ a^3 + 3b, \ a^4 + 4b, \ldots$$

Applications

63. Depreciation A construction company purchases a bulldozer for $160,000. Each year the value of the bulldozer depreciates by 20% of its value in the preceding year. Let V_n be the value of the bulldozer in the nth year. (Let $n = 1$ be the year the bulldozer is purchased.)
  (a) Find a formula for V_n.
  (b) In what year will the value of the bulldozer be less than $100,000?

64. Family Tree A person has two parents, four grandparents, eight great-grandparents, and so on. How many ancestors does a person have 15 generations back?

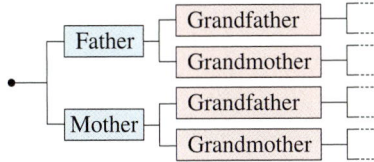

65. Bouncing Ball A ball is dropped from a height of 80 ft. The elasticity of this ball is such that it rebounds three-fourths of the distance it has fallen. How high does the ball rebound on the fifth bounce? Find a formula for how high the ball rebounds on the nth bounce.

66. Bacteria Culture A culture initially has 5000 bacteria, and its size increases by 8% every hour. How many bacteria

are present at the end of 5 hours? Find a formula for the number of bacteria present after n hours.

67. Mixing Coolant A truck radiator holds 5 gal and is filled with water. A gallon of water is removed from the radiator and replaced with a gallon of antifreeze; then, a gallon of the mixture is removed from the radiator and again replaced by a gallon of antifreeze. This process is repeated indefinitely. How much water remains in the tank after this process is repeated 3 times? 5 times? n times?

68. Musical Frequencies The frequencies of musical notes (measured in cycles per second) form a geometric sequence. Middle C has a frequency of 256, and the C that is an octave higher has a frequency of 512. Find the frequency of C two octaves below middle C.

69. Bouncing Ball A ball is dropped from a height of 9 ft. The elasticity of the ball is such that it always bounces up one-third the distance it has fallen.
  (a) Find the total distance the ball has traveled at the instant it hits the ground the fifth time.
  (b) Find a formula for the total distance the ball has traveled at the instant it hits the ground the nth time.

70. Geometric Savings Plan A very patient woman wishes to become a billionaire. She decides to follow a simple scheme: She puts aside 1 cent the first day, 2 cents the second day, 4 cents the third day, and so on, doubling the number of cents each day. How much money will she have at the end of 30 days? How many days will it take this woman to realize her wish?

71. St. Ives The following is a well-known children's rhyme:

> As I was going to St. Ives
> I met a man with seven wives;
> Every wife had seven sacks;
> Every sack had seven cats;
> Every cat had seven kits;
> Kits, cats, sacks, and wives,
> How many were going to St. Ives?

Assuming that the entire group is actually going to St. Ives, show that the answer to the question in the rhyme is a partial sum of a geometric sequence, and find the sum.

72. Drug Concentration A certain drug is administered once a day. The concentration of the drug in the patient's blood-stream increases rapidly at first, but each successive dose has less effect than the preceding one. The total amount of the drug (in mg) in the bloodstream after the nth dose is given by

$$\sum_{k=1}^{n} 50\left(\tfrac{1}{2}\right)^{k-1}$$

(a) Find the amount of the drug in the bloodstream after $n = 10$ days.

(b) If the drug is taken on a long-term basis, the amount in the bloodstream is approximated by the infinite series $\sum_{k=1}^{\infty} 50\left(\frac{1}{2}\right)^{k-1}$. Find the sum of this series.

73. Bouncing Ball A certain ball rebounds to half the height from which it is dropped. Use an infinite geometric series to approximate the total distance the ball travels, after being dropped from 1 m above the ground, until it comes to rest.

74. Bouncing Ball If the ball in Exercise 73 is dropped from a height of 8 ft, then 1 s is required for its first complete bounce—from the instant it first touches the ground until it next touches the ground. Each subsequent complete bounce requires $1/\sqrt{2}$ as long as the preceding complete bounce. Use an infinite geometric series to estimate the time interval from the instant the ball first touches the ground until it stops bouncing.

75. Geometry The midpoints of the sides of a square of side 1 are joined to form a new square. This procedure is repeated for each new square. (See the figure.)

(a) Find the sum of the areas of all the squares.

(b) Find the sum of the perimeters of all the squares.

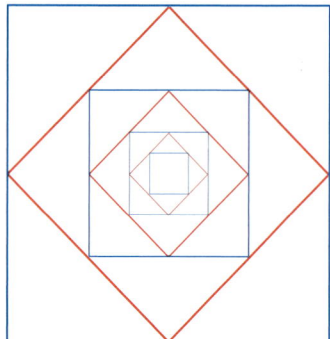

76. Geometry A circular disk of radius R is cut out of paper, as shown in figure (a). Two disks of radius $\frac{1}{2}R$ are cut out of paper and placed on top of the first disk, as in figure (b), and then four disks of radius $\frac{1}{4}R$ are placed on these two disks (figure (c)). Assuming that this process can be repeated indefinitely, find the total area of all the disks.

 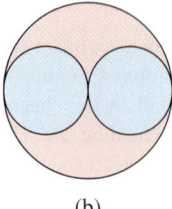 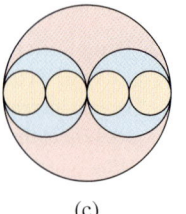

(a)   (b)   (c)

77. Geometry A yellow square of side 1 is divided into nine smaller squares, and the middle square is colored blue as shown in the figure. Each of the smaller yellow squares is in turn divided into nine squares, and each middle square is colored blue. If this process is continued indefinitely, what is the total area colored blue?

 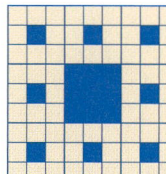 

Discovery • Discussion

78. Arithmetic or Geometric? The first four terms of a sequence are given. Determine whether these terms can be the terms of an arithmetic sequence, a geometric sequence, or neither. Find the next term if the sequence is arithmetic or geometric.

(a) $5, -3, 5, -3, \ldots$ **(b)** $\frac{1}{3}, 1, \frac{5}{3}, \frac{7}{3}, \ldots$

(c) $\sqrt{3}, 3, 3\sqrt{3}, 9, \ldots$ **(d)** $1, -1, 1, -1, \ldots$

(e) $2, -1, \frac{1}{2}, 2, \ldots$ **(f)** $x-1, x, x+1, x+2, \ldots$

(g) $-3, -\frac{3}{2}, 0, \frac{3}{2}, \ldots$ **(h)** $\sqrt{5}, \sqrt[3]{5}, \sqrt[6]{5}, 1, \ldots$

79. Reciprocals of a Geometric Sequence If $a_1, a_2, a_3, \ldots$ is a geometric sequence with common ratio r, show that the sequence

$$\frac{1}{a_1}, \frac{1}{a_2}, \frac{1}{a_3}, \ldots$$

is also a geometric sequence, and find the common ratio.

80. Logarithms of a Geometric Sequence If $a_1, a_2, a_3, \ldots$ is a geometric sequence with a common ratio $r > 0$ and $a_1 > 0$, show that the sequence

$$\log a_1, \log a_2, \log a_3, \ldots$$

is an arithmetic sequence, and find the common difference.

81. Exponentials of an Arithmetic Sequence If $a_1, a_2, a_3, \ldots$ is an arithmetic sequence with common difference d, show that the sequence

$$10^{a_1}, 10^{a_2}, 10^{a_3}, \ldots$$

is a geometric sequence, and find the common ratio.

DISCOVERY PROJECT

Finding Patterns

The ancient Greeks studied triangular numbers, square numbers, pentagonal numbers, and other **polygonal numbers**, like those shown in the figure.

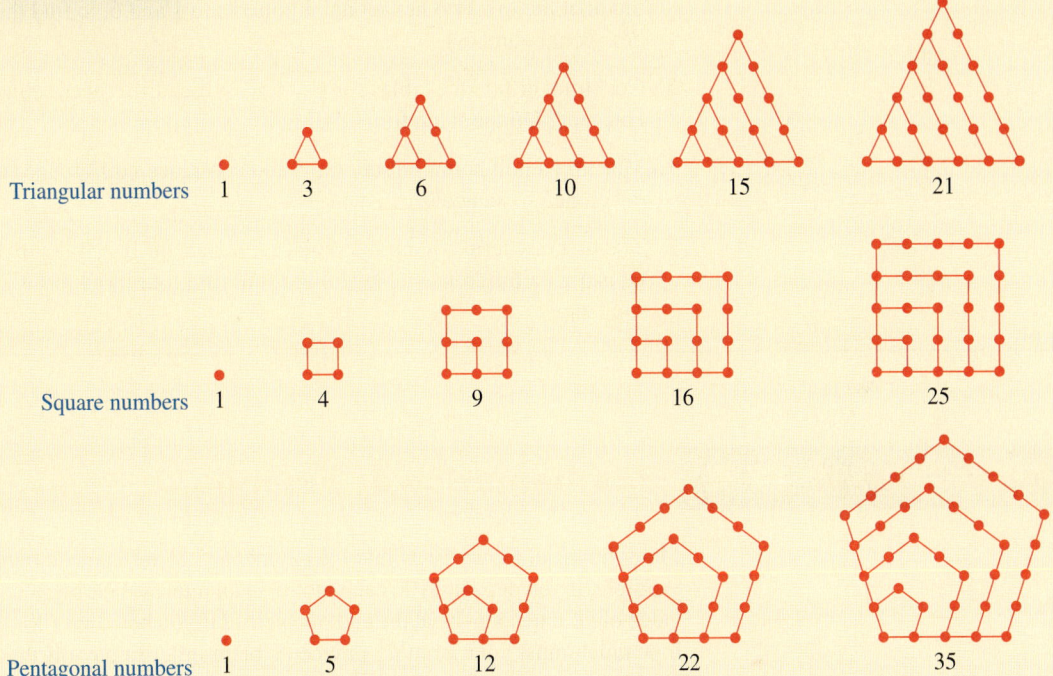

Triangular numbers 1   3   6   10   15   21

Square numbers 1   4   9   16   25

Pentagonal numbers 1   5   12   22   35

To find a pattern for such numbers, we construct a **first difference sequence** by taking differences of successive terms; we repeat the process to get a **second difference sequence, third difference sequence**, and so on. For the sequence of triangular numbers T_n we get the following **difference table**:

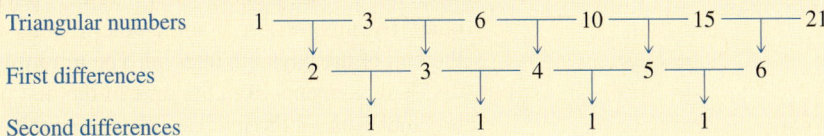

Triangular numbers   1   3   6   10   15   21

First differences   2   3   4   5   6

Second differences   1   1   1   1

We stop at the second difference sequence because it's a constant sequence. Assuming that this sequence will continue to have constant value 1, we can work backward from the bottom row to find more terms of the first difference sequence, and from these, more triangular numbers.

If a sequence is given by a polynomial function and if we calculate the first differences, the second differences, the third differences, and so on, then eventually we get a constant sequence. For example, the triangular numbers are given by the polynomial $T_n = \frac{1}{2}n^2 + \frac{1}{2}n$ (see the margin note on the next page); the second difference sequence is the constant sequence $1, 1, 1, \ldots$.

The formula for the *n*th triangular number can be found using the formula for the sum of the first *n* whole numbers (Example 2, Section 11.5). From the definition of T_n we have

$$T_n = 1 + 2 + \cdots + n$$
$$= \frac{n(n+1)}{2}$$
$$= \tfrac{1}{2}n^2 + \tfrac{1}{2}n$$

1. Construct a difference table for the square numbers and the pentagonal numbers. Use your table to find the tenth pentagonal number.

2. From the patterns you've observed so far, what do you think the second difference would be for the *hexagonal numbers*? Use this, together with the fact that the first two hexagonal numbers are 1 and 6, to find the first eight hexagonal numbers.

3. Construct difference tables for $C_n = n^3$. Which difference sequence is constant? Do the same for $F_n = n^4$.

4. Make up a polynomial of degree 5 and construct a difference table. Which difference sequence is constant?

5. The first few terms of a polynomial sequence are 1, 2, 4, 8, 16, 31, 57, ... Construct a difference table and use it to find four more terms of this sequence.

11.4   Mathematics of Finance

Many financial transactions involve payments that are made at regular intervals. For example, if you deposit $100 each month in an interest-bearing account, what will the value of your account be at the end of 5 years? If you borrow $100,000 to buy a house, how much must your monthly payments be in order to pay off the loan in 30 years? Each of these questions involves the sum of a sequence of numbers; we use the results of the preceding section to answer them here.

The Amount of an Annuity

An **annuity** is a sum of money that is paid in regular equal payments. Although the word *annuity* suggests annual (or yearly) payments, they can be made semiannually, quarterly, monthly, or at some other regular interval. Payments are usually made at the end of the payment interval. The **amount of an annuity** is the sum of all the individual payments from the time of the first payment until the last payment is made, together with all the interest. We denote this sum by A_f (the subscript *f* here is used to denote *final* amount).

Example 1   Calculating the Amount of an Annuity

An investor deposits $400 every December 15 and June 15 for 10 years in an account that earns interest at the rate of 8% per year, compounded semiannually. How much will be in the account immediately after the last payment?

Solution   We need to find the amount of an annuity consisting of 20 semiannual payments of $400 each. Since the interest rate is 8% per year, compounded semiannually, the interest rate per time period is $i = 0.08/2 = 0.04$. The first payment is in the account for 19 time periods, the second for 18 time periods, and so on.

When using interest rates in calculators, remember to convert percentages to decimals. For example, 8% is 0.08.

The last payment receives no interest. The situation can be illustrated by the time line in Figure 1.

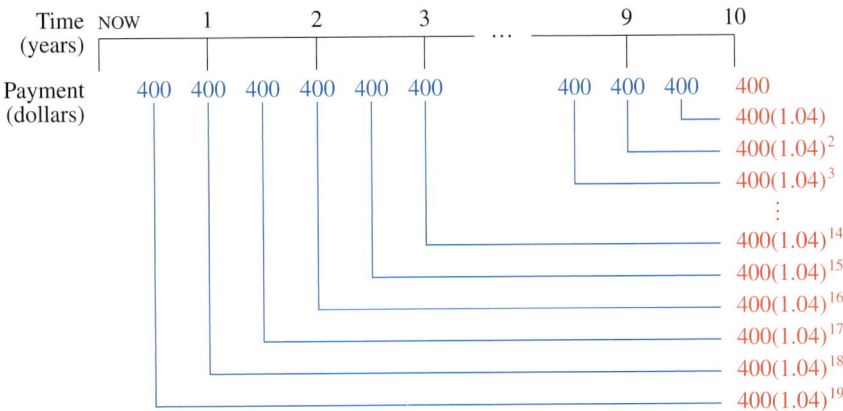

Figure 1

The amount A_f of the annuity is the sum of these 20 amounts. Thus

$$A_f = 400 + 400(1.04) + 400(1.04)^2 + \cdots + 400(1.04)^{19}$$

But this is a geometric series with $a = 400$, $r = 1.04$, and $n = 20$, so

$$A_f = 400 \frac{1 - (1.04)^{20}}{1 - 1.04} \approx 11{,}911.23$$

Thus, the amount in the account after the last payment is \$11,911.23. ∎

In general, the regular annuity payment is called the **periodic rent** and is denoted by R. We also let i denote the interest rate per time period and n the number of payments. *We always assume that the time period in which interest is compounded is equal to the time between payments.* By the same reasoning as in Example 1, we see that the amount A_f of an annuity is

$$A_f = R + R(1 + i) + R(1 + i)^2 + \cdots + R(1 + i)^{n-1}$$

Since this is the nth partial sum of a geometric sequence with $a = R$ and $r = 1 + i$, the formula for the partial sum gives

$$A_f = R \frac{1 - (1 + i)^n}{1 - (1 + i)} = R \frac{1 - (1 + i)^n}{-i} = R \frac{(1 + i)^n - 1}{i}$$

Amount of an Annuity

The amount A_f of an annuity consisting of n regular equal payments of size R with interest rate i per time period is given by

$$A_f = R \frac{(1 + i)^n - 1}{i}$$

Example 2   Calculating the Amount of an Annuity

How much money should be invested every month at 12% per year, compounded monthly, in order to have $4000 in 18 months?

Solution   In this problem $i = 0.12/12 = 0.01$, $A_f = 4000$, and $n = 18$. We need to find the amount R of each payment. By the formula for the amount of an annuity,

$$4000 = R\,\frac{(1 + 0.01)^{18} - 1}{0.01}$$

Solving for R, we get

$$R = \frac{4000(0.01)}{(1 + 0.01)^{18} - 1} \approx 203.928$$

Thus, the monthly investment should be $203.93. ∎

The Present Value of an Annuity

If you were to receive $10,000 five years from now, it would be worth much less than getting $10,000 right now. This is because of the interest you could accumulate during the next five years if you invested the money now. What smaller amount would you be willing to accept *now* instead of receiving $10,000 in five years? This is the amount of money that, together with interest, would be worth $10,000 in five years. The amount we are looking for here is called the *discounted value* or *present value*. If the interest rate is 8% per year, compounded quarterly, then the interest per time period is $i = 0.08/4 = 0.02$, and there are $4 \times 5 = 20$ time periods. If we let PV denote the present value, then by the formula for compound interest (Section 4.1) we have

$$10,000 = PV(1 + i)^n = PV(1 + 0.02)^{20}$$

so

$$PV = 10,000(1 + 0.02)^{-20} \approx 6729.713$$

Thus, in this situation, the present value of $10,000 is $6729.71. This reasoning leads to a general formula for present value:

$$PV = A(1 + i)^{-n}$$

Similarly, the **present value of an annuity** is the amount A_p that must be invested now at the interest rate i per time period in order to provide n payments, each of amount R. Clearly, A_p is the sum of the present values of each individual payment (see Exercise 22). Another way of finding A_p is to note that A_p is the present value of A_f:

$$A_p = A_f(1 + i)^{-n} = R\,\frac{(1 + i)^n - 1}{i}(1 + i)^{-n} = R\,\frac{1 - (1 + i)^{-n}}{i}$$

The Present Value of an Annuity

The **present value** A_p of an annuity consisting of n regular equal payments of size R and interest rate i per time period is given by

$$A_p = R\,\frac{1 - (1 + i)^{-n}}{i}$$

Example 3   Calculating the Present Value of an Annuity

A person wins $10,000,000 in the California lottery, and the amount is paid in yearly installments of half a million dollars each for 20 years. What is the present value of his winnings? Assume that he can earn 10% interest, compounded annually.

Solution   Since the amount won is paid as an annuity, we need to find its present value. Here $i = 0.1$, $R = \$500,000$, and $n = 20$. Thus

$$A_p = 500,000 \, \frac{1 - (1 + 0.1)^{-20}}{0.1} \approx 4,256,781.859$$

This means that the winner really won only $4,256,781.86 if it were paid immediately.   ∎

Installment Buying

When you buy a house or a car by installment, the payments you make are an annuity whose present value is the amount of the loan.

Example 4   The Amount of a Loan

A student wishes to buy a car. He can afford to pay $200 per month but has no money for a down payment. If he can make these payments for four years and the interest rate is 12%, what purchase price can he afford?

Solution   The payments the student makes constitute an annuity whose present value is the price of the car (which is also the amount of the loan, in this case). Here we have $i = 0.12/12 = 0.01$, $R = 200$, $n = 12 \times 4 = 48$, so

$$A_p = R \, \frac{1 - (1 + i)^{-n}}{i} = 200 \, \frac{1 - (1 + 0.01)^{-48}}{0.01} \approx 7594.792$$

Thus, the student can buy a car priced at $7594.79.   ∎

When a bank makes a loan that is to be repaid with regular equal payments R, then the payments form an annuity whose present value A_p is the amount of the loan. So, to find the size of the payments, we solve for R in the formula for the amount of an annuity. This gives the following formula for R.

Installment Buying

If a loan A_p is to be repaid in n regular equal payments with interest rate i per time period, then the size R of each payment is given by

$$R = \frac{iA_p}{1 - (1 + i)^{-n}}$$

Example 5   Calculating Monthly Mortgage Payments

A couple borrows $100,000 at 9% interest as a mortgage loan on a house. They expect to make monthly payments for 30 years to repay the loan. What is the size of each payment?

Solution   The mortgage payments form an annuity whose present value is $A_p = \$100,000$. Also, $i = 0.09/12 = 0.0075$, and $n = 12 \times 30 = 360$. We are looking for the amount R of each payment. From the formula for installment buying, we get

$$R = \frac{iA_p}{1 - (1+i)^{-n}}$$

$$= \frac{(0.0075)(100,000)}{1 - (1 + 0.0075)^{-360}} \approx 804.623$$

Thus, the monthly payments are $804.62.   ■

We now illustrate the use of graphing devices in solving problems related to installment buying.

Example 6   Calculating the Interest Rate from the Size of Monthly Payments

A car dealer sells a new car for $18,000. He offers the buyer payments of $405 per month for 5 years. What interest rate is this car dealer charging?

Solution   The payments form an annuity with present value $A_p = \$18,000$, $R = 405$, and $n = 12 \times 5 = 60$. To find the interest rate, we must solve for i in the equation

$$R = \frac{iA_p}{1 - (1+i)^{-n}}$$

A little experimentation will convince you that it's not possible to solve this equation for i algebraically. So, to find i we use a graphing device to graph R as a function of the interest rate x, and we then use the graph to find the interest rate corresponding to the value of R we want ($405 in this case). Since $i = x/12$, we graph the function

$$R(x) = \frac{\dfrac{x}{12}(18,000)}{1 - \left(1 + \dfrac{x}{12}\right)^{-60}}$$

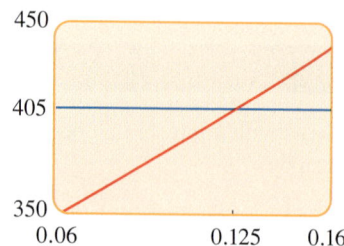

Figure 2

in the viewing rectangle $[0.06, 0.16] \times [350, 450]$, as shown in Figure 2. We also graph the horizontal line $R(x) = 405$ in the same viewing rectangle. Then, by moving the cursor to the point of intersection of the two graphs, we find that the corresponding x-value is approximately 0.125. Thus, the interest rate is about $12\frac{1}{2}\%$.   ■

11.4 Exercises

1. Annuity Find the amount of an annuity that consists of 10 annual payments of $1000 each into an account that pays 6% interest per year.

2. Annuity Find the amount of an annuity that consists of 24 monthly payments of $500 each into an account that pays 8% interest per year, compounded monthly.

3. Annuity Find the amount of an annuity that consists of 20 annual payments of $5000 each into an account that pays interest of 12% per year.

4. Annuity Find the amount of an annuity that consists of 20 semiannual payments of $500 each into an account that pays 6% interest per year, compounded semiannually.

5. Annuity Find the amount of an annuity that consists of 16 quarterly payments of $300 each into an account that pays 8% interest per year, compounded quarterly.

6. Saving How much money should be invested every quarter at 10% per year, compounded quarterly, in order to have $5000 in 2 years?

7. Saving How much money should be invested monthly at 6% per year, compounded monthly, in order to have $2000 in 8 months?

8. Annuity What is the present value of an annuity that consists of 20 semiannual payments of $1000 at the interest rate of 9% per year, compounded semiannually?

9. Funding an Annuity How much money must be invested now at 9% per year, compounded semiannually, to fund an annuity of 20 payments of $200 each, paid every 6 months, the first payment being 6 months from now?

10. Funding an Annuity A 55-year-old man deposits $50,000 to fund an annuity with an insurance company. The money will be invested at 8% per year, compounded semiannually. He is to draw semiannual payments until he reaches age 65. What is the amount of each payment?

11. Financing a Car A woman wants to borrow $12,000 in order to buy a car. She wants to repay the loan by monthly installments for 4 years. If the interest rate on this loan is $10\frac{1}{2}$% per year, compounded monthly, what is the amount of each payment?

12. Mortgage What is the monthly payment on a 30-year mortgage of $80,000 at 9% interest? What is the monthly payment on this same mortgage if it is to be repaid over a 15-year period?

13. Mortgage What is the monthly payment on a 30-year mortgage of $100,000 at 8% interest per year, compounded monthly? What is the total amount paid on this loan over the 30-year period?

14. Mortgage A couple can afford to make a monthly mortgage payment of $650. If the mortgage rate is 9% and the

couple intends to secure a 30-year mortgage, how much can they borrow?

15. Mortgage A couple secures a 30-year loan of $100,000 at $9\frac{3}{4}$% per year, compounded monthly, to buy a house.
(a) What is the amount of their monthly payment?
(b) What total amount will they pay over the 30-year period?
(c) If, instead of taking the loan, the couple deposits the monthly payments in an account that pays $9\frac{3}{4}$% interest per year, compounded monthly, how much will be in the account at the end of the 30-year period?

16. Financing a Car Jane agrees to buy a car for a down payment of $2000 and payments of $220 per month for 3 years. If the interest rate is 8% per year, compounded monthly, what is the actual purchase price of her car?

17. Financing a Ring Mike buys a ring for his fiancee by paying $30 a month for one year. If the interest rate is 10% per year, compounded monthly, what is the price of the ring?

18. Interest Rate Janet's payments on her $12,500 car are $420 a month for 3 years. Assuming that interest is compounded monthly, what interest rate is she paying on the car loan?

19. Interest Rate John buys a stereo system for $640. He agrees to pay $32 a month for 2 years. Assuming that interest is compounded monthly, what interest rate is he paying?

20. Interest Rate A man purchases a $2000 diamond ring for a down payment of $200 and monthly installments of $88 for 2 years. Assuming that interest is compounded monthly, what interest rate is he paying?

21. Interest Rate An item at a department store is priced at $189.99 and can be bought by making 20 payments of $10.50. Find the interest rate, assuming that interest is compounded monthly.

Discovery • Discussion

22. Present Value of an Annuity (a) Draw a time line as in Example 1 to show that the present value of an annuity is the sum of the present values of each payment, that is,

$$A_p = \frac{R}{1+i} + \frac{R}{(1+i)^2} + \frac{R}{(1+i)^3} + \cdots + \frac{R}{(1+i)^n}$$

(b) Use part (a) to derive the formula for A_p given in the text.

23. An Annuity That Lasts Forever An **annuity in perpetuity** is one that continues forever. Such annuities are useful in setting up scholarship funds to ensure that the award continues.

(a) Draw a time line (as in Example 1) to show that to set up an annuity in perpetuity of amount R per time period, the amount that must be invested now is

$$A_p = \frac{R}{1+i} + \frac{R}{(1+i)^2} + \frac{R}{(1+i)^3} + \cdots + \frac{R}{(1+i)^n} + \cdots$$

where i is the interest rate per time period.

(b) Find the sum of the infinite series in part (a) to show that

$$A_p = \frac{R}{i}$$

(c) How much money must be invested now at 10% per year, compounded annually, to provide an annuity in perpetuity of $5000 per year? The first payment is due in one year.

(d) How much money must be invested now at 8% per year, compounded quarterly, to provide an annuity in perpetuity of $3000 per year? The first payment is due in one year.

24. Amortizing a Mortgage   When they bought their house, John and Mary took out a $90,000 mortgage at 9% interest, repayable monthly over 30 years. Their payment is $724.17 per month (check this using the formula in the text). The

bank gave them an **amortization schedule**, which is a table showing how much of each payment is interest, how much goes toward the principal, and the remaining principal after each payment. The table below shows the first few entries in the amortization schedule.

Payment number	Total payment	Interest payment	Principal payment	Remaining principal
1	724.17	675.00	49.17	89,950.83
2	724.17	674.63	49.54	89,901.29
3	724.17	674.26	49.91	89,851.38
4	724.17	673.89	50.28	89,801.10

After 10 years they have made 120 payments and are wondering how much they still owe, but they have lost the amortization schedule.

(a) How much do John and Mary still owe on their mortgage? [*Hint:* The remaining balance is the present value of the 240 remaining payments.]

(b) How much of their next payment is interest and how much goes toward the principal? [*Hint:* Since 9% ÷ 12 = 0.75%, they must pay 0.75% of the remaining principal in interest each month.]

11.5   Mathematical Induction

There are two aspects to mathematics—discovery and proof—and both are of equal importance. We must discover something before we can attempt to prove it, and we can only be certain of its truth once it has been proved. In this section we examine the relationship between these two key components of mathematics more closely.

Conjecture and Proof

Let's try a simple experiment. We add more and more of the odd numbers as follows:

$$1 = 1$$
$$1 + 3 = 4$$
$$1 + 3 + 5 = 9$$
$$1 + 3 + 5 + 7 = 16$$
$$1 + 3 + 5 + 7 + 9 = 25$$

What do you notice about the numbers on the right side of these equations? They are in fact all perfect squares. These equations say the following:

The sum of the first 1 odd number is 1^2.

The sum of the first 2 odd numbers is 2^2.

The sum of the first 3 odd numbers is 3^2.

The sum of the first 4 odd numbers is 4^2.

The sum of the first 5 odd numbers is 5^2.

Consider the polynomial
$$p(n) = n^2 - n + 41$$
Here are some values of $p(n)$:

$$p(1) = 41 \quad p(2) = 43$$
$$p(3) = 47 \quad p(4) = 53$$
$$p(5) = 61 \quad p(6) = 71$$
$$p(7) = 83 \quad p(8) = 97$$

All the values so far are prime numbers. In fact, if you keep going, you will find $p(n)$ is prime for all natural numbers up to $n = 40$. It may seem reasonable at this point to conjecture that $p(n)$ is prime for *every* natural number n. But out conjecture would be too hasty, because it is easily seen that $p(41)$ is *not* prime. This illustrates that we cannot be certain of the truth of a statement no matter how many special cases we check. We need a convincing argument—a *proof*—to determine the truth of a statement.

This leads naturally to the following question: Is it true that for every natural number n, the sum of the first n odd numbers is n^2? Could this remarkable property be true? We could try a few more numbers and find that the pattern persists for the first 6, 7, 8, 9, and 10 odd numbers. At this point, we feel quite sure that this is always true, so we make a *conjecture*:

The sum of the first n odd numbers is n^2.

Since we know that the nth odd number is $2n - 1$, we can write this statement more precisely as

$$1 + 3 + 5 + \cdots + (2n - 1) = n^2$$

It's important to realize that this is still a conjecture. We cannot conclude by checking a finite number of cases that a property is true for all numbers (there are infinitely many). To see this more clearly, suppose someone tells us he has added up the first trillion odd numbers and found that they do *not* add up to 1 trillion squared. What would you tell this person? It would be silly to say that you're sure it's true because you've already checked the first five cases. You could, however, take out paper and pencil and start checking it yourself, but this task would probably take the rest of your life. The tragedy would be that after completing this task you would still not be sure of the truth of the conjecture! Do you see why?

Herein lies the power of mathematical proof. A **proof** is a clear argument that demonstrates the truth of a statement beyond doubt.

Mathematical Induction

Let's consider a special kind of proof called **mathematical induction**. Here is how it works: Suppose we have a statement that says something about all natural numbers n. Let's call this statement P. For example, we could consider the statement

P: For every natural number n, the sum of the first n odd numbers is n^2.

Since this statement is about *all* natural numbers, it contains infinitely many statements; we will call them $P(1), P(2), \ldots$.

$P(1)$:   The sum of the first 1 odd number is 1^2.

$P(2)$:   The sum of the first 2 odd numbers is 2^2.

$P(3)$:   The sum of the first 3 odd numbers is 3^2.

How can we prove all of these statements at once? Mathematical induction is a clever way of doing just that.

The crux of the idea is this: Suppose we can prove that whenever one of these statements is true, then the one following it in the list is also true. In other words,

For every k, if $P(k)$ is true, then $P(k + 1)$ is true.

This is called the **induction step** because it leads us from the truth of one statement to the next. Now, suppose that we can also prove that

$P(1)$ is true.

The induction step now leads us through the following chain of statements:

$$P(1) \text{ is true, so } P(2) \text{ is true.}$$
$$P(2) \text{ is true, so } P(3) \text{ is true.}$$
$$P(3) \text{ is true, so } P(4) \text{ is true.}$$

So we see that if both the induction step and $P(1)$ are proved, then statement P is proved for all n. Here is a summary of this important method of proof.

Principle of Mathematical Induction

For each natural number n, let $P(n)$ be a statement depending on n. Suppose that the following two conditions are satisfied.

1. $P(1)$ is true.

2. For every natural number k, if $P(k)$ is true then $P(k + 1)$ is true.

Then $P(n)$ is true for all natural numbers n.

To apply this principle, there are two steps:

Step 1 Prove that $P(1)$ is true.

Step 2 Assume that $P(k)$ is true and use this assumption to prove that $P(k + 1)$ is true.

Notice that in Step 2 we do not prove that $P(k)$ is true. We only show that *if* $P(k)$ is true, *then* $P(k + 1)$ is also true. The assumption that $P(k)$ is true is called the **induction hypothesis**.

©1979 National Council of Teachers of Mathematics. Used by permission. Courtesy of Andrejs Dunkels, Sweden.

We now use mathematical induction to prove that the conjecture we made at the beginning of this section is true.

Example 1   A Proof by Mathematical Induction

Prove that for all natural numbers n,

$$1 + 3 + 5 + \cdots + (2n - 1) = n^2$$

Solution   Let $P(n)$ denote the statement $1 + 3 + 5 + \cdots + (2n - 1) = n^2$.

Step 1   We need to show that $P(1)$ is true. But $P(1)$ is simply the statement that $1 = 1^2$, which is of course true.

Step 2   We assume that $P(k)$ is true. Thus, our induction hypothesis is

$$1 + 3 + 5 + \cdots + (2k - 1) = k^2$$

We want to use this to show that $P(k + 1)$ is true, that is,

$$1 + 3 + 5 + \cdots + (2k - 1) + [2(k + 1) - 1] = (k + 1)^2$$

[Note that we get $P(k + 1)$ by substituting $k + 1$ for each n in the statement $P(n)$.] We start with the left side and use the induction hypothesis to obtain the right side of the equation:

$$1 + 3 + 5 + \cdots + (2k - 1) + [2(k + 1) - 1]$$

This equals k^2 by the induction hypothesis.

$$= [1 + 3 + 5 + \cdots + (2k - 1)] + [2(k + 1) - 1] \quad \text{Group the first } k \text{ terms}$$

$$= k^2 + [2(k + 1) - 1] \quad \text{Induction hypothesis}$$

$$= k^2 + [2k + 2 - 1] \quad \text{Distributive Property}$$

$$= k^2 + 2k + 1 \quad \text{Simplify}$$

$$= (k + 1)^2 \quad \text{Factor}$$

Thus, $P(k + 1)$ follows from $P(k)$ and this completes the induction step.

Having proved Steps 1 and 2, we conclude by the Principle of Mathematical Induction that $P(n)$ is true for all natural numbers n. ■

Example 2   A Proof by Mathematical Induction

Prove that for every natural number n,

$$1 + 2 + 3 + \cdots + n = \frac{n(n + 1)}{2}$$

Solution   Let $P(n)$ be the statement $1 + 2 + 3 + \cdots + n = n(n + 1)/2$. We want to show that $P(n)$ is true for all natural numbers n.

Step 1   We need to show that $P(1)$ is true. But $P(1)$ says that

$$1 = \frac{1(1 + 1)}{2}$$

and this statement is clearly true.

Archivo Inconographico, S.A./Corbis

Blaise Pascal (1623–1662) is considered one of the most versatile minds in modern history. He was a writer and philosopher as well as a gifted mathematician and physicist. Among his contributions that appear in this book are Pascal's triangle and the Principle of Mathematical Induction.

Pascal's father, himself a mathematician, believed that his son should not study mathematics until he was 15 or 16. But at age 12, Blaise insisted on learning geometry, and proved most of its elementary theorems himself. At 19, he invented the first mechanical adding machine. In 1647, after writing a major treatise on the conic sections, he abruptly abandoned mathematics because he felt his intense studies were contributing to his ill health. He devoted himself instead to frivolous recreations such as gambling, but this only served to pique his interest in probability. In 1654 he miraculously survived a carriage accident in which his horses ran off a bridge. Taking this to be a sign from God, he entered a monastery, where he pursued theology and philosophy, writing his famous *Pensées*. He also continued his mathematical research. He valued faith and intuition more than reason as the source of truth, declaring that "the heart has its own reasons, which reason cannot know."

Step 2 Assume that $P(k)$ is true. Thus, our induction hypothesis is

$$1 + 2 + 3 + \cdots + k = \frac{k(k + 1)}{2}$$

We want to use this to show that $P(k + 1)$ is true, that is,

$$1 + 2 + 3 + \cdots + k + (k + 1) = \frac{(k + 1)[(k + 1) + 1]}{2}$$

So, we start with the left side and use the induction hypothesis to obtain the right side:

$1 + 2 + 3 + \cdots + k + (k + 1)$

$= [1 + 2 + 3 + \cdots + k] + (k + 1)$ *Group the first k terms*

$= \dfrac{k(k + 1)}{2} + (k + 1)$ *Induction hypothesis*

$= (k + 1)\left(\dfrac{k}{2} + 1\right)$ *Factor k + 1*

$= (k + 1)\left(\dfrac{k + 2}{2}\right)$ *Common denominator*

$= \dfrac{(k + 1)[(k + 1) + 1]}{2}$ *Write k + 2 as k + 1 + 1*

Thus, $P(k + 1)$ follows from $P(k)$ and this completes the induction step.

Having proved Steps 1 and 2, we conclude by the Principle of Mathematical Induction that $P(n)$ is true for all natural numbers n. ∎

Formulas for the sums of powers of the first n natural numbers are important in calculus. Formula 1 in the following box is proved in Example 2. The other formulas are also proved using mathematical induction (see Exercises 4 and 7).

Sums of Powers

0. $\displaystyle\sum_{k=1}^{n} 1 = n$ **1.** $\displaystyle\sum_{k=1}^{n} k = \frac{n(n + 1)}{2}$

2. $\displaystyle\sum_{k=1}^{n} k^2 = \frac{n(n + 1)(2n + 1)}{6}$ **3.** $\displaystyle\sum_{k=1}^{n} k^3 = \frac{n^2(n + 1)^2}{4}$

It might happen that a statement $P(n)$ is false for the first few natural numbers, but true from some number on. For example, we may want to prove that $P(n)$ is true for $n \geq 5$. Notice that if we prove that $P(5)$ is true, then this fact, together with the induction step, would imply the truth of $P(5), P(6), P(7), \ldots$. The next example illustrates this point.

Example 3   **Proving an Inequality by Mathematical Induction**

Prove that $4n < 2^n$ for all $n \geq 5$.

Solution Let $P(n)$ denote the statement $4n < 2^n$.

Step 1 $P(5)$ is the statement that $4 \cdot 5 < 2^5$, or $20 < 32$, which is true.

Step 2 Assume that $P(k)$ is true. Thus, our induction hypothesis is

$$4k < 2^k$$

We want to use this to show that $P(k + 1)$ is true, that is,

$$4(k + 1) < 2^{k+1}$$

We get $P(k + 1)$ by replacing k by $k + 1$ in the statement $P(k)$.

So, we start with the left side of the inequality and use the induction hypothesis to show that it is less than the right side. For $k \geq 5$, we have

$$
\begin{aligned}
4(k + 1) &= 4k + 4 \\
&< 2^k + 4 \qquad \textit{Induction hypothesis} \\
&< 2^k + 4k \qquad \textit{Because } 4 < 4k \\
&< 2^k + 2^k \qquad \textit{Induction hypothesis} \\
&= 2 \cdot 2^k \\
&= 2^{k+1} \qquad \textit{Property of exponents}
\end{aligned}
$$

Thus, $P(k + 1)$ follows from $P(k)$ and this completes the induction step.

Having proved Steps 1 and 2, we conclude by the Principle of Mathematical Induction that $P(n)$ is true for all natural numbers $n \geq 5$. ∎

11.5 Exercises

1–12 ■ Use mathematical induction to prove that the formula is true for all natural numbers n.

1. $2 + 4 + 6 + \cdots + 2n = n(n + 1)$

2. $1 + 4 + 7 + \cdots + (3n - 2) = \dfrac{n(3n - 1)}{2}$

3. $5 + 8 + 11 + \cdots + (3n + 2) = \dfrac{n(3n + 7)}{2}$

4. $1^2 + 2^2 + 3^2 + \cdots + n^2 = \dfrac{n(n + 1)(2n + 1)}{6}$

5. $1 \cdot 2 + 2 \cdot 3 + 3 \cdot 4 + \cdots + n(n + 1)$
$$= \dfrac{n(n + 1)(n + 2)}{3}$$

6. $1 \cdot 3 + 2 \cdot 4 + 3 \cdot 5 + \cdots + n(n + 2)$
$$= \dfrac{n(n + 1)(2n + 7)}{6}$$

7. $1^3 + 2^3 + 3^3 + \cdots + n^3 = \dfrac{n^2(n + 1)^2}{4}$

8. $1^3 + 3^3 + 5^3 + \cdots + (2n - 1)^3 = n^2(2n^2 - 1)$

9. $2^3 + 4^3 + 6^3 + \cdots + (2n)^3 = 2n^2(n + 1)^2$

10. $\dfrac{1}{1 \cdot 2} + \dfrac{1}{2 \cdot 3} + \dfrac{1}{3 \cdot 4} + \cdots + \dfrac{1}{n(n + 1)} = \dfrac{n}{(n + 1)}$

11. $1 \cdot 2 + 2 \cdot 2^2 + 3 \cdot 2^3 + 4 \cdot 2^4 + \cdots + n \cdot 2^n$
$$= 2[1 + (n - 1)2^n]$$

12. $1 + 2 + 2^2 + \cdots + 2^{n-1} = 2^n - 1$

13. Show that $n^2 + n$ is divisible by 2 for all natural numbers n.

14. Show that $5^n - 1$ is divisible by 4 for all natural numbers n.

15. Show that $n^2 - n + 41$ is odd for all natural numbers n.

16. Show that $n^3 - n + 3$ is divisible by 3 for all natural numbers n.

17. Show that $8^n - 3^n$ is divisible by 5 for all natural numbers n.

18. Show that $3^{2n} - 1$ is divisible by 8 for all natural numbers n.

19. Prove that $n < 2^n$ for all natural numbers n.

20. Prove that $(n + 1)^2 < 2n^2$ for all natural numbers $n \geq 3$.

21. Prove that if $x > -1$, then $(1 + x)^n \geq 1 + nx$ for all natural numbers n.

22. Show that $100n \leq n^2$ for all $n \geq 100$.

23. Let $a_{n+1} = 3a_n$ and $a_1 = 5$. Show that $a_n = 5 \cdot 3^{n-1}$ for all natural numbers n.

24. A sequence is defined recursively by $a_{n+1} = 3a_n - 8$ and $a_1 = 4$. Find an explicit formula for a_n and then use mathematical induction to prove that the formula you found is true.

25. Show that $x - y$ is a factor of $x^n - y^n$ for all natural numbers n.

$$[\text{Hint: } x^{k+1} - y^{k+1} = x^k(x - y) + (x^k - y^k)y]$$

26. Show that $x + y$ is a factor of $x^{2n-1} + y^{2n-1}$ for all natural numbers n.

27–31 ■ F_n denotes the nth term of the Fibonacci sequence discussed in Section 11.1. Use mathematical induction to prove the statement.

27. F_{3n} is even for all natural numbers n.

28. $F_1 + F_2 + F_3 + \cdots + F_n = F_{n+2} - 1$

29. $F_1^2 + F_2^2 + F_3^2 + \cdots + F_n^2 = F_n F_{n+1}$

30. $F_1 + F_3 + \cdots + F_{2n-1} = F_{2n}$

31. For all $n \geq 2$,

$$\begin{bmatrix} 1 & 1 \\ 1 & 0 \end{bmatrix}^n = \begin{bmatrix} F_{n+1} & F_n \\ F_n & F_{n-1} \end{bmatrix}$$

32. Let a_n be the nth term of the sequence defined recursively by

$$a_{n+1} = \frac{1}{1 + a_n}$$

and $a_1 = 1$. Find a formula for a_n in terms of the Fibonacci numbers F_n. Prove that the formula you found is valid for all natural numbers n.

33. Let F_n be the nth term of the Fibonacci sequence. Find and prove an inequality relating n and F_n for natural numbers n.

34. Find and prove an inequality relating $100n$ and n^3.

Discovery • Discussion

35. True or False? Determine whether each statement is true or false. If you think the statement is true, prove it. If you think it is false, give an example where it fails.

(a) $p(n) = n^2 - n + 11$ is prime for all n.

(b) $n^2 > n$ for all $n \geq 2$.

(c) $2^{2n+1} + 1$ is divisible by 3 for all $n \geq 1$.

(d) $n^3 \geq (n + 1)^2$ for all $n \geq 2$.

(e) $n^3 - n$ is divisible by 3 for all $n \geq 2$.

(f) $n^3 - 6n^2 + 11n$ is divisible by 6 for all $n \geq 1$.

36. All Cats Are Black? What is wrong with the following "proof" by mathematical induction that all cats are black? Let $P(n)$ denote the statement: In any group of n cats, if one is black, then they are all black.

Step 1 The statement is clearly true for $n = 1$.

Step 2 Suppose that $P(k)$ is true. We show that $P(k + 1)$ is true.

Suppose we have a group of $k + 1$ cats, one of whom is black; call this cat "Midnight." Remove some other cat (call it "Sparky") from the group. We are left with k cats, one of whom (Midnight) is black, so by the induction hypothesis, all k of these are black. Now put Sparky back in the group and take out Midnight. We again have a group of k cats, all of whom—except possibly Sparky—are black. Then by the induction hypothesis, Sparky must be black, too. So all $k + 1$ cats in the original group are black.

Thus, by induction $P(n)$ is true for all n. Since everyone has seen at least one black cat, it follows that all cats are black.

Midnight   Sparky

11.6   # The Binomial Theorem

An expression of the form $a + b$ is called a **binomial**. Although in principle it's easy to raise $a + b$ to any power, raising it to a very high power would be tedious. In this section we find a formula that gives the expansion of $(a + b)^n$ for any natural number n and then prove it using mathematical induction.

Expanding $(a + b)^n$

To find a pattern in the expansion of $(a + b)^n$, we first look at some special cases:

$$(a + b)^1 = a + b$$

$$(a + b)^2 = a^2 + 2ab + b^2$$

$$(a + b)^3 = a^3 + 3a^2b + 3ab^2 + b^3$$

$$(a + b)^4 = a^4 + 4a^3b + 6a^2b^2 + 4ab^3 + b^4$$

$$(a + b)^5 = a^5 + 5a^4b + 10a^3b^2 + 10a^2b^3 + 5ab^4 + b^5$$

$$\vdots$$

The following simple patterns emerge for the expansion of $(a + b)^n$:

1. There are $n + 1$ terms, the first being a^n and the last b^n.

2. The exponents of a decrease by 1 from term to term while the exponents of b increase by 1.

3. The sum of the exponents of a and b in each term is n.

For instance, notice how the exponents of a and b behave in the expansion of $(a + b)^5$.

The exponents of a decrease:

$$(a + b)^5 = a^{⑤} + 5a^{④}b^1 + 10a^{③}b^2 + 10a^{②}b^3 + 5a^{①}b^4 + b^5$$

The exponents of b increase:

$$(a + b)^5 = a^5 + 5a^4b^{①} + 10a^3b^{②} + 10a^2b^{③} + 5a^1b^{④} + b^{⑤}$$

With these observations we can write the form of the expansion of $(a + b)^n$ for any natural number n. For example, writing a question mark for the missing coefficients, we have

$$(a + b)^8 = a^8 + \text{?}\,a^7b + \text{?}\,a^6b^2 + \text{?}\,a^5b^3 + \text{?}\,a^4b^4 + \text{?}\,a^3b^5 + \text{?}\,a^2b^6 + \text{?}\,ab^7 + b^8$$

To complete the expansion, we need to determine these coefficients. To find a pattern, let's write the coefficients in the expansion of $(a + b)^n$ for the first few values of n in a triangular array as shown in the following array, which is called **Pascal's triangle**.

$(a + b)^0$						1					
$(a + b)^1$					1		1				
$(a + b)^2$				1		2		1			
$(a + b)^3$			①		③		3		1		
$(a + b)^4$		1		④		⑥		④		1	
$(a + b)^5$	1		5		10		⑩		5		1

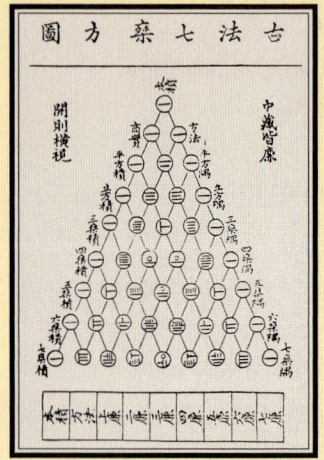

Pascal's triangle appears in this Chinese document by Chu Shikie, dated 1303. The title reads "The Old Method Chart of the Seven Multiplying Squares." The triangle was rediscovered by Pascal (see page 858).

The row corresponding to $(a + b)^0$ is called the zeroth row and is included to show the symmetry of the array. The key observation about Pascal's triangle is the following property.

Key Property of Pascal's Triangle

Every entry (other than a 1) is the sum of the two entries diagonally above it.

From this property it's easy to find any row of Pascal's triangle from the row above it. For instance, we find the sixth and seventh rows, starting with the fifth row:

$$(a + b)^5 \qquad\qquad 1 \quad\; 5 \quad\;\; 10 \quad\;\; 10 \quad\; 5 \quad\; 1$$
$$(a + b)^6 \qquad\; 1 \quad\; 6 \quad\; 15 \quad\; 20 \quad\; 15 \quad\; 6 \quad\; 1$$
$$(a + b)^7 \quad 1 \quad\; 7 \quad\; 21 \quad\; 35 \quad\; 35 \quad\; 21 \quad\; 7 \quad\; 1$$

To see why this property holds, let's consider the following expansions:

$$(a + b)^5 = a^5 + 5a^4b + 10a^3b^2 + 10a^2b^3 + 5ab^4 + b^5$$

$$(a + b)^6 = a^6 + 6a^5b + 15a^4b^2 + 20a^3b^3 + 15a^2b^4 + 6ab^5 + b^6$$

We arrive at the expansion of $(a + b)^6$ by multiplying $(a + b)^5$ by $(a + b)$. Notice, for instance, that the circled term in the expansion of $(a + b)^6$ is obtained via this multiplication from the two circled terms above it. We get this term when the two terms above it are multiplied by b and a, respectively. Thus, its coefficient is the sum of the coefficients of these two terms. We will use this observation at the end of this section when we prove the Binomial Theorem.

Having found these patterns, we can now easily obtain the expansion of any binomial, at least to relatively small powers.

Example 1   Expanding a Binomial Using Pascal's Triangle

Find the expansion of $(a + b)^7$ using Pascal's triangle.

Solution   The first term in the expansion is a^7, and the last term is b^7. Using the fact that the exponent of a decreases by 1 from term to term and that of b increases by 1 from term to term, we have

$$(a + b)^7 = a^7 + \,?\, a^6b + \,?\, a^5b^2 + \,?\, a^4b^3 + \,?\, a^3b^4 + \,?\, a^2b^5 + \,?\, ab^6 + b^7$$

The appropriate coefficients appear in the seventh row of Pascal's triangle. Thus

$$(a + b)^7 = a^7 + 7a^6b + 21a^5b^2 + 35a^4b^3 + 35a^3b^4 + 21a^2b^5 + 7ab^6 + b^7 \quad \blacksquare$$

Example 2   Expanding a Binomial Using Pascal's Triangle

Use Pascal's triangle to expand $(2 - 3x)^5$.

Solution   We find the expansion of $(a + b)^5$ and then substitute 2 for a and $-3x$ for b. Using Pascal's triangle for the coefficients, we get

$$(a + b)^5 = a^5 + 5a^4b + 10a^3b^2 + 10a^2b^3 + 5ab^4 + b^5$$

Substituting $a = 2$ and $b = -3x$ gives

$$(2 - 3x)^5 = (2)^5 + 5(2)^4(-3x) + 10(2)^3(-3x)^2 + 10(2)^2(-3x)^3 + 5(2)(-3x)^4 + (-3x)^5$$

$$= 32 - 240x + 720x^2 - 1080x^3 + 810x^4 - 243x^5 \qquad \blacksquare$$

The Binomial Coefficients

Although Pascal's triangle is useful in finding the binomial expansion for reasonably small values of n, it isn't practical for finding $(a + b)^n$ for large values of n. The reason is that the method we use for finding the successive rows of Pascal's triangle is recursive. Thus, to find the 100th row of this triangle, we must first find the preceding 99 rows.

We need to examine the pattern in the coefficients more carefully to develop a formula that allows us to calculate directly any coefficient in the binomial expansion. Such a formula exists, and the rest of this section is devoted to finding and proving it. However, to state this formula we need some notation.

The product of the first n natural numbers is denoted by $n!$ and is called ***n* factorial**:

$$4! = 1 \cdot 2 \cdot 3 \cdot 4 = 24$$
$$7! = 1 \cdot 2 \cdot 3 \cdot 4 \cdot 5 \cdot 6 \cdot 7 = 5040$$
$$10! = 1 \cdot 2 \cdot 3 \cdot 4 \cdot 5 \cdot 6 \cdot 7 \cdot 8 \cdot 9 \cdot 10$$
$$= 3{,}628{,}800$$

$$\boxed{n! = 1 \cdot 2 \cdot 3 \cdot \cdots \cdot (n - 1) \cdot n}$$

We also define 0! as follows:

$$\boxed{0! = 1}$$

This definition of 0! makes many formulas involving factorials shorter and easier to write.

The Binomial Coefficient

Let n and r be nonnegative integers with $r \le n$. The **binomial coefficient** is denoted by $\binom{n}{r}$ and is defined by

$$\binom{n}{r} = \frac{n!}{r!(n - r)!}$$

Example 3   Calculating Binomial Coefficients

(a) $\displaystyle \binom{9}{4} = \frac{9!}{4!(9 - 4)!} = \frac{9!}{4!5!} = \frac{1 \cdot 2 \cdot 3 \cdot 4 \cdot 5 \cdot 6 \cdot 7 \cdot 8 \cdot 9}{(1 \cdot 2 \cdot 3 \cdot 4)(1 \cdot 2 \cdot 3 \cdot 4 \cdot 5)}$

$$= \frac{6 \cdot 7 \cdot 8 \cdot 9}{1 \cdot 2 \cdot 3 \cdot 4} = 126$$

(b) $\displaystyle \binom{100}{3} = \frac{100!}{3!(100 - 3)!} = \frac{1 \cdot 2 \cdot 3 \cdot \cdots \cdot 97 \cdot 98 \cdot 99 \cdot 100}{(1 \cdot 2 \cdot 3)(1 \cdot 2 \cdot 3 \cdot \cdots \cdot 97)}$

$$= \frac{98 \cdot 99 \cdot 100}{1 \cdot 2 \cdot 3} = 161{,}700$$

(c) $\dbinom{100}{97} = \dfrac{100!}{97!(100-97)!} = \dfrac{1 \cdot 2 \cdot 3 \cdots \cdots 97 \cdot 98 \cdot 99 \cdot 100}{(1 \cdot 2 \cdot 3 \cdots \cdots 97)(1 \cdot 2 \cdot 3)}$

$$= \frac{98 \cdot 99 \cdot 100}{1 \cdot 2 \cdot 3} = 161{,}700 \qquad \blacksquare$$

Although the binomial coefficient $\binom{n}{r}$ is defined in terms of a fraction, all the results of Example 3 are natural numbers. In fact, $\binom{n}{r}$ is always a natural number (see Exercise 50). Notice that the binomial coefficients in parts (b) and (c) of Example 3 are equal. This is a special case of the following relation, which you are asked to prove in Exercise 48.

$$\binom{n}{r} = \binom{n}{n-r}$$

To see the connection between the binomial coefficients and the binomial expansion of $(a+b)^n$, let's calculate the following binomial coefficients:

$\dbinom{5}{2} = \dfrac{5!}{2!(5-2)!} = 10$

$\dbinom{5}{0} = 1 \qquad \dbinom{5}{1} = 5 \qquad \dbinom{5}{2} = 10 \qquad \dbinom{5}{3} = 10 \qquad \dbinom{5}{4} = 5 \qquad \dbinom{5}{5} = 1$

These are precisely the entries in the fifth row of Pascal's triangle. In fact, we can write Pascal's triangle as follows.

$$\binom{0}{0}$$

$$\binom{1}{0} \quad \binom{1}{1}$$

$$\binom{2}{0} \quad \binom{2}{1} \quad \binom{2}{2}$$

$$\binom{3}{0} \quad \binom{3}{1} \quad \binom{3}{2} \quad \binom{3}{3}$$

$$\binom{4}{0} \quad \binom{4}{1} \quad \binom{4}{2} \quad \binom{4}{3} \quad \binom{4}{4}$$

$$\binom{5}{0} \quad \binom{5}{1} \quad \binom{5}{2} \quad \binom{5}{3} \quad \binom{5}{4} \quad \binom{5}{5}$$

$$\cdot \qquad \cdot \qquad \cdot \qquad \cdot \qquad \cdot \qquad \cdot \qquad \cdot$$

$$\binom{n}{0} \quad \binom{n}{1} \quad \binom{n}{2} \quad \cdot \quad \cdot \quad \cdot \quad \binom{n}{n-1} \quad \binom{n}{n}$$

To demonstrate that this pattern holds, we need to show that any entry in this version of Pascal's triangle is the sum of the two entries diagonally above it. In other words, we must show that each entry satisfies the key property of Pascal's triangle. We now state this property in terms of the binomial coefficients.

Key Property of the Binomial Coefficients

For any nonnegative integers r and k with $r \leq k$,

$$\binom{k}{r-1} + \binom{k}{r} = \binom{k+1}{r}$$

Notice that the two terms on the left side of this equation are adjacent entries in the kth row of Pascal's triangle and the term on the right side is the entry diagonally below them, in the $(k+1)$st row. Thus, this equation is a restatement of the key property of Pascal's triangle in terms of the binomial coefficients. A proof of this formula is outlined in Exercise 49.

The Binomial Theorem

We are now ready to state the Binomial Theorem.

The Binomial Theorem

$$(a+b)^n = \binom{n}{0}a^n + \binom{n}{1}a^{n-1}b + \binom{n}{2}a^{n-2}b^2 + \cdots + \binom{n}{n-1}ab^{n-1} + \binom{n}{n}b^n$$

We prove this theorem at the end of this section. First, let's look at some of its applications.

Example 4   Expanding a Binomial Using the Binomial Theorem

Use the Binomial Theorem to expand $(x+y)^4$.

Solution   By the Binomial Theorem,

$$(x+y)^4 = \binom{4}{0}x^4 + \binom{4}{1}x^3y + \binom{4}{2}x^2y^2 + \binom{4}{3}xy^3 + \binom{4}{4}y^4$$

Verify that

$$\binom{4}{0} = 1 \qquad \binom{4}{1} = 4 \qquad \binom{4}{2} = 6 \qquad \binom{4}{3} = 4 \qquad \binom{4}{4} = 1$$

It follows that

$$(x+y)^4 = x^4 + 4x^3y + 6x^2y^2 + 4xy^3 + y^4$$

Example 5 Expanding a Binomial Using the Binomial Theorem

Use the Binomial Theorem to expand $\left(\sqrt{x} - 1 \right)^8$.

Solution We first find the expansion of $(a + b)^8$ and then substitute $\sqrt{x}$ for a and -1 for b. Using the Binomial Theorem, we have

$$(a + b)^8 = \binom{8}{0}a^8 + \binom{8}{1}a^7b + \binom{8}{2}a^6b^2 + \binom{8}{3}a^5b^3 + \binom{8}{4}a^4b^4$$
$$+ \binom{8}{5}a^3b^5 + \binom{8}{6}a^2b^6 + \binom{8}{7}ab^7 + \binom{8}{8}b^8$$

Verify that

$$\binom{8}{0} = 1 \quad \binom{8}{1} = 8 \quad \binom{8}{2} = 28 \quad \binom{8}{3} = 56 \quad \binom{8}{4} = 70$$
$$\binom{8}{5} = 56 \quad \binom{8}{6} = 28 \quad \binom{8}{7} = 8 \quad \binom{8}{8} = 1$$

So

$$(a + b)^8 = a^8 + 8a^7b + 28a^6b^2 + 56a^5b^3 + 70a^4b^4 + 56a^3b^5$$
$$+ 28a^2b^6 + 8ab^7 + b^8$$

Performing the substitutions $a = x^{1/2}$ and $b = -1$ gives

$$\left(\sqrt{x} - 1 \right)^8 = (x^{1/2})^8 + 8(x^{1/2})^7(-1) + 28(x^{1/2})^6(-1)^2 + 56(x^{1/2})^5(-1)^3$$
$$+ 70(x^{1/2})^4(-1)^4 + 56(x^{1/2})^3(-1)^5 + 28(x^{1/2})^2(-1)^6$$
$$+ 8(x^{1/2})(-1)^7 + (-1)^8$$

This simplifies to

$$\left(\sqrt{x} - 1 \right)^8 = x^4 - 8x^{7/2} + 28x^3 - 56x^{5/2} + 70x^2 - 56x^{3/2} + 28x - 8x^{1/2} + 1 \quad \blacksquare$$

The Binomial Theorem can be used to find a particular term of a binomial expansion without having to find the entire expansion.

General Term of the Binomial Expansion

The term that contains a^r in the expansion of $(a + b)^n$ is

$$\binom{n}{n - r}a^rb^{n-r}$$

Example 6   Finding a Particular Term in a Binomial Expansion

Find the term that contains x^5 in the expansion of $(2x + y)^{20}$.

Solution   The term that contains x^5 is given by the formula for the general term with $a = 2x$, $b = y$, $n = 20$, and $r = 5$. So, this term is

$$\binom{20}{15} a^5 b^{15} = \frac{20!}{15!(20 - 15)!} (2x)^5 y^{15} = \frac{20!}{15!5!} 32 x^5 y^{15} = 496{,}128 x^5 y^{15} \quad \blacksquare$$

Example 7   Finding a Particular Term in a Binomial Expansion

Find the coefficient of x^8 in the expansion of $\left(x^2 + \dfrac{1}{x} \right)^{10}$.

Solution   Both x^2 and $1/x$ are powers of x, so the power of x in each term of the expansion is determined by both terms of the binomial. To find the required coefficient, we first find the general term in the expansion. By the formula we have $a = x^2$, $b = 1/x$, and $n = 10$, so the general term is

$$\binom{10}{10 - r} (x^2)^r \left(\frac{1}{x} \right)^{10-r} = \binom{10}{10 - r} x^{2r} (x^{-1})^{10-r} = \binom{10}{10 - r} x^{3r - 10}$$

Thus, the term that contains x^8 is the term in which

$$3r - 10 = 8$$

$$r = 6$$

So the required coefficient is

$$\binom{10}{10 - 6} = \binom{10}{4} = 210 \quad \blacksquare$$

Proof of the Binomial Theorem

We now give a proof of the Binomial Theorem using mathematical induction.

■ **Proof**   Let $P(n)$ denote the statement

$$(a + b)^n = \binom{n}{0} a^n + \binom{n}{1} a^{n-1} b + \binom{n}{2} a^{n-2} b^2 + \cdots + \binom{n}{n - 1} ab^{n-1} + \binom{n}{n} b^n$$

Step 1   We show that $P(1)$ is true. But $P(1)$ is just the statement

$$(a + b)^1 = \binom{1}{0} a^1 + \binom{1}{1} b^1 = 1a + 1b = a + b$$

which is certainly true.

Step 2   We assume that $P(k)$ is true. Thus, our induction hypothesis is

$$(a + b)^k = \binom{k}{0} a^k + \binom{k}{1} a^{k-1} b + \binom{k}{2} a^{k-2} b^2 + \cdots + \binom{k}{k - 1} ab^{k-1} + \binom{k}{k} b^k$$

We use this to show that $P(k + 1)$ is true.

$$(a + b)^{k+1} = (a + b)[(a + b)^k]$$

$$= (a + b)\left[\binom{k}{0}a^k + \binom{k}{1}a^{k-1}b + \binom{k}{2}a^{k-2}b^2 + \cdots + \binom{k}{k-1}ab^{k-1} + \binom{k}{k}b^k\right] \quad \text{Induction hypothesis}$$

$$= a\left[\binom{k}{0}a^k + \binom{k}{1}a^{k-1}b + \binom{k}{2}a^{k-2}b^2 + \cdots + \binom{k}{k-1}ab^{k-1} + \binom{k}{k}b^k\right]$$

$$+ b\left[\binom{k}{0}a^k + \binom{k}{1}a^{k-1}b + \binom{k}{2}a^{k-2}b^2 + \cdots + \binom{k}{k-1}ab^{k-1} + \binom{k}{k}b^k\right] \quad \text{Distributive Property}$$

$$= \binom{k}{0}a^{k+1} + \binom{k}{1}a^k b + \binom{k}{2}a^{k-1}b^2 + \cdots + \binom{k}{k-1}a^2 b^{k-1} + \binom{k}{k}ab^k$$

$$+ \binom{k}{0}a^k b + \binom{k}{1}a^{k-1}b^2 + \binom{k}{2}a^{k-2}b^3 + \cdots + \binom{k}{k-1}ab^k + \binom{k}{k}b^{k+1} \quad \text{Distributive Property}$$

$$= \binom{k}{0}a^{k+1} + \left[\binom{k}{0} + \binom{k}{1}\right]a^k b + \left[\binom{k}{1} + \binom{k}{2}\right]a^{k-1}b^2$$

$$+ \cdots + \left[\binom{k}{k-1} + \binom{k}{k}\right]ab^k + \binom{k}{k}b^{k+1} \quad \text{Group like terms}$$

Using the key property of the binomial coefficients, we can write each of the expressions in square brackets as a single binomial coefficient. Also, writing the first and last coefficients as $\binom{k+1}{0}$ and $\binom{k+1}{k+1}$ (these are equal to 1 by Exercise 46) gives

$$(a + b)^{k+1} = \binom{k+1}{0}a^{k+1} + \binom{k+1}{1}a^k b + \binom{k+1}{2}a^{k-1}b^2 + \cdots + \binom{k+1}{k}ab^k + \binom{k+1}{k+1}b^{k+1}$$

But this last equation is precisely $P(k + 1)$, and this completes the induction step.

Having proved Steps 1 and 2, we conclude by the Principle of Mathematical Induction that the theorem is true for all natural numbers n. ■

11.6 Exercises

1–12 ■ Use Pascal's triangle to expand the expression.

1. $(x + y)^6$

2. $(2x + 1)^4$

3. $\left(x + \dfrac{1}{x}\right)^4$

4. $(x - y)^5$

5. $(x - 1)^5$

6. $(\sqrt{a} + \sqrt{b})^6$

7. $(x^2 y - 1)^5$

8. $(1 + \sqrt{2})^6$

9. $(2x - 3y)^3$

10. $(1 + x^3)^3$

11. $\left(\dfrac{1}{x} - \sqrt{x}\right)^5$

12. $\left(2 + \dfrac{x}{2}\right)^5$

13–20 ■ Evaluate the expression.

13. $\binom{6}{4}$

14. $\binom{8}{3}$

15. $\binom{100}{98}$

16. $\binom{10}{5}$

17. $\binom{3}{1}\binom{4}{2}$

18. $\binom{5}{2}\binom{5}{3}$

19. $\binom{5}{0} + \binom{5}{1} + \binom{5}{2} + \binom{5}{3} + \binom{5}{4} + \binom{5}{5}$

20. $\dbinom{5}{0} - \dbinom{5}{1} + \dbinom{5}{2} - \dbinom{5}{3} + \dbinom{5}{4} - \dbinom{5}{5}$

21–24 ■ Use the Binomial Theorem to expand the expression.

21. $(x + 2y)^4$

22. $(1 - x)^5$

23. $\left(1 + \dfrac{1}{x}\right)^6$

24. $(2A + B^2)^4$

25. Find the first three terms in the expansion of $(x + 2y)^{20}$.

26. Find the first four terms in the expansion of $(x^{1/2} + 1)^{30}$.

27. Find the last two terms in the expansion of $(a^{2/3} + a^{1/3})^{25}$.

28. Find the first three terms in the expansion of

$$\left(x + \dfrac{1}{x}\right)^{40}$$

29. Find the middle term in the expansion of $(x^2 + 1)^{18}$.

30. Find the fifth term in the expansion of $(ab - 1)^{20}$.

31. Find the 24th term in the expansion of $(a + b)^{25}$.

32. Find the 28th term in the expansion of $(A - B)^{30}$.

33. Find the 100th term in the expansion of $(1 + y)^{100}$.

34. Find the second term in the expansion of

$$\left(x^2 - \dfrac{1}{x}\right)^{25}$$

35. Find the term containing x^4 in the expansion of $(x + 2y)^{10}$.

36. Find the term containing y^3 in the expansion of $(\sqrt{2} + y)^{12}$.

37. Find the term containing b^8 in the expansion of $(a + b^2)^{12}$.

38. Find the term that does not contain x in the expansion of

$$\left(8x + \dfrac{1}{2x}\right)^8$$

39–42 ■ Factor using the Binomial Theorem.

39. $x^4 + 4x^3y + 6x^2y^2 + 4xy^3 + y^4$

40. $(x - 1)^5 + 5(x - 1)^4 + 10(x - 1)^3 +$
$10(x - 1)^2 + 5(x - 1) + 1$

41. $8a^3 + 12a^2b + 6ab^2 + b^3$

42. $x^8 + 4x^6y + 6x^4y^2 + 4x^2y^3 + y^4$

43–44 ■ Simplify using the Binomial Theorem.

43. $\dfrac{(x + h)^3 - x^3}{h}$

44. $\dfrac{(x + h)^4 - x^4}{h}$

45. Show that $(1.01)^{100} > 2$.

[*Hint*: Note that $(1.01)^{100} = (1 + 0.01)^{100}$ and use the Binomial Theorem to show that the sum of the first two terms of the expansion is greater than 2.]

46. Show that $\dbinom{n}{0} = 1$ and $\dbinom{n}{n} = 1$.

47. Show that $\dbinom{n}{1} = \dbinom{n}{n-1} = n$.

48. Show that $\dbinom{n}{r} = \dbinom{n}{n-r}$ for $0 \le r \le n$.

49. In this exercise we prove the identity

$$\dbinom{n}{r-1} + \dbinom{n}{r} = \dbinom{n+1}{r}$$

(a) Write the left side of this equation as the sum of two fractions.

(b) Show that a common denominator of the expression you found in part (a) is $r!(n - r + 1)!$.

(c) Add the two fractions using the common denominator in part (b), simplify the numerator, and note that the resulting expression is equal to the right side of the equation.

50. Prove that $\binom{n}{r}$ is an integer for all n and for $0 \le r \le n$.
[*Suggestion:* Use induction to show that the statement is true for all n, and use Exercise 49 for the induction step.]

Discovery • Discussion

51. Powers of Factorials Which is larger, $(100!)^{101}$ or $(101!)^{100}$? [*Hint*: Try factoring the expressions. Do they have any common factors?]

52. Sums of Binomial Coefficients Add each of the first five rows of Pascal's triangle, as indicated. Do you see a pattern?

$$1 + 1 = \boxed{?}$$

$$1 + 2 + 1 = \boxed{?}$$

$$1 + 3 + 3 + 1 = \boxed{?}$$

$$1 + 4 + 6 + 4 + 1 = \boxed{?}$$

$$1 + 5 + 10 + 10 + 5 + 1 = \boxed{?}$$

Based on the pattern you have found, find the sum of the nth row:

$$\binom{n}{0} + \binom{n}{1} + \binom{n}{2} + \cdots + \binom{n}{n}$$

Prove your result by expanding $(1 + 1)^n$ using the Binomial Theorem.

53. Alternating Sums of Binomial Coefficients   Find the sum

$$\binom{n}{0} - \binom{n}{1} + \binom{n}{2} - \cdots + (-1)^n\binom{n}{n}$$

by finding a pattern as in Exercise 52. Prove your result by expanding $(1 - 1)^n$ using the Binomial Theorem.

11   Review

Concept Check

1. (a) What is a sequence?

(b) What is an arithmetic sequence? Write an expression for the nth term of an arithmetic sequence.

(c) What is a geometric sequence? Write an expression for the nth term of a geometric sequence.

2. (a) What is a recursively defined sequence?

(b) What is the Fibonacci sequence?

3. (a) What is meant by the partial sums of a sequence?

(b) If an arithmetic sequence has first term a and common difference d, write an expression for the sum of its first n terms.

(c) If a geometric sequence has first term a and common ratio r, write an expression for the sum of its first n terms.

(d) Write an expression for the sum of an infinite geometric series with first term a and common ratio r. For what values of r is your formula valid?

4. (a) Write the sum $\displaystyle\sum_{k=1}^{n} a_k$ without using Σ-notation.

(b) Write $b_1 + b_2 + b_3 + \cdots + b_n$ using Σ-notation.

5. Write an expression for the amount A_f of an annuity consisting of n regular equal payments of size R with interest rate i per time period.

6. State the Principle of Mathematical Induction.

7. Write the first five rows of Pascal's triangle. How are the entries related to each other?

8. (a) What does the symbol $n!$ mean?

(b) Write an expression for the binomial coefficient $\binom{n}{r}$.

(c) State the Binomial Theorem.

(d) Write the term that contains a^r in the expansion of $(a + b)^n$.

Exercises

1–6 ■ Find the first four terms as well as the tenth term of the sequence with the given nth term.

1. $a_n = \dfrac{n^2}{n + 1}$

2. $a_n = (-1)^n \dfrac{2^n}{n}$

3. $a_n = \dfrac{(-1)^n + 1}{n^3}$

4. $a_n = \dfrac{n(n + 1)}{2}$

5. $a_n = \dfrac{(2n)!}{2^n n!}$

6. $a_n = \binom{n + 1}{2}$

7–10 ■ A sequence is defined recursively. Find the first seven terms of the sequence.

7. $a_n = a_{n-1} + 2n - 1, \quad a_1 = 1$

8. $a_n = \dfrac{a_{n-1}}{n}, \quad a_1 = 1$

9. $a_n = a_{n-1} + 2a_{n-2}, \quad a_1 = 1, a_2 = 3$

10. $a_n = \sqrt{3a_{n-1}}, \quad a_1 = \sqrt{3}$

11–14 ■ The nth term of a sequence is given.

(a) Find the first five terms of the sequence.

(b) Graph the terms you found in part (a).

(c) Determine if the series is arithmetic or geometric. Find the common difference or the common ratio.

11. $a_n = 2n + 5$

12. $a_n = \dfrac{5}{2^n}$

13. $a_n = \dfrac{3^n}{2^{n+1}}$

14. $a_n = 4 - \dfrac{n}{2}$

15–22 ■ The first four terms of a sequence are given. Determine whether they can be the terms of an arithmetic sequence, a geometric sequence, or neither. If the sequence is arithmetic or geometric, find the fifth term.

15. $5, 5.5, 6, 6.5, \ldots$

16. $1, -\frac{3}{2}, 2, -\frac{5}{2}, \ldots$

17. $\sqrt{2}, 2\sqrt{2}, 3\sqrt{2}, 4\sqrt{2}, \ldots$ **18.** $\sqrt{2}, 2, 2\sqrt{2}, 4, \ldots$

19. $t - 3, t - 2, t - 1, t, \ldots$ **20.** $t^3, t^2, t, 1, \ldots$

21. $\frac{3}{4}, \frac{1}{2}, \frac{1}{3}, \frac{2}{9}, \ldots$

22. $a, 1, \frac{1}{a}, \frac{1}{a^2}, \ldots$

23. Show that $3, 6i, -12, -24i, \ldots$ is a geometric sequence, and find the common ratio. (Here $i = \sqrt{-1}$.)

24. Find the nth term of the geometric sequence $2, 2 + 2i, 4i, -4 + 4i, -8, \ldots$ (Here $i = \sqrt{-1}$.)

25. The sixth term of an arithmetic sequence is 17, and the fourth term is 11. Find the second term.

26. The 20th term of an arithmetic sequence is 96, and the common difference is 5. Find the nth term.

27. The third term of a geometric sequence is 9, and the common ratio is $\frac{3}{2}$. Find the fifth term.

28. The second term of a geometric sequence is 10, and the fifth term is $\frac{1250}{27}$. Find the nth term.

29. A teacher makes \$32,000 in his first year at Lakeside School, and gets a 5% raise each year.
(a) Find a formula for his salary A_n in his nth year at this school.
(b) List his salaries for his first 8 years at this school.

30. A colleague of the teacher in Exercise 29, hired at the same time, makes \$35,000 in her first year, and gets a \$1200 raise each year.
(a) What is her salary A_n in her nth year at this school?
(b) Find her salary in her eighth year at this school, and compare it to the salary of the teacher in Exercise 29 in his eighth year.

31. A certain type of bacteria divides every 5 s. If three of these bacteria are put into a petri dish, how many bacteria are in the dish at the end of 1 min?

32. If $a_1, a_2, a_3, \ldots$ and $b_1, b_2, b_3, \ldots$ are arithmetic sequences, show that $a_1 + b_1, a_2 + b_2, a_3 + b_3, \ldots$ is also an arithmetic sequence.

33. If $a_1, a_2, a_3, \ldots$ and $b_1, b_2, b_3, \ldots$ are geometric sequences, show that $a_1 b_1, a_2 b_2, a_3 b_3, \ldots$ is also a geometric sequence.

34. (a) If $a_1, a_2, a_3, \ldots$ is an arithmetic sequence, is the sequence $a_1 + 2, a_2 + 2, a_3 + 2, \ldots$ arithmetic?
(b) If $a_1, a_2, a_3, \ldots$ is a geometric sequence, is the sequence $5a_1, 5a_2, 5a_3, \ldots$ geometric?

35. Find the values of x for which the sequence $6, x, 12, \ldots$ is
(a) arithmetic   (b) geometric

36. Find the values of x and y for which the sequence $2, x, y, 17, \ldots$ is
(a) arithmetic   (b) geometric

37–40 ■ Find the sum.

37. $\displaystyle\sum_{k=3}^{6} (k + 1)^2$

38. $\displaystyle\sum_{i=1}^{4} \frac{2i}{2i - 1}$

39. $\displaystyle\sum_{k=1}^{6} (k + 1)2^{k-1}$

40. $\displaystyle\sum_{m=1}^{5} 3^{m-2}$

41–44 ■ Write the sum without using sigma notation. Do not evaluate.

41. $\displaystyle\sum_{k=1}^{10} (k - 1)^2$

42. $\displaystyle\sum_{j=2}^{100} \frac{1}{j - 1}$

43. $\displaystyle\sum_{k=1}^{50} \frac{3^k}{2^{k+1}}$

44. $\displaystyle\sum_{n=1}^{10} n^2 2^n$

45–48 ■ Write the sum using sigma notation. Do not evaluate.

45. $3 + 6 + 9 + 12 + \cdots + 99$

46. $1^2 + 2^2 + 3^2 + \cdots + 100^2$

47. $1 \cdot 2^3 + 2 \cdot 2^4 + 3 \cdot 2^5 + 4 \cdot 2^6 + \cdots + 100 \cdot 2^{102}$

48. $\dfrac{1}{1 \cdot 2} + \dfrac{1}{2 \cdot 3} + \dfrac{1}{3 \cdot 4} + \cdots + \dfrac{1}{999 \cdot 1000}$

49–54 ■ Determine whether the expression is a partial sum of an arithmetic or geometric sequence. Then find the sum.

49. $1 + 0.9 + (0.9)^2 + \cdots + (0.9)^5$

50. $3 + 3.7 + 4.4 + \cdots + 10$

51. $\sqrt{5} + 2\sqrt{5} + 3\sqrt{5} + \cdots + 100\sqrt{5}$

52. $\frac{1}{3} + \frac{2}{3} + 1 + \frac{4}{3} + \cdots + 33$

53. $\displaystyle\sum_{n=0}^{6} 3(-4)^n$

54. $\displaystyle\sum_{k=0}^{8} 7(5)^{k/2}$

55. The first term of an arithmetic sequence is $a = 7$, and the common difference is $d = 3$. How many terms of this sequence must be added to obtain 325?

56. The sum of the first three terms of a geometric series is 52, and the common ratio is $r = 3$. Find the first term.

57. A person has two parents, four grandparents, eight great-grandparents, and so on. What is the total number of a person's ancestors in 15 generations?

58. Find the amount of an annuity consisting of 16 annual payments of $1000 each into an account that pays 8% interest per year, compounded annually.

59. How much money should be invested every quarter at 12% per year, compounded quarterly, in order to have $10,000 in one year?

60. What are the monthly payments on a mortgage of $60,000 at 9% interest if the loan is to be repaid in

    (a) 30 years?   **(b)** 15 years?

61–64 ■ Find the sum of the infinite geometric series.

61. $1 - \frac{2}{5} + \frac{4}{25} - \frac{8}{125} + \cdots$

62. $0.1 + 0.01 + 0.001 + 0.0001 + \cdots$

63. $1 + \dfrac{1}{3^{1/2}} + \dfrac{1}{3} + \dfrac{1}{3^{3/2}} + \cdots$

64. $a + ab^2 + ab^4 + ab^6 + \cdots$

65–67 ■ Use mathematical induction to prove that the formula is true for all natural numbers n.

65. $1 + 4 + 7 + \cdots + (3n - 2) = \dfrac{n(3n - 1)}{2}$

66. $\dfrac{1}{1 \cdot 3} + \dfrac{1}{3 \cdot 5} + \dfrac{1}{5 \cdot 7} + \cdots + \dfrac{1}{(2n - 1)(2n + 1)}$

    $= \dfrac{n}{2n + 1}$

67. $\left(1 + \dfrac{1}{1}\right)\left(1 + \dfrac{1}{2}\right)\left(1 + \dfrac{1}{3}\right) \cdots \left(1 + \dfrac{1}{n}\right) = n + 1$

68. Show that $7^n - 1$ is divisible by 6 for all natural numbers n.

69. Let $a_{n+1} = 3a_n + 4$ and $a_1 = 4$. Show that $a_n = 2 \cdot 3^n - 2$ for all natural numbers n.

70. Prove that the Fibonacci number F_{4n} is divisible by 3 for all natural numbers n.

71. Find and prove an inequality that relates 2^n and $n!$.

72–75 ■ Evaluate the expression.

72. $\dbinom{5}{2}\dbinom{5}{3}$

73. $\dbinom{10}{2} + \dbinom{10}{6}$

74. $\displaystyle\sum_{k=0}^{5} \dbinom{5}{k}$

75. $\displaystyle\sum_{k=0}^{8} \dbinom{8}{k}\dbinom{8}{8 - k}$

76–77 ■ Expand the expression.

76. $(1 - x^2)^6$

77. $(2x + y)^4$

78. Find the 20th term in the expansion of $(a + b)^{22}$.

79. Find the first three terms in the expansion of $(b^{-2/3} + b^{1/3})^{20}$.

80. Find the term containing A^6 in the expansion of $(A + 3B)^{10}$.

11 Test

1. Find the first four terms and the tenth term of the sequence whose nth term is $a_n = n^2 - 1$.

2. A sequence is defined recursively by $a_{n+2} = a_n^2 - a_{n+1}$, with $a_1 = 1$ and $a_2 = 1$. Find a_5.

3. An arithmetic sequence begins $2, 5, 8, 11, 14, \ldots$.
   (a) Find the common difference d for this sequence.
   (b) Find a formula for the nth term a_n of the sequence.
   (c) Find the 35th term of the sequence.

4. A geometric sequence begins $12, 3, 3/4, 3/16, 3/64, \ldots$.
   (a) Find the common ratio r for this sequence.
   (b) Find a formula for the nth term a_n of the sequence.
   (c) Find the tenth term of the sequence.

5. The first term of a geometric sequence is 25, and the fourth term is $\frac{1}{5}$.
   (a) Find the common ratio r and the fifth term.
   (b) Find the partial sum of the first eight terms.

6. The first term of an arithmetic sequence is 10 and the tenth term is 2.
   (a) Find the common difference and the 100th term of the sequence.
   (b) Find the partial sum of the first ten terms.

7. Let $a_1, a_2, a_3, \ldots$ be a geometric sequence with initial term a and common ratio r. Show that $a_1^2, a_2^2, a_3^2, \ldots$ is also a geometric sequence by finding its common ratio.

8. Write the expression without using sigma notation, and then find the sum.

   (a) $\displaystyle\sum_{n=1}^{5} (1 - n^2)$
   (b) $\displaystyle\sum_{n=3}^{6} (-1)^n 2^{n-2}$

9. Find the sum.

   (a) $\dfrac{1}{3} + \dfrac{2}{3^2} + \dfrac{2^2}{3^3} + \dfrac{2^3}{3^4} + \cdots + \dfrac{2^9}{3^{10}}$

   (b) $1 + \dfrac{1}{2^{1/2}} + \dfrac{1}{2} + \dfrac{1}{2^{3/2}} + \cdots$

10. Use mathematical induction to prove that, for all natural numbers n,

$$1^2 + 2^2 + 3^2 + \cdots + n^2 = \frac{n(n + 1)(2n + 1)}{6}$$

11. Expand $(2x + y^2)^5$.

12. Find the term containing x^3 in the binomial expansion of $(3x - 2)^{10}$.

13. A puppy weighs 0.85 lb at birth, and each week he gains 24% in weight. Let a_n be his weight in pounds at the end of his nth week of life.
   (a) Find a formula for a_n.
   (b) How much does the puppy weigh when he is six weeks old?
   (c) Is the sequence $a_1, a_2, a_3, \ldots$ arithmetic, geometric, or neither?

Focus on Modeling

Modeling with Recursive Sequences

Many real-world processes occur in stages. Population growth can be viewed in stages—each new generation represents a new stage in population growth. Compound interest is paid in stages—each interest payment creates a new account balance. Many things that change continuously are more easily measured in discrete stages. For example, we can measure the temperature of a continuously cooling object in one-hour intervals. In this *Focus* we learn how recursive sequences are used to model such situations. In some cases, we can get an explicit formula for a sequence from the recursion relation that defines it by finding a pattern in the terms of the sequence.

Recursive Sequences as Models

Suppose you deposit some money in an account that pays 6% interest compounded monthly. The bank has a definite rule for paying interest: At the end of each month the bank adds to your account $\frac{1}{2}$% (or 0.005) of the amount in your account at that time. Let's express this rule as follows:

$$
\boxed{\text{amount at the end of this month}} = \boxed{\text{amount at the end of last month}} + 0.005 \times \boxed{\text{amount at the end of last month}}
$$

Using the Distributive Property, we can write this as

$$
\boxed{\text{amount at the end of this month}} = 1.005 \times \boxed{\text{amount at the end of last month}}
$$

To model this statement using algebra, let A_0 be the amount of the original deposit, A_1 the amount at the end of the first month, A_2 the amount at the end of the second month, and so on. So A_n is the amount at the end of the nth month. Thus

$$A_n = 1.005A_{n-1}$$

We recognize this as a recursively defined sequence—it gives us the amount at each stage in terms of the amount at the preceding stage.

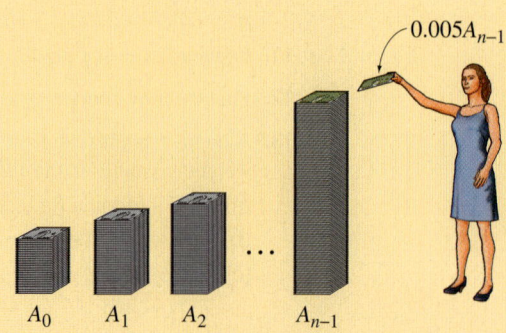

To find a formula for A_n, let's find the first few terms of the sequence and look for a pattern.

$$A_1 = 1.005A_0$$

$$A_2 = 1.005A_1 = (1.005)^2A_0$$

$$A_3 = 1.005A_2 = (1.005)^3A_0$$

$$A_4 = 1.005A_3 = (1.005)^4A_0$$

We see that in general, $A_n = (1.005)^nA_0$.

Example 1 Population Growth

A certain animal population grows by 2% each year. The initial population is 5000.

(a) Find a recursive sequence that models the population P_n at the end of the nth year.

(b) Find the first five terms of the sequence P_n.

(c) Find a formula for P_n.

Solution

(a) We can model the population using the following rule:

| population at the end of this year | = 1.02 × | population at the end of last year |

Algebraically we can write this as the recursion relation

$$P_n = 1.02P_{n-1}$$

(b) Since the initial population is 5000, we have

$$P_0 = 5000$$

$$P_1 = 1.02P_0 = (1.02)5000$$

$$P_2 = 1.02P_1 = (1.02)^25000$$

$$P_3 = 1.02P_2 = (1.02)^35000$$

$$P_4 = 1.02P_3 = (1.02)^45000$$

(c) We see from the pattern exhibited in part (b) that $P_n = (1.02)^n5000$. (Note that P_n is a geometric sequence, with common ratio $r = 1.02$.) ∎

Example 2 Daily Drug Dose

A patient is to take a 50-mg pill of a certain drug every morning. It is known that the body eliminates 40% of the drug every 24 hours.

(a) Find a recursive sequence that models the amount A_n of the drug in the patient's body after each pill is taken.

(b) Find the first four terms of the sequence A_n.

(c) Find a formula for A_n.

(d) How much of the drug remains in the patient's body after 5 days? How much will accumulate in his system after prolonged use?

Solution

(a) Each morning 60% of the drug remains in his system plus he takes an additional 50 mg (his daily dose).

$$\boxed{\text{amount of drug this morning}} = 0.6 \times \boxed{\text{amount of drug yesterday morning}} + 50 \text{ mg}$$

We can express this as a recursion relation

$$A_n = 0.6A_{n-1} + 50$$

(b) Since the initial dose is 50 mg, we have

$$A_0 = 50$$
$$A_1 = 0.6A_0 + 50 = 0.6(50) + 50$$
$$A_2 = 0.6A_1 + 50 = 0.6[0.6(50) + 50] + 50$$
$$= 0.6^2(50) + 0.6(50) + 50$$
$$= 50(0.6^2 + 0.6 + 1)$$
$$A_3 = 0.6A_2 + 50 = 0.6[0.6^2(50) + 0.6(50) + 50] + 50$$
$$= 0.6^3(50) + 0.6^2(50) + 0.6(50) + 50$$
$$= 50(0.6^3 + 0.6^2 + 0.6 + 1)$$

(c) From the pattern in part (b), we see that

$$A_n = 50(1 + 0.6 + 0.6^2 + \cdots + 0.6^n)$$
$$= 50\left(\frac{1 - 0.6^{n+1}}{1 - 0.6}\right)$$

Sum of a geometric sequence:
$$S_n = a\left(\frac{1 - r^{n+1}}{1 - r}\right)$$

$$= 125(1 - 0.6^{n+1})$$

Simplify

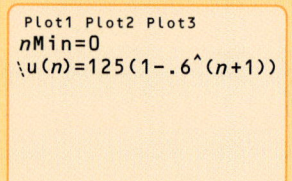

Plot1 Plot2 Plot3
nMin=0
\u(n)=125(1-.6^(n+1))

Enter sequence

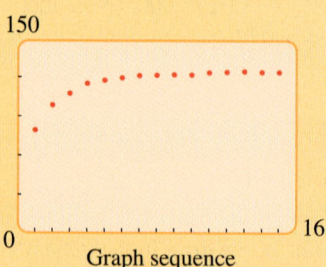

Graph sequence

Figure 1

(d) To find the amount remaining after 5 days, we substitute $n = 5$ and get
$$A_5 = 125(1 - 0.6^{5+1}) \approx 119 \text{ mg}.$$
To find the amount remaining after prolonged use, we let n become large. As n gets large, 0.6^n approaches 0. That is, $0.6^n \to 0$ as $n \to \infty$ (see Section 4.1). So as $n \to \infty$,

$$A_n = 125(1 - 0.6^{n+1}) \to 125(1 - 0) = 125$$

Thus, after prolonged use the amount of drug in the patient's system approaches 125 mg (see Figure 1, where we have used a graphing calculator to graph the sequence).

Problems

1. **Retirement Accounts**   Many college professors keep retirement savings with TIAA, the largest annuity program in the world. Interest on these accounts is compounded and credited *daily*. Professor Brown has $275,000 on deposit with TIAA at the start of 2006, and receives 3.65% interest per year on his account.
    (a) Find a recursive sequence that models the amount A_n in his account at the end of the nth day of 2006.
    (b) Find the first eight terms of the sequence A_n, rounded to the nearest cent.
    (c) Find a formula for A_n.

2. **Fitness Program**   Sheila decides to embark on a swimming program as the best way to maintain cardiovascular health. She begins by swimming 5 min on the first day, then adds $1\frac{1}{2}$ min every day after that.
    (a) Find a recursive formula for the number of minutes T_n that she swims on the nth day of her program.
    (b) Find the first 6 terms of the sequence T_n.
    (c) Find a formula for T_n. What kind of sequence is this?
    (d) On what day does Sheila attain her goal of swimming at least 65 min a day?
    (e) What is the total amount of time she will have swum after 30 days?

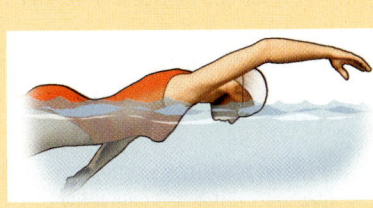

3. **Monthly Savings Program**   Alice opens a savings account paying 3% interest per year, compounded monthly. She begins by depositing $100 at the start of the first month, and adds $100 at the end of each month, when the interest is credited.
    (a) Find a recursive formula for the amount A_n in her account at the end of the nth month. (Include the interest credited for that month and her monthly deposit.)
    (b) Find the first 5 terms of the sequence A_n.
    (c) Use the pattern you observed in (b) to find a formula for A_n. [*Hint:* To find the pattern most easily, it's best *not* to simplify the terms *too* much.]
    (d) How much has she saved after 5 years?

4. **Stocking a Fish Pond**   A pond is stocked with 4000 trout, and through reproduction the population increases by 20% per year. Find a recursive sequence that models the trout population P_n at the end of the nth year under each of the following circumstances. Find the trout population at the end of the fifth year in each case.
    (a) The trout population changes only because of reproduction.
    (b) Each year 600 trout are harvested.
    (c) Each year 250 additional trout are introduced into the pond.
    (d) Each year 10% of the trout are harvested and 300 additional trout are introduced into the pond.

5. **Pollution**   A chemical plant discharges 2400 tons of pollutants every year into an adjacent lake. Through natural runoff, 70% of the pollutants contained in the lake at the beginning of the year are expelled by the end of the year.
    (a) Explain why the following sequence models the amount A_n of the pollutant in the lake at the end of the nth year that the plant is operating.

$$A_n = 0.30A_{n-1} + 2400$$

(b) Find the first five terms of the sequence A_n.

(c) Find a formula for A_n.

(d) How much of the pollutant remains in the lake after 6 years? How much will remain after the plant has been operating a long time?

(e) Verify your answer to part (d) by graphing A_n with a graphing calculator, for $n = 1$ to $n = 20$.

6. Annual Savings Program   Ursula opens a one-year CD that yields 5% interest per year. She begins with a deposit of $5000. At the end of each year when the CD matures, she reinvests at the same 5% interest rate, also adding 10% to the value of the CD from her other savings. (So for example, after the first year her CD has earned 5% of $5000 in interest, for a value of $5250 at maturity. She then adds 10%, or $525, bringing the total value of her renewed CD to $5775.)

(a) Find a recursive formula for the amount U_n in her CD when she reinvests at the end of the nth year.

(b) Find the first 5 terms of the sequence U_n. Does this appear to be a geometric sequence?

(c) Use the pattern you observed in (b) to find a formula for U_n.

(d) How much has she saved after 10 years?

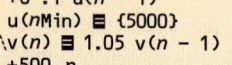

```
Plot1 Plot2 Plot3
\u(n) ⊟ 1.05 u(n − 1)
+0 .1 u(n − 1)
u(nMin) ⊟ {5000}
\v(n) ⊟ 1.05 v(n − 1)
+500 n
v(nMin) ⊟ {5000}
```

Entering the sequences

n	$u(n)$	$v(n)$
0	5000	5000
1	5750	5750
2	6612.5	7037.5
3	7604.4	8889.4
4	8745	11334
5	10057	14401
6	11565	18121
$n=0$		

Table of values
of the sequences

 7. Annual Savings Program   Victoria opens a one-year CD with a 5% annual interest yield at the same time as her friend Ursula in Problem 6. She also starts with an initial deposit of $5000. However, Victoria decides to add $500 to her CD when she reinvests at the end of the first year, $1000 at the end of the second, $1500 at the end of the third, and so on.

(a) Explain why the recursive formula displayed below gives the amount V_n in her CD when she reinvests at the end of the nth year.

$$V_n = 1.05V_{n-1} + 500n$$

(b) Using the $\texttt{Seq}$ ("sequence") mode on your graphing calculator, enter the sequences U_n and V_n as shown in the figure to the left. Then use the TABLE command to compare the two sequences. For the first few years, Victoria seems to be accumulating more savings than Ursula. Scroll down in the table to verify that Ursula eventually pulls ahead of Victoria in the savings race. In what year does this occur?

 8. Newton's Law of Cooling   A tureen of soup at a temperature of 170 °F is placed on a table in a dining room in which the thermostat is set at 70 °F. The soup cools according to the following rule, a special case of Newton's Law of Cooling: Each minute, the temperature of the soup declines by 3% of the difference between the soup temperature and the room temperature.

(a) Find a recursive sequence that models the soup temperature T_n at the nth minute.

(b) Enter the sequence T_n in your graphing calculator, and use the TABLE command to find the temperature at 10-min increments from $n = 0$ to $n = 60$. (See Problem 7(b).)

(c) Graph the sequence T_n. What temperature will the soup be after a long time?

9. Logistic Population Growth   Simple exponential models for population growth do not take into account the fact that when the population increases, survival becomes harder for each individual because of greater competition for food and other resources.

We can get a more accurate model by assuming that the birth rate is proportional to the size of the population, but the death rate is proportional to the square of the population. Using this idea, researchers find that the number of raccoons R_n on a certain island is modeled by the following recursive sequence:

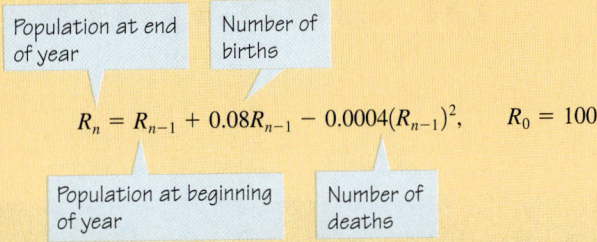

$$R_n = R_{n-1} + 0.08R_{n-1} - 0.0004(R_{n-1})^2, \qquad R_0 = 100$$

Here n represents the number of years since observations began, R_0 is the initial population, 0.08 is the annual birth rate, and 0.0004 is a constant related to the death rate.

(a) Use the ┌ TABLE ┐ command on a graphing calculator to find the raccoon population for each year from $n = 1$ to $n = 7$.

(b) Graph the sequence R_n. What happens to the raccoon population as n becomes large?

12

Limits: A Preview of Calculus

Chapter Overview

In this chapter we study the central idea underlying calculus—the concept of *limit*. Calculus is used in modeling numerous real-life phenomena, particularly situations that involve change or motion. To understand the basic idea of limits let's consider two fundamental examples.

To find the area of a polygonal figure we simply divide it into triangles and add the areas of the triangles, as in the figure to the left. However, it is much more difficult to find the area of a region with curved sides. One way is to approximate the area by inscribing polygons in the region. The figure illustrates how this is done for a circle.

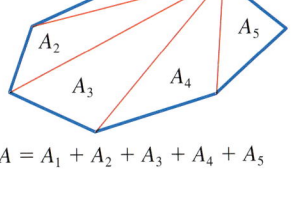

$$A = A_1 + A_2 + A_3 + A_4 + A_5$$

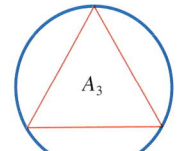 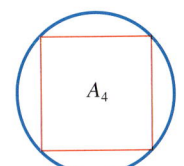 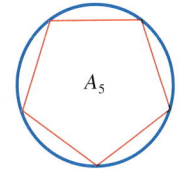 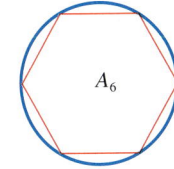 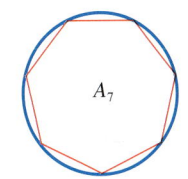 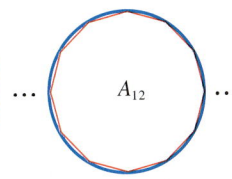

If we let A_n be the area of the inscribed regular polygon with n sides, then we see that as n increases A_n gets closer and closer to the area of the circle. We say that the area A of the circle is the *limit* of the areas A_n and write

$$\text{area} = \lim_{n \to \infty} A_n$$

If we can find a pattern for the areas A_n, then we may be able to determine the limit A exactly. In this chapter we use a similar idea to find areas of regions bounded by graphs of functions.

In Chapter 2 we learned how to find the average rate of change of a function. For example, to find average speed we divide the total distance traveled by the total time. But how can we find *instantaneous* speed—that is, the speed at a given instant? We can't divide the total distance traveled by the total time, because in an instant the total distance traveled is zero and the total time spent traveling is zero! But we can find the *average* rate of change on smaller and smaller intervals, zooming in on the instant we want. For example, suppose $f(t)$ gives the distance a car has traveled at time t. To find the speed of the car at exactly 2:00 P.M., we first find the average speed on an interval from 2 to a little after 2, that is, on the interval $[2, 2 + h]$. We know that the average speed on this interval is $[f(2 + h) - f(2)]/h$. By finding this average speed for smaller and smaller values of h (letting h go to zero), we zoom in on the instant we want. We can write

$$\text{instantaneous speed} = \lim_{h \to 0} \frac{f(2 + h) - f(2)}{h}$$

If we find a pattern for the average speed, we can evaluate this limit exactly.

The ideas in this chapter have wide-ranging applications. The concept of "instantaneous rate of change" applies to any varying quantity, not just speed. The concept of "area under the graph of a function" is a very versatile one. Indeed, numerous phenomena, seemingly unrelated to area, can be interpreted as area under the graph of a function. We explore some of these in *Focus on Modeling*, page 929.

12.1   Finding Limits Numerically and Graphically

In this section we use tables of values and graphs of functions to answer the question, What happens to the values $f(x)$ of a function f as the variable x approaches the number a?

Definition of Limit

We begin by investigating the behavior of the function f defined by

$$f(x) = x^2 - x + 2$$

for values of x near 2. The following table gives values of $f(x)$ for values of x close to 2 but not equal to 2.

x	$f(x)$
1.0	2.000000
1.5	2.750000
1.8	3.440000
1.9	3.710000
1.95	3.852500
1.99	3.970100
1.995	3.985025
1.999	3.997001

x	$f(x)$
3.0	8.000000
2.5	5.750000
2.2	4.640000
2.1	4.310000
2.05	4.152500
2.01	4.030100
2.005	4.015025
2.001	4.003001

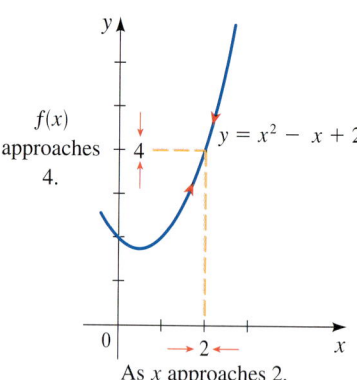

Figure 1

From the table and the graph of f (a parabola) shown in Figure 1 we see that when x is close to 2 (on either side of 2), $f(x)$ is close to 4. In fact, it appears that we can make the values of $f(x)$ as close as we like to 4 by taking x sufficiently close to 2. We express this by saying "the limit of the function $f(x) = x^2 - x + 2$ as x approaches 2 is equal to 4." The notation for this is

$$\lim_{x \to 2} (x^2 - x + 2) = 4$$

In general, we use the following notation.

Definition of the Limit of a Function

We write

$$\lim_{x \to a} f(x) = L$$

and say

"the limit of $f(x)$, as x approaches a, equals L"

if we can make the values of $f(x)$ arbitrarily close to L (as close to L as we like) by taking x to be sufficiently close to a, but not equal to a.

Roughly speaking, this says that the values of $f(x)$ get closer and closer to the number L as x gets closer and closer to the number a (from either side of a) but $x \neq a$.

An alternative notation for $\lim_{x \to a} f(x) = L$ is

$$f(x) \to L \qquad \text{as} \qquad x \to a$$

which is usually read "$f(x)$ approaches L as x approaches a." This is the notation we used in Section 3.6 when discussing asymptotes of rational functions.

Notice the phrase "but $x \neq a$" in the definition of limit. This means that in finding the limit of $f(x)$ as x approaches a, we never consider $x = a$. In fact, $f(x)$ need not even be defined when $x = a$. The only thing that matters is how f is defined *near a*.

Figure 2 shows the graphs of three functions. Note that in part (c), $f(a)$ is not defined and in part (b), $f(a) \neq L$. But in each case, regardless of what happens at a, $\lim_{x \to a} f(x) = L$.

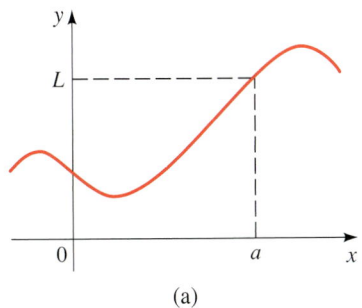

(a)

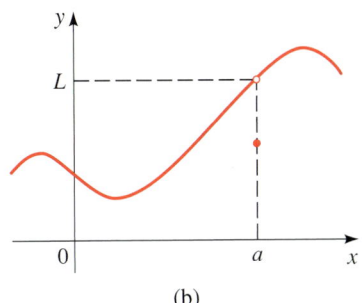

(b)

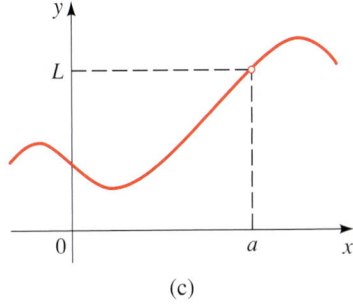

(c)

Figure 2

$\lim_{x \to a} f(x) = L$ in all three cases

Estimating Limits Numerically and Graphically

In Section 12.2 we will develop techniques for finding exact values of limits. For now, we use tables and graphs to estimate limits of functions.

Example 1   Estimating a Limit Numerically and Graphically

Guess the value of $\lim\limits_{x \to 1} \dfrac{x-1}{x^2-1}$. Check your work with a graph.

Solution   Notice that the function $f(x) = (x-1)/(x^2-1)$ is not defined when $x = 1$, but this doesn't matter because the definition of $\lim_{x \to a} f(x)$ says that we consider values of x that are close to a but not equal to a. The following tables give values of $f(x)$ (correct to six decimal places) for values of x that approach 1 (but are not equal to 1).

$x < 1$	$f(x)$	$x > 1$	$f(x)$
0.5	0.666667	1.5	0.4000000
0.9	0.526316	1.1	0.476190
0.99	0.502513	1.01	0.497512
0.999	0.500250	1.001	0.499750
0.9999	0.500025	1.0001	0.499975

On the basis of the values in the two tables, we make the guess that

$$\lim_{x \to 1} \frac{x-1}{x^2-1} = 0.5$$

As a graphical verification we use a graphing device to produce Figure 3. We see that when x is close to 1, y is close to 0.5. If we use the ZOOM and TRACE features to get a closer look, as in Figure 4, we notice that as x gets closer and closer to 1, y becomes closer and closer to 0.5. This reinforces our conclusion. ∎

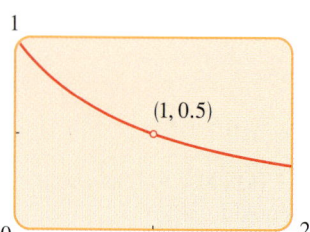

Figure 3

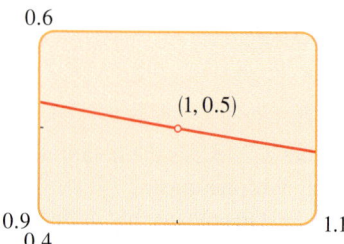

Figure 4

Example 2   Finding a Limit from a Table

Find $\lim\limits_{t \to 0} \dfrac{\sqrt{t^2+9}-3}{t^2}$.

Solution   The table in the margin lists values of the function for several values of t near 0. As t approaches 0, the values of the function seem to approach $0.1666666\ldots$, and so we guess that

$$\lim_{t \to 0} \frac{\sqrt{t^2+9}-3}{t^2} = \frac{1}{6}$$

∎

What would have happened in Example 2 if we had taken even smaller values of t? The table in the margin shows the results from one calculator; you can see that something strange seems to be happening.

If you try these calculations on your own calculator, you might get different values, but eventually you will get the value 0 if you make t sufficiently small. Does this mean that the answer is really 0 instead of $\frac{1}{6}$? No, the value of the limit is $\frac{1}{6}$, as we will show in the next section. The problem is that the calculator gave false values because $\sqrt{t^2+9}$ is very close to 3 when t is small. (In fact, when t is sufficiently small, a calculator's value for $\sqrt{t^2+9}$ is $3.000\ldots$ to as many digits as the calculator is capable of carrying.)

Something similar happens when we try to graph the function of Example 2 on a graphing device. Parts (a) and (b) of Figure 5 show quite accurate graphs of this func-

t	$\dfrac{\sqrt{t^2+9}-3}{t^2}$
± 1.0	0.16228
± 0.5	0.16553
± 0.1	0.16662
± 0.05	0.16666
± 0.01	0.16667

t	$\dfrac{\sqrt{t^2+9}-3}{t^2}$
± 0.0005	0.16800
± 0.0001	0.20000
± 0.00005	0.00000
± 0.00001	0.00000

tion, and when we use the $\boxed{\text{TRACE}}$ feature, we can easily estimate that the limit is about $\frac{1}{6}$. But if we zoom in too far, as in parts (c) and (d), then we get inaccurate graphs, again because of problems with subtraction.

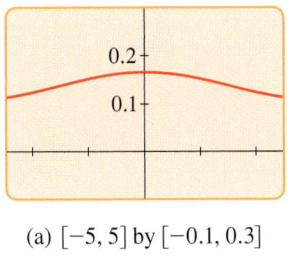

(a) $[-5, 5]$ by $[-0.1, 0.3]$

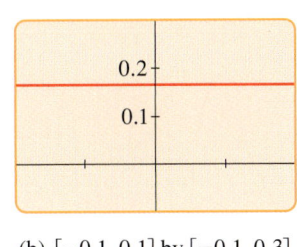

(b) $[-0.1, 0.1]$ by $[-0.1, 0.3]$

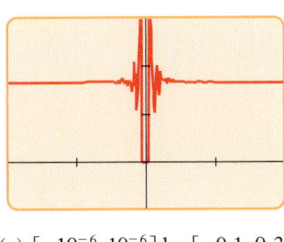

(c) $[-10^{-6}, 10^{-6}]$ by $[-0.1, 0.3]$

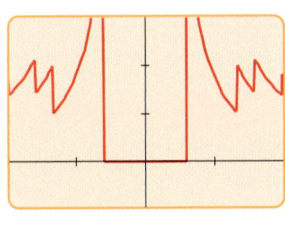

(d) $[-10^{-7}, 10^{-7}]$ by $[-0.1, 0.3]$

Figure 5

Limits That Fail to Exist

Functions do not necessarily approach a finite value at every point. In other words, it's possible for a limit not to exist. The next three examples illustrate ways in which this can happen.

Example 3 **A Limit That Fails to Exist (A Function with a Jump)**

The Heaviside function H is defined by

$$H(t) = \begin{cases} 0 & \text{if } t < 0 \\ 1 & \text{if } t \geq 0 \end{cases}$$

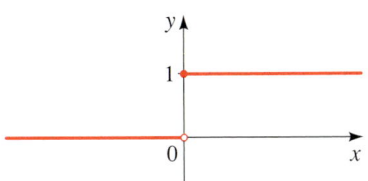

Figure 6

[This function is named after the electrical engineer Oliver Heaviside (1850–1925) and can be used to describe an electric current that is switched on at time $t = 0$.] Its graph is shown in Figure 6. Notice the "jump" in the graph at $x = 0$.

As t approaches 0 from the left, $H(t)$ approaches 0. As t approaches 0 from the right, $H(t)$ approaches 1. There is no single number that $H(t)$ approaches as t approaches 0. Therefore, $\lim_{t \to 0} H(t)$ does not exist. ∎

Example 4 **A Limit That Fails to Exist (A Function That Oscillates)**

Find $\lim_{x \to 0} \sin \dfrac{\pi}{x}$.

Solution   The function $f(x) = \sin(\pi/x)$ is undefined at 0. Evaluating the function for some small values of x, we get

$$f(1) = \sin \pi = 0 \qquad\qquad f(\tfrac{1}{2}) = \sin 2\pi = 0$$
$$f(\tfrac{1}{3}) = \sin 3\pi = 0 \qquad\qquad f(\tfrac{1}{4}) = \sin 4\pi = 0$$
$$f(0.1) = \sin 10\pi = 0 \qquad f(0.01) = \sin 100\pi = 0$$

Similarly, $f(0.001) = f(0.0001) = 0$. On the basis of this information we might be tempted to guess that

$$\lim_{x \to 0} \sin \frac{\pi}{x} \overset{?}{=} 0$$

🚫 but this time our guess is wrong. Note that although $f(1/n) = \sin n\pi = 0$ for any integer n, it is also true that $f(x) = 1$ for infinitely many values of x that approach 0. (See the graph in Figure 7.)

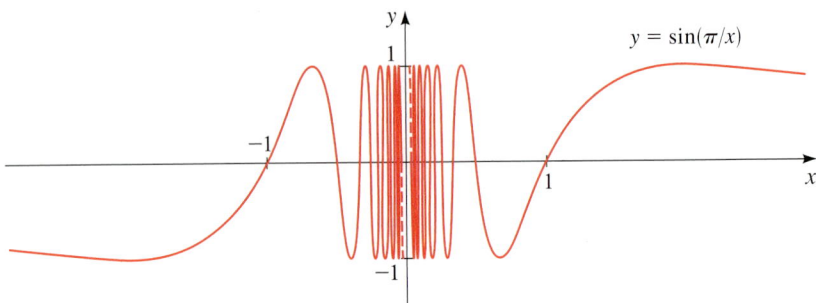

Figure 7

The broken lines indicate that the values of $\sin(\pi/x)$ oscillate between 1 and -1 infinitely often as x approaches 0. Since the values of $f(x)$ do not approach a fixed number as x approaches 0,

$$\lim_{x\to 0} \sin \frac{\pi}{x} \quad \text{does not exist} \qquad \blacksquare$$

🚫 Example 4 illustrates some of the pitfalls in guessing the value of a limit. It is easy to guess the wrong value if we use inappropriate values of x, but it is difficult to know when to stop calculating values. And, as the discussion after Example 2 shows, sometimes calculators and computers give incorrect values. In the next two sections, however, we will develop foolproof methods for calculating limits.

Example 5 A Limit That Fails to Exist (A Function with a Vertical Asymptote)

Find $\lim_{x\to 0} \dfrac{1}{x^2}$ if it exists.

Solution   As x becomes close to 0, x^2 also becomes close to 0, and $1/x^2$ becomes very large. (See the table in the margin.) In fact, it appears from the graph of the function $f(x) = 1/x^2$ shown in Figure 8 that the values of $f(x)$ can be made arbitrarily large by taking x close enough to 0. Thus, the values of $f(x)$ do not approach a number, so $\lim_{x\to 0}(1/x^2)$ does not exist.

x	$\dfrac{1}{x^2}$
± 1	1
± 0.5	4
± 0.2	25
± 0.1	100
± 0.05	400
± 0.01	10,000
± 0.001	1,000,000

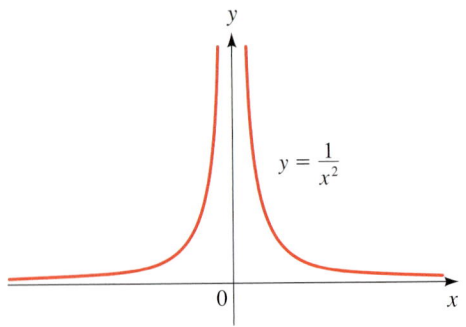

Figure 8 $\blacksquare$

To indicate the kind of behavior exhibited in Example 5, we use the notation

$$\lim_{x \to 0} \frac{1}{x^2} = \infty$$

 This does not mean that we are regarding ∞ as a number. Nor does it mean that the limit exists. It simply expresses the particular way in which the limit does not exist: $1/x^2$ can be made as large as we like by taking x close enough to 0. Notice that the line $x = 0$ (the y-axis) is a vertical asymptote in the sense we described in Section 3.6.

One-Sided Limits

We noticed in Example 3 that $H(t)$ approaches 0 as t approaches 0 from the left and $H(t)$ approaches 1 as t approaches 0 from the right. We indicate this situation symbolically by writing

$$\lim_{t \to 0^-} H(t) = 0 \qquad \text{and} \qquad \lim_{t \to 0^+} H(t) = 1$$

The symbol "$t \to 0^-$" indicates that we consider only values of t that are less than 0. Likewise, "$t \to 0^+$" indicates that we consider only values of t that are greater than 0.

Definition of a One-Sided Limit

We write

$$\lim_{x \to a^-} f(x) = L$$

and say the "left-hand limit of $f(x)$ as x approaches a" [or the "limit of $f(x)$ as x approaches a from the left"] is equal to L if we can make the values of $f(x)$ arbitrarily close to L by taking x to be sufficiently close to a and x less than a.

Notice that this definition differs from the definition of a two-sided limit only in that we require x to be less than a. Similarly, if we require that x be greater than a, we get "the **right-hand limit of $f(x)$ as x approaches a** is equal to L" and we write

$$\lim_{x \to a^+} f(x) = L$$

Thus, the symbol "$x \to a^+$" means that we consider only $x > a$. These definitions are illustrated in Figure 9.

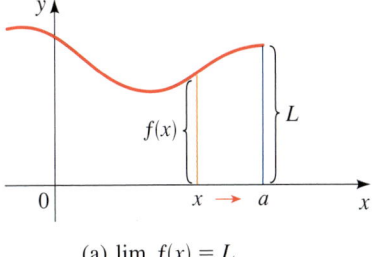

(a) $\lim_{x \to a^-} f(x) = L$

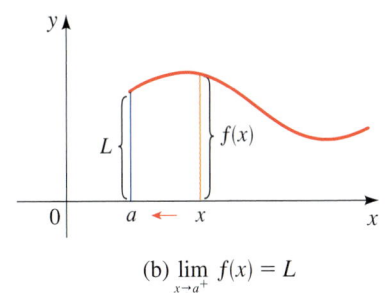

(b) $\lim_{x \to a^+} f(x) = L$

Figure 9

By comparing the definitions of two-sided and one-sided limits, we see that the following is true.

$$\lim_{x \to a} f(x) = L \quad \text{if and only if} \quad \lim_{x \to a^-} f(x) = L \quad \text{and} \quad \lim_{x \to a^+} f(x) = L$$

Thus, if the left-hand and right-hand limits are different, the (two-sided) limit does not exist. We use this fact in the next two examples.

Example 6   Limits from a Graph

The graph of a function g is shown in Figure 10. Use it to state the values (if they exist) of the following:

(a) $\lim\limits_{x \to 2^-} g(x)$, $\lim\limits_{x \to 2^+} g(x)$, $\lim\limits_{x \to 2} g(x)$

(b) $\lim\limits_{x \to 5^-} g(x)$, $\lim\limits_{x \to 5^+} g(x)$, $\lim\limits_{x \to 5} g(x)$

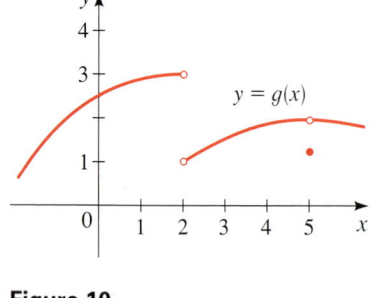

$y = g(x)$

Figure 10

Solution

(a) From the graph we see that the values of $g(x)$ approach 3 as x approaches 2 from the left, but they approach 1 as x approaches 2 from the right. Therefore

$$\lim_{x \to 2^-} g(x) = 3 \quad \text{and} \quad \lim_{x \to 2^+} g(x) = 1$$

Since the left- and right-hand limits are different, we conclude that $\lim_{x \to 2} g(x)$ does not exist.

(b) The graph also shows that

$$\lim_{x \to 5^-} g(x) = 2 \quad \text{and} \quad \lim_{x \to 5^+} g(x) = 2$$

This time the left- and right-hand limits are the same, and so we have

$$\lim_{x \to 5} g(x) = 2$$

Despite this fact, notice that $g(5) \neq 2$.   ∎

Example 7   A Piecewise-Defined Function

Let f be the function defined by

$$f(x) = \begin{cases} 2x^2 & \text{if } x < 1 \\ 4 - x & \text{if } x \geq 1 \end{cases}$$

Graph f, and use the graph to find the following:

(a) $\lim\limits_{x \to 1^-} f(x)$ (b) $\lim\limits_{x \to 1^+} f(x)$ (c) $\lim\limits_{x \to 1} f(x)$

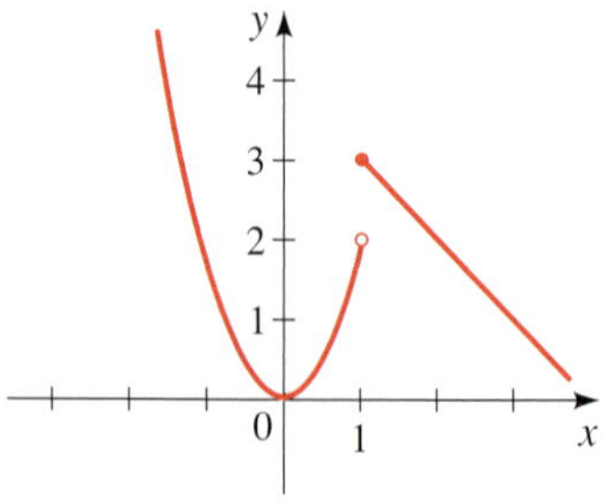

Figure 11

Solution   The graph of f is shown in Figure 11. From the graph we see that the values of $f(x)$ approach 2 as x approaches 1 from the left, but they approach 3 as x approaches 1 from the right. Thus, the left- and right-hand limits are not equal. So we have

(a) $\lim\limits_{x \to 1^-} f(x) = 2$ (b) $\lim\limits_{x \to 1^+} f(x) = 3$ (c) $\lim\limits_{x \to 1} f(x)$ does not exist.   ∎

12.1 Exercises

1–6 ■ Complete the table of values (to five decimal places) and use the table to estimate the value of the limit.

1. $\lim_{x \to 4} \dfrac{\sqrt{x} - 2}{x - 4}$

x	3.9	3.99	3.999	4.001	4.01	4.1
$f(x)$						

2. $\lim_{x \to 2} \dfrac{x - 2}{x^2 + x - 6}$

x	1.9	1.99	1.999	2.001	2.01	2.1
$f(x)$						

3. $\lim_{x \to 1} \dfrac{x - 1}{x^3 - 1}$

x	0.9	0.99	0.999	1.001	1.01	1.1
$f(x)$						

4. $\lim_{x \to 0} \dfrac{e^x - 1}{x}$

x	-0.1	-0.01	-0.001	0.001	0.01	0.1
$f(x)$						

5. $\lim_{x \to 0} \dfrac{\sin x}{x}$

x	± 1	± 0.5	± 0.1	± 0.05	± 0.01
$f(x)$					

6. $\lim_{x \to 0^+} x \ln x$

x	0.1	0.01	0.001	0.0001	0.00001
$f(x)$					

 7–12 ■ Use a table of values to estimate the value of the limit. Then use a graphing device to confirm your result graphically.

7. $\lim_{x \to -4} \dfrac{x + 4}{x^2 + 7x + 12}$

8. $\lim_{x \to 1} \dfrac{x^3 - 1}{x^2 - 1}$

9. $\lim_{x \to 0} \dfrac{5^x - 3^x}{x}$

10. $\lim_{x \to 0} \dfrac{\sqrt{x + 9} - 3}{x}$

11. $\lim_{x \to 1} \left(\dfrac{1}{\ln x} - \dfrac{1}{x - 1} \right)$

12. $\lim_{x \to 0} \dfrac{\tan 2x}{\tan 3x}$

13. For the function f whose graph is given, state the value of the given quantity, if it exists. If it does not exist, explain why.

(a) $\lim_{x \to 1^-} f(x)$ **(b)** $\lim_{x \to 1^+} f(x)$ **(c)** $\lim_{x \to 1} f(x)$

(d) $\lim_{x \to 5} f(x)$ **(e)** $f(5)$

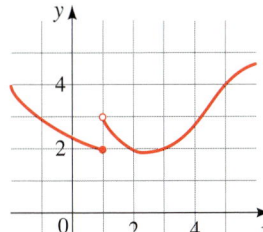

14. For the function f whose graph is given, state the value of the given quantity, if it exists. If it does not exist, explain why.

(a) $\lim_{x \to 0} f(x)$ **(b)** $\lim_{x \to 3^-} f(x)$ **(c)** $\lim_{x \to 3^+} f(x)$

(d) $\lim_{x \to 3} f(x)$ **(e)** $f(3)$

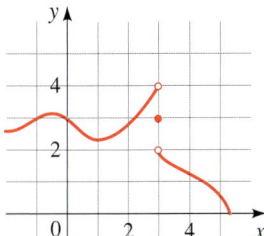

15. For the function g whose graph is given, state the value of the given quantity, if it exists. If it does not exist, explain why.

(a) $\lim_{t \to 0^-} g(t)$ **(b)** $\lim_{t \to 0^+} g(t)$ **(c)** $\lim_{t \to 0} g(t)$

(d) $\lim_{t \to 2^-} g(t)$ **(e)** $\lim_{t \to 2^+} g(t)$ **(f)** $\lim_{t \to 2} g(t)$

(g) $g(2)$ **(h)** $\lim_{t \to 4} g(t)$

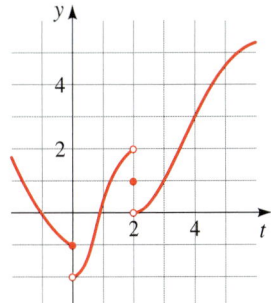

16. State the value of the limit, if it exists, from the given graph of f. If it does not exist, explain why.

(a) $\displaystyle\lim_{x\to 3} f(x)$ (b) $\displaystyle\lim_{x\to 1} f(x)$ (c) $\displaystyle\lim_{x\to -3} f(x)$

(d) $\displaystyle\lim_{x\to 2^-} f(x)$ (e) $\displaystyle\lim_{x\to 2^+} f(x)$ (f) $\displaystyle\lim_{x\to 2} f(x)$

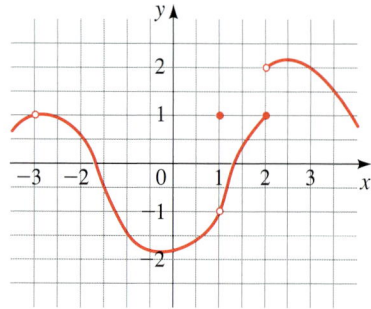

17–22 ■ Use a graphing device to determine whether the limit exists. If the limit exists, estimate its value to two decimal places.

17. $\displaystyle\lim_{x\to 1} \frac{x^3 + x^2 + 3x - 5}{2x^2 - 5x + 3}$ **18.** $\displaystyle\lim_{x\to 2} \frac{x^3 + 6x^2 - 5x + 1}{x^3 - x^2 - 8x + 12}$

19. $\displaystyle\lim_{x\to 0} \ln(\sin^2 x)$ **20.** $\displaystyle\lim_{x\to 0} \frac{x^2}{\cos 5x - \cos 4x}$

21. $\displaystyle\lim_{x\to 0} \cos \frac{1}{x}$ **22.** $\displaystyle\lim_{x\to 0} \frac{1}{1 + e^{1/x}}$

23–26 ■ Graph the piecewise-defined function and use your graph to find the values of the limits, if they exist.

23. $f(x) = \begin{cases} x^2 & \text{if } x \le 2 \\ 6 - x & \text{if } x > 2 \end{cases}$

(a) $\displaystyle\lim_{x\to 2^-} f(x)$ (b) $\displaystyle\lim_{x\to 2^+} f(x)$ (c) $\displaystyle\lim_{x\to 2} f(x)$

24. $f(x) = \begin{cases} 2 & \text{if } x < 0 \\ x + 1 & \text{if } x \ge 0 \end{cases}$

(a) $\displaystyle\lim_{x\to 0^-} f(x)$ (b) $\displaystyle\lim_{x\to 0^+} f(x)$ (c) $\displaystyle\lim_{x\to 0} f(x)$

25. $f(x) = \begin{cases} -x + 3 & \text{if } x < -1 \\ 3 & \text{if } x \ge -1 \end{cases}$

(a) $\displaystyle\lim_{x\to -1^-} f(x)$ (b) $\displaystyle\lim_{x\to -1^+} f(x)$ (c) $\displaystyle\lim_{x\to -1} f(x)$

26. $f(x) = \begin{cases} 2x + 10 & \text{if } x \le -2 \\ -x + 4 & \text{if } x > -2 \end{cases}$

(a) $\displaystyle\lim_{x\to -2^-} f(x)$ (b) $\displaystyle\lim_{x\to -2^+} f(x)$ (c) $\displaystyle\lim_{x\to -2} f(x)$

Discovery • Discussion

27. A Function with Specified Limits   Sketch the graph of an example of a function f that satisfies all of the following conditions.

$$\lim_{x\to 0^-} f(x) = 2 \qquad \lim_{x\to 0^+} f(x) = 0$$

$$\lim_{x\to 2} f(x) = 1 \qquad f(0) = 2 \qquad f(2) = 3$$

How many such functions are there?

28. Graphing Calculator Pitfalls

(a) Evaluate $h(x) = (\tan x - x)/x^3$ for $x = 1, 0.5, 0.1, 0.05, 0.01,$ and 0.005.

(b) Guess the value of $\displaystyle\lim_{x\to 0} \frac{\tan x - x}{x^3}$.

(c) Evaluate $h(x)$ for successively smaller values of x until you finally reach 0 values for $h(x)$. Are you still confident that your guess in part (b) is correct? Explain why you eventually obtained 0 values.

(d) Graph the function h in the viewing rectangle $[-1, 1]$ by $[0, 1]$. Then zoom in toward the point where the graph crosses the y-axis to estimate the limit of $h(x)$ as x approaches 0. Continue to zoom in until you observe distortions in the graph of h. Compare with your results in part (c).

12.2   Finding Limits Algebraically

In Section 12.1 we used calculators and graphs to guess the values of limits, but we saw that such methods don't always lead to the correct answer. In this section, we use algebraic methods to find limits exactly.

Limit Laws

We use the following properties of limits, called the *Limit Laws*, to calculate limits.

Limit Laws

Suppose that c is a constant and that the following limits exist:

$$\lim_{x \to a} f(x) \qquad \text{and} \qquad \lim_{x \to a} g(x)$$

Then

1. $\lim\limits_{x \to a} [f(x) + g(x)] = \lim\limits_{x \to a} f(x) + \lim\limits_{x \to a} g(x)$ Limit of a Sum

2. $\lim\limits_{x \to a} [f(x) - g(x)] = \lim\limits_{x \to a} f(x) - \lim\limits_{x \to a} g(x)$ Limit of a Difference

3. $\lim\limits_{x \to a} [cf(x)] = c \lim\limits_{x \to a} f(x)$ Limit of a Constant Multiple

4. $\lim\limits_{x \to a} [f(x)g(x)] = \lim\limits_{x \to a} f(x) \cdot \lim\limits_{x \to a} g(x)$ Limit of a Product

5. $\lim\limits_{x \to a} \dfrac{f(x)}{g(x)} = \dfrac{\lim\limits_{x \to a} f(x)}{\lim\limits_{x \to a} g(x)}$ if $\lim\limits_{x \to a} g(x) \neq 0$ Limit of a Quotient

These five laws can be stated verbally as follows:

Limit of a Sum   **1.** The limit of a sum is the sum of the limits.

Limit of a Difference   **2.** The limit of a difference is the difference of the limits.

Limit of a Constant Multiple   **3.** The limit of a constant times a function is the constant times the limit of the function.

Limit of a Product   **4.** The limit of a product is the product of the limits.

Limit of a Quotient   **5.** The limit of a quotient is the quotient of the limits (provided that the limit of the denominator is not 0).

It's easy to believe that these properties are true. For instance, if $f(x)$ is close to L and $g(x)$ is close to M, it is reasonable to conclude that $f(x) + g(x)$ is close to $L + M$. This gives us an intuitive basis for believing that Law 1 is true.

If we use Law 4 (Limit of a Product) repeatedly with $g(x) = f(x)$, we obtain the following Law 6 for the limit of a power. A similar law holds for roots.

Limit Laws

6. $\lim\limits_{x \to a} [f(x)]^n = \left[\lim\limits_{x \to a} f(x) \right]^n$ where n is a positive integer   Limit of a Power

7. $\lim\limits_{x \to a} \sqrt[n]{f(x)} = \sqrt[n]{\lim\limits_{x \to a} f(x)}$ where n is a positive integer   Limit of a Root

[If n is even, we assume that $\lim_{x \to a} f(x) > 0$.]

In words, these laws say:

Limit of a Power   **6.** The limit of a power is the power of the limit.

Limit of a Root   **7.** The limit of a root is the root of the limit.

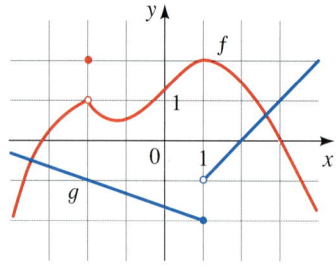

Figure 1

Example 1   Using the Limit Laws

Use the Limit Laws and the graphs of f and g in Figure 1 to evaluate the following limits, if they exist.

(a) $\lim\limits_{x \to -2} \left[f(x) + 5g(x) \right]$

(b) $\lim\limits_{x \to 1} \left[f(x)g(x) \right]$

(c) $\lim\limits_{x \to 2} \dfrac{f(x)}{g(x)}$

(d) $\lim\limits_{x \to 1} \left[f(x) \right]^3$

Solution

(a) From the graphs of f and g we see that

$$\lim_{x \to -2} f(x) = 1 \qquad \text{and} \qquad \lim_{x \to -2} g(x) = -1$$

Therefore, we have

$$\lim_{x \to -2} \left[f(x) + 5g(x) \right] = \lim_{x \to -2} f(x) + \lim_{x \to -2} \left[5g(x) \right] \qquad \text{\textcolor{blue}{Limit of a Sum}}$$

$$= \lim_{x \to -2} f(x) + 5 \lim_{x \to -2} g(x) \qquad \text{\textcolor{blue}{Limit of a Constant Multiple}}$$

$$= 1 + 5(-1) = -4$$

(b) We see that $\lim_{x \to 1} f(x) = 2$. But $\lim_{x \to 1} g(x)$ does not exist because the left- and right-hand limits are different:

$$\lim_{x \to 1^-} g(x) = -2 \qquad \lim_{x \to 1^+} g(x) = -1$$

So we can't use Law 4 (Limit of a Product). The given limit does not exist, since the left-hand limit is not equal to the right-hand limit.

(c) The graphs show that

$$\lim_{x \to 2} f(x) \approx 1.4 \qquad \text{and} \qquad \lim_{x \to 2} g(x) = 0$$

Because the limit of the denominator is 0, we can't use Law 5 (Limit of a Quotient). The given limit does not exist because the denominator approaches 0 while the numerator approaches a nonzero number.

(d) Since $\lim_{x \to 1} f(x) = 2$, we use Law 6 to get

$$\lim_{x \to 1} \left[f(x) \right]^3 = \left[\lim_{x \to 1} f(x) \right]^3 \qquad \text{\textcolor{blue}{Limit of a Power}}$$

$$= 2^3 = 8$$

∎

Applying the Limit Laws

In applying the Limit Laws, we need to use four special limits.

Some Special Units

1. $\lim\limits_{x \to a} c = c$

2. $\lim\limits_{x \to a} x = a$

3. $\lim\limits_{x \to a} x^n = a^n$ where n is a positive integer

4. $\lim\limits_{x \to a} \sqrt[n]{x} = \sqrt[n]{a}$ where n is a positive integer and $a > 0$

Special Limits 1 and 2 are intuitively obvious—looking at the graphs of $y = c$ and $y = x$ will convince you of their validity. Limits 3 and 4 are special cases of Limit Laws 6 and 7 (Limits of a Power and of a Root).

Example 2 Using the Limit Laws

Evaluate the following limits and justify each step.

(a) $\lim\limits_{x \to 5} (2x^2 - 3x + 4)$ (b) $\lim\limits_{x \to -2} \dfrac{x^3 + 2x^2 - 1}{5 - 3x}$

Solution

(a) $\lim\limits_{x \to 5} (2x^2 - 3x + 4) = \lim\limits_{x \to 5} (2x^2) - \lim\limits_{x \to 5} (3x) + \lim\limits_{x \to 5} 4$ Limits of a Difference and Sum

$\qquad\qquad\qquad\qquad\quad = 2 \lim\limits_{x \to 5} x^2 - 3 \lim\limits_{x \to 5} x + \lim\limits_{x \to 5} 4$ Limit of a Constant Multiple

$\qquad\qquad\qquad\qquad\quad = 2(5^2) - 3(5) + 4$ Special Limits 3, 2, and 1

$\qquad\qquad\qquad\qquad\quad = 39$

(b) We start by using Law 5, but its use is fully justified only at the final stage when we see that the limits of the numerator and denominator exist and the limit of the denominator is not 0.

$$\lim_{x \to -2} \frac{x^3 + 2x^2 - 1}{5 - 3x} = \frac{\lim\limits_{x \to -2} (x^3 + 2x^2 - 1)}{\lim\limits_{x \to -2} (5 - 3x)}$$ Limit of a Quotient

$$= \frac{\lim\limits_{x \to -2} x^3 + 2 \lim\limits_{x \to -2} x^2 - \lim\limits_{x \to -2} 1}{\lim\limits_{x \to -2} 5 - 3 \lim\limits_{x \to -2} x}$$ Limits of Sums, Differences, and Constant Multiples

$$= \frac{(-2)^3 + 2(-2)^2 - 1}{5 - 3(-2)}$$ Special Limits 3, 2, and 1

$$= -\frac{1}{11}$$

If we let $f(x) = 2x^2 - 3x + 4$, then $f(5) = 39$. In Example 2(a), we found that $\lim_{x \to 5} f(x) = 39$. In other words, we would have gotten the correct answer by substituting 5 for x. Similarly, direct substitution provides the correct answer in part (b). The functions in Example 2 are a polynomial and a rational function, respectively, and similar use of the Limit Laws proves that direct substitution always works for such functions. We state this fact as follows.

Limits by Direct Substitution

If f is a polynomial or a rational function and a is in the domain of f, then

$$\lim_{x \to a} f(x) = f(a)$$

Functions with this direct substitution property are called **continuous at** a. You will learn more about continuous functions when you study calculus.

Bill Sanederson/SPL/Photo Researchers, Inc.

Sir Isaac Newton (1642–1727) is universally regarded as one of the giants of physics and mathematics. He is well known for discovering the laws of motion and gravity and for inventing the calculus, but he also proved the Binomial Theorem and the laws of optics, and developed methods for solving polynomial equations to any desired accuracy. He was born on Christmas Day, a few months after the death of his father. After an unhappy childhood, he entered Cambridge University, where he learned mathematics by studying the writings of Euclid and Descartes.

During the plague years of 1665 and 1666, when the university was closed, Newton thought and wrote about ideas that, once published, instantly revolutionized the sciences. Imbued with a pathological fear of criticism, he published these writings only after many years of encouragement from Edmund Halley (who discovered the now-famous comet) and other colleagues.

Newton's works brought him enormous fame and prestige. Even poets were moved to praise; Alexander Pope wrote:

Nature and Nature's Laws
 lay hid in Night.
God said, "Let Newton be"
 and all was Light.

(continued)

Example 3   Finding Limits by Direct Substitution

Evaluate the following limits.

(a) $\lim_{x \to 3} (2x^3 - 10x - 8)$ (b) $\lim_{x \to -1} \dfrac{x^2 + 5x}{x^4 + 2}$

Solution

(a) The function $f(x) = 2x^3 - 10x - 12$ is a polynomial, so we can find the limit by direct substitution:

$$\lim_{x \to 3} (2x^3 - 10x - 12) = 2(3)^3 - 10(3) - 8 = 16$$

(b) The function $f(x) = (x^2 + 5x)/(x^4 + 2)$ is a rational function, and $x = -1$ is in its domain (because the denominator is not zero for $x = -1$). Thus, we can find the limit by direct substitution:

$$\lim_{x \to -1} \frac{x^2 + 5x}{x^4 + 2} = \frac{(-1)^2 + 5(-1)}{(-1)^4 + 2} = -\frac{4}{3}$$ ∎

Finding Limits Using Algebra and the Limit Laws

As we saw in Example 3, evaluating limits by direct substitution is easy. But not all limits can be evaluated this way. In fact, most of the situations in which limits are useful requires us to work harder to evaluate the limit. The next three examples illustrate how we can use algebra to find limits.

Example 4   Finding a Limit by Canceling a Common Factor

Find $\lim_{x \to 1} \dfrac{x - 1}{x^2 - 1}$.

Solution   Let $f(x) = (x - 1)/(x^2 - 1)$. We can't find the limit by substituting $x = 1$ because $f(1)$ isn't defined. Nor can we apply Law 5 (Limit of a Quotient) because the limit of the denominator is 0. Instead, we need to do some preliminary algebra. We factor the denominator as a difference of squares:

$$\frac{x - 1}{x^2 - 1} = \frac{x - 1}{(x - 1)(x + 1)}$$

The numerator and denominator have a common factor of $x - 1$. When we take the limit as x approaches 1, we have $x \neq 1$ and so $x - 1 \neq 0$. Therefore, we can cancel the common factor and compute the limit as follows:

$$\lim_{x \to 1} \frac{x - 1}{x^2 - 1} = \lim_{x \to 1} \frac{x - 1}{(x - 1)(x + 1)} \qquad \text{Factor}$$

$$= \lim_{x \to 1} \frac{1}{x + 1} \qquad \text{Cancel}$$

$$= \frac{1}{1 + 1} = \frac{1}{2} \qquad \text{Let } x \to 1$$

This calculation confirms algebraically the answer we got numerically and graphically in Example 1 in Section 12.1. ∎

Example 5 Finding a Limit by Simplifying

Evaluate $\lim\limits_{h\to 0} \dfrac{(3+h)^2 - 9}{h}$.

Solution We can't use direct substitution to evaluate this limit, because the limit of the denominator is 0. So we first simplify the limit algebraically.

$$\lim_{h\to 0} \frac{(3+h)^2 - 9}{h} = \lim_{h\to 0} \frac{(9 + 6h + h^2) - 9}{h} \qquad \text{Expand}$$

$$= \lim_{h\to 0} \frac{6h + h^2}{h} \qquad \text{Simplify}$$

$$= \lim_{h\to 0} (6 + h) \qquad \text{Cancel } h$$

$$= 6 \qquad \text{Let } h \to 0 \qquad\blacksquare$$

Example 6 Finding a Limit by Rationalizing

Find $\lim\limits_{t\to 0} \dfrac{\sqrt{t^2 + 9} - 3}{t^2}$.

Solution We can't apply Law 5 (Limit of a Quotient) immediately, since the limit of the denominator is 0. Here the preliminary algebra consists of rationalizing the numerator:

$$\lim_{t\to 0} \frac{\sqrt{t^2 + 9} - 3}{t^2} = \lim_{t\to 0} \frac{\sqrt{t^2 + 9} - 3}{t^2} \cdot \frac{\sqrt{t^2 + 9} + 3}{\sqrt{t^2 + 9} + 3} \qquad \text{Rationalize numerator}$$

$$= \lim_{t\to 0} \frac{(t^2 + 9) - 9}{t^2(\sqrt{t^2 + 9} + 3)} = \lim_{t\to 0} \frac{t^2}{t^2(\sqrt{t^2 + 9} + 3)}$$

$$= \lim_{t\to 0} \frac{1}{\sqrt{t^2 + 9} + 3} = \frac{1}{\sqrt{\lim\limits_{t\to 0}(t^2 + 9)} + 3} = \frac{1}{3 + 3} = \frac{1}{6}$$

This calculation confirms the guess that we made in Example 2 in Section 12.1. ∎

Using Left- and Right-Hand Limits

Some limits are best calculated by first finding the left- and right-hand limits. The following theorem is a reminder of what we discovered in Section 12.1. It says that *a two-sided limit exists if and only if both of the one-sided limits exist and are equal.*

$$\lim_{x\to a} f(x) = L \qquad \text{if and only if} \qquad \lim_{x\to a^-} f(x) = L = \lim_{x\to a^+} f(x)$$

When computing one-sided limits, we use the fact that the Limit Laws also hold for one-sided limits.

Example 7   Comparing Right and Left Limits

Show that $\lim\limits_{x \to 0} |x| = 0$.

Solution   Recall that

$$|x| = \begin{cases} x & \text{if } x \geq 0 \\ -x & \text{if } x < 0 \end{cases}$$

Since $|x| = x$ for $x > 0$, we have

$$\lim_{x \to 0^+} |x| = \lim_{x \to 0^+} x = 0$$

For $x < 0$, we have $|x| = -x$ and so

$$\lim_{x \to 0^-} |x| = \lim_{x \to 0^-} (-x) = 0$$

Therefore

$$\lim_{x \to 0} |x| = 0$$

The result of Example 7 looks plausible from Figure 2.

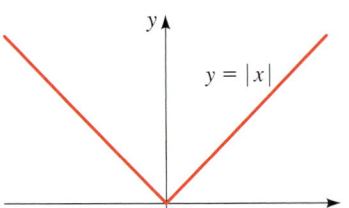

$y = |x|$

Figure 2

Example 8   Comparing Right and Left Limits

Prove that $\lim\limits_{x \to 0} \dfrac{|x|}{x}$ does not exist.

Solution   Since $|x| = x$ for $x > 0$ and $|x| = -x$ for $x < 0$, we have

$$\lim_{x \to 0^+} \frac{|x|}{x} = \lim_{x \to 0^+} \frac{x}{x} = \lim_{x \to 0^+} 1 = 1$$

$$\lim_{x \to 0^-} \frac{|x|}{x} = \lim_{x \to 0^-} \frac{-x}{x} = \lim_{x \to 0^-} (-1) = -1$$

Since the right-hand and left-hand limits exist and are different, it follows that $\lim_{x \to 0} |x|/x$ does not exist. The graph of the function $f(x) = |x|/x$ is shown in Figure 3 and supports the limits that we found.

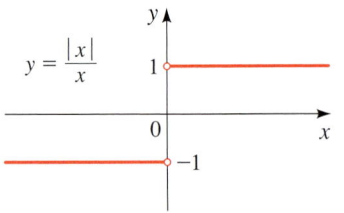

$y = \dfrac{|x|}{x}$

Figure 3

Example 9   The Limit of a Piecewise-Defined Function

Let

$$f(x) = \begin{cases} \sqrt{x - 4} & \text{if } x > 4 \\ 8 - 2x & \text{if } x < 4 \end{cases}$$

Determine whether $\lim\limits_{x \to 4} f(x)$ exists.

Solution   Since $f(x) = \sqrt{x - 4}$ for $x > 4$, we have

$$\lim_{x \to 4^+} f(x) = \lim_{x \to 4^+} \sqrt{x - 4} = \sqrt{4 - 4} = 0$$

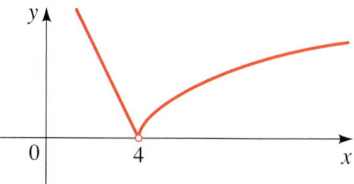

Figure 4

Since $f(x) = 8 - 2x$ for $x < 4$, we have

$$\lim_{x \to 4^-} f(x) = \lim_{x \to 4^-} (8 - 2x) = 8 - 2 \cdot 4 = 0$$

The right- and left-hand limits are equal. Thus, the limit exists and

$$\lim_{x \to 4} f(x) = 0$$

The graph of f is shown in Figure 4.   ■

12.2   Exercises

1. Suppose that

$$\lim_{x \to a} f(x) = -3 \qquad \lim_{x \to a} g(x) = 0 \qquad \lim_{x \to a} h(x) = 8$$

Find the value of the given limit. If the limit does not exist, explain why.

(a) $\lim_{x \to a} [f(x) + h(x)]$ (b) $\lim_{x \to a} [f(x)]^2$

(c) $\lim_{x \to a} \sqrt[3]{h(x)}$ (d) $\lim_{x \to a} \dfrac{1}{f(x)}$

(e) $\lim_{x \to a} \dfrac{f(x)}{h(x)}$ (f) $\lim_{x \to a} \dfrac{g(x)}{f(x)}$

(g) $\lim_{x \to a} \dfrac{f(x)}{g(x)}$ (h) $\lim_{x \to a} \dfrac{2f(x)}{h(x) - f(x)}$

2. The graphs of f and g are given. Use them to evaluate each limit, if it exists. If the limit does not exist, explain why.

(a) $\lim_{x \to 2} [f(x) + g(x)]$ (b) $\lim_{x \to 1} [f(x) + g(x)]$

(c) $\lim_{x \to 0} [f(x)g(x)]$ (d) $\lim_{x \to -1} \dfrac{f(x)}{g(x)}$

(e) $\lim_{x \to 2} x^3 f(x)$ (f) $\lim_{x \to 1} \sqrt{3 + f(x)}$

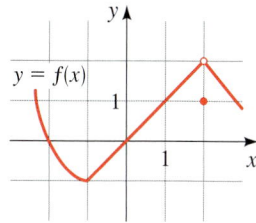

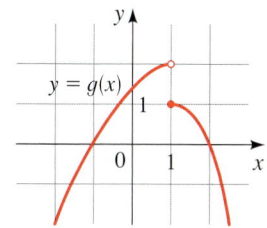

3–8 ■ Evaluate the limit and justify each step by indicating the appropriate Limit Law(s).

3. $\lim_{x \to 4} (5x^2 - 2x + 3)$

4. $\lim_{x \to 3} (x^3 + 2)(x^2 - 5x)$

5. $\lim_{x \to -1} \dfrac{x - 2}{x^2 + 4x - 3}$

6. $\lim_{x \to 1} \left(\dfrac{x^4 + x^2 - 6}{x^4 + 2x + 3} \right)^2$

7. $\lim_{t \to -2} (t + 1)^9 (t^2 - 1)$

8. $\lim_{u \to -2} \sqrt{u^4 + 3u + 6}$

9–20 ■ Evaluate the limit, if it exists.

9. $\lim_{x \to 2} \dfrac{x^2 + x - 6}{x - 2}$

10. $\lim_{x \to -4} \dfrac{x^2 + 5x + 4}{x^2 + 3x - 4}$

11. $\lim_{x \to 2} \dfrac{x^2 - x + 6}{x + 2}$

12. $\lim_{x \to 1} \dfrac{x^3 - 1}{x^2 - 1}$

13. $\lim_{t \to -3} \dfrac{t^2 - 9}{2t^2 + 7t + 3}$

14. $\lim_{h \to 0} \dfrac{\sqrt{1 + h} - 1}{h}$

15. $\lim_{h \to 0} \dfrac{(2 + h)^3 - 8}{h}$

16. $\lim_{x \to 2} \dfrac{x^4 - 16}{x - 2}$

17. $\lim_{x \to 7} \dfrac{\sqrt{x + 2} - 3}{x - 7}$

18. $\lim_{h \to 0} \dfrac{(3 + h)^{-1} - 3^{-1}}{h}$

19. $\lim_{x \to -4} \dfrac{\dfrac{1}{4} + \dfrac{1}{x}}{4 + x}$

20. $\lim_{t \to 0} \left(\dfrac{1}{t} - \dfrac{1}{t^2 + t} \right)$

21–24 ■ Find the limit and use a graphing device to confirm your result graphically.

21. $\lim_{x \to 1} \dfrac{x^2 - 1}{\sqrt{x} - 1}$

22. $\lim_{x \to 0} \dfrac{(4 + x)^3 - 64}{x}$

23. $\lim_{x \to -1} \dfrac{x^2 - x - 2}{x^3 - x}$

24. $\lim_{x \to 1} \dfrac{x^8 - 1}{x^5 - x}$

25. (a) Estimate the value of

$$\lim_{x \to 0} \dfrac{x}{\sqrt{1 + 3x} - 1}$$

by graphing the function $f(x) = x/(\sqrt{1 + 3x} - 1)$.

(b) Make a table of values of $f(x)$ for x close to 0 and guess the value of the limit.

(c) Use the Limit Laws to prove that your guess is correct.

26. (a) Use a graph of

$$f(x) = \dfrac{\sqrt{3 + x} - \sqrt{3}}{x}$$

to estimate the value of $\lim_{x \to 0} f(x)$ to two decimal places.

(b) Use a table of values of $f(x)$ to estimate the limit to four decimal places.

(c) Use the Limit Laws to find the exact value of the limit.

27–32 ■ Find the limit, if it exists. If the limit does not exist, explain why.

27. $\lim\limits_{x \to -4} |x + 4|$

28. $\lim\limits_{x \to -4^-} \dfrac{|x + 4|}{x + 4}$

29. $\lim\limits_{x \to 2} \dfrac{|x - 2|}{x - 2}$

30. $\lim\limits_{x \to 1.5} \dfrac{2x^2 - 3x}{|2x - 3|}$

31. $\lim\limits_{x \to 0^-} \left(\dfrac{1}{x} - \dfrac{1}{|x|} \right)$

32. $\lim\limits_{x \to 0^+} \left(\dfrac{1}{x} - \dfrac{1}{|x|} \right)$

33. Let

$$f(x) = \begin{cases} x - 1 & \text{if } x < 2 \\ x^2 - 4x + 6 & \text{if } x \ge 2 \end{cases}$$

(a) Find $\lim_{x \to 2^-} f(x)$ and $\lim_{x \to 2^+} f(x)$.

(b) Does $\lim_{x \to 2} f(x)$ exist?

(c) Sketch the graph of f.

34. Let

$$h(x) = \begin{cases} x & \text{if } x < 0 \\ x^2 & \text{if } 0 < x \le 2 \\ 8 - x & \text{if } x > 2 \end{cases}$$

(a) Evaluate each limit, if it exists.

(i) $\lim\limits_{x \to 0^+} h(x)$ (iv) $\lim\limits_{x \to 2^-} h(x)$

(ii) $\lim\limits_{x \to 0} h(x)$ (v) $\lim\limits_{x \to 2^+} h(x)$

(iii) $\lim\limits_{x \to 1} h(x)$ (vi) $\lim\limits_{x \to 2} h(x)$

(b) Sketch the graph of h.

Discovery • Discussion

35. Cancellation and Limits

(a) What is wrong with the following equation?

$$\frac{x^2 + x - 6}{x - 2} = x + 3$$

(b) In view of part (a), explain why the equation

$$\lim_{x \to 2} \frac{x^2 + x - 6}{x - 2} = \lim_{x \to 2} (x + 3)$$

is correct.

36. The Lorentz Contraction In the theory of relativity, the Lorentz contraction formula

$$L = L_0 \sqrt{1 - v^2/c^2}$$

expresses the length L of an object as a function of its velocity v with respect to an observer, where L_0 is the length of the object at rest and c is the speed of light. Find $\lim_{v \to c^-} L$ and interpret the result. Why is a left-hand limit necessary?

37. Limits of Sums and Products

(a) Show by means of an example that $\lim_{x \to a} [f(x) + g(x)]$ may exist even though neither $\lim_{x \to a} f(x)$ nor $\lim_{x \to a} g(x)$ exists.

(b) Show by means of an example that $\lim_{x \to a} [f(x)g(x)]$ may exist even though neither $\lim_{x \to a} f(x)$ nor $\lim_{x \to a} g(x)$ exists.

12.3   Tangent Lines and Derivatives

In this section we see how limits arise when we attempt to find the tangent line to a curve or the instantaneous rate of change of a function.

The Tangent Problem

A *tangent line* is a line that *just* touches a curve. For instance, Figure 1 shows the parabola $y = x^2$ and the tangent line t that touches the parabola at the point $P(1, 1)$. We will be able to find an equation of the tangent line t as soon as we know its slope m. The difficulty is that we know only one point, P, on t, whereas we need two points to compute the slope. But observe that we can compute an approximation to m by

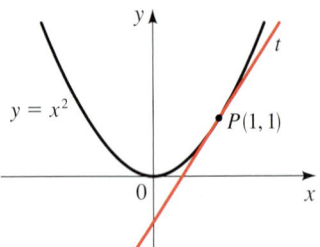

Figure 1

Example 1   Finding a Tangent Line to a Hyperbola

Find an equation of the tangent line to the hyperbola $y = 3/x$ at the point $(3, 1)$.

Solution   Let $f(x) = 3/x$. Then the slope of the tangent line at $(3, 1)$ is

$$m = \lim_{x \to 3} \frac{f(x) - f(3)}{x - 3} \qquad \text{Definition of } m$$

$$= \lim_{x \to 3} \frac{\dfrac{3}{x} - 1}{x - 3} \qquad f(x) = \frac{3}{x}$$

$$= \lim_{x \to 3} \frac{3 - x}{x(x - 3)} \qquad \begin{array}{l}\text{Multiply numerator}\\ \text{and denominator by } x\end{array}$$

$$= \lim_{x \to 3} \left(-\frac{1}{x} \right) \qquad \text{Cancel } x - 3$$

$$= -\frac{1}{3} \qquad \text{Let } x \to 3$$

Therefore, an equation of the tangent at the point $(3, 1)$ is

$$y - 1 = -\tfrac{1}{3}(x - 3)$$

which simplifies to

$$x + 3y - 6 = 0$$

The hyperbola and its tangent are shown in Figure 6.   ∎

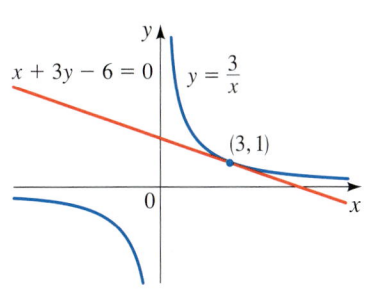

Figure 6

There is another expression for the slope of a tangent line that is sometimes easier to use. Let $h = x - a$. Then $x = a + h$, so the slope of the secant line PQ is

$$m_{PQ} = \frac{f(a + h) - f(a)}{h}$$

See Figure 7 where the case $h > 0$ is illustrated and Q is to the right of P. If it happened that $h < 0$, however, Q would be to the left of P.

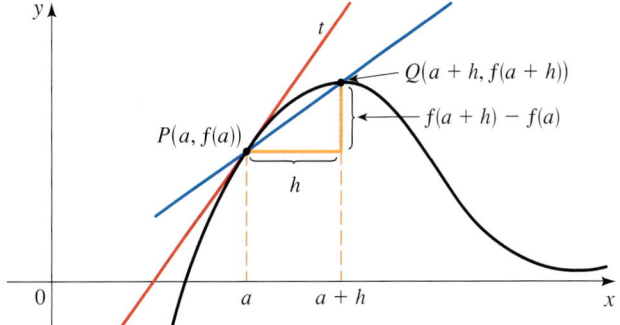

Figure 7

Notice that as x approaches a, h approaches 0 (because $h = x - a$), and so the expression for the slope of the tangent line becomes

$$m = \lim_{h \to 0} \frac{f(a + h) - f(a)}{h}$$

Example 2   Finding a Tangent Line

Find an equation of the tangent line to the curve $y = x^3 - 2x + 3$ at the point $(1, 2)$.

Solution   If $f(x) = x^3 - 2x + 3$, then the slope of the tangent line where $a = 1$ is

$$m = \lim_{h \to 0} \frac{f(1 + h) - f(1)}{h} \qquad \text{Definition of } m$$

$$= \lim_{h \to 0} \frac{[(1 + h)^3 - 2(1 + h) + 3] - [1^3 - 2(1) + 3]}{h} \qquad f(x) = x^3 - 2x + 3$$

$$= \lim_{h \to 0} \frac{1 + 3h + 3h^2 + h^3 - 2 - 2h + 3 - 2}{h} \qquad \text{Expand numerator}$$

$$= \lim_{h \to 0} \frac{h + 3h^2 + h^3}{h} \qquad \text{Simplify}$$

$$= \lim_{h \to 0} (1 + 3h + h^2) \qquad \text{Cancel } h$$

$$= 1 \qquad \text{Let } h \to 0$$

So an equation of the tangent line at $(1, 2)$ is

$$y - 2 = 1(x - 1) \qquad \text{or} \qquad y = x + 1 \qquad ■$$

Derivatives

We have seen that the slope of the tangent line to the curve $y = f(x)$ at the point $(a, f(a))$ can be written as

$$\lim_{h \to 0} \frac{f(a + h) - f(a)}{h}$$

It turns out that this expression arises in many other contexts as well, such as finding velocities and other rates of change. Because this type of limit occurs so widely, it is given a special name and notation.

Definition of a Derivative

The **derivative of a function f at a number a**, denoted by $f'(a)$, is

$$f'(a) = \lim_{h \to 0} \frac{f(a + h) - f(a)}{h}$$

if this limit exists.

Example 3 Finding a Derivative at a Point

Find the derivative of the function $f(x) = 5x^2 + 3x - 1$ at the number 2.

Solution According to the definition of a derivative, with $a = 2$, we have

$$f'(2) = \lim_{h \to 0} \frac{f(2 + h) - f(2)}{h} \qquad \text{Definition of } f'(2)$$

$$= \lim_{h \to 0} \frac{[5(2 + h)^2 + 3(2 + h) - 1] - [5(2)^2 + 3(2) - 1]}{h} \qquad f(x) = 5x^2 + 3x - 1$$

$$= \lim_{h \to 0} \frac{20 + 20h + 5h^2 + 6 + 3h - 1 - 25}{h} \qquad \text{Expand}$$

$$= \lim_{h \to 0} \frac{23h + 5h^2}{h} \qquad \text{Simplify}$$

$$= \lim_{h \to 0} (23 + 5h) \qquad \text{Cancel } h$$

$$= 23 \qquad \text{Let } h \to 0 \qquad\blacksquare$$

We see from the definition of a derivative that the number $f'(a)$ is the same as the slope of the tangent line to the curve $y = f(x)$ at the point $(a, f(a))$. So the result of Example 2 shows that the slope of the tangent line to the parabola $y = 5x^2 + 3x - 1$ at the point $(2, 25)$ is $f'(2) = 23$.

Example 4 Finding a Derivative

Let $f(x) = \sqrt{x}$.
(a) Find $f'(a)$.
(b) Find $f'(1)$, $f'(4)$, and $f'(9)$.

Solution
(a) We use the definition of the derivative at a:

$$f'(a) = \lim_{h \to 0} \frac{f(a + h) - f(a)}{h} \qquad \text{Definition of derivative}$$

$$= \lim_{h \to 0} \frac{\sqrt{a + h} - \sqrt{a}}{h} \qquad f(x) = \sqrt{x}$$

$$= \lim_{h \to 0} \frac{\sqrt{a + h} - \sqrt{a}}{h} \cdot \frac{\sqrt{a + h} + \sqrt{a}}{\sqrt{a + h} + \sqrt{a}} \qquad \text{Rationalize numerator}$$

$$= \lim_{h \to 0} \frac{(a + h) - a}{h(\sqrt{a + h} + \sqrt{a})} \qquad \text{Difference of squares}$$

$$= \lim_{h \to 0} \frac{h}{h(\sqrt{a + h} + \sqrt{a})} \qquad \text{Simplify numerator}$$

$$= \lim_{h \to 0} \frac{1}{\sqrt{a + h} + \sqrt{a}}$$ *Cancel h*

$$= \frac{1}{\sqrt{a} + \sqrt{a}} = \frac{1}{2\sqrt{a}}$$ *Let h → 0*

(b) Substituting $a = 1$, $a = 4$, and $a = 9$ into the result of part (a), we get

$$f'(1) = \frac{1}{2\sqrt{1}} = \frac{1}{2} \qquad f'(4) = \frac{1}{2\sqrt{4}} = \frac{1}{4} \qquad f'(9) = \frac{1}{2\sqrt{9}} = \frac{1}{6}$$

These values of the derivative are the slopes of the tangent lines shown in Figure 8.

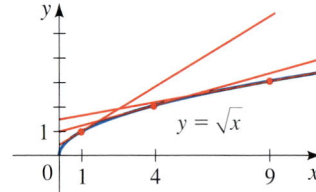

Figure 8

Instantaneous Rates of Change

In Section 2.3 we defined the average rate of change of a function f between the numbers a and x as

$$\text{average rate of change} = \frac{\text{change in } y}{\text{change in } x} = \frac{f(x) - f(a)}{x - a}$$

Suppose we consider the average rate of change over smaller and smaller intervals by letting x approach a. The limit of these average rates of change is called the instantaneous rate of change.

Instantaneous Rate of Change

If $y = f(x)$, the **instantaneous rate of change of y with respect to x at $x = a$** is the limit of the average rates of change as x approaches a:

$$\text{instantaneous rate of change} = \lim_{x \to a} \frac{f(x) - f(a)}{x - a} = f'(a)$$

Notice that we now have two ways of interpreting the derivative:

- $f'(a)$ is the slope of the tangent line to $y = f(x)$ at $x = a$
- $f'(a)$ is the instantaneous rate of change of y with respect to x at $x = a$

In the special case where $x = t = $ time and $s = f(t) = $ displacement (directed distance) at time t of an object traveling in a straight line, the instantaneous rate of change is called the **instantaneous velocity**.

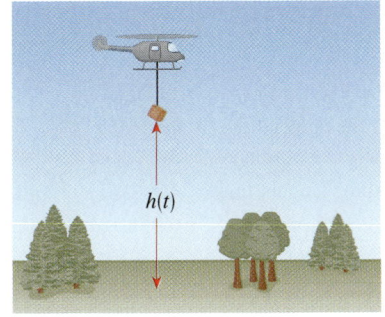

Example 5 Instantaneous Velocity of a Falling Object

If an object is dropped from a height of 3000 ft, its distance above the ground (in feet) after t seconds is given by $h(t) = 3000 - 16t^2$. Find the object's instantaneous velocity after 4 seconds.

Solution After 4 s have elapsed, the height is $h(4) = 2744$ ft. The instantaneous velocity is

$$h'(4) = \lim_{t \to 4} \frac{h(t) - h(4)}{t - 4} \qquad \text{Definition of } h'(4)$$

$$= \lim_{t \to 4} \frac{3000 - 16t^2 - 2744}{t - 4} \qquad h(t) = 3000 - 16t^2$$

$$= \lim_{t \to 4} \frac{256 - 16t^2}{t - 4} \qquad \text{Simplify}$$

$$= \lim_{t \to 4} \frac{16(4 - t)(4 + t)}{t - 4} \qquad \text{Factor numerator}$$

$$= \lim_{t \to 4} -16(4 + t) \qquad \text{Cancel } t - 4$$

$$= -16(4 + 4) = -128 \text{ ft/s} \qquad \text{Let } t \to 4$$

The negative sign indicates that the height is *decreasing* at a rate of 128 ft/s. ∎

Example 6 Estimating an Instantaneous Rate of Change

Let $P(t)$ be the population of the United States at time t. The table in the margin gives approximate values of this function by providing midyear population estimates from 1996 to 2004. Interpret and estimate the value of $P'(2000)$.

t	$P(t)$
1996	269,667,000
1998	276,115,000
2000	282,192,000
2002	287,941,000
2004	293,655,000

t	$\dfrac{P(t) - P(2000)}{t - 2000}$
1996	3,131,250
1998	3,038,500
2002	2,874,500
2004	2,865,750

Here we have estimated the derivative by averaging the slopes of two secant lines. Another method is to plot the population function and estimate the slope of the tangent line when $t = 2000$.

Solution The derivative $P'(2000)$ means the rate of change of P with respect to t when $t = 2000$, that is, the rate of increase of the population in 2000.

According to the definition of a derivative, we have

$$P'(2000) = \lim_{t \to 2000} \frac{P(t) - P(2000)}{t - 2000}$$

So we compute and tabulate values of the difference quotient (the average rates of change) as shown in the table in the margin. We see that $P'(2000)$ lies somewhere between 3,038,500 and 2,874,500. (Here we are making the reasonable assumption that the population didn't fluctuate wildly between 1996 and 2004.) We estimate that the rate of increase of the U.S. population in 2000 was the average of these two numbers, namely

$$P'(2000) \approx 2.96 \text{ million people/year} \qquad ∎$$

12.3　Exercises

1–6 ■ Find the slope of the tangent line to the graph of f at the given point.

1. $f(x) = 3x + 4$ 　at $(1, 7)$

2. $f(x) = 5 - 2x$ 　at $(-3, 11)$

3. $f(x) = 4x^2 - 3x$ 　at $(-1, 7)$

4. $f(x) = 1 + 2x - 3x^2$ 　at $(1, 0)$

5. $f(x) = 2x^3$ 　at $(2, 16)$

6. $f(x) = \dfrac{6}{x + 1}$ 　at $(2, 2)$

7–12 ■ Find an equation of the tangent line to the curve at the given point. Graph the curve and the tangent line.

7. $y = x + x^2$ 　at $(-1, 0)$

8. $y = 2x - x^3$ 　at $(1, 1)$

9. $y = \dfrac{x}{x - 1}$ 　at $(2, 2)$

10. $y = \dfrac{1}{x^2}$ 　at $(-1, 1)$

11. $y = \sqrt{x + 3}$ 　at $(1, 2)$

12. $y = \sqrt{1 + 2x}$ 　at $(4, 3)$

13–18 ■ Find the derivative of the function at the given number.

13. $f(x) = 1 - 3x^2$ 　at 2

14. $f(x) = 2 - 3x + x^2$ 　at -1

15. $g(x) = x^4$ 　at 1

16. $g(x) = 2x^2 + x^3$ 　at 1

17. $F(x) = \dfrac{1}{\sqrt{x}}$ 　at 4

18. $G(x) = 1 + 2\sqrt{x}$ 　at 4

19–22 ■ Find $f'(a)$, where a is in the domain of f.

19. $f(x) = x^2 + 2x$

20. $f(x) = -\dfrac{1}{x^2}$

21. $f(x) = \dfrac{x}{x + 1}$

22. $f(x) = \sqrt{x - 2}$

23. (a) If $f(x) = x^3 - 2x + 4$, find $f'(a)$.

(b) Find equations of the tangent lines to the graph of f at the points whose x-coordinates are 0, 1, and 2.

(c) Graph f and the three tangent lines.

24. (a) If $g(x) = 1/(2x - 1)$, find $g'(a)$.

(b) Find equations of the tangent lines to the graph of g at the points whose x-coordinates are -1, 0, and 1.

(c) Graph g and the three tangent lines.

Applications

25. Velocity of a Ball 　If a ball is thrown into the air with a velocity of 40 ft/s, its height (in feet) after t seconds is given by $y = 40t - 16t^2$. Find the velocity when $t = 2$.

26. Velocity on the Moon 　If an arrow is shot upward on the moon with a velocity of 58 m/s, its height (in meters) after t seconds is given by $H = 58t - 0.83t^2$.

(a) Find the velocity of the arrow after one second.

(b) Find the velocity of the arrow when $t = a$.

(c) At what time t will the arrow hit the moon?

(d) With what velocity will the arrow hit the moon?

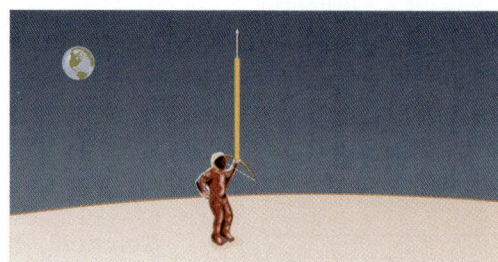

27. Velocity of a Particle 　The displacement s (in meters) of a particle moving in a straight line is given by the equation of motion $s = 4t^3 + 6t + 2$, where t is measured in seconds. Find the velocity of the particle s at times $t = a$, $t = 1$, $t = 2$, $t = 3$.

28. Inflating a Balloon 　A spherical balloon is being inflated. Find the rate of change of the surface area $\left(S = 4\pi r^2\right)$ with respect to the radius r when $r = 2$ ft.

29. Temperature Change 　A roast turkey is taken from an oven when its temperature has reached $185°F$ and is placed on a table in a room where the temperature is $75°F$. The graph shows how the temperature of the turkey decreases and eventually approaches room temperature. By measuring the slope of the tangent, estimate the rate of change of the temperature after an hour.

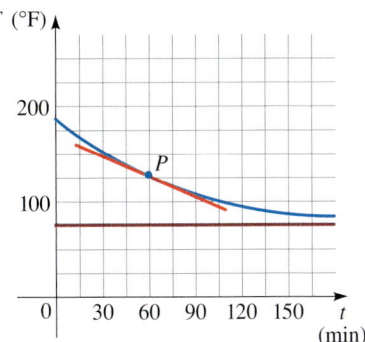

30. Heart Rate   A cardiac monitor is used to measure the heart rate of a patient after surgery. It compiles the number of heartbeats after t minutes. When the data in the table are graphed, the slope of the tangent line represents the heart rate in beats per minute.

t (min)	36	38	40	42	44
Heartbeats	2530	2661	2806	2948	3080

(a) Find the average heart rates (slopes of the secant lines) over the time intervals $[40, 42]$ and $[42, 44]$.

(b) Estimate the patient's heart rate after 42 minutes by averaging the slopes of these two secant lines.

31. Water Flow   A tank holds 1000 gallons of water, which drains from the bottom of the tank in half an hour. The values in the table show the volume V of water remaining in the tank (in gallons) after t minutes.

t (min)	5	10	15	20	25	30
V (gal)	694	444	250	111	28	0

(a) Find the average rates at which water flows from the tank (slopes of secant lines) for the time intervals $[10, 15]$ and $[15, 20]$.

(b) The slope of the tangent line at the point $(15, 250)$ represents the rate at which water is flowing from the tank after 15 minutes. Estimate this rate by averaging the slopes of the secant lines in part (a).

32. World Population Growth   The table gives the world's population in the 20th century.

Year	Population (in millions)	Year	Population (in millions)
1900	1650	1960	3040
1910	1750	1970	3710
1920	1860	1980	4450
1930	2070	1990	5280
1940	2300	2000	6080
1950	2560		

Estimate the rate of population growth in 1920 and in 1980 by averaging the slopes of two secant lines.

Discovery • Discussion

33. Estimating Derivatives from a Graph   For the function g whose graph is given, arrange the following numbers in increasing order and explain your reasoning.

$$0 \qquad g'(-2) \qquad g'(0) \qquad g'(2) \qquad g'(4)$$

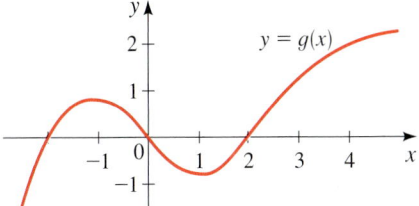

34. Estimating Velocities from a Graph   The graph shows the position function of a car. Use the shape of the graph to explain your answers to the following questions.

(a) What was the initial velocity of the car?

(b) Was the car going faster at B or at C?

(c) Was the car slowing down or speeding up at A, B, and C?

(d) What happened between D and E?

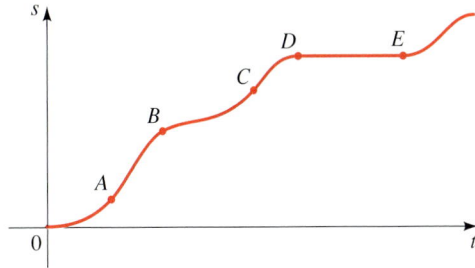

DISCOVERY PROJECT

Designing a Roller Coaster

Suppose you are asked to design the first ascent and drop for a new roller coaster. By studying photographs of your favorite coasters, you decide to make the slope of the ascent 0.8 and the slope of the drop -1.6. You then connect these two straight stretches $y = L_1(x)$ and $y = L_2(x)$ with part of a parabola

$$y = f(x) = ax^2 + bx + c$$

where x and $f(x)$ are measured in feet. For the track to be smooth there can't be abrupt changes in direction, so you want the linear segments L_1 and L_2 to be tangent to the parabola at the transition points P and Q, as shown in the figure.

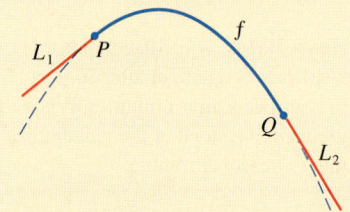

1. To simplify the equations, you decide to place the origin at P. As a consequence, what is the value of c?

2. Suppose the horizontal distance between P and Q is 100 ft. To ensure that the track is smooth at the transition points, what should the values of $f'(0)$ and $f'(100)$ be?

3. If $f(x) = ax^2 + bx + c$, show that $f'(x) = 2ax + b$.

4. Use the results of problems 2 and 3 to determine the values of a and b. That is, find a formula for $f(x)$.

 5. Plot L_1, f, and L_2 to verify graphically that the transitions are smooth.

6. Find the difference in elevation between P and Q.

12.4   Limits at Infinity; Limits of Sequences

In this section we study a special kind of limit called a *limit at infinity*. We examine the limit of a function $f(x)$ as x becomes large. We also examine the limit of a sequence a_n as n becomes large. Limits of sequences will be used in Section 12.5 to help us find the area under the graph of a function.

Limits at Infinity

Let's investigate the behavior of the function f defined by

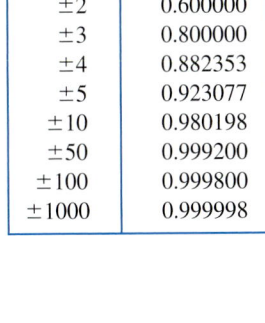

x	$f(x)$
0	-1.000000
± 1	0.000000
± 2	0.600000
± 3	0.800000
± 4	0.882353
± 5	0.923077
± 10	0.980198
± 50	0.999200
± 100	0.999800
± 1000	0.999998

$$f(x) = \frac{x^2 - 1}{x^2 + 1}$$

as x becomes large. The table in the margin gives values of this function correct to six decimal places, and the graph of f has been drawn by a computer in Figure 1.

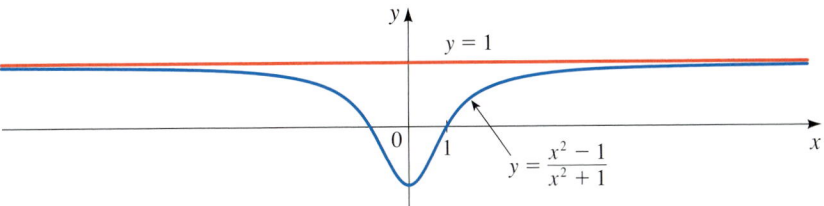

Figure 1

As x grows larger and larger, you can see that the values of $f(x)$ get closer and closer to 1. In fact, it seems that we can make the values of $f(x)$ as close as we like to 1 by taking x sufficiently large. This situation is expressed symbolically by writing

$$\lim_{x \to \infty} \frac{x^2 - 1}{x^2 + 1} = 1$$

In general, we use the notation

$$\lim_{x \to \infty} f(x) = L$$

to indicate that the values of $f(x)$ become closer and closer to L as x becomes larger and larger.

Limit at Infinity

Let f be a function defined on some interval (a, ∞). Then

$$\lim_{x \to \infty} f(x) = L$$

means that the values of $f(x)$ can be made arbitrarily close to L by taking x sufficiently large.

Another notation for $\lim_{x \to \infty} f(x) = L$ is

$$f(x) \to L \qquad \text{as} \qquad x \to \infty$$

The symbol ∞ does not represent a number. Nevertheless, we often read the expression $\lim_{x \to \infty} f(x) = L$ as

Limits at infinity are also discussed in Section 3.6.

"the limit of $f(x)$, as x approaches infinity, is L"

or   "the limit of $f(x)$, as x becomes infinite, is L"

or   "the limit of $f(x)$, as x increases without bound, is L"

Geometric illustrations are shown in Figure 2. Notice that there are many ways for the graph of f to approach the line $y = L$ (which is called a *horizontal asymptote*) as we look to the far right.

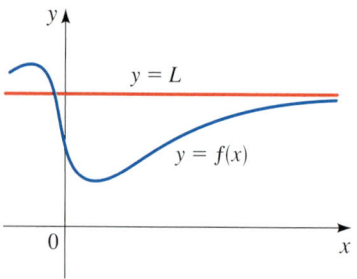
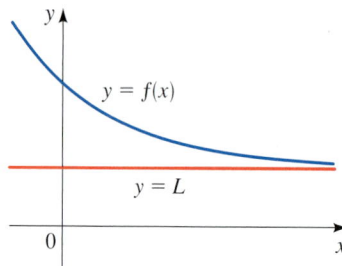
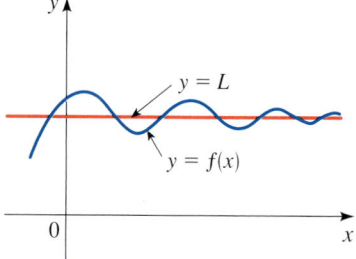

Figure 2
Examples illustrating $\lim\limits_{x \to \infty} f(x) = L$

Referring back to Figure 1, we see that for numerically large negative values of x, the values of $f(x)$ are close to 1. By letting x decrease through negative values without bound, we can make $f(x)$ as close as we like to 1. This is expressed by writing

$$\lim_{x \to -\infty} \frac{x^2 - 1}{x^2 + 1} = 1$$

The general definition is as follows.

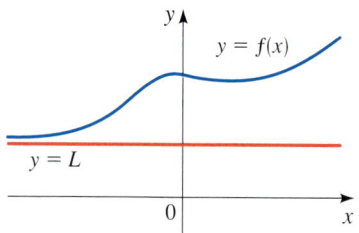

Limit at Negative Infinity

Let f be a function defined on some interval $(-\infty, a)$. Then

$$\lim_{x \to -\infty} f(x) = L$$

means that the values of $f(x)$ can be made arbitrarily close to L by taking x sufficiently large negative.

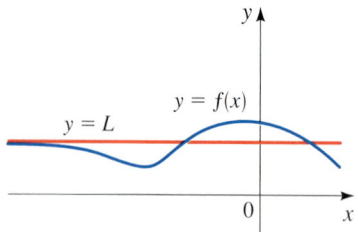

Again, the symbol $-\infty$ does not represent a number, but the expression $\lim\limits_{x \to -\infty} f(x) = L$ is often read as

"the limit of $f(x)$, as x approaches negative infinity, is L"

The definition is illustrated in Figure 3. Notice that the graph approaches the line $y = L$ as we look to the far left.

Figure 3
Examples illustrating $\lim\limits_{x \to -\infty} f(x) = L$

Horizontal Asymptote

The line $y = L$ is called a **horizontal asymptote** of the curve $y = f(x)$ if either

$$\lim_{x \to \infty} f(x) = L \qquad \text{or} \qquad \lim_{x \to -\infty} f(x) = L$$

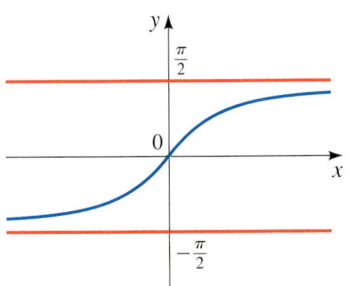

Figure 4

$y = \tan^{-1} x$

For instance, the curve illustrated in Figure 1 has the line $y = 1$ as a horizontal asymptote because

$$\lim_{x \to \infty} \frac{x^2 - 1}{x^2 + 1} = 1$$

As we discovered in Section 7.4, an example of a curve with two horizontal asymptotes is $y = \tan^{-1} x$ (see Figure 4). In fact,

$$\lim_{x \to -\infty} \tan^{-1} x = -\frac{\pi}{2} \qquad \text{and} \qquad \lim_{x \to \infty} \tan^{-1} x = \frac{\pi}{2}$$

so both of the lines $y = -\pi/2$ and $y = \pi/2$ are horizontal asymptotes. (This follows from the fact that the lines $x = \pm\pi/2$ are vertical asymptotes of the graph of tan.)

Example 1   Limits at Infinity

Find $\lim\limits_{x \to \infty} \dfrac{1}{x}$ and $\lim\limits_{x \to -\infty} \dfrac{1}{x}$.

We first investigated horizontal asymptotes and limits at infinity for rational functions in Section 3.6.

Solution   Observe that when x is large, $1/x$ is small. For instance,

$$\frac{1}{100} = 0.01 \qquad \frac{1}{10,000} = 0.0001 \qquad \frac{1}{1,000,000} = 0.000001$$

In fact, by taking x large enough, we can make $1/x$ as close to 0 as we please. Therefore

$$\lim_{x \to \infty} \frac{1}{x} = 0$$

Similar reasoning shows that when x is large negative, $1/x$ is small negative, so we also have

$$\lim_{x \to -\infty} \frac{1}{x} = 0$$

It follows that the line $y = 0$ (the x-axis) is a horizontal asymptote of the curve $y = 1/x$. (This is a hyperbola; see Figure 5.)   ∎

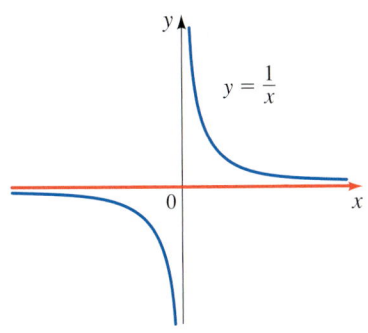

Figure 5

$\lim\limits_{x \to \infty} \dfrac{1}{x} = 0, \ \lim\limits_{x \to -\infty} \dfrac{1}{x} = 0$

The Limit Laws that we studied in Section 12.2 also hold for limits at infinity. In particular, if we combine Law 6 (Limit of a Power) with the results of Example 1, we obtain the following important rule for calculating limits.

> If k is any positive integer, then
>
> $$\lim_{x \to \infty} \frac{1}{x^k} = 0 \qquad \text{and} \qquad \lim_{x \to -\infty} \frac{1}{x^k} = 0$$

Example 2   Finding a Limit at Infinity

Evaluate $\lim\limits_{x \to \infty} \dfrac{3x^2 - x - 2}{5x^2 + 4x + 1}$.

Solution   To evaluate the limit at infinity of a rational function, we first divide both the numerator and denominator by the highest power of x that occurs in the denominator. (We may assume that $x \neq 0$ since we are interested only in large values of x.) In this case, the highest power of x in the denominator is x^2, so we have

$$\lim_{x \to \infty} \frac{3x^2 - x - 2}{5x^2 + 4x + 1} = \lim_{x \to \infty} \frac{3 - \dfrac{1}{x} - \dfrac{2}{x^2}}{5 + \dfrac{4}{x} + \dfrac{1}{x^2}}$$ *Divide numerator and denominator by x^2*

$$= \frac{\lim\limits_{x \to \infty} \left(3 - \dfrac{1}{x} - \dfrac{2}{x^2} \right)}{\lim\limits_{x \to \infty} \left(5 + \dfrac{4}{x} + \dfrac{1}{x^2} \right)}$$ *Limit of a Quotient*

$$= \frac{\lim\limits_{x \to \infty} 3 - \lim\limits_{x \to \infty} \dfrac{1}{x} - 2 \lim\limits_{x \to \infty} \dfrac{1}{x^2}}{\lim\limits_{x \to \infty} 5 + 4 \lim\limits_{x \to \infty} \dfrac{1}{x} + \lim\limits_{x \to \infty} \dfrac{1}{x^2}}$$ *Limits of Sums and Differences*

$$= \frac{3 - 0 - 0}{5 + 0 + 0} = \frac{3}{5}$$ *Let $x \to \infty$*

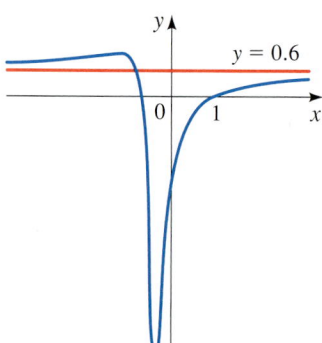

Figure 6

A similar calculation shows that the limit as $x \to -\infty$ is also $\frac{3}{5}$. Figure 6 illustrates the results of these calculations by showing how the graph of the given rational function approaches the horizontal asymptote $y = \frac{3}{5}$.   ■

Example 3   A Limit at Negative Infinity

Use numerical and graphical methods to find $\lim\limits_{x \to -\infty} e^x$.

Solution   From the graph of the natural exponential function $y = e^x$ in Figure 7 and the corresponding table of values, we see that

$$\lim_{x \to -\infty} e^x = 0$$

It follows that the line $y = 0$ (the x-axis) is a horizontal asymptote.

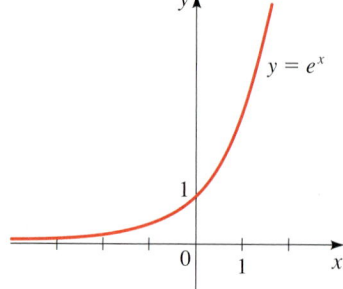

x	e^x
0	1.00000
-1	0.36788
-2	0.13534
-3	0.04979
-5	0.00674
-8	0.00034
-10	0.00005

Figure 7

■

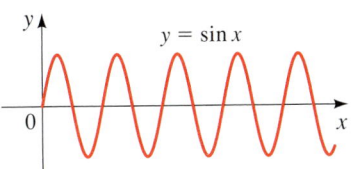

Figure 8

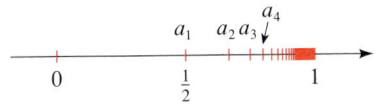

Figure 9

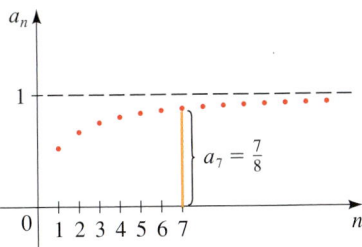

Figure 10

Example 4   **A Function with No Limit at Infinity**

Evaluate $\lim\limits_{x \to \infty} \sin x$.

Solution   From the graph in Figure 8 and the periodic nature of the sine function, we see that, as x increases, the values of $\sin x$ oscillate between 1 and -1 infinitely often and so they don't approach any definite number. Therefore, $\lim_{x \to \infty} \sin x$ does not exist.   ∎

Limits of Sequences

In Section 11.1 we introduced the idea of a sequence of numbers $a_1, a_2, a_3, \dots$. Here we are interested in their behavior as n becomes large. For instance, the sequence defined by

$$a_n = \frac{n}{n+1}$$

is pictured in Figure 9 by plotting its terms on a number line and in Figure 10 by plotting its graph. From Figure 9 or 10 it appears that the terms of the sequence $a_n = n/(n+1)$ are approaching 1 as n becomes large. We indicate this by writing

$$\lim_{n \to \infty} \frac{n}{n+1} = 1$$

Definition of the Limit of a Sequence

A sequence $a_1, a_2, a_3, \dots$ has the **limit** L and we write

$$\lim_{n \to \infty} a_n = L \qquad \text{or} \qquad a_n \to L \text{ as } n \to \infty$$

if the nth term a_n of the sequence can be made arbitrarily close to L by taking n sufficiently large. If $\lim_{n \to \infty} a_n$ exists, we say the sequence **converges** (or is **convergent**). Otherwise, we say the sequence **diverges** (or is **divergent**).

This definition is illustrated by Figure 11.

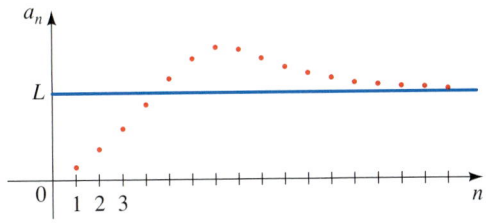

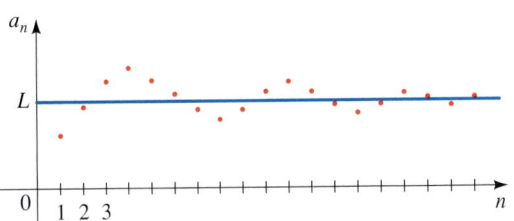

Figure 11

Graphs of two sequences with $\lim\limits_{n \to \infty} a_n = L$

If we compare the definitions of $\lim_{n \to \infty} a_n = L$ and $\lim_{x \to \infty} f(x) = L$, we see that the only difference is that n is required to be an integer. Thus, the following is true.

If $\lim\limits_{x \to \infty} f(x) = L$ and $f(n) = a_n$ when n is an integer, then $\lim\limits_{n \to \infty} a_n = L$.

In particular, since we know that $\lim_{x\to\infty}(1/x^k) = 0$ when k is a positive integer, we have

$$\lim_{n\to\infty}\frac{1}{n^k} = 0 \qquad \text{if } k \text{ is a positive integer}$$

Note that the Limit Laws given in Section 12.2 also hold for limits of sequences.

Example 5 Finding the Limit of a Sequence

Find $\lim_{n\to\infty}\dfrac{n}{n+1}$.

Solution The method is similar to the one we used in Example 2: Divide the numerator and denominator by the highest power of n and then use the Limit Laws.

$$\lim_{n\to\infty}\frac{n}{n+1} = \lim_{n\to\infty}\frac{1}{1+\dfrac{1}{n}} \qquad \text{Divide numerator and denominator by } n$$

$$= \frac{\lim_{n\to\infty} 1}{\lim_{n\to\infty} 1 + \lim_{n\to\infty}\dfrac{1}{n}} \qquad \text{Limits of a Quotient and a Sum}$$

$$= \frac{1}{1+0} = 1 \qquad \text{Let } n\to\infty$$

This result shows that the guess we made earlier from Figures 9 and 10 was correct.

Therefore, the sequence $a_n = n/(n+1)$ is convergent. ∎

Example 6 A Sequence That Diverges

Determine whether the sequence $a_n = (-1)^n$ is convergent or divergent.

Solution If we write out the terms of the sequence, we obtain

$$-1, 1, -1, 1, -1, 1, -1, \ldots$$

The graph of this sequence is shown in Figure 12. Since the terms oscillate between 1 and -1 infinitely often, a_n does not approach any number. Thus, $\lim_{n\to\infty}(-1)^n$ does not exist; that is, the sequence $a_n = (-1)^n$ is divergent. ∎

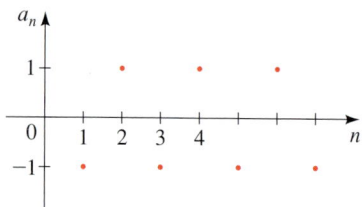

Figure 12

Example 7 Finding the Limit of a Sequence

Find the limit of the sequence given by

$$a_n = \frac{15}{n^3}\left[\frac{n(n+1)(2n+1)}{6}\right]$$

Solution Before calculating the limit, let's first simplify the expression for a_n. Because $n^3 = n\cdot n\cdot n$, we place a factor of n beneath each factor in the numerator that contains an n:

$$a_n = \frac{15}{6}\cdot\frac{n}{n}\cdot\frac{n+1}{n}\cdot\frac{2n+1}{n} = \frac{5}{2}\cdot 1\cdot\left(1+\frac{1}{n}\right)\left(2+\frac{1}{n}\right)$$

Now we can compute the limit:

$$\lim_{n\to\infty} a_n = \lim_{n\to\infty} \frac{5}{2}\left(1 + \frac{1}{n}\right)\left(2 + \frac{1}{n}\right)$$ Definition of a_n

$$= \frac{5}{2}\lim_{n\to\infty}\left(1 + \frac{1}{n}\right)\lim_{n\to\infty}\left(2 + \frac{1}{n}\right)$$ Limit of a Product

$$= \frac{5}{2}(1)(2) = 5$$ Let $n \to \infty$ ∎

12.4 Exercises

1–2 ■ **(a)** Use the graph of f to find the following limits.

  (i) $\lim_{x\to\infty} f(x)$

  (ii) $\lim_{x\to\infty} f(x)$

(b) State the equations of the horizontal asymptotes.

1.
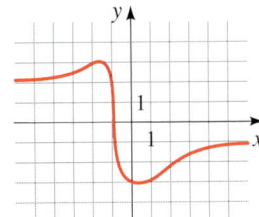

2.
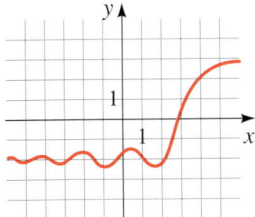

3–14 ■ Find the limit.

3. $\lim_{x\to\infty} \dfrac{6}{x}$

4. $\lim_{x\to\infty} \dfrac{3}{x^4}$

5. $\lim_{x\to\infty} \dfrac{2x + 1}{5x - 1}$

6. $\lim_{x\to\infty} \dfrac{2 - 3x}{4x + 5}$

7. $\lim_{x\to-\infty} \dfrac{4x^2 + 1}{2 + 3x^2}$

8. $\lim_{x\to-\infty} \dfrac{x^2 + 2}{x^3 + x + 1}$

9. $\lim_{t\to\infty} \dfrac{8t^3 + t}{(2t - 1)(2t^2 + 1)}$

10. $\lim_{r\to\infty} \dfrac{4r^3 - r^2}{(r + 1)^3}$

11. $\lim_{x\to\infty} \dfrac{x^4}{1 - x^2 + x^3}$

12. $\lim_{t\to\infty} \left(\dfrac{1}{t} - \dfrac{2t}{t - 1}\right)$

13. $\lim_{x\to-\infty} \left(\dfrac{x - 1}{x + 1} + 6\right)$

14. $\lim_{x\to\infty} \cos x$

15–18 ■ Use a table of values to estimate the limit. Then use a graphing device to confirm your result graphically.

15. $\lim_{x\to-\infty} \dfrac{\sqrt{x^2 + 4x}}{4x + 1}$

16. $\lim_{x\to\infty} \left(\sqrt{9x^2 + x} - 3x\right)$

17. $\lim_{x\to\infty} \dfrac{x^5}{e^x}$

18. $\lim_{x\to\infty} \left(1 + \dfrac{2}{x}\right)^{3x}$

19–30 ■ If the sequence is convergent, find its limit. If it is divergent, explain why.

19. $a_n = \dfrac{1 + n}{n + n^2}$

20. $a_n = \dfrac{5n}{n + 5}$

21. $a_n = \dfrac{n^2}{n + 1}$

22. $a_n = \dfrac{n - 1}{n^3 + 1}$

23. $a_n = \dfrac{1}{3^n}$

24. $a_n = \dfrac{(-1)^n}{n}$

25. $a_n = \sin(n\pi/2)$

26. $a_n = \cos n\pi$

27. $a_n = \dfrac{3}{n^2}\left[\dfrac{n(n + 1)}{2}\right]$

28. $a_n = \dfrac{5}{n}\left(n + \dfrac{4}{n}\left[\dfrac{n(n + 1)}{2}\right]\right)$

29. $a_n = \dfrac{24}{n^3}\left[\dfrac{n(n + 1)(2n + 1)}{6}\right]$

30. $a_n = \dfrac{12}{n^4}\left[\dfrac{n(n + 1)}{2}\right]^2$

Applications

31. Salt Concentration

(a) A tank contains 5000 L of pure water. Brine that contains 30 g of salt per liter of water is pumped into the tank at a rate of 25 L/min. Show that the concentration of salt after t minutes (in grams per liter) is

$$C(t) = \dfrac{30t}{200 + t}$$

(b) What happens to the concentration as $t \to \infty$?

32. Velocity of a Raindrop   The downward velocity of a falling raindrop at time t is modeled by the function

$$v(t) = 1.2(1 - e^{-8.2t})$$

(a) Find the terminal velocity of the raindrop by evaluating $\lim_{t \to \infty} v(t)$. (Use the result of Example 3.)

(b) Graph $v(t)$, and use the graph to estimate how long it takes for the velocity of the raindrop to reach 99% of its terminal velocity.

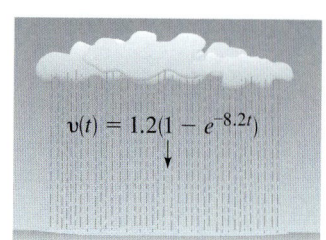

$$v(t) = 1.2(1 - e^{-8.2t})$$

Discovery • Discussion

33. The Limit of a Recursive Sequence

(a) A sequence is defined recursively by $a_1 = 0$ and

$$a_{n+1} = \sqrt{2 + a_n}$$

Find the first ten terms of this sequence correct to eight decimal places. Does this sequence appear to be convergent? If so, guess the value of the limit.

(b) Assuming the sequence in part (a) is convergent, let $\lim_{n \to \infty} a_n = L$. Explain why $\lim_{n \to \infty} a_{n+1} = L$ also, and therefore

$$L = \sqrt{2 + L}$$

Solve this equation to find the exact value of L.

12.5   Areas

We have seen that limits are needed to compute the slope of a tangent line or an instantaneous rate of change. Here we will see that they are also needed to find the area of a region with a curved boundary. The problem of finding such areas has consequences far beyond simply finding area. (See *Focus on Modeling*, page 929.)

The Area Problem

One of the central problems in calculus is the *area problem*: Find the area of the region S that lies under the curve $y = f(x)$ from a to b. This means that S, illustrated in Figure 1, is bounded by the graph of a function f (where $f(x) \geq 0$), the vertical lines $x = a$ and $x = b$, and the x-axis.

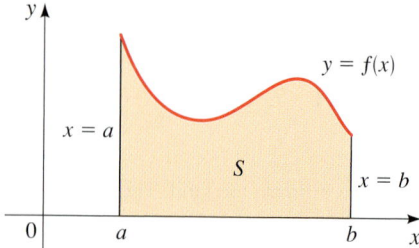

Figure 1

In trying to solve the area problem, we have to ask ourselves: What is the meaning of the word *area*? This question is easy to answer for regions with straight sides.

For a rectangle, the area is defined as the product of the length and the width. The area of a triangle is half the base times the height. The area of a polygon is found by dividing it into triangles (as in Figure 2) and adding the areas of the triangles.

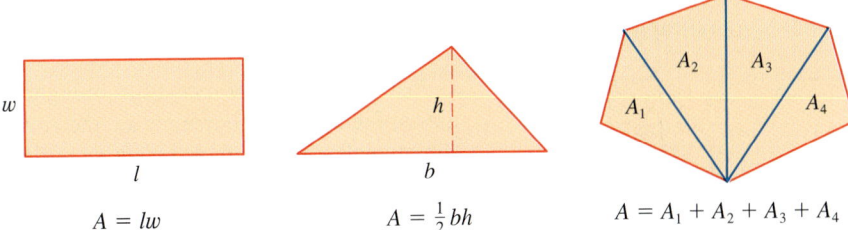

Figure 2

$$A = lw \qquad\qquad A = \tfrac{1}{2}bh \qquad\qquad A = A_1 + A_2 + A_3 + A_4$$

However, it is not so easy to find the area of a region with curved sides. We all have an intuitive idea of what the area of a region is. But part of the area problem is to make this intuitive idea precise by giving an exact definition of area.

Recall that in defining a tangent we first approximated the slope of the tangent line by slopes of secant lines and then we took the limit of these approximations. We pursue a similar idea for areas. We first approximate the region S by rectangles, and then we take the limit of the areas of these rectangles as we increase the number of rectangles. The following example illustrates the procedure.

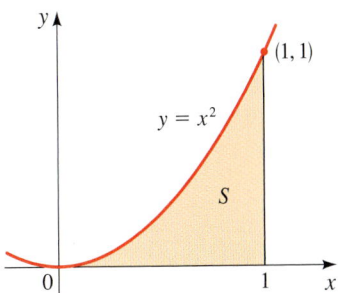

Figure 3

Example 1   Estimating an Area Using Rectangles

Use rectangles to estimate the area under the parabola $y = x^2$ from 0 to 1 (the parabolic region S illustrated in Figure 3).

Solution We first notice that the area of S must be somewhere between 0 and 1 because S is contained in a square with side length 1, but we can certainly do better than that. Suppose we divide S into four strips S_1, S_2, S_3, and S_4 by drawing the vertical lines $x = \frac{1}{4}$, $x = \frac{1}{2}$, and $x = \frac{3}{4}$ as in Figure 4(a). We can approximate each strip by a rectangle whose base is the same as the strip and whose height is the same as the right edge of the strip (see Figure 4(b)). In other words, the heights of these rectangles are the values of the function $f(x) = x^2$ at the right endpoints of the subintervals $\left[0, \frac{1}{4}\right]$, $\left[\frac{1}{4}, \frac{1}{2}\right]$, $\left[\frac{1}{2}, \frac{3}{4}\right]$, and $\left[\frac{3}{4}, 1\right]$.

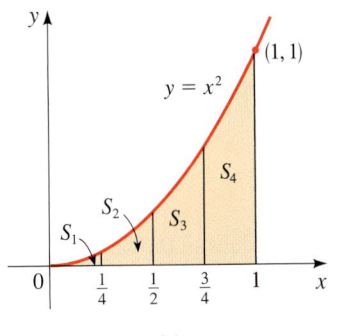
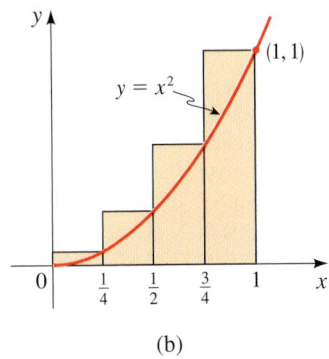

Figure 4   (a)   (b)

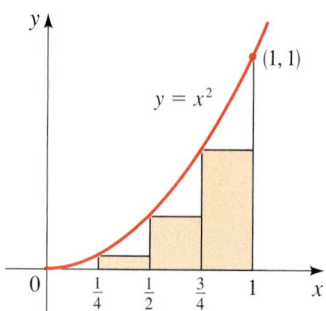

Figure 5

Each rectangle has width $\frac{1}{4}$ and the heights are $\left(\frac{1}{4}\right)^2$, $\left(\frac{1}{2}\right)^2$, $\left(\frac{3}{4}\right)^2$, and 1^2. If we let R_4 be the sum of the areas of these approximating rectangles, we get

$$R_4 = \tfrac{1}{4}\cdot\left(\tfrac{1}{4}\right)^2 + \tfrac{1}{4}\cdot\left(\tfrac{1}{2}\right)^2 + \tfrac{1}{4}\cdot\left(\tfrac{3}{4}\right)^2 + \tfrac{1}{4}\cdot1^2 = \tfrac{15}{32} = 0.46875$$

From Figure 4(b) we see that the area A of S is less than R_4, so

$$A < 0.46875$$

Instead of using the rectangles in Figure 4(b), we could use the smaller rectangles in Figure 5 whose heights are the values of f at the left endpoints of the subintervals. (The leftmost rectangle has collapsed because its height is 0.) The sum of the areas of these approximating rectangles is

$$L_4 = \tfrac{1}{4}\cdot0^2 + \tfrac{1}{4}\cdot\left(\tfrac{1}{4}\right)^2 + \tfrac{1}{4}\cdot\left(\tfrac{1}{2}\right)^2 + \tfrac{1}{4}\cdot\left(\tfrac{3}{4}\right)^2 = \tfrac{7}{32} = 0.21875$$

We see that the area of S is larger than L_4, so we have lower and upper estimates for A:

$$0.21875 < A < 0.46875$$

We can repeat this procedure with a larger number of strips. Figure 6 shows what happens when we divide the region S into eight strips of equal width. By computing the sum of the areas of the smaller rectangles (L_8) and the sum of the areas of the larger rectangles (R_8), we obtain better lower and upper estimates for A:

$$0.2734375 < A < 0.3984375$$

So one possible answer to the question is to say that the true area of S lies somewhere between 0.2734375 and 0.3984375.

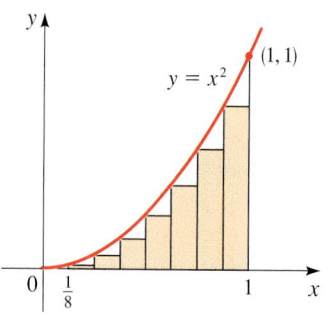 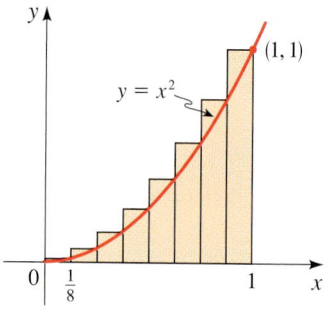

Figure 6

Approximating S with eight rectangles

(a) Using left endpoints

(b) Using right endpoints

We could obtain better estimates by increasing the number of strips. The table in the margin shows the results of similar calculations (with a computer) using n rectangles whose heights are found with left endpoints (L_n) or right endpoints (R_n). In particular, we see by using 50 strips that the area lies between 0.3234 and 0.3434. With 1000 strips we narrow it down even more: A lies between 0.3328335 and 0.3338335. A good estimate is obtained by averaging these numbers: $A \approx 0.3333335$. ∎

n	L_n	R_n
10	0.2850000	0.3850000
20	0.3087500	0.3587500
30	0.3168519	0.3501852
50	0.3234000	0.3434000
100	0.3283500	0.3383500
1000	0.3328335	0.3338335

From the values in the table it looks as if R_n is approaching $\frac{1}{3}$ as n increases. We confirm this in the next example.

Example 2 The Limit of Approximating Sums

For the region S in Example 1, show that the sum of the areas of the upper approximating rectangles approaches $\frac{1}{3}$, that is,

$$\lim_{n \to \infty} R_n = \frac{1}{3}$$

Solution R_n is the sum of the areas of the n rectangles shown in Figure 7. Each rectangle has width $1/n$, and the heights are the values of the function $f(x) = x^2$ at the points $1/n, 2/n, 3/n, \ldots, n/n$. That is, the heights are $(1/n)^2, (2/n)^2, (3/n)^2, \ldots, (n/n)^2$. Thus

$$R_n = \frac{1}{n}\left(\frac{1}{n}\right)^2 + \frac{1}{n}\left(\frac{2}{n}\right)^2 + \frac{1}{n}\left(\frac{3}{n}\right)^2 + \cdots + \frac{1}{n}\left(\frac{n}{n}\right)^2$$

$$= \frac{1}{n} \cdot \frac{1}{n^2}(1^2 + 2^2 + 3^2 + \cdots + n^2)$$

$$= \frac{1}{n^3}(1^2 + 2^2 + 3^2 + \cdots + n^2)$$

Here we need the formula for the sum of the squares of the first n positive integers:

$$1^2 + 2^2 + 3^2 + \cdots + n^2 = \frac{n(n+1)(2n+1)}{6}$$

Putting the preceding formula into our expression for R_n, we get

$$R_n = \frac{1}{n^3} \cdot \frac{n(n+1)(2n+1)}{6} = \frac{(n+1)(2n+1)}{6n^2}$$

Thus, we have

$$\lim_{n \to \infty} R_n = \lim_{n \to \infty} \frac{(n+1)(2n+1)}{6n^2}$$

$$= \lim_{n \to \infty} \frac{1}{6}\left(\frac{n+1}{n}\right)\left(\frac{2n+1}{n}\right)$$

$$= \lim_{n \to \infty} \frac{1}{6}\left(1 + \frac{1}{n}\right)\left(2 + \frac{1}{n}\right)$$

$$= \frac{1}{6} \cdot 1 \cdot 2 = \frac{1}{3} \qquad \blacksquare$$

It can be shown that the lower approximating sums also approach $\frac{1}{3}$, that is,

$$\lim_{n \to \infty} L_n = \frac{1}{3}$$

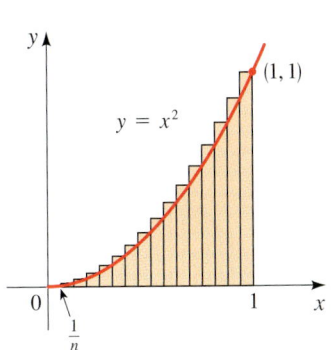

$y = x^2$

$(1, 1)$

Figure 7

This formula was discussed in Section 11.5.

From Figures 8 and 9 it appears that, as n increases, both R_n and L_n become better and better approximations to the area of S. Therefore, we *define* the area A to be the limit of the sums of the areas of the approximating rectangles, that is,

$$A = \lim_{n \to \infty} R_n = \lim_{n \to \infty} L_n = \tfrac{1}{3}$$

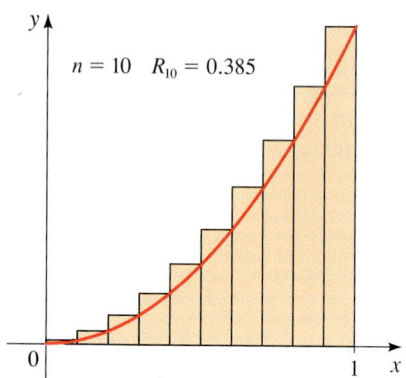
$n = 10$ $R_{10} = 0.385$

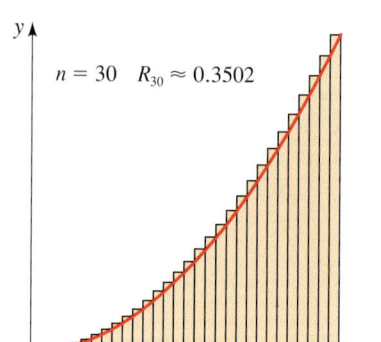
$n = 30$ $R_{30} \approx 0.3502$

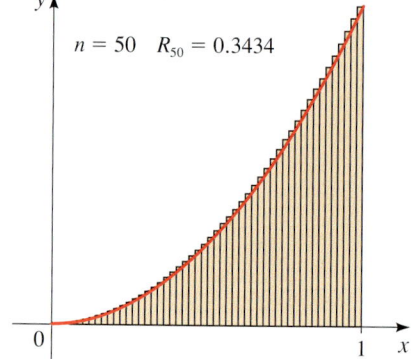
$n = 50$ $R_{50} = 0.3434$

Figure 8

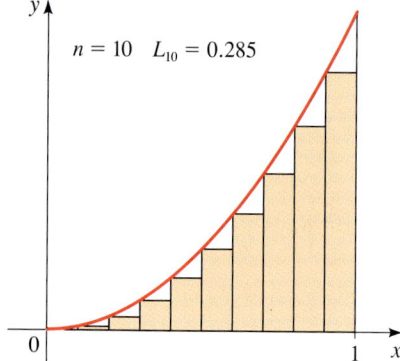
$n = 10$ $L_{10} = 0.285$

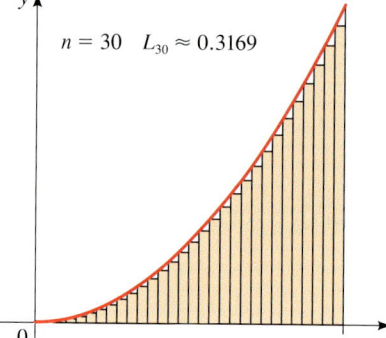
$n = 30$ $L_{30} \approx 0.3169$

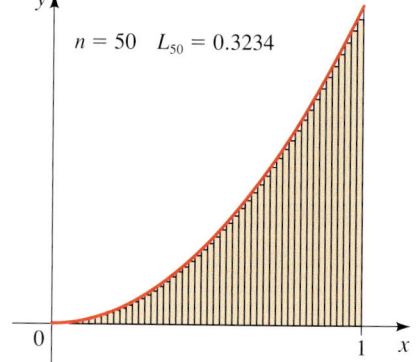
$n = 50$ $L_{50} = 0.3234$

Figure 9

Definition of Area

Let's apply the idea of Examples 1 and 2 to the more general region S of Figure 1. We start by subdividing S into n strips $S_1, S_2, \ldots, S_n$ of equal width as in Figure 10.

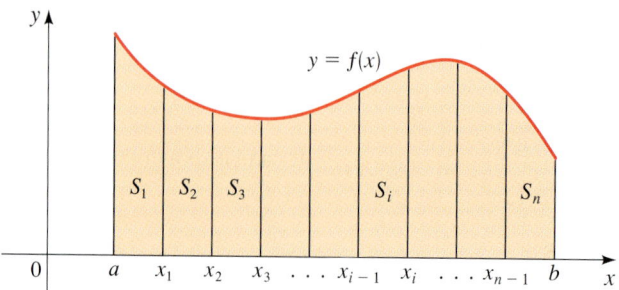

Figure 10

The width of the interval $[a, b]$ is $b - a$, so the width of each of the n strips is

$$\Delta x = \frac{b - a}{n}$$

These strips divide the interval $[a, b]$ into n subintervals

$$[x_0, x_1], \quad [x_1, x_2], \quad [x_2, x_3], \quad \ldots, \quad [x_{n-1}, x_n]$$

where $x_0 = a$ and $x_n = b$. The right endpoints of the subintervals are

$$x_1 = a + \Delta x, \quad x_2 = a + 2\,\Delta x, \quad x_3 = a + 3\,\Delta x, \quad \ldots, \quad x_k = a + k\,\Delta x, \quad \ldots$$

Let's approximate the kth strip S_k by a rectangle with width Δx and height $f(x_k)$, which is the value of f at the right endpoint (see Figure 11). Then the area of the kth rectangle is $f(x_k)\Delta x$. What we think of intuitively as the area of S is approximated by the sum of the areas of these rectangles, which is

$$R_n = f(x_1)\Delta x + f(x_2)\Delta x + \cdots + f(x_n)\Delta x$$

Figure 12 shows this approximation for $n = 2, 4, 8,$ and 12.

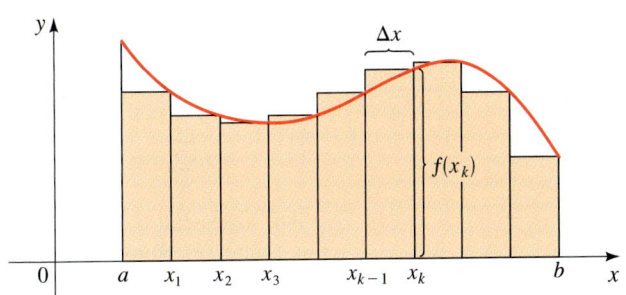

Figure 11

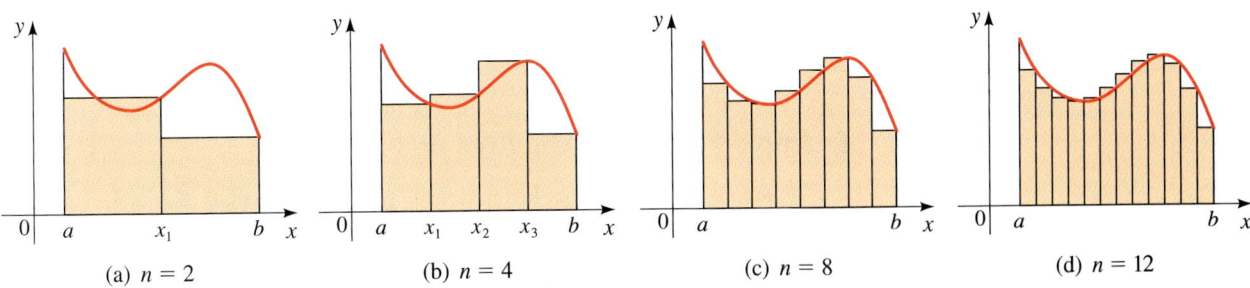

(a) $n = 2$ (b) $n = 4$ (c) $n = 8$ (d) $n = 12$

Figure 12

Notice that this approximation appears to become better and better as the number of strips increases, that is, as $n \to \infty$. Therefore, we define the area A of the region S in the following way.

Definition of Area

The **area** A of the region S that lies under the graph of the continuous function f is the limit of the sum of the areas of approximating rectangles:

$$A = \lim_{n \to \infty} R_n = \lim_{n \to \infty} \left[f(x_1)\Delta x + f(x_2)\Delta x + \cdots + f(x_n)\Delta x \right]$$

Using sigma notation, we write this as follows:

$$A = \lim_{n \to \infty} \sum_{k=1}^{n} f(x_k)\Delta x$$

In using this formula for area, remember that Δx is the width of an approximating rectangle, x_k is the right endpoint of the kth rectangle, and $f(x_k)$ is its height. So

Width: $\qquad \Delta x = \dfrac{b - a}{n}$

Right endpoint: $\qquad x_k = a + k\,\Delta x$

Height: $\qquad f(x_k) = f(a + k\,\Delta x)$

When working with sums, we will need the following properties from Section 11.1:

$$\sum_{k=1}^{n} (a_k \pm b_k) = \sum_{k=1}^{n} a_k \pm \sum_{k=1}^{n} b_k \qquad \sum_{k=1}^{n} ca_k = c \sum_{k=1}^{n} a_k$$

We will also need the following formulas for the sums of the powers of the first n natural numbers from Section 11.5.

$$\sum_{k=1}^{n} c = nc \qquad\qquad \sum_{k=1}^{n} k = \frac{n(n + 1)}{2}$$

$$\sum_{k=1}^{n} k^2 = \frac{n(n + 1)(2n + 1)}{6} \qquad\qquad \sum_{k=1}^{n} k^3 = \frac{n^2(n + 1)^2}{4}$$

Example 3   Finding the Area under a Curve

Find the area of the region that lies under the parabola $y = x^2$, $0 \le x \le 5$.

Solution   The region is graphed in Figure 13. To find the area, we first find the dimensions of the approximating rectangles at the nth stage.

Width: $\qquad \Delta x = \dfrac{b - a}{n} = \dfrac{5 - 0}{n} = \dfrac{5}{n}$

Right endpoint: $\qquad x_k = a + k\,\Delta x = 0 + k\left(\dfrac{5}{n}\right) = \dfrac{5k}{n}$

Height: $\qquad f(x_k) = f\left(\dfrac{5k}{n}\right) = \left(\dfrac{5k}{n}\right)^2 = \dfrac{25k^2}{n^2}$

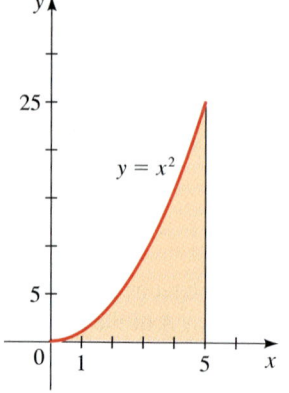

Figure 13

Now we substitute these values into the definition of area:

$$A = \lim_{n \to \infty} \sum_{k=1}^{n} f(x_k) \Delta x \qquad \text{Definition of area}$$

$$= \lim_{n \to \infty} \sum_{k=1}^{n} \frac{25k^2}{n^2} \cdot \frac{5}{n} \qquad f(x_k) = \frac{25k^2}{n^2}, \Delta x = \frac{5}{n}$$

$$= \lim_{n \to \infty} \sum_{k=1}^{n} \frac{125k^2}{n^3} \qquad \text{Simplify}$$

$$= \lim_{n \to \infty} \frac{125}{n^3} \sum_{k=1}^{n} k^2 \qquad \text{Factor } \frac{125}{n^3}$$

We can also calculate the limit by writing

$$\frac{125}{n^3} \cdot \frac{n(n+1)(2n+1)}{6}$$

$$= \frac{125}{6} \left(\frac{n}{n} \right) \left(\frac{n+1}{n} \right) \left(\frac{2n+1}{n} \right)$$

as in Example 2.

$$= \lim_{n \to \infty} \frac{125}{n^3} \cdot \frac{n(n+1)(2n+1)}{6} \qquad \text{Sum of squares formula}$$

$$= \lim_{n \to \infty} \frac{125(2n^2 + 3n + 1)}{6n^2} \qquad \text{Cancel } n \text{ and expand numerator}$$

$$= \lim_{n \to \infty} \frac{125}{6} \left(2 + \frac{3}{n} + \frac{1}{n^2} \right) \qquad \text{Divide numerator and denominator by } n^2$$

$$= \frac{125}{6}(2 + 0 + 0) = \frac{125}{3} \qquad \text{Let } n \to \infty$$

Thus, the area of the region is $\frac{125}{3} \approx 41.7$. ∎

Example 4   Finding the Area under a Curve

Find the area of the region that lies under the parabola $y = 4x - x^2$, $1 \leq x \leq 3$.

Solution   We start by finding the dimensions of the approximating rectangles at the nth stage.

Width: $\qquad \Delta x = \dfrac{b - a}{n} = \dfrac{3 - 1}{n} = \dfrac{2}{n}$

Right endpoint: $\qquad x_k = a + k\,\Delta x = 1 + k\left(\dfrac{2}{n} \right) = 1 + \dfrac{2k}{n}$

Figure 14 shows the region whose area is computed in Example 4.

Height: $\qquad f(x_k) = f\left(1 + \dfrac{2k}{n} \right) = 4\left(1 + \dfrac{2k}{n} \right) - \left(1 + \dfrac{2k}{n} \right)^2$

$$= 4 + \frac{8k}{n} - 1 - \frac{4k}{n} - \frac{4k^2}{n^2}$$

$$= 3 + \frac{4k}{n} - \frac{4k^2}{n^2}$$

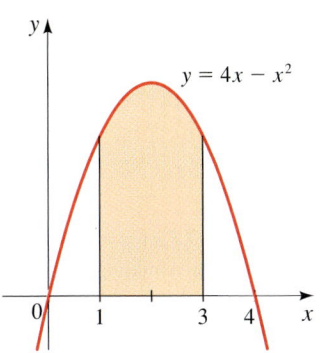

$y = 4x - x^2$

Figure 14

Thus, according to the definition of area, we get

$$A = \lim_{n \to \infty} \sum_{k=1}^{n} f(x_k)\,\Delta x = \lim_{n \to \infty} \sum_{k=1}^{n} \left(3 + \frac{4k}{n} - \frac{4k^2}{n^2} \right)\left(\frac{2}{n} \right)$$

$$= \lim_{n \to \infty} \left(\sum_{k=1}^{n} 3 + \frac{4}{n} \sum_{k=1}^{n} k - \frac{4}{n^2} \sum_{k=1}^{n} k^2 \right)\left(\frac{2}{n} \right)$$

$$= \lim_{n \to \infty} \left(\frac{2}{n} \sum_{k=1}^{n} 3 + \frac{8}{n^2} \sum_{k=1}^{n} k - \frac{8}{n^3} \sum_{k=1}^{n} k^2 \right)$$

$$= \lim_{n \to \infty} \left(\frac{2}{n}(3n) + \frac{8}{n^2}\left[\frac{n(n+1)}{2} \right] - \frac{8}{n^3}\left[\frac{n(n+1)(2n+1)}{6} \right] \right)$$

$$= \lim_{n \to \infty} \left(6 + 4 \cdot \frac{n}{n} \cdot \frac{n+1}{n} - \frac{4}{3} \cdot \frac{n}{n} \cdot \frac{n+1}{n} \cdot \frac{2n+1}{n} \right)$$

$$= \lim_{n \to \infty} \left[6 + 4\left(1 + \frac{1}{n} \right) - \frac{4}{3}\left(1 + \frac{1}{n} \right)\left(2 + \frac{1}{n} \right) \right]$$

$$= 6 + 4 \cdot 1 - \frac{4}{3} \cdot 1 \cdot 2 = \frac{22}{3} \qquad \blacksquare$$

12.5   Exercises

1. (a) By reading values from the given graph of f, use five rectangles to find a lower estimate and an upper estimate for the area under the given graph of f from $x = 0$ to $x = 10$. In each case, sketch the rectangles that you use.

   (b) Find new estimates using ten rectangles in each case.

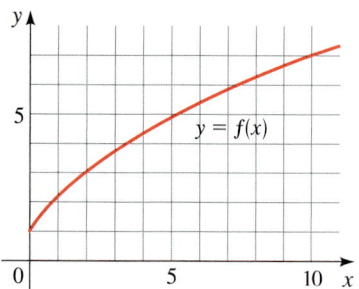

2. (a) Use six rectangles to find estimates of each type for the area under the given graph of f from $x = 0$ to $x = 12$.

   (i) L_6 (using left endpoints)

   (ii) R_6 (using right endpoints)

   (b) Is L_6 an underestimate or an overestimate of the true area?

   (c) Is R_6 an underestimate or an overestimate of the true area?

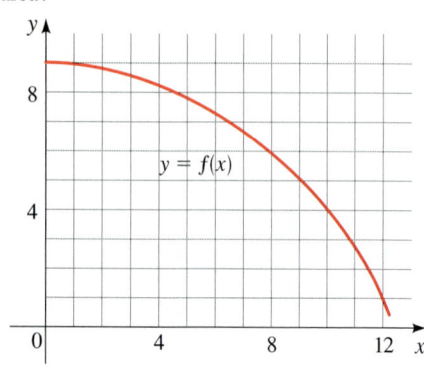

3–6 ■ Approximate the area of the shaded region under the graph of the given function by using the indicated rectangles. (The rectangles have equal width.)

3. $f(x) = \frac{1}{2}x + 2$

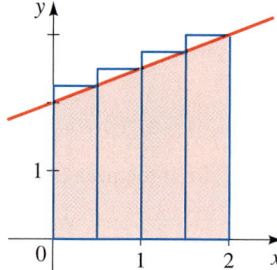

4. $f(x) = 4 - x^2$

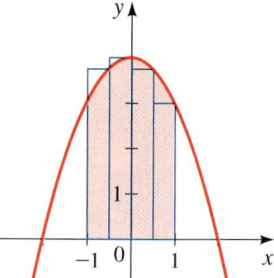

5. $f(x) = \dfrac{4}{x}$

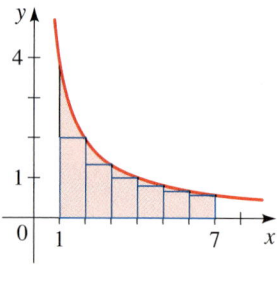

6. $f(x) = 9x - x^3$

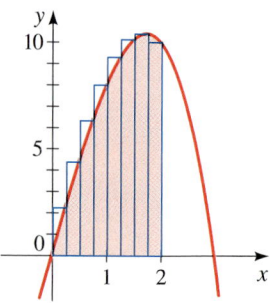

7. (a) Estimate the area under the graph of $f(x) = 1/x$ from $x = 1$ to $x = 5$ using four approximating rectangles and right endpoints. Sketch the graph and the rectangles. Is your estimate an underestimate or an overestimate?

   (b) Repeat part (a) using left endpoints.

8. (a) Estimate the area under the graph of $f(x) = 25 - x^2$ from $x = 0$ to $x = 5$ using five approximating rectangles and right endpoints. Sketch the graph and the rectangles. Is your estimate an underestimate or an overestimate?

   (b) Repeat part (a) using left endpoints.

9. (a) Estimate the area under the graph of $f(x) = 1 + x^2$ from $x = -1$ to $x = 2$ using three rectangles and right endpoints. Then improve your estimate by using six rectangles. Sketch the curve and the approximating rectangles.

   (b) Repeat part (a) using left endpoints.

10. (a) Estimate the area under the graph of $f(x) = e^{-x}$, $0 \le x \le 4$, using four approximating rectangles and taking the sample points to be

     (i) right endpoints

     (ii) left endpoints

     In each case, sketch the curve and the rectangles.

   (b) Improve your estimates in part (a) by using eight rectangles.

11–12 ■ Use the definition of area as a limit to find the area of the region that lies under the curve. Check your answer by sketching the region and using geometry.

11. $y = 3x$, $0 \le x \le 5$ 12. $y = 2x + 1$, $1 \le x \le 3$

13–18 ■ Find the area of the region that lies under the graph of f over the given interval.

13. $f(x) = 3x^2$, $0 \le x \le 2$

14. $f(x) = x + x^2$, $0 \le x \le 1$

15. $f(x) = x^3 + 2$, $0 \le x \le 5$

16. $f(x) = 4x^3$, $2 \le x \le 5$

17. $f(x) = x + 6x^2$, $1 \le x \le 4$

18. $f(x) = 20 - 2x^2$, $2 \le x \le 3$

Discovery • Discussion

19. **Approximating Area with a Calculator**   When we approximate areas using rectangles as in Example 1, then the more rectangles we use the more accurate the answer.

The following TI-83 program finds the approximate area under the graph of f on the interval $[a, b]$ using n rectangles. To use the program, first store the function f in Y_1. The program prompts you to enter N, the number of rectangles, and A and B, the endpoints of the interval.

(a) Approximate the area under the graph of $f(x) = x^5 + 2x + 3$ on $[1, 3]$ using 10, 20, and 100 rectangles.

(b) Approximate the area under the graph of f on the given interval using 100 rectangles.

   (i) $f(x) = \sin x$, on $[0, \pi]$

   (ii) $f(x) = e^{-x^2}$, on $[-1, 1]$

```
PROGRAM:AREA
:Prompt N
:Prompt A
:Prompt B
:(B-A)/N→D
:0→S
:A→X
:For (K,1,N)
:X+D→X
:S+Y₁→S
:End
:D*S→S
:Disp "AREA IS"
:Disp S
```

20. **Regions with Straight Versus Curved Boundaries**
Write a short essay that explains how you would find the area of a polygon, that is, a region bounded by straight line segments. Then explain how you would find the area of a region whose boundary is curved, as we did in this section. What is the fundamental difference between these two processes?

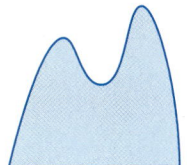

---

12 Review

Concept Check

1. Explain in your own words what is meant by the equation

$$\lim_{x \to 2} f(x) = 5$$

Is it possible for this statement to be true and yet $f(2) = 3$? Explain.

2. Explain what it means to say that

$$\lim_{x \to 1^-} f(x) = 3 \quad \text{and} \quad \lim_{x \to 1^+} f(x) = 7$$

In this situation is it possible that $\lim_{x \to 1} f(x)$ exists? Explain.

3. Describe several ways in which a limit can fail to exist. Illustrate with sketches.

4. State the following Limit Laws.
  (a) Sum Law
  (b) Difference Law
  (c) Constant Multiple Law
  (d) Product Law
  (e) Quotient Law
  (f) Power Law
  (g) Root Law

5. Write an expression for the slope of the tangent line to the curve $y = f(x)$ at the point $(a, f(a))$.

6. Define the derivative $f'(a)$. Discuss two ways of interpreting this number.

7. If $y = f(x)$, write expressions for the following.
  (a) The average rate of change of y with respect to x between the numbers a and x.
  (b) The instantaneous rate of change of y with respect to x at $x = a$.

8. Explain the meaning of the equation

$$\lim_{x \to \infty} f(x) = 2$$

Draw sketches to illustrate the various possibilities.

9. (a) What does it mean to say that the line $y = L$ is a horizontal asymptote of the curve $y = f(x)$? Draw curves to illustrate the various possibilities.
  (b) Which of the following curves have horizontal asymptotes?
    (i) $y = x^2$ (iv) $y = \tan^{-1} x$
    (ii) $y = 1/x$ (v) $y = e^x$
    (iii) $y = \sin x$ (vi) $y = \ln x$

10. (a) What is a convergent sequence?
  (b) What does $\lim_{n \to \infty} a_n = 3$ mean?

11. Suppose S is the region that lies under the graph of $y = f(x)$, $a \le x \le b$.
  (a) Explain how this area is approximated using rectangles.
  (b) Write an expression for the area of S as a limit of sums.

Exercises

 1–6 ■ Use a table of values to estimate the value of the limit. Then use a graphing device to confirm your result graphically.

1. $\lim\limits_{x \to 2} \dfrac{x - 2}{x^2 - 3x + 2}$

2. $\lim\limits_{t \to -1} \dfrac{t + 1}{t^3 - t}$

3. $\lim\limits_{x \to 0} \dfrac{2^x - 1}{x}$

4. $\lim\limits_{x \to 0} \dfrac{\sin 2x}{x}$

5. $\lim\limits_{x \to 1^+} \ln \sqrt{x - 1}$

6. $\lim\limits_{x \to 0^-} \dfrac{\tan x}{|x|}$

7. The graph of f is shown in the figure. Find each limit or explain why it does not exist.
  (a) $\lim\limits_{x \to 2^+} f(x)$ **(b)** $\lim\limits_{x \to -3^+} f(x)$
  (c) $\lim\limits_{x \to -3^-} f(x)$ **(d)** $\lim\limits_{x \to -3} f(x)$

  (e) $\lim\limits_{x \to 4} f(x)$ **(f)** $\lim\limits_{x \to \infty} f(x)$
  (g) $\lim\limits_{x \to -\infty} f(x)$ **(h)** $\lim\limits_{x \to 0} f(x)$

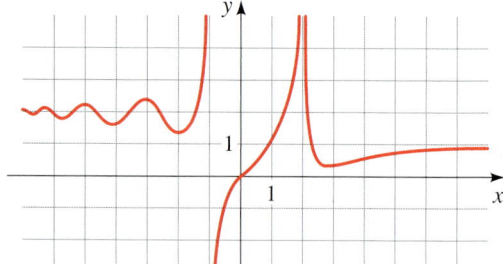

8. Let

$$f(x) = \begin{cases} 2 & \text{if } x < -1 \\ x^2 & \text{if } -1 \le x \le 2 \\ x + 2 & \text{if } x > 2 \end{cases}$$

Find each limit or explain why it does not exist.
  (a) $\lim\limits_{x \to -1^-} f(x)$ **(b)** $\lim\limits_{x \to -1^+} f(x)$
  (c) $\lim\limits_{x \to -1} f(x)$ **(d)** $\lim\limits_{x \to 2^-} f(x)$

(e) $\lim\limits_{x \to 2^+} f(x)$ **(f)** $\lim\limits_{x \to 2} f(x)$

(g) $\lim\limits_{x \to 0} f(x)$ **(h)** $\lim\limits_{x \to 3} (f(x))^2$

9–20 ■ Use the Limit Laws to evaluate the limit, if it exists.

9. $\lim\limits_{x \to 2} \dfrac{x + 1}{x - 3}$

10. $\lim\limits_{t \to 1} (t^3 - 3t + 6)$

11. $\lim\limits_{x \to 3} \dfrac{x^2 + x - 12}{x - 3}$

12. $\lim\limits_{x \to -2} \dfrac{x^2 - 4}{x^2 + x - 2}$

13. $\lim\limits_{u \to 0} \dfrac{(u + 1)^2 - 1}{u}$

14. $\lim\limits_{z \to 9} \dfrac{\sqrt{z} - 3}{z - 9}$

15. $\lim\limits_{x \to 3^-} \dfrac{x - 3}{|x - 3|}$

16. $\lim\limits_{x \to 0} \left(\dfrac{1}{x} + \dfrac{2}{x^2 - 2x} \right)$

17. $\lim\limits_{x \to \infty} \dfrac{2x}{x - 4}$

18. $\lim\limits_{x \to \infty} \dfrac{x^2 + 1}{x^4 - 3x + 6}$

19. $\lim\limits_{x \to \infty} \cos^2 x$

20. $\lim\limits_{t \to -\infty} \dfrac{t^4}{t^3 - 1}$

21–24 ■ Find the derivative of the function at the given number.

21. $f(x) = 3x - 5$, at 4   **22.** $g(x) = 2x^2 - 1$, at -1

23. $f(x) = \sqrt{x}$, at 16   **24.** $f(x) = \dfrac{x}{x + 1}$, at 1

25–28 ■ **(a)** Find $f'(a)$.   **(b)** Find $f'(2)$ and $f'(-2)$.

25. $f(x) = 6 - 2x$

26. $f(x) = x^2 - 3x$

27. $f(x) = \sqrt{x + 6}$

28. $f(x) = \dfrac{4}{x}$

29–30 ■ Find an equation of the tangent line shown in the figure.

29.

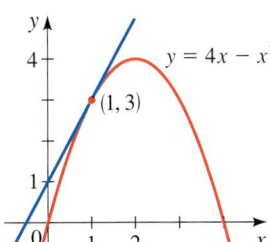

30.

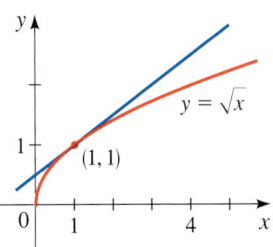

31–34 ■ Find an equation of the line tangent to the graph of f at the given point.

31. $f(x) = 2x$, at $(3, 6)$

32. $f(x) = x^2 - 3$, at $(2, 1)$

33. $f(x) = \dfrac{1}{x}$, at $\left(2, \dfrac{1}{2} \right)$

34. $f(x) = \sqrt{x + 1}$, at $(3, 2)$

35. A stone is dropped from the roof of a building 640 ft above the ground. Its height (in feet) after t seconds is given by $h(t) = 640 - 16t^2$.
   (a) Find the velocity of the stone when $t = 2$.
   (b) Find the velocity of the stone when $t = a$.
   (c) At what time t will the stone hit the ground?
   (d) With what velocity will the stone hit the ground?

36. If a gas is confined in a fixed volume, then according to Boyle's Law the product of the pressure P and the temperature T is a constant. For a certain gas, $PT = 100$, where P is measured in lb/in^2 and T is measured in kelvins (K).
   (a) Express P as a function of T.
   (b) Find the instantaneous rate of change of P with respect to T when $T = 300$ K.

37–42 ■ If the sequence is convergent, find its limit. If it is divergent, explain why.

37. $a_n = \dfrac{n}{5n + 1}$

38. $a_n = \dfrac{n^3}{n^3 + 1}$

39. $a_n = \dfrac{n(n + 1)}{2n^2}$

40. $a_n = \dfrac{n^3}{2n + 6}$

41. $a_n = \cos\left(\dfrac{n\pi}{2} \right)$

42. $a_n = \dfrac{10}{3^n}$

43–44 ■ Approximate the area of the shaded region under the graph of the given function by using the indicated rectangles. (The rectangles have equal width.)

43. $f(x) = \sqrt{x}$

44. $f(x) = 4x - x^2$

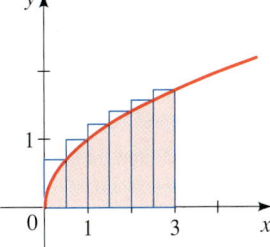

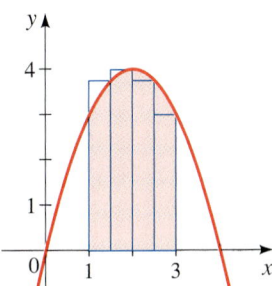

45–48 ■ Use the limit definition of area to find the area of the region that lies under the graph of f over the given interval.

45. $f(x) = 2x + 3$, $\quad 0 \le x \le 2$

46. $f(x) = x^2 + 1$, $\quad 0 \le x \le 3$

47. $f(x) = x^2 - x$, $\quad 1 \le x \le 2$

48. $f(x) = x^3$, $\quad 1 \le x \le 2$

12   Test

1. (a) Use a table of values to estimate the limit

$$\lim_{x \to 0} \frac{x}{\sin 2x}$$

(b) Use a graphing calculator to confirm your answer graphically.

2. For the piecewise-defined function f whose graph is shown, find:

(a) $\lim\limits_{x \to -1^-} f(x)$ (b) $\lim\limits_{x \to -1^+} f(x)$ (c) $\lim\limits_{x \to -1} f(x)$

(d) $\lim\limits_{x \to 0^-} f(x)$ (e) $\lim\limits_{x \to 0^+} f(x)$ (f) $\lim\limits_{x \to 0} f(x)$

(g) $\lim\limits_{x \to 2^-} f(x)$ (h) $\lim\limits_{x \to 2^+} f(x)$ (i) $\lim\limits_{x \to 2} f(x)$

$$f(x) = \begin{cases} 1 & \text{if } x < -1 \\ 0 & \text{if } x = -1 \\ x^2 & \text{if } -1 < x \le 2 \\ 4 - x & \text{if } 2 < x \end{cases}$$

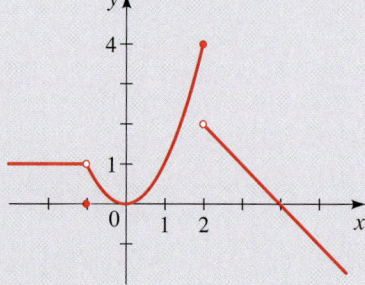

3. Evaluate the limit, if it exists.

(a) $\lim\limits_{x \to 2} \dfrac{x^2 + 2x - 8}{x - 2}$ (b) $\lim\limits_{x \to 2} \dfrac{x^2 - 2x - 8}{x + 2}$ (c) $\lim\limits_{x \to 2} \dfrac{1}{x - 2}$

(d) $\lim\limits_{x \to 2} \dfrac{x - 2}{|x - 2|}$ (e) $\lim\limits_{x \to 4} \dfrac{\sqrt{x} - 2}{x - 4}$ (f) $\lim\limits_{x \to \infty} \dfrac{2x^2 - 4}{x^2 + x}$

4. Let $f(x) = x^2 - 2x$. Find:

(a) $f'(x)$ (b) $f'(-1), f'(1), f'(2)$

5. Find the equation of the line tangent to the graph of $f(x) = \sqrt{x}$ at the point where $x = 9$.

6. Find the limit of the sequence.

(a) $a_n = \dfrac{n}{n^2 + 4}$ (b) $a_n = \sec n\pi$

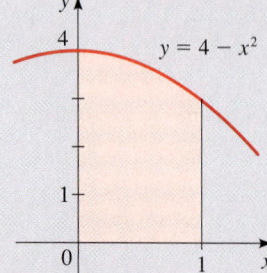

7. The region sketched in the figure in the margin lies under the graph of $f(x) = 4 - x^2$, above the interval $0 \le x \le 1$.

(a) Approximate the area of the region with five rectangles, equally spaced along the x-axis, using right endpoints to determine the heights of the rectangles.

(b) Use the limit definition of area to find the exact value of the area of the region.

Focus on Modeling
Interpretations of Area

The area under the graph of a function is used to model many quantities in physics, economics, engineering, and other fields. That is why the area problem is so important. Here we will show how the concept of work (Section 8.5) is modeled by area. Several other applications are explored in the problems.

Recall that the work W done in moving an object is the product of the force F applied to the object and the distance d that the object moves:

$$W = Fd \qquad \text{work} = \text{force} \times \text{distance}$$

This formula is used if the force is *constant*. For example, suppose you are pushing a crate across a floor, moving along the positive x-axis from $x = a$ to $x = b$, and you apply a constant force $F = k$. The graph of F as a function of the distance x is shown in Figure 1(a). Notice that the work done is $W = Fd = k(b - a)$, which is the area under the graph of F (see Figure 1(b)).

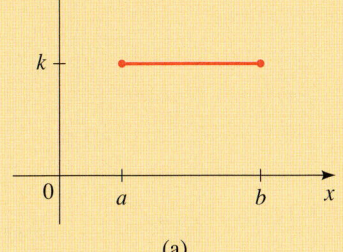 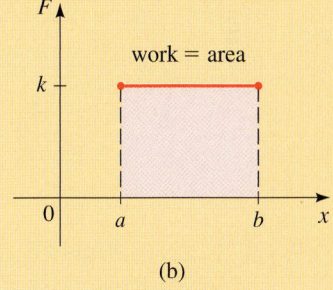

Figure 1

A constant force F

(a)   (b)

But what if the force is *not* constant? For example, suppose the force you apply to the crate varies with distance (you push harder at certain places than you do at others). More precisely, suppose that you push the crate along the x-axis in the positive direction, from $x = a$ to $x = b$, and at each point x between a and b you apply a force $f(x)$ to the crate. Figure 2 shows a graph of the force f as a function of the distance x.

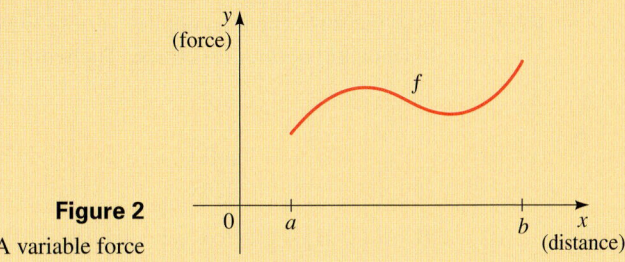

Figure 2

A variable force

How much work was done? We can't apply the formula for work directly because the force is not constant. So let's divide the interval $[a, b]$ into n subintervals with endpoints $x_0, x_1, \ldots, x_n$ and equal width Δx as shown in Figure 3(a) on the next page. The force at the right endpoint of the interval $[x_{k-1}, x_k]$ is $f(x_k)$. If n is large, then Δx is small, so the values of f don't change very much over the interval $[x_{k-1}, x_k]$. In other words f is almost constant on the interval, and so the work W_k that is done in moving the crate from x_{k-1} to x_k is approximately

$$W_k \approx f(x_k)\Delta x$$

Thus, we can approximate the work done in moving the crate from $x = a$ to $x = b$ by

$$W \approx \sum_{k=1}^{n} f(x_k)\Delta x$$

It seems that this approximation becomes better as we make n larger (and so make the interval $[x_{k-1}, x_k]$ smaller). Therefore, we define the work done in moving an object from a to b as the limit of this quantity as $n \to \infty$:

$$W = \lim_{n\to\infty} \sum_{k=1}^{n} f(x_k)\Delta x$$

Notice that this is precisely the area under the graph of f between $x = a$ and $x = b$ as defined in Section 12.5. See Figure 3(b).

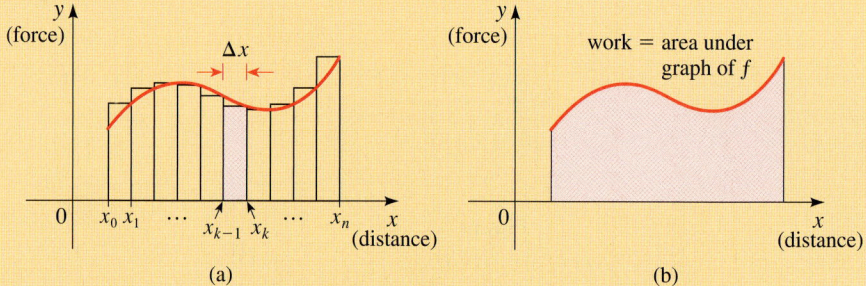

Figure 3
Approximating work

(a)   (b)

Example   The Work Done by a Variable Force

A man pushes a crate along a straight path a distance of 18 ft. At a distance x from his starting point, he applies a force given by $f(x) = 340 - x^2$. Find the work done by the man.

Solution   The graph of f between $x = 0$ and $x = 18$ is shown in Figure 4. Notice how the force the man applies varies—he starts by pushing with a force of 340 lb, but steadily applies less force. The work done is the area under the graph of f on

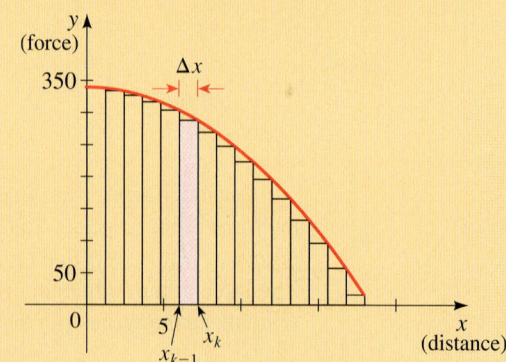

Figure 4

the interval $[0, 18]$. To find this area, we start by finding the dimensions of the approximating rectangles at the nth stage.

Width:
$$\Delta x = \frac{b - a}{n} = \frac{18 - 0}{n} = \frac{18}{n}$$

Right endpoint:
$$x_k = a + k\,\Delta x = 0 + k\left(\frac{18}{n}\right) = \frac{18k}{n}$$

Height:
$$f(x_k) = f\left(\frac{18k}{n}\right) = 340 - \left(\frac{18k}{n}\right)^2$$
$$= 340 - \frac{324k^2}{n^2}$$

Thus, according to the definition of work we get

$$W = \lim_{n \to \infty} \sum_{k=1}^{n} f(x_k)\,\Delta x = \lim_{n \to \infty} \sum_{k=1}^{n} \left(340 - \frac{324k^2}{n^2}\right)\left(\frac{18}{n}\right)$$

$$= \lim_{n \to \infty} \left(\frac{18}{n} \sum_{k=1}^{n} 340 - \frac{(18)(324)}{n^3} \sum_{k=1}^{n} k^2\right)$$

$$= \lim_{n \to \infty} \left(\frac{18}{n}\,340n - \frac{5832}{n^3}\left[\frac{n(n+1)(2n+1)}{6}\right]\right)$$

$$= \lim_{n \to \infty} \left(6120 - 972 \cdot \frac{n}{n} \cdot \frac{n+1}{n} \cdot \frac{2n+1}{n}\right)$$

$$= 6120 - 972 \cdot 1 \cdot 1 \cdot 2 = 4176$$

So the work done by the man in moving the crate is 4176 ft-lb.   ∎

Problems

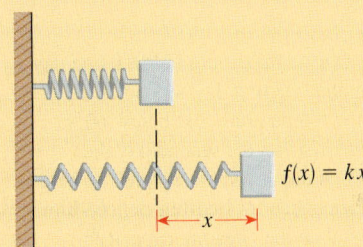

1. **Work Done by a Winch**   A motorized winch is being used to pull a felled tree to a logging truck. The motor exerts a force of $f(x) = 1500 + 10x - \frac{1}{2}x^2$ lb on the tree at the instant when the tree has moved x ft. The tree must be moved a distance of 40 ft, from $x = 0$ to $x = 40$. How much work is done by the winch in moving the tree?

2. **Work Done by a Spring**   Hooke's law states that when a spring is stretched, it pulls back with a force proportional to the amount of the stretch. The constant of proportionality is a characteristic of the spring known as the **spring constant**. Thus, a spring with spring constant k exerts a force $f(x) = kx$ when it is stretched a distance x.

   A certain spring has spring constant $k = 20$ lb/ft. Find the work done when the spring is pulled so that the amount by which it is stretched increases from $x = 0$ to $x = 2$ ft.

3. **Force of Water**   As any diver knows, an object submerged in water experiences pressure, and as depth increases, so does the water pressure. At a depth of x ft, the water pressure is $p(x) = 62.5x$ lb/ft^2. To find the force exerted by the water on a surface, we multiply the pressure by the area of the surface:

$$\text{force} = \text{pressure} \times \text{area}$$

$f(x) = kx$

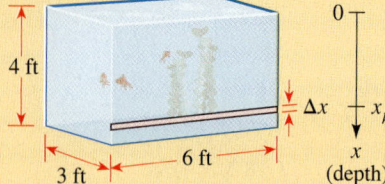

Suppose an aquarium that is 3 ft wide, 6 ft long, and 4 ft high is full of water. The bottom of the aquarium has area $3 \times 6 = 18$ ft^2, and it experiences water pressure of $p(4) = 62.5 \times 4 = 250$ lb/ft^2. Thus, the total force exerted by the water on the bottom is $250 \times 18 = 4500$ lb.

The water also exerts a force on the sides of the aquarium, but this is not as easy to calculate because the pressure increases from top to bottom. To calculate the force on one of the 4 ft by 6 ft sides, we divide its area into n thin horizontal strips of width Δx, as shown in the figure. The area of each strip is

$$\text{length} \times \text{width} = 6 \, \Delta x$$

If the bottom of the kth strip is at the depth x_k, then it experiences water pressure of approximately $p(x_k) = 62.5 x_k$ lb/ft^2—the thinner the strip, the more accurate the approximation. Thus, on each strip the water exerts a force of

$$\text{pressure} \times \text{area} = 62.5 x_k \times 6 \, \Delta x = 375 x_k \, \Delta x \text{ lb}$$

(a) Explain why the total force exerted by the water on the 4 ft by 6 ft sides of the aquarium is

$$\lim_{n \to \infty} \sum_{k=1}^{n} 375 x_k \, \Delta x$$

where $\Delta x = 4/n$ and $x_k = 4k/n$.

(b) What area does the limit in part (a) represent?

(c) Evaluate the limit in part (a) to find the force exerted by the water on one of the 4 ft by 6 ft sides of the aquarium.

(d) Use the same technique to find the force exerted by the water on one of the 4 ft by 3 ft sides of the aquarium.

NOTE Engineers use the technique outlined in this problem to find the total force exerted on a dam by the water in the reservoir behind the dam.

4. **Distance Traveled by a Car**   Since distance $=$ speed $\times$ time, it is easy to see that a car moving, say, at 70 mi/h for 5 h will travel a distance of 350 mi. But what if the speed varies, as it usually does in practice?

(a) Suppose the speed of a moving object at time t is $v(t)$. Explain why the distance traveled by the object between times $t = a$ and $t = b$ is the area under the graph of v between $t = a$ and $t = b$.

(b) The speed of a car t seconds after it starts moving is given by the function $v(t) = 6t + 0.1t^3$ ft/s. Find the distance traveled by the car from $t = 0$ to $t = 5$ s.

5. **Heating Capacity**   If the outdoor temperature reaches a maximum of 90 °F one day and only 80 °F the next, then we would probably say that the first day was hotter than the second. Suppose, however, that on the first day the temperature was below 60 °F for most of the day, reaching the high only briefly, whereas on the second day the temperature stayed above 75 °F all the time. Now which day is the hotter one? To better measure how hot a particular day is, scientists use the concept of **heating degree-hour**. If the temperature is a constant D degrees for t hours, then the "heating capacity" generated over this period is Dt heating degree-hours:

$$\text{heating degree-hours} = \text{temperature} \times \text{time}$$

If the temperature is not constant, then the number of heating degree-hours equals the area under the graph of the temperature function over the time period in question.

(a) On a particular day, the temperature (in °F) was modeled by the function $D(t) = 61 + \frac{6}{5}t - \frac{1}{25}t^2$, where t was measured in hours since midnight. How many heating degree-hours were experienced on this day, from $t = 0$ to $t = 24$?

(b) What was the maximum temperature on the day described in part (a)?

(c) On another day, the temperature (in °F) was modeled by the function $E(t) = 50 + 5t - \frac{1}{4}t^2$. How many heating degree-hours were experienced on this day?

(d) What was the maximum temperature on the day described in part (c)?

(e) Which day was "hotter"?

Answers to Odd-Numbered Exercises and Chapter Tests

Chapter 1

Section 1.1 ■ page 10

1. (a) 50 (b) 0, -10, 50 (c) 0, -10, 50, $\frac{22}{7}$, 0.538, 1.2$\bar{3}$,
$-\frac{1}{3}$ (d) $\sqrt{7}$, $\sqrt[3]{2}$ **3.** Commutative Property for addition
5. Associative Property for addition **7.** Distributive Property
9. Commutative Property for multiplication
11. $3 + x$ **13.** $4A + 4B$ **15.** $3x + 3y$ **17.** $8m$
19. $-5x + 10y$ **21.** (a) $\frac{17}{30}$ (b) $\frac{9}{20}$ **23.** (a) 3 (b) $\frac{25}{72}$
25. (a) $\frac{8}{3}$ (b) 6 **27.** (a) $<$ (b) $>$ (c) $=$ **29.** (a) False
(b) True **31.** (a) False (b) True **33.** (a) $x > 0$
(b) $t < 4$ (c) $a \geq \pi$ (d) $-5 < x < \frac{1}{3}$ (e) $|p - 3| \leq 5$
35. (a) $\{1, 2, 3, 4, 5, 6, 7, 8\}$ (b) $\{2, 4, 6\}$
37. (a) $\{1, 2, 3, 4, 5, 6, 7, 8, 9, 10\}$ (b) $\{7\}$
39. (a) $\{x \mid x \leq 5\}$ (b) $\{x \mid -1 < x < 4\}$

41. $-3 < x < 0$

43. $2 \leq x < 8$

45. $x \geq 2$

47. $(-\infty, 1]$

49. $(-2, 1]$

51. $(-1, \infty)$

53. (a) $[-3, 5]$ (b) $(-3, 5]$
55.

57.

59.

61. (a) 100 (b) 73 **63.** (a) 2 (b) -1 **65.** (a) 12
(b) 5 **67.** 5 **69.** (a) 15 (b) 24 (c) $\frac{67}{40}$ **71.** (a) $\frac{7}{9}$
(b) $\frac{13}{45}$ (c) $\frac{19}{33}$ **73.** Distributive Property
75. (a) Yes, no (b) 6 ft

Section 1.2 ■ page 21

1. $5^{-1/2}$ **3.** $\sqrt[3]{4^2}$ **5.** $5^{3/5}$ **7.** $\sqrt[5]{a^2}$ **9.** (a) -9 (b) 9
(c) 1 **11.** (a) 4 (b) $\frac{1}{81}$ (c) 16 **13.** (a) 4 (b) 2 (c) $\frac{1}{2}$

15. (a) $\frac{2}{3}$ (b) $-\frac{1}{4}$ (c) $-\frac{1}{2}$ **17.** (a) $\frac{3}{2}$ (b) 4 (c) -4
19. 5 **21.** 14 **23.** $7\sqrt{2}$ **25.** $3\sqrt[5]{3}$ **27.** a^4 **29.** $6x^7y^5$
31. $16x^{10}$ **33.** $4/b^2$ **35.** $64r^7s$ **37.** $648y^7$ **39.** $\dfrac{x^3}{y}$
41. $\dfrac{y^2z^9}{x^5}$ **43.** $\dfrac{s^3}{q^7r^6}$ **45.** $|x|$ **47.** $2x^2$ **49.** $|ab^3|$
51. $2|x|$ **53.** $x^{13/15}$ **55.** $\dfrac{-1}{9a^{5/4}}$ **57.** $16b^{9/10}$ **59.** $\dfrac{1}{c^{2/3}d}$
61. $y^{1/2}$ **63.** $\dfrac{32x^{12}}{y^{16/15}}$ **65.** $\dfrac{x^{15}}{y^{15/2}}$ **67.** $\dfrac{4a^2}{3b^{1/3}}$ **69.** $\dfrac{3t^{25/6}}{s^{1/2}}$
71. (a) 6.93×10^7 (b) 7.2×10^{12} (c) 2.8536×10^{-5}
(d) 1.213×10^{-4} **73.** (a) 319,000 (b) 272,100,000
(c) 0.00000002670 (d) 0.000000009999
75. (a) 5.9×10^{12} mi (b) 4×10^{-13} cm (c) 3.3×10^{19}
molecules **77.** 1.3×10^{-20} **79.** 1.429×10^{19}
81. 7.4×10^{-14} **83.** (a) $\dfrac{\sqrt{10}}{10}$ (b) $\dfrac{\sqrt{2x}}{x}$ (c) $\dfrac{\sqrt{3x}}{3}$
85. (a) $\dfrac{2\sqrt[3]{x^2}}{x}$ (b) $\dfrac{\sqrt[4]{y}}{y}$ (c) $\dfrac{xy^{3/5}}{y}$
87. (a) Negative (b) Positive (c) Negative (d) Negative
(e) Positive (f) Negative **89.** 2.5×10^{13} mi
91. 1.3×10^{21} L **93.** 4.03×10^{27} molecules
95. (a) 28 mi/h (b) 167 ft **97.** (a) 17.707 ft/s
(b) 1328.0 ft^3/s

Section 1.3 ■ page 31

1. Trinomial; x^2, $-3x$, 7; 2 **3.** Monomial; -8; 0
5. Four terms; $-x^4$, x^3, $-x^2$, x; 4 **7.** $7x + 5$
9. $5x^2 - 2x - 4$ **11.** $x^3 + 3x^2 - 6x + 11$ **13.** $9x + 103$
15. $-t^4 + t^3 - t^2 - 10t + 5$ **17.** $x^{3/2} - x$
19. $21t^2 - 29t + 10$ **21.** $3x^2 + 5xy - 2y^2$
23. $1 - 4y + 4y^2$ **25.** $4x^4 + 12x^2y^2 + 9y^4$
27. $2x^3 - 7x^2 + 7x - 5$ **29.** $x^4 - a^4$ **31.** $a - 1/b^2$
33. $1 + 3a^3 + 3a^6 + a^9$ **35.** $2x^4 + x^3 - x^2 + 3x - 2$
37. $1 - x^{2/3} + x^{4/3} - x^2$ **39.** $3x^4y^4 + 7x^3y^5 - 6x^2y^3 - 14xy^4$

41. $x^2 - y^2 - 2yz - z^2$ **43.** $2x(-x^2 + 8)$
45. $(y - 6)(y + 9)$ **47.** $xy(2x - 6y + 3)$
49. $(x - 1)(x + 3)$ **51.** $(2x - 5)(4x + 3)$
53. $(3x + 4)(3x + 8)$ **55.** $(3a - 4)(3a + 4)$
57. $(3x + y)(9x^2 - 3xy + y^2)$ **59.** $(x + 6)^2$
61. $(x + 4)(x^2 + 1)$ **63.** $(2x + 1)(x^2 - 3)$
65. $(x + 1)(x^2 + 1)$ **67.** $x^{1/2}(x + 1)(x - 1)$
69. $(x^2 + 3)(x^2 + 1)^{-1/2}$ **71.** $6x(2x^2 + 3)$
73. $(x - 4)(x + 2)$ **75.** $(2x + 3)(x + 1)$
77. $(3x + 2)(2x - 3)$ **79.** $(5s - t)^2$
81. $(2x - 5)(2x + 5)$ **83.** $4ab$
85. $(x + 3)(x - 3)(x + 1)(x - 1)$
87. $(2x + 5)(4x^2 - 10x + 25)$
89. $(x^2 - 2y)(x^4 + 2x^2y + 4y^2)$
91. $x(x + 1)^2$ **93.** $(y + 2)(y - 2)(y - 3)$
95. $(2x^2 + 1)(x + 2)$ **97.** $3(x - 1)(x + 2)$
99. $(a + 2)(a - 2)(a + 1)(a - 1)$
101. $2(x^2 + 4)^4(x - 2)^3(7x^2 - 10x + 8)$
103. $(x^2 + 3)^{-4/3}(\frac{1}{3}x^2 + 3)$
105. (d) $(a + b + c)(a + b - c)(a - b + c)(b - a + c)$

Section 1.4 ■ page 41

1. $\mathbb{R}$ **3.** $x \neq 4$ **5.** $x \geq -3$ **7.** $\dfrac{x + 2}{2(x - 1)}$

9. $\dfrac{1}{x + 2}$ **11.** $\dfrac{x + 2}{x + 1}$ **13.** $\dfrac{y}{y - 1}$ **15.** $\dfrac{x(2x + 3)}{2x - 3}$

17. $\dfrac{1}{4(x - 2)}$ **19.** $\dfrac{x + 3}{3 - x}$ **21.** $\dfrac{1}{t^2 + 9}$ **23.** $\dfrac{x + 4}{x + 1}$

25. $\dfrac{(2x + 1)(2x - 1)}{(x + 5)^2}$ **27.** $x^2(x + 1)$ **29.** $\dfrac{x}{yz}$

31. $\dfrac{3(x + 2)}{x + 3}$ **33.** $\dfrac{3x + 7}{(x - 3)(x + 5)}$ **35.** $\dfrac{1}{(x + 1)(x + 2)}$

37. $\dfrac{3x + 2}{(x + 1)^2}$ **39.** $\dfrac{u^2 + 3u + 1}{u + 1}$ **41.** $\dfrac{2x + 1}{x^2(x + 1)}$

43. $\dfrac{2x + 7}{(x + 3)(x + 4)}$ **45.** $\dfrac{x - 2}{(x + 3)(x - 3)}$ **47.** $\dfrac{5x - 6}{x(x - 1)}$

49. $\dfrac{-5}{(x + 1)(x + 2)(x - 3)}$ **51.** $-xy$ **53.** $\dfrac{c}{c - 2}$

55. $\dfrac{3x + 7}{x^2 + 2x - 1}$ **57.** $\dfrac{y - x}{xy}$ **59.** 1 **61.** $\dfrac{-1}{a(a + h)}$

63. $\dfrac{-3}{(2 + x)(2 + x + h)}$ **65.** $\dfrac{1}{\sqrt{1 - x^2}}$

67. $\dfrac{(x + 2)^2(x - 13)}{(x - 3)^3}$ **69.** $\dfrac{x + 2}{(x + 1)^{3/2}}$ **71.** $\dfrac{2x + 3}{(x + 1)^{4/3}}$

73. $2 + \sqrt{3}$ **75.** $\dfrac{2(\sqrt{7} - \sqrt{2})}{5}$ **77.** $\dfrac{y\sqrt{3} - y\sqrt{y}}{3 - y}$

79. $\dfrac{-4}{3(1 + \sqrt{5})}$ **81.** $\dfrac{r - 2}{5(\sqrt{r} - \sqrt{2})}$ **83.** $\dfrac{1}{\sqrt{x^2 + 1} + x}$

85. True   **87.** False   **89.** False   **91.** True

93. (a) $\dfrac{R_1 R_2}{R_1 + R_2}$ **(b)** $\frac{20}{3} \approx 6.7$ ohms

Section 1.5 ■ page 55

1. (a) No   **(b)** Yes   **3. (a)** Yes   **(b)** No   **5.** 12   **7.** 18
9. -3 **11.** 12   **13.** $-\frac{3}{4}$ **15.** 30   **17.** $-\frac{1}{3}$ **19.** $\frac{13}{3}$ **21.** -2
23. $R = \dfrac{PV}{nT}$ **25.** $R_1 = \dfrac{RR_2}{R_2 - R}$ **27.** $x = \dfrac{2d - b}{a - 2c}$

29. $x = \dfrac{1 - a}{a^2 - a - 1}$ **31.** $r = \pm\sqrt{\dfrac{3V}{\pi h}}$

33. $b = \pm\sqrt{c^2 - a^2}$ **35.** $t = \dfrac{-V_0 \pm \sqrt{V_0^2 + 2gh}}{g}$

37. $-4, 3$ **39.** $3, 4$ **41.** $-\frac{3}{2}, \frac{5}{2}$ **43.** $-2, \frac{1}{3}$ **45.** $-1 \pm \sqrt{6}$
47. $-\frac{7}{2}, \frac{1}{2}$ **49.** $-2 \pm \dfrac{\sqrt{14}}{2}$ **51.** $0, \frac{1}{4}$ **53.** $-3, 5$

55. $\dfrac{-3 \pm \sqrt{5}}{2}$ **57.** $-\frac{3}{2}, 1$ **59.** $\dfrac{1 \pm \sqrt{5}}{4}$ **61.** $-\frac{9}{2}, \frac{1}{2}$

63. $\dfrac{-5 \pm \sqrt{13}}{2}$ **65.** $-\dfrac{\sqrt{6}}{2}, \dfrac{\sqrt{6}}{6}$ **67.** $-\frac{7}{5}$

69. 2   **71.** 1   **73.** No real solution
75. $-\frac{7}{5}, 2$ **77.** $-50, 100$ **79.** -4 **81.** 4   **83.** 3
85. $\pm 2\sqrt{2}, \pm\sqrt{5}$ **87.** No real solution
89. $\pm 3\sqrt{3}, \pm 2\sqrt{2}$ **91.** $-1, 0, 3$ **93.** 27, 729   **95.** $-\frac{3}{2}, \frac{3}{2}$
97. 3.99, 4.01   **99.** 4.24 s   **101. (a)** After 1 s and $1\frac{1}{2}$ s
(b) Never   **(c)** 25 ft   **(d)** After $1\frac{1}{4}$ s   **(e)** After $2\frac{1}{2}$ s
103. (a) 0.00055, 12.018 m   **(b)** 234.375 kg/m^3
105. (a) After 17 yr, on Jan. 1, 2019   **(b)** After 18.621 yr, on
Aug. 12, 2020   **107.** 50   **109.** 132.6 ft

Section 1.6 ■ page 68

1. $3n + 3$ **3.** $\dfrac{160 + s}{3}$ **5.** $0.025x$

7. $A = 3w^2$ **9.** $d = \frac{3}{4}s$ **11.** $\dfrac{25}{x + 3}$ **13.** 51, 52, 53

15. 19 and 36   **17.** $9000 at $4\frac{1}{2}\%$ and $3000 at 4%
19. 7.5%   **21.** $7400   **23.** $45,000   **25.** Plumber, 70 h;
assistant, 35 h   **27.** 40 years old   **29.** 9 pennies, 9 nickels,
9 dimes   **31.** 6.4 ft from the fulcrum   **33. (a)** 9 cm
(b) 5 in.   **35.** 45 ft   **37.** 120 ft by 120 ft   **39.** 25 ft by 35 ft
41. 60 ft by 40 ft   **43.** 120 ft   **45.** 4 in.   **47.** 18 ft   **49.** 5 m
51. 4   **53.** 18 g   **55.** 0.6 L   **57.** 35%   **59.** 37 min 20 s
61. 3 h   **63.** Irene 3 h, Henry $4\frac{1}{2}$ h   **65.** 4 h   **67.** 500 mi/h
69. 50 mi/h (or 240 mi/h)   **71.** 6 km/h   **73.** 2 ft by
6 ft by 15 ft   **75.** 13 in. by 13 in.   **77.** 2.88 ft   **79.** 16 mi; no
81. 7.52 ft   **83.** 18 ft   **85.** 4.55 ft

Section 1.7 ■ page 84

1. $\{\sqrt{2}, 2, 4\}$ **3.** $\{4\}$ **5.** $\{-2, -1, 2, 4\}$
7. $(4, \infty)$ **9.** $(-\infty, 2]$

11. $\left(-\infty, -\frac{1}{2}\right)$

13. $[1, \infty)$

15. $\left(\frac{16}{3}, \infty\right)$

17. $(-\infty, -18)$

19. $(-\infty, -1]$

21. $[-3, -1)$

23. $(2, 6)$

25. $\left[\frac{9}{2}, 5\right)$

27. $\left(\frac{15}{2}, \frac{21}{2}\right]$

29. $(-2, 3)$

31. $\left(-\infty, -\frac{7}{2}\right] \cup [0, \infty)$

33. $[-3, 6]$

35. $(-\infty, -1] \cup [\frac{1}{2}, \infty)$

37. $(-1, 4)$

39. $(-\infty, -3) \cup (6, \infty)$

41. $(-2, 2)$

43. $(-\infty, \infty)$

45. $(-2, 0) \cup (2, \infty)$

47. $(-\infty, -1) \cup [3, \infty)$

49. $\left(-\infty, -\frac{3}{2}\right)$

51. $(-\infty, 5) \cup [16, \infty)$

53. $(-2, 0) \cup (2, \infty)$

55. $[-2, -1) \cup (0, 1]$

57. $[-2, 0) \cup (1, 3]$

59. $\left(-3, -\frac{1}{2}\right) \cup (2, \infty)$

61. $(-\infty, -1) \cup (1, \infty)$

63. $[-4, 4]$

65. $\left(-\infty, -\frac{7}{2}\right) \cup \left(\frac{7}{2}, \infty\right)$

67. $[2, 8]$

69. $[1.3, 1.7]$

71. $(-4, 8)$

73. $(-6.001, -5.999)$

75. $\left[-\frac{1}{2}, \frac{3}{2}\right]$

77. $|x| < 3$

79. $|x - 7| \geq 5$ **81.** $|x| \leq 2$ **83.** $|x| > 3$
85. $|x - 1| \leq 3$ **87.** $-\frac{4}{3} \leq x \leq \frac{4}{3}$ **89.** $x < -2$ or $x > 7$

91. (a) $x \geq \dfrac{c}{a} + \dfrac{c}{b}$ **(b)** $\dfrac{a - c}{b} \leq x < \dfrac{2a - c}{b}$

93. $68 \leq F \leq 86$ **95.** More than 200 mi
97. Between 12,000 mi and 14,000 mi   **99.** Distances between 20,000 km and 100,000 km   **101.** Between 0 and 60 mi/h

103. (a) $T = 20 - \dfrac{h}{100}$ **(b)** From 20°C down to −30°C

105. 24   **107. (a)** $|x - 0.020| \leq 0.003$
(b) $0.017 \leq x \leq 0.023$

Section 1.8 ■ page 97

1.

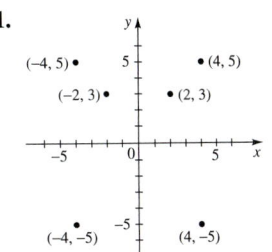

3. (a) $\sqrt{13}$ **(b)** $\left(\frac{3}{2}, 1\right)$ **5. (a)** 10   **(b)** $(1, 0)$
7. (a)   **9. (a)**

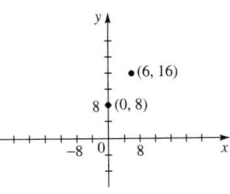
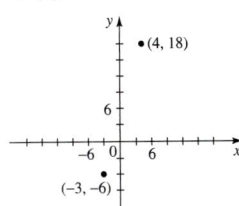

(b) 10   **(c)** $(3, 12)$ **(b)** 25   **(c)** $\left(\frac{1}{2}, 6\right)$
11. (a)   **13.** 24

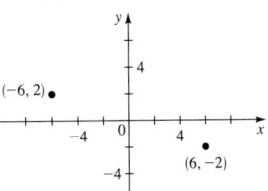
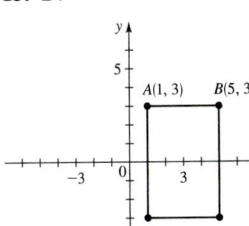

(b) $4\sqrt{10}$ **(c)** $(0, 0)$

15. Trapezoid, area = 9   **17.**

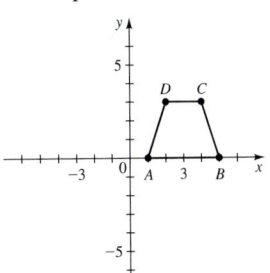
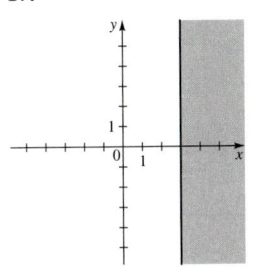

19.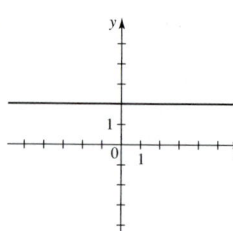

21.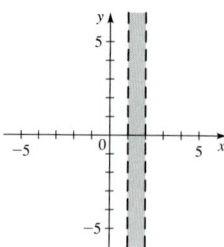

23.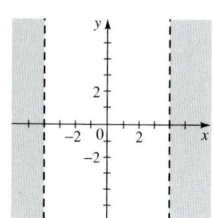

25.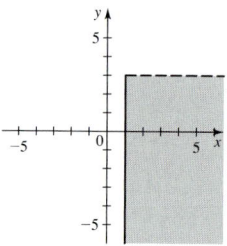

27. $A(6, 7)$ **29.** $Q(-1, 3)$
33. (b) 10   **37.** $(0, -4)$
39. $(2, -3)$

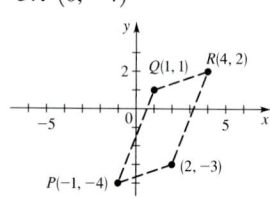

41. (a)   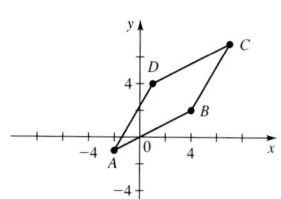   **(b)** $\left(\frac{5}{2}, 3\right), \left(\frac{5}{2}, 3\right)$

43. No, yes, yes   **45.** Yes, no, yes
47. x-intercepts 0, 4; y-intercept 0
49. x-intercepts -2, 2; y-intercepts -4, 4
51. x-intercept 4,
   y-intercept 4,
   no symmetry

53. x-intercept 3,
   y-intercept -6,
   no symmetry

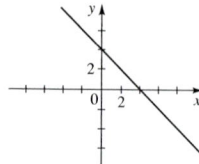   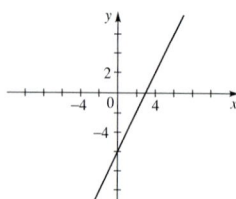

55. x-intercepts ± 1,
   y-intercept 1,
   symmetry about y-axis

57. x-intercept 0,
   y-intercept 0,
   symmetry about y-axis

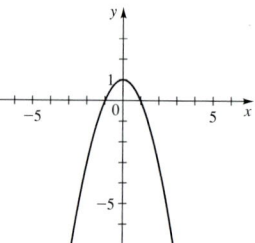

59. x-intercepts ± 3,
   y-intercept -9,
   symmetry about y-axis

61. No intercepts,
   symmetry about origin

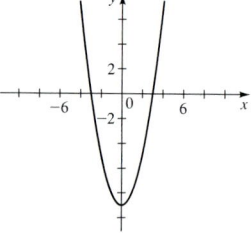

63. x-intercepts ± 2,
   y-intercept 2,
   symmetry about y-axis

65. x-intercept 4,
   y-intercepts -2, 2,
   symmetry about x-axis

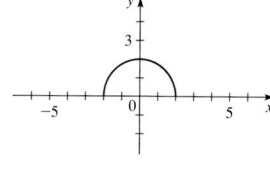

67. x-intercepts ± 2,
   y-intercept 16,
   symmetry about y-axis

69. x-intercepts ± 4,
   y-intercept 4,
   symmetry about y-axis

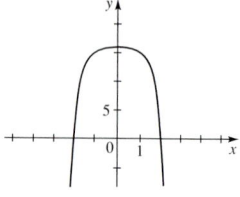

71. Symmetry about y-axis   **73.** Symmetry about origin
y-axis, and origin   **75.** Symmetry about origin

77.

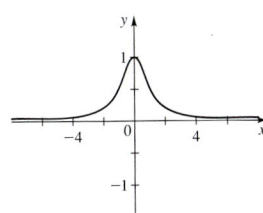

79.

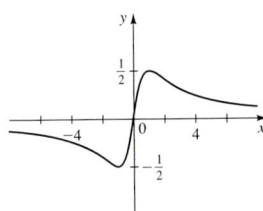

81. $(x - 2)^2 + (y + 1)^2 = 9$ **83.** $x^2 + y^2 = 65$
85. $(x - 7)^2 + (y + 3)^2 = 9$ **87.** $(x + 2)^2 + (y - 2)^2 = 4$
89. $(2, -5), 4$ **91.** $(\frac{1}{4}, -\frac{1}{4}), \frac{1}{2}$ **93.** $(\frac{3}{4}, 0), \frac{3}{4}$
95.

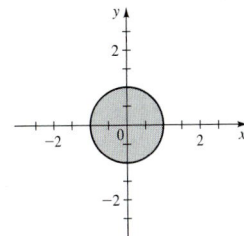

97. 12π

99. (a) 5 **(b)** 31; 25 **(c)** Points P and Q must either be on the same street or the same avenue. **101. (a)** 2 Mm, 8 Mm
(b) $-1.33, 7.33$; 2.40 Mm, 7.60 Mm

Section 1.9 ■ page 109

1. (c) **3.** (c) **5.** (c)
7.

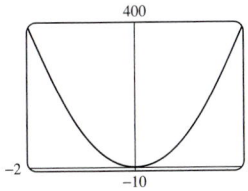

9.

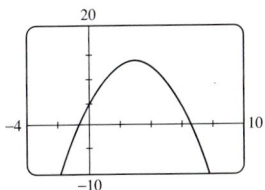

11.

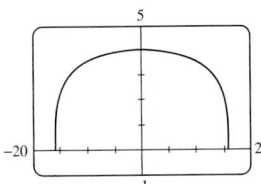

13.

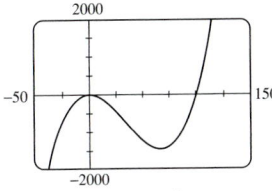

15.

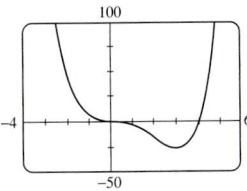

17.

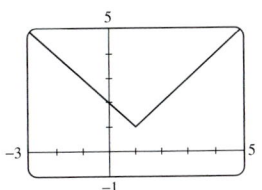

19.

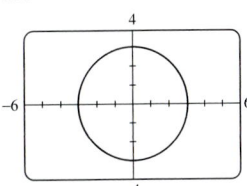

21.

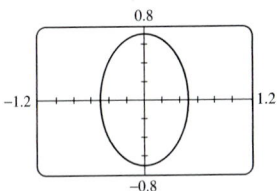

23. No **25.** Yes, 2 **27.** -4 **29.** $\frac{5}{14}$ **31.** $\pm 4\sqrt{2} \approx \pm 5.7$
33. $2.5, -2.5$ **35.** $5 + 2\sqrt[4]{5} \approx 7.99, 5 - 2\sqrt[4]{5} \approx 2.01$
37. 3.00, 4.00 **39.** 1.00, 2.00, 3.00 **41.** 1.62
43. $-1.00, 0.00, 1.00$ **45.** 2.55 **47.** $-2.05, 0, 1.05$
49. $[-2.00, 5.00]$ **51.** $(-\infty, 1.00] \cup [2.00, 3.00]$
53. $(-1.00, 0) \cup (1.00, \infty)$ **55.** $(-\infty, 0)$ **57.** 0, 0.01
59. (a) **(b)** 67 mi

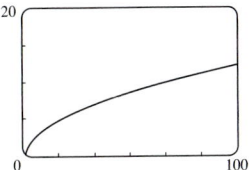

Section 1.10 ■ page 120

1. $\frac{1}{2}$ **3.** $\frac{1}{6}$ **5.** $-\frac{1}{2}$ **7.** $-\frac{9}{2}$ **9.** $-2, \frac{1}{2}, 3, -\frac{1}{4}$
11. $x + y - 4 = 0$ **13.** $3x - 2y - 6 = 0$
15. $x - y + 1 = 0$ **17.** $2x - 3y + 19 = 0$
19. $5x + y - 11 = 0$ **21.** $3x - y - 2 = 0$
23. $3x - y - 3 = 0$ **25.** $y = 5$ **27.** $x + 2y + 11 = 0$
29. $x = -1$ **31.** $5x - 2y + 1 = 0$
33. $x - y + 6 = 0$
35. (a) **(b)** $2x + y - 7 = 0$

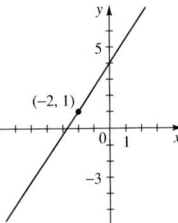

37. They all have the same slope.

39. They all have the same x-intercept.

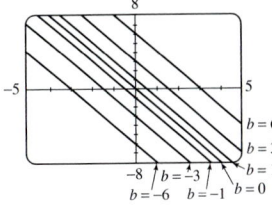

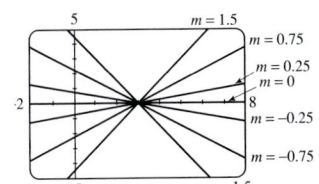

41. $-1, 3$

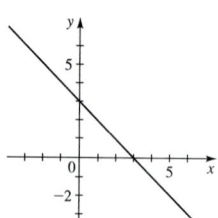

43. $-\frac{1}{3}, 0$

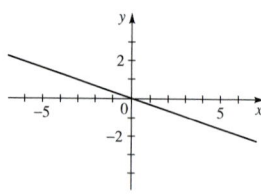

45. $\frac{3}{2}, 3$

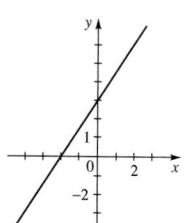

47. $0, 4$

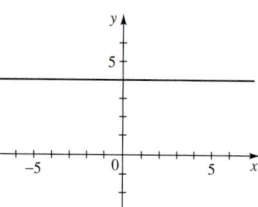

49. $\frac{3}{4}, -3$

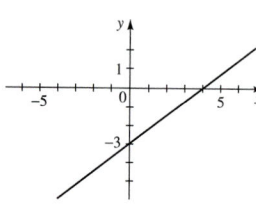

51. $-\frac{3}{4}, \frac{1}{4}$

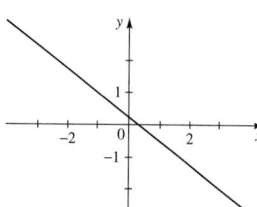

57. $x - y - 3 = 0$ **59. (b)** $4x - 3y - 24 = 0$
61. 16,667 ft   **63. (a)** 8.34; the slope represents the increase
in dosage for a one-year increase in age.   **(b)** 8.34 mg

65. (a)

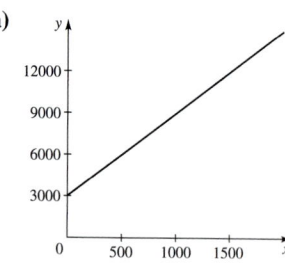

(b) The slope
represents production
cost per toaster; the
y-intercept represents
monthly fixed cost.

67. (a) $t = \frac{5}{24}n + 45$ **(b)** 76°F
69. (a) $P = 0.434d + 15$, where P is pressure in lb/in^2 and d is
depth in feet

(b)

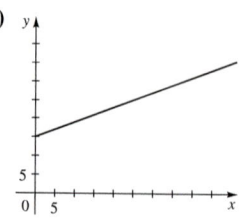

(c) The slope is the rate
of increase in water pressure, and the y-intercept
is the air pressure at the
surface.   **(d)** 196 ft

71. (a) $C = \frac{1}{4}d + 260$
(b) $635
(c) The slope represents
cost per mile.
(d) The y-intercept represents
monthly fixed cost.

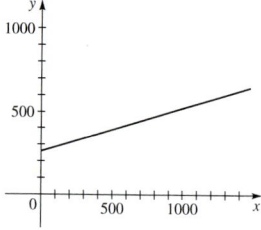

Section 1.11 ■ **page 127**
1. $T = kx$ **3.** $v = k/z$ **5.** $y = ks/t$ **7.** $z = k\sqrt{y}$
9. $V = klwh$ **11.** $R = k\dfrac{i}{Pt}$ **13.** $y = 7x$ **15.** $M = 15x/y$
17. $W = 360/r^2$ **19.** $C = 16lwh$ **21.** $s = 500/\sqrt{t}$
23. (a) $F = kx$ **(b)** 8   **(c)** 32 N
25. (a) $C = kpm$ **(b)** 0.125   **(c)** $57,500   **27. (a)** $P = ks^3$
(b) 0.012   **(c)** 324   **29.** 0.7 dB   **31.** 4   **33.** 5.3 mi/h
35. (a) $R = kL/d^2$ **(b)** $0.002916\overline{6}$ **(c)** $R \approx 137\ \Omega$
37. (a) 160,000   **(b)** 1,930,670,340   **39.** 36 lb
41. (a) $f = k/L$ **(b)** Halves it

Chapter 1 Review ■ **page 131**
1. Commutative Property for addition
3. Distributive Property
5. $-2 \le x < 6$
7. $[5, \infty)$
9. 6   **11.** $\frac{1}{72}$ **13.** $\frac{1}{6}$ **15.** 11   **17.** 4   **19.** $16x^3$ **21.** $12xy^8$
23. x^2y^2 **25.** $3x^{3/2}y^2$ **27.** $\dfrac{4r^{5/2}}{s^7}$ **29.** 7.825×10^{10}
31. 1.65×10^{-32} **33.** $3xy^2(4xy^2 - y^3 + 3x^2)$
35. $(x - 2)(x + 5)$ **37.** $(4t + 3)(t - 4)$
39. $(5 - 4t)(5 + 4t)$
41. $(x - 1)(x^2 + x + 1)(x + 1)(x^2 - x + 1)$
43. $x^{-1/2}(x - 1)^2$ **45.** $(x - 2)(4x^2 + 3)$
47. $\sqrt{x^2 + 2}(x^2 + x + 2)^2$ **49.** $6x^2 - 21x + 3$
51. $-7 + x$ **53.** $2x^3 - 6x^2 + 4x$
55. $\dfrac{3(x + 3)}{x + 4}$ **57.** $\dfrac{x + 1}{x - 4}$ **59.** $\dfrac{1}{x + 1}$
61. $-\dfrac{1}{2x}$ **63.** $3\sqrt{2} - 2\sqrt{3}$ **65.** 5   **67.** No solution
69. 2, 7   **71.** $-1, \frac{1}{2}$ **73.** $0, \pm\frac{5}{2}$ **75.** $\dfrac{-2 \pm \sqrt{7}}{3}$
77. -5 **79.** 3, 11   **81.** 20 lb raisins, 30 lb nuts
83. $\frac{1}{4}(\sqrt{329} - 3) \approx 3.78$ mi/h   **85.** 1 h 50 min
87. $(-3, \infty)$ **89.** $(-\infty, -6) \cup (2, \infty)$
91. $(-\infty, -2) \cup (2, 4]$ **93.** $[2, 8]$

95. $-1, 7$

97. $[1, 3]$

99. (a)

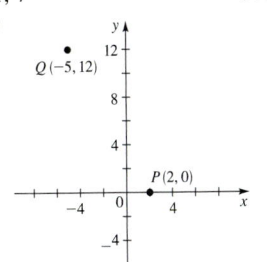

(b) $\sqrt{193}$

(c) $\left(-\frac{3}{2}, 6\right)$

(d) $y = -\frac{12}{7}x + \frac{24}{7}$

(e) $(x - 2)^2 + y^2 = 193$

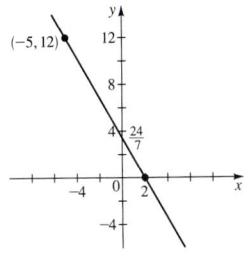

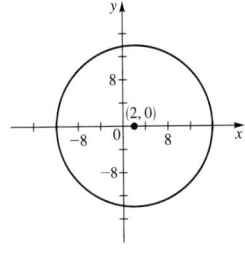

101.

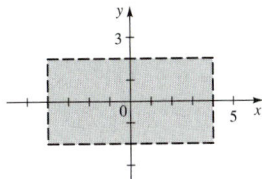

103. B **105.** $(x + 5)^2 + (y + 1)^2 = 26$

107. Circle, center $(-1, 3)$, radius 1   **109.** No graph

111. No symmetry

113. No symmetry

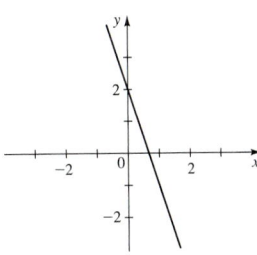

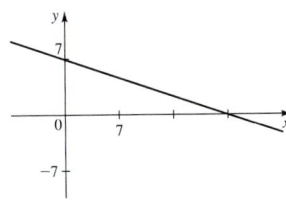

115. Symmetry about y-axis

117. No symmetry

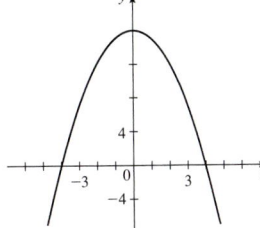

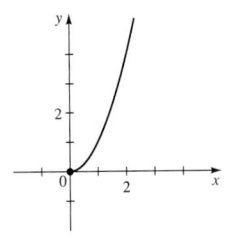

119.

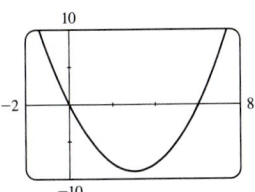

121.

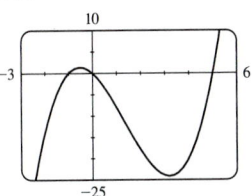

123. $2x - 3y - 16 = 0$ **125.** $3x + y - 12 = 0$

127. $x + 5y = 0$ **129.** $x^2 + y^2 = 169, 5x - 12y + 169 = 0$

131. (a) The slope represents the amount the spring lengthens for a one-pound increase in weight. The S-intercept represents the unstretched length of the spring.   **(b)** 4 in.

133. $M = 8z$ **135. (a)** $I = k/d^2$ **(b)** 64,000

(c) 160 candles   **137.** 11.0 mi/h

Chapter 1 Test ■ page 135

1. (a)

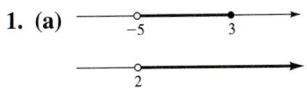

(b) $(-\infty, 3], [-1, 4)$ **(c)** 16   **2. (a)** 81   **(b)** -81 **(c)** $\frac{1}{81}$

(d) 25   **(e)** $\frac{9}{4}$ **(f)** $\frac{1}{8}$ **3. (a)** 1.86×10^{11} **(b)** 3.965×10^{-7}

4. (a) $6\sqrt{2}$ **(b)** $48a^5b^7$ **(c)** $\dfrac{x}{9y^7}$ **(d)** $\dfrac{x + 2}{x - 2}$ **(e)** $\dfrac{1}{x - 2}$

(f) $-(x + y)$ **5.** $5\sqrt{2} + 2\sqrt{10}$ **6. (a)** $11x - 2$

(b) $4x^2 + 7x - 15$ **(c)** $a - b$ **(d)** $4x^2 + 12x + 9$

(e) $x^3 + 6x^2 + 12x + 8$ **7. (a)** $(2x - 5)(2x + 5)$

(b) $(2x - 3)(x + 4)$ **(c)** $(x - 3)(x - 2)(x + 2)$

(d) $x(x + 3)(x^2 - 3x + 9)$ **(e)** $3x^{-1/2}(x - 1)(x - 2)$

(f) $xy(x - 2)(x + 2)$ **8. (a)** 6   **(b)** 1   **(c)** $-3, 4$

(d) $-1 \pm \dfrac{\sqrt{2}}{2}$ **(e)** No real solution   **(f)** $\pm 1, \pm\sqrt{2}$

(g) $\frac{2}{3}, \frac{22}{3}$ **9.** 120 mi   **10.** 50 ft by 120 ft

11. (a) $[-4, 3)$

(b) $(-2, 0) \cup (1, \infty)$

(c) $(1, 7)$

(d) $(-1, 4]$

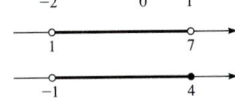

12. Between $41°F$ and $50°F$ **13.** $0 \le x \le 6$

14. (a) $-2.94, -0.11, 3.05$ **(b)** $[-1, 3]$

15. (a) $S(3, 6)$ **(b)** 18

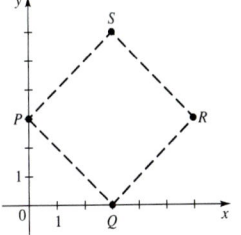

16. (a)

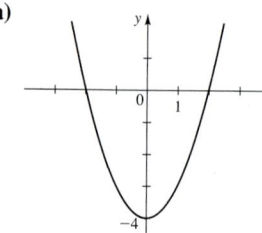

(b) x-intercepts $-2, 2$
y-intercept -4
(c) Symmetric about y-axis

17. (a)

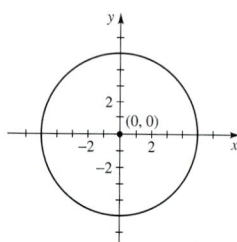

(b) $\sqrt{89}$ **(c)** $\left(1, \frac{7}{2}\right)$ **(d)** $\frac{5}{8}$ **(e)** $y = -\frac{8}{5}x + \frac{51}{10}$
(f) $(x - 1)^2 + \left(y - \frac{7}{2}\right)^2 = \frac{89}{4}$

18. (a) $(0,0), 5$

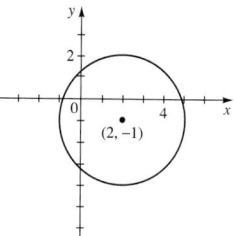

(b) $(2, -1), 3$

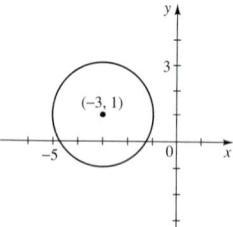

(c) $(-3, 1), 2$

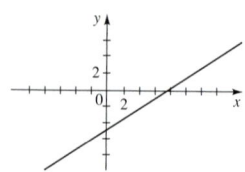

19. $y = \frac{2}{3}x - 5$

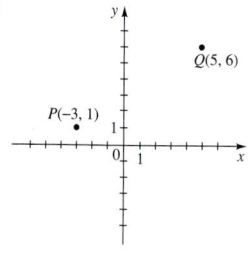

slope $\frac{2}{3}$; y-intercept -5

20. (a) $3x + y - 3 = 0$
(b) $2x + 3y - 12 = 0$
21. (a) $4°C$ **(b)**

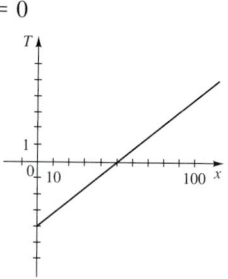

(c) The slope is the rate of change in temperature, the x-intercept is the depth at which the temperature is $0°C$, and the T-intercept is the temperature at ground level.
22. (a) $M = kwh^2/L$ **(b)** 400 **(c)** $12,000$ lb

Focus on Problem Solving ■ page 141

1. 37.5 mi/h **3.** 150 mi **5.** $427, 3n + 1$
7. 75 s **9.** The same amount
11. 2π **13.** 8.49 **15.** 7
19. The North Pole is one such point. There are infinitely many others near the South Pole.
21. π **23.** $1^3 + 12^3 = 9^3 + 10^3 = 1729$
27. Infinitely far
29.

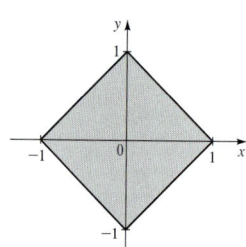

Chapter 2

Section 2.1 ■ page 155

1. $f(x) = 2(x + 3)$
3. $f(x) = (x - 5)^2$
5. Subtract 4, then divide by 3
7. Square, then add 2
9.

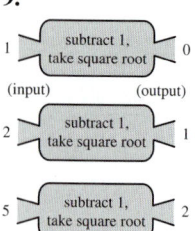

11.

x	f(x)
−1	8
0	2
1	0
2	2
3	8

13. 3, −3, 2, 2a + 1, −2a + 1, 2a + 2b + 1

15. $-\dfrac{1}{3}, -3, \dfrac{1}{3}, \dfrac{1-a}{1+a}, \dfrac{2-a}{a}$, undefined

17. −4, 10, −2, $3\sqrt{2}$, $2x^2 + 7x + 1$, $2x^2 − 3x − 4$

19. 6, 2, 1, 2, 2|x|, $2(x^2 + 1)$ **21.** 4, 1, 1, 2, 3

23. 8, $-\frac{3}{4}$, −1, 0, −1   **25.** $x^2 + 4x + 5$, $x^2 + 6$

27. $x^2 + 4$, $x^2 + 8x + 16$ **29.** 3a + 2, 3(a + h) + 2, 3

31. 5, 5, 0   **33.** $\dfrac{a}{a+1}, \dfrac{a+h}{a+h+1}, \dfrac{1}{(a+h+1)(a+1)}$

35. $3 − 5a + 4a^2$, $3 − 5a − 5h + 4a^2 + 8ah + 4h^2$,
$−5 + 8a + 4h$

37. $(-\infty, \infty)$ **39.** $[-1, 5]$ **41.** $\{x \mid x \neq 3\}$

43. $\{x \mid x \neq \pm 1\}$ **45.** $[5, \infty)$ **47.** $(-\infty, \infty)$ **49.** $\left[\frac{5}{2}, \infty\right)$

51. $[-2, 3) \cup (3, \infty)$ **53.** $(-\infty, 0] \cup [6, \infty)$ **55.** $(4, \infty)$

57. $\left(\frac{1}{2}, \infty\right)$ **59. (a)** $C(10) = 1532.1$, $C(100) = 2100$

(b) The cost of producing 10 yd and 100 yd

(c) $C(0) = 1500$

61. (a) $D(0.1) = 28.1$, $D(0.2) = 39.8$ **(b)** 41.3 mi

(c) 235.6 mi   **63. (a)** $v(0.1) = 4440$, $v(0.4) = 1665$

(b) Flow is faster near central axis.

(c)

r	v(r)
0	4625
0.1	4440
0.2	3885
0.3	2960
0.5	0

65. (a) 8.66 m, 6.61 m, 4.36 m

(b) It will appear to get shorter.

67. (a) $90, $105, $100, $105   **(b)** Total cost of an order, including shipping

69. (a) $F(x) = \begin{cases} 15(40 − x) & \text{if } 0 < x < 40 \\ 0 & \text{if } 40 \leq x \leq 65 \\ 15(x − 65) & \text{if } x > 65 \end{cases}$

(b) $150, $0, $150   **(c)** Fines for violating the speed limits

71.

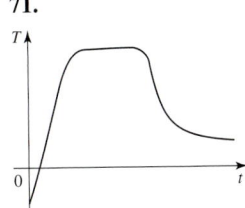

73.

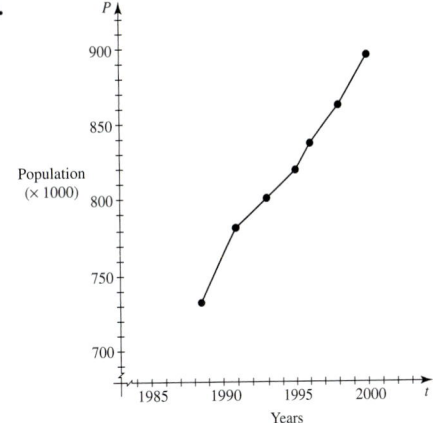

Section 2.2 ■ page 167

1.

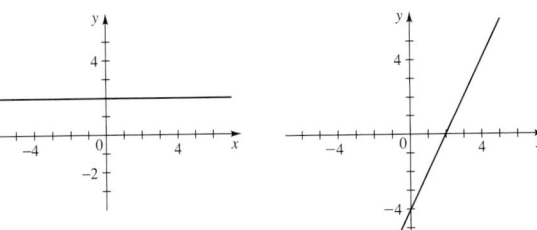

3.

5.

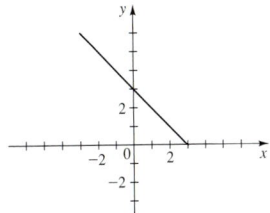

7.

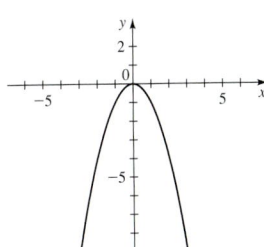

9.

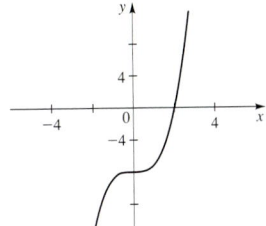

11.

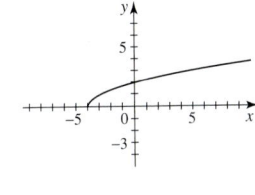

13.

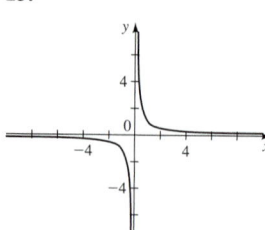

15.

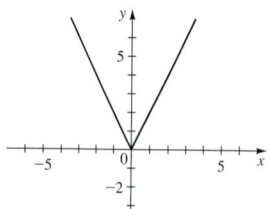

17.

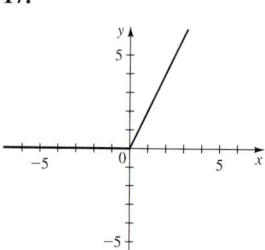

19.

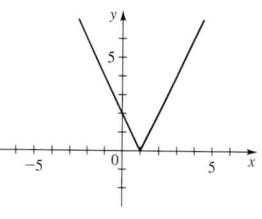

21.

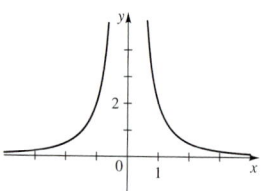

23. (a) $1, -1, 3, 4$ **(b)** Domain $[-3, 4]$, range $[-1, 4]$

25. (a) $f(0)$ **(b)** $g(-3)$ **(c)** $-2, 2$

27. (a)

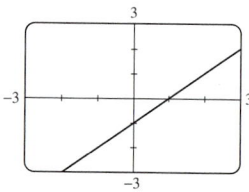

(b) Domain $(-\infty, \infty)$, range $(-\infty, \infty)$

29. (a)

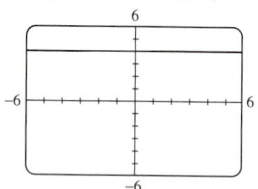

(b) Domain $(-\infty, \infty)$, range $\{4\}$

31. (a)

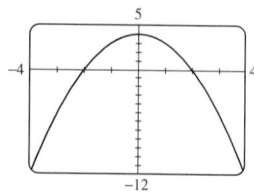

(b) Domain $(-\infty, \infty)$, range $(-\infty, 4]$

33. (a)

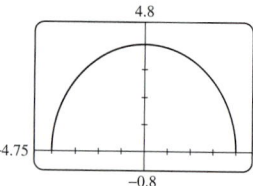

(b) Domain $[-4, 4]$, range $[0, 4]$

35. (a)

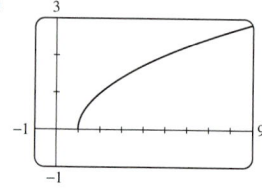

(b) Domain $[1, \infty)$, range $[0, \infty)$

37.

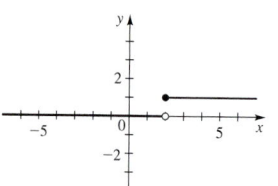

39.

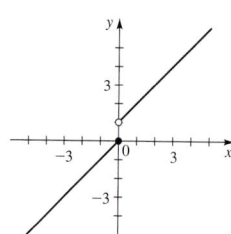

41.

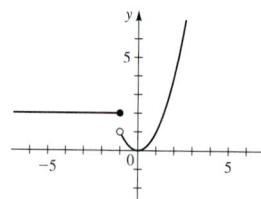

43.

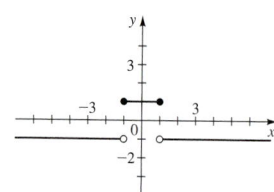

45.

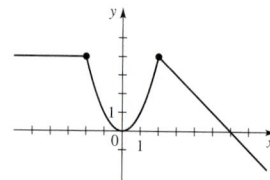

47.

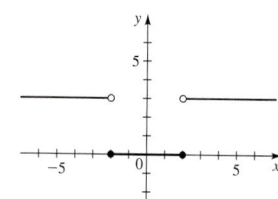

49.

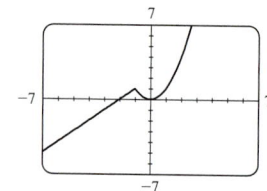

51.

53. $f(x) = \begin{cases} -2 & \text{if } x < -2 \\ x & \text{if } -2 \leq x \leq 2 \\ 2 & \text{if } x > 2 \end{cases}$

55. (a) Yes **(b)** No **(c)** Yes **(d)** No
57. Function, domain $[-3, 2]$, range $[-2, 2]$
59. Not a function **61.** Yes **63.** No **65.** No
67. Yes **69.** Yes **71.** Yes
73. (a)   **(b)**

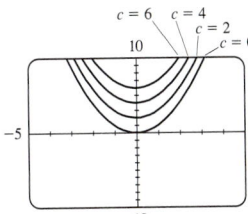
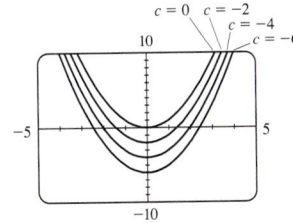

(c) If $c > 0$, then the graph of $f(x) = x^2 + c$ is the same as the graph of $y = x^2$ shifted upward c units. If $c < 0$, then the graph of $f(x) = x^2 + c$ is the same as the graph of $y = x^2$ shifted downward c units.

75. (a)   **(b)**

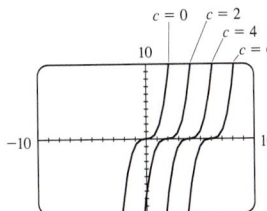
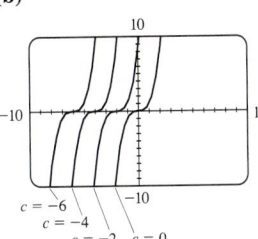

(c) If $c > 0$, then the graph of $f(x) = (x - c)^3$ is the same as the graph of $y = x^3$ shifted right c units. If $c < 0$, then the graph of $f(x) = (x - c)^3$ is the same as the graph of $y = x^3$ shifted left c units.

77. (a)   **(b)**

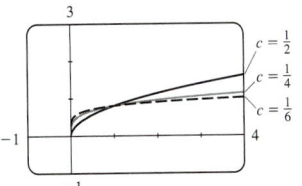
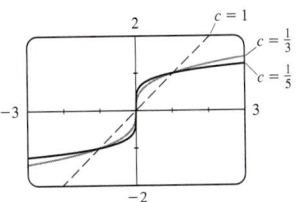

(c) Graphs of even roots are similar to $\sqrt{x}$; graphs of odd roots are similar to $\sqrt[3]{x}$. As c increases, the graph of $y = \sqrt[c]{x}$ becomes steeper near 0 and flatter when $x > 1$.
79. $f(x) = -\frac{7}{6}x - \frac{4}{3}, -2 \leq x \leq 4$
81. $f(x) = \sqrt{9 - x^2}, -3 \leq x \leq 3$
83. This person's weight increases as he grows, then continues to increase; the person then goes on a crash diet (possibly) at age 30, then gains weight again, the weight gain eventually leveling off.

85. A won the race. All runners finished. Runner B fell, but got up again to finish second. **87. (a)** 5 s **(b)** 30 s **(c)** 17 s
89.

$C(x) = \begin{cases} 2 & 0 < x \leq 1 \\ 2.2 & 1 < x \leq 1.1 \\ 2.4 & 1.1 < x \leq 1.2 \\ \vdots \\ 4.0 & 1.9 < x < 2.0 \end{cases}$

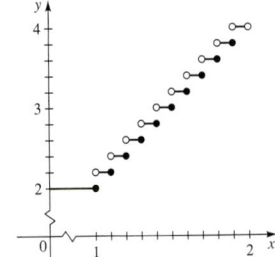

Section 2.3 ■ page 179

1. (a) $[-1, 1], [2, 4]$ **(b)** $[1, 2]$ **3. (a)** $[-2, -1], [1, 2]$
(b) $[-3, -2], [-1, 1], [2, 3]$
5. (a)   **7. (a)**

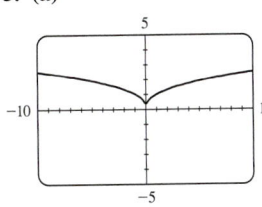
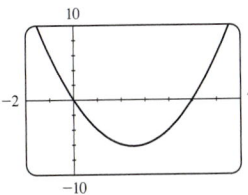

(b) Increasing on $[0, \infty)$;   **(b)** Increasing on $[2.5, \infty)$;
decreasing on $(-\infty, 0]$ decreasing on $(-\infty, 2.5]$
9. (a)   **11. (a)**

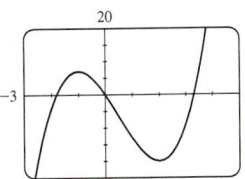
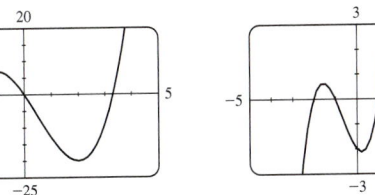

(b) Increasing on $(-\infty, -1]$,   **(b)** Increasing on
$[2, \infty)$; decreasing on $[-1, 2]$ $(-\infty, -1.55], [0.22, \infty)$;
                                          decreasing on $[-1.55, 0.22]$
13. $\frac{2}{3}$ **15.** $-\frac{4}{5}$ **17.** 3 **19.** 5 **21.** 60 **23.** $12 + 3h$
25. $-\frac{1}{a}$ **27.** $\frac{-2}{a(a + h)}$ **29. (a)** $\frac{1}{2}$
31. (a) Increasing on $[0, 150], [300, 365]$; decreasing on $[150, 300]$ **(b)** -0.25 ft/day **33. (a)** 245 persons/yr
(b) -328.5 persons/yr **(c)** 1997–2001 **(d)** 2001–2006
35. (a) 7.2 units/yr **(b)** 8 units/yr **(c)** -55 units/yr
(d) 2000–2001, 2001–2002

Section 2.4 ■ page 190

1. (a) Shift downward 5 units **(b)** Shift right 5 units
3. (a) Shift left $\frac{1}{2}$ unit **(b)** Shift up $\frac{1}{2}$ unit **5. (a)** Reflect in the x-axis and stretch vertically by a factor of 2 **(b)** Reflect in the x-axis and shrink vertically by a factor of $\frac{1}{2}$ **7. (a)** Shift right 4 units and upward $\frac{3}{4}$ unit **(b)** Shift left 4 units and downward $\frac{3}{4}$ unit **9. (a)** Shrink horizontally by a factor of $\frac{1}{4}$

(b) Stretch horizontally by a factor of 4   **11.** $g(x) = (x - 2)^2$
13. $g(x) = |x + 1| + 2$ **15.** $g(x) = -\sqrt{x + 2}$
17. (a) 3   **(b)** 1   **(c)** 2   **(d)** 4
19. (a)

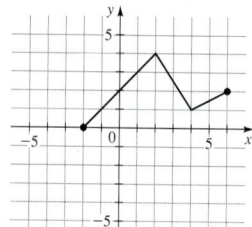

(b)

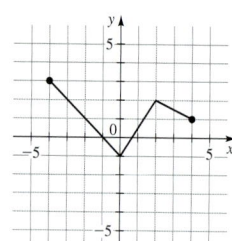

(c)

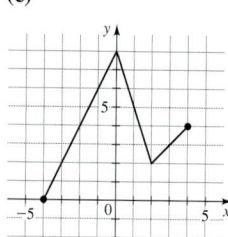

(d)

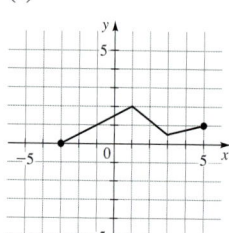

(e)

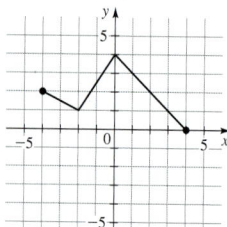

(f)

21. (a)

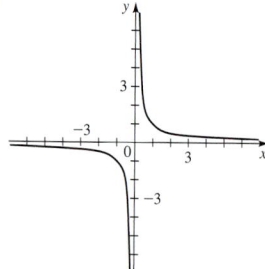

(b) (i)

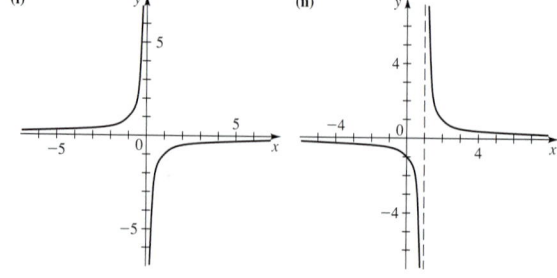

(ii)

(iii)

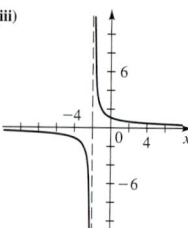

(iv)

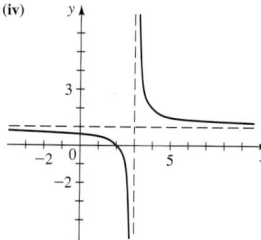

23. (a) Shift left 2 units   **(b)** Shift up 2 units
25. (a) Stretch vertically by a factor of 2
(b) Shift right 2 units, then shrink vertically by a factor of $\frac{1}{2}$
27. $g(x) = (x - 2)^2 + 3$ **29.** $g(x) = -5\sqrt{x + 3}$
31. $g(x) = 0.1|x - \frac{1}{2}| - 2$

33.

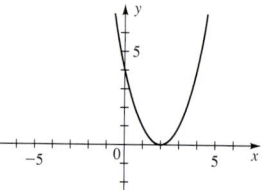

35.

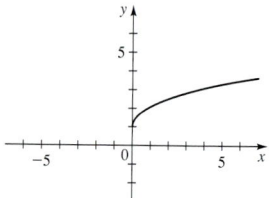

37.

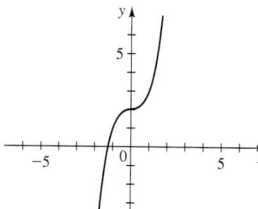

39.

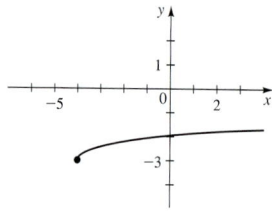

41.

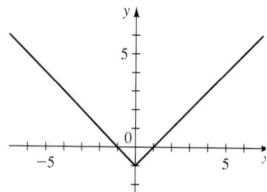

43.

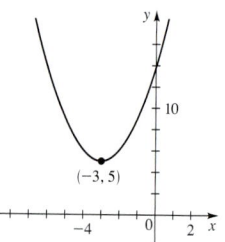

(−3, 5)

45.

47.

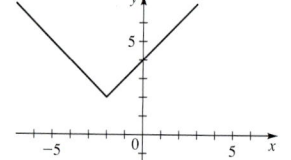

49.

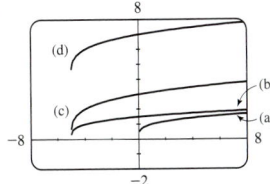

For part (b) shift the graph in (a) left 5 units; for part (c) shift the graph in (a) left 5 units and stretch vertically by a factor of 2; for part (d) shift the graph in (a) left 5 units, stretch vertically by a factor of 2, and then shift upward 4 units.

51.

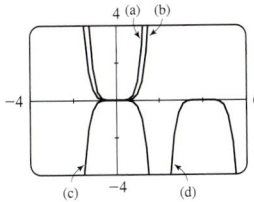

For part (b) shrink the graph in (a) vertically by a factor of $\frac{1}{3}$; for part (c) shrink the graph in (a) vertically by a factor of $\frac{1}{3}$ and reflect in the x-axis; for part (d) shift the graph in (a) right 4 units, shrink vertically by a factor of $\frac{1}{3}$, and then reflect in the x-axis.

53. (a)

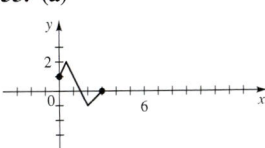

(b)

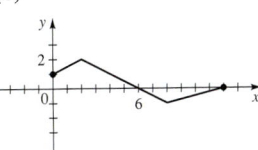

55. (a)

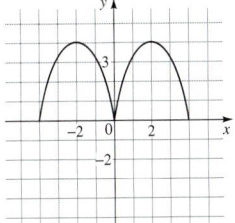

(b)

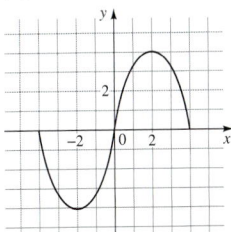

57.

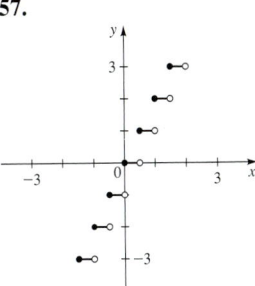

59.

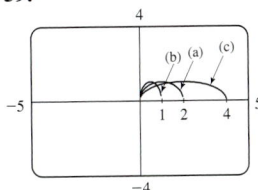

61. Even

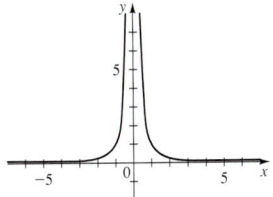

63. Neither

65. Odd

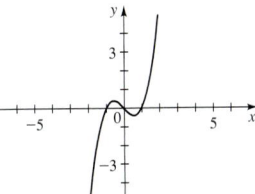

67. Neither
69. To obtain the graph of g, reflect in the x-axis the part of the graph of f that is below the x-axis.
71. (a)   **(b)**

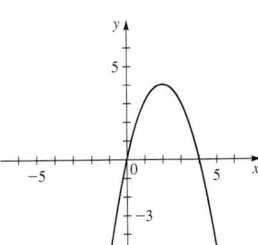   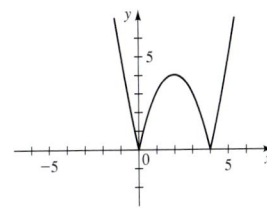

73. (a) Shift up 4 units, shrink vertically by a factor of 0.01
(b) Shift right 10 units; $g(t) = 4 + 0.01(t - 10)^2$

Section 2.5 ■ page 200

1. (a) $(3, 4)$ **(b)** 4 **3. (a)** $(1, -3)$ **(b)** -3
5. (a) $f(x) = (x - 3)^2 - 9$ **7. (a)** $f(x) = 2\left(x + \frac{3}{2}\right)^2 - \frac{9}{2}$
(b) Vertex $(3, -9)$ **(b)** Vertex $\left(-\frac{3}{2}, -\frac{9}{2}\right)$
x-intercepts 0, 6 x-intercepts 0, -3,
y-intercept 0 y-intercept 0
(c)   **(c)**

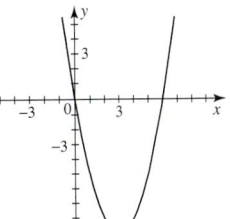

9. (a) $f(x) = (x + 2)^2 - 1$ **(b)** Vertex $(-2, -1)$
x-intercepts $-1, -3$, y-intercept 3
(c)

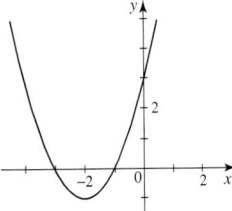

11. (a) $f(x) = -(x - 3)^2 + 13$ **(b)** Vertex $(3, 13)$;
x-intercepts $3 \pm \sqrt{13}$; y-intercept 4
(c)

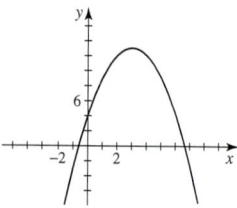

13. (a) $f(x) = 2(x + 1)^2 + 1$ **(b)** Vertex $(-1, 1)$;
no x-intercept; y-intercept 3
(c)

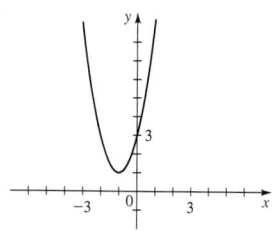

15. (a) $f(x) = 2(x - 5)^2 + 7$ **(b)** Vertex $(5, 7)$;
no x-intercept; y-intercept 57
(c)

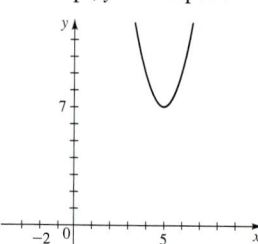

17. (a) $f(x) = -4(x + 2)^2 + 19$ **(b)** Vertex $(-2, 19)$;
x-intercepts $-2 \pm \frac{1}{2}\sqrt{19}$; y-intercept 3
(c)

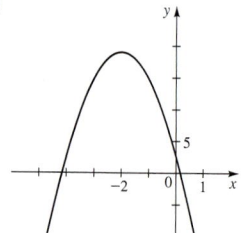

19. (a) $f(x) = -(x - 1)^2 + 1$ **21. (a)** $f(x) = (x + 1)^2 - 2$
(b) **(b)**

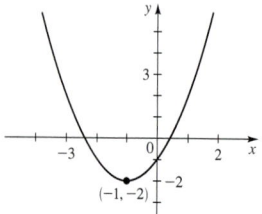

(c) Maximum $f(1) = 1$ **(c)** Minimum $f(-1) = -2$

23. (a) $f(x) = -(x + \frac{3}{2})^2 + \frac{21}{4}$
(b)

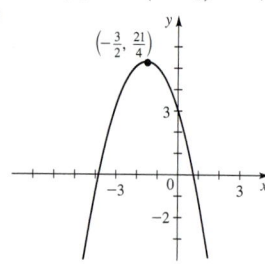

(c) Maximum $f(-\frac{3}{2}) = \frac{21}{4}$
25. (a) $g(x) = 3(x - 2)^2 + 1$
(b)

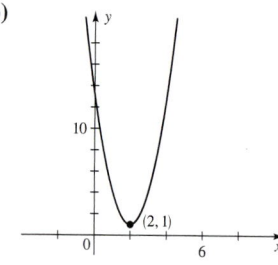

(c) Minimum $g(2) = 1$
27. (a) $h(x) = -(x + \frac{1}{2})^2 + \frac{5}{4}$
(b)

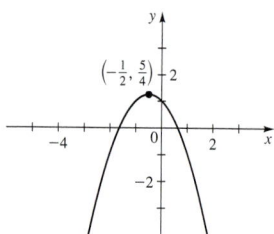

(c) Maximum $h(-\frac{1}{2}) = \frac{5}{4}$
29. Minimum $f(-\frac{1}{2}) = \frac{3}{4}$
31. Maximum $f(-3.5) = 185.75$
33. Minimum $f(0.6) = 15.64$
35. Minimum $h(-2) = -8$
37. Maximum $f(-1) = \frac{7}{2}$ **39.** $f(x) = 2x^2 - 4x$
41. $(-\infty, \infty), (-\infty, 1]$ **43.** $(-\infty, \infty), (-\frac{23}{2}, \infty)$
45. (a) -4.01 **(b)** -4.011025
47. Local maximum 2; local minimums $-1, 0$
49. Local maximums 0, 1; local minimums $-2, -1$
51. Local maximum ≈ 0.38 when $x \approx -0.58$;
local minimum ≈ -0.38 when $x \approx 0.58$
53. Local maximum ≈ 0 when $x = 0$;
local minimum ≈ -13.61 when $x \approx -1.71$;
local minimum ≈ -73.32 when $x \approx 3.21$
55. Local maximum ≈ 5.66 when $x \approx 4.00$
57. Local maximum ≈ 0.38 when $x \approx -1.73$;

local minimum ≈ -0.38 when $x \approx 1.73$ **59.** 25 ft
61. \$4,000, 100 units   **63.** 30 times   **65.** 50 trees per acre
67. 20 mi/h   **69.** $r \approx 0.67$ cm

Section 2.6 ■ page 210

1. $A(w) = 3w^2, w > 0$ **3.** $V(w) = \frac{1}{2}w^3, w > 0$
5. $A(x) = 10x - x^2, 0 < x < 10$
7. $A(x) = (\sqrt{3}/4)x^2, x > 0$ **9.** $r(A) = \sqrt{A/\pi}, A > 0$
11. $S(x) = 2x^2 + 240/x, x > 0$ **13.** $D(t) = 25t, t \geq 0$
15. $A(b) = b\sqrt{4 - b}, 0 < b < 4$
17. $A(h) = 2h\sqrt{100 - h^2}, 0 < h < 10$
19. (b) $p(x) = x(19 - x)$ **(c)** 9.5, 9.5   **21.** $-12, -12$
23. (b) $A(x) = x(2400 - 2x)$ **(c)** 600 ft by 1200 ft
25. (a) $f(w) = 8w + 7200/w$ **(b)** Width along road is 30 ft,
length is 40 ft   **(c)** 15 ft to 60 ft
27. (a) $R(p) = -3000p^2 + 57{,}000p$ **(b)** \$19   **(c)** \$9.50
29. (a) $A(x) = 15x - \left(\dfrac{\pi + 4}{8}\right)x^2$ **(b)** Width ≈ 8.40 ft,

height of rectangular part ≈ 4.20 ft
31. (a) $A(x) = x^2 + 48/x$ **(b)** Height ≈ 1.44 ft,
width ≈ 2.88 ft   **33. (a)** $A(x) = 2x + 200/x$
(b) 10 m by 10 m
35. (a) $E(x) = 14\sqrt{25 + x^2} + 10(12 - x)$
(b) To point C, 5.1 mi from point B

Section 2.7 ■ page 219

1. $(f + g)(x) = x^2 + x - 3, (-\infty, \infty)$;
$(f - g)(x) = -x^2 + x - 3, (-\infty, \infty)$;
$(fg)(x) = x^3 - 3x^2, (-\infty, \infty)$;
$\left(\dfrac{f}{g}\right)(x) = \dfrac{x - 3}{x^2}, (-\infty, 0) \cup (0, \infty)$
3. $(f + g)(x) = \sqrt{4 - x^2} + \sqrt{1 + x}, [-1, 2]$;
$(f - g)(x) = \sqrt{4 - x^2} - \sqrt{1 + x}, [-1, 2]$;
$(fg)(x) = \sqrt{-x^3 - x^2 + 4x + 4}, [-1, 2]$;
$\left(\dfrac{f}{g}\right)(x) = \sqrt{\dfrac{4 - x^2}{1 + x}}, (-1, 2]$
5. $(f + g)(x) = \dfrac{6x + 8}{x^2 + 4x}, x \neq -4, x \neq 0$;
$(f - g)(x) = \dfrac{-2x + 8}{x^2 + 4x}, x \neq -4, x \neq 0$;
$(fg)(x) = \dfrac{8}{x^2 + 4x}, x \neq -4, x \neq 0$;
$\left(\dfrac{f}{g}\right)(x) = \dfrac{x + 4}{2x}, x \neq -4, x \neq 0$
7. $[0, 1]$ **9.** $(3, \infty)$

11.
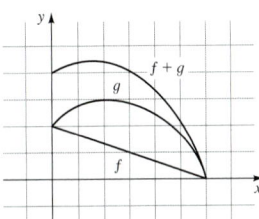

13.
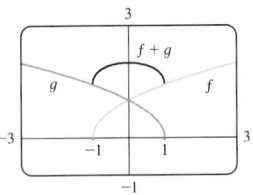

15.
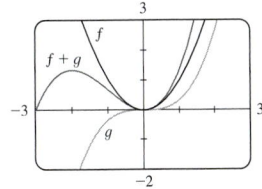

17. (a) 1   **(b)** -23 **19. (a)** -11 **(b)** -119
21. (a) $-3x^2 + 1$ **(b)** $-9x^2 + 30x - 23$
23. 4   **25.** 5   **27.** 4
29. $(f \circ g)(x) = 8x + 1, (-\infty, \infty)$;
$(g \circ f)(x) = 8x + 11, (-\infty, \infty)$;
$(f \circ f)(x) = 4x + 9, (-\infty, \infty)$;
$(g \circ g)(x) = 16x - 5, (-\infty, \infty)$
31. $(f \circ g)(x) = (x + 1)^2, (-\infty, \infty)$;
$(g \circ f)(x) = x^2 + 1, (-\infty, \infty)$; $(f \circ f)(x) = x^4, (-\infty, \infty)$;
$(g \circ g)(x) = x + 2, (-\infty, \infty)$
33. $(f \circ g)(x) = \dfrac{1}{2x + 4}, x \neq -2$; $(g \circ f)(x) = \dfrac{2}{x} + 4, x \neq 0$;
$(f \circ f)(x) = x, x \neq 0, (g \circ g)(x) = 4x + 12, (-\infty, \infty)$
35. $(f \circ g)(x) = |2x + 3|, (-\infty, \infty)$;
$(g \circ f)(x) = 2|x| + 3, (-\infty, \infty)$; $(f \circ f)(x) = |x|, (-\infty, \infty)$;
$(g \circ g)(x) = 4x + 9, (-\infty, \infty)$
37. $(f \circ g)(x) = \dfrac{2x - 1}{2x}, x \neq 0$;
$(g \circ f)(x) = \dfrac{2x}{x + 1} - 1, x \neq -1$;
$(f \circ f)(x) = \dfrac{x}{2x + 1}, x \neq -1, x \neq -\frac{1}{2}$;
$(g \circ g)(x) = 4x - 3, (-\infty, \infty)$
39. $(f \circ g)(x) = \sqrt[12]{x}, [0, \infty)$; $(g \circ f)(x) = \sqrt[12]{x}, [0, \infty)$;
$(f \circ f)(x) = \sqrt[9]{x}, (-\infty, \infty); (g \circ g)(x) = \sqrt[16]{x}, [0, \infty)$
41. $(f \circ g \circ h)(x) = \sqrt{x - 1} - 1$
43. $(f \circ g \circ h)(x) = (\sqrt{x} - 5)^4 + 1$
45. $g(x) = x - 9, f(x) = x^5$
47. $g(x) = x^2, f(x) = x/(x + 4)$
49. $g(x) = 1 - x^3, f(x) = |x|$
51. $h(x) = x^2, g(x) = x + 1, f(x) = 1/x$
53. $h(x) = \sqrt[3]{x}, g(x) = 4 + x, f(x) = x^9$
55. $R(x) = 0.15x - 0.000002x^2$

57. (a) $g(t) = 60t$ **(b)** $f(r) = \pi r^2$
(c) $(f \circ g)(t) = 3600\pi t^2$ **59.** $A(t) = 16\pi t^2$
61. (a) $f(x) = 0.9x$ **(b)** $g(x) = x - 100$
(c) $f \circ g(x) = 0.9x - 90$, $g \circ f(x) = 0.9x - 100$, $f \circ g$: first rebate, then discount, $g \circ f$: first discount, then rebate, $g \circ f$ is the better deal

Section 2.8 ■ page 230

1. No   **3.** Yes   **5.** No   **7.** Yes   **9.** Yes   **11.** No
13. No   **15.** No   **17. (a)** 2   **(b)** 3   **19.** 1
31. $f^{-1}(x) = \frac{1}{2}(x - 1)$ **33.** $f^{-1}(x) = \frac{1}{4}(x - 7)$
35. $f^{-1}(x) = 2x$ **37.** $f^{-1}(x) = (1/x) - 2$
39. $f^{-1}(x) = (5x - 1)/(2x + 3)$
41. $f^{-1}(x) = \frac{1}{5}(x^2 - 2)$, $x \geq 0$
43. $f^{-1}(x) = \sqrt{4 - x}$, $x \leq 4$ **45.** $f^{-1}(x) = (x - 4)^3$
47. $f^{-1}(x) = x^2 - 2x$, $x \geq 1$ **49.** $f^{-1}(x) = \sqrt[4]{x}$
51. (a)   **(b)**

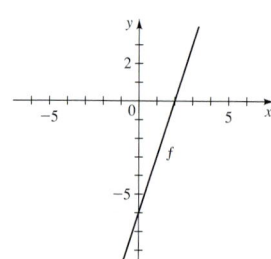   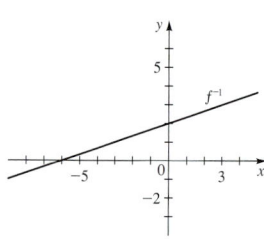

(c) $f^{-1}(x) = \frac{1}{3}(x + 6)$
53. (a)   **(b)**

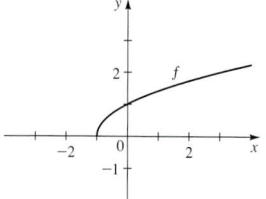   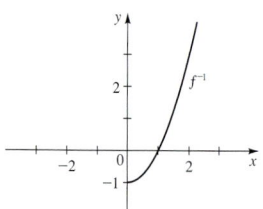

(c) $f^{-1}(x) = x^2 - 1$, $x \geq 0$
55. Not one-to-one   **57.** One-to-one

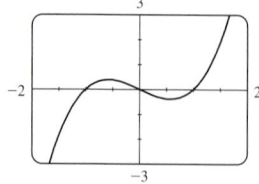   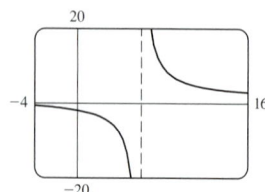

59. Not one-to-one

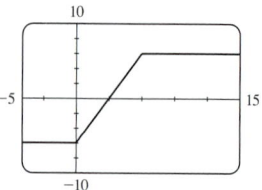

61. (a) $f^{-1}(x) = x - 2$
(b)

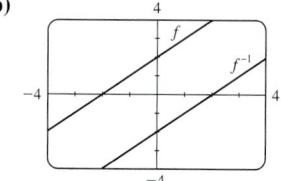

63. (a) $g^{-1}(x) = x^2 - 3$, $x \geq 0$
(b)

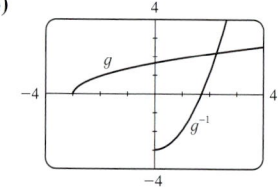

65. $x \geq 0$, $f^{-1}(x) = \sqrt{4 - x}$
67. $x \geq -2$, $h^{-1}(x) = \sqrt{x} - 2$
69.

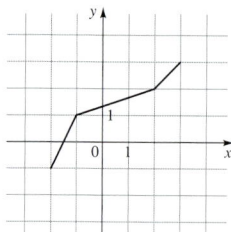

71. (a) $f(x) = 500 + 80x$ **(b)** $f^{-1}(x) = \frac{1}{80}(x - 500)$, the number of hours worked as a function of the fee   **(c)** 9; if he charges $1220, he worked 9 h

73. (a) $v^{-1}(t) = \sqrt{0.25 - \dfrac{t}{18,500}}$ **(b)** 0.498; at a distance 0.498 from the central axis, the velocity is 30
75. (a) $F^{-1}(x) = \frac{5}{9}(x - 32)$; the Celsius temperature when the Fahrenheit temperature is x **(b)** $F^{-1}(86) = 30$; when the temperature is 86°F, it is 30°C
77. (a) $f(x) = \begin{cases} 0.1x & \text{if } 0 \leq x \leq 20,000 \\ 2000 + 0.2(x - 20,000) & \text{if } x > 20,000 \end{cases}$

(b) $f^{-1}(x) = \begin{cases} 10x & \text{if } 0 \leq x \leq 2000 \\ 10,000 + 5x & \text{if } x > 2000 \end{cases}$
If you pay x euros in taxes, your income is $f^{-1}(x)$.
(c) $f^{-1}(10,000) = 60,000$ **79.** $f^{-1}(x) = \frac{1}{2}(x - 7)$. A pizza costing x dollars has $f^{-1}(x)$ toppings.

Chapter 2 Review ■ page 234

1. 6, 2, 18, $a^2 - 4a + 6$, $a^2 + 4a + 6$, $x^2 - 2x + 3$,
$4x^2 - 8x + 6$, $2x^2 - 8x + 10$ **3. (a)** $-1, 2$ **(b)** $[-4, 5]$
(c) $[-4, 4]$ **(d)** Increasing on $[-4, -2]$ and $[-1, 4]$; decreasing on $[-2, -1]$ and $[4, 5]$ **(e)** No **5.** Domain $[-3, \infty)$,
range $[0, \infty)$ **7.** $(-\infty, \infty)$ **9.** $[-4, \infty)$
11. $\{x \mid x \neq -2, -1, 0\}$ **13.** $(-\infty, -1] \cup [1, 4]$
15.

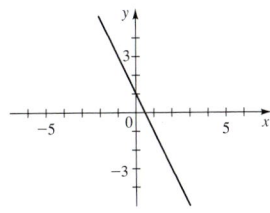

17.

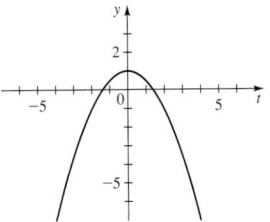

19.

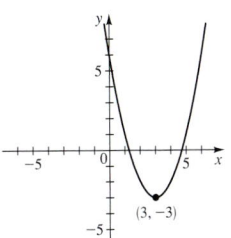

21.

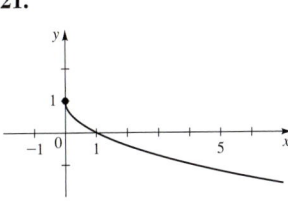

23.

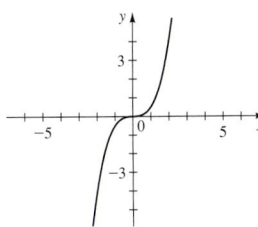

25.

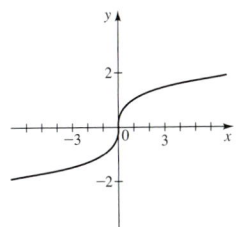

27.

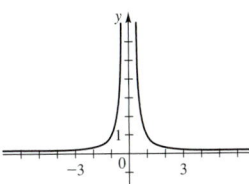

29.

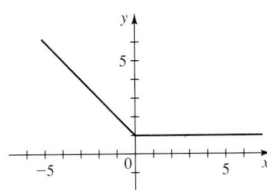

31.

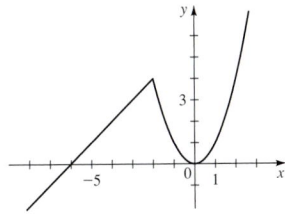

33. (iii)

35.

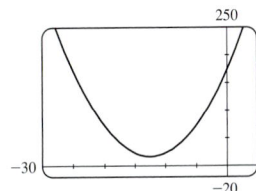

37.

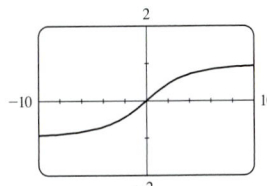

39. $[-2.1, 0.2] \cup [1.9, \infty)$ **41.** 5 **43.** $\dfrac{-1}{3(3 + h)}$

45.

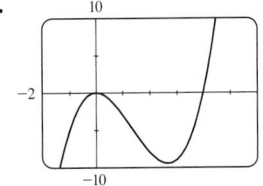

Increasing on $(-\infty, 0]$, $[2.67, \infty)$; decreasing on $[0, 2.67]$

47. (a) Shift upward 8 units **(b)** Shift left 8 units
(c) Stretch vertically by a factor of 2, then shift upward 1 unit **(d)** Shift right 2 units and downward 2 units
(e) Reflect in y-axis **(f)** Reflect in y-axis, then in x-axis
(g) Reflect in x-axis **(h)** Reflect in line $y = x$
49. (a) Neither **(b)** Odd **(c)** Even **(d)** Neither
51. $f(x) = (x + 2)^2 - 3$ **53.** $g(-1) = -7$ **55.** 68 ft
57. Local maximum ≈ 3.79 when $x \approx 0.46$; local minimum ≈ 2.81 when $x \approx -0.46$

59.

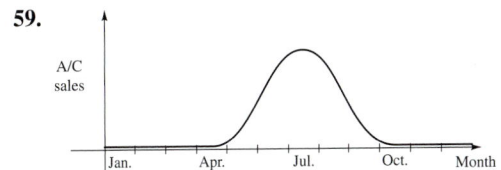

61. (a) $A(x) = 5\sqrt{3}x - \dfrac{\sqrt{3}}{2}x^2$ **(b)** 5 cm by $\dfrac{5\sqrt{3}}{2}$ cm
63. (a) $(f + g)(x) = x^2 - 6x + 6$
(b) $(f - g)(x) = x^2 - 2$
(c) $(fg)(x) = -3x^3 + 13x^2 - 18x + 8$
(d) $(f/g)(x) = (x^2 - 3x + 2)/(4 - 3x)$
(e) $(f \circ g)(x) = 9x^2 - 15x + 6$
(f) $(g \circ f)(x) = -3x^2 + 9x - 2$
65. $(f \circ g)(x) = -3x^2 + 6x - 1, (-\infty, \infty)$;
$(g \circ f)(x) = -9x^2 + 12x - 3, (-\infty, \infty)$;
$(f \circ f)(x) = 9x - 4, (-\infty, \infty)$;
$(g \circ g)(x) = -x^4 + 4x^3 - 6x^2 + 4x, (-\infty, \infty)$
67. $(f \circ g \circ h)(x) = 1 + \sqrt{x}$
69. Yes **71.** No **73.** No
75. $f^{-1}(x) = \dfrac{x + 2}{3}$ **77.** $f^{-1}(x) = \sqrt[3]{x} - 1$

79. (a), (b)

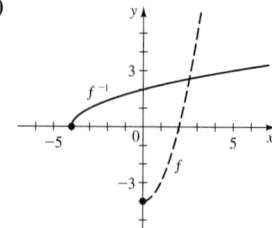

(c) $f^{-1}(x) = \sqrt{x + 4}$

Chapter 2 Test ■ page 237

1. (a) and **(b)** are graphs of functions, **(a)** is one-to-one
2. (a) $2/3$, $\sqrt{6}/5$, $\sqrt{a}/(a - 1)$ **(b)** $[-1, 0) \cup (0, \infty)$
3. 5
4. (a)   **(b)**

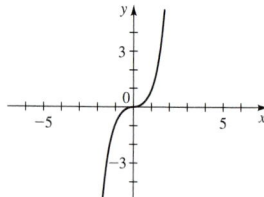   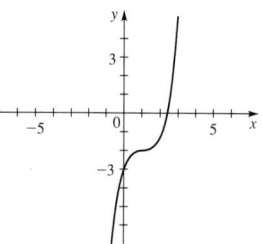

5. (a) Shift right 3 units, then shift upward 2 units
(b) Reflect in y-axis
6. (a) $f(x) = 2(x - 2)^2 + 5$
(b)   **(c)** $f(2) = 5$

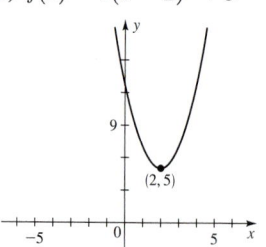

7. (a) $-3, 3$ **(b)**

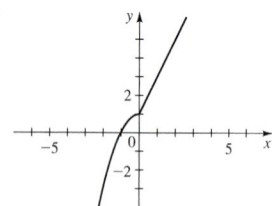

8. (a) $A(x) = -3x^2 + 900x$ **(b)** 150 ft
9. (a) $(f \circ g)(x) = (x - 3)^2 + 1$ **(b)** $(g \circ f)(x) = x^2 - 2$
(c) 2   **(d)** 2   **(e)** $(g \circ g \circ g)(x) = x - 9$
10. (a) $f^{-1}(x) = 3 - x^2, x \geq 0$

(b)

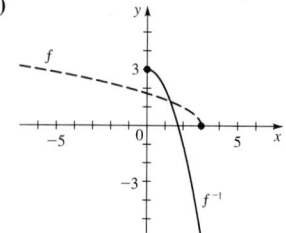

11. (a) Domain $[0, 6]$, range $[1, 7]$
(b)

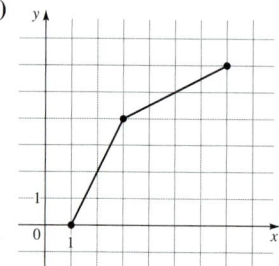

(c) $\frac{5}{4}$
12. (a)   **(b)** No

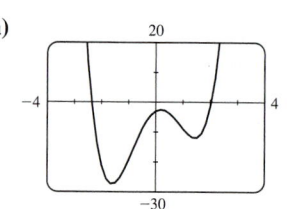

(c) Local minimum ≈ -27.18 when $x \approx -1.61$;
local maximum ≈ -2.55 when $x \approx 0.18$;
local minimum ≈ -11.93 when $x \approx 1.43$
(d) $[-27.18, \infty)$ **(e)** Increasing on
$[-1.61, 0.18] \cup [1.43, \infty)$; decreasing on
$(-\infty, -1.61] \cup [0.18, 1.43]$

Focus on Modeling ■ page 243

1. (a)

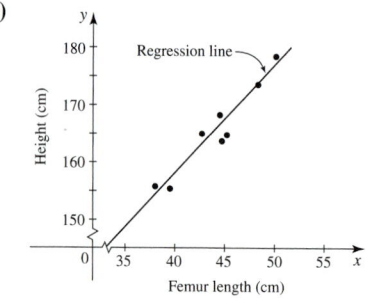

(b) $y = 1.8807x + 82.65$ **(c)** 191.7 cm

3. (a)

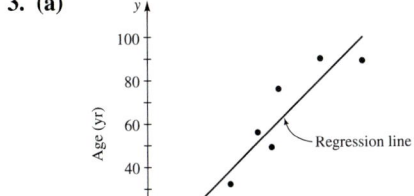

(b) $y = 6.451x - 0.1523$ **(c)** 116 years

5. (a)

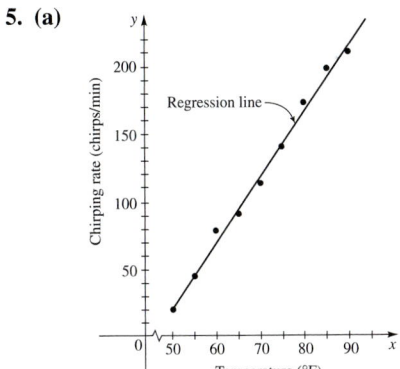

(b) $y = 4.857x - 220.97$ **(c)** 265 chirps/min

7. (a)

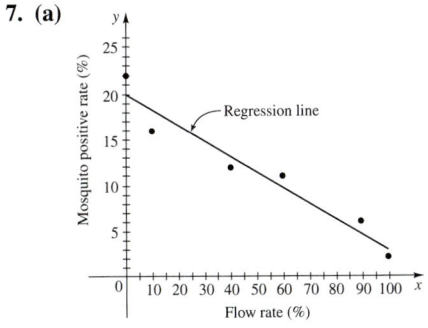

(b) $y = -0.168x + 19.89$ **(c)** 8.13%

9. (a)

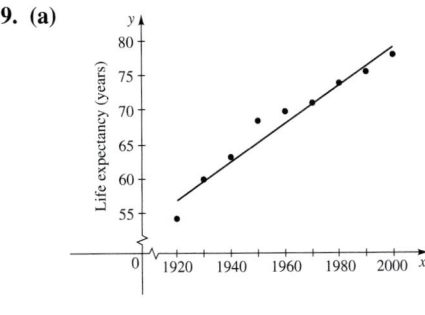

(b) $y = 0.2708x - 462.9$ **(c)** 78.2 years

11. (a) $y = -0.1729x + 64.717,\ y = -0.269x + 78.667$
(b) 2039

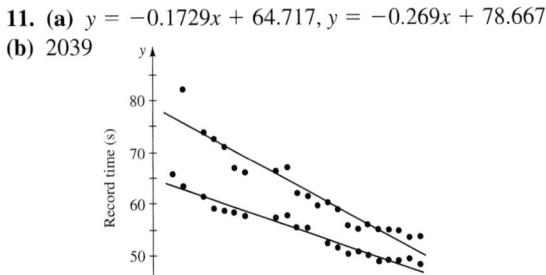

Chapter 3

Section 3.1 ■ page 262

1. (a)   **(b)**

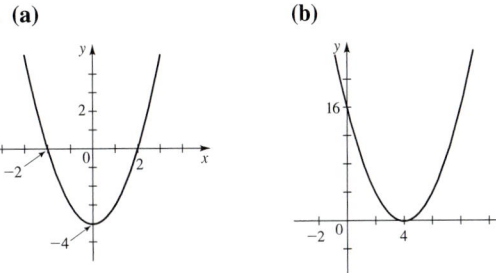

(c)   **(d)**

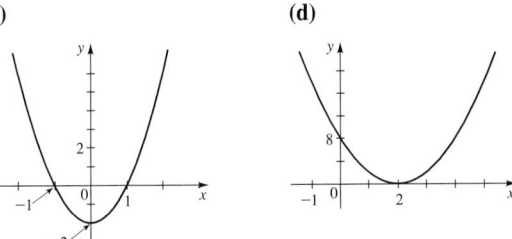

3. (a)   **(b)**

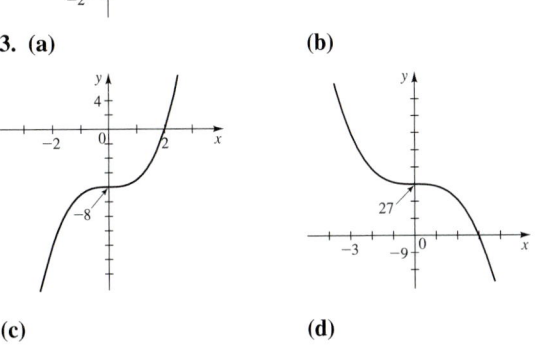

(c)   **(d)**

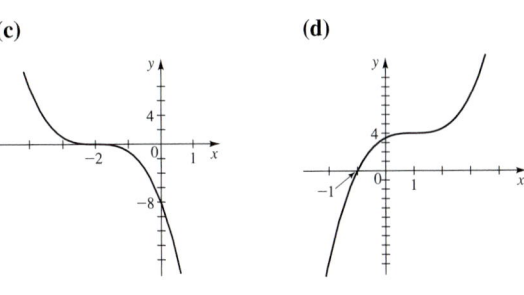

5. III   **7.** V   **9.** VI

11.

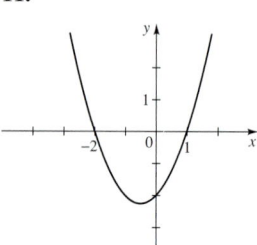

13.

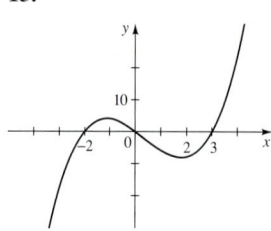

15.

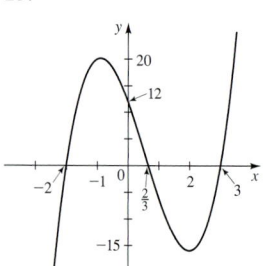

17.

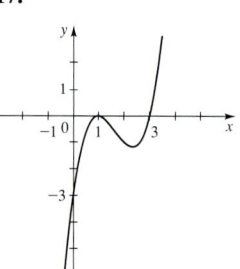

19.

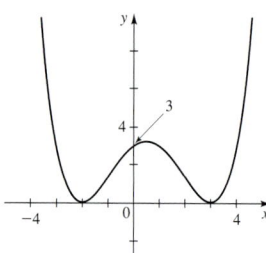

21.

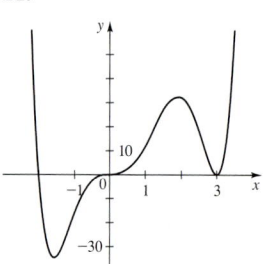

23. $P(x) = x(x + 2)(x - 3)$

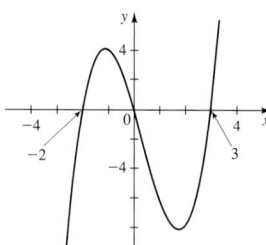

25. $P(x) = -x(x + 3)(x - 4)$

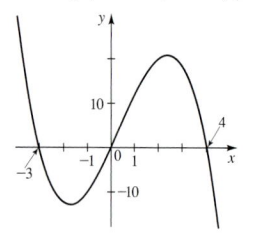

27. $P(x) = x^2(x - 1)(x - 2)$

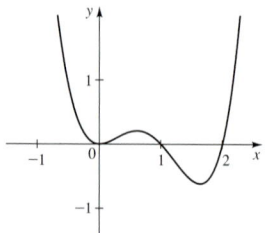

29. $P(x) = (x + 1)^2(x - 1)$

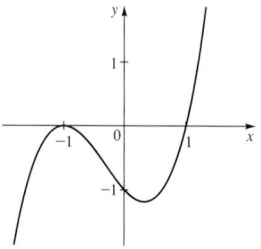

31. $P(x) = (2x - 1)(x + 3)(x - 3)$

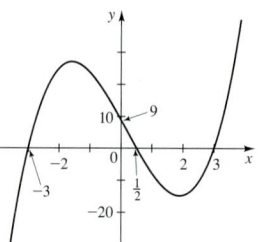

33. $P(x) = (x - 2)^2(x^2 + 2x + 4)$

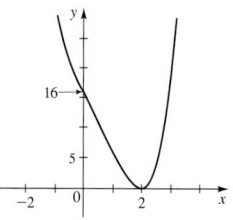

35. $P(x) = (x^2 + 1)(x + 2)(x - 2)$

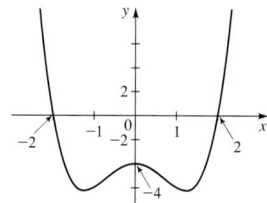

37. $y \to \infty$ as $x \to \infty, y \to -\infty$ as $x \to -\infty$
39. $y \to \infty$ as $x \to \pm\infty$
41. $y \to \infty$ as $x \to \infty, y \to -\infty$ as $x \to -\infty$
43. (a) x-intercepts 0, 4; y-intercept 0   (b) $(2, 4)$
45. (a) x-intercepts -2, 1; y-intercept -1
(b) $(-1, -2), (1, 0)$
47.

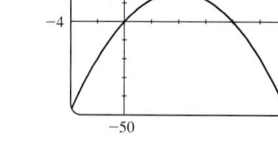

local maximum $(4, 16)$

49.

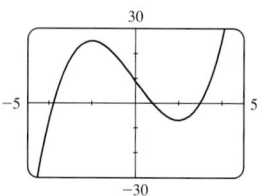

local maximum $(-2, 25)$,
local minimum $(2, -7)$

51.

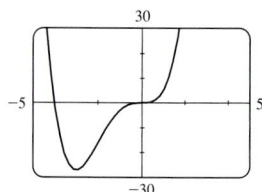

local minimum $(-3, -27)$

3. (a)

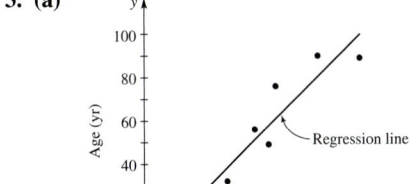

(b) $y = 6.451x - 0.1523$ **(c)** 116 years

5. (a)

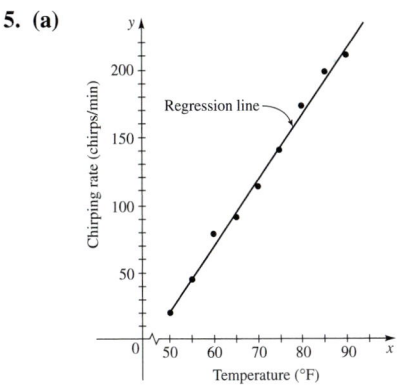

(b) $y = 4.857x - 220.97$ **(c)** 265 chirps/min

7. (a)

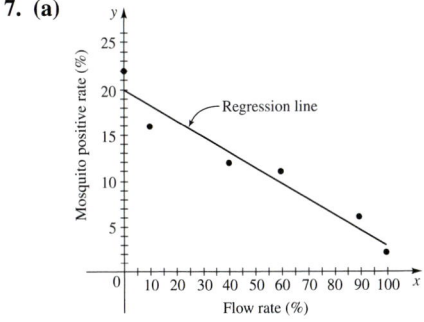

(b) $y = -0.168x + 19.89$ **(c)** 8.13%

9. (a)

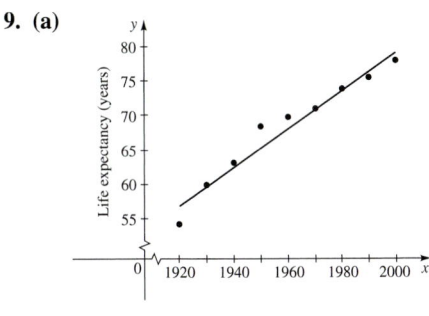

(b) $y = 0.2708x - 462.9$ **(c)** 78.2 years

11. (a) $y = -0.1729x + 64.717, \ y = -0.269x + 78.667$
(b) 2039

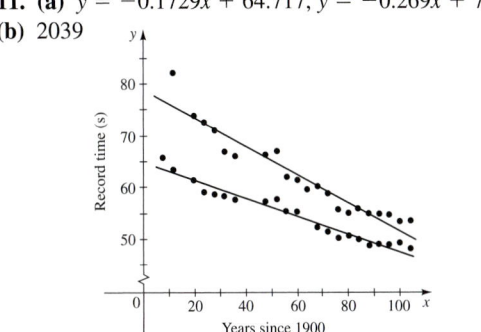

Chapter 3

Section 3.1 ■ page 262

1. (a)   **(b)**

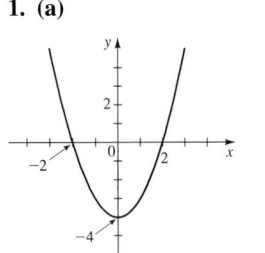   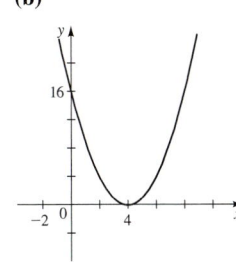

(c)   **(d)**

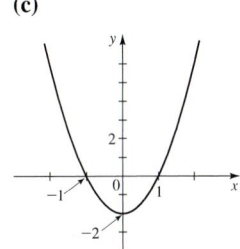   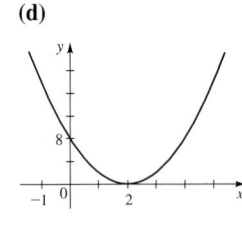

3. (a)   **(b)**

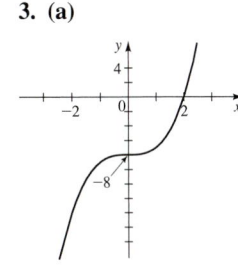   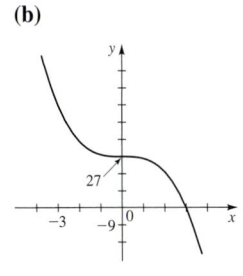

(c)   **(d)**

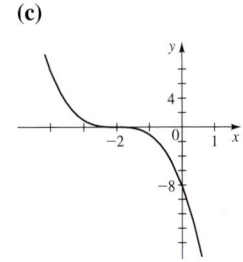   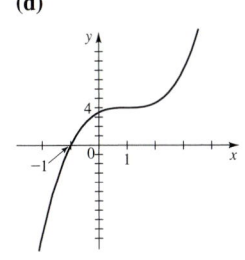

5. III   **7.** V   **9.** VI

11.

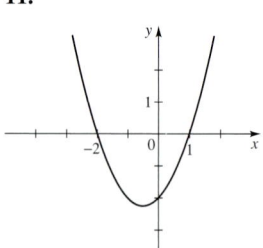

13.

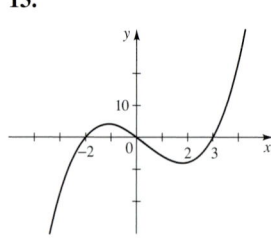

15.

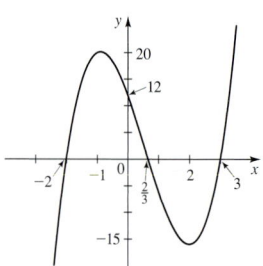

17.

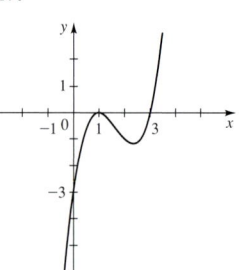

19.

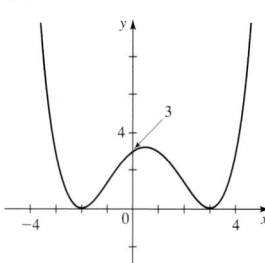

21.

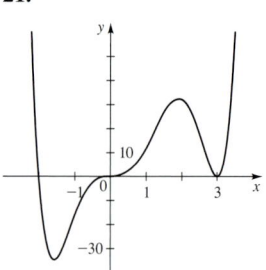

23. $P(x) = x(x + 2)(x - 3)$

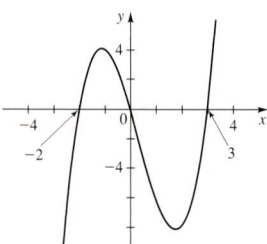

25. $P(x) = -x(x + 3)(x - 4)$

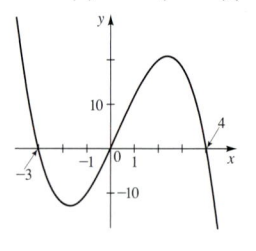

27. $P(x) = x^2(x - 1)(x - 2)$

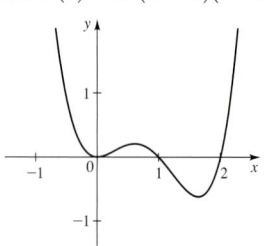

29. $P(x) = (x + 1)^2(x - 1)$

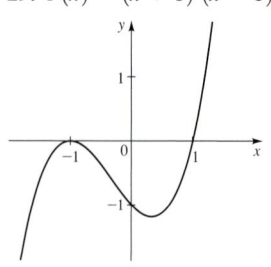

31. $P(x) = (2x - 1)(x + 3)(x - 3)$

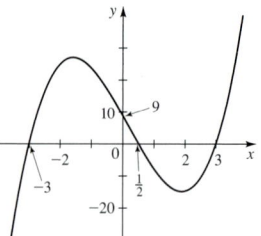

33. $P(x) = (x - 2)^2(x^2 + 2x + 4)$

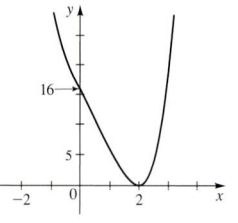

35. $P(x) = (x^2 + 1)(x + 2)(x - 2)$

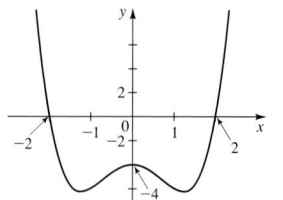

37. $y \to \infty$ as $x \to \infty, y \to -\infty$ as $x \to -\infty$
39. $y \to \infty$ as $x \to \pm\infty$
41. $y \to \infty$ as $x \to \infty, y \to -\infty$ as $x \to -\infty$
43. **(a)** x-intercepts 0, 4; y-intercept 0   **(b)** $(2, 4)$
45. **(a)** x-intercepts -2, 1; y-intercept -1
(b) $(-1, -2), (1, 0)$
47.

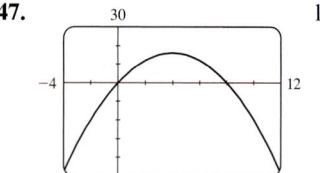

local maximum $(4, 16)$

49.

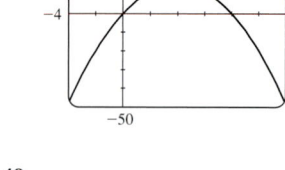

51.

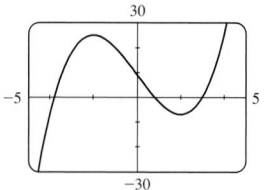

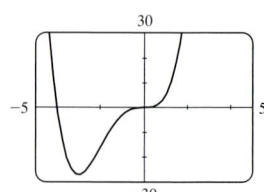

local maximum $(-2, 25)$,
local minimum $(2, -7)$

local minimum $(-3, -27)$

53.

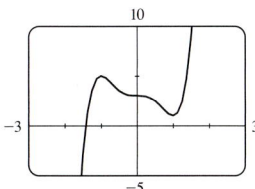

local maximum $(-1, 5)$,
local minimum $(1, 1)$

55. One local maximum, no local minimum
57. One local maximum, one local minimum
59. One local maximum, two local minima
61. No local extrema
63. One local maximum, two local minima
65.

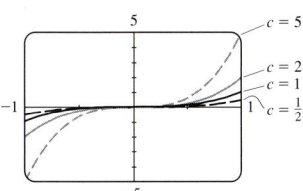

$c = 5$ Increasing the value of c
stretches the graph vertically.

$c = 2$
$c = 1$
$c = \frac{1}{2}$

67.

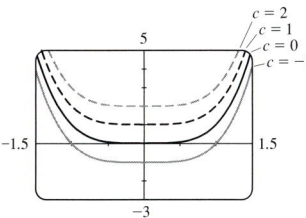

$c = 2$
$c = 1$ Increasing the value of c
$c = 0$ moves the graph up.
$c = -1$

69.

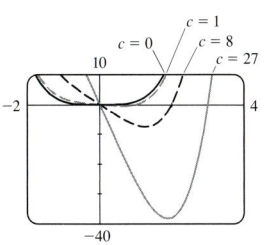

$c = 1$
$c = 0$ $c = 8$
      $c = 27$ Increasing the value of c
causes a deeper dip in the
graph in the fourth quadrant and
moves the positive x-intercept to
the right.

71. (a)

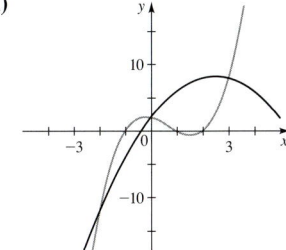

(b) Three   **(c)** $(0, 2), (3, 8), (-2, -12)$

73. (d) $P(x) = P_O(x) + P_E(x)$, where $P_O(x) = x^5 + 6x^3 - 2x$
and $P_E(x) = -x^2 + 5$
75. (a) Two local extrema

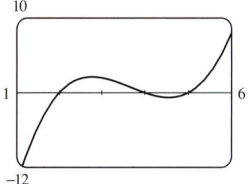

77. (a) 26 blenders   **(b)** No; $3276.22
79. (a) $V(x) = 4x^3 - 120x^2 + 800x$ **(b)** $0 < x < 10$
(c) Maximum volume ≈ 1539.6 cm^3

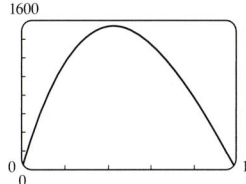

Section 3.2 ■ page 270

1. $(x + 3)(3x - 4) + 8$ **3.** $(2x - 3)(x^2 - 1) - 3$
5. $(x^2 + 3)(x^2 - x - 3) + (7x + 11)$
7. $x + 1 + \dfrac{-11}{x + 3}$

9. $2x - \frac{1}{2} + \dfrac{-\frac{15}{2}}{2x - 1}$ **11.** $2x^2 - x + 1 + \dfrac{4x - 4}{x^2 + 4}$

*In answers 13–36, the first polynomial given is the
quotient and the second is the remainder.*
13. $x - 2, -16$ **15.** $2x^2 - 1, -2$ **17.** $x + 2, 8x - 1$
19. $3x + 1, 7x - 5$ **21.** $x^4 + 1, 0$ **23.** $x - 2, -2$
25. $3x + 23, 138$ **27.** $x^2 + 2, -3$ **29.** $x^2 - 3x + 1, -1$
31. $x^4 + x^3 + 4x^2 + 4x + 4, -2$ **33.** $2x^2 + 4x, 1$
35. $x^2 + 3x + 9, 0$ **37.** -3 **39.** 12 **41.** -7 **43.** -483
45. 2159 **47.** $\frac{7}{3}$ **49.** -8.279 **55.** $-1 \pm \sqrt{6}$
57. $x^3 - 3x^2 - x + 3$ **59.** $x^4 - 8x^3 + 14x^2 + 8x - 15$
61. $-\frac{3}{2}x^3 + 3x^2 + \frac{15}{2}x - 9$ **63.** $(x + 1)(x - 1)(x - 2)$
65. $(x + 2)^2(x - 1)^2$

Section 3.3 ■ page 279

1. $\pm 1, \pm 3$ **3.** $\pm 1, \pm 2, \pm 4, \pm 8, \pm \frac{1}{2}$
5. $\pm 1, \pm 7, \pm \frac{1}{2}, \pm \frac{7}{2}, \pm \frac{1}{4}, \pm \frac{7}{4}$ **7. (a)** $\pm 1, \pm \frac{1}{5}$ **(b)** $-1, 1, \frac{1}{5}$
9. (a) $\pm 1, \pm 3, \pm \frac{1}{2}, \pm \frac{3}{2}$ **(b)** $-\frac{1}{2}, 1, 3$ **11.** $-2, 1$
13. $-1, 2$ **15.** 2 **17.** $-1, 2, 3$ **19.** -1 **21.** $\pm 1, \pm 2$
23. $1, -1, -2, -4$ **25.** $\pm 2, \pm \frac{3}{2}$ **27.** -2 **29.** $-1, -\frac{1}{2}, \frac{1}{2}$
31. $-\frac{3}{2}, \frac{1}{2}, 1$ **33.** $-\frac{5}{2}, -1, \frac{3}{2}$ **35.** $-1, \frac{1}{2}, 2$ **37.** $-3, -2, 1, 3$
39. $-1, -\frac{1}{3}, 2, 5$ **41.** $-2, -1 \pm \sqrt{2}$
43. $-1, 4, \dfrac{3 \pm \sqrt{13}}{2}$ **45.** $3, \dfrac{1 \pm \sqrt{5}}{2}$ **47.** $\frac{1}{2}, \dfrac{1 \pm \sqrt{3}}{2}$

49. $-1, -\frac{1}{2}, -3 \pm \sqrt{10}$

51. (a) $-2, 2, 3$ **(b)**

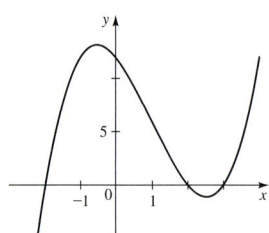

53. (a) $-\frac{1}{2}, 2$ **(b)**

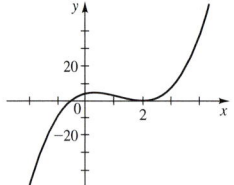

55. (a) $-1, 2$ **(b)**

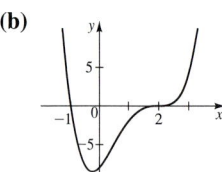

57. (a) $-1, 2$ **(b)**

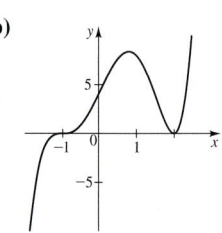

59. 1 positive, 2 or 0 negative; 3 or 1 real
61. 1 positive, 1 negative; 2 real **63.** 2 or 0 positive, 0 negative; 3 or 1 real (since 0 is a zero but is neither positive nor negative) **69.** $3, -2$ **71.** $3, -1$ **73.** $-2, \frac{1}{2}, \pm 1$ **75.** $\pm \frac{1}{2}, \pm \sqrt{5}$
77. $-2, 1, 3, 4$ **83.** $-2, 2, 3$ **85.** $-\frac{3}{2}, -1, 1, 4$
87. $-1.28, 1.53$ **89.** -1.50 **93.** 11.3 ft **95. (a)** It began to snow again. **(b)** No **(c)** Just before midnight on Saturday night **97.** 2.76 m **99.** 88 in. (or 3.21 in.)

Section 3.4 ■ page 289

1. Real part 5, imaginary part -7 **3.** Real part $-\frac{2}{3}$, imaginary part $-\frac{5}{3}$ **5.** Real part 3, imaginary part 0 **7.** Real part 0, imaginary part $-\frac{2}{3}$ **9.** Real part $\sqrt{3}$, imaginary part 2
11. $5 - i$ **13.** $3 + 5i$ **15.** $6 - i$ **17.** $2 - 2i$
19. $-19 + 4i$ **21.** $-\frac{1}{4} + \frac{1}{2}i$ **23.** $-4 + 8i$ **25.** $30 + 10i$
27. $-33 - 56i$ **29.** $27 - 8i$ **31.** $-i$ **33.** $\frac{8}{5} + \frac{1}{5}i$
35. $-5 + 12i$ **37.** $-4 + 2i$ **39.** $2 - \frac{4}{3}i$ **41.** $-i$
43. $-i$ **45.** 1 **47.** $5i$ **49.** -6

51. $(3 + \sqrt{5}) + (3 - \sqrt{5})i$ **53.** 2 **55.** $-i\sqrt{2}$ **57.** $\pm 3i$

59. $2 \pm i$ **61.** $-\frac{1}{2} \pm \frac{\sqrt{3}}{2}i$ **63.** $\frac{1}{2} \pm \frac{1}{2}i$ **65.** $-\frac{3}{2} \pm \frac{\sqrt{3}}{2}i$

67. $\dfrac{-6 \pm \sqrt{6}\,i}{6}$ **69.** $1 \pm 3i$

Section 3.5 ■ page 298

1. (a) $0, \pm 2i$ **(b)** $x^2(x - 2i)(x + 2i)$
3. (a) $0, 1 \pm i$ **(b)** $x(x - 1 - i)(x - 1 + i)$
5. (a) $\pm i$ **(b)** $(x - i)^2(x + i)^2$
7. (a) $\pm 2, \pm 2i$ **(b)** $(x - 2)(x + 2)(x - 2i)(x + 2i)$
9. (a) $-2, 1 \pm i\sqrt{3}$
(b) $(x + 2)(x - 1 - i\sqrt{3})(x - 1 + i\sqrt{3})$
11. (a) $\pm 1, \frac{1}{2} \pm \frac{1}{2}i\sqrt{3}, -\frac{1}{2} \pm \frac{1}{2}i\sqrt{3}$
(b) $(x - 1)(x + 1)(x - \frac{1}{2} - \frac{1}{2}i\sqrt{3})(x - \frac{1}{2} + \frac{1}{2}i\sqrt{3}) \times$
$(x + \frac{1}{2} - \frac{1}{2}i\sqrt{3})(x + \frac{1}{2} + \frac{1}{2}i\sqrt{3})$

In answers 13–30, the factored form is given first, then the zeros are listed with the multiplicity of each in parentheses.
13. $(x - 5i)(x + 5i)$; $\pm 5i\,(1)$
15. $[x - (-1 + i)][x - (-1 - i)]$; $-1 + i\,(1), -1 - i\,(1)$
17. $x(x - 2i)(x + 2i)$; $0\,(1), 2i\,(1), -2i\,(1)$
19. $(x - 1)(x + 1)(x - i)(x + i)$; $1\,(1), -1\,(1), i\,(1), -i\,(1)$
21. $16(x - \frac{3}{2})(x + \frac{3}{2})(x - \frac{3}{2}i)(x + \frac{3}{2}i)$;
$\frac{3}{2}\,(1), -\frac{3}{2}\,(1), \frac{3}{2}i\,(1), -\frac{3}{2}i\,(1)$
23. $(x + 1)(x - 3i)(x + 3i)$; $-1\,(1), 3i\,(1), -3i\,(1)$
25. $(x - i)^2(x + i)^2$; $i\,(2), -i\,(2)$
27. $(x - 1)(x + 1)(x - 2i)(x + 2i)$; $1\,(1), -1\,(1),$
$2i\,(1), -2i\,(1)$
29. $x(x - i\sqrt{3})^2(x + i\sqrt{3})^2$; $0\,(1), i\sqrt{3}\,(2), -i\sqrt{3}\,(2)$
31. $P(x) = x^2 - 2x + 2$ **33.** $Q(x) = x^3 - 3x^2 + 4x - 12$
35. $P(x) = x^3 - 2x^2 + x - 2$
37. $R(x) = x^4 - 4x^3 + 10x^2 - 12x + 5$
39. $T(x) = 6x^4 - 12x^3 + 18x^2 - 12x + 12$

41. $-2, \pm 2i$ **43.** $1, \dfrac{1 \pm i\sqrt{3}}{2}$ **45.** $2, \dfrac{1 \pm i\sqrt{3}}{2}$

47. $-\frac{3}{2}, -1 \pm i\sqrt{2}$ **49.** $-2, 1, \pm 3i$ **51.** $1, \pm 2i, \pm i\sqrt{3}$
53. 3 (multiplicity 2), $\pm 2i$ **55.** $-\frac{1}{2}$ (multiplicity 2), $\pm i$
57. 1 (multiplicity 3), $\pm 3i$ **59. (a)** $(x - 5)(x^2 + 4)$
(b) $(x - 5)(x - 2i)(x + 2i)$
61. (a) $(x - 1)(x + 1)(x^2 + 9)$
(b) $(x - 1)(x + 1)(x - 3i)(x + 3i)$
63. (a) $(x - 2)(x + 2)(x^2 - 2x + 4)(x^2 + 2x + 4)$

(b) $(x - 2)(x + 2)[x - (1 + i\sqrt{3})][x - (1 - i\sqrt{3})] \times$
$[x + (1 + i\sqrt{3})][x + (1 - i\sqrt{3})]$
65. (a) 4 real **(b)** 2 real, 2 imaginary **(c)** 4 imaginary

Section 3.6 ■ page 312

1. (a) $-3, -19, -199, -1999; 5, 21, 201, 2001; 1.2500,$
$1.0417, 1.0204, 1.0020; 0.8333, 0.9615, 0.9804, 0.9980$
(b) $r(x) \to -\infty$ as $x \to 2^-; r(x) \to \infty$ as $x \to 2^+$
(c) Horizontal asymptote $y = 1$
3. (a) $-22, -430, -40,300, -4,003,000;$
$-10, -370, -39,700, -3,997,000;$
$0.3125, 0.0608, 0.0302, 0.0030;$
$-0.2778, -0.0592, -0.0298, -0.0030$
(b) $r(x) \to -\infty$ as $x \to 2^-; r(x) \to -\infty$ as $x \to 2^+$
(c) Horizontal asymptote $y = 0$
5. x-intercept 1, y-intercept $-\frac{1}{4}$ **7.** x-intercepts $-1, 2$;
y-intercept $\frac{1}{3}$ **9.** x-intercepts $-3, 3$; no y-intercept
11. x-intercept 3, y-intercept 3, vertical $x = 2$;
horizontal $y = 2$ **13.** x-intercepts $-1, 1$; y-intercept $\frac{1}{4}$;
vertical $x = -2, x = 2$; horizontal $y = 1$
15. Vertical $x = -2$; horizontal $y = 0$ **17.** Vertical $x = 3$,
$x = -2$; horizontal $y = 1$ **19.** Horizontal $y = 0$
21. Vertical $x = -6, x = 1$; horizontal $y = 0$
23. Vertical $x = 1$
25.

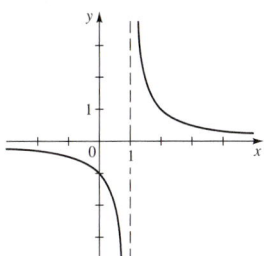

27.

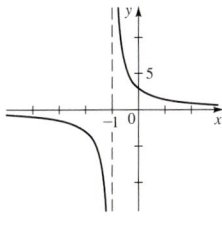

29.

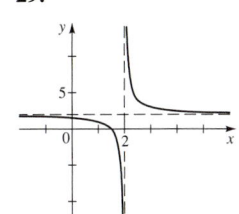

31.

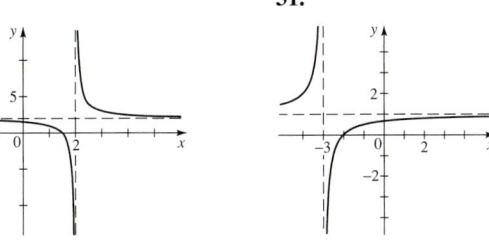

33.

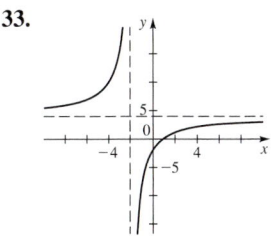

x-intercept 1
y-intercept -2
vertical $x = -2$
horizontal $y = 4$

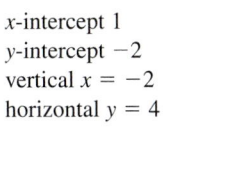

35.

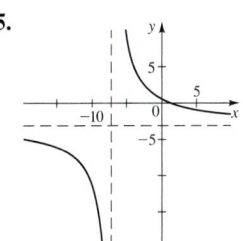

x-intercept $\frac{4}{3}$
y-intercept $\frac{4}{7}$
vertical $x = -7$
horizontal $y = -3$

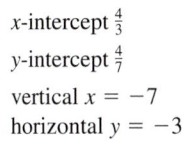

37.

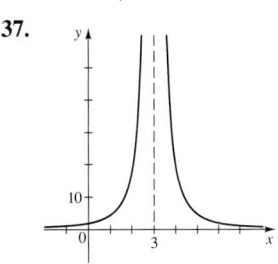

y-intercept 2
vertical $x = 3$
horizontal $y = 0$

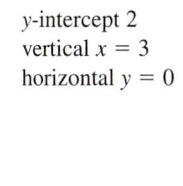

39.

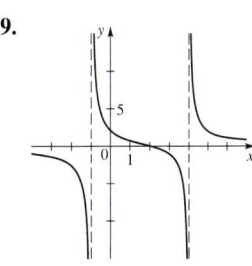

x-intercept 2
y-intercept 2
vertical $x = -1, x = 4$
horizontal $y = 0$

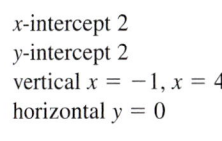

41.

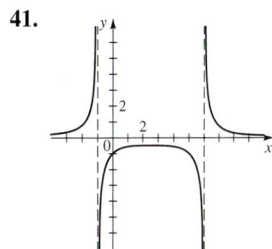

y-intercept -1
vertical $x = -1, x = 6$
horizontal $y = 0$

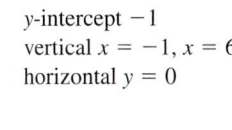

43.

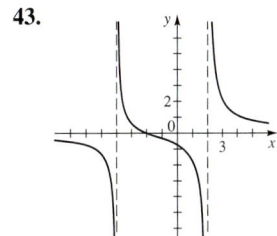

x-intercept -2
y-intercept $-\frac{3}{4}$
vertical $x = -4, x = 2$
horizontal $y = 0$

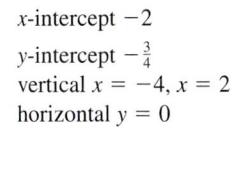

45.

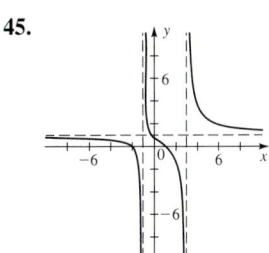

x-intercepts $-2, 1$
y-intercept $\frac{2}{3}$
vertical $x = -1, x = 3$
horizontal $y = 1$

47.
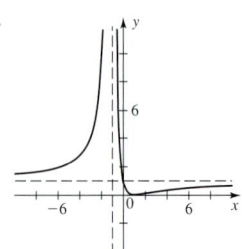
x-intercept 1
y-intercept 1
vertical $x = -1$
horizontal $y = 1$

49.
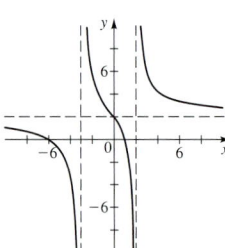
x-intercepts -6, 1
y-intercept 2
vertical $x = -3$, $x = 2$
horizontal $y = 2$

51.
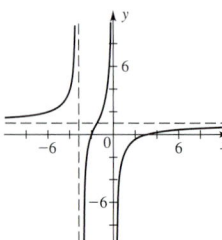
x-intercepts -2, 3
vertical $x = -3$, $x = 0$
horizontal $y = 1$

53.
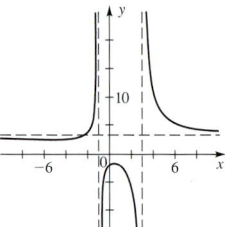
y-intercept -2
vertical $x = -1$, $x = 3$
horizontal $y = 3$

55.
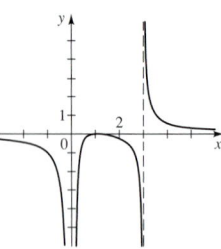
x-intercept 1
vertical $x = 0$, $x = 3$
horizontal $y = 0$

57.
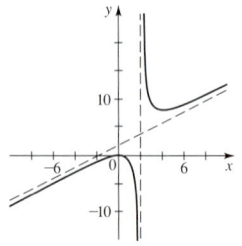
slant $y = x + 2$
vertical $x = 2$

59.
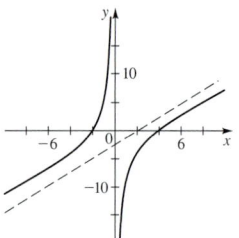
slant $y = x - 2$
vertical $x = 0$

61.
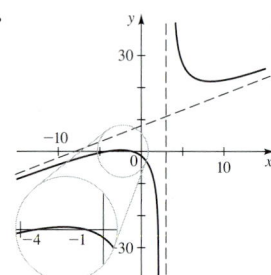
slant $y = x + 8$
vertical $x = 3$

63.
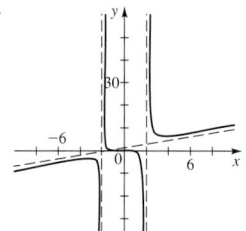
slant $y = x + 1$,
vertical $x = 2$, $x = -2$

65.
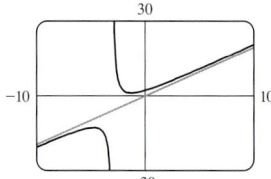
vertical $x = -3$

67.
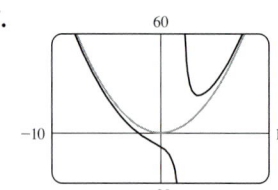
vertical $x = 2$

69.
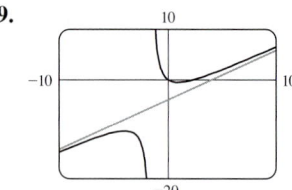
vertical $x = -1.5$
x-intercepts 0, 2.5
y-intercept 0, local
maximum $(-3.9, -10.4)$
local minimum $(0.9, -0.6)$
end behavior: $y = x - 4$

71.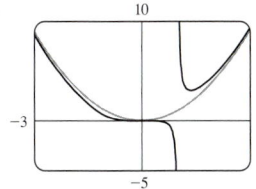

vertical $x = 1$
x-intercept 0
y-intercept 0
local minimum $(1.4, 3.1)$
end behavior: $y = x^2$

73.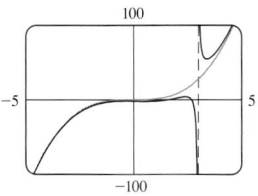

vertical $x = 3$
x-intercepts 1.6, 2.7
y-intercept -2
local maxima
$(-0.4, -1.8), (2.4, 3.8),$
local minima
$(0.6, -2.3), (3.4, 54.3)$
end behavior $y = x^3$

75. (a)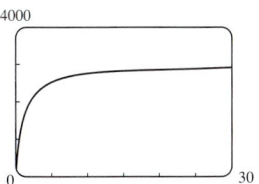

(b) It levels off at 3000.

77. (a) 2.50 mg/L   **(b)** It decreases to 0.   **(c)** 16.61 h

79.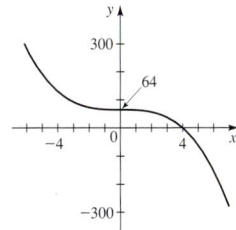

If the speed of the train approaches the speed of sound, then the pitch increases indefinitely (a sonic boom).

Chapter 3 Review ■ page 316

1.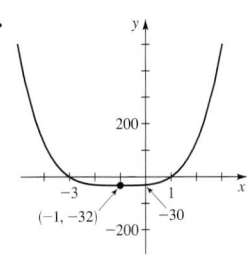

3.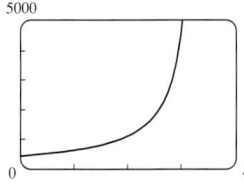

5.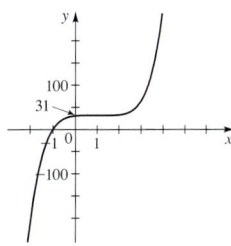

7.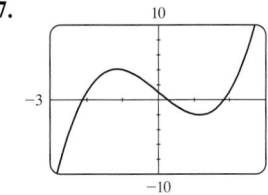

x-intercepts $-2.1, 0.3, 1.9$
y-intercept 1
local maximum $(-1.2, 4.1)$
local minimum $(1.2, -2.1)$
$y \to \infty$ as $x \to \infty$
$y \to -\infty$ as $x \to -\infty$

9.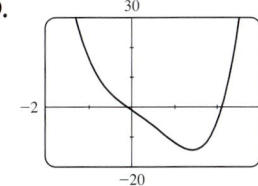

x-intercepts $-0.1, 2.1$
y-intercept -1
local minimum $(1.4, -14.5)$
$y \to \infty$ as $x \to \infty$
$y \to \infty$ as $x \to -\infty$

11. (a) $S = 13.8x(100 - x^2)$ **(b)** $0 \le x \le 10$
(c) 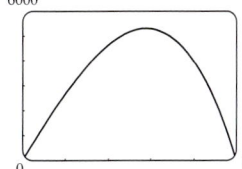   **(d)** 5.8 in.

In answers 13–20, the first polynomial given is the quotient and the second is the remainder.
13. $x - 1, 3$ **15.** $x^2 + 3x + 23, 94$
17. $x^3 - 5x^2 + 17x - 83, 422$
19. $2x - 3, 12$
21. 3   **25.** 8
27. (a) $\pm 1, \pm 2, \pm 3, \pm 6, \pm 9, \pm 18$
(b) 2 or 0 positive, 3 or 1 negative
29. (a) $-4, 0, 4$
(b)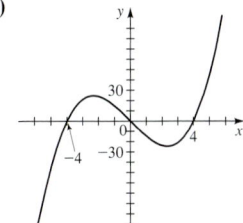

31. (a) $-2, 0$ (multiplicity 2), 1 **(b)**

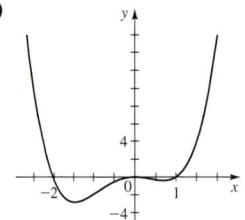

33. (a) $-2, -1, 2, 3$ **(b)**

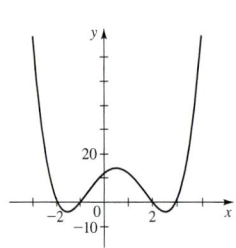

35. (a) $-\frac{1}{2}, 1$ **(b)**

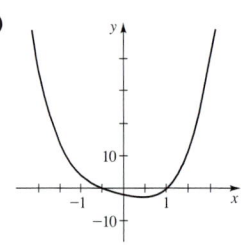

37. $3 + i$ **39.** $8 - i$ **41.** $\frac{6}{5} + \frac{8}{5}i$ **43.** i **45.** 2
47. $4x^3 - 18x^2 + 14x - 12$ **49.** No; since the complex
conjugates of imaginary zeros will also be zeros, the
polynomial would have 8 zeros, contradicting the requirement
that it have degree 4. **51.** $-3, 1, 5$ **53.** $-1 \pm 2i, -2$
(multiplicity 2) **55.** $\pm 2, 1$ (multiplicity 3)
57. $\pm 2, \pm 1 \pm i\sqrt{3}$
59. $1, 3, \dfrac{-1 \pm i\sqrt{7}}{2}$ **61.** $x = -0.5, 3$ **63.** $x \approx -0.24, 4.24$

65.

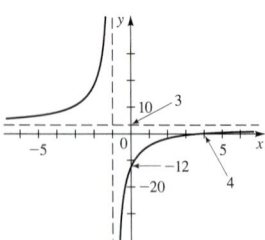

67.

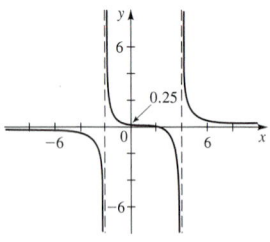

69.

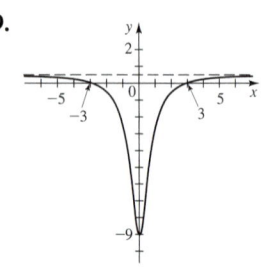

71.

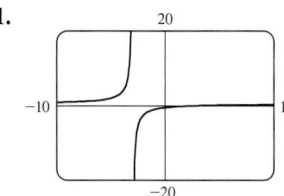

x-intercept 3
y-intercept -0.5
vertical $x = -3$
horizontal $y = 0.5$
no local extrema

73.

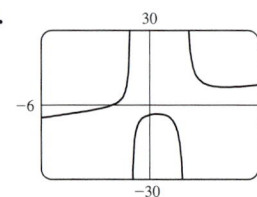

x-intercept -2
y-intercept -4
vertical $x = -1, x = 2$
slant $y = x + 1$
local maximum
$(0.425, -3.599)$
local minimum
$(4.216, 7.175)$

75. $(-2, -28), (1, 26), (2, 68), (5, 770)$

Chapter 3 Test ■ **page 319**
1.

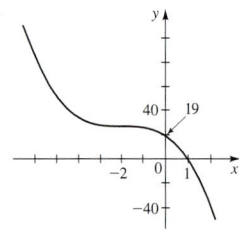

2. (a) $x^3 + 2x^2 + 2, 9$ **(b)** $x^3 + 2x^2 + \frac{1}{2}, \frac{15}{2}$
3. (a) $\pm 1, \pm 3, \pm\frac{1}{2}, \pm\frac{3}{2}$ **(b)** $2(x - 3)\left(x - \frac{1}{2}\right)(x + 1)$
(c) $-1, \frac{1}{2}, 3$
(d)

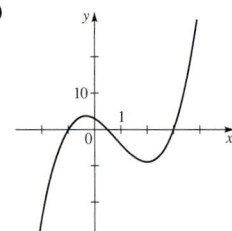

4. (a) $7 + i$ **(b)** $-1 - 5i$ **(c)** $18 + i$ **(d)** $\frac{6}{25} - \frac{17}{25}i$
(e) 1 **(f)** $6 - 2i$ **5.** $3, -1 \pm i$
6. $(x - 1)^2(x - 2i)(x + 2i)$ **7.** $x^4 + 2x^3 + 10x^2 + 18x + 9$
8. (a) 4, 2, or 0 positive; 0 negative
(c) 0.17, 3.93

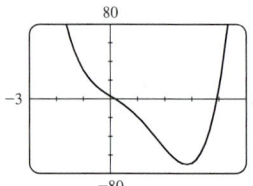

(d) Local minimum $(2.8, -70.3)$ **9. (a)** r, u **(b)** s **(c)** s

(d)

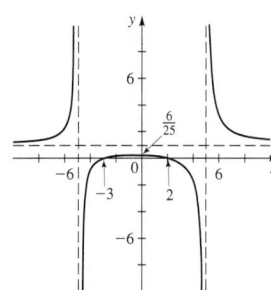

(e) $x^2 - 2x - 5$

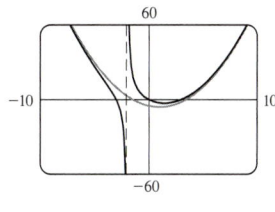

Focus on Modeling ■ page 323

1. (a) $y = -0.275428x^2 + 19.7485x - 273.5523$

(b)

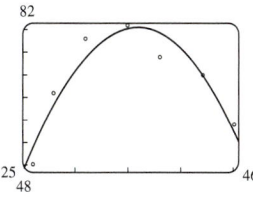

(c) 35.85 lb/in² **3. (a)** $y = 0.00203708x^3 - 0.104521x^2 + 1.966206x + 1.45576$

(b)

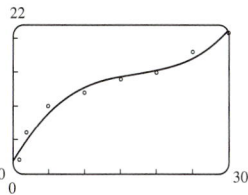

(c) 43 vegetables **(d)** 2.0 s **5. (a)** Degree 2

(b) $y = -16.0x^2 + 51.8429x + 4.20714$

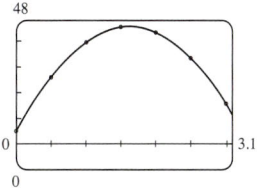

(c) 0.3 s and 2.9 s **(d)** 46.2 ft

Chapter 4

Section 4.1 ■ page 336

1. 2.000, 7.103, 77.880, 1.587 **3.** 0.885, 0.606, 0.117, 1.837

5.

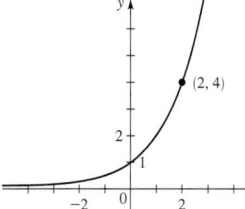

7.

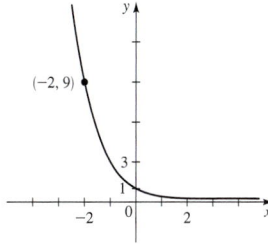

9.

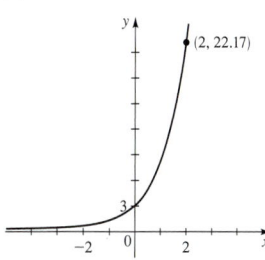

11.

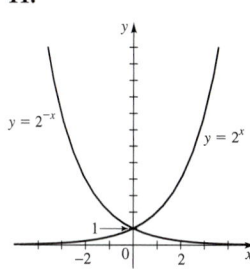

13.

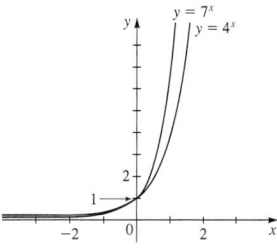

15. $f(x) = 3^x$ **17.** $f(x) = \left(\frac{1}{4}\right)^x$ **19.** III **21.** I **23.** II

25. $\mathbb{R}, (-\infty, 0), y = 0$ **27.** $\mathbb{R}, (-3, \infty), y = -3$

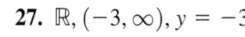

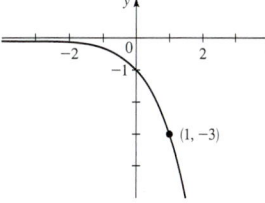

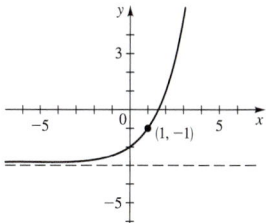

29. $\mathbb{R}, (4, \infty), y = 4$ **31.** $\mathbb{R}, (0, \infty), y = 0$

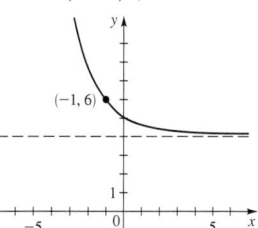

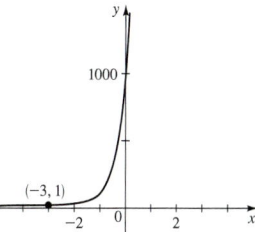

33. $\mathbb{R}, (-\infty, 0), y = 0$ **35.** $\mathbb{R}, (-1, \infty), y = -1$

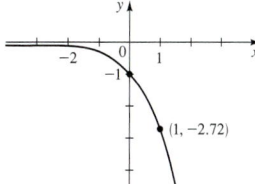

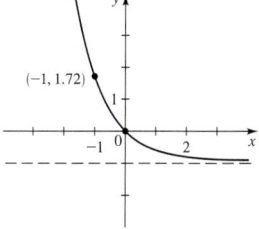

37. $\mathbb{R}$, $(0, \infty)$, $y = 0$ **39.** $y = 3(2^x)$

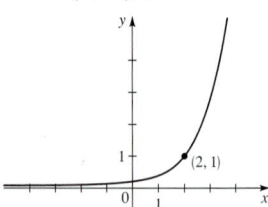

41. (a)

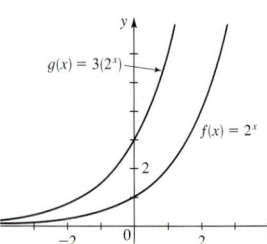

(b) The graph of g is steeper than that of f.

51. (a)

(i)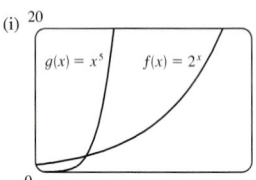

(ii)

(iii)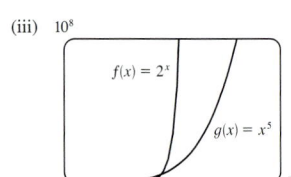

The graph of f ultimately increases much more quickly than g.
(b) 1.2, 22.4

53.

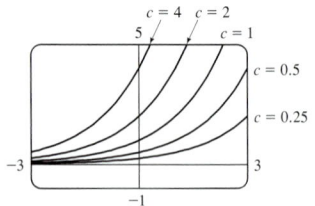

The larger the value of c, the more rapidly the graph increases.

45.

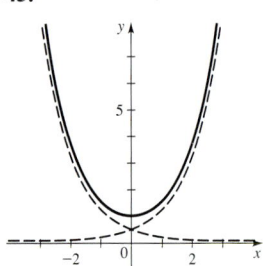

55.

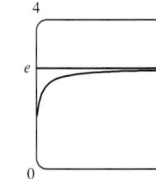

59.

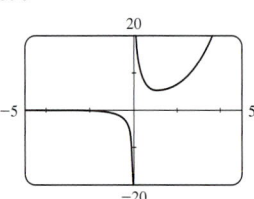

61. Local minimum $\approx (0.27, 1.75)$ **63. (a)** Increasing on $(-\infty, 1.00]$, decreasing on $[1.00, \infty)$ **(b)** $(-\infty, 0.37]$
65. (a) 13 kg   **(b)** 6.6 kg
67. (a) 0   **(b)** 50.6 ft/s, 69.2 ft/s
(c)   **(d)** 80 ft/s

69. (a) 100   **(b)** 482, 999, 1168   **(c)** 1200   **71.** 1.6 ft
73. $5203.71, $5415.71, $5636.36, $5865.99, $6104.98,
$6353.71   **75. (a)** $16,288.95   **(b)** $26,532.98
(c) $43,219.42   **77. (a)** $4,615.87   **(b)** $4,658.91
(c) $4,697.04   **(d)** $4,703.11   **(e)** $4,704.68   **(f)** $4,704.93
(g) $4,704.94   **79. (i)**   **81. (a)** $7,678.96   **(b)** $67,121.04

57. (a)

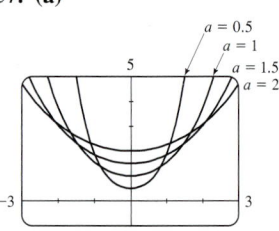

(b) The larger the value of a, the wider the graph.

vertical asymptote $x = 0$
horizontal asymptote $y = 0$, left side only

Section 4.2 ■ page 349

1.

Logarithmic form	Exponential form
$\log_8 8 = 1$	$8^1 = 8$
$\log_8 64 = 2$	$8^2 = 64$
$\log_8 4 = \frac{2}{3}$	$8^{2/3} = 4$
$\log_8 512 = 3$	$8^3 = 512$
$\log_8 \frac{1}{8} = -1$	$8^{-1} = \frac{1}{8}$
$\log_8 \frac{1}{64} = -2$	$8^{-2} = \frac{1}{64}$

3. (a) $5^2 = 25$ **(b)** $5^0 = 1$ **5. (a)** $8^{1/3} = 2$ **(b)** $2^{-3} = \frac{1}{8}$
7. (a) $e^x = 5$ **(b)** $e^5 = y$ **9. (a)** $\log_5 125 = 3$
(b) $\log_{10} 0.0001 = -4$ **11. (a)** $\log_8 \frac{1}{8} = -1$
(b) $\log_2 \frac{1}{8} = -3$ **13. (a)** $\ln 2 = x$ **(b)** $\ln y = 3$
15. (a) 1 **(b)** 0 **(c)** 2 **17. (a)** 2 **(b)** 2 **(c)** 10
19. (a) -3 **(b)** $\frac{1}{2}$ **(c)** -1 **21. (a)** 37 **(b)** 8 **(c)** $\sqrt{5}$
23. (a) $-\frac{2}{3}$ **(b)** 4 **(c)** -1 **25. (a)** 32 **(b)** 4
27. (a) 5 **(b)** 27 **29. (a)** 100 **(b)** 25 **31. (a)** 2 **(b)** 4
33. (a) 0.3010 **(b)** 1.5465 **(c)** -0.1761 **35. (a)** 1.6094
(b) 3.2308 **(c)** 1.0051 **37.** $y = \log_5 x$ **39.** $y = \log_9 x$
41. II **43.** III **45.** VI
47.

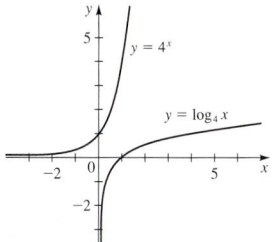

49. $(4, \infty), \mathbb{R}, x = 4$

51. $(-\infty, 0), \mathbb{R}, x = 0$

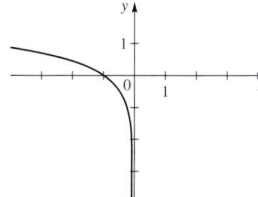

53. $(0, \infty), \mathbb{R}, x = 0$

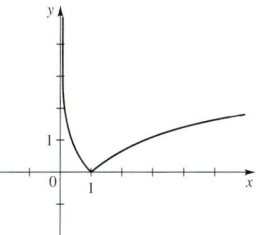

55. $(0, \infty), \mathbb{R}, x = 0$

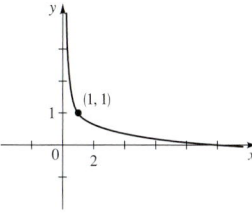

57. $(0, \infty), [0, \infty), x = 0$

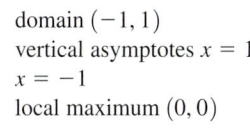

59. $(-3, \infty)$ **61.** $(-\infty, -1) \cup (1, \infty)$ **63.** $(0, 2)$
65.

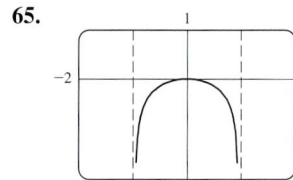

domain $(-1, 1)$
vertical asymptotes $x = 1$,
$x = -1$
local maximum $(0, 0)$

67.

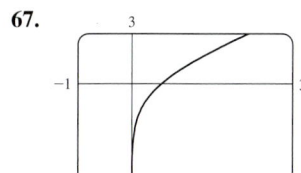

domain $(0, \infty)$
vertical asymptote $x = 0$
no maximum or minimum

69.

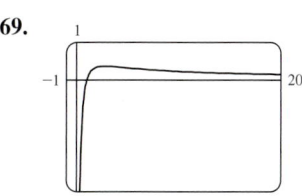

domain $(0, \infty)$
vertical asymptote $x = 0$
horizontal asymptote $y = 0$
local maximum $\approx (2.72, 0.37)$

71. The graph of f grows more slowly than g.
73. (a)

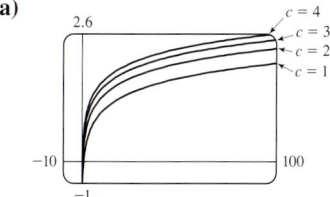

(b) The graph of
$f(x) = \log(cx)$ is
the graph of
$f(x) = \log(x)$
shifted upward
$\log c$ units.

75. (a) $(1, \infty)$ **(b)** $f^{-1}(x) = 10^{2^x}$
77. (a) $f^{-1}(x) = \log_2\left(\dfrac{x}{1 - x}\right)$ **(b)** $(0, 1)$
79. 2602 yr **81.** 11.5 yr, 9.9 yr, 8.7 yr **83.** 5.32, 4.32

Section 4.3 ■ page 356
1. $\frac{3}{2}$ **3.** 2 **5.** 3 **7.** 3 **9.** 200 **11.** 4 **13.** $1 + \log_2 x$
15. $\log_2 x + \log_2(x - 1)$ **17.** $10 \log 6$ **19.** $\log_2 A + 2 \log_2 B$
21. $\log_3 x + \frac{1}{2}\log_3 y$ **23.** $\frac{1}{3}\log_5(x^2 + 1)$ **25.** $\frac{1}{2}(\ln a + \ln b)$
27. $3 \log x + 4 \log y - 6 \log z$
29. $\log_2 x + \log_2(x^2 + 1) - \frac{1}{2}\log_2(x^2 - 1)$
31. $\ln x + \frac{1}{2}(\ln y - \ln z)$ **33.** $\frac{1}{4}\log(x^2 + y^2)$
35. $\frac{1}{2}[\log(x^2 + 4) - \log(x^2 + 1) - 2 \log(x^3 - 7)]$
37. $3 \ln x + \frac{1}{2}\ln(x - 1) - \ln(3x + 4)$ **39.** $\log_3 160$
41. $\log_2(AB/C^2)$ **43.** $\log\left(\dfrac{x^4(x - 1)^2}{\sqrt[3]{x^2 + 1}}\right)$ **45.** $\ln(5x^2(x^2 + 5)^3)$
47. $\log\left(\sqrt[3]{2x + 1}\,\sqrt{(x - 4)/(x^4 - x^2 - 1)}\right)$
49. 2.321928 **51.** 2.523719 **53.** 0.493008 **55.** 3.482892
57.

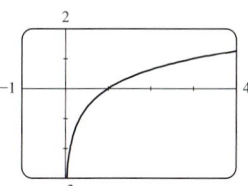

63. (a) $P = c/W^k$ **(b)** 1866, 64
65. (a) $M = -2.5 \log B + 2.5 \log B_0$

Section 4.4 ■ page 366

1. 1.3979   **3.** −0.9730   **5.** −0.5850   **7.** 1.2040
9. 0.0767   **11.** 0.2524   **13.** 1.9349   **15.** −43.0677
17. 2.1492   **19.** 6.2126   **21.** −2.9469   **23.** −2.4423
25. 14.0055   **27.** ±1   **29.** $0, \frac{4}{3}$ **31.** $\ln 2 \approx 0.6931, 0$
33. $\frac{1}{2} \ln 3 \approx 0.5493$ **35.** $e^{10} \approx 22026$ **37.** 0.01
39. $\frac{95}{3}$ **41.** $3 - e^2 \approx -4.3891$ **43.** 5   **45.** 5
47. $\frac{13}{12}$ **49.** 6   **51.** $\frac{3}{2}$ **53.** $1/\sqrt{5} \approx 0.4472$ **55.** 2.21
57. 0.00, 1.14   **59.** −0.57   **61.** 0.36
63. $2 < x < 4$ or $7 < x < 9$ **65.** $\log 2 < x < \log 5$
67. (a) \$6435.09   **(b)** 8.24 yr   **69.** 6.33 yr   **71.** 8.15 yr
73. 8.30%   **75.** 13 days   **77. (a)** 7337   **(b)** 1.73 yr
79. (a) $P = P_0 e^{-kh}$ **(b)** 56.47 kPa
81. (a) $t = -\frac{5}{13}\ln(1 - \frac{13}{60}I)$ **(b)** 0.218 s

Section 4.5 ■ page 379

1. (a) 500   **(b)** 45%   **(c)** 1929   **(d)** 6.66 h
3. (a) $n(t) = 18,000e^{0.08t}$ **(b)** 34,137

(c)

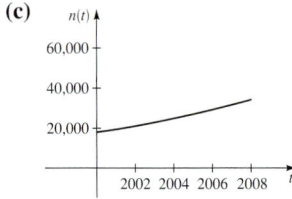

5. (a) $n(t) = 112,000e^{0.04t}$ **(b)** About 142,000
(c) 2008   **7. (a)** 20,000   **(b)** $n(t) = 20,000e^{0.1096t}$
(c) About 48,000   **(d)** 2010   **9. (a)** $n(t) = 8600e^{0.1508t}$
(b) About 11,600   **(c)** 4.6 h   **11. (a)** 2029
(b) 2049   **13.** 22.85 h   **15. (a)** $n(t) = 10e^{-0.0231t}$
(b) 1.6 g   **(c)** 70 yr   **17.** 18 yr   **19.** 149 h
21. 3560 yr   **23. (a)** 210°F   **(b)** 153°F
(c) 28 min   **25. (a)** 137°F   **(b)** 116 min   **27. (a)** 2.3
(b) 3.5   **(c)** 8.3   **29. (a)** 10^{-3} M   **(b)** 3.2×10^{-7} M
31. $4.8 \le \text{pH} \le 6.4$ **33.** $\log 20 \approx 1.3$
35. Twice as intense   **37.** 8.2   **39.** 6.3×10^{-3} W/m²
41. (b) 106 dB

Chapter 4 Review ■ page 383

1. $\mathbb{R}, (0, \infty), y = 0$ **3.** $\mathbb{R}, (3, \infty), y = 3$

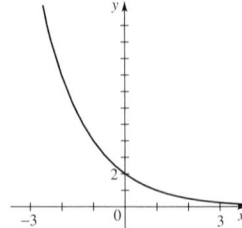 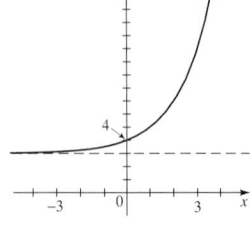

5. $(1, \infty), \mathbb{R}, x = 1$ **7.** $(0, \infty), \mathbb{R}, x = 0$

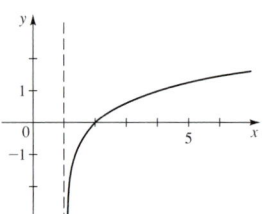 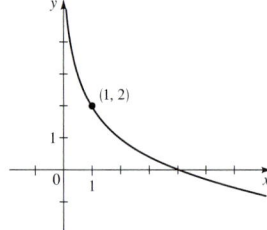

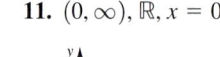

9. $\mathbb{R}, (-1, \infty), y = -1$ **11.** $(0, \infty), \mathbb{R}, x = 0$

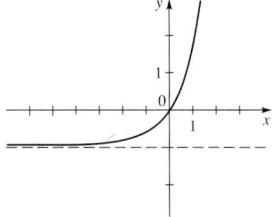 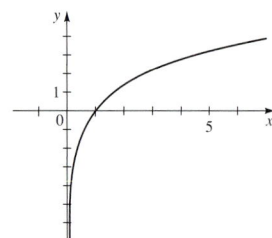

13. $\left(-\infty, \frac{1}{2}\right)$ **15.** $(-\infty, -2) \cup (2, \infty)$ **17.** $2^{10} = 1024$
19. $10^y = x$ **21.** $\log_2 64 = 6$ **23.** $\log 74 = x$ **25.** 7
27. 45   **29.** 6   **31.** −3   **33.** $\frac{1}{2}$ **35.** 2   **37.** 92
39. $\frac{2}{3}$ **41.** $\log A + 2 \log B + 3 \log C$
43. $\frac{1}{2}[\ln(x^2 - 1) - \ln(x^2 + 1)]$
45. $2 \log_5 x + \frac{3}{2} \log_5(1 - 5x) - \frac{1}{2}\log_5(x^3 - x)$

47. $\log 96$ **49.** $\log_2\left(\dfrac{(x - y)^{3/2}}{(x^2 + y^2)^2}\right)$ **51.** $\log\left(\dfrac{x^2 - 4}{\sqrt{x^2 + 4}}\right)$

53. −15   **55.** $\frac{1}{3}(5 - \log_5 26) \approx 0.99$ **57.** $\frac{4}{3}\ln 10 \approx 3.07$
59. 3   **61.** −4, 2   **63.** 0.430618   **65.** 2.303600

67.

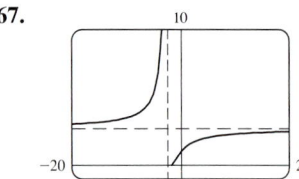

vertical asymptote
$x = -2$
horizontal asymptote
$y = 2.72$
no maximum or minimum

69.

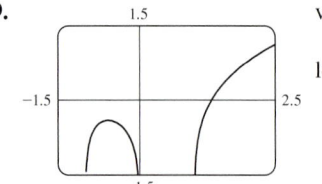

vertical asymptotes
$x = -1, x = 0, x = 1$
local maximum
$\approx (-0.58, -0.41)$

71. 2.42   **73.** $0.16 < x < 3.15$ **75.** Increasing on $(-\infty, 0]$
and $[1.10, \infty)$, decreasing on $[0, 1.10]$ **77.** 1.953445
79. $\log_4 258$ **81. (a)** \$16,081.15   **(b)** \$16,178.18
(c) \$16,197.64   **(d)** \$16,198.31   **83. (a)** $n(t) = 30e^{0.15t}$
(b) 55   **(c)** 19 yr   **85. (a)** 9.97 mg   **(b)** 1.39×10^5 yr

87. **(a)** $n(t) = 150e^{-0.0004359t}$ **(b)** 97.0 mg **(c)** 2520 yr
89. **(a)** $n(t) = 1500e^{0.1515t}$ **(b)** 7940 **91.** 7.9, basic **93.** 8.0

Chapter 4 Test ■ page 385

1.

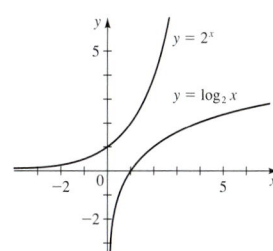

2. $(-1, \infty)$, $\mathbb{R}$, $x = -1$

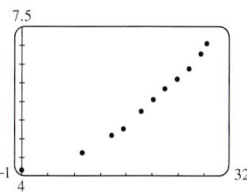

3. **(a)** $\frac{3}{2}$ **(b)** 3 **(c)** $\frac{2}{3}$ **(d)** 2
4. $\frac{1}{3}[\log(x + 2) - 4 \log x - \log(x^2 + 4)]$

5. $\ln\left(\dfrac{x\sqrt{3 - x^4}}{(x^2 + 1)^2}\right)$ **6.** **(a)** 4.32 **(b)** 0.77 **(c)** 5.39 **(d)** 2

7. **(a)** $n(t) = 1000e^{2.07944t}$ **(b)** 22,627 **(c)** 1.3 h
(d)

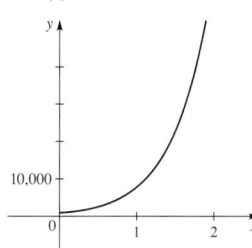

8. **(a)** $A(t) = 12{,}000\left(1 + \dfrac{0.056}{12}\right)^{12t}$
(b) \$14,195.06 **(c)** 9.249 yr
9. **(a)**

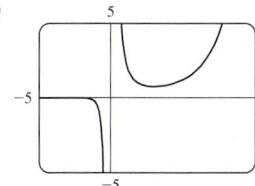

(b) $x = 0$, $y = 0$
(c) Local minimum $\approx$ $(3.00, 0.74)$
(d) $(-\infty, 0) \cup [0.74, \infty)$
(e) -0.85, 0.96, 9.92

Focus on Modeling ■ page 393

1. **(a)**

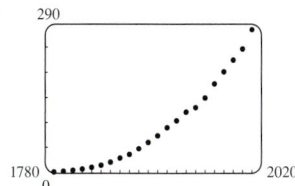

(b) $y = ab^t$, where $a = 1.180609 \times 10^{-15}$, $b = 1.0204139$, and y is the population in millions in the year t **(c)** 515.9 million
(d) 207.8 million **(e)** No

3. **(a)** Yes **(b)** Yes, the scatter plot appears linear.

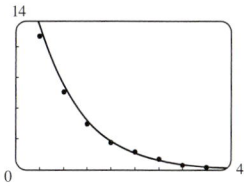

(c) $\ln E = 4.494411 + 0.0970921464t$, where t is years since 1970 and E is expenditure in billions of dollars
(d) $E = 89.51543173e^{at}$, where $a = 0.0970921464$
(e) 3948.2 billion dollars
5. **(a)** $I_0 = 22.7586444$, $k = 0.1062398$
(b)

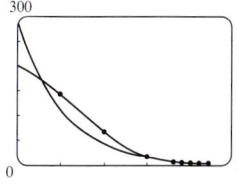

(c) 47.3 ft

7. **(a)** $y = ab^t$, where $a = 301.813054$, $b = 0.819745$, and t is the number of years since 1970
(b) $y = at^4 + bt^3 + ct^2 + dt + e$, where $a = -0.002430$, $b = 0.135159$, $c = -2.014322$, $d = -4.055294$, $e = 199.092227$, and t is the number of years since 1970
(c) From the graphs we see that the fourth-degree polynomial is a better model.

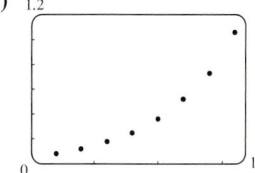

(d) 202.8, 27.8; 184.0, 43.5

9. **(a)**

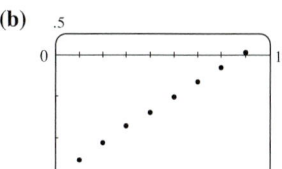

(b)

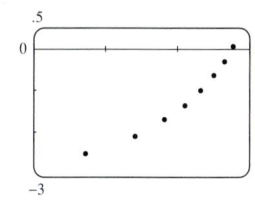

(c) Exponential function
(d) $y = ab^x$ where $a = 0.057697$ and $b = 1.200236$
11. **(a)** $y = \dfrac{c}{1 + ae^{-bx}}$, where $a = 49.10976596$, $b = 0.4981144989$, and $c = 500.855793$ **(b)** 10.58 days

Chapter 5

Section 5.1 ■ page 406

7. $-\frac{4}{5}$ **9.** $-2\sqrt{2}/3$ **11.** $3\sqrt{5}/7$ **13.** $P\left(\frac{4}{5}, \frac{3}{5}\right)$

15. $P\left(-\sqrt{5}/3, \frac{2}{3}\right)$ **17.** $P\left(-\sqrt{2}/3, -\sqrt{7}/3\right)$

19. $t = \pi/4, \left(\sqrt{2}/2, \sqrt{2}/2\right); t = \pi/2, (0, 1);$
$t = 3\pi/4, \left(-\sqrt{2}/2, \sqrt{2}/2\right); t = \pi, (-1, 0);$
$t = 5\pi/4, \left(-\sqrt{2}/2, -\sqrt{2}/2\right); t = 3\pi/2, (0, -1);$
$t = 7\pi/4, \left(\sqrt{2}/2, -\sqrt{2}/2\right); t = 2\pi, (1, 0)$

21. $(0, 1)$ **23.** $\left(-\sqrt{3}/2, \frac{1}{2}\right)$ **25.** $\left(\frac{1}{2}, -\sqrt{3}/2\right)$

27. $\left(-\frac{1}{2}, \sqrt{3}/2\right)$ **29.** $\left(-\sqrt{2}/2, -\sqrt{2}/2\right)$

31. (a) $\left(-\frac{3}{5}, \frac{4}{5}\right)$ (b) $\left(\frac{3}{5}, -\frac{4}{5}\right)$ (c) $\left(-\frac{3}{5}, -\frac{4}{5}\right)$ (d) $\left(\frac{3}{5}, \frac{4}{5}\right)$

33. (a) $\pi/4$ (b) $\pi/3$ (c) $\pi/3$ (d) $\pi/6$

35. (a) $2\pi/7$ (b) $2\pi/9$ (c) $\pi - 3 \approx 0.14$
(d) $2\pi - 5 \approx 1.28$ **37.** (a) $\pi/3$ (b) $\left(-\frac{1}{2}, \sqrt{3}/2\right)$

39. (a) $\pi/4$ (b) $\left(-\sqrt{2}/2, \sqrt{2}/2\right)$

41. (a) $\pi/3$ (b) $\left(-\frac{1}{2}, -\sqrt{3}/2\right)$

43. (a) $\pi/4$ (b) $\left(-\sqrt{2}/2, -\sqrt{2}/2\right)$

45. (a) $\pi/6$ (b) $\left(-\sqrt{3}/2, -\frac{1}{2}\right)$

47. (a) $\pi/3$ (b) $\left(\frac{1}{2}, \sqrt{3}/2\right)$ **49.** (a) $\pi/3$
(b) $\left(-\frac{1}{2}, -\sqrt{3}/2\right)$ **51.** $(0.5, 0.8)$ **53.** $(0.5, -0.9)$

Section 5.2 ■ page 416

1. $t = \pi/4, \sin t = \sqrt{2}/2, \cos t = \sqrt{2}/2; t = \pi/2, \sin t = 1,$
$\cos t = 0; t = 3\pi/4, \sin t = \sqrt{2}/2, \cos t = -\sqrt{2}/2;$
$t = \pi, \sin t = 0, \cos t = -1; t = 5\pi/4,$
$\sin t = -\sqrt{2}/2, \cos t = -\sqrt{2}/2; t = 3\pi/2, \sin t = -1,$
$\cos t = 0; t = 7\pi/4, \sin t = -\sqrt{2}/2, \cos t = \sqrt{2}/2;$
$t = 2\pi, \sin t = 0, \cos t = 1$ **3.** (a) $\sqrt{3}/2$ (b) $-1/2$
(c) $-\sqrt{3}$ **5.** (a) $-1/2$ (b) $-1/2$ (c) $-1/2$
7. (a) $-\sqrt{2}/2$ (b) $-\sqrt{2}/2$ (c) $\sqrt{2}/2$
9. (a) $\sqrt{3}/2$ (b) $2\sqrt{3}/3$ (c) $\sqrt{3}/3$
11. (a) -1 (b) 0 (c) 0 **13.** (a) 2 (b) $-2\sqrt{3}/3$ (c) 2
15. (a) $-\sqrt{3}/3$ (b) $\sqrt{3}/3$ (c) $-\sqrt{3}/3$
17. (a) $\sqrt{2}/2$ (b) $-\sqrt{2}$ (c) -1
19. (a) -1 (b) 1 (c) -1 **21.** (a) 0 (b) 1 (c) 0
23. $\sin 0 = 0, \cos 0 = 1, \tan 0 = 0, \sec 0 = 1,$
others undefined   **25.** $\sin \pi = 0, \cos \pi = -1, \tan \pi = 0,$
$\sec \pi = -1$, others undefined   **27.** $\frac{4}{5}, \frac{3}{5}, \frac{4}{3}$
29. $-\sqrt{11}/4, \sqrt{5}/4, -\sqrt{55}/5$ **31.** $\sqrt{13}/7, -6/7, -\sqrt{13}/6$
33. $-\frac{12}{13}, -\frac{5}{13}, \frac{12}{5}$ **35.** $\frac{21}{29}, -\frac{20}{29}, -\frac{21}{20}$ **37.** (a) 0.8 (b) 0.84147
39. (a) 0.9 (b) 0.93204 **41.** (a) 1 (b) 1.02964
43. (a) -0.6 (b) -0.57482 **45.** Negative
47. Negative   **49.** II   **51.** II   **53.** $\sin t = \sqrt{1 - \cos^2 t}$
55. $\tan t = (\sin t)/\sqrt{1 - \sin^2 t}$ **57.** $\sec t = -\sqrt{1 + \tan^2 t}$
59. $\tan t = \sqrt{\sec^2 t - 1}$ **61.** $\tan^2 t = (\sin^2 t)/(1 - \sin^2 t)$
63. $\cos t = -\frac{4}{5}, \tan t = -\frac{3}{4}, \csc t = \frac{5}{3}, \sec t = -\frac{5}{4}, \cot t = -\frac{4}{3}$
65. $\sin t = -2\sqrt{2}/3, \cos t = \frac{1}{3}, \tan t = -2\sqrt{2},$
$\csc t = -\frac{3}{4}\sqrt{2}, \cot t = -\sqrt{2}/4$

67. $\sin t = -\frac{3}{5}, \cos t = \frac{4}{5}, \csc t = -\frac{5}{3}, \sec t = \frac{5}{4}, \cot t = -\frac{4}{3}$
69. $\cos t = -\sqrt{15}/4, \tan t = \sqrt{15}/15, \csc t = -4,$
$\sec t = -4\sqrt{15}/15, \cot t = \sqrt{15}$
71. Odd   **73.** Odd   **75.** Even   **77.** Neither
79. $y(0) = 4, y(0.25) = -2.828, y(0.50) = 0,$
$y(0.75) = 2.828, y(1.00) = -4, y(1.25) = 2.828$
81. (a) 0.49870 amp   (b) -0.17117 amp

Section 5.3 ■ page 429

1.

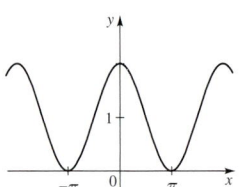

3.

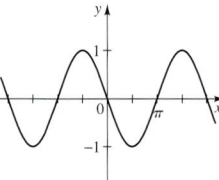

5.

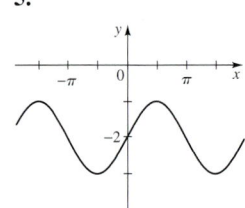

7.

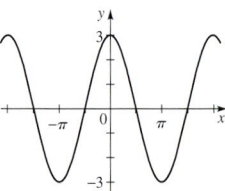

9.

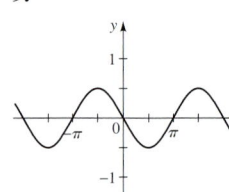

11.

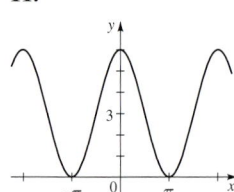

13.

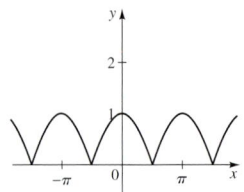

15.

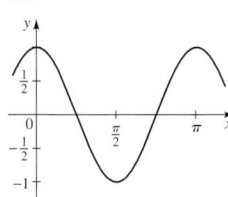

17. $3, 2\pi/3$

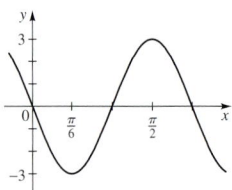

19. $10, 4\pi$

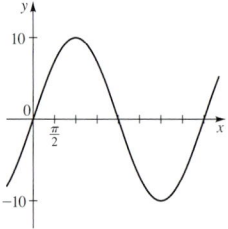

21. $\frac{1}{3}, 6\pi$

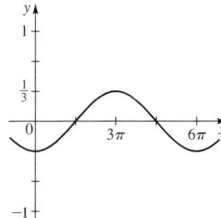

39. $1, 2\pi/3, -\pi/3$

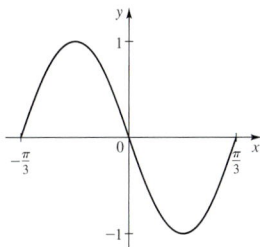

23. 2, 1

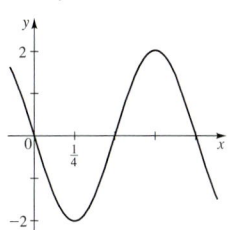

25. $\frac{1}{2}, 2$

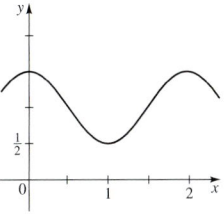

41. (a) $4, 2\pi, 0$ **(b)** $y = 4 \sin x$
43. (a) $\frac{3}{2}, \frac{2\pi}{3}, 0$ **(b)** $y = \frac{3}{2} \cos 3x$
45. (a) $\frac{1}{2}, \pi, -\frac{\pi}{3}$ **(b)** $y = -\frac{1}{2} \cos 2(x + \pi/3)$
47. (a) $4, \frac{3}{2}, -\frac{1}{2}$ **(b)** $y = 4 \sin \frac{4\pi}{3}(x + \frac{1}{2})$

49.

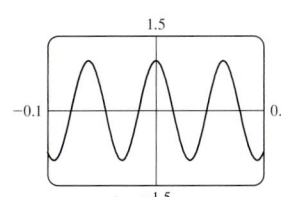

51.

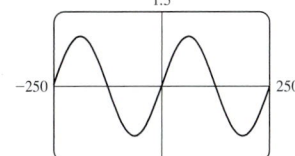

27. $1, 2\pi, \pi/2$

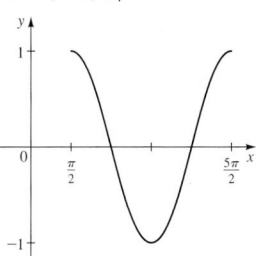

29. $2, 2\pi, \pi/6$

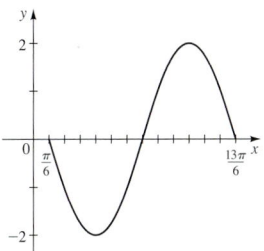

53.

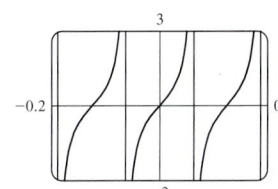

55.

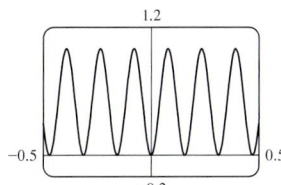

31. $4, \pi, -\pi/2$

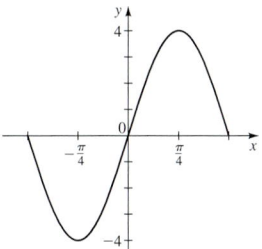

33. $5, 2\pi/3, \pi/12$

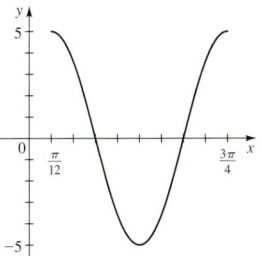

57.

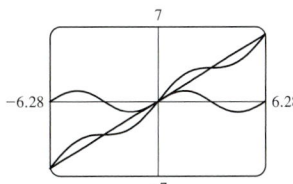

59.

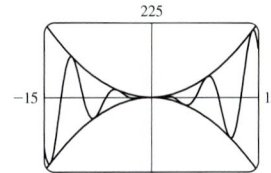

$y = x^2 \sin x$ is a sine curve that lies between the graphs of $y = x^2$ and $y = -x^2$

35. $\frac{1}{2}, \pi, \pi/6$

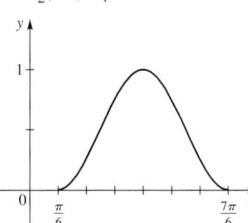

37. $3, 2, -\frac{1}{2}$

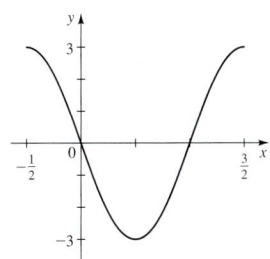

61.

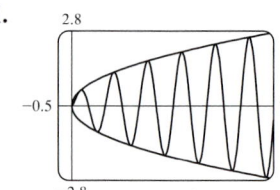

$y = \sqrt{x} \sin 5\pi x$ is a sine curve that lies between the graphs of $y = \sqrt{x}$ and $y = -\sqrt{x}$

63.

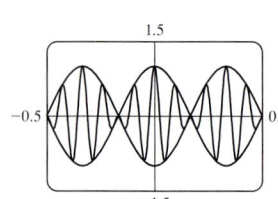

$y = \cos 3\pi x \cos 21\pi x$ is a cosine curve that lies between the graphs of $y = \cos 3\pi x$ and $y = -\cos 3\pi x$

65. Maximum value 1.76 when $x \approx 0.94$, minimum value -1.76 when $x \approx -0.94$ (The same maximum and minimum values occur at infinitely many other values of x.)
67. Maximum value 3.00 when $x \approx 1.57$, minimum value -1.00 when $x \approx -1.57$ (The same maximum and minimum values occur at infinitely many other values of x.)
69. 1.16   **71.** 0.34, 2.80
73. (a) Odd   **(b)** $0, \pm 2\pi, \pm 4\pi, \pm 6\pi, \ldots$
(c)

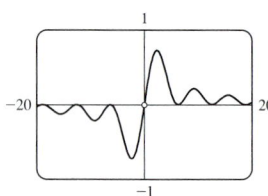

(d) $f(x)$ approaches 0
(e) $f(x)$ approaches 0

75. (a) 20 s   **(b)** 6 ft
77. (a) $\frac{1}{80}$ min   **(b)** 80
(c)

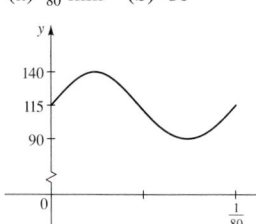

(d) $\frac{140}{90}$; it is higher than normal

Section 5.4 ■ page 441
1. II   **3.** VI   **5.** IV
7. π

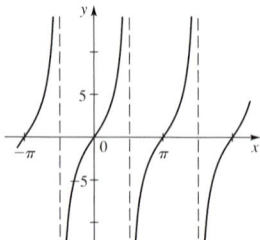

9. π

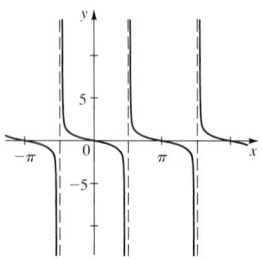

11. π

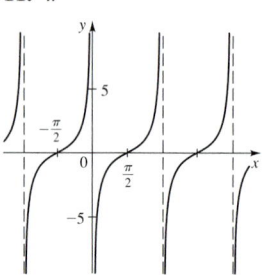

13. 2π

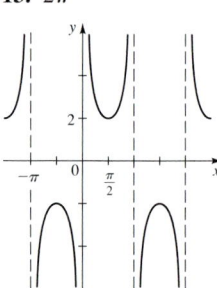

15. 2π

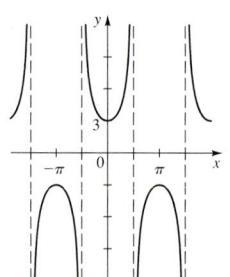

17. π

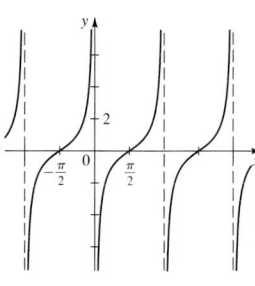

19. 2π

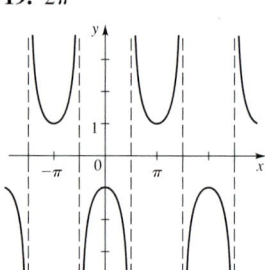

21. π

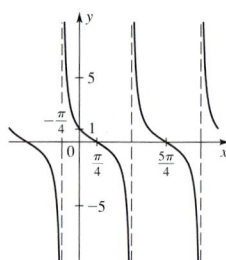

23. 2π

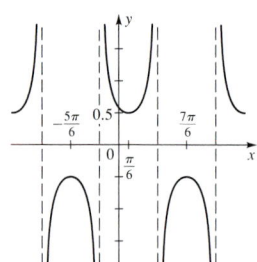

25. $\pi/2$

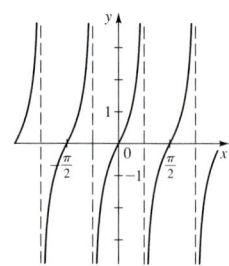

27. 4

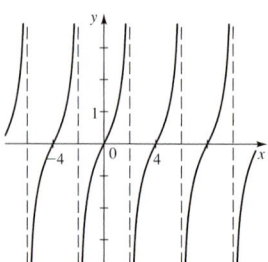

29. π

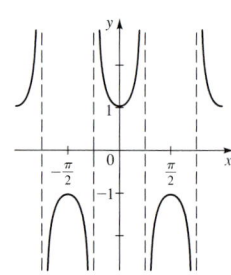

31. π

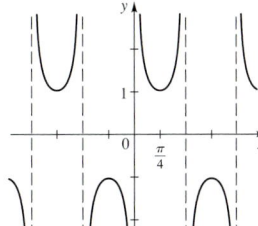

33. $\frac{1}{3}$

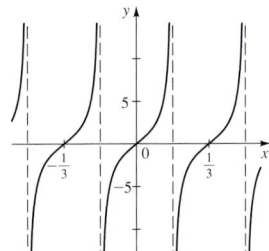

35. $\frac{4}{3}$

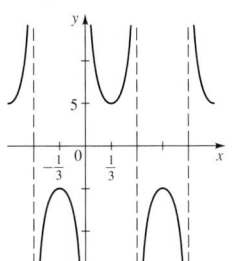

37. $\pi/2$

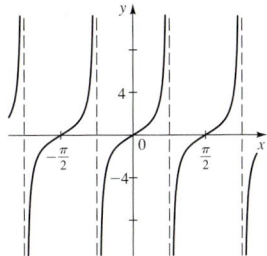

39. $\pi/2$

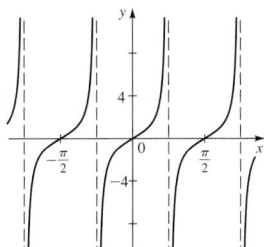

41. $\pi/2$

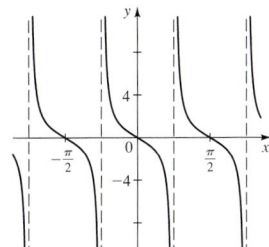

43. 2

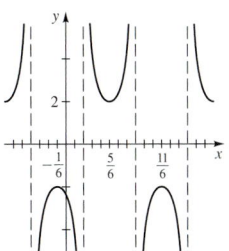

45. $2\pi/3$

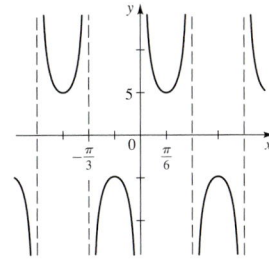

47. $3\pi/2$

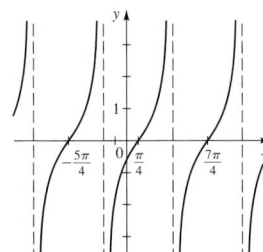

49. 2

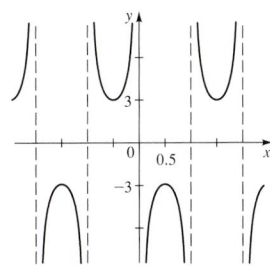

51. $\pi/2$

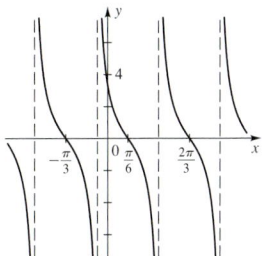

55. **(a)** 1.53 mi, 3.00 mi, 18.94 mi
(b)

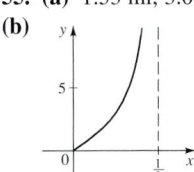

(c) $d(t)$ approaches ∞

Section 5.5 ■ **page 451**

1. (a) 2, $2\pi/3$, $3/(2\pi)$ **(b)**

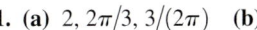
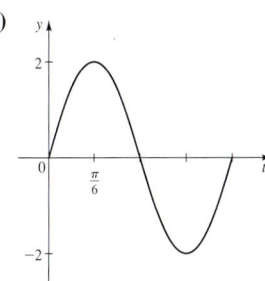

3. (a) 1, $20\pi/3$, $3/(20\pi)$ **(b)**

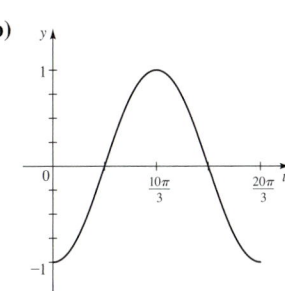

5. (a) $\frac{1}{4}$, $4\pi/3$, $3/(4\pi)$ **(b)**

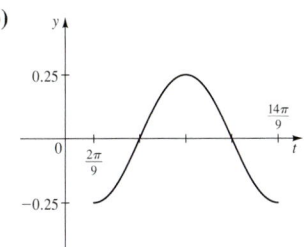

7. (a) $5, 3\pi, 1/(3\pi)$ **(b)**

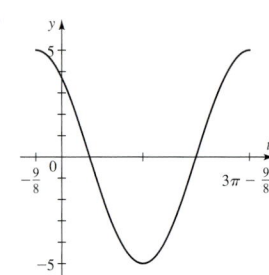

9. $y = 10 \sin\left(\dfrac{2\pi}{3}t\right)$ **11.** $y = 6 \sin(10t)$

13. $y = 60 \cos(4\pi t)$ **15.** $y = 2.4 \cos(1500\pi t)$

17. (a) $y = 2e^{-1.5t} \cos 6\pi t$ **(b)**

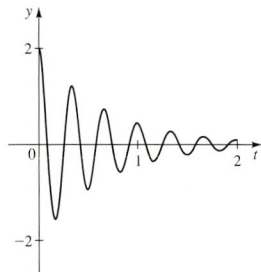

19. (a) $y = 100e^{-0.05t} \cos \dfrac{\pi}{2} t$

(b)

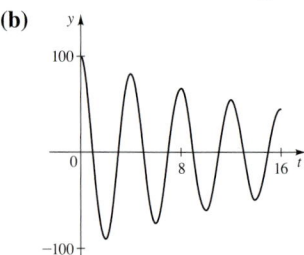

21. (a) $y = 7e^{-10t} \sin 12t$ **(b)**

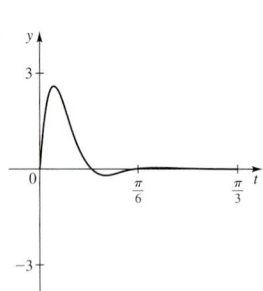

23. (a) $y = 0.3e^{-0.2t} \sin(40\pi t)$

(b)

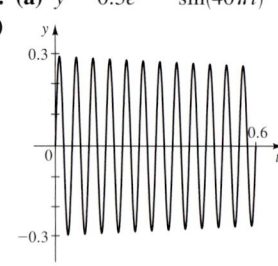

25. (a) 10 cycles per minute

(b) **(c)** 0.4 m

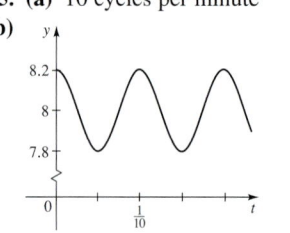

27. (a) 8900 **(b)** about 3.14 yr **29.** $d(t) = 5 \sin(5\pi t)$

31. $y = 21 \sin\left(\dfrac{\pi}{6}t\right)$

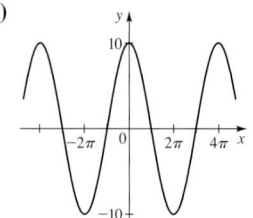

33. $y = 5 \cos(2\pi t)$ **35.** $y = 11 + 10 \sin\left(\dfrac{\pi t}{10}\right)$

37. $y = 3.8 + 0.2 \sin\left(\dfrac{\pi}{5}t\right)$

39. $E(t) = 310 \cos(200\pi t)$, 219.2 V

41. (a) 45 V **(b)** 40 **(c)** 40 **(d)** $E(t) = 45 \cos(80\pi t)$

43. $f(t) = e^{-0.9t} \sin \pi t$ **45.** $e = \dfrac{1}{3} \ln 4 \approx 0.46$

Chapter 5 Review ■ page 455

1. (b) $\frac{1}{2}, -\sqrt{3}/2, -\sqrt{3}/3$ **3. (a)** $\pi/3$ **(b)** $\left(-\frac{1}{2}, \sqrt{3}/2\right)$
(c) $\sin t = \sqrt{3}/2$, $\cos t = -\frac{1}{2}$, $\tan t = -\sqrt{3}$, $\csc t = 2\sqrt{3}/3$,
$\sec t = -2$, $\cot t = -\sqrt{3}/3$ **5. (a)** $\pi/4$
(b) $\left(-\sqrt{2}/2, -\sqrt{2}/2\right)$ **(c)** $\sin t = -\sqrt{2}/2$, $\cos t = -\sqrt{2}/2$,
$\tan t = 1$, $\csc t = -\sqrt{2}$, $\sec t = -\sqrt{2}$, $\cot t = 1$
7. (a) $\sqrt{2}/2$ **(b)** $-\sqrt{2}/2$ **9. (a)** 0.89121 **(b)** 0.45360
11. (a) 0 **(b)** Undefined **13. (a)** Undefined **(b)** 0
15. (a) $-\sqrt{3}/3$ **(b)** $-\sqrt{3}$ **17.** $(\sin t)/(1 - \sin^2 t)$
19. $(\sin t)/\sqrt{1 - \sin^2 t}$
21. $\tan t = -\frac{5}{12}$, $\csc t = \frac{13}{5}$, $\sec t = -\frac{13}{12}$, $\cot t = -\frac{12}{5}$
23. $\sin t = 2\sqrt{5}/5$, $\cos t = -\sqrt{5}/5$,
$\tan t = -2$, $\sec t = -\sqrt{5}$ **25.** $(16 - \sqrt{17})/4$ **27.** 3
29. (a) $10, 4\pi, 0$ **31. (a)** $1, 4\pi, 0$

(b) **(b)**

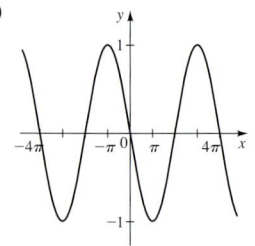

33. (a) $3, \pi, 1$
(b)

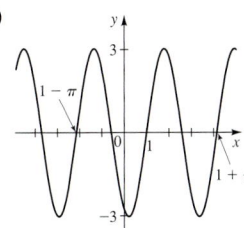

35. (a) $1, 4, -\frac{1}{3}$
(b)

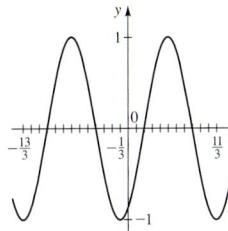

37. $y = 5 \sin 4x$ **39.** $y = \frac{1}{2} \sin 2\pi\left(x + \frac{1}{3}\right)$
41. π **43.** π

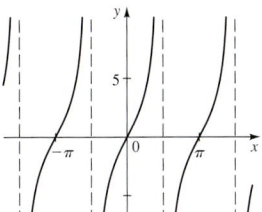

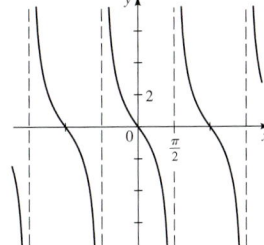

45. π **47.** 2π

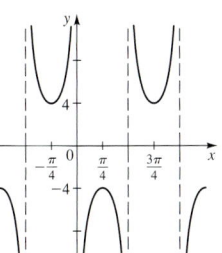

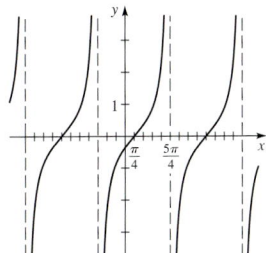

49. (a)

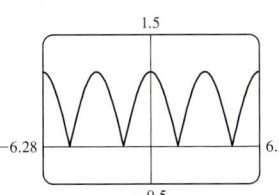

51. (a)

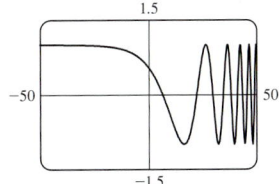

(b) Period π **(b)** Not periodic
(c) Even   **(c)** Neither
53. (a)   **55.**

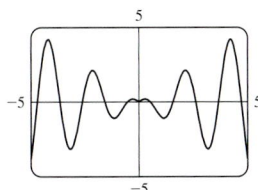

(b) Not periodic
(c) Even

$y = x \sin x$ is a sine function whose graph lies between those of $y = x$ and $y = -x$

57.

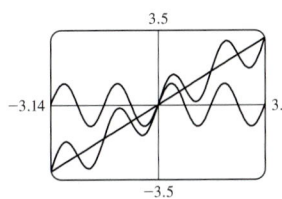

The graphs are related by graphical addition.

59. $1.76, -1.76$ **61.** $0.30, 2.84$
63. (a) Odd   **(b)** $0, \pm\pi, \pm2\pi, \ldots$
(c)

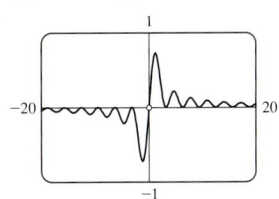

(d) $f(x)$ approaches 0
(e) $f(x)$ approaches 0

65. $y = 50 \cos(16\pi t)$ **67.** $y = 4 \cos\left(\frac{\pi}{6} t\right)$

Chapter 5 Test ■ page 458

1. $y = -\frac{5}{6}$ **2. (a)** $\frac{4}{5}$ **(b)** $-\frac{3}{5}$ **(c)** $-\frac{4}{3}$ **(d)** $-\frac{5}{3}$
3. (a) $-\frac{1}{2}$ **(b)** $-\sqrt{2}/2$ **(c)** $\sqrt{3}$ **(d)** -1
4. $\tan t = -(\sin t)/\sqrt{1 - \sin^2 t}$ **5.** $-\frac{2}{15}$
6. (a) $5, \pi/2, 0$ **7. (a)** $2, 4\pi, \pi/3$
(b)

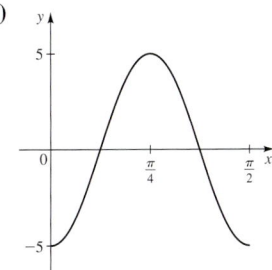

(b)

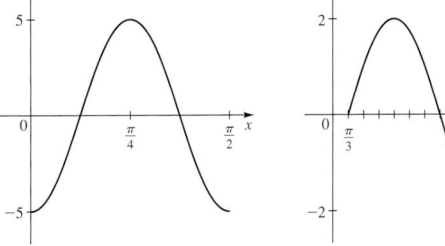

8. π **9.** $\pi/2$

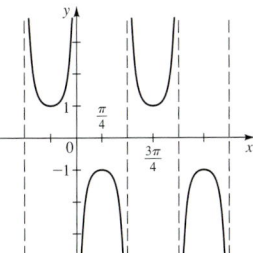

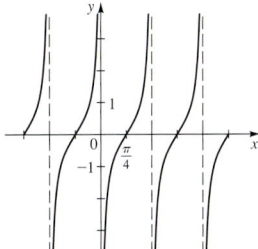

10. $y = 2 \sin 2(x + \pi/3)$
11. (a)

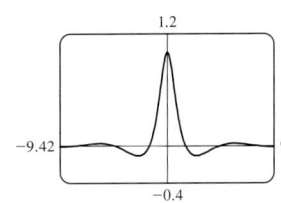

(b) Even
(c) Minimum value -0.11 when $x \approx \pm2.54$, maximum value 1 when $x = 0$

12. $y = 5 \sin(4\pi t)$

13. $y = 16e^{-0.1t} \cos 24\pi t$

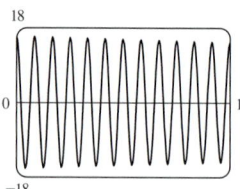

Focus on Modeling ■ **page 463**

1. (a) and **(c)**

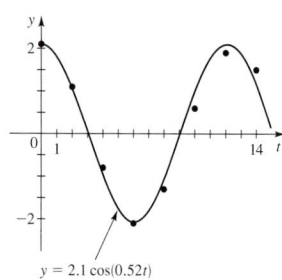

$y = 2.1 \cos(0.52t)$

(b) $y = 2.1 \cos(0.52t)$

(d) $y = 2.05 \sin(0.50t + 1.55) - 0.01$

(e) The formula of (d) reduces to
$y = 2.05 \cos(0.50t - 0.02) - 0.01$.
Same as (b), correct to one decimal.

3. (a) and **(c)**

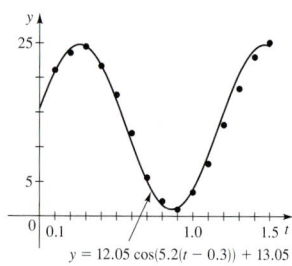

$y = 12.05 \cos(5.2(t - 0.3)) + 13.05$

(b) $y = 12.05 \cos(5.2(t - 0.3)) + 13.05$

(d) $y = 11.72 \sin(5.05t + 0.24) + 12.96$

(e) The formula of (d) reduces to
$y = 11.72 \cos(5.05(t - 0.26)) + 12.96$.
Close, but not identical, to (b).

5. (a) and **(c)**

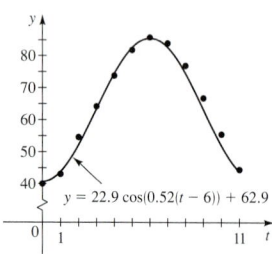

$y = 22.9 \cos(0.52(t - 6)) + 62.9$

(b) $y = 22.9 \cos(0.52(t - 6)) + 62.9$,
where y is temperature (°F) and t is months (January = 0)

(d) $y = 23.4 \sin(0.48t - 1.36) + 62.2$

7. (a) and **(c)**

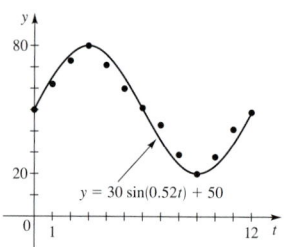

$y = 30 \sin(0.52t) + 50$

(b) $y = 30 \sin(0.52t) + 50$ where y is the owl population in
year t **(d)** $y = 25.8 \sin(0.52t - 0.02) + 50.6$

9. (a) and **(c)**

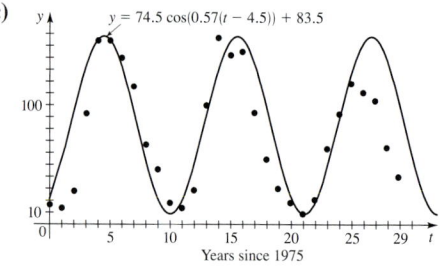

$y = 74.5 \cos(0.57(t - 4.5)) + 83.5$

Years since 1975

(b) $y = 74.5 \cos(0.57(t - 4.5)) + 83.5$, where y is the average
daily sunspot count, and t is the years since 1975

(d) $y = 67.65 \sin(0.62t - 1.65) + 74.5$

Chapter 6

Section 6.1 ■ **page 474**

1. $2\pi/5 \approx 1.257$ rad   **3.** $-\pi/4 \approx -0.785$ rad

5. $-5\pi/12 \approx -1.309$ rad   **7.** $6\pi \approx 18.850$ rad

9. $8\pi/15 \approx 1.676$ rad   **11.** $\pi/24 \approx 0.131$ rad

13. $210°$ **15.** $-225°$ **17.** $540/\pi \approx 171.9°$

19. $-216/\pi \approx 68.8°$ **21.** $18°$ **23.** $-24°$

25. $410°, 770°, -310°, -670°$

27. $11\pi/4, 19\pi/4, -5\pi/4, -13\pi/4$

29. $7\pi/4, 15\pi/4, -9\pi/4, -17\pi/4$ **31.** Yes   **33.** Yes

35. Yes   **37.** $13°$ **39.** $30°$ **41.** $280°$ **43.** $5\pi/6$

45. π **47.** $\pi/4$ **49.** $55\pi/9 \approx 19.2$ **51.** 4 **53.** 4 mi

55. 2 rad $\approx 114.6°$ **57.** $36/\pi \approx 11.459$ m

59. (a) 35.45 **(b)** 25 **61.** 50 m^2 **63.** 4 m

65. 6 cm^2 **67.** 13.9 mi   **69.** 330π mi ≈ 1037 mi

71. 1.6 million mi   **73.** 1.15 mi   **75.** 360π in$^2 \approx 1130.97$ in^2

77. $32\pi/15$ ft/s ≈ 6.7 ft/s   **79. (a)** 2000π rad/min

(b) $50\pi/3$ ft/s ≈ 52.4 ft/s   **81.** 39.3 mi/h   **83.** 2.1 m/s

85. (a) 10π cm ≈ 31.4 cm   **(b)** 5 cm   **(c)** 3.32 cm

(d) 86.8 cm^3

Section 6.2 ■ **page 484**

1. $\sin\theta = \frac{4}{5}$, $\cos\theta = \frac{3}{5}$,
$\tan\theta = \frac{4}{3}$, $\csc\theta = \frac{5}{4}$, $\sec\theta = \frac{5}{3}$, $\cot\theta = \frac{3}{4}$

3. $\sin\theta = \frac{40}{41}$, $\cos\theta = \frac{9}{41}$, $\tan\theta = \frac{40}{9}$, $\csc\theta = \frac{41}{40}$, $\sec\theta = \frac{41}{9}$,
$\cot\theta = \frac{9}{40}$

5. $\sin \theta = 2\sqrt{13}/13$, $\cos \theta = 3\sqrt{13}/13$, $\tan \theta = \frac{2}{3}$,
$\csc \theta = \sqrt{13}/2$, $\sec \theta = \sqrt{13}/3$, $\cot \theta = \frac{3}{2}$
7. (a) $3\sqrt{34}/34$, $3\sqrt{34}/34$ **(b)** $\frac{3}{5}, \frac{3}{5}$ **(c)** $\sqrt{34}/5$, $\sqrt{34}/5$
9. $\frac{25}{2}$ **11.** $13\sqrt{3}/2$ **13.** 16.51658
15. $x = 28 \cos \theta$, $y = 28 \sin \theta$
17. $\cos \theta = \frac{4}{5}$, $\tan \theta = \frac{3}{4}$, $\csc \theta = \frac{5}{3}$, $\sec \theta = \frac{5}{4}$, $\cot \theta = \frac{4}{3}$

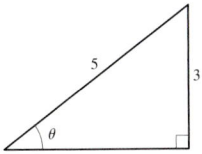

19. $\sin \theta = \sqrt{2}/2$, $\cos \theta = \sqrt{2}/2$, $\tan \theta = 1$,
$\csc \theta = \sqrt{2}$, $\sec \theta = \sqrt{2}$

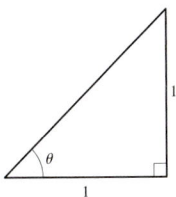

21. $\sin \theta = 3\sqrt{5}/7$, $\cos \theta = \frac{2}{7}$, $\tan \theta = 3\sqrt{5}/2$,
$\csc \theta = 7\sqrt{5}/15$, $\cot \theta = 2\sqrt{5}/15$

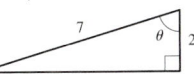

23. $(1 + \sqrt{3})/2$ **25.** 1 **27.** $\frac{1}{2}$
29.

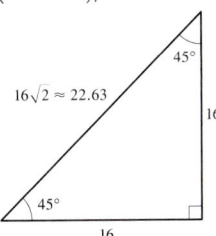

31.

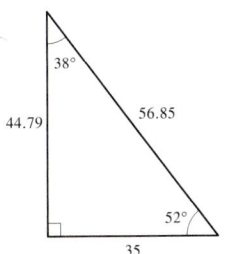

33.

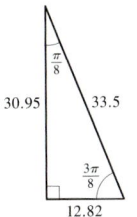

35.

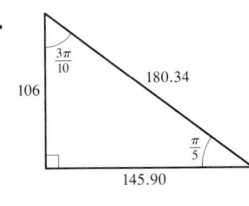

37. $\sin \theta \approx 0.45$, $\cos \theta \approx 0.89$, $\tan \theta = 0.50$, $\csc \theta \approx 2.24$,
$\sec \theta \approx 1.12$, $\cot \theta = 2.00$ **39.** 230.9 **41.** 63.7
43. $x = 10 \tan \theta \sin \theta$ **45.** 1026 ft **47. (a)** 2100 mi
(b) No **49.** 19 ft **51.** 38.7° **53.** 345 ft **55.** 415 ft, 152 ft
57. 2570 ft **59.** 5808 ft **61.** 91.7 million mi
63. 3960 mi **65.** 0.723 AU

Section 6.3 ■ page 495

1. (a) 30° **(b)** 30° **(c)** 30° **3. (a)** 45° **(b)** 90° **(c)** 75°
5. (a) $\pi/4$ **(b)** $\pi/6$ **(c)** $\pi/3$ **7. (a)** $2\pi/7$ **(b)** 0.4π

(c) 1.4 **9.** $\frac{1}{2}$ **11.** $-\sqrt{2}/2$ **13.** $-\sqrt{3}$ **15.** 1 **17.** $-\sqrt{3}/2$
19. $\sqrt{3}/3$ **21.** $\sqrt{3}/2$ **23.** -1 **25.** $\frac{1}{2}$ **27.** 2 **29.** -1
31. Undefined **33.** III **35.** IV
37. $\tan \theta = -\sqrt{1 - \cos^2\theta}/\cos \theta$ **39.** $\cos \theta = \sqrt{1 - \sin^2\theta}$
41. $\sec \theta = -\sqrt{1 + \tan^2\theta}$
43. $\cos \theta = -\frac{4}{5}$, $\tan \theta = -\frac{3}{4}$, $\csc \theta = \frac{5}{3}$, $\sec \theta = -\frac{5}{4}$,
$\cot \theta = -\frac{4}{3}$
45. $\sin \theta = -\frac{3}{5}$, $\cos \theta = \frac{4}{5}$, $\csc \theta = -\frac{5}{3}$, $\sec \theta = \frac{5}{4}$, $\cot \theta = -\frac{4}{3}$
47. $\sin \theta = \frac{1}{2}$, $\cos \theta = \sqrt{3}/2$, $\tan \theta = \sqrt{3}/3$,
$\sec \theta = 2\sqrt{3}/3$, $\cot \theta = \sqrt{3}$
49. $\sin \theta = 3\sqrt{5}/7$, $\tan \theta = -3\sqrt{5}/2$, $\csc \theta = 7\sqrt{5}/15$,
$\sec \theta = -\frac{7}{2}$, $\cot \theta = -2\sqrt{5}/15$
51. (a) $\sqrt{3}/2$, $\sqrt{3}$ **(b)** $\frac{1}{2}$, $\sqrt{3}/4$ **(c)** $\frac{3}{4}$, 0.88967
53. 19.1 **55.** 66.1° **57.** $(4\pi/3) - \sqrt{3} \approx 2.46$
61. (b)

θ	20°	60°	80°	85°
h	1922	9145	29,944	60,351

63. (a) $A(\theta) = 400 \sin \theta \cos \theta$
(b)

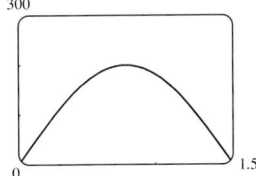

(c) width = depth $\approx$ 14.14 in.
65. (a) $9\sqrt{3}/4$ ft $\approx$ 3.897 ft, $\frac{9}{6}$ ft $\approx$ 0.5625 ft
(b) 23.982 ft, 3.462 ft
67. (a)

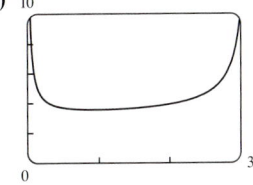

(b) 0.946 rad or 54°

69. 42°

Section 6.4 ■ page 506

1. 318.8 **3.** 24.8 **5.** 44° **7.** $\angle C = 114°$, $a \approx 51$, $b \approx 24$
9. $\angle A = 44°$, $\angle B = 68°$, $a \approx 8.99$
11. $\angle C = 62°$, $a \approx 200$, $b \approx 242$

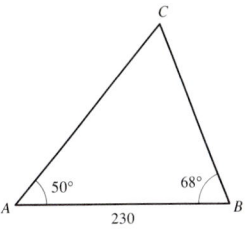

13. $\angle B = 85°$, $a \approx 5$, $c \approx 9$

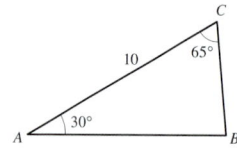

15. $\angle A = 100°$, $a \approx 89$, $c \approx 71$

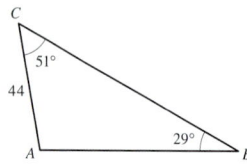

17. $\angle B \approx 30°$, $\angle C \approx 40°$, $c \approx 19$ **19.** No solution
21. $\angle A_1 \approx 125°$, $\angle C_1 \approx 30°$, $a_1 \approx 49$;
$\angle A_2 \approx 5°$, $\angle C_2 \approx 150°$, $a_2 \approx 5.6$
23. No solution   **25.** $\angle A_1 \approx 57.2°$, $\angle B_1 \approx 93.8°$, $b_1 \approx 30.9$;
$\angle A_2 \approx 122.8°$, $\angle B_2 \approx 28.2°$, $b_2 \approx 14.6$
27. (a) $91.146°$ **(b)** $14.427°$ **31. (a)** 1018 mi
(b) 1017 mi   **33.** 219 ft   **35.** 55.9 m   **37.** 175 ft
39. 192 m   **41.** 0.427 AU, 1.119 AU

Section 6.5 ■ page 513

1. 28.9 **3.** 47 **5.** $29.89°$ **7.** 15
9. $\angle A \approx 39.4°$, $\angle B \approx 20.6°$, $c \approx 24.6$
11. $\angle A \approx 48°$, $\angle B \approx 79°$, $c \approx 3.2$
13. $\angle A \approx 50°$, $\angle B \approx 73°$, $\angle C \approx 57°$
15. $\angle A_1 \approx 83.6°$, $\angle C_1 \approx 56.4°$, $a_1 \approx 193$;
$\angle A_2 \approx 16.4°$, $\angle C_2 \approx 123.6$, $a_2 \approx 54.9$
17. No such triangle   **19.** 2 **21.** 25.4 **23.** $89.2°$
25. 24.3 **27.** 54 **29.** 26.83 **31.** 5.33 **33.** 40.77
35. 3.85 cm^2 **37.** 2.30 mi   **39.** 23.1 mi   **41.** 2179 mi
43. (a) 62.6 mi   **(b)** S $18.2°$ E   **45.** $96°$ **47.** 211 ft
49. 3835 ft   **51.** $\$165,554$

Chapter 6 Review ■ page 516

1. (a) $\pi/3$ **(b)** $11\pi/6$ **(c)** $-3\pi/4$ **(d)** $-\pi/2$
3. (a) $450°$ **(b)** $-30°$ **(c)** $405°$ **(d)** $(558/\pi)° \approx 177.6°$
5. 8 m   **7.** 82 ft   **9.** 0.619 rad $\approx 35.4°$ **11.** $18{,}151$ ft^2
13. 300π rad/min ≈ 942.5 rad/min, 7539.8 in./min $=$
628.3 ft/min   **15.** $\sin \theta = 5/\sqrt{74}$, $\cos \theta = 7/\sqrt{74}$, $\tan \theta = \frac{5}{7}$,
$\csc \theta = \sqrt{74}/5$, $\sec \theta = \sqrt{74}/7$, $\cot \theta = \frac{7}{5}$
17. $x \approx 3.83$, $y \approx 3.21$ **19.** $x \approx 2.92$, $y \approx 3.11$
21.

23. $a = \cot \theta$, $b = \csc \theta$ **25.** 48 m   **27.** 1076 mi
29. $-\sqrt{2}/2$ **31.** 1 **33.** $-\sqrt{3}/3$ **35.** $-\sqrt{2}/2$
37. $2\sqrt{3}/3$ **39.** $-\sqrt{3}$
41. $\sin \theta = \frac{12}{13}$, $\cos \theta = -\frac{5}{13}$, $\tan \theta = -\frac{12}{5}$,
$\csc \theta = \frac{13}{12}$, $\sec \theta = -\frac{13}{5}$, $\cot \theta = -\frac{5}{12}$ **43.** $60°$

45. $\tan \theta = -\sqrt{1 - \cos^2\theta}/\cos \theta$
47. $\tan^2\theta = \sin^2\theta/(1 - \sin^2\theta)$
49. $\sin \theta = \sqrt{7}/4$, $\cos \theta = \frac{3}{4}$, $\csc \theta = 4\sqrt{7}/7$, $\cot \theta = 3\sqrt{7}/7$
51. $\cos \theta = -\frac{4}{5}$, $\tan \theta = -\frac{3}{4}$, $\csc \theta = \frac{5}{3}$,
$\sec \theta = -\frac{5}{4}$, $\cot \theta = -\frac{4}{3}$
53. $-\sqrt{5}/5$ **55.** 1 **57.** 5.32 **59.** 148.07 **61.** 77.82
63. 77.3 mi   **65.** 3.9 mi   **67.** 32.12

Chapter 6 Test ■ page 520

1. $11\pi/6$, $-3\pi/4$ **2.** $240°$, $-74.5°$
3. (a) 240π rad/min ≈ 753.98 rad/min
(b) $12{,}063.7$ ft/min $= 137$ mi/h   **4. (a)** $\sqrt{2}/2$
(b) $\sqrt{3}/3$ **(c)** 2 **(d)** 1 **5.** $(26 + 6\sqrt{13})/39$
6. $a = 24 \sin \theta$, $b = 24 \cos \theta$ **7.** $(4 - 3\sqrt{2})/4$
8. $-\frac{13}{12}$ **9.** $\tan \theta = -\sqrt{\sec^2\theta - 1}$ **10.** 19.6 ft
11. 9.1 **12.** 250.5 **13.** 8.4 **14.** 19.5 **15. (a)** 15.3 m^2
(b) 24.3 m   **16. (a)** $129.9°$ **(b)** 44.9 **17.** 554 ft

Focus on Modeling ■ page 523

1. 1.41 mi   **3.** 14.3 m   **5. (c)** 2349.8 ft
7.

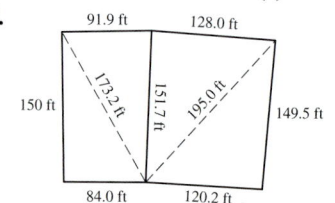

Chapter 7

Section 7.1 ■ page 533

1. $\sin t$ **3.** $\tan \theta$ **5.** -1 **7.** $\csc u$ **9.** $\tan \theta$ **11.** 1
13. $\cos y$ **15.** $\sin^2x$ **17.** $\sec x$ **19.** $2 \sec u$
21. $\cos^2x$ **23.** $\cos \theta$ **25.** LHS $= \sin \theta \dfrac{\cos \theta}{\sin \theta} =$ RHS
27. LHS $= \cos u \dfrac{1}{\cos u} \cot u =$ RHS
29. LHS $= \dfrac{\sin y}{\cos y} \sin y = \dfrac{1 - \cos^2y}{\cos y} = \sec y - \cos y =$ RHS
31. LHS $= \sin B + \cos B \dfrac{\cos B}{\sin B}$
$= \dfrac{\sin^2B + \cos^2B}{\sin B} = \dfrac{1}{\sin B} =$ RHS
33. LHS $= -\dfrac{\cos \alpha}{\sin \alpha} \cos \alpha - \sin \alpha = \dfrac{-\cos^2\alpha - \sin^2\alpha}{\sin \alpha}$
$= \dfrac{-1}{\sin \alpha} =$ RHS
35. LHS $= \dfrac{\sin \theta}{\cos \theta} + \dfrac{\cos \theta}{\sin \theta} = \dfrac{\sin^2\theta + \cos^2\theta}{\cos \theta \sin \theta}$
$= \dfrac{1}{\cos \theta \sin \theta} =$ RHS

37. LHS $= 1 - \cos^2\beta = \sin^2\beta =$ RHS

39. LHS $= \dfrac{(\sin x + \cos x)^2}{(\sin x + \cos x)(\sin x - \cos x)} = \dfrac{\sin x + \cos x}{\sin x - \cos x}$

$= \dfrac{(\sin x + \cos x)(\sin x - \cos x)}{(\sin x - \cos x)(\sin x - \cos x)} =$ RHS

41. LHS $= \dfrac{\frac{1}{\cos t} - \cos t}{\frac{1}{\cos t}} \cdot \dfrac{\cos t}{\cos t} = \dfrac{1 - \cos^2 t}{1} =$ RHS

43. LHS $= \dfrac{1}{\cos^2 y} = \sec^2 y =$ RHS

45. LHS $= \cot x \cos x + \cot x - \csc x \cos x - \csc x$

$= \dfrac{\cos^2 x}{\sin x} + \dfrac{\cos x}{\sin x} - \dfrac{\cos x}{\sin x} - \dfrac{1}{\sin x} = \dfrac{\cos^2 x - 1}{\sin x}$

$= \dfrac{-\sin^2 x}{\sin x} =$ RHS

47. LHS $= \sin^2 x\left(1 + \dfrac{\cos^2 x}{\sin^2 x}\right) = \sin^2 x + \cos^2 x =$ RHS

49. LHS $= 2(1 - \sin^2 x) - 1 = 2 - 2\sin^2 x - 1 =$ RHS

51. LHS $= \dfrac{1 - \cos\alpha}{\sin\alpha} \cdot \dfrac{1 + \cos\alpha}{1 + \cos\alpha}$

$= \dfrac{1 - \cos^2\alpha}{\sin\alpha(1 + \cos\alpha)} = \dfrac{\sin^2\alpha}{\sin\alpha(1 + \cos\alpha)} =$ RHS

53. LHS $= \dfrac{\sin^2\theta}{\cos^2\theta} - \dfrac{\sin^2\theta\cos^2\theta}{\cos^2\theta}$

$= \dfrac{\sin^2\theta(1 - \cos^2\theta)}{\cos^2\theta} = \dfrac{\sin^2\theta\sin^2\theta}{\cos^2\theta} =$ RHS

55. LHS $= \dfrac{\sin x - 1}{\sin x + 1} \cdot \dfrac{\sin x + 1}{\sin x + 1} = \dfrac{\sin^2 x - 1}{(\sin x + 1)^2} =$ RHS

57. LHS $= \dfrac{\sin^2 t + 2\sin t\cos t + \cos^2 t}{\sin t\cos t}$

$= \dfrac{\sin^2 t + \cos^2 t}{\sin t\cos t} + \dfrac{2\sin t\cos t}{\sin t\cos t} = \dfrac{1}{\sin t\cos t} + 2$

$=$ RHS

59. LHS $= \dfrac{1 + \frac{\sin^2 u}{\cos^2 u}}{1 - \frac{\sin^2 u}{\cos^2 u}} \cdot \dfrac{\cos^2 u}{\cos^2 u} = \dfrac{\cos^2 u + \sin^2 u}{\cos^2 u - \sin^2 u} =$ RHS

61. LHS $= \dfrac{\sec x}{\sec x - \tan x} \cdot \dfrac{\sec x + \tan x}{\sec x + \tan x}$

$= \dfrac{\sec x(\sec x + \tan x)}{\sec^2 x - \tan^2 x} =$ RHS

63. LHS $= (\sec v - \tan v) \cdot \dfrac{\sec v + \tan v}{\sec v + \tan v}$

$= \dfrac{\sec^2 v - \tan^2 v}{\sec v + \tan v} =$ RHS

65. LHS $= \dfrac{\sin x + \cos x}{\frac{1}{\cos x} + \frac{1}{\sin x}} = \dfrac{\sin x + \cos x}{\frac{\sin x + \cos x}{\cos x\sin x}}$

$= (\sin x + \cos x)\dfrac{\cos x\sin x}{\sin x + \cos x} =$ RHS

67. LHS $= \dfrac{\frac{1}{\sin x} - \frac{\cos x}{\sin x}}{\frac{1}{\cos x} - 1} \cdot \dfrac{\sin x\cos x}{\sin x\cos x} = \dfrac{\cos x(1 - \cos x)}{\sin x(1 - \cos x)}$

$= \dfrac{\cos x}{\sin x} =$ RHS

69. LHS $= \dfrac{\sin^2 u}{\cos^2 u} - \dfrac{\sin^2 u\cos^2 u}{\cos^2 u} = \dfrac{\sin^2 u}{\cos^2 u}(1 - \cos^2 u) =$ RHS

71. LHS $= (\sec^2 x - \tan^2 x)(\sec^2 x + \tan^2 x) =$ RHS

73. RHS $= \dfrac{\sin\theta - \frac{1}{\sin\theta}}{\cos\theta - \frac{\cos\theta}{\sin\theta}} = \dfrac{\frac{\sin^2\theta - 1}{\sin\theta}}{\frac{\cos\theta\sin\theta - \cos\theta}{\sin\theta}}$

$= \dfrac{\cos^2\theta}{\cos\theta(\sin\theta - 1)} =$ LHS

75. LHS $= \dfrac{-\sin^2 t + \tan^2 t}{\sin^2 t} = -1 + \dfrac{\sin^2 t}{\cos^2 t} \cdot \dfrac{1}{\sin^2 t}$

$= -1 + \sec^2 t =$ RHS

77. LHS $= \dfrac{\sec x - \tan x + \sec x + \tan x}{(\sec x + \tan x)(\sec x - \tan x)}$

$= \dfrac{2\sec x}{\sec^2 x - \tan^2 x} =$ RHS

79. LHS $= \tan^2 x + 2\tan x\cot x + \cot^2 x = \tan^2 x + 2 + \cot^2 x$

$= (\tan^2 x + 1) + (\cot^2 x + 1) =$ RHS

81. LHS $= \dfrac{\frac{1}{\cos u} - 1}{\frac{1}{\cos u} + 1} \cdot \dfrac{\cos u}{\cos u} =$ RHS

83. LHS $= \dfrac{(\sin x + \cos x)(\sin^2 x - \sin x\cos x + \cos^2 x)}{\sin x + \cos x}$

$= \sin^2 x - \sin x\cos x + \cos^2 x =$ RHS

85. LHS $= \dfrac{1 + \sin x}{1 - \sin x} \cdot \dfrac{1 + \sin x}{1 + \sin x} = \dfrac{(1 + \sin x)^2}{1 - \sin^2 x}$

$= \dfrac{(1 + \sin x)^2}{\cos^2 x} = \left(\dfrac{1 + \sin x}{\cos x}\right)^2 =$ RHS

87. LHS $= \left(\dfrac{\sin x}{\cos x} + \dfrac{\cos x}{\sin x}\right)^4 = \left(\dfrac{\sin^2 x + \cos^2 x}{\sin x\cos x}\right)^4$

$= \left(\dfrac{1}{\sin x\cos x}\right)^4 =$ RHS

89. $\tan\theta$ **91.** $\tan\theta$ **93.** $3\cos\theta$

95.

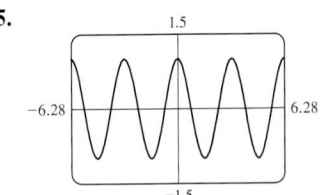

Yes

97.

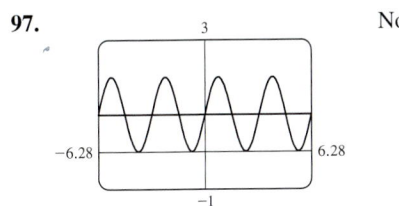

No

Section 7.2 ■ page 539

1. $\dfrac{\sqrt{6} + \sqrt{2}}{4}$ **3.** $\dfrac{\sqrt{2} - \sqrt{6}}{4}$

5. $2 - \sqrt{3}$ **7.** $-\dfrac{\sqrt{6} + \sqrt{2}}{4}$

9. $\sqrt{3} - 2$ **11.** $-\dfrac{\sqrt{6} + \sqrt{2}}{4}$

13. $\sqrt{2}/2$ **15.** $\frac{1}{2}$ **17.** $\sqrt{3}$

19. LHS $= \dfrac{\sin\left(\frac{\pi}{2} - u\right)}{\cos\left(\frac{\pi}{2} - u\right)} = \dfrac{\sin \frac{\pi}{2} \cos u - \cos \frac{\pi}{2} \sin u}{\cos \frac{\pi}{2} \cos u + \sin \frac{\pi}{2} \sin u}$

$= \dfrac{\cos u}{\sin u} = $ RHS

21. LHS $= \dfrac{1}{\cos\left(\frac{\pi}{2} - u\right)} = \dfrac{1}{\cos \frac{\pi}{2} \cos u + \sin \frac{\pi}{2} \sin u}$

$= \dfrac{1}{\sin u} = $ RHS

23. LHS $= \sin x \cos \frac{\pi}{2} - \cos x \sin \frac{\pi}{2} = $ RHS

25. LHS $= \sin x \cos \pi - \cos x \sin \pi = $ RHS

27. LHS $= \dfrac{\tan x - \tan \pi}{1 + \tan x \tan \pi} = $ RHS

29. LHS $= \cos x \cos \frac{\pi}{6} - \sin x \sin \frac{\pi}{6} + \sin x \cos \frac{\pi}{3} - \cos x \sin \frac{\pi}{3}$

$= \dfrac{\sqrt{3}}{2} \cos x - \frac{1}{2} \sin x + \frac{1}{2} \sin x - \dfrac{\sqrt{3}}{2} \cos x = $ RHS

31. LHS $= \sin x \cos y + \cos x \sin y$

$\qquad - (\sin x \cos y - \cos x \sin y) = $ RHS

33. LHS $= \dfrac{1}{\tan(x - y)} = \dfrac{1 + \tan x \tan y}{\tan x - \tan y}$

$= \dfrac{1 + \frac{1}{\cot x}\frac{1}{\cot y}}{\frac{1}{\cot x} - \frac{1}{\cot y}} \cdot \dfrac{\cot x \cot y}{\cot x \cot y} = $ RHS

35. LHS $= \dfrac{\sin x}{\cos x} - \dfrac{\sin y}{\cos y} = \dfrac{\sin x \cos y - \cos x \sin y}{\cos x \cos y} = $ RHS

37. LHS $= \dfrac{\sin x \cos y + \cos x \sin y - (\sin x \cos y - \cos x \sin y)}{\cos x \cos y - \sin x \sin y + \cos x \cos y + \sin x \sin y}$

$= \dfrac{2 \cos x \sin y}{2 \cos x \cos y} = $ RHS

39. LHS $= \sin((x + y) + z)$

$= \sin(x + y) \cos z + \cos(x + y) \sin z$

$= \cos z [\sin x \cos y + \cos x \sin y]$

$\qquad + \sin z [\cos x \cos y - \sin x \sin y] = $ RHS

41. $2 \sin\left(x + \dfrac{5\pi}{6}\right)$

43. $5\sqrt{2} \sin\left(2x + \dfrac{7\pi}{4}\right)$

45. (a) $f(x) = \sqrt{2} \sin\left(x + \dfrac{\pi}{4}\right)$

(b)

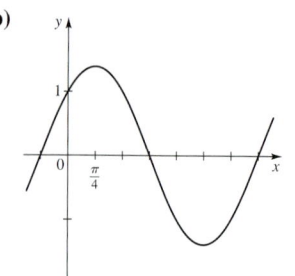

49. $\tan \gamma = \frac{17}{6}$

51. (a)

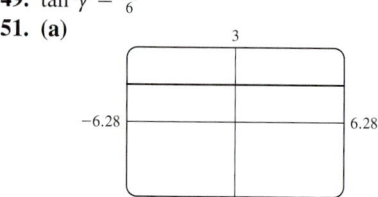

$\sin^2\left(x + \dfrac{\pi}{4}\right) + \sin^2\left(x - \dfrac{\pi}{4}\right) = 1$

53. $\pi/2$

55. (b) $k = 10\sqrt{3}, \phi = \pi/6$

Section 7.3 ■ page 548

1. $\frac{120}{169}, \frac{119}{169}, \frac{120}{119}$ **3.** $-\frac{24}{25}, \frac{7}{25}, -\frac{24}{7}$ **5.** $\frac{24}{25}, \frac{7}{25}, \frac{24}{7}$

7. $-\frac{3}{5}, \frac{4}{5}, -\frac{3}{4}$ **9.** $\frac{1}{2}\left(\frac{3}{4} - \cos 2x + \frac{1}{4} \cos 4x\right)$

11. $\frac{1}{16}(1 - \cos 2x - \cos 4x + \cos 2x \cos 4x)$

13. $\frac{1}{32}\left(\frac{3}{4} - \cos 4x + \frac{1}{4} \cos 8x\right)$ **15.** $\frac{1}{2}\sqrt{2 - \sqrt{3}}$ **17.** $\sqrt{2} - 1$

19. $-\frac{1}{2}\sqrt{2 + \sqrt{3}}$ **21.** $\sqrt{2} - 1$ **23.** $\frac{1}{2}\sqrt{2 + \sqrt{3}}$

25. $-\frac{1}{2}\sqrt{2 - \sqrt{2}}$ **27. (a)** $\sin 36°$ **(b)** $\sin 6\theta$

29. (a) $\cos 68°$ **(b)** $\cos 10\theta$ **31. (a)** $\tan 4°$ **(b)** $\tan 2\theta$

35. $\sqrt{10}/10, 3\sqrt{10}/10, \frac{1}{3}$

37. $\sqrt{(3 + 2\sqrt{2})/6}, \sqrt{(3 - 2\sqrt{2})/6}, 3 + 2\sqrt{2}$

39. $\sqrt{6}/6, -\sqrt{30}/6, -\sqrt{5}/5$ **41.** $\frac{1}{2}(\sin 5x - \sin x)$

43. $\frac{1}{2}(\sin 5x + \sin 3x)$ **45.** $\frac{3}{2}(\cos 11x + \cos 3x)$

47. $2 \sin 4x \cos x$ **49.** $2 \sin 5x \sin x$ **51.** $-2 \cos \frac{9}{2}x \sin \frac{5}{2}x$

53. $(\sqrt{2} + \sqrt{3})/2$ **55.** $\frac{1}{4}(\sqrt{2} - 1)$ **57.** $\sqrt{2}/2$

59. LHS $= \cos(2 \cdot 5x) = $ RHS

61. LHS $= \sin^2 x + 2 \sin x \cos x + \cos^2 x$

$= 1 + 2 \sin x \cos x = $ RHS

63. LHS $= \dfrac{2 \sin 2x \cos 2x}{\sin x} = \dfrac{2(2 \sin x \cos x)(\cos 2x)}{\sin x} = $ RHS

65. LHS $= \dfrac{2(\tan x - \cot x)}{(\tan x + \cot x)(\tan x - \cot x)} = \dfrac{2}{\tan x + \cot x}$

$= \dfrac{2}{\frac{\sin x}{\cos x} + \frac{\cos x}{\sin x}} \cdot \dfrac{\sin x \cos x}{\sin x \cos x} = \dfrac{2 \sin x \cos x}{\sin^2 x + \cos^2 x}$

$= 2 \sin x \cos x = $ RHS

67. LHS $= \tan(2x + x) = \dfrac{\tan 2x + \tan x}{1 - \tan 2x \tan x}$

$= \dfrac{\frac{2 \tan x}{1 - \tan^2 x} + \tan x}{1 - \frac{2 \tan x}{1 - \tan^2 x} \tan x}$

$= \dfrac{2 \tan x + \tan x(1 - \tan^2 x)}{1 - \tan^2 x - 2 \tan x \tan x} =$ RHS

69. LHS $= (\cos^2 x + \sin^2 x)(\cos^2 x - \sin^2 x)$
$= \cos^2 x - \sin^2 x =$ RHS

71. LHS $= \dfrac{2 \sin 3x \cos 2x}{2 \cos 3x \cos 2x} = \dfrac{\sin 3x}{\cos 3x} =$ RHS

73. LHS $= \dfrac{2 \sin 5x \cos 5x}{2 \sin 5x \cos 4x} =$ RHS

75. LHS $= \dfrac{2 \sin\left(\frac{x+y}{2}\right) \cos\left(\frac{x-y}{2}\right)}{2 \cos\left(\frac{x+y}{2}\right) \cos\left(\frac{x-y}{2}\right)} = \dfrac{\sin\left(\frac{x+y}{2}\right)}{\cos\left(\frac{x+y}{2}\right)} =$ RHS

81. LHS $= \dfrac{(\sin x + \sin 5x) + (\sin 2x + \sin 4x) + \sin 3x}{(\cos x + \cos 5x) + (\cos 2x + \cos 4x) + \cos 3x}$

$= \dfrac{2 \sin 3x \cos 2x + 2 \sin 3x \cos x + \sin 3x}{2 \cos 3x \cos 2x + 2 \cos 3x \cos x + \cos 3x}$

$= \dfrac{\sin 3x(2 \cos 2x + 2 \cos x + 1)}{\cos 3x(2 \cos 2x + 2 \cos x + 1)} =$ RHS

83. (a)

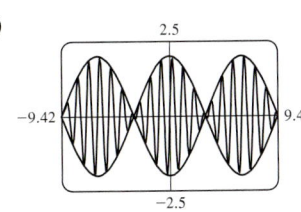

$\dfrac{\sin 3x}{\sin x} - \dfrac{\cos 3x}{\cos x} = 2$

85. (a)

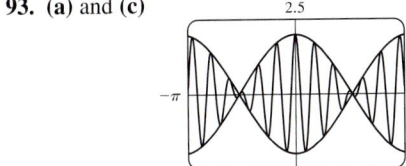

(c)

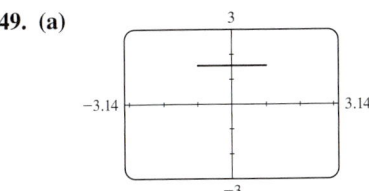

The graph of $y = f(x)$ lies between the two other graphs.

87. (a) $P(t) = 8t^4 - 8t^2 + 1$ **(b)** $Q(t) = 16t^5 - 20t^3 + 5t$
93. (a) and (c)

The graph of f lies between the graphs of $y = 2 \cos t$ and $y = -2 \cos t$. Thus, the loudness of the sound varies between $y = \pm 2 \cos t$.

Section 7.4 ■ page 557

1. (a) $\pi/6$ **(b)** $\pi/3$ **(c)** Not defined
3. (a) $\pi/4$ **(b)** $\pi/4$ **(c)** $-\pi/4$
5. (a) $\pi/2$ **(b)** 0 **(c)** π **7. (a)** $\pi/6$
(b) $-\pi/6$ **(c)** Not defined
9. (a) 0.13889 **(b)** 2.75876
11. (a) 0.88998 **(b)** Not defined   **13.** $\frac{1}{4}$ **15.** 5
17. $\pi/3$ **19.** $-\pi/6$ **21.** $-\pi/3$ **23.** $\sqrt{3}/3$ **25.** $\frac{1}{2}$
27. $\pi/3$ **29.** $\frac{4}{5}$ **31.** $\frac{12}{13}$ **33.** $\frac{13}{5}$ **35.** $\sqrt{5}/5$ **37.** $\frac{24}{25}$ **39.** 1
41. $\sqrt{1 - x^2}$ **43.** $x/\sqrt{1 - x^2}$ **45.** $\dfrac{1 - x^2}{1 + x^2}$ **47.** 0
49. (a)

Conjecture: $y = \pi/2$ for $-1 \le x \le 1$
51. (a) 0.28 **(b)** $(-3 + \sqrt{17})/4$
53. (a) $h = 2 \tan \theta$ **(b)** $\theta = \tan^{-1}(h/2)$
55. (a) $\theta = \sin^{-1}(h/680)$ **(b)** $\theta = 0.826$ rad
57. (a) $54.1°$ **(b)** $48.3°, 32.2°, 24.5°$. The function $\sin^{-1}$ is undefined for values outside the interval $[-1, 1]$.

Section 7.5 ■ page 568

1. $(2k + 1)\pi$ **3.** $\dfrac{\pi}{6} + 2k\pi, \dfrac{5\pi}{6} + 2k\pi$ **5.** $\dfrac{5\pi}{6} + k\pi$

7. $\dfrac{\pi}{3} + k\pi, \dfrac{2\pi}{3} + k\pi$ **9.** $\dfrac{(2k + 1)\pi}{4}$

11. $\dfrac{\pi}{3} + k\pi, \dfrac{2\pi}{3} + k\pi$ **13.** $\dfrac{\pi}{2} + k\pi, \dfrac{7\pi}{6} + 2k\pi, \dfrac{11\pi}{6} + 2k\pi$

15. $-\dfrac{\pi}{3} + k\pi$ **17.** $\dfrac{\pi}{2} + k\pi$ **19.** $\dfrac{\pi}{3} + 2k\pi, \dfrac{5\pi}{3} + 2k\pi$

21. $\dfrac{3\pi}{2} + 2k\pi$ **23.** No solution

25. $\dfrac{7\pi}{18} + \dfrac{2k\pi}{3}, \dfrac{11\pi}{18} + \dfrac{2k\pi}{3}$

27. $\dfrac{1}{4}\left(\dfrac{\pi}{3} + 2k\pi\right), \dfrac{1}{4}\left(-\dfrac{\pi}{3} + 2k\pi\right)$

29. $\dfrac{1}{2}\left(\dfrac{\pi}{6} + k\pi\right)$ **31.** $4k\pi$ **33.** $4\left(\dfrac{2\pi}{3} + k\pi\right)$ **35.** $\dfrac{k\pi}{3}$

37. $\dfrac{\pi}{6} + 2k\pi, \dfrac{2\pi}{3} + 2k\pi, \dfrac{5\pi}{6} + 2k\pi, \dfrac{4\pi}{3} + 2k\pi$

39. $\dfrac{\pi}{8} + \dfrac{k\pi}{2}, \dfrac{3\pi}{8} + \dfrac{k\pi}{2}$ **41.** $\dfrac{\pi}{9}, \dfrac{5\pi}{9}, \dfrac{7\pi}{9}, \dfrac{11\pi}{9}, \dfrac{13\pi}{9}, \dfrac{17\pi}{9}$

43. $\dfrac{\pi}{6}, \dfrac{3\pi}{4}, \dfrac{5\pi}{6}, \dfrac{7\pi}{4}$ **45.** $\dfrac{\pi}{3}, \dfrac{2\pi}{3}, \dfrac{4\pi}{3}, \dfrac{5\pi}{3}$ **47.** $0, \dfrac{2\pi}{3}, \dfrac{4\pi}{3}$

49. (a) $1.15928 + 2k\pi,\ 5.12391 + 2k\pi$
(b) $1.15928,\ 5.12391$
51. (a) $1.36944 + 2k\pi,\ 4.91375 + 2k\pi$
(b) $1.36944,\ 4.91375$
53. (a) $0.46365 + k\pi,\ 2.67795 + k\pi$
(b) $0.46365,\ 2.67795,\ 3.60524,\ 5.81954$
55. (a) $0.33984 + 2k\pi,\ 2.80176 + 2k\pi$
(b) $0.33984,\ 2.80176$

57. $\big((2k + 1)\pi, -2\big)$ **59.** $\left(\dfrac{\pi}{3} + k\pi, \sqrt{3}\right)$

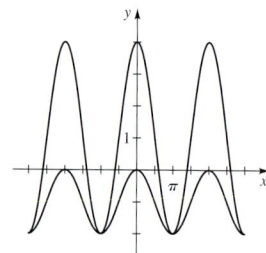 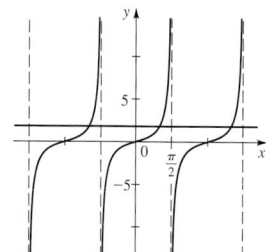

61. $\dfrac{\pi}{8}, \dfrac{3\pi}{8}, \dfrac{5\pi}{8}, \dfrac{7\pi}{8}, \dfrac{9\pi}{8}, \dfrac{11\pi}{8}, \dfrac{13\pi}{8}, \dfrac{15\pi}{8}$

63. $\dfrac{\pi}{9}, \dfrac{2\pi}{9}, \dfrac{7\pi}{9}, \dfrac{8\pi}{9}, \dfrac{13\pi}{9}, \dfrac{14\pi}{9}$ **65.** $\dfrac{\pi}{2}, \dfrac{7\pi}{6}, \dfrac{3\pi}{2}, \dfrac{11\pi}{6}$ **67.** 0

69. $\dfrac{k\pi}{2}$ **71.** $\dfrac{\pi}{9} + \dfrac{2k\pi}{3}, \dfrac{\pi}{2} + k\pi, \dfrac{5\pi}{9} + \dfrac{2k\pi}{3}$

73. $0, \pm0.95$ **75.** 1.92
77. ±0.71 **79.** $0.94721°$ or $89.05279°$ **81.** $44.95°$
83. (a) 34th day (February 3rd), 308th day (November 4th)
(b) 275 days   **85. (b)** $1.047 \approx 60°$

Chapter 7 Review ■ page 571

1. LHS $= \sin\theta\left(\dfrac{\cos\theta}{\sin\theta} + \dfrac{\sin\theta}{\cos\theta}\right) = \cos\theta + \dfrac{\sin^2\theta}{\cos\theta}$

$= \dfrac{\cos^2\theta + \sin^2\theta}{\cos\theta} =$ RHS

3. LHS $= (1 - \sin^2 x)\csc x - \csc x$

$= \csc x - \sin^2 x \csc x - \csc x$

$= -\sin^2 x \cdot \dfrac{1}{\sin x} =$ RHS

5. LHS $= \dfrac{\cos^2 x}{\sin^2 x} - \dfrac{\tan^2 x}{\sin^2 x} = \cot^2 x - \dfrac{1}{\cos^2 x} =$ RHS

7. LHS $= \dfrac{\cos x}{\frac{1}{\cos x}(1 - \sin x)} = \dfrac{\cos x}{\frac{1}{\cos x} - \frac{\sin x}{\cos x}} =$ RHS

9. LHS $= \sin^2 x \dfrac{\cos^2 x}{\sin^2 x} + \cos^2 x \dfrac{\sin^2 x}{\cos^2 x} = \cos^2 x + \sin^2 x =$ RHS

11. LHS $= \dfrac{2\sin x \cos x}{1 + 2\cos^2 x - 1} = \dfrac{2\sin x \cos x}{2\cos^2 x} = \dfrac{2\sin x}{2\cos x} =$ RHS

13. LHS $= \dfrac{1 - \cos x}{\sin x} = \dfrac{1}{\sin x} - \dfrac{\cos x}{\sin x} =$ RHS

15. LHS $= \frac{1}{2}[\cos((x + y) - (x - y))$
$\qquad\qquad - \cos((x + y) + (x - y))]$
$= \frac{1}{2}(\cos 2y - \cos 2x)$
$= \frac{1}{2}[1 - 2\sin^2 y - (1 - 2\sin^2 x)]$
$= \frac{1}{2}(2\sin^2 x - 2\sin^2 y) =$ RHS

17. LHS $= 1 + \dfrac{\sin x}{\cos x} \cdot \dfrac{1 - \cos x}{\sin x} = 1 + \dfrac{1 - \cos x}{\cos x}$

$= 1 + \dfrac{1}{\cos x} - 1 =$ RHS

19. LHS $= \cos^2\frac{x}{2} - 2\sin\frac{x}{2}\cos\frac{x}{2} + \sin^2\frac{x}{2}$
$= 1 - \sin\!\left(2 \cdot \frac{x}{2}\right) =$ RHS

21. LHS $= \dfrac{2\sin x \cos x}{\sin x} - \dfrac{2\cos^2 x - 1}{\cos x}$

$= 2\cos x - 2\cos x + \dfrac{1}{\cos x} =$ RHS

23. LHS $= \dfrac{\tan x + \tan\frac{\pi}{4}}{1 - \tan x \tan\frac{\pi}{4}} =$ RHS

25. (a) 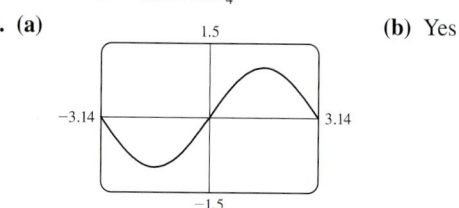   **(b)** Yes

27. (a) 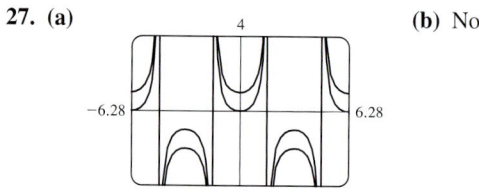   **(b)** No

29. (a) 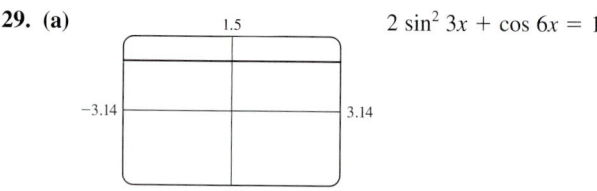 $2\sin^2 3x + \cos 6x = 1$

31. $0, \pi$ **33.** $\dfrac{\pi}{6}, \dfrac{5\pi}{6}$ **35.** $\dfrac{\pi}{3}, \dfrac{5\pi}{3}$ **37.** $\dfrac{2\pi}{3}, \dfrac{4\pi}{3}$

39. $\dfrac{\pi}{3}, \dfrac{2\pi}{3}, \dfrac{3\pi}{4}, \dfrac{4\pi}{3}, \dfrac{5\pi}{3}, \dfrac{7\pi}{4}$ **41.** $\dfrac{\pi}{6}, \dfrac{\pi}{2}, \dfrac{5\pi}{6}, \dfrac{7\pi}{6}, \dfrac{3\pi}{2}, \dfrac{11\pi}{6}$ **43.** $\dfrac{\pi}{6}$

45. 1.18 **47. (a)** $63.4°$ **(b)** No   **(c)** $90°$ **49.** $\frac{1}{2}\sqrt{2 + \sqrt{3}}$

51. $\sqrt{2} - 1$ **53.** $\sqrt{2}/2$ **55.** $\sqrt{2}/2$ **57.** $\dfrac{\sqrt{2} + \sqrt{3}}{4}$

59. $2\dfrac{\sqrt{10} + 1}{9}$ **61.** $\frac{2}{3}(\sqrt{2} + \sqrt{5})$ **63.** $\sqrt{(3 + 2\sqrt{2})}/6$

65. $\pi/3$ **67.** $\frac{1}{2}$ **69.** $2/\sqrt{21}$ **71.** $\frac{7}{9}$ **73.** $x/\sqrt{1+x^2}$

75. $\theta = \cos^{-1}\left(\dfrac{x}{3}\right)$ **77. (a)** $\theta = \tan^{-1}\left(\dfrac{10}{x}\right)$ **(b)** 286.4 ft

Chapter 7 Test ■ page 574

1. (a) LHS $= \dfrac{\sin\theta}{\cos\theta}\sin\theta + \cos\theta = \dfrac{\sin^2\theta + \cos^2\theta}{\cos\theta} = $ RHS

(b) LHS $= \dfrac{\tan x}{1-\cos x}\cdot\dfrac{1+\cos x}{1+\cos x} = \dfrac{\tan x(1+\cos x)}{1-\cos^2 x}$

$= \dfrac{\frac{\sin x}{\cos x}(1+\cos x)}{\sin^2 x} = \dfrac{1}{\sin x}\cdot\dfrac{1+\cos x}{\cos x} = $ RHS

(c) LHS $= \dfrac{2\tan x}{\sec^2 x} = \dfrac{2\sin x}{\cos x}\cdot\cos^2 x = 2\sin x\cos x = $ RHS

2. $\tan\theta$ **3. (a)** $\frac{1}{2}$ **(b)** $\dfrac{\sqrt{2}+\sqrt{6}}{4}$ **(c)** $\frac{1}{2}\sqrt{2-\sqrt{3}}$

4. $(10-2\sqrt{5})/15$ **5. (a)** $\frac{1}{2}(\sin 8x - \sin 2x)$
(b) $-2\cos\frac{7}{2}x\sin\frac{3}{2}x$ **6.** -2

7.
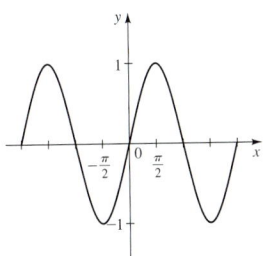
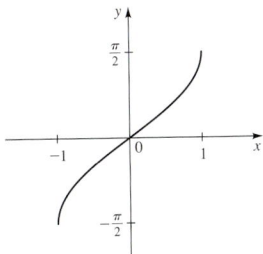
Domain $\mathbb{R}$ Domain $[-1,1]$

8. (a) $\theta = \tan^{-1}\dfrac{x}{4}$ **(b)** $\theta = \cos^{-1}\dfrac{3}{x}$

9. (a) $\dfrac{2\pi}{3},\dfrac{4\pi}{3}$ **(b)** $\dfrac{\pi}{6},\dfrac{\pi}{2},\dfrac{5\pi}{6},\dfrac{3\pi}{2}$
10. 0.57964, 2.56195, 3.72123, 5.70355 **11.** $\frac{40}{41}$

Focus on Modeling ■ page 578

1. (a) $y = -5\sin\left(\dfrac{\pi}{2}t\right)$ **(b)**
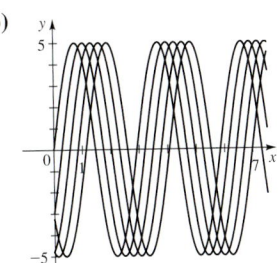

(c) $v = \pi/4$ Yes, it is a traveling wave.

3. $y(x,t) = 2.7\sin(0.68x - 4.10t)$
5. $y(x,t) = 0.6\sin(\pi x)\cos(40\pi t)$
7. (a) 1, 2, 3, 4

(b) 5:
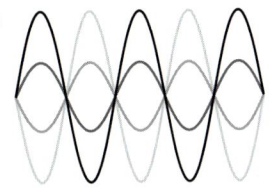

6:
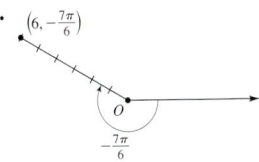

(c) 880π
(d) $y(x,t) = \sin t\cos(880\pi t)$;
$y(x,t) = \sin(2t)\cos(880\pi t)$;
$y(x,t) = \sin(3t)\cos(880\pi t)$;
$y(x,t) = \sin(4t)\cos(880\pi t)$

Chapter 8

Section 8.1 ■ page 586

1.
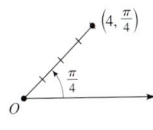
3.

5.

7.
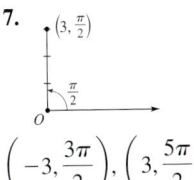

$\left(-3,\dfrac{3\pi}{2}\right),\left(3,\dfrac{5\pi}{2}\right)$

9.
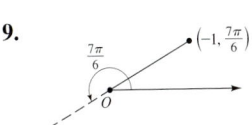
$\left(-1,-\dfrac{5\pi}{6}\right),\left(1,\dfrac{\pi}{6}\right)$

11.
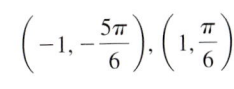
$(-5,2\pi),(5,\pi)$

13. Q **15.** Q **17.** P **19.** P
21. $(3\sqrt{2},3\pi/4)$ **23.** $\left(-\dfrac{5}{2},-\dfrac{5\sqrt{3}}{2}\right)$ **25.** $(2\sqrt{3},2)$
27. $(1,-1)$ **29.** $(-5,0)$ **31.** $(3\sqrt{6},-3\sqrt{2})$
33. $(\sqrt{2},3\pi/4)$ **35.** $(4,\pi/4)$ **37.** $(5,\tan^{-1}\frac{4}{3})$
39. $(6,\pi)$ **41.** $\theta = \pi/4$ **43.** $r = \tan\theta\sec\theta$

45. $r = 4 \sec \theta$ **47.** $x^2 + y^2 = 49$ **49.** $x = 6$

51. $x^2 + y^2 = \dfrac{y}{x}$ **53.** $y - x = 1$

55. $x^2 + y^2 = (x^2 + y^2 - x)^2$ **57.** $x = 2$

59. $y = \pm\sqrt{3}x$

Section 8.2 ■ page 594

1. VI   **3.** II   **5.** I

7. Symmetric about $\theta = \pi/2$

9. Symmetric about the polar axis

11. Symmetric about $\theta = \pi/2$

13. All three types of symmetry

15. 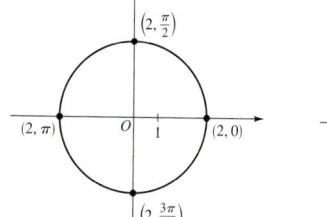   **17.**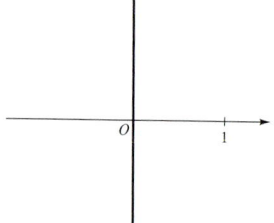

19. 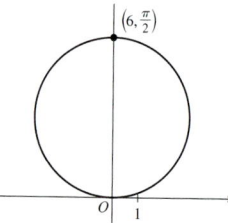   **21.**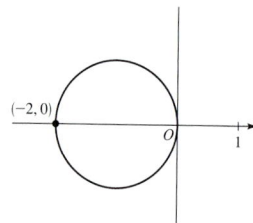

23. 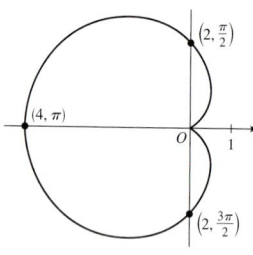   **25.**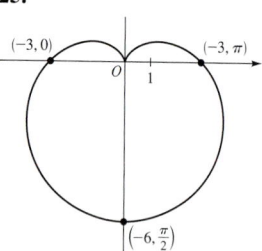

27. 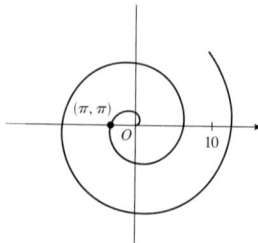   **29.**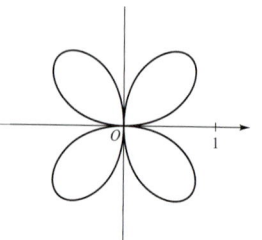

31. 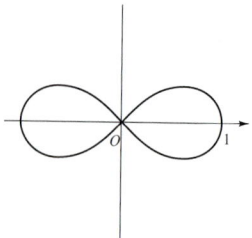   **33.**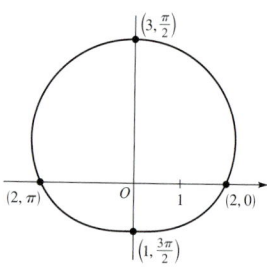

35. 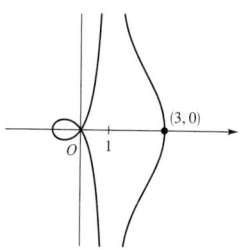   **37.** $0 \le \theta \le 4\pi$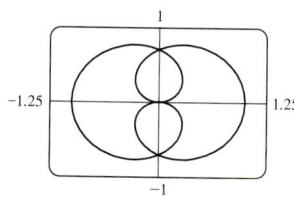

39. $0 \le \theta \le 4\pi$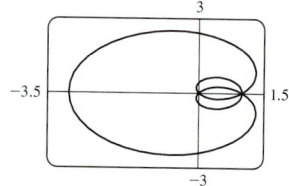

41. The graph of $r = 1 + \sin n\theta$ has n loops.

43. IV   **45.** III

47. 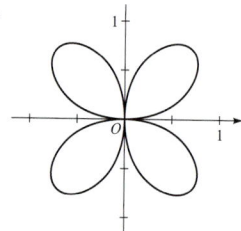   **49.**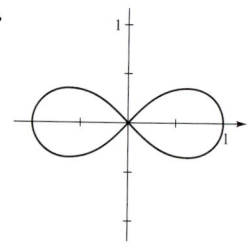

51. $\left(\dfrac{a}{2}, \dfrac{b}{2}\right), \dfrac{\sqrt{a^2 + b^2}}{2}$

53. (a) Elliptical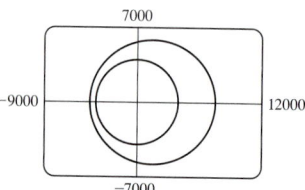

(b) π; 540 mi

Section 8.3 ■ page 603

1. 4

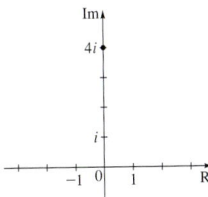

3. 2

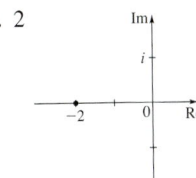

5. $\sqrt{29}$

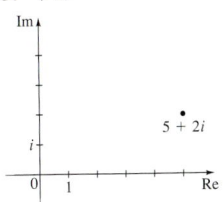

7. 2

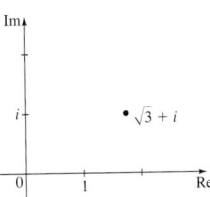

9. 1

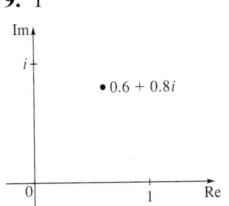

11.

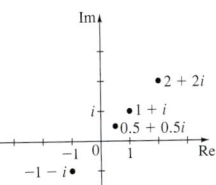

13.

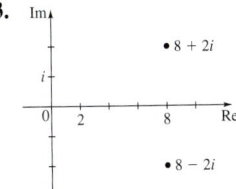

15.

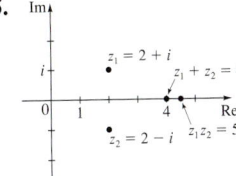

17.

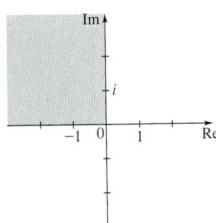

19.

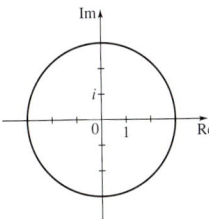

21.

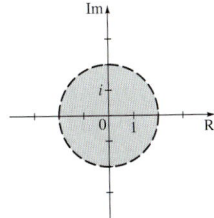

23.

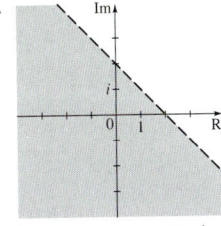

25. $\sqrt{2}\left(\cos\dfrac{\pi}{4} + i\sin\dfrac{\pi}{4}\right)$ **27.** $2\left(\cos\dfrac{7\pi}{4} + i\sin\dfrac{7\pi}{4}\right)$

29. $4\left(\cos\dfrac{11\pi}{6} + i\sin\dfrac{11\pi}{6}\right)$ **31.** $3\left(\cos\dfrac{3\pi}{2} + i\sin\dfrac{3\pi}{2}\right)$

33. $5\sqrt{2}\left(\cos\dfrac{\pi}{4} + i\sin\dfrac{\pi}{4}\right)$ **35.** $8\left(\cos\dfrac{11\pi}{6} + i\sin\dfrac{11\pi}{6}\right)$

37. $20(\cos\pi + i\sin\pi)$ **39.** $5[\cos(\tan^{-1}\tfrac{4}{3}) + i\sin(\tan^{-1}\tfrac{4}{3})]$

41. $3\sqrt{2}\left(\cos\dfrac{3\pi}{4} + i\sin\dfrac{3\pi}{4}\right)$ **43.** $8\left(\cos\dfrac{\pi}{6} + i\sin\dfrac{\pi}{6}\right)$

45. $\sqrt{5}[\cos(\tan^{-1}\tfrac{1}{2}) + i\sin(\tan^{-1}\tfrac{1}{2})]$

47. $2\left(\cos\dfrac{\pi}{4} + i\sin\dfrac{\pi}{4}\right)$

49. $z_1 z_2 = \cos\dfrac{4\pi}{3} + i\sin\dfrac{4\pi}{3}$ $\dfrac{z_1}{z_2} = \cos\dfrac{2\pi}{3} + i\sin\dfrac{2\pi}{3}$

51. $z_1 z_2 = 15\left(\cos\dfrac{3\pi}{2} + i\sin\dfrac{3\pi}{2}\right)$

$\dfrac{z_1}{z_2} = \dfrac{3}{5}\left(\cos\dfrac{7\pi}{6} - i\sin\dfrac{7\pi}{6}\right)$

53. $z_1 z_2 = 8(\cos 150° + i\sin 150°)$
$z_1/z_2 = 2(\cos 90° + i\sin 90°)$

55. $z_1 z_2 = 100(\cos 350° + i\sin 350°)$
$z_1/z_2 = \tfrac{4}{25}(\cos 50° + i\sin 50°)$

57. $z_1 = 2\left(\cos\dfrac{\pi}{6} + i\sin\dfrac{\pi}{6}\right)$

$z_2 = 2\left(\cos\dfrac{\pi}{3} + i\sin\dfrac{\pi}{3}\right)$

$z_1 z_2 = 4\left(\cos\dfrac{\pi}{2} + i\sin\dfrac{\pi}{2}\right)$

$\dfrac{z_1}{z_2} = \cos\dfrac{\pi}{6} - i\sin\dfrac{\pi}{6}$

$\dfrac{1}{z_1} = \dfrac{1}{2}\left(\cos\dfrac{\pi}{6} - i\sin\dfrac{\pi}{6}\right)$

59. $z_1 = 4\left(\cos\dfrac{11\pi}{6} + i\sin\dfrac{11\pi}{6}\right)$

$z_2 = \sqrt{2}\left(\cos\dfrac{3\pi}{4} + i\sin\dfrac{3\pi}{4}\right)$

$z_1 z_2 = 4\sqrt{2}\left(\cos\dfrac{7\pi}{12} + i\sin\dfrac{7\pi}{12}\right)$

$\dfrac{z_1}{z_2} = 2\sqrt{2}\left(\cos\dfrac{13\pi}{12} + i\sin\dfrac{13\pi}{12}\right)$

$\dfrac{1}{z_1} = \dfrac{1}{4}\left(\cos\dfrac{11\pi}{6} - i\sin\dfrac{11\pi}{6}\right)$

61. $z_1 = 5\sqrt{2}\left(\cos\dfrac{\pi}{4} + i\sin\dfrac{\pi}{4}\right)$

$z_2 = 4(\cos 0 + i\sin 0)$

$z_1 z_2 = 20\sqrt{2}\left(\cos\dfrac{\pi}{4} + i\sin\dfrac{\pi}{4}\right)$

$\dfrac{z_1}{z_2} = \dfrac{5\sqrt{2}}{4}\left(\cos\dfrac{\pi}{4} + i\sin\dfrac{\pi}{4}\right)$

$\dfrac{1}{z_1} = \dfrac{\sqrt{2}}{10}\left(\cos\dfrac{\pi}{4} - i\sin\dfrac{\pi}{4}\right)$

63. $z_1 = 20(\cos \pi + i \sin \pi)$

$z_2 = 2\left(\cos \dfrac{\pi}{6} + i \sin \dfrac{\pi}{6}\right)$

$z_1 z_2 = 40\left(\cos \dfrac{7\pi}{6} + i \sin \dfrac{7\pi}{6}\right)$

$\dfrac{z_1}{z_2} = 10\left(\cos \dfrac{5\pi}{6} + i \sin \dfrac{5\pi}{6}\right)$

$\dfrac{1}{z_1} = \tfrac{1}{20}(\cos \pi - i \sin \pi)$

65. -1024 **67.** $512(-\sqrt{3} + i)$

69. -1 **71.** 4096

73. $8(-1 + i)$ **75.** $\dfrac{1}{2048}(-\sqrt{3} - i)$

77. $2\sqrt{2}\left(\cos \dfrac{\pi}{12} + i \sin \dfrac{\pi}{12}\right),$

$2\sqrt{2}\left(\cos \dfrac{13\pi}{12} + i \sin \dfrac{13\pi}{12}\right)$

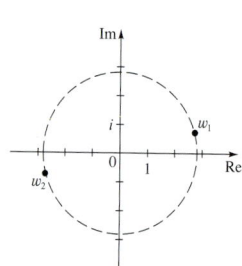

79. $3\left(\cos \dfrac{3\pi}{8} + i \sin \dfrac{3\pi}{8}\right),$

$3\left(\cos \dfrac{7\pi}{8} + i \sin \dfrac{7\pi}{8}\right),$

$3\left(\cos \dfrac{11\pi}{8} + i \sin \dfrac{11\pi}{8}\right),$

$3\left(\cos \dfrac{15\pi}{8} + i \sin \dfrac{15\pi}{8}\right)$

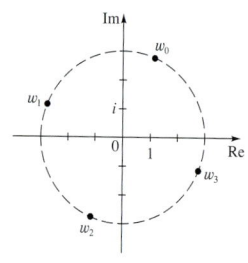

81. $\pm 1, \pm i, \pm \dfrac{\sqrt{2}}{2} \pm \dfrac{\sqrt{2}}{2}i$

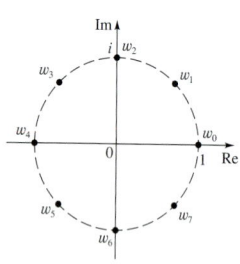

83. $\dfrac{\sqrt{3}}{2} + \dfrac{1}{2}i, -\dfrac{\sqrt{3}}{2} + \dfrac{1}{2}i, -i$

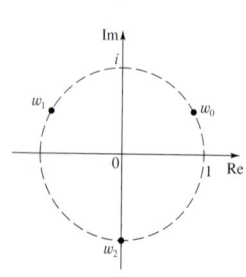

85. $\pm \dfrac{\sqrt{2}}{2} \pm \dfrac{\sqrt{2}}{2}i$

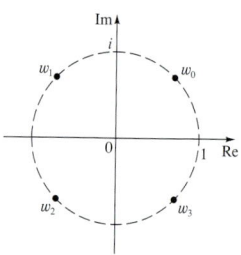

87. $\pm \dfrac{\sqrt{2}}{2} \pm \dfrac{\sqrt{2}}{2}i$

89. $2\left(\cos \dfrac{\pi}{18} + i \sin \dfrac{\pi}{18}\right), 2\left(\cos \dfrac{13\pi}{18} + i \sin \dfrac{13\pi}{18}\right),$

$2\left(\cos \dfrac{25\pi}{18} + i \sin \dfrac{25\pi}{18}\right)$

91. $2^{1/6}\left(\cos \dfrac{5\pi}{12} + i \sin \dfrac{5\pi}{12}\right), 2^{1/6}\left(\cos \dfrac{13\pi}{12} + i \sin \dfrac{13\pi}{12}\right),$

$2^{1/6}\left(\cos \dfrac{21\pi}{12} + i \sin \dfrac{21\pi}{12}\right)$

Section 8.4 ■ page 615

1. 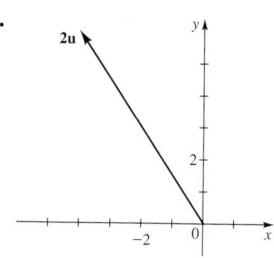   **3.**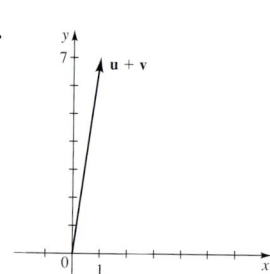

5.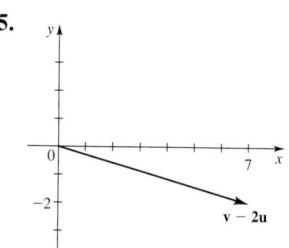

7. $\langle 3, 3 \rangle$ **9.** $\langle 3, -1 \rangle$ **11.** $\langle 5, 7 \rangle$

13. $\langle -4, -3 \rangle$ **15.** $\langle 0, 2 \rangle$

17. $\langle 4, 14 \rangle, \langle -9, -3 \rangle, \langle 5, 8 \rangle, \langle -6, 17 \rangle$

19. $\langle 0, -2 \rangle, \langle 6, 0 \rangle, \langle -2, -1 \rangle, \langle 8, -3 \rangle$

21. $4\mathbf{i}, -9\mathbf{i} + 6\mathbf{j}, 5\mathbf{i} - 2\mathbf{j}, -6\mathbf{i} + 8\mathbf{j}$

23. $\sqrt{5}, \sqrt{13}, 2\sqrt{5}, \tfrac{1}{2}\sqrt{13}, \sqrt{26}, \sqrt{10}, \sqrt{5} - \sqrt{13}$

25. $\sqrt{101}, 2\sqrt{2}, 2\sqrt{101}, \sqrt{2}, \sqrt{73}, \sqrt{145}, \sqrt{101} - 2\sqrt{2}$

27. $20\sqrt{3}\mathbf{i} + 20\mathbf{j}$ **29.** $-\dfrac{\sqrt{2}}{2}\mathbf{i} - \dfrac{\sqrt{2}}{2}\mathbf{j}$

31. $4 \cos 10°\mathbf{i} + 4 \sin 10°\mathbf{j} \approx 3.94\mathbf{i} + 0.69\mathbf{j}$

33. 5, 53.13°   **35.** 13, 157.38°   **37.** 2, 60°
39. $15\sqrt{3}, -15$ **41.** $2\mathbf{i} - 3\mathbf{j}$ **43. (a)** $40\mathbf{j}$
(b) $425\mathbf{i}$ **(c)** $425\mathbf{i} + 40\mathbf{j}$ **(d)** 427 mi/h, N 84.6° E
45. 794 mi/h, N 26.6° W   **47. (a)** $10\mathbf{i}$ **(b)** $10\mathbf{i} + 17.32\mathbf{j}$
(c) $20\mathbf{i} + 17.32\mathbf{j}$ **(d)** 26.5 mi/h, N 49.1° E
49. (a) $22.8\mathbf{i} + 7.4\mathbf{j}$ **(b)** 7.4 mi/h, 22.8 mi/h
51. (a) $\langle 5, -3 \rangle$ **(b)** $\langle -5, 3 \rangle$
53. (a) $-4\mathbf{j}$ **(b)** $4\mathbf{j}$ **55. (a)** $\langle -7.57, 10.61 \rangle$
(b) $\langle 7.57, -10.61 \rangle$
57. $\mathbf{T}_1 \approx -56.5\mathbf{i} + 67.4\mathbf{j}, \mathbf{T}_2 \approx 56.5\mathbf{i} + 32.6\mathbf{j}$

Section 8.5 ■ page 624

1. (a) 2   **(b)** 45°   **3. (a)** 13   **(b)** 56°
5. (a) -1 **(b)** 97°   **7. (a)** $5\sqrt{3}$ **(b)** 30°
9. Yes   **11.** No   **13.** Yes   **15.** 9   **17.** -5 **19.** $-\frac{12}{5}$
21. -24 **23. (a)** $\langle 1, 1 \rangle$ **(b)** $\mathbf{u}_1 = \langle 1, 1 \rangle, \mathbf{u}_2 = \langle -3, 3 \rangle$
25. (a) $\langle -\frac{1}{2}, \frac{3}{2} \rangle$ **(b)** $\mathbf{u}_1 = \langle -\frac{1}{2}, \frac{3}{2} \rangle, \mathbf{u}_2 = \langle \frac{3}{2}, \frac{1}{2} \rangle$
27. (a) $\langle -\frac{18}{5}, \frac{24}{5} \rangle$ **(b)** $\mathbf{u}_1 = \langle -\frac{18}{5}, \frac{24}{5} \rangle, \mathbf{u}_2 = \langle \frac{28}{5}, \frac{21}{5} \rangle$
29. -28 **31.** 25   **39.** 16 ft-lb   **41.** 8660 ft-lb
43. 1164 lb   **45.** 23.6°

Chapter 8 Review ■ page 627

1. (a)

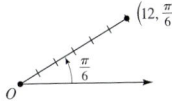

(b) $(6\sqrt{3}, 6)$

3. (a)

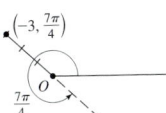

(b) $\left(\dfrac{-3\sqrt{2}}{2}, \dfrac{3\sqrt{2}}{2} \right)$

5. (a)

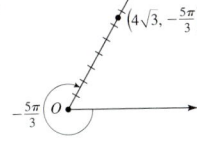

(b) $(2\sqrt{3}, 6)$

7. (a)

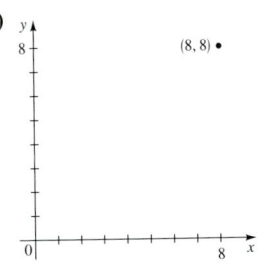

(b) $\left(8\sqrt{2}, \dfrac{\pi}{4} \right)$

(c) $\left(-8\sqrt{2}, \dfrac{5\pi}{4} \right)$

9. (a)

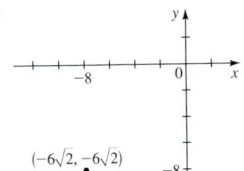

(b) $\left(12, \dfrac{5\pi}{4} \right)$

(c) $\left(-12, \dfrac{\pi}{4} \right)$

11. (a)

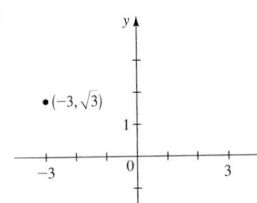

(b) $\left(2\sqrt{3}, \dfrac{5\pi}{6} \right)$

(c) $\left(-2\sqrt{3}, -\dfrac{\pi}{6} \right)$

13. (a) $r = \dfrac{4}{\cos \theta + \sin \theta}$

(b)

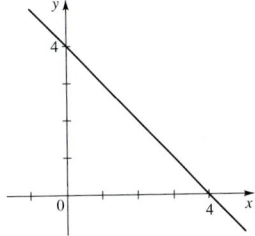

15. (a) $r = 4(\cos \theta + \sin \theta)$

(b)

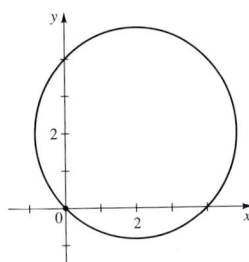

17. (a)

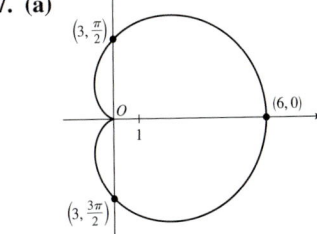

(b) $(x^2 + y^2 - 3x)^2 = 9(x^2 + y^2)$

19. (a)

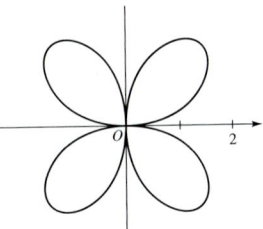

(b) $(x^2 + y^2)^3 = 16x^2 y^2$

21. (a)

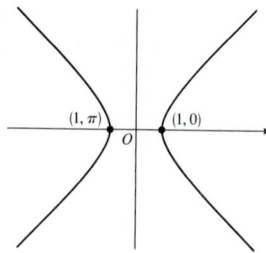

(b) $x^2 - y^2 = 1$

23. (a)

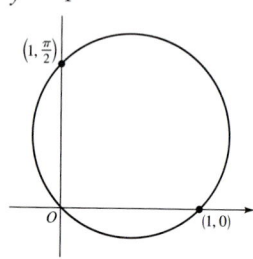

(b) $x^2 + y^2 = x + y$

25. $0 \le \theta \le 6\pi$ **27.** $0 \le \theta \le 6\pi$

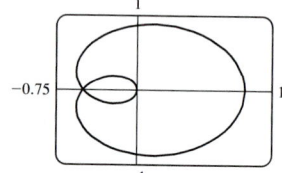

29. (a)

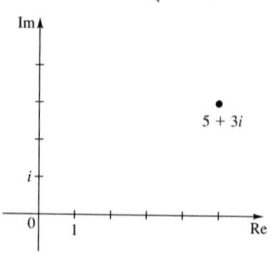

(b) $4\sqrt{2}, \dfrac{\pi}{4}$ **(c)** $4\sqrt{2}\left(\cos \dfrac{\pi}{4} + i \sin \dfrac{\pi}{4}\right)$

31. (a)

(b) $\sqrt{34}, \tan^{-1}\left(\frac{3}{5}\right)$ **(c)** $\sqrt{34}\left[\cos\left(\tan^{-1}\frac{3}{5}\right) + i \sin\left(\tan^{-1}\frac{3}{5}\right)\right]$

33. (a)

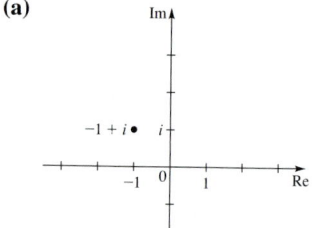

(b) $\sqrt{2}, \dfrac{3\pi}{4}$ **(c)** $\sqrt{2}\left(\cos \dfrac{3\pi}{4} + i \sin \dfrac{3\pi}{4}\right)$

35. $8\left(-1 + i\sqrt{3}\right)$ **37.** $-\frac{1}{32}\left(1 + i\sqrt{3}\right)$

39. $\pm 2\sqrt{2}(1 - i)$ **41.** $\pm 1, \pm\frac{1}{2} \pm \dfrac{\sqrt{3}}{2}i$

43. $\sqrt{13}, \langle 6, 4 \rangle, \langle -10, 2 \rangle, \langle -4, 6 \rangle, \langle -22, 7 \rangle$

45. $3\mathbf{i} - 4\mathbf{j}$ **47.** $(10, -2)$ **49. (a)** $(4.8\mathbf{i} + 0.4\mathbf{j}) \times 10^4$

(b) 4.8×10^4 lb, N 85.2° E   **51.** 5, 25, 60

53. $2\sqrt{2}, 8, 0$ **55.** Yes   **57.** No, 45°   **59. (a)** $17\sqrt{37}/37$

(b) $\langle \frac{102}{37}, -\frac{17}{37} \rangle$ **(c)** $\mathbf{u}_1 = \langle \frac{102}{37}, -\frac{17}{37} \rangle, \mathbf{u}_2 = \langle \frac{9}{37}, \frac{54}{37} \rangle$ **61.** -6

Chapter 8 Test ■ page 629

1. (a) $\left(-4\sqrt{2}, -4\sqrt{2}\right)$ **(b)** $\left(4\sqrt{3}, 5\pi/6\right), \left(-4\sqrt{3}, 11\pi/6\right)$

2. (a)   circle

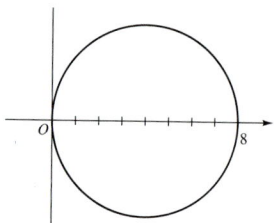

(b) $(x - 4)^2 + y^2 = 16$

3. (a)

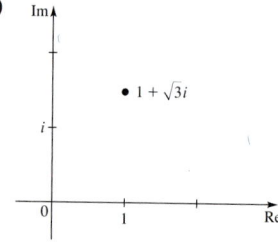

(b) $2\left(\cos \dfrac{\pi}{3} + i \sin \dfrac{\pi}{3}\right)$ **(c)** -512 **4.** $-8, \sqrt{3} + i$

5. $-3i, 3\left(\pm\dfrac{\sqrt{3}}{2} + \dfrac{1}{2}i\right)$

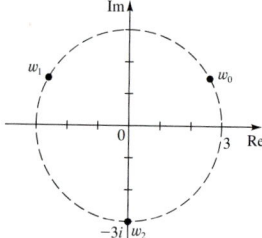

6. (a) $-6\mathbf{i} + 10\mathbf{j}$ **(b)** $2\sqrt{34}$

7. (a) $\langle 19, -3 \rangle$ **(b)** $5\sqrt{2}$ **(c)** 0 **(d)** Yes
8. (a)

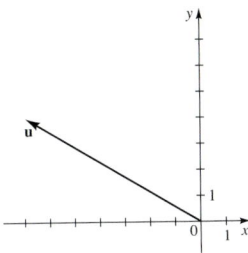

(b) $8, \dfrac{5\pi}{6}$ **9. (a)** $14\mathbf{i} + 6\sqrt{3}\,\mathbf{j}$ **(b)** 17.4 mi/h, N 53.4° E
10. (a) $45°$ **(b)** $\sqrt{26}/2$ **(c)** $\tfrac{5}{2}\mathbf{i} - \tfrac{1}{2}\mathbf{j}$ **11.** 90

Focus on Modeling ■ page 632

1. (a) $R = 18/\pi \approx 5.73$ **(b)** 691.2 mi
3. (a) $x \approx -12.23$, $y \approx 6.27$ **(b)** $x \approx 3.76$, $y \approx 8.43$
(c) $x \approx 15.12$, $y \approx -3.85$ **(d)** $x \approx -4.31$, $y \approx -2.42$
5. (a) 1.14 **(b)** 1.73 **(c)** 36.81 **7. (a)** 1.48
(b) 1.21 **(c)** 1.007

Chapter 9

Section 9.1 ■ page 642

1. $(3, 1)$ **3.** $(4, 16), (-3, 9)$ **5.** $(2, -2), (-2, 2)$
7. $(-25, 5), (-25, -5)$ **9.** $(1, 2)$ **11.** $(-3, 4), (3, 4)$
13. $(-2, -1), (-2, 1), (2, -1), (2, 1)$
15. $(-1, \sqrt{2}), (-1, -\sqrt{2}), (\tfrac{1}{2}, \sqrt{\tfrac{7}{2}}), (\tfrac{1}{2}, -\sqrt{\tfrac{7}{2}})$ **17.** $(-2, 3)$
19. $(2, 4), (-\tfrac{5}{2}, \tfrac{7}{4})$ **21.** $(0, 0), (1, -1), (-2, -4)$ **23.** $(4, 0)$
25. $(-2, -2)$ **27.** $(6, 2), (-2, -6)$ **29.** No solution
31. $(\sqrt{5}, 2), (\sqrt{5}, -2), (-\sqrt{5}, 2), (-\sqrt{5}, -2)$
33. $(3, -\tfrac{1}{2}), (-3, -\tfrac{1}{2})$ **35.** $(\tfrac{1}{5}, \tfrac{1}{3})$ **37.** $(-0.33, 5.33)$
39. $(2.00, 20.00), (-8.00, 0)$ **41.** $(-4.51, 2.17), (4.91, -0.97)$
43. $(1.23, 3.87), (-0.35, -4.21)$
45. $(-2.30, -0.70), (0.48, -1.19)$ **47.** 12 cm by 15 cm
49. 15, 20 **51.** $(400.50, 200.25), 447.77$ m **53.** $(12, 8)$

Section 9.2 ■ page 649

1. $(2, 2)$

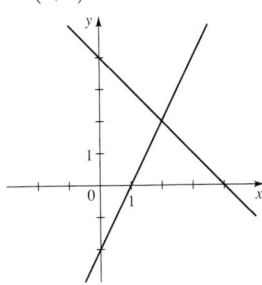

3. No solution

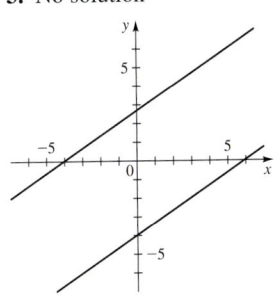

5. Infinitely many solutions

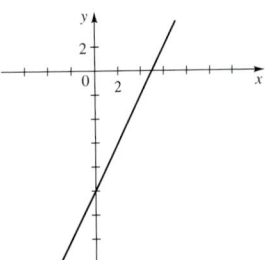

7. $(2, 2)$ **9.** $(3, -1)$ **11.** $(2, 1)$ **13.** $(3, 5)$ **15.** $(1, 3)$
17. $(10, -9)$ **19.** No solution **21.** No solution
23. $(x, \tfrac{1}{3}x - \tfrac{5}{3})$ **25.** $(x, 3 - \tfrac{3}{2}x)$ **27.** $(-3, -7)$
29. $(x, 5 - \tfrac{5}{6}x)$ **31.** $(5, 10)$ **33.** No solution

35. $(3.87, 2.74)$ **37.** $(61.00, 20.00)$ **39.** $\left(-\dfrac{1}{a-1}, \dfrac{1}{a-1}\right)$

41. $\left(\dfrac{1}{a+b}, \dfrac{1}{a+b}\right)$ **43.** 22, 12 **45.** 5 dimes, 9 quarters
47. Plane's speed 120 mi/h, wind speed 30 mi/h
49. Run 5 mi/h, cycle 20 mi/h **51.** 200 g of A, 40 g of B
53. 25%, 10% **55.** \$16,000 at 10%, \$32,000 at 6% **57.** 25

Section 9.3 ■ page 657

1. Linear **3.** Nonlinear **5.** $(1, 3, 2)$
7. $(4, 0, 3)$ **9.** $\left(5, 2, -\tfrac{1}{2}\right)$
11. $\begin{cases} x - 2y - z = 4 \\ \quad -y - 4z = 4 \\ 2x + \quad y + z = 0 \end{cases}$ **13.** $\begin{cases} 2x - y + 3z = 2 \\ x + 2y - z = 4 \\ \quad 3y + 7z = 14 \end{cases}$
15. $(1, 2, 1)$ **17.** $(5, 0, 1)$ **19.** $(0, 1, 2)$ **21.** $(1 - 3t, 2t, t)$
23. No solution **25.** No solution **27.** $(3 - t, -3 + 2t, t)$
29. $\left(2 - 2t, -\tfrac{2}{3} + \tfrac{4}{3}t, t\right)$ **31.** $(1, -1, 1, 2)$
33. \$30,000 in short-term bonds, \$30,000 in intermediate-term
bonds, \$40,000 in long-term bonds **35.** Impossible
37. 250 acres corn, 500 acres wheat, 450 acres soybeans

Section 9.4 ■ page 673

1. 3×2 **3.** 2×1 **5.** 1×3
7. (a) Yes **(b)** Yes **(c)** $\begin{cases} x = -3 \\ y = 5 \end{cases}$

9. (a) Yes **(b)** No **(c)** $\begin{cases} x + 2y + 8z = 0 \\ \quad y + 3z = 2 \\ \quad\quad 0 = 0 \end{cases}$

11. (a) No **(b)** No **(c)** $\begin{cases} x \quad\quad = 0 \\ \quad 0 = 0 \\ y + 5z = 1 \end{cases}$

13. (a) Yes **(b)** Yes **(c)** $\begin{cases} x + 3y - \quad w = 0 \\ \quad z + 2w = 0 \\ \quad\quad 0 = 1 \\ \quad\quad 0 = 0 \end{cases}$

15. $(1, 1, 2)$ **17.** $(1, 0, 1)$ **19.** $(-1, 0, 1)$ **21.** $(-1, 5, 0)$
23. $(10, 3, -2)$ **25.** No solution   **27.** $(2 - 3t, 3 - 5t, t)$
29. No solution   **31.** $(-2t + 5, t - 2, t)$
33. $x = -\frac{1}{2}s + t + 6, y = s, z = t$ **35.** $(-2, 1, 3)$
37. $(-9, 2, 0)$ **39.** $(0, -3, 0, -3)$ **41.** $(-1, 0, 0, 1)$
43. $\left(\frac{7}{4} - \frac{7}{4}t, -\frac{7}{4} + \frac{3}{4}t, \frac{9}{4} + \frac{3}{4}t, t\right)$
45. $x = \frac{1}{3}s - \frac{2}{3}t, y = \frac{1}{3}s + \frac{1}{3}t, z = s, w = t$
47. 2 VitaMax, 1 Vitron, 2 VitaPlus   **49.** 5-mile run,
2-mile swim, 30-mile cycle   **51.** Impossible

Section 9.5 ■ page 684

1. No   **3.** $\begin{bmatrix} 1 & 3 \\ 1 & 5 \end{bmatrix}$ **5.** $\begin{bmatrix} 3 & 6 \\ 12 & -3 \\ 3 & 0 \end{bmatrix}$ **7.** Impossible

9. $\begin{bmatrix} 5 & 2 & 1 \\ 7 & 10 & -7 \end{bmatrix}$ **11.** $\begin{bmatrix} -1 & -\frac{1}{2} \\ 1 & 2 \end{bmatrix}$ **13.** No solution

15. $\begin{bmatrix} 0 & -5 \\ -25 & -20 \\ -10 & 10 \end{bmatrix}$ **17.** $\begin{bmatrix} 5 & -2 & 5 \\ 1 & 1 & 0 \end{bmatrix}$ **19.** $\begin{bmatrix} -1 & -3 & -5 \\ -1 & 3 & -6 \end{bmatrix}$

21. $\begin{bmatrix} 13 & -\frac{7}{2} & 15 \\ 3 & 1 & 3 \end{bmatrix}$ **23.** $\begin{bmatrix} -14 & -8 & -30 \\ -6 & 10 & -24 \end{bmatrix}$

25. Impossible   **27.** $\begin{bmatrix} 3 & \frac{1}{2} & 5 \\ 1 & -1 & 3 \end{bmatrix}$ **29.** $\begin{bmatrix} 28 & 21 & 28 \end{bmatrix}$

31. $\begin{bmatrix} -1 \\ 8 \\ -1 \end{bmatrix}$ **33.** $\begin{bmatrix} 8 & -335 \\ 0 & 343 \end{bmatrix}$ **35.** Impossible

37. Impossible   **39.** $x = 2, y = -1$
41. $x = 1, y = -2$ **43.** $\begin{bmatrix} 2 & -5 \\ 3 & 2 \end{bmatrix}\begin{bmatrix} x \\ y \end{bmatrix} = \begin{bmatrix} 7 \\ 4 \end{bmatrix}$

45. $\begin{bmatrix} 3 & 2 & -1 & 1 \\ 1 & 0 & -1 & 0 \\ 0 & 3 & 1 & -1 \end{bmatrix}\begin{bmatrix} x_1 \\ x_2 \\ x_3 \\ x_4 \end{bmatrix} = \begin{bmatrix} 0 \\ 5 \\ 4 \end{bmatrix}$

47. Only ACB is defined. $ACB = \begin{bmatrix} -3 & -21 & 27 & -6 \\ -2 & -14 & 18 & -4 \end{bmatrix}$

49. (a) $\begin{bmatrix} 4{,}690 & 1{,}690 & 13{,}210 \end{bmatrix}$ **(b)** Total revenue in Santa
Monica, Long Beach, and Anaheim, respectively.
51. (a) $\begin{bmatrix} 105{,}000 & 58{,}000 \end{bmatrix}$ **(b)** The first entry is the total
amount (in ounces) of tomato sauce produced, and the second
entry is the total amount (in ounces) of tomato paste produced.
53.

(a) $\begin{bmatrix} 1 & 0 & 1 & 0 & 1 & 1 \\ 0 & 3 & 0 & 1 & 2 & 1 \\ 1 & 2 & 0 & 0 & 3 & 0 \\ 1 & 3 & 2 & 3 & 2 & 0 \\ 0 & 3 & 0 & 0 & 2 & 1 \\ 1 & 2 & 0 & 1 & 3 & 1 \end{bmatrix}$

(b) $\begin{bmatrix} 2 & 1 & 2 & 1 & 2 & 2 \\ 1 & 3 & 1 & 2 & 3 & 2 \\ 2 & 3 & 1 & 1 & 3 & 1 \\ 2 & 3 & 3 & 3 & 3 & 1 \\ 1 & 3 & 1 & 1 & 3 & 2 \\ 2 & 3 & 1 & 2 & 3 & 2 \end{bmatrix}$

(c) $\begin{bmatrix} 2 & 3 & 2 & 3 & 2 & 2 \\ 3 & 0 & 3 & 2 & 1 & 2 \\ 2 & 1 & 3 & 3 & 0 & 3 \\ 2 & 0 & 1 & 0 & 1 & 3 \\ 3 & 0 & 3 & 3 & 1 & 2 \\ 2 & 1 & 3 & 2 & 0 & 2 \end{bmatrix}$

(d) $\begin{bmatrix} 3 & 3 & 3 & 3 & 3 & 3 \\ 3 & 0 & 3 & 3 & 0 & 3 \\ 3 & 0 & 3 & 3 & 0 & 3 \\ 3 & 0 & 0 & 0 & 0 & 3 \\ 3 & 0 & 3 & 3 & 0 & 3 \\ 3 & 0 & 3 & 3 & 0 & 3 \end{bmatrix}$

(e) The letter E

Section 9.6 ■ page 697

5. $\begin{bmatrix} 1 & -2 \\ -\frac{3}{2} & \frac{7}{2} \end{bmatrix}$ **7.** $\begin{bmatrix} 2 & -3 \\ -3 & 5 \end{bmatrix}$ **9.** $\begin{bmatrix} 13 & 5 \\ -5 & -2 \end{bmatrix}$

11. No inverse   **13.** $\begin{bmatrix} 1 & 2 \\ -\frac{1}{2} & \frac{2}{3} \end{bmatrix}$ **15.** $\begin{bmatrix} -4 & -4 & 5 \\ 1 & 1 & -1 \\ 5 & 4 & -6 \end{bmatrix}$

17. No inverse   **19.** $\begin{bmatrix} -\frac{9}{2} & -1 & 4 \\ 3 & 1 & -3 \\ \frac{7}{2} & 1 & -3 \end{bmatrix}$

21. $\begin{bmatrix} 0 & 0 & -2 & 1 \\ -1 & 0 & 1 & 1 \\ 0 & 1 & -1 & 0 \\ 1 & 0 & 0 & -1 \end{bmatrix}$ **23.** $x = 8, y = -12$

25. $x = 126, y = -50$ **27.** $x = -38, y = 9, z = 47$
29. $x = -20, y = 10, z = 16$ **31.** $x = 3, y = 2, z = 1$
33. $x = 3, y = -2, z = 2$ **35.** $x = 8, y = 1, z = 0, w = 3$

37. $\begin{bmatrix} 7 & 2 & 3 \\ 10 & 3 & 5 \end{bmatrix}$ **39.** $\frac{1}{2a}\begin{bmatrix} 1 & 1 \\ -1 & 1 \end{bmatrix}$

41. $\begin{bmatrix} 1 & -\dfrac{1}{x} \\ -\dfrac{1}{x} & \dfrac{2}{x^2} \end{bmatrix}$; inverse does not exist for $x = 0$

43. $\frac{1}{2}\begin{bmatrix} 1 & e^{-x} & 0 \\ e^{-x} & -e^{-2x} & 0 \\ 0 & 0 & 1 \end{bmatrix}$; inverse exists for all x

45. $\begin{bmatrix} \cos x & -\sin x \\ \sin x & \cos x \end{bmatrix}$; inverse exists for all x

47. (a) $\begin{bmatrix} 0 & 1 & -1 \\ -2 & \frac{3}{2} & 0 \\ 1 & -\frac{3}{2} & 1 \end{bmatrix}$ **(b)** 1 oz A, 1 oz B, 2 oz C

(c) 2 oz A, 0 oz B, 1 oz C   **(d)** No

49. (a) $\begin{cases} x + y + 2z = 675 \\ 2x + y + z = 600 \\ x + 2y + z = 625 \end{cases}$

(b) $\begin{bmatrix} 1 & 1 & 2 \\ 2 & 1 & 1 \\ 1 & 2 & 1 \end{bmatrix} \begin{bmatrix} x \\ y \\ z \end{bmatrix} = \begin{bmatrix} 675 \\ 600 \\ 625 \end{bmatrix}$ **(c)** $A^{-1} = \begin{bmatrix} -\frac{1}{4} & \frac{3}{4} & -\frac{1}{4} \\ -\frac{1}{4} & -\frac{1}{4} & \frac{3}{4} \\ \frac{3}{4} & -\frac{1}{4} & -\frac{1}{4} \end{bmatrix}$

He earns \$125 on a standard set, \$150 or a deluxe set, and \$200 on a leather-bound set.

Section 9.7 ■ page 713

1. 6   **3.** -4 **5.** Does not exist   **7.** $\frac{1}{8}$ **9.** 20, 20
11. $-12, 12$ **13.** 0, 0   **15.** 4, has an inverse
17. -6, has an inverse   **19.** 5000, has an inverse
21. -4, has an inverse   **23.** -18 **25.** 120   **27. (a)** -2
(b) -2 **(c)** Yes   **29.** $(-2, 5)$ **31.** $(0.6, -0.4)$
33. $(4, -1)$ **35.** $(4, 2, -1)$ **37.** $(1, 3, 2)$ **39.** $(0, -1, 1)$
41. $\left(\frac{189}{29}, -\frac{108}{29}, \frac{88}{29}\right)$ **43.** $\left(\frac{1}{2}, \frac{1}{4}, \frac{1}{4}, -1\right)$ **45.** $abcde$
47. 0, 1, 2   **49.** 1, -1

51. 21   **53.** $\frac{63}{2}$ **57. (a)** $\begin{cases} 100a + 10b + c = 25 \\ 225a + 15b + c = 33\frac{3}{4} \\ 1600a + 40b + c = 40 \end{cases}$

(b) $y = -0.05x^2 + 3x$

Section 9.8 ■ page 720

1. $\dfrac{A}{x-1} + \dfrac{B}{x+2}$ **3.** $\dfrac{A}{x-2} + \dfrac{B}{(x-2)^2} + \dfrac{C}{x+4}$

5. $\dfrac{A}{x-3} + \dfrac{Bx+C}{x^2+4}$ **7.** $\dfrac{Ax+B}{x^2+1} + \dfrac{Cx+D}{x^2+2}$

9. $\dfrac{A}{x} + \dfrac{B}{2x-5} + \dfrac{C}{(2x-5)^2} + \dfrac{D}{(2x-5)^3}$
$+ \dfrac{Ex+F}{x^2+2x+5} + \dfrac{Gx+H}{(x^2+2x+5)^2}$

11. $\dfrac{1}{x-1} - \dfrac{1}{x+1}$ **13.** $\dfrac{1}{x-1} - \dfrac{1}{x+4}$

15. $\dfrac{2}{x-3} - \dfrac{2}{x+3}$ **17.** $\dfrac{1}{x-2} - \dfrac{1}{x+2}$

19. $\dfrac{3}{x-4} - \dfrac{2}{x+2}$ **21.** $\dfrac{-\frac{1}{2}}{2x-1} + \dfrac{\frac{3}{2}}{4x-3}$

23. $\dfrac{2}{x-2} + \dfrac{3}{x+2} - \dfrac{1}{2x-1}$ **25.** $\dfrac{2}{x+1} - \dfrac{1}{x} + \dfrac{1}{x^2}$

27. $\dfrac{1}{2x+3} - \dfrac{3}{(2x+3)^2}$ **29.** $\dfrac{2}{x} - \dfrac{1}{x^3} - \dfrac{2}{x+2}$

31. $\dfrac{4}{x+2} - \dfrac{4}{x-1} + \dfrac{2}{(x-1)^2} + \dfrac{1}{(x-1)^3}$

33. $\dfrac{3}{x+2} - \dfrac{1}{(x+2)^2} - \dfrac{1}{(x+3)^2}$ **35.** $\dfrac{x+1}{x^2+3} - \dfrac{1}{x}$

37. $\dfrac{2x-5}{x^2+x+2} + \dfrac{5}{x^2+1}$ **39.** $\dfrac{1}{x^2+1} - \dfrac{x+2}{(x^2+1)^2} + \dfrac{1}{x}$

41. $x^2 + \dfrac{3}{x-2} - \dfrac{x+1}{x^2+1}$ **43.** $A = \dfrac{a+b}{2}, B = \dfrac{a-b}{2}$

Section 9.9 ■ page 726

1.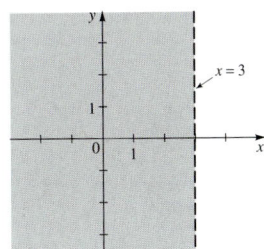

3.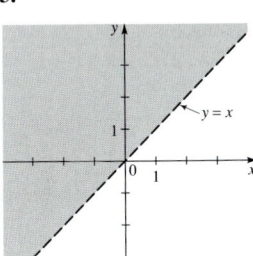

5.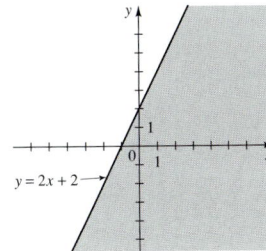

7.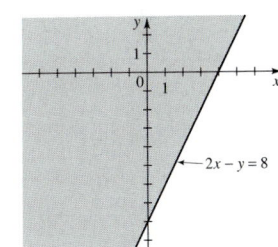

9.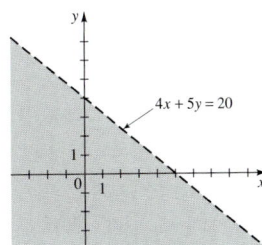

11.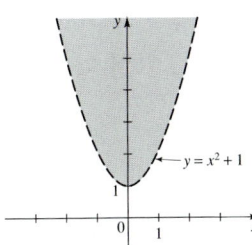

13.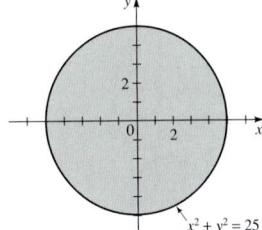

15. $y \le \frac{1}{2}x - 1$ **17.** $x^2 + y^2 > 4$

19.

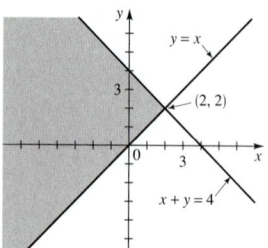

not bounded

21.

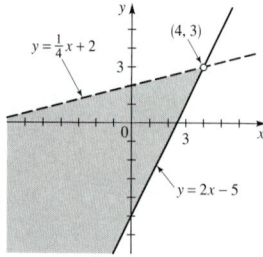

not bounded

39.

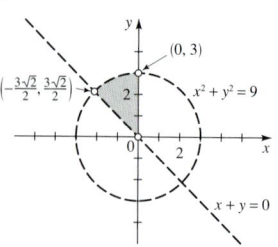

bounded

41.

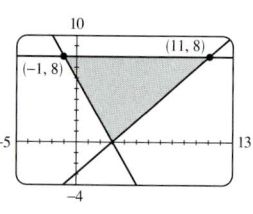

23.

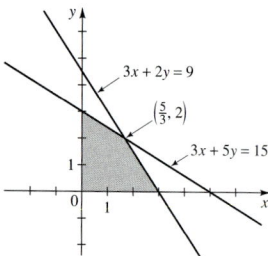

bounded

25.

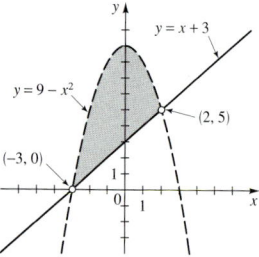

bounded

43.

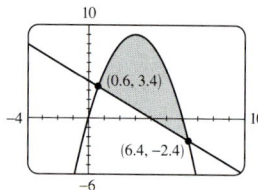

27.

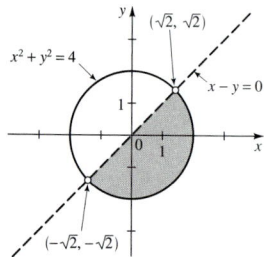

bounded

29.

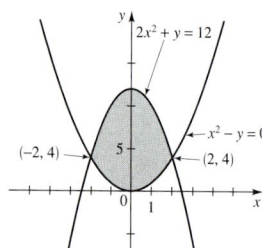

bounded

45. x = number of fiction books
y = number of nonfiction books
$$\begin{cases} x + y \le 100 \\ 20 \le y, x \ge y \\ x \ge 0, y \ge 0 \end{cases}$$

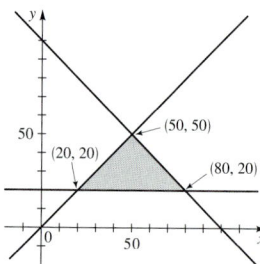

31.

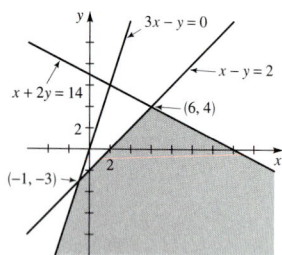

not bounded

33.

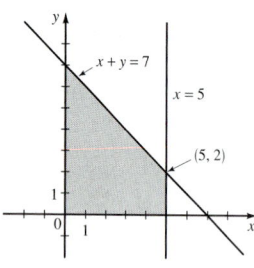

bounded

47. x = number of standard packages
y = number of deluxe packages
$$\begin{cases} \frac{1}{4}x + \frac{5}{8}y \le 80 \\ \frac{3}{4}x + \frac{3}{8}y \le 90 \\ x \ge 0, y \ge 0 \end{cases}$$

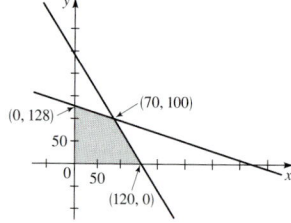

35.

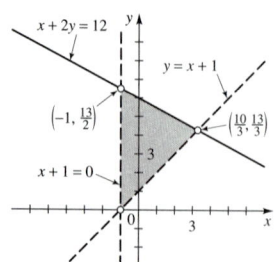

bounded

37.

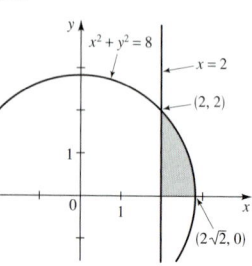

bounded

Chapter 9 Review ■ page 728

1. $(2, 1)$ **3.** $\left(-\frac{1}{2}, \frac{7}{4}\right), (2, -2)$

5. $(2, 1)$

7. x = any number
$y = \frac{2}{7}x - 4$

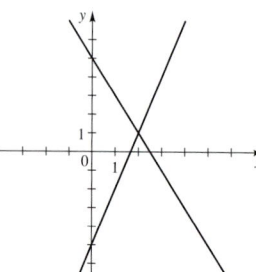

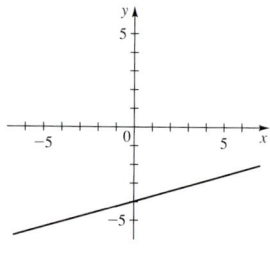

9. No solution

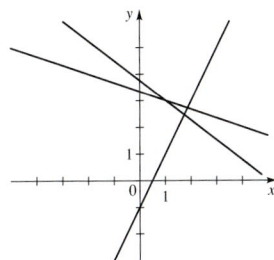

11. $(-3, 3), (2, 8)$ **13.** $\left(\frac{16}{7}, -\frac{14}{3}\right)$ **15.** $(21.41, -15.93)$
17. $(11.94, -1.39), (12.07, 1.44)$ **19. (a)** 2×3

(b) Yes **(c)** No **(d)** $\begin{cases} x + 2y = -5 \\ \quad\quad y = \quad 3 \end{cases}$

21. (a) 3×4 **(b)** Yes **(c)** Yes **(d)** $\begin{cases} x \quad + 8z = \quad 0 \\ \quad y + 5z = -1 \\ \quad\quad\quad 0 = \quad 0 \end{cases}$

23. (a) 3×4 **(b)** No **(c)** No **(d)** $\begin{cases} \quad y - 3z = 4 \\ x + \quad y \quad\quad = 7 \\ x + 2y + \quad z = 2 \end{cases}$

25. $(1, 1, 2)$ **27.** No solution **29.** $(-8, -7, 10)$
31. No solution **33.** $(1, 0, 1, -2)$
35. $x = -4t + 1, y = -t - 1, z = t$
37. $x = 6 - 5t, y = \frac{1}{2}(7 - 3t), z = t$
39. $\left(-\frac{4}{3}t + \frac{4}{3}, \frac{5}{3}t - \frac{2}{3}, t\right)$ **41.** $(s + 1, 2s - t + 1, s, t)$
43. No solution **45.** $(1, t + 1, t, 0)$
47. \$3000 at 6%, \$6000 at 7% **49.** \$11,250 in bank A,
\$22,500 in bank B, \$26,250 in bank C **51.** Impossible

53. $\begin{bmatrix} 4 & 18 \\ 4 & 0 \\ 2 & 2 \end{bmatrix}$ **55.** $\begin{bmatrix} 10 & 0 & -5 \end{bmatrix}$ **57.** $\begin{bmatrix} -\frac{7}{2} & 10 \\ 1 & -\frac{9}{2} \end{bmatrix}$

59. $\begin{bmatrix} 30 & 22 & 2 \\ -9 & 1 & -4 \end{bmatrix}$ **61.** $\begin{bmatrix} -\frac{1}{2} & \frac{11}{2} \\ \frac{15}{4} & -\frac{3}{2} \\ -\frac{1}{2} & 1 \end{bmatrix}$

65. $\frac{1}{3}\begin{bmatrix} -1 & -3 \\ -5 & 2 \end{bmatrix}$ **67.** $\begin{bmatrix} \frac{7}{2} & -2 \\ 0 & 8 \end{bmatrix}$ **69.** $\begin{bmatrix} 2 & -2 & 6 \\ -4 & 5 & -9 \end{bmatrix}$

71. $1, \begin{bmatrix} 9 & -4 \\ -2 & 1 \end{bmatrix}$ **73.** 0, no inverse

75. $-1, \begin{bmatrix} 3 & 2 & -3 \\ 2 & 1 & -2 \\ -8 & -6 & 9 \end{bmatrix}$ **77.** $24, \begin{bmatrix} 1 & 0 & 0 & -\frac{1}{4} \\ 0 & \frac{1}{2} & 0 & -\frac{1}{4} \\ 0 & 0 & \frac{1}{3} & -\frac{1}{4} \\ 0 & 0 & 0 & \frac{1}{4} \end{bmatrix}$

79. $(65, 154)$ **81.** $\left(-\frac{1}{12}, \frac{1}{12}, \frac{1}{12}\right)$
83. $\left(\frac{1}{5}, \frac{9}{5}\right)$ **85.** $\left(-\frac{87}{26}, \frac{21}{26}, \frac{3}{2}\right)$
87. 11 **89.** $\frac{2}{x - 5} + \frac{1}{x + 3}$ **91.** $\frac{-4}{x} + \frac{4}{x - 1} + \frac{-2}{(x - 1)^2}$
93. $\frac{-1}{x} + \frac{x + 2}{x^2 + 1}$ **95.** $x + y^2 \le 4$

97.

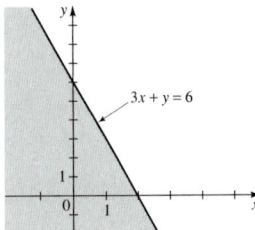

99.

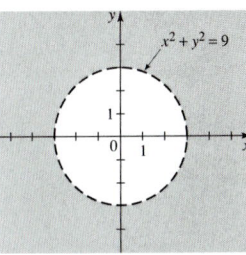

101.

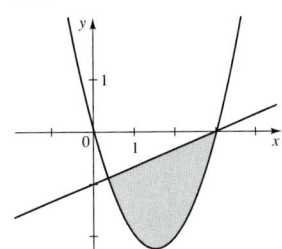

103.

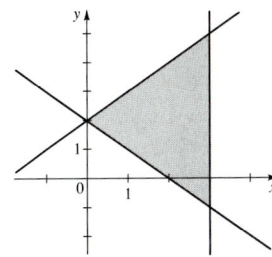

105.

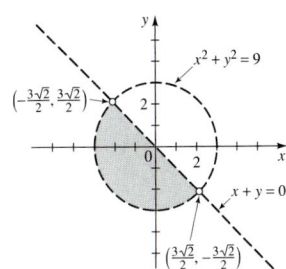

107.

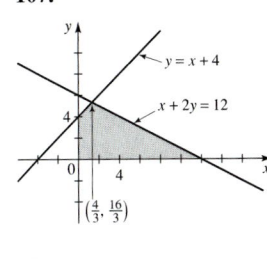

bounded   bounded
109. $x = \dfrac{b + c}{2}, y = \dfrac{a + c}{2}, z = \dfrac{a + b}{2}$ **111.** 2, 3

Chapter 9 Test ■ page 733

1. (a) Linear **(b)** $(-2, 3)$
2. (a) Nonlinear **(b)** $(1, -2), \left(\frac{5}{3}, 0\right)$
3. $(-0.55, -0.78), (0.43, -0.29), (2.12, 0.56)$
4. Wind 60 km/h, airplane 300 km/h **5. (a)** Row-echelon
form **(b)** Reduced row-echelon form **(c)** Neither
6. (a) $\left(\frac{5}{2}, \frac{5}{2}, 0\right)$ **(b)** No solution **7.** $\left(-\frac{3}{5} + \frac{2}{5}t, \frac{1}{5} + \frac{1}{5}t, t\right)$
8. Coffee \$1.50, juice \$1.75, donut \$0.75
9. (a) Incompatible dimensions **(b)** Incompatible dimensions

(c) $\begin{bmatrix} 6 & 10 \\ 3 & -2 \\ -3 & 9 \end{bmatrix}$ **(d)** $\begin{bmatrix} 36 & 58 \\ 0 & -3 \\ 18 & 28 \end{bmatrix}$ **(e)** $\begin{bmatrix} 2 & -\frac{3}{2} \\ -1 & 1 \end{bmatrix}$

(f) B is not square **(g)** B is not square **(h)** -3

10. (a) $\begin{bmatrix} 4 & -3 \\ 3 & -2 \end{bmatrix}\begin{bmatrix} x \\ y \end{bmatrix} = \begin{bmatrix} 10 \\ 30 \end{bmatrix}$ **(b)** $(70, 90)$

11. $|A| = 0, |B| = 2, B^{-1} = \begin{bmatrix} 1 & -2 & 0 \\ 0 & \frac{1}{2} & 0 \\ 3 & -6 & 1 \end{bmatrix}$

12. $(5, -5, -4)$

13. (a) $\dfrac{1}{x-1} + \dfrac{1}{(x-1)^2} - \dfrac{1}{x+2}$ (b) $-\dfrac{1}{x} + \dfrac{x+2}{x^2+3}$

14. (a)

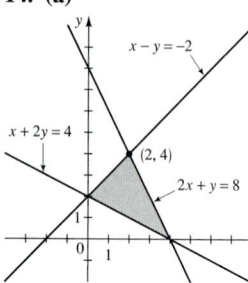

(b)

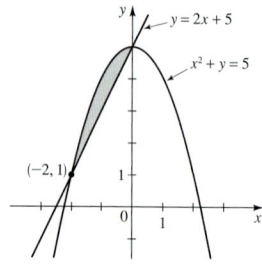

Focus on Modeling ■ **page 739**

1. 198, 195

3.

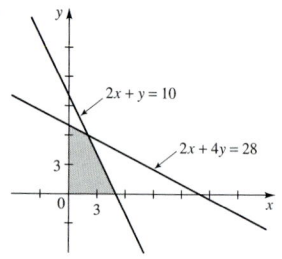 maximum 161
minimum 135

5. 3 tables, 34 chairs

7. 30 grapefruit crates, 30 orange crates

9. 15 Pasadena to Santa Monica, 3 Pasadena to El Toro, 0 Long Beach to Santa Monica, 16 Long Beach to El Toro

11. 90 standard, 40 deluxe

13. $7500 in municipal bonds, $2500 in bank certificates, $2000 in high-risk bonds

15. 4 games, 32 educational, 0 utility

Chapter 10

Section 10.1 ■ page 751

1. III   **3.** II   **5.** VI

Order of answers: focus; directrix; focal diameter

7. $F(1, 0)$; $x = -1$; 4   **9.** $F\left(0, \frac{9}{4}\right)$; $y = -\frac{9}{4}$; 9

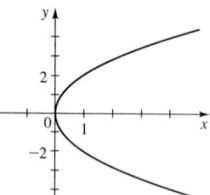   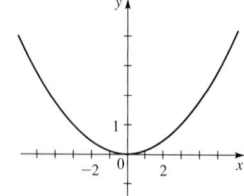

11. $F\left(0, \frac{1}{20}\right)$; $y = -\frac{1}{20}$; $\frac{1}{5}$ **13.** $F\left(-\frac{1}{32}, 0\right)$; $x = \frac{1}{32}$; $\frac{1}{8}$

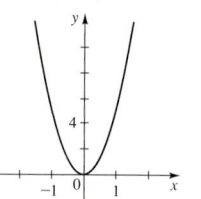   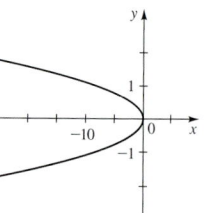

15. $F\left(0, -\frac{3}{2}\right)$; $y = \frac{3}{2}$; 6   **17.** $F\left(-\frac{5}{12}, 0\right)$; $x = \frac{5}{12}$; $\frac{5}{3}$

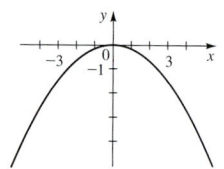   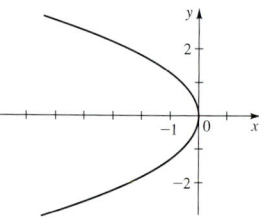

19.   **21.**

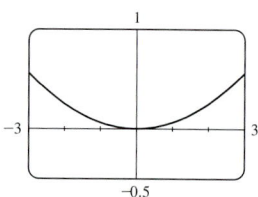   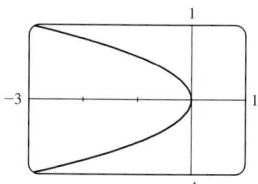

23.

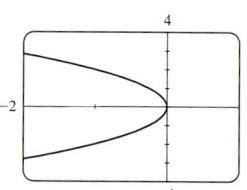

25. $x^2 = 8y$ **27.** $y^2 = -32x$

29. $y^2 = -8x$ **31.** $x^2 = 40y$

33. $y^2 = 4x$ **35.** $x^2 = 20y$

37. $x^2 = 8y$ **39.** $y^2 = -16x$

41. $y^2 = -3x$ **43.** $x = y^2$

45. $x^2 = -4\sqrt{2}\,y$

47. (a) $x^2 = -4py$, $p = \frac{1}{2}$, 1, 4, and 8

(b) The closer the directrix to the vertex, the steeper the parabola.

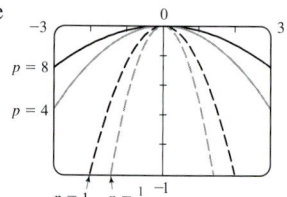

49. (a) $y^2 = 12x$ (b) $8\sqrt{15} \approx 31$ cm   **51.** $x^2 = 600y$

Section 10.2 ■ page 759

1. II   **3.** I

Order of answers: vertices; foci; eccentricity; major axis and minor axis

5. $V(\pm 5, 0)$; $F(\pm 4, 0)$; $\frac{4}{5}$; 10, 6

7. $V(0, \pm 3)$; $F(0, \pm\sqrt{5})$; $\sqrt{5}/3$; 6, 4

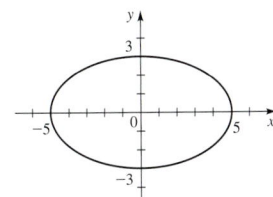

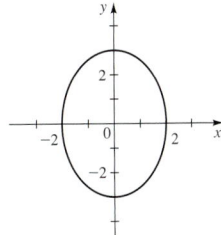

9. $V(\pm 4, 0)$; $F(\pm 2\sqrt{3}, 0)$; $\sqrt{3}/2$; 8, 4

11. $V(0, \pm\sqrt{3})$; $F(0, \pm\sqrt{3/2})$; $1/\sqrt{2}$; $2\sqrt{3}$, $\sqrt{6}$

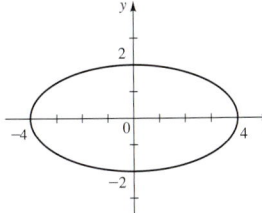

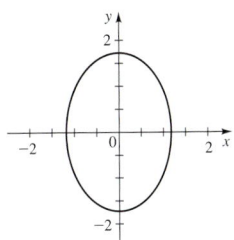

13. $V(\pm 1, 0)$; $F(\pm\sqrt{3}/2, 0)$; $\sqrt{3}/2$; 2, 1

15. $V(0, \pm\sqrt{2})$; $F(0, \pm\sqrt{3/2})$; $\sqrt{3}/2$; $2\sqrt{2}$, $\sqrt{2}$

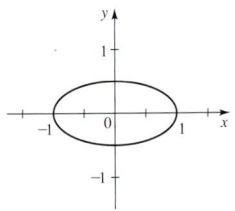

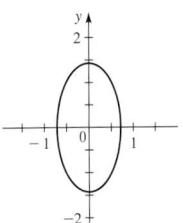

17. $V(0, \pm 1)$; $F(0, \pm 1/\sqrt{2})$; $1/\sqrt{2}$; 2, $\sqrt{2}$

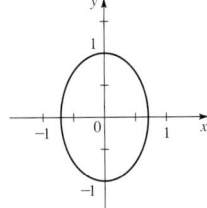

19. $\dfrac{x^2}{25} + \dfrac{y^2}{16} = 1$ **21.** $\dfrac{x^2}{4} + \dfrac{y^2}{8} = 1$

23. $\dfrac{x^2}{256} + \dfrac{y^2}{48} = 1$

25.

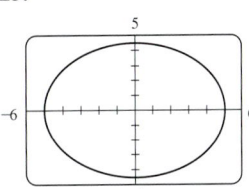

27.

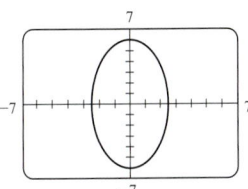

29. $\dfrac{x^2}{25} + \dfrac{y^2}{9} = 1$ **31.** $x^2 + \dfrac{y^2}{4} = 1$

33. $\dfrac{x^2}{9} + \dfrac{y^2}{13} = 1$ **35.** $\dfrac{x^2}{100} + \dfrac{y^2}{91} = 1$

37. $\dfrac{x^2}{25} + \dfrac{y^2}{5} = 1$ **39.** $\dfrac{64x^2}{225} + \dfrac{64y^2}{81} = 1$

41. $(0, \pm 2)$ **43.** $(\pm 1, 0)$

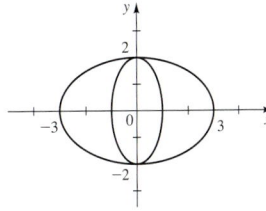

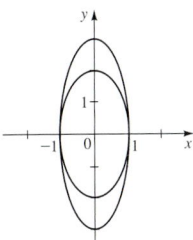

45. (a)

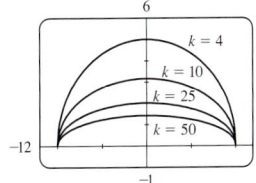

(b) Common major axes and vertices; eccentricity increases as k increases.

47. $\dfrac{x^2}{2.2500 \times 10^{16}} + \dfrac{y^2}{2.2491 \times 10^{16}} = 1$

49. $\dfrac{x^2}{1,455,642} + \dfrac{y^2}{1,451,610} = 1$ **51.** $5\sqrt{39}/2 \approx 15.6$ in.

Section 10.3 ■ page 768

1. III   **3.** II

Order of answers: vertices; foci; asymptotes

5. $V(\pm 2, 0)$; $F(\pm 2\sqrt{5}, 0)$; $y = \pm 2x$

7. $V(0, \pm 1)$; $F(0, \pm\sqrt{26})$; $y = \pm\frac{1}{5}x$

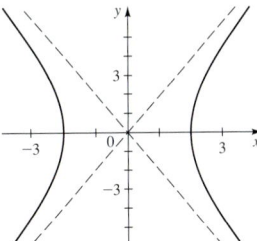

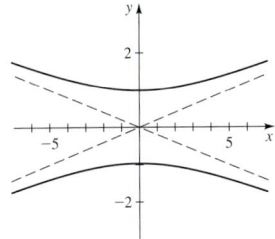

9. $V(\pm 1, 0)$; $F(\pm\sqrt{2}, 0)$; $y = \pm x$

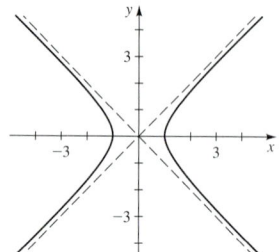

11. $V(0, \pm 3)$; $F(0, \pm\sqrt{34})$; $y = \pm\frac{3}{5}x$

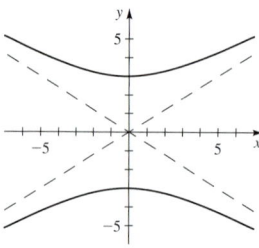

13. $V(\pm 2\sqrt{2}, 0)$; $F(\pm\sqrt{10}, 0)$; $y = \pm\frac{1}{2}x$

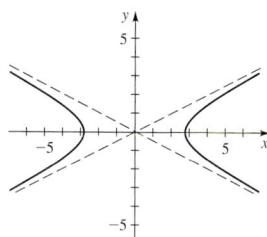

15. $V(0, \pm\frac{1}{2})$; $F(0, \pm\sqrt{5}/2)$; $y = \pm\frac{1}{2}x$

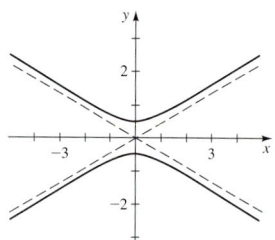

17. $\dfrac{x^2}{4} - \dfrac{y^2}{12} = 1$ **19.** $\dfrac{y^2}{16} - \dfrac{x^2}{16} = 1$

21. $\dfrac{x^2}{9} - \dfrac{4y^2}{9} = 1$

23.

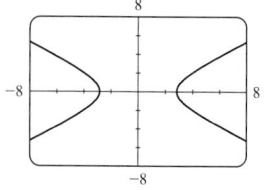

25.

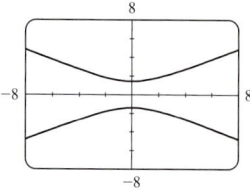

27. $\dfrac{x^2}{9} - \dfrac{y^2}{16} = 1$ **29.** $y^2 - \dfrac{x^2}{3} = 1$

31. $x^2 - \dfrac{y^2}{25} = 1$ **33.** $\dfrac{5y^2}{64} - \dfrac{5x^2}{256} = 1$

35. $\dfrac{x^2}{16} - \dfrac{y^2}{16} = 1$ **37.** $\dfrac{x^2}{9} - \dfrac{y^2}{16} = 1$

39. (b) $x^2 - y^2 = c^2/2$

43. (b)

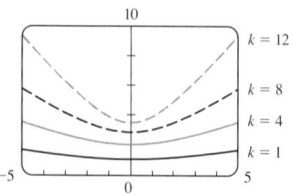

As k increases, the asymptotes get steeper.

45. $x^2 - y^2 = 2.3 \times 10^{19}$

Section 10.4 ■ page 781

1. Center $C(2, 1)$; foci $F(2 \pm \sqrt{5}, 1)$; vertices $V_1(-1, 1)$, $V_2(5, 1)$; major axis 6, minor axis 4

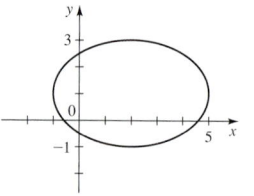

3. Center $C(0, -5)$; foci $F_1(0, -1)$, $F_2(0, -9)$; vertices $V_1(0, 0)$, $V_2(0, -10)$; major axis 10, minor axis 6

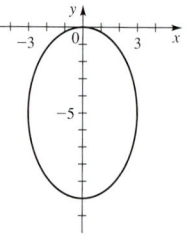

5. Vertex $V(3, -1)$; focus $F(3, 1)$; directrix $y = -3$

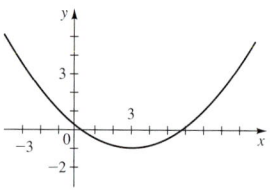

7. Vertex $V(-\frac{1}{2}, 0)$; focus $F(-\frac{1}{2}, -\frac{1}{16})$; directrix $y = \frac{1}{16}$

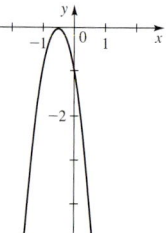

9. Center $C(-1, 3)$; foci $F_1(-6, 3)$, $F_2(4, 3)$; vertices $V_1(-4, 3)$, $V_2(2, 3)$; asymptotes $y = \pm\frac{4}{3}(x + 1) + 3$

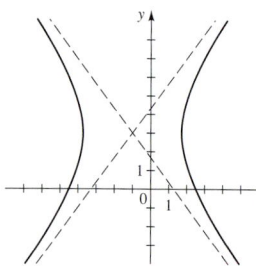

11. Center $C(-1, 0)$; foci $F(-1, \pm\sqrt{5})$; vertices $V(-1, \pm 1)$; asymptotes $y = \pm\frac{1}{2}(x + 1)$

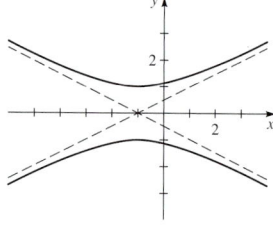

13. $x^2 = -\frac{1}{4}(y - 4)$ **15.** $\dfrac{(x - 5)^2}{25} + \dfrac{y^2}{16} = 1$

17. $(y - 1)^2 - x^2 = 1$

19. Ellipse; $C(2, 0)$; $F(2, \pm\sqrt{5})$; $V(2, \pm 3)$; major axis 6, minor axis 4

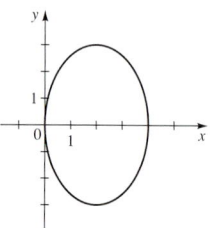

21. Hyperbola; $C(1, 2)$; $F_1(-\frac{3}{2}, 2)$, $F_2(\frac{7}{2}, 2)$; $V(1 \pm \sqrt{5}, 2)$; asymptotes $y = \pm\frac{1}{2}(x - 1) + 2$

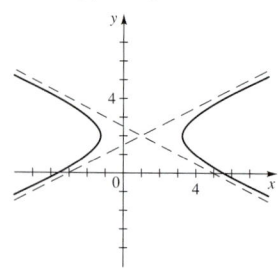

Section 10.2 ■ page 759

1. II   **3.** I

Order of answers: vertices; foci; eccentricity; major axis and minor axis

5. $V(\pm5,0)$; $F(\pm4,0)$; $\frac{4}{5}$; 10, 6

7. $V(0,\pm3)$; $F(0,\pm\sqrt{5})$; $\sqrt{5}/3$; 6, 4

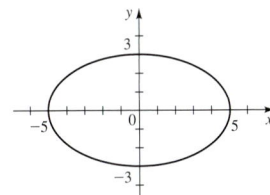
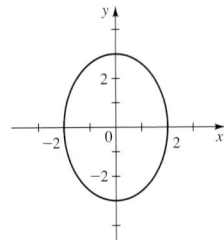

9. $V(\pm4,0)$; $F(\pm2\sqrt{3},0)$; $\sqrt{3}/2$; 8, 4

11. $V(0,\pm\sqrt{3})$; $F(0,\pm\sqrt{3/2})$; $1/\sqrt{2}$; $2\sqrt{3}$, $\sqrt{6}$

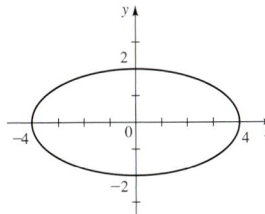
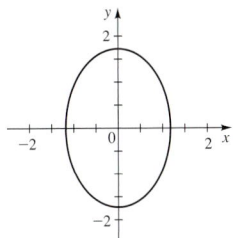

13. $V(\pm1,0)$; $F(\pm\sqrt{3}/2,0)$; $\sqrt{3}/2$; 2, 1

15. $V(0,\pm\sqrt{2})$; $F(0,\pm\sqrt{3/2})$; $\sqrt{3}/2$; $2\sqrt{2}$, $\sqrt{2}$

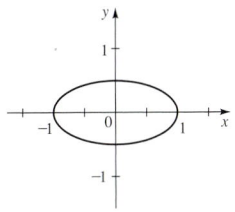
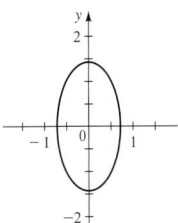

17. $V(0,\pm1)$; $F(0,\pm1/\sqrt{2})$; $1/\sqrt{2}$; 2, $\sqrt{2}$

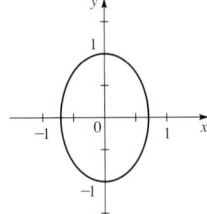

19. $\frac{x^2}{25}+\frac{y^2}{16}=1$ **21.** $\frac{x^2}{4}+\frac{y^2}{8}=1$

23. $\frac{x^2}{256}+\frac{y^2}{48}=1$

25.
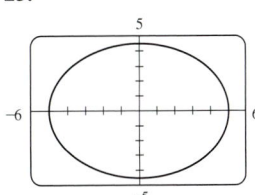

27.
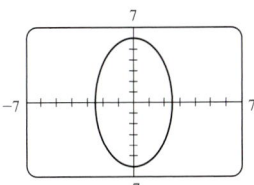

29. $\frac{x^2}{25}+\frac{y^2}{9}=1$ **31.** $x^2+\frac{y^2}{4}=1$

33. $\frac{x^2}{9}+\frac{y^2}{13}=1$ **35.** $\frac{x^2}{100}+\frac{y^2}{91}=1$

37. $\frac{x^2}{25}+\frac{y^2}{5}=1$ **39.** $\frac{64x^2}{225}+\frac{64y^2}{81}=1$

41. $(0,\pm2)$ **43.** $(\pm1,0)$

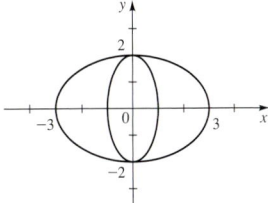
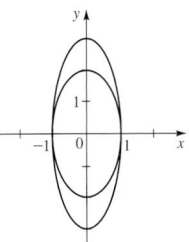

45. (a)
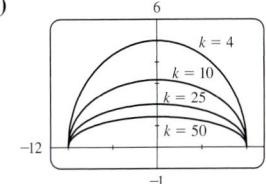
(b) Common major axes and vertices; eccentricity increases as k increases.

47. $\frac{x^2}{2.2500\times10^{16}}+\frac{y^2}{2.2491\times10^{16}}=1$

49. $\frac{x^2}{1,455,642}+\frac{y^2}{1,451,610}=1$ **51.** $5\sqrt{39}/2\approx15.6$ in.

Section 10.3 ■ page 768

1. III   **3.** II

Order of answers: vertices; foci; asymptotes

5. $V(\pm2,0)$; $F(\pm2\sqrt{5},0)$; $y=\pm2x$

7. $V(0,\pm1)$; $F(0,\pm\sqrt{26})$; $y=\pm\frac{1}{5}x$

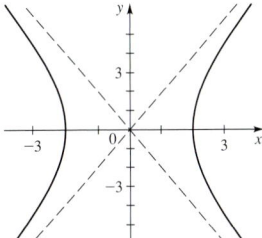
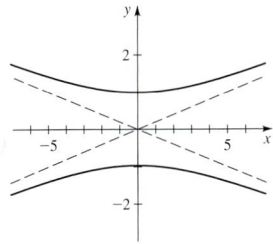

9. $V(\pm 1, 0)$; $F(\pm \sqrt{2}, 0)$; $y = \pm x$

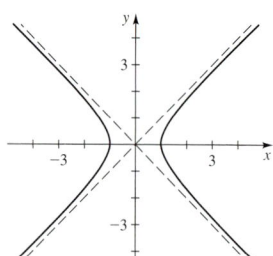

11. $V(0, \pm 3)$; $F(0, \pm \sqrt{34})$; $y = \pm \frac{3}{5}x$

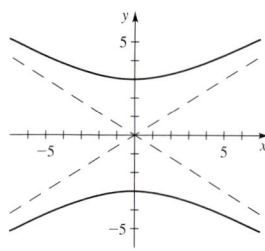

13. $V(\pm 2\sqrt{2}, 0)$; $F(\pm \sqrt{10}, 0)$; $y = \pm \frac{1}{2}x$

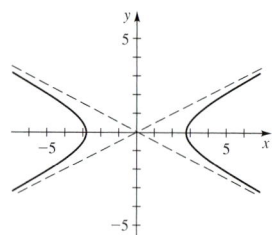

15. $V(0, \pm \frac{1}{2})$; $F(0, \pm \sqrt{5}/2)$; $y = \pm \frac{1}{2}x$

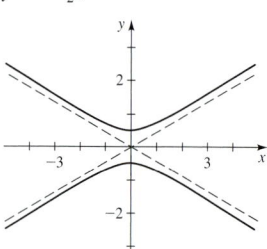

17. $\dfrac{x^2}{4} - \dfrac{y^2}{12} = 1$ **19.** $\dfrac{y^2}{16} - \dfrac{x^2}{16} = 1$

21. $\dfrac{x^2}{9} - \dfrac{4y^2}{9} = 1$

23.

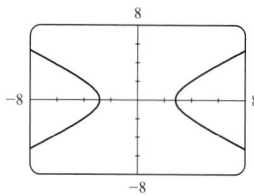

25.

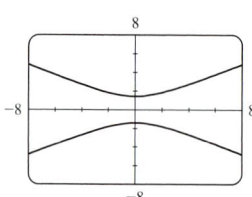

27. $\dfrac{x^2}{9} - \dfrac{y^2}{16} = 1$ **29.** $y^2 - \dfrac{x^2}{3} = 1$

31. $x^2 - \dfrac{y^2}{25} = 1$ **33.** $\dfrac{5y^2}{64} - \dfrac{5x^2}{256} = 1$

35. $\dfrac{x^2}{16} - \dfrac{y^2}{16} = 1$ **37.** $\dfrac{x^2}{9} - \dfrac{y^2}{16} = 1$

39. (b) $x^2 - y^2 = c^2/2$

43. (b)

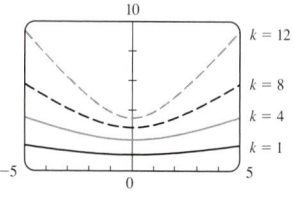

As k increases, the asymptotes get steeper.

45. $x^2 - y^2 = 2.3 \times 10^{19}$

Section 10.4 ■ page 781

1. Center $C(2, 1)$; foci $F(2 \pm \sqrt{5}, 1)$; vertices $V_1(-1, 1)$, $V_2(5, 1)$; major axis 6, minor axis 4

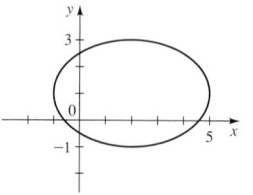

3. Center $C(0, -5)$; foci $F_1(0, -1)$, $F_2(0, -9)$; vertices $V_1(0, 0)$, $V_2(0, -10)$; major axis 10, minor axis 6

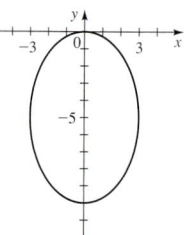

5. Vertex $V(3, -1)$; focus $F(3, 1)$; directrix $y = -3$

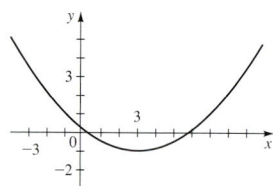

7. Vertex $V(-\frac{1}{2}, 0)$; focus $F(-\frac{1}{2}, -\frac{1}{16})$; directrix $y = \frac{1}{16}$

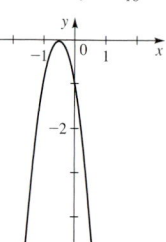

9. Center $C(-1, 3)$; foci $F_1(-6, 3)$, $F_2(4, 3)$; vertices $V_1(-4, 3)$, $V_2(2, 3)$; asymptotes $y = \pm \frac{4}{3}(x + 1) + 3$

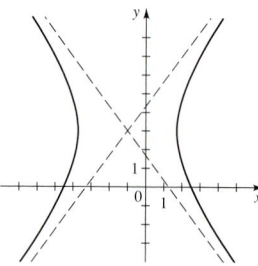

11. Center $C(-1, 0)$; foci $F(-1, \pm \sqrt{5})$; vertices $V(-1, \pm 1)$; asymptotes $y = \pm \frac{1}{2}(x + 1)$

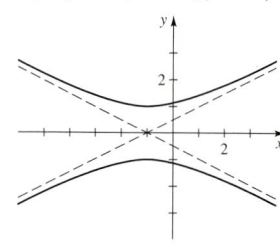

13. $x^2 = -\frac{1}{4}(y - 4)$ **15.** $\dfrac{(x - 5)^2}{25} + \dfrac{y^2}{16} = 1$

17. $(y - 1)^2 - x^2 = 1$

19. Ellipse; $C(2, 0)$; $F(2, \pm \sqrt{5})$; $V(2, \pm 3)$; major axis 6, minor axis 4

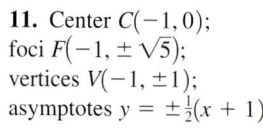

21. Hyperbola; $C(1, 2)$; $F_1(-\frac{3}{2}, 2)$, $F_2(\frac{7}{2}, 2)$; $V(1 \pm \sqrt{5}, 2)$; asymptotes $y = \pm \frac{1}{2}(x - 1) + 2$

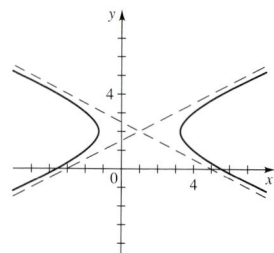

23. Ellipse; $C(3, -5)$;
$F(3 \pm \sqrt{21}, -5)$;
$V_1(-2, -5)$, $V_1(8, -5)$;
major axis 10, minor axis 4

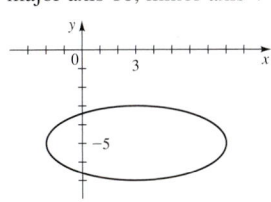

25. Hyperbola; $C(3, 0)$;
$F(3, \pm 5)$; $V(3, \pm 4)$;
asymptotes $y = \pm\frac{4}{3}(x - 3)$

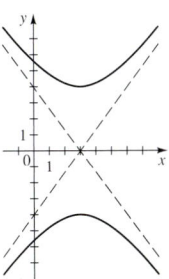

27. Degenerate conic
(pair of lines),
$y = \pm\frac{1}{2}(x - 4)$

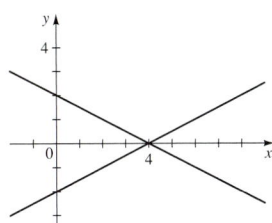

29. Point $(1, 3)$

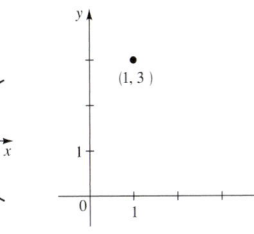

31.

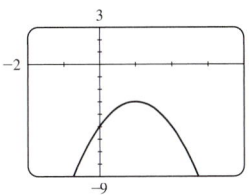

33.

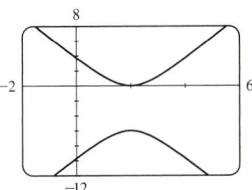

35. (a) $F < 17$ **(b)** $F = 17$ **(c)** $F > 17$

37. (a)

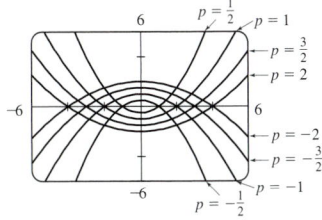

(c) The parabolas become narrower.

39. $\dfrac{(x + 150)^2}{18{,}062{,}500} + \dfrac{y^2}{18{,}040{,}000} = 1$

Section 10.5 ■ page 790

1. $(\sqrt{2}, 0)$ **3.** $(0, -2\sqrt{3})$ **5.** $(1.6383, 1.1472)$

7. $X^2 + \sqrt{3}XY + 2 = 0$

9. $7Y^2 - 48XY - 7X^2 - 40X - 30Y = 0$ **11.** $X^2 - Y^2 = 2$

13. (a) Hyberbola
(b) $X^2 - Y^2 = 16$
(c) $\phi = 45°$

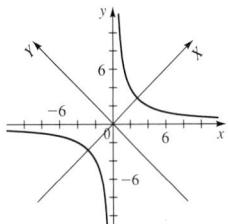

15. (a) Parabola
(b) $Y = \sqrt{2}X^2$
(c) $\phi = 45°$

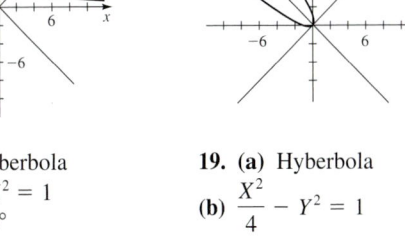

17. (a) Hyberbola
(b) $Y^2 - X^2 = 1$
(c) $\phi = 30°$

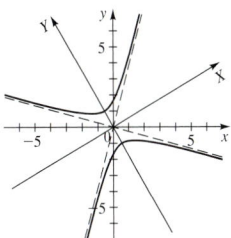

19. (a) Hyberbola
(b) $\dfrac{X^2}{4} - Y^2 = 1$
(c) $\phi \approx 53°$

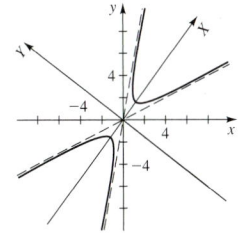

21. (a) Hyberbola
(b) $3X^2 - Y^2 = 2\sqrt{3}$
(c) $\phi = 30°$

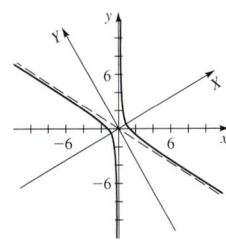

23. (a) Hyberbola
(b) $(X - 1)^2 - 3Y^2 = 1$
(c) $\phi = 60°$

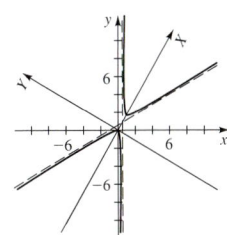

25. (a) Ellipse
(b) $X^2 + \dfrac{(Y + 1)^2}{4} = 1$
(c) $\phi \approx 53°$

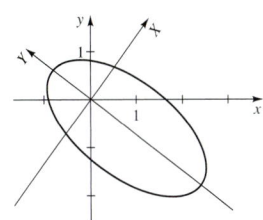

27. (a) Parabola
(b)

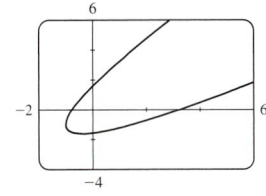

29. (a) Hyperbola
(b)

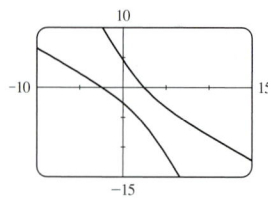

31. (a) $(X - 5)^2 - Y^2 = 1$
(b) XY-coordinates: $C(5, 0)$; $V_1(6, 0), V_2(4, 0)$; $F(5 \pm \sqrt{2}, 0)$;
xy-coordinates: $C(4, 3)$; $V_1\left(\frac{24}{5}, \frac{18}{5}\right), V_2\left(\frac{16}{5}, \frac{12}{5}\right)$;
$F_1\left(4 + \frac{4}{5}\sqrt{2}, 3 + \frac{3}{5}\sqrt{2}\right), F_2\left(4 - \frac{4}{5}\sqrt{2}, 3 - \frac{3}{5}\sqrt{2}\right)$
(c) $Y = \pm(X - 5)$; $7x - y - 25 = 0, x + 7y - 25 = 0$
33. $X = x \cos \phi + y \sin \phi$; $Y = -x \sin \phi + y \cos \phi$

Section 10.6 ■ page 799

1. $r = 6/(3 + 2 \cos \theta)$ **3.** $r = 2/(1 + \sin \theta)$
5. $r = 20/(1 + 4 \cos \theta)$ **7.** $r = 10/(1 + \sin \theta)$
9. II   **11.** VI   **13.** IV
15. (a) 3, hyperbola
(b)

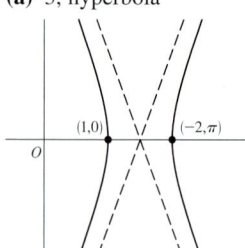

17. (a) 1, parabola
(b)

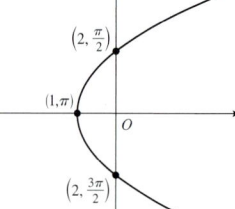

19. (a) $\frac{1}{2}$, ellipse
(b)

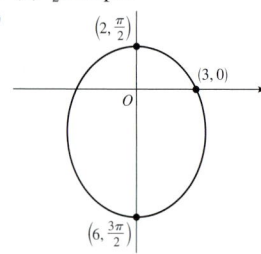

21. (a) $\frac{5}{2}$, hyperbola
(b)

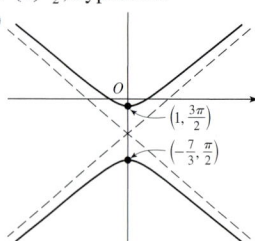

23. (a) $e = \frac{3}{4}$, directrix $x = -\frac{1}{3}$
(b) $r = \dfrac{1}{4 - 3 \cos\left(\theta - \dfrac{\pi}{3}\right)}$

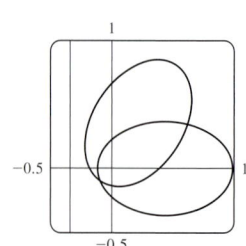

25. The ellipse is nearly circular when e is close to 0 and becomes more elongated as $e \to 1^-$. At $e = 1$, the curve becomes a parabola.

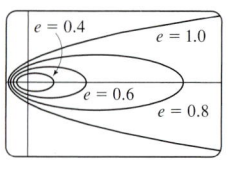

27. (b) $r = (1.49 \times 10^8)/(1 - 0.017 \cos \theta)$ **29.** 0.25

Section 10.7 ■ page 807

1. (a)

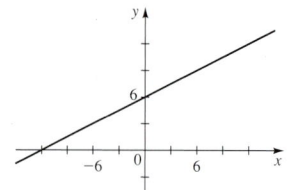

3. (a)

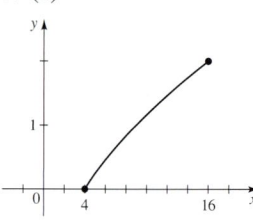

(b) $x - 2y + 12 = 0$
5. (a)

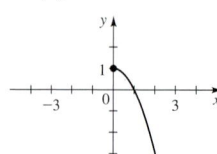

(b) $x = (y + 2)^2$
7. (a)

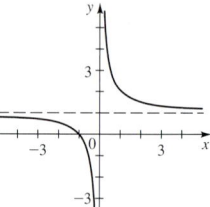

(b) $x = \sqrt{1 - y}$
9. (a)

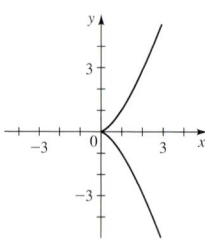

(b) $y = \dfrac{1}{x} + 1$
11. (a)

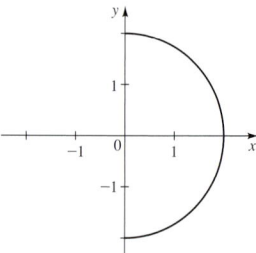

(b) $x^3 = y^2$
13. (a)

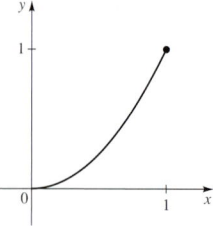

(b) $x^2 + y^2 = 4, x \geq 0$
15. (a)

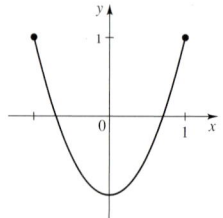

(b) $y = x^2, 0 \leq x \leq 1$
(b) $y = 2x^2 - 1, -1 \leq x \leq 1$

17. (a)

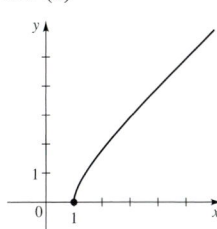

19. (a)

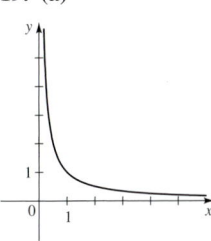

(b) $x^2 - y^2 = 1, x \geq 1, y \geq 0$ **(b)** $xy = 1, x \geq 0$

21. (a)

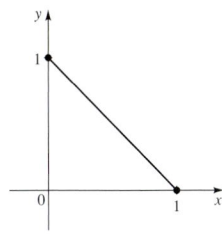

(b) $x + y = 1, 0 \leq x \leq 1$ **23.** $x = 4 + t, y = -1 + \frac{1}{2}t$
25. $x = 6 + t, y = 7 + t$ **27.** $x = a\cos t, y = a\sin t$
31.

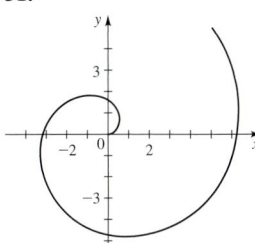

33.

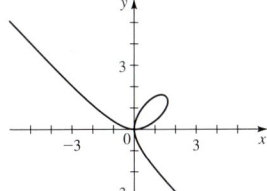

37.

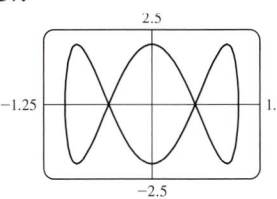

39.

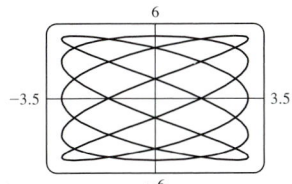

41.

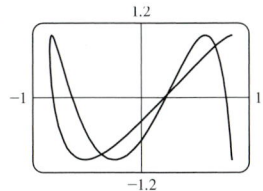

43. (a) $x = 2^{t/12}\cos t, y = 2^{t/12}\sin t$
(b)

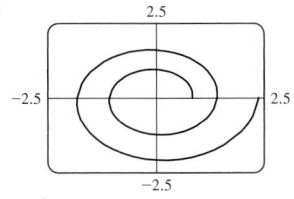

45. (a) $x = \dfrac{4\cos t}{2 - \cos t}, y = \dfrac{4\sin t}{2 - \cos t}$

(b)

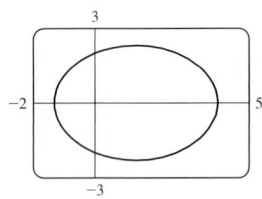

47. III   **49.** II
51.

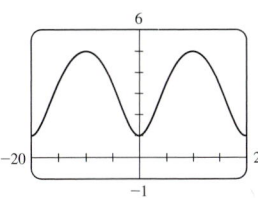

53. (b) $x^{2/3} + y^{2/3} = a^{2/3}$

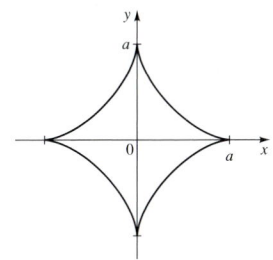

55. $x = a(\sin\theta\cos\theta + \cot\theta), y = a(1 + \sin^2\theta)$
57. (a) $x = a\sec\theta, y = b\sin\theta$
(b)

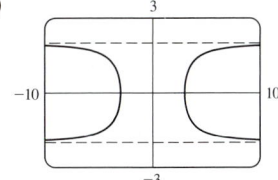

59. $y = a - a\cos\left(\dfrac{x + \sqrt{2ay - y^2}}{a}\right)$

61. (b)

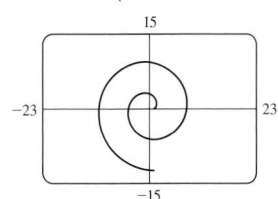

Chapter 10 Review ■ page 810

1. $V(0,0); F(1,0); x = -1$ **3.** $V(0,0); F(0,-2); y = 2$

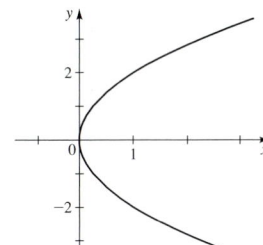
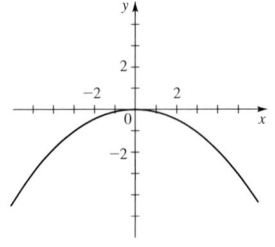

5. $V(-2, 2)$; $F\left(-\frac{7}{4}, 2\right)$; $x = -\frac{9}{4}$

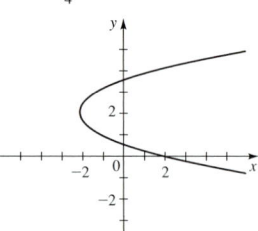

7. $V(-2, -3)$; $F(-2, -2)$; $y = -4$

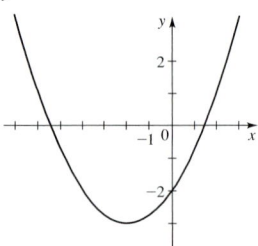

9. $C(0, 0)$; $V(0, \pm 5)$; $F(0, \pm 4)$; axes 10, 6

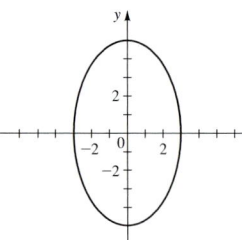

11. $C(0, 0)$; $V(\pm 4, 0)$; $F(\pm 2\sqrt{3}, 0)$; axes 8, 4

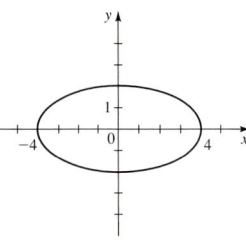

13. $C(3, 0)$; $V(3, \pm 4)$; $F(3, \pm\sqrt{7})$; axes 8, 6

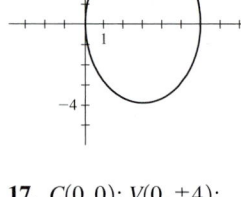

15. $C(0, 2)$; $V(\pm 3, 2)$; $F(\pm\sqrt{5}, 2)$; axes 6, 4

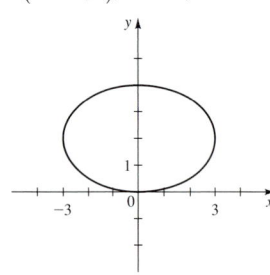

17. $C(0, 0)$; $V(0, \pm 4)$; $F(0, \pm 5)$; asymptotes $y = \pm\frac{4}{3}x$

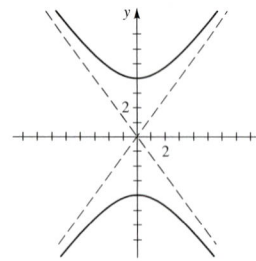

19. $C(0, 0)$; $V(\pm 4, 0)$; $F(\pm 2\sqrt{6}, 0)$; asymptotes $y = \pm\dfrac{1}{\sqrt{2}}x$

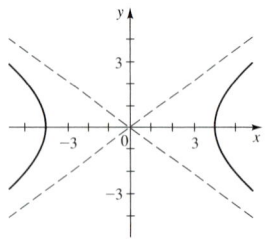

21. $C(-4, 0)$; $V_1(-8, 0)$, $V_2(0, 0)$; $F(-4 \pm 4\sqrt{2}, 0)$; asymptotes $y = \pm(x + 4)$

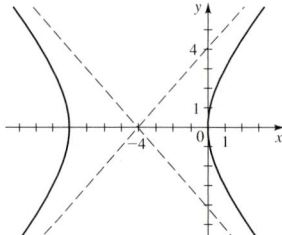

23. $C(-3, -1)$; $V(-3, -1 \pm \sqrt{2})$; $F(-3, -1 \pm 2\sqrt{5})$; asymptotes $y = \frac{1}{3}x$, $y = -\frac{1}{3}x - 2$

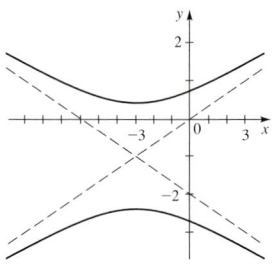

25. $y^2 = 8x$ **27.** $\dfrac{y^2}{16} - \dfrac{x^2}{9} = 1$

29. $\dfrac{(x-4)^2}{16} + \dfrac{(y-2)^2}{4} = 1$

31. Parabola; $F(0, -2)$; $V(0, 1)$

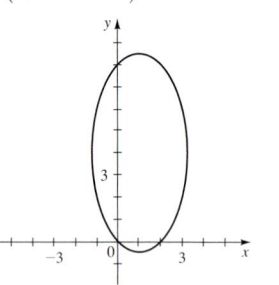

33. Hyperbola; $F(0, \pm 12\sqrt{2})$; $V(0, \pm 12)$

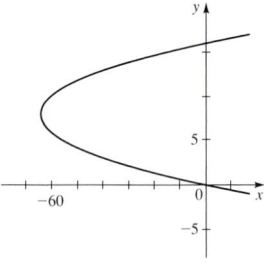

35. Ellipse; $F(1, 4 \pm \sqrt{15})$; $V(1, 4 \pm 2\sqrt{5})$

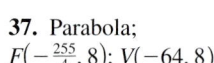

37. Parabola; $F\left(-\frac{255}{4}, 8\right)$; $V(-64, 8)$

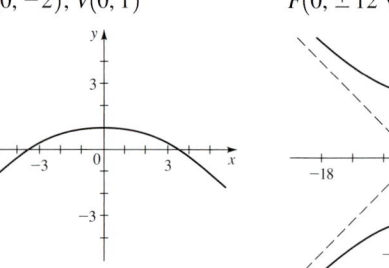

39. Ellipse; $F(3, -3 \pm 1/\sqrt{2})$; $V_1(3, -4)$, $V_2(3, -2)$

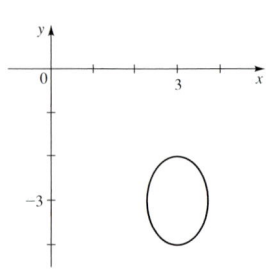

41. Has no graph   **43.** $x^2 = 4y$

45. $\dfrac{y^2}{4} - \dfrac{x^2}{16} = 1$ **47.** $\dfrac{(x-1)^2}{3} + \dfrac{(y-2)^2}{4} = 1$

49. $\dfrac{4(x-7)^2}{225} + \dfrac{(y-2)^2}{100} = 1$

51. (a) 91,419,000 mi   **(b)** 94,581,000 mi

53. (a)

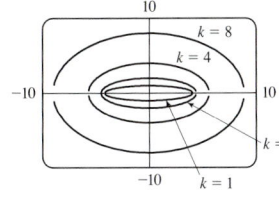

55. (a) Hyperbola
(b) $3X^2 - Y^2 = 1$
(c) $\phi = 45°$

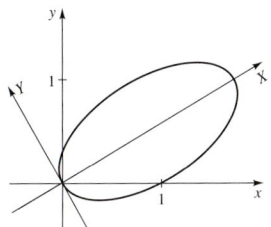

57. (a) Ellipse
(b) $(X-1)^2 + 4Y^2 = 1$
(c) $\phi = 30°$

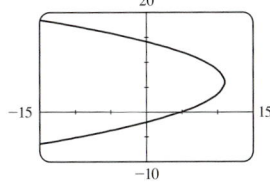

59. Ellipse

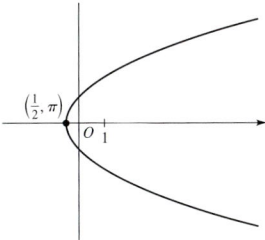

61. Parabola

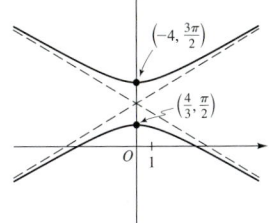

63. (a) $e = 1$, parabola
(b)

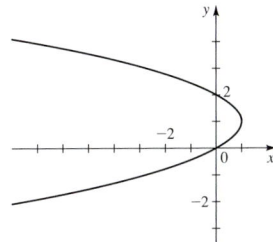

65. (a) $e = 2$, hyperbola

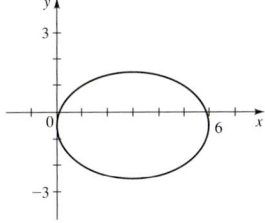

67. (a)

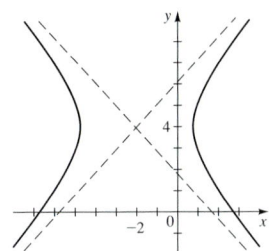

(b) $x = 2y - y^2$

69. (a)

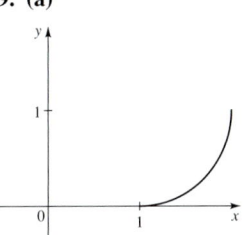

71.

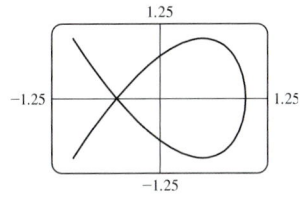

(b) $(x-1)^2 + (y-1)^2 = 1$, $1 \le x \le 2$, $0 \le y \le 1$
73. $x = \frac{1}{2}(1 + \cos\theta)$, $y = \frac{1}{2}(\sin\theta + \tan\theta)$

Chapter 10 Test ■ page 814

1. $F(0, -3)$, $y = 3$ **2.** $V(\pm 4, 0)$; $F(\pm 2\sqrt{3}, 0)$; 8, 4

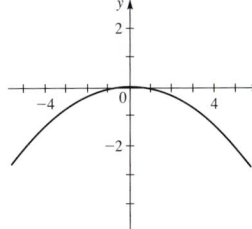

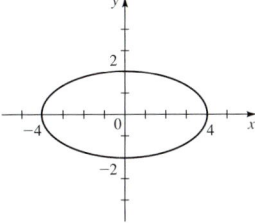

3. $V(0, \pm 3)$; $F(0, \pm 5)$; $y = \pm\frac{3}{4}x$

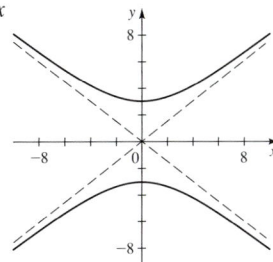

4. $y^2 = -x$ **5.** $\dfrac{x^2}{16} + \dfrac{(y-3)^2}{9} = 1$ **6.** $(x-2)^2 - \dfrac{y^2}{3} = 1$

7. $\dfrac{(x-3)^2}{9} + \dfrac{(y+\frac{1}{2})^2}{4} = 1$ **8.** $\dfrac{(x+2)^2}{8} - \dfrac{(y-4)^2}{9} = 1$

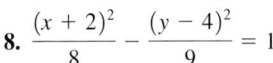

9. $(y + 4)^2 = -2(x - 4)$

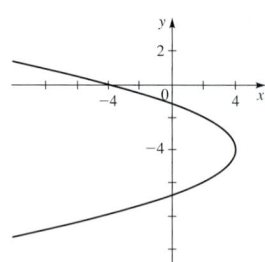

10. $\dfrac{y^2}{9} - \dfrac{x^2}{16} = 1$ **11.** $x^2 - 4x - 8y + 20 = 0$

12. $\frac{3}{4}$ in.   **13. (a)** Ellipse   **(b)** $\dfrac{X^2}{3} + \dfrac{Y^2}{18} = 1$

(c) $\phi \approx 27°$

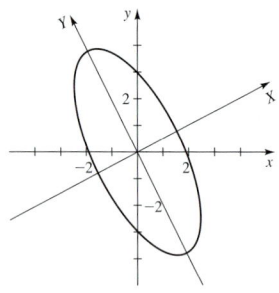

(d) $\left(-3\sqrt{2/5}, 6\sqrt{2/5}\right), \left(3\sqrt{2/5}, -6\sqrt{2/5}\right)$

14. (a) $r = \dfrac{1}{1 + 0.5\cos\theta}$ **(b)** Ellipse

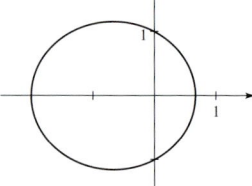

15. (a)   **(b)** $\dfrac{(x - 3)^2}{9} + \dfrac{y^2}{4} = 1, x \geq 3$

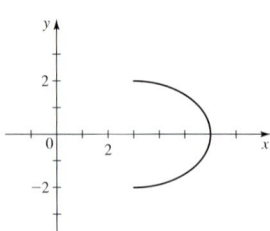

Focus on Modeling ■ page 818

1. $y = -\left(\dfrac{g}{2v_0^2 \cos^2\theta}\right)x^2 + (\tan\theta)x$

3. (a) 5.45 s   **(b)** 118.7 ft   **(c)** 5426.5 ft

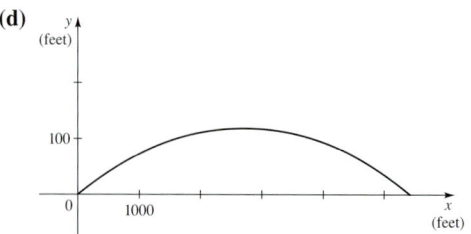

5. $\dfrac{v_0^2 \sin^2\theta}{2g}$ **7.** No, $\theta \approx 23°$

Chapter 11

Section 11.1 ■ page 830

1. 2, 3, 4, 5; 101   **3.** $\frac{1}{2}, \frac{1}{3}, \frac{1}{4}, \frac{1}{5}; \frac{1}{101}$ **5.** $-1, \frac{1}{4}, -\frac{1}{9}, \frac{1}{16}; \frac{1}{10,000}$
7. 0, 2, 0, 2; 2   **9.** 1, 4, 27, 256; 100^{100}
11. 3, 2, 0, -4, -12 **13.** 1, 3, 7, 15, 31   **15.** 1, 2, 3, 5, 8
17. (a) 7, 11, 15, 19, 23, 27, 31, 35, 39, 43
(b)

19. (a) 12, 6, 4, 3, $\frac{12}{5}$, 2, $\frac{12}{7}$, $\frac{3}{2}$, $\frac{4}{3}$, $\frac{6}{5}$
(b)

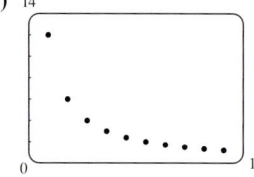

21. (a) 2, $\frac{1}{2}$, 2, $\frac{1}{2}$, 2, $\frac{1}{2}$, 2, $\frac{1}{2}$, 2, $\frac{1}{2}$
(b)

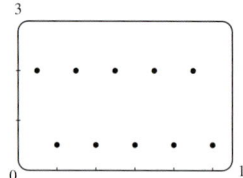

23. 2^n **25.** $3n - 2$ **27.** $(2n - 1)/n^2$ **29.** $1 + (-1)^n$
31. 1, 4, 9, 16, 25, 36   **33.** $\frac{1}{3}, \frac{4}{9}, \frac{13}{27}, \frac{40}{81}, \frac{121}{243}, \frac{364}{729}$
35. $\frac{2}{3}, \frac{8}{9}, \frac{26}{27}, \frac{80}{81}$; $S_n = 1 - \dfrac{1}{3^n}$
37. $1 - \sqrt{2}, 1 - \sqrt{3}, -1, 1 - \sqrt{5}$; $S_n = 1 - \sqrt{n + 1}$
39. 10   **41.** $\frac{11}{6}$ **43.** 8   **45.** 31   **47.** 385   **49.** 46,438
51. 22   **53.** $\sqrt{1} + \sqrt{2} + \sqrt{3} + \sqrt{4} + \sqrt{5}$
55. $\sqrt{4} + \sqrt{5} + \sqrt{6} + \sqrt{7} + \sqrt{8} + \sqrt{9} + \sqrt{10}$
57. $x^3 + x^4 + \cdots + x^{100}$ **59.** $\displaystyle\sum_{k=1}^{100} k$ **61.** $\displaystyle\sum_{k=1}^{10} k^2$

63. $\displaystyle\sum_{k=1}^{999} \frac{1}{k(k+1)}$ **65.** $\displaystyle\sum_{k=0}^{100} x^k$ **67.** $2^{(2^n-1)/2^n}$

69. **(a)** 2004.00, 2008.01, 2012.02, 2016.05, 2020.08, 2024.12
(b) $2149.16   **71.** **(a)** 35,700, 36,414, 37,142, 37,885, 38,643
(b) 42,665   **73.** **(b)** 6898   **75.** **(a)** $S_n = S_{n-1} + 2000$
(b) $38,000

Section 11.2 ■ page 837

1. **(a)** 5, 7, 9, 11, 13   **(b)** 2
(c)

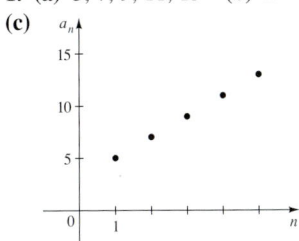

3. **(a)** $\frac{5}{2}, \frac{3}{2}, \frac{1}{2}, -\frac{1}{2}, -\frac{3}{2}$ **(b)** -1
(c)

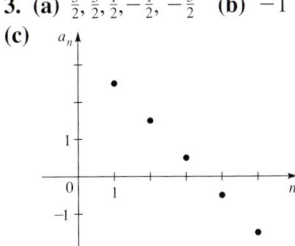

5. $a_n = 3 + 5(n-1), a_{10} = 48$
7. $a_n = \frac{5}{2} - \frac{1}{2}(n-1), a_{10} = -2$ **9.** Arithmetic, 3
11. Not arithmetic   **13.** Arithmetic, $-\frac{3}{2}$
15. Arithmetic, 1.7   **17.** 11, 18, 25, 32, 39; 7;
$a_n = 11 + 7(n-1)$ **19.** $\frac{1}{3}, \frac{1}{5}, \frac{1}{7}, \frac{1}{9}, \frac{1}{11}$; not arithmetic
21. $-4, 2, 8, 14, 20$; 6; $a_n = -4 + 6(n-1)$
23. 3, $a_5 = 14, a_n = 2 + 3(n-1), a_{100} = 299$
25. 5, $a_5 = 24, a_n = 4 + 5(n-1), a_{100} = 499$
27. 4, $a_5 = 4, a_n = -12 + 4(n-1), a_{100} = 384$
29. 1.5, $a_5 = 31, a_n = 25 + 1.5(n-1), a_{100} = 173.5$
31. $s, a_5 = 2 + 4s, a_n = 2 + (n-1)s, a_{100} = 2 + 99s$
33. $\frac{1}{2}$ **35.** $-100, -98, -96$ **37.** 30th   **39.** 100   **41.** 460
43. 1090   **45.** 20,301   **47.** 832.3   **49.** 46.75   **53.** Yes
55. 50   **57.** $1250   **59.** $403,500   **61.** 20   **63.** 78

Section 11.3 ■ page 844

1. **(a)** 5, 10, 20, 40, 80   **(b)** 2
(c)

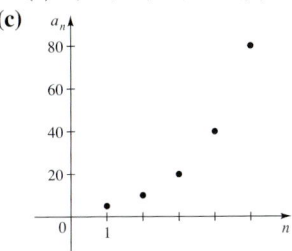

3. **(a)** $\frac{5}{2}, -\frac{5}{4}, \frac{5}{8}, -\frac{5}{16}, \frac{5}{32}$ **(b)** $-\frac{1}{2}$
(c)

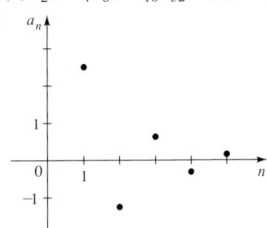

5. $a_n = 3 \cdot 5^{n-1}, a_4 = 375$ **7.** $a_n = \frac{5}{2}\left(-\frac{1}{2}\right)^{n-1}, a_4 = -\frac{5}{16}$
9. Geometric, 2   **11.** Geometric, $\frac{1}{2}$ **13.** Not geometric
15. Geometric, 1.1   **17.** 6, 18, 54, 162, 486; geometric,
common ratio 3; $a_n = 6 \cdot 3^{n-1}$ **19.** $\frac{1}{4}, \frac{1}{16}, \frac{1}{64}, \frac{1}{256}, \frac{1}{1024}$; geometric,
common ratio $\frac{1}{4}$; $a_n = \frac{1}{4}\left(\frac{1}{4}\right)^{n-1}$ **21.** 0, ln 5, 2 ln 5, 3 ln 5, 4 ln 5;
not geometric   **23.** 3, $a_5 = 162, a_n = 2 \cdot 3^{n-1}$
25. $-0.3, a_5 = 0.00243, a_n = (0.3)(-0.3)^{n-1}$
27. $-\frac{1}{12}, a_5 = \frac{1}{144}, a_n = 144\left(-\frac{1}{12}\right)^{n-1}$
29. $3^{2/3}, a_5 = 3^{11/3}, a_n = 3^{(2n+1)/3}$
31. $s^{2/7}, a_5 = s^{8/7}, a_n = s^{2(n-1)/7}$
33. $\frac{1}{2}$ **35.** $\frac{25}{4}$ **37.** 11th   **39.** 315   **41.** 441   **43.** 3280
45. $\frac{6141}{1024}$ **47.** $\frac{3}{2}$ **49.** $\frac{3}{4}$ **51.** $\frac{1}{648}$ **53.** $-\frac{1000}{117}$ **55.** $\frac{7}{9}$
57. $\frac{1}{33}$ **59.** $\frac{112}{999}$ **61.** 10, 20, 40
63. **(a)** $V_n = 160,000(0.80)^{n-1}$ **(b)** 4th year
65. 19 ft, $80\left(\frac{3}{4}\right)^n$ **67.** $\frac{64}{25}, \frac{1024}{625}, 5\left(\frac{4}{5}\right)^n$ **69.** **(a)** $17\frac{8}{9}$ ft
(b) $18 - \left(\frac{1}{3}\right)^{n-3}$ **71.** 2801   **73.** 3 m   **75.** **(a)** 2
(b) $8 + 4\sqrt{2}$ **77.** 1

Section 11.4 ■ page 853

1. $13,180.79   **3.** $360,262.21   **5.** $5,591.79   **7.** $245.66
9. $2,601.59   **11.** $307.24   **13.** $733.76, $264,153.60
15. **(a)** $859.15   **(b)** $309,294.00
(c) $1,841,519.29   **17.** $341.24   **19.** 18.16%   **21.** 11.68%

Section 11.5 ■ page 859

1. Let $P(n)$ denote the statement
$2 + 4 + \cdots + 2n = n(n+1)$.

Step 1 $P(1)$ is true since $2 = 1(1+1)$.
Step 2 Suppose $P(k)$ is true. Then

$$2 + 4 + \cdots + 2k + 2(k+1)$$
$$= k(k+1) + 2(k+1) \qquad \text{Induction hypothesis}$$
$$= (k+1)(k+2)$$

So $P(k+1)$ follows from $P(k)$. Thus, by the Principle of
Mathematical Induction, $P(n)$ holds for all n.

3. Let $P(n)$ denote the statement
$$5 + 8 + \cdots + (3n+2) = \frac{n(3n+7)}{2}.$$

Step 1 $P(1)$ is true since $5 = \dfrac{1(3 \cdot 1 + 7)}{2}$.

Step 2 Suppose $P(k)$ is true. Then

$$5 + 8 + \cdots + (3k + 2) + [3(k + 1) + 2]$$

$$= \frac{k(3k + 7)}{2} + (3k + 5) \qquad \text{Induction hypothesis}$$

$$= \frac{3k^2 + 13k + 10}{2}$$

$$= \frac{(k + 1)[3(k + 1) + 7]}{2}$$

So $P(k + 1)$ follows from $P(k)$. Thus, by the Principle of Mathematical Induction, $P(n)$ holds for all n.

5. Let $P(n)$ denote the statement

$$1 \cdot 2 + 2 \cdot 3 + \cdots + n(n + 1) = \frac{n(n + 1)(n + 2)}{3}.$$

Step 1 $P(1)$ is true since $1 \cdot 2 = \dfrac{1 \cdot (1 + 1) \cdot (1 + 2)}{3}$.

Step 2 Suppose $P(k)$ is true. Then

$$1 \cdot 2 + 2 \cdot 3 + \cdots + k(k + 1) + (k + 1)(k + 2)$$

$$= \frac{k(k + 1)(k + 2)}{3} + (k + 1)(k + 2) \qquad \text{Induction hypothesis}$$

$$= \frac{(k + 1)(k + 2)(k + 3)}{3}$$

So $P(k + 1)$ follows from $P(k)$. Thus, by the Principle of Mathematical Induction, $P(n)$ holds for all n.

7. Let $P(n)$ denote the statement

$$1^3 + 2^3 + \cdots + n^3 = \frac{n^2(n + 1)^2}{4}.$$

Step 1 $P(1)$ is true since $1^3 = \dfrac{1^2 \cdot (1 + 1)^2}{4}$.

Step 2 Suppose $P(k)$ is true. Then

$$1^3 + 2^3 + \cdots + k^3 + (k + 1)^3$$

$$= \frac{k^2(k + 1)^2}{4} + (k + 1)^3 \qquad \text{Induction hypothesis}$$

$$= \frac{(k + 1)^2[k^2 + 4(k + 1)]}{4}$$

$$= \frac{(k + 1)^2(k + 2)^2}{4}$$

So $P(k + 1)$ follows from $P(k)$. Thus, by the Principle of Mathematical Induction, $P(n)$ holds for all n.

9. Let $P(n)$ denote the statement

$$2^3 + 4^3 + \cdots + (2n)^3 = 2n^2(n + 1)^2.$$

Step 1 $P(1)$ is true since $2^3 = 2 \cdot 1^2(1 + 1)^2$.

Step 2 Suppose $P(k)$ is true. Then

$$2^3 + 4^3 + \cdots + (2k)^3 + [2(k + 1)]^3$$

$$= 2k^2(k + 1)^2 + [2(k + 1)]^3 \qquad \text{Induction hypothesis}$$

$$= (k + 1)^2(2k^2 + 8k + 8)$$

$$= 2(k + 1)^2(k + 2)^2$$

So $P(k + 1)$ follows from $P(k)$. Thus, by the Principle of Mathematical Induction, $P(n)$ holds for all n.

11. Let $P(n)$ denote the statement

$$1 \cdot 2 + 2 \cdot 2^2 + \cdots + n \cdot 2^n = 2[1 + (n - 1)2^n].$$

Step 1 $P(1)$ is true since $1 \cdot 2 = 2[1 + 0]$.

Step 2 Suppose $P(k)$ is true. Then

$$1 \cdot 2 + 2 \cdot 2^2 + \cdots + k \cdot 2^k + (k + 1) \cdot 2^{k+1}$$

$$= 2[1 + (k - 1)2^k] + (k + 1) \cdot 2^{k+1} \qquad \text{Induction hypothesis}$$

$$= 2 + (k - 1)2^{k+1} + (k + 1) \cdot 2^{k+1}$$

$$= 2 + 2k2^{k+1} = 2(1 + k2^{k+1})$$

So $P(k + 1)$ follows from $P(k)$. Thus, by the Principle of Mathematical Induction, $P(n)$ holds for all n.

13. Let $P(n)$ denote the statement $n^2 + n$ is divisible by 2.

Step 1 $P(1)$ is true since $1^2 + 1$ is divisible by 2.

Step 2 Suppose $P(k)$ is true. Now

$$(k + 1)^2 + (k + 1) = k^2 + 2k + 1 + k + 1$$

$$= (k^2 + k) + 2(k + 1)$$

But $k^2 + k$ is divisible by 2 (by the induction hypothesis) and $2(k + 1)$ is clearly divisible by 2, so $(k + 1)^2 + (k + 1)$ is divisible by 2. So $P(k + 1)$ follows from $P(k)$. Thus, by the Principle of Mathematical Induction, $P(n)$ holds for all n.

15. Let $P(n)$ denote the statement $n^2 - n + 41$ is odd.

Step 1 $P(1)$ is true since $1^2 - 1 + 41$ is odd.

Step 2 Suppose $P(k)$ is true. Now

$$(k + 1)^2 - (k + 1) + 41 = (k^2 - k + 41) + 2k$$

But $k^2 - k + 41$ is odd (by the induction hypothesis) and $2k$ is clearly even, so their sum is odd. So $P(k + 1)$ follows from $P(k)$. Thus, by the Principle of Mathematical Induction, $P(n)$ holds for all n.

17. Let $P(n)$ denote the statement $8^n - 3^n$ is divisible by 5.

Step 1 $P(1)$ is true since $8^1 - 3^1$ is divisible by 5.

Step 2 Suppose $P(k)$ is true. Now

$$8^{k+1} - 3^{k+1} = 8 \cdot 8^k - 3 \cdot 3^k$$

$$= 8 \cdot 8^k - (8 - 5) \cdot 3^k = 8 \cdot (8^k - 3^k) + 5 \cdot 3^k$$

which is divisible by 5 because $8^k - 3^k$ is divisible by 5 (by the induction hypothesis) and $5 \cdot 3^k$ is clearly divisible by 5. So $P(k + 1)$ follows from $P(k)$. Thus, by the Principle of Mathematical Induction, $P(n)$ holds for all n.

19. Let $P(n)$ denote the statement $n < 2^n$.

Step 1 $P(1)$ is true since $1 < 2^1$.
Step 2 Suppose $P(k)$ is true. Then

$$k + 1 < 2^k + 1 \qquad \text{Induction hypothesis}$$

$$< 2^k + 2^k \qquad \text{Because } 1 < 2^k$$

$$= 2 \cdot 2^k = 2^{k+1}$$

So $P(k + 1)$ follows from $P(k)$. Thus, by the Principle of Mathematical Induction, $P(n)$ holds for all n.

21. Let $P(n)$ denote the statement $(1 + x)^n \geq 1 + nx$ for $x > -1$.

Step 1 $P(1)$ is true since $(1 + x)^1 \geq 1 + 1 \cdot x$.
Step 2 Suppose $P(k)$ is true. Then

$$(1 + x)^{k+1} = (1 + x)(1 + x)^k$$

$$\geq (1 + x)(1 + kx) \qquad \text{Induction hypothesis}$$

$$= 1 + (k + 1)x + kx^2$$

$$\geq 1 + (k + 1)x$$

So $P(k + 1)$ follows from $P(k)$. Thus, by the Principle of Mathematical Induction, $P(n)$ holds for all n.

23. Let $P(n)$ denote the statement $a_n = 5 \cdot 3^{n-1}$.

Step 1 $P(1)$ is true since $a_1 = 5 \cdot 3^0 = 5$.
Step 2 Suppose $P(k)$ is true. Then

$$a_{k+1} = 3 \cdot a_k \qquad \text{Definition of } a_{k+1}$$

$$= 3 \cdot 5 \cdot 3^{k-1} \qquad \text{Induction hypothesis}$$

$$= 5 \cdot 3^k$$

So $P(k + 1)$ follows from $P(k)$. Thus, by the Principle of Mathematical Induction, $P(n)$ holds for all n.

25. Let $P(n)$ denote the statement $x - y$ is a factor of $x^n - y^n$.

Step 1 $P(1)$ is true since $x - y$ is a factor of $x^1 - y^1$.
Step 2 Suppose $P(k)$ is true. Now

$$x^{k+1} - y^{k+1} = x^{k+1} - x^k y + x^k y - y^{k+1}$$

$$= x^k(x - y) + (x^k - y^k)y$$

But $x^k(x - y)$ is clearly divisible by $x - y$ and $(x^k - y^k)y$ is divisible by $x - y$ (by the induction hypothesis), so their sum is divisible by $x - y$. So $P(k + 1)$ follows from $P(k)$. Thus, by the Principle of Mathematical Induction, $P(n)$ holds for all n.

27. Let $P(n)$ denote the statement F_{3n} is even.

Step 1 $P(1)$ is true since $F_{3 \cdot 1} = 2$, which is even.
Step 2 Suppose $P(k)$ is true. Now, by the definition of the Fibonacci sequence

$$F_{3(k+1)} = F_{3k+3} = F_{3k+2} + F_{3k+1}$$

$$= F_{3k+1} + F_{3k} + F_{3k+1}$$

$$= F_{3k} + 2 \cdot F_{3k+1}$$

But F_{3k} is even (by the induction hypothesis) and $2 \cdot F_{3k+1}$ is clearly even, so $F_{3(k+1)}$ is even. So $P(k + 1)$ follows from $P(k)$. Thus, by the Principle of Mathematical Induction, $P(n)$ holds for all n.

29. Let $P(n)$ denote the statement $F_1^2 + F_2^2 + \cdots + F_n^2 = F_n \cdot F_{n+1}$.

Step 1 $P(1)$ is true since $F_1^2 = F_1 \cdot F_2$ (because $F_1 = F_2 = 1$).
Step 2 Suppose $P(k)$ is true. Then

$$F_1^2 + F_2^2 + \cdots + F_k^2 + F_{k+1}^2$$

$$= F_k \cdot F_{k+1} + F_{k+1}^2 \qquad \text{Induction hypothesis}$$

$$= F_{k+1}(F_k + F_{k+1}) \qquad \begin{array}{l}\text{Definition of the}\\ \text{Fibonacci sequence}\end{array}$$

$$= F_{k+1} \cdot F_{k+2}$$

So $P(k + 1)$ follows from $P(k)$. Thus, by the Principle of Mathematical Induction, $P(n)$ holds for all n.

31. Let $P(n)$ denote the statement $\begin{bmatrix} 1 & 1 \\ 1 & 0 \end{bmatrix}^n = \begin{bmatrix} F_{n+1} & F_n \\ F_n & F_{n-1} \end{bmatrix}$.

Step 1 $P(2)$ is true since $\begin{bmatrix} 1 & 1 \\ 1 & 0 \end{bmatrix}^2 = \begin{bmatrix} 2 & 1 \\ 1 & 1 \end{bmatrix} = \begin{bmatrix} F_3 & F_2 \\ F_2 & F_1 \end{bmatrix}$.

Step 2 Suppose $P(k)$ is true. Then

$$\begin{bmatrix} 1 & 1 \\ 1 & 0 \end{bmatrix}^{k+1} = \begin{bmatrix} 1 & 1 \\ 1 & 0 \end{bmatrix}^k \begin{bmatrix} 1 & 1 \\ 1 & 0 \end{bmatrix}$$

$$= \begin{bmatrix} F_{k+1} & F_k \\ F_k & F_{k-1} \end{bmatrix} \begin{bmatrix} 1 & 1 \\ 1 & 0 \end{bmatrix} \qquad \text{Induction hypothesis}$$

$$= \begin{bmatrix} F_{k+1} + F_k & F_{k+1} \\ F_k + F_{k-1} & F_k \end{bmatrix}$$

$$= \begin{bmatrix} F_{k+2} & F_{k+1} \\ F_{k+1} & F_k \end{bmatrix} \qquad \begin{array}{l}\text{Definition of the}\\ \text{Fibonacci sequence}\end{array}$$

So $P(k + 1)$ follows from $P(k)$. Thus, by the Principle of Mathematical Induction, $P(n)$ holds for all $n \geq 2$.

33. Let $P(n)$ denote the statement $F_n \geq n$.

Step 1 $P(5)$ is true since $F_5 \geq 5$ (because $F_5 = 5$).
Step 2 Suppose $P(k)$ is true. Now

$$F_{k+1} = F_k + F_{k-1} \qquad \text{Definition of the Fibonacci sequence}$$

$$\geq k + F_{k-1} \qquad \text{Induction hypothesis}$$

$$\geq k + 1 \qquad \text{Because } F_{k-1} \geq 1$$

So $P(k + 1)$ follows from $P(k)$. Thus, by the Principle of Mathematical Induction, $P(n)$ holds for all $n \geq 5$.

Section 11.6 ■ page 868

1. $x^6 + 6x^5y + 15x^4y^2 + 20x^3y^3 + 15x^2y^4 + 6xy^5 + y^6$

3. $x^4 + 4x^2 + 6 + \dfrac{4}{x^2} + \dfrac{1}{x^4}$

5. $x^5 - 5x^4 + 10x^3 - 10x^2 + 5x - 1$

7. $x^{10}y^5 - 5x^8y^4 + 10x^6y^3 - 10x^4y^2 + 5x^2y - 1$

9. $8x^3 - 36x^2y + 54xy^2 - 27y^3$

11. $\dfrac{1}{x^5} - \dfrac{5}{x^{7/2}} + \dfrac{10}{x^2} + \dfrac{10}{x^{1/2}} + 5x - x^{5/2}$

13. 15   **15.** 4950   **17.** 18

19. 32   **21.** $x^4 + 8x^3y + 24x^2y^2 + 32xy^3 + 16y^4$

23. $1 + \dfrac{6}{x} + \dfrac{15}{x^2} + \dfrac{20}{x^3} + \dfrac{15}{x^4} + \dfrac{6}{x^5} + \dfrac{1}{x^6}$

25. $x^{20} + 40x^{19}y + 760x^{18}y^2$ **27.** $25a^{26/3} + a^{25/3}$

29. $48{,}620x^{18}$ **31.** $300a^2b^{23}$ **33.** $100y^{99}$

35. $13{,}440x^4y^6$ **37.** $495a^8b^8$ **39.** $(x + y)^4$

41. $(2a + b)^3$ **43.** $3x^2 + 3xh + h^2$

Chapter 11 Review ■ page 870

1. $\frac{1}{2}, \frac{4}{3}, \frac{9}{4}, \frac{16}{5}, \frac{100}{11}$ **3.** $0, \frac{1}{4}, 0, \frac{1}{32}, \frac{1}{500}$

5. $1, 3, 15, 105; 654{,}729{,}075$

7. $1, 4, 9, 16, 25, 36, 49$ **9.** $1, 3, 5, 11, 21, 43, 85$

11. (a) $7, 9, 11, 13, 15$

(b) 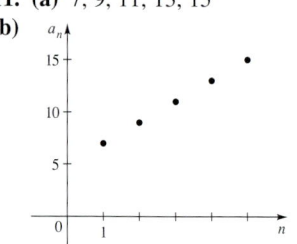   **(c)** Arithmetic, common difference 2

13. (a) $\frac{3}{4}, \frac{9}{8}, \frac{27}{16}, \frac{81}{32}, \frac{243}{64}$

(b) 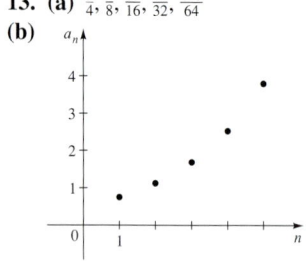   **(c)** Geometric, common ratio $\frac{3}{2}$

15. Arithmetic, 7   **17.** Arithmetic, $5\sqrt{2}$

19. Arithmetic, $t + 1$ **21.** Geometric, $\frac{4}{27}$

23. $2i$ **25.** 5   **27.** $\frac{81}{4}$ **29. (a)** $A_n = 32{,}000(1.05)^{n-1}$

(b) $32{,}000, \$33{,}600, \$35{,}280, \$37{,}044, \$38{,}896.20,$

$\$40{,}841.01, \$42{,}883.06, \$45{,}027.21$ **31.** 12,288

35. (a) 9   **(b)** $\pm 6\sqrt{2}$ **37.** 126   **39.** 384

41. $0^2 + 1^2 + 2^2 + \cdots + 9^2$

43. $\dfrac{3}{2^2} + \dfrac{3^2}{2^3} + \dfrac{3^3}{2^4} + \cdots + \dfrac{3^{50}}{2^{51}}$ **45.** $\displaystyle\sum_{k=1}^{33} 3k$ **47.** $\displaystyle\sum_{k=1}^{100} k2^{k+2}$

49. Geometric; 4.68559   **51.** Arithmetic, $5050\sqrt{5}$

53. Geometric, 9831   **55.** 13

57. 65,534   **59.** \$2390.27

61. $\frac{5}{7}$ **63.** $\frac{1}{2}(3 + \sqrt{3})$

65. Let $P(n)$ denote the statement

$$1 + 4 + 7 + \cdots + (3n - 2) = \frac{n(3n - 1)}{2}.$$

Step 1 $P(1)$ is true since $1 = \dfrac{1(3 \cdot 1 - 1)}{2}$.

Step 2 Suppose $P(k)$ is true. Then

$$1 + 4 + 7 + \cdots + (3k - 2) + [3(k + 1) - 2]$$

$$= \frac{k(3k - 1)}{2} + [3k + 1] \quad \text{Induction hypothesis}$$

$$= \frac{3k^2 - k + 6k + 2}{2}$$

$$= \frac{(k + 1)(3k + 2)}{2}$$

$$= \frac{(k + 1)[3(k + 1) - 1]}{2}$$

So $P(k + 1)$ follows from $P(k)$. Thus, by the Principle of Mathematical Induction, $P(n)$ holds for all n.

67. Let $P(n)$ denote the statement
$\left(1 + \frac{1}{1}\right)\left(1 + \frac{1}{2}\right) \cdots \left(1 + \frac{1}{n}\right) = n + 1.$

Step 1 $P(1)$ is true since $\left(1 + \frac{1}{1}\right) = 1 + 1.$
Step 2 Suppose $P(k)$ is true. Then

$$\left(1 + \frac{1}{1}\right)\left(1 + \frac{1}{2}\right) \cdots \left(1 + \frac{1}{k}\right)\left(1 + \frac{1}{k + 1}\right)$$

$$= (k + 1)\left(1 + \frac{1}{k + 1}\right) \quad \text{Induction hypothesis}$$

$$= (k + 1) + 1$$

So $P(k + 1)$ follows from $P(k)$. Thus, by the Principle of Mathematical Induction, $P(n)$ holds for all n.

69. Let $P(n)$ denote the statement $a_n = 2 \cdot 3^n - 2$.

Step 1 $P(1)$ is true since $a_1 = 2 \cdot 3^1 - 2 = 4$.
Step 2 Suppose $P(k)$ is true. Then

$$a_{k+1} = 3a_k + 4$$

$$= 3(2 \cdot 3^k - 2) + 4 \quad \text{Induction hypothesis}$$

$$= 2 \cdot 3^{k+1} - 2$$

So $P(k + 1)$ follows from $P(k)$. Thus, by the Principle of Mathematical Induction, $P(n)$ holds for all n.

71. Let $P(n)$ denote the statement $n! > 2^n$ for $n \geq 4$.

Step 1 $P(4)$ is true since $4! > 2^4$.

Step 2 Suppose $P(k)$ is true. Then

$$(k + 1)! = k!(k + 1)$$
$$> 2^k(k + 1) \qquad \textcolor{orange}{\text{Induction hypothesis}}$$
$$> 2^{k+1} \qquad \textcolor{orange}{\text{Because } k + 1 > 2}$$

So $P(k + 1)$ follows from $P(k)$. Thus, by the Principle of Mathematical Induction, $P(n)$ holds for all $n \geq 4$.
73. 255 **75.** 12,870
77. $16x^4 + 32x^3y + 24x^2y^2 + 8xy^3 + y^4$
79. $b^{-40/3} + 20b^{-37/3} + 190b^{-34/3}$

Chapter 11 Test ■ page 873

1. 0, 3, 8, 15; 99 **2.** −1 **3. (a)** 3
(b) $a_n = 2 + (n − 1)3$ **(c)** 104
4. (a) $\frac{1}{4}$ **(b)** $a_n = 12(\frac{1}{4})^{n-1}$ **(c)** $3/4^8$
5. (a) $\frac{1}{5}, \frac{1}{25}$ **(b)** $\dfrac{5^8 - 1}{12,500}$ **6. (a)** $-\frac{8}{9}, -78$ **(b)** 60
8. (a) $(1 − 1^2) + (1 − 2^2) + (1 − 3^2) + (1 − 4^2) + (1 − 5^2) = −50$
(b) $(-1)^3 2^1 + (-1)^4 2^2 + (-1)^5 2^3 + (-1)^6 2^4 = 10$
9. (a) $\frac{58,025}{59,049}$ **(b)** $2 + \sqrt{2}$
10. Let $P(n)$ denote the statement

$$1^2 + 2^2 + \cdots + n^2 = \frac{n(n + 1)(2n + 1)}{6}.$$

Step 1 $P(1)$ is true since $1^2 = \dfrac{1(1 + 1)(2 \cdot 1 + 1)}{6}$.

Step 2 Suppose $P(k)$ is true. Then

$$1^2 + 2^2 + \cdots + k^2 + (k + 1)^2$$
$$= \frac{k(k + 1)(2k + 1)}{6} + (k + 1)^2 \qquad \textcolor{orange}{\text{Induction hypothesis}}$$
$$= \frac{k(k + 1)(2k + 1) + 6(k + 1)^2}{6}$$
$$= \frac{(k + 1)[k(2k + 1) + 6(k + 1)]}{6}$$
$$= \frac{(k + 1)(2k^2 + 7k + 6)}{6}$$
$$= \frac{(k + 1)[(k + 1) + 1][2(k + 1) + 1]}{6}$$

So $P(k + 1)$ follows from $P(k)$. Thus, by the Principle of Mathematical Induction, $P(n)$ holds for all n.
11. $32x^5 + 80x^4y^2 + 80x^3y^4 + 40x^2y^6 + 10xy^8 + y^{10}$
12. $\binom{10}{3}(3x)^3(-2)^7 = -414,720x^3$
13. (a) $a_n = (0.85)(1.24)^n$
(b) 3.09 lb **(c)** Geometric

Focus on Modeling ■ page 877

1. (a) $A_n = 1.0001A_{n-1}, A_0 = 275,000$ **(b)** $A_0 = 275,000,$
$A_1 = 275,027.50, A_2 = 275,055.00, A_3 = 275,082.51,$
$A_4 = 275,110.02, A_5 = 275,137.53, A_6 = 275,165.04,$
$A_7 = 275,192.56$ **(c)** $A_n = 1.0001^n(275,000)$
3. (a) $A_n = 1.0025A_{n-1} + 100, A_0 = 100$ **(b)** $A_0 = 100,$
$A_1 = 200.25, A_2 = 300.75, A_3 = 401.50, A_4 = 502.51$
(c) $A_n = 100[(1.0025^{n+1} - 1)/0.0025]$ **(d)** \$603.76
5. (b) $A_0 = 2400, A_1 = 3120, A_2 = 3336, A_3 = 3400.8,$
$A_4 = 3420.2$ **(c)** $A_n = 3428.6(1 - 0.3^{n+1})$
(d) 3427.8 tons, 3428.6 tons
(e)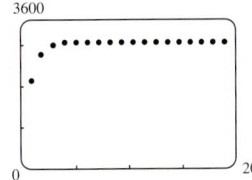

7. (b) In the 35th year
9. (a) $R_1 = 104, R_2 = 108, R_3 = 112, R_4 = 116, R_5 = 120,$
$R_6 = 124, R_7 = 127$
(b) It approaches 200.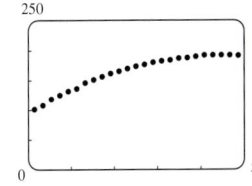

Chapter 12

Section 12.1 ■ page 889

1. $\frac{1}{4}$ **3.** $\frac{1}{3}$ **5.** 1 **7.** −1 **9.** 0.51 **11.** $\frac{1}{2}$ **13. (a)** 2
(b) 3 **(c)** Does not exist **(d)** 4 **(e)** Not defined
15. (a) −1 **(b)** −2 **(c)** Does not exist **(d)** 2 **(e)** 0
(f) Does not exist **(g)** 1 **(h)** 3 **17.** −8 **19.** Does not
exist **21.** Does not exist
23. (a) 4 **(b)** 4 **(c)** 4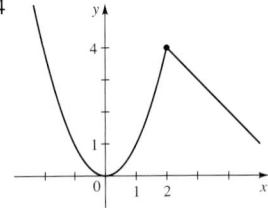

25. (a) 4 **(b)** 3 **(c)** Does not exist
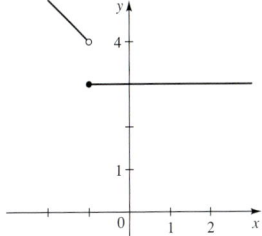

Section 12.2 ■ page 897

1. (a) 5 **(b)** 9 **(c)** 2 **(d)** $-\frac{1}{3}$ **(e)** $-\frac{3}{8}$ **(f)** 0
(g) Does not exist **(h)** $-\frac{6}{11}$ **3.** 75 **5.** $\frac{1}{2}$ **7.** -3 **9.** 5
11. 2 **13.** $\frac{6}{5}$ **15.** 12 **17.** $\frac{1}{6}$ **19.** $-\frac{1}{16}$
21. 4   **23.** $-\frac{3}{2}$

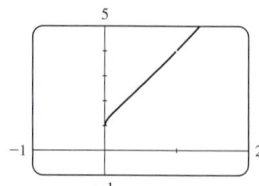   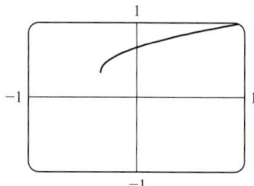

25. (a) 0.667

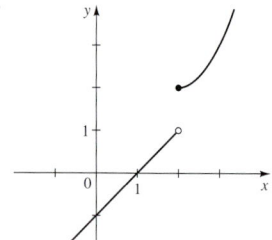

(b) 0.667

x	$f(x)$
0.1	0.71339
0.01	0.67163
0.001	0.66717
0.0001	0.66672

x	$f(x)$
-0.1	0.61222
-0.01	0.66163
-0.001	0.66617
-0.0001	0.66662

(c) $\frac{2}{3}$
27. 0 **29.** Does not exist **31.** Does not exist
33. (a) 1, 2 **(b)** Does not exist
(c)

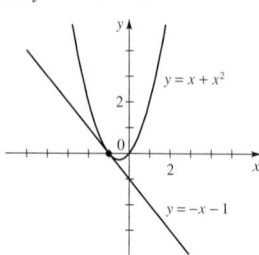

Section 12.3 ■ page 906

1. 3 **3.** -11 **5.** 24
7. $y = -x - 1$ **9.** $y = -x + 4$

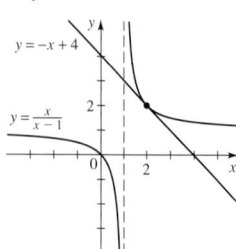

11. $y = \frac{1}{4}x + \frac{7}{4}$

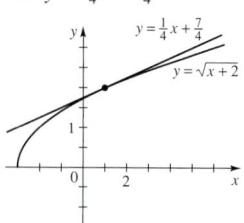

13. $f'(2) = -12$ **15.** $g'(1) = 4$ **17.** $F'(4) = -\frac{1}{16}$
19. $f'(a) = 2a + 2$ **21.** $f'(a) = \dfrac{1}{(a + 1)^2}$
23. (a) $f'(a) = 3a^2 - 2$
(b) $y = -2x + 4, \ y = x + 2, \ y = 10x - 12$
(c)

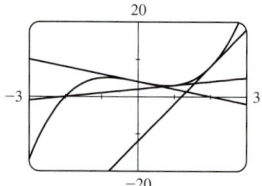

25. -24 ft/s **27.** $12a^2 + 6$ m/s, 18 m/s, 54 m/s, 114 m/s
29. 0.75°/min
31. (a) -38.3 gal/min, -27.8 gal/min **(b)** -33.3 gal/min

Section 12.4 ■ page 915

1. (a) $-1, 2$ **(b)** $y = -1, y = 2$ **3.** 0 **5.** $\frac{2}{5}$ **7.** $\frac{4}{3}$ **9.** 2
11. Does not exist **13.** 7 **15.** $-\frac{1}{4}$ **17.** 0 **19.** 0
21. Divergent **23.** 0 **25.** Divergent **27.** $\frac{3}{2}$ **29.** 8
31. (b) 30 g/L

Section 12.5 ■ page 924

1. (a) 40, 52

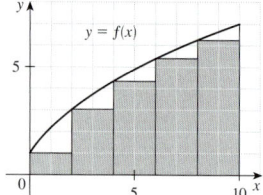   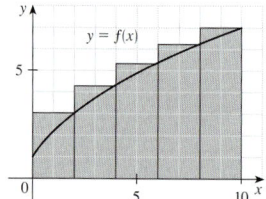

(b) 43.2, 49.2
3. 5.25 **5.** $\frac{223}{35}$
7. (a) $\frac{77}{60}$, underestimate   **(b)** $\frac{25}{12}$, overestimate

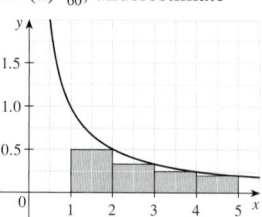   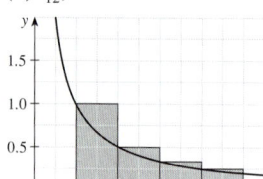

9. (a) 8, 6.875 **(b)** 5, 5.375

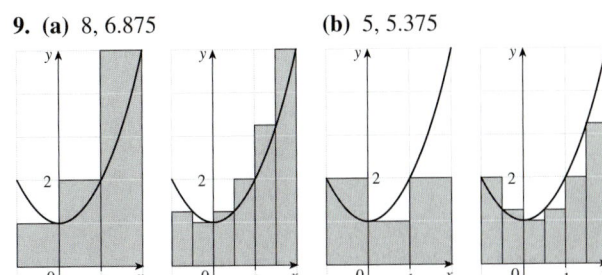

11. 37.5 **13.** 8 **15.** 166.25 **17.** 133.5

Chapter 12 Review ■ page 925

1. 1 **3.** 0.69 **5.** Does not exist **7. (a)** Does not exist
(b) 2.4 **(c)** 2.4 **(d)** 2.4 **(e)** 0.5 **(f)** 1 **(g)** 2 **(h)** 0
9. −3 **11.** 7 **13.** 2 **15.** −1 **17.** 2 **19.** Does not exist
21. $f'(4) = 3$ **23.** $f'(16) = \frac{1}{8}$ **25. (a)** $f'(a) = -2$
(b) −2, −2 **27. (a)** $f'(a) = 1/(2\sqrt{a+6})$
(b) $1/(4\sqrt{2})$, 1/4 **29.** $y = 2x + 1$ **31.** $y = 2x$
33. $y = -\frac{1}{4}x + 1$ **35. (a)** −64 ft/s **(b)** −32a ft/s
(c) $\sqrt{40} \approx 6.32$ s **(d)** −202.4 ft/s **37.** $\frac{1}{5}$ **39.** $\frac{1}{2}$
41. Divergent **43.** 3.83 **45.** 10 **47.** $\frac{5}{6}$

Chapter 12 Test ■ page 928

1. (a) $\frac{1}{2}$ **(b)**

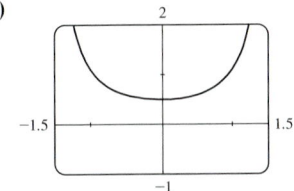

2. (a) 1 **(b)** 1 **(c)** 1 **(d)** 0 **(e)** 0 **(f)** 0 **(g)** 4 **(h)** 2
(i) Does not exist **3. (a)** 6 **(b)** −2 **(c)** Does not exist
(d) Does not exist **(e)** $\frac{1}{4}$ **(f)** 2 **4. (a)** $f'(x) = 2x - 2$
(b) −4, 0, 2 **5.** $y = \frac{1}{6}x + \frac{3}{2}$ **6. (a)** 0 **(b)** Does not exist
7. (a) $\frac{89}{25}$ **(b)** $\frac{11}{3}$

Focus on Modeling ■ page 931

1. $57{,}333\frac{1}{3}$ ft-lb **3. (b)** Area under the graph of
$p(x) = 375x$ between $x = 0$ and $x = 4$ **(c)** 3000 lb
(d) 1500 lb **5. (a)** 1625.28 heating-degree hours
(b) 70°F **(c)** 1488 heating-degree hours **(d)** 75°F
(e) The day in part (a)

Index

Index

Photo Credits

This page constitutes an extension of the copyright page. We have made every effort to trace the ownership of all copyrighted material and to secure permission from copyright holders. In the event of any question arising as to the use of any material, we will be pleased to make the necessary corrections in future printings. Thanks are due to the following authors, publishers, and agents for permission to use the material indicated.

Chapter 1
xxx © Bob Krist/Corbis
38 Courtesy of NASA
49 Public domain
75 The British Museum
100 Courtesy of NASA
103 National Portrait Gallery
112 Public domain
138 Stanford University News Service
141 © Bettmann/Corbis

Chapter 2
146 © Galen Rowell/Corbis
165 Stanford University News Service
188 The Granger Collection
241(top) © Alexandr Satinsky/AFP/Getty Images
241 (bottom) © Eric & David Hosking/Corbis
245 © Ed Kashi/Corbis
246 © Phil Schermeister/Corbis

Chapter 3
248 © J. L. Amos/SuperStock
256 Courtesy of Ford Motor Co.
273 Public domain
288 Public domain
294 © Corbis
321 © Ted Wood/The Image Bank/Getty Images

Chapter 4
326 © Theo Allofs/The Image Bank/Getty Images
331 Gary McMichael/Photo Researchers, Inc.
344 (left) © Bettmann/Corbis
344 (right) © Hulton-Deutsch Collection/Corbis
346 Public domain
374 © Joel W. Rogers/Corbis
378 © Roger Ressmeyer/Corbis
387 © Chabruken/The Image Bank/Getty Images

Chapter 5
398 © Robin Smith/Stone/Getty Images
432 © Jeffrey Lepore/Photo Researchers, Inc.
465 © SOHO/ESA/NASA/Photo Researchers, Inc.

Chapter 6
466 Gregory D. Dimijian, M.D.
504 © Alan Oddie/PhotoEdit
525 British Library

Chapter 7
526 © Tony Friedkin/Sony Pictures Classico/ZUMA/Corbis
562 © Getty Images

Chapter 8
580 Courtesy of NASA
605 (left) Courtesy of David Dewey
605 (right) © Stephen Gerard/Photo Researchers, Inc.
609 © Bill Ross/Corbis
626 © J. L. Amos/SuperStock

Chapter 9
634 © Art Wolfe/Stone/Getty Images
656 Courtesy of NASA

672 Courtesy of Caltech
678 The Archives of the National Academy of Sciences
683 © E. O. Hoppé/Corbis
688 © Art Wolfe/Stone/Getty Images
692 © The Granger Collection
696 © Volvox/Index Stock
705 © Baldwin H. Ward and Kathryn C. Ward/Corbis
707 © The Granger Collection

Chapter 10
742 © PhotoDisc/Getty Images
746 © Roger Ressmeyer/Corbis
771 (top left) © Nik Wheeler/Corbis
771 (top middle) © Architect of the Capitol
771 (top right) © Walter Schmid/Stone/Getty Images
771 (bottom left) © Richard T. Nowitz/Corbis
771 (bottom middle) © Courtesy of Chamber of Commerce, St. Louis, MO
771 (bottom right) © O. Alamany and E. Vicens/Corbis
774 www.skyscrapers.com
775 © Bob Krist/Corbis
802 © Bettmann/Corbis
817 © The Granger Collection

Chapter 11
820 © Susan Van Etten/PhotoEdit
826 © The Granger Collection
829 © Royalty-free/Corbis
840 © Michael Ng
858 © Archivo Iconografico, S.A./Corbis

Chapter 12
880 © Karl Ronstrom/Reuters/Landov
894 © Bill Sanderson/SPL/Photo Researchers, Inc.

SEQUENCES AND SERIES

Arithmetic

$$a, a + d, a + 2d, a + 3d, a + 4d, \ldots$$

$$a_n = a + (n - 1)d$$

$$S_n = \sum_{k=1}^{n} a_k = \frac{n}{2}[2a + (n - 1)d] = n\left(\frac{a + a_n}{2}\right)$$

Geometric

$$a, ar, ar^2, ar^3, ar^4, \ldots$$

$$a_n = ar^{n-1}$$

$$S_n = \sum_{k=1}^{n} a_k = a\frac{1 - r^n}{1 - r}$$

If $|r| < 1$, then the sum of an infinite geometric series is

$$S = \frac{a}{1 - r}$$

THE BINOMIAL THEOREM

$$(a + b)^n = \binom{n}{0}a^n + \binom{n}{1}a^{n-1}b + \cdots + \binom{n}{n-1}ab^{n-1} + \binom{n}{n}b^n$$

FINANCE

Compound interest

$$A = P\left(1 + \frac{r}{n}\right)^{nt}$$

where A is the amount after t years, P is the principal, r is the interest rate, and the interest is compounded n times per year.

Amount of an annuity

$$A_f = R\frac{(1 + i)^n - 1}{i}$$

where A_f is the final amount, i is the interest rate per time period, and there are n payments of size R.

Present value of an annuity

$$A_p = R\frac{1 - (1 + i)^{-n}}{i}$$

where A_p is the present value, i is the interest rate per time period, and there are n payments of size R.

Installment buying

$$R = \frac{iA_p}{1 - (1 + i)^{-n}}$$

where R is the size of each payment, i is the interest rate per time period, A_p is the amount of the loan, and n is the number of payments.

CONIC SECTIONS

Circles

$$(x - h)^2 + (y - k)^2 = r^2$$

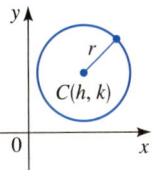

Parabolas

$$x^2 = 4py \qquad\qquad y^2 = 4px$$

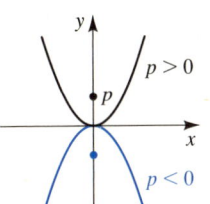 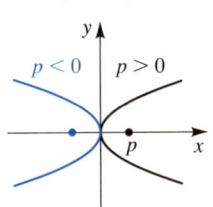

Focus $(0, p)$, directrix $y = -p$ Focus $(p, 0)$, directrix $x = -p$

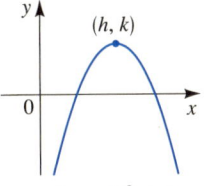 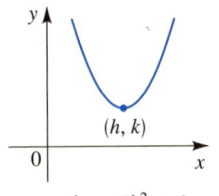

$y = a(x - h)^2 + k,$ $y = a(x - h)^2 + k,$
$a < 0, \quad h > 0, \quad k > 0$ $a > 0, \quad h > 0, \quad k > 0$

Ellipses

$$\frac{x^2}{a^2} + \frac{y^2}{b^2} = 1 \qquad\qquad \frac{x^2}{b^2} + \frac{y^2}{a^2} = 1$$

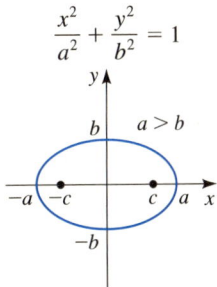 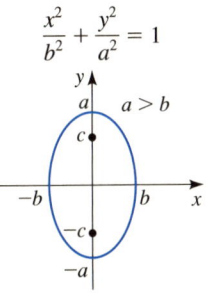

Foci $(\pm c, 0)$, $c^2 = a^2 - b^2$ Foci $(0, \pm c)$, $c^2 = a^2 - b^2$

Hyperbolas

$$\frac{x^2}{a^2} - \frac{y^2}{b^2} = 1 \qquad\qquad -\frac{x^2}{b^2} + \frac{y^2}{a^2} = 1$$

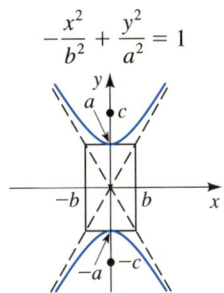

Foci $(\pm c, 0)$, $c^2 = a^2 + b^2$ Foci $(0, \pm c)$, $c^2 = a^2 + b^2$

ANGLE MEASUREMENT

π radians $= 180°$

$1° = \dfrac{\pi}{180}$ rad $\qquad 1$ rad $= \dfrac{180°}{\pi}$

$s = r\theta \qquad A = \frac{1}{2}r^2\theta \quad (\theta \text{ in radians})$

To convert from degrees to radians, multiply by $\dfrac{\pi}{180}$.

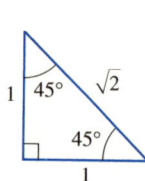

To convert from radians to degrees, multiply by $\dfrac{180}{\pi}$.

TRIGONOMETRIC FUNCTIONS OF REAL NUMBERS

$\sin t = y \qquad\qquad \csc t = \dfrac{1}{y}$

$\cos t = x \qquad\qquad \sec t = \dfrac{1}{x}$

$\tan t = \dfrac{y}{x} \qquad\qquad \cot t = \dfrac{x}{y}$

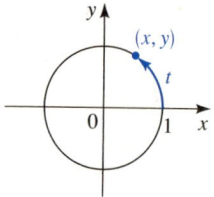

TRIGONOMETRIC FUNCTIONS OF ANGLES

$\sin \theta = \dfrac{y}{r} \qquad\qquad \csc \theta = \dfrac{r}{y}$

$\cos \theta = \dfrac{x}{r} \qquad\qquad \sec \theta = \dfrac{r}{x}$

$\tan \theta = \dfrac{y}{x} \qquad\qquad \cot \theta = \dfrac{x}{y}$

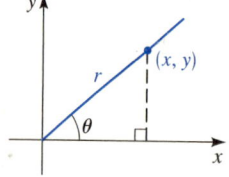

RIGHT ANGLE TRIGONOMETRY

$\sin \theta = \dfrac{\text{opp}}{\text{hyp}} \qquad\qquad \csc \theta = \dfrac{\text{hyp}}{\text{opp}}$

$\cos \theta = \dfrac{\text{adj}}{\text{hyp}} \qquad\qquad \sec \theta = \dfrac{\text{hyp}}{\text{adj}}$

$\tan \theta = \dfrac{\text{opp}}{\text{adj}} \qquad\qquad \cot \theta = \dfrac{\text{adj}}{\text{opp}}$

SPECIAL VALUES OF THE TRIGONOMETRIC FUNCTIONS

θ	radians	$\sin \theta$	$\cos \theta$	$\tan \theta$
0°	0	0	1	0
30°	$\pi/6$	$1/2$	$\sqrt{3}/2$	$\sqrt{3}/3$
45°	$\pi/4$	$\sqrt{2}/2$	$\sqrt{2}/2$	1
60°	$\pi/3$	$\sqrt{3}/2$	$1/2$	$\sqrt{3}$
90°	$\pi/2$	1	0	—
180°	π	0	-1	0
270°	$3\pi/2$	-1	0	—

SPECIAL TRIANGLES

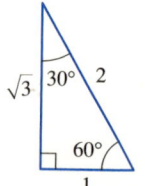

GRAPHS OF THE TRIGONOMETRIC FUNCTIONS

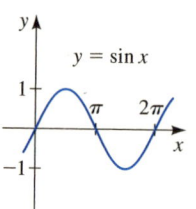
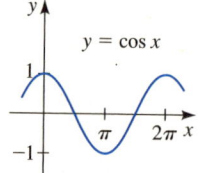
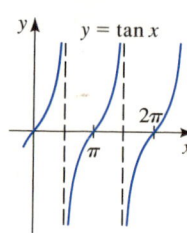

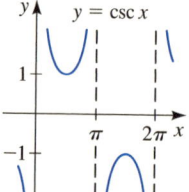
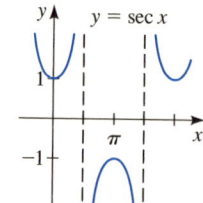
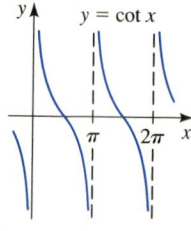

SINE AND COSINE CURVES

$y = a \sin k(x - b) \quad (k > 0) \qquad\qquad y = a \cos k(x - b) \quad (k > 0)$

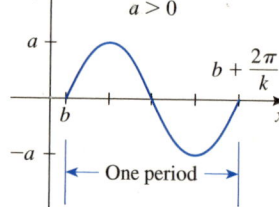
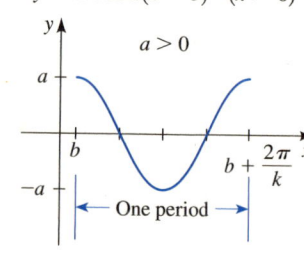

amplitude: $|a|$ period: $2\pi/k$ phase shift: b

GRAPHS OF THE INVERSE TRIGONOMETRIC FUNCTIONS

$y = \sin^{-1}x \qquad\qquad y = \cos^{-1}x \qquad\qquad y = \tan^{-1}x$

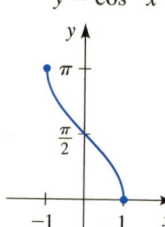
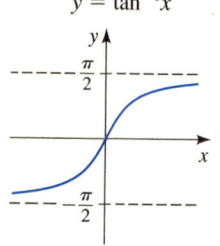